AF358633

New Developments in Boron Chemistry: From Oxidoborates to Hydrido(hetero)borane Derivatives – in Celebration of Professor John D. Kennedy's 80th Birthday

New Developments in Boron Chemistry: From Oxidoborates to Hydrido(hetero)borane Derivatives – in Celebration of Professor John D. Kennedy's 80th Birthday

Editors

Michael A. Beckett
Igor B. Sivaev

Basel • Beijing • Wuhan • Barcelona • Belgrade • Novi Sad • Cluj • Manchester

Editors
Michael A. Beckett
School of Natural Sciences
Bangor University
Bangor
UK

Igor B. Sivaev
A.N. Nesmeyanov Institute of
Organoelement Compounds
Russian Academy of Sciences
Moscow
Russia

Editorial Office
MDPI
St. Alban-Anlage 66
4052 Basel, Switzerland

This is a reprint of articles from the Special Issue published online in the open access journal *Molecules* (ISSN 1420-3049) (available at: https://www.mdpi.com/journal/molecules/special_issues/0406200G4).

For citation purposes, cite each article independently as indicated on the article page online and as indicated below:

Lastname, A.A.; Lastname, B.B. Article Title. *Journal Name* **Year**, *Volume Number*, Page Range.

ISBN 978-3-7258-0666-9 (Hbk)
ISBN 978-3-7258-0665-2 (PDF)
doi.org/10.3390/books978-3-7258-0665-2

Contents

About the Editors

Michael A. Beckett

Prof. Dr. Michael A. Beckett was born in 1957 in Exeter in the UK. He graduated with BSc (1978) and PhD (1981) degrees from the School of Chemical sciences at the University of East Anglia, Norwich, UK. After graduating, he held postdoctoral positions within the UK at Leeds University, QMC London, Exeter University, and Staffordshire University. He was appointed a Lecturer in Inorganic Chemistry at Bangor University in 1988. He remained at Bangor University throughout his academic career and was promoted to Professor in 2016. He became a Fellow of the Royal Society of Chemistry (FRSC) in 1997. He was Head of the School of Chemistry over the period of 2009–2018. He has been an Emeritus Professor of Inorganic Chemistry since his retirement in 2019. He has researched the co-ordination chemistry of heavier main group elements, metalloborane chemistry, organoborate chemistry and Lewis acidity measurement, crystal engineering, and the self-assembly of polyborate anions.

Igor B. Sivaev

Prof. Dr. Igor B. Sivaev was born in 1964 in Kivertsy (Volyn region, Ukraine). After graduation from the Mendeleev Moscow Institute of Chemical Technology (1987) as a Chemical Engineer, he studied the chemistry of polyhedral boron hydrides in the group of Prof. Nikolai Kuznetsov at the Kurnakov Institute of General and Inorganic Chemistry, Russian Academy of Sciences. Later, Igor pursued his PhD degree at the Department of Organic Chemistry, Uppsala University, Sweden, under the guidance of Prof. Stefan Sjöberg (2000) and joined the group of Prof. Vladimir Bregadze at the Nesmeyanov Institute of Organoelement Compounds, Russian Academy of Sciences. In 2014, Igor received a Doctor of Chemical Sciences degree (habilitation) at the Nesmeyanov Institute of Organoelement Compounds, Russian Academy of Sciences, where currently he is the Head of the Laboratory of Organoboron and Organoaluminium Compounds. His scientific interests are the chemistry and application of polyhedral boron hydrides, including boranes, carboranes, and metallacarboranes.

Preface

Professor John D. Kennedy, born 28th March 1943, is a leading and influential British inorganic chemist who made key contributions to the area of borane and metallaborane chemistry. He received his BSc (1965) and PhD (1968) from University College London, and his doctoral studies were supervised by Prof. A.G. Davies. He then spent a 3-year period at the State University of New York in Albany as a research associate with Prof. H.G. Kuivila. He returned to the UK in 1971 as a lecturer in organic chemistry at King's College London. He was then appointed as a research associate at the City of London Polytechnic with Prof. W. McFarlane.

In 1975, he moved from London to the University of Leeds, where he started to work on borane and metallaborane chemistry as a research fellow with Prof. N.N. Greenwood. This partnership resulted in 100 influential publications. Soon after his arrival in Leeds, John Kennedy also established himself as an independent researcher and was promoted to Professor in 2000. He has published more than 400 papers and has an h-index of 44. John Kennedy retired in 2008 and now holds the title Emeritus Professor of Inorganic Chemistry. The 'metallaborane' group at Leeds oversaw significant developments in metallaborane, heteroborane, metallaheteroborane, and carborane chemistry. In these studies, metals from both p and d blocks of the periodic table were incorporated into borane/heteroborane clusters. This research allowed significant advances to be made in macropolyhedral chemistry, 'disobedient skeletons', supramolecular chemistry, and cluster fluxional processes. At the same time, John Kennedy developed an active collaboration with the Czech boron group at the Institute of Inorganic Chemistry of the Czech Academy of Sciences.

Professor John Kennedy has always played an active role in the life of the boron community. He co-founded annual UK meetings for young boron chemists, known as the INTRABORON. He was a member of the international advisory boards for the international boron conferences IMEBORON and EUROBORON. In 1993, he received the J.E. Purkyně Medal from the Academy of Sciences of the Czech Republic, and he was awarded the Royal Society of Chemistry 1999 Silver Medal for Main-Group Chemistry.

It was Professor Kennedy's 80th birthday in March of 2023. Towards the end of summer 2022, we agreed to be Editors of a Special Issue of *Molecules* to celebrate his birthday. An invitation for articles and reviews was made, with the title of the Special Issue crafted to enable all aspects of boron chemistry to be considered. Many manuscripts were received during 2023. We (and John) were delighted by the response, which highlights the high esteem that Professor John Kennedy is held in by his colleagues and friends. These articles are collected together here in this reprint. Thank you to everyone who contributed.

Michael A. Beckett and Igor B. Sivaev

Editors

 *molecules* 

Review

Towards the Application of Purely Inorganic Icosahedral Boron Clusters in Emerging Nanomedicine

Francesc Teixidor, Rosario Núñez * and Clara Viñas *

Institut de Ciència de Materials de Barcelona, ICMAB-CSIC, 08193 Bellaterra, Spain; teixidor@icmab.es
* Correspondence: rosario@icmab.es (R.N.); clara@icmab.es (C.V.)

Abstract: Traditionally, drugs were obtained by extraction from medicinal plants, but more recently also by organic synthesis. Today, medicinal chemistry continues to focus on organic compounds and the majority of commercially available drugs are organic molecules, which can incorporate nitrogen, oxygen, and halogens, as well as carbon and hydrogen. Aromatic organic compounds that play important roles in biochemistry find numerous applications ranging from drug delivery to nanotechnology or biomarkers. We achieved a major accomplishment by demonstrating experimentally/theoretically that boranes, carboranes, as well as metallabis(dicarbollides), exhibit global 3D aromaticity. Based on the stability–aromaticity relationship, as well as on the progress made in the synthesis of derivatized clusters, we have opened up new applications of boron icosahedral clusters as key components in the field of novel healthcare materials. In this brief review, we present the results obtained at the Laboratory of Inorganic Materials and Catalysis (LMI) of the Institut de Ciència de Materials de Barcelona (ICMAB-CSIC) with icosahedral boron clusters. These 3D geometric shape clusters, the semi-metallic nature of boron and the presence of *exo*-cluster hydrogen atoms that can interact with biomolecules through non-covalent hydrogen and dihydrogen bonds, play a key role in endowing these compounds with unique properties in largely unexplored (bio)materials.

Keywords: carboranes; metallabis(dicarbollide); BNCT; proton therapy; PBFR; COSAN; FESAN; PET; SPECT; antimicrobial; luminescence; bioimaging; photodinamic therapy (PDT)

Citation: Teixidor, F.; Núñez, R.; Viñas, C. Towards the Application of Purely Inorganic Icosahedral Boron Clusters in Emerging Nanomedicine. *Molecules* **2023**, *28*, 4449. https:// doi.org/10.3390/molecules28114449

Academic Editor: Radosław Podsiadły

Received: 27 April 2023
Revised: 21 May 2023
Accepted: 24 May 2023
Published: 30 May 2023

1. Introduction

Boron was isolated in Penzance (Cornwall, England) in 1808 by the English chemist Humphry Davy [1], but boron as an element was identified by Jöns Jakob Berzelius in 1824 [2]. Boron is extracted as borate salts of different cations from minerals (Kaliborite, Karlite, Kernita and Kurnakovite, among others) [3,4]. Turkey (deposits existing in Kırka, Emet, Bigadiç, and Kestelek) has the largest world boron reserves, followed by the United States ("Death Valley" desert in California) and Russia at the second position [5], being Turkey the major country in boron production from 2010 to 2022 [6].

Borax ($Na_2B_4O_7.10H_2O$) was one of the first minerals to be exchanged in the times of the Ancient World. In the Egypt of the phaaraohs, the deceased were embalmed with mummification salts, being those containing borate the most reliable for preservation. Boric acid (H_3BO_3), which was produced from borax by the Dutch chemist William Homberg in 1702, has been widely used for topical administration since the 18th century due to its strong bactericidal and fungicidal activity [7].

Boron, which is located to the left of carbon on the periodic table, possesses and forms stable compounds with a wide variety of elements. Natural boron is composed of two stable isotopes ^{10}B and ^{11}B, the latter of which make up about 80% of natural boron. Boron, like carbon, can bond with itself, forming B-B bonds that give rise to boranes and heteroboranes (being the most known carboranes and metallacarboranes). These boron clusters form 3D aromatic [8–10], polyhedral structures with triangular faces in which the bonds that hold the cluster together are tricentric bonds with two electrons (3c-2e).

William Lipscomb received the Nobel Prize in Chemistry in 1976 for his studies on the 3c-2e bonding of borane structures [11]. These 3D molecular structures of boron clusters possess extraordinary chemical, biological, thermal, and photochemical stability that make them have unique applications in (nano)materials not possible with other elements, including carbon [12–17].

The traditional use of organic chemistry as the basis for all aspects of contemporary biomedical chemistry has provided truly miraculous results. Nowadays, most commercial drugs are purely organic molecules, but nitrogen, oxygen, phosphorus, sulfur, and halogens, all neighbors of carbon to the right, are part of a wide variety of the active principles of medicines. In the middle of the 20th century [18–21], the first investigations of boron compounds for their use in medicine were directed mainly towards the treatment of cancer by the therapy called BNCT (Boron Neutron Caption Therapy), but currently, a vibrant and growing research is being developed to employ boron-containing compounds in medicinal chemistry and chemical biology [22–33].

This mini-review focuses on the large research activity of the Inorganic Materials and Catalysis Laboratory (LMI) at the Institut de Ciència de Materials de Barcelona (ICMAB-CSIC) [34] with icosahedral boron clusters, which due to their geometric shape and the semi-metal nature of boron provide these compounds with unique properties in (bio)materials largely unexplored.

2. Characteristics of Icosahedral Neutral Carboranes and Anionic Metallabis(Dicarbollides)

2.1. Icosahedral Closo-Borane and Heteroborane Clusters

Figure 1 shows the inorganic icosahedral *closo*-dodecaborate ($[B_{10}H_{12}]^{2-}$), the dicarba-*closo*-dodecaborane (*closo* $C_2B_{10}H_{12}$), which exists in three isomeric forms that are named based on the positioning of the two CH vertices: 1,2- or *ortho*-, 1,7- or *meta*-, and 1,12- or *para*-carborane and, the sandwich metallabis(dicarbollides) $[M(C_2B_9H_{11})_2]^-$ (M = Co^{3+}, Fe^{3+}). Five different conformations can be found in the metallabis(dicarbollides): *cisoid*-1, *gauche*-1, *transoid*, *gauche*-2 and *cisoid*-2. However, *cisoid*-1 and *cisoid*-2, as well as *gauche*-1 and *gauche*-2, are equivalent in the non-substituted or symmetrically disubstituted clusters [35].

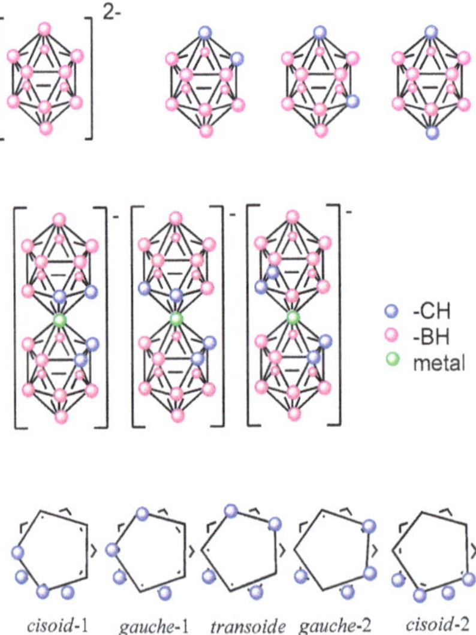

Figure 1. Representation of icosahedral borane, carboranes at the top as well as metallabis(dicarbollides) with the schematic representations of the different conformers of the most known metallabis(*o*-dicarbollide) in the middle and at the bottom, respectively.

Teixidor and Viñas believed that one of the main reasons for the lack of knowledge and poor application of these boron compounds is the lack of synthetic processes for their functionalization. Without these processes, the chemistry of boron is marginalized when its possibilities are enormous, and in many cases, complementary to the organic chemistry compounds.

2.2. Towards the Derivatization of the Icosahedral Boron Clusters

The neutral icosahedral *closo* $C_2B_{10}H_{12}$ carboranes have the potential for the incorporation of a large number of substituents at its 12 vertices (2 C-H and 10 B-H). The reactivity of the B-H vertices depends on the distance of each B-H vertex to the C-H ones. Most reactions that occur at the boron vertices do not affect the carbon vertices, and vice versa. Consequently, *o*-carborane offers the possibility to develop chemistry of neutral *closo*-carboranes at the C vertices, at the B vertices, as well as in both C and B vertices (Figure 2) [35].

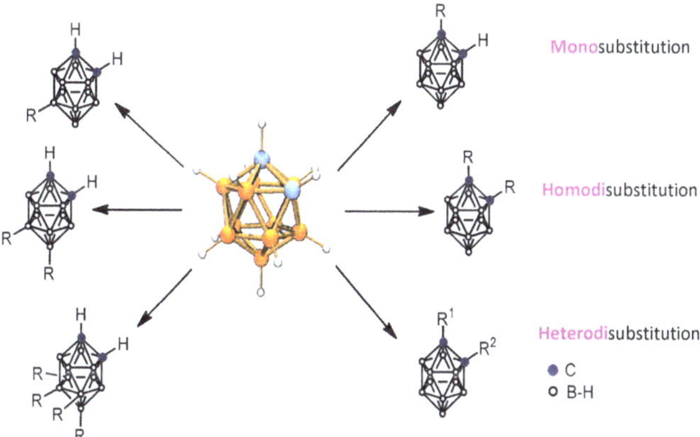

Figure 2. Schematic representation of some substituted *closo ortho*-carboranes at the B vertices (left), at the C vertices (right). Blue circles represent the C atoms, while orange ones the B atoms and white ones are the B atoms or B-H vertices (H atoms omitted for clarity).

Since 1982, Teixidor and Viñas have put emphasis on improving protocols of syntheses because their main objective was the application of icosahedral boron clusters [36], and clusters' derivatization was a necessary and key step to proceed on their use in (bio)materials [37]. Recently, several reviews summarizing the different synthetic procedures to achieve the substitution at the cluster vertices of the icosahedral boron clusters appeared [38–47].

3. Focusing on the Synthesis of Icosahedral Neutral Carborane and Anionic Metallabis(Dicarbollide) Derivatives for Medicinal Application

Teixidor and Viñas group carried out remarkable work in the synthesis of icosahedral carborane and metallacarborane derivatives as well as in their characterization with the objective of finding their application in different fields.

Endo and co-workers, based on the similarities between the phenyl group and the carborane cluster, pioneered the design of new drugs by substituting phenyl groups in compounds with known biological activity with icosahedral carborane groups [48–53]. The concept of 3D aromaticity has already been applied in boron cluster chemistry to relate the limited number of valence electrons in the clusters to their stability [54,55]. Recently [8–10], the 3D global aromaticity of the icosahedral boranes, carboranes, and cobaltabis(dicarbollides) was related to the more familiar 2D aromaticity abiding by Hückel's rule, indicating that both were two sides of the same coin. Then, in 2014 [7], grounded on the relationship between stability and aromaticity, new perspectives for applying icosahedral boron clusters as key components in the field of new biomaterials for healthcare were opened by Teixidor and Viñas group. The highlight is the development of potentiometric sensors for the detection of drugs [56–59], biosensors [60], and X-ray contrast agents for highly radiopaque vertebroplasty cement [61], among others.

Special emphasis is given to fostering advances in the application of boron compounds for the Boron Neutron Capture Therapy (BNCT) treatment of cancer due to the inherent property of the boron element itself (with 20% of ^{10}B). ^{10}B has a large neutron capture section opening up the application of icosahedral boron clusters to the treatment of cancer by the BNCT reaction between a thermal neutron and ^{10}B resulting in the generation of an α particle and 7Li nucleus (Scheme 1). Additionally, the 3D aromatic icosahedral boron clusters offer the possibility of holding twelve substituents covering the entire 3D space.

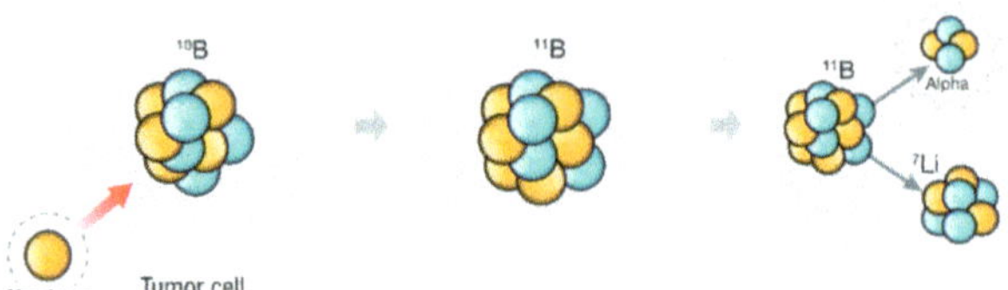

Scheme 1. Representation of the Boron Neutron Capture Therapy reaction (BNCT). Blue circles represents protons while yellow ones neutrons.

The twelve vertices of the cluster can be functionalized, and these small molecules can be converted into multifunctional scaffolds by themselves (Figure 3) [62–65] or bonded to inhibitors of kinases receptor molecules (Figure 4) [66–69] or in anchoring onto structures of nanocarriers (dendrimers [28,70], polymers [14], nanoparticles [36,71–75]) leading to payloads with high boron density (Figure 3). The objective was to synthesize anionic and water-soluble high-boron-containing molecules, which can incorporate in their scaffold either inhibitors of enzymes receptor (Figure 4) [66–69] and/or metal cores (Figure 5) for their use as multifunctional nanocarriers able to act as anticancer drugs by multi-therapy treatment [73–77].

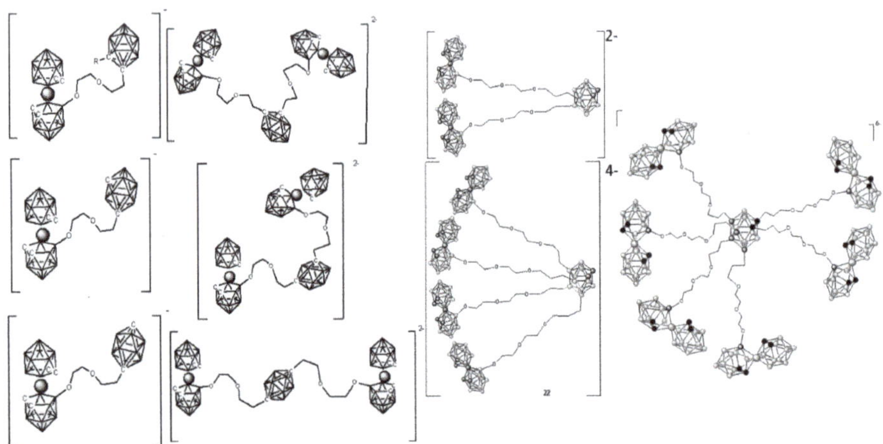

Figure 3. Neutral carborane clusters (*ortho-*, *meta-*, and *para-*) can be functionalized at B, at C, or at both C and B vertices to achieve water-soluble polyanionic high boron content molecules [62–65].

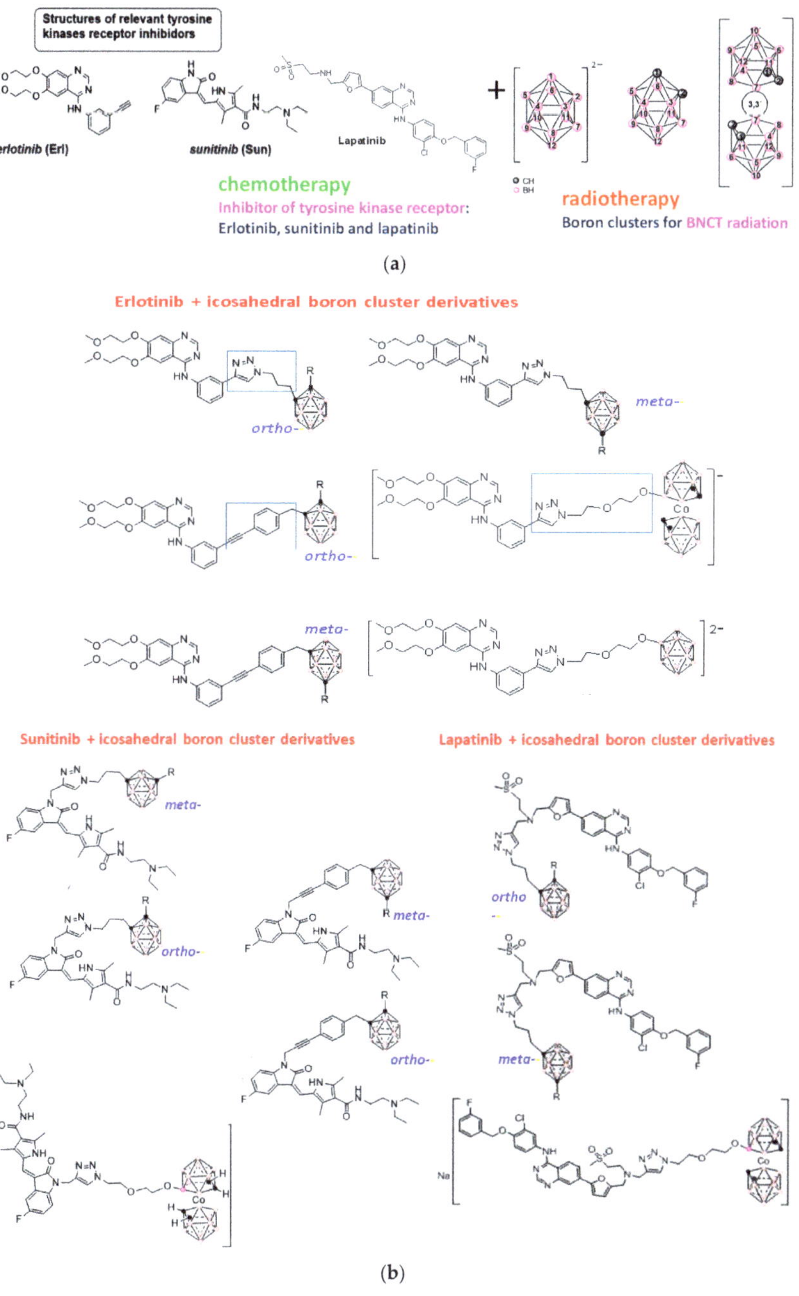

Figure 4. (**a**) Schematic representation of the two components (inhibitors of kinases receptor + icosahedral boron clusters) of the designed hybrid molecules with potential dual action. (**b**) The newly synthesized neutral and anionic boron clusters, which contain quinazoline molecules with potential dual action (chemotherapy + radiotherapy) result in significant clinical benefits [66–68]. The black, circles represents C atoms or C_c-H vertices, the pink circles represent B–H vertices and the purple ones Boron atoms.

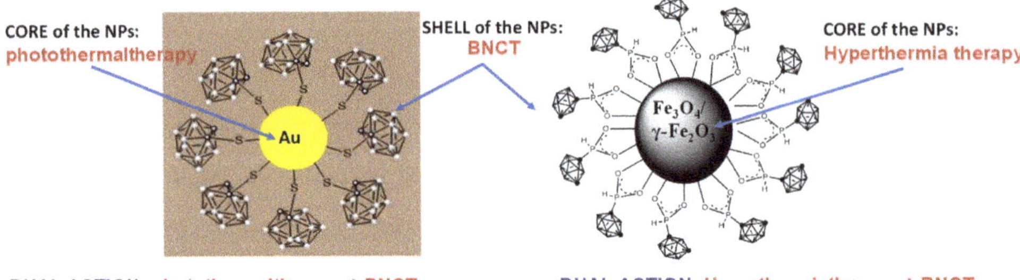

Figure 5. Neutral carboranes onto nanoparticles (gold NPs and magnetic NPs) as vehicles for cancer treatment [71,74]. Both families offer the possibility of dual action (photothermaltherapy + BNCT or hyperthermaltherapy + BNCT), which may result in significant clinical benefits [73].

4. Testing the Icosahedral Neutral Carboranes and Anionic Metallabis(Dicarbollides) in BNCT Cancer Treatment

4.1. Boron Neutron Capture Therapy

Regarding carboranes for BNCT, the research has focused on the development of new multifunctional hybrid (carboranyl + anilinoquinazolines) nanocarriers [66,67] and carborane-magnetic nanoparticles [73]. These (bio)materials exhibit desirable in vitro antitumor activities against preclinical rat glioblastoma F98, colorectal HT29, glioblastoma A172 cancer cell lines, and human brain endothelial hCMEC/D3 cell line.

Importantly, thermal neutrons irradiation in BNCT for 15 min reduced by 2.5 the number of cultured A172 glioblastoma cells after the treatment with carborane-magnetic nanoparticles (Figure 6a) and the systemic administration of carborane-magnetic nanoparticles in mice was well tolerated with no major signs of toxicity. The dual treatment by combining tyrosine kinase inhibition and BNCT irradiation for minutes on HT-29 cells after incubation with carboranyl + anilinoquinazoline hybrids provided better outcomes than *p*-Boronophenylalanine (BPA) [68]. The attractive profile of developed hybrids makes them interesting agents for combined therapy (Figure 6b).

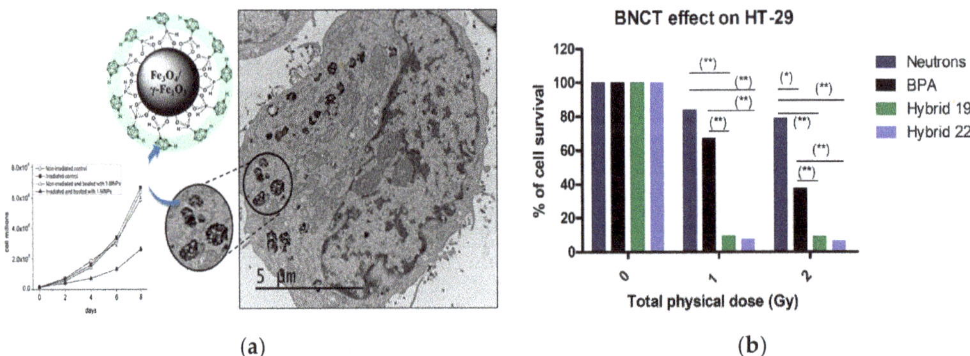

Figure 6. (a) CryoTEM image of glioblastoma A172 cancer cell after carborane-magnetic nanoparticles uptake and proliferation curves of A172 cells re-plated one day after BNCT treatment. BNCT studies were carried out by incubating A172 cells for 24 h with carborane-magnetic nanoparticles (20 µg/mL boron). The amount of internalized boron measured by ICP-MS was 133 ± 25 µg/g, corresponding to a ^{10}B concentration of 26 ± 5 µg/g [73]. (b) Effect on HT-29-cell survival without or with hybrids or BPA treatment post-neutron-irradiation (1 and 2 Gy). Compounds were studied at doses equivalent to 10.0 ppm of ^{10}B for 1 h of incubation. (*) $p < 0.05$; (**) $p < 0.01$. Reproduced from Ref. [61] with permission from Wiley & Sons.

4.2. Metallabis(Dicarbollides) Chemical and Physico-Chemical Properties and Cytotoxicity

Regarding the icosahedral metallacarboranes, the anionic metallabis(dicarbollides), $[3,3\text{-}M(1,2\text{-}C_2B_9H_{11})_2]^-$, (abbreviated as $[o\text{-}COSAN]^-$ and $[o\text{-}FESAN]^-$ for M = Co, Fe, respectively), which are inert to biochemical reactions, have attracted much attention in biology [35]. The 3D aromatic Na[o-COSAN] forms hydrogen and dihydrogen bonds that participate in its self-assembling, water solubility, and aggregates' formation [76]. The Na[o-COSAN] possesses the ability to readily cross cell membranes (Figure 7a) [77–79], is not cytotoxic against mammalian cells (HEK 293, HeLa, THP-1, 3T3), *D. discoideum* amoeba cells, and bacteria (*E. coli* and *Klebsiella*), but is cytostatic, and cells recover following its removal [79]. Furthermore, our studies on glioma-initiating cells (GIC7 and PG88) also supported Na[o-COSAN] cytostatic properties when cells were morphologically recovered 43 h after washing off the compound and increasing in the G2/M subpopulation. Additionally, the study showed that mesenchymal PG88 cells that are more resistant than proneural GIC7 cells to conventional radiotherapy have a lower EC_{50} Na[o-COSAN] and a higher uptake of the compound compared to GIC7 cells, suggesting a new resource to fight against resistant glioblastoma cells [80].

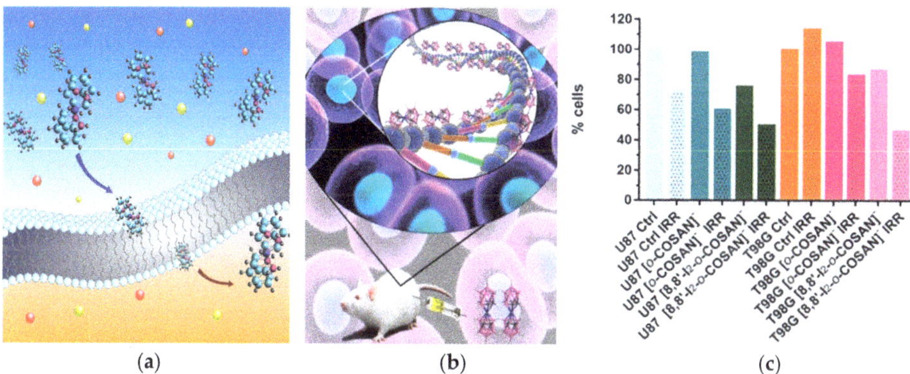

(a)   (b)   (c)

Figure 7. Schematic representation of (**a**) the $[o\text{-}COSAN]^-$ ability to readily cross cell membranes [65] and (**b**) $[o\text{-}COSAN]^-$ interaction with DNA [71]. (**c**) Percentage of viable U87 and T98G cells 24 h after neutron irradiation.

4.3. Synchrotron-Based Fourier-Transform Infrared Micro-Spectroscopy (SR-FTIRM) Studies

Having performed experiments in a round-bottom flask on a chemical scale, which showed that $[o\text{-}COSAN]^-$ and some of its halogenated derivatives interact with biomolecules (amino acids [56,57], proteins [81,82], *ds*-DNA [60,83] (Figure 7b) and glucose [84]), we wanted to go a step ahead by observing these interactions in vitro experiments by using SR-FTIRM. The round-bottom flask changed to a cell, and the solutions to the cell's physiological components. The first chemical scale studies between $[o\text{-}COSAN]^-$ anions and the biomolecules were done individually for each type, whereas the cell study incorporates the effect of all biomolecules interacting simultaneously. This study meant a step ahead to understand and detect that this anion modifies biomolecules (proteins, DNA, and lipids) and concentrates in the cell nucleus after their cellular uptake [68]. The small Na[o-COSAN] molecule, localized close to the cell's nucleus, induces proteins' conformational changes and spectral changes of the DNA region (Figure 7b) in both GIC cell lines, similar to the changes induced by other metal-based compounds like cisplatin that disrupt the double helix base pairing, suggesting that Na[o-COSAN] is a promising agent for BNCT of glioblastoma.

Consequently, in vitro tests in U87 and T98G cells conclude that the amount of ^{10}B inside the cells is enough for BNCT irradiation. BNCT becomes more effective on T98G after their incubation with $Na[8,8'\text{-}I_2\text{-}o\text{-}COSAN]$, whereas no apparent cell-killing effect was observed for untreated cells.

All this led to the following conclusions: These small molecules, particularly [8,8′-I$_2$-o-COSAN]$^-$, are serious candidates for BNCT now that the facilities of accelerator-based neutron sources are more accessible, providing an alternative treatment for resistant glioblastoma (Figure 7c) [85].

Then, in vivo experiments with Na[o-COSAN] and Na[8,8′-I$_2$-o-COSAN] were performed on *Caenorhabditis elegans (C. elegans)* at the L4-stage and their embryos. LD50 values for both cobaltabis(dicarbollides) in L4 *C. elegans* were found to be close to the IC$_{50}$ determined for T98G in vitro after 72 h (Figure 8) [86].

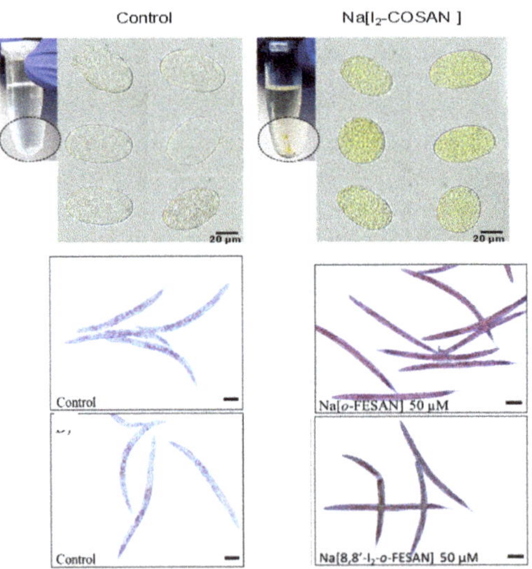

Figure 8. Optical microscopy images of *C. elegans*. (**Top**): The embryos and, (**bottom**): At the L4-stage before (control) and after incubation with Na[8,8′-I$_2$-o-COSAN] 200 µM [86].

Finally, in vivo evaluation in mammalian mice models were run trying to understand the ability of [o-COSAN]$^-$ to target the tumor cells, as well as to cross the blood–brain barrier. After intravenous administration, biodistribution studies of Na[o-COSAN] in BALB/c CrSlc mice (female, 5 weeks old) were run. Anionic [o-COSAN]$^-$ was distributed into many organs but mainly accumulated in the reticuloendothelial system (RES), including liver and spleen (Figure 9) [83].

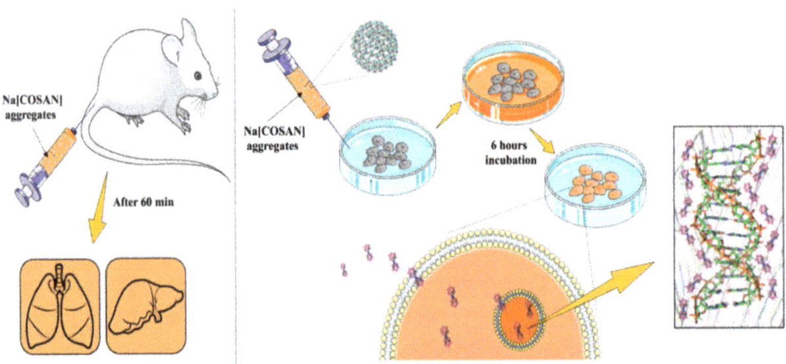

Figure 9. Schematic representation of the in vitro (**right**) and in vivo (**left**) studies of Na[o-COSAN] [85].

4.4. Contrast Agents

Furthermore, Na[8-I-*o*-COSAN] can be labeled with contrast agents, such as [124]I and ^{125}I, for in vivo markers by positron emission tomography (PET) and single photon emission computed tomography (SPECT) nuclear imaging techniques making these clusters very good scaffolds as theranostic agents (Figure 10) [87].

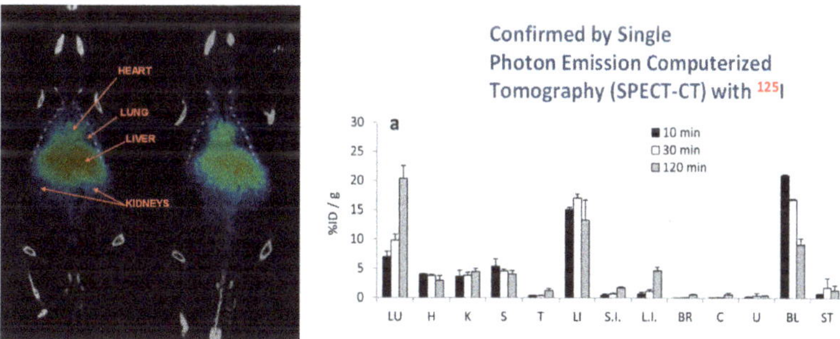

Figure 10. Imaging experiments by in vivo PET-CT and SPECT-CT with Na[8-I-*o*-COSAN] after nuclear interchange of I natural by [124]I and ^{125}I, respectively. Reproduced from Ref. [85]. with permission from the Royal Society of Chemistry.

The synthesis of these unprecedented radiolabeling Na[8-I-*o*-COSAN] anionic derivatives with either [125]I (gamma emitter) or [124]I (positron emitter) was achieved via palladium-catalyzed isotopic exchange reaction (Scheme 2a) following our previously reported synthesis of ^{125}I carborane derivatives (2-I-*p*-, 3-I-*o*-, 9-I-*o*-, 9-I-*m*-carborane, 1-phenyl-3-I-*o*-carborane, and 1,2-diphenyl-3-I-*o*-carborane) with some modifications (Scheme 2b) [87].

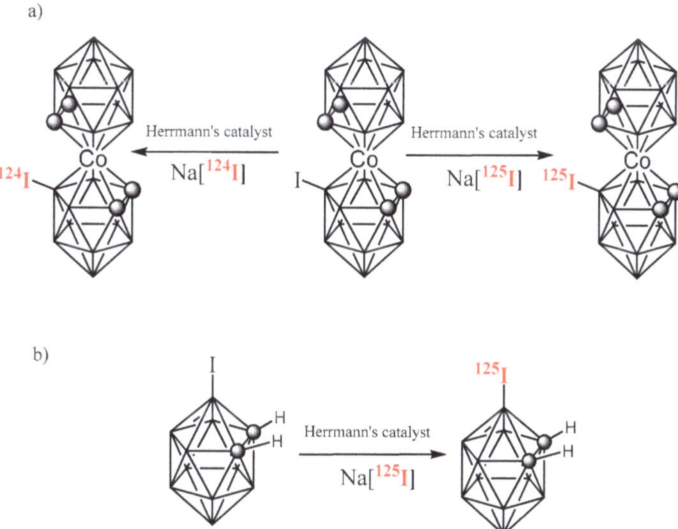

Scheme 2. Isotopic exchange: (**a**) Between [124I]iodide, [125I]iodide, and the anionic [8-I-*o*-COSAN]⁻ cluster, (**b**) between [125I]iodide and the neutral 3-iodo-*o*-carborane cluster by using Herrmann's catalyst (the organopalladium compound made by reaction of tris(*o*-tolyl)phosphine with palladium(II) acetate).

Recently, the sodium salt of the anionic [*o*-FESAN]$^-$ isotopically 100% ^{57}Fe was synthesized with the objective of treating glioblastoma cancer with Na[3,3′-^{57}Fe(1,2-C$_2$B$_9$H$_{11}$)$_2$] because the compound offers the possibility of dual-action (radiation + drug combinations) to improve clinical benefits and reduce healthy tissues toxicity. After [*o*-^{57}FESAN]$^-$ uptake by U87 glioblastoma cells, [*o*-^{57}FESAN]$^-$ was found to be within the cells with 29% of its uptake in the nuclear fraction, which is a particularly desirable target because the nucleus is the cell control center in which DNA and transcription machinery reside. The multi-therapies activity through irradiation with potential for glioblastoma treatment by the Mossbauer effect of [3,3′-^{57}Fe(1,2-C$_2$B$_9$H$_{11}$)$_2$]$^-$ was demonstrated (Figure 11) [88].

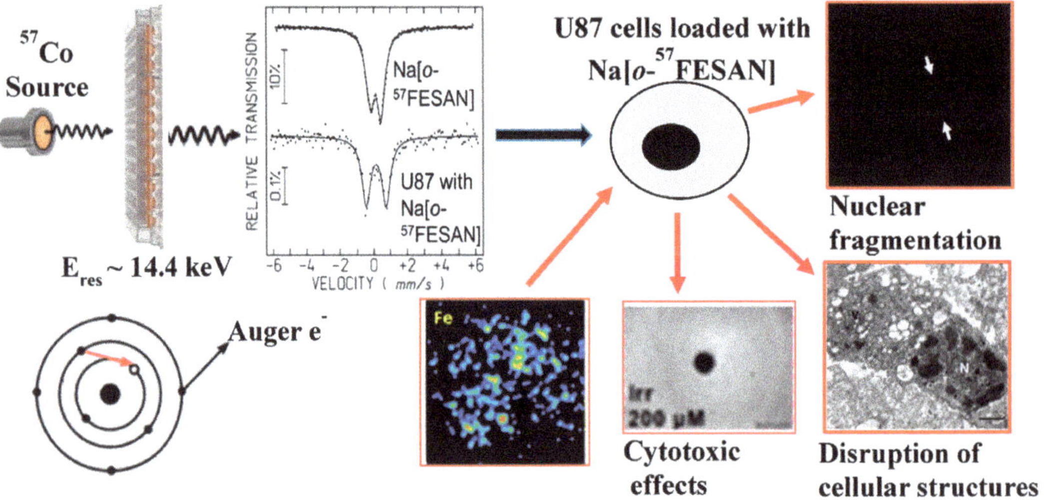

Figure 11. Mössbauer spectrum and 2D elemental maps of Fe distribution indicate that [*o*-^{57}FESAN]$^-$ is present inside U87 cells, an important requisite for selective energy deposition by Mössbauer absorption. Reproduced from [88] with permission from the Royal Society of Chemistry.

4.5. Proton Therapy Based on Boron

Proton therapy is an effective radiation treatment technique used in medicine, which consists of irradiating diseased tissue, most often to treat cancer, with a beam of protons [89,90]. Scheme 3 represents the Proton Boron Fusion Reaction (PBFR) between an energetic proton and ^{11}B resulting in the generation of three α particles. The viability of applying the proton boron fusion (PBF) reaction to the proton therapy to improve its effectiveness has been studied by using the Monte Carlo method [91–93] and experimentally using mercaptoundecahydro-closo-dodecaborate (abbreviated as BSH, which chemical formula is [SH-1-*closo*-B$_{12}$H$_{11}$]$^-$) [94]. Recently [95], taking advantage of the high ^{11}B isotope content in metallabis(dicarbollides), we tested, for the first time, metallacarboranes for the PBFR as a way to improve proton therapy with the [*o*-FESAN]$^-$ in the U87 glioblastoma cells. A simple calculation indicates that the use of PBFR would require 1/12 of isotopically natural molecules with respect to BNCT. Furthermore, in an ideal situation, BNCT can be used synchronously on the existing ^{10}B and Mössbauer on ^{57}Fe, resulting in multi therapies with only one compound. Results from the cellular damage response obtained suggest that PBFR radiation therapy, when applied to boron-rich compounds, is a promising modality to fight against resistant tumors.

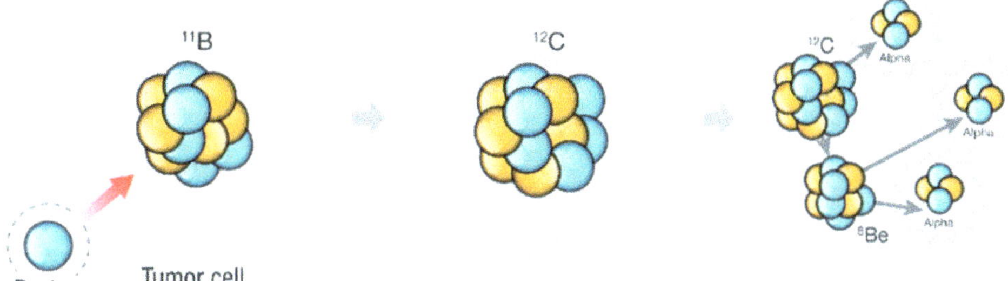

Scheme 3. Representation of the Proton Boron Fusion Reaction (PBFR). Blue circles represent protons while the yellow ones represent neutrons.

4.6. Antimicrobial Activity

In 2013, we started studying the physical–chemical properties and biological evaluation of the sodium salt of the small inorganic metallabis(dicarbollide) molecules ([*o*-COSAN]$^-$ and [*o*-FESAN]$^-$) and their derivatives [8-R(CH$_2$CH$_2$O)$_2$-*o*-COSAN)]$^-$ (R = -OOCCH$_3$; -OCH$_3$; -OCH$_2$CH$_3$) against pure cultures of 16 pathogenic bacterial strains (isolated from animals and humans as well as control strains) and 3 strains of *Candida* spp. as promising antimicrobial agents to tackle bacterial infections [96]. It is important to emphasize that the methicillin-resistant strain of *Staphylococcus aureus* (MRSA), the polyresistant strains of *Pseudomonas aeruginosa*, as well as of *Candida* spp., are sensitive to the compounds Na[8-CH$_3$(CH$_2$CH$_2$O)$_2$-*o*-COSAN)] and Na[8-CH$_3$CH$_2$(CH$_2$CH$_2$O)$_2$-*o*-COSAN)]. Recently, a review of the increasing evidence that boron cluster compounds are promising antimicrobial (antibacterial and antifungal) agents appeared [97]. Lately, with the objective to establish a structure–activity relationship, which clearly supports the antimicrobial activity of the pristine metallabis(dicarbollide) complexes, we tested the small molecules Na[*o*-COSAN], Na[*o*-FESAN], Na[*m*-COSAN)], Na[*m*-FESAN)], the di-iodinated derivatives Na[8,8′-I$_2$-*o*-COSAN], Na[8,8′-I$_2$-*o*-FESAN] and polyanionic species incorporating one or two cobaltabis(dicarbollide) anions with activity against four *Gram-positive* bacteria (two *Enterococcus faecalis* strains and two of *Staphylococcus aureus* including Multi-Resistant Staphylococcus Aureus (MRSA) strains), five *Gram-negative* bacteria (three strains of *Escherichia coli* and two of *Pseudomonas aeruginosa*), and three *Candida albicans* strains that have been responsible for human infections [98,99]. We demonstrated an antimicrobial effect against *Candida* species (Minimum Inhibitory Concentration (MIC) of 2 and 3 nM for Na[8,8′-I$_2$-*o*-COSAN] and Na[*m*-COSAN], respectively), and against *Gram-positive* and *Gram-negative* bacteria, including multi-resistant MRSA strains (MIC of 6 nM for Na[8,8′-I$_2$-*o*-COSAN]). The selectivity index (abbreviated as SI and, calculated as the ratio IC$_{50}$/MIC) for antimicrobial activity of Na[*o*-COSAN] and Na[8,8′-I$_2$-*o*-COSAN] compounds is very high (165 and 1180, respectively), which reveals that these small anionic metallacarborane molecules may be useful to tackle antibiotic-resistant bacteria because it is considered that an SI $\geq$ 10 is acceptable for a selective bioactive sample.

Furthermore, we demonstrated that the outer membrane of *Gram-negative* bacteria establishes an impermeable barrier for some of these metallabis(dicarbollide) small molecules (Scheme 4). Nonetheless, the addition of two iodine groups in the structure of the parent Na[*o*-COSAN] had an improved effect (3–7 times) against *Gram-negative* bacteria. It is important to emphasize that the most active metallabis(dicarbollides) (*meta*-isomers Na[*m*-COSAN)], Na[*m*-FESAN)] and the di-iodinated derivatives Na[8,8′-I$_2$-*o*-COSAN], Na[8,8′-I$_2$-*o*-FESAN]) are both *transoid* conformers in opposite to the Na[*o*-COSAN] that is *cisoid* conformer (see Figure 1), which represent structures with particular physical–chemical properties that make these small molecules more permeable to this barrier.

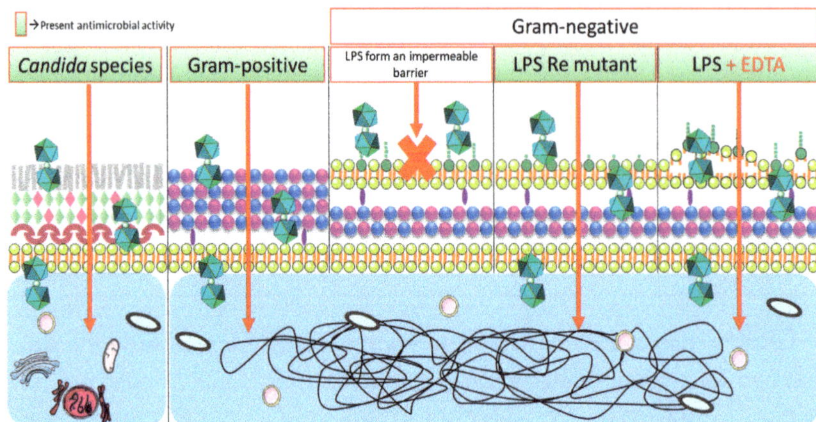

Scheme 4. Graphical representation of the transport of small anionic metallabis(dicarbollide) molecules through microbiological membranes of *Candida* sp. (**left**), Gram-positive (**center**), and Gram-negative (**right**) [100].

The fact that these small molecules cross the mammalian membrane and have antimicrobial properties but low toxicity for mammalian cells (high selectivity index SI) represents a promising tool to treat infectious intracellular bacteria as there is an urgent need for new antibiotics discovery and development. This achievement represents a relevant advance in the field.

5. Boron Clusters-Based Dyes as Theranostic Agents for Diagnosis and Therapy

Today, one of the most important tools in predicting disease is diagnosis. Molecular imaging is a remarkable diagnostic tool in vitro and in vivo that could provide crucial biological information regarding a targeted disease and can thus help to establish a particular treatment or therapy [100,101]. Moreover, the development of theranostic systems to integrate imaging and therapy is an efficient strategy for real-time tracking of the pharmacokinetics and biodistribution of a drug. Current imaging modalities include optics (e.g., fluorescence, Raman, photoacoustics), X-ray, magnetic resonance, radionuclides, and mass spectrometry [102]. Among them, Fluorescence Bioimaging is a common modality for cell and tissue visualization, being of special interest in preclinical research on theranostic agents. In this context, each fluorophore has its benefits and drawbacks, which requires the continued search for new fluorescent probes to meet stringent necessities for applications in terms of sensitive and selective use for bioimaging applications.

Moreover, imaging-guided BNCT is a challenge as it allows us to know the accurate position of the boron-containing compound in the body as well as the accumulation in the tumor. Therefore, it is an important issue to label the boron-containing compound with a fluorescence tracer in order to have relevant information for both diagnosis and therapy [103–107]. In particular, the near-infrared (NIR) boron carriers are of great interest due to their deeper penetration into the living body and their ability to avoid interference from body tissues [108,109].

Dr. Núñez has been a staff member of the LMI group since 2001. Over this time, she has developed synthetic strategies for the functionalization of a great variety of scaffolds, i.e., star-shape molecules and dendrimers [28,110], octasilsesquioxanes [111–114], carbon-based materials [115,116], among others [117], with icosahedral boron clusters and studied their properties. In 2007, Dr. Núñez reported a set of blue emissive Fréchet-type aryl ether core molecules peripherally functionalized with *closo*-carborane and *nido*-carborane clusters [118,119]. It was then demonstrated that the maximum wavelength and emission intensity depend on the $C_{cluster}$ substituent (Me or Ph), the solvent polarity, and the nature of the cluster (*closo* or *nido*). This work was the beginning of her immersion in the field of luminescence. Since then, her main interest has been the development of photoluminescent boron cluster-based organic π-conjugated dyes [120–128], revealing that incorporation of neutral and anionic boron clusters into the structure of well-known fluorophores is an attractive chemical strategy to modulate and improve their photoluminescence properties [15,120–128]. Her current research interest is more focused on new boron-based molecules and materials as theranostic agents for diagnosis (bioimaging) and boron carriers for BNCT.

A set of BODIPY-anionic boron cluster conjugates bearing dianionic $[B_{12}H_{12}]^{2-}$ and monoanionic, $[o\text{-COSAN}]^-$ and $[o\text{-FESAN}]^-$ clusters were designed and synthesized to be used as fluorescent cell probes and BNCT anticancer agents (**1–5** in Figure 12a) [129]. These conjugates were readily synthesized from the meso-(4-hydroxyphenyl)-4,4-difluoro-4-bora-3a,4a-diaza-s-indacene (BODIPY) by ring-opening reaction of the corresponding boron clusters derivatives. The luminescent properties of the BODIPY were not significantly altered by the linking of the anionic boron clusters, showing emission fluorescent quantum yields (Φ_F) in the range of 3–6%. Moreover, the cytotoxicity and cellular uptake of these compounds were analyzed in vitro at different concentrations of B (5, 50, and 100 µg B/mL) using HeLa cells. None of the compounds showed cytotoxicity at the lowest concentration (5 µg B/mL). Compound bearing $[B_{12}H_{12}]^{2-}$ and Na^+ as cation were non cytotoxic at any concentration, while the other compounds showed toxicity at the highest concentrations after 24h. Remarkably, all the compounds were successfully internalized by HeLa cells, exhibiting a strong cytoplasmic stain (Figure 12b). The internalization efficiency for all the compounds was assessed at the lowest concentration (5 µg B/mL), in which they are not cytotoxic. The exceptional cellular uptake and intracellular boron release, together with their fluorescent and biocompatibility properties, highlight the suitability of these boron cluster-containing dyes, especially $[o\text{-COSAN}]^-$ derivative, as potential candidates for cell labeling agents towards medical diagnosis in clinical biopsies. Moreover, the excellent cellular uptake, along with the boron-rich content of our conjugates, make them good candidates as boron carriers for BNCT.

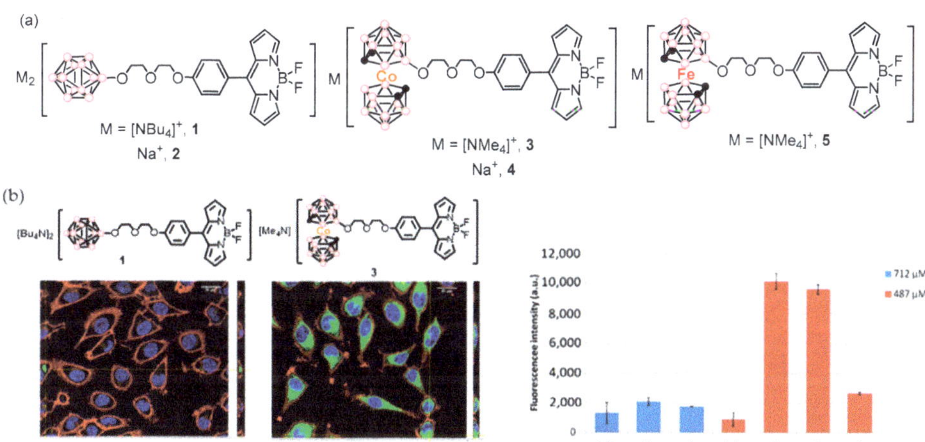

Figure 12. (**a**) Molecular structures of boron cluster-BODIPY conjugates **1–5** (pink circles in the cluster are B or B-H and black circles in the cluster are C-H. (**b**) Intracellular localization of BODIPY dyads **1** and **3** in HeLa cells obtained by confocal laser scanning microscopy (left). Cellular uptake comparison between BODIPY dye and BODIPY-boron clusters at 100 µg/mL of B for each compound (CLSM) (left). Mean values and SD from three independent experiments. a.u.: arbitrary units (right). Reprinted (adapted) with permission from Ref. [129] *Bioconjugate Chem.* **2018**, *29*, 1763–1773. Copyright 2018, American Chemical Society.

Our interest in the development of new boron delivery systems to be used for biological applications led us to prepare a family of fluorescent organotin compounds that have shown excellent properties as nucleoli and cytoplasmic markers in vitro [130]. These organotin compounds are based on 4-hydroxy-N′-((2-hydroxynaphthalen-1-yl)methylene)benzohydra zidato that was derivatized to contain two different boron clusters, $[B_{12}H_{12}]^{2-}$ and $[o$-COSAN$]^-$ following the oxonium ring opening reaction (**6–9** in Figure 13a). These compounds showed photoluminescence properties in solution with Φ_F values in the range from 24% to 49%. Remarkably, linking these anionic boron clusters to tin complexes improved their solubility in cell media, which resulted in better cell internalization and higher cellular uptake, as they do not aggregate either on the cell surface or in the extracellular media. Mouse melanoma B16F10 cells were incubated with 10 μg/mL of the different compounds for 2 h and then analyzed by confocal laser microscopy. Noticeably different staining effect was observed depending on the type of boron cluster attached to the organotin complexes. Compound **8** bearing the $[o$-COSAN$]^-$ anion and two phenyl rings coordinated to the Sn showed an important fluorescence in the cytoplasm, whereas that bearing $[B_{12}H_{12}]^{2-}$ (**6**) produced extraordinary nucleoli and cytoplasmic staining (Figure 13b). The remarkable fluorescence staining properties of these organotin compounds in B16F10 cells make them excellent candidates for in vitro fluorescent bioimaging.

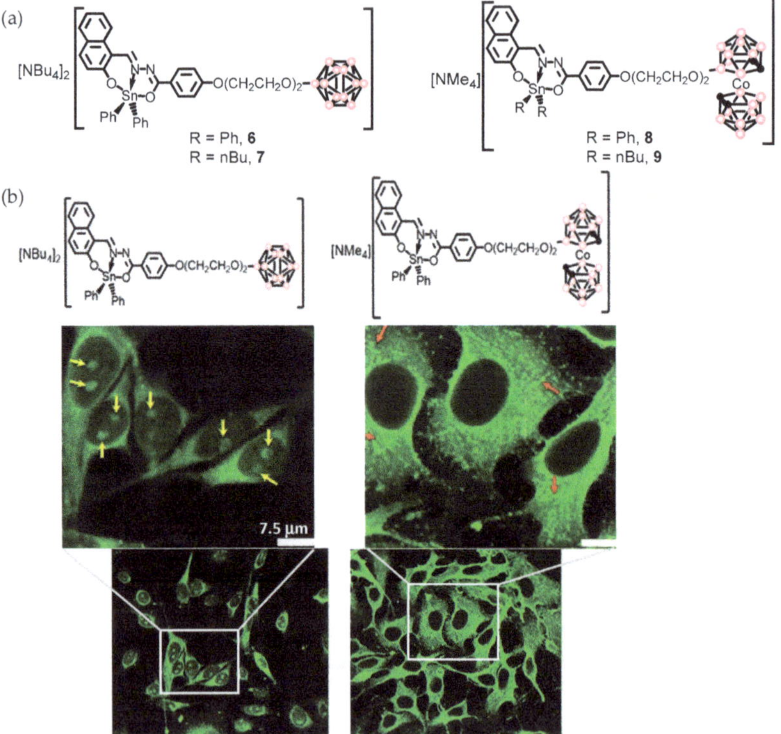

Figure 13. (**a**) Molecular structures of Tin complexes containing anionic boron clusters **6-9** (pink circles in the cluster are B or B-H, black circles in the cluster are C-H). (**b**) Cellular uptake of organotin compounds **6** and **8** bearing anionic $[B_{12}H_{12}]^{2-}$ (left) and $[o$-COSAN$]^-$ (right) by confocal laser scanning microscopy (CLSM). The yellow and red arrows show the internalization of our compounds in the cytoplasm and nucleoli [130]. Reprinted/adapted with permission from Ref. [130] Copyright 2018, Wiley.

Apart from previous BODIPY derivatives bearing anionic boron clusters, our group has also developed a family of neutral BODIPY-carboranyl conjugates which have been synthesized following Sonogashira or Heck cross-coupling reactions in which properly functionalized *ortho*- and *meta*-carborane clusters have been linked to light-emitting BOD-IPY or aza-BODIPY cores [131–133]. Figure 14 illustrates three different BODIPY-carboranyl systems with Ph-*ortho*-carborane (**10–11**) and Ph-*meta*-carborane (**12**), as examples. Due to their fluorescence properties, these fluorophores were studied in vitro as fluorescent probes. HeLa cells were incubated for 30 min with this set of BODIPYs, which presented very different behavior regarding cellular uptake and subcellular distribution (Figure 14) [132]. The differences seem to originate from their diverse static dipole moments and partition coefficients, which depend on the type of cluster isomer (*o*- or *m*-) linked to the BODIPY and that modulates the ability of these molecules to interact with the lipophilic microenvironments in cells. It can be highlighted that the *m*-carborane derivative with higher lipophilicity was much better internalized by cells than their *ortho* analogs. Confocal images of HeLa cells incubated with **12** (Figure 14) clearly indicate that **12** is accumulated in the cytoplasm of the cell. This evidence provides a molecular design strategy for improving the prospective applications of BODIPY-carboranyl dyads as potential fluorescence in vitro bioimaging agents and boron carriers for BNCT, suggesting that *m*-isomers are potentially better theranostic agents than *o*-isomers.

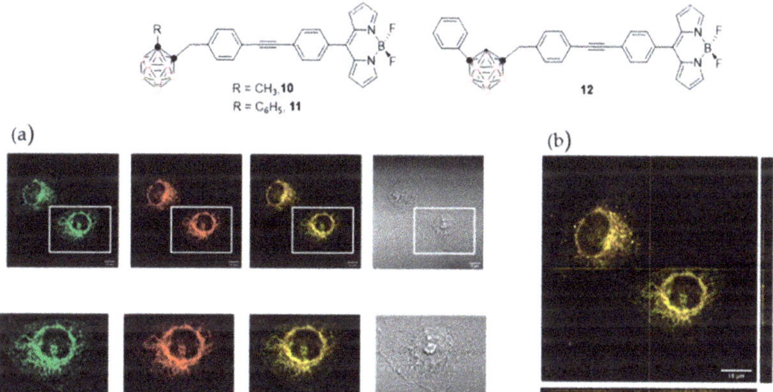

Figure 14. Molecular structures of the representative compounds: **10**, **11**, and **12** (top) ((pink circles in the cluster are B-H and black circles in the cluster are C). Confocal images of 10 μM **12** in live HeLa cells after 30 min incubation: (**a**) Panels from left to right show the confocal green channel (λ_{exc} = 486 nm, λ_{em} = 500 nm), red channel (λ_{exc} = 535 nm, λ_{em} = 610 nm), merged and bright channels, respectively. Scale bars represent 10 and 15 μm. (**b**) z-stack visualization [132]. Reprinted/adapted with permission from Ref. [132] Copyright 2020, Wiley.

Another type of well-known fluorophores are anthracene derivatives that exhibit excellent luminescence properties that make them perfect scaffolds for optical applications. Our group has developed efficient blue light-emitting materials by combining the properties of anthracene and *m*-carborane [134]. Three different *m*-carborane-anthracene dyads, in which the carborane is non-iodinated, mono-iodinated, or di-iodinated at B atoms, and the anthracene fragment is linked to one $C_{cluster}$ atom through a CH_2 spacer, were prepared. All of them exhibited exceptional fluorescence properties with high quantum yields ($\Phi_F \sim 100\%$) in solution with maximum emission of around 415 nm, confirming that simply linking the *m*-carborane fragment to one fluorophore produces a significant enhancement of the fluorescence emission in the target compound. Notably, the three conjugates exhibited good fluorescence efficiencies in aggregate state with Φ_F in the range 19–23%, indicating that our dyads are extremely good emitters in solution, while maintaining the emission properties in the aggregate state. Moreover, their cytotoxicity and cellular uptake in HeLa cells were evaluated. None of the compounds showed cytotoxicity at different concentrations for HeLa cells. Confocal microscopy studies confirmed that, although all compounds were internalized by cells via endocytosis, exhibiting high fluorescence emission intensity, the one with two iodo atoms is the one with a higher cellular uptake. This suggested that the presence of iodo units leads to a more efficient transport across the plasma membrane and a better internalization of the compounds. Figure 15 shows the autofluorescence of HeLa cells and fluorescence emitted by Hela cells incubated with the diiodinated antracece-*m*-carborane. We then conclude that the di-iodinated compound is an excellent candidate as a fluorescent dye for bioimaging studies in fixed cells, and due to the high boron content and exceptional cellular uptake, it could be used as a potential anticancer agent for BNCT.

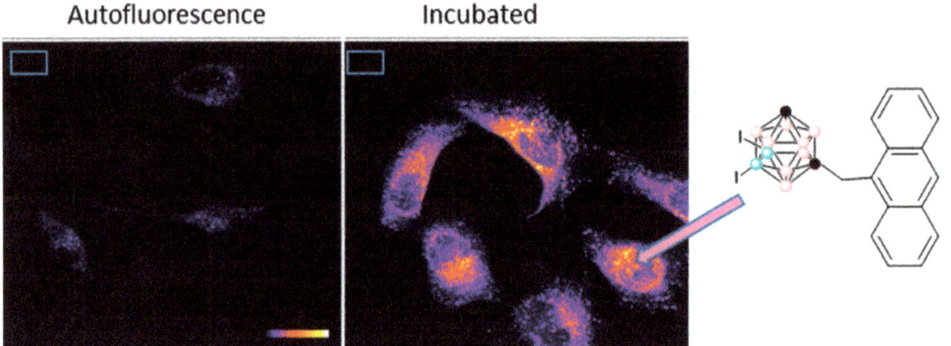

Figure 15. Fluorescence intensity emitted by HeLa cells incubated 4 h with 10 μM of diiodinated antracene-*m*-carborane. Image obtained with the confocal laser scanning microscope [134] (pink circles in the cluster are B-H, black circles in the cluster are C or C-H and blue circles in the cluster are B).

Besides previous luminescent materials, our group has also prepared carbon-based nanomaterials, which consist of graphene oxide (GO) functionalized on the surface by monoiodinated cobaltabis(dicarbollide) (GO-I-COSAN) for in vivo bioimaging. This GO-I-COSAN has been synthesized using the cobaltabis(dicarbollide) containing a B-I group and an amino group (I-COSAN) that is subsequently labeled with radioactive ^{124}I (Scheme 5) for its use in positron emission tomography (PET) [135]. After incubation of HeLa cells with different concentrations of GO-I-COSAN for 48 h, the results indicated that the nanomaterial was not cytotoxic, with cell mortality lower than 10%. Remarkably, internalization of the nanomaterial by cells was clearly confirmed by transmission electron microscopy (TEM), which showed that the GO-I-COSAN was accumulated in the cytoplasm without causing changes in either the size or morphology of the cells. Further in vivo studies using *C. elegans* indicated that GO-I-COSAN was ingested by the worms, showing no significant damage and very low toxicity, which supports the results observed in vitro. Radioisotopic labeling of I-COSAN using a palladium-catalyzed isotopic exchange reaction with Na[^{124}I]I and its subsequent functionalization onto GO was performed successfully, leading to the formation of the radioactive nanocomposite GO-[^{124}I]I-COSAN (Scheme 5). The radiolabeled nanomaterial was injected into the mice, and PET images at different times were taken (Figure 16), which revealed no activity in the thyroid and stomach even at long times, indicating that iodide did not detach from the material. GO-[^{124}I]I-COSAN presented a favorable biodistribution profile, with long residence time on blood, mainly accumulated in the liver and slightly in the lung, and progressive elimination via the gastrointestinal tract. It is noteworthy that the high boron content of this material paves the way toward theranostics because it benefits traceable boron delivery for BNCT.

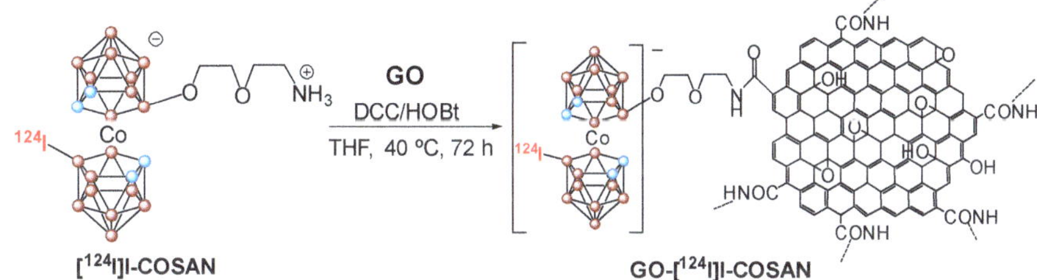

Scheme 5. Synthesis of the graphene oxide covalently bonded to the radiolabeled COSAN (pink circles in the cluster are B or B-H and blue circles in the cluster are C-H). Reprinted/adapted with permission from Ref. [135] Copyright 2021, American Chemical Society.

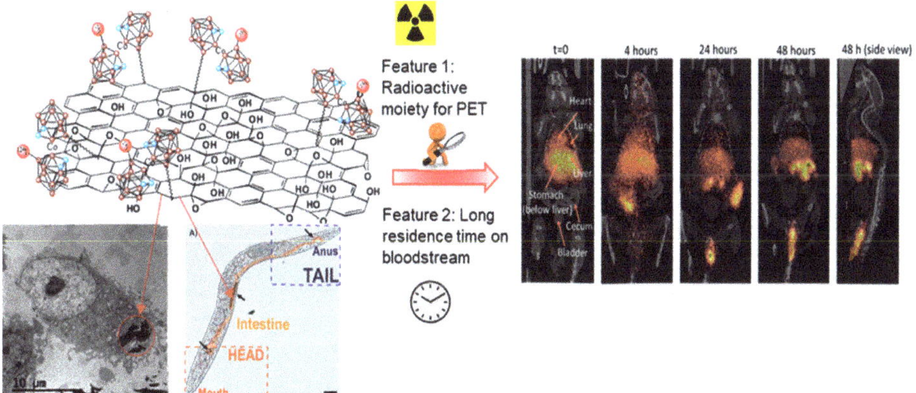

Figure 16. TEM image of GO-[^{124}I]I-COSAN internalized by HeLA cells, *C. elegans* incubated with GO-[^{124}I]I-COSAN and PET images of GO-[^{124}I]I-COSAN in the mice at different times (pink circles in the cluster are B or B-H, blue circles in the cluster are C-H and red circles exocluster ar the ^{124}I.). Reprinted/adapted with permission from Ref. [135] Copyright 2021, American Chemical Society.

Our group, in collaboration with S. Draper's group in Dublin, has also reported the preparation of transition metal-carborane photosensitizers by Sonogashira cross-coupling of (4-ethynylbenzyl)methyl-*o*-carborane with halogenated Ru(II)- or Ir(III)-phenanthroline complexes [136]. The resulting carboranyl-containing complexes (RuCB, IrCB, RuCB2, and IrCB2 in Figure 17) exhibited phosphorescence emission with maxima between 630 and 665 nm and lifetimes of 2.53, 0.38, 1.83, and 0.19 µs, respectively. All of them produce singlet oxygen with quantum yields (Φ_Δ) of 52%, 25%, 20%, and 10%, respectively, which suggests their use as triplet photosensitizers for photodynamic therapy (PDT). The subcellular uptake of all complexes was explored in SKBR-3 cells. Their localization and intensities were different depending on the number of carborane moieties and the nature of the transition metal centers. Complex IrCB was the best internalized with a clear accumulation in the cytoplasm. On the other hand, RuCB was hard to observe in the confocal microscopy images, but further microscopy experiments performed at a higher laser power showed that, in fact, RuCB was internalized. RuCB2 formed aggregates mainly located at the plasma membrane, whereas IrCB2 was poorly detected inside the cell (Figure 18). All of them showed the absence of dark toxicity under photodynamic therapy (PDT) conditions. Despite significant differences in the photophysical activities and cellular internalization of RuCB and IrCB, irradiation (λ_{ex} 405 nm; 3 min; mean intensity 55 µW) of both killed ~50% of SKBR-3 cells at 10 µM.

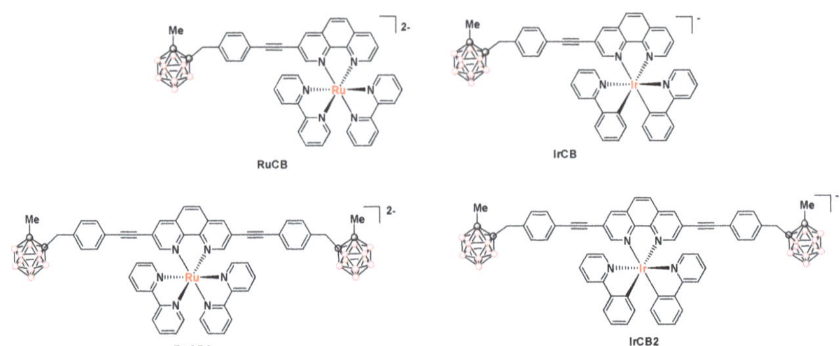

Figure 17. Structures of RuCB, IrCB, RuCB2, and IrCB2 (pink circles in the cluster are B-H and black circles in the cluster are C).

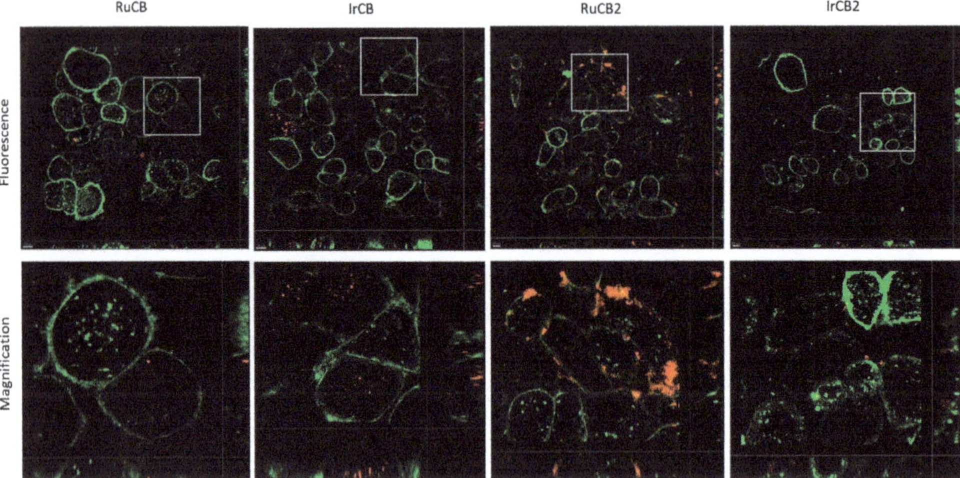

Figure 18. Orthogonal projection of z-stacks of live SKBR-3 cells incubated with 10 μM RuCB, IrCB, RuCB2, or IrCB2 for 4 h and observed under CLSM. (**Top row**): To analyze the localization of the compounds, fluorescence mode was used. Compound luminescence emission was detected in the range of 614–760 nm (red) by exciting the cells using the λ_{ex} 405 nm laser. Wheat Germ Agglutinin (WGA) fluorescence emission (membrane) was detected in the range of 496–579 nm (green) by exciting the cells using the λ_{ex} 488 nm laser. (**Bottom row**): Magnification of the selected areas (square boxes). Scale bar, 5 μm [136].

6. Conclusions

The progress in the synthesis of icosahedral boron clusters and their derivatives, the improvements in particles technology, the advances in medical imaging and computing, and the fact that new irradiation facilities are becoming available at hospitals makes radiotherapies such as BNCT and PBFR viable choices for new cancer medical therapies especially indicated for tumors resistant to chemotherapy and conventional radiotherapy. All this evidence promises to make BNCT and PBFR cutting-edge technology readily more accessible in the near future.

The fact that the icosahedral metallabis(dicarbollide) clusters reported in this review cross the mammalian membrane and have antimicrobial properties but low toxicity for mammalian cells (high selectivity index, SI) represents a promising tool to treat infectious intracellular bacteria. As there is an urgent need for antibiotic discovery and development, these small anionic molecules represent relevant and promising antimicrobial agents to tackle bacterial infections.

This review also gathers several families of boron clusters-based fluorophores with luminescent properties as potential theranostic agents for bioimaging and BNCT. Among them are a series of BODIPYs functionalized with either neutral or anionic boron clusters, a set of anthracene-*m*-carborane dyads, and a family of tin complexes linked to anionic boron clusters. All of them showed excellent fluorescence emission and high cellular uptake. The preparation and study of GO functionalized with radiolabeled cobaltabis(dicarbollide) for PET are described. To end, a set of Ru(II) and Ir(III)-phenanthroline photosensitizers bearing one or two Me-*o*-carborane cages, as well as the in vitro studies for PDT are reported.

Author Contributions: C.V. and F.T. developed the idea. C.V. and R.N. wrote the draft. C.V., F.T. and R.N. assembled the article and gave it in its final form. All authors have read and agreed to the published version of the manuscript.

Funding: This work research was funded by the European Union's Horizon 2020 grant number 768686, the Spanish Ministerio de Economía y Competitividad grant number PID2019-106832RB-I00 and, the Generalitat de Catalunya grant number 2017SGR1720.

Institutional Review Board Statement: Not applicable.

Informed Consent Statement: Not applicable.

Data Availability Statement: Not applicable.

Acknowledgments: This mini review is dedicated to Emeritus at University of Leeds John D. Kennedy on his 80th anniversary for his great contribution to the field of boron' clusters. F.T. and C.V. fondly remember the long and interesting evenings of discussion about B.-H.M. agostic bonds that we had in the early 1990s.

Conflicts of Interest: The authors declare no conflict of interest.

References

1. Davy, H. An account of some new analytical researches on the nature of certain bodies, particularly the alkalies, phosphorus, sulphur, carbonaceous matter, and the acids hitherto undecomposed: With some general observations on chemical theory. *Philos. Trans. Royal Soc.* **1809**, *99*, 39–104.
2. Berzelius, J. Undersökning af flusspatssyran och dess märkvärdigaste föreningar. *Proc. R. Soc.* **1824**, *12*, 46–98.
3. Garrett, D.E. *Borates: Handbook of Deposits, Processing, Properties, and Use*; Academic Press: San Diego, CA, USA, 1998.
4. Hobbs, D.Z.; Campbell, T.T.; Block, F.E. *Methods Used in Preparing Boron*; Bureau of Mines, US Department of the Interior: Pittsburgh, PA, USA, 1964.
5. World Boron Reserves as of 2022, by Major Countries. Available online: https://www.statista.com/statistics/264982/world-boron-reserves-by-major-countries/ (accessed on 26 April 2023).
6. Major Countries in Boron Production from 2010 to 2022. Available online: https://www.statista.com/statistics/264981/major-countries-in-boron-production/#statisticContainer (accessed on 26 April 2023).
7. Viñas, C. The Uniqueness of Boron as a novel challenging element for drugs in pharmacology, medicine and for smart biomaterials. *Future Med. Chem.* **2013**, *5*, 617–619. [CrossRef]
8. Poater, J.; Sola, M.; Viñas, C.; Teixidor, F. Hückel's Rule of Aromaticity Categorizes Aromatic closo Boron Hydride Clusters. *Chem. Eur. J.* **2016**, *22*, 7437–7443. [CrossRef]
9. Poater, J.; Solà, M.; Viñas, C.; Teixidor, F. Aromaticity and Three-Dimensional Aromaticity: Two sides of the Same Coin? *Angew. Chem. Int. Ed.* **2014**, *53*, 12191–12195. [CrossRef]
10. Poater, J.; Viñas, C.; Bennour, I.; Escayola-Gordils, S.; Solà, M.; Teixidor, F. Too Persistent to Give Up: Aromaticity in Boron Clusters Survives Radical Structural Changes. *J. Am. Chem. Soc.* **2020**, *142*, 9396–9407. [CrossRef]
11. The Nobel Prize in Chemistry 1976. Available online: www.nobelprize.org/prizes/chemistry/1976/summary/ (accessed on 26 April 2023).
12. Mukherjee, S.; Thilagar, P. Stimuli and shape responsive 'boron-containing' luminescent organic materials. *J. Mat. Chem. C* **2016**, *4*, 2647–2662. [CrossRef]
13. Lu, S.I.; Hamerton, I. Recent developments in the chemistry of halogen-free flame retardant polymers. *Prog. Polym. Sci.* **2002**, *27*, 1661–1712. [CrossRef]
14. Nuñez, R.; Romero, I.; Teixidor, F.; Viñas, C. Icosahedral boron clusters: A perfect tool for the enhancement of polymer features. *Chem. Soc. Rev.* **2016**, *45*, 5147–5173. [CrossRef] [PubMed]

15. Nuñez, R.; Tarrés, M.; Ferrer-Ugalde, A.; de Biani, F.F.; Teixidor, F. Electrochemistry and Photoluminescence of Icosahedral Carboranes, Boranes, Metallacarboranes, and Their Derivatives. *Chem. Rev.* **2016**, *116*, 14307–14378. [CrossRef] [PubMed]
16. Chujo, Y.; Tanaka, K. New Polymeric Materials Based on Element-Blocks. *Bull. Chem. Soc. Jpn.* **2015**, *88*, 633–643. [CrossRef]
17. Brand, R.; Lunkenheimer, P.; Loidl, A. Relaxation dynamics in plastic crystals. *J. Chem. Phys.* **2002**, *116*, 10386–10401. [CrossRef]
18. Plesek, J. Potential applications of the boron cluster compounds. *Chem. Rev.* **1992**, *92*, 269–278. [CrossRef]
19. Barth, R.F.; Soloway, A.H.; Fairchild, R.G.; Brugger, R.M. Boron Neutron-Capture Therapy for Cancer—Realities and Prospects. *Cancer* **1992**, *70*, 2995–3007. [CrossRef] [PubMed]
20. Hawthorne, M.F. The role of chemistry in the development of boron neutron-capture therapy of cancer. *Angew. Chem. Int. Ed.* **1993**, *32*, 950–984. [CrossRef]
21. Hawthone, M.F.; Maderna, A. Applications of radiolabeled boron clusters to the diagnosis and treatment of cancer. *Chem. Rev.* **1999**, *99*, 3421–3434. [CrossRef]
22. Valliant, J.F.; Guenther, K.J.; King, A.S.; Morel, P.; Schaffer, P.; Sogbein, O.O.; Stephenson, K.A. The medicinal chemistry of carboranes. *Coord. Chem. Soc.* **2002**, *232*, 173–230. [CrossRef]
23. Sivaev, I.; Bregadze, V.V. Polyhedral Boranes for Medical Applications: Current Status and Perspectives. *Eur. J. Inorg. Chem.* **2009**, *11*, 1433–1450. [CrossRef]
24. Leśnikowski, Z.J. Challenges and Opportunities for the Application of Boron Clusters in Drug Design. *J. Med. Chem.* **2016**, *59*, 7738–7758. [CrossRef]
25. Cerecetto, H.; Couto, M. Contemporary Diagnostic and Therapeutic Approaches. In *Glioma*; Omerhodzic, I., Arnautovic, K., Eds.; IntechOpen Ltd.: London, UK, 2019.
26. Hosmane, N.S. *Boron Science, New Technologies and Applications*, 1st ed.; CRC Press Taylor&Francis Group: Boca Raton, FL, USA, 2012. [CrossRef]
27. Axtell, J.C.; Saleh, L.M.A.; Qian, E.A.; Wixtrom, A.I.; Spokoyny, A.M. Synthesis and Applications of Perfunctionalized Boron Clusters. *Inorg. Chem.* **2018**, *57*, 2333–2350. [CrossRef]
28. Viñas, C.; Núñez, R.; Bennour, I.; Teixidor, F. Periphery Decorated and Core Initiated Neutral and Polyanionic Borane Large Molecules: Forthcoming and Promising Properties for Medicinal Applications. *Curr. Med. Chem.* **2019**, *26*, 5036–5076. [CrossRef]
29. Gozzi, M.; Schwarze, B.; Hey-Hawkins, E. Preparing (Metalla)carboranes for Nanomedicine. *ChemMedChem* **2021**, *16*, 1533–1565. [CrossRef] [PubMed]
30. Wolfgang, A.; Sauerwein, G.; Sancey, L.; Hey-Hawkins, E.; Kellert, M.; Panza, L.; Imperio, D.; Balcerzyk, M.; Rizzo, G.; Scalco, E.; et al. Theranostics in Boron Neutron Capture Therapy. *Life* **2021**, *11*, 330. [CrossRef]
31. Murphy, N.; McCarthy, E.; Dwyer, R.; Farràs, P. Boron clusters as breast càncer therapeutics. *J. Inorg. Biochem.* **2021**, *218*, 111412. [CrossRef]
32. Marfavi, A.; Kavianpour, P.; Rendina, L. Carboranes in drug discovery, chemical biology and molecular imaging. *Nat. Rev. Chem.* **2022**, *6*, 486–504. [CrossRef]
33. Hey-Hawkins, E.; Viñas, C. *Boron-Based Compounds: Potential and Emerging Applications in Medicine*; John Wiley & Sons, Ltd.: Oxford, UK, 2018.
34. ICMAB-CSIC. Available online: https://lmi.icmab.es/ (accessed on 26 April 2023).
35. Grimes, R.N. *Carboranes*, 3rd ed.; Elsevier Inc.: New York, NY, USA, 2016.
36. Teixidor, F.; Viñas, C.; Demonceau, A.; Nuñez, R. Boron clusters: Do they receive the deserved interest? *Pure Appl. Chem.* **2003**, *75*, 1305–1313. [CrossRef]
37. Olid, D.; Nuñez, R.; Viñas, C.; Teixidor, F. Methods to produce B-C, B-P, B-N and B-S bonds in boron clústers. *Chem. Soc. Rev.* **2013**, *42*, 3318–3336. [CrossRef] [PubMed]
38. Sivaev, I.B.; Bregadze, V.I.; Sjoberg, S. Chemistry of closo-dodecaborate anion $[B_{12}H_{12}](2-)$: A review. *Collect. Czechoslov. Chem. Commun.* **2002**, *67*, 679–727. [CrossRef]
39. Sivaev, I.B.; Bregadze, V.I. Chemistry of cobalt bis(dicarbollides). A review. *Collect. Czechoslov. Chem. Commun.* **1999**, *64*, 783–805. [CrossRef]
40. Sivaev, I.B.; Bregadze, V.I. Chemistry of nickel and iron bis(dicarbollides). A review. *Collect. Czechoslov. Chem. Commun.* **2002**, *614*, 27–36. [CrossRef]
41. Fisher, S.P.; Tomich, A.W.; Lovera, S.O.; Kleinsasser, J.F.; Guo, J.; Asay, M.J.; Nelson, H.M.; Lavallo, V. Nonclassical Applications of closo-Carborane Anions: From Main Group Chemistry and Catalysis to Energy Storage. *Chem. Rev.* **2019**, *119*, 8262–8290. [CrossRef] [PubMed]
42. Fisher, S.P.; Tomich, A.W.; Guo, J.; Lavallo, V. Teaching an old dog new tricks: New directions in fundamental and applied closo-carborane anion chemistry. *Chem. Commun.* **2019**, *55*, 1684–1701. [CrossRef]
43. Zhang, X.; Yan, H. Transition metal-induced B-H functionalization of o-carborane. *Coord. Chem. Rev.* **2019**, *378*, 466–482. [CrossRef]
44. Quan, Y.J.; Xie, Z.W. Controlled functionalization of o-carborane via transition metal catalyzed B-H activation. *Chem. Soc. Rev.* **2019**, *48*, 3660–3673. [CrossRef]
45. Qiu, Z.Z.; Xie, Z.W. A Strategy for Selective Catalytic B-H Functionalization of o-Carboranes. *Acc. Chem. Res.* **2021**, *54*, 4065–4079. [CrossRef]

46. Qiu, Z.Z.; Xie, Z.W. Functionalization of o-carboranes via carboryne intermediates. *Chem. Soc. Rev.* **2022**, *51*, 3164–3180. [CrossRef] [PubMed]

47. Dziedzic, R.M.; Spokoyny, A.M. Metal-catalyzed cross-coupling chemistry with polyhedral boranes. *Chem. Commun.* **2019**, *55*, 430–442. [CrossRef] [PubMed]

48. Endo, Y.; Iijima, T.; Yamakoshi, Y.; Yamaguchi, M.; Fukasawa, H.; Shudo, K. Potent Estrogenic Agonists Bearing Dicarba-closo-dodecaborane as a Hydrophobic Pharmacophore. *J. Med. Chem.* **1999**, *42*, 1501–1504. [CrossRef]

49. Endo, Y.; Iijima, T.; Yamakoshi, Y.; Kubo, A.; Itai, A. Structure-activity Study of Estrogenic Agonist Bearing Dicarba-closo-dodecaborane. Effect of Geometry and Separation Distance of Hydroxyl Groups at the Ends of Molecules. *Bioorg. Med. Chem. Lett.* **1999**, *9*, 3313–3318. [CrossRef]

50. Hawthorne, M.F. *Advances in Boron Chemistry*; Special Publication No. 201; Siebert, W., Ed.; Royal Society of Chemistry: London, UK, 1997; p. 261.

51. Endo, Y.; Yamamoto, K.; Kagechika, H. Utility of boron clusters for drug design. Relation between estrogen receptor binding affinity and hydrophobicity of phenols bearing various types of carboranyl groups. *Bioorg. Med. Chem. Lett.* **2003**, *13*, 4089–4092. [CrossRef]

52. Issa, F.; Kassiou, M.; Rendina, L.M. Boron in drug discovery: Carboranes as unique pharmacophores in biologically active compounds. *Chem. Rev.* **2011**, *111*, 5701–5722. [CrossRef] [PubMed]

53. Scholz, M.; Hey-Hawkins, E. Carbaboranes as pharmacophores: Properties, synthesis, application strategies. *Chem. Rev.* **2011**, *111*, 7035–7062. [CrossRef] [PubMed]

54. King, R.B. Three-dimensional aromaticity in polyhedral boranes and related molecules. *Chem. Rev.* **2001**, *101*, 1119–1152. [CrossRef] [PubMed]

55. Junqueira, G.M.A. Remarkable aromaticity of cobaltbis(dicarbollide) derivatives: A NICS study. *Theor. Chem. Acc.* **2018**, *137*, 92. [CrossRef]

56. Stoica, A.; Viñas, C.; Teixidor, F. Cobaltabisdicarbollide anion receptor for enantiomer-selective membrane electrode. *Chem Comun.* **2009**, *33*, 4988–4990. [CrossRef]

57. Stoica, A.; Viñas, C.; Teixidor, F. Application of the cobaltabisdicarbollide anion to the development of ion selective PVC membrane electrodes for tuberculosis drug analysis. *Chem. Commun.* **2008**, *48*, 6492–6494. [CrossRef]

58. Stoica, A.; Kleber, C.; Viñas, C.; Teixidor, F. Ion selective electrodes for protonable nitrogen containing analytes: Metallacarboranes as active membrane components. *Electrochim. Acta* **2013**, *113*, 94–98. [CrossRef]

59. Halima, H.B.; Baraket, A.; Viñas, C.; Zine, N.; Bausells, J.; Jaffrezic-Renault, N.; Teixidor, F.; Errachid, A. Selective Antibody-Free Sensing Membranes for Picogram Tetracycline Detection. *Biosensors* **2023**, *13*, 71. [CrossRef]

60. Garcia-Mendiola, T.; Bayon-Pizarro, V.; Zaulet, A.; Fuentes, I.; Pariente, F.; Teixidor, F.; Viñas, C.; Lorenzo, E. Metallacarboranes as tunable redox potential electrochemical indicators for screening of gene mutation. *Chem. Sci.* **2016**, *7*, 5786–5797. [CrossRef]

61. Pepiol, A.; Teixidor, F.; Saralidze, K.; van der Mare, C.; Willems, P.; Voss, L.; Knetsch, M.L.W.; Vinas, C.; Koole, L.H. A highly radiopaque vertebroplasty cement using tetraiodinated o-carborane additive. *Biomaterials* **2011**, *32*, 6389–6398. [CrossRef]

62. Farràs, P.; Cioran, A.M.; Sícha, V.; Teixidor, F.; Stíbr, B.; Gruner, B.; Viñas, C. Metallacarboranes as Building Blocks for Polyanionic Polyarmed Aryl-Ether Materials. *Inorg. Chem.* **2008**, *47*, 9497–9508. [CrossRef]

63. Teixidor, F.; Sillanpää, R.; Pepiol, A.; Lupu, M.; Viñas, C. Synthesis of Globular Precursors. *Chem. Eur. J.* **2015**, *21*, 12778–12786. [CrossRef] [PubMed]

64. Teixidor, F.; Pepiol, A.; Viñas, C. Synthesis of periphery-decorated and core-initiated borane polyanionic macromolecules. *Chem. Eur. J.* **2015**, *21*, 10650–10653. [CrossRef]

65. Teixidor, F.; Viñas, C. *Handbook of Boron Science. With Applications in Organometallics, Catalysis, Materials and Medicine*; Halogenated Icosahedral Carboranes: A Platform for Remarkable Applications; Hosmane, N., Eagling, R., Eds.; World Scientific Publishing: London, UK, 2019; Volume 1.

66. Couto, M.; Alamón, C.; García, M.F.; Kovacs, M.; Trias, E.; Nievas, S.; Pozzi, E.; Curotto, P.; Thorp, S.; Dagrosa, M.A.; et al. Closo-carboranyl-and Metallacarboranyl(1,2,3)-triazolyl-decorated Lapatinib-scaffold for Cancer Therapy Combining Tyrosine Kinase Inhibition and Boron Neutron Capture Therapy. *Cells* **2020**, *9*, 1408. [CrossRef] [PubMed]

67. Couto, M.; Alamón, C.; Nievas, S.; Perona, M.; Dagrosa, M.A.; Teixidor, F.; Cabral, P.; Viñas, C.; Cerecetto, H. Bimodal Therapeutic Agent against Glioblastoma, one the most Lethal Cancer. *Chem. Eur. J.* **2020**, *26*, 14335–14340. [CrossRef]

68. Alamón, C.; Dávila, B.; García, M.F.; Sánche, C.; Kovacs, M.; Trias, E.; Barbeito, L.; Gabay, M.; Zeineh, N.; Gavish, M.; et al. Sunitinib containing carborane pharmacophore with ability to inhibit tyrosine kinases receptors, FLT3, KIT, and PDGFR-β, exhibits powerful in vivo anti-glioblastoma activity. *Cancers* **2020**, *12*, 3423. [CrossRef]

69. Teixeira, R.G.; Marques, F.; Robaloc, M.P.; Fontrodona, X.; Garcia, M.H.; Geninatti Crich, S.; Viñas, C.; Valente, A. Ruthenium carboranyl complexes with 2,2′-bipyridine derivatives for potential bimodal therapy application. *RSC Adv.* **2020**, *10*, 16266–16276. [CrossRef] [PubMed]

70. Viñas, C.; Teixidor, F.; Nuñez, R. Boron clusters-based metallodendrimers. *Inorg. Chim. Acta* **2014**, *409*, 12–25. [CrossRef]

71. Cioran, A.M.; Musteti, A.D.; Teixidor, F.; Krpetić, Z.; Prior, I.A.; He, Q.; Kiely, C.J.; Brust, M.; Viñas, C. Mercaptocarborane-Capped Gold Nanoparticles: Electron Pools and Ion Traps with Switchable Hydrophilicity. *J. Am. Chem. Soc.* **2012**, *134*, 212–221. [CrossRef]

72. Oleshkevich, E.; Teixidor, F.; Rosell, A.; Viñas, C. Merging Icosahedral Boron Clusters and Magnetic Nanoparticles: Aiming toward Multifunctional Nanohybrid Materials. *Inorg. Chem.* **2018**, *57*, 462–470. [CrossRef]

73. Oleshkevich, E.; Morancho, A.; Galenkamp, K.M.O.; Grayston, A.; Gennatti Crich, S.; Alberti, D.; Protti, N.; Comella, J.X.; Teixidor, F.; Rosell, A.; et al. Combining magnetic nanoparticles and icosahedral boron clusters in biocompatible inorganic nanohybrids for cancer therapy. *Nanomedicine* **2019**, *20*, 101986. [CrossRef]

74. Grzelczak, M.P.; Danks, S.P.; Klipp, R.C.; Belic, D.; Zaulet, Z.; Kunstmann-Olsen, C.; Bradley, D.F.; Tsukuda, T.; Viñas, C.; Teixidor, F.; et al. Ion Transport across Biological Membranes by Carborane-Capped Gold Nanoparticles. *ACS Nano* **2017**, *11*, 12492–12499. [CrossRef]

75. Saha, A.; Oleshkevich, E.; Viñas, C.; Teixidor, F. Biomimetic Inspired Core–Canopy Quantum Dots: Ions Trapped in Voids Induce Kinetic Fluorescence Switching. *Adv. Mat.* **2017**, *29*, 1704238. [CrossRef]

76. Bauduin, P.; Prevost, S.; Farràs, P.; Teixidor, F.; Diat, O.; Zemb, T. A Theta-Shaped Amphiphilic Cobaltabisdicarbollide Anion: Transition From Monolayer Vesicles to Micelles. *Angew. Chem. Int. Ed.* **2011**, *50*, 5298–5300. [CrossRef]

77. Verdiá, C.; Alcaraz, A.; Aguilella, V.M.; Cioran, A.M.; Tachikawa, S.; Nakamura, H.; Teixidor, F.; Viñas, C. Amphiphilic COSAN and I2-COSAN crossing synthetic lipid membranes: Planar bilayers and liposomes. *Chem Commun* **2014**, *50*, 6700–6703. [CrossRef] [PubMed]

78. Tarrés, M.; Canetta, E.; Viñas, C.; Teixidor, F.; Harwood, A.J. Imaging in living cells using νB–H Raman spectroscopy: Monitoring COSAN uptake. *Chem. Commun.* **2014**, *50*, 3370–3372. [CrossRef] [PubMed]

79. Tarrés, M.; Canetta, E.; Paul, E.; Forbes, J.; Azzouni, K.; Teixidor, F.; Viñas, C.; Harwood, A.J. Biological interaction of living cells with COSAN-based synthetic vesicles. *Sci. Rep.* **2015**, *5*, 7804. [CrossRef] [PubMed]

80. Pedrosa, L.; Martinez-Rovira, I.; Yousef, I.; Diao, D.; Teixidor, F.; Stanzani, E.; Martinez-Soler, F.; Tortosa, A.; Sierra, A.; Gonzalez, J.J.; et al. Synchrotron-Based Fourier-Transform Infrared Micro-Spectroscopy (SR-FTIRM) Fingerprint of the Small Anionic Molecule Cobaltabis(dicarbollide) Uptake in Glioma Stem Cells. *Int. J. Mol. Sci.* **2021**, *22*, 9937. [CrossRef]

81. Fuentes, I.; Pujols, J.; Viñas, C.; Ventura, S.; Teixidor, F. Dual Binding Mode of Metallacarborane Produces a Robust Shield on Proteins. *Chem. Eur. J.* **2019**, *25*, 12820–12829. [CrossRef] [PubMed]

82. Goszczynski, T.M.; Fink, K.; Kowalski, K.; Lesnikowski, Z.J.; Boratynski, J. Interactions of Boron Clusters and their Derivatives with Serum Albumin. *Sci. Rep.* **2017**, *7*, 9800. [CrossRef]

83. Fuentes, I.; García-Mendiola, T.; Sato, S.; Pita, M.; Nakamura, H.; Lorenzo, E.; Teixidor, F.; Marques, F.; Viñas, C. Metallacarboranes on the Road to Anticancer Therapies: Cellular Uptake, DNA Interaction, and Biological Evaluation of Cobaltabisdicarbollide [COSAN]⁻. *Chem. Eur. J.* **2018**, *24*, 17239–17254. [CrossRef] [PubMed]

84. Merhi, T.; Jonchere, A.; Girard, L.; Diat, O.; Nuez, M.; Viñas, C.; Bauduin, P. Highlights on the Binding of Cobalta-Bis-(Dicarbollide) with Glucose Units. *Chem. Eur. J.* **2020**, *26*, 13935–13947. [CrossRef] [PubMed]

85. Gona, K.B.; Zaulet, A.; Gomez-Vallejo, V.; Teixidor, F.; Llop, J.; Viñas, C. COSAN as a molecular imaging platform: Synthesis and "in vivo" imaging. *Chem. Commun.* **2014**, *50*, 11415–11417. [CrossRef]

86. Nuez-Martinez, M.; Pinto, C.I.G.; Guerreiro, J.F.; Mendes, F.; Marques, F.; Muñoz-Juan, A.; Xavier, J.A.M.; Laromaine, A.; Bitonto, V.; Protti, N.; et al. Cobaltabis(dicarbollide) ([o-COSAN]−) as Multifunctional Chemotherapeutics: A Prospective Application in Boron Neutron Capture Therapy (BNCT) for Glioblastoma. *Cancers* **2021**, *13*, 6367. [CrossRef] [PubMed]

87. Winberg, K.J.; Barbera, G.; Eriksson, L.; Teixidor, F.; Tolmachev, V.; Viñas, C.; Sjöberg, S. High yield [¹²⁵I] iodide-labeling of iodinated carboranes by palladiumcatalyzed isotopic exchange. *J. Organomet. Chem.* **2003**, *680*, 188–192. [CrossRef]

88. Buades, A.B.; Pereira, L.C.J.; Vieira, B.J.C.; Cerdeira, A.C.; Waerenborgh, J.C.; Pinheiro, T.; Matos, A.P.A.; Pinto, C.G.; Guerreiro, J.F.; Mendes, F.; et al. The Mossbauer effect using Fe-57-ferrabisdicarbollide ([o-(57)FESAN](-)): A glance into the potential of a low-dose approach for glioblastoma radiotherapy. *Inorg. Chem. Front.* **2022**, *9*, 1490–1503. [CrossRef]

89. Paganetti, H. *Proton Therapy Physics*, 2nd ed.; Book series in Medical Physics and Biomedical Engineering; CRC Press: Boca Raton, FL, USA, 2020.

90. Yoon, D.-K.; Jung, J.-Y.; Suh, T.S. Application of proton boron fusion reaction to radiation therapy: A Monte Carlo simulation study. *Appl. Phys. Lett.* **2014**, *105*, 223507. [CrossRef]

91. Yoon, D.-K.; Naganawa, N.; Kimura, M.; Choi, M.-G.; Kim, M.-S.; Kim, Y.-J.; Law, M.W.-M.; Djeng, S.-K.; Shin, H.-B.; Choe, B.-Y.; et al. Application of proton boron fusion to proton therapy: Experimental verification to detect the alpha particles. *Appl. Phys. Lett.* **2019**, *115*, 223701. [CrossRef]

92. Shin, H.-B.; Yoon, D.-K.; Jung, J.-Y.; Kim, M.-S.; Suh, T.S. Prompt gamma ray imaging for verification of proton boron fusion therapy: A Monte Carlo study. *Phys. Med.* **2016**, *32*, 1271. [CrossRef]

93. Jung, J.-Y.; Yoon, D.-K.; Barraclough, B.; Lee, H.C.; Suh, T.S.; Lu, B. Comparison between proton boron fusion therapy (PBFT) and boron neutron capture therapy (BNCT): A Monte Carlo study. *Oncotarget* **2017**, *8*, 39774. [CrossRef]

94. Cirrone, G.A.P.; Manti, L.; Margarone, D.; Petringa, G.; Giuffrida, L.; Minopoli, A.; Picciotto, A.; Russo, G.; Cammarata, F.; Pisciotta, P.; et al. First experimental proof of Proton Boron Capture Therapy (PBCT) to enhance proton therapy effectiveness. *Sci. Rep.* **2018**, *8*, 1141. [CrossRef] [PubMed]

95. Nuez-Martınez, M.; Queralt-Martın, M.; Muñoz-Juan, A.; Aguilella, V.M.; Laromaine, A.; Teixidor, F.; Viñas, C.; Pinto, C.G.; Pinheiro, T.; Guerreiro, J.F.; et al. Boron clusters (ferrabisdicarbollides) shaping the future as radiosensitizers for multimodal (chemo/radio/PBFR) therapy of glioblastoma. *J. Mater. Chem. B* **2022**, *10*, 9727–9934. [CrossRef] [PubMed]

96. Popova, T.; Zaulet, A.; Teixidor, F.; Alexandrova, R.; Viñas, C. Investigations on antimicrobial activity of cobaltabisdicarbollides. *J. Organomet. Chem.* **2013**, *747*, 229–234. [CrossRef]

97. Fin, K.; Uchman, M. Boron cluster compounds as new chemical leads for antimicrobial Therapy. *Coord. Chem. Rev.* **2021**, *431*, 213684.
98. Romero, I.; Martinez-Medina, M.; Camprubi-Font, C.; Bennour, I.; Moreno, D.; Martinez-Martinez, L.; Teixidor, F.; Fox, M.A.; Viñas, C. Metallacarborane Assemblies as Effective Antimicrobial Agents, Including a Highly Potent Anti-MRSA Agent. *Organometallics* **2020**, *39*, 4253–4264. [CrossRef]
99. Bennour, I.; Ramos, M.N.; Nuez-Martínez, M.; Xavier, J.A.M.; Buades, A.B.; Sillanpää, R.; Teixidor, F.; Choquesillo-Lazarte, D.; Romero, I.; Martinez-Medina, M.; et al. Water soluble organometallic small molecules as promising antibacterial agents: Synthesis, physical–chemical properties and biological evaluation to tackle bacterial infections. *Dalton Trans.* **2022**, *51*, 7188–7209. [CrossRef]
100. Smith, B.R.; Gambhir, S.S. Nanomaterials for in vivo imaging. *Chem. Rev.* **2017**, *117*, 901–986. [CrossRef]
101. Fan, W.P.; Yung, B.; Huang, P.; Chen, X.Y. Nanotechnology for multimodal synergistic cancer therapy. *Chem. Rev.* **2017**, *117*, 13566–13638. [CrossRef]
102. Kunjachan, S.; Ehling, J.; Storm, G.; Kiessling, F.; Lammers, T. Noninvasive imaging of nanomedicines and nanotheranostics: Principles, progress, and prospects. *Chem. Rev.* **2015**, *15*, 10907–10937. [CrossRef]
103. Genady, A.R.; Ioppolo, J.A.; Azaam, M.M.; El-Zaria, M.E. New functionalized mercaptoundecahydrododecaborate derivatives for potential application in boron neutron capture therapy: Synthesis, characterization and dynamic visualization in cells. *Eur. J. Med. Chem.* **2015**, *93*, 574–583. [CrossRef]
104. Gao, D.; Aly, S.M.; Karsenti, P.-L.; Brisard, G.; Harvey, P.D. Application of the boron center for the design of a covalently bonded closely spaced triad of porphyrinfullerene mediated by dipyrromethane. *Dalton Trans.* **2017**, *46*, 6278–6290. [CrossRef]
105. Nakase, I.; Katayama, M.; Hattori, Y.; Ishimura, M.; Inaura, S.; Fujiwara, D.; Takatani-Nakase, T.; Fujii, I.; Futakie, S.; Kirihata, M. Intracellular target delivery of cell-penetrating peptide-conjugated dodecaborate for boron neutron capture therapy (BNCT). *Chem. Commun.* **2019**, *55*, 13955. [CrossRef] [PubMed]
106. Zhao, F.; Hu, K.; Shao, C.; Jin, G. Original boron cluster covalent with poly-zwitterionic BODIPYs for boron neutron capture therapy agent. *Polym. Test.* **2021**, *100*, 107269. [CrossRef]
107. Ruan, Z.; Liu, L.; Fu, L.; Xing, T.; Yan, L. An amphiphilic block copolymer conjugated with carborane and a NIR fluorescent probe for potential imaging-guided BNCT therapy. *Polym. Chem.* **2016**, *7*, 4411. [CrossRef]
108. Kalot, G.; Godard, A.; Busser, B.; Pliquett, J.; Broekgaarden, M.; Motto-Ros, V.; Wegner, K.D.; Resch-Genger, U.; Koster, U.; Denat, F.; et al. Aza-BODIPY: A new vector for enhanced theranostic boron neutron capture therapy Applications. *Cells* **2020**, *9*, 1953. [CrossRef] [PubMed]
109. Cabrera-González, J.; Xochitiotzi-Flores, E.; Viñas, C.; Teixidor, F.; García-Ortega, H.; Farfán, N.; Santillan, R.; Parella, T.; Núñez, R. High-Boron-Content Porphyrin-Cored Aryl Ether Dendrimers: Controlled Synthesis, Characterization, and Photophysical Properties. *Inorg. Chem.* **2015**, *54*, 5021–5031. [CrossRef]
110. Ferrer-Ugalde, A.; Juárez-Pérez, E.J.; Teixidor, F.; Viñas, C.; Núñez, R. Synthesis, Characterization, and Thermal Behavior of Carboranyl–Styrene Decorated Octasilsesquioxanes: Influence of the Carborane Clusters on Photoluminescence. *Chem. Eur. J.* **2013**, *19*, 17021–17030. [CrossRef]
111. Cabrera-González, J.; Sánchez-Arderiu, V.; Viñas, C.; Parella, T.; Teixidor, F.; Núñez, R. Redox active metallacarborane-decorated octasilsesquioxanes. Electrochemical and thermal properties. *Inorg. Chem.* **2016**, *55*, 11630–11634. [CrossRef]
112. Cabrera-González, J.; Ferrer-Ugalde, A.; Bhattacharyya, S.; Chaari, M.; Teixidor, F.; Gierschner, J.; Núñez, R. Fluorescent carborane vinylstilbene functionalised octasilsesquioxanes: Synthesis, structural, thermal and photophysical properties. *J. Mater. Chem. C* **2017**, *5*, 10211–10219. [CrossRef]
113. Cabrera-González, J.; Teixidor, F.; Viñas, C.; Núñez, R. Blue emitting star-shaped and octasilsesquioxane-based polyanions bearing boron clusters. Photophysical and thermal properties. *Molecules* **2020**, *25*, 1210. [CrossRef]
114. Cabana, L.; González-Campo, A.; Ke, X.; Van Tendeloo, G.; Núñez, R.; Tobias, G. Efficient Chemical Modification of Carbon Nanotubes with Metallacarboranes. *Chem. Eur. J.* **2015**, *21*, 16792–16795. [CrossRef]
115. Cabrera-González, J.; Cabana, L.; Ballesteros, B.; Tobias, G.; Núñez, R. Highly Dispersible and Stable Anionic Boron Cluster–Graphene Oxide Nanohybrids. *Chem. Eur. J.* **2016**, *22*, 5096–5101. [CrossRef] [PubMed]
116. Ferrer-Ugalde, A.; González-Campo, A.; Planas, J.G.; Viñas, C.; Teixidor, F.; Sáez, I.M.; Núñez, R. Tuning the Liquid Crystallinity of Cholesteryl-o-Carborane Dyads: Synthesis, Structure, Photoluminescence, and Mesomorphic Properties. *Crystals* **2021**, *11*, 133. [CrossRef]
117. Lerouge, F.; Viñas, C.; Teixidor, F.; Núñez, R.; Abreu, A.; Xochitiotzi, E.; Santillán, R.; Farfán, N. High Boron Content Carboranyl-Functionalized Aryl Ether Derivatives Displaying Photoluminescent Properties. *Dalton Trans.* **2007**, 1898–1903. [CrossRef] [PubMed]
118. Lerouge, F.; Ferrer-Ugalde, A.; Viñas, C.; Teixidor, F.; Núñez, R.; Abreu, A.; Xochitiotzi, E.; Santillán, R.; Farfán, N. Synthesis and Fluorescence Emission of Neutral and Anionic Di- and Tetra-Carboranyl Compounds. *Dalton Trans.* **2011**, *40*, 7541–7550. [CrossRef] [PubMed]
119. Ferrer-Ugalde, A.; Juárez-Pérez, E.J.; Teixidor, F.; Viñas, C.; Sillanpää, R.; Pérez-Inestrosa, E.; Núñez, R. Synthesis and Characterization of New Fluorescent Styrene-Containing Carborane Derivatives: The Singular Quenching Role of a Phenyl Substituent. *Chem. Eur. J.* **2012**, *18*, 544. [CrossRef]

120. Ferrer-Ugalde, A.; González-Campo, A.; Viñas, C.; Rodríguez-Romero, J.; Santillan, R.; Farfán, N.; Sillanpää, R.; Sousa-Pedrares, A.; Núñez, R.; Teixidor, F. Fluorescence of new o-carborane compounds with different fluorophores: Can it be tuned? *Chem. Eur. J.* **2014**, *20*, 9940–9951. [CrossRef]

121. Cabrera-González, J.; Viñas, C.; Haukka, M.; Bhattacharyya, S.; Gierschner, J.; Núñez, R. Photoluminescence in Carborane-Stilbene Triads: A Structural, Spectroscopic, and Computational Study. *Chem. Eur. J.* **2016**, *22*, 13588–13598. [CrossRef]

122. Ferrer-Ugalde, A.; Cabrera-González, J.; Juárez-Pérez, E.J.; Viñas, C.; Teixidor, F.; Pérez-Inestrosa, E.; Montenegro, J.M.; Sillanpää, R.; Haukka, M.; Núñez, R. Carborane-stilbene dyads: Influence of substituents and cluster isomers on the photoluminescent properties. *Dalton Trans.* **2017**, *46*, 2091–2104. [CrossRef]

123. Chaari, M.; Kelemen, Z.; Giner-Planas, J.; Teixidor, F.; Choquesillo-Lazarte, D.; Ben Salah, A.; Viñas, C.; Núñez, R. Photoluminescence in *m*-carborane–anthracene triads: A combined experimental and computational study. *J. Mater. Chem. C* **2018**, *6*, 11336–11347. [CrossRef]

124. Chaari, M.; Cabrera-González, J.; Kelemen, Z.; Ferrer-Ugalde, A.; Viñas, C.; Choquecillo-Lazarte, D.; Ben Salah, A.; Teixidor, F.; Núñez, R. Luminescence properties of carborane-containing distyrylaromatic systems. *J. Organomet. Chem.* **2018**, *865*, 206–213. [CrossRef]

125. Chaari, M.; Kelemen, Z.; Choquesillo-Lazarte, D.; Teixidor, F.; Viñas, C.; Núñez, R. Anthracene–styrene-substituted *m*-carborane derivatives: Insights into the electronic and structural effects of substituents on photoluminescence. *Inorg. Chem Front.* **2020**, *7*, 2370. [CrossRef]

126. Parejo, L.; Chaari, M.; Santiago, S.; Guirado, G.; Teixidor, F.; Núñez, R.; Hernando, J. Reversibly Switchable Fluorescent Molecular Systems Based on Metallacarborane–Perylenediimide Conjugates. *Chem. Eur. J.* **2021**, *27*, 270–280. Available online: https://ddd.uab.cat/record/266062 (accessed on 26 April 2023). [CrossRef] [PubMed]

127. Soldevila-Sanmartín, J.; Ruiz, E.; Choquesillo-Lazarte, D.; Light, M.E.; Viñas, C.; Teixidor, F.; Núñez, R.; Pons, J.; Planas, J.G. Tuning the architectures and luminescence properties of Cu(I) compounds of phenyl and carboranyl pyrazoles: The impact of 2D versus 3D aromatic moieties in the ligand backbone. *J. Mater. Chem. C* **2021**, *9*, 7643–7657. [CrossRef]

128. Sinha, S.; Kelemen, Z.; Hümpfner, E.; Ratera, I.; Malval, J. –P.; Piers Jurado, J.; Viñas, C.; Teixidor, F.; Núñez, R. o-Carborane-based fluorophores as efficient luminescent systems both as solids and as water-dispersible nanoparticles. *Chem Comm.* **2022**, *58*, 4016–4019. [CrossRef]

129. Chaari, M.; Gaztelumendi, N.; Cabrera-González, J.; Peixoto-Moledo, P.; Viñas, C.; Xochitiotzi-Flores, E.; Farfan, N.; Ben Salah, A.; Nogués, C.; Núñez, R. Fluorescent BODIPY-anionic boron cluster conjugates as potential agents for cell tracking. *Bioconjugate Chem.* **2018**, *29*, 1763–1773. [CrossRef]

130. Cabrera-González, J.; Muñoz-Flores, B.M.; Viñas, C.; Chávez-Reyes, A.; Dias, H.V.R.; Jiménez-Pérez, V.M.; Núñez, R. Organotin dyes bearing anionic boron clusters as cell-staining fluorescent probes. *Chem. Eur. J.* **2018**, *4*, 5601–5612. [CrossRef]

131. Bellomo, C.; Chaari, M.; Cabrera-González, J.; Blangetti, M.; Lombardi, C.; Deagostino, A.; Viñas, C.; Gaztelumendi, N.; Nogués, C.; Núñez, R.; et al. Carborane-BODIPY dyads: New photoluminescent materials through an efficient Heck coupling. *Chem. Eur. J.* **2018**, *24*, 15622–15630. [CrossRef]

132. Labra-Vázquez, P.; Flores-Cruz, R.; Galindo-Hernández, A.; Cabrera-González, J.; Guzmán-Cedillo, C.; Jiménez-Sánchez, A.; Lacroix, P.G.; Santillan, R.; Farfán, N.; Núñez, R. Tuning the cell uptake and subcellular distribution in BODIPY–carboranyl dyads: An experimental and theoretical study. *Chem. Eur. J.* **2020**, *26*, 16530–16540. [CrossRef]

133. Bellomo, C.; Zanetti, D.; Cardano, F.; Sinha, S.; Chaari, M.; Fin, A.; Maranzana, A.; Núñez, R.; Blangetti, M.; Prandi, C. Red light-emitting Carborane-BODIPY dyes: Synthesis and properties of visible-light tuned fluorophores with enhanced boron content. *Dyes Pigm.* **2021**, *194*, 109644. [CrossRef]

134. Chaari, M.; Kelemen, Z.; Choquesillo-Lazarte, D.; Teixidor, F.; Gaztelumendi, N.; Viñas, C.; Nogues, C.; Núñez, R. Efficient blue light emitting materials based on m-carborane-anthracene dyads. Structure, photophysics and bioimaging studies. *Biomat. Sci.* **2019**, *7*, 5324–5337. [CrossRef] [PubMed]

135. Ferrer-Ugalde, A.; Sandoval, S.; Reddy Pulagam, K.; Muñoz-Juan, A.; Laromaine, A.; Llop, J.; Tobias, G.; Núñez, R. Radiolabeled cobaltabis(dicarbollide) anion graphene oxide nanocomposites for in vivo bioimaging and boron delivery. *ACS Appl. Nano Mater.* **2021**, *4*, 1613–1625. [CrossRef]

136. Conway-Kenny, R.; Ferrer-Ugalde, A.; Careta, O.; Cui, X.; Zhao, J.; Nogués, C.; Núñez, R.; Cabrera-González, J.; Draper, S.M. Ru(II) and Ir(III) phenanthroline-based photosensitisers bearing o-carborane: PDT agents with boron carriers for potential BNCT. *Biomater. Sci.* **2021**, *9*, 5691–5702. [CrossRef] [PubMed]

MDPI

Review

Chemistry of Carbon-Substituted Derivatives of Cobalt Bis(dicarbollide)(1⁻) Ion and Recent Progress in Boron Substitution †

Lucia Pazderová [1], Ece Zeynep Tüzün [1,2], Dmytro Bavol [1], Miroslava Litecká [1], Lukáš Fojt [3] and Bohumír Grűner [1,*]

[1] Institute of Inorganic Chemistry of the Czech Academy of Sciences, 250 68 Řež, Czech Republic; pazderova@iic.cas.cz (L.P.); tuzun@iic.cas.cz (E.Z.T.); bavol@iic.cas.cz (D.B.); litecka@iic.cas.cz (M.L.)
[2] Department of Inorganic Chemistry, Faculty of Natural Science, Charles University, Hlavova 2030/8, 128 43 Prague, Czech Republic
[3] Institute of Biophysics of the Czech Academy of Sciences, Královopolská 135, 612 00 Brno, Czech Republic; fojtlukas@gmail.com
* Correspondence: gruner@iic.cas.cz
† This paper is dedicated to John D. Kennedy on the occasion of his 80th birthday in recognition of his outstanding contributions to metallaborane chemistry and his founder's role in the famous ACPC, which successfully continues after many decades.

Abstract: The cobalt bis(dicarbollide)(1⁻) anion (**1**⁻), [(1,2-C₂B₉H₁₁)₂-3,3′-Co(III)](1⁻), plays an increasingly important role in material science and medicine due to its high chemical stability, 3D shape, aromaticity, diamagnetic character, ability to penetrate cells, and low cytotoxicity. A key factor enabling the incorporation of this ion into larger organic molecules, biomolecules, and materials, as well as its capacity for "tuning" interactions with therapeutic targets, is the availability of synthetic routes that enable easy modifications with a wide selection of functional groups. Regarding the modification of the dicarbollide cage, syntheses leading to substitutions on boron atoms are better established. These methods primarily involve ring cleavage of the ether rings in species containing an oxonium oxygen atom connected to the B(8) site. These pathways are accessible with a broad range of nucleophiles. In contrast, the chemistry on carbon vertices has remained less elaborated over the previous decades due to a lack of reliable methods that permit direct and straightforward cage modifications. In this review, we present a survey of methods based on metalation reactions on the acidic C-H vertices, followed by reactions with electrophiles, which have gained importance in only the last decade. These methods now represent the primary trends in the modifications of cage carbon atoms. We discuss the scope of currently available approaches, along with the stereochemistry of reactions, chirality of some products, available types of functional groups, and their applications in designing unconventional drugs. This content is complemented with a report of the progress in physicochemical and biological studies on the parent cobalt bis(dicarbollide) ion and also includes an overview of recent syntheses and emerging applications of boron-substituted compounds.

Keywords: borane; carborane; dicarbollide; cobalt bis(dicarbollide); lithiation; metalation

Citation: Pazderová, L.; Tüzün, E.Z.; Bavol, D.; Litecká, M.; Fojt, L.; Grűner, B. Chemistry of Carbon-Substituted Derivatives of Cobalt Bis(dicarbollide)(1⁻) Ion and Recent Progress in Boron Substitution. *Molecules* **2023**, *28*, 6971. https://doi.org/10.3390/molecules28196971

Academic Editors: Michael A. Beckett and Igor B. Sivaev

Received: 21 August 2023
Revised: 27 September 2023
Accepted: 3 October 2023
Published: 7 October 2023

1. Introduction

The 3,3-*commo-closo*-bis(-1,2-dicarba-3-cobalt dodecaborane)(**1**⁻) anion, often abbreviated using the trivial name "cobalt bis(dicarbollide) anion" or "COSAN" (**1**⁻), was discovered through the pioneering work of M.F. Hawthorne in 1965. M. F. Hawthorne found that the open face of the deprotonated 7,8-dicarba-nido-undecaborate ion (dicarbollide ion) [C₂B₉H₁₁]²⁻ is iso-electronic with the cyclopentadiene ion, providing six electrons for bonding with a transition metal in a η⁵-fashion, similar to organic metallocenes [1,2]. Indeed, the cobalt bis(dicarbollide) ion possesses fully occupied bonding inner orbitals with an 18-electron structure, akin to organometallic ferrocene, making it the most stable compound within the series of metalla bis(dicarbollide) ions. However, significant differences arise, starting with Co(III) as the central atom, its oxidation state, ionic charge of the cluster, 3D architecture, overall size, presence of B-H and C-H sites in the molecule, and dipole moments. This compound can be synthesized through the deboronation of ortho-carborane, followed by deprotonation and treatment with cobalt chloride [2–5].

The chemistry of this ion has consistently garnered interest and, consequently, has been the subject of constant interest and has been the topic of several previous reviews and book chapters by, for example, Bregadze and Sivaev [6], Dash and Hosmane [7], and Grimes [8]. Its unique electronic structure, characterized by the presence of hydridic BH vertices, charge delocalization throughout the structure, and aromaticity, contributes to its high stability. The ion is marked by its ionic negative charge, diamagnetic character, low nucleophilicity, and very high thermal, chemical, and radiolytic stability. Its versatility is evident in various applications, including nuclear waste treatment [9,10], conducting polymers and coordination polymers [11], electrochemistry, ion-selective electrodes [11–14], catalysts, photocatalysts [15,16], and more. The ion (1^-) also exhibits hydrogen and dihydrogen bonding capabilities, which facilitate self-assembly [17–19], water solubility [20], and the formation of micelles and vesicles [21–23]. Moreover, it shows resistance to catabolism [24], easy penetration into cells, low toxicity in both in vitro and in vivo models [25–29], and the potential for unconventional types of interactions [30]. Therefore, this ion represents an attractive candidate for medicinal applications, such as inhibitors of therapeutically relevant enzymes like HIV Protease [31], cancer-associated Carbonic Anhydrase IX (CA-IX) [32], NO synthase [33], kinases [34], antibacterial drugs [35,36], and antimycotic agents [37], as well as a boron delivery platform for Boron Neutron Capture Therapy (BNCT) of cancer [38], modifications and labeling of DNA [39], and the development of electrochemical, photoredox, and radiochemical probes in diagnostics [24,40–43].

The derivatization of the cobalt bis(dicarbollide) anion can be achieved at both the carbon and boron sites of the cluster. However, the synthetic methods for these derivatives differ [6–8]. There has been greater emphasis on studying substitution at the boron atoms due to the availability of convenient synthetic pathways, generally proceeding via the Electrophile Induced Nucleophilic Substitution (EINS) mechanism [8]. Functionalization at the carbon atoms has been accessible only through indirect synthetic routes involving metal insertion into modified dicarbollide anions as ligands. In contrast to dicarba-*closo*-dodecaboranes (carboranes) [44] and the 1-carba-*closo*-dodecaborane ion [45], the CH sites in the cobalt bis(dicarbollide) anion appear less susceptible to metalation reactions, and, thus, direct modifications of carbon became available much later, generally over the course of the last decade. Although some early reports of successful lithiation can be found in the literature [46], reproducibility under the described reaction conditions was poor or rather inaccessible. Therefore, chemists working in this area almost lost hope, after many unsuccessful attempts, that direct synthesis via metalation and reactions with electrophiles could be a feasible method for the successful derivatization of the cage.

Interest was reborn after the first successful report on the synthesis of compounds containing carbon–silicon and carbon–phosphorus bonds by the group of Teixidor and Viñas [47–49]. Following the development of approaches and reaction conditions allowing for reliable lithiation and subsequent reactions, the direct synthesis of carbon-modified cobalt bis(dicarbollide) anions containing carbon–carbon, carbon–silicon, carbon–phosphorus or carbon–halogen bonds, and various terminal groups became accessible, forming the main focus of this review. Since the last comprehensive review on cobalt bis(dicarbollide) derivatives in 2017, when Dash and Hosmane et al. [7] published an overview of cobalt bis(dicarbollide) derivatives functionalized on both carbon and boron atoms, significant progress has been made in this field. The purpose of our review is, therefore, to summarize the main achievements in cobalt bis(dicarbollide) ion chemistry published since the last review and to provide a comprehensive overview of the developments in this area over recent years.

2. General Properties of the Cobalt Bis(dicarbollide) Ion

Cobalt bis(dicarbollide) is a low-spin diamagnetic (d^6) anionic complex composed of a Co(III) atom at its center, coordinated to two $[C_2B_9H_{11}]^{2-}$ ligands. The electrons are fully delocalized in the inner bonding orbitals, and the ion is considered an aromatic species [50–53]. While two known isomeric forms exist, $[(1,2-C_2B_9H_{11})_2-3,3'-Co(III)]^-$ and $[(1,7-C_2B_9H_{11})_2-2,2'-Co(III)]^-$, the number of potential isomers can reach 45 [9]. The latter isomer, with non-adjacent carbon atoms in the open $\{C_2B_3\}$ face of the dicarbollide ligands, corresponds to the thermodynamically more stable species [54]. Isomerization into the meta-isomer requires high energy and temperature, but the presence of substituents may reduce the activation barrier for isomerization [8].

The structure of the cobalt bis(dicarbollide) ion resembles a rod-like "peanut" shape with nanometer dimensions. The distance between terminal H(10) atoms in apical positions is approximately 1.2 nm long, and the cage is 0.72 nm wide [32]. Both dicarbollide ligands can rotate around the *z*-axis in a solution. The surface is covered with B-H bonds, with partial negative charges localized on the hydrogen atoms. Four C-H groups are slightly acidic, though however, the determination of the corresponding pK_a values is pending. XRD structures and chemical calculations indicate the presence of several discrete conformers in solid-state, including *transoid-(staggered)*, *gauche-(gauche)*, and *cisoid-(eclipsed)* geometries as the main energetically favorable forms (Figure 1). The fast rotation of both ligand planes around the central cobalt atom occurs in a solution. Density Functional Theory (DFT) calculations suggest that the energetic separation of the three rotamers, around 11 kJ mol^{-1}, and barriers for their interconversion, approximately 41 kJ mol^{-1} [54,55], are low. The rapid rotation of ligand planes in metalla bis(dicarbollide) ions is also supported by experiments; the most recent paper on a similar iron sandwich compound was published in 2022 [56]. Low energy interconversion precludes the isolation of the rotamers from the solution, however, these can be seen in solid state structures. The Cs$^+$ salt of the parent ion typically adopts a *transoid* conformation in solid-state structures, although structures of many other salts with alkyl ammonium cations in the Cambridge Structural Database (CCDC) show a *cisoid* arrangement. Most derivatives of ion **1**$^-$ in the CCDC usually adopt *cisoid* or *gauche* conformations with a wide variety of torsion angles. Only derivatives with heavier halogens and those heterosubstituted with Ph and I adopt a *transoid* conformation, attributed to the formation of B-H-Hal or B-H-Ar bonds [57,58]. As reported in the literature, a search of the CCDC Database revealed that approximately 80% of the deposited structures adopt a *cisoid* conformation, 17% are *gauche*, and only 5% are *transoid* arrangements of the carbon atoms [59]. Conformational mobility can be partly or fully restrained by the presence of two (or more) bulky substituents on the ligand planes sandwiching the cobalt ion [60] or due to a bridge substituent (*ansa*-substitution) interlinking the boron or carbon positions.

The cobalt bis(dicarbollide) ion has a high surface area to charge ratio, with maximum electron density located on the B(8), B(9), and B(12) skeletal vertices [61,62]. No recent computational analysis had been reported until 2005 [54]. A DFT analysis showed that the electronegativity difference of B and C increases the electron density around the carbons of the dicarbollide ligand, and the Highest Occupied Molecular Orbital (HOMO) of cobalt bis(dicarbollide) exhibits the expected predominant d(metal) and pronounced ligand character, respectively [54].

Cobalt bis(dicarbollide) can be reduced to its Co(II) dianion by sodium amalgam or metallic cesium and, subsequently, oxidized back into its Co(III) state [46]. Its ability to undergo a Co(II/III) redox process makes it suitable for use in electrochemical and electronic applications [63]. Its ability to undergo a reversible 1e$^-$ oxidation 1e$^-$ reduction cycle was first reported in 1971, suggesting an Electron transfer, Chemical reaction, and Electron transfer (ECE-type) mechanism supported by cyclic voltammetry [64].

^{11}B NMR is a comprehensive method for the identification of B and C-substituted carboranes and cobaltacarboranes. The paramagnetic Co(II) complex $[Co(C_2B_9H_{11})_2]^{2-}$ has a wide ^{11}B NMR range (from +132 to −66 ppm), while the diamagnetic Co(III) complex has peaks typically in the +6 to −23 ppm region [46]. The sp^n hybridization of dicarbollide ligands is expected to result in larger coupling constants compared to ligands of pure p-character in their bonding orbitals, such as in a {C$_5$H$_5$}$^-$ ion [65]. Despite distinguishable B(8) substitution, signals of the 18 BH vertices can often overlap, particularly those observed for B(4, 7, 9, and 12), which may complicate the assignment under practical circumstances. Bűhl et al. demonstrated that the ^{11}B NMR shifts in the spectra of cobalt bis(dicarbollide) and its derivatives can be predicted using DFT calculations within a relatively small deviation [54]. Theoretical nucleus-induced chemical shift (NICS) studies showed that cobalt bis(dicarbollide) derivatives possess high aromaticity, providing excellent stability and reactivity [50].

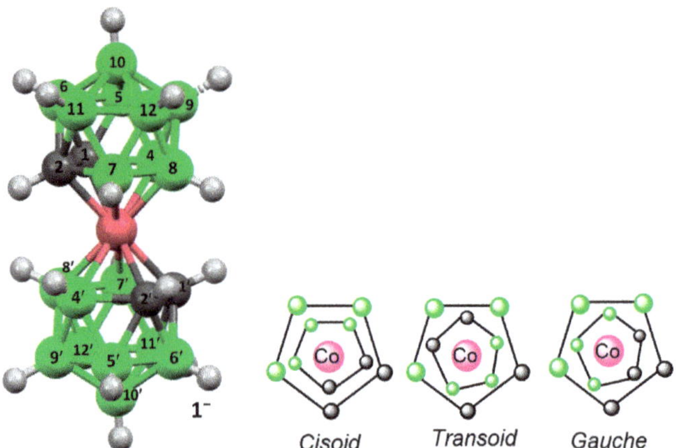

Figure 1. Schematic presentation of the molecular structure of cobalt bis(dicarbollide) ion (**1$^-$**) (from XRD structure of Cs$^+$ salt) presented in ref. [61] and three main energetically favored conformations [54]. Boron atoms are drawn in green, carbon in gray, and cobalt in rose color.

3. Carbon-Substituted Derivatives

In the chemistry of the neutral 1,2-dicarba-*closo*-dodecaborane, it is well-established that the properties of functional groups on carbon and boron atoms can differ significantly. For instance, the pK_a of functions such as NH$_2$, SH, and COOH, as example, attached to C(1) and B(9) positions in 1,2-dicarba-*closo*-dodecaborate can vary appreciably [66]. In general, groups on carbon atoms are notably more acidic than those bound to boron. A similar trend can be expected for groups on B and C atoms of the cobalt bis(dicarbollide) cage, although a systematic and sufficiently quantitative set of physicochemical data is not yet available.

Concerning amine derivatives, optimized structures and relative proton affinities have been calculated based on B3LYP and BP86 functionals along with conformational changes using BP86/6-31G* quantum-chemical computations [67]. The calculations were performed over the whole series of seven isomeric C(1), B(4), B(5), B(6), B(8), B(9), and B(10) substituted amines in *transoid*-conformation, as well as 11 possible isomers of C(1), C(2), B(4), B(5), B(6), B(7), B(8), B(9), B(10), B(11), and B(12) -NH$_2$ derivatives of the cobalt bis(dicarbollide) ion. The results clearly demonstrate a significant increase in the acidity of amino functions when present on carbon atoms. However, in a *transoid*-conformation, the B(6)-NH$_2$ group appears more acidic than the C(1)-NH$_2$ group. Conversely, when considering *cisoid*-arrangements, the latter becomes the most acidic. The energy difference between proton affinity in *cisoid*-conformation is 178 kJ mol^{-1} between C(1)-NH$_2$ and the most basic B(8)-NH$_2$.

Available experimental evidence supports this theory and shows that NH$_2$ groups bonded to boron atoms readily undergo protonation to -NH$_3$ under slightly acidic conditions or even at or below pH 7, producing stable zwitterionic compounds. However, the protonation of C-substituted amines with 3 M HCl proceeds differently and results in the respective hydrochlorides -NH$_2$·HCl instead, which is characterized by the formation of double salts. This feature has also been observed in several unpublished solid-state structures of amines that have recently been crystallized [68]; one example being [1,1′-(HCl·H$_2$N-CH$_2$-C$_2$B$_9$H$_{10}$)$_2$-3,3′-Co(III)]Me$_4$N (**2⁻**), schematically depicted in Figure 2. As a practical consequence, these compounds tend to behave as anions in solution.

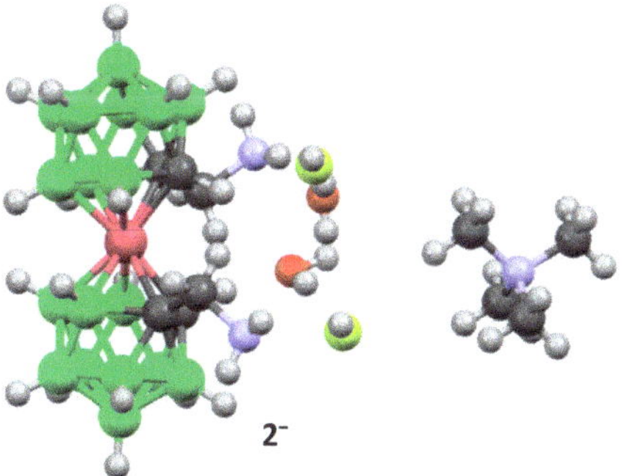

Figure 2. Schematic presentation of the molecular structure of the double salt of the formulation [1,1′-(HCl·H$_2$N-CH$_2$-C$_2$B$_9$H$_{10}$)$_2$-3,3′-Co(III)]Me$_4$N based on unpublished structural data [68]. Boron atoms are drawn in green, carbon in gray, cobalt in rose, chlorine in green-yellow, nitrogen in blue, and oxygen in red color.

Recent demands for fine-tuning interactions of cobalt bis(dicarbollide) derivatives with biological targets in medicinal chemistry have sparked increased interest in the synthesis and use of carbon-substituted derivatives as a viable solution, offering better control over the distance between functional groups and the cage. Indeed, C-substitution allows for the introduction of one or two substitution groups, which may be identical, or eventually different. When combined with boron substitution, this opens up a broad range of geometrical and functional possibilities. In practical terms, these derivatives have proven advantageous, for example, in the design of CA-IX inhibitors, demonstrating clear benefits which proved to have clearly better properties compared to boron-substituted derivatives, particularly in terms of greater flexibility in modifications and improved inhibitory properties (as discussed in Section 5.2.1 below) [32].

3.1. Indirect Substitutions via Modified 1,2-Dicarba-closo-dodecaborate

Indirect substitution pathways were extensively covered in previous reviews and books [6–8]. The first known methods for obtaining C-substituted derivatives involved the modifications of *ortho*-carborane followed by the degradation of specific derivatives into modified dicarbollide ions. This was followed by deprotonation with a strong base (typically NaH or KO^{t}Bu) in THF or DME and metal insertion via a reaction with anhydrous CoCl$_2$. Generally, this approach primarily leads to disubstituted compounds, typically obtained as a mixture of two diastereoisomers. The first is represented by the asymmetric (*racemic*) form, and the second corresponds to the form that has a plane of symmetry (*meso*) (see Section 3.3 on stereochemistry below). The substituents are then located in two ligand planes up and down the cobalt atom. This method can also be employed in the synthesis of monosubstituted compounds, provided that a mixture of modified and parent dicarbollide ligand is reacted with cobalt(II) chloride (or other Co(II) salts). However, its use is quite uneconomical since the reaction kinetics of the parent dicarbollide ion is faster than that of the substituted ligand. Therefore, a considerable amount of the parent sandwich ion **1⁻** forms in these reactions, along with some disubstituted products that start to form when the parent dicarbollide ion is spent.

Derivatives substituted on carbon atoms with two or four alkyls or phenyls were prepared using this approach. Additionally, a series of bridged *ansa*-derivatives were synthesized, where two carboranes or dicarbollide anions were interconnected via a chain, eventually containing heteroatoms, into which a cobalt atom was inserted (Figure 3). This category also includes Hawthorne's famous *ansa*-pyrazolyl Venus flytrap system designed for easy labeling with [60]Co(III) for radioimaging [69]. In addition, Hawthorne's group also synthesized bridged derivatives on cage carbons of cobalt bis(dicarbollide) with -propyl-, -butyl-, -pentyl-, and -TosN(CH$_2$CH$_2$) spacers via deboronation, deprotonation, and cobalt insertion steps with yields ranging from 21% to 61% [70]. They also similarly prepared -COC- and -CSC- bridged derivatives, with the latter being oxidizable into its corresponding sulfone derivative using meta-chloroperoxybenzoic acid in dichloromethane [71,72]. More recent studies by Nabakka et al. and Viñas et al. focused on preparing cobalt bis(dicarbollide) substituted with ethylene glycol units, which proved advantageous in partitioning ^{137}Cs, ^{90}Sr, and [152]Eu from simulated nuclear waste [72–74]. Recently, carbon-substituted dimethylsulfanyl derivatives [60] and some alkyl-sulfonamides alkylsulfonamides [75] were prepared using the indirect approach.

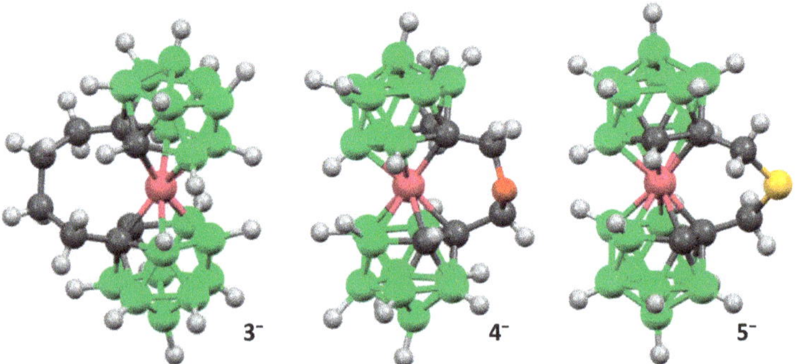

Figure 3. Schematic presentation of the molecular structures of (**3⁻**) *rac*-form of the butylene bridged derivative [70], (**4⁻**) *rac*-CH$_2$OCH$_2$- bridged [71], and (**5⁻**) *meso*-thioether bridged [72] cobalt bis(dicarbollide). Boron atoms are drawn in green, carbon in gray, cobalt in rose, sulfur in yellow, and oxygen in red color.

3.2. Direct Modifications

Since the cesium salt of cobalt bis(dicarbollide) has become widely available commercially, direct lithiation reactions offer advantages over the above indirect methods by allowing the synthesis of either mono- or disubstituted compounds in a single step. However, it is worth noting that the metalation with RLi requires strictly controlled low-temperature conditions, although the exact reason for this requirement remains unclear. Considering similar, but neutral, ferrocene, the acidity of the C-H group is significantly enhanced compared to common aromatic molecules, allowing for easy lithiation [76]. In the case of cobalt bis(dicarbollide), a similar effect can probably be anticipated, however, corresponding systematic studies are not yet available.

It is currently uncertain as to what extent C-Li interactions display an ionic character, as seen in the case of the [CB$_{11}$H$_{12}$]⁻ ion, which has been shown to stabilize a "naked" Li⁺ cation in a solution [77]. Complications in reacting lithiated species could be connected with the abstraction of Li⁺ by other species present in the solution, the aggregation of the RLi reagent depending on the specific solvent [78], or unspecific interactions of Li⁺ complexes in a solution with other sites on the boron cage. Some observations indicate that C-Li vertices can, under certain reaction conditions, follow unexpected pathways, such as nucleophilic reactions with R present in RLi, resulting in C-R substitution [68]. An example of a 1-butyl derivative isolated from a reaction with BuLi in DME is presented in Figure 4. Other examples include nucleophilic substitution on a carbonyl group observed in reactions with

N-(ω-bromoalkyl)phthalimides (see Section 3.7.3 below) [79] or the preferential formation of a C-Br bond in a reaction with BrCN (see Section 3.10) [80].

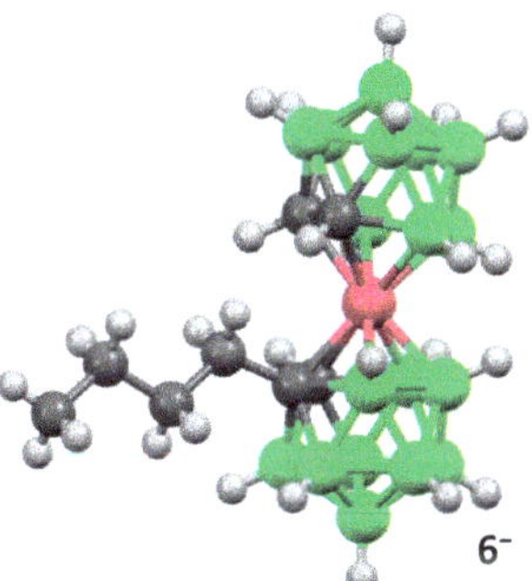

Figure 4. Drawing of the molecular structure of the butyl-substituted derivative (**6⁻**) based on XRD analysis [68]. Boron atomsare drawn in green, carbon in gray, and cobalt in rose color.

On the other hand, metalation, and subsequent reactions with electrophiles, offer advantages in terms of direct and rapid syntheses, the possibility of generating mono-substituted compounds, and the capability to introduce substituents that would interfere during metal insertion (due to coordination to cobalt), which may react with dicarbollide ligands or decompose under the highly alkaline conditions that are necessary when using indirect methods.

3.3. Stereochemistry of Substitution

In principle, there are four CH vertices available for metalation in cobalt bis(dicarbollide) ions. However, experimental trials and compounds reported in the literature [46,81] suggest that it is challenging to achieve more than disubstituted products, even when using a larger excess (up to six equivalents) of RLi and an appropriate ratio of reagents. Although small quantities, up to approximately 20%, of tri- and tetra-substituted species may form, these reactions do not typically go to completion. Several factors could contribute to this limitation, including the steric bulk of the Li^+ (Solvent)$_x$ moiety on the carbon vertex, the potential involvement of B-H interactions, or electronic repulsion of the lithiated C-vertices, which carry a partial negative charge.

In contrast to ferrocene chemistry [79], the *ortho*-metalation on the cobalt bis(dicarbollide) ion appears to be difficult, at least based on current knowledge. Indirect experimental evidence indicates that, in cases of disubstitution, typically two nonadjacent vertices are metalated and subsequently substituted, resulting in the formation of one or two diastereoisomers out of three possible options (Figure 5). The exact outcome depends (Scheme 1) on the reagent and reaction conditions. Most often, a mixture of *rac-* and *meso*-forms is obtained. Typically, the asymmetric *rac*-form, characterized by the highest spatial separation of the substituted carbon atoms, is the most abundant product. From a study published by Juarez-Perez et al. on the silylation of carbon vertices, it follows that the presence of the *meso*-form can increase with temperature [48]. Solvent choice can also favor the prevalence of *rac*-form, e.g., the replacement of dimethoxy ethane (DME) with diethoxy ethane (DEE) led to a higher ratio of the *rac*-form [80,81]. Some reagents, such as CO_2, $CH_3OCH_2CH_2OCH_2Cl$ (MEMCl) [46] in reaction with lithiated parent cobalt bis(dicarbollide), or trimethylene oxide, in case of the ion **1⁻** rigidified with an oxygen bridge, produced essentially pure *rac-* isomer [82]. This is likely due to increased steric strain around the reaction sites induced by the coordination of Li^+ to a specific solvent, or steric requirements of the initial substituent entering the carbon site and/or its coordination ability. In the latter case, the preference was connected with the inherent *cisoid*-arrangement of carbons in the bridged molecule [82], which apparently resulted in higher steric strain around the carbon atoms.

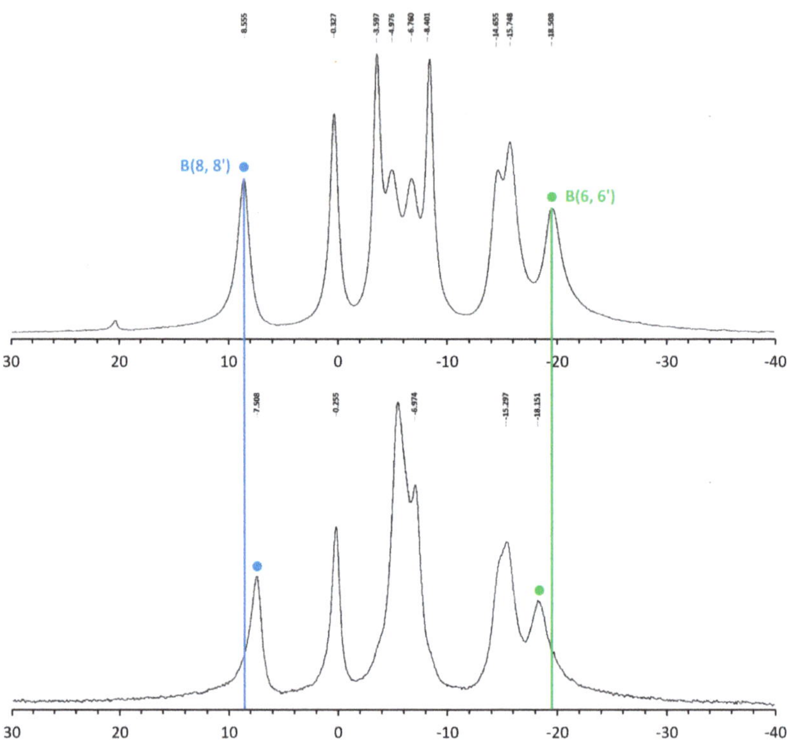

Figure 5. ^{11}B {^{1}H} NMR spectra of *rac-* (up) and *meso-*forms (down) of disulfonamidoethyl of **1**$^-$.

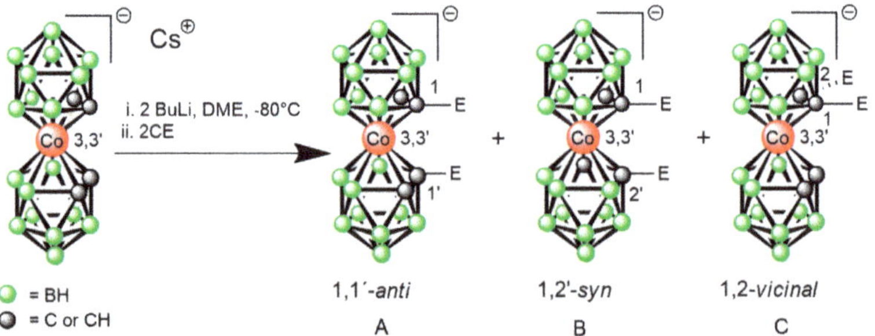

Scheme 1. Possible stereoisomers of carbon disubstituted species, E means a general substituent bind by simple bond. (**A**) *meso-*, (**B**) *rac-*, and (**C**) *vicinal*-isomer.

The presence of two (or three) isomers can be discerned from the ^{11}B, ^{1}H, and ^{13}C spectra [46,81]. An example of differences in ^{11}B NMR signals between *rac-* and *meso-*forms is illustrated in Figure 5. It can be seen that both forms exhibit very similar spectral patterns with the same number of nine boron signals. Only the peaks in the lowest and highest field corresponding to B(8,8′) and B(6,6′) may be, depending on the specific substituent, mutually and distinguishably shifted. Other ^{11}B signals for particular boron vertices exhibit only minor differences in chemical shifts. These signals are often overlapping, and overlap especially in the analysis of unseparated mixtures of diastereoisomers.

The pseudosymmetric spectral pattern in the ^{11}B -NMR of the *rac-*form is notably unexpected. Even so, the spectrum reflects the asymmetry of each ligand but corresponds to the magnetic equivalency of both dicarbollide cages in a solution. This results in the signal averaging in the NMR time scale due to time-averaged rotational disorder. The ^{11}B NMR spectrum of the symmetric *vicinal*-isomer consists of two shifted symmetric patterns of dicarbollide ligands in the spectrum; each part shows six boron signals of relative intensities 1:1:2:2:2:1 [46]. Each isomer typically shows one CH signal with an intensity of two in the corresponding ^{1}H and ^{13}C spectra, usually differing in chemical shifts. The ratios of isomers can be reliably analyzed using ion-pair reverse-phase (IP-RP) chromatography on RP columns modified with octyl or phenyl groups, which often allows for the baseline separation of the peaks of respective stereoisomers [83]. Compared to normal-phase chromatography, the separations on the RP column often provide better results when considering the preparative separations of diastereoisomers [75].

3.4. Chirality

Substitutions on the cobalt bis(dicarbollide) ion can readily induce asymmetry in the cage, leading to chirality. While mainly electroneutral compounds with chirality induced by asymmetric bridging units between boron atoms were previously described and resolved into enantiomers [83–86], carbon substitutions offer a broader range of possibilities for preparing chiral species.

Monosubstituted compounds with the general formulation $[(1\text{-}X\text{-}1,2\text{-}C_2B_9H_{10})(1',2'\text{-}C_2B_9H_{11})\text{-}3,3'\text{-}Co(III)]^-$ and $[(1\text{-}Y^+\text{-}1,2\text{-}C_2B_9H_{10})(1',2'\text{-}C_2B_9H_{11})\text{-}3,3'\text{-}Co(III)]^0$ are inherently chiral species displaying planar chirality. Trisubstituted derivatives can also exhibit the same type of asymmetry. In the case of disubstitutions with identical groups, of the substitutions with short or rigid bridges, and out of the three possible stereoisomers (Scheme 1, A: *rac*-, B: *meso*-, and C: *vicinal*-), only the *rac*-stereoisomer is inherently asymmetric and represents a prochiral species with axial chirality. Chiral molecules can also be obtained when two different substituents are present on carbon atoms in the *meso*-and *vicinal*-forms or when the second group is located on a boron atom.

Enantioseparations of these chiral ionic cobalt bis(dicarbollide) derivatives using chiral HPLC have been challenging, in particular, due to the unspecific interactions of the anions with silica-gel-based Chiral Stationary Phases (CSPs). Horáček and Kučera recently achieved the successful resolution of this issue by demonstrating that the addition of strong electrolytes to the mobile phase can effectively mitigate these effects [87]. The separation of approximately 16 enantiomeric pairs of chiral ionic, carbon-substituted, cobalt bis(dicarbollide) derivatives on different modified cellulose and cyclodextrin stationary phases was recently described using HPLC and SFC chromatographic techniques [82,88–90]. The majority of resolved compounds belong to the type with simple functional groups with preserved rotation of the ligand planes around the cobalt atom. Surprisingly, these compounds could be resolved with similar or even better selectivity and resolution compared to anions rigidified with bridges between $B(8,8')$ boron positions, indicating that the conformational mobility of the cage has little effect on the separation of enantiomers.

3.5. Derivatives and Structural Blocks Prepared through Metalation Reactions

The carbon-substituted compounds in question are categorized and discussed in the following sections based on the chemical origin of the terminal groups attached to the cage. This classification encompasses scenarios where these groups are either separated by a spacer group or not. Particularly in the context of medicinal applications, mostly compounds containing a linker have been studied due to the search for optimal interactions with biological targets.

Notably, compounds with functional groups directly attached to the cage carbon atoms, although they have received less attention, are of significant importance due to their distinct chemical and physicochemical properties. Indeed, even the terminal groups bound to the cage via a methylene spacer show chemically different properties compared to those with ethylene and longer pendant groups [32,91]. In general, the synthesis and isolation of compounds containing groups directly attached to carbon atoms pose various challenges due to the proximal steric and electronic effects of the bulky cobalt bis(dicarbollide) ion.

Currently, the first members of the series are the subject of extensive investigation in our laboratories.

3.5.1. Alkyl Derivatives

The first description of deprotonation and nucleophilic substitution reactions in carbon atoms was presented by Miller-Chamberlin et al., who utilized butyllithium lithiation followed by reactions with methylene iodide, hexyl bromide, or the glycolic reagent MEMCl. These reactions resulted in the alkylation of one to four cage carbon atoms [46]. The reactions were conducted in tetrahydrofuran (THF) at ambient temperature or at $-30\,°C$ for NMR experiments. Despite detailed NMR studies supporting the synthetic results of the cage deprotonation with BuLi and a thorough characterization of the products, later attempts by different research groups to reproduce these syntheses under the described conditions failed. The NMR results indicated the formation of purple or blue diamagnetic species resulting from the metalation of the C-H sites. Consequently, the reduction of the cobalt atom to Co(II) with BuLi was ruled out. The deprotonated sites kinetically determined the stereochemistry of the substitution reactions. The lithiated form was observed to undergo deuterium exchange with D_2O. In the case of disubstituted compounds, unseparated mixtures of *rac*- and *meso*-isomers were formed in a ratio of 2.5:1 for methyl and hexyl species, while the pure *rac*-diastereoisomer of the di-(MEM) derivative $[1,1'\text{-}(CH_3OCH_2CH_2OCH_2)_2\text{-}(1,2\text{-}C_2B_9H_{10})_2\text{-}3,3'\text{-}Co(III)]^-$ emerged as the sole product (**7**⁻, Figure 6). Larger ratios of BuLi exceeding two equivalents were reported to cause the degradation of the cage, leading to the release of metallic cobalt. The lithiation of boron-substituted mono- and dibromo/iodo-cobalt bis(dicarbollide) (followed by a reaction with MeI) did not yield any methylated derivatives. This can be explained by the interactions of large-sized and electron-rich Cl and Br atoms with the acidic C-H hydrogens in the *transoid* conformation preferred by the halide derivatives [46]. These compounds were synthesized for the purpose of their use in radionuclide extraction.

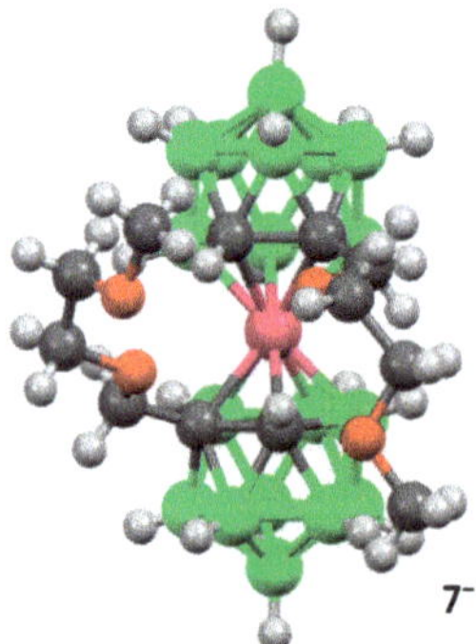

Figure 6. Schematic presentation of the molecular structure of the *rac*-isomer of di-(MEM)⁻ derivative presented in ref. [46]. Cesium atoms are omitted for clarity. Boron atoms are drawn in green, carbon in gray, cobalt in rose, and oxygen in red color.

3.5.2. Hydroxyalkyl Derivatives and Their Respective Esters

Reactions of lithiated cobalt bis(1,2-dicarbollide)(**1**⁻) in the presence of paraformaldehyde, ethylene oxide, or trimethylene oxide led to the substitution of **1**⁻ at the carbon atoms, resulting in the high-yield formation of monosubstituted alkylhydroxy derivatives $[(1\text{-}HO(CH_2)_n\text{-}1,2\text{-}C_2B_9H_{10})(1',2'\text{-}C_2B_9H_{11})\text{-}3,3'\text{-}Co(III)]$ (n = 1–3) (**8**⁻ **a, b, c**), along with disubstituted products of general formulation $[(HO(CH_2)_n\text{-}1,2\text{-}C_2B_9H_{10})2\text{-}3,3'\text{-}Co(III)]^-$ (n = 1–3) (**10**⁻ **a, b,** and **c**) (Scheme 2) [81]. Disubstituted compounds are, in fact, mixtures of 1,1'-*rac*- and 1,2'-*meso*-diastereoisomers. In the case of the longest pendant group in **10c**⁻, the 1,2-*vicinal*-isomer also formed. Initially, only the *rac*-isomer could be isolated in pure form through multiple fractional crystallizations from dichloromethane layered by hexane, in the case of shorter chain compounds **10a⁻** and **10b⁻**. Later, the *meso*-isomer of **10b⁻** and **10c⁻** was also isolated through a combination of crystallization and chromatography [32]. All these alkylhydroxy derivatives have been further utilized as structural blocks in the synthesis of various other derivatives in previous (see Section 5) and ongoing drug design.

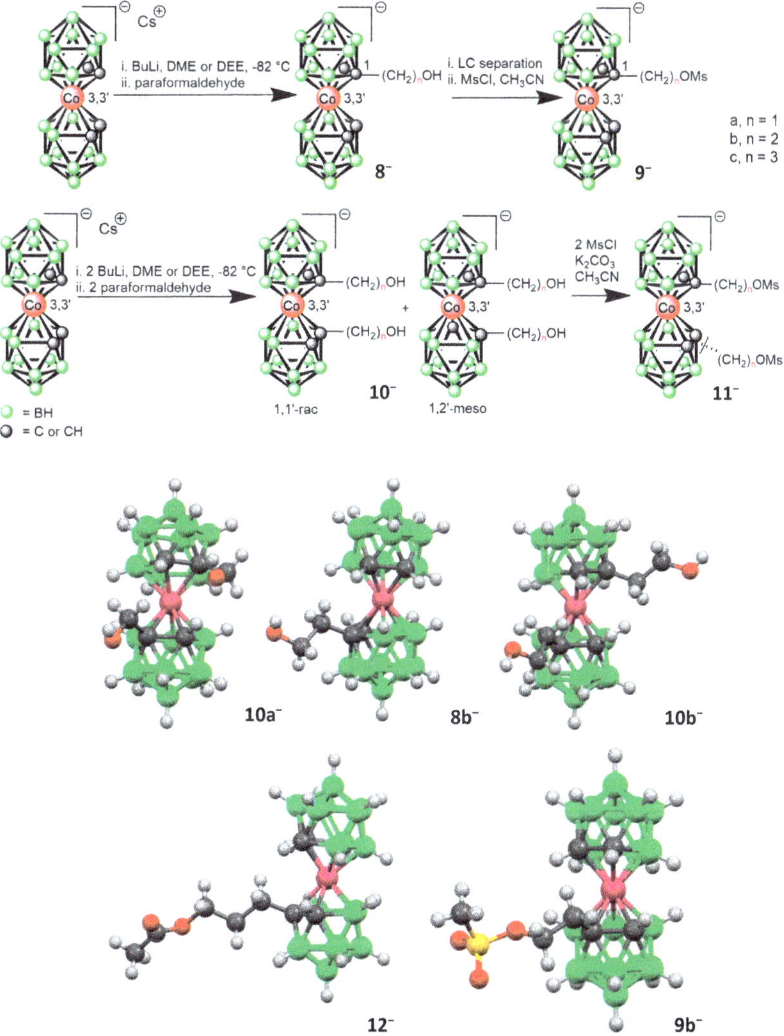

Scheme 2. Synthesis of mono- and disubstituted hydroxylalkyl and the respective mesyl esters of cobalt bis(dicarbollide) from the reaction of hydroxyalkyl derivatives with methylsulfonyl chloride; i. Cs_2CO_3 or K_2CO_3, CH_3CN or CH_3CN/toluene; ii. MsCl or TsCl, 45 or 80 °C; ii. MeOH, aq. Me_4NCl, drying; the schematic presentation of the molecular structures of compounds **10a⁻**, **8b⁻**, **10b⁻**, **12⁻**, and **9b⁻** [75,81]. Boron atoms are drawn in green, carbon in gray, and cobalt in rose, oxygen in red, and sulfur in dark yellow color.

These alcohols readily undergo esterification reactions with chlorides and anhydrides of carboxylic acids (see the acetyl ester in Scheme 2). In addition, esters of toluensulfonic or methylsulfonic acid, of the general formula $[1\text{-}X\text{-}O\text{-}(CH_2)_n\text{-}(1,2\text{-}C_2B_9H_{10})(1',2'\text{-}C_2B_9H_{11})\text{-}3,3'\text{-}Co(III))]Me_4N$ (where n = 1–3 and X = $-SO_2CH_3$ or $-SO_2(-C_6H_4\text{-}4\text{-}CH_3)$ and diesters

$[1,1'-(X-O-(CH_2)_n)_2-(1,2-C_2B_9H_{10})_2-3,3'-Co(III)]Me_4N$ (n = 1 and 2), are easily available in good to excellent yields using reactions of trimethylammonium salts of alkylhydroxy derivatives with methylsulfonyl or *p*-toluenesulfonyl chloride. Lower conversions were observed only for starting compounds with short methylene linkers [75]. The presence of Me_3NH^+ in the starting salt seems essential for attaining good yields of all compounds, due to the catalytic effect of the tertiary amines, a well-known concept in organic chemistry [92]. The Ms and Ts esters were further employed as structural blocks in the synthesis of the respective amines, azides, and sulfamides.

The low-temperature lithiation and reaction with trimethylene oxide were recently studied using the cobalt bis(dicarbollide) cage, which was rigidified by a polar oxygen atom positioned between boron positions (B8,8′), lying in the plane intersecting the bonds between the carbon atoms in dicarbollide ligands located across the central atom [82]. Therefore, the starting anion had a restricted rotation of the ligand planes that corresponded to the symmetric C_{2h} Point Group. The reaction procedure led to the high yield formation of mixtures of hydroxypropyl $[\mu-8,8'-O-(1-HOC_3H_6-(1,2-C_2B_9H_9)(1',2'-C_2B_9H_{10})-3,3'-Co(III))]^-$ (**13⁻**) and dihydroxypropyl $[\mu-8,8'-O-(1,1'-(HOC_3H_6)_2-(1,2-C_2B_9H_9)_2-3,3'-Co(III))]^-$ (**14⁻**) derivatives, with one product prevailing depending on the initial ratios of BuLi used in the reaction. The compounds were characterized using a combination of spectroscopic methods, X-ray crystallography, and chemical analysis. Dissimilarly to the parent cage, only the *racemic*-isomer of dihydroxypropyl derivative formed when the conformation mobility of the cobalt bis(dicarbollide) cage was restricted. As also follows from XRD analysis, the molecular structures of **13⁻** and **14⁻** are completely asymmetric due to substituent sites located on carbon atoms C1 and C(1,1′), respectively. Both types of this heterosubstitution are inherently chiral, and the enantiomers were resolved using analytical HPLC techniques on chiral stationary phases. The molecular structures of these compounds are depicted in Figure 7.

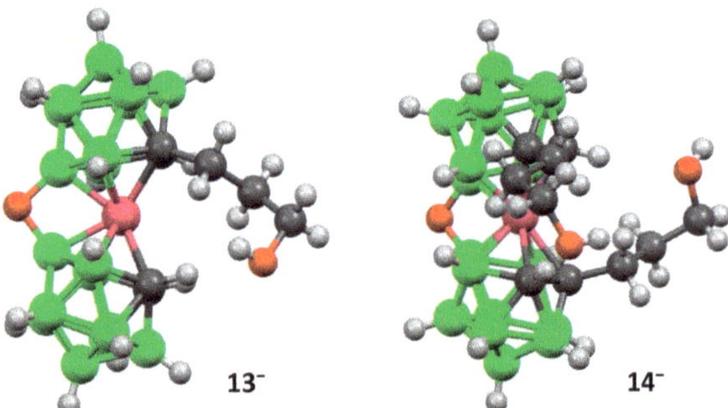

Figure 7. Schematic presentation of the ms of the anions heterosubstituted with one (**13⁻**) or two (**14⁻**) terminal hydroxyalkyl groups, and the oxygen bridge presented in ref. [82]. Cations are omitted for clarity. Boron atoms are drawn in green, carbon in gray, cobalt in rose, and oxygen in red color.

The Me_3NH^+ salts of the hydroxypropyl derivatives were easily converted to their corresponding methylsulfonyl (Ms) esters of formulae $[\mu-8,8'-O-[(1-(CH_3SO_2O-C_3H_6-1,2-C_2B_9H_9)(1,2-C_2B_9H_{10})-3,3'-Co(III)]^-$ and $[\mu-8,8'-O-(1,1'-(CH_3SO_2O-C_3H_6)_2-(1,2-C_2B_9H_9)_2-3,3'-Co(III))]^-$.

Recently, the feasibility of the reaction of lithiated cobalt bis(1,2-dicarbollide) with ethylene oxide was confirmed by Śmiałkowski et al. for the production of compounds containing substituents on both boron and carbon sites (2023) [93]. The starting compounds were B(8,8′) substituted derivatives that contained either two $-OEtOSi(CH_3)_2C(CH_3)_3$ groups or a phosphorothioate bridge that were substituted with a trityl-protected hydroxyethyl group at the sulfur atom (see Section 4). In both cases, the lithiation followed by reactions with ethylene oxide provided the expected type of substitution, with ethoxy groups on carbon atoms. Although low yields of the respective products were obtained, the results represent a rare example of simultaneous substitutions with terminal hydroxy groups bonded to the cage via an alkyl chain on two boron positions, along with the presence of another group on one or two carbons. The products substituted with four groups on two C- and two B-sites are reported to be a mixture of stereoisomers and enantiomers, and these mixtures could not be separated.

3.5.3. Carboxylic Acids and Esters

The first report on the synthesis of C-dicarboxylic acid appeared in the literature as early as in 1997 [94]. However, the paper primarily focused on polymer synthesis and lacked sufficient details on the synthesis and characterization of the precursor acids. The polyamide polymer backbone incorporating cobalt bis(dicarbollide) anion (15^-) was tested as a cation exchanger for the separation of $^{137}Cs^+$ and $^{90}Sr(II)$ from radioactive waste (Figure 8) [95]. This was paralleled by other studies of this team on radionuclide partitioning, which were conducted using extraction agents based on carbon alkylated [96] or boron halogenated [97] cobalt bis(dicarbollide) ions.

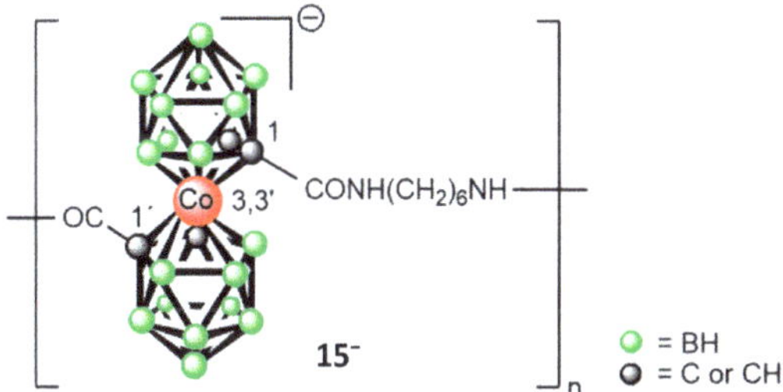

Figure 8. Condensation polymerization product of cobalt bis(dicarbollide) dicarboxylic acid [1-(CONH(CH$_2$)$_6$NH)-1'-(CO)$_2$-(1,2-C$_2$B$_9$H$_{10}$)$_2$-3,3'-Co(III)]$_n^-$ (15^-) [94,98].

Later, a detailed study was published on the reactions of lithiated cobalt bis(1,2-dicarbollide)(1^-) with carbon dioxide in DME that led to the substitution of 1^- on the C-atoms by carboxy function(s) [99]. This reaction pathway resulted, according to conditions, in good yields of the monosubstituted and disubstituted carboxylic acids of formulations [(1-HOOC-1,2-C$_2$B$_9$H$_{10}$)(1',2'-C$_2$B$_9$H$_{11}$)-3,3'-Co(III)]$^-$ (16^-) and [(HOOC)$_2$-(1,2-C$_2$B$_9$H$_{10}$)$_2$-3,3'-Co(III)]$^-$ (17^- a, b), respectively. The 1,1'-*rac*-isomer ($17a^-$) forms in this reaction as the main product, along with minor quantities of 1,2'-*meso*-isomer ($17b^-$, up to 10% by HPLC). Only the 1,1'-*rac*-isomer could be isolated in its pure form. The stereochemistry of these species was supported by the geometry optimizations and calculations of ^{11}B NMR shifts at the GIAO-DFT level. Scheme 3 presents a more recent XRD structure of the main 1,1'-*rac*-diastereoisomer that was measured and refined in the meantime [68]. This fully confirms its asymmetric structure. In addition, compounds where the carboxylic function is separated from the cage with different pendant groups were reported in this article. The first has a formula [(1-HOOC-CH$_2$-1,2-C$_2$B$_9$H$_{10}$)(1',2'-C$_2$B$_9$H$_{11}$)-3,3'-Co(III)]$^-$ ($19a^-$) and resulted from a lithiation followed by a reaction with BrCH$_2$COOEt and the hydrolysis of the respective ethyl ester (18^-). The second compound, which contains an ethylene pendant group, [(1-HOOC-(CH$_2$)$_2$-1,2-C$_2$B$_9$H$_{10}$)(1',2'-C$_2$B$_9$H$_{11}$)-3,3'-Co(III)]$^-$ ($19b^-$) between the cage and carboxylic group, was prepared by the oxidation of a hydroxypropyl derivative of the ion 1^-. *Para*-nitrophenyl esters were also prepared (20^-) as building blocks with the aim of the easy formation of amidic bonds between the boron cage and organic primary amino functions (Scheme 4). Their use was exemplified with model organic amines; butylamide and benzylamide [(1-RNHOC-(CH$_2$)$_n$-1,2-C$_2$B$_9$H$_{10}$)(1',2'-C$_2$B$_9$H$_{11}$)-3,3'-Co(III)]$^-$ (21^- a, b) (n = 1, 2; R = Bu, R = Bn) were isolated and characterized. In addition, the feasibility of the synthesis of compounds containing two cobalt bis(dicarbollide) cages, inter-connected via an amidic bond, was also demonstrated.

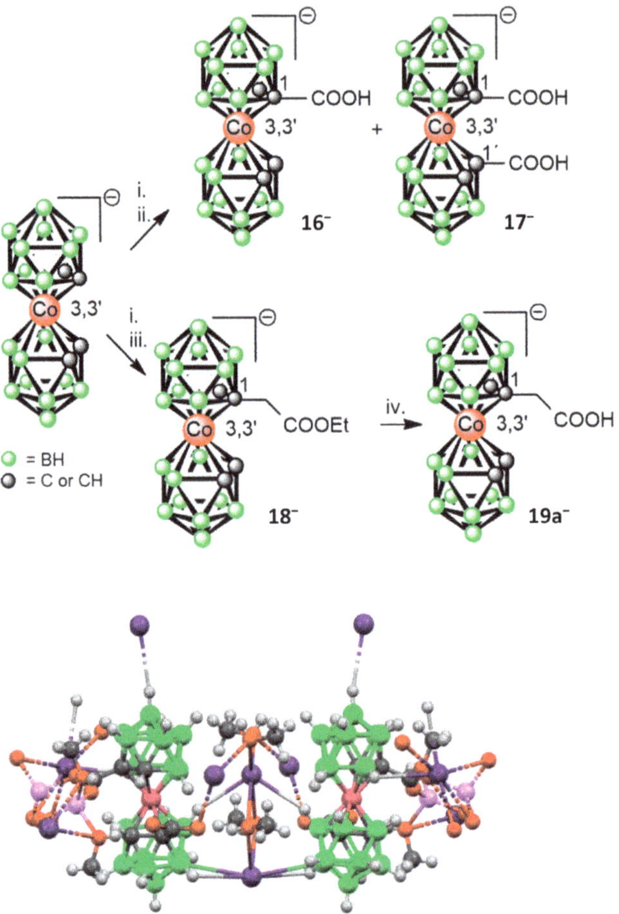

Scheme 3. Synthesis of the carboxylic acids **16⁻**, **17a⁻**, and **19a⁻** from **1⁻** by direct lithiation reactions. Reaction conditions: i. DME, −82 °C, n-BuLi; ii. CO₂(s); iii. BrCH₂COOEt at −82 °C; iv. aq. NaOH in EtOH, reflux. BH and CH hydrogen atoms are omitted for clarity. Schematic drawing of the molecular structure of the double Cs⁺/Li⁺ salt of dicarboxylic acid **17a⁻** based on unpublished crystallographic data [68]. Boron atoms are drawn in green, carbon in gray, cobalt in rose, oxygen in red, and Li⁺ and Cs⁺ cations in magenta and dark violet color, respectively. The compound was isolated by the crystallization of the reaction mixture from aqueous MeOH.

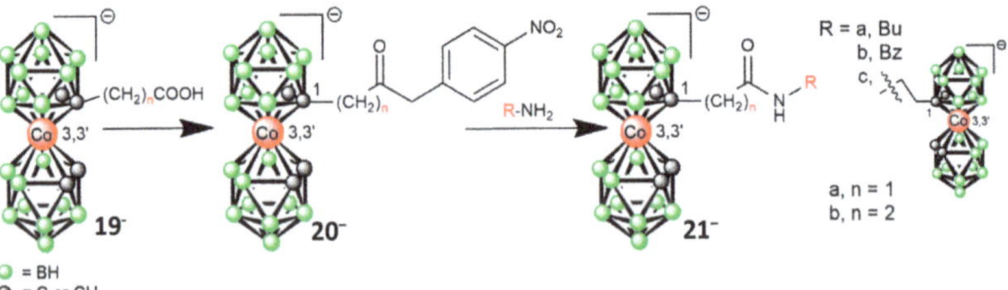

Scheme 4. Active esters of cobalt bis(dicarbollide)(**1⁻**) carboxylic acids allow for incorporation of this ion into functional molecules by amidic bonds [99].

3.6. Nitrogen Containing Compounds

3.6.1. Nitriles

Three compounds with terminal nitrile groups have been reported in the literature [80]. These were synthesized through reactions of Ms esters containing longer linkers with KCN in a polar solvent, such as DMF, resulting in the respective alkylnitriles in good to excellent yields (Scheme 5). Compounds with ethylene and propylene connectors between the nitrile group and the cage, with a general formula of $[(1\text{-}NC(CH_2)_n\text{-}1,2\text{-}C_2B_9H_{10})(1',2'\text{-}C_2B_9H_{11})\text{-}3,3'\text{-}Co(III)]^-$ (n = 2 and 3, Me$_4$N**22b** and Me$_4$N**22c**), were isolated as Me$_4$N$^+$ salts after chromatography. Using this procedure, a C,C-disubstituted member of the series of formula $[1,1'\text{-}(NC(CH_2)_2\text{-}1,2\text{-}C_2B_9H_{10})_2\text{-}3,3'\text{-}Co(III)]^-$ (Me$_4$N**23**) was also prepared that comprised two ethylnitrile groups in a *rac*-arrangement. The starting Ms ester was prepared from the pure *rac*-diastereoisomer of the corresponding alcohol, isolated through crystallization [25,81]. On the other hand, the synthesis of the first members of the series containing a short methylene connection failed, apparently due to the steric effects of the cage and ionic repulsion. Interestingly, procedures involving bromoacetonitrile or cyanogen bromide resulted in substitutions with bromine instead (see Section 3.10 on halogen derivatives). These compounds serve as widely applicable building blocks, and their reactivity was tested in dipolar cycloaddition reactions, resulting in the formation of the tetrazole motif.

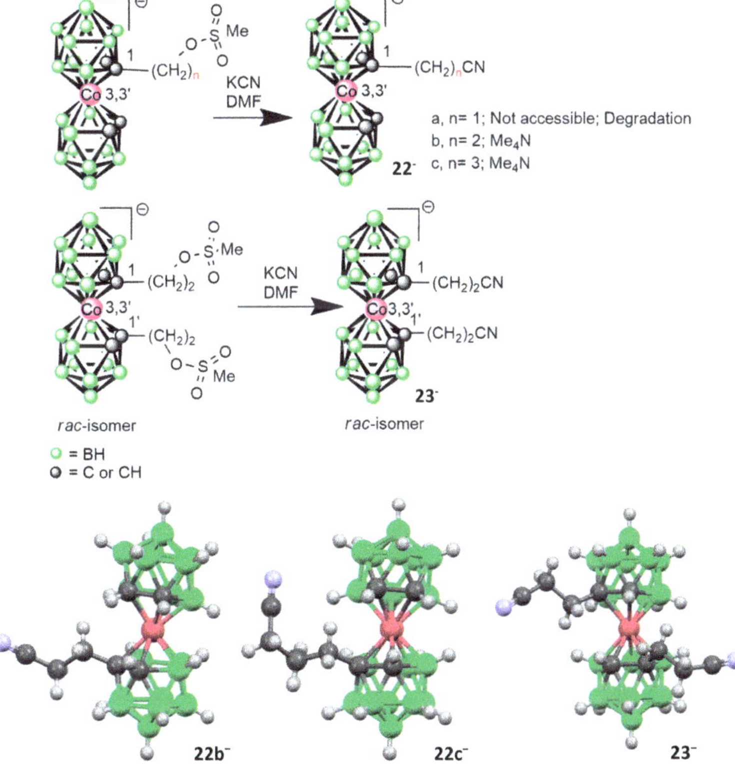

Scheme 5. Synthesis of alkylnitrile derivatives of the cobalt bis(dicarbollide) ion starting from mesyl esters, which were prepared according to ref. [91]. Schematic figures of molecular structures of three nitrile derivatives, determined using XRD and reported in ref. [80], are depicted at the bottom. Cations and solvent molecules are omitted for clarity. Boron atoms are drawn in green, carbon in gray, cobalt in rose, and nitrogen in blue color.

3.6.2. Azides

Monosubstituted propylazide and *rac*-isomer of dialkylazide [μ-8,8'-O- [(1-(N₃C₃H₆-1,2-C₂B₉H₉)(1,2-C₂B₉H₁₀)-3,3'-Co(III)]⁻ and dipropylhydroxy [μ-8,8'-O-(1,1'-(N₃C₃H₆-1,2-C₂B₉H₉)₂-3,3'-Co(III)]⁻ ($\mathbf{26^-}$, $\mathbf{27^-}$) were successfully prepared by the reaction of the corresponding Ms esters $\mathbf{24^-}$ and $\mathbf{25^-}$ with NaN₃ in dry dimethyl formamide (DMF) (Scheme 6). Only the azides derived from the bridge compound with conformationally restricted geometry have been reported thus far. These compounds were then studied as building blocks suitable for Huisgen–Sharpless dipolar [2 + 3] cycloadditions with organic alkynes [82]

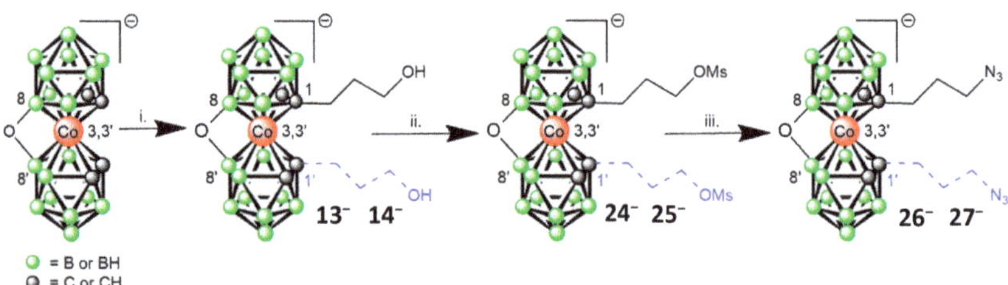

Scheme 6. Synthesis of C-alkylazides of the oxygen bridged cobalt bis(dicarbollide) ion. Reaction conditions: i. BuLi, trimethylene oxide, DME, −78 °C, then Me₃NH⁺; ii. MsCl, K₂CO₃, CH₃CN, 45 °C, then Me₄N⁺; iii. NaN₃, 45 °C [82].

3.6.3. Amines

C-substituted alkylamine derivatives of the cobalt bis(dicarbollide) ion are readily available when starting from primary hydroxyalkyl derivatives with varying lengths of the aliphatic spacer between the C(1) or C(1) and C(1') cage positions. The reaction pathways involve two steps: the conversion of hydroxyalkyl derivatives to methylsulfonyl or p-toluenesulfonyl esters and their subsequent reactions with ammonia or primary and secondary amines in toluene. These reactions resulted in compounds with primary, secondary, and tertiary alkylamine groups attached to the carbon C(1) cage positions. These synthetic routes also enable the easy disubstitution of the cobalt bis(dicarbollide) with two amine functions ($\mathbf{28^-}$). Less satisfactory results were obtained only for groups attached to the cage via a methylene linker. In particular, the dimesyl ester ($\mathbf{10a^-}$) was shown to react with ammonia and primary amines in a different pathway, producing bridge structures consisting of three atoms of the formula [(μ-(-CH₂NRCH₂)-(1,2-C₂B₉H₁₀)₂-3,3'-Co(III))]⁻ (R = H or Bu) ($\mathbf{29^-}$**a, b**). In our opinion, this synthetic route is the most reliable and convenient method for the synthesis of C-substituted amines, although three reaction steps are necessary. An example of disubstituted compounds is shown in Scheme 7 [99].

In principle, alkylamines are accessible from reactions of halogenoalkyl amines with protected amino functions (using phthalimido, *t*-Bu-carbonyl, or other protective groups). However, this reaction pathway is compromised by a nucleophilic substitution that may simultaneously proceed on the carbonyl group (if present in the particular protective group). Thus, only one successful reaction is described in the literature, which consists of the treatment of the lithiated anion $\mathbf{1^-}$ with bromomethyl phthalimide. This produces, after deprotection, methylene amine in a low yield of up to 18% (Scheme 8) [79].

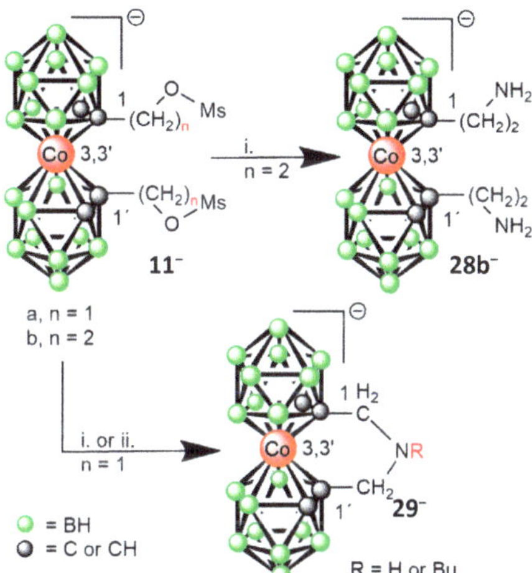

Scheme 7. Diethylamine substituted cobalt bis(dicarbollide) (**34⁻**) and bridge derivatives (**35⁻ a, b**) from the reactions of monosubstituted methylsulfonyl and p-toluenesulfonyl esters of cobalt bis(dicarbollide) with excess of ammonia, primary and secondary amines; i. NH₃ in excess, heating; ii. nBuNH₂, 60 °C; R = H or n-butyl [99].

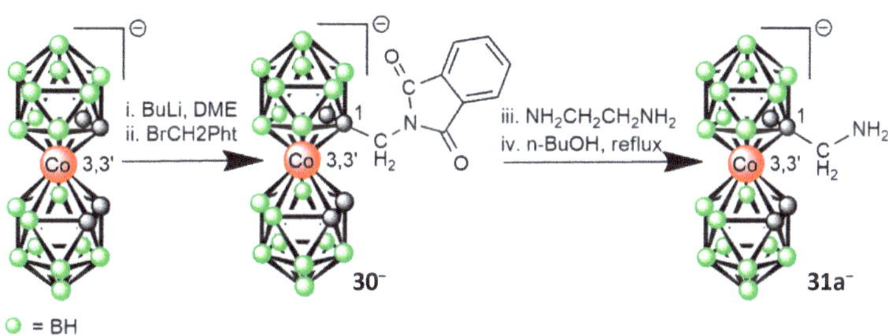

Scheme 8. Synthesis of aminomethyl (**32⁻**) cobalt bis(dicarbollide) by low temperature reaction of bromomethyl phthalimide with lithiated Cs**1** and subsequent cleavage of the protective group (**31⁻**). Alternatively, NaBH₄ in *i*-PrOH-water can be used, followed by hydrolysis by 3 M HCl in glacial acetic acid [79].

3.7. Nitrogen-Containing Heterocyclic Compounds

3.7.1. Triazines

El Anwar et al. reported a series of compounds heterosubstituted with 1*H*-1,2,3-triazole ring and an oxygen bridge between boron sites, which was synthesized starting from azidopropyl derivatives with conformationally restricted geometry by the presence of an oxygen bridge [82]. The main aim of this study has been the testing and development of reaction conditions suitable for the Huisgen–Sharpless [2 + 3] cycloaddition reactions of azides with alkynes. This would allow for an easy and reliable merging of cobalt bis(dicarbollide) ion with complex organic functional molecules and bioconjugations. The reactions of the boronated alkylazides with model alkynes such as phenylacetylene, ethynylsilane, trimethylsilane, propargyl amine, dimethylpropargyl malonate, propargyl glycine, and ethynylpyrene were tested. The optimized procedure comprised the use of CuI (5 mol%) together with *N,N*-diisopropylethylamine (DIPEA) in anhydrous ethanol at 37 °C over a period from 20 h to 5 days. In general, using these conditions, reactions of **36⁻** with most of the alkynes provided the expected products of the general formula $[\mu\text{-}8,8'\text{-}O\text{-}(1\text{-}(4\text{-}R\text{-}1,2,3\text{-triazolyl})\text{-}(1,2\text{-}C_2B_9H_9)(1',2'\text{-}C_2B_9H_{10})\text{-}3,3'\text{-Co(III)})]^-$ (**33⁻** to **36⁻**)

in good yields. Also, the feasibility of reliable disubstitutions was studied starting from diazidopropyl derivative **27⁻**. This was supported by the successful synthesis of compounds [μ-8,8′-O-[(1-(4-R-1,2,3-triazolyl-1,2-C$_2$B$_9$H$_9$)$_2$-3,3′-Co(III)]⁻ **38⁻** to **40⁻** (Scheme 9). Difficulties encountered due to the tedious isolations and the removal of side products could be seen in the case of **35⁻** and **40⁻**. This was reflected in the low yields of both latter compounds (12%). Drawing based on an unpublished XRD structure of tetrazole derivative **37⁻** is shown in Scheme 9, bottom. All these compounds are inherently chiral. This has been verified by the enantioseparation of a large part of these anionic species using HPLC techniques on chiral stationary phases [82].

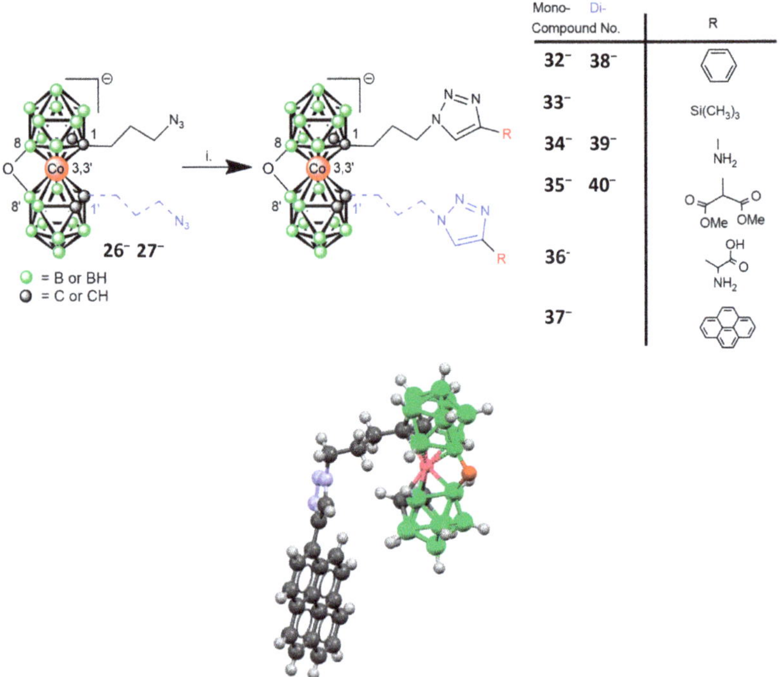

Scheme 9. Conjugation of azides with conformationally restricted geometry with organic functional groups via copper-catalyzed (2 + 3) cycloaddition reactions. Reaction conditions: CuI, DIPEA, alkyne, 37 °C, and the schematic drawing of the molecular structure of product **37⁻** containing pyrene substituted triazine ring (in bottom), which is based on unpublished XRD data [68]. Boron atoms are drawn in green, carbon in gray, cobalt in rose, oxygen in red and nitrogen in blue color.

3.7.2. Tetrazoles

El Anwar et al. (2020) used an advanced organocatalysis approach in the cycloaddition reactions of nitriles with azide ion [80]. Thus, L-proline catalyzed reactions of nitriles **22⁻b, c** with NaN$_3$ in DMF provided substitution of the cobalt bis(dicarbollide) ion with a tetrazole ring and compounds of the general formula [(1-(tetrazol-5-yl)-(CH$_2$)$_n$-1,2-C$_2$B$_9$H$_{10}$)(1′,2′-C$_2$B$_9$H$_{11}$)-3,3′-Co(III)]⁻ (**41⁻a, b**, Scheme 10). Prolonged heating at 130 °C for 4–11 days was needed, however, the products were obtained in good yields and purity after necessary purifications using chromatography and crystallization. In addition, disubstituted compound comprising two tetrazole rings per cobalt bis(dicarbollide) cage of formula [(1-(tetrazol-5-yl)-C$_2$H$_4$)-1,2-C$_2$B$_9$H$_{10}$)$_2$-3,3′-Co(III)]⁻ (**42⁻**, Scheme 4) was prepared in good yield, starting from *rac*-form of the di(ethylnitrile) **23⁻**. Alternatively, cycloadditions using a previously known isonitrilium derivative [8-R-CN-(1,2-C$_2$B$_9$H$_{10}$)(1′,2′-C$_2$B$_9$H$_{11}$)-3,3′-Co(III)]⁰ (**43a, b** R = Me, Ph) bonded via a cage boron atom B(8), were studied for comparison. The reactions with NaN_3 proceeded readily even without a catalyst in CH_3CN at room temperature, and afforded products of the formulae [(8-(5-Me-tetrazol-1-yl)-1,2-$C_2B_9H_{10}$)(1′,2′-$C_2B_9H_{11}$)-3,3′-Co(III)]$^-$ (**44a**$^-$) and [(8-(5-Ph-tetrazol-1-yl)-1,2-$C_2B_9H_{10}$)(1′,2′-$C_2B_9H_{11}$)-3,3′-Co(III)]$^-$ (**44b**$^-$) which were obtained in good isolated yields. On the other hand, the corresponding reactions with organic azides could be accomplished in neither case, despite a variety of tested catalysts and conditions. The molecular structures of two compounds corresponding to both types of tetrazole derivatives Me$_3$NH**41b** and Me$_3$NH**44a** were determined using single-crystal X-ray diffraction analysis and are presented in Scheme 10, bottom. The 5-substituted tetrazole species belongs to key structural motifs that mimic carboxylic functions and provide additional interactions and hydrogen bonding, which has been widely introduced into drug design [100]. The structural modifications of the cobalt bis(dicarbollide) ion with a tetrazole ring may help to further tune its properties as a hydrophobic pharmacophore.

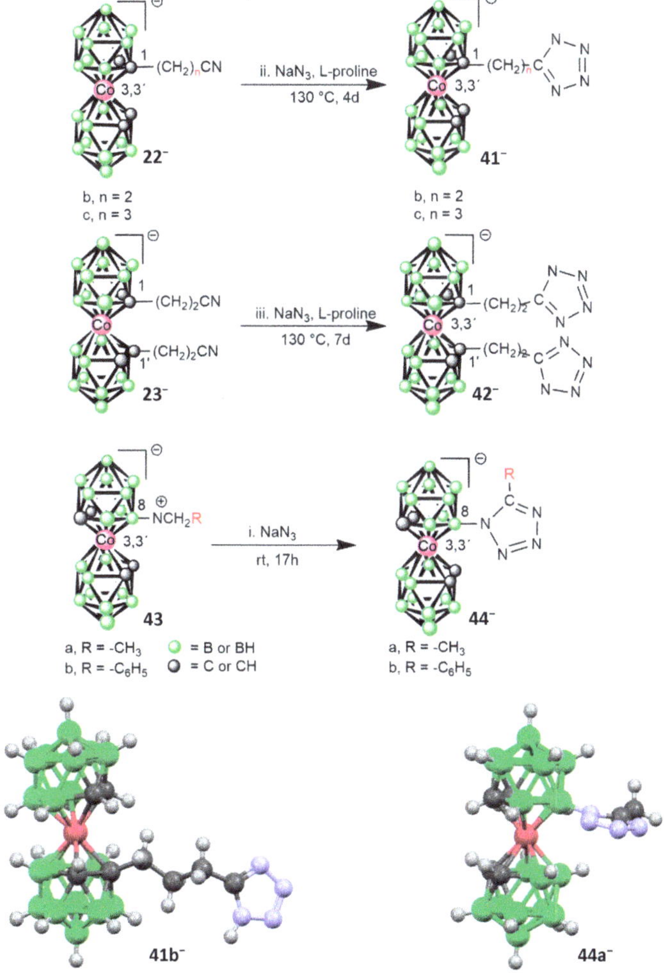

Scheme 10. Formation of tetrazoles from nitrilium and isonitrilium derivatives of cobalt bis(dicarbollide) and sodium azide. The drawings of the molecular structures in the bottom part are based on structural data from ref. [80]. Boron atoms are drawn in green, carbon in gray, cobalt in rose, and nitrogen in blue color.

3.7.3. Isoindolones and Related Species

Grűner et al. reported that the low-temperature lithiation of cobalt bis(dicarbollide) ion and subsequent reactions with N-(ω-bromoalkyl)phthalimides, Br-$(CH_2)_n$-$N(CO)_2NC_6H_4$ (where n = 2 and 3) give compounds substituted at cage carbon atoms by tricyclic isoindolone moieties, with five- (**45a⁻**) or six-member (**45b⁻**) lateral oxazine rings as the predominant products. In addition, two minor products (**46⁻**) were isolated that correspond to unusual cyclic substitution. Anion **46⁻** contains a bicyclic benzofuranone ring attached by a quaternary carbon in a bridging manner, and anion **46⁻** consists of a ketobenzoic acid amide substituent with a lateral azetidine ring [79]. The structures of all the isolated cyclic products were determined using single-crystal X-ray crystallography. As depicted in Scheme 11, the first step in the main reaction mechanism consists of the nucleophilic addition of the lithiated carbon of the metallacarborane to the carbonyl moiety present in the phthalimide group. This is followed by the intramolecular cyclization process of the alkylbromide end of the molecule, which results in the formation of a tricyclic heterocycle of the isoindolone type. Only the later stages of the reaction correspond to an SN_2 mechanism in which the nucleophilic alcoholate oxygen reacts with the alkylbromide end and forms the lateral ring in the isoindolone scaffold (see Scheme 11). The formation of a stable pentagonal or hexagonal ring apparently contributes significantly to the observed cyclization pathways. The mechanism that rationalizes the formation of the isoindolone rings thus closely resembles that observed previously for other metalated species such as phenyllithium [101] or lithiated 1-phenyl-1,2-carborane [102].

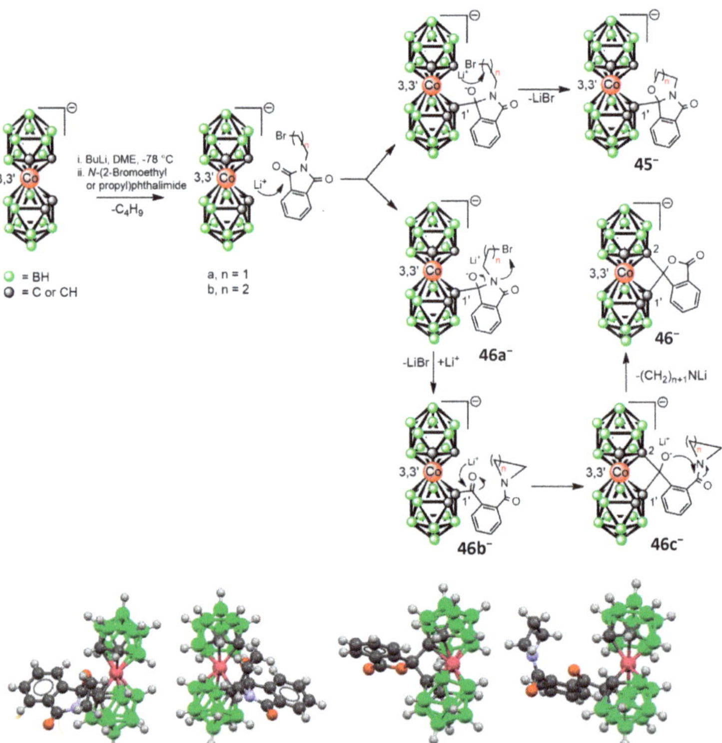

Scheme 11. Proposed mechanism for the formation of the main cyclic products **45⁻** and **46⁻** via low temperature reaction of bromoethyl or bromopropyl phthalimide with lithiated Cs1 and the schematic images of molecular structures of the compounds determined by XRD and presented in ref. [79], cations and solvent atoms are omitted for clarity. Boron atomsare drawn in green, carbon in gray, cobalt in rose, nitrogen in blue, and oxygen in red color.

3.8. Silicon Derivatives

Teixidor's group reported the synthesis of bridged and non-bridged dimethylsilyl derivatives of cobalt bis(dicarbollide) anion after lithiation at low temperatures [48,103]. Dimethylsilyl-bridged derivative can be obtained using Me_2SiCl_2 or even Me_2SiHCl in high yield due to the loss of H_2 from the acidic proton of the cage carbon and hydridic Si-H, which results in bridged substitution between carbon sites, compound $[1,1'-\mu-(Me_2Si)-(1,2-C_2B_9H_{10})_2-3,3'-Co(III)]^-$ (**48⁻**). The non-bridged mono-substituted derivatives can be obtained in lower yields compared to their bridged derivatives (Scheme 12). The non-bridged mono-substituted derivative $[(1-Me_3Si-1,2-C_2B_9H_{10})(1',2'-C_2B_9H_{11})-3,3'-Co(III)]^-$ (**47⁻**) and disilylated compound $[1,1'-(Me_3Si)_2-(1,2-C_2B_9H_{10})_2-3,3'-Co(III)]^-$ (**51⁻**) could be obtained via the reaction of the lithiated cobalt bis(dicarbollide) ion with Me_3SiCl [48].

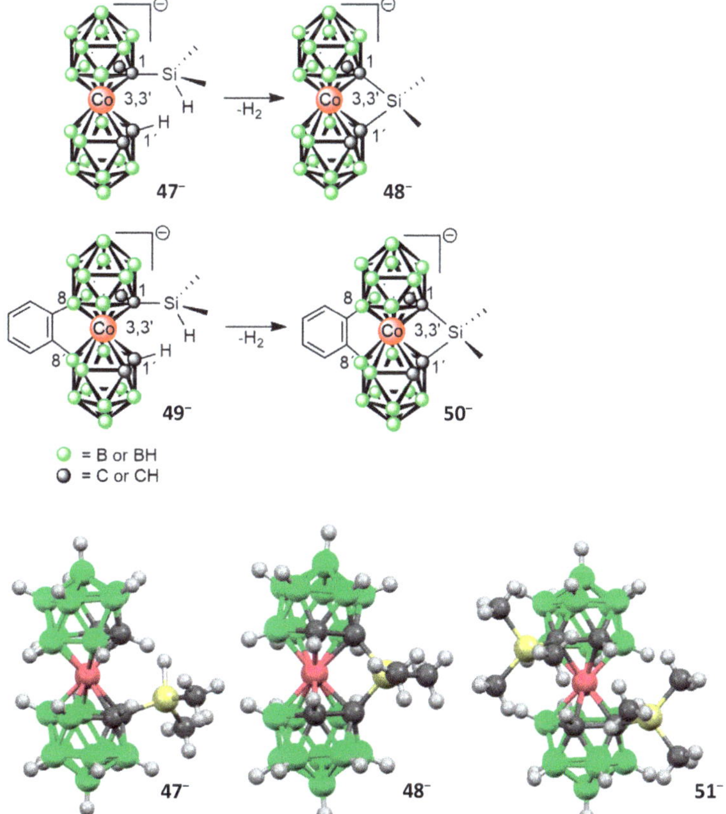

Scheme 12. The formation of the bridge due to the elimination of the hydrogen molecule from cage-C-H and hydridic Si-H is shown on top; and the schematic drawings of the molecular structures of the trimethylsilyl, bridged dimethylsilyl, and bis(trimethylsilyl) derivatives determined using XRD [48] are presented on bottom. Boron atoms are drawn in green, carbon in gray, cobalt in rose, and silicon in pale yellow color.

In addition, both of the silyl reagents gave the *meso*-substituted doubly bridged dimethyl silyl-cobalt bis(dicarbollide) when the reaction was started from lithiated 8,8'-μ-C_6H_4-1,2-cobalt bis(dicarbollide) (**49⁻**). The preference of *meso*-form in (**50⁻**; Scheme 12) is explained in terms of the presence of a phenylene bridge, which does not allow for the inclination of the dihedral angle that would enable the attaining of optimal bond lengths in the C(1)-Si-C(1′) bridging group. This represents a unique example when *meso*-isomer forms preferentially. In this case, the singly bound analogue could not be isolated due to H$_2$ loss from the structure. While the intermediate from the rotation-restricted phenylene derivative could have never been isolated, it was possible to isolate the mono-silylated intermediate of the parent ion, cobalt bis(dicarbollide), [1,1′-μ-SiMeH-(1,2-C$_2$B$_9$H$_{10}$)$_2$-3,3′-Co(III)]$^-$, which might be due to the availability of rotation in the structure.

DFT calculations supported the experimental outcome that the *rac*-isomers are more stable than *meso*- which leads to the preferential formation of *rac*-isomer and would also allow for the chiral separation of the enantiomers [48]. It was demonstrated that temperature may play a role in the stereoselectivity of these reactions. Relatively high temperatures, such as −40 °C, favored the formation of *meso*-derivative and, in turn, a higher ratio of *rac*-isomer formed at −78 °C. The [1,1′-μ-SiMeH-(1,2-C$_2$B$_9$H$_{10}$)$_2$-3,3′-Co(III)]$^-$ derivative was used in the regiospecific hydrosilylation reactions of vinyl-terminated dendrimers, producing species decorated with up to eight boron clusters on the periphery [48].

3.9. Phosphorylated Derivatives

The synthesis of compounds containing two diphenylphosphine groups was prepared via the direct reaction of lithiated anion **1**$^-$ with diphenyl chlorophosphine in DME at a low temperature (Figure 9).

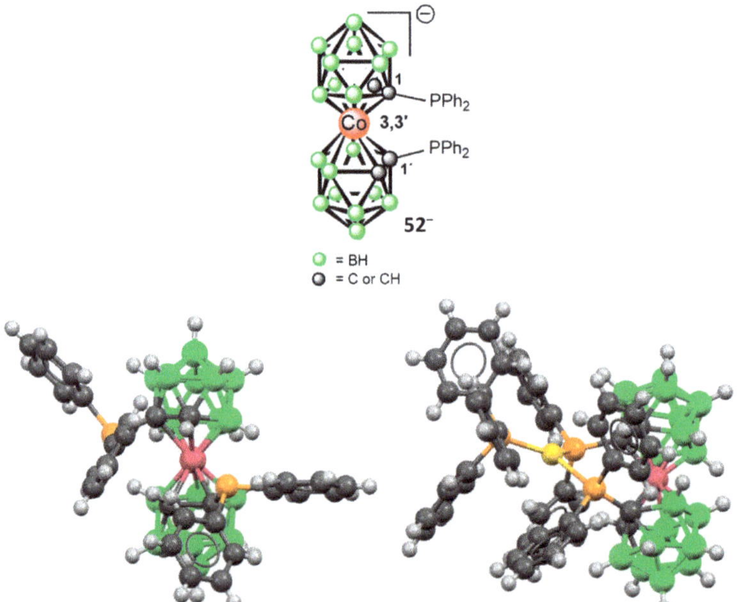

Figure 9. Schematic presentation of the bis(diphenylphosphine) derivative *rac*-(PPh$_2$)$_2$-cobalt bis(dicarbollide) (**52**$^-$) [47]; and molecular structures of *rac*-bis(PPh$_3$) ligand **52**$^-$ and its gold complex PPh$_3$Au.**52** drawn from XRD data presented in ref. [47]. Boron atoms are depicted in green, carbon in gray, cobalt in rose, phosphorus in orange, and gold in yellow color.

The compound [1,1′-(PPh$_2$)$_2$-(1,2-C$_2$B$_9$H$_{10}$)$_2$-3,3′-Co(III)]$^-$ **52**$^-$, with two diphenylphosphine groups, has been denoted as a versatile ligand available for metal complexation and has been denoted here as an alternative of organic BINAP ligand or chiral ferrocenyl phosphines with asymmetric substitution [47]. Indeed, this derivative was prepared in *rac*-form and can potentially serve as a versatile chiral auxiliary for metal complexation in asymmetric catalysis. The coordination of rhodium, palladium, silver, and gold with phosphine groups was described, which resulted in the formation of an *ansa*-arrangement of the complexes that contain the triatomic -Ph$_2$P-ML-Ph$_2$- bridging group. Three XRD structures of these complexes were determined using X-ray crystallography. The XRD structure of the PPh$_3$Au.**52** complex is shown in Figure 9, bottom. The two metal coordination sites have been shown to be less separated than in the case of the corresponding phosphinoferrocenes and more mutually separated than in BINAP. Therefore, the plane-intersecting PPh$_2$ has a different relative orientation with respect to the rotation axis.

Later, the reaction of the lithiated anion **1**$^-$ with phosphoryl chloride was studied that provided, after hydrolysis, substitution at two carbon sites, which was denoted as a bridged derivative of the formula [1,1'-μ-HO(O)P-(1,2-C$_2$B$_9$H$_{10}$)$_2$]$^-$ (**53**$^-$). Thus, the product was characterized using XRD as *rac*-isomer (Scheme 13), bottom.

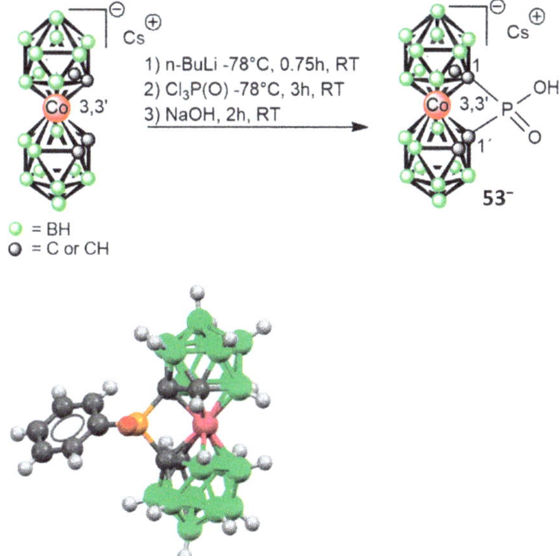

Scheme 13. Bridge-substituted phosphine oxide derivative **53**$^-$ of cobalt bis(dicarbollide) and its molecular structures are drawn from XRD data presented in ref. [48]. Boron atoms are drawn in green, carbon in gray, cobalt in rose, phosphorus in orange, and oxygen in red color.

Grűner et al. developed another series of phosphorus derivatives containing a phosphoric group that was separated from the cage by a propylene pendant group. A compound substituted with one group was prepared by a reaction of hydroxypropyl derivative Me$_3$NH**8c** with NaH and one equivalent of POCl$_3$ [81]. This resulted, after hydrolysis, in a phosphorylated compound of the formula [(1-(HO)$_2$P(O)OC$_3$H$_6$)-1,2-C$_2$B$_9$H$_{10}$)(1',2'-C$_2$B$_9$H$_{11}$)-3,3'-Co(III)]$^-$, with Me$_3$NH**54** as the main product. The corresponding reaction of the hydroxypropyl derivative with half of the equivalent produces a high yield of phosphoric acid diester (Me$_3$NH)$_2$**55**, in which structure two cages are connected via a propylene linker(s) to the central phosphoric acid moiety. The calcium salt Ca(**56**)$_2$ of the bridged ion [(μ-(HO)P(O)(OC$_3$H$_6$)$_2$-(1,2-C$_2$B$_9$H$_{10}$)$_2$-3,3'-Co(III)]$^-$ was isolated from the reaction of Me$_3$NH**10c** with NaH and one equivalent of POCl$_3$, followed by hydrolysis and the addition of CaCl$_2$. All new compounds were characterized using multinuclear NMR spectroscopy and mass spectrometry (Scheme 14).

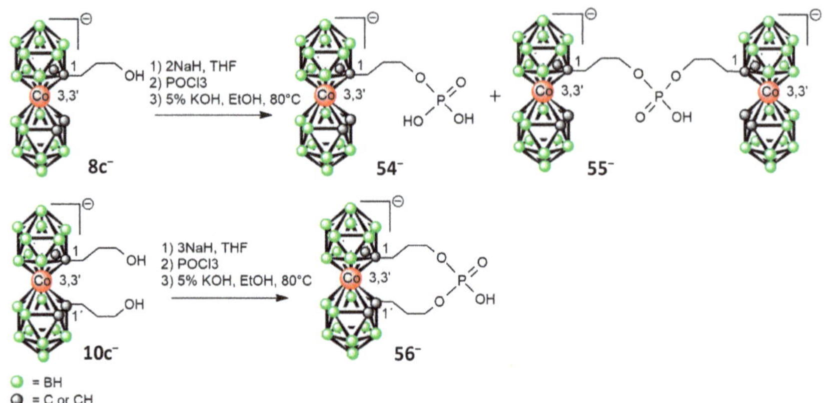

Scheme 14. Preparation of phosphoric acid derivatives [81].

3.10. Halogen Derivatives

The treatment of lithiated cobalt bis(dicarbollide) with bromoalkylnitriles resulted in mono- and di-bromo derivatives, unexpectedly. Hence, the reaction with 1-bromoacetonitrile gave low yields of monosubstituted product of the formulation $[(1\text{-Br-}1,2\text{-}C_2B_9H_{10})(1',2'\text{-}C_2B_9H_{11})\text{-}3,3'\text{-Co(III)}]^-$ (**57**$^-$). A similar low-temperature reaction (-78 °C, Scheme 15) of the cobalt bis(dicarbollide) ion lithiated by two equivalents of n-BuLi in DME with BrCN proceeded smoothly with high conversion to C,C-dibromide $[Br_2\text{-}(1,2\text{-}C_2B_9H_{10})_2\text{-}3,3'\text{-Co(III)}]^-$ (**58**$^-$), along with a smaller quantity of monobrominated compound **57**$^-$ [80]. Thus, in both instances, the bromine atom acts as a preferred reagent, possibly due to its properties lying on the borderline between soft and hard electrophiles. It was assumed in this study that the lithiated large cobalt bis(dicarbollide) ion might be, due to its bulky structure and low electron density, considered to act as a soft nucleophile that would react preferentially with soft electrophiles rather than with hard ones like CN^+ or $^+CH_2CN$. The disubstituted compound was separated using chromatography combined with crystallization and characterized as a mixture of all three possible diastereoisomers, which were present in an approximate ratio of 5:1:1 according to NMR. Pure *rac*-isomer $[1,1'\text{-Br}_2\text{-}(1,2\text{-}C_2B_9H_{10})_2\text{-}3,3'\text{-Co(III)}]^-$ was then isolated by repeating the chromatography on an RP-C18 column using 55% aqueous MeOH as the mobile phase.

In addition to all the carbon-substituted cobalt bis(dicarbollide) derivatives mentioned in the above sections, derivatives with terminal sulfamide, sulfonamide, and phthalimide groups were prepared (please see details in Section 5.2) focusing on medicinal applications.

The carbon-substituted derivatives currently form a portfolio of structural blocks applicable in diverse areas of chemical research. They form viable alternatives compared to more elaborated chemistry on boron vertices and offer new opportunities in tuning the spatial arrangement of substituents, dipole moments, interactions with medicinal targets, and in the development of chiral platforms. This has already been shown by the applications of carbon-substituted derivatives in drug design. Several types of new compounds addressing different therapeutic targets have already been prepared and studied, often showing improved properties when compared with related boron-substituted analogues (see Section 5) devoted to medicinal applications.

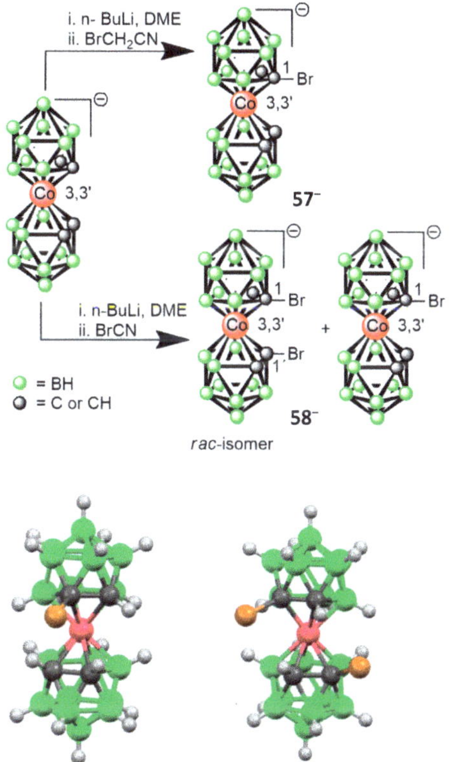

Scheme 15. Synthesis of bromo derivatives of the cobalt bis(dicarbollide) ion and schematic drawings of their molecular structures determined using XRD (**57⁻**, **58⁻**) and presented in ref. [80]. Boron atoms are drawn in green, carbon in gray, cobalt in rose, and bromine in brown color.

4. Recent Progress in Boron Substitution

The demand for new derivatives of cobalt bis(dicarbollide) ion continues to drive ongoing synthesis efforts. Over the past 30 years, significant progress has been achieved in synthesizing derivatives through ring-opening reactions of cyclic oxonium derivatives when treated with various nucleophiles. This particular reaction, initially invented by Plešek [104,105], provides a highly efficient method for introducing functional groups onto the B(8) atom of the cage. It is sometimes referred to as "boron-click" due to its effectiveness in producing various types of derivatives with functional groups attached to the boron atoms via a flexible six-atom spacer [106–109]. These reactions have been the subject of review articles published in 2008, 2012 [110,111], and 2021 [112]. For some terminal groups, it has been observed that the six-atom spacer can undergo cleavage to form a shorter chain consisting of a B-OCH₂CH₂OH group [113]. This offers new opportunities for tuning the distance between the functional group and the cage. Another type of oxonium compound of this kind consists of a diatomic (-O⁺(CH₃)-CH₂-) bridge in (B8,8') (**60**) positions of the cage. Cleaving this bridge with nitrogen nucleophiles results in a short methylene linker present in the B(8)-CH₂-Nu unit [114]. However, the use of this building block has certain limitations due to demethylation, which, in some circumstances, proceeds preferentially. The latter pathway occurs with oxygen nucleophiles and some bulky amines. Due to their high versatility, these methods have been extensively investigated since they represent an easy solution for introducing a wide variety of functional groups and bioorganic molecules to boron atoms (Scheme 16) [112,115]. Another widely applicable method, which was also initiated by Plešek et al., consists of a similar cleavage of a monoatomic bromonium or iodonium bridging moiety between (B8,8′) positions (**61**) [116]. This cleavage results in the formation of a B(8)-Nu bond along with a B(8′)-Hal substitution (**61**) [104,117,118].

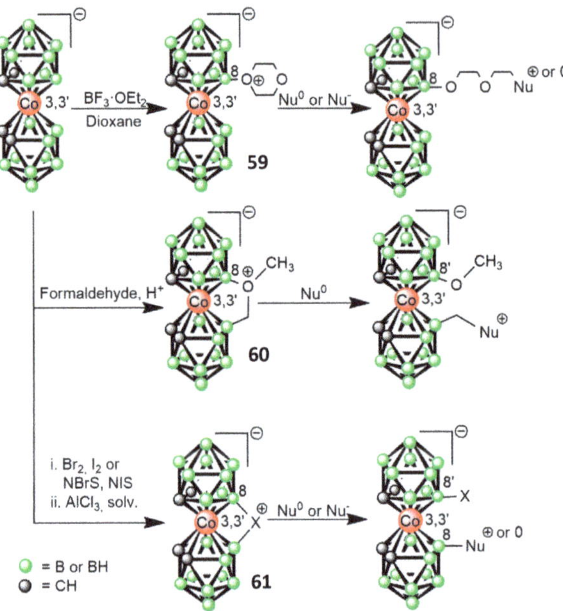

Scheme 16. General synthetic routes for easy introduction of variety group(s) and molecule(s) to boron atom(s).

From the perspective of tuning the spatial arrangements of substituents on the cage, especially for interactions with medicinal targets and materials science, recent results on compounds substituted at "unusual" boron sites B(10), B(4), and B(9) are particularly important.

Safronov, Hawthorne et al. described disubstituted iododerivative in apical positions B(10,10′) accessible through the degradation of the corresponding 8-I-dicarba-*closo*-dodecaborane followed by the insertion of the cobalt atom [119].

Pichaandi et al. described two approaches for synthesizing rod-like polymers containing metal bis(dicarbollide) anions connected by organic linkers. In the first synthetic route, *nido*-bis(carborane) precursors with acetylene or 1,4-dibutyl-2,5-diethynylbenzene linkers were used. The synthesis involved the formation of the cobalt bis(dicarbollide) dianion through the reaction of *nido*-bis(carboranes) with n-butyllithium, followed by metal insertion using cobalt(II) acetylacetonate. These precursors resulted in the formation of oligomers **62⁻a, b, c** (Figure 10). In the second synthetic route, the synthesis started with diiodo-substituted cobaltacarborane, and a zinc derivative of the 1,4-dibutyl-2,5-diethynylbenzene linker was prepared. The reaction between the complex and the zinc derivative with a palladium(0) catalyst resulted in the formation of a mixture of oligomers and polymers **63⁻** (Figure 10) [120]. The mentioned synthetic strategy was also later used in the synthesis of bis(ferrocenylethynyl)cobalt bis(dicarbollide) anion **64⁻** (Figure 10). In this case, the organozinc derivative of the 1,4-dibutyl-2,5-diethynylbenzene was replaced with an ethynylferrocenyl zinc reagent. This reagent was used in a Pd-catalyzed Kumada cross-coupling reaction with B(8,8′)-diiodo bis(dicarbollide) [121].

Shmal'ko, Sivaev, Bregadze et al. reported the synthesis of phenyl disubstituted anion **1⁻** in position B(6,6′) of the cage (**65⁻**), which was prepared from the corresponding 3-Ph-dicarba-*closo*-dodecaborane derivative (Figure 11) [122]. Additionally, a wider series of (9,9′, 12,12′) tetrasubstituted cobalt bis(dicarbollide) derivatives containing bromo, alkyl, and phenyl groups were prepared using these pathways [123]. These compounds have substituents at unusual sites located in the upper skeletal pentagonal plane, which is non-adjacent to the cobalt atom.

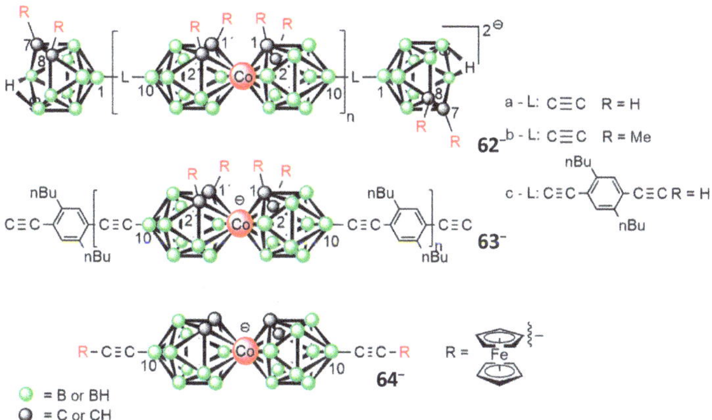

Figure 10. Structural formula(s) of compound(s) {7,8-R_2-$C_2B_9H_8$-[1,1′,2,2′-(R)$_4$-10-L-(1,2-$C_2B_9H_8$)$_2$-3,3′-Co(III)]$_n$}$^{2-}$ **62**$^-$, {[1,1′,2,2′-(R)$_4$-10-((2,5-C_4H_9)$_2$-Ph)-C_2)-10′-C_2-(1,2-$C_2B_9H_8$)$_2$-3,3′-Co(III)]}$_n^-$ (**63**$^-$), [10,10′-(R-C_2)$_2$-(1,2-$C_2B_9H_{10}$)$_2$-3,3′-Co(III)]$^-$ (**64**$^-$).

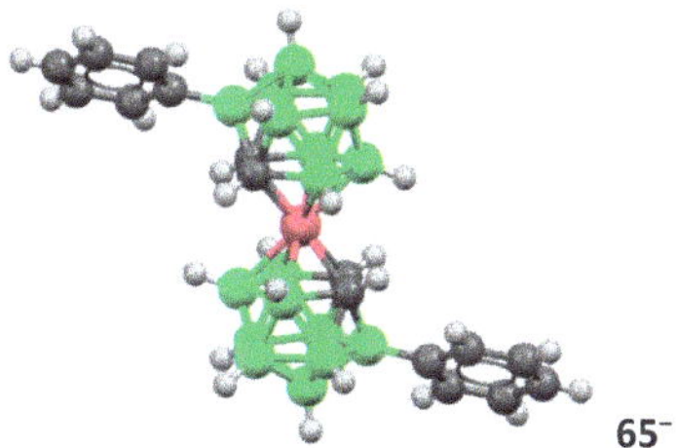

Figure 11. Drawing of the molecular structure of **65**$^-$ reported in ref. [122]. Boron atoms are drawn in green, carbon in gray, and cobalt in rose color.

From an electrochemical perspective, Nar et al. reported the synthesis and characterization of a new unsymmetrical zinc phthalocyanine with a cobalt bis(dicarbollide) unit **66**$^-$ (Figure 12). In the synthesis of this compound (Pc), a statistical condensation approach was employed using two different precursors. Phthalonitrile derivatives 4,5-bis(3-diethylaminophenoxy)-phthalonitrile and 4-hydroxyphenoxyphthalonitrile were used as starting materials. The metal-mediated cyclotetramerization of these compounds with $ZnCl_2$ produced a mixture of symmetrical and unsymmetrical analogs of phthalocyanine, which were separated using column chromatography. The compound **66**$^-$ was then obtained by reacting Pc with the oxonium derivative of cobalt bis(dicarbollide) in acetone under reflux. By introducing diethylaminophenoxy moieties to the phthalocyanine fragments, high solubilities in organic solvents were achieved, enabling electropolymerization on the electrode surface [124].

The synthesis of alkoxy derivatives via the alkylation of hydroxy derivatives of cobalt bis(dicarbollide) remains limited [99,125,126]. Stogniy et al. performed the reactions of monohydroxy derivatives of cobalt bis(dicarbollide) with methyl, ethyl, or propyl iodide in the presence of sodium hydride, resulting in the formation of alkoxy derivatives [8-R-O-(1,2-$C_2B_9H_{10}$)$_2$-3,3′-Co(III)]$^-$ (R = Me, Et, Pr) (**67**$^-$ **a**, **b**, **c**; Figure 13) in high yields. Similarly, it was possible to obtain symmetrically substituted 8,8′-dialkoxy derivatives (**68**$^-$ **a**, **b**, **c**; Figure 13) [8,8′-(R-O)$_2$-(1,2-$C_2B_9H_{10}$)(1′,2′-$C_2B_9H_{10}$)-3,3′-Co(III)]$^-$ (R = Me, Et, Pr) either directly or using a two-step synthesis depending on the amount of NaH [126].

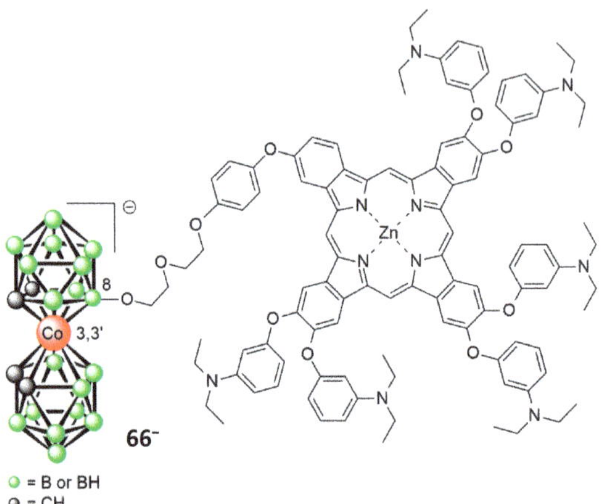

Figure 12. Chemical structure(s) of 2,3,9,10,16,17-Hexakis(3-dietylaminophenoxy)-23-[3,3′-Co(1,2-$C_2B_9H_{10})_2$]phthalocyaninato Zinc(II) (**66**⁻).

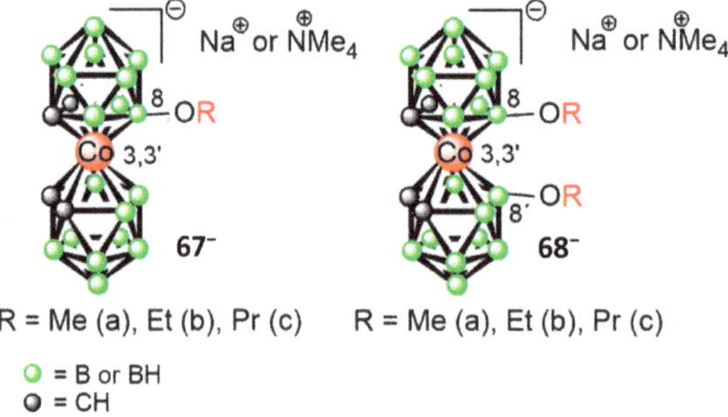

Figure 13. Structural formulas of compounds of various alkoxy derivatives of cobalt bis(dicarbollide) (**67**⁻, **68**⁻).

Śmiałkowski et al. (2023) focused on the oligofunctionalization of cobalt bis(1,2-dicarbollide) using hydroxyalkyl ligands to maximize the distance between functional groups and the cluster core. The study focused on functionalizing cobalt bis(1,2-dicarbollide) at boron atoms 8 and 8′ proceeding via the direct alkylation of hydroxy groups. The cobalt bis(1,2-dicarbollide) was converted into formula [8,8′-$(OH)_2$-(1,2-$C_2B_9H_{10})_2$-3,3′-Co(III)]⁻ using aqueous sulfuric acid, following the procedure described by Plešek et al. [127]. The authors employed different alkylating agents (4-trityloxybutyl 4-methylbenzenesulfonate and 3-bromopropoxy)(tert-butyl)dimethylsilane) for protective substitution of OH groups, however, the complete alkylation of both hydroxyl groups was not achieved, resulting in a mixture of mono- and disubstituted products (**69**⁻**a, b, 70**⁻**a, b;** Scheme 17) [93], which had to be separated.

The same research group investigated the functionalization of boron atoms 8 and 8′ in 8,8′-O,O-[cobalt bis(1,2-dicarbollide)] phosphorothioate (**72**⁻) using S-alkylation. The synthesis of compound **72**⁻ involved a two-step procedure (Figure 14). Initially, they converted the 8,8′-dihydroxy-derivative into 8,8′-bridged 8,8′-O,O-[cobalt bis(1,2-dicarbollide)] H-phosphonate acid ester (**71**⁻) using tris(1H-imidazol-1-yl)phosphine and the in situ hydrolysis of the resultant imidazolide. Subsequently, they treated H-phosphonate acid ester **72**⁻ with elemental sulfur in anhydrous methanol and with a strong organic base present in the reaction mixture [93].

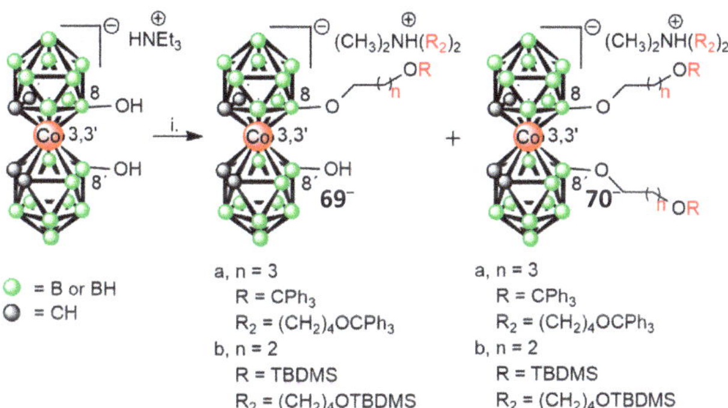

Scheme 17. Alkylation of $[8,8'\text{-}(OH)_2\text{-}(1,2\text{-}C_2B_9H_{10})_2\text{-}3,3'\text{-}Co(III)]^-$; i. NaH, alkylating agents in DMF, heating [93].

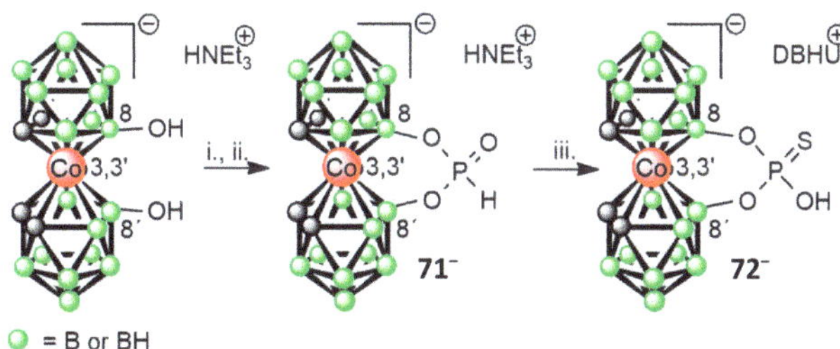

Figure 14. Preparation of H-phosphonate (71^-) and thiophosphate (72^-) esters of $[8,8'\text{-}(OH)_2\text{-}(1,2\text{-}C_2B_9H_{10})_2\text{-}3,3'\text{-}Co(III)]^-$; i. PCl_3, imidazole, Et_3N in THF; ii. H_2O; iii. S_8, DBU in MeOH. DBU = 1,8-diazabicyclo(5.4.0)undec-7-en [93].

Furthermore, the *S*-alkylation of compound 72^- could be performed using both linear and branched alkylating agents with hydroxyl functions protected by a trityl group. The nucleophilic properties of sulfur played a beneficial role in this alkylation reaction, resulting in high yields of the alkylated products 73^- **a**, **b**, and 74^- (Figure 15). Interestingly, derivatives of bridged 8,8'-μ-O,O-[cobalt bis(1,2-dicarbollide)] phosphate, which have an O,O-phosphate bridge instead of a phosphorothioate group, have not been previously described. This distinction in reactivity could be attributed to the "metallacarborane effect", which leads to the lower nucleophilicity of the phosphorus center and oxygen compared to sulfur [93].

In the study of Sardo et al., bridged thiophosphate 75^- and its oxo-analogue 76^- (Figure 16) were synthesized by reacting the B(8,8')-dihydroxy cobalt bis(dicarbollide) derivative with the corresponding dichloride esters in pyridine at room temperature. The remarkable stability of O-(4-nitrophenyl)phosphorothioate and phosphate esters of cobalt bis(dicarbollide) ion under alkali conditions was observed. The study demonstrated that metallacarboranes exhibit a significantly decreased reactivity of phosphorus-bridged metallacarborane esters when compared with organic analogues. This reduction in reactivity is attributed to the inductive effect and electrostatic shielding of the phosphorus atom by the metallacarborane, which weakens the electrophilicity of the phosphorus and hinders the approach of negatively charged nucleophiles. These effects, along with steric hindrance and the limited flexibility of the six-membered ring (considering the involvement of phosphorus, oxygens, boron, and cobalt atoms), are referred to as the metallacarborane effect. The metallacarborane effect has a significant influence on the reactivity of phosphate and phosphorothioate functional groups bonded via a bridge, while the thio- effect has a minor role in this respect [128].

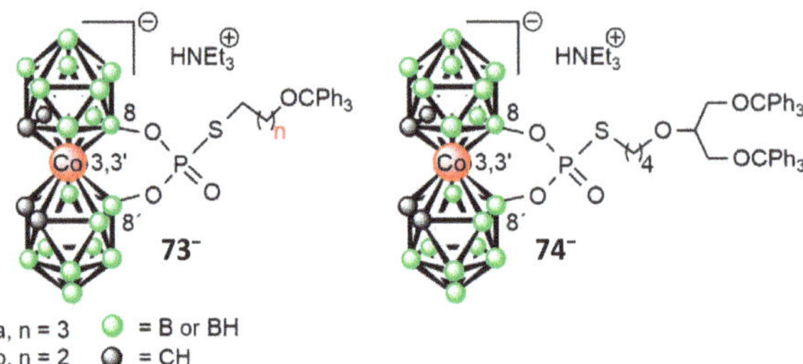

Figure 15. Structural formulas of *S*-alkylated phosphorothioates compounds 8,8′-bridged [8,8′-μ-O$_2$P(O)S(CH$_2$)$_n$OCPh$_3$-(1,2-C$_2$B$_9$H$_{10}$)$_2$-3,3′-Co(III)]HNEt$_3$ (**73**$^-$) and 8,8′-bridged {8,8′-μ-O$_2$P(O)S[(CH$_2$)$_4$OCH(CH$_2$OCPh$_3$)$_2$]-(1,2-C$_2$B$_9$H$_{10}$)$_2$-3,3′-Co(III)}HNEt$_3$ (**74**$^-$) [93].

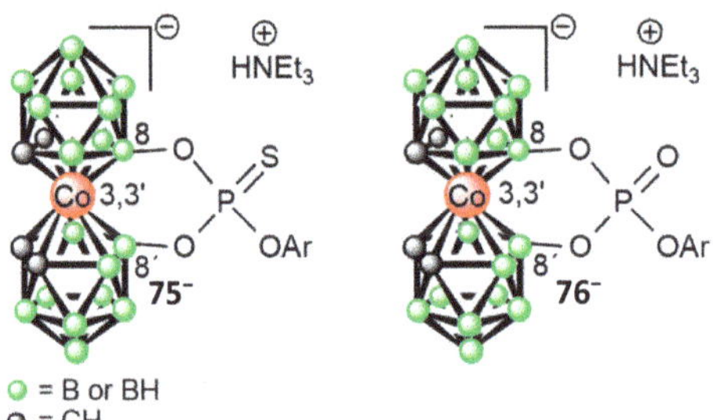

Figure 16. Chemical structures of [8,8′-μ-O$_2$P(S)OC$_6$H$_4$NO$_2$-(1,2-C$_2$B$_9$H$_{10}$)$_2$-3,3′-Co(III)]$^-$ (**75**$^-$) and [8,8′-μ-O$_2$P(O)OC$_6$H$_4$NO$_2$-(1,2-C$_2$B$_9$H$_{10}$)$_2$-3,3′-Co(III)]$^-$ (**76**$^-$); Ar = 4-nitrophenyl [128].

The synthesis of 8,8′-bis(methylsulfanyl) derivatives of cobalt bis(dicarbollide) **77**$^-$ was achieved through two different strategies (Figure 17). The first method involved the substitution of hydrogen atoms in the parent cobalt bis(dicarbollide) molecule. A dithiol derivative was obtained via the reaction of the parent molecule with carbon disulfide, followed by acidic hydrolysis. The alkylation of the dithiol derivative with various alkylating agents provided alkylthio derivatives. However, this method could not be extended to other transition-metal bis(dicarbollides) or to the preparation of complexes with asymmetrically substituted dicarbollide ligands [129–132]. The second strategy involved the synthesis of the corresponding carborane ligand in the first step followed by metal insertion. This approach allows for the synthesis of transition-metal bis(dicarbollide) complexes with substituents at different positions of the dicarbollide ligands. A symmetrically substituted methylsulfanyl derivative of cobalt bis(dicarbollide) was synthesized by reacting the corresponding carborane ligand with CoCl$_2$ in a strongly alkaline solution. The products were isolated in good yields [133,134].

Methylsulfanyl derivatives of carboranes have been shown to form stable tetra- and pentacarbonyltungsten complexes [135,136]. Timofeev et al. hypothesized that the reaction between the 8,8′-bis(methylsulfanyl) derivatives of cobalt bis(dicarbollide) and (MeCN)$_3$W(CO)$_3$ results in the formation of a chelate complex {(CO)$_4$W[μ-8,8′-(MeS)$_2$-(1,2-C$_2$B$_9$H$_{10}$)$_2$-3,3′-Co(III)]}$^-$, involving the rotation of the dicarbollide ligands from *transoid*- to *cisoid*-conformation. However, the reaction yielded a mixture of pentacarbonyl complexes (**78**$^-$, **79**$^-$; Figure 18) instead [137]. The 8,8′-bis(methylsulfanyl) derivatives of cobalt bis(dicarbollide) can also form chelate complexes with various transition metals such as copper, silver, palladium, and rhodium. This complexation process causes a conformational change from the *transoid*- to the *cisoid*-form. The transition between these conformations is reversible, and it can be reversed by removing the external transition metal using stronger ligands or precipitating agents. This reversible conformational change may offer potential for the design of coordination-driven molecular switches based on transition metal bis(dicarbollide) complexes [60].

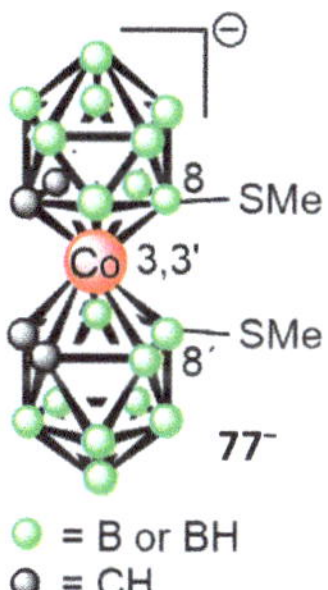

Figure 17. Chemical structures of $[8,8'\text{-}(MeS)_2\text{-}(1,2\text{-}C_2B_9H_{10})_2\text{-}3,3'\text{-}Co(III)]^-$ ($\mathbf{77^-}$).

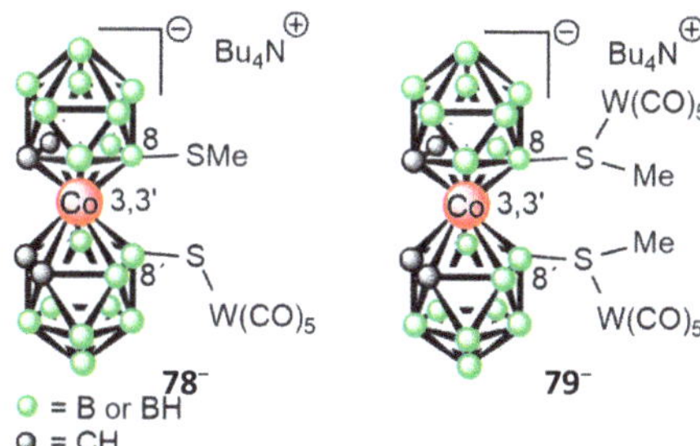

Figure 18. Chemical structures of $\{(CO)_5W[\mu\text{-}8,8'\text{-}(MeS)_2\text{-}(1,2\text{-}C_2B_9H_{10})_2\text{-}3,3'\text{-}Co(III)]\}^-$ ($\mathbf{78^-}$) and $\{[(CO)_5W]_2[\mu\text{-}8,8'\text{-}(MeS)_2\text{-}(1,2\text{-}C_2B_9H_{10})_2\text{-}3,3'\text{-}Co(III)]\}^-$ ($\mathbf{79^-}$).

Amidines are nitrogen analogues of carboxylic acids and esters and they are also important organic compounds widely used in the synthesis of various heterocycles and as pharmacophores in biologically active compounds. Bogdanova et al. studied the nucleophilic addition reactions of primary and secondary amines to the highly activated $B\text{-}N^+\equiv C\text{-}R$ triple bond of the propionitrilium derivative **80**. Several new amidines, based on metallacarboranes, were successfully synthesized (**81, 82**; Figure 19). It was observed that the reactions with primary amines produced mixtures of *E*- and *Z*-isomers, while the reactions with secondary amines selectively yielded *E*-isomers [138].

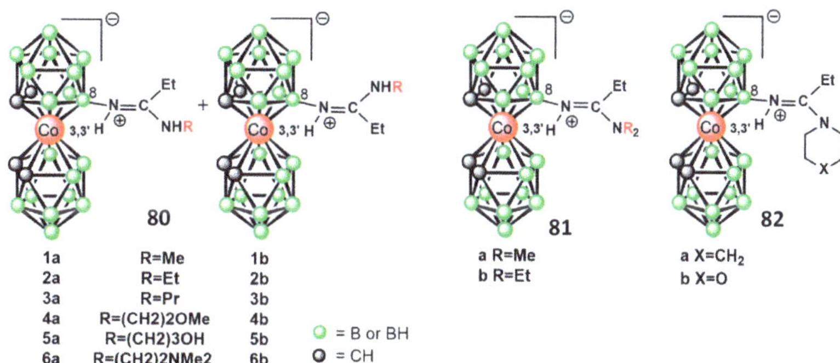

1a	R=Me	1b	
2a	R=Et	2b	
3a	R=Pr	3b	
4a	R=(CH2)2OMe	4b	
5a	R=(CH2)3OH	5b	
6a	R=(CH2)2NMe2	6b	

a R=Me	a X=CH₂
b R=Et	b X=O

Figure 19. Schematic structures of $[8\text{-}EtC(NHR)\text{=}HN\text{-}(1,2\text{-}C_2B_9H_{10})_2\text{-}3,3'\text{-}Co(III)]$ (**80**), $[8\text{-}EtC(NR_2)\text{=}HN\text{-}(1,2\text{-}C_2B_9H_{10})_2\text{-}3,3'\text{-}Co(III)]$ (**81**), and $[8\text{-}EtC(NC_4H_8X)\text{=}HN\text{-}(1,2\text{-}C_2B_9H_{10})_2\text{-}3,3'\text{-}Co(III)]$ (**82**).

4.1. Cobalt Bis(dicarbollide) as Extraction Agent

The field of radionuclide extraction using the cobalt bis(dicarbollide) ion was covered in two book chapters by Rais et al. (2004) [9], Grűner et al. (2012) [10], and a ligand designed for f-elements by Leoncini et al. (2017) [139]. Halogenated cobalt bis(dicarbollide) and their derivatives play a significant role in the extraction and separation of radioactive isotopes from nuclear waste, especially in the processes related to spent nuclear fuel (SNF) reprocessing and waste management. Due to their remarkable selectivity for cesium, strontium (with PEGs as synergists), and even lanthanides/minor actinides (either in synergic mixtures or after chemical modification), they have been shown to serve as a valuable tool for reducing nuclear waste volume and assisting in environmental remediation during nuclear accidents. Ongoing research in the field of radiochemistry focuses on cobalt bis(dicarbollide), with the aim of enhancing its extraction efficiency and optimizing its performance for various applications in nuclear waste treatment and management [140–142].

In 2017, Leoncini et al. published a review paper on extraction agents, summarizing developments in the ligand design, optimization, and extraction properties for the liquid–liquid extraction of f-elements from nuclear waste [139]. The following discussion will primarily highlight scientific papers published after the aforementioned review work.

A study conducted by Shishkin et al. (2020) reported the use of different extractive agents in the separation of elements present in weakly acidic raffinate. The study demonstrates that the extractive agent chlorinated cobalt dicarbollide (CCD) and di-2-ethylhexyl phosphoric acid (HDEHP) can effectively extract americium and europium from the solution, while Cs, Zr, U, Mo, and Fe are also extracted. Strontium and chromium remain in the raffinate. This extractive system also facilitates the separation of rare earths (REE) and transplutonium elements (TPE) with good separation factors (β). Urea is found to enhance the separation of REE and TPE in the extractive system. A scheme (see below; Chart 1) for the extraction of TPE from weakly acidic raffinate using the CCD + HDEHP extractive agent is presented, which reduces the volume of solutions to be evaporated and minimizes the use of complexating agents and auxiliary substances [143].

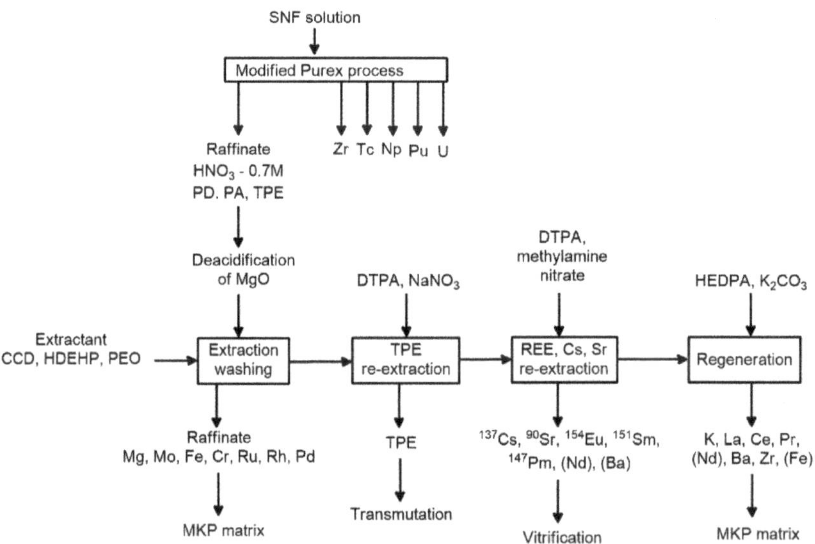

Chart 1. The flow sheet for extraction of TPE from weakly acidic raffinate.

In 2022, this research group presented the efficiency of a synergic mixture comprising CCD, HDEHP, and polyethylene oxide (PEO) in separating cesium, strontium, REE, and TPE from weakly acidic raffinate. Adjusting the raffinate's acidity with magnesium oxide enhances the extraction of cesium, strontium, americium, and europium. The nitric acid concentration affects the distribution coefficients of REE and TPE. The use of diethylene-triaminepentaacetic acid (DTPA) and sodium nitrate allows for the effective separation of REE and TPE. The addition of methylamine nitrate enables the separation of radiotoxic lanthanides. The regeneration of the extractant is possible using solutions of potassium carbonate and HEDPA. The deep partitioning scheme offers advantages in element separation and minimizes vitrified HLW volume [144].

The kinetics of Cs^+ and Sr(II) extraction and back-extraction were investigated using hexachlorinated cobalt bis(dicarbollide)/polyethylene glycol-400/FS-13 and a solvent system consisting of sulfone FS-13 (83^-, **84**; Figure 20) in batch experiments. Cs^+ extraction was relatively fast, reaching a steady state within 10 s, while stripping took about 50 s. Surprisingly, Sr(II) exhibited slower extraction kinetics compared to its back-extraction, contrary to common trends. In micromixer settler experiments, both Cs^+ and Sr(II) extraction and stripping were improved using guanidine carbonate and diethylenetriamine pentaacetic acid (DTPA). The microbore-based helix provided short residence times, achieving quantitative extraction and back-extraction of Cs^+, showing significantly faster kinetics than batch mixing. The mini-settler used in the experiments demonstrated effective settling, despite the role reversal of phases. Overall, the study showcased the kinetics of Cs^+ and Sr(II) extraction, emphasizing the advantages of microchannel mixers over batch mixing methods [145].

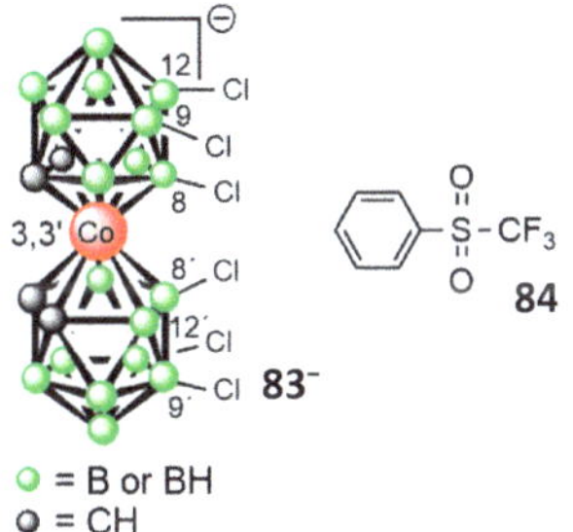

Figure 20. Chemical structures of $[8,8',9,9',12,12'-Cl_6-(1,2-C_2B_9H_{10})_2-3,3'-Co(III)]^-$ (83^-) and phenyl trifluoromethyl sulfone (**84**).

4.2. Cobalt Bis(dicarbollide) as Potentiometric Membrane Sensors

The sensitivity of novel sensors to single-, double-, and triple-charged ions was studied by Khaydukova et al., along with their selectivity. The sensor sensitivity varied depending on the type of ionophore used, either TODGA or CMPO (N,N,N',N'-tetraoctyl diglycolamide or di-phenyl-N,N-di-i-sobutylcarbamoylmethylene phoshine oxide). In the first case, the ionophore was either covalently bound to the cage or used in a synergic mixture with chlorinated cobalt bis(dicarbollide) ion in the later studied system (85^-**a**, **b**, Figure 21). For single-charged ions, CD-bonded sensors showed sensitivity to K^+. Double-charged ions exhibited good sensitivity to Ca(II) with TODGA-based sensors. The highest sensitivity was observed for Pb(II) ions, but sensitivity to Zn(II), Cd(II), and Cu(II) species remained low. Triple-charged ions (REEs) displayed sensitivity correlating with atomic number, particularly with TODGA-based sensors. The study also explored the effects of plasticizers and chlorinated cobalt bis(dicarbollide) concentration on sensitivity and selectivity. Sensors with bonded ligands were found to be suitable for multisensor systems for the simultaneous determination of several cations [146].

Chaudhury et al. (2014) investigated the electro-driven selective transport of Cs^+ from low acidic/neutral and high-level waste solutions through ion-exchange membranes. The main aim was to reach high Cs^+/Na^+ selectivity. The study, which measured potential drop and membrane resistance, demonstrated that cellulose-triacetate-based polymer inclusion membranes proved to have advanced properties compared to conventional types of membranes. The polymer inclusion membrane containing hexachlorinated cobalt bis(dicarbollide) efficiently transported Cs^+ in an electric field with high selectivity, making it energy-efficient and eco-friendly for nuclear waste treatment [147]. In 2018, the same research group published another study on the electro-driven transport of Cs^+ using chlorinated cobaltbis(dicarbollide) loaded membranes. The authors show that hollow-fiber-supported liquid membrane (HFSLM) can effectively treat large volumes of waste solutions with high Cs^+ selectivity, outperforming PIM in terms of transport rate and decontamination factor. The stability of the membrane against carrier leaching is also confirmed. Furthermore, the text provides current profiles and efficiencies for Cs+ transport from both neutral and acidic feed solutions. It emphasizes the time needed for complete Cs^+ transport in highly acidic solutions. The successful transport of Cs^+ from environmental samples is also reported, demonstrating the selective removal of Cs^+ over Na^+, K^+, and calcium(II) [148].

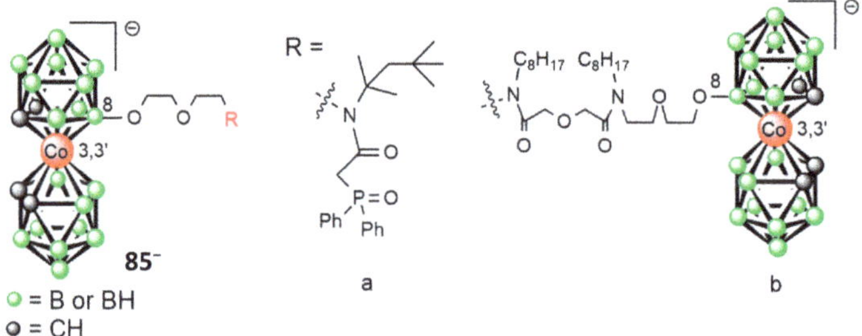

Figure 21. Chemical structures of [8-*R*(C$_2$H$_4$O)$_2$-(1,2-C$_2$B$_9$H$_{10}$)(1′,2′-C$_2$B$_9$H$_{11}$)-3,3′-Co(III)]$^-$ (**95**$^-$) (2-(diphenylphosphoryl)acetyl)(2,4,4-trimethylpentan-2-yl) (**a**) and *N-R-N,N′*-dioctyl diglycolamide[8-(C$_2$H$_4$O)$_2$-(1,2-C$_2$B$_9$H$_{10}$)(1′,2′-C$_2$B$_9$H$_{11}$)-3,3′-Co(III)]$^-$ (**b**).

Further information on the significance of metallacarboranes for producing potentiometric membrane sensors, to monitor amine-containing drugs and other nitrogen-containing molecules, is reported in a mini-review from Stoica et al. published in 2022 [14].

5. Recent Medicinal Applications

Polyhedral boron clusters have garnered significant attention as unconventional inorganic pharmacophores [66,149–158]. In recent years, there has been a growing focus on exploring the applications of cobalt bis(dicarbollide) ion in medicinal chemistry, particularly in the design of biologically active compounds with antiviral [31,159], antibacterial [35], and anticancer properties [32,160–162] and carriers for BNCT [38,163]. This is a binary treatment method for cancer that relies on the selective accumulation of boron compounds in tumor cells, followed by irradiation with a thermal neutron flux. The irradiation leads to the selective destruction of tumor cells while minimizing damage to surrounding normal tissues. To achieve the successful development of BNCT, it is crucial to reach the selective delivery of a large load of ^{10}B nucleus to tumor cells and ensure high boron accumulation (by uptake), while maintaining a low concentration of boron in the surrounding normal tissues [69,164–166].

To enhance the functionality of parent anion, the nucleophilic cleavage of a cyclic ether ring is employed as the building block for conjugation with biomolecules. Additionally, the regioselective click reaction, specifically the [3 + 2] cycloaddition of alkynes to azides [108,167], is a useful method for the modification of biomolecules with boron clusters. The resulting boron-containing triazole linkage mimics the stability and properties of the peptide bond [108]. The following paragraphs discuss the biomedical versatility of cobalt bis(dicarbolide) anion derivatives.

5.1. Recent Studies on the Parent Cobalt Bis(dicarbollide) Ion

Advancements in the emerging medicinal chemistry of the cobalt bis(dicarbollide) ion have triggered recent interest in its interactions with cells and the phenomena that occur in aqueous solutions. Indeed, the solution of the behavior of this ion is unusual and parallels that of other bulky surface-active inorganic ions denoted as chaotropes or even superchaotropes [168,169]. Understanding these properties plays a key role in the practical aspects of drug design and the pharmaceutical formulations of active compounds.

Moreover, from the perspective of practical synthesis and biological properties, the effects of charge, interactions with cations, solubility of the particular salt in selected solvents, cage solvation, and the tendency to self-assemble correspond to factors that have not been fully understood and warrant further investigation.

Self-assembly in an aqueous solution, host–guest properties, and the interactions of boron cluster compounds, inclusive of cobalt bis(dicarbollide) ion with biomolecules, are the subject of a recent review article published by Cebula et al. (2022) [170]. Here, we briefly describe and comment on the recent results of physicochemical and biological studies.

In 2006, Matějíček et al. made a significant step forward in understanding the physicochemical phenomena in aqueous solutions of the ion 1^-. The aggregation of the sodium salt of cobalt bis(dicarbollide) (Na1) and other metallacarboranes in aqueous solutions was described [171]. Recent studies on the Na1 revealed a two-step aggregation process [172–174]. The first step involves the formation of small aggregates with a pentamer structure, driven by favorable enthalpy and positive entropy. The second step leads to larger aggregates with more counterion binding and exothermic enthalpy. The proposed model suggests that pentamers are stabilized by two embedded Na^+ counterions, a hypothesis supported by quantum chemistry calculations [175].

Water-soluble cobalt bis(dicarbollide) salts, including H1, Li1, K1, and Cs1, exhibit an aggregation tendency, with similar critical aggregation concentration (CAC) values and aggregation numbers, suggesting a common aggregation mechanism. However, the aggregation behavior of Cs1 differs due to the large size of the cesium cation, making it unable to fit into the space between cobalt bis(dicarbollide) clusters. Molecular dynamics simulations were used to investigate the aggregation behavior of Na1 in explicit water. While the simulation snapshots showed changes in rotational isomerism, the fixed *cisoid*-cobalt bis(dicarbollide) clusters model provided more realistic results with roughly uniform pentamers forming around CAC. The simulation results are in agreement with the experimental evidence, and the presence of pentamers as the most probable pattern of the cobalt bis(dicarbollide) aggregates in solutions around CAC was confirmed. Additional investigation is required to gain a comprehensive understanding of the counterions' role in the aggregates of various cobalt bis(dicarbollide) salts and other metallacarborane analogues [175].

The study by Merhi et al. (2020) delved into how the cobalt bis(dicarbollide) interacts with an octyl-glucopyranoside surfactant (C8G1), leading to the formation of mixed aggregates. Depending on the concentrations of both substances, various types of assemblies were formed. At low cobalt bis(dicarbollide) content, cobalt bis(dicarbollide) vesicles and monomers coexisted. The addition of C8G1 in a monomeric form disrupted cobalt bis(dicarbollide) vesicles, forming cobalt bis(dicarbollide)-C8G1 nano-assemblies through hydrophobic interactions. At low cobalt bis(dicarbollide) content and high C8G1 concentrations, the ion adsorbed to the surface of C8G1 micelles, stabilizing them and reducing the critical micellar concentration of C8G1. However, with high cobalt bis(dicarbollide) content, the ion disrupted C8G1 micelles and penetrated the micellar surface. At higher cobalt bis(dicarbollide)/C8G1 ratios, assemblies similar to cobalt bis(dicarbollide) micelles containing solubilized C8G1 were formed. The study unveiled the superchaotropic behavior of cobalt bis(dicarbollide), where it spontaneously adsorbed onto C8G1 micelles, akin to other nanometric ions. This behavior was more pronounced at high C8G1 concentrations, superseding the hydrophobic effect observed at low concentrations. Cobalt bis(dicarbollide) and its derivatives demonstrated surfactant properties and hydrophobic characteristics, enabling them to cross biological membranes, including cell membranes, with potential applications in the pharmaceutical field [176].

Zaulet et al. (2018) explored the self-assembly of cobalt bis(dicarbollide) in water and the role of intermolecular interactions in forming aggregates. Salts of cobalt bis(dicarbollide) with alkali metal cations of Na^+, K^+, and Li^+ of cobalt bis(dicarbollide) were prepared using cation exchange, and their crystal structures revealed the presence of dihydrogen bonds. These dihydrogen bonds were responsible for the formation of stable supramolecular entities in water, avoiding direct contact with surrounding water molecules. The intermolecular B-H_B $^{\delta-}\ldots^{\delta+}H_C$-C interactions played a crucial role in the self-assembly process. The coordination of counter ions also influenced the self-assembly, as evidenced by the formation of a layered 2D structure in the crystal structure of $[Na(H_2O)_4][Co(C_2B_9H_{11})_2]$ [177].

Fernandez-Alvarez et al. (2019) focused on examining how presence of cobalt bis (dicarbollide) ion affect solution behavior and morphology of star-like polyelectrolyte micelles with a fixed number of arms. The two hydrophobic counterions consisted of cobalt bis(dicarbollide) anion, and the polycationic copolymeric micelles, which had polycationic star-like structures with frozen cores. By conducting computer simulations and experiments, they observed that the gradual addition of cobalt bis(dicarbollide) caused the micelles to precipitate at specific ratios, irrespective of the micelle's ionization level. In contrast, the addition of an indifferent electrolyte like NaCl did not produce the same effect. The simulations revealed that hydrophobic counterions altered the ionization profile of micelles, causing the formation of compact domains. The precipitation mechanisms differed based on ionization levels: less ionized stars collapsed into a single globule, while fully ionized stars formed an infinite gel network with counterion pearls. An NMR analysis confirmed the collapsed domains due to hydrophobic counterions. The findings could have applications in pH-controlled drug delivery with hydrophobic ionic solutes in polyelectrolyte-based nanostructures [178].

Traditionally, bulky monovalent cations and alkylammonium salt are used to precipitate the cobalt bis(dicarbollide) anion, but divalent cations are of particular biological relevance. There is a study by Zaulet et al. (2021) that explores the use of biocompatible divalent cations, such as Ca(II), Mg(II), and Fe(II), to isolate the cobalt bis(dicarbollide) anion. Ca(II) and Mg(II) are classified as hard Lewis acids, which means they form aqua ions when water is present during the synthesis process. In this study, all solid Ca(II) and Mg(II) salts examined were found to contain coordinated water molecules, as detected by IR and TGA/DSC analyses. On the other hand, Fe(II) is considered a medium-hard Lewis acid, while Fe(III) is classified as a hard Lewis acid. When placed in acetone, Fe(II) and Fe(III) ions do not coordinate with the cobalt bis(dicarbollide) anion but instead become solvated by acetone [153].

Rak et al. (2010), for the first time, studied the deaggregation of the cobalt bis(dicarbollide) ion in aqueous solutions on a panel of hydrophobic HIV-Protease and NO synthase inhibitors. A broad series of modified cyclodextrins (CD), classical surfactants, and amphiphilic copolymers were tested with the aim of finding general biocompatible excipients. They found several hints such as Pluronic F-127, DIMEB, or PVP, which may serve as suitable excipients. The strong interaction of metallacarorane derivatives with Human Serum Albumin (HSA) was found to compete with the solubilizing action of these species. However, it was found that HSA can itself act as a solubilization agent [179].

In the study of Goszczyński et al. (2017), they examined the fluorescence quenching mechanism, binding constants, and binding modes of Bovine Serum Albumin (BSA) when exposed to various boron clusters. The results showed that metallacarboranes caused substantial fluorescence quenching, utilizing both dynamic and static mechanisms. Metallacarboranes displayed strong binding constants, primarily engaging in hydrophobic interactions. At low stoichiometry, they first exhibited specific interactions with BSA within the hydrophobic cavity. However, at high stoichiometry, they also displayed non-specific interactions with the protein surface. Cobalt bis(dicarbollide) ion had a significant impact on BSA, inducing changes in BSA's α-helix content and forming stable complexes that affected BSA's hydrodynamic size. These findings offered valuable contributions to the development of novel bioactive compounds incorporating boron clusters [180].

Assaf et al. (2019) studied in detail the supramolecular chemistry of the series of parent isomeric cobalt bis(dicarbollide) ions, and seven simply substituted derivatives with cyclodextrins in an aqueous solution, using NMR and UV-visible spectroscopy, MS, electrochemistry, and isothermal titration calorimetry techniques. The compounds were found to be strongly bound to cyclodextrins [181], in particular into β-CD and γ-CD cavities. The binding constants of the inclusion complexes reached values in the micromolar range and thus exceeded those of highly hydrophobic nanodiamonds and were comparable with perhalogenated dodecaborate clusters [168]. The translocation of several cobalt bis(dicarbollide) derivatives through the lipid bilayer was shown by using fluorescence monitoring in supramolecular tandem membrane assays. No damage of the bilayer was observed.

The research of Abdelgawwad et al. (2021) aimed to explore the light-induced stabilization of cobalt bis(dicarbollide) isomers in water, with potential applications in light-switchable surfactants for drug delivery. *Transoid*-cobalt bis(dicarbollide) showed higher stability in the excited state than *cisoid*-cobalt bis(dicarbollide). Photoexcitation in water promoted rotation to the *transoid*-form, creating a fast photoinduced rotation. The study also revealed non-radiative decay mechanisms affecting luminescence in 3D fluorescence experiments. Overall, the research provided insights into cobalt bis(dicarbollide)'s photochemical behavior and its potential for light-responsive surfactants in drug delivery [182].

A recent study by Chazapi, Diat, and Bauduin (2023) demonstrated that sodium salt of cobalt bis(dicarbollide) ion can act, due to its surfactant properties, as an efficient aqueous solubilizer of medium-chain alcohols ($0.6 < \log P < 1.5$) and hydrophobic organic compounds, when the sodium salt is present in a solution in a monomeric form [183]. This relates to concentrations lying below its critical micelle formation. Mechanistically, the solubilization corresponds to the two-dimensional anisotropic growth of cobalt bis(dicarbollide)/butanol co-assemblies, whereas solubilization by the surfactant occurs via an isotropic swelling of micelles. Such co-assemblies with 2-butanol efficiently solubilize more hydrophobic compounds and dyes with $\log P_{ov}$ values from 0.6 to approximately 5.6. This seems to be an important result, which suggests good prospects for use in several fields of contemporary applications, in particular in formulations of lipophilic cobalt bis(dicarbollide) drugs, e.g., those dicluster inhibitors of HIV-Protease [165], lipophilic organic compounds such as aromatics, terpenoids, aldehydes, and ketones, and ethers used as fragrances, pharmaceutical active species, and dyes.

The penetration of a parent cobalt bis(dicarbollide) ion and its dihalogenated derivatives through lipid membranes was studied by Rokitskaya, Bregadze et al. (2017). The rates of penetration increased with the molecular weight of the halogen and the molecular volume of the anion. Thus, the parent ion and its fluorinated derivative exhibited the slowest transmembrane penetration compared to the compounds modified with heavier halogens. The authors expressed the hypothesis that this is connected with different conformations of the adsorbed species, which complicates the permeation due to the rotational component of the transmembrane diffusion [184].

Barba-Bon, Nau et al. (2022) demonstrated that both the isomers of parent cobalt bis(dicarbollide) ion and its halogenated and methylated derivatives are selective and highly efficient molecular carriers of impermeable hydrophilic oligopeptides through both artificial and cellular membranes. At the studied low micromolar concentrations of the carrier, no damage to the membrane was observed. The compounds were shown to transport both arginine- and lysine-rich peptides. Neither low-molecular-weight analytes, such as amino acids, nor neurotransmitters as well as neutral and anionic cargos (phalloidin and BSA) were co-transported in these experiments. U-tube experiments and electrophysiology establish that the transport is mediated by a molecular carrier mechanism and exclude alternative uptake pathways, such as channel or pore formation. It has been shown that the oligopeptides are delivered into the cytosol and nucleus by direct penetration [185].

Fink et al. (2023) conducted a comprehensive study to investigate the interaction between cobalt bis(dicarbollide) and DNA using various techniques. UV-Vis absorption spectroscopy, circular dichroism, linear dichroism, viscosity measurements, and differential scanning calorimetry all indicated that cobalt bis(dicarbollide) did not strongly interact with DNA. These methods showed no significant changes in the DNA absorption spectra, secondary structure, or intercalation of cobalt bis(dicarbollide) into DNA, suggesting no strong binding. Additionally, three other methods (isothermal titration calorimetry, equilibrium dialysis, and NMR) were used to detect potential cobalt bis(dicarbollide)-DNA complexes, but no strong interaction between cobalt bis(dicarbollide) and DNA was found. Regarding toxicity, cobalt bis(dicarbollide) showed moderate hemolytic activity towards anuclear cells (like red blood cells) and also demonstrated toxicity towards nuclear cells (various cell lines) at similar concentrations. The toxicity towards nuclear cells could potentially involve cell membrane disruption, but further investigation is needed to confirm this hypothesis. Thus, the study's overall findings suggest that cobalt bis(dicarbollide) does not strongly bind to DNA and exhibits moderate toxicity towards both anuclear and nuclear cells [26].

5.2. Application of Carbon-Substituted Compounds

5.2.1. Anticancer Compounds

Inhibitors of Carbonic Anhydrase IX Enzyme

Carbonic anhydrase IX (CA-IX) is a transmembrane zinc metalloenzyme that regulates pH in hypoxic tumors and promotes tumor cell survival. Its expression is associated with the occurrence of metastases and poor patient prognosis [186]. Carboranes and cobalt bis(dicarbollide) ions may act as highly potent inhibitors of this enzyme if the cage is substituted with a zinc binding group attached via an aliphatic pendant group. The progress in the field of boron cluster inhibitors forms the topic of a recently published review [32].

The first series of highly specific and selective cobalt bis(dicarbollide)(1^-) inhibitors of CA-IX, substituted either on the boron or on carbon sites by alkylsulfamide group(s), was reported by Grüner, Hajdúch, and Řezáčová et al. in 2019 [25]. A key step forward that enabled the rational design of inhibitors in this research was connected with the preparative availability of C-aminoalkyl-substituted metallacarboranes with tunable linker length. The synthesis started with a three-step process based on cobalt bis(dicarbollide) described in Section 3.6.3, which we consider as the most reliable reaction pathways leading to C-alkylamines. In only one case, a different strategy was used consisting of indirect cobalt insertion i.e., cobalt insertion into $[7\text{-}H_3NCH_2\text{-}C_2B_9H_{11}]$ that resulted in corresponding metallacarborane amine $[(1,1'\text{-}(NH_2CH_2\text{-}1,2\text{-}C_2B_9H_{10})_2\text{-}3,3'\text{-}Co(III)]^-$. The *rac*-isomer of the di(methyleneamino)-substituted sandwich was obtained as the only isolatable product, as verified by NMR and its X-ray structure (see Figure 2). The primary alkylamino groups were then converted to corresponding sulfamides upon heating with sulfamide in dioxane in the presence of K_2CO_3 (Scheme 18), providing inhibitors with one sulfamido group of formula $[(1\text{-}H_2NSO_2NH\text{-}(CH_2)_n\text{-}1,2\text{-}C_2B_9H_{10})(1,2\text{-}C_2B_9H_{11})\text{-}3,3'\text{-}Co(III)]^-$ (**86$^-$a, b, c**, n = 1–3). In addition, four compounds containing two alkylsulfamido groups were prepared: $[rac\text{-}1,1'\text{-}(NH_2CH_2\text{-}1,2\text{-}C_2B_9H_{10})_2\text{-}3,3'\text{-}Co(III)]\text{-}$, $[rac\text{-}1,1'\text{-}(NH_2\text{-}S(O)_2NH\text{-}C_2H_4\text{-}1,2\text{-}C_2B_9H_{10})_2\text{-}3,3'\text{-}Co(III)]^-$ (**88$^-$**), $[meso\text{-}1,2'\text{-}(NH_2\text{-}S(O)_2NH\text{-}C_2H_4\text{-}1,2\text{-}C_2B_9H_{10})_2\text{-}3,3'\text{-}Co(III)]^-$ (**89$^-$**) and $[meso\text{-}1,2'\text{-}(NH_2\text{-}S(O)_2NH\text{-}C_3H_6\text{-}1,2\text{-}C_2B_9H_{10})_2\text{-}3,3'\text{-}Co(III)]^-$ (**90$^-$**). The interactions of these compounds with the active site of CA-IX were explored on the atomic level using protein crystallography.

The inhibitory profile of the compounds was tested in vitro using the stopped-flow carbon dioxide hydration assay. Two selected derivatives **86$^-$** and **88$^-$**, that showed *sub*-nanomolar or picomolar K_i values and high selectivity for the tumor-specific CA-IX over cytosolic isoform CAII, were the subject of subsequent detailed biological tests. Both derivatives proved to have a time-dependent effect on the growth of multicellular spheroids of HT-29 and HCT116 colorectal cancer cells, and, in addition, facilitated the penetration and/or accumulation of doxorubicin into spheroids. The results from in vivo tests demonstrated that both inhibitors displayed low toxicity and showed promising pharmacokinetics and a significant inhibitory effect on tumor growth in syngeneic breast 4T1 and colorectal HT-29 cancer xenotransplants in mice. The results from preclinical studies indicate the promising potential of these compounds to be developed into drug-like forms and used in cancer therapy [25,32].

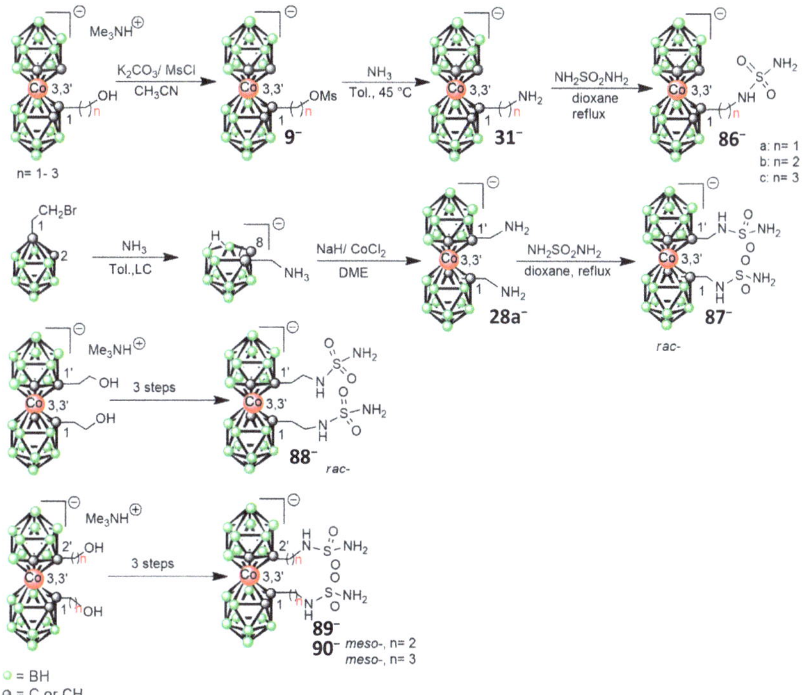

Scheme 18. Synthetic procedures used for the synthesis of the pilot series of inhibitors based on modifications of the cobalt bis(dicarbollide) ion at carbon atoms with alkylsulfamido group [25,32].

More recently, the synthetic routes to compounds substituted with sulfonamido groups have been developed. The sulfamido group is known to bind to the zinc atom in the active site more tightly than sulfamide, typically resulting in improved inhibitory activity compared to corresponding isostructural compounds with the sulfamide group. Even in the case of cobalt bis(dicarbollide) inhibitors, the substitution for the sulfonamide function resulted in an increase in inhibition activity by approximately one order of magnitude [75].

Several methods for the direct substitution of the cobalt bis(dicarbollide) cage with one alkylsulfonamido group were tested. These included the conversion of the respective mesyl esters to corresponding derivatives containing thiourea derivatives with the formation of the corresponding derivative $[1\text{-}X\text{-}(CH_2)_n\text{-}(1,2\text{-}C_2B_9H_{10})(1',2'\text{-}C_2B_9H_{11})\text{-}3,3'\text{-}Co(III)]^-$ $(X = \text{-}SC(NH_2)_2, n = 2$ and $3)$. The terminal thiourea group was then converted by oxidative chlorination in the second step to the corresponding sulfamoyl chloride group. Mild conditions were used using NaClO in a two-phase system at low temperatures and high dilution. However, even then it was impossible to avoid cage chlorination at the two most electron-rich boron $(B8,8')$ sites. These sulfamoyl chlorides were then subsequently converted, by reaction with aqueous NH_4OH, to the sulfonamides of formula $[1\text{-}H_2NSO_2\text{-}(CH_2)_n\text{-}8,8'\text{-}Cl_2\text{-}(1,2\text{-}C_2B_9H_9)(1',2'\text{-}C_2B_9H_{10})\text{-}3,3'\text{-}Co(III)]^-$. Two compounds comprising ethylene and propylene connectors were prepared [75].

The primary synthetic routes, however, involved the use of an indirect approach. Thus, a cobalt atom was inserted into a substituted and NaH deprotonated 11-vertex *nido*-species of general formula $[1\text{-}H_2N_S(O)_2\text{-}(CH_2)_n\text{-}7,8\text{-}C_2B_9H_{11}]^-$ (**92**$^-$**a, b, c, d**, n = 1–4). The insertion of the Co(III) central atom provided the respective di(sulfonamidoalkyl) cobalt bis(dicarbollide) ions of formula $[1,1'\text{-}(NH_2\text{-}S(O)_2\text{-}(CH_2)_n)_2\text{-}(1,2\text{-}C_2B_9H_{10})_2\text{-}3,3'\text{-}Co(III)]^-$ (**93**$^-$**a, b, c, d–104**$^-$**a, b, c**) in moderate (for n = 1 and 4) or excellent yields (for n = 2 and 3). As follows from their origin, it was possible to synthesize mainly the disubstituted derivatives. The sandwich compounds were obtained as a mixture of *meso-* and *rac-*stereoisomers. Both isomers could be readily separated using preparative flash chromatography on a C18-RP column using a mobile phase consisting of aqueous methanol. Also, compounds from the monosubstituted series of formula $[1\text{-}H_2NSO_2\text{-}(CH_2)_n\text{-}(1,2\text{-}C_2B_9H_{10})(1,2\text{-}C_2B_9H_{11})\text{-}3,3'\text{-}Co(III)]^-$ (**95**$^-$) (Scheme 19, bottom) could be prepared using a similar indirect method of reacting the sulfonamide-substituted eleven-vertex ligand together with the parent unsubstituted $[C_2B_9H_{12}]^-$ ion, however, only in a low yield [75].

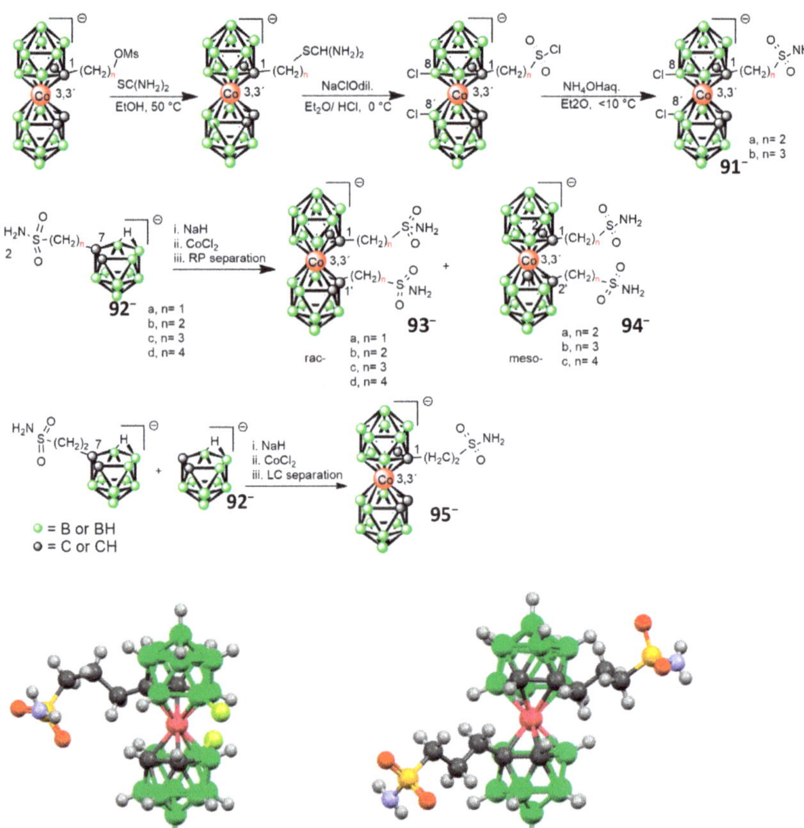

Scheme 19. Synthetic methods used for the preparation of the series of alkylsulfonamido-substituted inhibitors. Schematic presentation of molecular structures of 8,8′-dichloro-sulfonamidopropyl and di(sulfonamidopropyl) cobalt bis(dicarbollide) ions, **91b**$^-$ and **93c**$^-$, are shown in the bottom, from refs. [32,75]. Boron atoms are drawn in green, carbon in gray, cobalt in rose, sulfur in yellow, nitrogen in blue, oxygen in red, and chlorine in yellow-green color.

The inhibitory profile of the compounds was tested in vitro using the stopped-flow carbon dioxide hydration assay. To evaluate selectivity, the inhibition of two CA isoforms, the cancer-associated CA-IX and the widespread cytosolic CA-II isoform, was studied (Figure 22). The selectivity index (SI), expressed as a ratio of the two K_i values, is presented in Table 1. The whole series of compounds proved to serve as potent inhibitors of both CA isoforms, however, the values for CA-IX were significantly lower, which resulted in high selectivity [32,75].

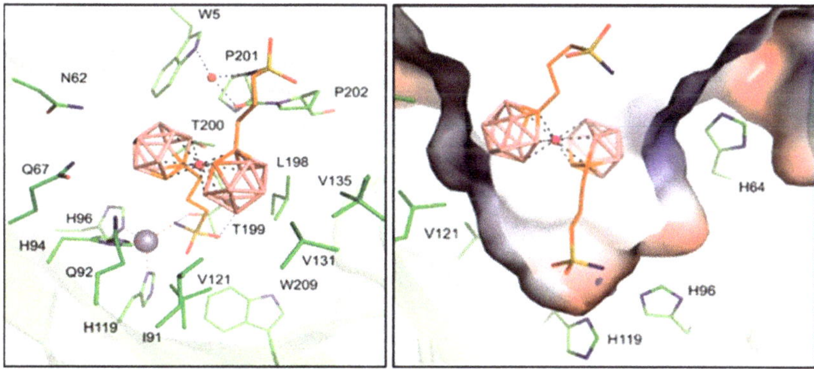

Figure 22. The binding mode of the compound **93c⁻** in the active site of CA-IX-mimic enzyme, according to refs. [75]. The boron atoms of the cage are drawn in light-pink, carbon in orange, cobalt in rose, sulfur in yellow, nitrogen in blue, oxygen in red, and zinc in the active site in gray color.

Table 1. In vitro inhibition of selected carbonic anhydrase isoenzymes [75].

Compound	Linker, Stereochemistry or Substitution	K_i (CA-II) [nM]	K_i (CA-IX) [nM]	Selectivity Index [a]
95⁻1	Mono, C_2	133.10 ± 15.35	0.92 ± 0.28	144.2
91a⁻	Mono, C_2, Cl_2	29.85 ± 4.46	0.89 ± 0.15	33.5
91b⁻	Mono, C_3, Cl_2	8.10 ± 0.86	0.10 ± 0.02	82.7
93a⁻	C_1, *rac-*	41.09 ± 6.09	0.86 ± 0.12	47.6
93b⁻	C_2, *rac-*	164.90 ± 19.86	0.86 ± 0.15	191.5
94a⁻	C_2, *meso-*	74.17 ± 12.67	0.37 ± 0.05	199.9
93c⁻	C_3, *rac-*	11.36 ± 1.25	0.02 ± 0.003	668.2
94b⁻	C_3, *meso-*	26.76 ± 2.45	0.60 ± 0.12	44.5
93d⁻	C_4, *rac-*	10.07 ± 0.61	0.29 ± 0.01	35.2
94c⁻	C_4, *meso-*	31.57 ± 3.69	0.06 ± 0.01	574.0

[a] Selectivity index is the ratio between K_i (CA-II) and K_i (CA-IX).

Interestingly, comparing K_i values for **91a⁻** and **91b⁻**, the chlorination of the cage at B(8,8′) sites improved the inhibition of CA-II, rendering the activity towards CA-IX on the same level. However, this results in decreased selectivity towards CA-IX (Table 1).

A three-atomic pendant group appears to be of optimal length for CA-IX inhibition over the whole series. Thus, the most active inhibitor *rac-***93c⁻** selectively inhibits CA-IX with a picomolar K_i value and also shows the highest S_I value 668. This is one order of magnitude better than for the corresponding *meso-*isoform. On the other hand, isomeric couples for other lengths of the linker do not show significant differences in their S_I. Therefore, no conclusion about the effect of stereochemistry on CA-IX inhibition could be drawn over the whole series of inhibitors [32].

1,8-Naphthalimides Derivatives, Analogues of Mitonafide, and Pinafide

1,8-Naphthalimides are compounds well known for their high antitumor activities with mechanisms proceeding via intercalation into DNA. Conjugates containing naphthalimide moiety attached to carbon and boron atoms of the cobalt bis(dicarbollide) ion, analogues of organic drugs mitonafide, were synthesized and studied, along with a panel of related carborane derivatives. The biological evaluations included in vitro cytotoxicity, type of cell death, cell cycle, and ROS production testing. The boron cluster compounds showed a significant effect on the proliferation of cancer cells, although a different kind of activity was observed compared to organic drug leads, mitonafide, and pinafide. The most promising derivative from this series corresponded to a disubstituted cobalt bis(dicarbollide) ion containing the C(1) naphthalimide group attached to the cage via an ethylene linker (**96**⁻, Figure 23) along with a second aminoethyl substituent on C(1′) atom. Compounds without this auxiliary substitution or those containing two naphthalimide groups in the same cage positions proved to decrease in activity. The compound **96**⁻ was selected for detailed studies, which revealed cytotoxicity against HepG2 cells and the activation of cell apoptosis. The compound caused cell cycle arrest in HepG2 cells. Further investigations in HepG2 cells revealed that compound **96**⁻ can also induce ROS generation, particularly mitochondrial ROS (mtROS), which was reflected by an increased 8-oxo-dG level in DNA. The interactions of new compounds with ct-DNA were also studied by CD spectra and melting temperature. The results confirmed the weak intercalation of boronated compounds into DNA, which is in contrast to the organic naphthalimide derivatives that usually show significant intercalating action [162].

Figure 23. Schematic structure of the derivative **96**⁻ showing the most promising antiproliferative activity.

5.2.2. Antiparasitic Activity

Boronated mitonafide analogues, naphthalimide derivatives of carboranes and metallacarborane Co(III) cages **97**⁻–**101**⁻ (Figure 24), were prepared using reactions of C- and B-substituted amines with 3-nitro-1,8-naphthalic anhydride [56]. The biological tests of antihelmintic activity against Rhabditis sp. have shown that the cobalt bis(dicarbollide) conjugates **98**⁻, **99**⁻, attached via a carbon atom, showed the highest activity from the series. The lowest LC50 values and strongest nematicidal activity were observed for *N*-[(8-(3-oxa-pentoxy)-cobalt bis(dicarbollide)]-1,8-naphthalimide conjugate, in which the terminal group was attached via a bis(ethylenglycol) linker. Its LC_{50} value was as low as 0.148 mg mL⁻¹. Slightly lower activity with an LC50 value of 0.207 mg mL⁻¹ was observed for the carbon-bound conjugate corresponding to *N*-[2-cobalt bis(1,2-dicarbollide)ethyl)]-1,8-naphthalimide. Interestingly, compound **97**⁻, comprising a longer propylene linker between the modified group and the naphthalimide residue, exhibited significantly lower activity. These modified naphthalimide conjugates, **98**⁻ and **99**⁻, displayed considerably higher activity than the mitonafide and mebendazole, drugs approved for the treatment of this type of parasitic infection.

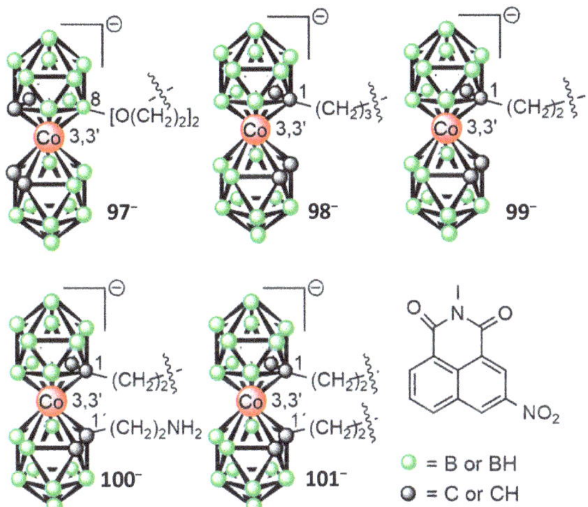

Figure 24. Chemical structures of [8-*R*-(1,2-C$_2$B$_9$H$_{10}$)(1,2-C$_2$B$_9$H$_{11}$)-3,3′-Co(III)] (R = (OC$_2$H$_4$)$_2$ (**97**$^-$), [1-*R*-(1,2-C$_2$B$_9$H$_{10}$)(1,2-C$_2$B$_9$H$_{11}$)-3,3′-Co(III)] (R = (CH$_2$)$_3$ (**98**$^-$), (CH$_2$)$_2$ (**99**$^-$)), [1-(CH$_2$)$_2$-2′-(NH$_2$(CH$_2$)$_2$)-(1,2-C$_2$B$_9$H$_{10}$)$_2$-3,3′-Co(III)]$^-$ (**100**$^-$), and [1,1′-C$_2$H$_4$)$_2$-(1,2-C$_2$B$_9$H$_{10}$)$_2$-3,3′-Co(III)]$^-$ (**101**$^-$) [56].

5.3. Application of Boron-Substituted Compounds in Medicinal Chemistry

Comprehensive review articles about the potential of boron-substituted cobalt bis (dicarbollide) ions in medicinal chemistry and novel healthcare materials were published in 2023 by Teixidor et al. [150], 2022 by Chen et al. [187], and in 2022 by Das et al. [188]. Therefore, the following chapters will focus exclusively on selected recent articles, highlighting the overall use of cobalt bis(dicarbollide) as an unconventional pharmacophore in biomedicine.

5.3.1. Antimicrobial Active Compounds and Antibiofilm Agents

The rapidly emerging resistance of bacterial pathogens to "classical" antibiotics presents a current and highly challenging issue in contemporary clinical practice. In this context, the use of unconventional hydrophobic pharmacophores based on boron clusters may potentially introduce a viable alternative. The possibility of using boron clusters in combating resistant forms of bacterial pathogens has been denoted by Plešek in his review published three decades ago [189]. His idea was based on an assumption that boron polyhedra are abiotic man-made compounds, which contain an unnatural type of bonding, and their surface composed of B-H bonds provides different interactions. From that time, knowledge about the mechanism of interactions of the cluster boron compounds with biological targets and cells increased appreciably. However, understanding the effects on the mechanism of resistance in particular types of pathogens, remains still limited. Some awareness of this idea can be seen in a recent paper published by Zheng et al. concerning the treatment of antibacterial infections [190] or a study by Řezáčová et al. and Kožíšek et al. showing that bis(dicarbollide) derivatives proved to have promising inhibitory action towards resistant mutants of viral HIV-Pr enzyme [31,159].

Considering antibacterial compounds, amines of the cobalt bis(dicarbollide) ion were recently proved to have antibiotic properties towards Gram-positive bacteria, in particular to Methicillin-resistant Staphylococcus aureus (MRSA), which is of high current concern. This is a significant pathogen that poses a threat to public health, particularly in healthcare settings. The emergence of drug-resistant strains has reduced the effectiveness of conventional treatments. This topic was recently covered by a review article published by Fink and Uchman [35]. Thus, here we focus only on several selected recent articles.

Microorganisms can exist in two forms: suspension cells or biofilms. Biofilms, which are communities of cells embedded in an extracellular matrix attached to a surface, can cause problematic infections. Biofilm formation often increases microbial resistance to antibiotics, necessitating the search for new antibiofilm compounds [191–194]. Cobalt bis(dicarbollide) and its derivatives have shown promising antimicrobial and pharmacological properties. While their antimicrobial activity has been studied, their antibiofilm activity remains largely unexplored [195–198]. Previous studies have primarily unlocked the antimicrobial activity of cobalt bis(dicarbollide) and its derivatives, focusing on Gram-positive bacteria and certain fungi, with limited efficacy observed against Gram-negative bacteria [196–198]. In the study by Vaňková et al., the antimicrobial and antibiofilm activities of cobalt bis(dicarbollide) derivatives (Na1, 8-NH$_3$-1, and 8-PhNH$_2$-1 were investigated [199]. They showed effective antimicrobial properties against Gram-positive bacteria but limited activity against Gram-negative bacteria and Candida parapsilosis. All three compounds inhibited the growth of the filamentous fungus T. cutaneum. Na1 exhibited strong antibiofilm activity against gram-positive bacteria, while 8-NH$_3$-1 and 8-PhNH$_2$-1 supported biofilm formation in P. aeruginosa. Na1 disrupted the biofilm structure, inhibiting biofilm formation by at least 80%. Moreover, Na1 had low cytotoxicity, making it a potential treatment for biofilm-associated infections. Further research is needed to explore their therapeutic applications [195].

Popova et al. focused on the preparation, characterization, and antimicrobial properties of alkylammonium derivatives. The authors conducted the first investigation into the in vitro antimicrobial activity of compound **102** and its derivatives decorated with ethyl or salicylic ester groups (Scheme 20). They discovered that these derivatives possess antimicrobial activity similar to the antibiotic thiamphenicol against both bacterial and Candida spp. strains. Compound **102a** exhibited the non-selective growth inhibition of Gram-positive bacteria and fungi. Compound **102b** selectively inhibited Trichosporon cutaneum, while compound **102e** showed significant activity against filamentous fungi. Other derivatives were not effective. The increase in substituent bulkiness correlated with hydrophobicity and stability. The exact mechanism and selectivity remain unknown. Compounds **102a**, **102b**, and **102e** show promise as inhibitors of opportunistic pathogens [197].

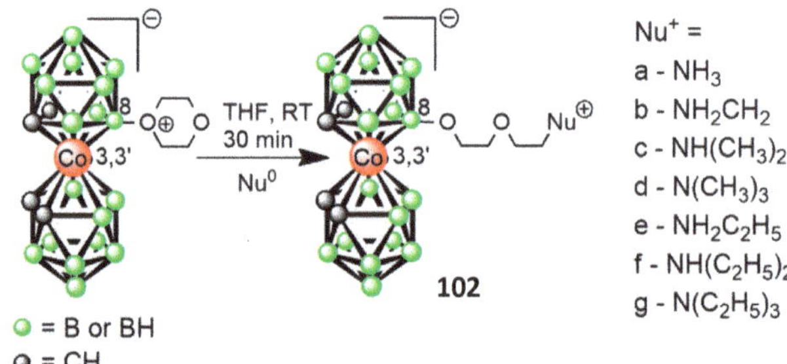

Scheme 20. Preparation of [8-*Nu*(C$_2$H$_4$O)$_2$-(1,2-C$_2$B$_9$H$_{10}$)(1′,2′-C$_2$B$_9$H$_{11}$)-3,3′-Co(III)] (**102**).

New conjugates of cobalt bis(dicarbollide) with curcumin were synthesized using the "click" methodology. The antibacterial activity of the synthesized curcumin derivatives was assessed. The compounds showed no activity against Gram-negative bacteria. However, variations in MIC values were observed for Gram-positive bacteria. Bacillus cereus was susceptible to all compounds, while Staphylococcus aureus and Enterococcus faecalis exhibited higher sensitivity to derivative **103e⁻**. Compound **103e⁻** also inhibited the growth of Aspergillus fumigatus, while the other samples had MIC values above 1000 mg L^{-1}. In the presence of curcumin and derivatives **103a⁻** and **103b⁻** (Scheme 21), a decrease in the growth density of Candida albicans was observed. Based on the results, compound **103b⁻** exhibited the highest activity against Gram-positive bacteria, A. fumigatus, and a decrease in the growth density of C. albicans, followed by curcumin, derivative **103a$^-$**. These findings suggest the potential of these conjugates as antibacterial agents [200].

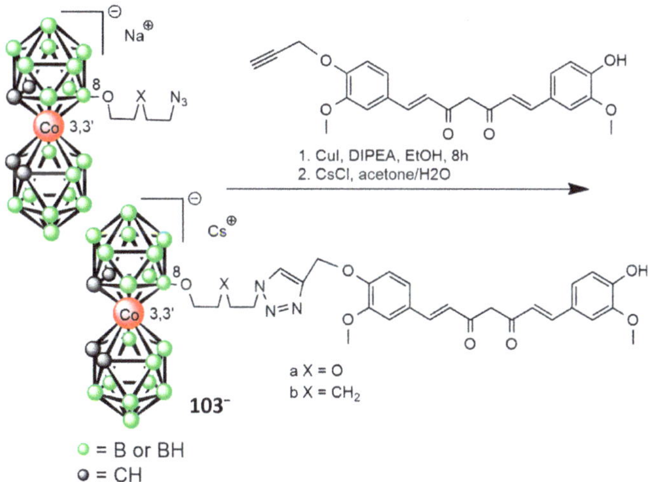

Scheme 21. Preparation of [8-OC$_4$H$_8$X-4-((4-((1*E*,6*E*)-7-(4-hydroxy-3-methoxyphenyl)-3,5-dioxohepta-1,6-dien-1-yl)-2-methoxyphenoxy)methyl)-1H-1,2,3-triazole-(1,2-C$_2$B$_9$H$_{10}$)(1′,2′-C$_2$B$_9$H$_{11}$)-3,3′-Co(III)]$^-$ (**103$^-$**).

In 2020, Romero et al. reported on the synthesis, characterization, and antimicrobial testing of cobalt bis(dicarbollide) derivatives Na**104**, K$_2$**105**, Na**106**, and K$_2$**107** (Scheme 22). The compounds were obtained through a ring-opening reaction of the cyclic oxonium zwitterion with various organic and inorganic carboxylates as nucleophiles. Microdilution tests revealed that the compounds lacked significant antibacterial effects on Gram-negative bacteria. However, they exhibited potent antibacterial activity against Gram-positive bacteria and had moderate antifungal activity against Candida albicans. The study assessed four Gram-positive bacteria strains, among them the life-threatening superbug MRSA, known for its resistance to various antimicrobial drugs. The cobalt bis(dicarbollide) derivative Na**104** displayed exceptional inhibitory effects on MRSA, with a standard minimum inhibitory concentration of 1 mg L^{-1} and a minimum bactericidal concentration of 2 mg L^{-1}. These results suggest that Na**104** holds promise as a potent antibacterial agent against MRSA [201].

Swietnicki et al. (2021) conducted research on cobalt bis(dicarbollide) derivatives for antibacterial activity against Yersinia enterocolitica and Pseudomonas aeruginosa (Scheme 23). They utilized an iodonium bridge-opening reaction for synthesis, yielding predominantly O-linked derivatives. Compounds **1$^-$**, **108$^-$a**, **b**, **c**, **d**, and **109b** were most effective against Yersinia, with short aliphatic chains, being Pseudomonas, showing lower susceptibility. The compounds showed unique chemistry and induced resistance in Yersinia. They acted in a bacteriostatic manner, affecting cell division. *N*-linked boron derivatives were more potent in mammalian cell toxicity studies, and compound **108d$^-$** had the lowest mortality rate in zebrafish toxicity tests at 20 μmol L^{-1} concentration [202].

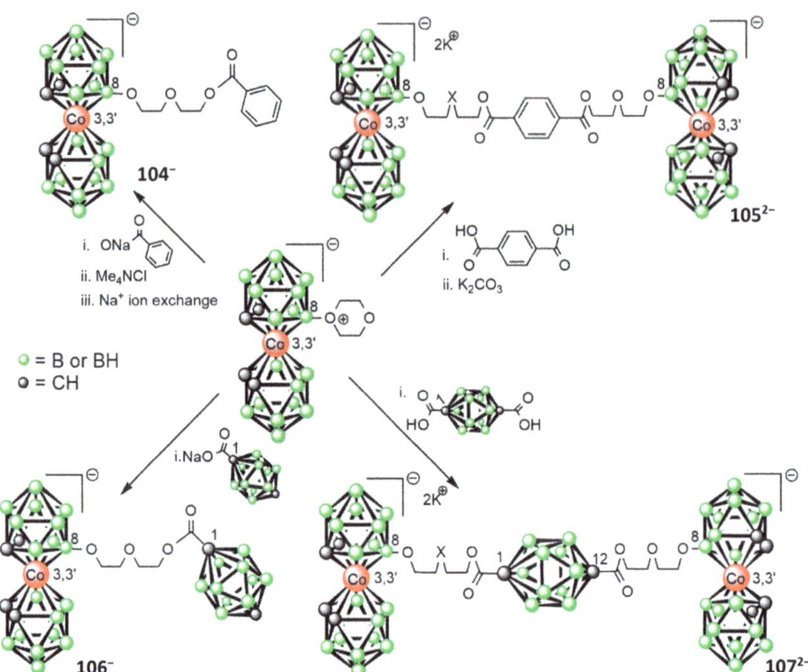

Scheme 22. Reaction scheme of [8-O$_2$C$_4$H$_8$-(1,2-C$_2$B$_9$H$_{10}$)(1′,2′-C$_2$B$_9$H$_{11}$)-3,3′-Co(III)] with sodium benzoate (**104$^-$**), terephtalic acid (**105^{2-}**), *p*-carborane-1-carboxylic acid (**106$^-$**), and *p*-carborane-1,12-dicarboxylic acid (**107^{2-}**).

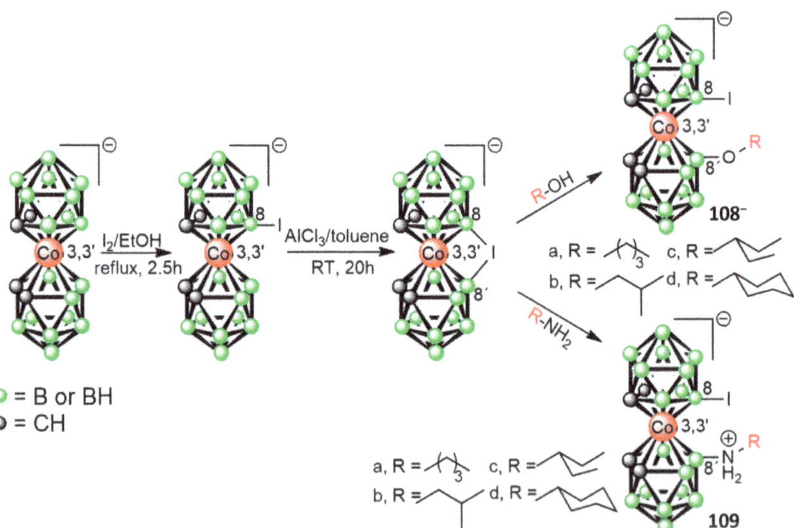

Scheme 23. Synthetic road for preparation of [8-I-8′-O*R*-(1,2-C$_2$B$_9$H$_{10}$)$_2$-3,3′-Co(III)]$^-$ (**108$^-$**) and [8-I-8′-NH$_2$*R*-(1,2-C$_2$B$_9$H$_{10}$)$_2$-3,3′-Co(III)] (**109**).

The synthesis of a series of cobalt bis(dicarbollide) derivatives containing an iodine atom at position B(8) and an organic substituent at B(8′) is described by Kubiński et al. (2022). These derivatives fall into two types: anionic products linked through oxygen

(B(8′)−O bond) and zwitterionic products linked through a nitrogen atom (B(8′)−N). The reactions involve the opening of the iodonium bridge with selected nucleophiles to yield desired bifunctional derivatives. The compounds' antimicrobial activity was tested against Gram-positive and Gram-negative bacteria, as well as Candida albicans. The unmodified cobalt bis(dicarbollide) anion showed good activity against Gram-positive bacteria and certain fungi but not against Gram-negative bacteria. Synthesized metallacarborane derivatives displayed strong to moderate antimicrobial activity, comparable to some systemic drugs. Compounds **109a**, **109b**, and **109e** exhibited the strongest antibacterial properties (Scheme 24). They were also effective against Candida albicans, with the potential to overcome drug resistance. Some compounds showed synergy with amphotericin B. The compounds had low toxicity towards mammalian cells, making them promising for antiviral and anticancer therapies. Modifications to the parental compound improved safety in vivo [37].

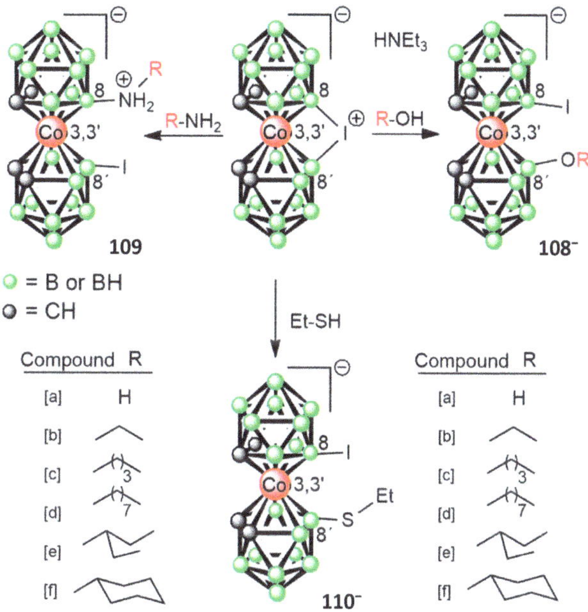

Scheme 24. Synthesis of [8-NH₂*R*-8′-I-(1,2-C₂B₉H₁₀)₂-3,3′-Co(III)] (**109**), [8-I-8′-O*R*-(1,2-C₂B₉H₁₀)₂-3,3′-Co(III)]⁻ (**108⁻**), and [8-I-8′-SEt-(1,2-C₂B₉H₁₀)₂-3,3′-Co(III)]⁻ (**110⁻**).

To address the treatment of MRSA, Kosenko et al. and Zheng et al. described the synthesis [125] and antimicrobial properties of **111⁻**, a cobalt bis(1,2-dicarbollide) alkoxy derivative, as a potential therapeutic agent (Figure 25). Compound **111⁻** demonstrated potent anti-MRSA activity, effectively reducing the number of MRSA colonies and completely eradicating the bacteria at suitable concentrations. It showed specific activity against MRSA and did not significantly affect other drug-resistant bacteria. The concentration-dependent anti-MRSA effect of **111⁻** was observed, with the complete eradication of MRSA at a concentration as low as 8 µg mL⁻¹. **111⁻** exhibited the fastest killing kinetics among the reported metallacarboranes. Importantly, multiple treatments with **111⁻** did not induce drug resistance in MRSA, unlike vancomycin. **111⁻** also showed excellent antibiofilm activity, inhibiting MRSA biofilm formation at sub-MIC concentrations. The mechanism of action of **111⁻** involved damaging the MRSA cell wall/membrane, which was confirmed using microscopy and staining techniques. The increase in reactive oxygen species (ROS) induced by **111⁻** contributed to cell membrane damage. **111⁻** exhibited excellent biocompatibility with mammalian cells and negligible cytotoxicity. These findings highlight the potential of **111⁻** as a nonantibiotic therapeutic agent for the treatment of MRSA infections [190].

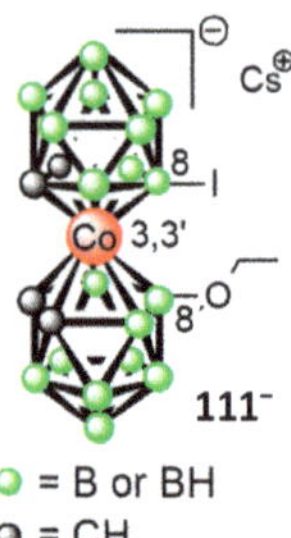

Figure 25. Chemical structure and [8-I-8′-OEt-(1,2-C$_2$B$_9$H$_{10}$)$_2$-3,3′-Co(III)]⁻ (**111⁻**).

Kosenko et al. investigated an iodonium-bridged derivative of **1⁻** for its reactivity with nucleophilic reagents, such as alcohols and amines, leading to disubstituted derivatives. The reaction of the starting derivative with propargyl alcohol yielded terminal alkyne, which underwent a Sonogashira reaction with 5-iodo-2′deoxyuridine to form cobalt bis(dicarbollide) and 5-ethynyl-2′-deoxyuridine conjugate **112⁻** (Scheme 25). Further treatment of **112⁻** resulted in an intramolecular cyclization product, furo [2,3-*d*]pyrimidin-2(3*H*)-one conjugate **113⁻**. The cytotoxicity of **112⁻** and **113⁻** was moderate, with the fetal lung fibroblast MRC-5 cells being the most sensitive. The compounds did not show antiviral activity against tested DNA and RNA viruses. Another attempt to synthesize a 5-ethynyl-2′-deoxyuridine conjugate with a similar spacer length but with a compensating charge did not yield the desired product [203].

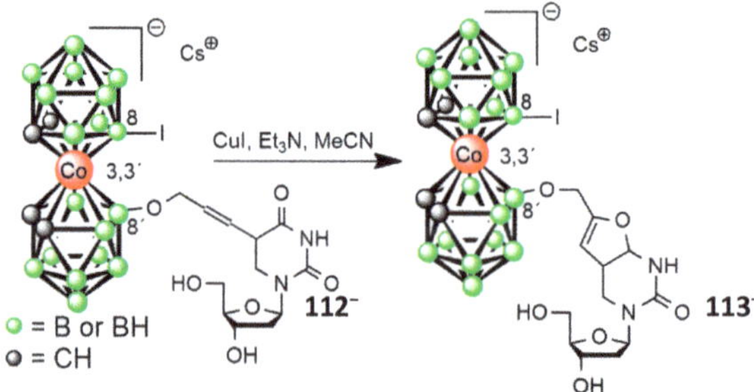

Scheme 25. Reaction scheme of cobalt bis(dicarbollide) with 5-ethynyl-2′-deoxyuridine and further intramolecular cyclization (**113⁻**).

5.3.2. Anticancer Compounds

Cholesterol-Containing Compounds for Anticancer Therapy

Considering anticancer compounds, cholesterol-containing compounds have gained attention for anticancer therapy. Liposomes, known for their biocompatibility, biodegradability, and low immunogenicity, are widely used as nanocarriers for hydrophobic and hydrophilic molecules. They are already actively used in medicine for the transport of some anticancer drugs, such as doxorubicin [204] or paclitaxel [205]. Cancer cells require an increased demand for cholesterol to build their membranes, making boron-containing cholesterols a promising strategy for selective boron delivery using liposomes. Recently, cobalt bis(dicarbollide) conjugates with cholesterol have been synthesized, forming stable and non-toxic liposomes. The conjugates were prepared by performing a copper(I)-catalyzed cycloaddition reaction. Cobalt bis(dicarbollide) derivatives were used with terminal azido groups, which were then reacted with an alkyne compound. The resulting triazoles contained cobalt bis(dicarbollide) at position 1 and cholesterol at position 4. Compounds **114**$^-$ (Figure 26) were tested for their anti-cancer effects using an MTT assay on various cell lines, including MCF7, HCT116, A549, and WI38. The cytotoxicity of the compounds, compared to cisplatin, was assessed. They demonstrated low toxicity (IC50 > 200 μmol L^{-1}) and exhibited consistent antiproliferative activity across different cell lines, irrespective of the spacer structure. This suggests their suitability for BNCT without significant toxicity concerns [206].

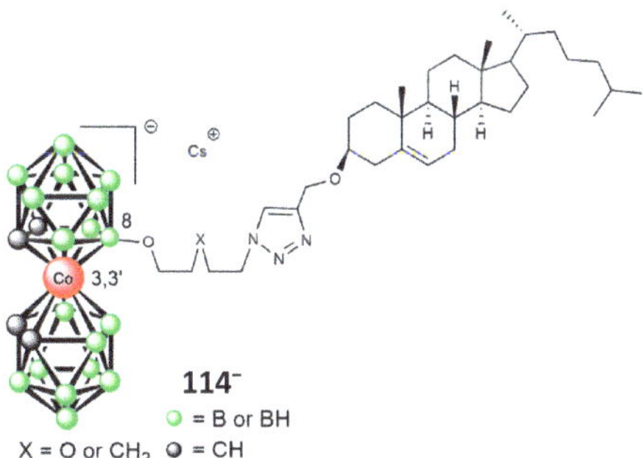

Figure 26. Chemical structure of [8-(OC$_4$H$_8$-X-N$_3$(3β-(2-azido-ethyl)cholest-5-ene))-(1,2-C$_2$B$_9$H$_{10}$)(1′,2′-C$_2$B$_9$H$_{11}$)-3,3′-Co(III)]$^-$ (**114**$^-$).

Druzina et al. synthesized cholesterol derivatives of cobalt bis(dicarbollide) with different spacers, namely hydrophilic (CH$_2$CH$_2$O)$_2$ in compound **115a**$^-$ and lipophilic (CH$_2$)$_5$ in compound **115b**$^-$ (Scheme 26). These derivatives were obtained through nucleophilic ring-opening reactions using oxonium derivatives of cobalt bis(dicarbollide) and a modified cholesterol derivative. The cytotoxicity of compounds **115a**$^-$ and **115b**$^-$ was evaluated against glioblastoma cells (U-87 MG) and human embryo fibroblasts (FECH-15). Compound **115a**$^-$ showed lower cytotoxicity towards normal cells compared to tumor cells, with an IC50 value of 1.56 mg mL^{-1} for normal cells and 0.06 mg mL^{-1} for tumor cells. Compound **115b**$^-$ exhibited similar cytotoxicity towards both cells, with IC50 values of 39.5 mg mL^{-1} and 1.56 mg mL^{-1}, respectively. The selective cytotoxicity index (CC50) for compounds **115a**$^-$ and **115b**$^-$ indicated their potential as antitumor agents. These cholesterol cobalt bis(dicarbollide) conjugates could be promising for BNCT and as independent antitumor agents. They also synthetized cholesterol-cobalt bis(dicarbollide) conjugates through the "click" cycloaddition of 3β-(2-azido-ethyl)cholest-5-ene and acetylene-functionalized cobalt bis(dicarbollides). These boronated cholesterols hold potential for application as drug delivery systems for Boron Neutron Capture Therapy in cancer treatment [164].

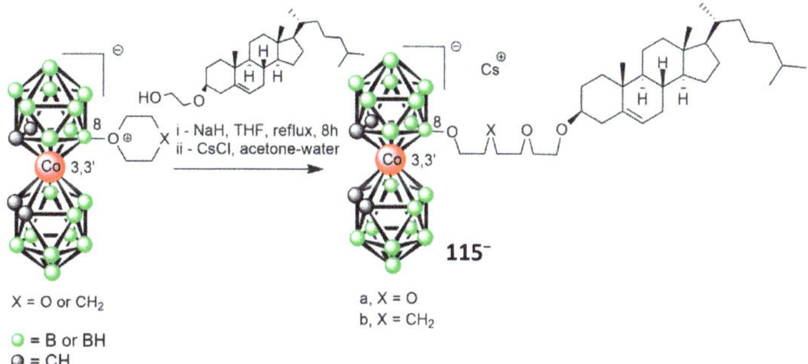

Scheme 26. Preparation of [8-(O$_2$C$_4$H$_8$-X-(3β-(2-azido-ethyl)cholest-5-ene))-(1,2-C$_2$B$_9$H$_{10}$)(1′,2′-C$_2$B$_9$H$_{11}$)-3,3′-Co(III)]$^-$ (**115**$^-$).

Furthermore, Druzina et al. reported on the synthesis of derivatives of cobalt bis(dicarbollide) terminal alkynes with charge-compensated groups (**116**; Scheme 27). They describe a reaction in which the oxonium derivative of cobalt bis(dicarbollide) reacts with tertiary amines to form corresponding ammonium salts. The authors successfully prepared terminal alkynes with quaternary ammonium groups based on cobalt bis(dicarbollide) using *N*,*N*-dimethylprop-2-yn-1-amine, resulting in novel derivatives with high yields. These compounds were further used in the synthesis of boronated cholesterol derivatives utilizing click reactions between terminal alkynes on metallacarborane and an organic azido-cholesterol derivative (**117**; Figure 27). The reactions yielded novel boron conjugates in high yields. The crystal structures of the alkynes prepared from 1,4-dioxane and tetrahydropyran derivatives of cobalt bis(dicarbollide) were determined. The synthesized compounds will be used to prepare boronated liposomes for delivering boron clusters to cancer cells in future BNCT experiments [108].

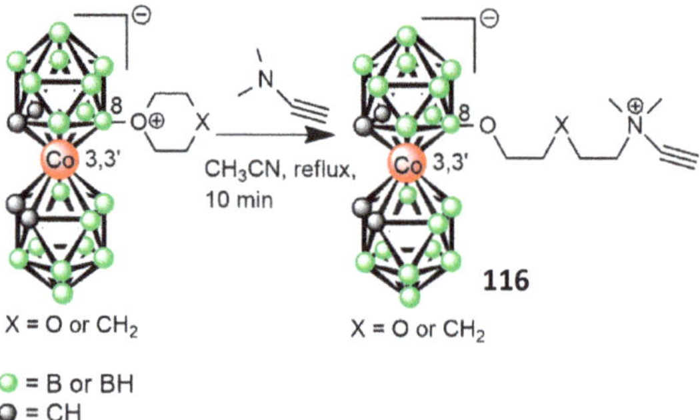

Scheme 27. Preparation of [8-OC₄H₈-*X*-NC₄H₇-(1,2-C₂B₉H₁₀)(1′,2′-C₂B₉H₁₁)-3,3′-Co(III)] (**116**).

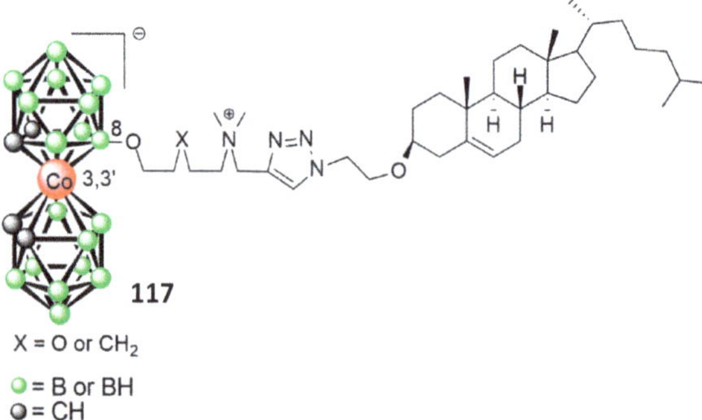

Figure 27. Chemical structure of [8-(OC₄H₈-*X*-NC₅H₉N₃-(3*β*-(2-azido-ethyl)cholest-5-ene))-(1,2-C₂B₉H₁₀)(1′,2′-C₂B₉H₁₁)-3,3′-Co(III)] (**117**).

The work of Dubey et al. (2021) explored the use of different linkers in liposomal formulations and their effects on various properties. The linkers tested include triazole, polyethylene glycol, benzene, and hydrocarbon chains (Figure 28). **118f⁻** is identified as the most hydrophilic, while **118e⁻** is the most hydrophobic among the compounds studied. The liposomal formulations are prepared using lipid thin-film hydration and extrusion, resulting in stable and uniformly sized particles with high zeta potential, indicating their physical stability. Encapsulation efficiency is found to be excellent for most formulations, with **118b⁻**, **118d⁻**, and **118f⁻** showing optimal hydrophobicity and the absence of benzene rings leading to high efficiency. The liposomes exhibit minimal drug release at physiological pH, suggesting sustained drug availability for tumor tissues through the enhanced permeability and retention (EPR) effect [207].

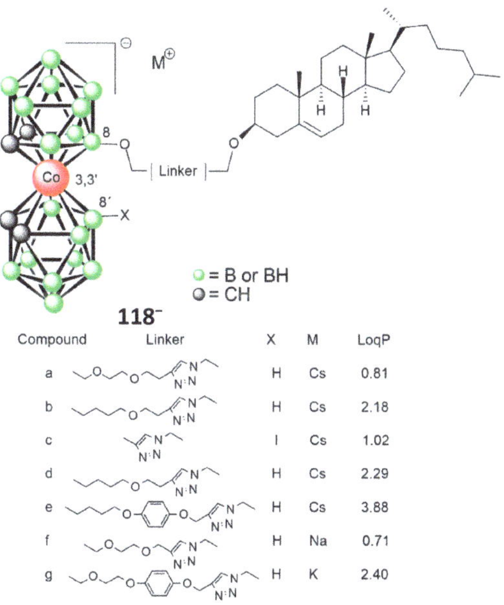

Figure 28. Chemical structure of [8-(OCH$_2$-*Linker*-(3β-(2-azido-ethyl)cholest-5-ene))-(1,2-C$_2$B$_9$H$_{10}$)(1',2'-C$_2$B$_9$H$_{11}$)-3,3'-Co(III)]⁻ (**118⁻**) [207].

Curcumin-Containing Compounds for Anticancer Therapy

In the synthesis of cobalt bis(dicarbollide)-curcumin conjugates, a series of compounds were designed using a "boron click" reaction. The reaction involved the nucleophilic ring-opening of cyclic oxonium derivatives of cobalt bis(dicarbollide) with the OH-group of curcumin. Different oxonium cycles and curcumin were used to obtain the boronated curcumin derivatives of cobalt bis(dicarbollide). Cyclic esters of cobalt bis(dicarbollide) provided spacers with varying hydrophilicity (–(CH$_2$CH$_2$O)$_2$– spacer) or lipophilicity (–(CH2)4-5– spacer) between the boron cage and the biological macromolecule. The choice of spacers allowed for flexibility and biocompatibility in the synthesized compounds. Monoanionic and dianionic products were isolated as potassium salts with good yields. The viability of human tumor cell lines (HCT116 and K562) and non-malignant human skin fibroblasts (hFB-hTERT6) was assessed to evaluate the effect of prepared conjugates. Curcumin and doxorubicin were used as reference compounds. Curcumin exhibited cytotoxicity against all cell lines, with slightly lower toxicity towards normal fibroblasts compared to malignant cells. However, the cobalt bis(dicarbollide)-curcumin conjugates **119⁻a, b**, and **120²⁻a, b** were found to be inactive (Figure 29). To determine if the lack of cytotoxicity was due to poor cell penetration, intracellular drug accumulation was examined using flow cytometry. It was confirmed that all the conjugates entered HCT116 cells in a time-dependent manner. Conjugate **119a⁻** showed the highest penetration, while compounds **120a²⁻** and **120b²⁻** required longer incubation times for complete accumulation. The compounds did not affect cell viability. These findings indicate that compounds **119⁻a, b**, and **120²⁻a, b** can penetrate cells without causing cytotoxicity [109].

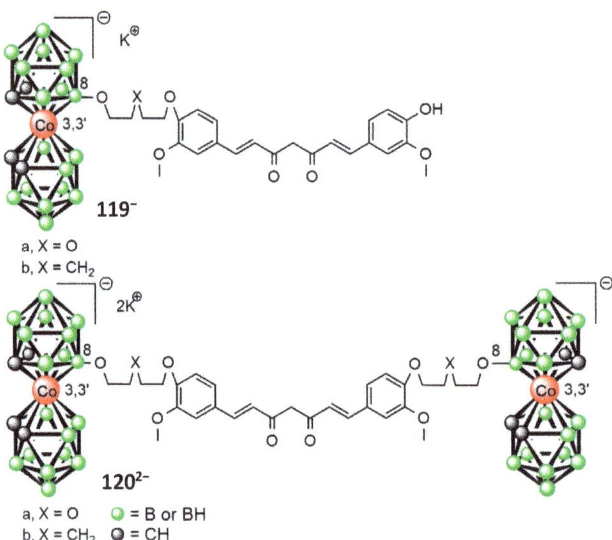

Figure 29. Chemical structure of [8-OC$_2$H$_4$-X-C$_2$H$_4$O-((1E,6E)-1-(4-hydroxy-3-methoxyphenyl)-7-(3-methoxyphenyl)hepta-1,6-diene-3,5-dione)-(1,2-C$_2$B$_9$H$_{10}$)(1′,2′-C$_2$B$_9$H$_{11}$)-3,3′-Co(III)]$^-$ (**119$^-$**) and {(((1E,6E)-1,7-bis(3-methoxyphenyl)hepta-1,6-diene-3,5-dione)-[8-OC$_2$H$_4$-X-C$_2$H$_4$O-(1,2-C$_2$B$_9$H$_{10}$) (1′,2′-C$_2$B$_9$H$_{11}$)-3,3′-Co(III)]$_2$}$^{2-}$ (**120^{2-}**).

Chlorin-Containing Compounds for Anticancer Therapy

Determining boron content in blood and tissue samples is crucial for BNCT planning and application. Ex vivo methods, like inductively coupled plasma mass spectroscopy (ICP-MS), atomic emission spectrometry (ICP-AES) [109,208,209], and prompt gamma-ray spectroscopy (PGRS) [210–212], are commonly used for boron accumulation assessment. Fluorescent boron-containing compounds offer a non-invasive approach to the quantitative assessment of boron content in biological tissues, as their fluorescence intensity corresponds to the boron amount [213]. Chlorin e_6-cobalt bis(dicarbollide) conjugates (CCDC) have shown selective accumulation in lung adenocarcinoma cells suitable for BNCT. However, accurately assessing the boron content using conjugate fluorescence remains a challenge [214,215]. To address this, mathematical simulations based on pharmacokinetic models were used to provide a preliminary assessment of medication doses and compound content in various organs. In the case of studying new boron-containing compounds for BNCT, the development of pharmacokinetic models is of significant interest [216]. The study of Volovetsky et al. demonstrates a direct correlation between boron concentration and CCDC fluorescence in different tissues, indicating the stability of the compound. The authors successfully achieved a contrasting accumulation of boron between tumor and muscle tissues using the chlorin e_6-cobalt bis(dicarbollide) conjugate (**121$^-$**; Figure 30, top). This highlights the potential of fluorescent methods for the non-invasive determination of boron content in living organisms. Furthermore, a mathematical model was developed that accurately describes the accumulation and distribution of CCDC in tissues. The model's high fit with the experimental data suggests its reliability and usefulness in understanding and predicting the behavior of boron accumulation and distribution in various organs [217].

Fedotova et al. discussed the combination of BNCT and photodynamic therapy (PDT) for treating head and neck tumors [218]. BNCT involves using non-toxic isotopes ^{10}B and thermal neutrons to target cancer cells, while PDT uses chlorine derivatives as photosensitizers to absorb light and induce therapy [215,219]. Researchers aim to create a theranostic conjugate by adding boron clusters to the chlorin–cyclen conjugate (**122$^-$**; Figure 30, bottom), increasing boron atom concentration for more efficient BNCT [220].

The text describes two approaches for preparing boron-containing chlorin conjugates, which comprise a chlorin–cyclen conjugate that reacts with a nitrile derivative of bis(1,2-dicarbollide) cobalt, resulting in the addition of one boron cluster; or, in the second case, the chlorin–cyclen conjugate reacts with a dioxane derivative of bis(1,2-dicarbollide) cobalt, yielding a mixture of mono- and di-substituted derivatives. Interestingly, when using nitrile derivatives, only one product was obtained, while the dioxane derivative allowed for the introduction of a second boron cluster. This difference can be attributed to the steric effects and the length of the spacer group between the macrocycle and boron polyhedron [218].

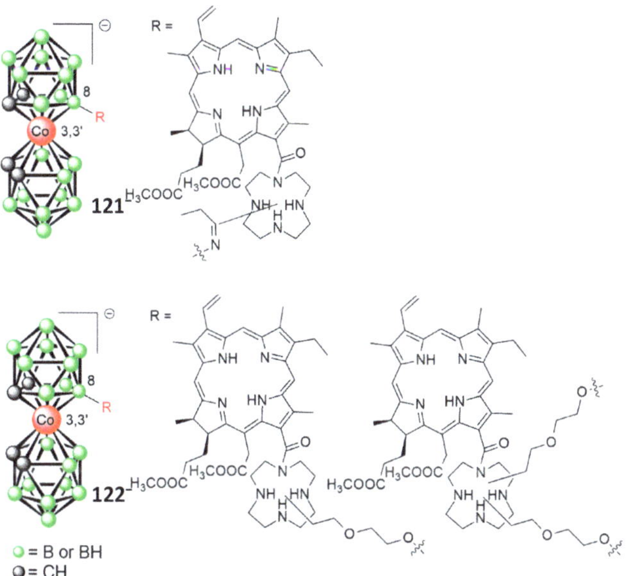

Figure 30. Chemical structure of chlorin–cyclen cobalt bis(dicarbollide) derivatives.

Coumarin-Containing Compounds

Coumarins are natural biologically active compounds with versatile biomedical applications (e.g., anticancer, antimicrobial, and anticoagulant activity). That is why they are given increased attention [221]. In 2020, Kosenko et al. performed a synthesis of coumarin derivatives attached to cobalt bis(dicarbollide) (**123⁻**, **124⁻**, **125⁻**; Figure 31). The authors used nucleophilic ring cleavage reactions with oxonium derivatives to prepare these compounds. They found that reactions with cobalt bis(dicarbollide) resulted in conjugates attached at either the C-3 or C-7 positions of coumarin. The new compounds showed good yields and fluorescent properties. The authors also attempted to attach the boron cluster at other positions of the coumarin ring, which has been found to be challenging [222]. Serdyukov et al. (2021) reported on a coumarin conjugate with cobalt bis(dicarbollide) as well (**126⁻**; Figure 31). The synthesis involved the ring cleavage of cyclic oxonium derivative of cobalt bis(dicarbollide) ion by coumarin moiety. The reaction of charge-compensated cobalt bis(dicarbollide) derivative with 7-diethylamino-4-hydroxycoumarin resulted in an anionic product. A new anionic boron conjugate **126⁻** was obtained in the form of a potassium salt. The lipophilicity of the synthesized compound was determined using a $\log D_{7.4}$ measurement, indicating that this conjugate seems promising for medicinal applications [223].

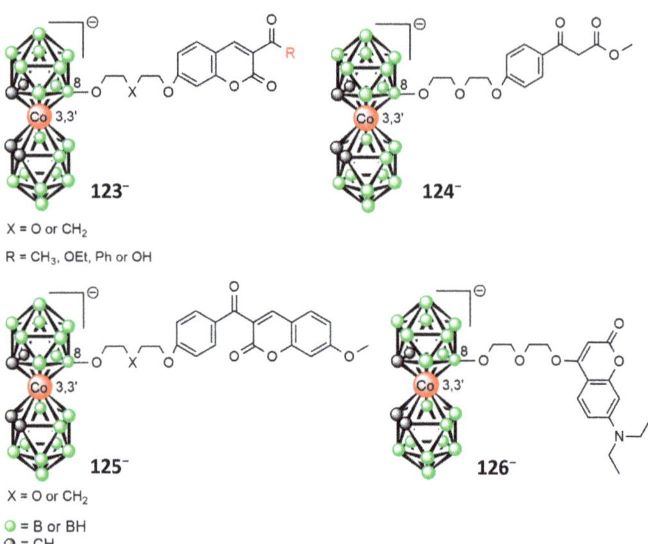

Figure 31. Chemical structure of [8-OC$_2$H$_4$-X-C$_2$H$_4$O-(2H-chromen-2-one)-CO-R-(1,2-C$_2$B$_9$H$_{10}$)(1′,2′-C$_2$B$_9$H$_{11}$)-3,3′-Co(III)]$^-$ (**123**$^-$), [8-O$_3$C$_4$H$_8$-(methyl-3-oxo-3-phenylpropanoate)-(1,2-C$_2$B$_9$H$_{10}$)(1′,2′-C$_2$B$_9$H$_{11}$)-3,3′-Co(III)]$^-$ (**124**$^-$), [8-OC$_2$H$_4$-X-C$_2$H$_4$O-(3-benzoyl-7-methoxy-2H-chromen-2-one)-(1,2-C$_2$B$_9$H$_{10}$)(1′,2′-C$_2$B$_9$H$_{11}$)-3,3′-Co(III)]$^-$ (**125**$^-$), and [8-O$_3$C$_4$H$_8$-(7-(diethylamino)-2H-chromen-2-one)(1,2-C$_2$B$_9$H$_{10}$)(1′,2′-C$_2$B$_9$H$_{11}$)-3,3′-Co(III)]$^-$ (**126**$^-$).

Glioblastoma and Neuroblastoma

Glioblastoma (GBM) is a highly aggressive primary brain tumor with limited treatment options and a poor prognosis. Standard treatments, such as surgical resection followed by chemoradiotherapy, are met with limited success due to the development of therapy resistance and tumor recurrence. Over the years, BNCT has been considered as a potential alternative for the treatment of GBM [224–228]. To further explore the potential of **1**$^-$ for BNCT, Serdyukov et al. employed Synchrotron Radiation-Fourier Transform Infrared (SR-FTIRM) micro-spectroscopy. This advanced technique allows for the non-destructive analysis of key molecular structures within cells, providing valuable diagnostic information. By utilizing the characteristic ν B-H frequency range of 2.600–2.500 cm^{-1}, specific to boron clusters, SR-FTIRM enabled the detection and localization of **1**$^-$ within cells [223]. The study of Nuez-Martinez et al. investigates the uptake and effects of Na**1** in glioma-initiating cells (GICs). It demonstrates that Na**1** enters GICs and induces changes in cell phenotype, particularly in radio-resistant mesenchymal cells. The study highlights the potential of Na**1** as a therapeutic compound for glioblastoma treatment, especially in resistant cases of GBM. The use of SR-FTIRM provided insight into drug distribution and radio-resistance. The high uptake and rapid clearance of Na**1** make it a promising candidate for BNCT. Further research is needed to understand the underlying mechanisms and optimize its delivery [224].

In another study by Nuez-Martinez et al., the potential of **1**$^-$ for both chemotherapy and BNCT in GBM was investigated using in vitro and in vivo models. Studies using spheroids derived from the U87 and T98G cell lines revealed that T98G spheroids showed increased resistance to treatment compared to that of the U87 type, contrary to results observed in 2D monolayer cultures. This highlights the importance of employing 3D models for GBM studies. In vitro tests demonstrated that **1**$^-$ and [8,8′-I$_2$-**1**$^-$], at non-cytotoxic concentrations, effectively loaded sufficient levels of boron into GBM cells for successful boron neutron capture (BNC) reactions. T98G cells, known for their resistance to standard radiotherapies, exhibited enhanced sensitivity to neutron irradiation after incubation with

Na[8,8′-I$_2$-**1**$^-$] due to their higher boron uptake. In vivo tests using C. elegans nematodes and embryos confirmed the toxicity of these compounds. The compounds formed hybrids with C. elegans' eggs and arrested larvae development. Thus, [8,8′-I$_2$-**1**$^-$] and related compounds have the potential as candidates for the combination BNCT treatment of resistant variants of GBM [38].

The neuroblastoma (NB) is the most common solid extracranial tumor in children [229]. Current diagnosis involves invasive procedures and general anesthesia for young patients. Screening for NB can be based on monitoring the levels of catecholamine metabolites, homovanillic acid (HVA), and vanillylmandelic acid (VMA) (Figure 32). Various laboratory techniques are used for their analysis. Immobilized supramolecular systems are proposed to enhance the detection of NB metabolites through selective interactions. In this study, Shishanova et al. designed a cobalt bis(dicarbollide) derivative including an o-phenylenediamine unit (CB-oPD) that shows potential for specific interactions with carboxylate analytes (**127**; Figure 32). The objective was to evaluate electrochemical techniques for determining VMA and HVA, considering their structural similarity. Electropolymerized film based on CB-oPD is proposed as a "host" for the selective detection of the metabolites ("guests"). The focus is on monitoring VMA and HVA as individual species and as a mixture to aid in correct diagnosis. This approach is considered novel for the electrochemical detection of these metabolites [230–234]. The electrochemical recognition of neuroblastoma metabolites was investigated on a modified electrode surface using electrochemical impedance spectroscopy (EIS), potentiometry, and differential pulse voltammetry (DPV). The association constants indicate the higher affinity of the electrode surface for HVA (2.5×10^5) compared to VMA (4.3×10^4). The binding of the metabolites to the pCB-oPD-modified electrode allows for the differentiation of VMA and HVA in a mixture, offering practical applications [234].

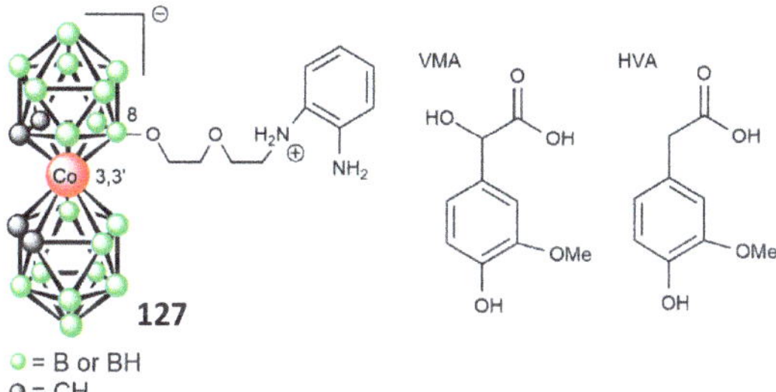

Figure 32. Chemical structure of [8-C$_4$H$_8$O$_2$NH$_2$-C$_6$H$_6$N-I-(1,2-C$_2$B$_9$H$_{10}$)(1′,2′-C$_2$B$_9$H$_{11}$)-3,3′-Co(III)] (**127**).

5.3.3. Nanocomposites for Bioimaging and Drug Delivery

Nanomaterials have shown great potential in the development of efficient in vivo imaging probes for noninvasive techniques like positron emission tomography (PET), magnetic resonance imaging, and optical imaging [234]. Carbon nanomaterials, such as graphene oxide (GO), have attracted attention due to their unique properties, high biocompatibility, and prolonged blood circulation times. GO can be functionalized and used as a platform for drug delivery systems, including anticancer agents [235–237]. The cobalt bis(dicarbollide) derivatives have been linked to macromolecules and nanoparticles, including GO, resulting in materials with improved properties and cellular uptake. The chemical modification of GO with cobalt bis(dicarbollide) anions has led to hybrid materials with improved dispersibility and thermal properties. Radiolabeling cobalt bis(dicarbollide) enables its visualization in vivo using PET. In the study of Ferrer-Ugalde et al., boron-enriched carbon-based materials tagged with the positron emitter iodine-124 (^{124}I) were developed as theranostic agents for whole-body imaging. The nanocomposite, **128**$^-$ (Figure 33), was prepared by functionalizing graphene oxide (GO) with a monoiodinated boron-cluster compound (I-**1**$^-$). To synthesize I-**1**$^-$, iodine crystals were added to a solution of [8-C$_4$H$_8$O$_2$NH$_3$-3,3'-Co(1,2-C$_2$B$_9$H$_{10}$)(1',2'-C$_2$B$_9$H$_{11}$)] in dichloromethane. To attach I-**1**$^-$ to GO, a covalent coupling reaction was performed using *N,N'*-dicyclohexylcarbodiimide (DCC) and 1-hydroxybenzotriazole (HOBt) as coupling reagents. The resulting **128**$^-$ exhibited exfoliated graphene derivatives and a graphitic structure. The biocompatibility of **128**$^-$ was evaluated in healthy cells using cytotoxicity studies. The results showed low cytotoxicity, with cell viability above 90% at different concentrations and incubation times. A TEM analysis confirmed the internalization of **128**$^-$, and live/dead tests confirmed that the majority of cells remained alive, even at the highest concentration. The positive control exhibited high cell mortality. The TEM analysis showed the presence of material aggregates in the cytoplasm without affecting cell structure. In vivo toxicity tests using C. elegans demonstrated the lack of toxicity and internalization of both **128**$^-$ and I-**1**$^-$. The radiolabeling of I-**1**$^-$ with iodine isotopes enabled PET imaging, revealing the accumulation of [^{124}I] **128**$^-$ in the liver, lungs, and heart, with long circulation time and elimination through the gastrointestinal tract. These results indicate the potential of **128**$^-$ as a theranostic agent for Boron Neutron Capture Therapy [238].

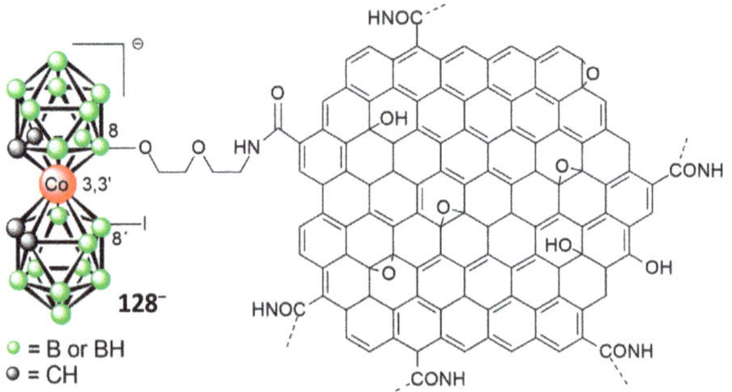

Figure 33. Chemical structure of [8-C$_4$H$_8$O$_2$NH-*GO*-8'-I-(1,2-C$_2$B$_9$H$_{10}$)$_2$-3,3'-Co(III)]$^-$ (**128**$^-$).

Pulagam et al. (2021) developed PEG–cobalt bis(dicarbollide)-AuNRs, a nanoconjugate with potential applications in BNCT. They prepared CTAB-stabilized gold nanorods (AuNRs) and modified them with mPEG (poly(ethylene glycol) methyl ether thiol) for biocompatibility and circulation enhancement. Cobalt bis(dicarbollide) derivatives were then conjugated onto the AuNR surface. The presence of cobalt bis(dicarbollide) ion on the nanoconjugates was confirmed through various analyses, including NMR, STEM-EDXS, and XPS. Despite gold's high cross-section for thermal neutrons, their calculations showed that the presence of gold as a boron carrier is not a limitation for BNCT applications. The nanoconjugates showed good stability, low cytotoxicity, and accumulated in the tumor in a mouse model of gastric adenocarcinoma. They exhibited promising results for combined Photothermal Therapy and BNCT in vitro, inducing localized thermal heating and cell damage under NIR and neutron irradiation. However, the accumulation of nanoconjugates in the tumor needs improvement for more effective BNCT treatment [239].

5.4. Electrochemistry

The electrochemistry of cobalt bis(dicarbollide) anion in non-aqueous media has already been reported in the early 1970s [64,240] and served predominantly as a characterization method for establishing the physicochemical properties of newly synthetized species. These works include, for example, studies on the B-substituted porphyrin derivatives and their spectroelectrochemistry [241], the electrochemical and spectroelectrochemical characterization of the phthalocyanines B-substituted potential BNCT treatment candidate [242], and the electrochemistry of different substituted metallacarboranes including the cobalt ones [243]. This progress was covered by a recent review article [11].

Growing attention to the use of the cobalt bis(dicarbollide) anion in biological applications led to the first attempts of electrochemistry in the aqueous milieu [11,12,14,153]. The authors report not only the reversible redox signal of the Co(III) central atom (as was reported previously in non-aqueous media [240]) but also the irreversible signal of the core boron cage (Figure 34) [63,82,244]. The electrochemical response for the parent cobalt bis(dicarbollide) anion (using a polished glassy carbon electrode in a phosphate buffer of pH = 8) is rather complicated. It consists of the reversible signal of the Co(III)/Co(II) redox process on the cathodic part of the voltammogram. On the anodic part, a set of at least three overlapping peaks could be observed in the bride potential range. This signal was ascribed to the electrooxidation of the dicarbollide ligands. Interestingly, in the case of the bridged samples, the set of overlapping peaks merged into one with a markedly higher current density response. This happens only in the case of a short bridge linker (as a single atom); with prolonging the linker chain, the signals tend to split again into a set of overlapping peaks [244]. This behavior could be explained by the presence of the three main rotamers of the parent cobalt bis(dicarbollide) anion, where the short bridging linker does not allow the transition between the different states. Study [63] compares the electrochemical behavior of B- and C-substituted derivatives, namely the alkylhydroxy and carboxy groups. Although the electronic properties of the C- and B-derivatives (with comparable ligand structure) markedly differ, this work does not find any significant difference between the electrochemical behavior of these derivatives. Interestingly, other studies conducted in aqueous media reported only the reversible signal of Co(II), for example, the original electrochemical work of the Teixidor group [12], and the use of bis(dicarbollide) ion as a transducer on an electrode for neuroblastoma markers sensing [245].

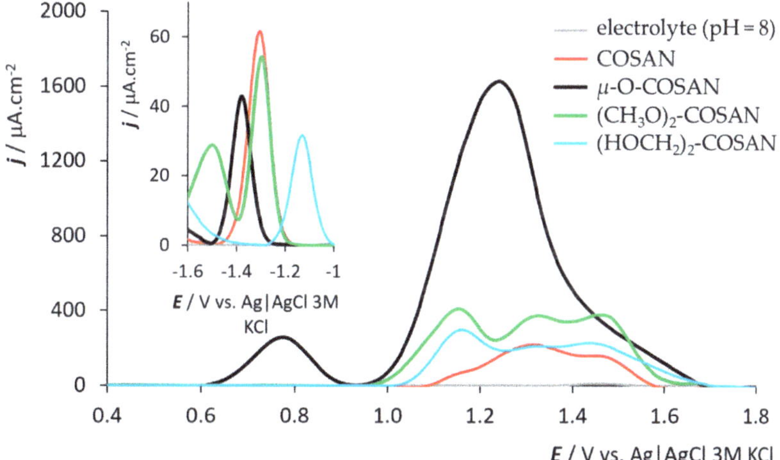

Figure 34. Example of DPV (differential pulse voltammogram) of some cobalt bis(dicarbollide) (COSAN) anions in phosphate buffer, pH = 8500 µmol L^{-1} concentrations of all samples, polished glassy carbon electrode. COSAN-*closo*-[(1,2-C$_2$B$_9$H$_{11}$)$_2$-3,3′-Co(III)]Cs; μ-O-COSAN–[μ-8,8′-O-(*ortho*-1,2-*closo*-C$_2$B$_9$H$_{10}$)$_2$-3,3′-Co(III)]Na.nH$_2$O; (CH$_3$O)$_2$-COSAN–[8,8′-(CH$_3$O)$_2$-(*ortho*-1,2-*closo*-C$_2$B$_9$H$_{10}$)$_2$-3,3′-Co(III)]Me$_4$N; and (HOCH$_2$)$_2$-COSAN–[1,1′-(HOCH$_2$)$_2$-1,2-C$_2$B$_9$H$_{10}$)$_2$-3,3′-Co(III)]Me$_3$NH [63,244].

6. Conclusions and Outlook

This review particularly and primarily focuses on the chemistry of the carbon-substituted derivatives of the cobalt bis(dicarbollide) cage, which proceeds via the direct metalation of C-H vertices followed by reactions with electrophiles. We have demonstrated that these methods can be used as a versatile solution for producing a large variety of available structural blocks that offer viable alternatives to boron-substituted compounds and can be used in diverse applications. As demonstrated, due to its flexibility, this approach already resulted in improved inhibitory properties of enzyme inhibitors designed to target cancer-associated CA-IX or compounds that proved to have better antiparasitic activity than the boron-substituted analogues. Furthermore, the methods outlined here also allow for an easy heterosubstitution on boron and carbon vertices.

In light of recent advancements in ferrocene chemistry, we can anticipate further progress in the understanding of the mechanism of metalation of the cobalt bis(dicarbollide) cage. This progress will likely be accompanied by further improvements in reaction conditions, including the use of auxiliary bases and reagents, aimed at enhancing the stereospecificity of these reactions. It is worth investigating the *ortho*-directive effect of certain groups located on carbons and the activation of B-H bonds in close proximity.

The chemistry of carbon atoms provides a straightforward means of inducing chirality into molecules. These compounds may, in principle, serve as chiral platforms resembling BINOL, BINAP, and chiral ferrocenes. Progress in the chemistry of chiral species may, thus, be anticipated, leading to the production of enantiomers of high enantiomeric excess when chiral additives are used in metalation reactions. This will be paralleled with the further development of analytical (and preparative) HPLC and other separation methods that would allow for the fast analysis of enantiomeric purity.

Author Contributions: Conceptualization, L.P., E.Z.T. and B.G.; methodology, D.B.; validation, L.P., M.L. and D.B.; data curation, D.B., L.F. and M.L.; writing—original draft preparation, L.P., E.Z.T., D.B., L.F. (paragraph on electrochemistry) and B.G.; writing—review and editing, L.P. and B.G.; visualization, D.B.; supervision, B.G.; funding acquisition, B.G. All authors have read and agreed to the published version of the manuscript.

Funding: This work was funded by the Czech Science Foundation, Project No. 21-14409S.

Acknowledgments: We wish to thank Zdeňka Růžičková, Department of General and Inorganic Chemistry, Faculty of Chemical Technology, University of Pardubice for providing the crystallographic data used in the preparation of Figure 2 and for drawing figures based on previously published data. The authors acknowledge the language editing by Michael. G. Londesborough. We appreciate the support from the Czech Science Foundation, Project No. 21-14409S.

Conflicts of Interest: The authors declare no conflict of interest.

References

1. Hawthorne, M.F. Chemistry of the polyhedral species derived from transition metals and carboranes. *Accounts Chem. Res.* **1968**, *1*, 281–288. [CrossRef]
2. Hawthorne, M.F.; Andrews, T.D. Carborane analogues of cobalticinium ion. *Chem. Commun.* **1965**, *19*, 443–444. [CrossRef]
3. Francis, J.N.; Hawthorn, M.F.; Jones, C.J. Chemistry of bis(.pi.-7,8-dicarballyl)metalates. Reaction between [(.pi.-7,8-B9C2H11)2Co]- and aryl diazonium salts. *J. Am. Chem. Soc.* **1972**, *94*, 4878–4881. [CrossRef]
4. Viñas, C.; Pedrajas, J.; Bertran, J.; Teixidor, F.; Kivekäs, R.; Sillanpää, R. Synthesis of cobaltabis(dicarbollyl) complexes incorporating ex-ocluster SR substituents and the improved synthesis of [3,3′-Co(1-R-2-R′-1,2-C2B9H9)2]- derivatives. *Inorg. Chem.* **1997**, *36*, 2482–2486. [CrossRef]
5. Hawthorne, M.F.; Young, D.C.; Andrews, T.D.; Howe, D.V.; Pilling, R.L.; Pitts, A.D.; Reintjes, M.; Warren, L.F., Jr.; Wegner, P.A. pi.-Dicarbollyl derivatives of the transition metals. Metallocene analogs. *J. Am. Chem. Soc.* **1968**, *90*, 879–896. [CrossRef]
6. Sivaev, I.B.; Bregadze, V.I. Chemistry of cobalt bis(dicarbollides). A review. *Collect. Czech. Chem. Commun.* **1999**, *64*, 783–805. [CrossRef]
7. Dash, B.P.; Satapathy, R.; Swain, B.R.; Mahanta, C.S.; Jena, B.B.; Hosmane, N.S. Cobalt bis(dicarbollide) anion and its derivatives. *J. Organomet. Chem.* **2017**, *849–850*, 170–194.
8. Grimes, R.N. Metallacarboranes of the transition and lanthanide elements. In *Carboranes*; Elsevier: Amsterdam, The Netherlands, 2016; pp. 711–903.
9. Rais, J.; Grűner, B. Extraction with Metal bis(Dicarbollide) Anions; Metal bis(dicarbollide) extractants and their applications in separation chemistry. In *Ion Exchange, Solvent Extraction*, 1st ed.; Marcus, Y., SenGupta, A.K., Eds.; Marcel Dekker: New York, NY, USA, 2004; Volume 17, pp. 243–334.
10. Grűner, B.; Rais, J.; Selucký, P.; Lucanikova, M. Recent progress in extraction agents based on cobalt bis(dicarbollides) for partitioning of radionuclides from high level nuclear waste. In *Chapter 19 in Boron Science, New Technologies and Applications*; Hosmane, N.S., Ed.; CRC Press: Boca Raton, FL, USA, 2012.
11. Núñez, R.; Tarrés, M.; Ferrer-Ugalde, A.; de Biani, F.F.; Teixidor, F. Electrochemistry and photoluminescence of icosahedral carboranes, boranes, metallacarboranes, and their derivatives. *Chem. Rev.* **2016**, *116*, 14307–14378.

12. Xavier, J.A.M.; Viñas, C.; Lorenzo, E.; García-Mendiola, T.; Teixidor, F. Potential application of metallacarboranes as an internal reference: An electrochemical comparative study to ferrocene. *Chem. Commun.* **2022**, *58*, 4196–4199. [CrossRef]

13. Xavier, J.A.M.; Fuentes, I.; Nuez-Martínez, M.; Kelemen, Z.; Andrio, A.; Viñas, C.; Compañ, V.; Teixidor, F. How to switch from a poor PEDOT:X oxygen evolution reaction (OER) to a good one. A study on dual redox reversible PEDOT:metallacarborane. *J. Mater. Chem.* **2022**, *10*, 16182–16192. [CrossRef]

14. Stoica, A.I.; Viñas, C.; Teixidor, F. History of cobaltabis(dicarbollide) in potentiometry, no need for ionophores to get an excellent selectivity. *Molecules* **2022**, *27*, 14.

15. Teixidor, F.; Viñas, C.; Planas, J.G.; Romero, I.; Núñez, R. Advances in the catalytic and photocatalytic behavior of carborane derived metal complexes. In *Advances in the Synthesis and Catalytic Applications of Boron Cluster: A Tribute to the Works of Professor Francesc Teixidor and Professor Clara Viñas*, 1st ed.; Dieguez, M., Núñez, R., Eds.; Academic Press: Cambridge, MA, USA, 2022; Volume 71, pp. 1–45.

16. Guerrero, I.; Viñas, C.; Romero, I.; Teixidor, F. A stand-alone cobalt bis(dicarbollide) photoredox catalyst epoxidates alkenes in water at extremely low catalyst load. *Green Chem.* **2021**, *23*, 10123–10131. [CrossRef]

17. Hardie, M.J.; Raston, C.L. Solid state supramolecular assemblies of charged supermolecules (Na[2.2.2]cryptate) and anionic carboranes with host cyclotriveratrylene. *Chem. Commun.* **2001**, *10*, 905–906.

18. Fox, M.A.; Hughes, A.K. Cage C-H center dot center dot center dot X interactions in solid-state structures of icosahedral carboranes. *Coord. Chem. Rev.* **2004**, *248*, 457–476. [CrossRef]

19. Brusselle, D.; Bauduin, P.; Girard, L.; Zaulet, A.; Viñas, C.; Teixidor, F.; Ly, I.; Diat, O. Lyotropic lamellar phase formed from monolayered theta-shaped carborane-cage amphiphiles. *Angew. Chem. Int. Edit.* **2013**, *52*, 12114–12118. [CrossRef] [PubMed]

20. Tarrés, M.; Viñas, C.; González-Cardoso, P.; Hänninen, M.M.; Sillanpää, R.; Ďord'ovič, V.; Uchman, M.; Teixidor, F.; Matejicek, P. Aqueous self-assembly and cation selectivity of cobaltabisdicarbollide dianionic dumbbells. *Chem. Eur. J.* **2014**, *20*, 6786–6794. [CrossRef]

21. Bauduin, P.; Prevost, S.; Farràs, P.; Teixidor, F.; Diat, O.; Zemb, T. A Theta-shaped amphiphilic cobaltabisdicarbollide anion: Transition from monolayer vesicles to micelles. *Angew. Chem. Int. Edit.* **2011**, *50*, 5298–5300. [CrossRef] [PubMed]

22. Viñas, C.; Tarres, M.; González-Cardoso, P.; Farràs, P.; Bauduin, P.; Teixidor, F. Surfactant behaviour of metallacarboranes. A study based on the electrolysis of water. *Dalton Trans.* **2014**, *43*, 5062–5068. [PubMed]

23. Ďord'ovič, V.; Tošner, Z.; Uchman, M.; Zhigunov, A.; Reza, M.; Ruokolainen, J.; Pramanik, G.; Cígler, P.; Kalíková, K.; Gradzielski, M.; et al. Stealth amphiphiles: Self-assembly of polyhedral boron clusters. *Langmuir* **2016**, *32*, 6713–6722. [CrossRef]

24. Hao, E.; Sibrian-Vazquez, M.; Serem, W.; Garno, J.C.; Fronczek, F.R.; Vicente, M.G.H. Synthesis, aggregation and cellular investigations of porphyrin–cobaltacarborane conjugates. *Chem. Eur. J.* **2007**, *13*, 9035–9042.

25. Grűner, B.; Brynda, J.; Das, V.; Šícha, V.; Štěpánková, J.; Nekvinda, J.; Holub, J.; Pospíšilová, K.; Fábry, M.; Pachl, P.; et al. Metallacarborane sulfamides: Unconventional, specific, and highly selective inhibitors of carbonic anhydrase IX. *J. Med. Chem.* **2019**, *62*, 9560–9575. [CrossRef] [PubMed]

26. Fink, K.; Cebula, J.; Tošner, Z.; Psurski, M.; Uchman, M.; Goszczyński, T.M. Cobalt bis(dicarbollide) is a DNA-neutral pharmacophore. *Dalton Trans.* **2023**, *52*, 10338–10347.

27. Chen, Y.; Barba-Bon, A.; Grűner, B.; Winterhalter, M.; Aksoyoglu, M.A.; Pangeni, S.; Ashjari, M.; Brix, K.; Salluce, G.; Folgar-Cameán, Y.; et al. Metallacarborane cluster anions of the cobalt bisdicarbollide-type as chaotropic carriers for transmembrane and intracellular delivery of cationic peptides. *J. Am. Chem. Soc.* **2023**, *145*, 13089–13098. [CrossRef]

28. Gan, L.; Nord, M.T.; Lessard, J.M.; Tufts, N.Q.; Chidambaram, A.; Light, M.E.; Huang, H.L.; Solano, E.; Fraile, J.; Suárez-García, F.; et al. Biomimetic photodegradation of glyphosate in carborane-functionalized nanoconfined spaces. *J. Am. Chem. Soc.* **2023**, *145*, 13730–13741.

29. Fuentes, I.; García-Mendiola, T.; Sato, S.; Pita, M.; Nakamura, H.; Lorenzo, E.; Teixidor, F.; Marques, F.; Viñas, C. Metallacarboranes on the road to anticancer therapies: Cellular uptake, DNA interaction, and biological evaluation of cobaltabisdicarbollide COSAN (-). *Chem.-Eur. J.* **2018**, *24*, 17239–17254.

30. Farràs, P.; Juárez-Pérez, E.J.; Lepšík, M.; Luque, R.; Núñez, R.; Teixidor, F. Metallacarboranes and their interactions: Theoretical insights and their applicability. *Chem. Soc. Rev.* **2012**, *41*, 3445–3463.

31. Řezáčová, P.; Cígler, P.; Matějíček, P.; Pokorná, J.; Grűner, B.; Konvalinka, J. Medicinal application of carboranes: Inhibition of HIV protease. In *Boron Science- New Technologies and Applications*; Hosmane, N.S., Ed.; CRC Press: Boca Raton, FL, USA, 2012; pp. 45–63.

32. Kugler, M.; Nekvinda, J.; Holub, J.; El Anwar, S.; Das, V.; Šícha, V.; Pospíšilová, K.; Fábry, M.; Král, V.; Brynda, J.; et al. Inhibitors of CA IX enzyme based on polyhedral boron compounds. *ChemBioChem* **2021**, *22*, 2741–2761. [PubMed]

33. Kaplánek, R.; Martásek, P.; Grűner, B.; Panda, S.; Rak, J.; Masters, B.S.S.; Král, V.; Roman, L.J. Nitric oxide synthases activation and inhibition by metallacarborane-cluster-based isoform-specific affectors. *J. Med. Chem.* **2012**, *55*, 9541–9548.

34. Couto, M.; Mastandrea, I.; Cabrera, M.; Cabral, P.; Teixidor, F.; Cerecetto, H.; Viñas, C. Small-molecule kinase-inhibitors-loaded boron cluster as hybrid agents for glioma-cell-targeting therapy. *Chem. Eur. J.* **2017**, *23*, 9233–9238. [CrossRef]

35. Fink, K.; Uchman, M. Boron cluster compounds as new chemical leads for antimicrobial therapy. *Coord. Chem. Rev.* **2021**, *431*, 213684.

36. Bennour, I.; Ramos, M.N.; Nuez-Martínez, M.; Xavier, J.A.M.; Buades, A.B.; Sillanpää, R.; Teixidor, F.; Choquesillo-Lazarte, D.; Romero, I.; Martinez-Medina, M.; et al. Water soluble organometallic small molecules as promising antibacterial agents: Synthesis, physical-chemical properties and biological evaluation to tackle bacterial infections. *Dalton Trans.* **2022**, *51*, 7188–7209. [PubMed]

37. Kubiński, K.; Masłyk, M.; Janeczko, M.; Goldeman, W.; Nasulewicz-Goldeman, A.; Psurski, M.; Martyna, A.; Boguszewska-Czubara, A.; Cebula, J.; Goszczyński, T.M. Metallacarborane derivatives as innovative anti-candida albicans agents. *J. Med. Chem.* **2022**, *65*, 13935–13945. [CrossRef] [PubMed]

38. Nuez-Martinez, M.; Pinto, C.I.G.; Guerreiro, J.F.; Mendes, F.; Marques, F.; Muñoz-Juan, A.; Xavier, J.A.M.; Laromaine, A.; Bitonto, V.; Protti, N.; et al. Cobaltabis(dicarbollide) (o-COSAN (-)) as multifunctional chemotherapeutics: A prospective application in boron neutron capture therapy (BNCT) for glioblastoma. *Cancers* **2021**, *13*, 22.

39. Olejniczak, A.B.; Nawrot, B.; Leśnikowski, Z.J. DNA modified with boron-metal cluster complexes M(C2B9H11)(2) synthesis, properties, and applications. *Int. J. Mol. Sci.* **2018**, *19*, 13.

40. Grimes, R.N. Boron clusters come of age. *J. Chem. Educ.* **2004**, *81*, 657. [CrossRef]

41. Masalles, C.; Borrós, S.; Viñas, C.; Teixidor, F. Surface layer formation on polypyrrole films. *Adv. Mater.* **2002**, *14*, 449–452. [CrossRef]

42. Hardie, M.J. The use of carborane anions in coordination polymers and extended solids. *J. Chem. Crystallogr.* **2007**, *37*, 69–80.

43. Stoica, A.I.; Viñas, C.; Teixidor, F. Cobaltabisdicarbollide anion receptor for enantiomer-selective membrane electrodes. *Chem. Commun.* **2009**, *33*, 4988–4990. [CrossRef]

44. Grimes, R.N. *Carboranes*, 3rd ed.; Academic Press: Cambridge, MA, USA; Elsevier Science Ltd: London, UK, 2016; pp. 1–1041.

45. Körrbe, S.; Schreiber, P.J.; Michl, J. Chemistry of the carba-closo-dodecaborate(-) anion, CB11H12. *Chem. Rev.* **2006**, *106*, 5208–5249.

46. Chamberlin, R.M.; Scott, B.L.; Melo, M.M.; Abney, K.D. Butyllithium deprotonation *vs.* alkali metal reduction of cobalt dicarbollide: A new synthetic route to C-substituted derivatives. *Inorg. Chem.* **1997**, *36*, 809–817.

47. Rojo, I.; Teixidor, F.; Viñas, C.; Kivekäs, R.; Sillanpää, R. Synthesis and coordinating ability of an anionic cobaltabisdicarbollide ligand geometrically analogous to BINAP. *Chem.-Eur. J.* **2004**, *10*, 5376–5385. [CrossRef]

48. Juárez-Pérez, E.J.; Viñas, C.; González-Campo, A.; Teixidor, F.; Sillanpää, R.; Kivekäs, R.; Núñez, R. Controlled direct synthesis of C-mono- and C-disubstituted derivatives of 3,3'-Co(1,2-C(2)B(9)H(11))(2) (-) with organosilane groups: Theoretical calculations compared with experimental results. *Chem.-Eur. J.* **2008**, *14*, 4924–4938. [CrossRef] [PubMed]

49. Farràs, P.; Teixidor, F.; Rojo, I.; Kivekäs, R.; Sillanpää, R.; González-Cardoso, P.; Viñas, C. Relaxed but highly compact diansa metallacyclophanes. *J. Am. Chem. Soc.* **2011**, *133*, 16537–16552. [CrossRef] [PubMed]

50. Junqueira, G.M.A. Remarkable aromaticity of cobalt bis(dicarbollide) derivatives: A NICS study. *Theor. Chem. Acc.* **2018**, *137*, 7. [CrossRef]

51. Poater, J.; Solà, M.; Viñas, C.; Teixidor, F. π aromaticity and three-dimensional aromaticity: Two sides of the same coin? *Angew. Chem.-Int. Edit.* **2014**, *53*, 12191–12195. [CrossRef]

52. Poater, J.; Viñas, C.; Olid, D.; Solà, M.; Teixidor, F. Aromaticity and extrusion of benzenoids linked to o-COSAN (-): Clar has the answer. *Angew. Chem. Int. Edit.* **2022**, *61*. [CrossRef]

53. Poater, J.; Viñas, C.; Solà, M.; Teixidor, F. 3D and 2D aromatic units behave like oil and water in the case of benzocarborane derivatives. *Nat. Commun.* **2022**, *13*, 8.

54. Bühl, M.; Hnyk, D.; Macháček, J. Computational study of structures and properties of metallaboranes: Cobalt bis(dicarbollide). *Chem.-Eur. J.* **2005**, *11*, 4109–4120.

55. Bühl, M.; Holub, J.; Hnyk, D.; Macháček, J. Computational studies of structures and properties of metallaboranes. 2. Transition-metal dicarbollide complexes. *Organometallics* **2006**, *25*, 2173–2181. [CrossRef]

56. Bogucka-Kocka, A.; Kołodziej, P.; Makuch-Kocka, A.; Różycka, D.; Rykowski, S.; Nekvinda, J.; Grűner, B.; Olejniczak, A.B. Nematicidal activity of naphthalimide–boron cluster conjugates. *Chem. Commun.* **2022**, *58*, 2528–2531.

57. Sivaev, I.B.; Kosenko, I.D. Rotational conformation of 8,8'-dihalogenated derivatives of cobalt bis(dicarbollide) in solution. *Russ. Chem. Bull.* **2021**, *70*, 753–756. [CrossRef]

58. Sivaev, I.B. Ferrocene and transition metal bis(dicarbollides) as platform for design of rotatory molecular switches. *Molecules* **2017**, *22*, 30.

59. Juárez-Pérez, E.J.; Núñez, R.; Viñas, C.; Sillanpää, R.; Teixidor, F. The role of C-H center dot center dot center dot H-B interactions in establishing rotamer configurations in metallabis(dicarbollide) systems. *Eur. J. Inorg. Chem.* **2010**, *16*, 2385–2392.

60. Anufriev, S.A.; Timofeev, S.V.; Anisimov, A.A.; Suponitsky, K.Y.; Sivaev, I.B. Bis(dicarbollide) complexes of transition metals as a platform for molecular switches. study of complexation of 8,8'-bis(methylsulfanyl) derivatives of cobalt and iron bis(dicarbollides). *Molecules* **2020**, *25*, 5745. [CrossRef] [PubMed]

61. Zalkin, A.; Hopkins, T.E.; Templeton, D.H. Crystal structure of Cs(B9C2H11)2CO. *Inorg. Chem.* **1967**, *6*, 1911–1915. [CrossRef]

62. Mortimer, M.D.; Knobler, C.B.; Hawthorne, M.F. Methylation of boron vertices of the cobalt dicarbollide anion. *Inorg. Chem.* **1996**, *35*, 5750–5751. [CrossRef]

63. Fojt, L.; Grűner, B.; Nekvinda, J.; Tüzün, E.Z.; Havran, L.; Fojta, M. Electrochemistry of cobalta bis(dicarbollide) ions substituted at carbon atoms with hydrophilic alkylhydroxy and carboxy groups. *Molecules* **2022**, *27*, 1761. [CrossRef]

64. Geiger, W.E.; Smith, D.E. Electrochemical indications of new oxidation states in transition-metal dicarbollide complexes. *J. Chem. Soc. D* **1971**, *1*, 8–9. [CrossRef]

65. Manning, M.J.; Knobler, C.B.; Hawthorne, M.F.; Do, Y. Dicarbollide complexes of thallium—Structural and B-11 NMR-studies. *Inorg. Chem.* **1991**, *30*, 3589–3591. [CrossRef]
66. Scholz, M.; Hey-Hawkins, E. Carbaboranes as pharmacophores: Properties, synthesis, and application strategies. *Chem. Rev.* **2011**, *111*, 7035–7062.
67. Oliva-Enrich, J.M.; Humbel, S.; Dávalos, J.Z.; Holub, J.; Hnyk, D. Proton affinities of amino group functionalizing 2D and 3D boron compounds. *Afinidad* **2018**, *75*, 260–266.
68. Růžičková, Z.; Litecká, M.; Pazderová, L.; Tüzün, E.; Grüner, B. *Cobalt Bis(dicarbollide) Ion with Functional Groups Directly Attached to Carbon Atoms*; Correspondence Grüner, B.; Institute of Inorganic Chemistry: Czech Republic, 2023; manuscript in preparation.
69. Hawthorne, M.F.; Maderna, A. Applications of radiolabeled boron clusters to the diagnosis and treatment of cancer. *Chem. Rev.* **1999**, *99*, 3421–3434. [CrossRef] [PubMed]
70. Gomez, F.A.; Johnson, S.E.; Knobler, C.B.; Hawthorne, M.F. Synthesis and structural characterization of metallacarboranes containing bridged dicarbollide ligands. *Inorg. Chem.* **1992**, *31*, 3558–3567. [CrossRef]
71. Harwell, D.E.; Nabakka, J.; Knobler, C.B.; Hawthorne, M.F. Synthesis and structural characterization of an ether-bridged cobalta-bis(dicarbollide)—a model for venus flytrap cluster reagents. *Can. J. Chem.* **1995**, *73*, 1044–1049. [CrossRef]
72. Nabakka, J.M.; Harwell, D.E.; Knobler, C.; Hawthorne, M.F. The synthesis and characterization of a thioether-bridged cobalta-bis(dicarbollide): A model for Venus flytrap cluster reagents. *Abstr. Pap. Am. Chem. Soc.* **1996**, *211*, 186.
73. Viñas, C.; Bertran, J.; Gomez, S.; Teixidor, F.; Dozol, J.F.; Rouquette, H.; Kivekäs, R.; Sillanpää, R. Aromatic substituted metallacarboranes as extractants of Cs-137 and Sr-90 from nuclear wastes. *J. Chem. Soc.-Dalton Trans.* **1998**, *17*, 2849–2853. [CrossRef]
74. Viñas, C.; Gomez, S.; Bertran, J.; Teixidor, F.; Dozol, J.F.; Rouquette, H. Cobaltabis(dicarbollide) derivatives as extractants for europium from nuclear wastes. *Chem. Commun.* **1998**, *2*, 191–192. [CrossRef]
75. Grüner, B.; Kugler, M.; El Anwar, S.; Holub, J.; Nekvinda, J.; Bavol, D.; Růžičková, Z.; Pospíšilová, K.; Fábry, M.; Král, V.; et al. Cobalt bis(dicarbollide) alkylsulfonamides: Potent and highly selective inhibitors of tumor specific carbonic anhydrase IX. *ChemPlusChem* **2021**, *86*, 352–363. [CrossRef]
76. Schaarschmidt, D.; Lang, H. Selective syntheses of planar-chiral ferrocenes. *Organometallics* **2013**, *32*, 5668–5704.
77. Kitazawa, Y.; Takita, R.; Yoshida, K.; Muranaka, A.; Matsubara, S.; Uchiyama, M. "Naked" lithium cation: Strongly activated metal cations facilitated by carborane anions. *J. Org. Chem.* **2017**, *82*, 1931–1935.
78. Reich, H.J. Role of organolithium aggregates and mixed aggregates in organolithium mechanisms. *Chem. Rev.* **2013**, *113*, 7130–7178. [PubMed]
79. Grüner, B.; Šícha, V.; Hnyk, D.; Londesborough, M.G.S.; Císařová, I. The synthesis and structural characterization of polycyclic derivatives of cobalt bis(dicarbollide)(1(-)). *Inorg. Chem.* **2015**, *54*, 3148–3158. [CrossRef] [PubMed]
80. El Anwar, S.; Růžičková, Z.; Bavol, D.; Fojt, L.; Grüner, B. Tetrazole ring substitution at carbon and boron sites of the cobalt bis(dicarbollide) ion available via dipolar cycloadditions. *Inorg. Chem.* **2020**, *59*, 17430–17442. [CrossRef]
81. Grüner, B.; Švec, P.; Šícha, V.; Padělková, Z. Direct and facile synthesis of carbon substituted alkylhydroxy derivatives of cobalt bis(1,2-dicarbollide), versatile building blocks for synthetic purposes. *Dalton Trans.* **2012**, *41*, 7498–7512. [CrossRef] [PubMed]
82. El Anwar, S.; Pazderová, L.; Bavol, D.; Bakardjiev, M.; Růžičková, Z.; Horáček, O.; Fojt, L.; Kučera, R.; Grüner, B. Structurally rigidified cobalt bis(dicarbollide) derivatives, a chiral platform for labelling of biomolecules and new materials. *Chem. Commun.* **2022**, *58*, 2572–2575.
83. Grüner, B.; Plzák, Z. High-performance liquid chromatographic separations of boron-cluster compounds. *J. Chromatogr. A* **1997**, *789*, 497–517. [CrossRef]
84. Plešek, J. The age of chiral deltahedral borane derivatives. *Inorg. Chim. Acta* **1999**, *289*, 45–50. [CrossRef]
85. Grüner, B.; Císařová, I.; Franken, A.; Plešek, J. Resolution of the 6,6'-mu-(CH3)(2)P-(1,7-(C2B9H10)(2))-2-Co bridged cobal-tacarborane to enantiomers pure by chiral HPLC, circular dichroism spectra and absolute configurations by X-ray diffraction. *Tetrahedron-Asymmetry* **1998**, *9*, 79–88. [CrossRef]
86. Horáková, H.; Grüner, B.; Vespalec, R. Emerging subject for chiral separation science: Cluster boron compounds. *Chirality* **2011**, *23*, 307–319. [CrossRef]
87. Horáček, O.; Papajová-Janetková, M.; Grüner, B.; Lochman, L.; Štěrbová-Kovaříková, P.; Vespalec, R.; Kučera, R. The first chiral HPLC separation of dicarba-nido-undecarborate anions and their chromatographic behavior. *Talanta* **2021**, *222*, 9.
88. Horáček, O.; Marvalová, J.; Stilcová, K.; Holub, J.; Grüner, B.; Kučera, R. Reversed-phase chromatography as an effective tool for the chiral separation of anionic and zwitterionic carboranes using polysaccharide-based chiral selectors. *J. Chromatogr. A* **2022**, *1672*, 463051. [PubMed]
89. Horáček, O.; Nováková, L.; Tüzün, E.; Grüner, B.; Švec, F.; Kučera, R. Advanced tool for chiral separations of anionic and zwitterionic (metalla)carboranes: Supercritical fluid chromatography. *Anal. Chem.* **2022**, *94*, 17551–17558. [PubMed]
90. Horáček, O.; Dhaubhadel, U.; Holub, J.; Grüner, B.; Armstrong, D.W.; Kučera, R. Employment of chiral columns with superficially porous particles in chiral separations of cobalt bis (dicarbollide) and nido-7,8-C2B9H12(1-) derivatives. *Chirality* **2023**, 1–15. [CrossRef]
91. Nekvinda, J.; Švehla, J.; Císařová, I.; Grüner, B. Chemistry of cobalt bis(1,2-dicarbollide) ion; the synthesis of carbon substituted alkylamino derivatives from hydroxyalkyl derivatives via methylsulfonyl or p-toluenesulfonyl esters. *J. Organomet. Chem.* **2015**, *798*, 112–120.

92. Carey, F.A.; Sundberg, R.J. *Advanced Organic Chemistry, Part B: Reaction and Synthesis*, 5th ed.; Springer: Berlin/Heidelberg, Germany, 2007.

93. Śmiałkowski, K.; Sardo, C.; Leśnikowski, Z.J. Metallacarborane synthons for molecular constructionoligofunctionalization of cobalt bis(1,2-dicarbollide) on boron and carbon atoms with extendable ligands. *Molecules* **2023**, *28*, 4118. [CrossRef]

94. Fino, S.A.; Benwitz, K.A.; Sullivan, K.M.; LaMar, D.L.; Stroup, K.M.; Giles, S.M.; Balaich, G.J.; Chamberlin, R.M.; Abney, K.D. Condensation polymerization of cobalt dicarbollide dicarboxylic acid. *Inorg. Chem.* **1997**, *36*, 4604–4606. [CrossRef] [PubMed]

95. Miller, R.L.; Pinkerton, A.B.; Hurlburt, P.K.; Abney, K.D. Efficient extraction of Cs and Sr into hydrocarbons using modified cobalt dicarbollide. *Abstr. Pap. Am. Chem. S* **1995**, *209*, 146.

96. Miller, R.L.; Pinkerton, A.B.; Hurlburt, P.K.; Abney, K.D. Extraction of cesium and strontium into hydrocarbon solvents using tetra-C-alkyl cobalt dicarbollide. *Solvent Extr. Ion Exc.* **1995**, *13*, 813–827. [CrossRef]

97. Hurlburt, P.K.; Miller, R.L.; Abney, K.D.; Foreman, T.M.; Butcher, R.J.; Kinkead, S.A. New synthetic routes to B-halogenated derivatives of cobalt dicarbollide. *Inorg. Chem.* **1995**, *34*, 5215–5219. [CrossRef]

98. Steckle, W.P.; Duke, J.R.; Jorgensen, B.S. Cobalt dicarbollide containing polymer resins for cesium and strontium uptake. In *Metal-Containing Polymeric Materials*; Springer: Boston, MA, USA, 1996; pp. 277–285.

99. Nekvinda, J.; Šícha, V.; Hnyk, D.; Grűner, B. Synthesis, characterisation and some chemistry of C- and B-substituted carboxylic acids of cobalt bis(dicarbollide). *Dalton Trans.* **2014**, *43*, 5106–5120. [CrossRef] [PubMed]

100. Bredael, K.; Geurs, S.; Clarisse, D.; De Bosscher, K.; D'Hooghe, M. Carboxylic acid bioisosteres in medicinal chemistry: Synthesis and properties. *J. Chem.* **2022**, *2022*, 21. [CrossRef]

101. Wharton, C.J.; Wrigglesworth, R. Synthesis and reactions of 2,3-dihydro-oxazolo 2,3-a isoindol-5(9BH)-ones. *J. Chem. Soc. Perkin Trans. 1* **1985**, 809–813. [CrossRef]

102. Vyakaranam, K.; Li, S.J.; Zheng, C.; Hosmane, N.S. Substituent effect on the carborane coupling reaction: Synthesis and crystal structure of 1-phenyl-2- 2,3-benzobicyclo(3,3,0)-1-oxo-4-oxa-7-aza-8-yl -1,2-dicarba- closo-dodecaborane(12). *Inorg. Chem. Commun.* **2001**, *4*, 180–182. [CrossRef]

103. Juarez-Perez, E.J.; Viñas, C.; Teixidor, F.; Núñez, R. First example of the formation of a Si-C bond from an intramolecular Si-H center dot center dot center dot H-C diyhydrogen interaction in a metallacarborane: A theoretical study. *J. Organomet. Chem.* **2009**, *694*, 1764–1770. [CrossRef]

104. Selucký, P.; Plešek, J.; Rais, J.; Kyrš, M.; Kadlecová, L. Extraction of fission-products into nitrobenzene with dicobalt tris-dicarbollide and ethyleneoxy-substituted cobalt bis- dicarbollide. *J. Radioanal. Nucl. Chem. Artic.* **1991**, *149*, 131–140. [CrossRef]

105. Plešek, J.; Heřmánek, S.; Franken, A.; Císařová, I.; Nachtigal, C. Dimethyl sulfate induced nucleophilic substitution of the bis(1,2-dicarbollido)-3-cobalt(1-) ate ion. Syntheses, properties and structures of its 8,8'-mu-sulfato, 8-phenyl and 8-dioxane derivatives. *Collect. Czech. Chem. Commun.* **1997**, *62*, 47–56. [CrossRef]

106. Plešek, J.; Grűner, B.; Heřmánek, S.; Báča, J.; Mareček, V.; Jänchenová, J.; Lhotský, A.; Holub, K.; Selucký, P.; Rais, J.; et al. Synthesis of functionalized cobaltacarboranes based on the closo-[(1,2-C2B9H11)2-3,3'-Co]− ion bearing polydentate ligands for separation of M3+ cations from nuclear waste solutions. Electrochemical and liquid–liquid extraction study of selective transfer of M3+ metal cations to an organic phase. Molecular structure of the closo-[(8-(2-CH3O C6H4 O)-(CH2CH2O)2-1,2-C2B9H10)-(1',2'-C2B9H11)-3,3'-Co]Na determined by X-ray diffraction analysis. *Polyhedron* **2002**, *21*, 975–986.

107. Sivaev, I.B.B.; Bregadze, V.I. *Boron Science: New Technologies and Applications*; Hosmane, N.S., Ed.; CRC Press: Boca Raton, FL, USA, 2012; pp. 624–637.

108. Druzina, A.A.; Kosenko, I.D.; Zhidkova, O.B.; Ananyev, I.V.; Timofeev, S.V.; Bregadze, V.I. Novel cobalt bis(dicarbollide) based on terminal alkynes and their click-reactions. *Eur. J. Inorg. Chem.* **2020**, *2020*, 2658–2665. [CrossRef]

109. Dezhenkova, L.G.; Druzina, A.A.; Volodina, Y.L.; Dudarova, N.V.; Nekrasova, N.A.; Zhidkova, O.B.; Grin, M.A.; Bregadze, V.I. Synthesis of cobalt bis(dicarbollide)-curcumin conjugates for potential use in boron neutron capture therapy. *Molecules* **2022**, *27*, 4658. [CrossRef]

110. Semioshkin, A.A.; Sivaev, I.B.; Bregadze, V.I. Cyclic oxonium derivatives of polyhedral boron hydrides and their synthetic applications. *Dalton Trans.* **2008**, *2008*, 977–992. [CrossRef]

111. Sivaev, I.B.; Bregadze, V.I. Cyclic Oxonium Derivatives as an Efficient Synthetic Tool for the Modification of Polyhedral Boron Hydrides. *Chem. Inform.* **2012**, *43*, 623–637. [CrossRef]

112. Druzina, A.A.; Shmalko, A.V.; Sivaev, I.B.; Bregadze, V.I. Cyclic oxonium derivatives of cobalt and iron bis(dicarbollides) and their use in organic synthesis. *Russ. Chem. Rev.* **2021**, *90*, 785–830.

113. Shmal'ko, A.V.; Stogniy, M.Y.; Kazakov, G.S.; Anufriev, S.A.; Sivaev, I.B.; Kovalenko, L.V.; Bregadze, V.I. Cyanide free contraction of disclosed 1,4-dioxane ring as a route to cobalt bis(dicarbollide) derivatives with short spacer between the boron cage and terminal functional group. *Dalton Trans.* **2015**, *44*, 9860–9871. [CrossRef]

114. Plešek, J.; Grűner, B.; Šícha, V.; Böhmer, V.; Císařová, I. The zwitterion [(8,8'-μ-CH2O(CH3)-(1,2-C2B9H10)-3,3'-Co]0 as a versatile building block for introduction of the cobalt bis(dicarbollide) ion into organic molecules. *Organometallics* **2012**, *31*, 1703–1715. [CrossRef]

115. Druzina, A.A.; Kosenko, I.D.; Zhidkova, O.B. Synthesis of novel conjugates of closo-dodecaborate derivatives with cholesterol. *INEOS OPEN* **2020**, *3*, 70–74. [CrossRef]

116. Plešek, J.; Štíbr, B.; Heřmánek, S. A 8,8'-mu-I-3-CO(1,2-C2B9H10)2 metallacarborane complex with a iodonium bridge—evidence for a bromonium analog. *Collect. Czech. Chem. Commun.* **1984**, *49*, 1492–1496. [CrossRef]

117. Kosenko, I.D.; Lobanova, I.A.; Starikova, Z.A.; Bregadze, V.I. Synthesis of new charge-compensated cobalt bis(1,2-dicarbollide) derivatives. *Russ. Chem. Bull.* **2013**, *62*, 1914–1918. [CrossRef]
118. Kosenko, I.D.; Lobanova, I.A.; Godovikov, I.A.; Starikova, Z.A.; Sivaev, I.B.; Bregadze, V.I. Mild C-H activation of activated aromatics with 8,8 '-mu-I-3,3 '-Co(1,2-C2B9H10)(2): Just mix them. *J. Organomet. Chem.* **2012**, *721*, 70–77. [CrossRef]
119. Safronov, A.V.; Sevryugina, Y.V.; Jalisatgi, S.S.; Kennedy, R.D.; Barnes, C.L.; Hawthorne, M.F. Unfairly forgotten member of the iodocarborane family: Synthesis and structural characterization of 8-iodo-1,2-dicarba-closo-dodecaborane, its precursors, and derivatives. *Inorg. Chem.* **2012**, *51*, 2629–2637. [CrossRef] [PubMed]
120. Pichaandi, K.R.; Safronov, A.V.; Sevryugina, Y.V.; Everett, T.A.; Jalisatgi, S.S.; Hawthorne, M.F. Rodlike polymers containing nickel and cobalt metal bis(dicarbollide) anions: Synthesis and characterization. *Organometallics* **2017**, *36*, 3823–3829. [CrossRef]
121. Pichaandi, K.R.; Nilakantan, L.; Safronov, A.V.; Sevryugina, Y.V.; Jalisatgi, S.S.; Hawthorne, M.F. Electronic interactions between ferrocenyl units facilitated by the cobalt bis(dicarbollide) anion linker: An experimental and DFT study. *Eur. J. Inorg. Chem.* **2018**, *2018*, 666–670. [CrossRef]
122. Shmal'ko, A.V.; Anufriev, S.A.; Anisimov, A.A.; Stogniy, M.Y.; Sivaev, I.B.; Bregadze, V.I. Synthesis of cobalt and nickel 6,6-diphenylbis(dicarbollides). *Russ. Chem. Bull.* **2019**, *68*, 1239–1247. [CrossRef]
123. Anufriev, S.A.; Sivaev, I.B.; Bregadze, V.I. Synthesis of 9,9',12,12'-substituted cobalt bis(dicarbollide) derivatives. *Russ. Chem. Bull.* **2015**, *64*, 712–717. [CrossRef]
124. Nar, I.; Atsay, A.; Gümrükçü, S.; Karazehir, T.; Hamuryudan, E. Low-symmetry phthalocyanine cobalt bis(dicarbollide) conjugate for hydrogen reduction. *Eur. J. Inorg. Chem.* **2018**, *2018*, 3878–3882. [CrossRef]
125. Kosenko, I.D.; Lobanova, I.A.; Ananyev, I.V.; Godovikov, I.A.; Chekulaeva, L.A.; Starikova, Z.A.; Qi, S.; Bregadze, V.I. Novel alkoxy derivatives of cobalt bis(1,2-dicarbollide). *J. Organomet. Chem.* **2014**, *769*, 72–79. [CrossRef]
126. Stogniy, M.Y.; Suponitsky, K.Y.; Chizhov, A.O.; Sivaev, I.B.; Bregadze, V.I. Synthesis of 8-alkoxy and 8,8'-dialkoxy derivatives of cobalt bis(dicarbollide). *J. Organomet. Chem.* **2018**, *865*, 138–144. [CrossRef]
127. Plešek, J.; Grűner, B.; Báča, J.; Fusek, J.; Císařová, I. Syntheses of the B(8)-hydroxy- and B(8,8')-dihydroxy-derivatives of the bis(1,2-dicarbollido)-3-cobalt(1-)ate ion by its reductive acetoxylation and hydroxylation: Molecular structure of [8,8'-μ-CH3C(O)2 (1,2-C2B9H10)2-3-Co]0 zwitterion determined by X-ray diffraction analysis. *J. Organomet. Chem.* **2002**, *649*, 181–190.
128. Sardo, C.; Janczak, S.; Leśnikowski, Z.J. Unusual resistance of cobalt bis dicarbollide phosphate and phosphorothioate bridged esters towards alkaline hydrolysis: The "metallacarborane effect". *J. Organomet. Chem.* **2019**, *896*, 70–76. [CrossRef]
129. Anufriev, S.A.; Erokhina, S.A.; Suponitsky, K.Y.; Godovikov, I.A.; Filippov, O.A.; Fabrizi de Biani, F.; Corsini, M.; Chizhov, A.O.; Sivaev, I.B. Methylsulfanyl-stabilized rotamers of cobalt bis(dicarbollide). *Eur. J. Inorg. Chem.* **2017**, *2017*, 4444–4451. [CrossRef]
130. Churchill, M.R.; Gold, K.; Francis, J.N.; Hawthorne, M.F. Preparation and crystallographic characterization of a bridged metallo-carborane complex containing a carbonium ion center: (B9C2H10)2CoS2CH. *J. Am. Chem. Soc.* **1969**, *91*, 1222–1223. [CrossRef]
131. Francis, J.N.; Hawthorn, M.F. Synthesis and reactions of novel bridged dicarbollide complexes having electron-deficient carbon atoms. *Inorg. Chem.* **1971**, *10*, 594.
132. Frank, R.; Ahrens, V.M.; Boehnke, S.; Beck-Sickinger, A.G.; Hey-Hawkins, E. Charge-compensated metallacarborane building blocks for conjugation with peptides. *ChemBioChem* **2016**, *17*, 308–317. [CrossRef]
133. Sivaev, I.B.; Stogniy, M.Y.; Anufriev, S.A.; Zakharova, M.V.; Bregadze, V.I. New sulfur derivatives of carboranes and metallacarboranes. *Phosphorus Sulfur Silicon Relat. Elem.* **2017**, *192*, 192–196. [CrossRef]
134. Anufriev, S.A.; Sivaev, I.B.; Suponitsky, K.Y.; Bregadze, V.I. Practical synthesis of 9-methylthio-7,8-nido-carborane [9-MeS-7,8-C2B9H11]-. Some evidences of BH···X hydride-halogen bonds in 9- XCH2(Me)S-7,8-C2B9H11 (X = Cl, Br, I). *J. Organomet. Chem.* **2017**, *849*, 315–323.
135. Timofeev, S.V.; Zakharova, M.V.; Mosolova, E.M.; Godovikov, I.A.; Ananyev, I.V.; Sivaev, I.B.; Bregadze, V.I. Tungsten carbonyl σ-complexes of nido-carborane thioethers. *J. Organomet. Chem.* **2012**, *721*, 92–96. [CrossRef]
136. Timofeev, S.V.; Zhidkova, O.B.; Mosolova, E.M.; Sivaev, I.B.; Godovikov, I.A.; Suponitsky, K.Y.; Starikova, Z.A.; Bregadze, V.I. Tungsten carbonyl σ-complexes with charge-compensated nido-carboranyl thioether ligands. *Dalton Trans.* **2015**, *44*, 6449–6456. [CrossRef] [PubMed]
137. Timofeev, S.V.; Anufriev, S.A.; Sivaev, I.B.; Bregadze, V.I. Synthesis of cobalt bis(8-methylthio-1,2-dicarbollide)- pentacarbonyl-tungsten complexes. *Russ. Chem. Bull.* **2018**, *67*, 570–572. [CrossRef]
138. Bogdanova, E.V.; Stogniy, M.Y.; Suponitsky, K.Y.; Sivaev, I.B.; Bregadze, V.I. Synthesis of boronated amidines by addition of amines to nitrilium derivative of cobalt bis(dicarbollide). *Molecules* **2021**, *26*, 16.
139. Leoncini, A.; Huskens, J.; Verboom, W. Ligands for f-element extraction used in the nuclear fuel cycle. *Chem. Soc. Rev.* **2017**, *46*, 7229–7273.
140. Logunov, M.V.; Voroshilov, Y.A.; Babain, V.A.; Skobtsov, A.S. Experience of mastering, industrial exploitation, and optimization of the integrated extraction–precipitation technology for fractionation of liquid high-activity wastes at mayak production association. *Radiochemistry* **2020**, *62*, 700–722. [CrossRef]
141. Herbst, R.S.; Law, J.D.; Todd, T.A.; Romanovskii, V.N.; Babain, V.A.; Esimantovski, V.M.; Zaitsev, B.N.; Smirnov, I.V. Development and testing of a cobalt dicarbollide based solvent extraction process for the separation of cesium and strontium from acidic tank waste. *Sep. Sci. Technol.* **2002**, *37*, 1807–1831. [CrossRef]

142. Grűner, B.; Kvíčalová, M.; Plešek, J.; Šícha, V.; Císařová, I.; Lučaníková, M.; Selucký, P. Cobalt bis(dicarbollide) ions functionalized by CMPO-like groups attached to boron by short bonds; efficient extraction agents for separation of trivalent f-block elements from highly acidic nuclear waste. *J. Organomet. Chem.* **2009**, *694*, 1678–1689.

143. Shishkin, D.N.; Petrova, N.K.; Goletskii, N.D. On the possibility of extractive fractionation of REEs and TPUs from weakly acid raffinate produced of irradiated fuel elements with a mixture of ChCD and D2EHPA in polar solvent. *Radiochemistry* **2020**, *62*, 31–36. [CrossRef]

144. Shishkin, D.N.; Petrova, N.K.; Goletskii, N.D.; Mamchich, M.V.; Ushanov, A.D.; Bizin, A.V. Study of the possibility of deep partitioning of the spent nuclear fuel reprocessing raffinate according to the scheme of a pilot demonstration center by extraction with a mixture of CCD, PEO, and HDEHP in a polar solvent. *Radiochemistry* **2022**, *64*, 294–299. [CrossRef]

145. Kumar, S.; Rao, R.V.S. Mass transfer studies in a micromixer-settler: Extraction of Cs and Sr with CCD-PEG-400 solvent from simulated acidic radwaste solutions. *J. Radioanal. Nucl. Chem.* **2021**, *329*, 351–357. [CrossRef]

146. Khaydukova, M.; Militsyn, D.; Karnaukh, M.; Grűner, B.; Selucký, P.; Babain, V.; Wilden, A.; Kirsanov, D.; Legin, A. Modified diamide and phosphine oxide extracting compounds as membrane components for cross-sensitive chemical sensors. *Chemosensors* **2019**, *7*, 41. [CrossRef]

147. Chaudhury, S.; Bhattacharyya, A.; Goswami, A. Electrodriven selective transport of cs+ using chlorinated cobalt dicarbollide in polymer inclusion membrane: A novel approach for cesium removal from simulated nuclear waste solution. *Environ. Sci. Technol.* **2014**, *48*, 12994–13000. [CrossRef]

148. Chaudhury, S.; Bhattacharyya, A.; Ansari, S.A.; Goswami, A. A new approach for selective Cs+ separation from simulated nuclear waste solution using electrodriven cation transport through hollow fiber supported liquid membranes. *J. Membr. Sci.* **2018**, *545*, 75–80. [CrossRef]

149. Issa, F.; Kassiou, M.; Rendina, L.M. Boron in Drug Discovery: Carboranes as unique pharmacophores in biologically active compounds. *Chem. Rev.* **2011**, *111*, 5701–5722.

150. Teixidor, F.; Núñez, R.; Viñas, C. Towards the application of purely inorganic icosahedral boron clusters in emerging nanomedicine. *Molecules* **2023**, *28*, 24.

151. Pinheiro, T.; Alves, L.C.; Corregidor, V.; Teixidor, F.; Viñas, C.; Marques, F. Metallacarboranes for proton therapy using research accelerators: A pilot study. *EPJ Tech. Instrum.* **2023**, *10*, 5. [CrossRef]

152. Messner, K.; Vuong, B.; Tranmer, G.K. The boron advantage: The evolution and diversification of boron's applications in medicinal chemistry. *Pharmaceuticals* **2022**, *15*, 264. [CrossRef] [PubMed]

153. Zaulet, A.; Nuez, M.; Sillanpää, R.; Teixidor, F.; Viñas, C. Towards purely inorganic clusters in medicine: Biocompatible divalent cations as counterions of cobaltabis(dicarbollide) and its iodinated derivatives. *J. Organomet. Chem.* **2021**, *950*, 121997. [CrossRef]

154. Leśnikowski, Z.J. Challenges and opportunities for the application of boron clusters in drug design. *J. Med. Chem.* **2016**, *59*, 7738–7758. [CrossRef]

155. Leśnikowski, Z.J. What are the current challenges with the application of boron clusters to drug design. *Expert Opin. Drug Discov.* **2021**, *16*, 481–483. [CrossRef]

156. Gabel, D. Boron clusters in medicinal chemistry: Perspectives and problems. *Pure Appl. Chem.* **2015**, *87*, 173–179. [CrossRef]

157. Gozzi, M.; Schwarze, B.; Hey-Hawkins, E. Preparing (metalla)carboranes for nanomedicine. *ChemMedChem* **2021**, *16*, 1533–1565.

158. Marfavi, A.; Kavianpour, P.; Rendina, L.M. Carboranes in drug discovery, chemical biology and molecular imaging. *Nat. Rev. Chem.* **2022**, *6*, 486–504.

159. Kožíšek, M.; Cígler, P.; Lepšík, M.; Fanfrlík, J.; Řezáčová, P.; Brynda, J.; Pokorná, J.; Plešek, J.; Grűner, B.; Grantz-Šašková, K.; et al. Inorganic polyhedral metallacarborane inhibitors of HIV protease: A new approach to overcoming antiviral resistance. *J. Med. Chem.* **2008**, *51*, 4839–4843. [CrossRef]

160. Murphy, N.; McCarthy, E.; Dwyer, R.; Farràs, P. Boron clusters as breast cancer therapeutics. *J. Inorg. Biochem.* **2021**, *218*, 11.

161. Bednarska-Szczepaniak, K.; Przelazły, E.; Kania, K.D.; Szwed, M.; Litecká, M.; Grűner, B.; Leśnikowski, Z.J. Interaction of adenosine, modified using carborane clusters, with ovarian cancer cells: A new anticancer approach against chemoresistance. *Cancers* **2021**, *13*, 48. [CrossRef] [PubMed]

162. Nekvinda, J.; Różycka, D.; Rykowski, S.; Wyszko, E.; Fedoruk-Wyszomirska, A.; Gurda, D.; Orlicka-Płocka, M.; Giel-Pietraszuk, M.; Kiliszek, A.; Rypniewski, W.; et al. Synthesis of naphthalimide-carborane and metallacarborane conjugates: Anticancer activity, DNA binding ability. *Bioorganic Chem.* **2020**, *94*, 16. [CrossRef]

163. Beck-Sickinger, A.G.; Becker, D.P.; Chepurna, O.; Das, B.; Flieger, S.; Hey-Hawkins, E.; Hosmane, N.; Jalisatgi, S.S.; Nakamura, H.; Patil, R.; et al. New boron delivery agents. *Cancer. Biother. Radiopharm.* **2023**, *38*, 160–172. [CrossRef]

164. Druzina, A.A.; Dudarova, N.V.; Zhidkova, O.B.; Razumov, I.A.; Solovieva, O.I.; Kanygin, V.V.; Bregadze, V.I. Synthesis and cytotoxicity of novel cholesterol-cobalt bis(dicarbollide) conjugates. *Mendeleev Commun.* **2022**, *32*, 354–356. [CrossRef]

165. Gozzi, M.; Schwarze, B.; Hey-Hawkins, E. Half- and mixed-sandwich metallacarboranes for potential applications in medicine. *Pure Appl. Chem.* **2019**, *91*, 563–573.

166. Sivaev, I.B.; Bregadze, V.V. Polyhedral boranes for medical applications: Current status and perspectives. *Eur. J. Inorg. Chem.* **2009**, *11*, 1433–1450.

167. Wojtczak, B.A.; Andrysiak, A.; Grűner, B.; Leśnikowski, Z.J. "Chemical Ligation": A versatile method for nucleoside modification with boron clusters. *Chem. Eur. J.* **2008**, *14*, 10675–10682. [CrossRef] [PubMed]

168. Assaf, K.I.; Nau, W.M. The chaotropic effect as an assembly motif in chemistry. *Angew. Chem.-Int. Edit.* **2018**, *57*, 13968–13981. [CrossRef]
169. Assaf, K.I.; Wilińska, J.; Gabel, D. Ionic boron clusters as superchaotropic anions: Implications for drug design. *Boron-Based Compd. Potential Emerg. Appl. Med.* **2018**, 109–125.
170. Cebula, J.; Fink, K.; Boratyński, J.; Goszczyński, T.M. Supramolecular chemistry of anionic boron clusters and its applications in biology. *Coord. Chem. Rev.* **2023**, *477*, 19.
171. Matějíček, P.; Cígler, P.; Procházka, K.; Král, V. Molecular assembly of metallacarboranes in water: Light scattering and microscopy study. *Langmuir* **2006**, *22*, 575–581.
172. Medoš, Z.; Bešter-Rogač, M. Two-step micellization model: The case of long-chain carboxylates in water. *Langmuir* **2017**, *33*, 7722–7731. [CrossRef]
173. Medoš, Z.; Friesen, S.; Buchner, R.; Bešter-Rogač, M. Interplay between aggregation number, micelle charge and hydration of catanionic surfactants. *Phys. Chem. Chem. Phys.* **2020**, *22*, 9998–10009. [CrossRef]
174. Woolley, E.M.; Burchfield, T.E. Model for thermodynamics of ionic surfactant solutions. 2. Enthalpies, heat capacities, and volumes. *J. Phys. Chem.* **1984**, *88*, 2155–2163. [CrossRef]
175. Medoš, Z.; Hleli, B.; Tošner, Z.; Ogrin, P.; Urbič, T.; Kogej, K.; Bešter-Rogač, M.; Matějíček, P. Counterion-induced aggregation of metallacarboranes. *J. Phys. Chem. C* **2022**, *126*, 5735–5742. [CrossRef]
176. Merhi, T.; Jonchère, A.; Girard, L.; Diat, O.; Nuez, M.; Viñas, C.; Bauduin, P. Highlights on the binding of cobalta-bis-(dicarbollide) with glucose units. *Chem. Eur. J.* **2020**, *26*, 13935–13947. [CrossRef] [PubMed]
177. Zaulet, A.; Teixidor, F.; Bauduin, P.; Diat, O.; Hirva, P.; Ofori, A.; Viñas, C. Deciphering the role of the cation in anionic cobaltabisdicarbollide clusters. *J. Organomet. Chem.* **2018**, *865*, 214–225.
178. Fernandez-Alvarez, R.; Nová, L.; Uhlík, F.; Kereïche, S.; Uchman, M.; Košovan, P.; Matějíček, P. Interactions of star-like polyelectrolyte micelles with hydrophobic counterions. *J. Colloid Interface Sci.* **2019**, *546*, 371–380.
179. Rak, J.; Kaplánek, R.; Král, V. Solubilization and deaggregation of cobalt bis(dicarbollide) derivatives in water by biocompatible excipients. *Bioorg. Med. Chem. Lett.* **2010**, *20*, 1045–1048. [CrossRef] [PubMed]
180. Goszczyński, T.M.; Fink, K.; Kowalski, K.; Leśnikowski, Z.J.; Boratyński, J. Interactions of boron clusters and their derivatives with serum albumin. *Sci. Rep.* **2017**, *7*, 12.
181. Assaf, K.I.; Begaj, B.; Frank, A.; Nilam, M.; Mougharbel, A.S.; Kortz, U.; Nekvinda, J.; Grüner, B.; Gabel, D.; Nau, W.M. High-affinity binding of metallacarborane cobalt bis(dicarbollide) anions to cyclodextrins and application to membrane translocation. *J. Org. Chem.* **2019**, *84*, 11790–11798. [CrossRef]
182. Abdelgawwad, A.M.A.; Xavier, J.A.M.; Roca-Sanjuán, D.; Viñas, C.; Teixidor, F.; Francés-Monerris, A. Light-induced on/off switching of the surfactant character of the o-cobaltabis(dicarbollide) anion with no covalent bond alteration. *Angew. Chem. Int. Edit.* **2021**, *60*, 25753–25757.
183. Chazapi, I.; Diat, O.; Bauduin, P. Aqueous solubilization of hydrophobic compounds by inorganic nano-ions: An unconventional mechanism. *J. Colloid Interface Sci.* **2023**, *638*, 561–568. [CrossRef]
184. Rokitskaya, T.I.; Kosenko, I.D.; Sivaev, I.B.; Antonenko, Y.N.; Bregadze, V.I. Fast flip-flop of halogenated cobalt bis(dicarbollide) anion in a lipid bilayer membrane. *Phys. Chem. Chem. Phys.* **2017**, *19*, 25122–25128. [CrossRef]
185. Barba-Bon, A.; Salluce, G.; Lostalé-Seijo, I.; Assaf, K.I.; Hennig, A.; Montenegro, J.; Nau, W.M. Boron clusters as broadband membrane carriers. *Nature* **2022**, *603*, 637–642. [CrossRef]
186. Langella, E.; Esposito, D.; Monti, S.M.; Supuran, C.T.; De Simone, G.; Alterio, V. A combined in silico and structural study opens new perspectives on aliphatic sulfonamides, a still poorly investigated class of ca inhibitors. *Biology* **2023**, *12*, 281. [CrossRef]
187. Chen, Y.; Du, F.K.; Tang, L.Y.; Xu, J.R.; Zhao, Y.S.; Wu, X.; Li, M.X.; Shen, J.; Wen, Q.L.; Cho, C.H.; et al. Carboranes as unique pharmacophores in antitumor medicinal chemistry. *Mol. Ther. Oncolytics.* **2022**, *24*, 400–416. [PubMed]
188. Das, B.C.; Nandwana, N.K.; Das, S.; Nandwana, V.; Shareef, M.A.; Das, Y.; Saito, M.; Weiss, L.M.; Almaguel, F.; Hosmane, N.S.; et al. Boron chemicals in drug discovery and development: Synthesis and medicinal perspective. *Molecules* **2022**, *27*, 2615. [CrossRef] [PubMed]
189. Plešek, J. Potential Applications of the Boron Cluster Compounds. *Chem. Rev.* **1992**, *92*, 269–278. [CrossRef]
190. Zheng, Y.K.; Liu, W.W.; Chen, Y.; Jiang, H.; Yan, H.; Kosenko, I.; Chekulaeva, L.; Sivaev, I.; Bregadze, V.; Wang, X.M. A highly potent antibacterial agent targeting methicillin-resistant staphylococcus aureus based on cobalt bis(1,2-dicarbollide) alkoxy derivative. *Organometallics* **2017**, *36*, 3484–3490.
191. Jefferson, K.K. What drives bacteria to produce a biofilm? *FEMS Microbiol. Lett.* **2004**, *236*, 163–173. [CrossRef]
192. Archer, G.L. Staphylococcus aureus: A Well-Armed Pathogen. *Clin. Infect. Dis.* **1998**, *26*, 1179–1181. [CrossRef]
193. Campodónico, V.L.; Gadjeva, M.; Paradis-Bleau, C.; Uluer, A.; Pier, G.B. Airway epithelial control of pseudomonas aeruginosa infection in cystic fibrosis. *Trends. Mol. Med.* **2008**, *14*, 120–133. [PubMed]
194. Thebault, P.; Lequeux, I.; Jouenne, T. Antibiofilm strategies. *J. Wound. Tech.* **2013**, *21*, 36–39.
195. Vaňková, E.; Lokočová, K.; Maťátková, O.; Křížová, I.; Masák, J.; Grüner, B.; Kaule, P.; Čermák, J.; Šícha, V. Cobalt bis-dicarbollide and its ammonium derivatives are effective antimicrobial and antibiofilm agents. *J. Organomet. Chem.* **2019**, *899*, 8. [CrossRef]
196. Kvasničková, E.; Masák, J.; Čejka, J.; Maťátková, O.; Šícha, V. Preparation, characterization, and the selective antimicrobial activity of N-alkylammonium 8-diethyleneglycol cobalt bis-dicarbollide derivatives. *J. Organomet. Chem.* **2017**, *827*, 23–31. [CrossRef]

197. Popova, T.; Zaulet, A.; Teixidor, F.; Alexandrova, R.; Viñas, C. Investigations on antimicrobial activity of cobaltabisdicarbollides. *J. Organomet. Chem.* **2013**, *747*, 229–234. [CrossRef]
198. Totani, T.; Aono, K.; Yamamoto, K.; Tawara, K. Synthesis and in vitro antimicrobial property of o-carborane derivatives. *J. Med. Chem.* **1981**, *24*, 1492–1499. [CrossRef]
199. Vaňková, E.; Lokočová, K.; Kašparová, P.; Hadravová, R.; Křížová, I.; Mat'átková, O.; Masák, J.; Šícha, V. Cobalt bis-dicarbollide enhances antibiotics action towards staphylococcus epidermidis planktonic growth due to cell envelopes disruption. *Pharmaceuticals* **2022**, *15*, 534. [CrossRef]
200. Druzina, A.A.; Grammatikova, N.E.; Zhidkova, O.B.; Nekrasova, N.A.; Dudarova, N.V.; Kosenko, I.D.; Grin, M.A.; Bregadze, V.I. Synthesis and antibacterial activity studies of the conjugates of curcumin with closo-dodecaborate and cobalt bis(dicarbollide) boron clusters. *Molecules* **2022**, *27*, 2920. [CrossRef]
201. Romero, I.; Martinez-Medina, M.; Camprubí-Font, C.; Bennour, I.; Moreno, D.; Martínez-Martínez, L.; Teixidor, F.; Fox, M.A.; Viñas, C. Metallacarborane assemblies as effective antimicrobial agents, including a highly potent anti-MRSA agent. *Organometallics* **2020**, *39*, 4253–4264. [CrossRef]
202. Swietnicki, W.; Goldeman, W.; Psurski, M.; Nasulewicz-Goldeman, A.; Boguszewska-Czubara, A.; Drab, M.; Sycz, J.; Goszczyński, T.M. Metallacarborane derivatives effective against pseudomonas aeruginosa and yersinia enterocolitica. *Int. J. Mol. Sci.* **2021**, *22*, 6762. [CrossRef] [PubMed]
203. Kosenko, I.; Ananyev, I.; Druzina, A.; Godovikov, I.; Laskova, J.; Bregadze, V.; Studzinska, M.; Paradowska, E.; Leśnikowski, Z.J.; Semioshkin, A. Disubstituted cobalt bis(1,2-dicarbollide)(-I) terminal alkynes: Synthesis, reactivity in the Sonogashira reaction and application in the synthesis of cobalt bis(1,2-dicarbollide)(-I) nucleoside conjugates. *J. Organomet. Chem.* **2017**, *849*, 142–149. [CrossRef]
204. Olusanya, T.O.B.; Haj Ahmad, R.R.; Ibegbu, D.M.; Smith, J.R.; Elkordy, A.A. Liposomal drug delivery systems and anticancer drugs. *Molecules* **2018**, *23*, 907–913. [CrossRef] [PubMed]
205. Huang, S.T.; Wang, Y.P.; Chen, Y.H.; Lin, C.T.; Li, W.S.; Wu, H.C. Liposomal paclitaxel induces fewer hematopoietic and cardiovascular complications than bioequivalent doses of Taxol. *Int. J. Oncol.* **2018**, *53*, 1105–1117. [CrossRef] [PubMed]
206. Erdelyi, K.E.; Antonets, A.A.; Zhidkova, O.B.; Druzina, A.A.; Nazarov, A.A.; Timofeev, S.V.; Sivaev, I.B.; Bregadze, V.I. Cobalt and iron bis(dicarbollide) conjugates with cholesterol: Synthesis and evaluation of antiproliferative activity. *Russ. Chem. Bull.* **2023**, *72*, 1059–1065. [CrossRef]
207. Dubey, R.D.; Sarkar, A.; Shen, Z.Y.; Bregadze, V.I.; Sivaev, I.B.; Druzina, A.A.; Zhidkova, O.B.; Shmal'ko, A.V.; Kosenko, I.D.; Sreejyothi, P.; et al. Effects of linkers on the development of liposomal formulation of cholesterol conjugated cobalt bis(dicarbollides). *J. Pharm. Sci.* **2021**, *110*, 1365–1373. [CrossRef]
208. Probst, T.U.; Berryman, N.G.; Lemmen, P.; Weissfloch, L.; Auberger, T.; Gabel, D.; Carlsson, J.; Larsson, B. Comparison of inductively coupled plasma atomic emission spectrometry and inductively coupled plasma mass spectrometry with quantitative neutron capture radiography for the determination of boron in biological samples from cancer therapy. *J. Anal. At. Spectrom.* **1997**, *12*, 1115–1122. [CrossRef]
209. Laakso, J.; Kulvik, M.; Ruokonen, I.; Vähätalo, J.; Zilliacus, R.; Färkkilä, M.; Kallio, M. Atomic emission method for total boron in blood during neutron-capture therapy. *Clin. Chem.* **2001**, *47*, 1796–1803. [CrossRef] [PubMed]
210. Linko, S.; Revitzer, H.; Zilliacus, R.; Kortesniemi, M.; Kouri, M.; Savolainen, S. Boron detection from blood samples by ICP-AES and ICP-MS during boron neutron capture therapy. *Scand. J. Clin. Lab. Investig.* **2008**, *68*, 696–702. [CrossRef]
211. Kobayashi, T.; Kanda, K. Microanalysis system of ppm-order 10B concentrations in tissue for neutron capture therapy by prompt gamma-ray spectrometry. *Nucl. Instrum. Methods Phys. Res.* **1983**, *204*, 525–531. [CrossRef]
212. Matsumoto, T.; Aoki, M.; Aizawa, O. Phantom experiment and calculation for in vivo 10boron analysis by prompt gamma ray spectroscopy. *Phys. Med. Biol.* **1991**, *36*, 329–338. [CrossRef] [PubMed]
213. Mukai, K.; Nakagawa, Y.; Matsumoto, K. Prompt gamma ray spectrometry for in vivo measurement of boron-10 concentration in rabbit brain tissue. *Neurol. Med. Chir.* **1995**, *35*, 855–860. [CrossRef]
214. Kashino, G.; Fukutani, S.; Suzuki, M.; Liu, Y.; Nagata, K.; Masunaga, S.I.; Maruhashi, A.; Tanaka, H.; Sakurai, Y.; Kinashi, Y.; et al. A simple and rapid method for measurement of b-10-para-boronophenylalanine in the blood for boron neutron capture therapy using fluorescence spectrophotometry. *J. Radiat. Res.* **2009**, *50*, 377–382. [CrossRef]
215. Efremenko, A.V.; Ignatova, A.A.; Grin, M.A.; Sivaev, I.B.; Mironov, A.F.; Bregadze, V.I.; Feofanov, A.V. Chlorin e(6) fused with a cobalt-bis(dicarbollide) nanoparticle provides efficient boron delivery and photoinduced cytotoxicity in cancer cells. *Photochem. Photobiol. Sci.* **2014**, *13*, 92–102. [CrossRef]
216. Efremenko, A.V.; Ignatova, A.A.; Borsheva, A.A.; Grin, M.A.; Bregadze, V.I.; Sivaev, I.B.; Mironov, A.F.; Feofanov, A.V. Cobalt bis(dicarbollide) versus closo-dodecaborate in boronated chlorin e(6) conjugates: Implications for photodynamic and boron-neutron capture therapy. *Photochem. Photobiol. Sci.* **2012**, *11*, 645–652.
217. Volovetsky, A.; Sukhov, V.; Balalaeva, I.; Dudenkova, V.; Shilyagina, N.; Feofanov, A.; Efremenko, A.; Grin, M.; Mironov, A.; Sivaev, I.; et al. Pharmacokinetics of chlorin e(6)-cobalt bis(dicarbollide) conjugate in balb/c mice with engrafted carcinoma. *Int. J. Mol. Sci.* **2017**, *18*, 2556. [CrossRef] [PubMed]
218. Fedotova, M.K.; Usachev, M.N.; Bogdanova, E.V.; Diachkova, E.; Vasil'ev, Y.; Bregadze, V.I.; Mironov, A.F.; Grin, M.A. Highly purified conjugates of natural chlorin with cobalt bis(dicarbollide) nanoclusters for PDT and BNCT therapy of cancer. *Bioengineering* **2022**, *9*, 5. [CrossRef]

219. Barth, R.F.; Vicente, M.H.; Harling, O.K.; Kiger, W.; Riley, K.J.; Binns, P.J.; Wagner, F.M.; Suzuki, M.; Aihara, T.; Kato, I.; et al. Current status of boron neutron capture therapy of high grade gliomas and recurrent head and neck cancer. *Radiat. Oncol.* **2012**, *7*, 21.

220. Grin, M.A.; Titeev, R.A.; Brittal, D.I.; Ulybina, O.V.; Tsiprovskiy, A.G.; Berzina, M.Y.; Lobanova, I.A.; Sivaev, I.B.; Bregadze, V.I.; Mironov, A.F. New conjugates of cobalt bis(dicarbollide) with chlorophyll a derivatives. *Mendeleev Commun.* **2011**, *21*, 84–86.

221. Al-Warhi, T.; Sabt, A.; Elkaeed, E.B.; Eldehna, W.M. Recent advancements of coumarin-based anticancer agents: An up-to-date review. *Bioorganic Chem.* **2020**, *103*, 15.

222. Kosenko, I.; Laskova, J.; Kozlova, A.; Semioshkin, A.; Bregadze, V.I. Synthesis of coumarins modified with cobalt bis (1,2-dicarbolide) and closo-dodecaborate boron clusters. *J. Organomet. Chem.* **2020**, *921*, 9. [CrossRef]

223. Serdyukov, A.; Kosenko, I.; Druzina, A.; Grin, M.; Mironov, A.F.; Bregadze, V.I.; Laskova, J. Anionic polyhedral boron clusters conjugates with 7-diethylamino-4-hydroxycoumarin. Synthesis and lipophilicity determination. *J. Organomet. Chem.* **2021**, *946–947*, 121905.

224. Nuez-Martínez, M.; Pedrosa, L.; Martinez-Rovira, I.; Yousef, I.; Diao, D.; Teixidor, F.; Stanzani, E.; Martínez-Soler, F.; Tortosa, A.; Sierra, À.; et al. Synchrotron-based fourier-transform infrared micro-spectroscopy (SR-FTIRM) fingerprint of the small anionic molecule cobaltabis(dicarbollide) uptake in glioma stem cells. *Int. J. Mol. Sci.* **2021**, *22*, 9937. [CrossRef] [PubMed]

225. Coghi, P.; Li, J.; Hosmane, N.S.; Zhu, Y. Next generation of boron neutron capture therapy (BNCT) agents for cancer treatment. *Med. Res. Rev.* **2023**, *43*, 1809–1830. [CrossRef]

226. Seneviratne, D.S.; Saifi, O.; Mackeyev, Y.; Malouff, T.; Krishnan, S. Next-generation boron drugs and rational translational studies driving the revival of BNCT. *Cells* **2023**, *12*, 1398.

227. Malouff, T.D.; Seneviratne, D.S.; Ebner, D.K.; Stross, W.C.; Waddle, M.R.; Trifiletti, D.M.; Krishnan, S. Boron neutron capture therapy: A review of clinical applications. *Front. Oncol.* **2021**, *11*, 351.

228. Dymova, M.A.; Taskaev, S.Y.; Richter, V.A.; Kuligina, E.V. Boron neutron capture therapy: Current status and future perspectives. *Cancer Commun.* **2020**, *40*, 406–421. [CrossRef] [PubMed]

229. Ishola, T.A.; Chung, D.H. Neuroblastoma. *Surg. Oncol.* **2007**, *16*, 149–156. [CrossRef]

230. Ross, J.A.; Davies, S.M. Screening for neuroblastoma: Progress and pitfalls. *Cancer Epidemiol. Biomarkers Prev.* **1999**, *8*, 189–194. [PubMed]

231. Maris, J.M. Recent advances in neuroblastoma. *N. Engl. J. Med.* **2010**, *362*, 2202–2211. [CrossRef]

232. Peaston, R.T.; Weinkove, C. Measurement of catecholamines and their metabolites. *Ann. Clin. Biochem.* **2004**, *41*, 17–38. [CrossRef]

233. Verly, I.R.N.; van Kuilenburg, A.B.P.; Abeling, N.G.G.M.; Goorden, S.M.I.; Fiocco, M.; Vaz, F.M.; van Noesel, M.M.; Zwaan, C.M.; Kaspers, G.L.; Merks, J.H.M.; et al. Catecholamines profiles at diagnosis: Increased diagnostic sensitivity and correlation with biological and clinical features in neuroblastoma patients. *Eur. J. Cancer* **2017**, *72*, 235–243. [CrossRef] [PubMed]

234. Smith, B.R.; Gambhir, S.S. Nanomaterials for in vivo imaging. *Chem. Rev.* **2017**, *117*, 901–986.

235. Hong, G.S.; Diao, S.O.; Antaris, A.L.; Dai, H.J. Carbon nanomaterials for biological imaging and nanomedicinal therapy. *Chem. Rev.* **2015**, *115*, 10816–10906. [PubMed]

236. Ji, D.K.; Ménard-Moyon, C.; Bianco, A. Physically-triggered nanosystems based on two-dimensional materials for cancer theranostics. *Adv. Drug Deliv. Rev.* **2019**, *138*, 211–232. [PubMed]

237. Wang, J.T.W.; Klippstein, R.; Martincic, M.; Pach, E.; Feldman, R.; Šefl, M.; Michel, Y.; Asker, D.; Sosabowski, J.K.; Kalbac, M.; et al. Neutron activated Sm-153 sealed in carbon nanocapsules for in vivo imaging and tumor radiotherapy. *ACS Nano* **2020**, *14*, 129–141. [CrossRef]

238. Ferrer-Ugalde, A.; Sandoval, S.; Pulagam, K.R.; Muñoz-Juan, A.; Laromaine, A.; Llop, J.; Tobias, G.; Núñez, R. Radiolabeled cobaltabis(dicarbollide) anion-graphene oxide nanocomposites for in vivo bioimaging and boron delivery. *ACS Appl. Nano Mater.* **2021**, *4*, 1613–1625. [CrossRef]

239. Pulagam, K.R.; Henriksen-Lacey, M.; Uribe, K.B.; Renero-Lecuna, C.; Kumar, J.; Charalampopoulou, A.; Facoetti, A.; Protti, N.; Gómez-Vallejo, V.; Baz, Z.; et al. In vivo evaluation of multifunctional gold nanorods for boron neutron capture and photothermal therapies. *ACS Appl. Mater. Interfaces* **2021**, *13*, 49589–49601. [CrossRef]

240. Morris, J.H.; Gysling, H.J.; Reed, D. Electrochemistry of boron compounds. *Chem. Rev.* **1985**, *85*, 51–76. [CrossRef]

241. Hao, E.; Zhang, M.; Wenbo, E.; Kadish, K.M.; Fronczek, F.R.; Courtney, B.H.; Vicente, M.G.H. Synthesis and spectroelectrochemistry of N-cobaltacarborane porphyrin conjugates. *Bioconjugate Chem.* **2008**, *19*, 2171–2181. [CrossRef] [PubMed]

242. Nar, I.; Gül, A.; Sivaev, I.B.; Hamuryudan, E. Cobaltacarborane functionalized phthalocyanines: Synthesis, photophysical, electrochemical and spectroelectrochemical properties. *Synth. Met.* **2015**, *210*, 376–385. [CrossRef]

243. Núñez, R.; Tutusaus, O.; Teixidor, F.; Viñas, C.; Sillanpää, R.; Kivekäs, R. Highly stable neutral and positively charged dicarbollide sandwich complexes. *Chem. Eur. J.* **2005**, *11*, 5637–5647. [CrossRef] [PubMed]

244. Fojt, L.; Grüner, B.; Šícha, V.; Nekvinda, J.; Vespalec, R.; Fojta, M. Electrochemistry of icosahedral cobalt bis(dicarbollide) ions and their carbon and boron substituted derivatives in aqueous phosphate buffers. *Electrochim. Acta* **2020**, *342*, 136112. [CrossRef]

245. Shishkanova, T.V.; Sinica, A. Electrochemically deposited cobalt bis(dicarbollide) derivative and the detection of neuroblastoma markers on the electrode surface. *J. Electroanal. Chem.* **2022**, *921*, 7. [CrossRef]

Review

Decaborane: From Alfred Stock and Rocket Fuel Projects to Nowadays [†]

Igor B. Sivaev

A.N. Nesmeyanov Institute of Organoelement Compounds, Russian Academy of Sciences, 28 Vavilov Str., 119334 Moscow, Russia; sivaev@ineos.ac.ru
[†] Dedicated to Professor John D. Kennedy on his 80th birthday and in recognition of his outstanding contributions to the chemistry of boranes and metallaboranes.

Abstract: The review covers more than a century of decaborane chemistry from the first synthesis by Alfred Stock to the present day. The main attention is paid to the reactions of the substitution of hydrogen atoms by various atoms and groups with the formation of *exo*-polyhedral boron–halogen, boron–oxygen, boron–sulfur, boron–nitrogen, boron–phosphorus, and boron–carbon bonds. Particular attention is paid to the chemistry of *conjucto*-borane *anti*-$[B_{18}H_{22}]$, whose structure is formed by two decaborane moieties with a common edge, the chemistry of which has been intensively developed in the last decade.

Keywords: decaborane; history; properties; derivatives

Citation: Sivaev, I.B. Decaborane: From Alfred Stock and Rocket Fuel Projects to Nowadays. *Molecules* **2023**, *28*, 6287. https://doi.org/10.3390/molecules28176287

Academic Editor: M. Concepción Gimeno

Received: 2 August 2023
Revised: 21 August 2023
Accepted: 24 August 2023
Published: 28 August 2023

1. Introduction

Decaborane $[B_{10}H_{14}]$ plays a central role in the chemistry of polyhedral boron hydrides. Decaborane is an essential boron reagent for the preparation of medium and higher carboranes $C_2B_nH_{n+2}$ (n = 8–10) [1] and the carba-*closo*-decaborate anions $[CB_9H_{10}]^-$ [2]. Until recently, the synthesis of the *closo*-dodecaborate $[B_{12}H_{12}]^{2-}$ [3,4] and the carba-*closo*-dodecaborate $[CB_{11}H_{12}]^-$ [5,6] anions was also based on the use of decaborane, and it is still used for the synthesis of the *closo*-decaborate anion $[B_{10}H_{10}]^{2-}$ [7,8]. In addition, decaborane can be used to prepare boron coatings [9–13], nanoparticles [14], microcrystals [15,16], boron nitride nanosheets [17], and nanotubes [18], as well as various metal boride thin films [19–24]. Recently, a decaborane-based fuel cell power source with a high energy density was developed [25]. The intensive development of the chemistry of decaborane is associated with the 1950s to the early 1960s, when the main types of its transformations were discovered and described. These early studies were reviewed in the 1960s by Hawthorne [26] and Zakharkin et al. [27]. This area was also partly elucidated in *Boron Hydride Chemistry* [28] and *Comprehensive Inorganic Chemistry I* [29]. Recent studies in the field of decaborane chemistry, deepening and expanding the previously described conclusions using modern instrumental methods, were briefly covered in *Comprehensive Inorganic Chemistry III* [30]. Therefore, the purpose of this review is to give the most complete picture of the current state of the chemistry of decaborane and its derivatives.

2. Synthesis, Structure, and General Properties

The formation of this ten-vertex cluster during the pyrolysis of diborane B_2H_6 was first described by Alfred Stock and co-workers more than 100 years ago [31,32]. The best yields of decaborane(14) were obtained by heating diborane to 120 °C for 47 h. The low volatility of decaborane allows it to be easily separated from other volatile boron hydrides while being volatile enough to be easily separated from non-volatile products. Decaborane is a colorless, air-stable, easily subliming, malodorous, crystalline solid that melts at 99.7 °C and boils with decomposition at 213 °C [33]. For a long time, interest in the chemistry of boron hydrides was mainly academic but was supported by the fact that boranes and some related compounds did not comply with the usual rules relating the chemical composition to the classical theory of valence. At the same time, various assumptions were made about the structure of $B_{10}H_{14}$, including linear [34] or naphthalene-like [35] structures.

Practical interest in boron hydrides, and decaborane in particular, arose shortly after World War II, when the United States government launched programs (Projects Hermes, Zip, and HEF (High-Energy Fuels)) [36,37] whose purpose was to develop borane-based aviation and rocket fuels capable of generating much higher energy than conventional kerosene-based fuels [38–40]. As a part of this program, two chemical companies, Callery Chemical Company and Olin-Mathieson Corporation, developed eight pilot and production plants and produced an array of borane-derived energetic products, including methyldecaborane (HEF-4), ethyldecaborane (HEF-3), and ethylacetylenedecaborane (HEF-5), to be tested as additives to propellants and explosives [41]. Amost at the same time, due to the development of various physical research methods, such as single-crystal X-ray diffraction, neutron diffraction, and gas phase electron diffraction, the molecular structure of $[B_{10}H_{14}]$ was determined [42–50]. The decaborane molecule was found to be shaped like a boat built from ten BH-units, with four additional BHB bridges decorating its bow and stern (Figure 1).

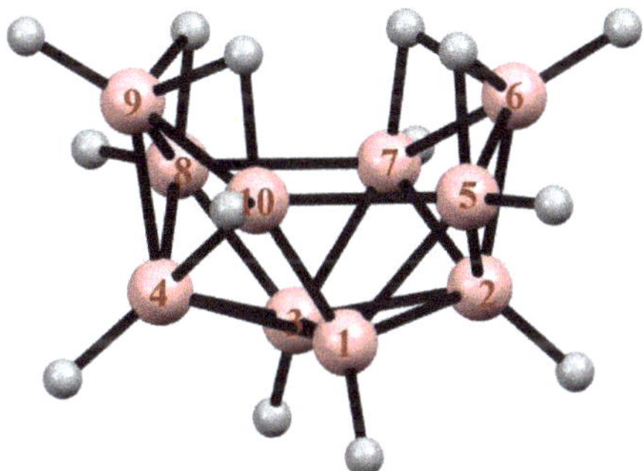

Figure 1. Structure and numbering of atoms in decaborane $B_{10}H_{14}$.

Shortly thereafter, the future Nobel Winner Lipscomb and his collaborators developed bond counting rules and topological principles that made it possible to describe bonding in boron hydrides. According to the topological formalism, the binding in the decaborane molecule can be described by a combination of four 3c-2e B-H-B bonds, six closed or fractional closed 3c-2e B-B-B bonds, and two 2c-2e B-B bonds [51–53]. Some time later, molecular orbital theory in the form of the extended Hückel theory was originated and applied to decaborane to provide an alternative to this topological approach [54–56]. Subsequently, both the logical basis and the parameters for these molecular orbitals were greatly improved using the more rigorous molecular self-consistent field (SCF) method [57,58]. More recently, the electronic structure of decaborane has been described in terms of the BadeR's theory "Atoms in Molecules" (AIM) [59].

A powerful tool for determining the structure of polyhedral boron hydrides is NMR spectroscopy, the practical birth of which coincided with a wave of interest in the chemistry of boron hydrides. Therefore, it is not surprising that decaborane was one of the first molecules to be investigated using NMR spectroscopy [60–62]. The subsequent development of the instrumental base and methods of NMR spectroscopy caused repeated studies [63–70]. The decaborane molecule has also been characterized by IR [59,71,72], Raman [59], electron [73–75], NQR [76,77], photoelectron [78], and electron energy loss [79] spectroscopy. The ionization potentials of decaborane and ^{11}B-enriched decaborane were determined to be 11.0 eV [80] and 10.26 eV [81], respectively. The dipole moment of decaborane was determined by measuring the dielectric constants of benzene, cyclohexane, and carbon disulfide solutions and varies from 3.17 D in carbon disulfide to 3.62 D in benzene [82]. The magnetic susceptibility of decaborane is $-116 \pm 1.5 \times 10^{-6}$ emu mol^{-1} [82]. The heat of formation of decaborane was determined to be -66.1 kJ/mol [83]. The heat capacity of decaborane has been measured, and the derived thermodynamic functions have been calculated [84,85]. The heats of melting and vaporization [85], as well as the vapor pressure of decaborane [85,86], were also determined. Pressure-induced room temperature transformations of decaborane up to 131 GPa were studied using in situ optical spectroscopy techniques [87].

The industrial production of boron hydrides, which involved more than 2000 people, was accompanied by various accidents, which led to the discovery of the high toxicity of decaborane and its derivatives [88–92]. With decaborane, intoxication, headaches, tremors, impaired coordination, confusion, anxiety, photophobia, and other symptoms are observed. Moreover, intoxication can occur from relatively small amounts of decaborane. Decaborane can be detected by its odor at or near its maximum acceptable concentration, but there is considerable olfactory fatigue. Repeated exposure to decaborane can cause severe damage to the nervous system [93]. The effects of decaborane on various animals have also been studied [93–111]. A number of studies were directed to study the mechanism of decaborane action on living organisms [112–130].

The production of decaborane, established in the 1950s, was based on the pyrolytic conversion of diborane proposed by Alfred Stock [33]. At the same time, attempts were made to find an alternative to this dangerous process, among which the use of the CW CO$_2$ laser is worth mentioning [131]. Almost at the same time, a convenient and effective method was proposed, which is based on the oxidation of sodium tetrahydroborate NaBH$_4$ to the octahydrotriborate anion [B$_3$H$_8$]$^-$, followed by its pyrolysis in diglyme at 105 °C to the tetradecahydro-*nido*-undecaborate anion [B$_{11}$H$_{14}$]$^-$ [132]. The subsequent mild oxidation of [B$_{11}$H$_{14}$]$^-$ gives decaborane [B$_{10}$H$_{14}$] (Scheme 1) [133–136]. Decaborane can also be obtained by the cage-opening of the *closo*-decaborate anion [B$_{10}$H$_{10}$]$^{2-}$ on protonation with strong acids such as sulfuric acid [137].

$$\text{NaBH}_4 \xrightarrow{\text{[O]}} \text{Na[B}_3\text{H}_8] \xrightarrow[\text{diglyme}]{105\ ^\circ\text{C}} \cdots \xrightarrow{\text{[O]}} \cdots$$

Scheme 1. Synthesis of decaborane(14) from sodium tetrahydroborate NaBH$_4$.

Decaborane(14) has an acidic character [138] and can be deprotonated with strong bases, such as sodium hydride [139], tetraalkylammonium hydroxides [140], diethylamine [140], triethylamine [140,141], methylenetriphenylphosphorane [141,142], or a Proton Sponge (PS) [143,144] to give the corresponding salts of the tridecahydro-*nido*-decaborate [B$_{10}$H$_{13}$]$^-$ anion (Scheme 2). The pK_a value of decaborane(14) in aqueous ethanol was found to vary from 2.41 to 3.21 depending on the water content [145].

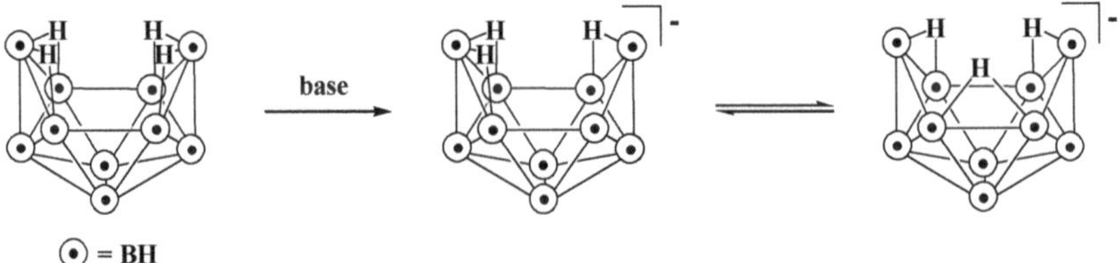

Scheme 2. Preparation of the [B$_{10}$H$_{13}$]$^-$ anion and its tautomeric forms.

The (Et₃NH)[B₁₀H₁₃] and (Et₄N)[B₁₀H₁₃] salts obtained by the deprotonation of decaborane with Et₃N and (Et₄N)OH, respectively, have been found to trigger the hypergolic reactivity of some polar aprotic organic solvents, such as tetrahydrofuran and ethyl acetate [146].

The solid state structures of (Et₃NH)[B₁₀H₁₃] [147], (BnNMe₃)[B₁₀H₁₃] [148], and (HPS)[B₁₀H₁₃] [144] were determined by single-crystal X-ray diffraction. The solid state structure of the [B₁₀H₁₃]⁻ anion (Figure 2) can be derived from the structure of B₁₀H₁₄ by μ-H(9,10) deprotonation.

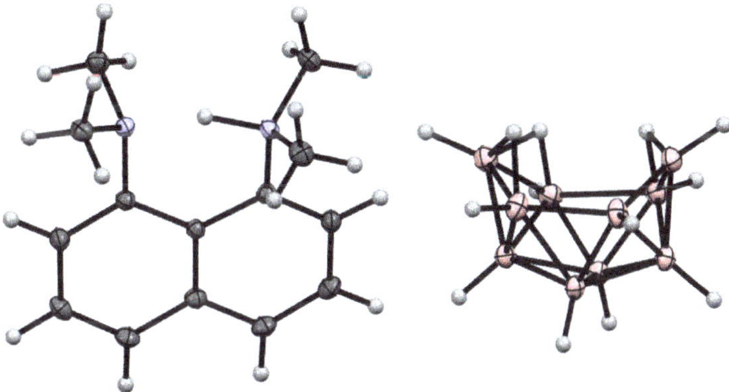

Figure 2. Solid state structures of the (HPS)⁺ cation (**left**) and of the [*nido*-B₁₀H₁₃]⁻ anion (**right**) in the crystal structure of (HPS)[B₁₀H₁₃].

In the solution, the [B₁₀H₁₃]⁻ anion exists as a mixture of symmetrical and unsymmetrical *H*-tautomers with different arrangements of bridging hydrogens (Figure 2) [143,149], with an interconversion ΔG‡ value of less than 7 kcal/mol [144].

Strong bases such as sodium hydride in ether solvents are able to remove two protons from decaborane(14) to form the [B₁₀H₁₂]²⁻ dianion [140,150,151]. The latter is unstable in solution and transforms into other decaborates and their derivatives [151]. According to quantum chemical calculations, the [B₁₀H₁₂]²⁻ anion has the C_2-symmetric structure with μ-B(5)HB(6) and μ-B(8)HB(9) bridging hydrogens [152].

The reduction of decaborane(14) with KBH₄ in water results in the formation of the [*arachno*-B₁₀H₁₄]²⁻ anion with a boron cage geometry near the same as that of the starting [*nido*-B₁₀H₁₄] (Scheme 3), which was isolated by precipitation from an aqueous solution in the form of rubidium, cesium, or tetramethylammonium salts [153,154]. The structure of the [B₁₀H₁₄]²⁻ anion was proposed using ¹¹B NMR spectroscopy [155]. The solid state structure of (Me₄N)₂[B₁₀H₁₄] was determined by single-crystal X-ray diffraction [154]. It was supposed that the reaction proceeds by hydride transfer with the formation of the [B₁₀H₁₅]⁻ anion. The latter is unstable in the solution and loses hydrogen to form [*nido*-B₁₀H₁₃]⁻ [156,157].

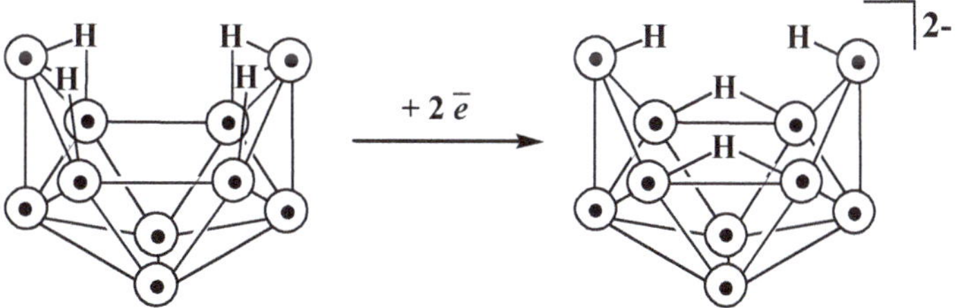

Scheme 3. Preparation of the [*arachno*-B₁₀H₁₄]²⁻ anion.

It should be borne in mind that decaborane itself has pronounced reducing properties. Due to this, the possibility of its use as a reducing agent in organic synthesis was studied. In particular, decaborane can be used for the reduction of acetals to ethers [158], the reductive esterification of aromatic aldehydes [159,160], and the reductive amination of acetals with aromatic amines [161]. Decaborane can also be used for chemoselective reduction aldehydes and ketones [162–164], the dehalogenation of α-halocarbonyl compounds [165], and the hydrogenation of alkenes or alkynes [166].

The bridging hydrogens in decaborane(14) were found to exchange rapidly for deuterium atoms with D_2O in 1,4-dioxane or acetonitrile to give $[\mu_4\text{-}B_{10}H_{10}D_4]$ [72,167,168]. The use of DCl in 1,4-dioxane makes it possible to obtain the decadeuterated decaborane $[\mu_4\text{-}5,6,7,8,9,10\text{-}B_{10}H_4D_{10}]$ [169]. The treatment of decaborane(14) with DCl in a carbon disulfide solution and in the presence of $AlCl_3$ results in the tetradeuterated decaborane $[1,2,3,4\text{-}B_{10}H_{10}D_4]$ [169,170]. If the reaction is carried out under heating in sealed ampoule, the product is the octadeuterated decaborane $[1,2,3,4,5,7,8,10\text{-}B_{10}H_6D_8]$ [72]. The tetradeuterated decaborane $[5,7,8,10\text{-}B_{10}H_{10}D_4]$ was obtained by the reaction of $[1,2,3,4,5,7,8,10\text{-}B_{10}H_6D_8]$ with HCl in a carbon disulfide solution and in the presence of $AlCl_3$ [72]. The octadeuterated decaborane $[\mu_4\text{-}1,2,3,4\text{-}B_{10}H_6D_8]$ was prepared by the reaction of $[1,2,3,4\text{-}B_{10}H_{10}D_4]$ with D_2O in acetonitrile [72], whereas the dodecadeuterated decaborane $[\mu_4\text{-}1,2,3,4,5,7,8,10\text{-}B_{10}H_2D_{12}]$ was obtained by heating $[\mu_4\text{-}B_{10}H_{10}D_4]$ with DCl in carbon disulfide in sealed ampoule in the presence of $AlCl_3$ [72]. Deuterated aromatic solvents can also act as a source of deuterium. For example, heating decaborane(14) in benzene-d_6 in the presence of $AlCl_3$ under reflux leads to the the tetradeuterated decaborane $[1,2,3,4\text{-}B_{10}H_{10}D_4]$, while the reaction of decaborane(14) with $AlCl_3$ in toluene-d_8 at 5 °C results in the dideuterated decaborane $[2,4\text{-}B_{10}H_{12}D_2]$ [171]. The reaction of $[1,2,3,4\text{-}B_{10}H_{10}D_4]$ with $AlCl_3$ in benzene leads to $[1,3\text{-}B_{10}H_{12}D_2]$ [172].

3. Halogen Derivatives

Stock first reported the preparation of halogen derivatives of decaborane(14) by the direct reaction of decaborane with halogens in a sealed tube [33]. These reactions were re-investigated in the 1960s. It was found that the reaction of decaborane(14) with 1 equiv. iodine at 110–120 °C leads to the formation of a mixture of 1- and 2-iodo derivatives of decaborane in a ratio of ~1:2 [173,174]. The resulting mixture of isomers can be separated by fractional crystallization from low-boiling alkanes (pentane, hexane, heptane) [173] or chromatographically [175]. The assignment of the substitution position was made based on the ^{11}B NMR spectra [174,176,177]; however, the 1-isomer was initially erroneously assigned as the 5-isomer [173,178]. The subsequent reaction of 2-iododecaborane with iodine at 110 °C results in a mixture of the 1,2- and 2,4-diiodo derivatives $[1,2\text{-}I_2\text{-}B_{10}H_{12}]$ and $[2,4\text{-}I_2\text{-}B_{10}H_{12}]$ in a nearly equal ratio, while the similar reaction of 1-iododecaborane produces mainly $[1,2\text{-}I_2\text{-}B_{10}H_{12}]$. The reaction of decaborane(14) with an excess of iodine was found to give a mixture of $[1,2\text{-}I_2\text{-}B_{10}H_{12}]$ and $[2,4\text{-}I_2\text{-}B_{10}H_{12}]$ in a ratio of ~1:2 [173]. The reaction of decaborane(14) with iodine or iodine chloride in carbon disulfide in the presence of $AlCl_3$ was found to give a mixture of the 1- and 2-iodo derivatives of decaborane in the same ratio of ~1:2 [179]. Later, the 1- and 2-iodo derivatives of decaborane were synthesized by the reaction of decaborane(14) with iodine chloride in refluxing dichloromethane in the presence of $AlCl_3$ and reliably characterized by NMR spectroscopy [180]. The solid state structures of $[1\text{-}I\text{-}B_{10}H_{13}]$ [181], $[2\text{-}I\text{-}B_{10}H_{13}]$ (Figure 3) [180], and $[2,4\text{-}I_2\text{-}B_{10}H_{12}]$ [182] were determined by single-crystal X-ray diffraction.

The 5- and 6-iodo derivatives of decaborane were prepared in an indirect way. The treatment of $[arachno\text{-}6,9\text{-}(Me_2S)_2\text{-}B_{10}H_{12}]$ with anhydrous HI in benzene under reflux conditions results in a mixture of the 5- and 6-iodo derivatives of decaborane $[5\text{-}I\text{-}B_{10}H_{13}]$ and $[6\text{-}I\text{-}B_{10}H_{13}]$ [174], which was separated by fraction crystallization from hexane [174] or chromatographically [175]. The reaction of $[arachno\text{-}6,9\text{-}(Et_2S)_2\text{-}B_{10}H_{12}]$ with anhydrous HI in benzene at room temperature was found to give the 5-iodo derivative $[5\text{-}I\text{-}B_{10}H_{13}]$ [183–185]. The reaction of $(NH_4)_2[closo\text{-}B_{10}H_{10}]$ with anhydrous HCl in a mixture of AlI_3 and 1-butyl-3-

methylimidazolium iodide (bmimI) at 70 °C proceeds with the boron cage opening and results in the 6-iodo derivative of decaborane [6-I-$B_{10}H_{13}$] [186]. The 6-iodo derivative isomerizes to the 5-iodo derivative [5-I-$B_{10}H_{13}$] in the presence of a catalytic amount of triethylamine in toluene at 60 °C. It is assumed that the isomerization occurs through the transformation of [6-I-$B_{10}H_{13}$] into the [6-I-$B_{10}H_{13}$]$^-$ anion, followed by its isomerization [187]. The 6-iodo derivative [6-I-$B_{10}H_{13}$] was also found to undergo photochemical isomerization to [5-I-$B_{10}H_{13}$] under UV-irradiation in a solution [187]. The solid state structures of [5-I-$B_{10}H_{13}$] [187] and [6-I-$B_{10}H_{13}$] [186] were determined by single-crystal X-ray diffraction (Figure 4). The 6-iodo derivative [6-I-$B_{10}H_{13}$] was also obtained by the reaction of $(NH_4)_2[closo\text{-}B_{10}H_{10}]$ with AlI_3, followed by the hydrolysis of the resulting intermediate [188].

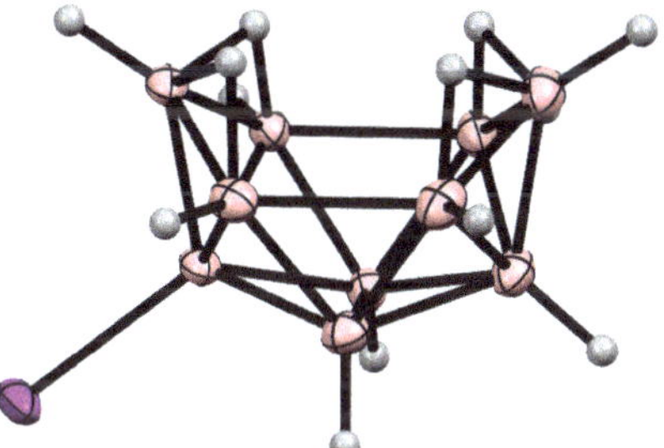

Figure 3. Solid state structure of the 2-iodo derivative of decaborane [2-I-$B_{10}H_{13}$].

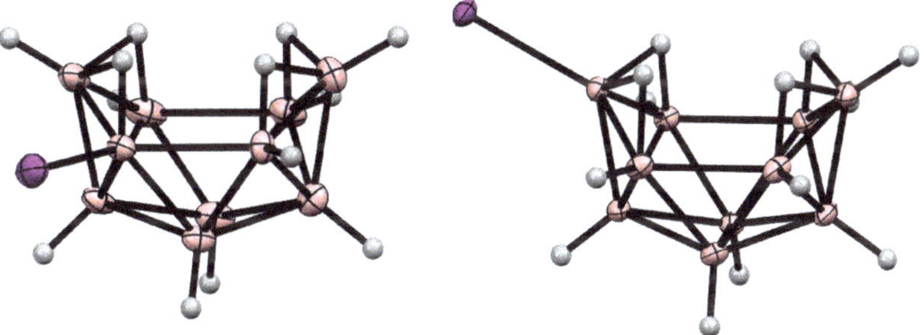

Figure 4. Solid state structures of [5-I-$B_{10}H_{13}$] (**left**) and [6-I-$B_{10}H_{13}$] (**right**).

The reactions of decaborane(14) with bromine in dichloromethane in the presence of $AlCl_3$ [189] or $AlBr_3$ [190], or in carbon disulfide in the presence of $AlCl_3$ [174], lead to the formation of a mixture of 1- and 2-bromo derivatives of decaborane [1-Br-$B_{10}H_{13}$] and [2-Br-$B_{10}H_{13}$], which can be separated by fractional crystallization from hexane [173,189,190] or chromatographically [175].

The 5- and 6-bromo derivatives of decaborane were also prepared in an indirect way. The treatment of [*arachno*-6,9-$(Me_2S)_2$-$B_{10}H_{12}$] with anhydrous HBr in benzene under reflux conditions results in a mixture of the 5- and 6-bromo derivatives of decaborane [5-Br-$B_{10}H_{13}$] and [6-Br-$B_{10}H_{13}$] in a ratio of ~1:4, which was separated by fraction crystallization from hexane [174] or chromatographically [175,190]. The reactions of [*arachno*-6,9-$(R_2S)_2$-$B_{10}H_{12}$] (R = Me, Et) with anhydrous HBr in benzene at room temperature were found to give the 5-bromo derivative [5-Br-$B_{10}H_{13}$] [183–185]. The reaction of $(NH_4)_2[closo\text{-}B_{10}H_{10}]$ with anhydrous HBr in a mixture of $AlBr_3$ and 1-butyl-3-methylimidazolium bromide (bmimBr) at 70 °C proceeds with the boron cage opening and results in the 6-bromo derivative of decaborane [6-Br-$B_{10}H_{13}$] [186]. The 6-bromo derivative isomerizes to the 5-bromo derivative [5-Br-$B_{10}H_{13}$] in the presence of a catalytic amount of triethylamine in toluene at 60 °C [187]. The solid state structures of [5-Br-$B_{10}H_{13}$] [187] and [6-Br-$B_{10}H_{13}$] [186] were determined by single-crystal X-ray diffraction (Figure 5). The 6-bromo derivative [6-Br-

$B_{10}H_{13}$] was also obtained by the reaction of $(NH_4)_2[closo\text{-}B_{10}H_{10}]$ with $AlBr_3$, followed by the hydrolysis of the resulting intermediate [188].

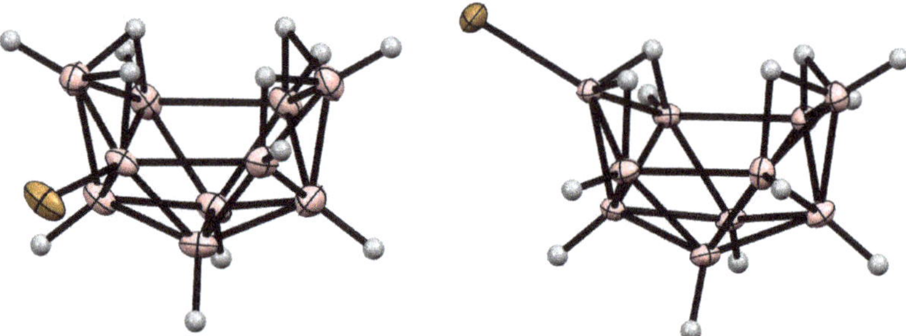

Figure 5. Solid state structures of $[5\text{-}Br\text{-}B_{10}H_{13}]$ (**left**) and $[6\text{-}Br\text{-}B_{10}H_{13}]$ (**right**).

The reaction of dimethylstannaundecaborane [$nido$-$Me_2SnB_{10}H_{12}$] with bromine in carbon disulfide leads to the oxidative removal of tin with the formation of the 5,10-dibromo derivative of decaborane [$5,10\text{-}Br_2\text{-}B_{10}H_{12}$] (Figure 6) [191].

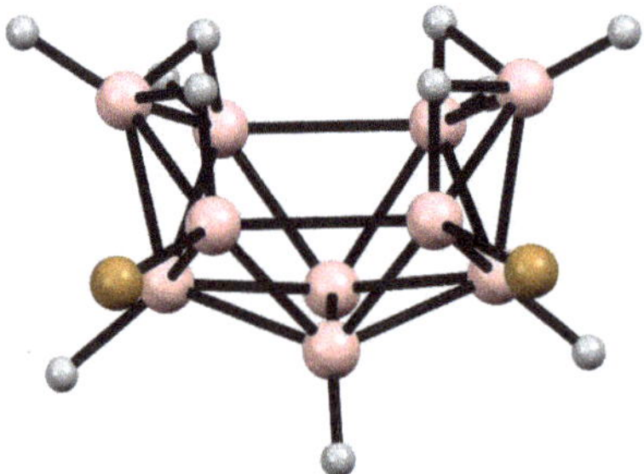

Figure 6. Solid state structure of the 5,10-dibromo derivative of decaborane [$5,10\text{-}Br_2\text{-}B_{10}H_{12}$].

The reaction of decaborane(14) with chlorine in dichloromethane in the presence of $AlCl_3$ results in a mixture of the 1- and 2-chloro derivatives of decaborane [$1\text{-}Cl\text{-}B_{10}H_{13}$] and [$2\text{-}Cl\text{-}B_{10}H_{13}$] [174,189]. Unexpectedly, a mixture of the 1- and 2-chloro derivatives of decaborane was obtained in the reaction of decaborane(14) with 1,1-difluoroethane in carbon disulfide in the presence of $AlCl_3$ [192]. The isomers were separated by fractional crystallization from pentane or hexane [174,189,192] or chromatographically [175], and the substitution position was assigned using ^{11}B NMR spectroscopy [174,193].

Similar to the corresponding iodo and bromo derivatives, the 5- and 6-chloro derivatives of decaborane were prepared in an indirect way. The treatment of [$arachno$-6,9-$(Me_2S)_2\text{-}B_{10}H_{12}$] with anhydrous HCl in benzene under reflux conditions results mainly in the 6-chloro derivative of decaborane [$6\text{-}Cl\text{-}B_{10}H_{13}$] with some amount of the 5-chloro isomer [174]. The reactions of [$arachno$-6,9-$(Et_2S)_2\text{-}B_{10}H_{12}$] with anhydrous HCl or $HgCl_2$ in benzene at room temperature were also found to give the 6-chloro derivative [$6\text{-}Cl\text{-}B_{10}H_{13}$] [183–185]. The reaction of $(NH_4)_2[closo\text{-}B_{10}H_{10}]$ with anhydrous HCl in a mixture of $AlCl_3$ and 1-butyl-3-methylimidazolium chloride (bmimCl) at 70 °C proceeds with the boron cage opening and results in the 6-chloro derivative of decaborane [$6\text{-}Cl\text{-}B_{10}H_{13}$] [186]. The 6-chloro derivative of decaborane can also be prepared by the reactions of $(NH_4)_2[closo\text{-}B_{10}H_{10}]$ with triflic acid in dichloromethane, respectively [186]. The 6-chloro derivative isomerizes to the 5-chloro derivative [$5\text{-}Cl\text{-}B_{10}H_{13}$] in the presence of a catalytic amount of triethylamine in toluene at 60 °C [187]. It was assumed that the isomerization occurs through the transformation of [$6\text{-}Cl\text{-}B_{10}H_{13}$] into the [$6\text{-}Cl\text{-}B_{10}H_{13}$]$^-$ anion, followed by its isomerization. The solid state structures of [$6\text{-}Cl\text{-}B_{10}H_{13}$], (HPS)[$6\text{-}Cl\text{-}B_{10}H_{12}$], and

(HPS)[5-Cl-$B_{10}H_{12}$] were determined by single-crystal X-ray diffraction (Figure 7) [186,187]. The B-H-B bridge deprotonation at the site adjacent to the halogenated boron atoms was revealed [187].

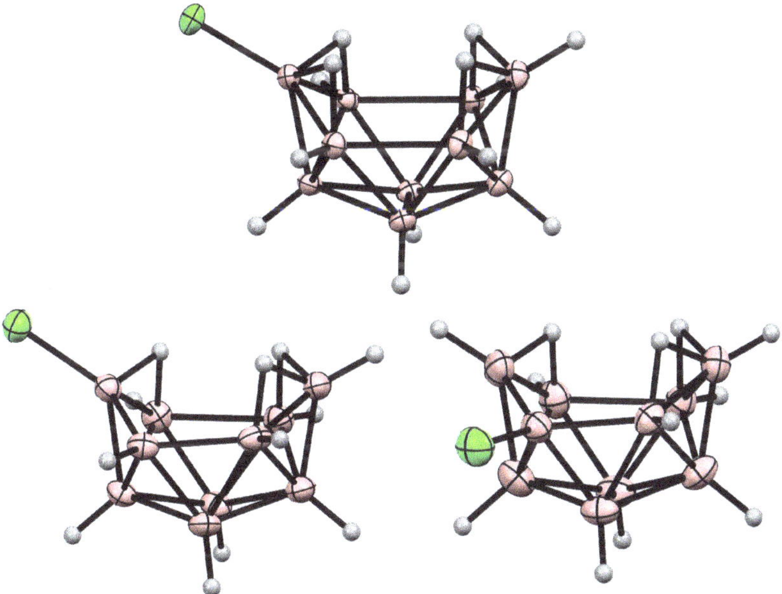

Figure 7. Solid state structures of [6-Cl-$B_{10}H_{13}$] (**top**) as well as the [6-Cl-$B_{10}H_{12}$]$^-$ (**bottom left**) and [5-Cl-$B_{10}H_{12}$]$^-$ (**bottom right**) anions in the crystal structures of the corresponding protonated Proton Sponge salts.

The 6-chloro derivative [6-Cl-$B_{10}H_{13}$] was also obtained by the reaction of $(NH_4)_2$[*closo*-$B_{10}H_{10}$] or $(Et_4N)_2$[*closo*-$B_{10}H_{10}$] with $AlCl_3$, followed by the hydrolysis of the resulting intermediate [188,194,195].

The 6-fluoro derivative of decaborane [6-F-$B_{10}H_{13}$] was first obtained by the reaction of [*arachno*-6,9-$(Et_2S)_2$-$B_{10}H_{12}$] with anhydrous HF in benzene at room temperature [183–185]. Later, the 6-fluoro derivative was prepared by the reaction of $(NH_4)_2$[*closo*-$B_{10}H_{10}$] with triflic acid in 1-fluoropentane [186]. The solid state structure of [6-F-$B_{10}H_{13}$] was determined by single-crystal X-ray diffraction (Figure 8) [186].

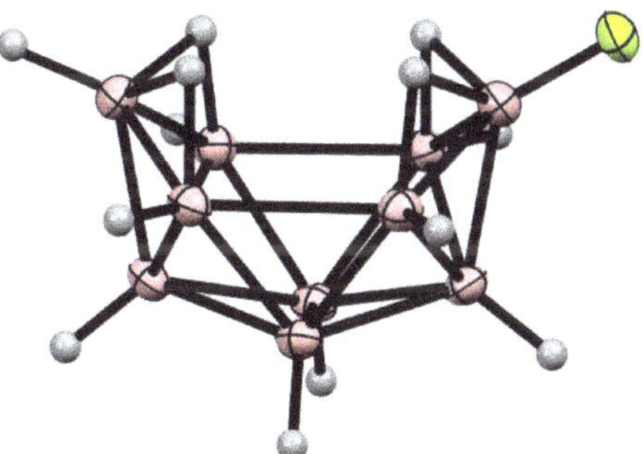

Figure 8. Solid state structure of the 6-fluoro derivative of decaborane [6-F-$B_{10}H_{13}$].

4. Derivatives with a B-O Bond

Due to the lack of electrophilic reagents for the introduction of oxygen substituents, the corresponding decaborane derivatives with substituents localized in the "bottom" of the decaborane basket have not yet been obtained. Alkoxy derivatives of decaborane [5-RO-$B_{10}H_{13}$] (R = Me, Et, Pr, Bu) were first obtained in low (13–20%) yields by trying to iodinate Na[$B_{10}H_{13}$] in the corresponding esters [196]. The phenoxy derivative [5-PhO-$B_{10}H_{13}$] was obtained in the same way, using anisole as a solvent [196]. The substitution position was determined using ^{11}B NMR spectroscopy [197]. It was assumed that their formation proceeds through the formation of oxonium derivatives of *arachno*-decaborane [R_2O-$B_{10}H_{13}$]$^-$, followed by the elimination of one alkyl group [26]. The reaction of Na[$B_{10}H_{13}$] with $SnCl_4$ in diethyl ether leads to a mixture of 6- and 5-alkoxy derivatives of decaborane [6-EtO-$B_{10}H_{13}$] and [5-EtO-$B_{10}H_{13}$] in a ratio varying from 85:15 to 70:30 depending on the reaction temperature. The isomers were separated using column chromatography on silica [198]. It should be noted that the direct reactions of decaborane(14) with alcohols and phenols ROH leads to its complete degradation to the corresponding trialkyl- or triarylborates $(RO)_3B$ [199]. The trimethylsiloxy derivative [6-Me_3SiO-$B_{10}H_{13}$] was obtained in a low yield (5–20%) from the reactions of Na[$B_{10}H_{13}$] and Na_2[$B_{10}H_{12}$] with Me_3SiCl in diethyl ether [198]. The report on the preparation of a trimethylsilyl derivative [Me_3Si-$B_{10}H_{13}$] under similar conditions [200] should apparently be considered erroneous.

The reactions of 5-bromo derivative [5-Br-$B_{10}H_{13}$] with alcohols ROH in the presence of $NaHCO_3$ in dichloromethane lead to 6-alkoxy derivatives [6-RO-$B_{10}H_{13}$] (R = Me, Et, *t*-Bu, *c*-Hx, CH_2CH_2SH, CH_2CH_2I, $CH_2CH_2OCH_2CH_2Cl$, $(CH_2)_3C{\equiv}CH$, $CH(CH_2CH{=}CH_2)_2$), while the reactions of 6-bromo derivative [6-Br-$B_{10}H_{13}$] with alcohols ROH in the presence of $NaHCO_3$ in dichloromethane lead to 5-alkoxy derivatives [5-RO-$B_{10}H_{13}$] (R = Me, *t*-Bu, cHx, CH_2CH_2SH, CH_2CH_2I, $CH_2CH_2N(CO)_2C_2H_4$, $CH_2CH_2OCH_2CH_2Cl$, $(CH_2)_3C{\equiv}CH$, $CH_2C{\equiv}CCH_3$, $CH(CH_2CH{=}CH_2)_2$). The reactions of [5-Br-$B_{10}H_{13}$] and [6-Br-$B_{10}H_{13}$] with 1,4-cyclohexyldiol lead to the compounds [μ-6,6′-($OC_6H_{10}O$)-($B_{10}H_{13}$)$_2$] and [μ-5,5′-($OC_6H_{10}O$)-($B_{10}H_{13}$)$_2$], respectively. The reactions of alcohols with [6-Br-$B_{10}H_{13}$] proceed quickly at room temperature, while those with [5-Br-$B_{10}H_{13}$] require heating (70 °C) to achieve completion. The reaction of [6-Br-$B_{10}H_{13}$] with ethanol was largely complete after 12 h at room temperature, but the reactions with 2-iodoethanol (~20 h), 2-bromoethanol (~40 h), 2-chloroethanol (~100 h), and 2-fluoroethanol (~125 h) all took increasingly longer times. The reactions with chloro- and iodo-derivatives of decaborane were found to proceed in a similar way; however, the reaction rate decreases in the halogen series I~Br > Cl [198]. The solid state structures of [5-MeO-$B_{10}H_{13}$] (Figure 9), [6-*t*-BuO-$B_{10}H_{13}$] (Figure 9), [5-$ClCH_2CH_2OCH_2CH_2O$-$B_{10}H_{13}$], [6-$ClCH_2CH_2OCH_2CH_2O$-$B_{10}H_{13}$], [5-$MeC{\equiv}CCH_2O$-$B_{10}H_{13}$], and [μ-6,6′-($OC_6H_{10}O$)-($B_{10}H_{13}$)$_2$] (Figure 9) were determined by single-crystal X-ray diffraction [201].

The 6-triflate derivative of decaborane [6-TfO-$B_{10}H_{13}$] was prepared by the reaction of Cs_2[*closo*-$B_{10}H_{10}$] with neat triflic acid at an ambient temperature [202,203]. In contrast, the reaction of $(NH_4)_2$[*closo*-$B_{10}H_{10}$] with triflic acid in 1-butyl-3-methylimidazolium triflate at 60 °C results in the 5-triflate derivative of decaborane [5-TfO-$B_{10}H_{13}$] [204]. It was found that the reaction proceeds through the formation of the 6-triflate derivative, which, upon heating, isomerizes into the 5-triflate derivative. In the presence of a catalytic amount of triethylamine, the isomerization of [6-TfO-$B_{10}H_{13}$] to [5-TfO-$B_{10}H_{13}$] proceeds even at room temperature [204]. The reactions of [5-TfO-$B_{10}H_{13}$] with methanol and 4-methoxyphenol in 1,2-dichloroethane at 70 °C result in the corresponding ethers [6-RO-$B_{10}H_{13}$] (R = Me, C_6H_4-4-OMe) [204]. The solid state structures of [6-TfO-$B_{10}H_{13}$] [202] and [5-TfO-$B_{10}H_{13}$] [204] were determined by single-crystal X-ray diffraction (Figure 10).

The reactions of Na_2[*closo*-$B_{10}H_{10}$] with alcohols ROH (R = Me, Et, *i*-Pr, Bu, Ph) in hexane in the presence of trimethylsilyl triflate lead to the corresponding 6-alkoxy derivatives of decaborane [6-RO-$B_{10}H_{13}$] [205]. The reaction with water under the same conditions results in the 6-trimethylsiloxy derivative [6-Me_3SiO-$B_{10}H_{13}$] [205].

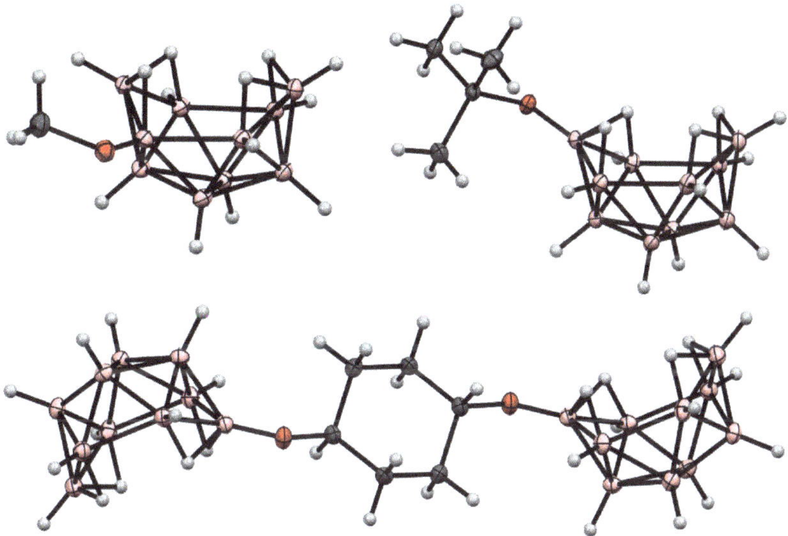

Figure 9. Solid state structures of [5-MeO-$B_{10}H_{13}$] (**top left**), [6-*t*-BuO-$B_{10}H_{13}$] (**top right**), and [μ-6,6′-(OC$_6$H$_{10}$O)-($B_{10}H_{13}$)$_2$] (**bottom**).

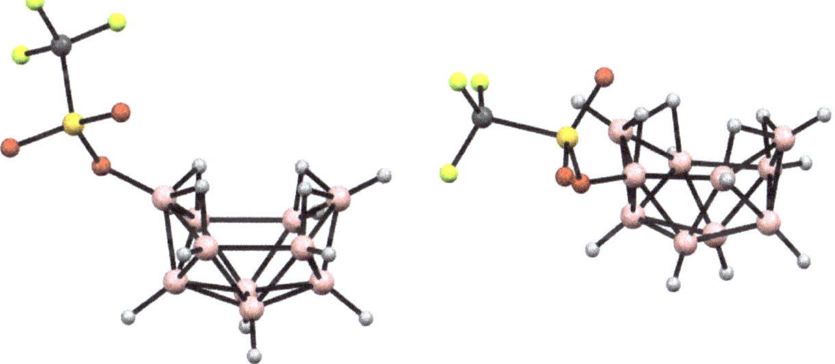

Figure 10. Solid state structures of [6-TfO-$B_{10}H_{13}$] (**left**) and [5-TfO-$B_{10}H_{13}$] (**right**).

In a similar way, the reaction of (NH$_4$)$_2$[*closo*-$B_{10}H_{10}$] with sulfuric acid produces the 6-hydroxy derivative of decaborane [6-HO-$B_{10}H_{13}$] [206]. The 6-hydroxy derivative was also obtained as a by-product of the reaction of [*arachno*-6,9-(Me$_2$S)$_2$-$B_{10}H_{12}$] with sulfuric acid in benzene [207].

The 6-acetoxy derivative of *nido*-decaborane [6-AcO-$B_{10}H_{13}$] was obtained by the reaction of [*arachno*-6,9-(Me$_2$S)$_2$-$B_{10}H_{12}$] with mercury acetate [208]. The 6-acetoxy derivative of the *arachno*-decaborate anion [*arachno*-6-AcO-$B_{10}H_{13}$]$^{2-}$ was obtained by the reaction of decaborane with 1-ethyl-3-methylimidazolium acetate (C$_2$mim)(OAc). The solid state structure of (C$_2$mim)$_2$[6-AcO-$B_{10}H_{13}$] was determined by single-crystal X-ray diffraction (Figure 11) [209].

The bis(decaboranyl) ether [μ-6,6′-O-($B_{10}H_{13}$)$_2$] was prepared by the reaction of [*arachno*-6,9-(R$_2$S)$_2$-$B_{10}H_{12}$] (R = Me, Et) with sulfuric acid in benzene [210,211]. Its structure was determined by ^{11}B NMR spectroscopy [183,211] and supported by single-crystal X-ray diffraction [212]. The bis(decaboranyl) ether [μ-6,6′-O-($B_{10}H_{13}$)$_2$] was also obtained by the dehydration of (H$_3$O)$_2$[*closo*-$B_{10}H_{10}$] [213].

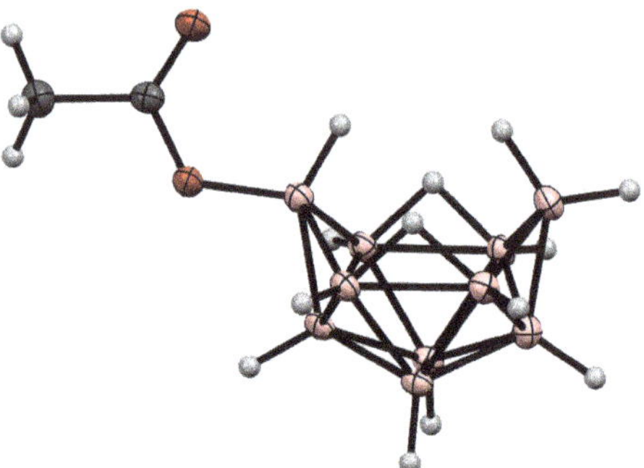

Figure 11. Structure of the [6-*arachno*-AcO-$B_{10}H_{13}$]$^{2-}$ anion in the crystal structure of (C_2mim)$_2$[6-AcO-$B_{10}H_{13}$].

The preparation of decaborane derivatives with amides [*arachno*-6,9-(MeRN(R′)CO)$_2$-$B_{10}H_{12}$] (R = H, R′ = H, Me; R = Me, R′ = H, Me), triphenylphosphine oxide [*arachno*-6,9-(Ph$_3$PO)$_2$-$B_{10}H_{12}$], and dimethylsulfoxide [*arachno*-6,9-(Me$_2$SO)$_2$-$B_{10}H_{12}$] has also been reported [214–217].

5. Derivatives with a B-S Bond

The reaction of decaborane(14) with sulfur in the presence of AlCl$_3$ at 120 °C results in a mixture of the mercapto derivatives [1-HS-$B_{10}H_{13}$], [2-HS-$B_{10}H_{13}$], and [1,2-(HS)$_2$-$B_{10}H_{12}$] [218,219]. The solid state structures of these mercapto derivatives were determined by single-crystal X-ray diffraction (Figure 12) [219].

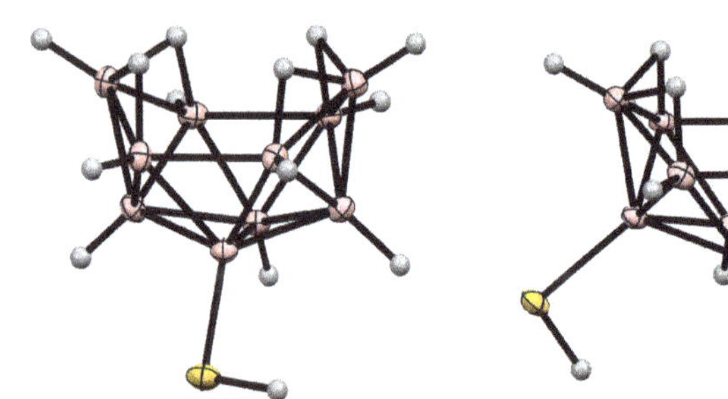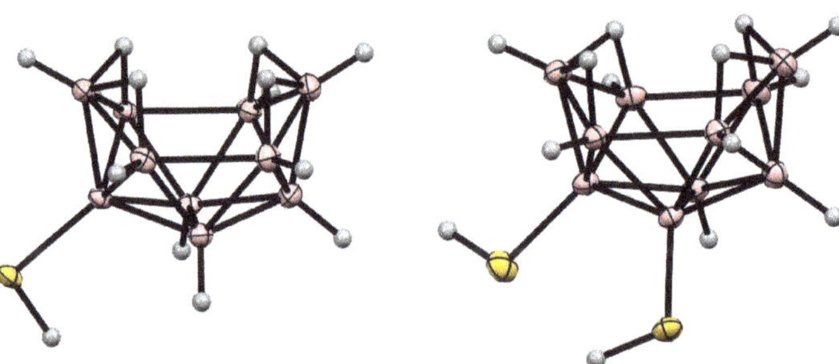

Figure 12. Solid state structures of the mercapto derivatives of decaborane [1-HS-$B_{10}H_{13}$] (**left**), [2-HS-$B_{10}H_{13}$] (**middle**), and [1,2-(HS)$_2$-$B_{10}H_{12}$] (**right**).

The reactions of Na$_2$[*closo*-$B_{10}H_{10}$] with thiols RSH (R = *i*-Pr, *i*-Bu, *c*-Hx, C_6H_4-*p*-Me, C_6H_4-*p*-F) in hexane in the presence of trimethylsilyl triflate led to the corresponding 6-alkyl- and 6-arylsulfides [6-RS-$B_{10}H_{13}$] [205]. It should be noted that the direct reactions of decaborane(14) with alkyl thiols RS lead to its complete degradation to the corresponding trialkylthioborates (RS)$_3$B [220].

Due to its use in the synthesis of carboranes, the 6,9-bis(dimethylsulfonium) derivative of the *arachno*-decaborate anion [*arachno*-6,9-(Me$_2$S)$_2$-$B_{10}H_{12}$], which is formed by refluxing decaborane with dimethyl sulfide in ether or benzene, is the most known decaborane derivative with a B-S bond [214,221,222]. The solid state structure of [*arachno*-6,9-(Me$_2$S)$_2$-$B_{10}H_{12}$] was determined by single-crystal X-ray diffraction [223]. The 6,9-bis(dimethylsulfonium) derivative was studied by X-ray and X-ray photoelectron spectroscopy [224–226], and its diamagnetic susceptibility was determined [227]. The Me$_2$S substituents in [*arachno*-6,9-(Me$_2$S)$_2$-$B_{10}H_{12}$] can be easily replaced by stronger Lewis bases [214,228]. A series of other bis(dialkylsulfonium) derivatives ([*arachno*-6,9-(RR′S)$_2$-

$B_{10}H_{12}$] (R = R′ = Et, Pr; RR′ = (CH$_2$)$_4$, (CH$_2$CH$_2$)$_2$S, (CH$_2$CH$_2$)$_2$O)) have been prepared in a similar manner [215,220,228,229].

The bis(diethylsulfonium) derivative [*arachno*-6,9-(Et$_2$S)$_2$-B$_{10}$H$_{12}$] can also be obtained by the reaction of (NH$_4$)$_2$[*closo*-B$_{10}$H$_{10}$] with anhydrous hydrogen chloride in diethylsulfide [230]. This approach has been extended to other salts of the *closo*-decaborate anion and other strong acids, including (H$_3$O)$_2$[*closo*-B$_{10}$H$_{10}$] [231,232].

The reactions of 2-halogen derivatives of decaborane [2-X-B$_{10}$H$_{13}$] (X = Cl, Br, I) with dimethylsulfide Me$_2$S give the corresponding 2-halogen-6,9-bis(dimethylsulfonium) derivatives [*arachno*-2-X-6,9-(Me$_2$S)$_2$-B$_{10}$H$_{11}$] [233,234]. In a similar way, the reactions of 5-halogen derivatives of decaborane [5-X-B$_{10}$H$_{13}$] (X = F, Br, I) with dialkylsulfides R$_2$S (R = Me, Et) result in the corresponding 5-halogen-6,9-bis(dialkylsulfonium) derivatives [*arachno*-5-X-6,9-(R$_2$S)$_2$-B$_{10}$H$_{11}$], while the reactions of the 6-chloro derivative proceed with the halogen displacement, giving [*arachno*-6,9-(R$_2$S)$_2$-B$_{10}$H$_{12}$] [185]. The solid state structure of [5-Br-6,9-(R$_2$S)$_2$-B$_{10}$H$_{11}$] was determined by single-crystal X-ray diffraction [235]. The reactions of 2,4-dichloro- and 1,2,4-trichloro derivatives of decaborane [*nido*-2,4-Cl$_2$-B$_{10}$H$_{12}$] and [*nido*-1,2,4-Cl$_3$-B$_{10}$H$_{11}$] with dimethylsulfide were found to proceed with the boron cage rearrangement, resulting in [*arachno*-1,7-Cl$_2$-6,9-(Me$_2$S)$_2$-B$_{10}$H$_{10}$] and [*arachno*-1,3,7-Cl$_3$-6,9-(Me$_2$S)$_2$-B$_{10}$H$_9$], respectively [236]. The solid state structures of [1,7-Cl$_2$-6,9-(Me$_2$S)$_2$-B$_{10}$H$_{10}$] and [1,3,7-Cl$_3$-6,9-(Me$_2$S)$_2$-B$_{10}$H$_9$] were determined by single-crystal X-ray diffraction [236].

The reaction of the 5-triflato derivative of decaborane [5-TfO-B$_{10}$H$_{13}$] with dimethylsulfide in toluene results in the 5-triflato-6,9-bis(dialkylsulfonium) derivative [*arachno*-5-TfO-6,9-B$_{10}$H$_{11}$(SMe$_2$)$_2$], while the similar reaction of the 5-triflato derivative [6-TfO-B$_{10}$H$_{13}$] proceeds with the substitution of the triflate group, giving [*arachno*-6,9-B$_{10}$H$_{12}$(SMe$_2$)$_2$] [204]. The solid state structure of [5-TfO-6,9-(Me$_2$S)$_2$-B$_{10}$H$_{11}$] was determined by single-crystal X-ray diffraction [204].

The 5-dimethylsulfonium derivative of decaborane [*nido*-5-Me$_2$S-B$_{10}$H$_{12}$] was obtained by heating [*arachno*-6,9-(Me$_2$S)$_2$-B$_{10}$H$_{12}$] in toluene or mesitylene, and its solid state structure was determined by single-crystal X-ray diffraction (Figure 13) [215,222,237,238]. It was found that the B-H-B bridge deprotonation occurs at the site adjacent to the substituted boron atom, and thus, the structure of [5-Me$_2$S-B$_{10}$H$_{12}$] is similar to the structure of the [6-Cl-B$_{10}$H$_{12}$]$^-$ anion [187].

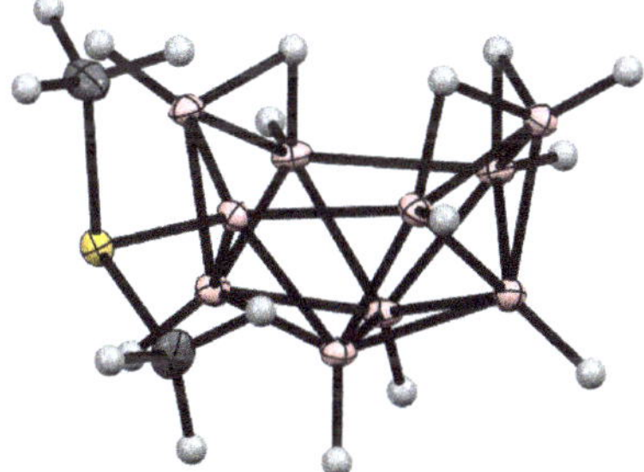

Figure 13. Solid state structure of [*nido*-5-Me$_2$S-B$_{10}$H$_{12}$].

The preparation of the bis(dimethylthioformamide) [*arachno*-6,9-(Me$_2$N(H)CS)$_2$-B$_{10}$H$_{12}$] [215] and the bis(di(alkyl/aryl)thiourea) [*arachno*-6,9-((RHN)$_2$CS)$_2$-B$_{10}$H$_{12}$] (R = Et, Ph) [217,239] derivatives has also been reported.

6. Derivatives with a B-N Bond

Heating decaborane(14) in acetonitrile under reflux proceeds with hydrogen elimination, resulting in the 6,9-bis(acetonitrilium) derivative of the *arachno*-decaborate anion [*arachno*-6,9-(MeC≡N)$_2$-B$_{10}$H$_{12}$] [240], which was the first structurally characterized decaborane derivative [217,241–243] (Figure 14). The 6,9-bis(acetonitrilium) derivative was studied by X-ray photoelectron spectroscopy [226], and its diamagnetic susceptibility was determined [227]. The 6,9-bis(propionitrilium) and 6,9-bis(benzonitrilium) derivatives [*arachno*-6,9-(RC≡N)$_2$-B$_{10}$H$_{12}$] (R = Et, Ph) were prepared in a similar way from decaborane(14) and propionitrile [244] or benzonitrile [217], respectively. The reaction of the 6,9-bis(acetonitrilium) derivative with diethylcyanamide in diethyl ether results in [*arachno*-6,9-(Et$_2$NC≡N)$_2$-B$_{10}$H$_{12}$] [214,245]. The reaction of the 2-bromo derivative of decaborane [2-Br-B$_{10}$H$_{13}$] gives [*arachno*-6,9-(MeC≡N)$_2$-2-Br-B$_{10}$H$_{11}$] [233].

Figure 14. Solid state structure of [*arachno*-6,9-(MeC≡N)$_2$-B$_{10}$H$_{12}$].

The 6,9-bis(acetonitrilium) derivative [*arachno*-6,9-(MeC≡N)$_2$-B$_{10}$H$_{12}$] reacts with *N*-nucleophiles (primary and secondary amines and hydrazine), giving the corresponding amidines [*arachno*-6,9-(RR'N(Me)C=HN)$_2$-B$_{10}$H$_{12}$] (R = H, R' = Et, Pr, Bu, Ph, NH$_2$, NHMe; R = R' = Et, Pr, Bu] [246–248]. The solid state structures of the amidines [6,9-(Bu$_2$N(Me)C=HN)$_2$-B$_{10}$H$_{12}$] and [6,9-(PhHN(Me)C=HN)$_2$-B$_{10}$H$_{12}$]·Et$_2$O were determined by single-crystal X-ray diffraction (Figure 15) [248]. It should be noted that the reactions with primary amines produce mainly the *ZE* isomers, whereas the reactions with secondary amines result only in the *EE* isomers.

The reaction of the 6,9-bis(acetonitrilium) derivative with methanol results in the formation of the corresponding imidate [*arachno*-6,9-(MeO(Me)C=HN)$_2$-B$_{10}$H$_{12}$] [248]. Nowadays, the addition of nucleophiles to the activated triple -C≡N- bond of nitrilium derivatives of various polyhedral boron hydrides has become a widely used method for their modification [249–258].

The reactions of the 6,9-bis(acetonitrilium) derivative with tertiary amines in refluxing benzene or toluene lead to the corresponding 6,9-bis(trialkylammonium) derivatives [*arachno*-6,9-(R$_3$N)$_2$-B$_{10}$H$_{12}$] (R = Me, Et) [214,245]. The 6,9-bis(ammonium) derivative [*arachno*-6,9-(H$_3$N)$_2$-B$_{10}$H$_{12}$] was prepared by the reaction of decaborane(14) with ammonia in benzene or toluene [259,260]. The solid state structures of [6,9-(H$_3$N)$_2$-B$_{10}$H$_{12}$] (Figure 16) [243,261], [6,9-(Me$_3$N)$_2$-B$_{10}$H$_{12}$] [262], and [6,9-(Et$_3$N)$_2$-B$_{10}$H$_{12}$] [263] were determined by single-crystal X-ray diffraction. [6,9-(H$_3$N)$_2$-B$_{10}$H$_{12}$] and [6,9-(Et$_3$N)$_2$-B$_{10}$H$_{12}$] were studied by X-ray photoelectron and X-ray fluorescence spectroscopy [225,226,264]. The diamagnetic susceptibilities of [6,9-(Me$_3$N)$_2$-B$_{10}$H$_{12}$] and [6,9-(Et$_3$N)$_2$-B$_{10}$H$_{12}$] were determined [227]. The thermal decomposition of the 6,9-bis(ammonium) derivative [*arachno*-6,9-(H$_3$N)$_2$-B$_{10}$H$_{12}$] was studied [260,265,266].

The reaction of decaborane(14) with diethylamine in cyclohexane results in the diethylammonium derivative of the *arachno*-decaborate anion (Et$_2$NH$_2$)[*arachno*-6-Et$_2$HN-B$_{10}$H$_{13}$] [267]. Similar alkylammonium derivatives (Me$_4$N)[*arachno*-6-RR'R''N-B$_{10}$H$_{13}$] (R = Et, R' = R'' = H; R = R' = Et, R'' = H; R = R' = R'' = Et; RR' = (CH$_2$)$_5$, R'' = H) were prepared by the reactions of Na[B$_{10}$H$_{13}$] with the corresponding amines, followed by precipitation with (Me$_4$N)Cl [267]. Heating (Et$_2$NH$_2$)[*arachno*-6-Et$_2$HN-B$_{10}$H$_{13}$] in THF under reflux gives the 6,9-bis(diethylammonium) derivative [*arachno*-6,9-(Et$_2$HN)$_2$-B$_{10}$H$_{12}$], whereas the similar reaction in acetonitrile leads to [*arachno*-6-Et$_2$HN-9-Et$_2$N(Me)C=HN-B$_{10}$H$_{12}$] [267].

The reactions of Na[*arachno*-6-Et$_2$HN-B$_{10}$H$_{13}$] with acetonitrile and dimethylsulfide in the presence of dry HCl result in [*arachno*-6-Et$_2$NH-9-MeC≡N-B$_{10}$H$_{12}$] and [*arachno*-6-Et$_2$NH-9-Me$_2$S-B$_{10}$H$_{12}$], respectively [267]. In a similar way, the reaction of (Me$_4$N)[*arachno*-6-Et$_3$N-B$_{10}$H$_{13}$] with acetonitrile leads to [*arachno*-6-Et$_3$N-9-MeC≡N-B$_{10}$H$_{12}$] [221]. The reactions of Na[*arachno*-6-Et$_2$HN-B$_{10}$H$_{13}$] with amines in THF produce the corresponding 6,9-bis(alkylammonium) derivatives [*arachno*-6-Et$_2$NH-9-RR′R″N-B$_{10}$H$_{12}$] (R = Et, R′ = R″ = H; R = R′ = Et, R″ = H; R = R′ = R″ = Me) [267].

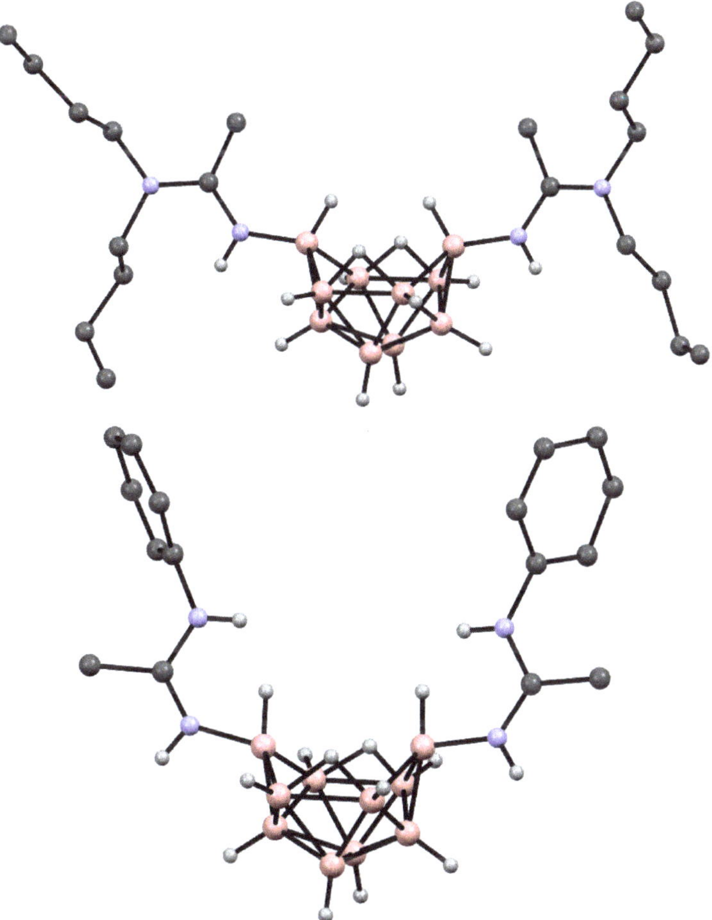

Figure 15. Solid state structures of the decaborane-based amidines [6,9-(Bu$_2$N(Me)C=HN)$_2$-B$_{10}$H$_{12}$] (**top**) and [6,9-(PhHN(Me)C=HN)$_2$-B$_{10}$H$_{12}$] (**bottom**).

The reactions of decaborane(14) with pyridines, quinolines, and isoquinoline lead to the corresponding [*arachno*-6,9-L$_2$-B$_{10}$H$_{12}$] (L = pyridine, 2-methylpyridine, 3-methylpyridine, 4-methylpyridine, 2-ethynylpyridine, 2-cyanopyridine, quinoline, 2-methylquinoline, 8-methylquinoline, isoquinoline) derivatives (Scheme 4) [214,236,268,269]. A more convenient way to prepare 6,9-bis(pyridinium) derivatives is the nucleophilic substitution of the dialkylsulfide groups in [*arachno*-6,9-(R$_2$S)$_2$-B$_{10}$H$_{12}$] (R = Me, Et). In this way, the [*arachno*-6,9-L$_2$-B$_{10}$H$_{12}$] (L = pyridine, 2-methylpyridine, 3-methylpyridine, 4-methylpyridine, 2,3-dimethylpyridine, 2,4-dimethylpyridine, 2,5-dimethylpyridine, 2,6-dimethylpyridine, 3,4-dimethylpyridine, 2-methyl-5-ethylpyridine, 2-phenylpyridine, 4-benzylpyridine, 4-styrylpyridine,

2-methoxypyridine, 4-methoxypyridine, 3-chloropyridine, 4-chloropyridine, 2-bromopyridine, 4-bromopyridine, 3,5-dibromopyridine, 3-cyanopyridine, 4-cyanopyridine, 4-acetylpyridine, quinoline) derivatives were synthesized (Scheme 4) [229,270–272].

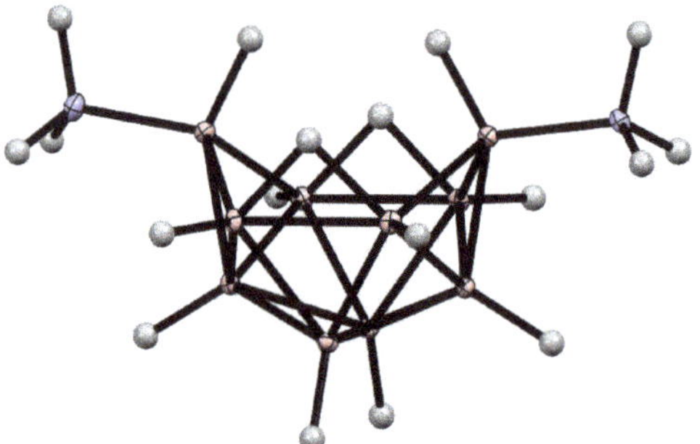

Figure 16. Solid state structure of [*arachno*-6,9-$(H_3N)_2$-$B_{10}H_{12}$].

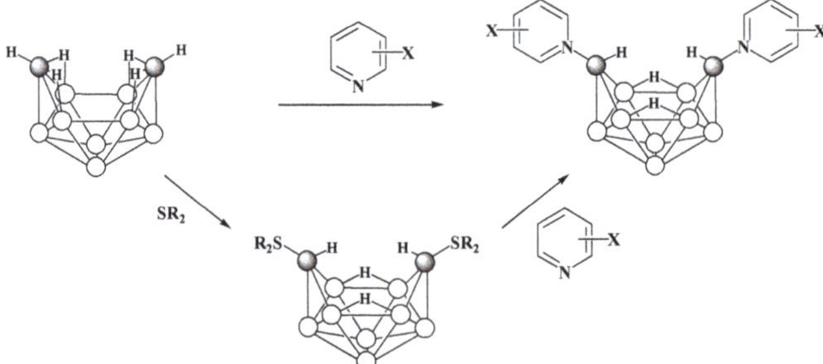

Scheme 4. Synthesis of 6,9-bis(pyridinium) derivatives [6,9-L_2-$B_{10}H_{12}$].

All compounds of this series are brightly colored from yellow to red, which is the reason for the interest in their study by UV and luminescent spectroscopy [229,271–273]. The 6,9-bis(pyridinium) derivative [6,9-Py_2-$B_{10}H_{12}$] was studied by X-ray photoelectron and X-ray fluorescence spectroscopy [225,226] and, its diamagnetic susceptibility was determined [227]. The thermal decomposition of the 6,9-bis(pyridinium) and 6,9-bis(quinolinium) derivatives was studied [274]. The solid state structures of [6,9-Py_2-$B_{10}H_{12}$] [275], [6,9-(HC≡C-*o*-$C_5H_4N)_2$-$B_{10}H_{12}$] [269], and [6,9-(N≡C-*o*-$C_5H_4N)_2$-$B_{10}H_{12}$]·CH_2Cl_2 [269] were determined by single-crystal X-ray diffraction (Figure 17).

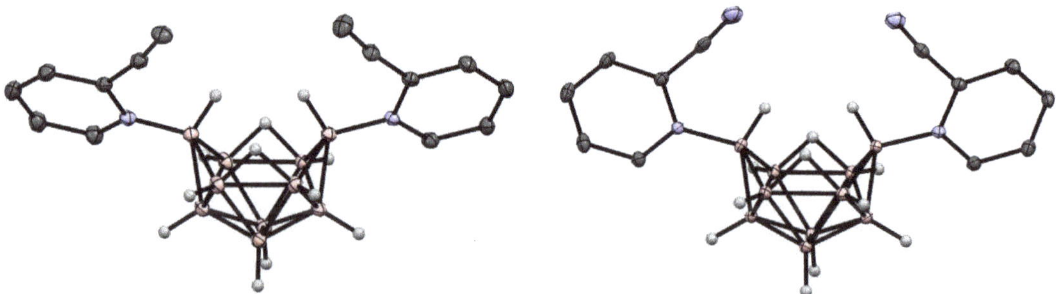

Figure 17. Solid state structures of [*arachno*-6,9-(HC≡C-*o*-$C_5H_4N)_2$-$B_{10}H_{12}$] (**left**) and [*arachno*-6,9-(N≡C-*o*-$C_5H_4N)_2$-$B_{10}H_{12}$] (**right**). Hydrogen atoms of organic substituents are omitted for clarity.

It should be noted that the reaction of decaborane(14) with pyridine at low temperatures was found to form the 6,6-bis(pyridinium) derivative [*arachno*-6,6-Py$_2$-B$_{10}$H$_{12}$], which, upon refluxing in dry degassed pyridine, converts into the more stable 6,9-isomer (Scheme 5) [268].

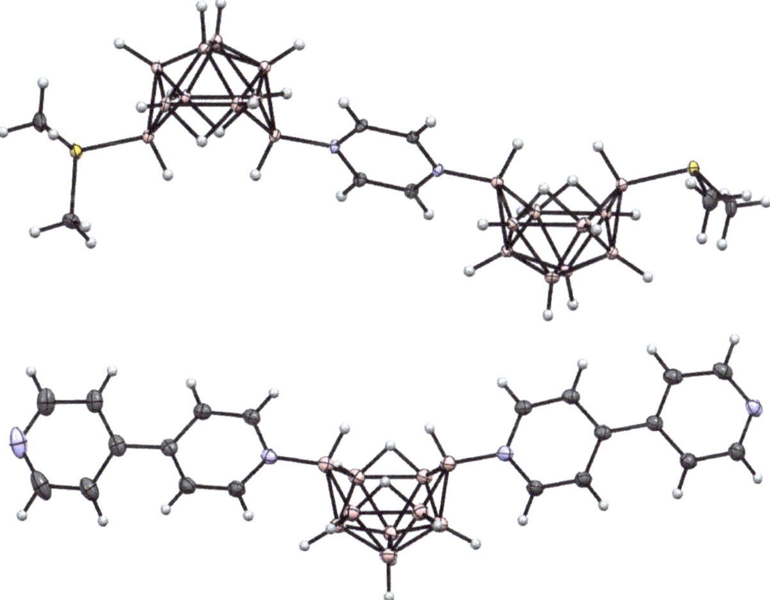

Scheme 5. Synthesis of [*arachno*-6,6-Py$_2$-B$_{10}$H$_{12}$] and its transformation to [*arachno*-6,9-Py$_2$-B$_{10}$H$_{12}$].

The reaction of [6,9-(Me$_2$S)$_2$-B$_{10}$H$_{12}$] with pyrazine in dichloromethane gives the pyrazine-bridged derivative [μ-6,6′-pyrazine-(9-Me$_2$S-B$_{10}$H$_{12}$)$_2$], whereas the reaction with 4,4′-bipyridine leads to the product of the substitution of the Me$_2$S groups with azaheterocycle [6,9-(NC$_5$H$_4$C$_5$H$_4$N)$_2$-B$_{10}$H$_{12}$] (Figure 18) [276].

Figure 18. Solid state structures of [μ-6,6′-pyrazine-(9-Me$_2$S-B$_{10}$H$_{12}$)$_2$] (**top**) and [6,9-(NC$_5$H$_4$C$_5$H$_4$N)$_2$-B$_{10}$H$_{12}$] (**bottom**).

The similar reactions of [6,9-(Me$_2$S)$_2$-B$_{10}$H$_{12}$] with 1,4-bis[β-(4-pyridyl)vinyl]benzene and 1,4-bis[β-(4-quinolyl)vinyl]benzene were found to produce mixtures of the corresponding bridged and terminal substituted derivatives (Scheme 6) [277].

The reactions of Na[*arachno*-6-Et$_2$HN-B$_{10}$H$_{13}$] and (Me$_4$N)[*arachno*-6-Et$_3$N-B$_{10}$H$_{13}$] with pyridine in THF produce the corresponding 6-alkylammonium-9-pyridinium derivatives [*arachno*-6-Et$_2$RN-9-Py-B$_{10}$H$_{12}$] (R = H, Et) [267,278].

The reactions of decaborane(14) with imidazoles in refluxing benzene result in the corresponding 6,9-bis(imidazolium) derivatives [*arachno*-6,9-(RIm)$_2$-Me$_2$S-B$_{10}$H$_{12}$] (R = H, Me, Et, Bu) (Figure 19). The hypergolic properties of the 6,9-bis(imidazolium) derivatives prepared were studied [279].

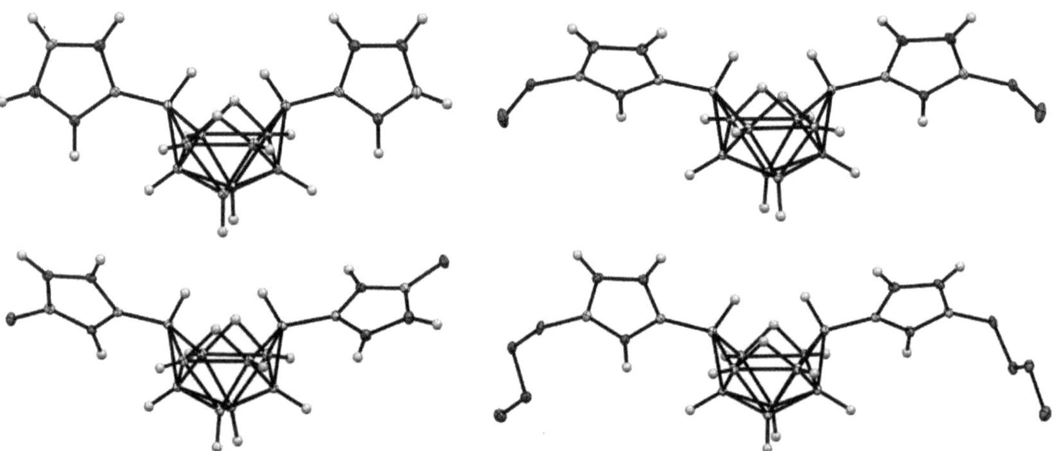

Scheme 6. Reactions of [*arachno*-6,9-(Me$_2$S)$_2$-B$_{10}$H$_{12}$] with 1,4-bis[β-(4-pyridyl)vinyl]benzene and 1,4-bis[β-(4-quinolyl)vinyl]benzene.

Figure 19. Solid state structures of [6,9-(HIm)$_2$-Me$_2$S-B$_{10}$H$_{12}$] (**top left**), [6,9-(MeIm)$_2$-Me$_2$S-B$_{10}$H$_{12}$] (**top right**), [6,9-(EtIm)$_2$-Me$_2$S-B$_{10}$H$_{12}$] (**bottom left**), and [6,9-(BuIm)$_2$-Me$_2$S-B$_{10}$H$_{12}$] (**top right**). Hydrogen atoms of alkyl groups are omitted for clarity.

The reactions of decaborane(14) with 2-isopropyl- and 2-methyl-5-(2-chloroethyl) tetrazoles in benzene result in the corresponding 6,9-bis(tetrazolium) derivatives [*arachno*-6,9-L$_2$-Me$_2$S-B$_{10}$H$_{12}$] [280].

The 6-isothiocyanato derivative [6-SCN-B$_{10}$H$_{13}$] was prepared by the reaction of [6,9-(R$_2$S)$_2$-B$_{10}$H$_{12}$] (R = Me, Et) with mercury isothiocyanate [208]. Alternatively, the 6-isothiocyanato derivative can be prepared by the reaction of decaborane(14) with NaSCN in 1,2-dimethoxyethane in the presence of dry HCl [281]. The solid state structure of [6-SCN-B$_{10}$H$_{13}$] was determined by single-crystal X-ray diffraction (Figure 20) [281].

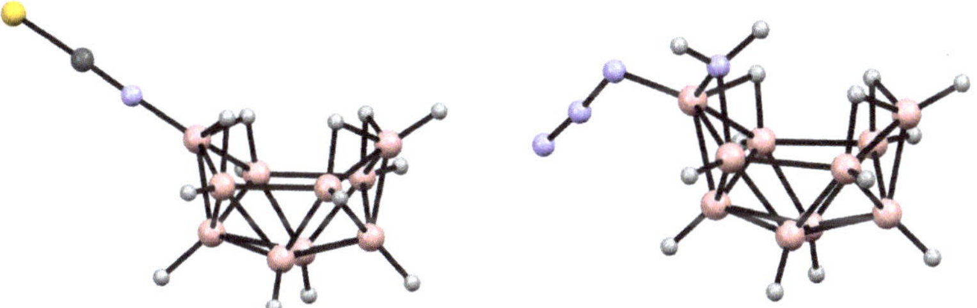

Figure 20. Solid state structures of [6-SCN-B$_{10}$H$_{13}$] (**left**) and [6-N$_3$-μ-5,6-NH$_2$-B$_{10}$H$_{11}$] (**right**).

The reaction of [6,9-(Me$_2$S)$_2$-B$_{10}$H$_{12}$] with excess HN$_3$ in toluene results in the 6-azido-μ-5,6-amino derivative [6-N$_3$-μ-5,6-NH$_2$-B$_{10}$H$_{11}$], the structure of which was determined by single-crystal X-ray diffraction (Figure 20) [282].

7. Derivatives with a B-P Bond

The reaction of decaborane(14) with triphenylphosphine in diethyl ether under reflux results in the 6,9-bis(triphenylphosphonium) derivative of the *arachno*-decaborate anion [*arachno*-6,9-(Ph$_3$P)$_2$-B$_{10}$H$_{12}$]; the same product can be prepared by the reaction of [6,9-(MeC≡N)$_2$-B$_{10}$H$_{12}$] with triphenylphosphine in hot acetonitrile [214,245,283]. The solid state structure of [*arachno*-6,9-(Ph$_3$P)$_2$-B$_{10}$H$_{12}$]·2DMF·H$_2$O was determined by single-crystal X-ray diffraction [284]. The 6,9-bis(triphenylphosphonium) derivative was also studied by X-ray emission and X-ray photoelectron spectroscopy [279,280], and its diamagnetic susceptibility was determined [227]. The reaction of [6,9-(Me$_2$S)$_2$-2-Br-B$_{10}$H$_{11}$] with triphenylphosphine in benzene results in the corresponding 6,9-bis(triphenylphosphonium) derivative [6,9-(Ph$_3$P)$_2$-2-Br-B$_{10}$H$_{11}$] [233].

The reaction of decaborane(14) with PhMe$_2$P at low temperatures (~200 K) gives a mixture of *exo,exo*- and *exo,endo*-isomers of [6,9-(PhMe$_2$P)$_2$-*arachno*-B$_{10}$H$_{12}$], which were separated chromatographically [285,286]. The structure of both isomers was confirmed by single-crystal X-ray diffraction (Figure 21) [286].

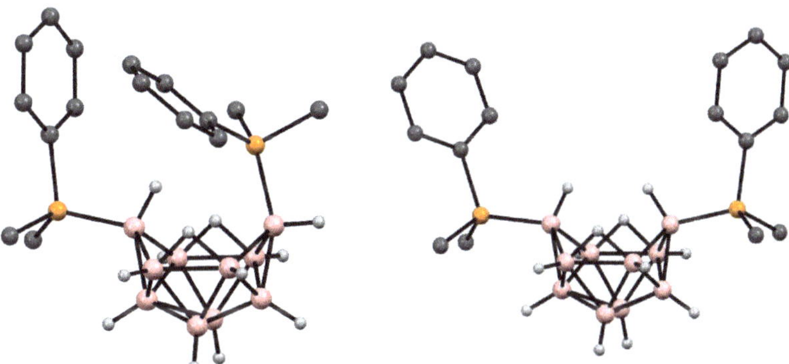

Figure 21. Solid state structures of *exo,endo*-[6,9-(PhMe$_2$P)$_2$-*arachno*-B$_{10}$H$_{12}$] (**left**) and *exo,exo*-[6,9-(PhMe$_2$P)$_2$-*arachno*-B$_{10}$H$_{12}$] (**right**). Hydrogen atoms of organic substituents are omitted for clarity.

Under similar conditions, the reaction of the 2,4-dichloro derivatives of decaborane with PhMe$_2$P solely produces the *exo,endo*-isomer of [6,9-(PhMe$_2$P)$_2$-2,4-Cl$_2$-*arachno*-B$_{10}$H$_{10}$], while the reaction of 2-bromo leads to mixtures of *exo,exo*- and *exo,endo*-isomers and [6,9-

(PhMe$_2$P)$_2$-2-Br-*arachno*-B$_{10}$H$_{11}$]. It is interesting to note that in the reaction of the 2-bromo derivative, it is precisely the 6,9-*exo,endo*-isomer that is formed, without any traces of the 9,6-*exo,endo*-isomer. The solid state structure of the *exo,endo*-isomer [6,9-(PhMe$_2$P)$_2$-2-Br-*arachno*-B$_{10}$H$_{12}$] was determined by single-crystal X-ray diffraction (Figure 22) [286].

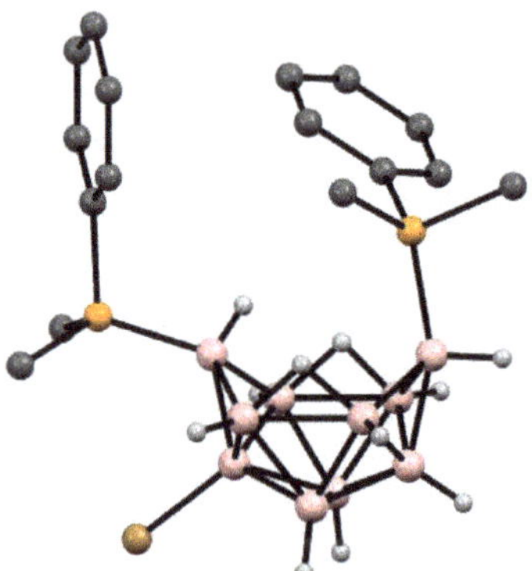

Figure 22. Solid state structure of *exo,endo*-[6,9-(PhMe$_2$P)$_2$-2-Br-*arachno*-B$_{10}$H$_{12}$]. Hydrogen atoms of organic substituents are omitted for clarity.

The reaction of decaborane(14) with triethylphosphine in benzene results in the 6,9-bis(triethylphosphonium) derivative [*arachno*-6,9-(Et$_3$P)$_2$-B$_{10}$H$_{12}$] [287], whereas the 6,9-bis(diphenylphosphonium) and 6,9-bis(phenylphosphonium) derivatives [*arachno*-6,9-(Ph$_2$HP)$_2$-B$_{10}$H$_{12}$] and [*arachno*-6,9-(PhH$_2$P)$_2$-B$_{10}$H$_{12}$] were prepared by the reactions of the corresponding phosphines with [*arachno*-6,9-(Et$_2$S)$_2$-B$_{10}$H$_{12}$] [217].

The phosphite, phosphinite, and thiophosphite derivatives of the *arachno*-decaborate anion [*arachno*-6,9-(R$_2$R'P)$_2$-B$_{10}$H$_{12}$] (R = R' = OMe, OEt, OPh; R = Ph, R' = OEt; R = OBu, R' = Ph; R = R' = SEt) were prepared by the direct reactions of decaborane(14) with the corresponding phosphorus compounds or via substitution of the Me$_2$S and MeCN groups in [*arachno*-6,9-(Me$_2$S)$_2$-B$_{10}$H$_{12}$] and [*arachno*-6,9-(MeC≡N)$_2$-B$_{10}$H$_{12}$], respectively [217,288–290].

The reaction of decaborane(14) with diphenylchlorophosphine in diethyl ether gives the 6,9-bis(chlorodiphenylphosphonium) derivative [*arachno*-6,9-(ClPh$_2$P)$_2$-B$_{10}$H$_{12}$], which, upon the treatment with dimethylamine in alcohols, lead to the corresponding phosphonites [6,9-(ROPh$_2$P)$_2$-B$_{10}$H$_{12}$] (R = Me, Et, CH$_2$CH$_2$OH) [290]. The reaction of [6,9-(ClPh$_2$P)$_2$-B$_{10}$H$_{12}$] with dimethylamine in aqueous solution water results in the bis(dimethylammonium) salt of the corresponding acid (Me$_2$NH$_2$)$_2$[6,9-(OPh$_2$P)$_2$-B$_{10}$H$_{12}$] [290]. The 6,9-bis-(hydroxy diphenylphosphonium) derivative [6,9-(HOPh$_2$P)$_2$-B$_{10}$H$_{12}$] was prepared by the reaction of [6,9-(ClPh$_2$P)$_2$-B$_{10}$H$_{12}$] with water in acetone [290].

The 6,9-bis(chlorodiphenylphosphonium) derivative reacts with ammonia, hydrazine, primary aliphatic amines, and ethylenimine in alcohols to form the corresponding 6,9-bis(aminodiphenylphosphonium) derivatives [6,9-(R'RNPh$_2$P)$_2$-B$_{10}$H$_{12}$] (R = H, R' =NH$_2$, Me, Bu; R = R' = CH$_2$CH$_2$) [290]. The reactions of decaborane(14) or [*arachno*-6,9-(R$_2$S)$_2$-B$_{10}$H$_{12}$] (R = Me, Et) with dimethylaminophosphines in refluxing benzene lead to the corresponding 6,9-bis(dimethylaminophosphonium) derivatives [6,9-(RR'(Me$_2$N)P)$_2$-B$_{10}$H$_{12}$] (R = R' = NMe$_2$; R = NMe$_2$, R' = Ph, Cl; R = R' = Ph; R = Ph, R' = Cl; RR' = OCH$_2$CH$_2$O) [217].

The reaction of [6,9-(ClPh$_2$P)$_2$-B$_{10}$H$_{12}$] with NaN$_3$ in ethanol results in the 6,9-bis-(azididi phenylphosphonium) derivative [6,9-(N$_3$Ph$_2$P)$_2$-B$_{10}$H$_{12}$] [290], which, upon the treatment with triphenylphosphine in refluxing benzene, gives the 6,9-bis(triphenylphosphineiminodiphenyl phosphonium) derivative [6,9-(Ph$_3$P=NPh$_2$P)$_2$- B$_{10}$H$_{12}$] [291].

The bifunctional derivatives [6,9-(XPh$_2$P)$_2$-B$_{10}$H$_{12}$] (X = Cl, OH, N$_3$) were used for the synthesis of decaborane-based polymers [291,292]. The chemistry of decaborane-based polymers is considered in detail in the review [293]. The formation of decaborane-based polymers along with a small amount of [6,9-(dppf)$_2$-B$_{10}$H$_{12}$] has also been reported in the reaction of decaborane(14) with 1,1′-bis(diphenylphosphino)ferrocene (dppf) [294].

The reaction of Na[B$_{10}$H$_{13}$] with tributylphosphine in acetonitrile in the presence of dry HCl leads to [*arachno*-6-Bu$_3$P-9-MeC≡N-B$_{10}$H$_{12}$] [221]. The reaction of Na[B$_{10}$H$_{13}$] with diphenylchlorophosphine in diethyl ether leads to the diphenylphosphine derivative [*nido*-µ-5,6-Ph$_2$P-B$_{10}$H$_{13}$] [295], whose structure was determined by single-crystal X-ray diffraction [296]. The same compound was reported to be formed in the reaction of the so-called "Grignard derivative" [B$_{10}$H$_{13}$MgI] formed by the treatment of decaborane(14) with MeMgI, with diphenylchlorophosphine in diethyl ether [297]. The diphenylphosphine derivative [*nido*-µ-5,6-Ph$_2$P-B$_{10}$H$_{13}$] is easily deprotonated with triethylamine or sodium hydroxide to form the corresponding salts [295,297]. The solid state structure of the triphenylmethylphosphonium salt (Ph$_3$PMe)[*arachno*-µ-6,9-Ph$_2$P-B$_{10}$H$_{12}$] was determined by single-crystal X-ray diffraction (Figure 23) [298].

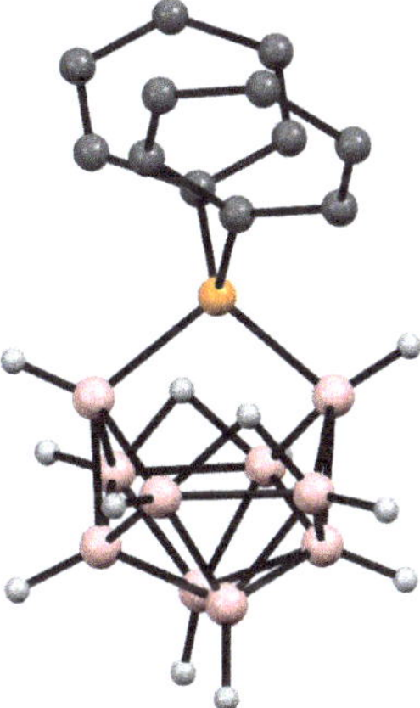

Figure 23. Solid state structure of the [*arachno*-µ-6,9-Ph$_2$P-B$_{10}$H$_{12}$]$^-$ anion. Hydrogen atoms of organic substituents are omitted for clarity.

It should be noted that the reactions of [6,9-(MeC≡N)$_2$-B$_{10}$H$_{12}$] with low-coordinated phosphorus compounds, such as phosphaalkynes RC≡P (R = *t*-Bu, Ad), do not lead to substitution o hydrogens but to the incorporation of phosphorus into the decaborane basket with the formation of 11-vertex phosphoboranes [*nido*-RC(H)=PB$_{10}$H$_{13}$] [299,300].

8. Derivatives with a B-As Bond

The 6,9-bis(trialkyl/arylarsonium) derivatives [6,9-(R$_3$As)$_2$-B$_{10}$H$_{12}$] (R = Et, Ph) were prepared by the reactions of decaborane(14) or [6,9-(MeC≡N)$_2$-B$_{10}$H$_{12}$] with the corresponding arsines in benzene or toluene [287]. The reaction of decaborane(14) with triethoxyarsine in benzene leads to [6,9-((EtO)$_3$As)$_2$-B$_{10}$H$_{12}$] [289].

9. Derivatives with a B-C Bond

Decaborane derivatives with a B-C bond are probably the most studied area of decaborane chemistry. Like the halogenation of decaborane, the direct alkylation reactions result in the substitution of hydrogen atoms at "the bottom" of the decaborane basket. The reaction of decaborane(14) with methyl bromide in carbon disulfide in the presence of AlCl$_3$ at 80 °C gives a mixture of the 2-methyl [2-Me-B$_{10}$H$_{13}$], 1,2- and 3,4-dimethyl [1,2-Me$_2$-B$_{10}$H$_{12}$] and [2,4-Me$_2$-B$_{10}$H$_{12}$], 1,2,3- and 1,2,4-trimethyl [1,2,3-Me$_3$-B$_{10}$H$_{11}$] and [1,2,4-Me$_3$-B$_{10}$H$_{11}$], 1,2,3,4- and 1,2,3,5(or 8)-tetramethyl [1,2,3,4-Me$_3$-B$_{10}$H$_{10}$], and [1,2,3,5(or 8)-Me$_4$-B$_{10}$H$_{10}$] derivatives, which were chromatographically separated [301]. The methylation of decaborane(14) was also studied using methyl chloride [302,303].

The reaction of decaborane(14) with neat methyl iodide in the presence of $AlCl_3$ at room temperature gives the 1,2,3,4-tetramethyl derivative [1,2,3,4-Me_4-$B_{10}H_{10}$], whereas the reaction at 120 °C leads to the octasubstituted product [1-I-2,3,4,5,6,7,8-Me_7-$B_{10}H_6$]. The similar octasubstituted derivative [1-TfO-2,3,4,5,6,7,8-Me_7-$B_{10}H_6$] was obtained by the reaction of decaborane(14) with TfOMe in the presence of a catalytic amount of triflic acid at 120 °C. The solid state structures of [1-I-2,3,4,5,6,7,8-Me_7-$B_{10}H_6$] and [1-TfO-2,3,4,5,6,7,8-Me_7-$B_{10}H_6$] were determined by single-crystal X-ray diffraction (Figure 24) [304]. The methyl derivatives prepared can be easily deprotonated with a Proton Sponge to give the corresponding salts [304].

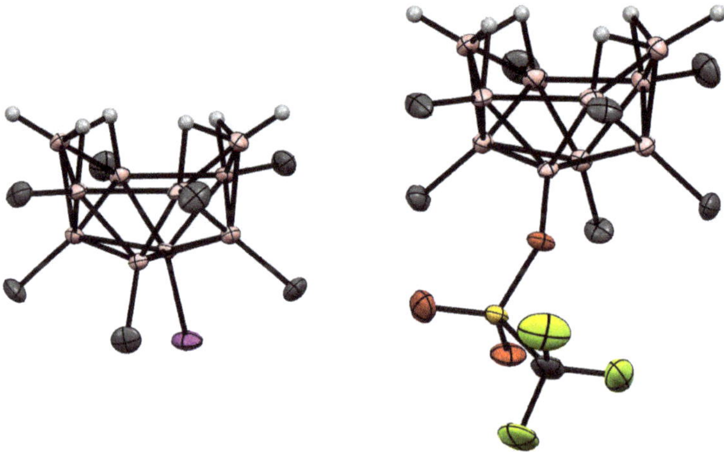

Figure 24. Solid state structures of [1-I-2,3,4,5,6,7,8-Me_7-$B_{10}H_6$] (**left**) and [1-TfO-2,3,4,5,6,7,8-Me_7-$B_{10}H_6$] (**right**). Hydrogen atoms of methyl groups are omitted for clarity.

The reaction of decaborane(14) with ethyl bromide in carbon disulfide in the presence of $AlCl_3$ under reflux gives a mixture of mono-, di-, and triethyl derivatives [305]. The monoethyl derivative of decaborane was prepared by the reaction of decaborane(14) with neat ethyl bromide in the presence of $AlCl_3$ [306]. The solid state structure of the 1-ethyl derivative of decaborane [1-Et-$B_{10}H_{13}$] was determined by single-crystal X-ray diffraction [307].

The reaction of decaborane(14) with MeLi in benzene followed by treatment with HCl has been reported to give a mixture of the 6-methyl-, 6,5(or 8)- and 6,9-dimethyl derivatives of decaborane [308]. The reaction with EtLi in benzene gives the 6-ethyl derivative [6-Et-$B_{10}H_{13}$] [308]. The reaction of decaborane(14) with MeMgI has been shown to proceed by two routes. The major reaction yields the so-called "Grignard derivative" [$B_{10}H_{13}MgI$] and methane, and the minor reaction produces the 6-methyl derivative of decaborane. The reaction of the "Grignard derivative" with dimethyl sulfate produces a mixture of the 5- and 6-methyl derivatives of decaborane [309]. In a similar way, the reaction of decaborane(14) with EtMgI produces the "Grignard derivative" as the main product and the 6-ethyl derivative of decaborane as a by-product. The reaction of the "Grignard derivative" with [Et_3O]BF_4 or diethyl sulfate gives the 5-ethyl derivative of decaborane [5-Et-$B_{10}H_{13}$] [309]. A series of alkyl derivatives [R-$B_{10}H_{13}$] (R = butyl, amyl, hexyl, cyclohexyl, heptyl, octyl) was prepared by the reactions of the "Grignard derivative" with the corresponding alkyl fluorides [310]. The 6-benzyl derivative of decaborane [6-Bn-$B_{10}H_{13}$] can be prepared by the reaction of the "Grignard derivative" with benzyl chloride or Na[$B_{10}H_{13}$] with benzyl bromide [309,311,312].

Later, these reactions were re-examined, and it was shown that the first stage of the reaction of decaborane(14) with the alkyllithium reagents RLi is deprotonation of decaborane with the formation of Li[$B_{10}H_{13}$]. The reaction with the second equivalent of RLi produces Li_2[*arachno*-6-R-$B_{10}H_{13}$], which, when treated with HCl, gives Li[*arachno*-6-R-$B_{10}H_{14}$] and then [*nido*-6-R-$B_{10}H_{13}$] (R = Me, *n*-Bu, *t*-Bu). The use of pre-prepared salts of the [$B_{10}H_{13}$]⁻ anion makes it possible to reduce the formation of by-products [237,313].

Another approach to the 6-alkyl derivatives of decaborane includes the reactions of [*arachno*-6,9-(Me_2S)$_2$-$B_{10}H_{12}$] with alkenes in dichloromethane, resulting in [*nido*-6-R-8-Me_2S-$B_{10}H_{11}$] (R = cyclohexyl, cyclohexenyl, hexyl, octyl, 2,3-dimethyl-1-butyl, 2,3-dimethyl-2-butyl, 2-methyl-2-butyl, (1R)-(+)-α-pinene, and (1S)-(−)-β-pinene), which can be reduced using Superhydride Li[Et_3BH] in THF to [6-R-$B_{10}H_{12}$]⁻ and then protonated with HCl/Et_2O to [6-R-$B_{10}H_{13}$] [222,237,313–315]. From the point of view of organic chemistry, these reactions can be considered as the hydroboration reactions. The solid state structures of [6-Chx-8-Me_2S-$B_{10}H_{11}$] [316], [6-Thx-8-Me_2S-$B_{10}H_{11}$] [237] (Figure 25), and [6-Thx-$B_{10}H_{13}$] [237] (Figure 26) were determined by single-crystal X-ray diffraction.

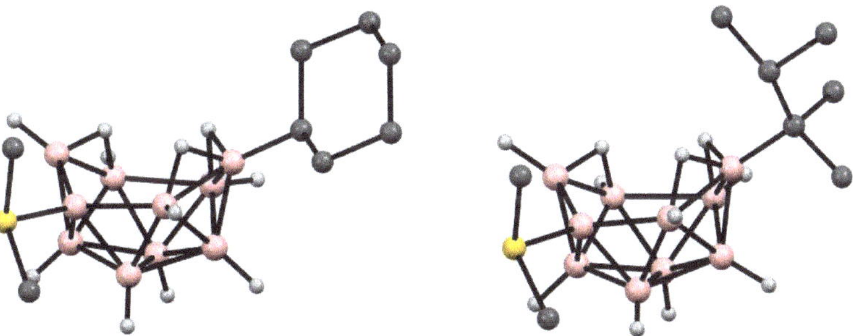

Figure 25. Solid state structures of [6-Chx-8-Me_2S-$B_{10}H_{11}$] (**left**) and [6-Thx-8-Me_2S-$B_{10}H_{11}$] (**right**). Hydrogen atoms of organic substituents are omitted for clarity.

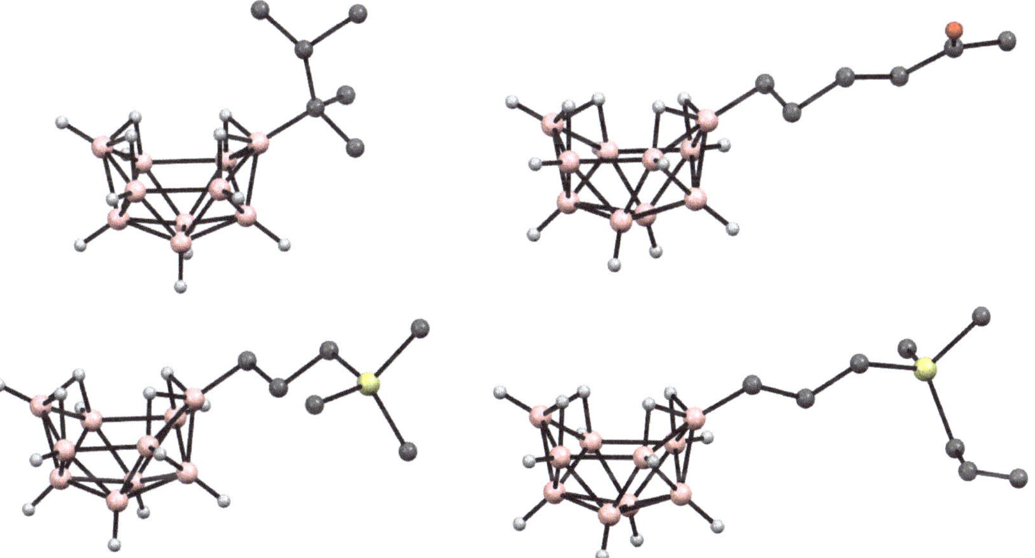

Figure 26. Solid state structures of [6-Thx-$B_{10}H_{13}$] (**top left**), [6-MeC(O)$CH_2CH_2CH_2CH_2$-$B_{10}H_{13}$] (**top right**), [6-$Me_3SiCH_2CH_2CH_2$-$B_{10}H_{13}$] (**bottom left**), and [6-H_2C=CHCH$_2$SiMe$_2$CH$_2$CH$_2$CH$_2$-$B_{10}H_{13}$] (**bottom right**). Hydrogen atoms of organic substituents are omitted for clarity.

Another convenient method for the synthesis of 6-alkyl derivatives of decaborane is based on the use of ionic liquids as a solvent. The reactions of decaborane(14) with terminal alkenes in biphasic ionic-liquid/toluene mixtures lead to the corresponding 6-alkyl derivatives [6-R-$B_{10}H_{13}$] (R = C_6H_{13}, C_8H_{17}, $C_{16}H_{33}$, CH(*i*-Pr)CH_2CHMe$_2$, $(CH_2)_2C_6H_5$, $(CH_2)_3C_6H_5$, $(CH_2)_6$Br, $(CH_2)_4$CH=CH$_2$, $(CH_2)_6$CH=CH$_2$, $(CH_2)_3$OC$_3$H$_7$, $(CH_2)_3$SiMe$_3$, $(CH_2)_4$COMe, $(CH_2)_6$OAc, $(CH_2)_3$OBn, $(CH_2)_3$OH, and $(CH_2)_3$Bpin, norbornenyl) [317–320]. The best results were observed for reactions with [bmim]X (1-butyl-3-methylimidazolium, X = Cl$^-$ or BF$_4$$^-$) and bmpyX (1-butyl-4-methylpyridinium, X = Cl$^-$ or BF$_4$$^-$). The reaction mechanism includes the ionic-liquid-promoted formation of the [$B_{10}H_{13}$]$^-$ anion, its addition to the alkene to form the [6-R-$B_{10}H_{12}$]$^-$ anion, and, finally, the protonation of the last one to form the final product [6-R-$B_{10}H_{13}$] [314]. The solid state structures of [6-Me$_3$Si(CH$_2$)$_3$-$B_{10}H_{13}$] and [6-MeC(O)(CH$_2$)$_4$-$B_{10}H_{13}$] were determined by single-crystal X-ray diffraction (Figure 26) [318]. The 6-cyclohexyl derivative of decaborane [*nido*-6-C_6H_{11}-$B_{10}H_{13}$] was obtained in a low yield from the reaction of Cs$_2$[*closo*-$B_{10}H_{10}$] with triflic acid in cyclohexane (Figure 27) [202,203]. In a similar way, the 6-hexyl derivative [*nido*-6-C_6H_{13}-$B_{10}H_{13}$] was isolated from the reaction of (NH$_4$)$_2$[*closo*-$B_{10}H_{10}$] with concentrated nitric acid in hexane [321].

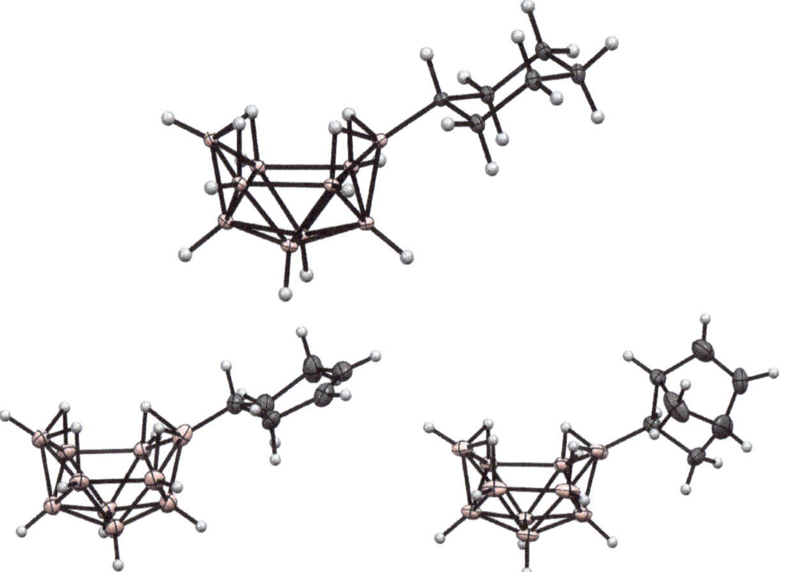

Figure 27. Solid state structures of [*nido*-6-C_6H_{11}-$B_{10}H_{13}$] (**top**), [6-(4′-cyclohexenyl)-$B_{10}H_{13}$] (**bottom left**) and [6-(5′-norbornenyl)-$B_{10}H_{13}$] (**bottom right**).

The Cp$_2$Ti(CO)$_2$-catalyzed reactions of decaborane(14) with terminal alkenes have been found to result in the high-yield formation of 6-alkyl derivatives of decaborane [6-R-$B_{10}H_{13}$] (R = C_6H_{13}, C_8H_{17}, $(CH_2)_3$SiMe$_3$) [322,323]. The reactions of decaborane(14) with equimolar amounts of bifunctional alkenes such as diallyldimethylsilane, 1,5-hexadiene, 1,4-cyclohexadiene, 1,5-cyclooctadiene, and 2,5-norbornadiene produce the corresponding decaborane derivatives with a double bond in the substituent [6-R-$B_{10}H_{13}$] (R = $(CH_2)_3$SiMe$_2$CH$_2$CH=CH$_2$, $(CH_2)_4$CH=CH$_2$, 4-cyclohexenyl, 5-cyclooctenyl, 5-norbornenyl) [322–325]. The solid state structures of [6-H$_2$C=CHCH$_2$SiMe$_2$(CH$_2$)$_3$-$B_{10}H_{13}$] (Figure 26) [322], [6-(4′-cyclohexenyl)-$B_{10}H_{13}$] (Figure 27) [325], and [6-(5′-norbornenyl)-$B_{10}H_{13}$] (Figure 27) [324] were determined by single-crystal X-ray diffraction.

The reactions of multifunctional alkenes with an excess amount of decaborane(14) produce the saturated linked-cage compounds with two ([μ-6,6′-Me$_2$Si-(6-(CH$_2$)$_3$-$B_{10}H_{13}$)$_2$], [μ-6,6′-(CH$_2$)$_6$-(B$_{10}H_{13}$)$_2$], [μ-6,6′-(1″,5″-cyclooctyl)-(B$_{10}H_{13}$)$_2$], and [μ-6,6′-(2″,5″-norbornyl)-

($B_{10}H_{13})_2$]) or four ([μ^4-6,6',6'',6'''-Si-(6-($CH_2)_3$-$B_{10}H_{13})_4$]) decaborane units [322,323,325]. The solid state structures of [μ-6,6'-($CH_2)_6$-($B_{10}H_{13})_2$], [μ-6,6'-(2'',5''-norbornyl)-($B_{10}H_{13})_2$], and [μ^4-6,6',6'',6'''-Si-(6-($CH_2)_3$-$B_{10}H_{13})_4$] were determined by single-crystal X-ray diffraction (Figure 28) [322,323,325].

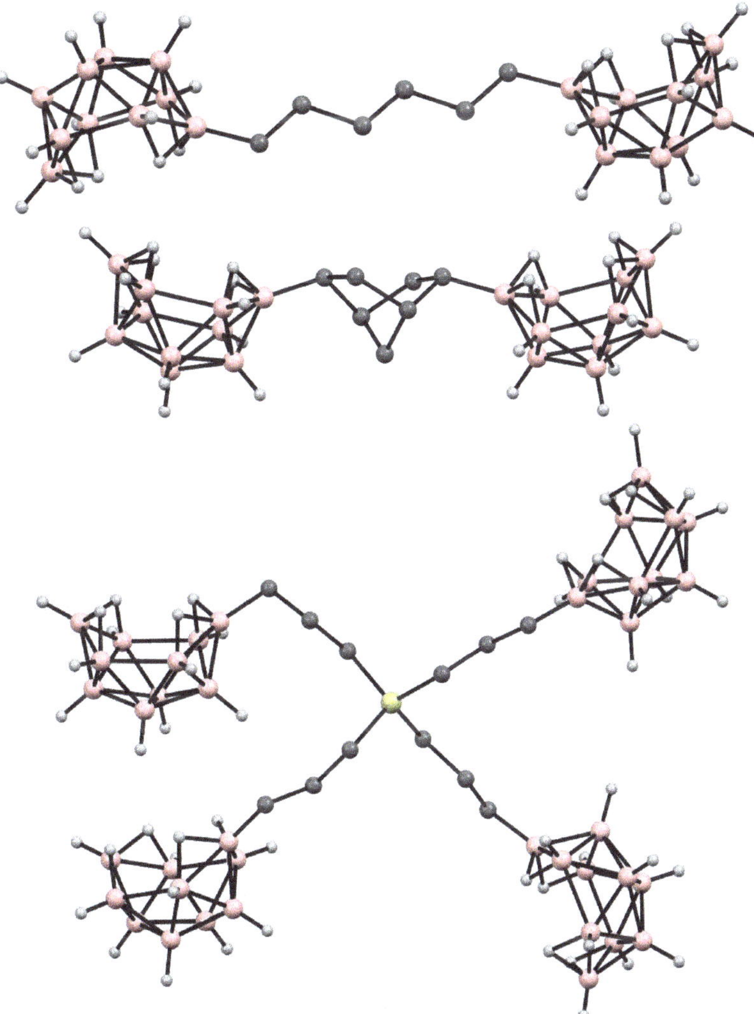

Figure 28. Solid state structures of [μ-6,6'-($CH_2)_6$-($B_{10}H_{13})_2$] (**top**), [μ-6,6'-(2'',5''-norbornyl)-($B_{10}H_{13})_2$] (**middle**), and [μ^4-6,6',6'',6'''-Si-(6-($CH_2)_3$-$B_{10}H_{13})_4$] (**bottom**). Hydrogen atoms of organic substituents are omitted for clarity.

The derivatives with the two decaborane units [μ-6,6'-(1'',5''-cyclooctyl)-($B_{10}H_{13})_2$] and [μ-6,6'-(2'',5''-norbornyl)-($B_{10}H_{13})_2$] can also be prepared by the titanium-catalyzed reactions of decaborane(14) with [6-(5'-cyclooctenyl)-$B_{10}H_{13}$] and [6-(5'-norbornenyl)-$B_{10}H_{13}$], respectively [325].

The 6-alkyl derivatives of decaborane with substituents containing double bonds in the side chain (hexenyl, norbornenyl, etc.) are used for the synthesis of decaborane-based polymers and boron-containing ceramics [319,324–332].

The reactions of decaborane(14) with terminal alkenes in the presence of catalytic amounts of $PtBr_2$ or H_2PtCl_6 lead to the 6,9-dialkyl derivatives *nido*-[6,9-R_2-$B_{10}H_{12}$] (R = C_2H_5, C_3H_7, C_4H_9, C_5H_{11}) [333]. The reactions of [5-TfO-$B_{10}H_{13}$] and [5-I-$B_{10}H_{13}$] with 1-pentene in the presence of a catalytic amount of $PtBr_2$ at 55 °C lead to the corresponding 6,9-dialkyl derivatives [6,9-$(C_5H_{11})_2$-5-TfO-$B_{10}H_{11}$] and [6,9-$(C_5H_{11})_2$-5-I-$B_{10}H_{11}$] [204]. The solid state structure of [6,9-$(C_5H_{11})_2$-5-I-$B_{10}H_{11}$] was determined by single-crystal X-ray diffraction (Figure 29) [204].

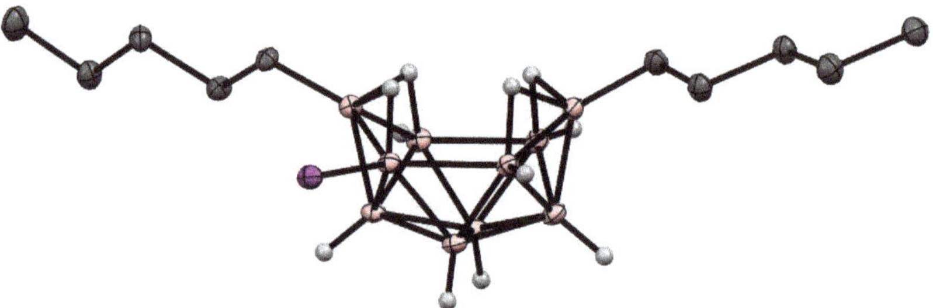

Figure 29. Solid state structure of [6,9-$(C_5H_{11})_2$-5-I-$B_{10}H_{11}$]. Hydrogen atoms of organic substituents are omitted for clarity.

The reactions of decaborane(14) with terminal alkynes in toluene in the presence of [Cp*IrCl$_2$]$_2$ or [(*p*-cymene)RuCl$_2$]$_2$ as catalysts lead to the corresponding 6,9-di(β-alkenyl) derivatives of decaborane [6,9-((*E*)-RCH=CH)$_2$-$B_{10}H_{12}$] (R = H, C_6H_{13}, C_6H_5, $(CH_2)_2Br$, $(CH_2)_3Cl$, $SiMe_3$) [334,335]. The solid state structures of [6,9-((*E*)-Br$(CH_2)_2$CH=CH)$_2$-$B_{10}H_{12}$] and [6,9-((*E*)-Me$_3$SiCH=CH)$_2$-$B_{10}H_{12}$] were determined by single-crystal X-ray diffraction (Figure 30) [334,335].

Figure 30. Solid state structures of [6,9-((*E*)-Br$(CH_2)_2$CH=CH)$_2$-$B_{10}H_{12}$] (**left**) and [6,9-((*E*)-Me$_3$SiCH=CH)$_2$-$B_{10}H_{12}$] (**right**).

In contrast to [(*p*-cymene)RuCl$_2$]$_2$, the reactions of decaborane(14) with terminal alkynes in the presence of [(*p*-cymene)RuI$_2$]$_2$ result in the 6,9-di(α-alkenyl) derivatives [6,9-(R(H$_2$C=)C)$_2$-$B_{10}H_{12}$] (R = C_6H_{13}, CH$_2$-*c*-C_6H_{11}, $(CH_2)_2Br$, $(CH_2)_3Cl$) [334,335]. The solid state structure of [6,9-(*c*-C_6H_{11}CH$_2$(H$_2$C=)C)$_2$-$B_{10}H_{12}$] was determined by single-crystal X-ray diffraction (Figure 31) [334,335].

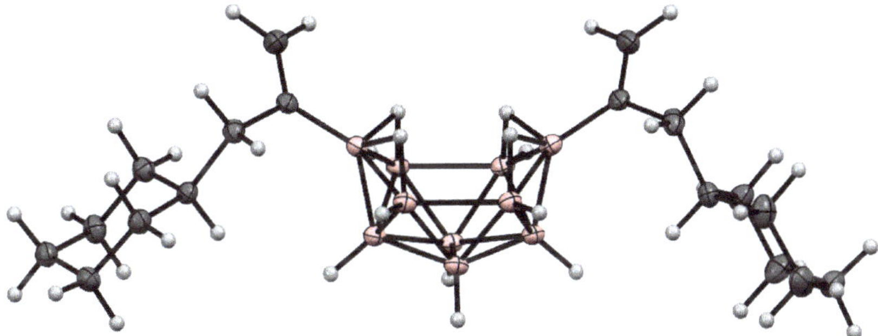

Figure 31. Solid state structure of [6,9-(*c*-C$_6$H$_{11}$CH$_2$(H$_2$C=)C)$_2$-B$_{10}$H$_{12}$]. Hydrogen atoms of organic substituents are omitted for clarity.

In a similar way, the reactions of 6-alkyldecaboranes [6-R-B$_{10}$H$_{13}$] with terminal alkynes in the presence of [Cp*IrCl$_2$]$_2$ give asymmetrically substituted 6-alkyl-9-alkenyl-derivatives [6-R-9-((*E*)-R′CH=CH)$_2$-B$_{10}$H$_{12}$] (R = (CH$_2$)$_3$SiMe$_3$, R′ = H, C$_6$H$_5$, C$_6$H$_4$-*m*-CH≡CH; CH$_2$CH=CH$_2$; R = C$_5$H$_{11}$, R′ = H). The solid state structure of [6-Me$_3$Si(CH$_2$)$_3$-9-(*E*)-*m*-HC≡CC$_6$H$_4$CH=CH-B$_{10}$H$_{12}$] was determined by single-crystal X-ray diffraction (Figure 32) [334,335].

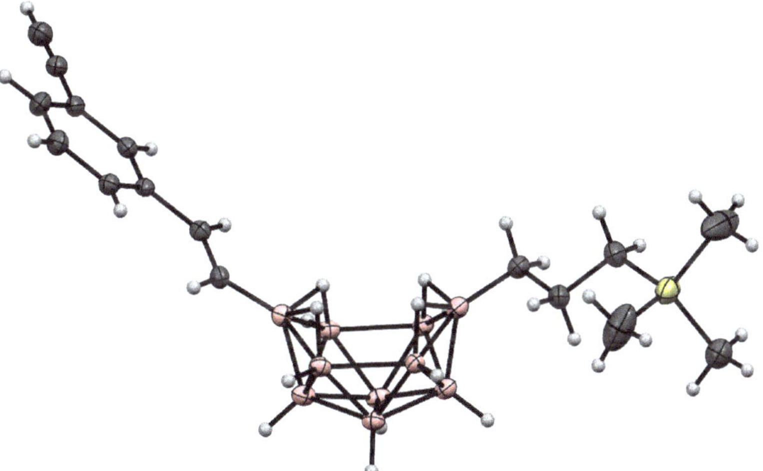

Figure 32. Solid state structure of [6-Me$_3$Si(CH$_2$)$_3$-9-(*E*)-*m*-HC≡CC$_6$H$_4$CH=CH-B$_{10}$H$_{12}$].

While [Cp*IrCl$_2$]$_2$ proved to be inactive for inducing the hydroboration of simple olefins, such as 1-pentene, by either decaborane or the 6-alkyl-decaboranes, it was found to catalyze the hydroboration of 6-alkyl-9-vinyldecaboranes [6-R-9-CH$_2$=CH-B$_{10}$H$_{12}$] (R = C$_5$H$_{11}$, (CH$_2$)$_3$SiMe$_3$) by 6-alkyl-decaboranes [6-R-B$_{10}$H$_{13}$] (R = C$_5$H$_{11}$, (CH$_2$)$_3$SiMe$_3$) to yield linked-cage products [9,9′-μ-CH$_2$CH$_2$-(6-R-B$_{10}$H$_{12}$)$_2$] (R = C$_5$H$_{11}$, (CH$_2$)$_3$SiMe$_3$) (Figure 33) [335].

The vinyl derivative [6-Me$_3$Si(CH$_2$)$_3$-9-CH$_2$=CH)$_2$-B$_{10}$H$_{12}$] was found to readily undergo both homo- and cross-metathesis reactions in the presence of Grubbs' II catalyst, giving the corresponding products [9,9′-μ-CH=CH-(6-Me$_3$Si(CH$_2$)$_3$-B$_{10}$H$_{12}$)$_2$] (Figure 34) and [6-Me$_3$Si(CH$_2$)$_3$-9-RCH=CH-B$_{10}$H$_{12}$] (R = C$_3$H$_7$, (CH$_2$)$_4$Br, CH$_2$SiMe$_3$) [335].

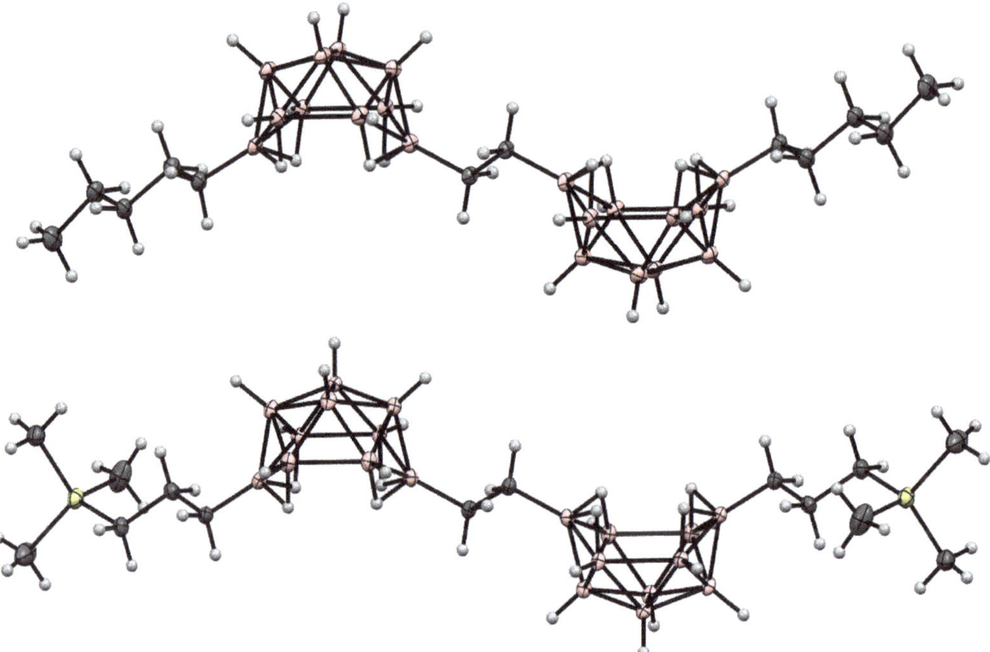

Figure 33. Solid state structures of [9,9′-μ-CH₂CH₂-(6-C₅H₁₁-B₁₀H₁₂)₂] (**top**) and [9,9′-μ-CH₂CH₂-(6-Me₃Si(CH₂)₃-B₁₀H₁₂)₂] (**bottom**).

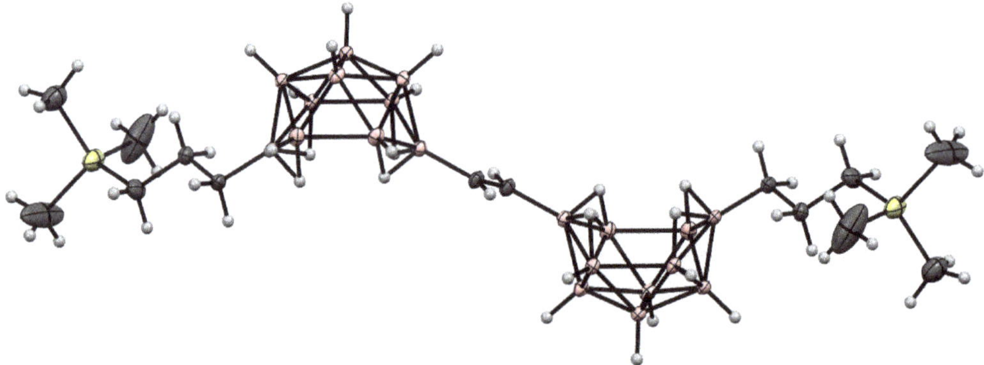

Figure 34. Solid state structure of [9,9′-μ-CH=CH-(6-Me₃Si(CH₂)₃-B₁₀H₁₂)₂].

Heating [*arachno*-6,9-(Me₂S)₂-B₁₀H₁₂] with silylated acetylenes Me₃SiC≡CR (R = Me, Bu, SiMe₃) leads to the corresponding trimethylsilyl alkenyl derivatives [*nido*-6-Me₃Si(R)C=CH-5-Me₂S-B₁₀H₁₁] [336,337]. The solid state structures of [6-Me₃Si(Me)C=CH-5-Me₂S-B₁₀H₁₁] [336], [6-Me₃Si(Bu)C=CH-5-Me₂S-B₁₀H₁₁] [337], and [6-(Me₃Si)₂C=CH-5-Me₂S-B₁₀H₁₁] [337] were determined by single-crystal X-ray diffraction (Figure 35).

The reaction of [6,9-(Me₂S)₂-B₁₀H₁₂] with the phosphaalkyne *t*-BuC≡P in refluxing benzene leads to [μ-6(C),6′(C),5′(P)-C(*t*-Bu)PH-(*nido*-8-Me₂S-B₁₀H₁₁)(*nido*-B₁₀H₁₂)], in which two decaborane units are linked by the C(*t*-Bu)PH-bridge (Figure 36) [338].

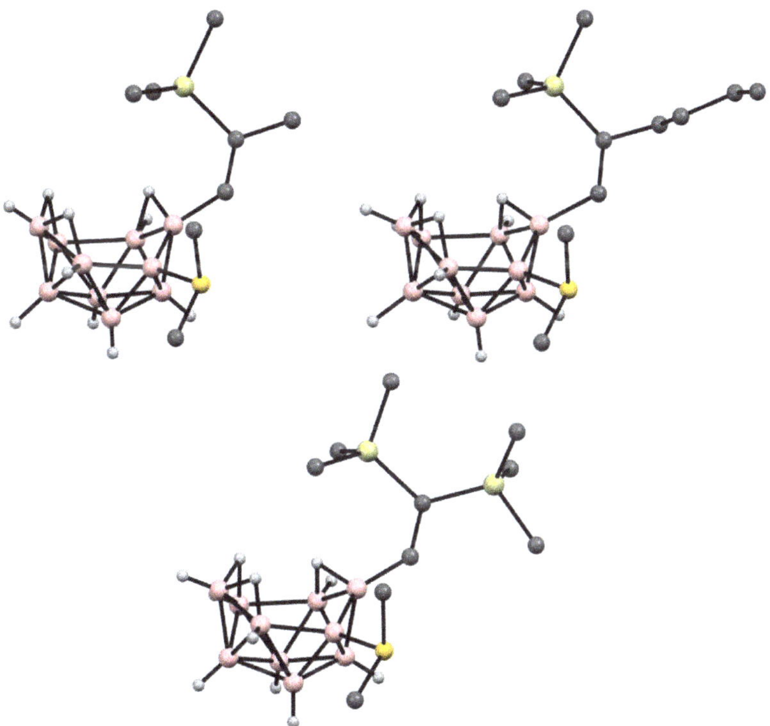

Figure 35. Solid state structures of [6-Me$_3$Si(Me)C=CH-5-Me$_2$S-B$_{10}$H$_{11}$] (**top left**), [6-Me$_3$Si(Bu)C=CH-5-Me$_2$S-B$_{10}$H$_{11}$] (**top right**), and [6-(Me$_3$Si)$_2$C=CH-5-Me$_2$S-B$_{10}$H$_{11}$] (**bottom**). Hydrogen atoms of organic substituents are omitted for clarity.

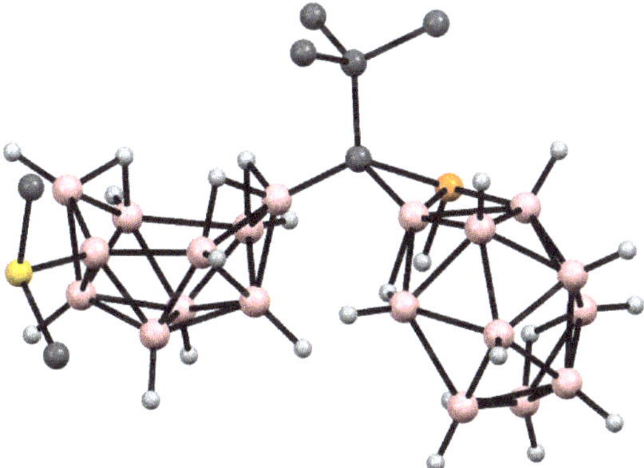

Figure 36. Solid state structure of [μ-6(*C*),6′(*C*),5′(*P*)-C(*t*-Bu)PH-(*nido*-8-Me$_2$S-B$_{10}$H$_{11}$)(*nido*-B$_{10}$H$_{12}$)]. Hydrogen atoms of organic substituents are omitted for clarity.

The derivatives [μ-(*exo*-6(*C*),*endo*-6(*N*)-CH=CH-*o*-C$_5$H$_4$N)-9(*N*)-HC≡C-*o*-C$_5$H$_4$N-*arachno*-B$_{10}$H$_{11}$] (Figure 34) [265] and [μ-(*exo*-6(*C*),*endo*-6(*N*)-(*closo*-1′,2′-C$_2$B$_{10}$H$_{10}$-2′-)-*o*-C$_5$H$_4$N)-μ-(*exo*-8(*C*),*exo*-9(*N*)-CH$_2$CH$_2$-*o*-C$_5$H$_4$N)-*arachno*-B$_{10}$H$_{10}$] (Figure 37) [339] were isolated in minor amounts as products of the intramolecular hydroboration of [*arachno*-6,9-(HC≡C-*o*-C$_5$H$_4$N)$_2$-B$_{10}$H$_{12}$] during its thermolysis in 1,2-dichloroethane. In the latter compound, the formation of the *ortho*-carborane fragment occurs as a result of the reaction of the acetylene group of the substituent with the second molecule of the decaborane derivative.

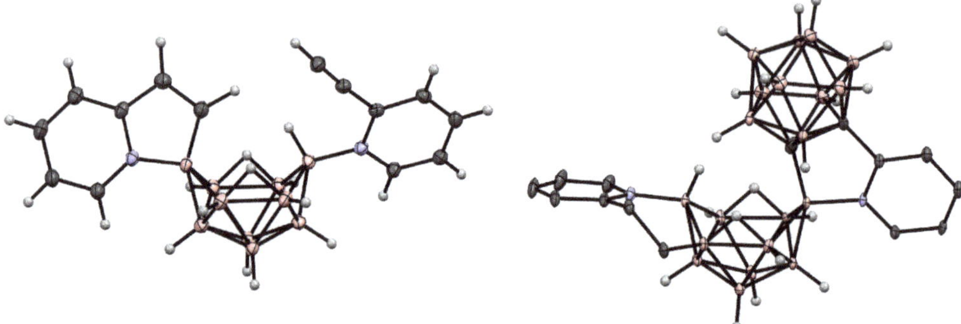

Figure 37. Solid state structures of [μ-(*exo*-6(C),*endo*-6(*N*)-CH=CH-*o*-C$_5$H$_4$N)-9(*N*)-HC≡C-*o*-C$_5$H$_4$N-*arachno*-B$_{10}$H$_{11}$] (**left**) and [μ-(*exo*-6(C),*endo*-6(*N*)-(*closo*-1′,2′-C$_2$B$_{10}$H$_{10}$-2′-)-*o*-C$_5$H$_4$N)-μ-(*exo*-8(C),*exo*-9(*N*)-CH$_2$CH$_2$-*o*-C$_5$H$_4$N)-*arachno*-B$_{10}$H$_{10}$] (**right**). Hydrogen atoms of organic substituents in the left structure are omitted for clarity.

The reactions of Cs$_2$[*closo*-B$_{10}$H$_{10}$] with triflic acid or (NH$_4$)$_2$[*closo*-B$_{10}$H$_{10}$] with sulfuric acid in the presence of aromatic hydrocarbons produce the corresponding 6-aryl derivatives of decaborane [*nido*-6-Ar-B$_{10}$H$_{13}$] (Ar = Ph, C$_6$H$_4$-4-Me, C$_6$H$_3$-3,5-Me$_2$, C$_6$H$_2$-2,4,6-Me$_3$, C$_6$H$_2$-2,4,6-^{i}Pr$_3$, C$_6$H$_4$Cl, C$_6$H$_4$CF$_3$) [202,203,321,340]. The solid state structures of [6-Ph-B$_{10}$H$_{13}$], [6-*p*-Tol-B$_{10}$H$_{13}$], and [6-(2′,4′,6′-^{i}Pr$_3$-C$_6$H$_2$-B$_{10}$H$_{13}$] were determined by single-crystal X-ray diffraction (Figure 38) [202,203,340].

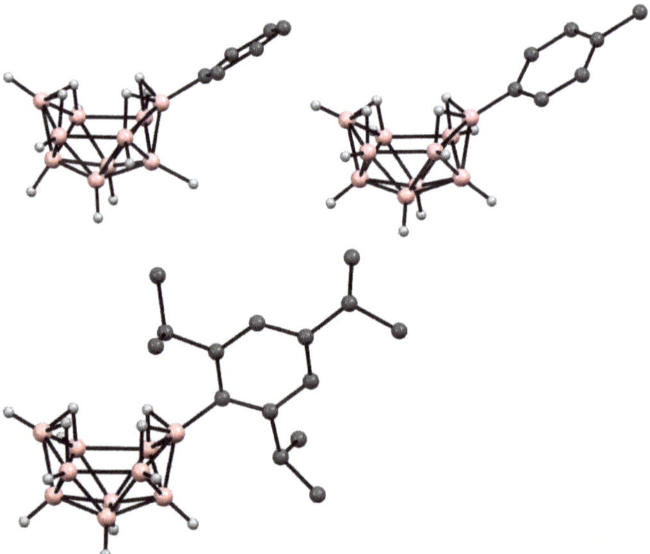

Figure 38. Solid state structures of [6-Ph-B$_{10}$H$_{13}$] (**top left**), [6-*p*-Tol-B$_{10}$H$_{13}$] (**top right**), and [6-(2′,4′,6′-^{i}Pr$_3$-C$_6$H$_2$-B$_{10}$H$_{13}$] (**bottom**). Hydrogen atoms of organic substituents are omitted for clarity.

The 6-phenyl derivative [6-Ph-B$_{10}$H$_{13}$] was also obtained by the reaction of decaborane(14) with PhLi, followed by the treatment with HCl in Et$_2$O [237] as well as by the solid state pyrolysis of [*nido*-Ph$_2$SnB$_{10}$H$_{12}$] at 95 °C [198].

The pyrolysis of decaborane(14) in benzene at 200 °C gives the 5-phenyl derivative [5-Ph-$B_{10}H_{13}$] as the main product together with some amounts of the 6-isomer and 5,8-diphenyl derivative [5,8-Ph_2-$B_{10}H_{12}$] [341]. The solid state structures of [5-Ph-$B_{10}H_{13}$] and [5,8-Ph_2-$B_{10}H_{12}$] were determined by single-crystal X-ray diffraction (Figure 39) [341].

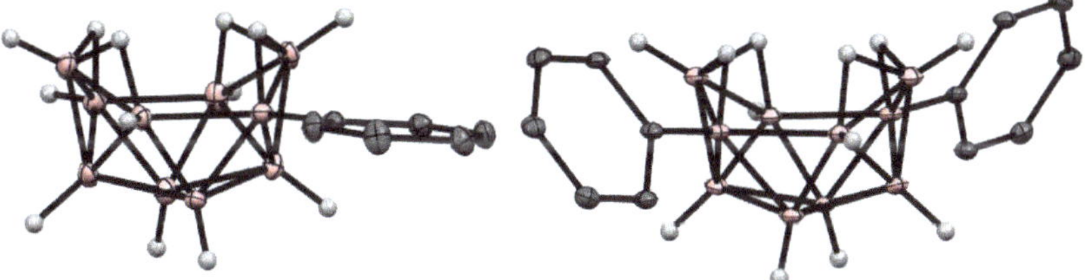

Figure 39. Solid state structures of [5-Ph-$B_{10}H_{13}$] (**left**) and [5,8-Ph_2-$B_{10}H_{12}$] (**right**). Hydrogen atoms of organic substituents are omitted for clarity.

The pyrolysis of decaborane(14) in toluene at 250 °C affords the novel microporous polymer named "activated borane", in which the decaborane clusters are interconnected by toluene moieties. Activated borane displays a high surface area of 774 m^2 g^{-1}, a thermal stability up to 1000 °C (under Ar), and a sorption capacity to emerging pollutants exceeding the capacity of commercial activated carbon [342].

Heating (HPS)[$B_{10}H_{13}$] in acetonitrile under reflux results in the formation of the bridged imino derivative (HPS)[*arachno*-μ-6(*C*),9(*N*)-MeC=NH-$B_{10}H_{12}$] (Figure 40) [343].

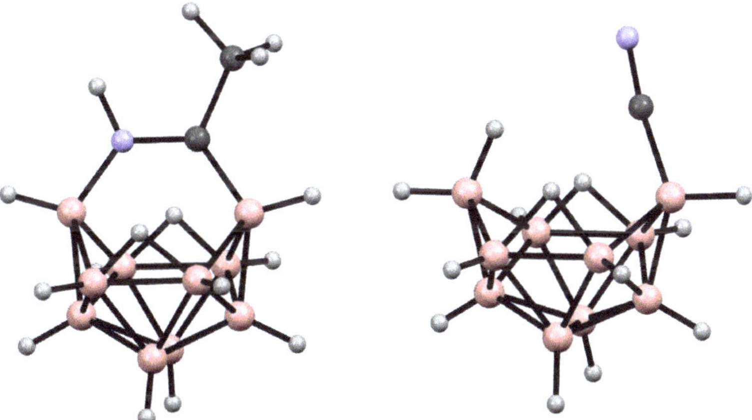

Figure 40. Solid state structures of the [*arachno*-μ-6(*C*),9(*N*)-MeC=NH-$B_{10}H_{12}$]$^-$ (**left**) and [*arachno*-*endo*-6-N≡C-$B_{10}H_{12}$]$^{2-}$ (**right**) anions.

The reaction of decaborane(14) with sodium cyanide in water followed by the addition of CsCl gives Cs$_2$[*arachno*-*endo*-6-N≡C-$B_{10}H_{13}$] [215]. The solid state structure of the trimethylphenylammonium salt (Me$_3$NPh)$_2$[*endo*-6-N≡C-$B_{10}H_{13}$] (Figure 40) [344] and lead complex {(Bipy)$_2$Pb[*endo*-6-N≡C-$B_{10}H_{13}$]} [345] were determined by single-crystal X-ray diffraction.

The reactions of decaborane(14) with sodium cyanide and dimethylsulfide or tetrahydrothiophene lead to the corresponding Na[*arachno*-6-N≡C-9-RR′S-$B_{10}H_{12}$] (R = R′ = Me, RR′ = (CH$_2$)$_4$). Na[6-N≡C-9-Me$_2$S-$B_{10}H_{12}$] can also be prepared by the reaction of [6,9-(Me$_2$S)$_2$-$B_{10}H_{12}$] with sodium cyanide in dimethylsulfide [215].

10. Derivatives with an *exo*-Polyhedral B-B Bond

Decaborane derivatives with an *exo*-polyhedral B-B bond are rare, since the reaction of decaborane(14) with boron hydrides usually leads to the completion of the polyhedral backbone with the formation of tetradecahydro-*nido*-undecaborate $[B_{11}H_{14}]^-$ [346] and dodecahydro-*closo*-dodecaborate $[B_{12}H_{12}]^{2-}$ [3] anions and their derivatives.

The reactions of decaborane(14) with sterically hindered (alkyl/arylimino)(2,2,6,6-tetramethylpiperidino)boranes lead to the corresponding 6-substituted aminoborane derivatives [*nido*-6-(RNH)$(C_5H_6Me_4NH)$B-$B_{10}H_{13}$] (R = *t*-Bu, C_6H_3-2,6-^{i}Pr$_2$) (Figure 41) [347].

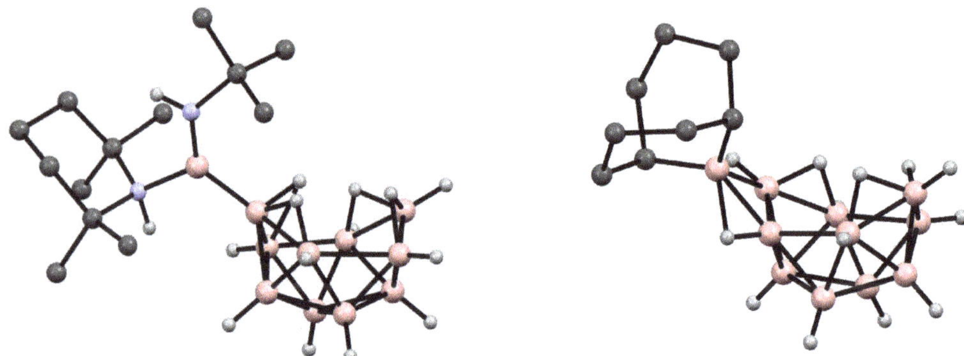

Figure 41. Solid state structures of [*nido*-6-(*t*-BuNH)$(C_5H_6Me_4NH)$B-$B_{10}H_{13}$] (**left**) and [μ-5,6-(9-BBN)-$B_{10}H_{13}$] (**right**). Hydrogen atoms of alkyl groups are omitted for clarity.

The reaction of Na[$B_{10}H_{13}$] with 9-bora[3.3.1]bicyclononane (9-Br-BBN) in dichloromethane results in the formation of [μ-5,6-(9-BBN)-$B_{10}H_{13}$], where the 9-BBN group appears in the role of an asymmetric bridge between the B(5) and B(6) positions of the decaborane basket (Figure 41). It is noteworthy that upon the deprotonation of [μ-5,6-(9-BBN)-$B_{10}H_{13}$] with a Proton Sponge in dichloromethane, the 9-BBN bridging group migrates from the B(6) atom to the B(10) atom, which leads to the formation of an 11-vertex *nido*-structure (HPS)[μ-7,7-CH$(CH_2CH_2CH_2)_2$CH-$B_{11}H_{12}$], being a formal derivative of the $[B_{11}H_{14}]^-$ anion [348].

Decaborane derivatives with an *exo*-polyhedral B-B bond also include isomeric *conjuncto*-decaboranes $[B_{10}H_{13}]_2$, which consist of two *nido*-B_{10} units linked by a direct B-B bond. These compounds were first identified as trace impurities in technical decaborane (14) [349]. In principle, there can be 11 different geometric isomers of *conjuncto*-decaborane $[B_{10}H_{13}]_2$, 4 of which are in the form of enantiomeric pairs. Therefore, various routes (photolysis [350,351], pyrolysis [350,351], γ-irradiation [352], high-energy electron bombardment [351], silent electrical discharge [353]) for synthesizing these compounds have been developed. All of them, as a rule, lead to the formation of mixtures with different isomeric compositions. The solid state structures of the 1,1′- [353], 1,2′- [354], 1,5′- [352], 1,6′- [238], 2,2′- [355,356], and 2,6′- [356] isomers have been determined by single-crystal X-ray diffraction, whereas the 2,5′- [351], 5,5′- [351], and 6,6′- [357] isomers have been characterized by NMR spectroscopy.

The asymmetric derivatives [*arachno*-1-(6′-*nido*-$B_{10}H_{13}$)-6,9-$(Me_2S)_2$-$B_{10}H_{11}$] and [*nido*-4-(2′-*nido*-$B_{10}H_{13}$)-5-Me_2S-$B_{10}H_{11}$] (Figure 42) were isolated from the reaction of [1,5′-(*nido*-$B_{10}H_{13}$)$_2$] with dimethylsulfide under reflux [238]. The first compound was also obtained by the thermolysis of [*arachno*-6,9-$(Me_2S)_2$-$B_{10}H_{12}$] in refluxing toluene [238].

Figure 42. Solid state structures of [*arachno*-1-(6′-*nido*-B$_{10}$H$_{13}$)-6,9-(Me$_2$S)$_2$-B$_{10}$H$_{11}$] (**left**) and [*nido*-4-(2′-*nido*-B$_{10}$H$_{13}$)-5-Me$_2$S-B$_{10}$H$_{11}$] (**right**). Hydrogen atoms of alkyl groups are omitted for clarity.

The isomeric tridecaboranyl species [*arachno*-1,5-(6′-*nido*-B$_{10}$H$_{13}$)$_2$-6,9-(Me$_2$S)$_2$-B$_{10}$H$_{10}$] (Figure 43) and [*arachno*-1,3-(6′-*nido*-B$_{10}$H$_{13}$)$_2$-6,9-(Me$_2$S)$_2$-B$_{10}$H$_{10}$] were isolated from the products of the thermolysis of [*arachno*-6,9-(Me$_2$S)$_2$-B$_{10}$H$_{12}$] in refluxing benzene [238,358].

Figure 43. Solid state structure of [*arachno*-1,5-(6′-*nido*-B$_{10}$H$_{13}$)$_2$-6,9-(Me$_2$S)$_2$-B$_{10}$H$_{10}$]. Hydrogen atoms of organic substituents are omitted for clarity.

Here, it is also worth mentioning an unusual structure, in which two hydrogen atoms at positions 5 and 6 of the decaborane basket are replaced by pentaborane moieties—5-

(*nido*-pentaboran-2-yl)-6-(*nido*-pentaboran-1-yl)-*nido*-decaborane, formed as a result of the long-term storage (23 years) of a sealed, under-vacuum pentaborane(9) sample under ambient lighting and temperature conditions (Figure 44) [339].

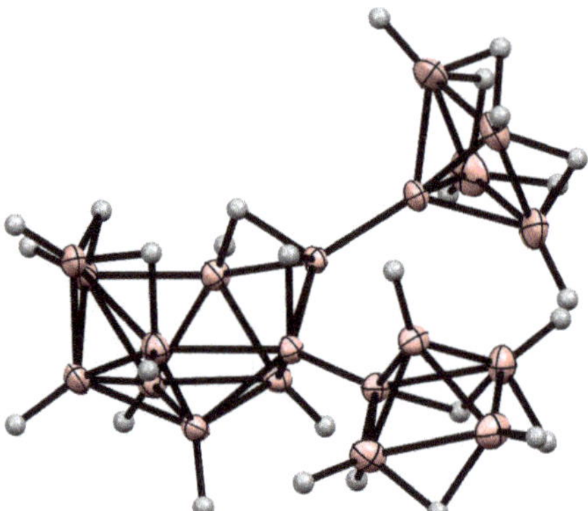

Figure 44. Structure of 5-(*nido*-pentaboran-2-yl)-6-(*nido*-pentaboran-1-yl)-*nido*-decaborane.

11. Decaborane-Related *conjucto*-Boranes [$B_{18}H_{22}$]

Another type of compound worth considering here are the *conjuncto*-boranes [$B_{18}H_{22}$], in which two decaborane baskets are connected by a common edge. Octadecaborane(22) [$B_{18}H_{22}$] in the form of a mixture of *syn*- and *anti*-isomers (Figure 45) is formed on the hydrolysis of $(H_3O)_2[trans\text{-}B_{20}H_{18}]\cdot nH_2O$ and can be separated by fractional crystallization [359]. More recently, a convenient method has been proposed for the synthesis of *anti*-[$B_{18}H_{22}$] by mild oxidation of $(Me_4N)[nido\text{-}B_9H_{12}]$ with iodine in toluene, which gives excellent yields (~80%) and thus provides a large-scale and safe route to this important polyborane cluster [360].

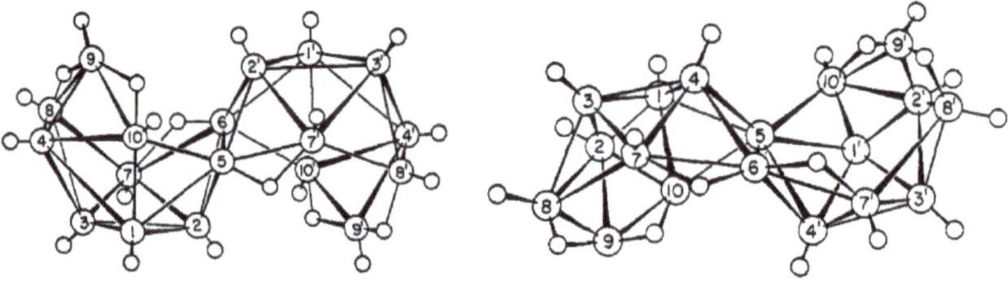

Figure 45. Structures and numbering of atoms in *syn*- (**left**) and *anti*- (**right**) isomers of [$B_{18}H_{22}$]. Reprinted with permission from Ref. [359]. Copyright (1968) the American Chemical Society.

The photophysics of both isomers have been studied by UV–vis spectroscopic techniques and quantum chemical calculations. In an air-saturated hexane solution, *anti*-[$B_{18}H_{22}$] shows fluorescence with a high quantum yield, $\Phi_F = 0.97$, and singlet oxygen $O_2(^1\Delta_g)$ production ($\Phi_\Delta \sim 0.008$). Conversely, the isomer *syn*-[$B_{18}H_{22}$] shows no measurable fluorescence, instead displaying a much faster, picosecond nonradiative decay of excited singlet states [361]. Due to this, *anti*-[$B_{18}H_{22}$] can be considered as a potential blue laser material. The photophysical properties of *anti*-[$B_{18}H_{22}$] can be tuned by the partial substitution of hydrogen atoms with various functional groups. Because of this, combined with its high stability [362–366], *anti*-[$B_{18}H_{22}$] is attracting increasing research interest, while the *syn*-[$B_{18}H_{22}$] isomer has received much less attention.

The reaction of *anti*-[$B_{18}H_{22}$] with chlorine generated in situ from *N*-chlorosuccinimide (NCS) with HCl/dioxane in dichloromethane leads to the 7-chloro derivative *anti*-[7-Cl-$B_{18}H_{21}$] (Figure 46) [367]. The reaction of *anti*-[$B_{18}H_{22}$] with $AlCl_3$ in tetrachloromethane results in a mixture of the 3,3′- and 3,4′-dichloro derivatives *anti*-[3,3′-Cl₂-B₁₈H₂₀] (Figure 46) and *anti*-[3,4′-Cl₂-B₁₈H₂₀] (Figure 45) together with minor amounts of the other isomeric dichloro derivatives *anti*-[4,4′-Cl₂-B₁₈H₂₀] (Figure 46), *anti*-[3,1′-Cl₂-B₁₈H₂₀] (Figure 46), and *anti*-[7,3′-Cl₂-B₁₈H₂₀], as well as the 3- and 4-chloro derivatives *anti*-[3-Cl-B₁₈H₂₁] and *anti*-[4-Cl-B₁₈H₂₁] and the 3,4,3′- and 3,4,4′-trichloro derivatives *anti*-[3,4,3′-Cl₃-B₁₈H₁₉] and *anti*-[3,4,4′-Cl₃-B₁₈H₁₉], which were separated chromatographically [368].

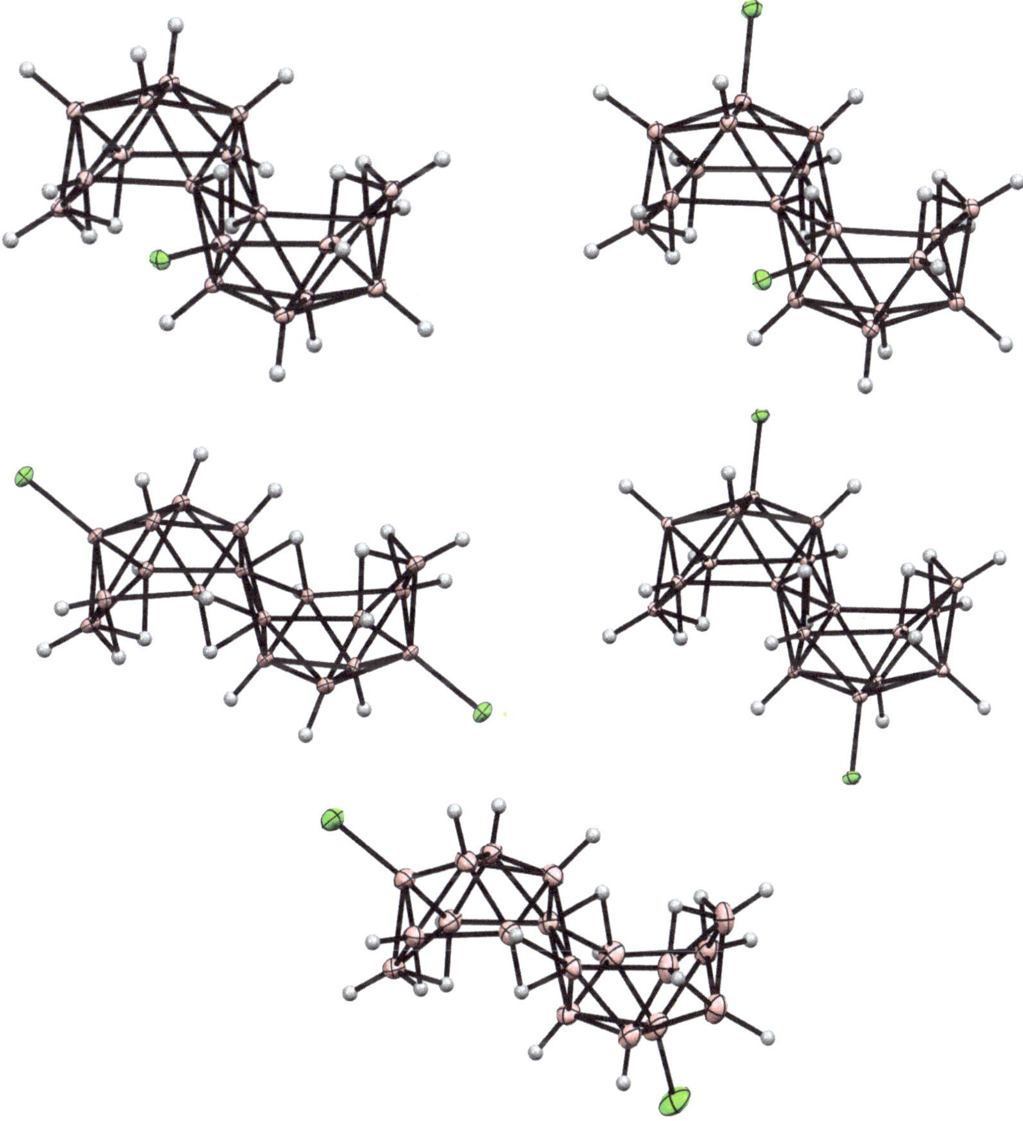

Figure 46. Solid state structures of *anti*-[7-Cl-B₁₈H₂₁] (**top left**), *anti*-[3,1′-Cl₂-B₁₈H₂₀] (**top right**), *anti*-[4,4′-Cl₂-B₁₈H₂₀] (**middle left**), *anti*-[3,3′-Cl₂-B₁₈H₂₀] (**middle right**), and *anti*-[3,4′-Cl₂-B₁₈H₂₀] (**bottom**).

The bromination of *anti*-[$B_{18}H_{22}$] with bromine in dichloromethane in the presence of $AlCl_3$ leads to the 4-bromo or 4,4′-dibromo derivatives *anti*-[4-Br-$B_{18}H_{21}$] or *anti*-[4,4′-Br_2-$B_{18}H_{20}$] depending on the reagent ratio (Figure 47) [369,370].

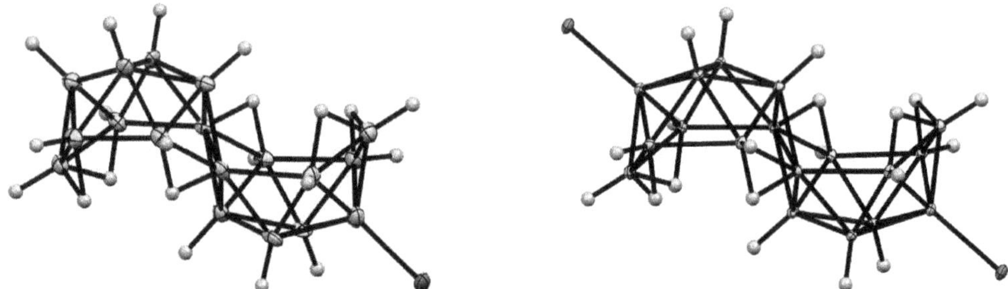

Figure 47. Solid state structures of *anti*-[4-Br-$B_{18}H_{21}$] (**left**) and *anti*-[4,4′-Br_2-$B_{18}H_{20}$] (**right**).

The reaction of *anti*-[$B_{18}H_{22}$] with iodine in ethanol leads to the 4-iodo derivative *anti*-[7-I-$B_{18}H_{21}$] (Figure 48) [359,371], while the reaction with I_2 or ICl in the presence of $AlCl_3$ in dichloromethane results in the 4-iodo- and 4,4′-diiodo derivatives *anti*-[4-I_2-$B_{18}H_{21}$] and *anti*-[4,4′-I_2-$B_{18}H_{20}$] (Figure 48), depending on the reagent ratio [367,371].

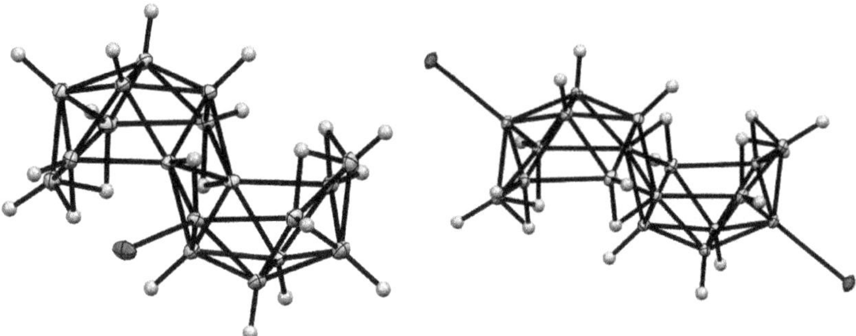

Figure 48. Solid state structures of *anti*-[7-I-$B_{18}H_{21}$] (**left**) and *anti*-[4,4′-I_2-$B_{18}H_{20}$] (**right**).

The iodine atom in *anti*-[7-I-$B_{18}H_{21}$] can be substituted by various nucleophiles: the reaction with trifluoroacetamide in toluene in the presence of K_3PO_4 gives the corresponding *N*-boronated amide *anti*-[7-CF_3CONH-$B_{18}H_{21}$]; the reactions with *t*-BuOK, 4-FC_6H_4OK, and 1-AdSK in toluene or tetrahydrofuran lead to the corresponding (thio)ethers *anti*-[7-RX-$B_{18}H_{21}$]. The reaction with potassium 2,6-dimethylthiophenolate in toluene results in the corresponding thioether *anti*-[7-(2′,6′-$Me_2C_6H_3S$)-$B_{18}H_{21}$], while the reaction in tetrahydrofuran produces *anti*-[7-(2′,6′-$Me_2C_6H_3S(CH_2)_4O$)-$B_{18}H_{21}$] [372]. The Pd-catalyzed reactions of *anti*-[7-I-$B_{18}H_{21}$] with CF_3CONH_2, *t*-BuOK, and 2,6-$Me_2C_6H_3OK$ in the presence of catalytic amounts of RuPhos Pd G4 and RuPhos in 1,4-dioxane lead to the corresponding derivatives with B-N and B-O bonds *anti*-[7-X-$B_{18}H_{21}$] [372].

The reaction of *anti*-[$B_{18}H_{22}$] with neat methyl iodide in the presence of $AlCl_3$ at room temperature results in the 3,3′,4,4′-tetramethyl derivative *anti*-[3,3′,4,4′-Me_4-$B_{18}H_{18}$] (Figure 49) as the main product, together with minor amounts of the 4,4′-dimethyl derivative *anti*-[4,4′-Me_2-$B_{18}H_{20}$] (Figure 49), the 3,4,4′- and 3,3′,4-trimethyl derivatives *anti*-[3,4,4′-Me_3-$B_{18}H_{19}$] (Figure 47) and *anti*-[3,3′,4-Me_3-$B_{18}H_{19}$], as well as the 1,3,3′,4,4′- and 3,3′,4,4′,8-pentamethyl derivatives *anti*-[1,3,3′,4,4′-Me_5-$B_{18}H_{17}$] and *anti*-[3,3′,4,4′,8-Me_5-$B_{18}H_{17}$] and the 1,1′,3,3′,4,4′-hexamethyl derivative *anti*-[1,1′,3,3′,4,4′-Me_6-$B_{18}H_{16}$] [373]. The similar reaction with ethyl iodide gives the 3,3′,4,4′-tetraethyl derivative *anti*-[3,3′,4,4′-Et_4-$B_{18}H_{18}$] (Figure 49) [373].

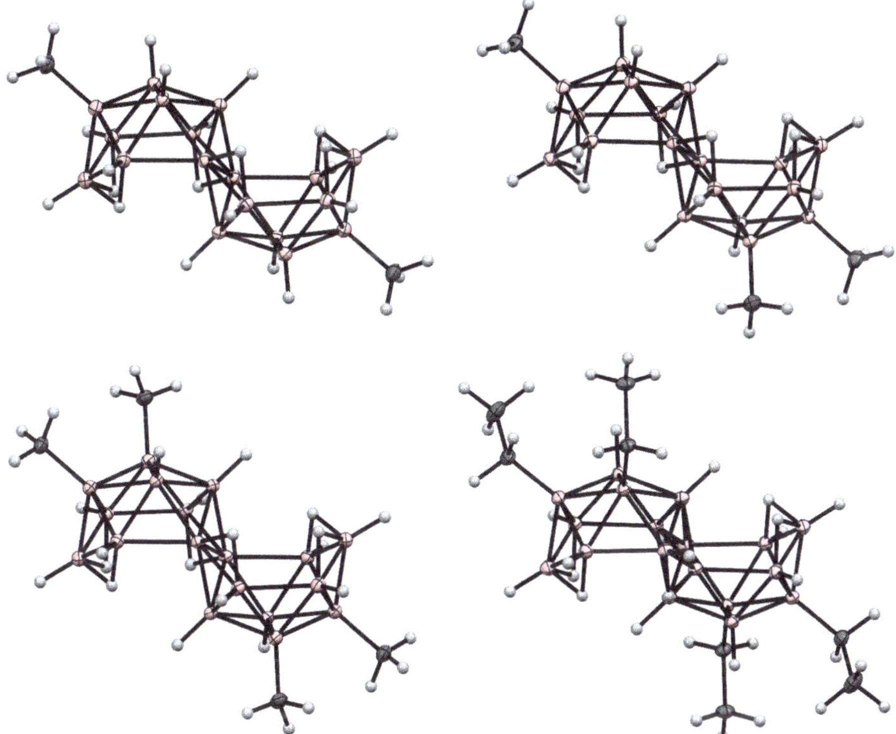

Figure 49. Solid state structures of *anti*-[4,4′-Me$_2$-B$_{18}$H$_{20}$] (**top left**), *anti*-[3,4,4′-Me$_3$-B$_{18}$H$_{19}$] (**top right**), *anti*-[3,3′,4,4′-Me$_4$-B$_{18}$H$_{18}$] (**bottom left**), and *anti*-[3,3′,4,4′-Et$_4$-B$_{18}$H$_{18}$] (**bottom right**).

The dichloroundecamethyl *anti*-[2,2′-Cl$_2$-1,1′,3,3′,4,4′,7,7′,8,8′,10′-Me$_{11}$-B$_{18}$H$_9$], dichlorododecamethyl *anti*-[2,2′-Cl$_2$-1,1′,3,3′,4,4′,7,7′,8,8′,10,10′-Me$_{12}$-B$_{18}$H$_8$] (Figure 50), and dichlorotridecamethyl *anti*-[2,2′-Cl$_2$-1,1′,3,3′,4,4′,7,7′,8,8′,9,10,10′-Me$_{13}$-B$_{18}$H$_7$] derivatives were obtained by the reaction of *anti*-[B$_{18}$H$_{22}$] with methyl iodide in the presence of AlCl$_3$ in dichloromethane at 55 °C [373,374].

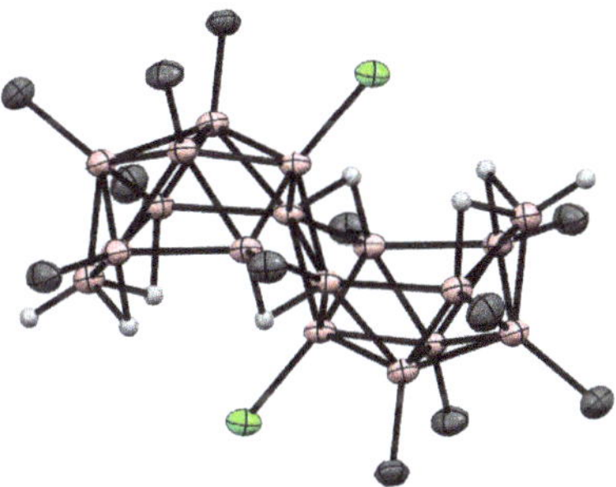

Figure 50. Solid state structure of *anti*-[2,2′-Cl$_2$-1,1′,3,3′,4,4′,7,7′,8,8′,10,10′-Me$_{12}$-B$_{18}$H$_8$]. Hydrogen atoms of organic substituents are omitted for clarity.

The reaction of *anti*-[B$_{18}$H$_{22}$] with elemental sulfur in the presence of AlCl$_3$ at 125 °C leads to the 4,4′-dimercapto derivative *anti*-[4,4′-(HS)$_2$-B$_{18}$H$_{20}$] (Figure 51) [375].

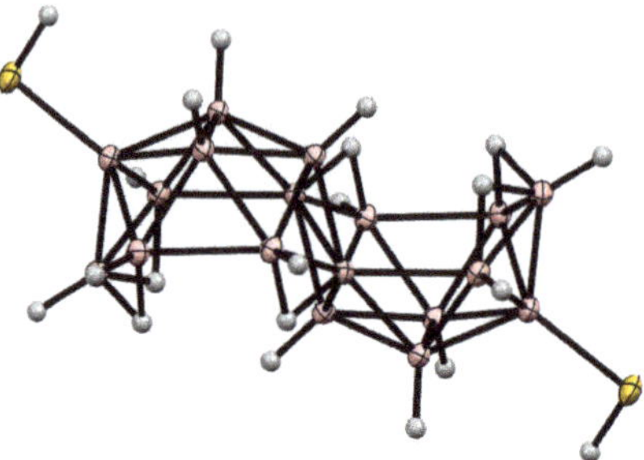

Figure 51. Solid state structure of *anti*-[4,4′-(HS)$_2$-B$_{18}$H$_{20}$].

The reaction of *anti*-[B$_{18}$H$_{22}$] with pyridine in refluxing chloroform or benzene unexpectedly results in a two-fold substitution in one of the B$_{10}$-baskets to form *nido-arachno*-[6′,9′-Py$_2$-B$_{18}$H$_{20}$] (Figure 52) together with some amount of *anti*-[8′-Py-B$_{18}$H$_{21}$] (Figure 52) and [3′,8′-Py$_2$-B$_{16}$H$_{18}$] as the main degradation product. In contrast to the thermochromic fluorescence of *nido-arachno*-[6′,9′-Py$_2$-B$_{18}$H$_{20}$] (from 620 nm brick red at room temperature to 585 nm yellow at 8 K), *anti*-[8′-Py-B$_{18}$H$_{21}$] exhibits no luminescence [376,377]. The 6′,9′-disubstituted derivatives with 4-picoline [377], isoquinoline [378], and 5-hydroxyisoquinoline [379] were prepared in a similar way.

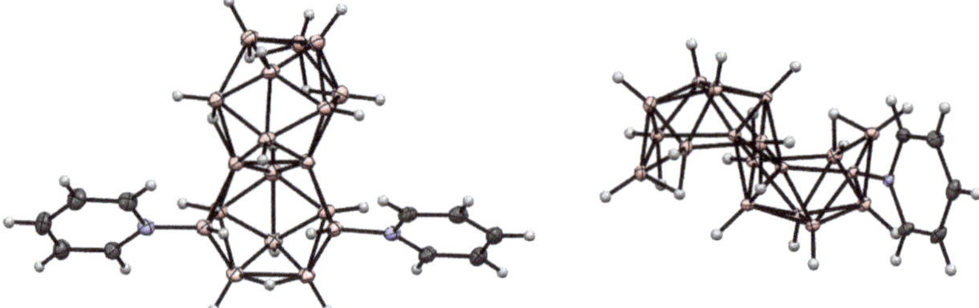

Figure 52. Solid state structures of *nido-arachno*-[6′,9′-Py$_2$-B$_{18}$H$_{20}$] (**left**) and *anti*-[8-Py-B$_{18}$H$_{21}$] (**right**).

The reaction of *anti*-[B$_{18}$H$_{22}$] with methyl isonitrile MeNC in benzene leads to *anti*-[7-{(MeNH)C$_3$N$_2$HMe$_2$}-B$_{18}$H$_{20}$], in which a reductive trimerization of MeNC gives an unusual imidazole-based carbene, {(MeNH)C$_3$N$_2$HMe$_2$}, that is stabilized by coordination to the macropolyhedral boron cluster (Figure 53) [380].

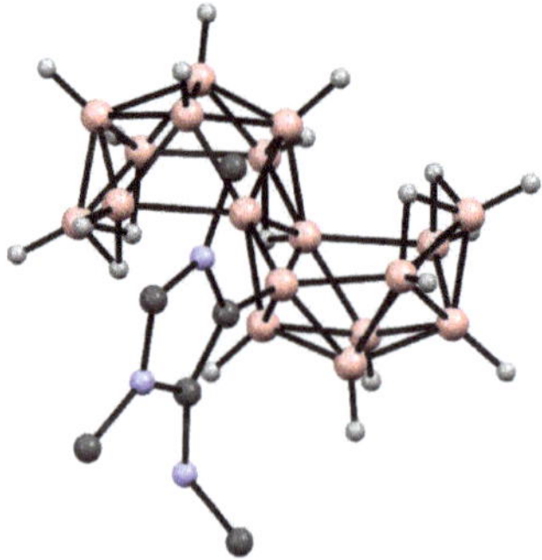

Figure 53. Solid state structure of *anti*-[7-{(MeNH)C$_3$N$_2$HMe$_2$}-B$_{18}$H$_{20}$]. Hydrogen atoms of organic substituents are omitted for clarity.

The reaction with *tert*-butyl isonitrile in 1,2-dichloroethane results in *anti*-[7-{(*t*-BuNHCH){*t*-BuNHC(CN)}CH$_2$}-B$_{18}$H$_{20}$], in which a reductive oligomerization of *t*-BuNC has given the complex polynitrogen base {(*t*-BuNHCH){*t*-BuNHC(CN)}CH$_2$:} formally as a zwitterionic carbene attached to the macropolyhedral boron cluster (Figure 54) [381].

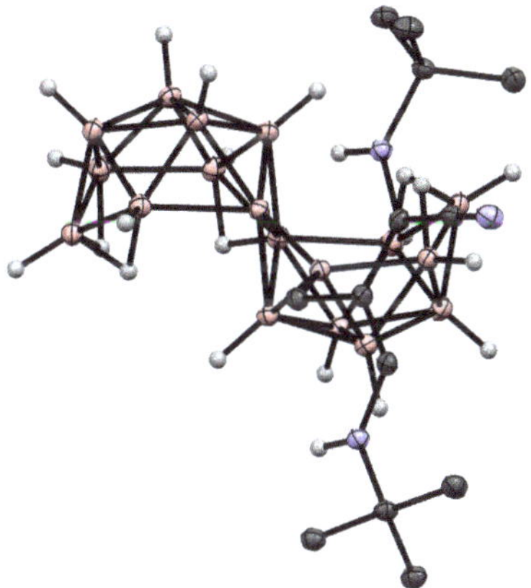

Figure 54. Solid state structure of *anti*-[7-{(*t*-BuNHCH){*t*-BuNHC(CN)}CH$_2$}-B$_{18}$H$_{20}$]. Hydrogen atoms of organic substituents are omitted for clarity.

The synthesis of few substituted derivatives of *syn*-[B$_{18}$H$_{22}$] was reported. Heating *syn*-[B$_{18}$H$_{22}$] with sulfur in the presence of anhydrous AlCl$_3$ at 125 °C results in a mixture of the isomeric mercapto derivatives *syn*-[1-HS-B$_{18}$H$_{21}$], *syn*-[3-HS-B$_{18}$H$_{21}$], and *syn*-[4-HS-B$_{18}$H$_{21}$], which were all separated by chromatography on silica (Figure 55) [382].

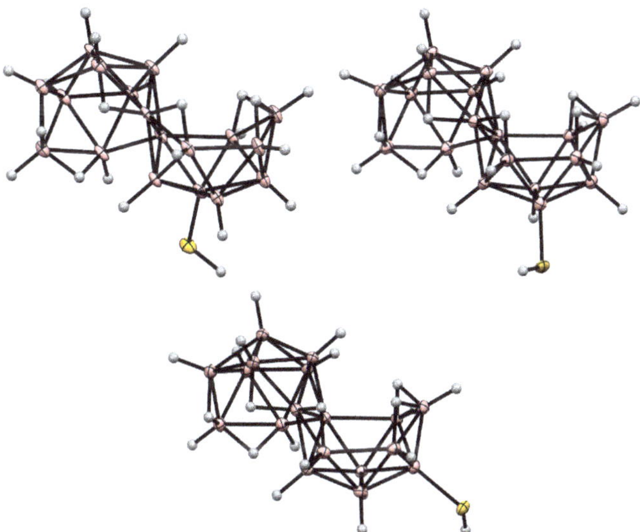

Figure 55. Solid state structures of *syn*-[1-HS-B$_{18}$H$_{21}$] (**top left**), *syn*-[3-HS-B$_{18}$H$_{21}$] (**top right**), and *syn*-[4-HS-B$_{18}$H$_{21}$] (**bottom**).

The 3- and 4-mercapto derivatives of *syn*-[B$_{18}$H$_{22}$], *syn*-[3-HS-B$_{18}$H$_{21}$], and *syn*-[4-HS-B$_{18}$H$_{21}$] were obtained as byproducts of thermolysis of nonathiaborane *arachno*-[SB$_8$H$_{12}$]

in boiling cyclohexane [383]. The mercapto derivatives obtained are brightly luminescent under UV irradiation, making these compounds rare examples of a luminescent derivative of *syn*-[$B_{18}H_{22}$] [382,383].

The deprotonation of *syn*-[$B_{18}H_{22}$] with NaH in 1,2-dimethoxyethane, followed by the reaction with iodine and dimethylsulfide under reflux, results in the 7-dimethylsulfonium derivative *syn*-[7-Me_2S-$B_{18}H_{20}$] (Figure 56) [384].

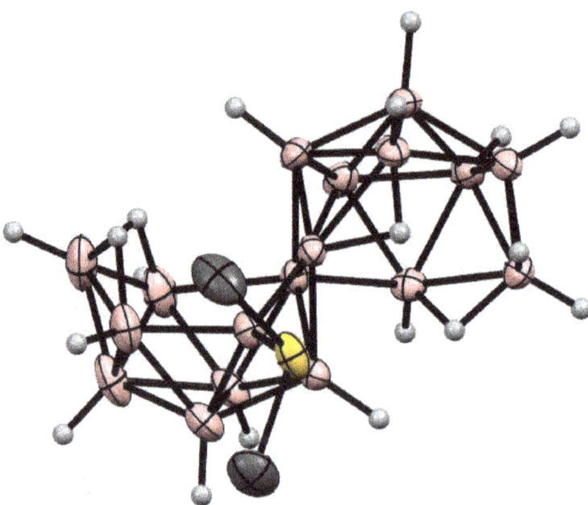

Figure 56. Solid state structure of *syn*-[7-Me_2S-$B_{18}H_{20}$]. Hydrogen atoms of organic substituents are omitted for clarity.

12. Conclusions

The purpose of this review was to give the most complete picture of the current state of the chemistry of decaborane and its derivatives. After its rapid development on the verge of the 1950s and 1960s, associated with the study of the chemistry of decaborane as a potential component of rocket fuels, the chemistry of decaborane was studied by many research groups. However, no comprehensive review elucidating it in full has appeared for more than 50 years, which certainly hindered the development of this important area of boron cluster chemistry. We would like to hope that this review will be useful both for young researchers just starting their way in boron chemistry and for researchers actively working in this field.

Funding: This work was supported by the Ministry of Science and Higher Education of the Russian Federation (075-03-2023-642).

Data Availability Statement: No new data were created.

Conflicts of Interest: The author declares no conflict of interests.

Sample Availability: Not applicable.

References

1. Grimes, R.N. *Carboranes*, 3rd ed.; Academic Press: London, UK, 2016. [CrossRef]
2. Shmal'ko, A.V.; Sivaev, I.B. Chemistry of carba-*closo*-decaborate anions [CB_9H_{10}]$^-$. *Russ. J. Inorg. Chem.* **2019**, *64*, 1726–1749. [CrossRef]
3. Sivaev, I.B.; Bregadze, V.I.; Sjöberg, S. Chemistry of *closo*-dodecaborate anion [$B_{12}H_{12}$]$^{2-}$: A review. *Collect. Czech. Chem. Commun.* **2002**, *67*, 679–727. [CrossRef]
4. Zhao, X.; Yang, Z.; Chen, H.; Wang, Z.; Zhou, X.; Zhang, H. Progress in three-dimensional aromatic-like *closo*-dodecaborate. *Coord. Chem. Rev.* **2021**, *444*, 214042. [CrossRef]
5. Körbe, S.; Schreiber, P.J.; Michl, J. Chemistry of the carba-*closo*-dodecaborate(-) anion, $CB_{11}H_{12}^-$. *Chem. Rev.* **2006**, *106*, 5208–5249. [CrossRef]
6. Douvris, C.; Michl, J. Update 1 of: Chemistry of the carba-*closo*-dodecaborate(-) anion, $CB_{11}H_{12}^-$. *Chem. Rev.* **2013**, *113*, R179–R233. [CrossRef]

7. Sivaev, I.B.; Prikaznov, A.V.; Naoufal, D. Fifty years of the *closo*-decaborate anion chemistry. *Collect. Czech. Chem. Commun.* **2010**, *75*, 1149–1199. [CrossRef]

8. Mahfouz, N.; Abi Ghaida, F.; El Hajj, Z.; Diab, M.; Floquet, S.; Mehdi, A.; Naoufal, D. Recent achievements on functionalization within *closo*-decahydrodecaborate $[B_{10}H_{10}]^{2-}$ clusters. *ChemistrySelect* **2022**, *7*, e202200770. [CrossRef]

9. Nakamura, K. Preparation and properties of amorphous boron films deposited by pyrolysis of decaborane in the molecular flow region. *J. Electrochem. Soc.* **1984**, *131*, 2691. [CrossRef]

10. Kim, Y.G.; Dowben, P.A.; Spencer, J.T.; Ramseyer, G.O. Chemical vapor deposition of boron and boron nitride from decaborane(14). *J. Vac. Sci. Technol. A* **1989**, *7*, 2796–2799. [CrossRef]

11. Perkins, F.K.; Rosenberg, R.A.; Lee, S.; Dowben, P.A. Synchrotron-radiation-induced deposition of boron and boron carbide films from boranes and carboranes: Decaborane. *J. Appl. Phys.* **1991**, *69*, 4103–4109. [CrossRef]

12. Saidoh, M.; Ogiwara, N.; Shimada, M.; Arai, T.; Hiratsuka, H.; Koike, T.; Shimizu, M.; Ninomiya, H.; Nakamura, H.; Jimbou, R. Initial boronization of JT-60U tokamak using decaborane. *Jpn. J. Appl. Phys.* **1993**, *32*, 3276–3281. [CrossRef]

13. Kodama, H.; Oyaidzu, M.; Sasaki, M.; Kimura, H.; Morimoto, Y.; Oya, Y.; Matsuyama, M.; Sagara, A.; Noda, N.; Okuno, K. Studies on structural and chemical characterization for boron coating films deposited by PCVD. *J. Nucl. Mater.* **2004**, *329–333*, 889–893. [CrossRef]

14. Bellott, B.J.; Noh, W.; Nuzzo, R.G.; Girolami, G.S. Nanoenergetic materials: Boron nanoparticles from the pyrolysis of decaborane and their functionalisation. *Chem. Commun.* **2009**, 3214–3215. [CrossRef]

15. Ekimov, E.A.; Zibrov, I.P.; Zoteev, A.V. Preparation of boron microcrystals via high-pressure, high-temperature pyrolysis of decaborane, $B_{10}H_{14}$. *Inorg. Mater.* **2011**, *47*, 1194–1198. [CrossRef]

16. Ekimov, E.A.; Lebed, J.B.; Sidorov, V.A.; Lyapin, S.G. High-pressure synthesis of crystalline boron in B-H system. *Sci. Technol. Adv. Mater.* **2011**, *12*, 055009. [CrossRef] [PubMed]

17. Chatterjee, S.; Luo, Z.; Acerce, M.; Yates, D.M.; Johnson, A.T.C.; Sneddon, L.G. Chemical vapor deposition of boron nitride nanosheets on metallic substrates via decaborane/ammonia reactions. *Chem. Mater.* **2011**, *23*, 4414–4416. [CrossRef]

18. Chatterjee, S.; Kim, M.J.; Zakharov, D.N.; Kim, S.M.; Stach, E.A.; Maruyama, B.; Sneddon, L.G. Syntheses of boron nitride nanotubes from borazine and decaborane molecular precursors by catalytic chemical vapor deposition with a floating nickel catalyst. *Chem. Mater.* **2012**, *24*, 2872–2879. [CrossRef]

19. Kher, S.S.; Spencer, J.T. Chemical vapor deposition precursor chemistry. 3. Formation and characterization of crystalline nickel boride thin films from the cluster-assisted deposition of polyhedral borane compounds. *Chem. Mater.* **1992**, *4*, 538–544. [CrossRef]

20. Kher, S.S.; Tan, Y.; Spencer, J.T. Chemical vapor deposition of metal borides. 4. The application of polyhedral boron clusters to the chemical vapor deposition formation of gadolinium boride thin-film materials. *Appl. Organomet. Chem.* **1996**, *10*, 297–304. [CrossRef]

21. Kher, S.S.; Romero, J.V.; Caruso, J.D.; Spencer, J.T. Chemical vapor deposition of metal borides. 6. The formation of neodymium boride thin film materials from polyhedral boron clusters and metal halides by chemical vapor deposition. *Appl. Organomet. Chem.* **2008**, *22*, 300–307. [CrossRef]

22. Tynell, T.; Aizawa, T.; Ohkubo, I.; Nakamura, K.; Mori, T. Deposition of thermoelectric strontium hexaboride thin films by a low pressure CVD method. *J. Cryst. Growth* **2016**, *449*, 10–14. [CrossRef]

23. Guélou, G.; Martirossyan, M.; Ogata, K.; Ohkubo, I.; Kakefuda, Y.; Kawamoto, N.; Kitagawa, Y.; Ueda, J.; Tanabe, S.; Maeda, K.; et al. Rapid deposition and thermoelectric properties of ytterbium boride thin films using hybrid physical chemical vapor deposition. *Materialia* **2018**, *1*, 244–248. [CrossRef]

24. Ohkubo, I.; Aizawa, T.; Nakamura, K.; Mori, T. Control of competing thermodynamics and kinetics in vapor phase thin-film growth of nitrides and borides. *Front. Chem.* **2021**, *9*, 642388. [CrossRef] [PubMed]

25. Zhang, Y.; Peng, Y.; Wan, Q.; Ye, D.; Wang, A.; Zhang, L.; Jiang, W.; Liu, Y.; Li, J.; Zhuang, X.; et al. Fuel cell power source based on decaborane with high energy density and low crossover. *Mater. Today Energy* **2023**, *32*, 101244. [CrossRef]

26. Hawthorne, M.F. Decaborane-14 and its derivatives. In *Advances in Inorganic Chemistry and Radiochemistry*; Academic Press: New York, NY, USA, 1963; Volume 5, pp. 307–347. [CrossRef]

27. Stanko, V.I.; Chapovskii, Y.A.; Brattsev, V.A.; Zakharkin, L.I. The chemistry of decaborane and its derivatives. *Russ. Chem. Rev.* **1965**, *34*, 424–439. [CrossRef]

28. Shore, S.G. *Nido*- and *arachno*-boron hydrides. In *Boron Hydride Chemistry*; Muetterties, E.L., Ed.; Academic Press: New York, NY, USA, 1975; pp. 79–174. [CrossRef]

29. Greenwood, N.N. Boron. In *Comprehensive Inorganic Chemistry*; Pergamon Press: Oxford, UK, 1973; Volume 1, pp. 818–837. [CrossRef]

30. Sivaev, I.B. Molecular boron clusters. In *Comprehensive Inorganic Chemistry*; Elsevier: Amsterdam, The Netherlands, 2023; Volume 1, pp. 740–777. [CrossRef]

31. Stock, A.; Friederici, K.; Priess, O. Borwasserstoffe. III. Feste Borwasserstoffe: Zur Kenntnis des B_2H_6. *Ber. Dtsch. Chem. Ges.* **1913**, *46*, 3353–3365. [CrossRef]

32. Stock, A.; Pohland, E. Borwasserstoffe. XII. Zur Kenntnis des $B_{10}H_{14}$. *Ber. Dtsch. Chem. Ges.* **1929**, *62*, 90–99. [CrossRef]

33. Stock, A. *Hydrides of Boron and Silicon*; Cornell University Press: Ithaca, NY, USA, 1933.

34. Stock, A. Borwasserstoffe. XI. Strukturformeln der Borhydride. *Ber. Dtsch. Chem. Ges.* **1926**, *59*, 2226–2229. [CrossRef]

35. Huggins, M.L. Boron hydrides. *J. Phys. Chem.* **1922**, *26*, 833–835. [CrossRef]

36. Clark, J.D. *Ignition! An Informal History of Liquid Rocket Propellants*; Rutgers University Press: New Brunswick, NJ, USA, 1972. [CrossRef]
37. *Boron High Energy Fuels: Hearings Before the Committee on Science and Astronautics U.S. House of Representatives. Eighty-Six Congress, First Session*; United States Government Printing Office: Washington, DC, USA, 1959.
38. Goodger, E. Unconventional fuels. *New Sci.* **1957**, 27–29.
39. Siegel, B.; Mack, J.L. The boron hydrides. *J. Chem. Ed.* **1957**, *34*, 314–317. [CrossRef]
40. Martin, D.R. The development of borane fuels. *J. Chem. Ed.* **1959**, *36*, 208–214. [CrossRef]
41. Gold, J.W.; Militscher, C.; Slauson, D.D. Pentaborane Disposal: Taming the Dragon. Available online: https://dokumen.tips/documents/pentaborane-taming-the-dragonpdf.html (accessed on 23 August 2023).
42. Kasper, J.S.; Lucht, C.M.; Harker, D. The structure of the decaborane molecule. *J. Am. Chem. Soc.* **1948**, *70*, 881. [CrossRef]
43. Kasper, J.S.; Lucht, C.M.; Harker, M. The crystal structure of decaborane, $B_{10}H_{14}$. *Acta Cryst.* **1950**, *3*, 436–455. [CrossRef]
44. Moore, E.B.; Dickerson, R.E.; Lipscomb, W.N. Least squares refinements of $B_{10}H_{14}$, B_4H_{10}, and B_5H_{11}. *J. Chem. Phys.* **1957**, *27*, 209–211. [CrossRef]
45. Brill, R.; Dietrich, H.; Dierks, H. Distribution of the bonding electrons in decaborane-14. *Angew. Chem. Int. Ed.* **1970**, *9*, 524–526. [CrossRef]
46. Brill, R.; Dietrich, H.; Dierks, H. Die Verteilung der Bindungselektronen im Dekaboran-molekül ($B_{10}H_{14}$). *Acta Cryst. B* **1971**, *27*, 2003–2018. [CrossRef]
47. Dietrich, H.; Scheringer, C. Refinement of the molecular charge distribution in decaborane(14). *Acta Cryst. B* **1978**, *34*, 54–63. [CrossRef]
48. Tippe, A.; Hamilton, W.C. Neutron diffraction study of decaborane. *Inorg. Chem.* **1969**, *8*, 464–470. [CrossRef]
49. Vilkov, L.V.; Mastryukov, V.S.; Akishin, P.A. An electron-diffraction study of the structure of the decaborane molecule in the vapor state. *J. Struct. Chem.* **1964**, *4*, 301–304. [CrossRef]
50. Mastryukov, V.S.; Dorofeeva, O.V.; Vilkov, L.V. Reexamination of the electron-diffraction data for the decaborane(14) molecule. *J. Struct. Chem.* **1975**, *16*, 110–112. [CrossRef]
51. Eberhardt, W.H.; Crawford, B.; Lipscomb, W.N. The valence structure of the boron hydrides. *J. Chem. Phys.* **1954**, *22*, 989–1001. [CrossRef]
52. Lipscomb, W.N. Valence in the boron hydrides. *J. Phys. Chem.* **1957**, *61*, 23–27. [CrossRef]
53. Lipscomb, W.N. *Boron Hydrides*; W. A. Benjamin, Inc.: New York, NY, USA, 1963.
54. Moore, E.B.; Lohr, L.L.; Lipscomb, W.N. Molecular orbitals in some boron compounds. *J. Chem. Phys.* **1961**, *35*, 1329–1334. [CrossRef]
55. Moore, E.B. Molecular orbitals in $B_{10}H_{14}$. *J. Chem. Phys.* **1962**, *37*, 675–677. [CrossRef]
56. Hoffmann, R.; Lipscomb, W.N. Boron hydrides: LCAO-MO and resonance studies. *J. Chem. Phys.* **1962**, *37*, 2872–2883. [CrossRef]
57. Laws, E.A.; Stevens, R.M.; Lipscomb, W.N. Self-consistent field study of decaborane(14). *J. Am. Chem. Soc.* **1972**, *94*, 4467–4474. [CrossRef]
58. Lipscomb, W.N. Advances in theoretical studies of boron hydrides and carboranes. In *Boron Hydride Chemistry*; Muetterties, E.L., Ed.; Academic Press: New York, NY, USA, 1975; pp. 39–78. [CrossRef]
59. Kononova, E.G.; Klemenkova, Z.S. The electronic structure of *nido*-$B_{10}H_{14}$ and [6-Ph-*nido*-6-CB_9H_{11}]$^-$ in terms of Bader's theory (AIM). *J. Mol. Struct.* **2013**, *1036*, 311–317. [CrossRef]
60. Shoolery, J.N. The relation of high resolution nuclear magnetic resonance spectra to molecular structures. *Discuss. Faraday Soc.* **1955**, *19*, 215–225. [CrossRef]
61. Williams, R.E.; Shapiro, I. Reinterpretation of nuclear magnetic resonance spectra of decaborane. *J. Chem. Phys.* **1958**, *29*, 677–678. [CrossRef]
62. Phillips, W.D.; Miller, H.C.; Muetterties, E.L. B^{11} Magnetic resonance study of boron compounds. *J. Am. Chem. Soc.* **1959**, *81*, 4496–4500. [CrossRef]
63. Pilling, R.L.; Tebbe, F.N.; Hawthorne, M.F.; Pier, E.A. Boron-11 nuclear magnetic resonance spectra of two boron hydride derivatives at 60 Mc./sec. *Proc. Chem. Soc.* **1964**, 402–403. [CrossRef]
64. Keller, P.C.; MacLean, D.; Schaeffer, R. Final assignment of B1 (3) N.M.R. resonance of decaborane. *Chem. Commun.* **1965**, 204. [CrossRef]
65. Williams, R.L.; Greenwood, N.N.; Morris, J.H. The nuclear magnetic resonance spectrum of decaborane-14. *Spectrochim. Acta* **1965**, *21*, 1579–1587. [CrossRef]
66. Bodner, G.M.; Sneddon, L.G. An assignment of the hydrogen-1 magnetic resonance spectrum of decaborane at 220 MHz. *Inorg. Chem.* **1970**, *9*, 1421–1423. [CrossRef]
67. Onak, T.; Marynick, D. Application of a ring current model to decaborane(14). A correlation of proton and boron-11 nuclear magnetic resonance chemical shifts. *Trans. Faraday Soc.* **1970**, *66*, 1843–1847. [CrossRef]
68. Greenwood, N.N.; Kennedy, J.D. N.M.R. evidence for a transition between isotropic and anisotropic thermal motion of *nido*-decaborane in an aromatic solvent. *J. Chem. Soc. Chem. Commun.* **1979**, 1099–1101. [CrossRef]
69. Colquhoun., I.J.; McFarlane, W. Heteronuclear two-dimensional nuclear magnetic resonance: Decaborane. *J. Chem. Soc. Dalton Trans.* **1981**, 2014–2016. [CrossRef]

70. Venable, T.L.; Hutton, W.C.; Grimes, R.N. Two-dimensional boron-11-boron-11 nuclear magnetic resonance spectroscopy as a probe of polyhedral structure: Application to boron hydrides, carboranes, metallaboranes, and metallacarboranes. *J. Am. Chem. Soc.* **1984**, *106*, 29–37. [CrossRef]
71. Keller, W.E.; Johnston, H.L. A note on the vibrational frequencies and the entropy of decaborane. *J. Chem. Phys.* **1952**, *20*, 1749–1751. [CrossRef]
72. Hanousek, F.; Štíbr, B.; Heřmánek, S.; Plešek, J. Chemistry of boranes. XXXI. Infrared spectra of decaborane and its deuterio derivatives. *Collect. Czech. Chem. Commun.* **1973**, *38*, 1312–1320. [CrossRef]
73. Pimentel, G.C.; Pitzer, K.S. The ultraviolet absorption and luminescence of decaborane. *J. Chem. Phys.* **1949**, *17*, 882–884. [CrossRef]
74. Pondy, P.R.; Beachell, H.C. Near infrared spectra of diborane, pentaborane, and decaborane. *J. Chem. Phys.* **1956**, *25*, 238–241. [CrossRef]
75. Haaland, A.; Eberhardt, W.H. Electronic spectrum of decaborane. *J. Chem. Phys.* **1962**, *36*, 2386–2392. [CrossRef]
76. Lötz, A.; Olliges, J.; Voitländer, J. The nuclear quadrupole double resonance spectrum of decaborane. *Chem. Phys. Lett.* **1982**, *93*, 560–563. [CrossRef]
77. Hiyama, Y.; Butler, L.G.; Brown, T.L. Boron-10 and boron-11 nuclear quadrupole resonance spectrum of decaborane[14]. *J. Magn. Res.* **1985**, *65*, 472–480. [CrossRef]
78. Lloyd, D.R.; Lynaugh, N.; Roberts, P.J.; Guest, M.F. Photoelectron studies of boron compouds. Part 5. Higher boron hydrides B_4H_{10}, B_5H_9 and $B_{10}H_{14}$. *J. Chem. Soc. Faraday Trans. 2* **1975**, *71*, 1382–1394. [CrossRef]
79. Hitchcock, A.P.; Wen, A.T.; Lee, S.; Glass, J.A.; Spencer, J.T.; Dowben, P.A. Inner-shell excitation of boranes and carboranes. *J. Phys. Chem.* **1993**, *97*, 8171–8181. [CrossRef]
80. Margrave, J.L. Ionization potentials of B_5H_9, B_5H_8I, $B_{10}H_{14}$, and $B_{10}H_{13}C_2H_5$ from electron impact studies. *J. Chem. Phys.* **1960**, *32*, 1889. [CrossRef]
81. Kaufman, J.J.; Koski, W.S.; Kuhns, L.J.; Law, R.W. Appearance and ionization potentials of selected fragments from decaborane, $B^{11}{}_{10}H_{14}{}^1$. *J. Am. Chem. Soc.* **1962**, *84*, 4198–4205. [CrossRef]
82. Bottei, R.S.; Laubengayer, A.W. The dipole moment and magnetic susceptibility of decaborane. *J. Phys. Chem.* **1962**, *66*, 1449–1451. [CrossRef]
83. Johnson, W.H.; Kilday, M.V.; Prosen, E.J. Heat of formation of decaborane. *J. Res. Natl. Bur. Stand. A Phys. Chem.* **1960**, *64*, 521–525. [CrossRef] [PubMed]
84. Kerr, E.C.; Hallett, N.C.; Johnston, H.L. Low temperature heat capacity of inorganic solids. VI. The heat capacity of decaborane, $B_{10}H_{14}$, from 14 to 305 K. *J. Am. Chem. Soc.* **1951**, *73*, 1117–1119. [CrossRef]
85. Furukawa, G.T.; Park, R.P. Heat capacity, heats of fusion and vaporization, and vapor pressure of decaborane ($B_{10}H_{14}$). *J. Res. Natl. Bur. Stand.* **1955**, *55*, 255–260. [CrossRef]
86. Miller, G.A. The vapor pressure of solid decaborane. *J. Phys. Chem.* **1963**, *67*, 1363–1364. [CrossRef]
87. Nakano, S.; Hemley, R.J.; Gregoryanz, E.A.; Goncharov, A.F.; Mao, H.-K. Pressure-induced transformations of molecular boron hydride. *J. Phys. Condens. Matter* **2002**, *14*, 10453. [CrossRef]
88. Rozendaal, H.M. Clinical observation on the toxicology of boron hydrides. *AMA Arch. Ind. Hyg. Occup. Med.* **1951**, *4*, 257–260. Available online: https://archive.org/details/sim_a-m-a-archives-of-industrial-health_1951-09_4_3/page/256/mode/2up (accessed on 31 July 2023).
89. Krackow, E.H. Toxicity and health hazards of boron hydrides. *AMA Arch. Ind. Hyg. Occup. Med.* **1953**, *8*, 335–339. Available online: https://archive.org/details/sim_a-m-a-archives-of-industrial-health_1953-10_8_4/page/334/mode/2up (accessed on 31 July 2023).
90. Cole, V.V.; Hill, D.L.; Oikemus, A.H. Problems in the study of decaborane and possible therapy of its poisoning. *AMA Arch. Ind. Hyg. Occup. Med.* **1954**, *10*, 158–161. Available online: https://archive.org/details/sim_a-m-a-archives-of-industrial-health_1954-08_10_2/page/158/mode/2up (accessed on 31 July 2023).
91. Lowe, H.; Freeman, G. Boron hydride (borane) intoxication in man. *AMA Arch. Ind. Hyg. Occup. Med.* **1957**, *12*, 523–533. Available online: https://archive.org/details/sim_a-m-a-archives-of-industrial-health_1957-12_16_6/page/522/mode/2up (accessed on 31 July 2023). [CrossRef]
92. Cordasco, E.M.; Cooper, R.W.; Murphy, J.V.; Anderson, C. Pulmonary aspects of some toxic experimental space fuels. *Dis. Chest* **1962**, *41*, 68–72. [CrossRef]
93. Merritt, J.H. Pharmacology and toxicology of propellant fuels: Boron hydrides. *Aeromed. Rev.* **1966**, *3*, 1–11.
94. Svirbely, J.L. Acute toxicity studies of decaborane and pentaborane by inhalation. *AMA Arch. Ind. Hyg. Occup. Med.* **1954**, *10*, 298–304. Available online: https://archive.org/details/sim_a-m-a-archives-of-industrial-health_1954-10_10_4/page/298/mode/2up (accessed on 31 July 2023).
95. Svirbely, J.L. Subacute toxicity of decaborane and pentaborane vapors. *AMA Arch. Ind. Hyg. Occup. Med.* **1954**, *10*, 305–311. Available online: https://archive.org/details/sim_a-m-a-archives-of-industrial-health_1954-10_10_4/page/304/mode/2up (accessed on 31 July 2023).
96. Svirbely, J.L. Toxicity tests of decaborane for laboratory animals. I. Acute toxicity studies. *AMA Arch. Ind. Hyg. Occup. Med.* **1955**, *11*, 132–137. Available online: https://archive.org/details/sim_a-m-a-archives-of-industrial-health_1955-02_11_2/page/132/mode/2up (accessed on 31 July 2023).

97. Svirbely, J.L. Toxicity tests of decaborane for laboratory animals. I. Effect of repeated doses. *AMA Arch. Ind. Hyg. Occup. Med.* **1955**, *11*, 138–141. Available online: https://archive.org/details/sim_a-m-a-archives-of-industrial-health_1955-02_11_2/page/138/mode/2up (accessed on 31 July 2023).

98. Walton, R.P.; Richardson, J.A.; Brodie, O.J. Cardiovascular actions of decaborane. *J. Pharmacol. Exp. Ther.* **1955**, *114*, 367–378.

99. Lamberti, J.M. Review of the toxicological properties of pentaborane, diborane, decaborane, and boric acid. *NASA Tech. Rep.* **1956**, NACA-RM-E56H13a. Available online: https://ntrs.nasa.gov/api/citations/19930090289/downloads/19930090289.pdf (accessed on 31 July 2023).

100. Roush, G. The toxicology of the boranes. *J. Occup. Med.* **1959**, *1*, 46–52. Available online: https://www.jstor.org/stable/44999044 (accessed on 31 July 2023). [CrossRef]

101. Fabre, R.; Chary, R.; Bocquet, P.; Jayot, R. Etat actuel de la toxicologie des hydrides de bore. *Arch. Mal. Prof. Med. Trav. Secur. Soc.* **1959**, *20*, 701–712.

102. Miller, D.F.; Tamas, A.; Robinson, L.; Merriweather, E. Observations on experimental boron hydride exposures. *Toxicol. Appl. Pharmacol.* **1960**, *2*, 430–440. [CrossRef]

103. Miller, D.F.; Tamas, A.A.; Robinson, L.; Merriweather, E. Cumulative effects of borane toxicity as revealed by a clinical test. *Tech. Rep. Aerospace Med. Res. Lab.* **1960**, WADD-60-604. Available online: https://web.archive.org/web/20181030113902/http://www.dtic.mil/dtic/tr/fulltext/u2/247355.pdf (accessed on 31 July 2023).

104. Delgado, J.M. Effects of decaborane on brain activity. *Tech. Rep. Aerospace Med. Res. Lab.* **1963**, AMRL-TDR-63-41. Available online: https://web.archive.org/web/20180726163136/http://www.dtic.mil/dtic/tr/fulltext/u2/411769.pdf (accessed on 31 July 2023).

105. Delgado, J.M.R.; Back, K.C.; Tamas, A.A. The effect of boranes on the monkey brain. *Arch. Int. Pharmacodyn. Thérap.* **1963**, *141*, 262–270.

106. Lalli, G. Sulla tossicità di alcuni propellenti per missile. *Ann. Geophys.* **1963**, *16*, 385–406. [CrossRef]

107. Merrit, J.A. Methylene blue in the treatment of decaborane toxicity. *Arch. Environ. Health* **1965**, *10*, 452–454. [CrossRef]

108. Reynolds, H.H.; Back, K.C. Effect of decaborane injection on operant behavior of monkeys. *Toxicol. Appl. Pharmacol.* **1966**, *8*, 197–209. [CrossRef]

109. Fairchild, M.D.; Sterman, M.B.; McRae, G.L. The effects of decaborane on cerebral electrical activity and locomotor behavior in the cat. *Tech. Rep. Aerosp. Med. Res. Lab.* **1972**, AMRL-TR-72-80. Available online: https://web.archive.org/web/20180724164936/http://www.dtic.mil/dtic/tr/fulltext/u2/756526.pdf (accessed on 31 July 2023).

110. Tadepalli, A.S.; Buckley, J.P. Cardiac and peripheral vascular effects of decaborane. *Toxicol. Appl. Pharmacol.* **1974**, *29*, 210–222. [CrossRef]

111. Dekaboran. *The MAK-Collection for Occupational Health and Safety*; Greim, G., Ed.; Wiley-VCH Verlag: Berlin, Germany, 2002; Volume 1. [CrossRef]

112. Merritt, J.H.; Schultz, E.J.; Wykes, A.A. Effect of decaborane on the norepinephrine content of rat brain. *Biochem. Pharmacol.* **1964**, *13*, 1364–1366. [CrossRef]

113. Oliverio, A. Release of cardiac noradrenaline by decaborane in the heart-lung preparation of guinea pig. *Biochem. Pharmacol.* **1965**, *14*, 1689–1692. [CrossRef]

114. von Euler, U.S.; Lishajko, F. Stereospecific catecholamine uptake in rabbit hearts depleted by decaborane. *Int. J. Neuropharmacol.* **1965**, *4*, 273–280. [CrossRef]

115. von Euler, U.S.; Lishajko, F. Catecholamine depletion and uptake in adrenergic nerve vesicles and in rabbit organs after decaborane. *Acta Physiol. Scand.* **1965**, *65*, 324–330. [CrossRef]

116. Byodeman, S.; von Euler, U.S. Neurotransmitter deficiency and reloading in noradrenaline depleted rabbits. *Acta Physiol. Scand.* **1966**, *66*, 134–140. [CrossRef]

117. Merritt, J.H.; Schultz, E.J. The effect of decaborane on the biosynthesis and metabolism of norepinephrine in the rat brain. *Life Sci.* **1966**, *5*, 27–32. [CrossRef]

118. Johnson, D.G. The effect of cold exposure on the catecholamine excretion of rats treated with decaborane. *Acta Physiol. Scand.* **1966**, *68*, 129–133. [CrossRef]

119. Merritt, J.H.; Sulkowski, T.S. Inhibition of aromatic L-amino acid decarboxylation by decaborane. *Biochem. Pharmacol.* **1967**, *16*, 369–373. [CrossRef]

120. Bhattacharya, I.C. Uptake of noradrenaline in the isolated perfused rat heart after depletion with decaborane. *Acta Physiol. Scand.* **1968**, *73*, 128–138. [CrossRef]

121. Medina, M.A.; Landez, J.H.; Foster, L.L. Inhibition of tissue histamine formation by decaborane. *J. Pharmacol. Exp. Ther.* **1969**, *169*, 132–137. Available online: https://jpet.aspetjournals.org/content/169/1/132 (accessed on 31 July 2023).

122. Scott, W.N.; Landez, J.H.; Cole, H.D. Effects of boranes upon tissues of the rat. I. Aspartate aminotransferase and lactic dehydrogenase. *Proc. Soc. Exp. Biol. Med.* **1970**, *134*, 348–352. [CrossRef]

123. Korty, P.; Scott, W.N. Effects of boranes upon tissues of the rat. II. Tissue amino acid content in rats on a normal diet. *Proc. Soc. Exp. Biol. Med.* **1970**, *135*, 629–632. [CrossRef]

124. Landez, J.H.; Scott, W.N. Effects of boranes upon tissues of the rat. III. Tissue amino acids in rats on a pyridoxine-deficient diet. *Proc. Soc. Exp. Biol. Med.* **1971**, *136*, 1389–1393. [CrossRef]

125. Malmfors, T.; von Euler, U.S. Depletion and repletion of noradrenaline in adrenergic nerves of the rat after decaborane treatment. *Experientia* **1971**, *27*, 417–419. [CrossRef]

126. Shahab, L.; Lishajko, F.; von Euler, U.S. Differentiated storage mechanisms for noradrenaline and dopamine in the rabbit heart. *Neuropharmacology* **1971**, *10*, 765–769. [CrossRef]

127. Menon, M.; Clark, W.G.; Aures, D. Effect of tremorine, oxotremorine and decaborane on brain histamine levels in rats. *Pharmacol. Res. Commun.* **1971**, *3*, 345–350. [CrossRef]

128. Schayer, R.W.; Reilly, M.A. Effect of decaborane on histamine formation in mice. *J. Pharmacol. Exp. Ther.* **1971**, *177*, 177–180. Available online: https://jpet.aspetjournals.org/content/177/1/177 (accessed on 31 July 2023).

129. Naeger, L.L.; Leibman, K.C. Mechanisms of decaborane toxicity. *Toxicol. Appl. Pharmacol.* **1972**, *22*, 517–527. [CrossRef]

130. Valerino, D.M.; Soliman, M.R.I.; Aurori, K.C.; Tripp, S.L.; Wykes, A.A.; Vesell, E.S. Studies on the interaction of several boron hydrides with liver microsomal enzymes. *Toxicol. Appl. Pharmacol.* **1974**, *29*, 358–366. [CrossRef]

131. Merritt, J.A.; Meyer, H.C.; Greenberg, R.I.; Tanton, G.A. The production of decaborane-14 from diborane by laser induced chemistry. *Propellants Explos. Pyrotech.* **1979**, *4*, 78–82. [CrossRef]

132. Dunks, G.B.; Palmer Ordonez, K. A one-step synthesis of $B_{11}H_{14}^-$ ion from $NaBH_4$. *Inorg. Chem.* **1978**, *17*, 1514–1516. [CrossRef]

133. Dunks, G.B.; Palmer Ordonez, K. A simplified preparation of $B_{10}H_{14}$ from $NaBH_4$. *Inorg. Chem.* **1978**, *17*, 2555–2556. [CrossRef]

134. Dunks, G.B.; Barker, K.; Hedaya, E.; Hefner, C.; Palmer-Ordonez, K.; Remec, P. Simplified synthesis of $B_{10}H_{14}$ from $NaBH_4$ via $B_{11}H_{14}^-$ ion. *Inorg. Chem.* **1981**, *20*, 1692–1697. [CrossRef]

135. Belov, P.P.; Storozhenko, P.A.; Voloshina, N.S.; Kuznetsova, M.G. Synthesis of decaborane by the reaction of sodium undecaborate with mild organic oxidants. *Russ. J. Appl. Chem.* **2018**, *90*, 1804–1809. [CrossRef]

136. Voloshina, N.S.; Belov, P.P.; Storozhenko, P.A.; Shebashova, N.M.; Kozlova, E.E.; Egorova, N.V.; Kuznetsova, M.G.; Gurkova, E.L. Specific features of oxidation of sodium tetradecahydroundecaborate to decaborane with manganese dioxide. *Russ. J. Appl. Chem.* **2020**, *93*, 807–812. [CrossRef]

137. Mongeot, H.; Atchekzai, H.R. Opening of the $B_{10}H_{10}^{2-}$ cage to give $B_{10}H_{14}$. *Z. Naturforsch. B* **1981**, *36*, 313–314. [CrossRef]

138. Guter, G.A.; Schaeffer, G.W. The strong acid behavior of decaborane. *J. Am. Chem. Soc.* **1956**, *78*, 3546. [CrossRef]

139. Norment, H.J. Unit cells and space groups for two etherates of sodium tridecahydrodecaborate(1-). *Acta Cryst.* **1959**, *12*, 695. [CrossRef]

140. Greenwood, N.N.; Sharrocks, D.N. Decaborane anions and the synthesis of polyhedral borane complexes of mercury(II) and cobalt(II). *J. Chem. Soc. A* **1969**, 2334–2338. [CrossRef]

141. Hawthorne, M.F.; Pitochelli, A.R.; Strahm, R.D.; Miller, J.J. The preparation and characterization of salts which contain the $B_{10}H_{13}$ anion. *J. Am. Chem. Soc.* **1960**, *82*, 1825–1829. [CrossRef]

142. Hawthorne, M.F. The reaction of phosphine methylenes with boron hydrides. *J. Am. Chem. Soc.* **1958**, *80*, 3480–3481. [CrossRef]

143. Onak, T.; Rosendo, H.; Siwapinyoyos, G.; Kubo, R.; Liauw, L. Reaction of 1,8-bis(dimethylamino)naphthalene, a highly basic and weakly nucleophilic amine, with several polyboranes and with boron trifluoride. *Inorg. Chem.* **1979**, *18*, 2943–2945. [CrossRef]

144. Pérez, S.; Sanz Miguel, P.J.; Macías, R. Decaborane anion tautomerism: Ion pairing and proton transfer control. *Dalton Trans.* **2018**, *47*, 5850–5859. [CrossRef] [PubMed]

145. Heřmánek, S.; Plotová, H.; Plešek, J. On the acidity characteristics of decaborane(14) and its benzyl derivatives in organic solvent-water systems. *Collect. Czech. Chem. Commun.* **1975**, *40*, 3593–3601. [CrossRef]

146. McCrary, P.D.; Barber, P.S.; Kelley, S.P.; Rogers, R.D. Nonaborane and decaborane cluster anions can enhance the ignition delay in hypergolic ionic liquids and induce hypergolicity in molecular solvents. *Inorg. Chem.* **2014**, *53*, 4770–4776. [CrossRef] [PubMed]

147. Sneddon, L.G.; Huffman, J.C.; Schaeffer, R.O.; Streib, W.E. Structure of the $B_{10}H_{13}^-$ ion. *J. Chem. Soc. Chem. Commun.* **1972**, 474–475. [CrossRef]

148. Wynd, A.J.; Welch, A.J. Structure of $[PhCH_2NMe_3]^+[B_{10}H_{13}]^-$. *Acta Cryst. C* **1989**, *45*, 615–617. [CrossRef]

149. Siedle, A.R.; Bodner, G.M.; Todd, L.J. Studies in boron hydrides—V: Assignment of the ^{11}B NMR spectrum of the tridecahydro decaborate(1−) ion. *J. Inorg. Nucl. Chem.* **1971**, *33*, 3671–3676. [CrossRef]

150. Wilks, P.H.; Carter, J.C. Preparation and properties of sodium decaboranate(12,2-). *J. Am. Chem. Soc.* **1966**, *88*, 3441. [CrossRef]

151. Bridges, A.N.; Gaines, D.F. The dianion of *nido*-decaborane(14), *nido*-dodecahydrodecaborate(2-), $[B_{10}H_{12}^{2-}]$, and its solution behavior. *Inorg. Chem.* **1995**, *34*, 4523–4524. [CrossRef]

152. Hofmann, M.; von Ragué Schleyer, P. Structures of *arachno*- and *hypho*-B_{10} clusters and stability of their possible Lewis base adducts ($[B_{10}H_{12}]^{2-}$, $[B_{10}H_{12}\cdot L]^{2-}$, $[B_{10}H_{12}\cdot 2L]^{2-}$, $[B_{10}H_{13}]^-$, $[B_{10}H_{13}\cdot L]^-$, $[B_{10}H_{12}\cdot 2L]$). An *ab initio*/IGLO/NMR investigation. *Inorg. Chem.* **1998**, *37*, 5557–5565. [CrossRef]

153. Muetterties, E.L. Chemistry of boranes. VI. Preparation and structure of $B_{10}H_{14}^{2-}$. *Inorg. Chem.* **1963**, *2*, 647–648. [CrossRef]

154. Kendall, D.S.; Lipscomb, W.N. Crystal structure of tetramethylammonium tetradecahydrodecaborate. Structure of the $B_{10}H_{14}^{2-}$ ion. *Inorg. Chem.* **1973**, *12*, 546–551. [CrossRef]

155. Lipscomb, W.N.; Wiersema, R.J.; Hawthorne, M.F. Structural ambiguity of the $B_{10}H_{14}^{2-}$ ion. *Inorg. Chem.* **1972**, *11*, 651–652. [CrossRef]

156. Schaeffer, R.; Tebbe, F. Formation of $B_{10}H_{15}^-$ as an intermediate in borohydride attack on decaborane-14. *Inorg. Chem.* **1964**, *3*, 1638–1640. [CrossRef]

157. Rietz, R.R.; Siedle, A.R.; Schaeffer, R.O.; Todd, L.J. High-resolution nuclear magnetic resonance study of the pentadecahydro-decaborate(1-) ion. *Inorg. Chem.* **1973**, *12*, 2100–2102. [CrossRef]

158. Zahkarkin, L.I.; Stanko, V.I.; Chapovskii, Y.A. Reactions of acetals and ortho-ethers with decaborane and diacetonitrile decaborane. *Bull. Acad. Sci. USSR Div. Chem. Sci.* **1962**, *11*, 1048–1049. [CrossRef]

159. Lee, S.H.; Park, Y.J.; Yoon, C.M. Reductive etherification of aromatic aldehydes with decaborane. *Tetrahedron Lett.* **1999**, *40*, 6049–6050. [CrossRef]
160. Funke, U.; Jiay, H.; Fischer, S.; Scheunemann, M.; Steinbach, J. One-step reductive etherification of 4-[^{18}F]fluoro-benzaldehyde with decaborane. *J. Label Compd. Radiopharm.* **2006**, *49*, 745–755. [CrossRef]
161. Park, E.S.; Lee, J.H.; Kim, S.J.; Yoon, C.M. One-pot reductive amination of acetals with aromatic amines using decaborane ($B_{10}H_{14}$) in methanol. *Synth. Commun.* **2003**, *33*, 3387–3396. [CrossRef]
162. Toyosuke, T.; Tsuneo, M.; Katsunori, K.; Yuichi, I. The reaction of decaborane with carbonyl compounds. *Bull. Chem. Soc. Jpn.* **1978**, *51*, 1259–1260. [CrossRef]
163. Bae, J.W.; Lee, S.H.; Jung, Y.J.; Yoon, C.-O.M.; Yoon, C.M. Reduction of ketones to alcohols using a decaborane/pyrrolidine/cerium(III) chloride system in methanol. *Tetrahedron Lett.* **2001**, *42*, 2137–2139. [CrossRef]
164. Lee, S.H.; Nam, M.H.; Cho, M.Y.; Yoo, B.W.; Rhee, H.J.; Yoon, C.M. Chemoselective reduction of aldehydes using decaborane in aqueous solution. *Synth. Commun.* **2006**, *36*, 2469–2474. [CrossRef]
165. Lee, S.H.; Jung, Y.J.; Cho, Y.J.; Yoon, C.-O.M.; Hwang, H.-J.; Yoon, C.M. Dehalogenation of α-halocarbonyls using decaborane as a transfer hydrogen agent in methanol. *Synth. Commun.* **2001**, *31*, 2251–2254. [CrossRef]
166. Lee, S.H.; Park, J.Y.; Yoon, C.M. Hydrogenation of alkenes or alkynes using decaborane in methanol. *Tetrahedron Lett.* **2000**, *41*, 887–889. [CrossRef]
167. Hawthorne, M.F.; Miller, J.J. Deuterium exchange of decaborane with deuterium oxide and deuterium chloride. *J. Am. Chem. Soc.* **1958**, *80*, 754. [CrossRef]
168. Miller, J.J.; Hawthorne, M.F. The course of base-catalyzed hydrogen exchange in decaborane. *J. Am. Chem. Soc.* **1959**, *81*, 4501–4503. [CrossRef]
169. Dupont, J.A.; Hawthorne, M.F. The nature of the electrophilic deuterium exchange reaction of decaborane with deuterium chloride. *J. Am. Chem. Soc.* **1962**, *84*, 1804–1808. [CrossRef]
170. Dupont, J.A.; Hawthorne, M.F. Deuterium exchange of decaborane with deuterium chloride under electrophilic conditions. *J. Am. Chem. Soc.* **1959**, *81*, 4998–4999. [CrossRef]
171. Dopke, J.A.; Gaines, D.F. Deuteration of decaborane(14) via exchange with deuterated aromatic solvents. *Inorg. Chem.* **1999**, *38*, 4896–4897. [CrossRef]
172. Gaines, D.F.; Beall, H. Hydrogen−deuterium exchange in decaborane(14): Mechanistic studies. *Inorg. Chem.* **2000**, *39*, 1812–1813. [CrossRef] [PubMed]
173. Hillman, M. The chemistry of decaborane. Iodination studies. *J. Am. Chem. Soc.* **1960**, *82*, 1096–1099. [CrossRef]
174. Sprecher, R.F.; Aufderheide, B.E.; Luther, G.W., III; Carter, J.C. Boron-11 nuclear magnetic resonance chemical shift assignments for monohalogenated decaborane(14) isomers. *J. Am. Chem. Soc.* **1974**, *96*, 4404–4410. [CrossRef]
175. Stuchlík, J. Gas chromatographic separation of 1- and 2-halogenodecaboranes. *J. Chromatogr. A* **1973**, *81*, 142–143. [CrossRef]
176. Schaeffer, R.; Shoolery, J.N.; Jones, R. Structures of halogen substituted boranes. *J. Am. Chem. Soc.* **1958**, *80*, 2670–2673. [CrossRef]
177. Williams, R.E.; Onak, T.P. Boron-11 nuclear magnetic resonance spectra (32.1 Mc.) of alkylated derivatives of dicarbahexaborane(8) and 1-iododecaborane(14). *J. Am. Chem. Soc.* **1964**, *86*, 3159–3160. [CrossRef]
178. Hillman, M. The chemistry of decaborane. II. Iodination in solvent. *J. Inorg. Nucl. Chem.* **1960**, *12*, 384–385. [CrossRef]
179. Wallbridge, M.H.G.; Williams, J.; Williams, R.L. Boron hydride derivatives. Part XI. Iodination of decaborane. *J. Chem. Soc. A* **1967**, 132–133. [CrossRef]
180. Safronov, A.V.; Sevryugina, Y.V.; Jalisatgi, S.S.; Kennedy, R.D.; Barnes, C.L.; Hawthorne, M.F. Unfairly forgotten member of the iodocarborane family: Synthesis and structural characterization of 8-iodo-1,2-dicarba-*closo*-dodecaborane, its precursors, and derivatives. *Inorg. Chem.* **2012**, *51*, 2629–2637. [CrossRef] [PubMed]
181. Sequeira, A.; Hamilton, W.C. Crystal and molecular structure of monoiododecaborane. *Inorg. Chem.* **1967**, *6*, 1281–1286. [CrossRef]
182. Schaeffer, R. The molecular structure of $B_{10}H_{12}I_2$. *J. Am. Chem. Soc.* **1957**, *79*, 2726–2728. [CrossRef]
183. Plešek, J.; Štíbr, B.; Heřmánek, S. Chemistry of boranes. VI. The reaction of bis-dialkylsulphido-dodecahydrodecaboranes with hydrohalogens. General preparation of 6- (or 5-) halogentridecahydrodecaboranes. *Collect. Czech. Chem. Commun.* **1966**, *31*, 4744–4745. [CrossRef]
184. Sedmera, P.; Hanousek, F.; Samek, Z. Chemistry of boranes. XIII. Determination of structure of some halogenedecaboranes and oxido-bis-tridecahydrodecaborane by means of ^{11}B NMR and IR spectra. *Collect. Czech. Chem. Commun.* **1968**, *33*, 2169–2175. [CrossRef]
185. Štíbr, B.; Plešek, J.; Heřmánek, S. Chemistry of boranes. XV. Synthesis, properties, reactions and mechanism of formation of 5(6)-halogenotridecahydrodecaboranes. *Collect. Czech. Chem. Commun.* **1969**, *34*, 194–205. [CrossRef]
186. Ewing, W.C.; Carroll, P.J.; Sneddon, L.G. Crystallographic characterizations and new high-yield synthetic routes for the complete series of 6-X-$B_{10}H_{13}$ halodecaboranes (X = F, Cl, Br, I) via superacid-induced cage-opening reactions of *closo*-$B_{10}H_{10}^{2-}$. *Inorg. Chem.* **2008**, *47*, 8580–8582. [CrossRef]
187. Ewing, W.C.; Carroll, P.J.; Sneddon, L.G. Efficient Syntheses of 5-X-$B_{10}H_{13}$ halodecaboranes *via* the photochemical (X = I) and/or base-catalyzed (X = Cl, Br, I) isomerization reactions of 6-X-$B_{10}H_{13}$. *Inorg. Chem.* **2010**, *49*, 1983–1994. [CrossRef] [PubMed]
188. Zakharkin, L.I.; Kalinin, V.N. Halogenation of decaborane in the presence of aluminum chloride. *Zh. Obshch. Khim.* **1966**, *36*, 2160–2162.

189. Bonnetot, B.; Miele, P.; Naoufal, D.; Mongeot, H. The interaction of the $[B_{10}H_{10}]^{2-}$ cage with Lewis acids and the formation of decaborane derivatives by cage-opening reactions. *Collect. Czech. Chem. Commun.* **1997**, *62*, 1273–1278. [CrossRef]
190. Stuchlík, J.; Heřmánek, S.; Plešek, J.; Štíbr, B. Chemistry of boranes. XVIII. A preparative separation of halogenodecaboranes. The isolation of 1-, 2-, 5-, and 6-bromotridecahydrodecaboranes. *Collect. Czech. Chem. Commun.* **1970**, *35*, 339–343. [CrossRef]
191. Dupont, T.J.; Loffredo, R.E.; Haltiwanger, R.C.; Turner, C.A.; Norman, A.D. Oxidative cleavage of dimethylstanna-undecaborane: Preparation and structural characterization of 5,10-dibromodecaborane(14). *Inorg. Chem.* **1978**, *17*, 2062–2067. [CrossRef]
192. Hillman, M.; Mangold, D.J. Chlorodecaborane. *Inorg. Chem.* **1965**, *4*, 1356–1357. [CrossRef]
193. Williams, R.E.; Pier, E. Chlorodecaboranes identified as 1-$ClB_{10}H_{13}$ and 2-$ClB_{10}H_{13}$ by 64.2-Mc. ^{11}B nuclear magnetic resonance spectra. *Inorg. Chem.* **1965**, *4*, 1357–1358. [CrossRef]
194. Mongeot, H.; Atchekzai, J.; Bonnetot, B.; Colombier, M. Preparation du 6-$B_{10}H_{13}Cl$ a partir de melanges $AlCl_3$-$(Et_4N)_2B_{10}H_{10}$. *Bull. Soc. Chim. Fr.* **1987**, 75–77.
195. Bonnetot, B.; Aboukhassib, A.; Mongeot, H. Study of the interaction of $AlCl_3$ with the $B_{10}H_{10}^{2-}$ cage in the solid state. *Inorg. Chim. Acta* **1989**, *156*, 183–187. [CrossRef]
196. Hawthorne, M.F.; Miller, J.J. The alkoxylation of decaborane. *J. Am. Chem. Soc.* **1960**, *82*, 500. [CrossRef]
197. Norman, A.D.; Rosell, S.L. Evidence of terminal ethoxy group substitution in ethoxydecaborane(14). *Inorg. Chem.* **1969**, *8*, 2818–2820. [CrossRef]
198. Loffredo, R.E.; Drullinger, L.F.; A. Slater, J.A.; Turner, C.A.; Norman, A.D. Preparation and properties of 6-ethoxy-, 6-phenyl-, and 6-trimethylsiloxydecaborane(14). *Inorg. Chem.* **1976**, *15*, 478–480. [CrossRef]
199. Beachell, H.C.; Schar, W.C. The reaction of decaborane with substituted alcohols. *J. Am. Chem. Soc.* **1958**, *80*, 2943–2945. [CrossRef]
200. Amberger, E.; Leidl, P. Synthese von $(CH_3)_3SiB_{10}H_{13}$. *J. Organomet. Chem.* **1969**, *18*, 345–347. [CrossRef]
201. Ewing, W.C.; Carroll, P.J.; Sneddon, L.G. Syntheses and surprising regioselectivity of 5- and 6-substituted decaboranyl ethers via the nucleophilic attack of alcohols on 6- and 5-halodecaboranes. *Inorg. Chem.* **2011**, *50*, 4054–4064. [CrossRef]
202. Hawthorne, M.F.; Mavunkal, I.J.; Knobler, C.B. Electrophilic reactions of protonated *closo*-$B_{10}H_{10}^{2-}$ with arenes, alkane C-H bonds, and triflate ion forming aryl, alkyl, and triflate *nido*-6-X-$B_{10}H_{13}$ derivatives. *J. Am. Chem. Soc.* **1992**, *114*, 4427–4429. [CrossRef]
203. Bondarev, O.; Sevryugina, Y.V.; Jalisatgi, S.S.; Hawthorne, M.F. Acid-induced opening of [*closo*-$B_{10}H_{10}]^{2-}$ as a new route to 6-Substituted *nido*-$B_{10}H_{13}$ decaboranes and related carboranes. *Inorg. Chem.* **2012**, *51*, 9935–9942. [CrossRef]
204. Berkeley, E.R.; Ewing, W.C.; Carroll, P.J.; Sneddon, L.G. Synthesis, structural characterization, and reactivity studies of 5-CF_3SO_3-$B_{10}H_{13}$. *Inorg. Chem.* **2014**, *53*, 5348–5358. [CrossRef]
205. Wang, Y.; Han, H.; Kang, J.-X.; Peng, J.; Lu, X.; Cao, H.-J.; Liu, Z.; Chen, X. Silylium ion-mediated cage-opening functionalization of *closo*-$B_{10}H_{10}^{2-}$ salts. *Chem. Commun.* **2022**, *58*, 11933–11936. [CrossRef] [PubMed]
206. Naoufal, D.; Kodeih, M.; Cornu, D.; Miele, P. New method of synthesis of 6-hydroxy-*nido*-decaborane 6-$(OH)B_{10}H_{13}$ by cage opening of *closo*-$[B_{10}H_{10}]^{2-}$. *J. Organomet. Chem.* **2005**, *690*, 2787–2789. [CrossRef]
207. Greenwood, N.N.; Hails, M.J.; Kennedy, J.D.; McDonald, W.S. Reactions of 6,6′-bis(*nido*-decaboranyl) oxide and 6-hydroxy-*nido*-decaborane with dihalogenobis(phosphine) complexes of nickel, palladium, and platinum, and some related chemistry; nuclear magnetic resonance investigations and the crystal and molecular structures of bis(dimethylphosphine)-di-μ-(2,3,4-η^3-*nido*-hexaboranyl)-diplatinum(*Pt–Pt*), [Pt_2(μ-η^3-$B_6H_9)_2(PMe_2Ph)_2$], and of 2,4-dichloro-1,1-bis(dimethylphenylphosphine)-*closo*-1-nickeladecaborane, [$(PhMe_2P)_2NiB_9H_7Cl_2$]. *J. Chem. Soc. Dalton Trans.* **1985**, 953–972. [CrossRef]
208. Štíbr, B.; Plešek, J.; Hanousek, F.; Heřmánek, S. Chemistry of boranes. XXIII. Reaction of 6,9-bis(dialkylsulfido)-dodecahydrodecaboranes with mercuric salts. *Collect. Czech. Chem. Commun.* **1971**, *36*, 1794–1799. [CrossRef]
209. Kelley, S.P.; Rachiero, G.P.; Titi, H.M.; Rogers, R.D. New reactions for old ions: Cage rearrangements, hydrolysis, and two-electron reduction of *nido*-decaborane in neat 1-ethyl-3-methylimidazolium acetate. *ACS Omega* **2018**, *3*, 8491–8496. [CrossRef] [PubMed]
210. Plešek, J.; Heřmánek, S.; Štíbr, B. Chemistry of boranes. VIII. Synthesis and reactions of 6,6′-oxido-bis-tridecahydrodecaborane. *Collect. Czech. Chem. Commun.* **1968**, *33*, 691–698. [CrossRef]
211. Kennedy, J.D.; Greenwood, N.N. A proton and boron-11 NMR study of icosaborane oxide, $B_{20}H_{16}O$. *Inorg. Chim. Acta* **1980**, *38*, 93–96. [CrossRef]
212. Greenwood, N.N.; McDonald, W.S.; Spalding, T.R. Crystal and molecular structure of 6,6′-bis(*nido*-decaboranyl) oxide $(B_{10}H_{13})_2O$. *J. Chem. Soc. Dalton Trans.* **1980**, 1251–1252. [CrossRef]
213. Bonnetot, B.; Tangi, A.; Colombier, M.; Mongeot, H. Dehydration of $(H_3O)_2B_{10}H_{10}$: An improved preparation of icosaborane oxide, $(B_{10}H_{13})_2O$. *Inorg. Chim. Acta* **1985**, *105*, L15–L16. [CrossRef]
214. Pace, R.J.; Williams, J.; Williams, R.L. Boron hydride derivatives. Part VII. The characterisation of some decaborane derivatives of the type, $B_{10}H_{12}$,2M. *J. Chem. Soc.* **1961**, 2196–2204. [CrossRef]
215. Knoth, W.H.; Muetterties, E.L. Chemistry of boranes. II: Decaborane derivatives based on the $B_{10}H_{12}$ structural unit. *J. Inorg. Nucl. Chem.* **1961**, *20*, 66–72. [CrossRef]
216. Fein, M.M.; Green, J.; Bobinski, J.; Cohen, M.S. Reaction products from decaborane and amides. *Inorg. Chem.* **1965**, *4*, 583–584. [CrossRef]
217. Cragg, R.H.; Fortuin, M.S.; Greenwood, N.N. Complexes of decaborane. Part I. Ultraviolet spectra of some bis-(ligand) complexes containing phosphorus and sulphur. *J. Chem. Soc. A* **1970**, 1817–1821. [CrossRef]
218. Janoušek, Z.; Plešek, J.; Plzák, Z. Open cage boranes and heteroborane thiols: Syntheses, structures, and some properties. *Collect. Czech. Chem. Commun.* **1979**, *44*, 2904–2907.

219. Bould, J.; Macháček, J.; Londesborough, M.G.S.; Macías, R.; Kennedy, J.D.; Bastl, Z.; Rupper, P.; Baše, T. Decaborane thiols as building blocks for self-assembled monolayers on metal surfaces. *Inorg. Chem.* **2012**, *51*, 1685–1694. [CrossRef]

220. Zakharkin, L.I.; Stanko, V.I.; Okhlobystin, O.Y. Reaction of decaborane and pentaborane with mercaptans and sulfides. *Bull. Acad. Sci. USSR Div. Chem. Sci.* **1961**, *10*, 1942–1943. [CrossRef]

221. Hawthorne, M.F.; Pilling, R.L.; Grimes, R.N. The mechanism of $B_{10}H_{10}{}^{-2}$ formation from $B_{10}H_{12}$(ligand)$_2$ species. *J. Am. Chem. Soc.* **1967**, *89*, 1067–1074. [CrossRef]

222. Beall, H.; Gaines, D.F. Reactions of 6,9-bis(dimethyl sulfide)-decaborane(14), 6,9-[(CH$_3$)$_2$S]$_2$B$_{10}$H$_{12}$: Mechanistic considerations. *Inorg. Chem.* **1998**, *37*, 1420–1422. [CrossRef]

223. Sands, D.E.; Zalkin, A. The crystal structure of $B_{10}H_{12}[S(CH_3)_2]_2$. *Acta Cryst.* **1962**, *15*, 410–417. [CrossRef]

224. Yumatov, V.D.; Murakhtanov, V.V.; Volkov, V.V.; Il'inchik, E.A.; Volkov, O.V. Comparative study of electronic structure of 6,9-bis(dimethyl sulfide)-*arachno*-decaborane(12) $B_{10}H_{12}[S(CH_3)]_2$ in a series of sulfide derivatives by the X-ray emission method. *Russ. J. Inorg. Chem.* **1998**, *43*, 1557–1561.

225. Yumatov, V.D.; Il'inchik, E.A.; Mazalov, L.N.; Volkov, O.V.; Volkov, V.V. X-Ray and X-ray photoelectron spectroscopy studies of the electronic structure of borane derivatives. *J. Struct. Chem.* **2001**, *42*, 281–295. [CrossRef]

226. Il'inchik, E.A.; Volkov, V.V.; Mazalov, L.N. X-ray photoelectron spectroscopy of boron compounds. *J. Struct. Chem.* **2005**, *46*, 523–534. [CrossRef]

227. Volkov, V.V.; Ikorskii, V.N.; Dunaev, S.T. Diamagnetism of compounds of the $B_{10}H_{12}L_2$ series. *Bull. Acad. Sci. USSR Div. Chem. Sci.* **1988**, *37*, 845–848. [CrossRef]

228. Zakharkin, L.I.; Stanko, V.I.; Klimova, A.I. Exchange reactions of decaborane complexes of the type $B_{10}H_{12}[X]_2$. *Bull. Acad. Sci. USSR Div. Chem. Sci.* **1964**, *13*, 857–858. [CrossRef]

229. Graybill, B.M.; Hawthorne, M.F. The nature of the colored 6,9-bis-pyridine decaborane molecule, $B_{10}H_{12}Py_2$. *J. Am. Chem. Soc.* **1961**, *83*, 2673–2676. [CrossRef]

230. Marshall, M.D.; Hunt, R.M.; Hefferan, G.T.; Adams, R.M.; Makhlouf, J.M. Opening the $B_{10}H_{10}{}^{2-}$ cage to produce $B_{10}H_{12}(Et_2S)_2$. *J. Am. Chem. Soc.* **1967**, *89*, 3361–3362. [CrossRef]

231. Guillevic, G.; Dazord, J.; Mongeot, H.; Cueilleron, J. Improved conversion of potassium tetrahydroborate into bis(dialkylsulfide)-decaborane(12), $B_{10}H_{12}(R_2S)_2$, via bis(tetraethylammonium) decahydrodecaborate, $(Et_4N)_2B_{10}H_{10}$. *J. Chem. Res. (S)* **1978**, 402.

232. Wang, G.-C.; Lu, Y.-X.; Huang, X.-Y.; Dai, L.-X. A new method for the synthesis of bis(diethylsulfide)decaborane. *Acta Chim. Sin.* **1981**, *39*, 251–254. Available online: http://sioc-journal.cn/Jwk_hxxb/EN/Y1981/V39/I3/251 (accessed on 31 July 2023).

233. Heying, T.L.; Naar-Colin, C. Some chemistry of substituted decaboranes. *Inorg. Chem.* **1964**, *3*, 282–285. [CrossRef]

234. Ahmad, R.; Crook, J.E.; Greenwood, N.N.; Kennedy, J.D. Synthesis, reactions, and nuclear magnetic resonance studies of some substituted *arachno*-decaborane and *arachno*-nonaborane derivatives, and the isolation of novel polyhedral diplatinaboranes. Crystal and molecular structure of $[Pt_2(PMe_2Ph)_2(\eta^3\text{-}B_2H_5)(\eta^3\text{-}B_6H_9)]$. *J. Chem. Soc. Dalton Trans.* **1986**, 2433–2442. [CrossRef]

235. V. Petřiček, V.; Cisařova, I.; Subrtova, V. Structure of σ(+)-5-bromo-6,9-bis(dimethylsulphido)-*nido*-decaborane(12), $C_4H_{23}B_{10}BrS_2$, determined with a twinned crystal. *Acta Cryst. C* **1983**, *39*, 1070–1072. [CrossRef]

236. Bould, J.; Dörfler, U.; Thornton-Pett, M.; Kennedy, J.D. A rearrangement of the 10-boron *nido/arachno* decaboranyl cluster. *Inorg. Chem. Commun.* **2001**, *4*, 544–546. [CrossRef]

237. Bridges, A.N.; Powell, D.R.; Dopke, J.A.; Desper, J.M.; Gaines, D.F. Monoalkyldecaborane(14) syntheses via nucleophilic alkylation and hydroboration. *Inorg. Chem.* **1998**, *37*, 503–509. [CrossRef]

238. Bould, J.; Dörfler, U.; Rath, N.P.; Barton, L.; Kilner, C.A.; Londesborough, M.G.S.; Ormsby, D.L.; Kennedy, J.D. Macropolyhedral boron-containing cluster chemistry. A synthetic approach via the auto-fusion of [6,9-(SMe$_2$)$_2$-*arachno*-B$_{10}$H$_{12}$]. *Dalton Trans.* **2006**, 3752–3765. [CrossRef]

239. Beachell, H.C.; Hoffman, D.E. The reaction of decaborane with amines and related compounds. *J. Am. Chem. Soc.* **1962**, *84*, 180–182. [CrossRef]

240. Schaeffer, R. A new type of substituted borane. *J. Am. Chem. Soc.* **1957**, *79*, 1006–1007. [CrossRef]

241. van der Maas Reddy, J.; Lipscomb, W.N. Molecular structure of $B_{10}H_{12}(CH_3CN)_2$. *J. Am. Chem. Soc.* **1959**, *81*, 754. [CrossRef]

242. van der Maas Reddy, J.; Lipscomb, W.N. Molecular structure of $B_{10}H_{12}(CH_3CN)_2$. *J. Chem. Phys.* **1959**, *31*, 610–616. [CrossRef]

243. Mebs, S.; Kalinowski, R.; Grabowsky, S.; Förster, D.; Kickbusch, R.; Justus, E.; Morgenroth, W.; Paulmann, C.; Luger, P.; Gabel, D.; et al. Real-space indicators for chemical bonding. Experimental and theoretical electron density studies of four deltahedral boranes. *Inorg. Chem.* **2011**, *50*, 90–103. [CrossRef] [PubMed]

244. Beall, H. Icosahedral carboranes. XVII. Simplified preparation of *o*-carborane. *Inorg. Chem.* **1972**, *11*, 637–638. [CrossRef]

245. Hawthorne, M.F.; Pitochelli, A.R. Displacement reactions on the $B_{10}H_{12}$ unit. *J. Am. Chem. Soc.* **1958**, *80*, 6685. [CrossRef]

246. Hawthorne, M.F.; Pitochelli, A.R. The reactions of bis-acetonitrile decaborane with amines. *J. Am. Chem. Soc.* **1959**, *81*, 5519. [CrossRef]

247. Fein, M.M.; Bobinski, J.; Paustian, J.E.; Grafstein, D.; Cohen, M.S. Reaction of decaborane and its derivatives. II. Addition reactions of 6,9-bis(acetonitrile)decaborane with hydrazine. *Inorg. Chem.* **1965**, *4*, 422. [CrossRef]

248. Froehner, G.; Challis, K.; Gagnon, K.; Getman, T.D.; Luck, R.L. A re-investigation of the reactions of amines and alcohols with 6,9-bis-(acetonitrile)decaborane. *Synth. React. Inorg. Metal-Org. Nano-Metal Chem.* **2007**, *36*, 777–785. [CrossRef]

249. Stogniy, M.Y.; Erokhina, S.A.; Sivaev, I.B.; Bregadze, V.I. Nitrilium derivatives of polyhedral boron compounds (boranes, carboranes, metallacarboranes): Synthesis and reactivity. *Phosphorus Sulfur Silicon Relat. Elem.* **2019**, *194*, 983–988. [CrossRef]

250. Zhdanov, A.P.; Nelyubin, A.V.; Klyukin, I.N.; Selivanov, N.A.; Bortnikov, E.O.; Grigoriev, M.S.; Zhizhin, K.Y.; Kuznetsov, N.T. Nucleophilic addition reaction of secondary amines to acetonitrilium *closo*-decaborate [2-$B_{10}H_9NCCH_3$]$^-$. *Russ. J. Inorg. Chem.* **2019**, *64*, 841–846. [CrossRef]

251. Stogniy, M.Y.; Erokhina, S.A.; Suponitsky, K.Y.; Anisimov, A.A.; Godovikov, I.A.; Sivaev, I.B.; Bregadze, V.I. Synthesis of novel carboranyl amidines. *J. Organomet. Chem.* **2020**, *909*, 121111. [CrossRef]

252. Bogdanova, E.V.; Stogniy, M.Y.; Chekulaeva, L.A.; Anisimov, A.A.; Suponitsky, K.Y.; Sivaev, I.B.; Grin, M.A.; Mironov, A.F.; Bregadze, V.I. Synthesis and reactivity of propionitrilium derivatives of cobalt and iron bis(dicarbollides). *New J. Chem.* **2020**, *44*, 15836–15848. [CrossRef]

253. El Anwar, S.; Růžičková, Z.; Bavol, D.; Fojt, L.; Grüner, B. Tetrazole ring substitution at carbon and boron sites of the cobalt bis(dicarbollide) ion available via dipolar cycloadditions. *Inorg. Chem.* **2020**, *59*, 17430–17442. [CrossRef] [PubMed]

254. Bogdanova, E.V.; Stogniy, M.Y.; Suponitsky, K.Y.; Sivaev, I.B.; Bregadze, V.I. Synthesis of boronated amidines by addition of amines to nitrilium derivative of cobalt bis(dicarbollide). *Molecules* **2021**, *26*, 6544. [CrossRef] [PubMed]

255. Voinova, V.V.; Selivanov, N.A.; Plyushchenko, I.V.; Vokuev, M.F.; Bykov, A.Y.; Klyukin, I.N.; Novikov, A.S.; Zhdanov, A.P.; Grigoriev, M.S.; Rodin, I.A.; et al. Fused 1,2-diboraoxazoles based on *closo*-decaborate anion—Novel members of diboroheterocycle class. *Molecules* **2021**, *26*, 248. [CrossRef] [PubMed]

256. Stogniy, M.Y.; Erokhina, S.A.; Suponitsky, K.Y.; Markov, V.Y.; Sivaev, I.B. Synthesis and crystal structures of nickel(II) and palladium(II) complexes with *o*-carboranyl amidine ligands. *Dalton Trans.* **2021**, *50*, 4967–4975. [CrossRef]

257. Nelyubin, A.V.; Selivanov, N.A.; Bykov, A.Y.; Klyukin, I.N.; Novikov, A.S.; Zhdanov, A.P.; Karpechenko, N.Y.; Grigoriev, M.S.; Zhizhin, K.Y.; Kuznetsov, N.T. Primary amine nucleophilic addition to nitrilium *closo*-dodecaborate [$B_{12}H_{11}NCCH_3$]$^-$: A simple and effective route to the new BNCT drug design. *Int. J. Mol. Sci.* **2021**, *22*, 13391. [CrossRef]

258. Laskova, J.; Ananiev, I.; Kosenko, I.; Serdyukov, A.; Stogniy, M.; Sivaev, I.; Grin, M.; Semioshkin, A.; Bregadze, V.I. Nucleophilic addition reactions to nitrilium derivatives [$B_{12}H_{11}NCCH_3$]$^-$ and [$B_{12}H_{11}NCCH_2CH_3$]$^-$. Synthesis and structures of *closo*-dodecaborate-based iminols, amides and amidines. *Dalton Trans.* **2022**, *51*, 3051–3059. [CrossRef]

259. Williams, J.; Williams, R.L.; Wright, J.C. Boron hydride derivatives. Part IX. The reaction of decaborane with ammonia. *J. Chem. Soc.* **1963**, 5816–5824. [CrossRef]

260. Rozenberg, A.S.; Neehiporenko, G.N.; Alekseev, A.P. Decomposition of nitrogen-hydrogen complexes of decaborane(12). 1. The kinetics of the thermal decomposition of decaborane(12) diammoniate. *Bull. Acad. Sci. USSR Div. Chem. Sci.* **1978**, *27*, 284–287. [CrossRef]

261. Baidina, I.A.; Podberezskaya, N.V.; Alekseev, V.I.; Volkov, V.V.; Borisov, S.V. Crystal structure of 6,9-bis-aminododecahydro-*nido*-decaborane $B_{10}H_{12}(NH_3)_2$. *J. Struct. Chem.* **1978**, *19*, 476–479. [CrossRef]

262. Polyanskaya, T.M.; Volkov, V.V. Crystal and molecular structure of 6,9-bis(trimethylamino)-*nido*-decaborane(12) $B_{10}H_{12}[N(CH_3)_3]_2$. *J. Struct. Chem.* **1989**, *30*, 629–634. [CrossRef]

263. Baidina, I.A.; Podberezskaya, N.V.; Volkov, V.V.; Borisov, S.V. Crystal structure of 6,9-bis-(triethylamino)-*nido*-decaborane $B_{10}H_{12}[(C_2H_5)_3N]_2$. *J. Struct. Chem.* **1978**, *19*, 479–482. [CrossRef]

264. Volkov, V.V.; Il'inchik, E.A.; Khudorozhko, G.F.; Yumatov, V.D.; Mazalov, L.N. Comparative study of $B_{10}H_{12}(NH_3)_2$ and $B_{10}H_{12}(NEt_3)_2$. *Bull. Acad. Sci. USSR Div. Chem. Sci.* **1985**, *34*, 2079–2085. [CrossRef]

265. Rozenberg, A.S. Decomposition of nitrogen-hydrogen complexes of decaborane(12). 2. An IR study of the mechanism of decomposition of decaborane(12) diammoniate. *Bull. Acad. Sci. USSR Div. Chem. Sci.* **1978**, *27*, 287–290. [CrossRef]

266. Isaenko, L.I.; Myakishev, K.G.; Posnaya, I.S.; Volkov, V.V. About the thermal stability of several derivatives of boron hydrides. *Izv. Sib. Otd. Akad. Nauk SSSR Ser. Khim.* **1982**, 73–78.

267. Graybill, B.M.; Pitochelli, A.R.; Hawthorne, M.F. The preparation and reactions of $B_{10}H_{13}$(Ligand) anions. *Inorg. Chem.* **1962**, *1*, 622–626. [CrossRef]

268. Cendrowski-Guillaume, S.M.; O'Loughlin, J.L.; Pelczer, I.; Spencer, J.T. Reactivity of decaborane(14) with pyridine: Synthesis and characterization of the first 6,6-substituted isomer of *nido*-$B_{10}H_{14}$, 6,6-($C_5H_5N)_2B_{10}H_{12}$, and application of ^{11}B-^{11}B double-quantum NMR spectroscopy. *Inorg. Chem.* **1995**, *34*, 3935–3941. [CrossRef]

269. Bould, J.; Laromaine, A.; Bullen, N.J.; Viñas, C.; Thornton-Pett, M.; Sillanpää, R.; Kivekäs, R.; Kennedy, J.D.; Teixidor, F. Borane reaction chemistry. Alkyne insertion reactions into boron-containing clusters. Products from the thermolysis of [6,9-(2-HC≡C–$C_5H_4N)_2$-*arachno*-$B_{10}H_{12}$]. *Dalton Trans.* **2008**, 1552–1563. [CrossRef]

270. Volkov, V.V.; Myakishev, K.G.; Potapova, O.G.; Polyanskaya, T.M.; Dunaev, S.T.; Il'inchik, E.A.; Baidina, I.A.; Hudorozhko, G.F. Synthesis and physico-chemical study of 6,9-bis-pyridino-*nido*-decaborane(14). *Izv. Sib. Otd. Akad. Nauk SSSR Ser. Khim.* **1988**, *3*, 20–27.

271. Il'inchik, E.A.; Volkov, V.V.; Dunaev, S.T. Structural effects in the electron absorption and luminescence spectra of decaborane(14) derivatives $B_{10}H_{12}L_2$. *J. Struct. Chem.* **1996**, *37*, 51–57. [CrossRef]

272. Volkov, V.V.; Il'inchik, E.A.; Volkov, O.V.; Yuryeva, O.P. The luminescence of cluster derivatives of boron hydrides, and some applied aspects. *Chem. Sustain. Develop.* **2000**, *8*, 185–191. Available online: https://sibran.ru/upload/iblock/fed/the_luminescence_of_cluster_derivatives_of_boron_hydrides_and_some_applied_aspects.PDF (accessed on 31 July 2023).

273. Volkov, V.V.; Il'inchik, E.A.; Yur'eva, O.P.; Volkov, O.V. Luminescence of the derivatives of boranes and adducts of decaborane(14) of the $B_{10}H_{12}[Py(X)]_2$ type. *J. Appl. Spectr.* **2000**, *67*, 864–870. [CrossRef]

274. Volkov, V.V.; Myakishev, K.G.; Dunaev, S.T. Thermal transformations of $B_{10}H_{12}(NH_3)_2$, $B_{10}H_{12}(C_5H_5N)_2$, and $B_{10}H_{12}(C_9H_7N)_2$. *Bull. Acad. Sci. USSR Div. Chem. Sci.* **1988**, *37*, 2234–2236. [CrossRef]

275. Polyanskaya, T.M.; Volkov, V.V.; Andrianov, V.I.; Il'inchik, E.A. Structures of two modifications of 6,9-bis-pyridine-*nido*-decaborane(12). *Bull. Acad. Sci. USSR Div. Chem. Sci.* **1989**, *38*, 1751–1754. [CrossRef]

276. Londesborough, M.G.S.; Price, C.; Thornton-Pett, M.; Clegg, W.; Kennedy, J.D. Two potential pyridine-borane oligomer and polymer building blocks. Structural characterisation of $[NC_5H_4 \cdot C_5H_4N \cdot B_{10}H_{12} \cdot NC_5H_4 \cdot C_5H_4N]$ and $[Me_2S \cdot B_{10}H_{12} \cdot NC_4H_4N \cdot B_{10}H_{12} \cdot SMe_2]$ by conventional and synchrotron X-ray methods. *Inorg. Chem. Commun.* **1999**, *2*, 298–300. [CrossRef]

277. Genady, A.R.; Fayed, T.A.; Gabel, D. Synthesis, characterization, and spectrophotometric studies of novel fluorescent *arachno* decaborane and nonaborane clusters containing aza-distyrylbenzene derivatives. *J. Organomet. Chem.* **2008**, *693*, 1065–1072. [CrossRef]

278. Hawthorne, M.F.; Pilling, R.L.; Vasavada, R.C. The mechanism of ligand exchange with $B_{10}H_{12}(ligand)_2$ species. *J. Am. Chem. Soc.* **1967**, *89*, 1075–1078. [CrossRef]

279. Rachiero, G.P.; Titi, H.M.; Rogers, R.D. Versatility and remarkable hypergolicity of *exo*-6, *exo*-9 imidazole-substituted *nido*-decaborane. *Chem. Commun.* **2017**, *53*, 7736–7739. [CrossRef]

280. Fetter, N.R. Reaction of decaborane with 2-isopropyl- and 2-methyl-5-(2-chloroethyl) tetrazole. *Chem. Ind.* **1959**, 1548.

281. Kendall, D.S.; Lipscomb, W.N. Molecular structure and two crystal structures of 6-isothiocyanodecaborane, 6-$B_{10}H_{13}NCS$. *Inorg. Chem.* **1973**, *12*, 2915–2919. [CrossRef]

282. Müller, J.; Paetzold, P.; Boese, R. The reaction of decaborane with hydrazoic acid: A novel access to azaboranes. *Heteroat. Chem.* **1990**, *1*, 461–465. [CrossRef]

283. Il'inchik, E.A.; Dunaev, S.T.; Myakishev, K.G.; Asanov, I.P. On solvatation and thermal transformations of $B_{10}H_{12}(PPh_3)_2$. *Zh. Neorg. Khim.* **1994**, *39*, 1071–1074.

284. Polyanskaya, T.M.; Yumatov, V.D.; Volkov, V.V. Molecular and electronic structure of 6,9-$(PPh_3)_2$-*arachno*-$B_{10}H_{12}$. *Dokl. Chem.* **2003**, *390*, 144–147. [CrossRef]

285. Fontaine, X.L.R.; Kennedy, J.D. Identification of the *endo,exo* isomer of 6,9-$(PMe_2Ph)_2$-*arachno*-$B_{10}H_{12}$ by nuclear magnetic resonance spectroscopy. *J. Chem. Soc. Dalton Trans.* **1987**, 1573–1575. [CrossRef]

286. Dörfler, U.; McGrath, T.D.; Cooke, P.A.; Kennedy, J.D.; Thornton-Pett, M. *exo,endo* and *exo,exo* isomers of 6,9-$(PMe_2Ph)_2$-*arachno*-$B_{10}H_{12}$ and its halogenated derivatives. Molecular structures of *exo,endo*- and *exo,exo*-6,9-$(PMe_2Ph)_2$-*arachno*-$B_{10}H_{12}$ and *exo*-6,*endo*-9-$(PMe_2Ph)_2$-2-Br-*arachno*-$B_{10}H_{11}$. *J. Chem. Soc. Dalton Trans.* **1997**, 4739–4746. [CrossRef]

287. Zakharkin, L.I.; Stanko, V.I. Complexes of decaborane with organic phosphorus and arsenic compounds. *Bull. Acad. Sci. USSR Div. Chem. Sci.* **1961**, *10*, 1936–1937. [CrossRef]

288. Polak, R.J.; Heying, T.L. The preparation of phosphite and phosphinite decaboranes. *J. Org. Chem.* **1962**, *27*, 1483–1484. [CrossRef]

289. Stanko, V.I.; Klimova, A.I.; Zakharkin, L.I. Complexes of decaborane with trialkyl-, triaryl-, trialkyltrithiophosphites and trialkyl-, trialkyltrithioarsenites. *Bull. Acad. Sci. USSR Div. Chem. Sci.* **1962**, *11*, 856–857. [CrossRef]

290. Schroeder, H.; Reiner, J.R.; Heying, T.L. Chemistry of decaborane-phosphorus compounds. I. Nucleophilic substitutions of bis-(chlorodiphenylphosphine)-decaborane. *Inorg. Chem.* **1962**, *1*, 618–621. [CrossRef]

291. Schroeder, H.; Reiner, J.R.; Knowles, T.A. Chemistry of decaborane-phosphorus compounds. III. Decaborane-14-phosphine polymers. *Inorg. Chem.* **1963**, *2*, 393–396. [CrossRef]

292. Seyferth, D.; Rees, W.S.; Haggerty, J.S.; Lightfoot, A. Preparation of boron-containing ceramic materials by pyrolysis of the decaborane(14)-derived $[B_{10}H_{12} \cdot Ph_2POPPh_2]_x$ polymer. *Chem. Mater.* **1989**, *1*, 45–52. [CrossRef]

293. Packirisamy, S. Decaborane(14)-based polymers. *Progr. Polym. Sci.* **1996**, *21*, 707–773. [CrossRef]

294. Donaghy, K.J.; Carroll, P.J.; Sneddon, L.G. Reactions of 1,1'-bis(diphenylphosphino)ferrocene with boranes, thiaboranes, and carboranes. *Inorg. Chem.* **1997**, *36*, 547–553. [CrossRef]

295. Schroeder, H. Chemistry of decaborane-phosphorus compounds. II. Synthesis and reactions of diphenylphosphinodecaborane-14. *Inorg. Chem.* **1963**, *2*, 390–393. [CrossRef]

296. Friedman, L.B.; Perry, S.L. Crystal and molecular structure of 5,6-μ-diphenylphosphino-decaborane(14). *Inorg. Chem.* **1973**, *12*, 288–293. [CrossRef]

297. Muetterties, E.L.; Aftandilian, V.D. Chemistry of boranes. IV. Phosphine derivatives of $B_{10}H_{14}$ and B_9H_{15}. *Inorg. Chem.* **1962**, *1*, 731–734. [CrossRef]

298. Thornton-Pett, M.; Beckett, M.A.; Kennedy, J.D. Polyhedral phosphaborane chemistry: Crystal and molecular structure of the diphenylphosphido-bridged *arachno*-decaboranyl cluster compound $[PMePh_3][6,9-\mu-(PPh_2)B_{10}H_{12}]$. *J. Chem. Soc. Dalton Trans.* **1986**, 303–308. [CrossRef]

299. Miller, R.W.; Spencer, J.T. Small heteroborane cluster systems. 7. Reaction of phosphaalkyne *t*-BuCP with bis-(acetonitrile) decaborane(12). A new synthetic route to a large phosphaborane cluster compound. *Polyhedron* **1996**, *15*, 3151–3155. [CrossRef]

300. Miller, R.W.; Spencer, J.T. Small heteroborane cluster systems. 8. Preparation of phosphaborane clusters from the reaction of polyhedral boranes with low-coordinate phosphorus compounds: Reaction chemistry of phosphaalkynes with decaborane(14). *Organometallics* **1996**, *15*, 4293–4300. [CrossRef]

301. Williams, R.L.; Dunstan, I.; Blay, N.J. Boron hydride derivatives. Part IV. Friedel–Crafts methylation of decaborane. *J. Chem. Soc.* **1960**, 5006. [CrossRef]

302. Obenland, C.O.; Newberry, J.R.; Schreiner, W.L.; Bartoszek, E.J. Friedel-Crafts methylation of decaborane. *Ind. Eng. Chem. Prod. Res. Dev.* **1965**, *4*, 281–283. [CrossRef]
303. Polak, R.J.; Goodspeed, N.C. Catalyst study in methylation of decaborane. *Ind. Eng. Chem. Prod. Res. Dev.* **1965**, *4*, 158–160. [CrossRef]
304. Holub, J.; Růžička, A.; Růžičková, Z.; Fanfrlík, J.; Hnyk, D.; Štíbr, B. Electrophilic methylation of decaborane(14): Selective synthesis of tetramethylated and heptamethylated decaboranes and their conjugated bases. *Inorg. Chem.* **2020**, *59*, 10540–10547. [CrossRef] [PubMed]
305. Blay, N.J.; Dunstan, I.; Williams, R.L. Boron hydride derivatives. Part III. Electrophilic substitution in pentaborane and decaborane. *J. Chem. Soc.* **1960**, 430–433. [CrossRef]
306. Cueilleron, J.; Guillot, P. Preparation de quelques composes du decaborane. *Bull. Chim. Fr.* **1960**, 2044–2052.
307. Perloff, A. The crystal structure of 1-ethyldecaborane. *Acta Cryst.* **1964**, *17*, 332–338. [CrossRef]
308. Dunstan, I.; Williams, R.L.; Blay, N.J. Boron hydride derivatives. Part V. Nucleophilic substitution in decaborane. *J. Chem. Soc.* **1960**, 5012–5015. [CrossRef]
309. Dunstan, I.; Blay, N.J.; Williams, R.L. Boron hydride derivatives. Part VI. Decaborane Grignard reagent. *J. Chem. Soc.* **1960**, 5016–5019. [CrossRef]
310. Gallaghan, J.; Siegel, B. Grignard synthesis of alkyl decaboranes. *J. Am. Chem. Soc.* **1959**, *81*, 504. [CrossRef]
311. Siegel, B.; Mack, J.L.; Lowe, J.U.; Gallaghan, J. Decaborane Grignard reagents. *J. Am. Chem. Soc.* **1958**, *80*, 4523–4526. [CrossRef]
312. Palchak, R.J.F.; Norman, J.H.; Williams, R.E. Decaborane, "6-benzyl" $B_{10}H_{13}$ chemistry. *J. Am. Chem. Soc.* **1961**, *83*, 3380–3384. [CrossRef]
313. Gaines, D.F.; Bridges, A.N. New routes to monoalkyl decaborane(14) derivatives. *Organometallics* **1993**, *12*, 2015–2016. [CrossRef]
314. Tolpin, E.I.; Mizusawa, E.; Becker, D.S.; Venzel, J. Synthesis and chemistry of 9-cyclohexyl-5(7)-(dimethyl sulfide)-*nido*-decaborane(11), $B_{10}H_{11}C_6H_{11}S(CH_3)_2$. *Inorg. Chem.* **1980**, *19*, 1182–1187. [CrossRef]
315. Millan, M.D.; Davis, J.H. Hydroboration of (1R)-(+)-α-pinene and (1S)-(−)-β-pinene with $B_{10}H_{12}(SMe_2)_2$: A straightforward approach to the preparation of optically active 6-(alkyl)-*nido*-$B_{10}H_{13}$ derivatives. *Tetrahedron Asymmetry* **1998**, *9*, 709–712. [CrossRef]
316. Mizusawa, E.; Rudnick, S.E.; Eriks, K. The crystal and molecular structure of 9-cyclohexyl-5(7)-(dimethyl sulfide)-*nido*-decaborane(11), $B_{10}H_{11}C_6H_{11}S(CH_3)_2$. *Inorg. Chem.* **1980**, *19*, 1188–1191. [CrossRef]
317. Kusari, U.; Li, Y.; Bradley, M.G.; Sneddon, L.G. Polyborane reactions in ionic liquids: New efficient routes to functionalized decaborane and *o*-carborane clusters. *J. Am. Chem. Soc.* **2004**, *126*, 8662–8663. [CrossRef] [PubMed]
318. Kusari, U.; Carroll, P.J.; Sneddon, L.G. Ionic-liquid-promoted decaborane olefin-hydroboration: A new efficient route to 6-R-$B_{10}H_{13}$ derivatives. *Inorg. Chem.* **2008**, *47*, 9203–9215. [CrossRef]
319. Yu, X.-H.; Cao, K.; Huang, Y.; Yang, J.; Li, J.; Chang, G. Platinum catalyzed sequential hydroboration of decaborane: A facile approach to poly(alkenyldecaborane) with decaborane in the mainchain. *Chem. Commun.* **2014**, *50*, 4585–4587. [CrossRef]
320. Boggio, P.; Toppino, A.; Geninatti Crich, S.; Alberti, D.; Marabello, D.; Medana, C.; Prandi, C.; Venturello, P.; Aime, S.; Deagostino, A. The hydroboration reaction as a key for a straightforward synthesis of new MRI-NCT agents. *Org. Biomol. Chem.* **2015**, *13*, 3288–3297. [CrossRef]
321. Naoufal, D.; Laila, Z.; Yazbeck, O.; Hamad, H.; Ibrahim, G.; Aoun, R.; Safa, A.; El Jamal, M. Kanj Synthesis, characterization and mechanism of formation of 6-substituted *nido*-$B_{10}H_{13}$ decaboranes by the opening reaction of *closo*-decahydrodecaborate $[B_{10}H_{10}]^{2-}$ cage. *Main Group Chem.* **2013**, *12*, 39–48. [CrossRef]
322. Pender, M.J.; Wideman, T.; Carroll, P.J.; Sneddon, L.G. Transition metal promoted reactions of boron hydrides. 15. Titanium-catalyzed decaborane–olefin hydroborations: One-step, high-yield syntheses of monoalkyldecaboranes. *J. Am. Chem. Soc.* **1998**, *120*, 9108–9109. [CrossRef]
323. Pender, M.J.; Carroll, P.J.; Sneddon, L.G. Transition-metal-promoted reactions of boron hydrides. 17. Titanium-catalyzed decaborane–olefin hydroborations. *J. Am. Chem. Soc.* **2001**, *123*, 12222–12231. [CrossRef] [PubMed]
324. Wei, X.; Carroll, P.J.; Sneddon, L.G. New routes to organodecaborane polymers via ruthenium-catalyzed ring-opening metathesis polymerization. *Organometallics* **2004**, *23*, 163–165. [CrossRef]
325. Wei, X.; Carroll, P.J.; Sneddon, L.G. Ruthenium-catalyzed ring-opening polymerization syntheses of poly(organodecaboranes): New single-source boron-carbide precursors. *Chem. Mater.* **2006**, *18*, 1113–1123. [CrossRef]
326. Pender, M.J.; Forsthoefel, K.M.; Sneddon, L.G. Molecular and polymeric precursors to boron carbide nanofibers, nanocylinders, and nanoporous ceramics. *Pure Appl. Chem.* **2003**, *75*, 1287–1294. [CrossRef]
327. Pender, M.J.; Sneddon, L.G. An efficient template synthesis of aligned boron carbide nanofibers using a single-source molecular precursor. *Chem. Mater.* **2000**, *12*, 280–283. [CrossRef]
328. Li, J.; Yang, H.; Liang, Y.; Chen, J.; Xie, M.; Zhou, L.; Xiong, Q.; Li, W. Ruthenium-catalyzed cascade ring-opening polymerization/hydroboration for the synthesis of low cross-linking poly(6-norbornenyldecaborane) and its thermal property. *Adv. Mater. Res.* **2014**, *1035*, 288–291.
329. Zhang, X.; Li, J.; Cao, K.; Yi, Y.; Yang, J.; Li, B. Synthesis and characterization of B-C polymer hollow microspheres from a new organodecaborane preceramic polymer. *RSC Adv.* **2015**, *5*, 86214–86218. [CrossRef]
330. Wang, J.; Gou, Y.; Jian, K.; Huang, J.; Wang, H. Boron carbide ceramic hollow microspheres prepared from poly(6-CH_2=$CH(CH_2)_4$-$B_{10}H_{13}$) precursor. *Mater. Design* **2016**, *109*, 408–414. [CrossRef]

331. Wang, J.; Gou, Y.; Zhang, Q.; Jian, K.; Chen, Z.; Wang, H. Linear organodecaborane block copolymer as a single-source precursor for porous boron carbide ceramics. *J. Eur. Ceram. Soc.* **2017**, *37*, 1937–1943. [CrossRef]

332. Li, J.; Cao, K.; Li, J.; Liu, M.; Zhang, S.; Yang, J.; Zhang, Z.; Li, B. Synthesis and ceramic conversion of a new organodecaborane preceramic polymer with high-ceramic-yield. *Molecules* **2018**, *23*, 2461. [CrossRef]

333. Mazighi, K.; Carroll, P.J.; Sneddon, L.G. Transition metal promoted reactions of boron hydrides. 13. Platinum catalyzed synthesis of 6,9-dialkyldecaboranes. *Inorg. Chem.* **1993**, *32*, 1963–1969. [CrossRef]

334. Chatterjee, S.; Carroll, P.J.; Sneddon, L.G. Metal-catalyzed decaborane-alkyne hydroboration reactions: Efficient routes to alkenyldecaboranes. *Inorg. Chem.* **2010**, *49*, 3095–3097. [CrossRef] [PubMed]

335. Chatterjee, S.; Carroll, P.J.; Sneddon, L.G. Iridium and ruthenium catalyzed syntheses, hydroborations, and metathesis reactions of alkenyl-decaboranes. *Inorg. Chem.* **2013**, *52*, 9119–9130. [CrossRef] [PubMed]

336. Ernest, R.L.; Quintana, W.; Rosen, R.; Carroll, P.J.; Sneddon, L.G. Reactions of decaborane(14) with silylated acetylenes. Synthesis of the new monocarbon carborane 9-Me$_2$S-7-[(Me$_3$Si)$_2$CH]CB$_{10}$H$_{11}$. *Organometallics* **1987**, *6*, 80–88. [CrossRef]

337. Burgos-Adorno, G.; Carroll, P.J.; Quintana, W. Synthesis and characterization of a new alkenyldecaborane and alkenyl monocarbon carboranes. *Inorg. Chem.* **1996**, *35*, 2568–2575. [CrossRef]

338. Meyer, F.; Paetzold, P.; Englert, U. Reaktion von Decaboran mit dem Phosphaalkin PCtBu. *Chem. Ber.* **1994**, *127*, 93–95. [CrossRef]

339. Bould, J.; Londesborough, M.G.S.; Ormsby, D.L.; MacBride, J.A.H.; Wade, K.; Kilner, C.A.; Clegg, W.; Teat, S.J.; Thornton-Pett, M.; Greatrex, R.; et al. Macropolyhedral boron-containing cluster chemistry: Models for intermediates en route to globular and discoidal megaloborane assemblies. Structures of [*nido*-B$_{10}$H$_{12}$(*nido*-B$_5$H$_8$)$_2$] and [(CH$_2$CH$_2$C$_5$H$_4$N)-*arachno*-B$_{10}$H$_{10}$(NC$_5$H$_4$-*closo*-C$_2$B$_{10}$H$_{10}$)] as determined by synchrotron X-ray diffraction analysis. *J. Organomet. Chem.* **2002**, *657*, 256–261. [CrossRef]

340. Kim, S.; Treacy, J.W.; Nelson, Y.A.; Gonzalez, J.A.M.; Gembicky, M.; Houk, K.N.; Spokoyny, A.M. Arene C–H borylation strategy enabled by a non-classical boron cluster-based electrophile. *Nat. Commun.* **2023**, *14*, 1671. [CrossRef]

341. Demel, J.; Kloda, M.; Lang, K.; Škoch, K.; Hynek, J.; Opravil, A.; Novotný, M.; Bould, J.; Ehn, M.; Londesborough, M.G.S. Direct Phenylation of *nido*-B$_{10}$H$_{14}$. *J. Org. Chem.* **2022**, *87*, 10034–10043. [CrossRef]

342. Bůžek, D.; Škoch, K.; Ondrušová, S.; Kloda, M.; Bavol, D.; Mahun, A.; Kobera, L.; Lang, K.; Londesborough, M.G.S.; Demel, J. "Activated Borane"—A porous borane cluster network as an effective adsorbent for removing organic pollutants. *Chem. Eur. J.* **2022**, *28*, e202201885. [CrossRef]

343. Wille, A.E.; Su, K.; Carroll, P.J.; Sneddon, L.G. New synthetic routes to azacarborane clusters: Nitrile insertion reactions of *nido*-5,6-C$_2$B$_8$H$_{11}^-$ and *nido*-B$_{10}$H$_{13}^-$. *J. Am. Chem. Soc.* **1996**, *118*, 6407–6421. [CrossRef]

344. Baše, K.; Alcock, N.W.; Howarth, O.W.; Powell, H.R.; Harrison, A.T.; Wallbridge, M.G.H. The structure of the *arachno*-[B$_{10}$H$_{13}$C≡N]$^{2-}$ anion: An example of *endo* substitution in the decaborane(14) framework. *J. Chem. Soc. Chem. Commun.* **1988**, 341–342. [CrossRef]

345. Orlova, A.M.; Sivaev, I.B.; Lagun, V.L.; Katser, S.B.; Solntsev, K.A.; Kuznetsov, N.T. Synthesis and structure of Pb(Bipy)$_2$(6-B$_{10}$H$_{13}$CN). *Koord. Khim.* **1996**, *22*, 119–124.

346. Sivaev, I.B. Chemistry of 11-vertex polyhedral boron hydrides (Review). *Russ. J. Inorg. Chem.* **2019**, *64*, 955–976. [CrossRef]

347. Geisberger, G.; Linti, G.; Nöth, H. Beiträge zur Chemie des Bors. 213. Reaktionen eines Amino-imino-borans mit Triboran(7) und Decaboran(14). *Chem. Ber.* **1992**, *125*, 2691–2699. [CrossRef]

348. Bridges, A.N.; Liu, J.; Kultyshev, R.G.; Gaines, D.F.; Shore, S.G. Partial insertion of the 9-BBN unit into the *nido*-B$_{10}$ framework: Preparation and structural characterization of (9-BBN)B$_{10}$H$_{13}$ and [(9-BBN)B$_{10}$H$_{12}$]$^-$. *Inorg. Chem.* **1998**, *37*, 3276–3283. [CrossRef]

349. Greenwood, N.N.; Kennedy, J.D.; Taylorson, D. Mass spectroscopic evidence for icosaborane(26). *J. Phys. Chem.* **1978**, *82*, 623–625. [CrossRef]

350. Greenwood, N.N.; Kennedy, J.D.; Spalding, T.R.; Taylorson, D. Isomers of icosaborane(26): Some synthetic routes and preliminary characterisations in the bis(*nido*-decaboranyl) system. *J. Chem. Soc. Dalton Trans.* **1979**, 840–846. [CrossRef]

351. Boocock, S.K.; Cheek, Y.M.; Greenwood, N.N.; Kennedy, J.D. A new route to isomers of icosaborane(26), B$_{20}$H$_{26}$. The use of 115.5-MHz ^{11}B and ^{11}B-{^{1}H} nuclear magnetic resonance spectroscopy for the comparison and characterisation of separated isomers and the identification of three further icosaboranes as 1,2'-, 2,5'-, and 5,5'(or 5,7')-(B$_{10}$H$_{13}$)$_2$. *J. Chem. Soc. Dalton Trans.* **1981**, 1430–1437. [CrossRef]

352. Brown, G.M.; Pinson, J.W.; Ingram, L.L. Crystal and molecular structure of 1,5'-bidecaboran(14)yl: A new borane from γ irradiation of decaborane(14). *Inorg. Chem.* **1979**, *18*, 1951–1956. [CrossRef]

353. Bould, J.; Clegg, W.; Kennedy, J.D.; Teat, S.J. Isomeric icosaboranes B$_{20}$H$_{26}$: The synchrotron structure of 1,1'-bis(*nido*-decaboranyl). *Acta Cryst. C* **2001**, *57*, 779–780. [CrossRef]

354. Barrett, S.A.; Greenwood, N.N.; Kennedy, J.D.; Thornton-Pett, M. The chemistry of isomers of icosaborane(26): Crystal and molecular structure of 1,2'-bi(*nido*-decaboranyl). *Polyhedron* **1985**, *4*, 1981–1984. [CrossRef]

355. Greenwood, N.N.; Kennedy, J.D.; McDonald, W.S.; Staves, J.; Taylorson, D. Isomers of B$_{20}$H$_{26}$: Structural characterisation by X-ray diffraction of 2,2'-bi(*nido*-decaboranyl). *J. Chem. Soc. Chem. Commun.* **1979**, 17–18. [CrossRef]

356. Boocock, S.K.; Greenwood, N.N.; Kennedy, J.D.; McDonald, W.S.; Staves, J. The chemistry of isomeric icosaboranes, B$_{20}$H$_{26}$. Molecular structures and physical characterization of 2,2'-bi(*nido*-decaboranyl) and 2,6'-bi(*nido*-decaboranyl). *J. Chem. Soc. Dalton Trans.* **1980**, 790–796. [CrossRef]

357. Boocock, S.K.; Greenwood, N.N.; Kennedy, J.D.; Taylorson, D. Isomers of B$_{20}$H$_{26}$: Elucidation of the structure of 6,6'-bi(*nido*-decaboranyl) by ^{11}B-{^{1}H} and ^{1}H-{^{11}B} n.m.r. spectroscopy. *J. Chem. Soc. Chem. Commun.* **1979**, 16–17. [CrossRef]

358. Bould, J.; Dörfler, U.; Clegg, W.; Teat, S.J.; Thornton-Pett, M.; Kennedy, J.D. Triple linking of the decaboranyl cluster. Structure of [(SMe$_2$)$_2$B$_{10}$H$_{10}$(B$_{10}$H$_{13}$)$_2$] as determined by synchrotron X-ray diffraction analysis. *Chem. Commun.* **2001**, 1788–1789. [CrossRef]

359. Olsen, F.P.; Vasavada, R.C.; Hawthorne, M.F. The chemistry of *n*-B$_{18}$H$_{22}$ and *i*-B$_{18}$H$_{22}$. *J. Am. Chem. Soc.* **1968**, *90*, 3946–3951. [CrossRef]

360. Li, Y.; Sneddon, L.G. Improved synthetic route to *n*-B$_{18}$H$_{22}$. *Inorg. Chem.* **2006**, *45*, 470–471. [CrossRef]

361. Londesborough, M.G.S.; Hnyk, D.; Bould, J.; Bould, J.; Serrano-Andres, L.; Sauri, V.; Oliva, J.M.; Kubat, P.; Polivka, T.; Lang, K. Distinct photophysics of the isomers of B$_{18}$H$_{22}$ explained. *Inorg. Chem.* **2012**, *51*, 1471–1479. [CrossRef]

362. Londesborough, M.G.S.; Dolanský, J.; Braborec, J.; Cerdán, L. Interaction of *anti*-B$_{18}$H$_{22}$ with light. In *Handbook of Boron Science with Applications in Organometallics, Catalysis, Materials and Medicine*; Hosmane, N.S., Eagling, R., Eds.; World Scientific Publishing: London, UK, 2019; Volume 3, pp. 115–136. [CrossRef]

363. Cerdan, L.; Frances-Monerris, A.; Roca-Sanjuan, D.; Bould, J.; Dolandky, J.; Fuciman, M.; Londesborough, M.G.S. Unveiling the role of upper excited electronic states in the photochemistry and laser performance of *anti*-B$_{18}$H$_{22}$. *J. Mater. Chem. C* **2020**, *8*, 12806–12818. [CrossRef]

364. Tan, C.; Zhang, B.; Chen, J.; Zhang, L.; Huang, X.; Meng, H. Study of hydrolysis kinetic of new laser material [*anti*-B$_{18}$H$_{22}$]. *Russ. J. Inorg. Chem.* **2019**, *64*, 1359–1364. [CrossRef]

365. Ševčik, J.; Urbanek, P.; Hanulikova, B.; Čapkova, T.; Urbanek, M.; Antoš, J.; Londesborough, M.G.S.; Bould, J.; Ghasemi, B.; Petřkovsky, L.; et al. The photostability of novel boron hydride blue emitters in solution and polystyrene matrix. *Materials* **2021**, *14*, 589. [CrossRef] [PubMed]

366. Capkova, T.; Hanulikova, B.; Sevcik, J.; Urbanek, P.; Antos, J.; Urbanek, M.; Kuritka, I. Incorporation of the new *anti*-octadecaborane laser dyes into thin polymer films: A temperature-dependent photoluminescence and infrared spectroscopy study. *Int. J. Mol. Sci.* **2022**, *23*, 8832. [CrossRef] [PubMed]

367. Anderson, K.P.; Rheingold, A.L.; Djurovich, P.I.; Soman, O.; Spokoyny, A.M. Synthesis and luminescence of monohalogenated B$_{18}$H$_{22}$ clusters. *Polyhedron* **2022**, *227*, 116099. [CrossRef]

368. Ehn, M.; Bavol, D.; Bould, J.; Strnad, V.; Litecká, M.; Lang, K.; Kirakci, K.; Clegg, W.; Waddell, P.G.; Londesborough, M.G.S. A window into the workings of *anti*-B$_{18}$H$_{22}$ luminescence—Blue-fluorescent isomeric pair 3,3′-Cl$_2$-B$_{18}$H$_{20}$ and 3,4′-Cl$_2$-B$_{18}$H$_{20}$ (and others). *Molecules* **2023**, *28*, 4505. [CrossRef]

369. Anderson, K.P.; Waddington, M.A.; Balaich, G.J.; Stauber, J.M.; Bernier, N.A.; Caram, J.R.; Djurovich, P.I.; Spokoyny, A.M. A molecular boron cluster-based chromophore with dual emission. *Dalton Trans.* **2020**, *49*, 16245–16251. [CrossRef] [PubMed]

370. Anderson, K.P.; Hua, A.S.; Plumley, J.B.; Ready, A.D.; Rheingold, A.L.; Peng, T.L.; Djurovich, I.; Kerestes, C.; Snyder, N.A.; Andrews, A.; et al. Benchmarking the dynamic luminescence properties and UV stability of B$_{18}$H$_{22}$-based materials. *Dalton Trans.* **2022**, *51*, 9223–9228. [CrossRef]

371. Londesborough, M.G.S.; Dolansky, J.; Bould, J.; Braborec, J.; Kirakci, K.; Lang, K.; Cisařova, I.; Kubat, P.; Roca-Sanjuan, D.; Frances-Monerris, A.; et al. Effect of iodination on the photophysics of the laser borane *anti*-B$_{18}$H$_{22}$: Generation of efficient photosensitizers of oxygen. *Inorg. Chem.* **2019**, *58*, 10248–10259. [CrossRef]

372. Anderson, K.P.; Djurovich, P.I.; Rubio, V.P.; Liang, A.; Spokoyny, A.M. Metal-catalyzed and metal-free nucleophilic substitution of 7-I-B$_{18}$H$_{21}$. *Inorg. Chem.* **2022**, *61*, 15051–15057. [CrossRef]

373. Bould, J.; Lang, K.; Kirakci, K.; Cerdan, L.; Roca-Sanjuan, D.; Frances-Monerris, A.; Clegg, W.; Waddell, P.G.; Fuciman, M.; Polivka, T.; et al. A series of ultra-efficient blue borane fluorophores. *Inorg. Chem.* **2020**, *59*, 17058–17070. [CrossRef]

374. Londesborough, M.G.S.; Lang, K.; Clegg, W.; Waddell, P.G.; Bould, J. Swollen polyhedral volume of the *anti*-B$_{18}$H$_{22}$ cluster via extensive methylation: Anti-B$_{18}$H$_8$Cl$_2$Me$_{12}$. *Inorg. Chem.* **2020**, *59*, 2651–2654. [CrossRef]

375. Sauri, V.; Oliva, J.M.; Hnyk, D.; Bould, J.; Braborec, J.; Merchan, M.; Kubat, P.; Cisařova, I.; Lang, K.; Londesborough, M.G.S. Tuning the Photophysical Properties of *anti*-B$_{18}$H$_{22}$: Efficient intersystem crossing between excited ainglet and triplet states in new 4,4′-(HS)$_2$-*anti*-B$_{18}$H$_{20}$. *Inorg. Chem.* **2013**, *52*, 9266–9274. [CrossRef]

376. Londesborough, M.G.S.; Dolansky, J.; Cerdan, L.; Lang, K.; Jelinek, T.; Oliva, J.M.; Hnyk, D.; Roca-Sanjuan, D.; Frances-Monerris, A.; Martinčik, J.; et al. Thermochromic fluorescence from B$_{18}$H$_{20}$(NC$_5$H$_5$)$_2$: An inorganic–organic composite luminescent compound with an unusual molecular geometry. *Adv. Opt. Mater.* **2017**, *5*, 1600694. [CrossRef]

377. Londesborough, M.G.S.; Dolansky, J.; Jelinek, T.; Kennedy, J.D.; Cisařova, I.; Kennedy, R.D.; Roca-Sanjuan, D.; Frances-Monerris, A.; Lang, K.; Clegg, W. Substitution of the laser borane *anti*-B$_{18}$H$_{22}$ with pyridine: A structural and photophysical study of some unusually structured macropolyhedral boron hydrides. *Dalton Trans.* **2018**, *47*, 1709–1725. [CrossRef] [PubMed]

378. Chen, J.; Xiong, L.; Zhang, L.; Huang, X.; Meng, H.; Tan, C. Synthesis, aggregation-induced emission of a new *anti*-B$_{18}$H$_{22}$-isoquinoline hybrid. *Chem. Phys. Lett.* **2020**, *747*, 137328. [CrossRef]

379. Xiong, L.; Zheng, Y.; Wang, H.; Yan, J.; Huang, X.; Meng, H.; Tan, C. A novel AIEE active *anti*-B$_{18}$H$_{22}$ derivative-based Cu^{2+} and Fe^{3+} fluorescence off-on-off sensor. *Methods Appl. Fluoresc.* **2022**, *10*, 035004. [CrossRef]

380. Jelinek, T.; Kennedy, J.D.; Štibr, B.; Thornton-Pett, M. Macropolyhedral boron-containing cluster chemistry. A reductive trimerisation of MeNC to give an imidazole-based carbene stabilized by coordination to boron in an eighteen-vertex cluster compound. *J. Chem. Soc. Chem. Commun.* **1994**, 1999–2000. [CrossRef]

381. Jelinek, T.; Kilner, C.A.; Štibr, B.; Thornton-Pett, M.; Kennedy, J.D. Macropolyhedral borane reaction chemistry: Reductive oligomerisation of ^{ter}BuNC by *anti*-B$_{18}$H$_{22}$ to give the boron-coordinated {(^{ter}BuNHCH){^{ter}BuNHC(CN)}CH$_2$:} carbene residue. *Inorg. Chem. Commun.* **2005**, *8*, 491–494. [CrossRef]

382. Patel, D.K.; Sooraj, B.S.; Kirakci, K.; Macháček, J.; Kučeráková, M.; Bould, J.; Dušek, M.; Frey, M.; Neumann, C.; Ghosh, S.; et al. Macropolyhedral *syn*-$B_{18}H_{22}$, the "forgotten" isomer. *J. Am. Chem. Soc.* **2023**, *145*, 17975–17986. [CrossRef] [PubMed]

383. Ehn, M.; Litecká, M.; Londesborough, M.G.S. Unexpected minor products from the thermal auto-fusion of *arachno*-SB_8H_{12}: Luminescent 4-(HS)-*syn*-$B_{18}H_{21}$ and 3-(HS)-*syn*-$B_{18}H_{21}$. *Inorg. Chem. Commun.* **2023**, *155*, 111021. [CrossRef]

384. Jelínek, T.; Grüner, B.; Císařová, I.; Štíbr, B.; Kennedy, J.D. Macropolyhedral boron-containing cluster chemistry: The reaction of *syn*-$B_{18}H_{22}$ with SMe_2 and I_2 in monoglyme: Structure of [7-(SMe_2)-*syn*-$B_{18}H_{20}$]. *Inorg. Chem. Commun.* **2007**, *10*, 125–128. [CrossRef]

Review

Recent Progress in Crystalline Borates with Edge-Sharing BO$_4$ Tetrahedra

Jing-Jing Li, Wei-Feng Chen, You-Zhao Lan and Jian-Wen Cheng *

Key Laboratory of the Ministry of Education for Advanced Catalysis Materials, Institute of Physical Chemistry, Zhejiang Normal University, Jinhua 321004, China
* Correspondence: jwcheng@zjnu.cn

Abstract: Crystalline borates have received great attention due to their various structures and wide applications. For a long time, the corner-sharing B–O unit is considered a basic rule in borate structural chemistry. The Dy$_4$B$_6$O$_{15}$ synthesized under high-pressure is the first oxoborate with edge-sharing [BO$_4$] tetrahedra, while the KZnB$_3$O$_6$ is the first ambient pressure borate with the edge-sharing [BO$_4$] tetrahedra. The edge-sharing connection modes greatly enrich the structural chemistry of borates and are expected to expand new applications in the future. In this review, we summarize the recent progress in crystalline borates with edge-sharing [BO$_4$] tetrahedra. We discuss the synthesis, fundamental building blocks, structural features, and possible applications of these edge-sharing borates. Finally, we also discuss the future perspectives in this field.

Keywords: borate; edge-sharing; fundamental building blocks; structural chemistry

1. Introduction

Borates show rich structural chemistry and have broad applications as birefringent materials and nonlinear optical (NLO) materials [1–31]. The famous KBe$_2$BO$_3$F$_2$ (KBBF), LiB$_3$O$_5$ (LBO), and β-BaB$_2$O$_4$ (β-BBO) crystals are used to generate ultraviolet (UV) or deep-UV lasers through cascaded frequency conversion in practical application [32–34]. α-BaB$_2$O$_4$ (α-BBO) is an excellent UV birefringent crystal with a wide transparency window from 190 nm to 3500 nm and a large birefringence of 0.15 at 266 nm [35]. To date, the number of synthetic borates and borate minerals are over 3900 in the documented literature [1]. Three types of B–O units of linear [BO$_2$], triangular [BO$_3$], and tetrahedral [BO$_4$] are observed in these borates in which linear [BO$_2$] with sp hybridized chemical bonds are extremely rare; only 0.1% of borates contain the linear [BO$_2$] configuration. M$_5$Ba$_2$(B$_{10}$O$_{17}$)$_2$(BO$_2$) (M = K, Rb) and NaRb$_6$(B$_4$O$_5$(OH)$_4$)$_3$(BO$_2$) are three typical examples; the former two compounds contain unusual [BO$_2$] with the traditional [BO$_3$] and [BO$_4$] units and exhibit suitable birefringence (Δn = 0.06) and transparency windows down to the deep-UV region (<190 nm) [36,37]. Theoretical analyses reveal that the [BO$_3$] and [BO$_4$] units have the smaller polarizability anisotropy compared with linear [BO$_2$]. While the latter one is the first noncentrosymmetric and chiral structure with the linear [BO$_2$] unit and displays a weak second-harmonic generation response (SHG) (0.1 $\times$ SiO$_2$) and wide transparency of about 21.2% at 200 nm [38].

In 2021, Pan and coworkers summarized the synthesis, fundamental building blocks (FBBs), symmetries, structure features, and functional properties of the reported anhydrous borates [1]. The FBBs of polynuclear borates are generally formed by corner-/edge-sharing [BO$_3$] and [BO$_4$] units. Cs$_3$B$_7$O$_{12}$ contains a large FBB with 63 boron atoms in which 35 (or 37) BO$_3$ triangles and 28 (or 26) BO$_4$ tetrahedra are linked to form thick anionic sheets stacked along the c direction [39]. Mg$_7$@[B$_{69}$O$_{108}$(OH)$_{18}$] contains 42 [BO$_3$] triangles and 27 [BO$_4$] tetrahedra; it exhibits a supramolecular framework with hexagonal snowflake-like channels; unique triple-helical ribbons are found in {B$_{69}$} FBBs [40]. This huge [B$_{69}$O$_{108}$(OH)$_{18}$] cluster represents the largest FBB in borates. The FBBs can further

Citation: Li, J.-J.; Chen, W.-F.; Lan, Y.-Z.; Cheng, J.-W. Recent Progress in Crystalline Borates with Edge-Sharing BO$_4$ Tetrahedra. *Molecules* **2023**, *28*, 5068. https://doi.org/10.3390/molecules28135068

Academic Editors: Michael A. Beckett and Igor B. Sivaev

Received: 1 June 2023
Revised: 25 June 2023
Accepted: 26 June 2023
Published: 28 June 2023

polymerize into 1D chains, 2D layers, and 3D networks [41–49]. For example, we obtained three alkali and alkaline earth-metal borates, namely $Ba_2B_{10}O_{16}(OH)_2\cdot(H_3BO_3)(H_2O)$, $Na_2B_{10}O_{17}\cdot H_2en$, and $Ca_2[B_5O_9]\cdot(OH)\cdot H_2O$ [41–43], in which pentaborates are used to construct a single-layered structure, 2D microporous layers, and a 3D network, respectively. $Ca_2[B_5O_9]\cdot(OH)\cdot H_2O$ is impressive with a dense net consisting of *pcu* B−O net and *dia* Ca−O net and exhibits a short UV cutoff edge below 200 nm and a strong SHG response of ~three times that of KH_2PO_4 (KDP) [43].

In 2002, Huppertz and coworkers reported the high-pressure synthesis of $Dy_4B_6O_{15}$; it is the first oxoborate with an edge-sharing BO_4 tetrahedra [50]. The edge-sharing $[BO_4]$ tetrahedra in $Dy_4B_6O_{15}$ changes the rule of corner-sharing $[BO_3]/[BO_4]$ units in borate structural chemistry. In addition, it is considered that the extreme synthetic conditions, such as high pressure, is necessary for edge-sharing borates. In 2010, the discovery of $KZnB_3O_6$ changed this view; $KZnB_3O_6$ represents the first ambient pressure edge-sharing $[BO_4]$-containing borate [51]. To date, edge-sharing $[BO_4]$-containing borates are still rare; less than 1% of borates contain edge-sharing BO_4 tetrahedra. Over the past decade, the synthesis, crystal structures, and properties of KBBF-like borates [52], fluorooxoborates [53,54], high-temperature borates [55], high-pressure borates [56], *f*-element borates [57], zincoborates [58,59], aluminoborates [60,61], borogermanates [62], hybrid *d*- or *p*-block metal borates [63], and hydrated borates with non-metal or transition-metal complex cations have been well reviewed [64]. Herein, we give a detailed summary of the recent progress in crystalline borates with edge-sharing BO_4 tetrahedra. These edge-sharing borates can be grouped into two types in terms of their synthetic method: (i) high pressure synthesis of borates with edge-sharing $[BO_4]$ tetrahedra and (ii) ambient pressure synthesis of borates with edge-sharing $[BO_4]$ tetrahedra. We discuss the synthesis, FBBs, structural features, potential applications, and future perspectives of edge-sharing borates.

2. High Pressure Synthesis of Borates with Edge-Sharing [BO₄] Tetrahedra

The existence of uncommon edge-sharing $[BO_4]$ tetrahedra disobeys Pauling's third rule. The borates containing the so-called edge-sharing $[B_2O_6]$ dimer were initially believed to be obtained only under extreme conditions, such as high temperature and high pressure. Since the first case of this species was discovered, multi-anvil high-pressure synthesis is the dominant route to obtain the new edge-sharing $[BO_4]$ tetrahedra-containing borates. Up to now, there are 26 high-pressure edge-sharing borates within the scope of discussion. Boron atoms tend to coordinate with four O atoms to form $[BO_4]$ tetrahedra under a high-pressure environment, as evidenced by most of these high-pressure compounds constructed merely from $[BO_4]$ tetrahedra. Even in $[BO_3]$-containing borates, such as high-pressure AB_3O_5, the proportion of the $[BO_3]$ triangle is only 1/3.

2.1. Rare Earth Borates

$RE_4B_6O_{15}$ (RE = Dy and Ho). $Dy_4B_6O_{15}$ is the first reported metal borate with edge-sharing $[BO_4]$ tetrahedra; it was obtained under high-temperature (1273 K) and high-pressure (8 Gpa) conditions by Huppertz et al. in 2002 [50]. Shortly after, isostructural Ho-analogues was prepared under the same extreme high-pressure condition in 2003 [65]. The $RE_4B_6O_{15}$ series crystallize in the monoclinic crystal system with the space group of $C2/c$ (no. 15); their structures exhibit corrugated $^2[B_6O_{15}]_\infty$ layers formed by the linkage of the adjacent $[B_{12}O_{35}]$ clusters (Figure 1b). The large $[B_{12}O_{35}]$ cluster, incorporating edge-sharing and corner-sharing $[BO_4]$ tetrahedra with the ratio of 8:4, can be considered as the FBB of $RE_4B_6O_{15}$ (Figure 1a). Furthermore, the interlayer rare earth ions connect these corrugated layers to form the final 3D structures (Figure 1c). The multi-anvil techniques, which can offer external pressures, accelerate the discovery of borates with unusual edge-sharing $[BO_4]$ tetrahedra and initiate the era of exploring such borates under multi-anvil high-pressure conditions.

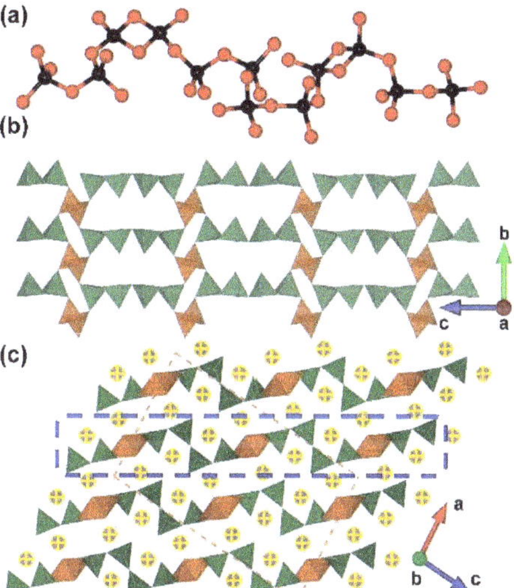

Figure 1. (a) The [B$_{12}$O$_{35}$] FBB; (b) the ^{2}[B$_6$O$_{15}$]$_\infty$ corrugated layer; (c) the total structure of RE$_4$B$_6$O$_{15}$ (RE = Dy and Ho) along [010] direction. Key: cross-centered purple ball, rare earth atom; black ball, B atom; red ball, O atom; orange/olive tetrahedron, edge/vertex-sharing [BO$_4$]; purple triangle, [BO$_3$].

α-RE$_2$B$_4$O$_9$ (RE = Sm, Eu, Gd, Tb, Dy, Ho and Y). α-RE$_2$B$_4$O$_9$ borates (RE = Sm, Eu, Gd, Tb, Dy, Ho and Y) are another rare earth borate series with edge-sharing [BO$_4$] tetrahedra reported in the period of 2002 to 2017 [66–69]. Similar to the RE$_4$B$_6$O$_{15}$ series, the α-RE$_2$B$_4$O$_9$ series crystallize in the same space group (*C*2/*c*, no. 15) in which all the incorporating boron atoms are four-coordinated. In these structures, the complex [B$_{20}$O$_{55}$] FBB is comprised with edge- and corner-sharing [BO$_4$] tetrahedra according to the ratio of 18:2 (Figure 2a and blue blanket in Figure 2b). With respect to the whole covalent B–O framework of RE$_4$B$_6$O$_{15}$, the ^{3}[B$_6$O$_{15}$]$_\infty$ network is formed by the linkage of [B$_{20}$O$_{55}$] FBBs, the rare earth cations located in the channels (Figure 2b).

La$_3$B$_6$O$_{13}$(OH). During the synthetic process, the replacement of the anhydrous boron source with boric acid, hydrated borates, or borates containing water molecules are sometimes obtained. La$_3$B$_6$O$_{13}$(OH) is the first SHG-active edge-sharing [BO$_4$] tetrahedra-containing borate [70]. This compound was obtained by a high-pressure/high-temperature condition at 6 GPa and 1673 K and was immediately identified as an NLO crystal by Huppertz et al. in 2020. It crystallizes in the chiral space group, *P*2$_1$ (no. 4), and presents a 2D ^{2}[B$_6$O$_{13}$(OH)]$_\infty$ layered structure with La ions located between the layers (Figure 3). The FBB of La$_3$B$_6$O$_{13}$(OH) features a 'sechser'-ring, which is constructed of one [B$_2$O$_6$], three vertex-sharing [BO$_4$], and one [BO$_3$(OH)] (Figure 3a). The [B$_6$O$_{16}$(OH)] FBBs are linked into a 2D ^{2}[B$_6$O$_{13}$(OH)]$_\infty$ layer along the *bc* plane, which further stack along [100] direction with La ions residing in the interlayer space (Figure 3b). Although La$_3$B$_6$O$_{13}$(OH) crystallizes in a noncentrosymmetric space group, its basic B–O units in the lattice are all the non-π-conjugated tetrahedral. La$_3$B$_6$O$_{13}$(OH) displays a relatively weak SHG effect. Compared to the non-π-conjugated [BO$_4$] tetrahedron with negligible hyperpolarization, the π-conjugated motifs represented by planar [BO$_3$] and [B$_3$O$_6$] in the borate system are superior NLO-active functional modules, and thus, the powder SHG response of La$_3$B$_6$O$_{13}$(OH) based on the Kurtz–Perry method is as weak as 2/3 times that of quartz.

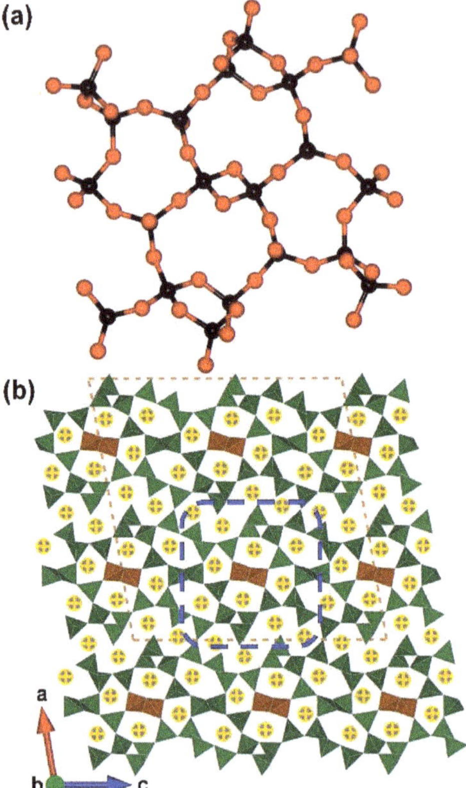

Figure 2. (**a**) The $[B_{20}O_{55}]$ FBB; (**b**) the total structure of $RE_2B_4O_9$ (RE = Sm, Eu, Gd, Tb, Dy, Ho, and Y) along [010] direction. Key: cross-centered purple ball, rare earth atom; black ball, B atom; red ball, O atom; orange/olive tetrahedron, edge/vertex-sharing $[BO_4]$.

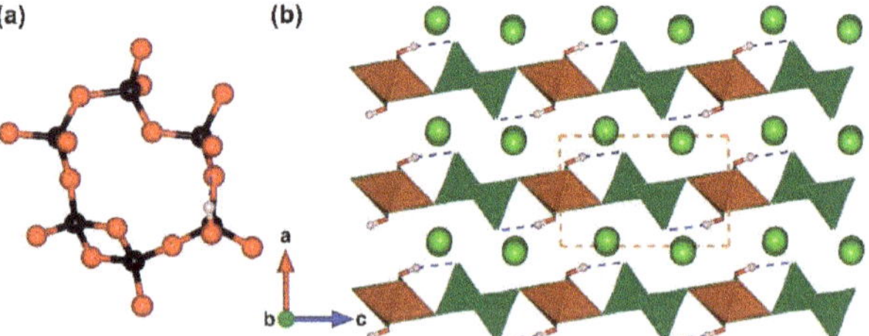

Figure 3. (**a**) The $[B_6O_{16}(OH)]$ FBB of $La_3B_6O_{13}(OH)$; (**b**) the total structure of $La_3B_6O_{13}(OH)$ along [010] direction. Key: green ball, La atom; black ball, B atom; red ball, O atom; small pink ball, H atom; orange/olive tetrahedron, edge/vertex-sharing $[BO_4]$.

2.2. Transition Metal Borates

TMB_2O_4 (TM = Ni, Fe and Co). Previous research on edge-sharing $[BO_4]$-containing borates mainly focus on lanthanide borates. Later, researchers achieved the combination of transition metal and edge-sharing $[BO_4]$ tetrahedra. From 2007 to 2010, a series of high-pressure transition metal borates, TMB_2O_4 (TM = Ni, Fe and Co), were discovered by Huppertz and coworkers [71–73]. All boron atoms in this species are four-coordinated, and the FBB is the simplest $[B_2O_6]$ cluster (Figure 4b). Each edge-shared $[B_2O_6]$ dimmer is linked to four surrounding [B$_2$O$_6$] units through μ_2-O atoms, resulting in a dense 2D ^{2}[B$_2$O$_4$]$_\infty$ layer with six-member rings (6 MRs) (Figure 4a). The stacking of ^{2}[B$_2$O$_4$]$_\infty$ layer along [100] direction is further linked by interlayer TM ions, which leads to the final structure of TMB$_2$O$_4$ (Figure 4c).

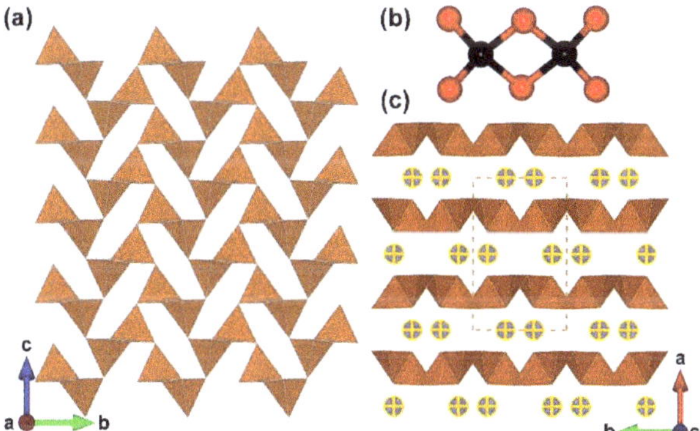

Figure 4. (a) The ^{2}[B$_2$O$_4$]$_\infty$ layer expanding in the *bc* plane; (b) the [B$_2$O$_6$] FBB; (c) the total structure of TMB$_2$O$_4$ (TM = Ni, Fe and Co) along [001] direction. Key: cross-centered purple ball, divalent transition metal atom; black ball, B atom; red ball, O atom; orange tetrahedron, edge-sharing [BO$_4$].

γ-HfB$_2$O$_5$. In 2021, the γ-phase of HfB$_2$O$_5$, which incorporates edge-sharing [BO$_4$] tetrahedra, was obtained under extreme pressure (120 GPa) by Huppertz [74]. γ-HfB$_2$O$_5$ crystallizes in the centrosymmetric monoclinic space group, $P2_1/c$ (no. 14). The tetravalent transition metal Hf^{4+} cation displays higher coordination numbers than divalent cations, and the FBB in γ-HfB$_2$O$_5$ is changed to [B$_3$O$_9$] with the additional one vertex-sharing [BO$_4$] (Figure 5a). Similar to the stuctures of TMB$_2$O$_4$ series, the structure of γ-HfB$_2$O$_5$ borate also shows layered sheets with Hf ions filling the interlayer space (Figure 5b). It is interesting to note that β-HfB$_2$O$_5$ was synthesized at 7.5 GPa in the multi-anvil press, upon further compression up to 120 GPa, a shrinkage of the cell parameters during the compression process was observed, and finally the β-phase is transformed to the γ-phase. The layer in β-HfB$_2$O$_5$ contains four MRs and eight MRs by the corner-sharing BO$_4$ tetrahedra, while γ-HfB$_2$O$_5$ contains ten MRs, including the edge-sharing BO$_4$ tetrahedra. Edge-sharing BO$_4$ tetrahedra in new phase γ-HfB$_2$O$_5$ shows exceptionally short B–O and B$\cdots$B distances. The coordination number of the Hf^{4+} cations in γ-HfB$_2$O$_5$ increased to nine in comparison to eight in its ambient pressure counterpart.

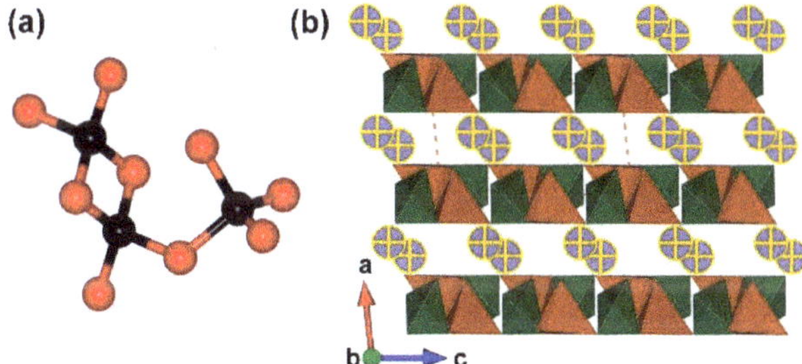

Figure 5. (a) The [B$_3$O$_9$] FBB; (b) the total structure of HfB$_2$O$_5$ along [010] direction. Key: cross-centered purple ball, Hf atom; black ball, B atom; red ball, O atom; orange tetrahedron, edge-sharing [BO$_4$].

M$_6$B$_{22}$O$_{39}$·H$_2$O (M = Fe and Co). The first two acentric edge-sharing [BO$_4$] tetrahedra-containing borates M$_6$B$_{22}$O$_{39}$·H$_2$O (M = Fe and Co) were prepared under the high-pressure

(6 GPa) and high-temperature (880 °C for Fe and 950 °C for Co) conditions in a Walker-type multi-anvil apparatus by Huppertz et al. in 2010 [75]. The $M_6B_{22}O_{39} \cdot H_2O$ series crystallize in a noncentrosymmetric orthorhombic space group, $Pmn2_1$ (no. 31). The unusually long B–O bond lengths as well as the short distances between the two boron cores are shown in the structure, which indicates the successful capture of intermediate states on the way to edge-sharing $[BO_4]$ tetrahedra. The structure of $M_6B_{22}O_{39} \cdot H_2O$ shows a 3D $[B_{22}O_{39}]_\infty$ anhydrous B–O framework with the Fe or Co ions and water molecules located in the structural channels (Figure 6a,c). Specifically, taking $Fe_6B_{22}O_{39} \cdot H_2O$ as an example, the B(11), O(2), O(15), and O(24) in the structure are not strictly in the same plane, and the B(11)-O(16) bond length (1.883(6) Å) is obviously longer than the common B–O distances (Figure 6b). The group of B(11) center and its three coordinated O(2,15,24) atoms as well as the neighboring O(16) can be regarded as a highly twisted polyhedron or the intermediate states between $[BO_3]$ tringle and $[BO_4]$ tetrahedron. The discovery of $M_6B_{22}O_{39} \cdot H_2O$ is helpful for understanding the dynamic process from the vertex-sharing $[BO_3] + [BO_4]$ model to edge-sharing $[BO_4] + [BO_4]$ model.

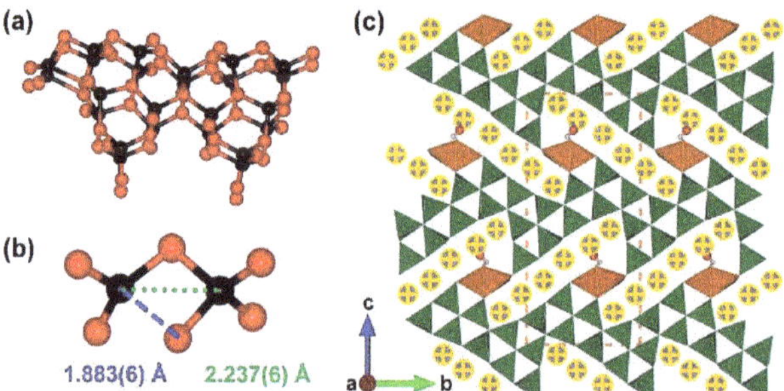

Figure 6. (a) The $[B_{24}O_{54}]$ FBB of $M_6B_{22}O_{39} \cdot H_2O$ (M = Fe and Co); (**b**) coordination spheres of boron atoms B(11) and B(8) in $Fe_6B_{22}O_{39} \cdot H_2O$; (**c**) the total structure of $M_6B_{22}O_{39} \cdot H_2O$ (M = Fe and Co) along [100] direction. Key: cross-centered purple ball, Fe/Co atom; black ball, B atom; red ball, O atom; small pink ball, H atom; orange/olive tetrahedron, edge/vertex-sharing $[BO_4]$.

$Co_7B_{24}O_{42}(OH)_2 \cdot 2H_2O$. Although the cobalt hydrated borate $Co_7B_{24}O_{42}(OH)_2 \cdot 2H_2O$ crystallizes in a centrosymmetric space group, *Pbam* (no. 55), it shares similar structural characteristics with $Co_6B_{22}O_{39} \cdot H_2O$. This species was prepared under high-pressure (6 GPa) and high-temperature (1153 K) conditions by Huppertz et al. in 2012 [76]. The complex FBB of $Co_7B_{24}O_{42}(OH)_2 \cdot 2H_2O$ is comprised of twenty-two corner- and two edge-sharing $[BO_4]$ tetrahedra with two hydroxy group locating in the mirror plane (Figure 7a). The structure of $Co_7B_{24}O_{42}(OH)_2 \cdot 2H_2O$ shows the $^3[B_{24}O_{42}(OH)_2]_\infty$ framework with Co ions and water molecules located in the structural channels (Figure 7b).

2.3. Borates with Monovalent or Divalent A-Site Cations

AB_3O_5 [A = K, NH_4, Rb, Tl and $Cs_{1-x}(H_3O)_x$ (x = 0.5–0.7)]. During the period of 2011 to 2014, the AB_3O_5 series [A = K, NH_4, Rb, Tl and $Cs_{1-x}(H_3O)_x$ (x = 0.5–0.7)] were synthesized under high-pressure/high-temperature conditions by Huppertz et al. [77–80]. KB_3O_5 is the first compound with various configurations, including corner-sharing $[BO_3]$, corner-sharing $[BO_4]$, and edge-sharing $[BO_4]$. The FBB of the isostructural AB_3O_5 contains two $[BO_3]$ triangles, four corner-sharing $[BO_4]$ tetrahedra, and two edge-sharing $[BO_4]$ tetrahedra (Figure 8a). It should be noted that the $[B_2O_6]$ rings in AB_3O_5 can be regarded as six connected nodes; the two endocyclic O atoms of each $[B_2O_6]$ ring are further connected with two corner-sharing $[BO_4]$ tetrahedra. The total structures of the AB_3O_5 series exhibit 3D B–O anionic skeletons with monovalent cations locating the structural channels (Figure 8b). Although the boron source in the synthesis of AB_3O_5 series are boric acid, only the $Cs_{1-x}(H_3O)_xB_3O_5$ ($x = 0.5$–0.7) phase successfully incorporates oxonium ions into its structure.

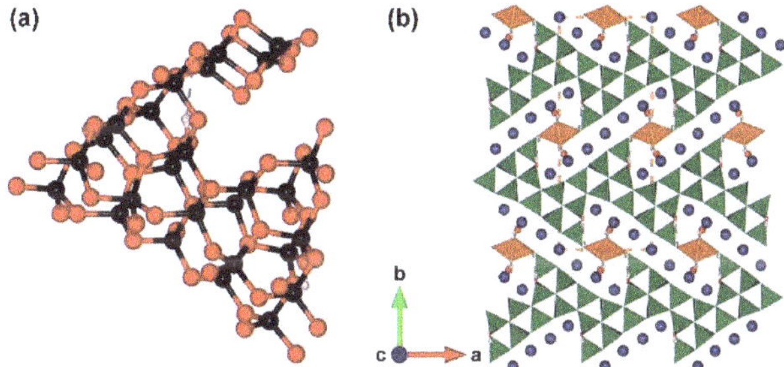

Figure 7. (a) The $[B_{24}O_{48}(OH)_2]$ FBB of $Co_7B_{24}O_{42}(OH)_2 \cdot 2H_2O$; (b) the total structure of $Co_7B_{24}O_{42}(OH)_2 \cdot 2H_2O$ along [001] direction. Key: navy ball, Co atom; black ball, B atom; red ball, O atom; small pink ball, H atom; orange/olive tetrahedron, edge/vertex-sharing $[BO_4]$.

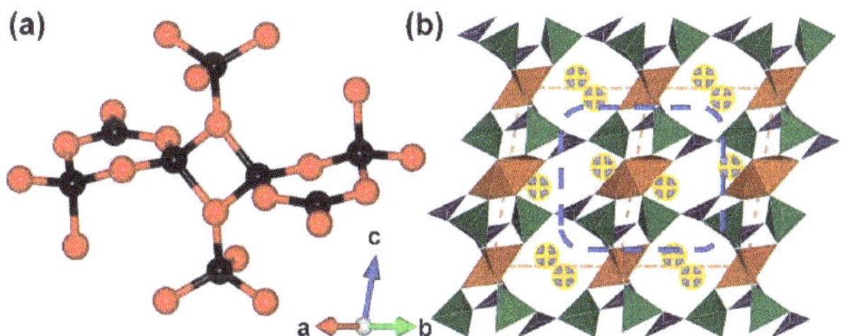

Figure 8. (a) The $[B_8O_{20}]$ FBB; (b) the total structure of AB_3O_5 along [110] direction. Key: cross-centered purple ball, monovalent cation; black ball, B atom; red ball, O atom; orange/olive tetrahedron, edge/vertex-sharing $[BO_4]$; purple triangle $[BO_3]$.

CsB_5O_8. CsB_5O_8 is another alkali metal borate prepared under high-pressure (6 Gpa) and high-temperature (1173 K) conditions in a Walker-type multianvil apparatus [81]. Structurally, CsB_5O_8 features a similar structure to the aforementioned AB_3O_5 series. The basic B–O building blocks of CsB_5O_8 are corner-sharing $[BO_3]$, corner-sharing $[BO_4]$, and edge-sharing $[BO_4]$; these units exhibit a ratio of 2:1:2, respectively (Figure 9a). The structure of CsB_5O_8 exhibits a 3D B–O covalent framework with Cs ions locating in the structural channels (Figure 9b).

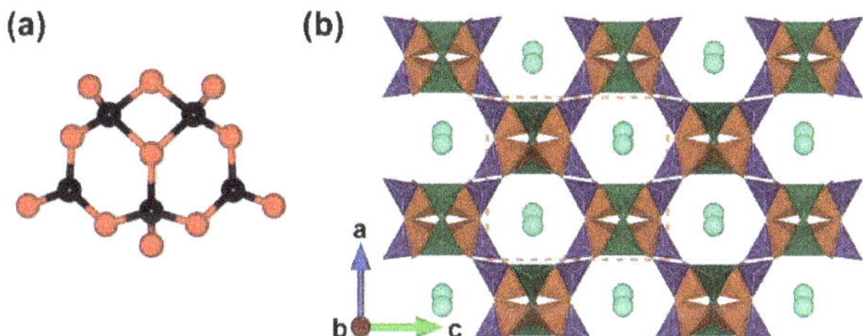

Figure 9. (a) The $[B_5O_{11}]$ FBB; (b) the total structure of CsB_5O_8 along [010] direction. Key: cyan ball, Cs atom; black ball, B atom; red ball, O atom; orange/olive tetrahedron, corner-/edge-sharing $[BO_4]$; purple tringle $[BO_3]$.

$NaBSi_3O_8$. In 2022, Gorelova et al. studied the high-pressure modification of $NaBSi_3O_8$, and revealed the transformation behaviors of $NaBSi_3O_8$ during continuous pressure increase [82]. Unexpectedly, above 27.8 GPa the crystal structure of $NaBSi_3O_8$ achieves the coexistence of the rare edge-sharing [BO$_4$] tetrahedra and earlier unknown edge-sharing [SiO$_5$] square pyramids. The structures under 16.2 and 27.8 Gpa are quite different. Both the 16.2 Gpa- and 27.8 Gpa-phases crystallize in the $P\bar{1}$. The Si atoms in the 16.2 Gpa-phase are all four-coordinated, and the corner-sharing [SiO$_4$] tetrahedra are incorporated into the 1D [Si$_3$O$_8$]$_\infty$ chains, while 1/3 Si atoms are five-coordinated in the 27.8 Gpa-phase. These [SiO$_5$] square pyramids are dimerized into [Si$_2$O$_8$] units (Figure 10a,c). SiO$_4$ tetrahedra undergo geometrical distortion leading to the formation of SiO$_5$ polyhedra due to the pressure-induced transformations. The [BO$_4$] tetrahedra in 16.2 Gpa-phase and the [B$_2$O$_6$] dimers in 27.8 Gpa-phase act as linkers and further stable the whole structures (Figure 10b,d).

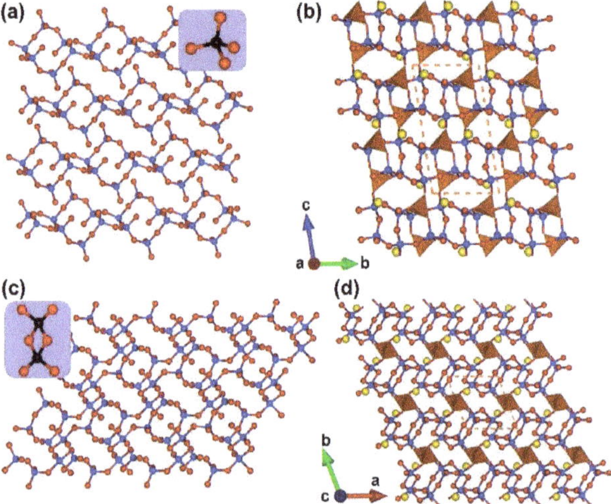

Figure 10. (**a**) The ^{2}[Si$_3$O$_8$]$_\infty$ pseudo layer and [BO$_4$] linkage in the structure of 16.2 Gpa-NaBSi$_3$O$_8$; (**b**) the view of the whole structure of 16.2 Gpa-NaBSi$_3$O$_8$ along [100] direction; (**c**) the ^{2}[Si$_3$O$_8$]$_\infty$ layer and [B$_2$O$_6$] linkage in the structure of 24.8 Gpa-NaBSi$_3$O$_8$; (**d**) the view of the whole structure of 24.8 Gpa-NaBSi$_3$O$_8$ along [001] direction. Key: yellow ball, Na atom; blue ball, Si atom; black ball, B atom; red ball, O atom; orange tetrahedron, edge-sharing [BO$_4$].

γ-BaB$_2$O$_4$. The α- and β-phases of barium metaborate are famously commercialized birefringent and nonlinear optical materials. Relevant theoretical studies offered various predicted phase of barium metaborate. In 2022, the third phase, γ-BaB$_2$O$_4$, was synthesized experimentally by Bekker et al. under conditions of 3 GPa and 1173 K [83]. γ-BaB$_2$O$_4$ crystallizes in a centrosymmetrical space group, $P2_1/n$ (no. 14). Its anionic B–O skeleton exhibits 1D chains, which is completely different from the isolated [B$_3$O$_6$] planar cluster in α- and β-phases. The [B$_4$O$_{10}$] FBB in γ-BaB$_2$O$_4$ is comprised of one [B$_2$O$_6$] ring and two additional [BO$_3$] triangles (Figure 11a); these [B$_4$O$_{10}$] FBBs further assemble into the ^{1}[B$_2$O$_4$]$_\infty$ chains (Figure 11b). Finally, the [BaO$_{10}$] polyhedra stable the ^{1}[B$_2$O$_4$]$_\infty$ chains in the lattice to form the total 3D structure of γ-BaB$_2$O$_4$ (Figure 11c). The calculated band gap is up to 7.045 eV, which implies transparency in the deep-UV region. The most intense band at a frequency of 853 cm^{-1} in the Raman spectra corresponds to the symmetric bending mode of the B–O–B–O ring in edge-sharing tetrahedra.

α-Ba$_3$[B$_{10}$O$_{17}$(OH)$_2$]. Apart from the extreme high pressure afforded by the multi-anvil high-pressure device, the hydrothermal reactor can also provide relatively high pressure. In 2019, Lii et al. reported the structures of α-Ba$_3$[B$_{10}$O$_{17}$(OH)$_2$], which were obtained through hydrothermal reactions at 773 K and 0.1 Gpa. α-Ba$_3$[B$_{10}$O$_{17}$(OH)$_2$]. The phase containing edge-sharing [BO$_4$] tetrahedra crystallizes in the monoclinic space group, $P2_1/n$ (no. 14), and presents a hydrated 3D B–O framework with Ba ions filling in the cavities

(Figure 12b) [84]. In terms of its FBB, the complex $[B_{20}O_{40}(OH)_4]$ can be divided into the double $[B_5O_{12}]$ (the blue dotted ellipse part) and $[B_{10}O_{18}(OH)_4]$ (the red dotted blanket) categories (Figure 12a). Unlike FBBs mentioned in other borates, the $[B_2O_4(OH)_2]$ units act as two connected nodes in the structure as the targeted $[B_2O_6]$ units are bounded to hydrogen atoms as terminal hydroxy groups. The aggregation of $[B_5O_{12}]$ clusters expanding in the *ac* plane leads to a corrugated layer, and the hydrated $[B_{10}O_{18}(OH)_4]$ clusters connect the adjacent antiparallel layers to form the $^3[B_{10}O_{17}(OH)_2]_\infty$ covalent skeleton.

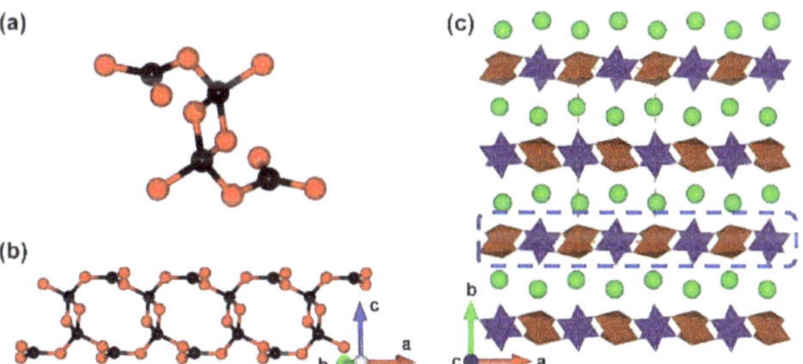

Figure 11. (**a**) The $[B_4O_{10}]$ FBB of γ-BaB$_2$O$_4$; (**b**) the view of 1D $^1[BO_2]_\infty$ chain in the structure; (**c**) the total structure of γ-BaB$_2$O$_4$ along [001] direction. Key: green ball, Ba atom; black ball, B atom; red ball, O atom; orange tetrahedron, edge-sharing $[BO_4]$; purple tringle $[BO_3]$.

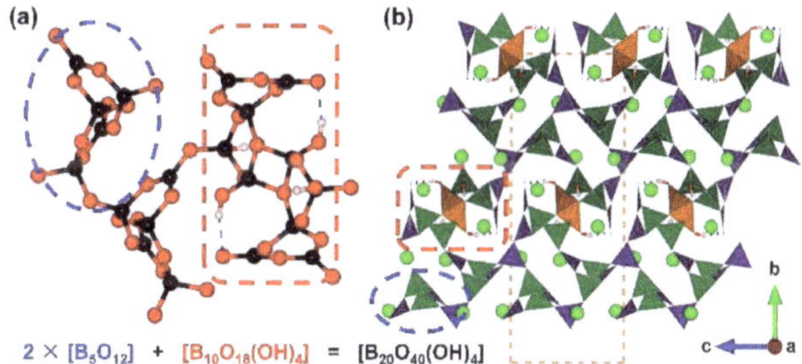

Figure 12. (**a**) The $[B_{20}O_{40}(OH)_4]$ FBB comprised with two $[B_5O_{12}]$ clusters and one $[B_{10}O_{18}(OH)_4]$ cluster; (**b**) the total structure of α-Ba$_3$[B$_{10}$O$_{17}$(OH)$_2$] along [100] direction. Key: green ball, Ba atom; black ball, B atom; red ball, O atom; small pink ball, H atom; orange/olive tetrahedron, edge/vertex-sharing $[BO_4]$; purple triangle $[BO_3]$.

3. Ambient Pressure Synthesis of Borates with Edge-Sharing $[BO_4]$ Tetrahedra

The edge-sharing $[BO_4]$ tetrahedra-containing borates obtained from classical high-temperature solution and cooling method make it possible to obtain this species more conveniently. More importantly, borates obtained under ambient pressure might incorporate more π-conjugated $[BO_3]$ units. Edge-sharing borates with high $[BO_3]$:$[BO_4]$ ratios, such as β-CsB$_9$O$_{14}$ (7:2) and Ba$_6$Zn$_6$(B$_3$O$_6$)$_6$(B$_6$O$_{12}$) (22:2), are identified as birefringent crystals.

KZnB$_3$O$_6$. The first case of borate containing edge-sharing $[BO_4]$ tetrahedra was synthesized under atmospheric pressure. KZnB$_3$O$_6$ was reported by Chen et al. and Wu et al. independently in 2010 [51,85]. KZnB$_3$O$_6$ crystallizes in the triclinic space group ($P\bar{1}$, no. 2) with a low symmetry. The $[B_6O_{12}]$ FBB features isolated B–O cluster containing four $[BO_3]$ tringles and two edge-sharing $[BO_4]$ tetrahedra (Figure 13a). The aligned repetition of isolated $[B_6O_{12}]$ clusters in the lattice gives a 2D $[B_6O_{12}]_\infty$ pseudo layer (see the green dotted blankets in Figure 13b,c), the pairs of edge-sharing $[ZnO_4]$ polyhedra connect the adjacent six $[B_6O_{12}]$ clusters to form the $^3[ZnB_3O_6]_\infty$ network with K cations filling the cavities. Later, KZnB$_3$O$_6$ was defined as highly thermally stable by Chen et al., and its unidirectional thermal expansion property was investigated. Its unique property is explained by the cooperative rotations of rigid groups $[B_6O_{12}]$ and $[Zn_2O_6]$ driven by anharmonic thermal vibrations of K ions [86–88]. The discovery of $KZnB_3O_6$ indicated that high pressure is not essential for obtaining edge-sharing $[BO_4]$ tetrahedra-containing borates, and subsequently, ambient pressure edge-sharing $[BO_4]$ tetrahedra-containing borates have been synthesized successfully one after another.

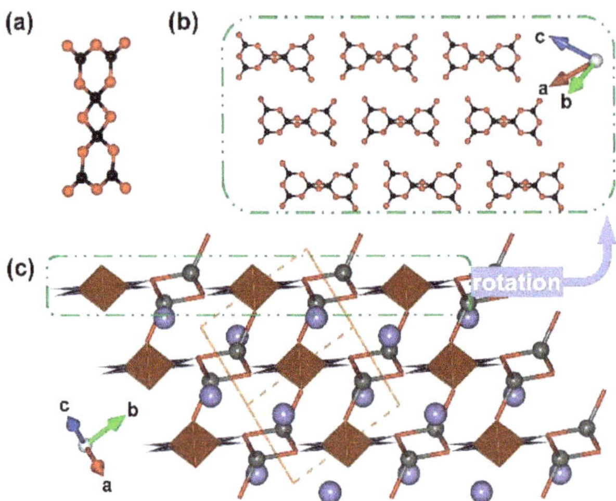

Figure 13. (a) The $[B_6O_{12}]$ FBB; (b) the 2D $[B_6O_{12}]_\infty$ pseudo layer; (c) the total structure of $KZnB_3O_6$ along [110] direction. Key: purple ball, K atom; grey ball, Zn atom; black ball, B atom; red ball, O atom; orange tetrahedron, edge-sharing $[BO_4]$; purple tringle $[BO_3]$.

$Ba_4Na_2Zn_4(B_3O_6)_2(B_{12}O_{24})$. $Ba_4Na_2Zn_4(B_3O_6)_2(B_{12}O_{24})$ is another edge-sharing $[BO_4]$ tetrahedra-containing borate obtained without an extreme pressure condition, as reported by Chen et al. in 2013 [89]. $Ba_4Na_2Zn_4(B_3O_6)_2(B_{12}O_{24})$ crystallizes in the triclinic space group, $P\bar{1}$ (no. 2); it features a complex sandwich-like layered structure. There are two kinds of FBBs in the structure of $Ba_4Na_2Zn_4(B_3O_6)_2(B_{12}O_{24})$, namely $[B_{12}O_{24}]$ and $[B_3O_6]$, respectively (Figure 14a,b). The aggregation of $[B_{12}O_{24}]$ FBBs and $[Zn(1)O_4]$ tetrahedra according to the stoichiometric ratio of 1:2 gives a layered $[Zn_2(B_{12}O_{24})]_\infty$ configuration expanding in the *ab* planes, while the second FBBs $[B_3O_6]$ are connected to $[Zn(2)O_4]$ to form $[Zn(B_3O_6)]_\infty$ sheets. The assembly of two $[Zn(B_3O_6)]_\infty$ sheets and one $[Zn(B_{12}O_{24})]_\infty$ layer leads to the formation of a complex $[Zn_4(B_3O_6)_2(B_{12}O_{24})]_\infty$ sandwiched structure. Split Na(1,2) atoms with the occupancy of 0.47:0.53 fill in the cavities of the sandwiched layers, and spherical coordinated Ba ions fill in the adjacent sandwiched layers (Figure 14c).

$Li_4Na_2CsB_7O_{20}$. The trimetallic borate $Li_4Na_2CsB_7O_{20}$ was reported by Pan et al. in 2019, and its expansion rate was investigated at the same time [90]. $Li_4Na_2CsB_7O_{20}$ crystallizes in a triclinic crystal system with the space group of $P\bar{1}$ (no. 2). With respect to its unique $[B_{14}O_{28}]$ FBB, the centered $[B_2O_6]$ ring acts as a four-connected node and further connects with one $[BO_3]$ tringle and one $[B_5O_{11}]$ cluster (Figure 15a). The total crystal structure of $Li_4Na_2CsB_7O_{20}$ displays a 3D configuration with monovalent alkali metal Li, Na, and Cs ions residing in the free spaces (Figure 15b). The temperature-dependent unit cell parameters were collected experimentally. as Additionally, the theoretical evaluation of thermal expansion along the principal axes indicate the highly anisotropic thermal expansion behavior of $Li_4Na_2CsB_7O_{20}$. The expansion rates for X_1, X_2, and X_3 were evaluated to be 3.51×10^{-6}, 17×10^{-6}, and 25.4×10^{-6} K^{-1}, respectively. This compound may be used as a thermal expansion material.

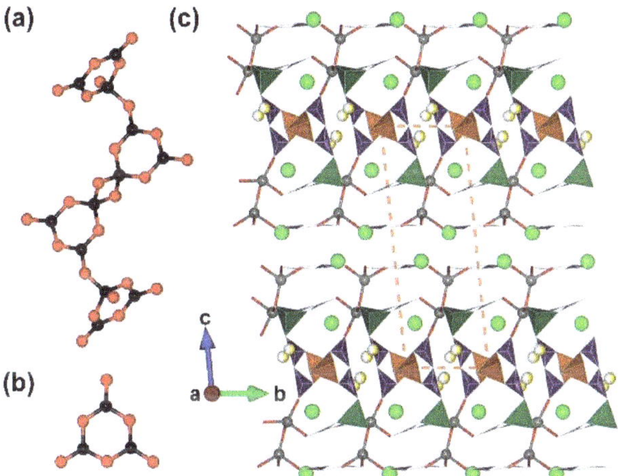

Figure 14. Two types of different FBBs occur in $Ba_4Na_2Zn_4(B_3O_6)_2(B_{12}O_{24})$: (**a**) $[B_{12}O_{24}]$ FBB; (**b**) $[B_3O_6]$ FBB; (**c**) the complex layered structure of $Ba_4Na_2Zn_4(B_3O_6)_2(B_{12}O_{24})$ along [100] direction. Key: green ball, Ba atom; yellow ball, Na atom; grey ball, Zn atom; black ball, B atom; red ball, O atom; olive/orange tetrahedron, corner-/edge-sharing $[BO_4]$; purple tringle $[BO_3]$.

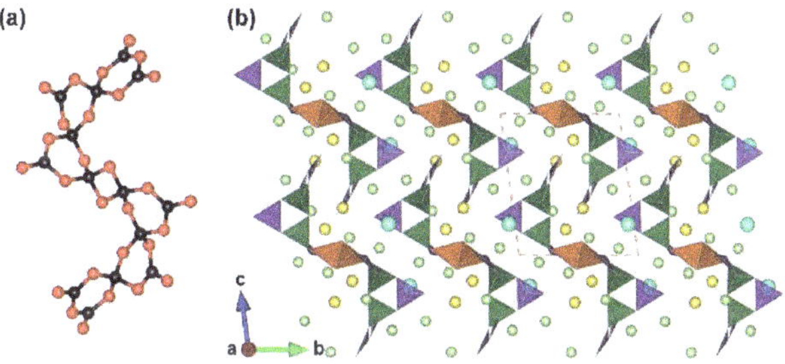

Figure 15. (**a**) The $[B_{14}O_{28}]$ FBB; (**b**) view of the whole crystal structure of $Li_4Na_2CsB_7O_{14}$ along [100] direction. Key: green ball, Li atom; yellow ball, Na atom; cyan ball, Cs atom; black ball, B atom; red ball, O atom; orange/olive tetrahedron, edge-/corner-sharing $[BO_4]$; purple tringle $[BO_3]$.

$BaAlBO_4$. In 2019, Pan et al. reported the synthesis and experimental and theoretical studies of an edge-sharing $[BO_4]$ tetrahedra-containing aluminum oxyborate, $BaAlBO_4$. $BaAlBO_4$ was synthesized via the high-temperature solution method under atmospheric pressure [91]. Single-crystal X-ray diffraction analysis reveals that $BaAlBO_4$ crystallizes in a monoclinic space group, $P2_1/c$ (no. 14). The crystal structure of $BaAlBO_4$ exhibits a 3D framework, which is comprised with $[AlO_4]$ tetrahedra, $[B_4O_{10}]$ clusters, and A-site Ba^{2+} cations filling the structural channels. The corner-sharing $[AlO_4]$ units in the *ab* plane give a 2D $^2[Al_2O_5]_\infty$ layer with six MRs (Figure 16b). The $[B_2O_6]$ rings connect with two $[BO_3]$ tringles end to end to form the isolated $[B_4O_{10}]$ cluster (Figure 16a), which can be considered as the FBB of $BaAlBO_4$. The combination of $[B_4O_{10}]$ clusters and the neighboring $^2[Al_2O_5]_\infty$ layer give the final 3D framework.

β-CsB_9O_{14}. In 2019, Pan et al. prepared the β-CsB_9O_{14} under ambient pressure. This compound is the first case of triple-layered borate with edge-sharing $[BO_4]$ tetrahedra [92]. Taking the $[B_6O_{12}]$ cluster in the $KZnB_3O_6$ as the prototype, the sandwich-like $[B_{18}O_{34}]$ FBB can be evolved from the combination of one $[B_6O_{12}]$ and two anti-parallel $[B_6O_{12}]$ double-ring units (Figure 17a). The further aggregation of $[B_{18}O_{34}]$ FBBs in the *bc* plane leads to the formation of corrugated layers with A-site Ba^{2+} cations residing in the channels; the whole structure of β-CsB_9O_{14} is formed by stacking of these triple-layered sheets along

[100] direction (Figure 17b). The B–O anionic skeleton of β-CsB$_9$O$_{14}$ possesses a high [BO$_3$]:[BO$_4$] (7:2) ratio; the layered structure as well as the well-aligned [BO$_3$] units in the lattice lead to a large optical anisotropy. The experimental and theoretical studies indicate that β-CsB$_9$O$_{14}$ can be identified as a potential deep-ultraviolet birefringent material with a wide band gap (>6.72 eV) and large birefringence (0.115 or 0.135 at 1064 nm).

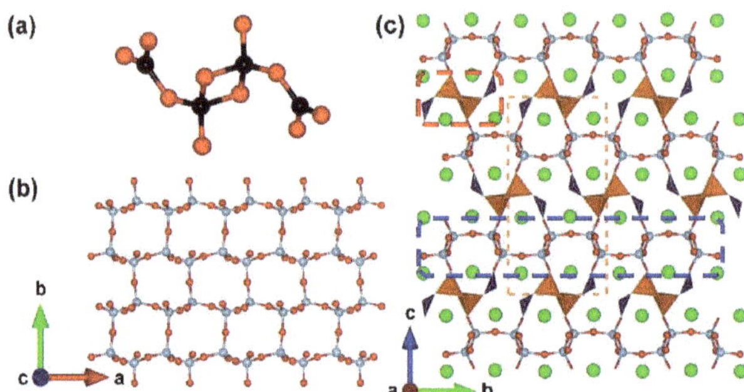

Figure 16. (**a**) The [B$_4$O$_{10}$] FBB of BaAlBO$_4$; (**b**) the 2D [Al$_2$O$_5$]$_\infty$ layer constructed by vertex-sharing [AlO$_4$] tetrahedra expanding in the ab plane; (**c**) the total structure of BaAlBO$_4$ along [100] direction. Key: green ball, Ba atom; light blue ball, Al atom; black ball, B atom; red ball, O atom; small pink ball, H atom; orange tetrahedron, edge-sharing [BO$_4$]; purple tringle [BO$_3$].

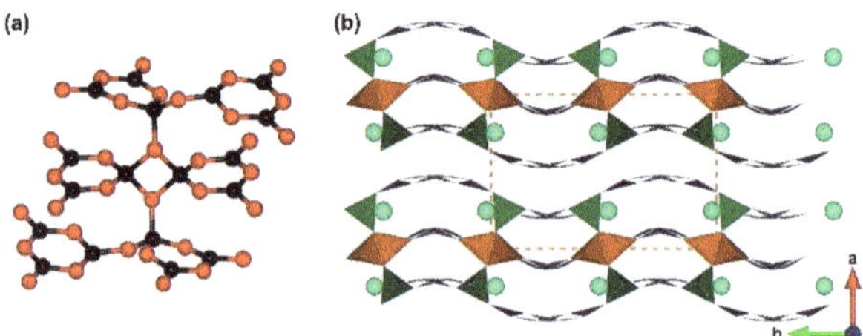

Figure 17. (**a**) The [B$_{18}$O$_{34}$] FBB of β-CsB$_9$O$_{14}$; (**b**) view of the whole crystal structure of β-CsB$_9$O$_{14}$ along [100] direction. Key: cyan ball, Cs atom; black ball, B atom; red ball, O atom; orange/green tetrahedron, edge/corner-sharing [BO$_4$]; purple tringle [BO$_3$].

Pb$_{2.28}$Ba$_{1.72}$B$_{10}$O$_{19}$. In 2021, an edge-sharing [BO$_4$]-containing borate, Pb$_{2.28}$Ba$_{1.72}$B$_{10}$O$_{19}$, was obtained under ambient pressure by Pan et al. [93]. It features a 3D B–O anionic framework. Pb$_{2.28}$Ba$_{1.72}$B$_{10}$O$_{19}$ crystallizes in a monoclinic crystal system with the space group of $C2/c$ (no. 15). Its asymmetric unit consists of one Pb atom, five B atoms, ten O atoms, and one common site of the Ba/Pb atom with the occupancy of 0.14:0.86. Unlike the [B$_2$O$_6$] basic ring in most of edge-sharing [BO$_4$]-containing borate with four exocyclic O atoms acting as connection nodes, the centered [B$_2$O$_6$] in [B$_{10}$O$_{24}$] FBB connects two [BO$_4$] tetrahedra by the two exocyclic μ_2-O atoms and two [B$_3$O$_8$] by two endocyclic μ_3-O atoms (Figure 18a). The whole [B$_{10}$O$_{19}$] anionic framework is assembled from [B$_{10}$O$_{24}$] FBBs and Pb and Ba ions located in the structural channels (Figure 18b).

K$_3$Sb$_4$BO$_{13}$. In 2021, Quarez et al. discovered the complete transformation of adjacent [BO$_3$] pairs into [B$_2$O$_6$] dimers in the α- and β-phase of K$_3$Sb$_4$BO$_{13}$ driven by cooling [94]. The [BO$_3$]-containing α-phase of K$_3$Sb$_4$BO$_{13}$ is obtained from the traditional high-temperature solution method, and the symmetry increasing from $P\bar{1}$ to $C2/c$ during the cooling process, accompanied with the transformation of two close [BO$_3$] tringles into edge-sharing [B$_2$O$_6$] units. The structures of α- and β-K$_3$Sb$_4$BO$_{13}$ display complex ^{2}[Sb$_4$O$_{13}$]$_\infty$ layers separated by [BO$_3$] pairs or edge-sharing [BO$_4$] tetrahedra (Figure 19a,b). The anti-parallel [BO$_3$] pair in the α-phase displays a short B$\cdots$B distance (3.083(6) Å) and an extremely long B$\cdots$O secondary bond (2.623(6) Å), while the coordination spheres of corresponding B atoms in the β-phase are distorted into tetrahedra (Figure 19c,d). The low temperature brings a lattice compression, which finally leads to B_2O_6 units, which shortens the $B\cdots B$ and $B\cdots O$ distances in each pair of adjacent BO_3 triangles units. Further studies show that B K-edge electron energy loss (EELS) spectroscopes provide a characteristic signal of the B_2O_6 units; the EELS method may widely use to identify edge-sharing B_2O_6 units more convenient in the future.

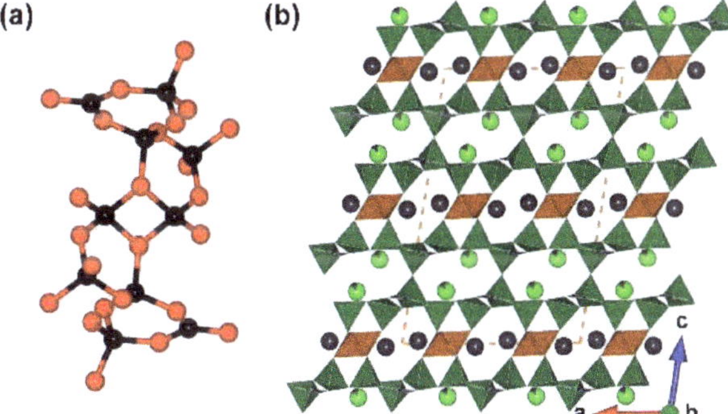

Figure 18. (**a**) The $[B_{10}O_{24}]$ FBB of $Pb_{2.28}Ba_{1.72}B_{10}O_{19}$; (**b**) the view of the whole structure of $Pb_{2.28}Ba_{1.72}B_{10}O_{19}$ along [010] direction. Key: grey ball, Pb atom; green ball, Ba atom; black ball, B atom; red ball, O atom; orange/olive tetrahedron, edge/vertex-sharing $[BO_4]$.

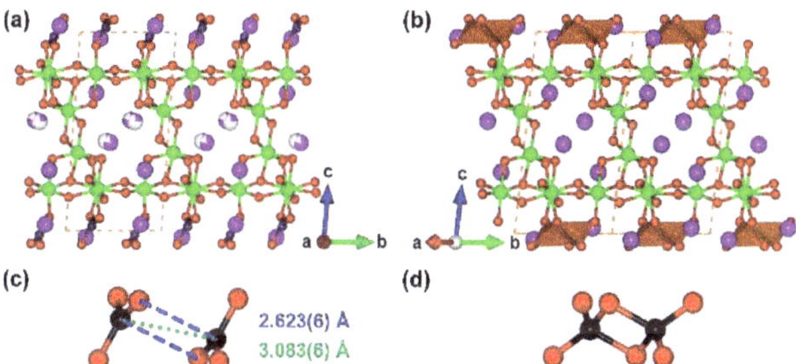

Figure 19. (**a**) The view of the whole structure of β-$K_3Sb_4BO_{13}$ along [100] direction; (**b**) the view of the whole structure of α-$K_3Sb_4BO_{13}$ along [100] direction; (**c**) the anti-parallel $[BO_3]$ pair in the β-$K_3Sb_4BO_{13}$; (**d**) the edge-sharing $[BO_4]$ tetrahedra in the α-$K_3Sb_4BO_{13}$. Key: purple ball, K atom; green ball, Sb atom; black ball, B atom; red ball, O atom; orange tetrahedron, edge-sharing $[BO_4]$; purple triangle, $[BO_3]$.

$Ba_6Zn_6(B_3O_6)_6(B_6O_{12})$. $Ba_6Zn_6(B_3O_6)_6(B_6O_{12})$ was reported by Mao et al. and Pan et al. independently in 2022 and identified as a potential birefringent crystal with a deep-ultraviolet absorption cut-off edge and strong optical anisotropy [95,96]. The structure of $Ba_6Zn_6(B_3O_6)_6(B_6O_{12})$ features a 2D $[ZnB_4O_8]_\infty$ network constructed by $[ZnO_4]$ tetrahedra and two kinds of B–O clusters ($[B_6O_{12}]$ and $[B_3O_6]$) with Ba cations located in the cavities (Figure 20). It should be noted that $Ba_6Zn_6(B_3O_6)_6(B_6O_{12})$ shows extremely low symmetry (space group P-1, no. 2), and its asymmetric unit includes six Ba atoms, six Zn atoms, six planar $[B_3O_6]$ clusters, and two $[B_3O_6]$ fragments (half of $[B_6O_{12}]$ cluster). To simplify the description of structure, we use B–O cluster-1 and B–O cluster-2 to represent the basic structural units (Figure 20a–c). In the sandwiched $[ZnB_4O_8]_\infty$ layers, the top and bottom of well-aligned $[B_6O_{12}]$ clusters are shielded by the anti-parallel $^2[Zn(B_3O_6)]_\infty$ sheets. The . . . A-A′-A . . . stacking sequence of $[ZnB_4O_8]_\infty$ along the [001] direction leads to the formation of the total covalent skeleton, and Ba ions act as counterions in the lattice. From the structural perspective, the uniformly arrangement of two kinds of B–O clusters and the high ratio of highly birefringence-active $[BO_3]$ tringles and $[BO_4]$ tetrahedra

(22:20) indicate that $Ba_6Zn_6(B_3O_6)_6(B_6O_{12})$ may have remarkable optical anisotropy. In addition, the dangling bonds of terminal in two kinds of B–O clusters are eliminated by the covalent $[ZnO_4]$ tetrahedra; thus, the short-wavelength absorption cut off edge has a blue shift. The basic physical properties of $Ba_6Zn_6(B_3O_6)_6(B_6O_{12})$ were also studied. The transmission/absorption spectra indicate that $Ba_6Zn_6(B_3O_6)_6(B_6O_{12})$ possesses a wide transparency window from 180 nm to 3405 nm. The difference of refractive indices based on a (001) wafer at 589.3 nm is as large as 0.14, which indicates that the birefringence of $Ba_6Zn_6(B_3O_6)_6(B_6O_{12})$ is even larger than the commercialized α-BaB_2O_4. Moreover, thermal analysis demonstrates that $Ba_6Zn_6(B_3O_6)_6(B_6O_{12})$ melts congruently. The acquirement of bulk crystals could be anticipated as is evidenced by the already grown sub-centimeter sized crystals.

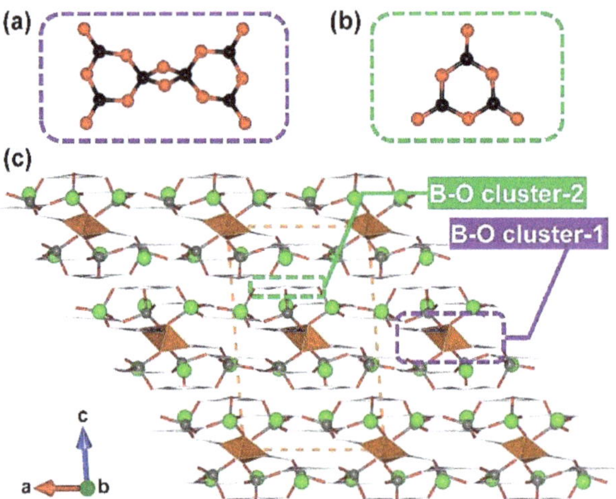

Figure 20. Two types of different FBBs occur in $Ba_6Zn_6(B_3O_6)_6(B_6O_{12})$: (**a**) $[B_6O_{12}]$ FBB; (**b**) $[B_3O_6]$ FBB; (**c**) the structure of $Ba_6Zn_6(B_3O_6)_6(B_6O_{12})$ along [010] direction. Key: green ball, Ba atom; grey ball, Zn atom; black ball, B atom; red ball, O atom; orange/green tetrahedron, edge-/corner-sharing $[BO_4]$; purple tringle $[BO_3]$.

4. Conclusions

The synthesis of edge-sharing borates greatly changes the rule of corner sharing B–O units in borate structures, and further work demonstrates that the extreme synthetic conditions, such as high pressure, are not necessary for edge-sharing borates. The crystalline borates with edge-sharing $[BO_4]$ tetrahedra continue to develop; about 34 new edge-sharing borates containing edge-sharing B_2O_6 unit have been found in recent years, among which three are crystallized in noncentrosymmetric space groups, only about 10% in the whole edge-sharing borates. This ratio is much smaller than 35% for the entire borate system, which may be attributable to the $[BO_4]$ units likely formed under the high-pressure environment [97]. Noncentrosymmetric edge-sharing borates are needed to better understand the NLO property in these types of structures. Fortunately, more π-conjugated $[BO_3]$ units are found under the ambient-pressure environment; the high $[BO_3]$ and $[BO_4]$ ratio in edge-sharing borates may be beneficial for the formation of noncentrosymmetric structures.

The signal of the B_2O_6 structural motif can be unambiguously assigned in the B K-edge EELS spectrum. Some of these edge-sharing borates exhibit interesting properties, such as unusual anisotropic thermal expansion behavior. It is curious to chemists whether edge-sharing BO_3/BO_4, BO_3/BO_3, or even face-sharing B–O units can be realized in the future. It is also expected that the synthesis of edge-sharing $[BO_3F]^{4-}$, $[BO_2F_2]^{3-}$, and $[BOF_3]^{2-}$ units in the future will greatly enrich the structural chemistry of crystalline fluorooxoborates.

Finally, we should better understand the structure–property relationships of these edge-sharing borates, which will help us to find more applications.

Funding: This research was funded by the National Natural Science Foundation of China (Grant 21975224).

Institutional Review Board Statement: Not applicable.

Informed Consent Statement: Not applicable.

Data Availability Statement: Not applicable.

Conflicts of Interest: The authors declare no conflict of interest.

References

1. Mutailipu, M.; Poeppelmeier, K.R.; Pan, S. Borates: A rich source for optical materials. *Chem. Rev.* **2021**, *121*, 1130–1202. [CrossRef] [PubMed]
2. Chen, J.; Hu, C.; Kong, F.; Mao, J. High-performance second-harmonic-generation (SHG) materials: New developments and new strategies. *Acc. Chem. Res.* **2021**, *54*, 2775–2783. [CrossRef] [PubMed]
3. Tran, T.T.; Yu, H.; Rondinelli, J.M.; Poeppelmeier, K.R.; Halasyamani, P.S. Deep ultraviolet nonlinear optical materials. *Chem. Mater.* **2016**, *28*, 5238–5258. [CrossRef]
4. Halasyamani, P.S.; Zhang, W. Viewpoint: Inorganic materials for UV and deep-UV nonlinear-optical applications. *Inorg. Chem.* **2017**, *56*, 12077–12085. [CrossRef]
5. Kang, L.; Lin, Z. Deep-ultraviolet nonlinear optical crystals: Concept development and materials discovery. *Light Sci. Appl.* **2022**, *11*, 201. [CrossRef]
6. Mutailipu, M.; Yang, Z.; Pan, S. Toward the enhancement of critical performance for deep-ultraviolet frequency-doubling crystals utilizing covalent tetrahedra. *Acc. Mater. Res.* **2021**, *2*, 282–291. [CrossRef]
7. Li, X.; Li, J.; Cheng, J.; Yang, G. Two acentric aluminoborates incorporated d^{10} cations: Syntheses, structures, and nonlinear optical properties. *Inorg. Chem.* **2023**, *62*, 1264–1271. [CrossRef]
8. Li, Y.; Zhou, Z.; Zhao, S.; Liang, F.; Ding, Q.; Sun, J.; Lin, Z.; Hong, M.; Luo, J. A deep-UV nonlinear optical borosulfate with incommensurate modulations. *Angew. Chem. Int. Ed.* **2021**, *60*, 11457–11463. [CrossRef]
9. Shen, Y.; Zhao, S.; Luo, J. The role of cations in second-order nonlinear optical materials based on π-conjugated $[BO_3]^{3-}$ groups. *Coord. Chem. Rev.* **2018**, *366*, 1–28. [CrossRef]
10. Bai, S.; Wang, D.; Liu, H.; Wang, Y. Recent advances of oxyfluorides for nonlinear optical applications. *Inorg. Chem. Front.* **2021**, *8*, 1637–1654. [CrossRef]
11. Lin, Z.; Yang, G. Oxo boron clusters and their open frameworks. *Eur. J. Inorg. Chem.* **2011**, *26*, 3857–3867. [CrossRef]
12. Dong, Y.; Chen, C.; Chen, J.; Cheng, J.; Li, J.; Yang, G. Two porous-layered borates built by $B_7O_{13}(OH)$ clusters and AlO_4/GaO_4 tetrahedra. *Cryst. Eng. Comm.* **2022**, *24*, 8027–8033. [CrossRef]
13. Peng, G.; Lin, C.; Fan, H.; Chen, K.; Li, B.; Zhang, G.; Ye, N. $Be_2(BO_3)(IO_3)$: The first anion-mixed van der waals member in the $KBe_2BO_3F_2$ family with a very strong second harmonic generation response. *Angew. Chem. Int. Ed.* **2021**, *60*, 17415–17418. [CrossRef]
14. Pan, Y.; Guo, S.; Liu, B.; Xue, H.; Guo, G. Second-order nonlinear optical crystals with mixed anions. *Coord. Chem. Rev.* **2018**, *374*, 464–469. [CrossRef]
15. Song, J.; Hu, C.; Xu, X.; Kong, F.; Mao, J. A facile synthetic route to a new SHG material with two types of parallel π-conjugated planar triangular units. *Angew. Chem. Int. Ed.* **2015**, *54*, 3679–3682. [CrossRef]
16. Chen, C.; Pan, R.; Li, X.; Qin, D.; Yang, G. Four inorganic-organic hybrid borates: From 2D layers to 3D oxoboron cluster organic frameworks. *Inorg. Chem.* **2021**, *60*, 18283–18290. [CrossRef]
17. Wu, C.; Jiang, X.; Lin, L.; Dan, W.; Lin, Z.; Huang, Z.; Humphrey, M.G.; Zhang, C. Strong SHG responses in a beryllium-free deep-UV-transparent hydroxyborate via covalent bond modification. *Angew. Chem. Int. Ed.* **2021**, *60*, 27151–27157. [CrossRef]
18. Zhang, Y.; Li, Q.; Chen, B.; Lan, Y.; Cheng, J.; Yang, G. $Na_3B_6O_{10}(HCOO)$: An ultraviolet nonlinear optical sodium borate-formate. *Inorg. Chem. Front.* **2022**, *9*, 5032–5038. [CrossRef]
19. Jin, C.; Li, F.; Li, X.; Lu, J.; Yang, Z.; Pan, S.; Mutailipu, M. Difluoro(oxalato)borates as short-wavelength optical crystals with bifunctional $[BF_2C_2O_4]$ units. *Chem. Mater.* **2022**, *34*, 7516–7525. [CrossRef]
20. Jin, C.; Li, F.; Cheng, B.; Qiu, H.; Yang, Z.; Pan, S.; Mutailipu, M. Double-modification oriented design of a deep-UV birefringent crystal functionalized by $[B_{12}O_{16}F_4(OH)_4]$ Clusters. *Angew. Chem. Int. Ed.* **2022**, *61*, e202203984. [CrossRef]
21. Wang, Q.; Yang, F.; Wang, X.; Zhou, J.; Ju, J.; Huang, L.; Gao, D.; Bi, J.; Zou, G. Deep-ultraviolet mixed-alkali-metal borates with induced enlarged birefringence derived from the structure rearrangement of the LiB_3O_5. *Inorg. Chem.* **2019**, *58*, 5949–5955. [CrossRef] [PubMed]
22. Liu, J.; Lee, M.-H.; Li, C.; Meng, X.; Yao, J. Growth, structure, and optical properties of a nonlinear optical niobium borate crystal $CsNbOB_2O_5$ with distorted NbO_5 square pyramids. *Inorg. Chem.* **2022**, *61*, 19302–19308. [CrossRef] [PubMed]
23. Huang, J.; Jin, C.; Xu, P.; Gong, P.; Lin, Z.; Cheng, J.; Yang, G. $Li_2CsB_7O_{10}(OH)_4$: A deep-ultraviolet nonlinear-optical mixed-alkaline borate constructed by unusual heptaborate anions. *Inorg. Chem.* **2019**, *58*, 1755–1758. [CrossRef] [PubMed]

24. Ma, W.; Zhang, J.; Yu, F.; Dai, B. NaK$_5$Zn$_2$(B$_5$O$_{10}$)$_2$ and β-K$_3$ZnB$_5$O$_{10}$: Two zincoborates with deep-UV cutoff edge. *Inorg. Chem.* **2022**, *61*, 16533–16538. [CrossRef]

25. Tian, H.; Wang, W.; Gao, Y.; Deng, T.; Wang, J.; Feng, Y.; Cheng, J. Facile assembly of an unusual lead borate with different cluster building units *via* a hydrothermal process. *Inorg. Chem.* **2013**, *52*, 6242–6244. [CrossRef]

26. Cao, G.; Wei, Q.; Cheng, J.; Cheng, L.; Yang, G. A zeolite CAN-type aluminoborate with gigantic 24-ring channels. *Chem. Commun.* **2016**, *52*, 1729–1732. [CrossRef]

27. Liu, Q.; Wu, Q.; Wang, T.; Kang, L.; Lin, Z.; Wang, Y.; Xia, M. Polymorphism of LiCdBO$_3$: Crystal structures, phase transitions and optical characterizations. *Chin. J. Struct. Chem.* **2023**, *42*, 100026. [CrossRef]

28. Pan, R.; Cheng, J.; Yang, B.; Yang, G. CsB$_x$Ge$_{6-x}$O$_{12}$ (x = 1): A zeolite sodalite-type borogermanate with a high Ge/B ratio by partial boron substitution. *Inorg. Chem.* **2017**, *56*, 2371–2374. [CrossRef]

29. Li, W.; Deng, J.; Pan, C. BO$_3$ Triangle and B@Zn$_2$O$_3$ cationic layer in the structure of the hybrid zinc acetate borate [ZnAc]·[ZnBO$_3$]. *Inorg. Chem.* **2021**, *60*, 1289–1293. [CrossRef]

30. Liu, Y.; Pan, R.; Cheng, J.; He, H.; Yang, B.; Zhang, Q.; Yang, G. A series of aluminoborates templated or supported by zinc-amine complexes. *Chem. Eur. J.* **2015**, *21*, 15732–15739. [CrossRef]

31. Yu, S.; Gu, X.; Deng, T.; Huang, J.; Cheng, J.; Yang, G. Centrosymmetric (Hdima)$_2$[Ge$_5$B$_3$O$_{15}$(OH)] and noncentrosymmetric Na$_4$Ga$_3$B$_4$O$_{12}$(OH): Solvothermal/surfactant-thermal synthesis of open-framework borogermanate and galloborate. *Inorg. Chem.* **2017**, *56*, 12695–12698. [CrossRef]

32. Wu, B.; Tang, D.; Ye, N.; Chen, C. Linear and nonlinear optical properties of the KBe$_2$BO$_3$F$_2$ (KBBF) crystal. *Opt. Mater.* **1996**, *5*, 105–109. [CrossRef]

33. Chen, C.; Wu, Y.; Jiang, A.; Wu, B.; You, G.; Li, R.; Lin, S. New nonlinear-optical crystal: LiB$_3$O$_5$. *J. Opt. Soc. Am. B* **1989**, *6*, 616–621. [CrossRef]

34. Chen, C.; Wu, B.; Jiang, A.; You, G. A new-type ultraviolet SHG crystal β-BaB$_2$O$_4$. *Sci. Sin. Ser. B* **1985**, *28*, 235–243.

35. Zhou, G.; Xu, J.; Chen, X.; Zhong, H.; Wang, S.; Xu, K.; Deng, P.; Gan, F. Growth and spectrum of a novel birefringent α-BaB$_2$O$_4$ crystal. *J. Cryst. Growth* **1998**, *191*, 517–519.

36. Huang, C.; Mutailipu, M.; Zhang, F.; Griffith, K.J.; Hu, C.; Yang, Z.; Griffin, J.M.; Poeppelmeier, K.R.; Pan, S. Expanding the chemistry of borates with functional [BO$_2$]$^-$ anions. *Nat. Commun.* **2021**, *12*, 2597. [CrossRef]

37. Zhang, Y.; Li, F.; Yang, R.; Yang, Y.; Zhang, F.; Yang, Z.; Pan, S. Rb$_5$Ba$_2$(B$_{10}$O$_{17}$)$_2$(BO$_2$): The formation of unusual functional [BO$_2$]$^-$ in borates with deep-ultraviolet transmission window. *Sci. China Chem.* **2022**, *65*, 719–725. [CrossRef]

38. Ding, F.; Griffith, K.J.; Zhang, W.; Cui, S.; Zhang, C.; Wang, Y.; Kamp, K.; Yu, H.; Halasyamani, P.S.; Yang, Z.; et al. NaRb$_6$(B$_4$O$_5$(OH)$_4$)$_3$(BO$_2$) featuring noncentrosymmetry, chirality, and the linear anionic group BO$_2$. *J. Am. Chem. Soc.* **2023**, *145*, 4928–4933. [CrossRef]

39. Nowogrocki, G.; Penin, N.; Touboul, M. Crystal structure of Cs$_3$B$_7$O$_{12}$ containing a new large polyanion with 63 boron atoms. *Solid State Sci.* **2003**, *5*, 795–803. [CrossRef]

40. Wang, J.; Yang, G. A novel supramolecular magnesoborate framework with snowflake-like channels built by unprecedented huge B$_{69}$ cluster cages. *Chem. Commun.* **2017**, *53*, 10398–10401. [CrossRef]

41. Zhao, W.; Zhang, Y.; Lan, Y.; Cheng, J.; Yang, G. Ba$_2$B$_{10}$O$_{16}$(OH)$_2$·(H$_3$BO$_3$)(H$_2$O): A possible deep-ultraviolet nonlinear-optical barium borate. *Inorg. Chem.* **2022**, *61*, 4246–4250. [CrossRef] [PubMed]

42. Wang, J.; Wei, Q.; Cheng, J.; He, H.; Yang, B.; Yang, G. Na$_2$B$_{10}$O$_{17}$·H$_2$en: A three dimensional open-framework layered borate co-templated by inorganic cations and organic amines. *Chem. Commun.* **2015**, *51*, 5066–5068. [CrossRef] [PubMed]

43. Wei, Q.; Cheng, J.; He, C.; Yang, G. An acentric calcium borate Ca$_2$[B$_5$O$_9$]·(OH)·H2O: Synthesis, structure, and nonlinear optical property. *Inorg. Chem.* **2014**, *53*, 11757–11763. [CrossRef]

44. Li, X.; Yang, G. LiB$_9$O$_{15}$·H$_2$dap·H$_2$O: A cotemplated acentric layer-pillared borate built by mixed oxoboron clusters. *Inorg. Chem.* **2021**, *60*, 16085–16089. [CrossRef] [PubMed]

45. Li, X.; Yang, G. Two mixed alkali-metal borates templated from cations to clusters. *Inorg. Chem.* **2022**, *61*, 10205–10210. [CrossRef]

46. Wang, E.; Huang, J.; Yu, S.; Lan, Y.; Cheng, J.; Yang, G. An ultraviolet nonlinear optic borate with 13-ring channels constructed from different building units. *Inorg. Chem.* **2017**, *56*, 6780–6783. [CrossRef]

47. Wei, Q.; Wang, J.; He, C.; Cheng, J.; Yang, G. Deep-ultraviolet nonlinear optics in a borate framework with 21-Ring channels. *Chem. Eur. J.* **2016**, *22*, 10759–10762. [CrossRef]

48. Wang, J.; Cheng, J.; Wei, Q.; He, H.; Yang, B.; Yang, G. NaB$_3$O$_5$·0.5H$_2$O and NH$_4$NaB$_6$O$_{10}$: Two cluster open frameworks with chiral quartz and achiral primitive cubic nets constructed from oxo boron cluster building units. *Eur. J. Inorg. Chem.* **2014**, *2014*, 4079–4083. [CrossRef]

49. Zhi, S.; Wang, Y.; Sun, L.; Cheng, J.; Yang, G. Linking 1D transition-metal coordination polymers and different inorganic boron oxides to construct a series of 3D inorganic-organic hybrid borates. *Inorg. Chem.* **2018**, *57*, 1350–1355. [CrossRef]

50. Huppertz, H.; von der Eltz, B. Multianvil high-pressure synthesis of Dy$_4$B$_6$O$_{15}$: The first oxoborate with edge-sharing BO$_4$ tetrahedra. *J. Am. Chem. Soc.* **2002**, *124*, 9376–9377. [CrossRef]

51. Jin, S.; Cai, G.; Wang, W.; He, M.; Wang, S.; Chen, X. Stable oxoborate with edge-sharing BO$_4$ tetrahedra synthesized under ambient pressure. *Angew. Chem. Int. Ed.* **2010**, *49*, 4967–4970. [CrossRef]

52. Ouyang, T.; Shen, Y.; Zhao, S. Accurate design and synthesis of nonlinear optical crystals employing KBe$_2$BO$_3$F$_2$ as structural template. *Chin. J. Struct. Chem.* **2023**, *42*, 100024. [CrossRef]

53. Su, H.; Yan, Z.; Hou, X.; Zhang, M. Fluorooxoborates: A precious treasure of deep-ultraviolet nonlinear optical materials. *Chin. J. Struct. Chem.* **2023**, *42*, 100027. [CrossRef]

54. Mutailipu, M.; Zhang, M.; Yang, Z.; Pan, S. Targeting the next generation of deep-ultraviolet nonlinear optical materials: Expanding from borates to borate fluorides to fluorooxoborates. *Acc. Chem. Res.* **2019**, *52*, 791–801. [CrossRef]

55. Leonyuk, N.I.; Maltsev, V.V.; Volkova, E.A. Crystal chemistry of high-temperature borates. *Molecules* **2020**, *25*, 2450. [CrossRef]

56. Huppertz, H. New synthetic discoveries *via* high-pressure solid-state chemistry. *Chem. Commun.* **2011**, *47*, 131–140. [CrossRef]

57. Silver, M.A.; Albrecht-Schmitt, T.E. Evaluation of *f*-element borate chemistry. *Coord. Chem. Rev.* **2016**, *323*, 36–51. [CrossRef]

58. Schubert, D.M. Hydrated zinc borates and their industrial use. *Molecules* **2019**, *24*, 2419. [CrossRef]

59. Chen, Y.; Zhang, M.; Mutailipu, M.; Poeppelmeier, K.R.; Pan, S. Research and development of zincoborates: Crystal growth, structural chemistry and physicochemical properties. *Molecules* **2019**, *24*, 2763. [CrossRef]

60. Jiao, J.; Zhang, M.; Pan, S. Aluminoborates as nonlinear optical materials. *Angew. Chem. Int. Ed.* **2022**, *61*, e202217037.

61. Li, Q.; Chen, W.; Lan, Y.; Cheng, J. Recent progress in ultraviolet and deep-ultraviolet nonlinear optical aluminoborates. *Chin. J. Struct. Chem.* **2023**, *42*, 100036. [CrossRef]

62. Zhang, J.; Kong, F.; Xu, X.; Mao, J. Crystal structures and second-order NLO properties of borogermanates. *J. Solid State Chem.* **2012**, *195*, 63–72. [CrossRef]

63. Xin, S.; Zhou, M.; Beckett, M.A.; Pan, C. Recent advances in crystalline oxidopolyborate complexes of *d*-block or *p*-block metals: Structural aspects, syntheses, and physical properties. *Molecules* **2021**, *26*, 3815. [CrossRef] [PubMed]

64. Beckett, M.A. Recent advances in crystalline hydrated borates with non-metal or transition-metal complex cations. *Coord. Chem. Rev.* **2016**, *323*, 2–14. [CrossRef]

65. Huppertz, H. High-pressure preparation, crystal structure, and properties of $RE_4B_6O_{15}$ (RE = Dy, Ho) with an extension of the "fundamental building block"-descriptors. *Z. Naturforsch.* **2003**, *58*, 278–290. [CrossRef]

66. Emme, H.; Huppertz, H. $Gd_2B_4O_9$: Ein weiteres oxoborat mit kanten-verknüpften BO_4-tetraedern. *Z. Anorg. Allg. Chem.* **2002**, *628*, 2165. [CrossRef]

67. Emme, H.; Huppertz, H. High-pressure preparation, crystal structure, and properties of α-$(RE)_2B_4O_9$ (RE = Eu, Gd, Tb, Dy): Oxoborates displaying a new type of structure with edge-sharing BO_4 tetrahedra. *Chem. Eur. J.* **2003**, *9*, 3623–3633. [CrossRef]

68. Emme, H.; Huppertz, H. High-pressure syntheses of α-$RE_2B_4O_9$ (RE = Sm, Ho), with a structure type displaying edge-sharing BO_4 tetrahedra. *Acta Crystallogr. C* **2005**, *61*, I29–I31. [CrossRef]

69. Schmitt, M.K.; Huppertz, H. High-pressure synthesis and crystal structure of α-$Y_2B_4O_9$. *Z. Naturforsch.* **2017**, *72*, 977–982. [CrossRef]

70. Fuchs, B.; Heymann, G.; Wang, X.F.; Tudi, A.; Bayarjargal, L.; Siegel, R.; Schmutzler, A.; Senker, J.; Joachim-Mrosko, B.; Saxer, A.; et al. $La_3B_6O_{13}(OH)$: The first acentric high-pressure borate displaying edge-sharing BO_4 tetrahedra. *Chem. Eur. J.* **2020**, *26*, 6851–6861. [CrossRef]

71. Knyrim, J.S.; Roeßner, F.; Jakob, S.; Johrendt, D.; Kinski, I.; Glaum, R.; Huppertz, H. Formation of edge-sharing BO_4 tetrahedra in the high-pressure borate HP-NiB_2O_4. *Angew. Chem. Int. Ed.* **2007**, *46*, 9097–9100. [CrossRef]

72. Neumair, S.C.; Glaum, R.; Huppertz, H. Synthesis and crystal structure of the high-pressure iron borate β-FeB_2O_4. *Z. Naturforsch.* **2009**, *64b*, 883–890. [CrossRef]

73. Neumair, S.C.; Kaindl, R.; Huppertz, H. Synthesis and crystal structure of the high-pressure cobalt borate HP-CoB_2O_4. *Z. Naturforsch.* **2010**, *65b*, 1311–1317. [CrossRef]

74. Pakhomova, A.; Fuchs, B.; Dubrovinsky, L.S.; Natalia Dubrovinskaia, N.; Huppertz, H. Polymorphs of the gadolinite-type borates ZrB_2O_5 and HfB_2O_5 under extreme pressure. *Chem. Eur. J.* **2021**, *27*, 6007–6014. [CrossRef]

75. Neumair, S.C.; Knyrim, J.S.; Oeckler, O.; Glaum, R.; Kaindl, R.; Stalder, R.; Huppertz, H. Intermediate states on the way to edge-sharing BO_4 tetrahedra in $M_6B_{22}O_{39} \cdot H_2O$ (M = Fe, Co). *Chem. Eur. J.* **2010**, *16*, 13659–13670. [CrossRef]

76. Neumair, S.C.; Kaindl, R.; Huppertz, H. The new high-pressure borate $Co_7B_{24}O_{42}(OH)_2 \cdot 2\,H_2O$—Formation of edge-sharing BO_4 tetrahedra in a hydrated borate. *J. Solid State Chem.* **2012**, *185*, 1–9. [CrossRef]

77. Neumair, S.C.; Vanicek, S.; Kaindl, R.; Többens, D.M.; Martineau, C.; Taulelle, F.; Senker, J.; Huppertz, H. HP-KB_3O_5 highlights the structural diversity of borates: Corner-sharing BO_3/BO_4 groups in combination with edge-sharing BO_4 tetrahedra. *Eur. J. Inorg. Chem.* **2011**, *2011*, 4147–4152. [CrossRef]

78. Sohr, G.; Neumair, S.C.; Huppertz, H. High-pressure synthesis and characterization of the alkali metal borate HP-RbB_3O_5. *Z. Naturforsch.* **2012**, *67b*, 1197–1204. [CrossRef]

79. Sohr, G.; Perfler, L.; Huppertz, H. The high-pressure thallium triborate HP-TlB_3O_5. *Z. Naturforsch.* **2014**, *69b*, 1260–1268. [CrossRef]

80. Sohr, G.; Neumair, S.C.; Heymann, G.; Wurst, K.; Schmedt auf der Gunne, J.; Huppertz, H. Oxonium ions substituting cesium ions in the structure of the new high-pressure borate HP-$Cs_{1-x}(H_3O)_xB_3O_5$ (x = 0.5–0.7). *Chem. Eur. J.* **2014**, *20*, 4316–4323. [CrossRef]

81. Sohr, G.; Többens, D.M.; Schmedt auf der Gunne, J.; Huppertz, H. HP-CsB_5O_8: Synthesis and characterization of an outstanding borate exhibiting the simultaneous linkage of all structural units of borates. *Chem. Eur. J.* **2014**, *20*, 17059–17067. [CrossRef] [PubMed]

82. Gorelova, L.; Pakhomova, A.; Aprilis, G.; Yin, Y.Q.; Laniel, D.; Winkler, B.; Krivovichev, S.; Pekov, I.; Dubrovinskaia, N.; Dubrovinsky, L. Edge-sharing BO_4 tetrahedra and penta-coordinated silicon in the high-pressure modification of $NaBSi_3O_8$. *Inorg. Chem. Front.* **2022**, *9*, 1735–1742. [CrossRef]

83. Tatyana, B.; Bekker, T.B.; Podborodnikov, I.V.; Sagatov, N.E.; Shatskiy, A.; Rashchenko, S.; Sagatova, D.N.; Davydov, A.; Litasov, K.D. γ-BaB$_2$O$_4$: High-pressure high-temperature polymorph of barium borate with edge-sharing BO$_4$ tetrahedra. *Inorg. Chem.* **2022**, *61*, 2340–2350.

84. Jen, I.-H.; Lee, Y.C.; Tsai, C.E.; Lii, K.H. Edge-sharing BO$_4$ tetrahedra in the structure of hydrothermally synthesized barium borate: α-Ba$_3$[B$_{10}$O$_{17}$(OH)$_2$]. *Inorg. Chem.* **2019**, *58*, 4085–4088. [CrossRef]

85. Wu, Y.; Yao, J.; Zhang, J.; Fu, P.Z.; Wu, Y. Potassium zinc borate, KZnB$_3$O$_6$. *Acta. Cryst. E* **2010**, *66*, i45. [CrossRef]

86. Yang, L.; Fan, W.; Li, Y.; Sun, H.; Wei, L.; Cheng, X.; Zhao, X. Theoretical insight into the structural stability of KZnB$_3$O$_6$ polymorphs with different BO$_x$ polyhedral networks. *Inorg. Chem.* **2012**, *51*, 6762–6770. [CrossRef]

87. Lou, Y.; Li, D.; Li, Z.; Jin, S.; Chen, X. Unidirectional thermal expansion in edge-sharing BO$_4$ tetrahedra contained KZnB$_3$O$_6$. *Sci. Rep.* **2015**, *5*, 10996. [CrossRef]

88. Lou, Y.; Li, D.; Li, Z.; Zhang, H.; Jin, S.; Chen, X. Unidirectional thermal expansion in KZnB$_3$O$_6$: Role of alkali metals. *Dalton Trans.* **2015**, *44*, 19763–19767. [CrossRef]

89. Chen, X.; Chen, Y.; Sun, C.; Chang, X.; Xiao, W. Synthesis, crystal structure, spectrum properties, and electronic structure of anew three-borate Ba$_4$Na$_2$Zn$_4$(B$_3$O$_6$)$_2$(B$_{12}$O$_{24}$) with two isolated types of blocks:3[3Δ] and 3[2Δ + 1T]. *J. Alloys Compd.* **2013**, *568*, 60–67. [CrossRef]

90. Mutailipu, M.; Zhang, M.; Li, H.; Fan, X.; Yang, Z.; Jin, S.; Wang, G.; Pan, S. Li$_4$Na$_2$CsB$_7$O$_{14}$: A new edge-sharing [BO$_4$]$^{5-}$ tetrahedra containing borate with high anisotropic thermal expansion. *Chem. Commun.* **2019**, *55*, 1295–1298. [CrossRef]

91. Guo, F.; Han, J.; Cheng, S.; Yu, S.; Yang, Z.; Pan, S. Transformation of the B$-$O units from corner-sharing to edge-sharing linkages in BaMBO$_4$ (M = Ga, Al). *Inorg. Chem.* **2019**, *58*, 8237–8244. [CrossRef]

92. Han, S.; Huang, C.; Tudi, A.; Hu, S.; Yang, Z.; Pan, S. β-CsB$_9$O$_{14}$: A triple-layered borate with edge-sharing BO$_4$ tetrahedra exhibiting a short cutoff edge and a large birefringence. *Chem.-Eur. J.* **2019**, *25*, 11614–11619. [CrossRef]

93. Guo, S.; Zhang, W.; Yang, R.; Zhang, M.; Yang, Z.; Pan, S. Pb$_{2.28}$Ba$_{1.72}$B$_{10}$O$_{19}$ featuring a three-dimensional B–O anionic network with edge-sharing [BO$_4$] obtained under ambient pressure. *Inorg. Chem. Front.* **2021**, *8*, 3716–3722. [CrossRef]

94. Quarez, E.; Gautron, E.; Paris, M.; Gajan, D.; Mevellec, J. Toward the coordination fingerprint of the edge-sharing BO$_4$. *Inorg. Chem.* **2021**, *60*, 2406–2413. [CrossRef]

95. Xie, W.; Fang, Z.; Mao, J. Ba$_6$Zn$_6$(B$_3$O$_6$)$_6$(B$_6$O$_{12}$): Barium zinc borate contains π-conjugated [B$_3$O$_6$]$^{3-}$ anions and [B$_6$O$_{12}$]$^{6-}$ anion with edge-sharing BO$_4$ tetrahedra. *Inorg. Chem.* **2022**, *61*, 18260–18266. [CrossRef]

96. Han, J.; Liu, K.; Chen, L.; Li, F.; Yang, Z.; Zhang, F.; Pan, S.; Mutailipu, M. Finding a deep-UV borate BaZnB$_4$O$_8$ with edge-sharing [BO$_4$] tetrahedra and strong optical anisotropy. *Chem. Eur. J.* **2023**, *29*, e202203000. [CrossRef]

97. Edwards, T.; Endo, T.; Walton, J.H.; Sen, S. Observation of the transition state for pressure-induced BO$_3$→BO$_4$ conversion in glass. *Science* **2014**, *345*, 1027–1029. [CrossRef]

Communication

Facile Attachment of Halides and Pseudohalides to Dodecaborate(2-) via Pd-catalyzed Cross-Coupling

Mahmoud K. Al-Joumhawy, Jui-Chi Chang, Fariba Sabzi and Detlef Gabel *

School of Science, Constructor University, 28759 Bremen, Germany
* Correspondence: dgabel@constructor.university

Abstract: Cross-coupling reactions with $[B_{12}H_{11}I]^{2-}$ as one partner have been used successfully for Kumada and Buchwald Hartwig couplings with Pd catalysis. Here, we found that the iodide could be substituted easily, and unexpectedly, with other halides such as Br and Cl, and with pseudohalides such as cyanide, azide, and isocyanate. We found that for Cl, Br, N_3, and NCO, tetrabutylammonium salts—or sodium salts—were successful halide sources, whereas for cyanide, CuCN was the only halide source that allowed a successful exchange. The azide could be reacted further in a click reaction with triazoles. While no substitution with fluoride occurred, tetrabutylammonium fluoride in the presence of water led to $[B_{12}H_{11}OH]^{2-}$. Yields were high to very high, and reaction times were short when using a microwave oven as a heating source.

Keywords: dodecaborate(2-); dodecahydrido-closo-dodecaborate(2-); Pd catalysis; halogen exchange; click reaction

Citation: Al-Joumhawy, M.K.; Chang, J.-C.; Sabzi, F.; Gabel, D. Facile Attachment of Halides and Pseudohalides to Dodecaborate(2-) via Pd-catalyzed Cross-Coupling. *Molecules* **2023**, *28*, 3245. https://doi.org/10.3390/molecules28073245

Academic Editor: Michael A. Beckett

Received: 13 March 2023
Revised: 30 March 2023
Accepted: 2 April 2023
Published: 5 April 2023

1. Introduction

Dodecaborate clusters are of increasing interest because of their special supramolecular properties: they are water soluble due to their inherent charge, and yet they can interact strongly with hydrophobic surfaces. This can be used in, e.g., their binding to the interior of cyclodextrins [1–4], or to the outside surface of cucurbiturils [5], and it also manifests itself in large retention factors in chromatography on hydrophilic matrices in water [6]. They can form ionic liquids even with Li^+ as the cation [7]. At the basis of these phenomena is the weak interaction of the cluster with water as a solvent [8]. While a proper choice of counterion allows good to excellent solubility in water, the solvent interacts only weakly and can therefore be removed easily. This phenomenon makes dodecaborate compounds the first example of superchaotropic agents [4,9].

Such properties are not found in organic compounds. Making full use of these properties requires that the boron cluster is attached to other (organic, organometallic, or inorganic) units. This requires that suitable functionalization methods allow the dodecaborate to react with other functional groups in other building blocks, so that a linkage between them can combine properties from both components. In the past, known types of reactions have been rather limited, and they were achieved mostly by attaching organic moieties to heteroatoms such as O [10,11], N [12–14], and S [2,15] (for an overview, see Scheme 1).

Recently, we have found that Pd-catalyzed cross-coupling with anilines and amides of monoiodo-dodecaborate ($[B_{12}H_{11}I]^{2-}$) leads to the formation of B–N-linkers between the dodecaborate and the organic moiety [16]. Previously, $[B_{12}H_{11}I]^{2-}$ was reacted under Kumada coupling conditions with Grignard reagents [17,18]. The procedure to introduce an iodine atom into the $[B_{12}H_{12}]^{2-}$ cluster could be improved considerably by using not molecular iodine, but N-iodosuccinimide, resulting in very clean stoichiometric iodinations [19]. The corresponding N-chloro- and N-bromosuccinimides result in the corresponding chloro- and bromododecaborates, again in clean reactions with well-controllable stoichiometry. The $[B_{12}H_{11}NH_3]^-$ was obtained before by reaction with hydroxylamine-O-sulfonic acid; using a Buchwald–Hartwig cross-coupling, we could obtain $[B_{12}H_{11}NH_3]^-$ from $[B_{12}H_{11}I]^{2-}$ with urea as source of nitrogen, followed by acid-catalyzed hydrolysis.

Previous work

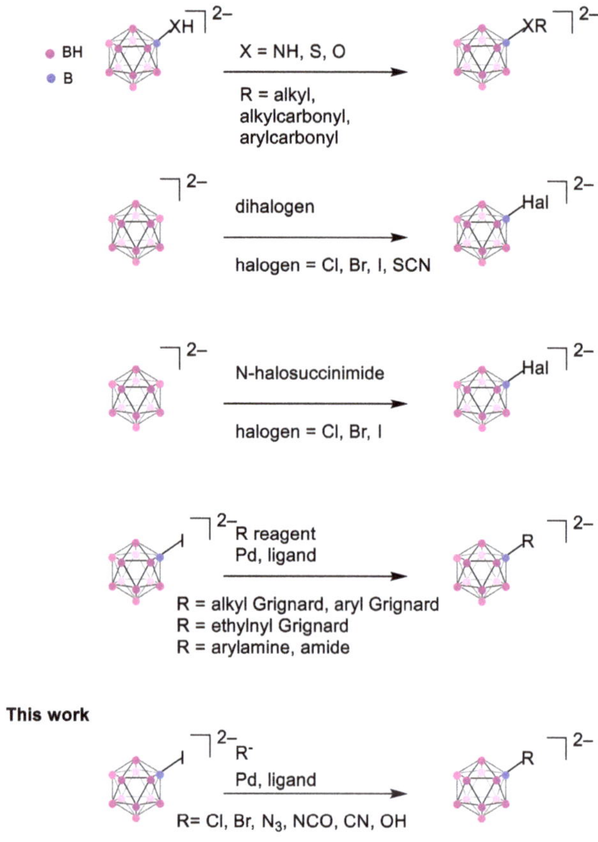

Scheme 1. Substitution reactions of $[B_{12}H_{12}]^{2-}$.

While the selective introduction of a single halogen atom is possible, the formation of $[B_{12}H_{11}OH]^{2-}$, described before [20], uses strongly acidic conditions; it therefore usually leads also to di-substituted products as byproducts [20]. Other dodecaborate substituents such as cyanide, azide, cyanate, and thiocyanate have not been described. The isocyanate derivative has been obtained before through a Curtius rearrangement of a carbonyl azide [21].

In the course of our continued exploration of the further reactivity of $[B_{12}H_{11}I]^{2-}$, especially with respect to other Pd-catalyzed reactions, reactions with apparently innocent salts, such as NaBr and tetrabutylammonium bromide, occurred. We were very surprised to find that an exchange of the iodide with other nucleophiles was possible, and we could obtain compounds which had not been obtained before. Several of these compounds can be reacted further, while others might be unwanted side products in further reactions.

2. Results

Our investigation started with an observation when we tried to run Pd-catalyzed cross-coupling reactions in a heterogeneous system of water and immiscible solvents. In order to increase the solubility of water-soluble anions, we wanted to resort to phase transfer catalysts such as Bu$_4$NBr to transfer the anion into the organic phase. To our surprise, even in the absence of any further reagent (except KOH, which would serve as a base for other reactions) we found conversion of [B$_{12}$H$_{11}$I]$^{2-}$ to the corresponding [B$_{12}$H$_{11}$Br]$^{2-}$. Halogen exchange is known to occur in aryl halides [22], but we had not observed this reaction when further, probably better, groups with good ability to interact with the Pd cation were present. We therefore checked further salts, first with the tetrabutylammonium cation, and then with Na$^+$ as the counterion.

Successful exchange reactions could be performed using Pd$_2$(dba)$_3$ as the Pd source and Davephos as the ligand in DMSO, as previously found for Buchwald–Hartwig type cross-couplings (Ref. [16]). For the reaction with the halides and pseudohalides Br$^-$, Cl$^-$, and N$_3$$^-$, the choice of counterion (Bu$_4$N$^+$ or Na$^+$) did not matter, and the corresponding derivatives were obtained in good yield. Depending on the workup, there might be a partial or complete exchange of the cation when starting with Bu$_4$N$^+$ salts of the cluster and using a sodium halide as a halide source.

Reactions were considerably faster, at identical temperatures, when conducted with microwave heating. This is in agreement with our previous experience with substitution reactions on dodecaborates [16,23]. We attribute this to the inherent dipole of the cluster and the ionic conduction, both of which are known to increase the energy transfer of microwave radiation into the solution.

We then checked whether KOH as base (required for Buchwald–Hartwig or Suzuki reactions) was necessary by leaving it out, and the reaction proceeded equally well without KOH. Scheme 2 shows the conditions and the scope.

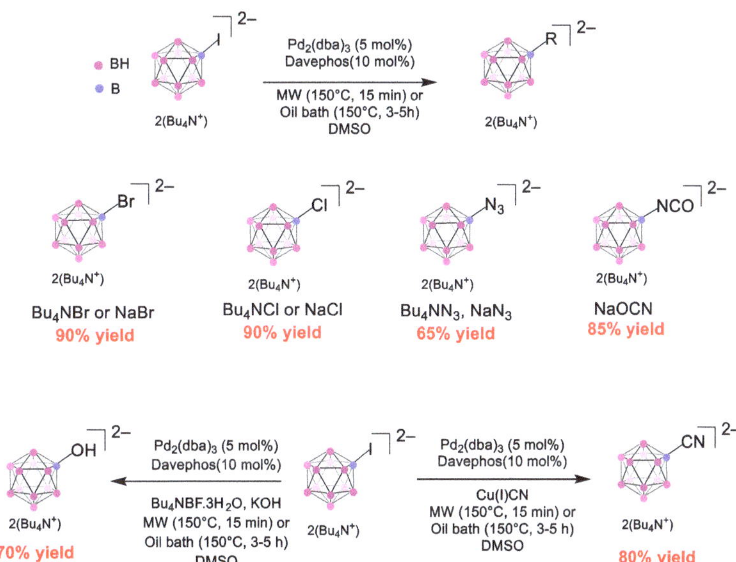

Scheme 2. Reaction of [B$_{12}$H$_{11}$I]$^{2-}$ to halides and pseudohalides.

We were able to obtain an isocyanate using NaNCO as a cyanate donor. This compound has been obtained before by the reaction of the CO derivative of dodecaborate [24] with sodium azide through a Curtius rearrangement [21]. On the basis of the ^{11}B NMR spectrum, we identified the N atom as the B-bonded atom, which resonates at -10 ppm in agreement with the product described in the literature [21]. For oxygen compounds bound to B, boron-11 resonates at positive ppm values (relative to BF$_3$ etherate) are observed [20,25], while the compound isolated here, as all other B-N cluster bonds, has a resonance at negative ppm values for the heteroatom-substituted B atom.

The analogous reaction with thiocyanate did not provide any product. This might, in part, be attributed to strong interactions between the sulfur of the thiocyanate and the Pd atom (observed for similar structures not containing boron [26]), preventing any further reactions. The thiocyanate had been obtained before by the reaction of the dirhodanide pseudohalogen with $[B_{12}H_{12}]^{2-}$ [27], in a reaction similar to other halogenation reactions.

When trying to exchange the iodide of $[B_{12}H_{11}I]^{2-}$ with F^-, using anhydrous tetrabutylammonium fluoride, or NaF, we did not succeed in obtaining any substitution with fluoride. When using the trihydrate of the Bu_4NF and KOH as a hydroxide source, a very clean reaction to $[B_{12}H_{11}OH]^{2-}$ occurred. Given the ease of preparing $[B_{12}H_{11}I]^{2-}$ in its pure monosubstituted form with N-iodosuccinimide and its smooth conversion to $[B_{12}H_{11}OH]^{2-}$ described here, this route might be preferable to the routes described in the past, which used aqueous acid.

The hydroxy derivative was a side product when using the common hydroxides (sodium, potassium, tetrabutylammonium—the latter either in water or in THF). With these reagents, only a little hydroxydodecaborate (20% at most) was formed. This also manifests itself in the observation that KOH is a good base for cross-coupling reactions, both here and in the Buchwald–Hartwig reaction, without the formation of hydroxydodecaborate [16]. This is in contrast with the reaction of aryl halides with KOH in water, which, under Pd catalysis, leads to phenols in excellent yields [28].

For CN^- as anion, neither tetrabutylammonium nor sodium as countercations gave the desired product. In addition, KCN, $Zn(CN)_2$, and $K_4[Fe(CN)_6]$ did not provide any cyano products. Only its copper(I) salt could be used successfully (Scheme 2).

Previously, the complete replacement of the iodine atoms on $[B_{12}I_{12}]^{2-}$ with CN^-, using Pd salts and prolonged high-temperature heating with microwave irradiation, has been reported by Kamin et al. [29]. A bulky Pd ligand, *t*BuPrettPhos, more space demanding than the Davephos used here, was needed for achieving reasonable yields.

Further reactions of the new compounds obtained here (see Scheme 3) are also of great interest.

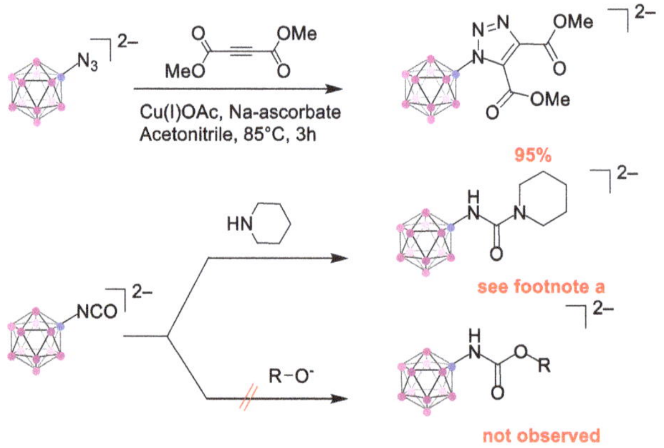

Scheme 3. Reactions of $[B_{12}H_{11}NCO]^{2-}$ and $[B_{12}H_{11}N_3]^{2-}$. (a): Ref. [21].

The azide obtained here for the first time invites one to perform click chemistry. While this has been described for carboranyl azides [30], such reactions have not yet been described for the dodecaborate cluster. As a prototype reaction, we used acetylene dimethylcarboxylate. The desired product was obtained in excellent yield after only short reaction times and a simple workup.

The isocyanate (described previously and used for reactions by others [21]) is known to react to the corresponding urea derivative with amines. We tested whether we could also cause a reaction with alcohols under basic conditions to the corresponding carbamates. To our surprise, alkoxides could not react to form the desired product. We attribute this to the large electron density donated by the cluster to the carbonyl carbon. We have seen previously that the cluster considerably increases the pK_a values of the –SH and the –NH$_2$ groups attached to it [15,31].

3. Discussion

The exchange reactions described here work only with $[B_{12}H_{11}I]^{2-}$ as the starting material. The bromide and the chloride did not react; we had already observed this for cross-coupling reactions with amides and anilines [16]. This is in contrast to carboranes, where the bromide can undergo reactions to the isocyanate and azide, and to the amides [30].

We speculate that the exchange with the halides and pseudohalides happens after the oxidative insertion of the Pd into the B–I bond, similar to the mechanisms proposed for the aryl halide exchange [22]. The iodide that had been bound originally to the cluster can exchange with the incoming halide or pseudohalide, and the resulting complex undergoes reductive elimination to the final product. While one would expect this to work in both directions, we found before that in contrast to $[B_{12}H_{11}I]^{2-}$, the chloro- or bromoderivatives were non-reactive in cross-coupling with Pd under Buchwald–Hartwig conditions [16]. Thus, even if the formation of the bromo- or chloro-dodecaborate might not be an energetically favored reaction, due to the low reactivity of the product under cross-coupling conditions, the oxidative insertion of Pd into the B–Br or B–Cl bond will not be possible, and the product will therefore accumulate. The fluoride might be too hard a nucleophile to bind well to the Pd (which is known to be a soft electrophile) and to replace the iodide in the complex.

Our results might also be of importance when salts are added as presumably non-reacting additives to cross-coupling reactions with $[B_{12}H_{11}I]^{2-}$, as the anions of the salts might react preferentially.

The surprising formation of $[B_{12}H_{11}OH]^{2-}$ offers a new route to this compound, which can act as a nucleophile and be alkylated or acylated [11].

The preparation of the isocyanate has been achieved before through a multistep reaction from the carbonyl-substituted cluster [21]. The method proposed here is an alternative, which might result in an easier preparation of this compound.

While the bromo- and chloro-dodecaborates will be largely chemically inert, the pseudohalogens azide, isocyanate, and cyanide can react further. The reaction of the isocyanate with amines is known [21]. Even more interesting for further reactions is the click reaction of the azide with alkynes. As the click reaction occurs under mild conditions, and there are plenty of potential alkyne reaction partners, their conjugation with the dodecaborate and its unique properties might open up new opportunities in cellular transport of attached biological effectors [32], in covalent stabilization of non-covalent networks [33], and in further applications.

4. Materials and Methods

Dodecahydrido-*closo*-dodecaborate was purchased as Cs$^+$ salt from BASF. It was converted to tetrabutylammonium salt by precipitation from an aqueous solution with Bu$_4$NCl dissolved in water. All other chemicals and solvents were from Sigma-Aldrich, St. Louis, MO, USA, or Carl Roth, Karlsruhe, Germany, and were used as received.

^{11}B NMR spectra were recorded on a JEOL 400 MHz spectrometer at 25 °C. Chemical shifts were referenced relative to external BF$_3$ etherate. MestReNova V10.0.2-15465 S3 software (Mestrelab Research, Santiago de Compostela, Spain) was used to visualize the spectra. Coupling constants (J) are reported in Hertz (Hz). NMR spectra of the new compounds described here are provided in the Supplementary Materials (Figures S1–S16).

Mass spectra were recorded on a Waters QTOF Premier spectrometer in negative mode, using acetonitrile as solvent for electrospray. The m/z values reported below are those of the most intense peaks. MS spectra of the new compounds described here are provided in the Supplementary Materials (Figures S17–S22).

Reagents and solvents were commercially available and used without further purification. The $[B_{12}H_{11}I]^{2-}$ $(Bu_4N)_2$ was prepared according to the literature procedure, with some modifications. Tetrabutylammonium bromide, tetrabutylammonium chloride, tetrabutylammonium fluoride trihydrate, copper cyanide, sodium bromide, sodium chloride, tetrabutylammonium azide, sodium azide, Davephos, and $Pd_2(dba)_3$ were purchased from Sigma-Aldrich and were used as received. The DMSO, dichloromethane, acetonitrile, and silica gel (Grade 60, 230–400 Mesh) were from Carl Roth. Celite (545 filter aid, not acid washed, powder) was from Fisher. All cross-coupling reactions were performed in an oven-dried 10 mL round-bottom flask. The thin-layer chromatography (TLC) AluSil plates were from MachereyNagel, Düren, Germany. The TLC samples for borane-containing compounds were stained with 1 wt.% $PdCl_2$ in 6 M HCl and were developed using a heat gun. An open-vessel microwave oven (CEM Discover, Model 908860, or HNZXIB, Model MCR-3) was used.

4.1. General Procedure

A dry 10 mL round-bottom flask fitted with condenser was charged with 1.0 equivalents of: $(Bu_4N)_2B_{12}H_{11}I$, $Pd_2(dba)_3$ (5 mol%); Davephos (10 mol%); 3–5 equivalents of tetrabutylammonium bromide or NaBr for synthesis of $(Bu_4N)_2B_{12}H_{11}Br$; Tetrabutylammonium chloride or NaCl for synthesis of $(Bu_4N)_2B_{12}H_{11}Cl$; Tetrabutylammonium azide or NaN_3 for synthesis of $(Bu_4N)_2B_{12}H_{11}N_3$; Tetrabutylammonium fluoride·$3H_2O$ and 5.0 equivalents of KOH were used for the synthesis of $(Bu_4N)_2B_{12}H_{11}OH$; and copper(I)cyanide was used for the synthesis of $(Bu_4N)_2B_{12}H_{11}CN$. Subsequently, 2.0 mL of anhydrous DMSO was added. The reaction flask was filled with N_2 and connected to a condenser. The round-bottom flask was placed in a CEM microwave oven and heated to 150 °C, at a maximum power of 300 W, for 15 min with stirring (high), or heated in an oil bath at 150 °C for 3–5 h until the starting $(Bu_4N)_2B_{12}H_{11}I$ was completely consumed, as judged by ^{11}B NMR and TLC. The mixture was cooled to room temperature and then filtered through a funnel filled with cotton, celite, and filter paper. The resulting solution was concentrated under reduced pressure and the crude product was subjected to silica gel chromatography using a gradient of acetonitrile (0–25%) in DCM.

The spectroscopic data of the compounds known from the literature (Br, Cl, OH) agree with the literature data.

$(Bu_4N)_2B_{12}H_{11}Br$: ^{11}B NMR (129 MHz, DMSO-d_6, δ): −19.30 (d, J = 128 Hz, 1B, B-H), −16.27 (d, J = 134 Hz, 5B, B-H), −14.71 (d, J = 136 Hz, 5B, B-H), −8.91 (s, 1B, B-Br). HRMS (ESI/TOF) m/z for $B_{12}H_{11}Br$ $[M]^{2-}$: 110.55 (found: 110.57).

$(Bu_4N)_2B_{12}H_{11}Cl$: ^{11}B NMR (129 MHz, DMSO-d_6, δ): −20.8 (d, J = 129 Hz, 1B, B-H), −17.2 (d, J = 129 Hz, 5B, B-H), −15.5 (d, J = 131 Hz, 5B, B-H), −3.7 (s, 1B, B-Cl). HRMS (ESI/TOF) m/z for $B_{12}H_{11}Cl$ $[M]^{2-}$: 88.08 (found: 88.10).

$(Bu_4N)_2B_{12}H_{11}N_3$: ^{11}B NMR (129 MHz, DMSO-d_6, δ): −19.61 (d, J = 129 Hz, 1B, B-H), −16.07 (d, J = 129 Hz, 5B, B-H), −16.43 (d, J = 131 Hz, 5B, B-H), −2.68 (s, 1B, B-N_3). HRMS (ESI/TOF) m/z for $B_{12}H_{11}N_3$ $[M]^{2-}$: 91.60 (found: 91.62).

$(Bu_4N)_2B_{12}H_{11}OH$: ^{11}B NMR (129 MHz, DMSO-d_6, δ): −25.81 (d, J = 129 Hz, 1B, B-H), −19.71 (d, J = 125 Hz, 5B, B-H), −17.08 (d, J = 124 Hz, 5B, B-H), 4.23 (s, 1B, B-OH). HRMS (ESI/TOF) m/z for $B_{12}H_{11}OH$ $[M]^{2-}$: 79.10 (found: 79.10).

$(Bu_4N)_2B_{12}H_{11}CN$: ^{11}B NMR (129 MHz, DMSO-d_6, δ): −22.86 (s, 1B, B-CN) −18.60 (d, J = 149 Hz, 1B, B-H), −16.43 (d, J = 123 Hz, 10B, B-H). HRMS (ESI/TOF) m/z for $B_{12}H_{11}N_3$ $[M]^{2-}$: 83.60 (found: 83.61).

$(Bu_4N)_2B_{12}H_{11}NCO$: ^{11}B NMR (129 MHz, DMSO-d_6, δ): −21.25 (d, J = 123.7 Hz, 1B, B-H), −18.72 (d, J = 154.14 Hz, 5B, B-H), −17.40 (d, J = 154.14 Hz, 5B, B-H), −9.75 (s, 1B, B-NCO). HRMS (ESI/TOF) m/z for $B_{12}H_{11}NCO$ $[M]^{2-}$: 91.59 (found 91.60).

4.2. Click Reaction

$(Bu_4N)_2B_{12}H_{11}N_3$, (1 mmol) was dissolved in CH_3CN; subsequently, Cu(I)OAc (0.2 mmol) and sodium ascorbate (0.3 mmol) were added. Then, dimethyl but-2-ynedioate

(2.0 mmol) was added to the mixture and the reaction mixture was stirred at 85 °C for 3 h. The reaction progress was monitored by TLC and NMR. After the complete consumption of the starting material, the reaction mixture was allowed to cool to room temperature, and the insoluble materials were removed via filtration. The desired product was collected in 95% yield by a short plug of silica gel using 1:1 DCM:Acetonitrile.

^{11}B NMR (129 MHz, Acetonitrile-d_3, δ): -5.4 (s, 1B), -16.4 (d, $J = 50$ Hz, 5B), -17.1 (d, $J = 48.9$ Hz, 5B), -18.4 (d, $J = 83$ Hz, 1B). ^{1}H NMR (400 MHz, Acetonitrile-d_3, δ): 0.95–0.98 (t, 24H), 1.31–1.40, 16H), 1.56–1.64 (m, 16H), 3.07, 3.09, 3.11 (m, 16H), 3.80 (s, 3H), 3.84 (s, 3H). HRMS (ESI/TOF) m/z for $B_{12}H_{17}C_6O_4N_3$ [M]$^{2-}$: 162.61 (found: 162.60).

5. Conclusions

We have found that $[B_{12}H_{11}I]^{2-}$ reacts with other halides and pseudohalides under cross-coupling conditions to yield compounds which have not been previously obtained, and to yield other products in a simple manner. Several of these compounds can be reacted further, thus considerably broadening the possibilities of linking the dodecaborate cluster to other fragments of interest for biology, drug development, and material sciences.

Supplementary Materials: The following supporting information can be downloaded at: https://www.mdpi.com/article/10.3390/molecules28073245/s1, Figure S1: ^{11}B{H}-NMR of (Bu$_4$N)$_2$B$_{12}$H$_{11}$Br; Figure S2: ^{11}B-NMR of (Bu$_4$N)$_2$B$_{12}$H$_{11}$Br; Figure S3: ^{11}B{H}-NMR of (Bu$_4$N)$_2$B$_{12}$H$_{11}$Cl; Figure S4: ^{11}B-NMR of (Bu$_4$N)$_2$B$_{12}$H$_{11}$Cl; Figure S5: ^{11}B{H}-NMR of (Bu$_4$N)$_2$B$_{12}$H$_{11}$OH; Figure S6: ^{11}B-NMR of (Bu$_4$N)$_2$B$_{12}$H$_{11}$OH; Figure S7: ^{11}B{H}-NMR of (Bu$_4$N)$_2$B$_{12}$H$_{11}$NCO; Figure S8: ^{11}B-NMR of (Bu$_4$N)$_2$B$_{12}$H$_{11}$NCO; Figure S9: ^{11}B{H}-NMR of (Bu$_4$N)$_2$B$_{12}$H$_{11}$CN; Figure S10: ^{11}B{H}-NMR of (Bu$_4$N)$_2$B$_{12}$H$_{11}$CN; Figure S11: ^{11}B{H}-NMR of (Bu$_4$N)$_2$B$_{12}$H$_{11}$N$_3$; Figure S12: ^{11}B{H}-NMR of (Bu$_4$N)$_2$B$_{12}$H$_{11}$N$_3$; Figure S13: ^{11}B{H}-NMR of (Bu$_4$N)$_2$B$_{12}$H$_{11}$-C$_6$H$_6$O$_4$; Figure S13: ^{11}B{H}-NMR of (Bu$_4$N)$_2$B$_{12}$H$_{11}$-C$_6$H$_6$O$_4$; Figure S15: ^{1}H-NMR of the click reaction product; Figure S16: MS spectrum of $[B_{12}H_{11}Cl]^{2-}$; Figure S17: MS spectrum of $[B_{12}H_{11}Br]^{2-}$; Figure S18: MS spectrum of $[B_{12}H_{11}OH]^{2-}$; Figure S19: MS spectrum of $[B_{12}H_{11}NCO]^{2-}$; Figure S20: MS spectrum of $[B_{12}H_{11}N_3]^{2-}$; Figure S21: MS spectrum of $[B_{12}H_{11}CN]^{2-}$; Figure S22: MS spectrum of $[B_{12}H_{11}$-C$_6$H$_6$O$_4]^{2-}$.

Author Contributions: Conceptualization: M.K.A.-J. and D.G.; experimental work: M.K.A.-J., F.S. and J.-C.C.; writing—original draft preparation, D.G.; writing—review and editing, M.K.A.-J., J.-C.C. and D.G. All authors have read and agreed to the published version of the manuscript.

Funding: This research was funded by DFG German Research Council, grant number GA 250/55-1

Institutional Review Board Statement: Not applicable.

Informed Consent Statement: Not applicable.

Data Availability Statement: All experimental procedures are described in the paper. The NMR and MS spectra are deposited in the Supporting Materials.

Conflicts of Interest: The authors declare no conflict of interest.

References

1. Marei, T.; Al-Joumhawy, M.K.; Alnajjar, M.A.; Nau, W.M.; Assaf, K.I.; Gabel, D. Binding affinity of aniline-substituted dodecaborates to cyclodextrins. *Chem. Commun.* **2022**, *58*, 2363–2366. [CrossRef] [PubMed]
2. Assaf, K.I.; Suckova, O.; Al Danaf, N.; von Glasenapp, V.; Gabel, D.; Nau, W.M. Dodecaborate-Functionalized Anchor Dyes for Cyclodextrin-Based Indicator Displacement Applications. *Org. Lett.* **2016**, *18*, 932–935. [CrossRef]
3. Assaf, K.I.; Gabel, D.; Zimmermann, W.; Nau, W.M. High-affinity host-guest chemistry of large-ring cyclodextrins. *Org. Biomol. Chem.* **2016**, *14*, 7702–7706. [CrossRef]
4. Assaf, K.I.; Ural, M.S.; Pan, F.; Georgiev, T.; Simova, S.; Rissanen, K.; Gabel, D.; Nau, W.M. Water Structure Recovery in Chaotropic Anion Recognition: High-Affinity Binding of Dodecaborate Clusters to gamma-Cyclodextrin. *Angew. Chem. Int. Ed.* **2015**, *54*, 6852–6856. [CrossRef] [PubMed]
5. Wang, W.; Wang, X.; Cao, J.; Liu, J.; Qi, B.; Zhou, X.; Zhang, S.; Gabel, D.; Nau, W.M.; Assaf, K.I.; et al. The chaotropic effect as an orthogonal assembly motif for multi-responsive dodecaborate-cucurbituril supramolecular networks. *Chem. Commun.* **2018**, *54*, 2098–2101. [CrossRef]

6. Fan, P.; Neumann, J.; Stolte, S.; Arning, J.; Ferreira, D.; Edwards, K.; Gabel, D. Interaction of dodecaborate cluster compounds on hydrophilic column materials in water. *J. Chromatogr. A* **2012**, *1256*, 98–104. [CrossRef] [PubMed]
7. Justus, E.; Rischka, K.; Wishart, J.F.; Werner, K.; Gabel, D. Trialkylammoniododecaborates: Anions for Ionic Liquids with Potassium, Lithium and Protons as Cations. *Chem. Eur. J.* **2008**, *14*, 1918–1923. [CrossRef]
8. Karki, K.; Gabel, D.; Roccatano, D. Structure and dynamics of dodecaborate clusters in water. *Inorg. Chem.* **2012**, *51*, 4894–4896. [CrossRef]
9. Assaf, K.I.; Nau, W.M. The Chaotropic Effect as an Assembly Motif in Chemistry. *Angew. Chem. Int. Ed.* **2018**, *57*, 13968–13981. [CrossRef]
10. Semioshkin, A.A.; Sivaev, I.B.; Bregadze, V.I. Cyclic oxonium derivatives of polyhedral boron hydrides and their synthetic applications. *Dalton Trans.* **2008**, 977–992. [CrossRef]
11. Peymann, T.; Lork, E.; Gabel, D. Hydroxoundecahydro-*closo*-dodecaborate(2-) as a nucleophile. Preparation and structural characterization of *O*-alkyl and *O*-acyl derivatives of hydroxoundecahydro-*closo*-dodecaborate(2-). *Inorg. Chem.* **1996**, *35*, 1355–1360. [CrossRef] [PubMed]
12. Hoffmann, S.; Justus, E.; Ratajski, M.; Lork, E.; Gabel, D. $B_{12}H_{11}$-containing guanidinium derivatives by reaction of carbodiimides with $H_3N-B_{12}H_{11}(1-)$. A new method for connecting boron clusters to organic compounds. *J. Organomet. Chem.* **2005**, *690*, 2757–2760. [CrossRef]
13. Sun, Y.; Zhang, J.; Zhang, Y.; Liu, J.; van der Veen, S.; Duttwyler, S. The closo-Dodecaborate Dianion Fused with Oxazoles Provides 3D Diboraheterocycles with Selective Antimicrobial Activity. *Chem. Eur. J.* **2018**, *24*, 10364–10371. [CrossRef]
14. Sivaev, I.B.; Bruskin, A.B.; Nesterov, V.V.; Antipin, M.Y.; Bregadze, V.I.; Sjöberg, S. Synthesis of Schiff bases derived from the ammoniaundecahydro-*closo*- dodecaborate (1–) anion, $[B_{12}H_{11}NH=CHR]^-$, and their reduction into monosubstituted amines $[B_{12}H_{11}NH_2CH_2R]^-$: A new route to water soluble agents for BNCT. *Inorg. Chem.* **1999**, *38*, 5887–5893. [CrossRef]
15. Gabel, D.; Moller, D.; Harfst, S.; Roesler, J.; Ketz, H. Synthesis of S-alkyl and S-acyl derivatives of mercaptoundecahydrododecaborate, a possible boron carrier for neutron capture therapy. *Inorg. Chem.* **1993**, *32*, 2276–2278. [CrossRef]
16. Al-Joumhawy, M.K.; Marei, T.; Shmalko, A.; Cendoya, P.; La Borde, J.; Gabel, D. B–N bond formation through palladium-catalyzed, microwave-assisted cross-coupling of nitrogen compounds with iodo-dodecaborate. *Chem. Commun.* **2021**, *57*, 10007–10010. [CrossRef]
17. Peymann, T.; Knobler, C.B.; Hawthorne, M.F. Synthesis of Alkyl and Aryl Derivatives of closo-$B_{12}H_{12}{}^{2-}$ by the Palladium-Catalyzed Coupling of closo-$B_{12}H_{11}I^{2-}$ with Grignard Reagents. *Inorg. Chem.* **1998**, *37*, 1544–1548. [CrossRef]
18. Himmelspach, A.; Finze, M.; Vöge, A.; Gabel, D. Cesium and Tetrabutylammonium Salt of the Ethynyl-closo-dodecaborate Dianion. *Z. Anorg. Allg. Chem.* **2012**, *638*, 512–519. [CrossRef]
19. Al-Joumhawy, M.; Cendoya, P.; Shmalko, A.; Marei, T.; Gabel, D. Improved synthesis of halo- and oxonium derivatives of dodecahydrido-closo-dodecaborate(2-). *J. Organomet. Chem.* **2021**, *949*, 121967. [CrossRef]
20. Semioshkin, A.A.; Petrovskii, P.V.; Sivaev, I.B.; Balandina, E.G.; Bregadze, V.I. Synthesis and NMR spectra of the hydroxyundecahydro-closo-dodecaborate $[B_{12}H_{11}OH]^{2-}$ and its acylated derivatives. *Russ. Chem. Bull.* **1996**, *45*, 683–686. [CrossRef]
21. Alam, F.; Soloway, A.H.; Barth, R.F.; Mafune, N.; Adams, D.M.; Knoth, W.H. Boron neutron capture therapy: Linkage of a boronated macromolecule to monoclonal antibodies directed against tumor-associated antigens. *J. Med. Chem.* **1989**, *32*, 2326–2330. [CrossRef]
22. Evano, G.; Nitelet, A.; Thilmany, P.; Dewez, D.F. Metal-Mediated Halogen Exchange in Aryl and Vinyl Halides: A Review. *Front. Chem.* **2018**, *6*, 114. [CrossRef]
23. Himmelspach, A.; Reiss, G.J.; Finze, M. Microwave-assisted Kumada-type cross-coupling reactions of iodinated carba-closo-dodecaborate anions. *Inorg. Chem.* **2012**, *51*, 2679–2688. [CrossRef]
24. Sivaev, I.B.; Bregadze, V.I.; Sjöberg, S. A mild synthesis of the $[B_{12}H_{11}CO]^-$ anion. *Russ. Chem. Bull.* **1998**, *47*, 193–194. [CrossRef]
25. Sivaev, I.B.; Kulikova, N.Y.; Nizhnik, E.A.; Vichuzhanin, M.V.; Starikova, Z.A.; Semioshkin, A.A.; Bregadze, V.I. Practical synthesis of 1,4-dioxane derivative of the closo-dodecaborate anion and its ring opening with acetylenic alkoxides. *J. Organomet. Chem.* **2008**, *693*, 519–525. [CrossRef]
26. Owen, G.R.; Vilar, R.; White, A.J.P.; Williams, D.J. Studies on the Reactivity of Isocyanates and Isothiocyanates with Palladium—Imidoyl Complexes. *Organometallics* **2003**, *22*, 4511–4521. [CrossRef]
27. Srebny, H.G.; Preetz, W. Darstellung und Charakterisierung von Thiocyanatderivaten der Hydroboratanionen $B_{10}H_{10}{}^{2-}$ und $B_{12}H_{12}{}^{2-}$. *Z. Anorg. Allg. Chem.* **1984**, *513*, 7–14. (In German) [CrossRef]
28. Anderson, K.W.; Ikawa, T.; Tundel, R.E.; Buchwald, S.L. The Selective Reaction of Aryl Halides with KOH: Synthesis of Phenols, Aromatic Ethers, and Benzofurans. *J. Am. Chem. Soc.* **2006**, *128*, 10694–10695. [CrossRef] [PubMed]
29. Kamin, A.A.; Juhasz, M.A. Exhaustive Cyanation of the Dodecaborate Dianion: Synthesis, Characterization, and X-ray Crystal Structure of $[B_{12}(CN)_{12}]^{2-}$. *Inorg. Chem.* **2020**, *59*, 189–192. [CrossRef]
30. Mu, X.; Hopp, M.; Dziedzic, R.M.; Waddington, M.A.; Rheingold, A.L.; Sletten, E.M.; Axtell, J.C.; Spokoyny, A.M. Expanding the Scope of Palladium-Catalyzed B–N Cross-Coupling Chemistry in Carboranes. *Organometallics* **2020**, *39*, 4380–4386. [CrossRef] [PubMed]
31. Justus, E.; Vöge, A.; Gabel, D. *N*-alkylation of ammonioundecahydro-*closo*-dodecaborate(1-) for the preparation of anions for ionic liquids. *Eur. J. Inorg. Chem.* **2008**, *2008*, 5245–5250. [CrossRef]

32. Barba-Bon, A.; Salluce, G.; Lostalé-Seijo, I.; Assaf, K.I.; Hennig, A.; Montenegro, J.; Nau, W.M. Boron clusters as broadband membrane carriers. *Nature* **2022**, *603*, 637–642. [CrossRef] [PubMed]
33. Zhang, Y.; Yang, L.; Wang, L.; Duttwyler, S.; Xing, H. A Microporous Metal-Organic Framework Supramolecularly Assembled from a CuII Dodecaborate Cluster Complex for Selective Gas Separation. *Angew. Chem. Int. Ed.* **2019**, *58*, 8145–8150. [CrossRef] [PubMed]

Communication

Room Temperature Reduction of Titanium Tetrachloride-Activated Nitriles to Primary Amines with Ammonia-Borane

P. Veeraraghavan Ramachandran * and Abdulkhaliq A. Alawaed

Department of Chemistry, Purdue University, 560 Oval Drive, West Lafayette, IN 47907, USA
* Correspondence: chandran@purdue.edu

Abstract: The reduction of a variety of aromatic and aliphatic nitriles, activated by a molar equivalent of titanium tetrachloride, has been achieved at room temperature using ammonia borane as a safe reductant. The corresponding methanamines were isolated in good to excellent yields following a simple acid-base workup.

Keywords: reduction; ammonia-borane; nitrile; primary amines; titanium tetrachloride; catalysis

1. Introduction

The amine moiety in organic molecules is considered extremely important due to their multifaceted functions, especially in life sciences [1] and industrial chemistry [2,3]. Their applications encompass agro, materials, dye, textile, pharma, surfactant, plastic, and paper industries, to name a few. Accordingly, their syntheses have been the subject of intense research for organic chemists [4,5]. Primary amines function as intermediates or end-products in organic synthesis and have received their due attention. While reductive amination using ammonia or ammonium salts can be envisioned for their synthesis, it is often very challenging to perform [6–8]. Another simple process for primary amines is readily achieved via the reduction of organonitriles (Scheme 1) [9].

(i) H$_2$/cat.
(ii) LiAlH$_4$
(iii) NaBH$_4$ + additives
(iv) B-THF, BMS

R−CN $\longrightarrow$ R$\diagup$NH$_2$

(v) catechol/pinacolborane+ catalyst
(vi) TH, amine-borane + catalysts
(vii) aminoboranes
(viii) silanes + catalyst

Scheme 1. Reduction of organonitriles to primary amines.

Catalytic hydrogenation (Scheme 1i) Of nitriles to amines depends on the reaction conditions [10] and often the intermediate imine-derivatives undergo side reactions to form secondary and/or tertiary amines. Among the hydride reducing agents, lithium aluminum hydride (LAH) can reduce nitriles to amines [11,12] (Scheme 1ii), whereas sodium borohydride (SBH) fails to achieve the reduction [13–15]. However, SBH with metal/metal salt additives [16–20], such as aluminum chloride, indium chloride [21], zinc chloride,

Citation: Ramachandran, P.V.; Alawaed, A.A. Room Temperature Reduction of Titanium Tetrachloride -Activated Nitriles to Primary Amines with Ammonia-Borane. *Molecules* **2023**, *28*, 60. https://doi.org/ 10.3390/molecules28010060

Academic Editors: Michael A. Beckett and Igor B. Sivaev

Received: 7 December 2022
Revised: 16 December 2022
Accepted: 18 December 2022
Published: 21 December 2022

R_2SeBr_2 [22], etc. has been utilized for this reduction (Scheme 1iii). Borane derivatives, such as borane-tetrahydrofuran (B-THF) and borane-dimethyl sulfide (BMS), have also been used for the reduction of nitriles [23,24] (Scheme 1iv). Oxoborane derivatives, such as catecholborane or pinacolborane, which are slow in their reactivity compared to borane or its alkyl derivatives have also been used to reduce nitriles by using transition metal activators as catalysts [25–28] (Scheme 1v). Several metal-catalyzed transfer hydrogenations (TH) using amine-boranes, including ammonia-borane to convert a nitrile to an amine have also been reported recently (Scheme 1vi) [29–34]. Diisopropylaminoborane generated via the dehydrogenation of the corresponding amine-borane has been shown to be effective for the reduction of nitriles to amines [35] (Scheme 1vii). The literature is also permeated with reports on the use of silane derivatives in the presence of activators for nitrile reduction [36–42] (Scheme 1viii).

Most of these procedures have several serious drawbacks, such as air- and moisture-sensitivity of the reagents, expensive nature of the reagents or catalysts, the formation of dialkylamines, etc. Efficient procedures for the reduction of nitriles are still being sought actively. As part of our program on amine-boranes, we recently described the reduction of ketones [43] as well as carboxylic acids [44] using ammonia borane (AB, **1a**) in diethyl ether (Et_2O), in the presence of sub-stoichiometric titanium tetrachloride as an activator (Scheme 2). While comparing the competitive reduction of a carboxylic acid and an organonitrile, under standard conditions, the latter was observed to be unreactive. Due to the importance of primary amines, we were eager to learn whether an organonitrile will succumb to the reaction under modified conditions. Accordingly, a systematic examination was initiated and reported herein is the facile conversion of nitriles to primary amines, at room temperature, with ammonia borane in the presence of a molar equivalent of titanium tetrachloride as the activator. Both aliphatic and aromatic nitriles underwent reduction, and the products were isolated, in good to excellent yields, using a simple acid-base workup.

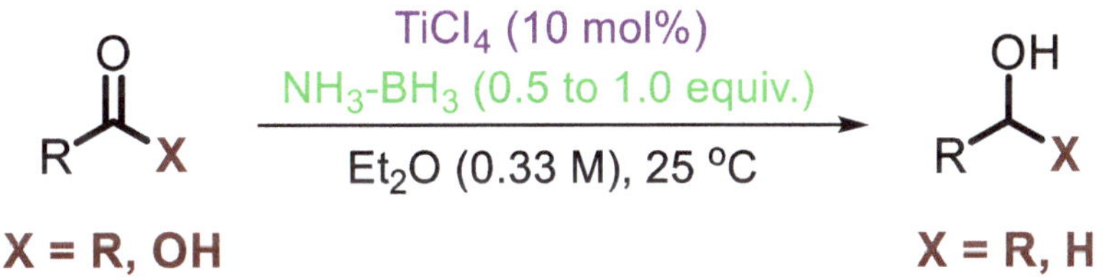

Scheme 2. Reduction of carbonyls activated by $TiCl_4$ using ammonia borane.

2. Results and Discussion

The successful hydroboration of alkenes [45] and alkynes [46] with **1a** in refluxing tetrahydrofuran (THF), as opposed to the lack of any reaction at room temperature [47], led us to attempt the hydroboration (reduction) of a representative nitrile, benzonitrile (**2a**) with **1a** in refluxing THF. Surprisingly, even after 20 h, the reaction was only ~24% complete with one equiv. and ~60% complete with two equiv. of the reducing agent **1a** (entries 2 and 3, Table 1). This prompted a logical modification of the reaction of **2a**, which was conducted, under catalyzed conditions, in diethyl ether at room temperature. Unlike the sub-stoichiometric (10%) catalyst loading reported for the reduction of ketones and acids [43,44], the reaction was now performed in the presence of increasing amounts of the activator titanium tetrachloride as well as differing stoichiometries of the reducing agent, **1a**. The reaction was followed by thin layer chromatography for the disappearance of **2a**. After several attempts, we were delighted to observe that using a molar ratio of 1:2:0.7 for **2a:1a**: activator led to complete conversion of 2a to benzylamine (3a) in 77% isolated yield within 3 h (entry 4, Table 1). No conversion of benzonitrile was observed in the absence of the catalyst, confirming the crucial necessity of $TiCl_4$ for this reduction (entry 1, Table 1). Remarkably, increasing the stoichiometry of $TiCl_4$ to 1 equiv. and of **1a** to 2 equiv. provided an increase in yield (95%), while decreasing the reaction time to an hour (entry 5, Table 1). Decreasing the reagent load of 1a to 1.5 equiv., however, resulted in an inefficient reaction and the yield decreased to 71% (entry 6, Table 1). Increasing the reaction time up to 24 h did not have any effect on the yield.

Table 1. Optimization of reaction conditions for the reduction of benzonitrile.

Entry	LA	R_3N-BH_3	LA: R_3N-BH_3	Solvent	Time, h	Product Yield [a], %
1	none	**1a**	0:2	Et_2O	24	NR [b]
2	none	**1a**	0:1	THF	20	24 [c]
3	none	**1a**	0:2	THF	20	60 [c]
4	$TiCl_4$	**1a**	0.7:2	Et_2O	3	77
5 [d]	**$TiCl_4$**	**1a**	**1:2**	**Et_2O**	**1**	**95**
6	$TiCl_4$	**1a**	1:1.5	Et_2O	24	71
7	$TiCl_4$	**1a**	0.7:2	CH_2Cl_2	3	71
8	$TiCl_4$	**1a**	0.7:2	THF	3	20
9	$TiCl_4$	**1a**	0.7:2	pentane	3	NR [b]
10	$TiBr_4$	**1a**	0.5:2	Et_2O	3	95
11	$HfCl_4$	**1a**	0.7:2	Et_2O	3	63
12	$BF_3.OEt_2$	**1a**	0.7:2	Et_2O	3	17
13	$AlCl_3$	**1a**	0.7:2	Et_2O	3	65
14	$FeCl_3$	**1a**	0.7:2	Et_2O	3	55
15	$TiCl_4$	**1b**	1:2	Et_2O	3	16
16	$TiCl_4$	**1c**	1:2	Et_2O	3	17
17	$TiCl_4$	**1d**	1:2	Et_2O	3	NR [b]
18	$TiCl_4$	**1e**	1:2	Et_2O	3	34

[a] isolated yield. [b] NR = no reaction. [c] determined as a mixture of **3a** and **2a** by PMR after workup. [d] optimal condition.

The conversions and reaction rates for amine formation from nitriles strongly depended on the reaction parameters, such as solvent, Lewis acid and its equivalences, as well as the amine-borane used. The effect of the solvent was exemplified by replacing Et_2O, with dichloromethane (CH_2Cl_2), THF, and pentane under similar conditions. These observations confirmed that Et_2O is the best solvent to effect the transformation effortlessly (entries 7–9 in Table 1).

Next, other common Lewis acids, such as $TiBr_4$, $HfCl_4$, $BF_3 \bullet Et_2O$, $AlCl_3$, and $FeCl_3$, were examined (entries 10–14, Table 1). Among these catalysts, $TiBr_4$ showed good catalytic activity (entry 10). However, due to the relatively higher cost of $TiBr_4$, $TiCl_4$ was used as the catalyst for subsequent studies.

The effect of the amine-borane reductant was then examined by incorporating amine-boranes of differing substitutions on the nitrogen, prepared in our laboratory [48,49], in place of **1a**. Thus, 1°-(*n*-propylamine-borane, **1b**) 2°-(dimethylamine-borane, **1c**), 3°-(triethylamine-borane, **1d**) and heteroaromatic (pyridine-borane, **1e**) were examined (Figure 1) and the results reveal that **1a** is the most efficient among all the amine-boranes tested (entries 15–18 in Table 1). Notably, when triethylamine-borane (**1d**) was used, no reduction was observed (entry 17, Table 1).

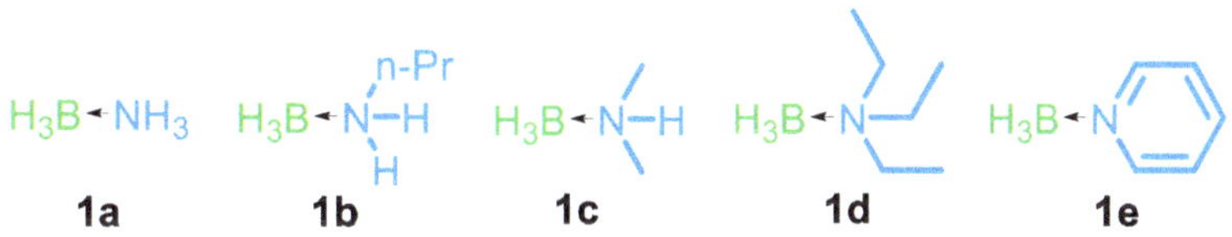

Figure 1. Amine-boranes examined for the reduction of organonitriles.

Having optimized the reaction conditions to achieve the reduction of benzonitrile in 95% yield, the scope of the methodology was studied with respect to the organonitrile partner (Figure 2). Initially, the effect of substitutions on the benzene ring at the *ortho*-, *meta*-, and *para*-positions was evaluated. Thus, *ortho*-chlorobenzonitrile (**2b**), *meta*-fluorobenzonitrile (**2c**), and *para*-fluoro-(**2d**), -chloro-(**2e**), and -bromo-(**2f**) benzonitriles *were converted to* the corresponding amines (**3b–3f**) in 70–72% yields, respectively. Additionally, 2,4-difluorobenzonitrile (**2g**) provided the desired benzylamine product **3g** in 77% yield. No dehalogenation product was observed in all these cases.

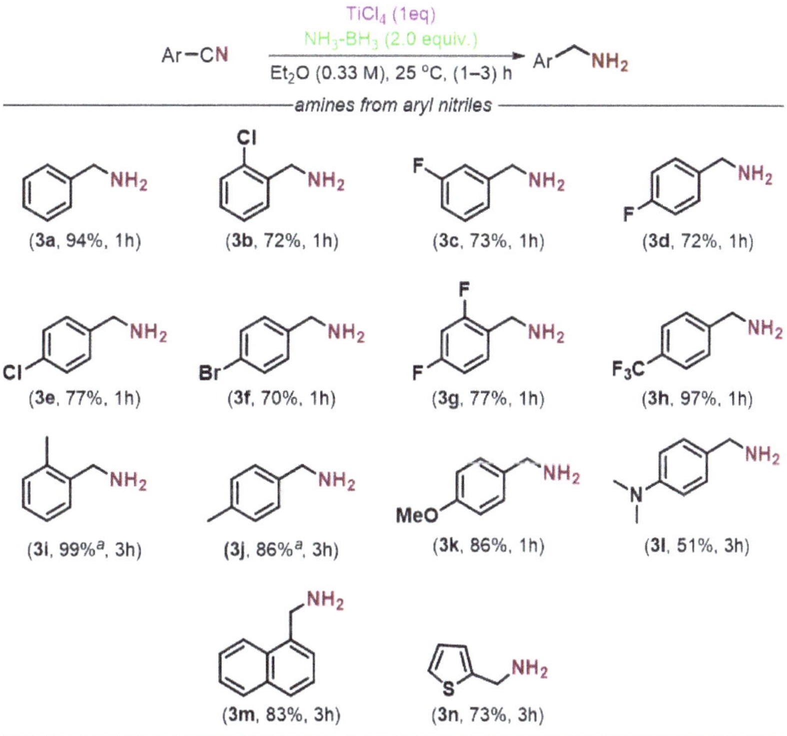

^{a}2.5 equiv. NH$_3$BH$_3$ was used.

Figure 2. Reduction of activated aromatic nitriles with ammonia borane.

Reductions of benzonitrile with an electron-deficient group on the aromatic ring, for example, 4-trifluoromethylbenzonitrile (**2h**) underwent the reduction efficiently to the corresponding amine **3h** in almost quantitative yield 97%, indicating that weak electron-withdrawing groups have no impact on this transformation. The electron-donating 2- and 4- methyl (**2i–2j**) did not inhibit the formation of amines (**3i–3j**), isolated in 99% and 86% yields in 3 h, respectively, although a slightly higher molar equivalent of **1a** (2.5 equiv.) was necessary for complete reduction. Furthermore, the reaction of increased electron-donating 4-methoxybenzonitrile was converted to the desired product methanamine 3k in good yield (86%). It should be noted that higher temperatures were required when diisopropylaminoborane reagent was used for reduction of **2k** [35]. However, *para-N, N*-dimethylaminobenzonitrile (**2l**) provided the corresponding aminobenzylamine (**3l**), *albeit* in diminished yield (51%) even when the reaction was extended to 24 h. The sluggish reactivity was attributed to the deleterious effect of the dimethylamino group, which might be exchanging borane with ammonia and rendering the reduction ineffective or due to the deactivation of the catalyst by complexation with the non-bonding electrons on nitrogen.

As a representative of a bulky aryl nitrile, 1-cyanonaphthalene (**2m**) was subjected to the new ammonia borane reduction under the optimized conditions when the corresponding 1-naphthylmethanamine (**3m**) was isolated in 83% yield. In addition, a representative heteroaromatic nitrile, 1-thiophenenitrile (**2n**) also proved highly amenable to the reaction conditions and afforded thiophenylmethanamine (**3n**) in 73% yield.

Reduction of alkyl nitriles is considered a challenge and numerous methodologies have failed to reduce aliphatic nitriles to primary amines, mainly due to a competitive deprotonation of the acidic α-proton prior to the reduction of the nitrile moiety. Accordingly, a series of straight chain and branched aliphatic nitriles were also included in the study. We were pleased to observe that the ammonia-borane/TiCl$_4$ reducing system is effective for the reduction of these nitriles as well (Figure 3). Our catalytic system does not induce deprotonation and, indeed, all the aliphatic nitriles were reduced, within 3 h, to their corresponding amines at room temperature in excellent yields. For example, acyclic octane- (**2o**) and dodecanenitrile (**2p**) were reduced to the amines **3o** and **3p**, respectively, in 89% and 92% yields. A branched nitrile, cyclohexanenitrile (**2q**) was also reduced, *albeit*, in a decreased 63% yield, to the corresponding methanamine **3q**. Additionally, substituted, and unsubstituted 2-phenylethanenitriles were examined and all of them provided the corresponding amines in >90% yields. Thus, the parent 2-phenylethanenitrile (**2r**), 2-(4-fluorophenyl)ethanenitrile (**2s**), 2-(4-methoxyphenyl)ethanenitrile (**2t**) and 2-(2′,4′-difluorophenyl)ethanenitrile (**2u**) with electron-neutral,-poor, and -rich substituents were converted to the corresponding amines **3r–3t** in 90%–98% yields. Gratifyingly, even a highly hindered ethanenitrile derivative, such as α,α-diphenylethanenitrile (**2v**), was reduced to the corresponding β,β-diphenylethylamine (**3v**) in quantitative yield 95%.

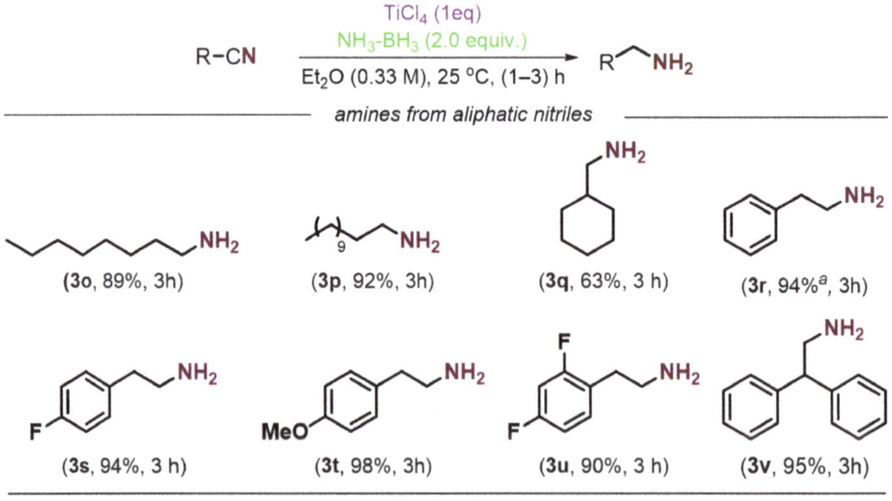

^{a}2.5 equiv. NH$_3$BH$_3$ was used.

Figure 3. Reduction of activated aliphatic nitriles with ammonia borane.

3. Materials and Methods

3.1. General Information

Ammonia-borane [50] and other amine-boranes used in this study were prepared according to our earlier published procedures [48,49]. Other reagents and solvents as well as the amines or amine-hydrochlorides to prepare the amine-boranes were purchased from Sigma-Aldrich or Oakwood chemicals. The nitriles, amines, sodium bicarbonate, and sodium borohydride were used as received. Anhydrous diethyl ether was prepared by distillation over sodium-benzophenone and anhydrous dichloromethane was prepared by distillation over calcium hydride and stored under nitrogen atmosphere. Thin layer chromatography (TLC) was performed on silica gel F60 plates and visualized under UV light or ceric ammonium molybdate solution. The structures of the product amines were confirmed by nuclear magnetic resonance (NMR) spectroscopy and measured in δ values in parts per million (ppm). ^{1}H NMR spectra of reduction products were recorded on a Bruker 400 MHz spectrometer at ambient temperature and calibrated against the residual solvent peak of $CDCl_3$ (δ = 7.26 ppm) as an internal standard. The ^{13}C NMR spectra were reported at 101 MHz (297 K) and calibrated using $CDCl_3$ (δ = 77.0 ppm) as an internal standard. Coupling constants (J) are given in hertz (Hz), and signal multiplicities are described of NMR data as s = singlet, d = doublet, t = triplet, dd = doublet of doublets, dt = doublet of triplets, qd = quartet of doublets, q = quartet, quint and p = pentet, m = multiplet, and br = broad. ^{11}B, ^{1}H (300 MHz), and ^{13}C NMR (75 MHz) spectra of synthesized amine-boranes were recorded at room temperature on a Varian INOVA or MERCURY 300 MHz NMR instrument. ^{11}B NMR spectra were recorded at 96 MHz and chemical shifts were reported relative to the external standard, BF_3:OEt_2 (δ = 0 ppm).

3.2. Experimental

3.2.1. General Procedure for the Preparation of Amines from Nitriles

The preparation of benzylamine from benzonitrile is typical. A 50 mL oven dried round bottom flask was charged with benzonitrile (1 mmol, 1 equiv.) and a magnetic stirring bar. The flask was sealed using a rubber septum. After purging the flask with nitrogen, dry diethyl ether (or other solvents) (3 mL) was added, and the solution was cooled to 0 °C with an ice bath. Subsequently, $TiCl_4$ (or other Lewis acids) (1 mmol, 1 equiv.) was added to the solution, dropwise via syringe if a liquid or by temporarily removing the septum (under a flow of nitrogen) if a solid. The septum was then carefully opened (under a flow of nitrogen) and ammonia borane (or other solid amine-borane) (2.0 mmol, 2.0 equiv.) was added slowly to the reaction mixture (liquid amine-boranes were added via a syringe). Upon complete addition, the reaction flask was again sealed with a septum. After stirring at 0 °C for 1 min, the reaction mixture was allowed to warm to room temperature, stirred and monitored by TLC for completion (disappearance of the starting nitrile), when the crude mixture was brought to 0 °C using an ice bath and quenched by the slow addition of cold 3 *M* HCl. The acidic solution was stirred for 30 min, made basic with 3 M NaOH to pH 11, transferred to a separatory funnel and extracted with diethyl ether (2 × 15 mL). The combined organic layers were washed with brine (1 × 3 mL), dried over anhydrous sodium sulfate, filtered through cotton, and concentrated under aspirator vacuum using a rotary evaporator. Any remaining traces of solvent were removed by subjecting to high vacuum for 30 min. The product amines were characterized using ^{1}H and ^{13}C NMR spectroscopy and compared with those reported in the literature and their references have been included. The spectra are available in Supporting Information. The results from the optimization experiments are shown in Table 1. Ammonia borane as the reductant and titanium chloride as the Lewis acid in diethyl ether solvent was established as the optimal procedure for subsequent reactions.

3.2.2. Characterization of Product Amines

Benzylamine (3a); The compound was prepared as described in the general procedure (colorless oil, yield = 100 mg, 94%); ^{1}**H NMR** (400 MHz, $CDCl_3$) δ 7.35–7.30 (m, 4H),

7.27–7.23 (m, 1H), 3.87 (s, 2H), 1.51 (s, 2H). **^{13}C NMR** (101 MHz, CDCl$_3$) δ143.2, 128.4, 127.0, 126.7, 46.4. Compound characterization is in accordance with previous reports [51].

2-Chlorobenzylamine (3b); The compound was prepared as described in the general procedure (colorless oil, yield = 102 mg, 72%); **^{1}H NMR** (400 MHz, CDCl$_3$) δ 7.37–7.30 (m, 2H), 7.27–7.13 (m, 3H), 3.90 (d, *J* = 2.5 Hz, 2H), 1.60 (s, 2H). **^{13}C NMR** (101 MHz, CDCl$_3$) δ140.5, 133.2, 129.4, 128.8, 128.1, 127.0, 44.5. Compound characterization is in accordance with previous reports [52].

3-Flourobenzylamine (3c); The compound was prepared as described in the general procedure (colorless oil, yield = 91 mg, 73%); **^{1}H NMR** (400 MHz, CDCl$_3$) δ 7.33–7.23 (m, 1H), 7.05 (dd, *J* = 17.0, 8.4 Hz, 2H), 6.92 (t, *J* = 7.8 Hz, 1H), 3.86 (s, 2H), 1.48 (s, 2H). **^{13}C NMR** (101 MHz, CDCl$_3$) δ161.7, 145.8, 129.9, 129.8, 122.4, 113.9, 113.7, 113.6, 113.4, 45.9. **^{19}F NMR** (376 MHz, CDCl$_3$) δ −114.9. Compound characterization is in accordance with previous reports [53].

4-Fluorobenzylamine (3d); The compound was prepared as described in the general procedure (yellow oil, yield = 90 mg, 72%); **^{1}H NMR** (400 MHz, CDCl$_3$) δ 7.27 (dd, *J* = 8.5, 5.5 Hz, 2H), 7.01 (t, *J* = 8.7 Hz, 2H), 3.84 (s, 2H), 1.44 (s, 2H). **^{13}C NMR** (101 MHz, CDCl$_3$) δ162.9, 160.5, 138.8, 128.6, 128.5, 115.3, 115.0, 45.7. **^{19}F NMR** (376 MHz, CDCl$_3$) δ −117.9. Compound characterization is in accordance with previous reports [53].

4-Chlorobenzylamine (3e); The compound was prepared as described in the general procedure (colorless oil, yield = 109 mg, 77%); **^{1}H NMR** (400 MHz, CDCl$_3$) δ 7.29 (d, *J* = 7.7 Hz, 2H), 7.25 (d, *J* = 7.0 Hz, 2H), 3.84 (s, 2H), 1.53 (s, 2H). **^{13}C NMR** (101 MHz, CDCl$_3$) δ141.5, 132.4, 128.5, 128.4, 45.7. Compound characterization is in accordance with previous reports [51].

4-Bromobenzylamine (3f); The compound was prepared as described in the general procedure (colorless oil, yield = 130 mg, 70%); **^{1}H NMR** (400 MHz, CDCl$_3$) δ 7.42 (d, *J* = 8.0 Hz, 2H), 7.16 (d, *J* = 8.2 Hz, 2H), 3.79 (s, 1H), 1.53 (s, 1H). **^{13}C NMR** (101 MHz, CDCl$_3$) δ 142.0, 131.4, 128.7, 120.4, 45.7. Compound characterization is in accordance with previous reports [52].

2,4-Diflourobenzylamine (3g); The compound was prepared as described in the general procedure (colorless oil, yield = 110 mg, 77%); **^{1}H NMR** (400 MHz, CDCl$_3$) δ 7.33–7.23 (m, 1H), 6.87–6.72 (m, 2H), 3.84 (s, 2H), 1.52 (s, 2H). **^{13}C NMR** (101 MHz, CDCl$_3$) δ163.2, 161.9, 159.6, 129.9, 129.8, 129.7, 126.1, 126.0, 111.1, 110.9, 110.8, 103.9, 103.7, 103.4, 39.95, 39.91. **^{19}F NMR** (376 MHz, CDCl$_3$) δ −113.9, −117.4.

(4-(Trifluoromethyl)phenyl)methanamine (3h); The compound was prepared as described in the general procedure (colorless oil, yield = 170 mg, 97%); **^{1}H NMR** (400 MHz, CDCl$_3$) δ 7.58 (d, *J* = 8.0 Hz, 2H), 7.42 (d, *J* = 8.1 Hz, 2H), 3.93 (s, 2H), 1.61 (s, 1H). **^{13}C NMR** ^{13}C NMR (101 MHz, CDCl$_3$) δ146.9, 128.5, 127.2, 125.35, 125.31, 122.8, 45.8. **^{19}F NMR** (376 MHz, CDCl$_3$) δ −63.9. Compound characterization is in accordance with previous reports [53].

o-Tolylmethanamine (3i); The compound was prepared as described in the general procedure (colorless oil, yield = 118 mg, 99%); **^{1}H NMR** (400 MHz, CDCl$_3$) δ 7.31 (d, *J* = 6.6 Hz, 1H), 7.24–7.14 (m, 3H), 3.86 (s, 2H), 2.34 (s, 3H), 1.69 (s, 1H). **^{13}C NMR** (101 MHz, CDCl$_3$) δ140.9, 135.4, 130.2, 127.0, 126.7, 126.1, 43.9, 18.7. Compound characterization is in accordance with previous reports [53].

4-Methylbenzyl amine (3j); The compound was prepared as described in the general procedure (colorless oil, yield = 104 mg, 86%); **^{1}H NMR** (400 MHz, CDCl$_3$) δ 7.19 (d, *J* = 8.0 Hz, 2H), 7.15 (d, *J* = 5.4 Hz, 2H), 3.81 (s, 2H), 2.33 (s, 3H), 1.54–1.44 (m, 2H). **^{13}C NMR** (101 MHz, CDCl$_3$) δ140.3, 136.2, 129.1, 126.9, 46.2, 21.0. Compound characterization is in accordance with previous reports [51].

4-Methoxybenzylamine (3k); The compound was prepared as described in the general procedure (Colorless oil, yield = 118 mg, 86%); **^{1}H NMR** (400 MHz, CDCl$_3$) δ 7.22 (d, *J* = 8.6 Hz, 2H), 6.87 (d, *J* = 8.6 Hz, 2H), 3.79 (s, 5H), 1.43 (s, 2H). **^{13}C NMR** (101 MHz, CDCl$_3$) δ158.4, 135.5, 128.2, 113.8, 55.2, 45.8. Compound characterization is in accordance with previous reports.

4-(Aminomethyl)-N,N-dimethylaniline (3l); The compound was prepared as described in the general procedure (yellow oil, yield = 77mg, 51%); **^{1}H NMR** (400 MHz, CDCl$_3$) δ 7.18 (d, *J* = 8.6 Hz, 2H), 6.72 (d, *J* = 8.7 Hz, 2H), 3.76 (s, 2H), 2.93 (s, 6H), 1.47 (s, 2H). **^{13}C NMR**

(101 MHz, CDCl$_3$) δ149.7, 131.5, 127.9, 112.8, 45.9, 40.7. Compound characterization is in accordance with previous reports [54].

Naphthalen-1-ylmethanamine (3m); The compound was prepared as described in the general procedure (yellow oil, yield = 130 mg, 83%); **^{1}H NMR** (400 MHz, CDCl$_3$) δ 8.07 (s, 1H), 7.89 (d, *J* = 7.8 Hz, 1H), 7.78 (d, *J* = 8.1 Hz, 1H), 7.59–7.43 (m, 4H), 4.32 (s, 2H), 1.59 (s, 2H). **^{13}C NMR** (101 MHz, CDCl$_3$) δ138.9, 133.8, 131.1, 128.8, 127.5, 126.1, 125.6, 125.5, 124.4, 123.1, 44.0. Compound characterization is in accordance with previous reports [53].

Thiophen-2-ylmethanamine (3n); The compound was prepared as described in the general procedure (yellow oil, yield = 82 mg, 73%); **^{1}H NMR** (400 MHz, CDCl$_3$) δ 7.18 (d, *J* = 5.0 Hz, 1H), 6.94 (t, *J* = 4.2 Hz, 1H), 6.90 (d, *J* = 2.9 Hz, 1H), 4.03 (s, 2H), 1.65 (s, 2H). **^{13}C NMR** (101 MHz, CDCl$_3$) δ 147.4, 126.7, 123.9, 123.5, 41.3.

Octan-1-amine (3o); The compound was prepared as described in the general procedure (colorless oil, yield = 115 mg, 89%); **^{1}H NMR** (400 MHz, CDCl$_3$) δ 2.65 (t, *J* = 8.0 Hz, 2H), 1.45–1.20 (m, 14H), 0.87 (t, *J* = 8.0 Hz, 3H). **^{13}C NMR** (101 MHz, CDCl$_3$) δ42.1, 33.8, 31.7, 29.4, 29.2, 26.8, 22.5, 14.0. Compound characterization is in accordance with previous reports [55].

Dodecan-1-amine (3p); The compound was prepared as described in the general procedure (colorless oil, yield = 170 mg, 92%); **^{1}H NMR** (400 MHz, CDCl$_3$) δ 2.66 (t, *J* = 6.9 Hz, 2H), 1.41 (s, 2H), 1.24 (s, 20H), 0.86 (t, *J* = 6.4 Hz, 3H). **^{13}C NMR** (101 MHz, CDCl$_3$) δ42.2, 33.8, 31.8, 29.54, 29.53, 29.41, 29.41, 29.24, 29.24, 26.8, 22.6, 14.0. Compound characterization is in accordance with previous reports [52].

Cyclohexylmethanamine (3q); The compound was prepared as described in the general procedure (colorless oil, yield = 71 mg, 63%); **^{1}H NMR** (400 MHz, CDCl$_3$) δ 2.49 (d, *J* = 6.3 Hz, 2H), 1.75–1.62 (m, 6H), 1.29–1.18 (m, 5H), 0.87 (t, *J* = 11.8 Hz, 2H). **^{13}C NMR** (101 MHz, CDCl$_3$) δ48.8, 41.2, 30.7, 26.6, 25.9. Compound characterization is in accordance with previous reports [52].

Phenethylamine (3r); The compound was prepared as described in the general procedure (yellow oil, yield = 114 mg, 94%); **^{1}H NMR** (400 MHz, CDCl$_3$) δ 7.34–7.25 (m, 2H), 7.23–7.18 (m, 3H), 2.97 (t, *J* = 7.1 Hz, 2H), 2.77 (t, *J* = 6.9 Hz, 2H), 2.13 (s, 2H). **^{13}C NMR** (101 MHz, CDCl$_3$) δ139.5, 128.7, 128.4, 126.1, 43.2, 39.5. Compound characterization is in accordance with previous reports [56].

2-(4-Fluorophenyl)ethan-1-amine (3s); The compound was prepared as described in the general procedure (yellow oil, yield = 130 mg, 94%); **^{1}H NMR** (400 MHz, CDCl$_3$) δ 7.14 (d, *J* = 13.4 Hz, 2H), 6.97 (d, *J* = 17.4 Hz, 2H), 2.93 (s, 2H), 2.71 (s, 2H). **^{13}C NMR** (101 MHz, CDCl$_3$) δ160.2, 135.4, 130.0, 115.2, 115.0, 43.5, 39.1. **^{19}F NMR** (376 MHz, CDCl$_3$) δ −118.8. Compound characterization is in accordance with previous reports [56].

2-(4-Methoxyphenyl)ethanamine (3t); The compound was prepared as described in the general procedure (yellow oil, yield = 148 mg, 98%); **^{1}H NMR** (400 MHz, CDCl$_3$) δ 7.10 (d, *J* = 8.3 Hz, 2H), 6.83 (d, *J* = 8.2 Hz, 2H), 3.77 (s, 3H), 2.91 (s, 2H), 2.68 (s, 2H), 1.32 (s, 2H). **^{13}C NMR** (101 MHz, CDCl$_3$) δ157.9, 131.7, 129.6, 114.4, 113.8, 55.1, 43.6, 39.0. Compound characterization is in accordance with previous reports [51].

2-(2,4-difluorophenyl)ethan-1-amine (3u); The compound was prepared as described in the general procedure (yellow oil, yield = 141 mg, 90%); **^{1}H NMR** (400 MHz, CDCl$_3$) δ 7.18–7.09 (m, 1H), 6.82–6.73 (m, 2H), 2.91 (t, *J* = 6.9 Hz, 2H), 2.72 (t, *J* = 6.9 Hz, 2H), 1.24 (s, 2H). **^{13}C NMR** (101 MHz, CDCl$_3$) δ160.3, 159.9, 131.5, 131.4, 131.3, 122.5, 122.3, 111.0, 110.8, 103.9, 103.6, 103.3, 42.3, 32.8. **^{19}F NMR** (376 MHz, CDCl$_3$) δ −114.8, −115.9.

2,2-Diphenylethan-1-amine (3v); The compound was prepared as described in the general procedure (colorless, yield = 187 mg, 95%); **^{1}H NMR** (400 MHz, CDCl$_3$) δ 7.33–7.20 (m, 10H), 3.99 (t, *J* = 7.6 Hz, 1H), 3.33 (d, *J* = 7.6 Hz, 2H), 1.22 (s, 2H). **^{13}C NMR** (101 MHz, CDCl$_3$) δ142.7, 128.5, 128.0, 126.4, 55.1, 47.0.

4. Conclusions

In conclusion, we have developed a simple protocol for the reduction of nitriles to afford primary amines using ammonia-borane as the reductant in the presence of one molar equivalent of TiCl$_4$ in diethyl ether at room temperature. A broad range of aromatic, heteroaromatic, benzylic, and aliphatic nitriles were efficiently reduced under this condition in moderate to very high yields. This reducing system affords negligible side products, and the workup of the reaction mixture is very simple. The reaction is believed to progress via the activation of the nitrile by titanium tetrachloride, followed by the hydroboration of the carbon nitrogen triple bond.

Supplementary Materials: The following supporting information can be downloaded at: https://www.mdpi.com/article/10.3390/molecules28010060/s1, NMR spectra of product amines.

Author Contributions: Conceptualization, supervision, resources, project administration, and funding acquisition, P.V.R. Both authors have contributed to methodology, validation, investigation, data curation, writing—original draft preparation, review and editing; visualization, A.A.A. All authors have read and agreed to the published version of the manuscript.

Funding: This research was funded by the Herbert C. Brown Center for Borane Research of Purdue University.

Institutional Review Board Statement: Not applicable.

Informed Consent Statement: Not applicable.

Data Availability Statement: All of the ^{1}H, $^{11=9}$F, and ^{13}C NMR spectra are available in the supporting information.

Acknowledgments: Helpful discussions with Henry J. Hamann are gratefully acknowledged.

Conflicts of Interest: The authors declare no conflict of interest.

Sample Availability: Samples of the compounds are not available from the authors.

References

1. Ricci, A. *Amino Group Chemistry—From Synthesis to the Life Sciences*; Wiley-VCH: Weinheim, Germany, 2008.
2. Llevot, A.; Dannecker, P.-K.; von Czapiewski, M.; Over, L.C.; Söyer, Z.; Meier, M.A.R. Renewability is not Enough: Recent Advances in the Sustainable Synthesis of Biomass-Derived Monomers and Polymers. *Chem. Eur. J.* **2016**, *22*, 11510–11521. [CrossRef] [PubMed]
3. Roose, P.; Eller, K.; Henkes, E.; Rossbacher, R.; Höke, H. Amines, Aliphatic. In *Ullmann's Encyclopedia of Industrial Chemistry*; Wiley-VCH: Weinheim, Germany, 2015.
4. Ricci, A. *Modern Amination Methods*; Wiley-VCH: Weinheim, Germany, 2000.
5. Ricci, A.; Bernardi, L. *Methodologies in Amine Synthesis: Challenges and Applications*; Wiley-VCH: Weinheim, Germany, 2021.
6. Afanasyev, O.I.; Kuchuk, E.; Usanov, D.L.; Chusov, D. Reductive Amination in the Synthesis of Pharmaceuticals. *Chem. Rev.* **2019**, *119*, 11857–11911. [CrossRef] [PubMed]
7. Irrgang, T.; Kempe, R. Transition-Metal-Catalyzed Reductive Amination Employing Hydrogen. *Chem. Rev.* **2020**, *120*, 9583–9674. [CrossRef]
8. Abdel-Magid, A.F.; Mehrman, S.J. A Review on the Use of Sodium Triacetoxyborohydride in the Reductive Amination of Ketones and Aldehydes. *Org. Process. Res. Dev.* **2006**, *10*, 971–1031. [CrossRef]
9. Das, S.; Maity, J.; Panda, T.K. Metal/Non-Metal Catalyzed Activation of Organic Nitriles. *Chem. Rec.* **2022**, *22*, e202200192. [CrossRef]
10. Monguchi, Y.; Mizuno, M.; Ichikawa, T.; Fujita, Y.; Murakami, E.; Hattori, T.; Maegawa, T.; Sawama, Y.; Sajiki, H. Catalyst-Dependent Selective Hydrogenation of Nitriles: Selective Synthesis of Tertiary and Secondary Amines. *J. Org. Chem.* **2017**, *82*, 10939–10944. [CrossRef]
11. Novakov, I.A.; Orlinson, B.S.; Mamutova, N.N.; Savel'Ev, E.N.; Potayonkova, E.A.; Pyntya, L.A.; Nakhod, M.A. Reduction of unsaturated adamantyl-containing nitriles with lithium aluminum hydride in 2-methyltetrahydrofuran. *Russ. J. Gen. Chem.* **2016**, *86*, 1255–1258. [CrossRef]
12. Brown, H.C.; Weissman, P.M.; Yoon, N.M. Selective Reductions. IX. Reaction of Lithium Aluminum Hydride with Selected Organic Compounds Containing Representative Functional Groups. *J. Am. Chem. Soc.* **1966**, *88*, 1458–1463. [CrossRef]
13. Chaikin, S.W.; Brown, W.G. Reduction of Aldehydes, Ketones and Acid Chlorides by Sodium Borohydride. *J. Am. Chem. Soc.* **1949**, *71*, 122–125. [CrossRef]
14. Brown, H.C.; Krishnamurthy, S. Forty years of hydride reductions. *Tetrahedron* **1979**, *35*, 567–607. [CrossRef]
15. Brown, H.C.; Ramachandran, P.V. Sixty years of hydride reductions. *Reduct. Org. Synth. Recent Adv. Pract. Appl.* **1996**, *641*, 1–30.
16. Satoh, T.; Suzuki, S.; Suzuki, Y.; Miyaji, Y.; Imai, Z. Reduction of organic compounds with sodium borohydride-transition metal salt systems: Reduction of organic nitrile, nitro and amide compounds to primary amines. *Tetrahedron Lett.* **1969**, *10*, 4555–4558. [CrossRef]

17. Bang, S.; Kim, J.; Jang, B.W.; Kang, S.-G.; Kwak, C.-H. Reduction of nitrile to primary amine using sodium borohydride: Synthesis and protonation of a cis-octahedral macrocyclic nickel(II) complex bearing two 2-aminoethyl pendant arms. *Inorg. Chim. Acta* **2016**, *444*, 176–180. [CrossRef]
18. Caddick, S.; Judd, D.B.; Lewis, A.K.K.; Reich, M.T.; Williams, M.R. A generic approach for the catalytic reduction of nitriles. *Tetrahedron* **2003**, *59*, 5417–5423. [CrossRef]
19. Liu, S.; Yang, Y.; Zhen, X.; Li, J.; He, H.; Feng, J.; Whiting, A. Enhanced reduction of C–N multiple bonds using sodium borohydride and an amorphous nickel catalyst. *Org. Biomol. Chem.* **2011**, *10*, 663–670. [CrossRef]
20. Periasamy, M.; Thirumalaikumar, M. Methods of enhancement of reactivity and selectivity of sodium borohydride for applications in organic synthesis. *J. Organomet. Chem.* **2000**, *609*, 137–151. [CrossRef]
21. Saavedra, J.Z.; Resendez, A.; Rovira, A.; Eagon, S.; Haddenham, D.; Singaram, B. Reaction of InCl$_3$ with Various Reducing Agents: InCl$_3$–NaBH$_4$-Mediated Reduction of Aromatic and Aliphatic Nitriles to Primary Amines. *J. Org. Chem.* **2011**, *77*, 221–228. [CrossRef]
22. Akabori, S.; Takanohashi, Y. Novel borane–selenium complex: Highly selective reduction of tertiary amides and nitriles to the corresponding amines with sodium borohydride–dialkylselenium dibromide. *J. Chem. Soc. Perkin Trans. 1* **1991**, *22*, 479–482. [CrossRef]
23. Brown, H.C.; Choi, Y.M.; Narasimhan, S. Improved procedure for borane-dimethyl sulfide reduction of nitriles. *Synthesis* **1981**, *8*, 605–606. [CrossRef]
24. Fowler, J.S.; MacGregor, R.R.; Ansari, A.N.; Atkins, H.L.; Wolf, A.P. Radiopharmaceuticals. 12. New rapid synthesis of carbon-11-labeled norepinephrine hydrochloride. *J. Med. Chem.* **1974**, *17*, 246–248. [CrossRef]
25. Geri, J.B.; Szymczak, N.K. A Proton-Switchable Bifunctional Ruthenium Complex That Catalyzes Nitrile Hydroboration. *J. Am. Chem. Soc.* **2015**, *137*, 12808–12814. [CrossRef] [PubMed]
26. Huang, Z.; Wang, S.; Zhu, X.; Yuan, Q.; Wei, Y.; Zhou, S.; Mu, X. Well-Defined Amidate-Functionalized N-Heterocyclic Carbene -Supported Rare-Earth Metal Complexes as Catalysts for Efficient Hydroboration of Unactivated Imines and Nitriles. *Inorg. Chem.* **2018**, *57*, 15069–15078. [CrossRef] [PubMed]
27. Kaithal, A.; Chatterjee, B.; Gunanathan, C. Ruthenium-Catalyzed Selective Hydroboration of Nitriles and Imines. *J. Org. Chem.* **2016**, *81*, 11153–11161. [CrossRef] [PubMed]
28. Pradhan, S.; Sankar, R.V.; Gunanathan, C. A Boron-Nitrogen Double Transborylation Strategy for Borane-Catalyzed Hydroboration of Nitriles. *J. Org. Chem.* **2022**, *87*, 12386–12396. [CrossRef]
29. Van der Waals, D.; Pettman, A.; Williams, J.M.J. Copper-catalysed reductive amination of nitriles and organic-group reductions using dimethylamine borane. *RSC Adv.* **2014**, *4*, 51845–51849. [CrossRef]
30. Sarkar, K.; Das, K.; Kundu, A.; Adhikari, D.; Maji, B. Phosphine-Free Manganese Catalyst Enables Selective Transfer Hydrogenation of Nitriles to Primary and Secondary Amines Using Ammonia–Borane. *ACS Catal.* **2021**, *11*, 2786–2794. [CrossRef]
31. Nixon, T.D.; Whittlesey, M.K.; Williams, J.M. Ruthenium-catalysed transfer hydrogenation reactions with dimethylamine borane. *Tetrahedron Lett.* **2011**, *52*, 6652–6654. [CrossRef]
32. Song, H.; Xiao, Y.; Zhang, Z.; Xiong, W.; Wang, R.; Guo, L.; Zhou, T. Switching Selectivity in Copper-Catalyzed Transfer Hydrogenation of Nitriles to Primary Amine-Boranes and Secondary Amines under Mild Conditions. *J. Org. Chem.* **2021**, *87*, 790–800. [CrossRef]
33. Hou, S.-F.; Chen, J.-Y.; Xue, M.; Jia, M.; Zhai, X.; Liao, R.-Z.; Tung, C.-H.; Wang, W. Cooperative Molybdenum-Thiolate Reactivity for Transfer Hydrogenation of Nitriles. *ACS Catal.* **2019**, *10*, 380–390. [CrossRef]
34. Lau, S.; Gasperini, D.; Webster, R.L. Amine–Boranes as Transfer Hydrogenation and Hydrogenation Reagents: A Mechanistic Perspective. *Angew. Chem. Int. Ed.* **2020**, *60*, 14272–14294. [CrossRef]
35. Haddenham, D.; Pasumansky, L.; DeSoto, J.; Eagon, S.; Singaram, B. Reductions of Aliphatic and Aromatic Nitriles to Primary Amines with Diisopropylaminoborane. *J. Org. Chem.* **2009**, *74*, 1964–1970. [CrossRef] [PubMed]
36. Cabrita, I.; Fernandes, A.C. A novel efficient and chemoselective method for the reduction of nitriles using the system silane/oxorhenium complexes. *Tetrahedron* **2011**, *67*, 8183–8186. [CrossRef]
37. Itazaki, M.; Nakazawa, H. Selective Double Addition Reaction of an E–H Bond (E = Si, B) to a C$\equiv$N Triple Bond of Organonitriles. *Molecules* **2018**, *23*, 2769. [CrossRef] [PubMed]
38. Ito, M.; Itazaki, M.; Nakazawa, H. Selective Double Hydrosilylation of Nitriles Catalyzed by an Iron Complex Containing Indium Trihalide. *ChemCatChem* **2016**, *8*, 3323–3325. [CrossRef]
39. Ito, M.; Itazaki, M.; Nakazawa, H. Selective Double Hydroboration and Dihydroborylsilylation of Organonitriles by an Iron–indium Cooperative Catalytic System. *Inorg. Chem.* **2017**, *56*, 13709–13714. [CrossRef] [PubMed]
40. Laval, S.; Dayoub, W.; Favre-Reguillon, A.; Berthod, M.; Demonchaux, P.; Mignani, G.; Lemaire, M. A mild and efficient method for the reduction of nitriles. *Tetrahedron Lett.* **2009**, *50*, 7005–7007. [CrossRef]
41. Sanagawa, A.; Nagashima, H. Hydrosilane Reduction of Nitriles to Primary Amines by Cobalt–Isocyanide Catalysts. *Org. Lett.* **2018**, *21*, 287–291. [CrossRef]
42. Gandhamsetty, N.; Jeong, J.; Park, J.; Park, S.; Chang, S. Boron-Catalyzed Silylative Reduction of Nitriles in Accessing Primary Amines and Imines. *J. Org. Chem.* **2015**, *80*, 7281–7287. [CrossRef]
43. Ramachandran, P.V.; Alawaed, A.A.; Hamann, H.J. TiCl$_4$-Catalyzed Hydroboration of Ketones with Ammonia Borane. *J. Org. Chem.* **2022**, *87*, 13259–13269. [CrossRef]

44. Ramachandran, P.V.; Alawaed, A.A.; Hamann, H.J. A Safer Reduction of Carboxylic Acids with Titanium Catalysis. *Org. Lett.* **2022**, *24*, 8481–8486. [CrossRef]
45. Ramachandran, P.V.; Drolet, M.P.; Kulkarni, A.S. A non-dissociative open-flask hydroboration with ammonia borane: Ready synthesis of ammonia–trialkylboranes and aminodialkylboranes. *Chem. Commun.* **2016**, *52*, 11897–11900. [CrossRef] [PubMed]
46. Ramachandran, P.V.; Drolet, M.P. Direct, high-yielding, one-step synthesis of vic-diols from aryl alkynes. *Tetrahedron Lett.* **2018**, *59*, 967–970. [CrossRef]
47. Hutchins, R.O.; Learn, K.; Nazer, B.; Pytlewski, D.; Pelter, A. Amine boranes as selective reducing and hydroborating agents. A review. *Org. Prep. Proceed. Int.* **1984**, *16*, 335–372. [CrossRef]
48. Ramachandran, P.V.; Kulkarni, A.S.; Zhao, Y.; Mei, J. Amine–boranes bearing borane-incompatible functionalities: Application to selective amine protection and surface functionalization. *Chem. Commun.* **2016**, *52*, 11885–11888. [CrossRef] [PubMed]
49. Ramachandran, P.V.; Kulkarni, A.S. Open-Flask Synthesis of Amine–Boranes via Tandem Amine–Ammonium Salt Equilibration–Metathesis. *Inorg. Chem.* **2015**, *54*, 5618–5620. [CrossRef] [PubMed]
50. Ramachandran, P.V.; Kulkarni, A.S. Water-promoted, safe and scalable preparation of ammonia borane. *Int. J. Hydrog. Energy* **2017**, *42*, 1451–1455. [CrossRef]
51. Amberchan, G.; Snelling, R.A.; Moya, E.; Landi, M.; Lutz, K.; Gatihi, R.; Singaram, B. Reaction of Diisobutylaluminum Borohydride, a Binary Hydride, with Selected Organic Compounds Containing Representative Functional Groups. *J. Org. Chem.* **2021**, *86*, 6207–6227. [CrossRef]
52. Ke, D.; Zhou, S. General Construction of Amine via Reduction of N=X (X = C, O, H) Bonds Mediated by Supported Nickel Boride Nanoclusters. *Int. J. Mol. Sci.* **2022**, *23*, 9337. [CrossRef]
53. Liu, Y.; He, S.; Quan, Z.; Cai, H.; Zhao, Y.; Wang, B. Mild palladium-catalysed highly efficient hydrogenation of C[triple bond, length as m-dash]N, C–NO$_2$, and C[double bond, length as m-dash]O bonds using H$_2$ of 1 atm in H$_2$O. *Green Chem.* **2019**, *21*, 830–838. [CrossRef]
54. Kita, Y.; Kuwabara, M.; Yamadera, S.; Kamata, K.; Hara, M. Effects of ruthenium hydride species on primary amine synthesis by direct amination of alcohols over a heterogeneous Ru catalyst. *Chem. Sci.* **2020**, *11*, 9884–9890. [CrossRef]
55. Liu, Y.; Quan, Z.; He, S.; Zhao, Z.; Wang, J.; Wang, B. Heterogeneous palladium-based catalyst promoted reduction of oximes to amines: Using H$_2$ at 1 atm in H$_2$O under mild conditions. *React. Chem. Eng.* **2019**, *4*, 1145–1152. [CrossRef]
56. Ishitani, H.; Furiya, Y.; Kobayashi, S. Enantioselective Sequential-Flow Synthesis of Baclofen Precursor via Asymmetric 1,4-Addition and Chemoselective Hydrogenation on Platinum/Carbon/Calcium Phosphate Composites. *Chem. Asian J.* **2020**, *15*, 1688–1691. [CrossRef] [PubMed]

Communication

1,8-Dihydroxy Naphthalene—A New Building Block for the Self-Assembly with Boronic Acids and 4,4′-Bipyridine to Stable Host–Guest Complexes with Aromatic Hydrocarbons

Chamila P. Manankandayalage, Daniel K. Unruh, Ryan Perry and Clemens Krempner *

Department of Chemistry & Biochemistry, Texas Tech University, P.O. Box 41061, Lubbock, TX 79409-1061, USA; chamilapriya1984@gmail.com (C.P.M.); ryan.perry@ttu.edu (R.P.)
* Correspondence: clemens.krempner@ttu.edu; Tel.: +1 806 834 3507

Abstract: The new Lewis acid–base adducts of general formula X(nad)B←NC$_5$H$_4$-C$_5$H$_4$N→B(nad)X [nad = 1,8-O$_2$C$_{10}$H$_6$, X = C$_6$H$_5$ (**2c**), 3,4,5-F$_3$-C$_6$H$_2$ (**2d**)] were synthesized in high yields via reactions of 1,8-dihydroxy naphthalene [nadH$_2$] and 4,4′-bipyridine with the aryl boronic acids C$_6$H$_5$B(OH)$_2$ and 3,4,5-F$_3$-C$_6$H$_2$B(OH)$_2$, respectively, and structurally characterized by multi-nuclear NMR spectroscopy and SCXRD. Self-assembled H-shaped Lewis acid–base adduct **2d** proved to be effective in forming thermally stable host–guest complexes, **2d** × solvent, with aromatic hydrocarbon solvents such as benzene, toluene, mesitylene, aniline, and *m-*, *p-*, and *o*-xylene. Crystallographic analysis of these solvent adducts revealed host–guest interactions to primarily occur via π···π contacts between the 4,4′-bipyridyl linker and the aromatic solvents, resulting in the formation of 1:1 and 1:2 host–guest complexes. Thermogravimetric analysis of the isolated complexes **2d** × solvent revealed their high thermal stability with peak temperatures associated with the loss of solvent ranging from 122 to 147 °C. **2d**, when self-assembled in an equimolar mixture of *m-*, *p-*, and *o*-xylene (1:1:1), preferentially binds to *o*-xylene. Collectively, these results demonstrate the ability of 1,8-dihydroxy naphthalene to serve as an effective building block in the selective self-assembly to supramolecular aggregates through dative covalent N→B bonds.

Keywords: boron; boronic acid ester; Lewis acid–base adduct; self-assembly; boronic acid; dative N→B bonding; host–guest complex; π–π stacking; aromatic hydrocarbons

Citation: Manankandayalage, C.P.; Unruh, D.K.; Perry, R.; Krempner, C. 1,8-Dihydroxy Naphthalene—A New Building Block for the Self-Assembly with Boronic Acids and 4,4′-Bipyridine to Stable Host–Guest Complexes with Aromatic Hydrocarbons. *Molecules* **2023**, *28*, 5394. https://doi.org/10.3390/molecules28145394

Academic Editors: Michael A. Beckett and Igor B. Sivaev

Received: 30 April 2023
Revised: 22 June 2023
Accepted: 7 July 2023
Published: 14 July 2023

1. Introduction

The last decades have witnessed increased interest in the rational design of supramolecular architectures based on the self-assembly of various suitable building blocks [1–12]. Amongst the various strategies, the utilization of intermolecular dative covalent N→B bond interactions in the design and self-assembly of well-defined supramolecular structures has become a promising synthetic approach with potential in host–guest chemistry, gas storage, and crystal engineering [13–16]. Examples of this recent development are trigonal planar catechol-based aryl boronic esters [17]. These strongly Lewis acidic building blocks, in combination with N-containing strongly Lewis basic ligands, have been demonstrated to facilitate self-assembly to give—through coordinate covalent N→B bonds—boronic ester-based supramolecular structures, such as coordination polymers, cage- and nanostructures, and gels [18–27]. Recently, the groups of Höpfl and Morales-Rojas [28] have shown that reactions of catechol-based aryl boronic esters with bipyridine linkers in the presence of various aromatic hydrocarbon solvents readily form self-assembled Lewis-type N→B adducts with aromatic hydrocarbons incorporated. It was hypothesized that upon N→B coordination, the aromatic bipyridine linker becomes increasingly electron-deficient, enabling the recognition and selective isolation of host–guest complexes with aromatic hydrocarbons [29]. However, catecholate-based boronic esters, despite their high Lewis acidity, readily hydrolyze, particularly when used in "wet" solvents. Our group has recently shown that 1,8-dihydroxy naphthalene-based aryl boronic esters are as strongly Lewis acidic as their catechol-based counterparts but significantly more hydrolytically stable (Scheme 1) [30]. Based on our recent findings, we envisioned that 1,8-dihydroxy naphthalene should be an effective building block in the self-assembly with aryl boronic acids and a suitable bipyridine linker to form thermally stable host–guest complexes with aromatic hydrocarbons.

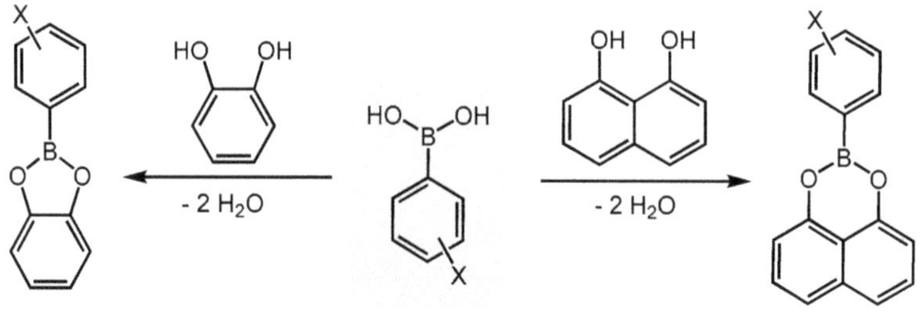

Scheme 1. Formation of catechol- and 1,8-dihydroxy naphthalene-based aryl boronic esters.

2. Results and Discussion

At the outset, we investigated the ability of our recently prepared boronic esters of general formula R-Bnad (**1a–e**), where nad = 1,8-naphthtalenediolate (**a**: R = mesityl; **b**: R = 2,6-dichlorophenyl; **c**: R = phenyl; **d**: R = 3,4,5-trifluorophenyl; **e**: R = pentafluorophenyl) to form stable Lewis acid–base adducts with 4,4′-bipyridine (Scheme 2). The reactions were performed in acetone-D_6 as the solvent and monitored by ^{11}B and ^{1}H NMR spectroscopy. However, the ^{1}H NMR spectra of all five reactions exclusively gave one set of signals, even when an excess of 4,4′-bipyridine or one of the respective boronic esters was used, indicating rapid equilibrium to occur at room temperature. The ^{11}B NMR spectra, on the other hand, showed very different spectral features (Figure 1). Thus, a 1:2 mixture of 4,4′-bipyridine and the bulkiest and least Lewis acidic ester **1a** (R = mesityl) showed a signal at around 29 ppm, which is close to the ^{11}B NMR chemical shift of **1a** (δ = 27 ppm in CDCl$_3$) [30], showing that the equilibrium strongly favors the individual acid and base components over adduct **2a**. For the system 4,4′-bipyridine/**1b** (1:2), two distinct signals at ca. 27 ppm and 7 ppm were observed; the former can be assigned to ester **1b**, while the latter is due to the desired Lewis acid–base adduct **2b**. In contrast, mixtures of 4,4′-bipyridine with **1c**, **1d**, and **1e**, respectively, showed up-field shifted signals at ca. 12 ppm (**1c**), 6 ppm (**1d**), and 4 ppm (**1e**), indicating the formation of the respective adducts **2c**, **2d**, and **2e**.

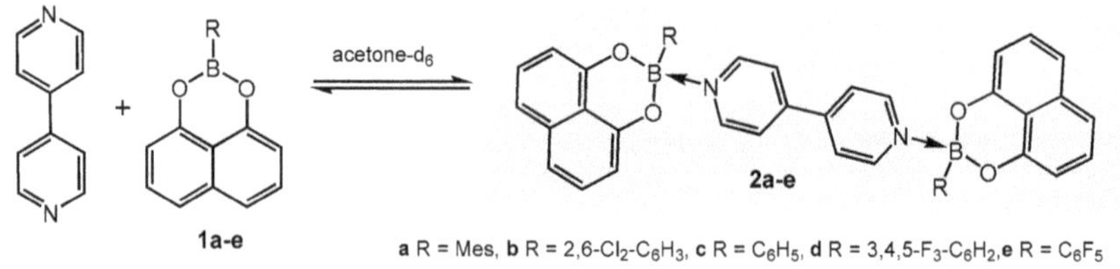

Scheme 2. Equilibrium between 4,4′-bipyridine/**1a–e** and the Lewis acid–base adducts **2a–e** in dilute acetone-d$_6$ at room temperature.

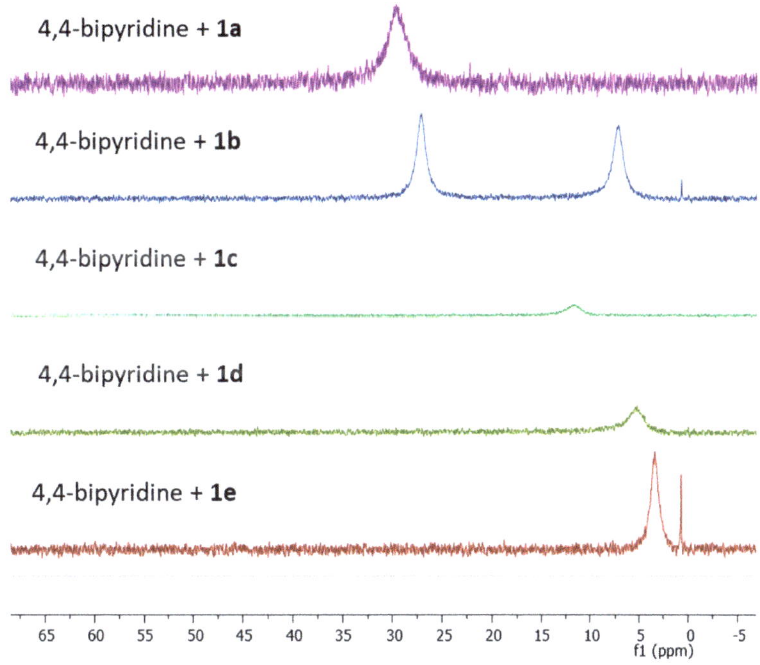

Figure 1. ^{11}B NMR spectroscopic study of the equilibrium reaction of the esters **1a–e** with 4,4′-bipyridine in acetone-D$_6$ at room temperature.

To confirm our structural assignments, efforts were undertaken to grow single crystals suitable for X-ray analysis from concentrated acetone solutions of 4,4′-bipyridine and **2a–e**. The results for **2a**, **2b**, and **2d** are shown in Figure 2 and reveal that two boronic ester units bind to a 4,4′-bipyridine moiety linked via dative N→B bonds. Compounds **2a** and **2b** can structurally be described as double-tweezers. **2a** exhibits an almost perfect double tweezer shaped structure with nearly planar bipyridine units, while **2b** is markedly twisted with a twisting angle of the bipyridine moiety of about 35°. Adduct **2d**, on the other hand, is composed of a more H-shaped structure with a nearly planar bipyridine unit (twisting angle ~10°). The bond parameters for all three adducts are similar, with B-N and B-O distances ranging from 1.61 to 1.67 Å and 1.45–1.47 Å, respectively. The central boron atoms in **2a**, **2b**, and **2d** have slightly distorted tetrahedral coordination environments, with O-B-O and N-B-O angles ranging from 111–116° to 102–105°. Unfortunately, it was not possible to obtain suitable single crystals of **2c** and **2e** from acetone as the solvent. Not only was **2e** highly soluble in acetone, but it also slowly degraded to a second product exhibiting a relatively sharp signal in the ^{11}B NMR with a chemical shift of ca. 0 ppm (see also Figure 1). This new product crystallized from acetone after a week and was identified by NMR spectroscopy and SCXRD as a spirocyclic 4,4′-bipyridinium borate (see supporting information). However, switching the solvent from acetone to benzene resulted in the formation of single crystalline material, which by X-ray analysis was identified as the phenyl derivative **2c** × benzene (Figure 2). Similar to **2d**, adduct **2c** × benzene has an H-shaped structure with a nearly planar bipyridine unit (twisting angle ~ 5°) and average B-O and B-N distances of 1.45 and 1.67 Å, respectively. Notably, **2c** × benzene co-crystallizes with two molecules of benzene, both being entrapped via π–π stacking with the 4,4′-bipyridine moiety and C-H ⋯ π contacts with both the phenyl and the naphthalene substituents.

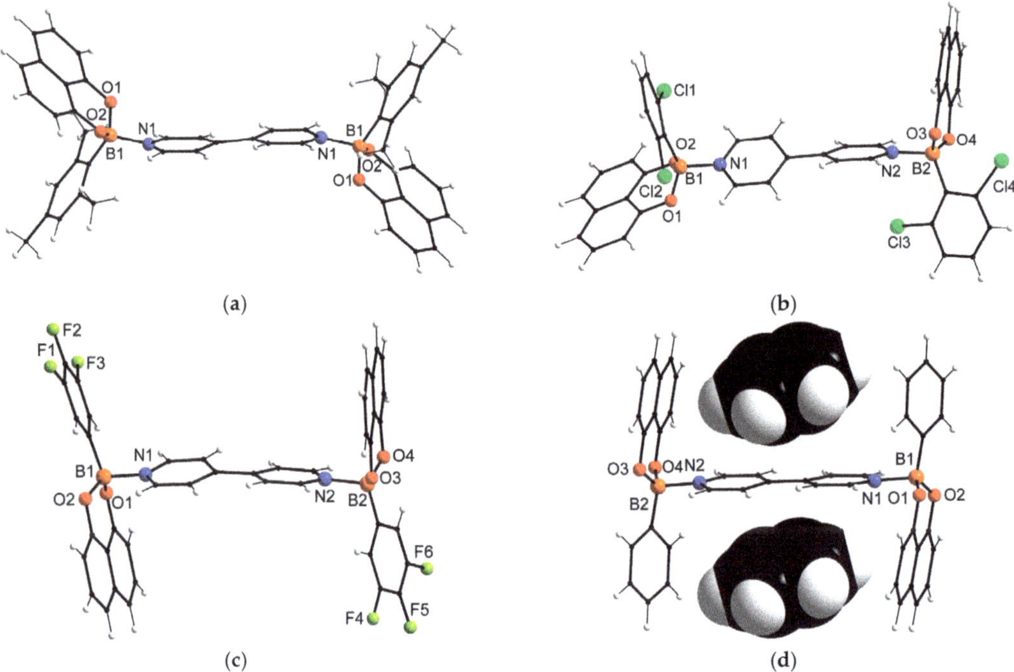

Figure 2. Solid-state structures of the 4,4′-bipyridine adducts **2a**, **2b**, **2d** (co-crystallizing acetone molecules omitted for clarity), and **2c** × benzene. Color codes: red = oxygen, orange = boron, green = fluorine, white = hydrogen, black = carbon, blue = nitrogen. Selected bond lengths [Å] and angles [°]: (**a**) **2a**, O1 B1 1.476(2), O2 B1 1.462(2), N1 B1 1.627(2), O2 B1 O1 110.9(1), O2 B1 N1 104.7(1), O1 B1 N1 100.3(1); (**b**) **2b**, O1 B1 1.448(2), O2 B1 1.449(2), O3 B2 1.453(2), O4 B2 1.442(2), N1 B1 1.611(2), N2 B2 1.650(2), O1 B1 O2 112.1(1), O1 B1 N1 104.9(1), O2 B1 N1 105.4(1), O4 B2 N2 106.2(1), O3 B2 N2 102.5(1), O4 B2 O3 116.2(1); (**c**) **2d**, B1 O2 1.449(2), B1 O1 1.453(2), B1 N1 1.675(2), B2 O4 1.447(2), B2 O3 1.452(2), B2 N2 1.669(2), O2 B1 N1 105.9(1), O1 B1 N1 104.5(1), O2 B1 O1 115.0(1), O4 B2 N2 104.4(1), O3 B2 N2 105.9(1), O4 B2 O3 115.2(1); (**d**) **2c** × benzene, O1 B1 1.455(2), O2 B1 1.456(2), O3 B2 1.458(2), O4 B2 1.453(2), N1 B1 1.673(2), N2 B2 1.670(2), O1 B1 O2 114.1(1), O1 B1 N1 104.8(1), O2 B1 N1 104.4(1), O4 B2 O3 114.1(1), O4 B2 N2 104.9(1), O3 B2 N2 104.5(1).

It was also possible to synthesize and isolate adducts **2c** and **2d** via a one-pot procedure in excellent yields starting from the respective boronic acids, without the need of preparing the respective esters **1c** and **1d** (Scheme 3). For example, when two equivalents of 1,8-dihydroxy naphthalene and one equivalent of 4,4′-bipyridine were reacted with two equivalents of the respective aryl boronic acid in acetone as a solvent, **2c** and **2d** were isolated in yields of 80% and 90%, respectively, as bright orange crystalline materials. Both compounds were fully characterized by ^{1}H, ^{13}C, ^{19}F, and ^{11}B NMR spectroscopy as well as by combustion analysis.

Encouraged by these results, the self-assembly with diboronic acids was investigated. Thus, upon reacting 1,8-dihydroxy naphthalene with 4,4′-bipyridine and 1,4-phenylene diboronic acid in a 2:1:1 ratio in acetone as the solvent, an orange crystalline material formed. However, once precipitated, the isolated material proved to be insoluble in common organic solvents, rendering its characterization by NMR spectroscopy impossible. Attempts to grow single crystals suitable for X-ray analysis failed as well. On the other hand, replacing 4,4′-bipyridine with trans-1,2-di(4-pyridyl)ethylene gave— upon reaction with 1,8-dihydroxy naphthalene and 1,4-phenylene diboronic acid in acetone—an orange microcrystalline precipitate, from which single crystals suitable for X-ray analysis could be obtained. The results revealed the compound to be polymeric Lewis acid–base adduct **3**. Polymer **3** is composed of diboronic ester units coordinated with the 4,4′-bipyridine moieties in a zig-zag fashion (Scheme 3, Figure 3a). Due to poor diffraction and multiple twinning of the single crystals measured, only the connectivity of the polymer could be confirmed.

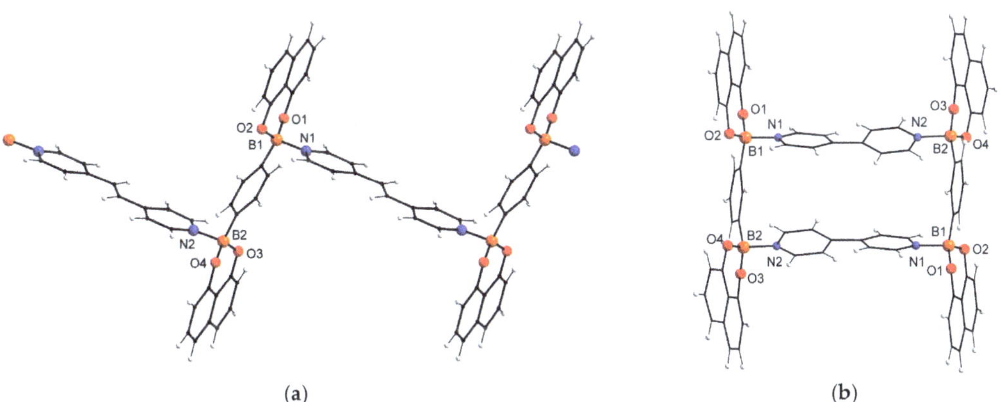

Scheme 3. One-pot synthesis of **2c**, **2d**, **3**, and **4**.

(a) (b)

Figure 3. Solid-state structures of **3** (**a**) (disordered DMSO molecules omitted for clarity) and **4** (**b**) (disordered acetone molecules omitted for clarity). Color codes: red = oxygen, orange = boron, white = hydrogen, black = carbon, blue = nitrogen. Selected bond lengths in [Å] and angles [°]: **4**, O1 B1 1.454(2), O2 B1 1.458(2), O3 B2 1.451(2), O4 B2 1.457(2), N1 B1 1.672(2), N2 B2 1.678(2), O1 B1 O2 114.9(1), O1 B1 N1 105.6(1), O2 B1 N1 105.3(1), O3 B2 O4 114.9(1), O3 B2 N2 105.1(1), O4 B2 N2 105.9(1).

In contrast, the reaction of 1,3-phenylene diboronic acid under otherwise identical conditions did not lead to the formation of a polymeric structure; instead, the rectangular dimer **4** (yield 94%) was formed as confirmed by single-crystal X-ray analysis (Scheme 3, Figure 3b). Both bipyridine units in **4** are markedly twisted, with twisting angles of about

$40°$. The average B-N and B-O distances are in the expected range of 1.675 and 1.455 Å, respectively.

Encouraged by the ability of **2c** × benzene to form a stable host–guest complex with two molecules of benzene, a detailed study of the inclusion properties of H-shaped **2d** was undertaken (Scheme 4). **2d** was selected because of its enhanced hydrolytic stability in solution and the ability of the fluorine substituents to significantly decrease the electron density at boron, thereby increasing the stability of the Lewis acid–base adducts to be formed. Suspensions of 1,8-dihydroxy naphthalene, 4,4′-bipyridine, and 3,4,5-trifluorophenyl boronic acid in the respective aromatic hydrocarbon solvent (i.e., benzene, toluene, *o*-xylene, *m*-xylene, *p*-xylene, mesitylene, aniline) were heated until clear solutions were obtained. Upon slowly cooling to room temperature, crystalline solids were obtained in isolated yields ranging from 34–81%. In addition to being characterized by ^{1}H NMR spectroscopy and combustion analysis, all compounds were analyzed by X-ray crystallography.

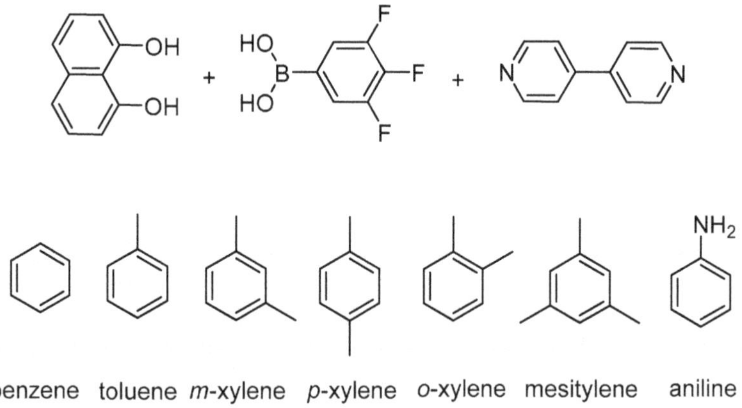

Scheme 4. Host–guest chemistry of **2c** with various aromatic hydrocarbon solvents.

The results of the X-ray analysis are shown in Figure 4 and reveal that **2d** is capable of forming various stable host–guest complexes with all aromatic hydrocarbon solvents used in this study via π–π stacking between the 4,4′-bipyridyl unit of **2d** and the respective aromatic solvent. Thus, in benzene as a solvent, **2d** × benzene is formed (Figure 4a), which is composed of three benzene molecules per two molecules of **2d**. Identical results were obtained with aniline as the solvent, generating host–guest complex **2d** × aniline (Figure 4b) with a **2d**/aniline ratio of 2:3. In contrast, carrying out the self-assembly reactions in toluene and *p*-xylene gave the host–guest complexes **2d** × toluene and **2d** × *p*-xylene (Figure 4c,d), respectively, which consist of one aromatic guest per one molecule of **2d** (1:1 complex). Interestingly, crystallization experiments in meta-xylene gave two crystallographically distinct single crystals of **2d** × *m*-xylene; the results for one revealed a classical 1:1 host–guest complex (Figure 4e), while X-ray analysis of the second one revealed a 1:2 complex (Figure 4f) with two additional unbound *m*-xylene molecules per three equivalents of the 1:2 complex. On the other hand, X-ray analysis of crystalline **2d** × *o*-xylene obtained from *o*-xylene as a solvent confirmed a 1:2 complex (Figure 4g). The X-ray analysis for **2d** × mesitylene (Figure 4h) also confirms a 1:2 host–guest complex but with two molecules of **2d** and two molecules of unbound mesitylene in the unit cell.

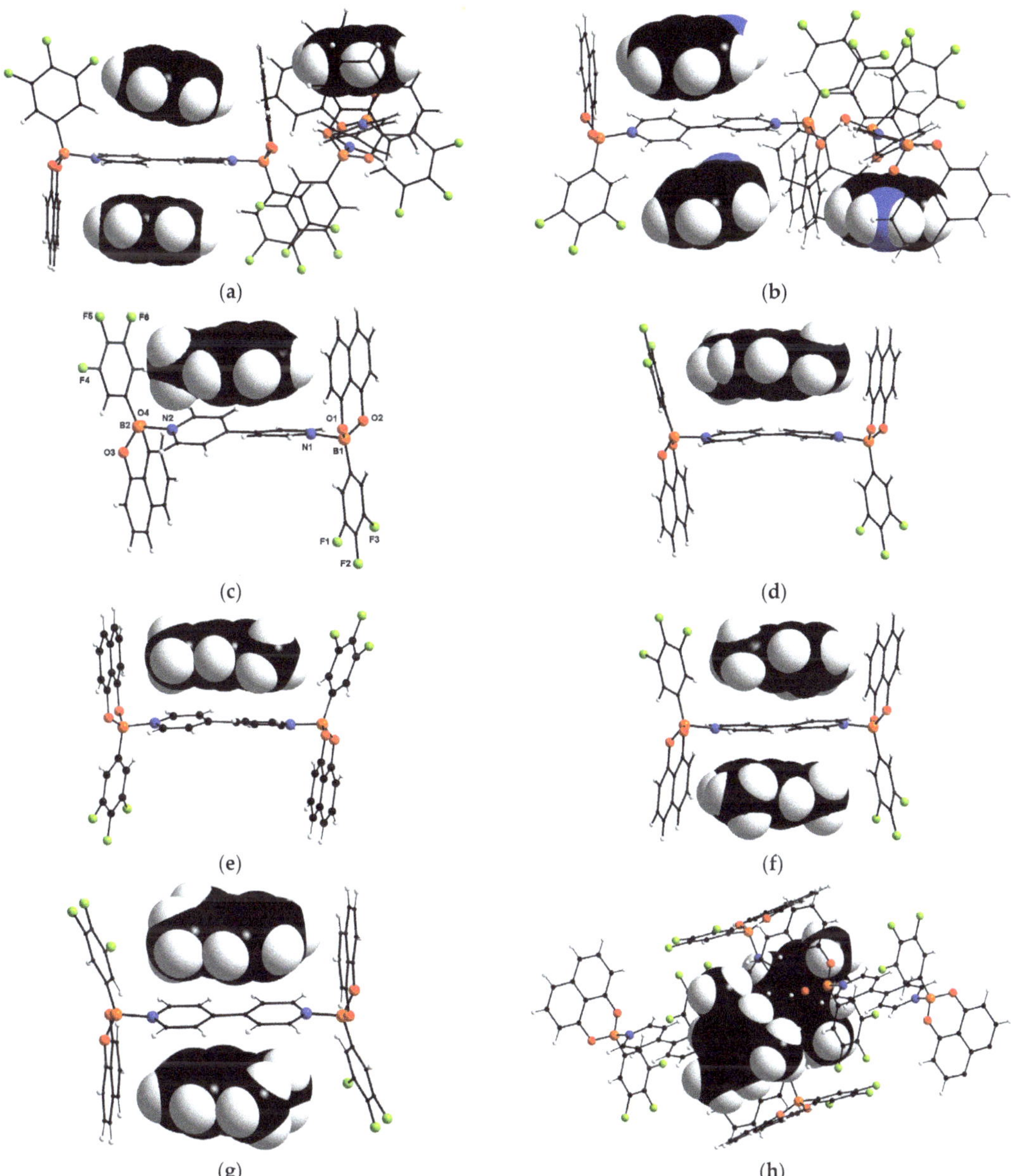

Figure 4. Solid-state structures of host–guest complexes of **2c** with various aromatic solvents (color codes: red = oxygen, orange = boron, white = hydrogen, black = carbon, blue = nitrogen, green = fluorine): (**a**) **2d** × benzene; (**b**) **2d** × aniline, (**c**) **2d** × toluene, (**d**) **2d** × *p*-xylene; (**e**) **2d** × *m*-xylene; (**f**) **2d** × *m*-xylene (unbound *m*-xylene omitted for clarity); (**g**) **2d** × *o*-xylene; (**h**) **2d** × mesitylene (unbound mesitylene omitted for clarity).

Table 1 compares the host–guest ratios of **2d** × solvent determined from the X-ray analysis of single crystals with those obtained from the ^{1}H NMR data of the bulk materials; the latter were dried in air for circa 24 h. It was found that crystals of **2d** × *o*-xylene and **2d**

× *m*-xylene, upon drying in air, lose guest molecules to form host–guest complexes with a formal 1:1 host–guest ratio. While in **2d** × aniline and **2d** × benzene, the 2:3 host–guest ratio remains unchanged, and **2d** × mesitylene contains significantly more mesitylene than what was found via X-ray analysis of single crystalline material. This seems to imply that **2d** × mesitylene may crystallize in various forms with different host–guest ratios.

Table 1. Molar **2d**/solvent ratios of **2d** × solvent from SCXRD and ^{1}H NMR data.

2d × Solvent	**2d/Solvent Ratio (from X-ray Data)**	**2d/Solvent Ratio (from ^{1}H NMR Data) [1]**
2d × benzene	2:3	1:1.5
2d × toluene	1:1	1:1
2d × *o*-xylene	1:2	1:1
2d × *m*-xylene [2]	3:6 [3] and 1:1	1:1
2d × *p*-xylene	1:1	1:1
2d × mesitylene	1:2 [4]	1:2.66
2d × aniline	2:3	1:1.5

[1] The isolated crystalline materials were dried in air for 24 h; [2] Two crystallographically distinct crystals were found; [3] Crystal contains two additional unbound *m*-xylene molecules; [4] Crystal contains two molecules of **2d** and two molecules of unbound mesitylene.

To study the thermal behavior of the host–guest complexes **2d** × solvent, a thermogravimetric analysis of the bulk materials was performed; the results are shown in Figure 5, Figures S30 and S31, and Table S1. Figure 5 (bottom) shows the TG curve of **2d**, revealing complete weight loss to occur at about 260 °C. This indicates that the acid–base adduct **2d** either sublimes or that the decomposition and evaporation of its individual acid and base components occur within the same temperature range. The TG graphs of all host–guest compounds exhibit a two-step weight loss. This is exemplarily shown for the solvent adducts **2d** × benzene, **2d** × toluene, **2d** × *o*-xylene, **2d** × *m*-xylene, and **2d** × *p*-xylene (Figure 5), for which the observed weight losses are consistent with the stoichiometry of the bulk materials established by ^{1}H NMR spectroscopy (Table 1). After the loss of the guest, the residual material likely consists of the solid phase established for solvent-free **2d**. In this series, for **2d** × benzene and **2d** × toluene, the peak temperatures (T_{peak}) associated with the liberation of the guest were found to be 131 °C and 147 °C, respectively, suggesting stronger host–guest interactions in **2d** × toluene, and consistent with the results of DSC analysis (Figures S32 and S33). For the xylene series, the peak temperatures for the loss of solvent are 122 °C (**2d** × *m*-xylene), 133 °C (**2d** × *o*-xylene), and 136 °C (**2d** × *p*-xylene). The slightly higher peak temperatures for **2d** × *o*-xylene and **2d** × *p*-xylene suggest somewhat stronger host–guest interactions compared to **2d** × *m*-xylene, which is supported by DSC measurements (Figures S34–S36).

In a related study, Hőpfl and Morales-Rojas recently disclosed the thermal properties of self-assembled host–guest complexes derived from reactions of 1,2-catechol and 3,4,5-trifluorphenylboronic acid with 4,4′-bipyridine in various aromatic hydrocarbon solvents [28]. Thermogravimetric analysis of these catechol-based host–guest complexes revealed the peak temperatures associated with the loss of solvent to be 65 °C (*m*-xylene), 71 °C (*o*-xylene), 76 °C (*p*-xylene), 121 °C (toluene), 139 °C (benzene), and 100 °C (mesitylene). Except for benzene, these values are significantly lower than what was measured for our 1,8-dihydroxy naphthalene-based host–guest complexes **2d** × solvent (Figure 5). The replacement of catechol with the 1,8-dihydroxy naphthalene ligand at the central boron appears to result in stronger π-type host–guest interactions between the 4,4′-bipyridine unit and the aromatic solvent. Unfortunately, Hőpfl and Morales-Rojas did not disclose thermodynamic data from DSC measurements for comparison [28].

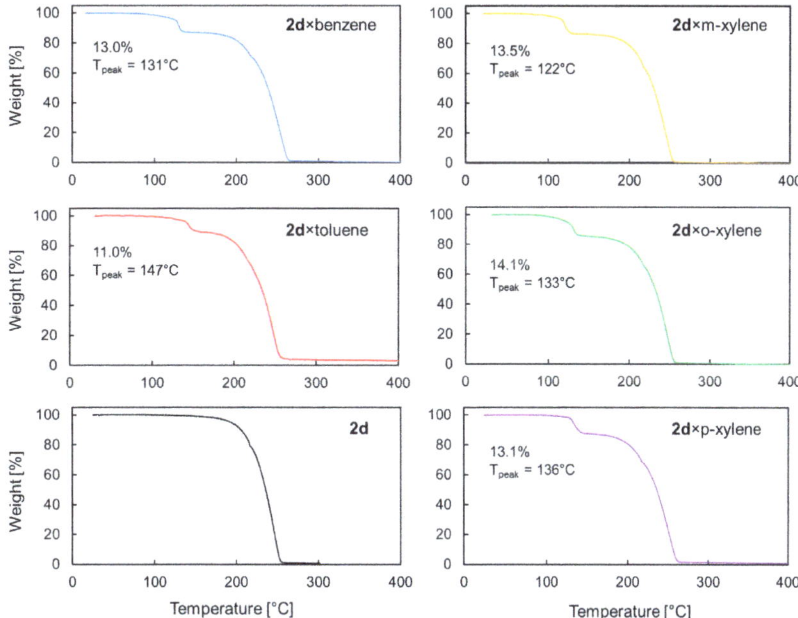

Figure 5. Thermogravimetric analysis of **2d** and its solvent adducts **2d** × solvent with corresponding weight loss [%] and peak temperature T_{peak} [°C] for the loss of solvent.

Based on the results from the host–guest chemistry with *m-*, *o-*, and *p-*xylene, we wondered whether self-assembled **2d** would selectively incorporate one of the three xylene isomers. Thus, 1,8-dihydroxy naphthalene, 4,4′-bipyridine, and 3,4,5-trifluoroboronic acid were reacted in an equimolar solvent mixture of *m-*, *p-*, and *o-*xylene (1:1:1). The obtained crystalline material was collected and analyzed by ^{1}H NMR spectroscopy using DMSO-D_6 as the solvent (see supporting information for more details). The results revealed a higher selectivity of **2d** for *o-*xylene (ca. 57%), with an *m-*, *p-*, to *o-*xylene ratio of about 2:1:4, respectively, which is similar to those of the respective catechol-based host–guest complexes reported by Höpfl and Morales-Rojas [28]. Considering the results from the TGA and DSC measurements, the fairly low uptake of *p-*xylene appears to be surprising but perhaps can be attributed to kinetic effects during the crystallization process.

Finally, to further broaden the scope of the host–guest approach, the modified bipyridine linker 1,2-bis(4-pyridyl)ethane was employed in reactions with 1,8-dihydroxy naphthalene and various boronic acids. Unfortunately, in none of the cases could host–guest complexes with aromatic hydrocarbon solvents be crystallized and structurally determined. Nonetheless, single crystals of the solvent-free double-tweezers **5** and **6** could be obtained from acetone solutions and were characterized by single-crystal X-ray crystallography (Figure 6a,b).

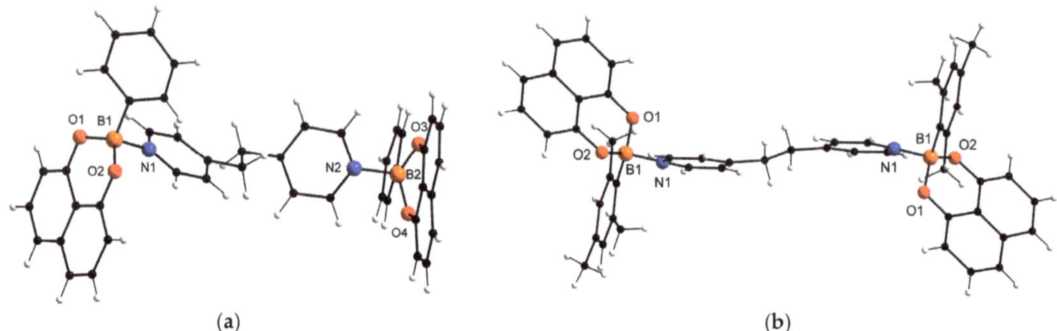

Figure 6. Solid-state structures of **5** and **6** (color codes: red = oxygen, orange = boron, white = hydrogen, black = carbon, blue = nitrogen). Selected bond lengths [Å] and angles [°]: (**a**) **5**, O1 B1 1.447(4), O2 C3 1.367(3), O2 B1 1.466(4), N1 B1 1.656(4), N2 B2 1.682(4), O1 B1 O2 114.4(3), O1 B1 N1 105.4(2), O2 B1 N1 104.7(2); (**b**) **6**, O1 B1 1.471(2), O2 B1 1.467(2), N1 B1 1.622(2), O2 B1 O1 112.3(1), O2 B1 N1 106.0(1), O1 B1 N1 99.7(1).

3. Conclusions

We have demonstrated the ability of 1,8-dihydroxy naphthalene to self-assemble with various aryl boronic acids and 4,4′-bipyridine as a linker to give double-tweezer and H-shaped Lewis acid–base adducts in very good isolated yields by means of N→B bond formation. In addition, a series of host–guest complexes, **2d** × solvent, was prepared via reactions of 4,4′-bipyridine, 1,8-dihydroxy naphthalene, and 3,4,5-trifluorophenyl boronic acid with various aromatic hydrocarbons. SCXRD analysis of these solvent adducts revealed host–guest interactions to primarily occur via π···π contacts between the 4,4′-bipyridyl linker and the aromatic solvents to give 1:1 and 1:2 host–guest complexes, which was confirmed by ^{1}H NMR spectroscopic analysis of the bulk samples. TGA measurements of the bulk samples showed well-defined weight losses at peak temperatures ranging from 122 to 147 °C, which in every case could be attributed to the complete loss of solvent (guest). Self-assembled Lewis acid–base adduct **2d**, when recrystallized from an equimolar mixture of *m*-, *p*-, and *o*-xylene (1:1:1), preferentially binds to *o*-xylene, generating **2d** × *o*-xylene as the major component, similar to what was reported for structurally related catechol-based host–guest complexes [28]. Collectively, the present contribution demonstrates the effectiveness of 1,8-dihydroxy naphthalene in constructing supramolecular structures via dative N-B bond formation.

Supplementary Materials: The following supporting information can be downloaded at: https://www.mdpi.com/article/10.3390/molecules28145394/s1. Figure S1: ^{1}H NMR spectrum of **2c** (acetone-d$_6$); Figure S2: ^{13}C NMR spectrum of **2c** (acetone-d$_6$); Figure S3: ^{11}B NMR spectrum of **2c** (acetone-d$_6$); Figure S4: ^{11}B NMR spectrum of **2c** (DMSO-d$_6$); Figure S5: ^{1}H NMR spectrum of **2c** × benzene in DMSO-d$_6$; Figure S6: ^{1}H NMR spectrum **2d** (acetone-d$_6$); Figure S7: ^{13}C NMR spectrum of **2d** (acetone-d$_6$); Figure S8: ^{19}F NMR spectrum of **2d** (acetone-d$_6$); Figure S9: ^{11}B NMR spectrum of **2d** (acetone-d$_6$); Figure S10: ^{1}H NMR spectrum of 4,4-bipyridinium-bis(1,8-naphthalenediolato)borate (DMSO-d$_6$); Figure S11: ^{11}B NMR spectrum of 4,4-bipyridinium-bis(1,8-naphthalenediolato)borate (DMSO-d$_6$); Figure S12: ^{13}C NMR spectrum of 4,4-bipyridinium-bis(1,8-naphthalenediolato)borate (DMSO-d$_6$); Figure S13: ^{13}C NMR-DEPT spectrum of 4,4-bipyridinium-bis(1,8-naphthalenediolato) borate (DMSO-d$_6$); Figure S14: ^{1}H NMR spectrum of 1,4-benzene-diboronic-acid-1,8-naphthalenediolate ester in DMSO-d$_6$; Figure S15: ^{13}C NMR spectrum of 1,4-benzene-diboronic-acid-1,8-naphthalenediolate ester in DMSO-d$_6$; Figure S16: ^{11}B NMR spectrum of 1,4-benzene-diboronic-acid-1,8-naphthalenediolate ester in DMSO-d$_6$; Figure S17: ^{11}B NMR spectrum of 1,4-benzene-diboronic-acid-1,8-naphthalenediolate ester in acetone-d$_6$; Figure S18: ^{1}H NMR spectrum of **4** in DMSO-d$_6$; Figure S19: ^{1}H NMR spectrum of **2d** × benzene in acetone-d$_6$; Figure S20: ^{1}H NMR spectrum of **2d** × toluene in DMSO-d$_6$; Figure S21: ^{1}H NMR spectrum of **2d** × *m*-xylene in DMSO-d$_6$; Figure S22: ^{1}H NMR spectrum of **2d** × *p*-xylene in DMSO-d$_6$; Figure S23: ^{1}H NMR spectrum of **2d** $\times$ *o*-xylene in DMSO-d$_6$; Figure S24: ^{1}H NMR spectrum of **2d** $\times$ mesitylene in DMSO-d$_6$; Figure S25: ^{1}H NMR spectrum of **2d** $\times$ aniline in DMSO-d$_6$; Figure S26: ^{1}H NMR spectrum (DMSO-d$_6$) of adduct **2d** $\times$ xylene isolated from a 1:1:1 mixture of *m*-, *p*-, and *o*-xylene; Figure S27: Aliphatic region of the ^{1}H NMR spectrum (DMSO-d$_6$) of adduct **2d** $\times$ xylene isolated from a 1:1:1 mixture of *m*-, *p*-, and *o*-xylene; Figure S28: ^{1}H NMR spectrum (DMSO-d$_6$) of **2d** $\times$ *m*-xylene after soaking in *o*-xylene for 24 h; Figure S29: ^{1}H NMR spectrum (DMSO-d$_6$) of **2d** $\times$ *p*-xylene after soaking in *o*-xylene for 24 h; Figure S30: Thermogravimetric analysis of **2d** $\times$ mesitylene with weight loss [%] and peak temperature T$_{peak}$ [$^\circ$C] for the loss of solvent; Figure S31: Thermogravimetric analysis of **2d** $\times$ aniline with weight loss [%] and peak temperature T$_{peak}$ [$^\circ$C] for the loss of solvent; Figure S32: DSC analysis of **2d** $\times$ benzene for the loss of solvent; Figure S33: DSC analysis of **2d** $\times$ toluene for the loss of solvent; Figure S34: DSC analysis of **2d** $\times$ p-xylene for the loss of solvent; Figure S35: DSC analysis of **2d** $\times$ o-xylene for the loss of solvent; Figure S36: DSC analysis of **2d** $\times$ m-xylene for the loss of solvent; Table S1. Selected information from the TGA analysis for **2d** $\times$ solvent; Table S2: Crystal data and structure refinement for **2d** $\times$ toluene; Table S3: Crystal data and structure refinement for **6**; Table S4: Crystal data and structure refinement for **2b**; Table S5: Crystal data and structure refinement for **2d** $\times$ benzene; Table S6: Crystal data and structure refinement for **2d** $\times$ p-xylene; Table S7: Crystal data and structure refinement for **2d** $\times$ *m*-xylene; Table S8: Crystal data and structure refinement for **2d** $\times$ aniline; Table S9: Crystal data and structure refinement for **5**; Table S10: Crystal data and structure refinement for **2d**; Table S11: Crystal data and structure refinement for **3**; Table S12: Crystal data and structure refinement for **2d** $\times$ *m*-xylene; Table S13: Crystal data and structure refinement for **2c**; Table S14: Crystal data and structure refinement for **2a**; Table S15: Crystal data and structure refinement for **2d** $\times$ mesitylene; Table S16: Crystal data and structure refinement for **2d** $\times$ o-xylene; Table S17: Crystal data and structure refinement for **4**. References [31–38] are cited in the supplementary materials.

Author Contributions: Conceptualization, C.K.; methodology, C.P.M. and R.P.; formal analysis, C.P.M., R.P. and D.K.U.; investigation, C.P.M. and R.P.; writing—review and editing, C.K.; supervision, C.K.; project administration, C.K.; funding acquisition, C.K. All authors have read and agreed to the published version of the manuscript.

Funding: This research was funded by the U.S. Department of Energy, Office of Science, Office of Basic Energy Sciences, Catalysis Science Program, under Award DE-SC0019094.

Institutional Review Board Statement: Not applicable.

Informed Consent Statement: Not applicable.

Data Availability Statement: Not applicable.

Conflicts of Interest: The authors declare no conflict of interest.

Sample Availability: Not applicable.

References

1. Desiraju, G.R. Supramolecular synthons in crystal engineering—A new organic synthesis. *Angew. Chem. Int. Ed. Engl.* **1995**, *34*, 2311–2327. [CrossRef]
2. Brunet, P.; Simard, M.; Wuest, J.D. Molecular tectonics. Porous hydrogen-bonded networks with unprecedented structural integrity. *J. Am. Chem. Soc.* **1997**, *119*, 2737–2738. [CrossRef]
3. Hosseini, M.W. Molecular tectonics: From simple tectons to complex molecular networks. *Acc. Chem. Res.* **2005**, *38*, 313–323. [CrossRef] [PubMed]
4. Mastalerz, M. Shape-Persistent Organic Cage Compounds by Dynamic Covalent Bond Formation. *Angew. Chem. Int. Ed.* **2010**, *49*, 5042–5053. [CrossRef]
5. Chakrabarty, R.; Mukherjee, P.S.; Stang, P.J. Supramolecular coordination: Self-assembly of finite two- and three-dimensional ensembles. *Chem. Rev.* **2011**, *111*, 6810–6918.
6. Smulders, M.M.J.; Riddell, I.A.; Browne, C.; Nitschke, J.R. Building on architectural principles for three-dimensional metallo-supramolecular construction. *Chem. Soc. Rev.* **2013**, *42*, 1728–1754. [CrossRef]
7. Goesten, M.G.; Kapteijn, F.; Gascon, J. Fascinating chemistry or frustrating unpredictability: Observations in crystal engineering of metal−organic frameworks. *CrystEngComm* **2013**, *15*, 9249–9257. [CrossRef]
8. Harris, K.; Fujita, D.; Fujita, M. Giant hollow MnL$_{2n}$ spherical complexes: Structure, functionalisation and applications. *Chem. Commun.* **2013**, *49*, 6703–6712. [CrossRef]

9. Song, M.; Sun, Z.; Han, C.; Tian, D.; Li, H.; Kim, J.S. Calixarene based chemosensors by means of click chemistry. *Chem. Asian J.* **2014**, *9*, 2344–2357. [CrossRef]

10. Durola, F.; Heitz, V.; Reviriego, F.; Roche, C.; Sauvage, J.-P.; Sour, A.; Trolez, Y. Cyclic [4]rotaxanes containing two parallel porphyrinic plates: Toward switchable molecular receptors and Compressors. *Acc. Chem. Res.* **2014**, *47*, 633–645. [CrossRef] [PubMed]

11. Zhang, G.; Mastalerz, M. Organic cage compounds—From shape-persistency to function. *Chem. Soc. Rev.* **2014**, *43*, 1934–1947. [CrossRef]

12. Li, H.; Yao, Z.-J.; Liu, D.; Jin, G.-X. Multi-component coordination-driven self-assembly toward heterometallic macrocycles and cages. *Coord. Chem. Rev.* **2015**, *293–294*, 139–157.

13. Torres-Huerta, A.; Velásquez-Hernández, M.; Martínez-Otero, D.; Höpfl, H.; Jancik, V. Structural induction via solvent variation in assemblies of triphenylboroxine and piperazine—Potential application as self-assembly molecular sponge. *Cryst. Growth Des.* **2017**, *17*, 2438–2452. [CrossRef]

14. Wang, W.; Wang, L.; Du, F.; Wang, G.-D.; Hou, L.; Zhu, Z.; Liu, B.; Wang, Y.-Y. Dative B←N bonds based crystalline organic framework with permanent porosity for acetylene storage and separation. *Chem. Sci.* **2023**, *14*, 533–539.

15. Stephens, A.J.; Scopelliti, R.; Tirani, F.F.; Solari, E.; Severin, K. Crystalline Polymers Based on Dative Boron-Nitrogen Bonds and the Quest for Porosity. *ACS Mater. Lett.* **2019**, *1*, 3–7. [CrossRef]

16. Bhandary, S.; Shukla, R.; Van Hecke, K. Effect of chemical substitution on the construction of boroxine-based supramolecular crystalline polymers featuring B←N dative bonds. *CrystEngComm* **2022**, *24*, 1695–1699. [CrossRef]

17. Madura, I.D.; Czerwinska, K.; Jakubczyk, M.; Pawelko, A.; Adamczyk-Wozniak, A.; Sporzynski, A. Weak C–H⋯O and Dipole–Dipole Interactions as Driving Forces in Crystals of Fluorosubstituted Phenylboronic Catechol Esters. *Cryst. Growth Des.* **2013**, *13*, 5344–5352. [CrossRef]

18. Luisier, N.; Scopelliti, R.; Severin, K. Supramolecular gels based on boronate esters and imidazolyl donors. *Soft Matter* **2016**, *12*, 588–593. [CrossRef] [PubMed]

19. Salazar-Mendoza, D.; Cruz-Huerta, J.; Höpfl, H.; Hernández-Ahuactzi, I.F.; Sanchez, M. Macrocycles and coordination polymers derived from self-complementary tectons based on N-containing boronic acids. *Cryst. Growth Des.* **2013**, *13*, 2441–2454. [CrossRef]

20. Ray, K.K.; Campillo-Alvarado, G.; Morales-Rojas, H.; Hopfl, H.; MacGillivray, L.R.; Tivanski, A.V. Semiconductor Cocrystals Based on Boron: Generated Electrical Response with π-Rich Aromatic Molecules. *Cryst. Growth Des.* **2020**, *20*, 3–8. [CrossRef]

21. Icli, B.; Solari, E.; Kilbas, B.; Scopelliti, R.; Severin, K. Multicomponent Assembly of Macrocycles and Polymers by Coordination of Pyridyl Ligands to 1,4-Bis(benzodioxaborole)benzene. *Chem. Eur. J.* **2012**, *18*, 14867–14874. [CrossRef]

22. Christinat, N.; Scopelliti, R.; Severin, K. Boron-based rotaxanes by multicomponent self-assembly. *Chem. Commun.* **2008**, 3660–3662. [CrossRef] [PubMed]

23. Hartwick, C.J.; Yelgaonkar, S.P.; Reinheimer, E.W.; Campillo-Alvarado, G.; MacGillivray, L.R. Self-Assembly of Diboronic Esters with U-Shaped Bipyridines: "Plugin-Socket" Assemblies. *Cryst. Growth Des.* **2021**, *21*, 4482–4487. [CrossRef] [PubMed]

24. Christinat, N.; Croisier, E.; Scopelliti, R.; Cascella, M.; Rothlisberger, U.; Severin, K. Formation of boronate ester polymers with efficient intrastrand charge-transfer transitions by three-component reactions. *Eur. J. Inorg. Chem.* **2007**, *2007*, 5177–5181. [CrossRef]

25. Herrera-España, A.D.; Campillo-Alvarado, G.; Román-Bravo, P.; Herrera-Ruiz, D.; Höpfl, H.; Morales-Rojas, H. Selective Isolation of Polycyclic Aromatic Hydrocarbons by Self-Assembly of a Tunable N→B Clathrate. *Cryst. Growth Des.* **2015**, *15*, 1572–1576. [CrossRef]

26. Dhara, A.; Beuerle, F. Reversible Assembly of a Supramolecular Cage Linked by Boron–Nitrogen Dative Bonds. *Chem. Eur. J.* **2015**, *21*, 17391–17396. [CrossRef]

27. Icli, B.; Sheepwash, E.; Riis-Johannessen, T.; Schenk, K.; Filinchuk, Y.; Scopelliti, R.; Severin, K. Dative boron–nitrogen bonds in structural supramolecular chemistry: Multicomponent assembly of prismatic organic cages. *Chem. Sci.* **2011**, *2*, 1719–1721.

28. Campillo-Alvarado, G.; Vargas-Olvera, E.C.; Hopfl, H.; Herrera-Espana, A.D.; Sanchez-Guadarrama, O.; Morales-Rojas, H.; MacGillivray, L.R.; Rodriguez-Molina, B.; Farfan, N. Self-Assembly of Fluorinated Boronic Esters and 4,4'-Bipyridine into 2:1 N→B Adducts and Inclusion of Aromatic Guest Molecules in the Solid State: Application for the Separation of o,m,p-Xylene. *Cryst. Growth Des.* **2018**, *18*, 2726–2743. [CrossRef]

29. Herrera-Espana, A.D.; Hoepfl, H.; Morales-Rojas, H. Boron-Nitrogen Double Tweezers Comprising Arylboronic Esters and Diamines: Self-Assembly in Solution and Adaptability as Hosts for Aromatic Guests in the Solid State. *ChemPlusChem* **2020**, *85*, 548–560. [CrossRef] [PubMed]

30. Manankandayalage, C.P.; Unruh, D.K.; Krempner, C. Boronic, diboronic and boric acid esters of 1,8-naphthalenediol–synthesis, structure and formation of boronium salts. *Dalton Trans.* **2020**, *49*, 4834–4842. [CrossRef]

31. *Bruker (2021) SADABS v2016/2*, Bruker AXS Inc.: Madison, WI, USA, 2015.

32. Sheldrick, G.M. Crystal structure refinement with SHELXL. *Acta Crystallogr. Sect. C Struct. Chem.* **2015**, *C71*, 3–8. [CrossRef] [PubMed]

33. Spek, A.L. Structure validation in chemical crystallography. *Acta Crystallogr. Sect. D Biol. Crystallogr.* **2009**, *D65*, 148–155. [CrossRef] [PubMed]

34. Spek, A.L. PLATON SQUEEZE: A tool for the calculation of the disordered solvent contribution to the calculated structure factors. *Acta Crystallogr. Sect. C Struct. Chem.* **2015**, *C71*, 9–18. [CrossRef]

35. CrysAlisPRO. *Oxford Diffraction*; Agilent Technologies UK Ltd.: Yarnton, UK, 2018.
36. *SCALE3 ABSPACK*, Oxford Diffraction Ltd.: Abingdon, UK, 2005.
37. Sheldrick, G.M. SHELXT–Integrated space-group and crystal-structure determination. *Acta Crystallogr. Sect. A Found. Adv.* **2015**, *A71*, 3–8. [CrossRef] [PubMed]
38. Dolomanov, O.V.; Bourhis, L.J.; Gildea, R.J.; Howard, J.A.K.; Puschmann, H. OLEX2: A complete structure solution, refinement and analysis program. *J. Appl. Crystallogr.* **2009**, *42*, 339–341. [CrossRef]

*molecules*

Article

New Boron Containing Acridines: Synthesis and Preliminary Biological Study

Anna A. Druzina [1,*], Nadezhda V. Dudarova [1], Ivan V. Ananyev [2], Anastasia A. Antonets [3], Dmitry N. Kaluzhny [4], Alexey A. Nazarov [3], Igor B. Sivaev [1,5] and Vladimir I. Bregadze [1,†]

[1]   A.N. Nesmeyanov Institute of Organoelement Compounds, Russian Academy of Sciences, 28 Vavilov Str., 119334 Moscow, Russia; nadezjdino_96@mail.ru (N.V.D.); sivaev@ineos.ac.ru (I.B.S.); bre@ineos.ac.ru (V.I.B.)

[2]   N.S. Kurnakov Institute of General and Inorganic Chemistry, Russian Academy of Sciences, 31 Leninskii pr., 119991 Moscow, Russia; i.ananyev@gmail.com

[3]   Department of Chemistry, M.V. Lomonosov Moscow State University, 1/3 Leninskie Gory, 119991 Moscow, Russia; antonets.anastasia.a@gmail.com (A.A.A.); nazarov@med.chem.msu.ru (A.A.N.)

[4]   V.A. Engelhardt Institute of Molecular Biology, Russian Academy of Sciences, 32 Vavilov Str., 11991 Moscow, Russia; uzhny@mail.ru

[5]   Basic Department of Chemistry of Innovative Materials and Technologies, G.V. Plekhanov Russian University of Economics, 36 Stremyannyi Line, 117997 Moscow, Russia

*   Correspondence: ilinova_anna@mail.ru; Tel.: +7-926-404-5566

†   Dedicated to John Kennedy on the occasion of his 80th Jubilee and in recognition of his outstanding contributions to borane and metallaborane chemistry.

Abstract: The synthesis of the first conjugates of acridine with cobalt bis(dicarbollide) are reported. A novel 9-azido derivative of acridine was prepared through the reaction of 9-methoxyacridine with $N_3CH_2CH_2NH_2$, and its solid-state molecular structure was determined via single-crystal X-ray diffraction. The azidoacridine was used in a copper (I)-catalyzed azide-alkyne cycloaddition reaction with cobalt bis(dicarbollide)-based terminal alkynes to give the target 1,2,3-triazoles. DNA interaction studies via absorbance spectroscopy showed the weak binding of the obtained conjugates with DNA. The antiproliferative activity (IC_{50}) of the boronated conjugates against a series of human cell lines was evaluated through an MTT assay. The results suggested that acridine derivatives of cobalt bis(dicarbollide) might serve as a novel scaffold for the future development of new agents for boron neutron capture therapy (BNCT).

Keywords: acridine; cobalt bis(dicarbollide); synthesis; DNA-interaction; antiproliferative activity

Citation: Druzina, A.A.; Dudarova, N.V.; Ananyev, I.V.; Antonets, A.A.; Kaluzhny, D.N.; Nazarov, A.A.; Sivaev, I.B.; Bregadze, V.I. New Boron Containing Acridines: Synthesis and Preliminary Biological Study. *Molecules* **2023**, *28*, 6636. https://doi.org/10.3390/molecules28186636

Academic Editor: Renhua Qiu

Received: 24 July 2023
Revised: 5 September 2023
Accepted: 14 September 2023
Published: 15 September 2023

1. Introduction

Boron neutron capture therapy (BNCT) is a binary therapeutic method based on the nuclear capture reaction that takes place when the stable isotope ^{10}B is irradiated with neutrons to release high-energy α-particles and ^7Li nuclei [1,2]. The most important requirements for BNCT agents are (1) low toxicity; (2) high tumor uptake (~20–35 μg ^{10}B) and low normal tissue uptake, with a sufficient tumor/normal tissue boron concentration ratio of >3:1; and (3) relatively rapid clearance from the blood and normal tissues, and persistence in the tumor for at least several hours during irradiation with a neutron flux [3–6]. Since α-particles have very short pathlengths in biological tissues (5–9 μm), their destructive effects are practically limited to cells that contain boron. In theory, α-particles can selectively destroy not only tumor cells, but also adjoining normal cells. Because BNCT primarily is a biologically, rather than physically, targeted type of particle radiation therapy, it offers the possibility to selectively destroy tumor cells without affecting the normal cells and tissues of an organism. The requirement, however, is that sufficient amounts of ^{10}B and thermal neutrons are delivered to the site of the tumor.

Therefore, polyhedral boron hydrides, such as cobalt bis(dicarbollide) $[3,3'\text{-Co}(1,2\text{-}C_2B_9H_{11})_2]^-$, containing eighteen boron atoms in the molecule and characterized by outstanding chemical and thermal stability and the availability of convenient modification methods [7,8], are good candidates for the design of BNCT agents. Cobalt bis(dicarbollide) as a sodium salt demonstrates good water solubility and low toxicity both in vitro [9,10] and in vivo [9,11]. In particular, it does not show acute toxicity when intravenously [9] or intraperitoneally [11] injected into wild-type mice. It was also found that cobalt bis(dicarbollide) can pass directly through lipid membranes [12–14] and accumulate in cells without disrupting membrane integrity [10].

Calculations have shown that the radiobiological effectiveness of the boron neutron capture reaction [^{10}B(^{1}n,^{4}He)^{7}Li] can be significantly enhanced when it occurs in the cell nucleus rather than in the cytoplasm or the cell membrane [15]. Boron accumulating in the cell nucleus is much more efficient in cell killing than the same amount of boron when it is uniformly distributed. Consequently, when the BNCT drug is localized in the cell nucleus, a lower concentration is required [16]. These data have implications for the choice of boron carriers in neutron capture therapy.

This leads to an interest in DNA-binding BNCT agents, such as DNA intercalators (acridine, phenanthridinium, naphthalimide, and others) [17–27]. The first boron-containing acridine was synthesized by Snyder and Konecky and contained two aryl dihydroxyboryl groups [28]. This compound was too toxic to be used as a BNCT agent, but it led to the synthesis of the first carborane-based acridines [29]. Although these latter compounds were less toxic and achieved higher concentrations in tumors compared with the normal brain and blood, these values were significantly lower than those found in the liver, kidney, and spleen. Later, it was proposed to introduce additional hydrophilic groups into the carborane part of the molecule in order to improve its bioavailability [30]. Recently, synthesis of a series of acridines modified with carborane clusters and the evaluation of their DNA-binding ability and cytotoxicity has been described [31]. Also recently, one example of boronated acridine has been reported in which the boron moiety is cobalt bis(dicarbollide). This compound was synthesized through the reaction of 9-aminoacridine with a 1,4-dioxane derivative of cobalt bis(dicarbollide) [32].

In this contribution, we describe the synthesis of a series of novel 9-aminoacridine derivatives with the cobalt bis(dicarbollide) moiety as potential candidates for BNCT via the Cu(I)-catalyzed 1,3-dipolar [3 + 2] cycloaddition reaction of alkynes to azides ("click" reaction) as well as the evaluation of their antiproliferative activity and DNA interaction.

2. Results and Discussion

2.1. Synthesis of N^9-Azidooacridine 2

9-Aminoacridine derivatives are an interesting group of heterocyclic compounds whose chemical structure causes them to have a variety of biological activities [33,34]. Thus, it was found that 9-aminoacridines can selectively accumulate in cell nuclei and other cellular organoids containing nucleic acids. Earlier studies on 9-aminoacridine have shown that this compound intercalates between the base stacks into a DNA double helix. It is assumed that the 9-aminoacridine with its 9-amino group lies in the minor groove and the 4- and 5-positions of the acridine ring are oriented toward the major groove [35]. It should also be noted that the literature provides information that the cytotoxicity of acridine derivatives is related to the presence and nature of various types of substituents at the 9-amino group of the acridine heterocycle. Thus, the substitution of one of the hydrogen atoms of the amino group in 9-aminoacridine leads to a decrease in the toxicity of substances [36].

We decided to synthesize boron-containing acridine derivatives with the introduction of the cobalt bis(dicarbollide) moiety through the "click" reaction. Among the methods for obtaining bioconjugates, the "click" reaction is widely used, leading to the formation of 1,2,3-triazoles [37–39]. Earlier, the "click" reaction was successfully used to obtain a wide range of conjugates of cobalt bis(dicarbollide) with chlorine e_6 [40], nucleosides [41], and cholesterol [42]. In the present work, we used the "click" reaction to obtain new conjugates of the cobalt bis(dicarbollide) with acridine. At the start of our study, only a few examples of known boron-containing acridines were represented by carborane derivatives [29,31], while the first example of conjugate of acridine with cobalt bis(dicarbollide) was reported very recently, and it was prepared through the direct reaction of 9-aminoacridine with the 1,4-dioxane derivative of cobalt bis(dicarbollide) [32].

Thus, we prepared the azido derivative of acridine which can be used for conjugation with acetylenic derivatives of cobalt bis(dicarbollide). The reaction of 9-methoxyacridine **1** with 2-azidoethanamine hydrochloride in acetonitrile in the presence of Et_3N upon reflux for 17 h results in N^9-azidoacridine hydrochloride **2**, which was isolated as a water-soluble pale yellow crystalline solid after crystallization from ethanol in 76% yield (Scheme 1).

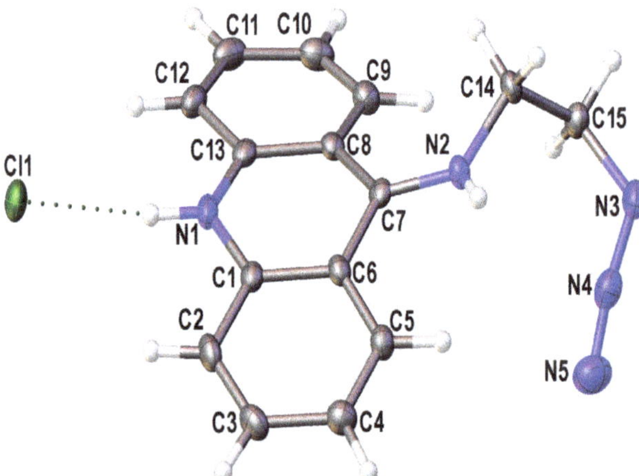

Scheme 1. Synthesis of N^9-substituted azidoacridine **2**.

The synthesized azido derivative of acridine **2** was characterized using the methods of ^1H and ^{13}C NMR spectroscopy, IR-spectroscopy, and high-resolution mass spectrometry (see Supplementary Materials). In the ^1H-NMR spectrum, the signals of the methylene groups are observed at 3.63 and 3.72 ppm, and the characteristic signals of the acridine core **2** are observed in the range of 6.96–7.61 ppm. The IR spectrum of compound **2** exhibits an absorption band characteristic of the N_3 group at 2065 cm^{-1}.

2.2. Single-Crystal X-ray Diffraction Studies of N^9-Azidoacridine 2

The solid-state structure of 9-azidoacridine **2** was determined through a single-crystal X-ray diffraction study (Figure 1). Crystals of **2** suitable for single-crystal X-ray analysis were grown via crystallization from ethanol.

Figure 1. The independent unit in crystal of **2** in the representation of non-hydrogen atoms as probability ellipsoids of atomic displacements ($p = 0.5$). The H-bond is shown by a dotted line.

The independent part of the unit cell of **2** contains an H-bonded contact ionic pair: the chloride anion and the protonated substituted acridine as a cation. According to a Cambridge Structural Database [43] search, the structure of **2** is one of many known examples of protonated acridine salts [44–48]. Note that in all these structures, the acridinium cation is nearly flat, while this is not the case for **2**, where the angle between mean-squared planes composed of the carbon atoms of two lateral hydrocarbon rings is 21.6 (2)°. Moreover, the N2 nitrogen atom in **2** is significantly shifted out of the plane of the central acridine cycle: the non-bonding N1...C7-N2 angle equals 167.0 (1)°. Considering the opposite directions of the displacement of the substituent and the lateral acridine cycles from the mean-square acridine plane, one can suppose the presence of steric repulsion between them. Indeed, there are several rather short H...H contacts within the cation; with the normalization of X-H bond lengths, the distances between the hydrogen atoms at the C5 and C9 atoms and the hydrogen atoms at the N2, C14, and C15 atoms are less than 2.05 Å.

The role of steric repulsion is supported by the geometry optimization performed for the isolated cation on the PBE0-D3/def2TZVP level. The relaxed structure is heavily distorted even without the influence of media effects: the mentioned interplane angle is 12.1° and the N1...C7-N2 angle equals 165.8°. Overall, both the relaxed isolated and crystal structures of cation **2** are quite similar (the r.m.s. difference for non-hydrogen atoms does not exceed 0.12 Å, Figure S1). What is noteworthy is that this conformation is indeed unfavorable for the acridine moiety: according to the calculations for the unsubstituted acridinium cation, the energy of the distorted conformation (constructed on the basis of that in **2** with the optimization of only hydrogen atoms) is 6.3 kcal·mol^{-1} higher than that in the fully relaxed structure. The electronic structure of the isolated cation **2** was then analyzed using the real space methods to determine the role of interatomic contacts in the (de)stabilization of the cation's conformation.

According to the analysis of non-covalent interactions based on the reduced density gradient (RDG) and sign $(\lambda_2) \cdot \rho$ (**r**) functions [49] (λ_2—intermediate eigenvalue of electron density Hessian), there are a number of regions having low RDG values, which indicate the presence of non-covalent interactions in the area between the substituent and the acridine moiety. In particular, the RDG/sign $(\lambda_2) \cdot \rho$ (**r**) plot (Figure S2) and the corresponding 3D isosurface plot (Figure 2, left) demonstrates the regions potentially having (1) rather strong non-covalent interactions between the H2N and H5 atoms and between the H9 and H14B atoms (the negative λ_2 sign corresponds to the accumulation of electron density), (2) weak Van der Waals interaction between the N5 and H5 atoms, and (3) forced interactions between the H9 and H15A atoms, between the H9 and N2 atoms, and between the N2 and C5 atoms, all formed due to steric effects (the positive λ_2 sign corresponds to the electronic charge depletion). The bonding nature of the dihydrogen H2N...H5 and H9...H14B contacts as well as of the weak N5...H5 contact is confirmed by the presence of (3, −1) critical points of electron density between corresponding topological atoms (Figure 2, right) that is the real space manifestation of preferred exchange–interaction channels [50,51]. Note that the net energy of these bonding non-covalent interactions estimated from the interatomic surface integrals of electron density [52] equals −7.1 kcal·mol^{-1}, which is in exceptionally good agreement with the loss of energy occurring upon the distortion of acridine moiety conformation (see above). The main contribution (−3.6 and −3.2 kcal·mol^{-1}) arises from the dihydrogen bonds, whereas the energy of C-H...N interaction is less than 0.3 kcal·mol^{-1} in magnitude. Thus, the steric repulsion between the substituent and acridine fragments leading to the pronounced distortion of the aromatic fragment is compensated by two dihydrogen bonds.

Finally, the crystal packing of **2** is expected, considering the presence of two strong proton donors in the cation. Namely, cation moieties are aggregated into the infinite chains by rather strong H-bonds between the N-H groups and chloride anions (N...Cl 3.131 and 3.156 Å, $\angle$ (NHCl) with the normalization of N-H bond lengths 171.3° and 156.4°, Figure 3). Due to their geometry, these H-bonds, together with weaker C-H...Cl contacts, can additionally stabilize the distorted structure of the cation. The abovementioned chains are bound together through numerous weaker interactions such as π...π stacking interactions between acridine moieties, C-H...Cl hydrogen bonds, and C-H...π interactions.

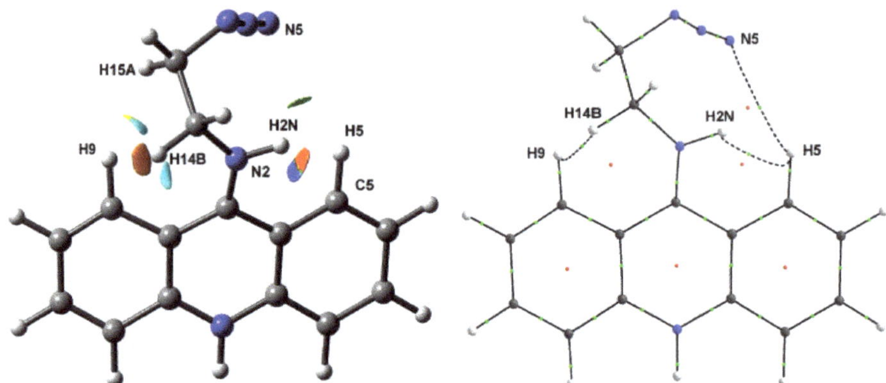

Figure 2. Real space electronic structure peculiarities in the area of substituent in the isolated optimized cation from **2**. On the left: the RDG isosurface (0.3) colored by the sign $(\lambda_2)\cdot\rho$ (**r**) function (from -0.0019 a.u. in red to 0.0019 a.u. in blue). On the right: the atomic connectivity graph according to the topological analysis of ρ (**r**); dashed curves denote non-covalent interactions; green and red points denote (3, -1) and (3, $+1$) critical points of ρ (**r**), correspondingly.

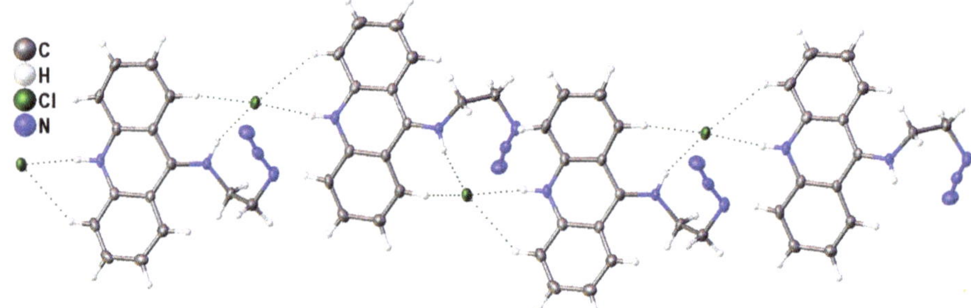

Figure 3. A fragment of infinite chain in the crystal of **2**. H-bonds are shown by dotted lines.

2.3. Synthesis of Cobalt Bis(dicarbollide)-Acridine Derivatives with 1,2,3-Triazoles **7–10**

One of the important goals of BNCT, as already mentioned, is the synthesis of compounds which provide higher accumulation of boron in tumor than the clinically used compounds. This can be achieved through using boron clusters with a high content of boron atoms in the molecule, such as cobalt bis(dicarbollide) [6,53]. In this contribution, the "click" reaction was proposed to combine the cobalt bis(dicarbollide) cluster, providing a high content of boron atoms in the molecule, and the acridine system, providing the delivery of the boron component to the cell nucleus due to intercalation between DNA base stacks. Moreover, through changing the type and the size of a spacer between cobalt bis(dicarbollide) cluster and acridine, it is possible to control, to some extent, the hydrophilic/hydrophobic balance of the compounds.

The obtained N^9-azidooacridine **2** was used for the synthesis of target conjugates of cobalt bis(dicarbollide) with acridine **7–10** (Scheme 2). The acetylenic derivatives of cobalt bis(dicarbollide) **3–6** were prepared through the nucleophilic cleavage reactions of the corresponding cyclic oxonium derivatives of cobalt bis(dicarbollide) with 2-propyn-1-ol and 3-butyn-1-ol [40,54]. The "click" reactions were carried out refluxing boron-containing acetylenic derivatives **3–6** with the N^9-azidoacridine **2** in ethanol for 2 h in the presence of CuI and diisopropylethylamine (DIPEA). The desired triazoles **7–10** were isolated as cesium salts via precipitation with CsCl from aqueous acetone followed by column chromatography on silica using a CH_2Cl_2-CH_3CN mixture (70–75%) as eluent.

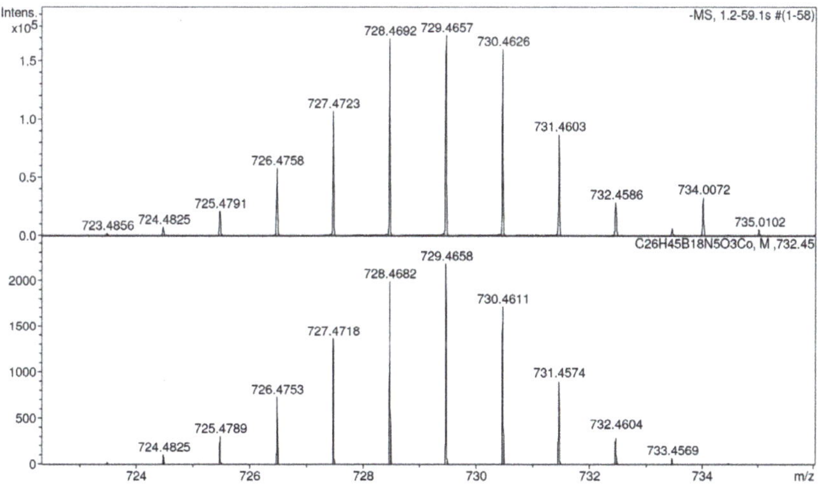

Scheme 2. Synthesis of conjugates of cobalt bis(dicarbollide) with acridine **7–10**.

The conjugation products were characterized via ^{1}H-, ^{11}B-, and ^{13}C-NMR; IR-spectroscopy; and high-resolution mass spectrometry (see Supplementary Materials). The purity of the obtained compounds was confirmed via chemical analysis.

In the ^{1}H-NMR spectra of the synthesized compounds **7–10**, the characteristic signals of the triazole CH hydrogens appear in the region of 7.83–7.99 ppm; the signals of the methylene groups next to the triazole cycle are observed in the range of 4.90–5.14 ppm and the characteristic signals of the aromatic system of acridine are observed in the range of 7.62–8.58 ppm for compounds **7–10**. For conjugates **7–10**, the signals of the CH_{carb} groups in the ^{1}H-NMR spectra appear as broad singlets in the ranges of 4.19–4.25 and 4.11–4.19 ppm. In the ^{13}C-NMR spectra, the signals of the triazole CH carbons for **7–10** are observed in the range of 123.2–124.0 ppm, whereas the signals of the second triazole carbons appear in the range of 139.8–140.2 ppm. In the ^{13}C-NMR spectra, the signals of CH$_{carb}$ groups appeared in the range of 46.4–53.9 ppm. The ^{11}B{^{1}H} NMR spectra of compounds **7–10** exhibit singlets between 23.4–23.8 ppm from the substituted boron atom.

The IR spectra of compounds **7–10** exhibit absorption bands characteristic of the BH groups at 2547–2553 cm^{-1} and the 1,2,3-triazole heterocycles at 1581–1593 cm^{-1}. The mass spectra of boron-containing conjugates **7–10** show negatively charged ions corresponding to [M-Cs]$^-$, which are consistent with the predicted isotope distributions; this confirms the molecular formulas of the obtained compounds (Figure 4 presents the mass spectrum of conjugate **7**).

Figure 4. Mass spectrum of conjugate **7** (found 729.4657 [M]$^-$, [C$_{26}$H$_{45}$B$_{18}$CoN$_5$O$_3$]$^-$, calculated 729.4658).

2.4. DNA Interaction Study

In order to test the ability of the compounds to bind DNA, spectral changes were tested through increasing the concentration of DNA in solution. Due to the low solubility of the compounds, a concentration of 5 µM was used. Calf thymus DNA at concentrations of up to 80 µM was added to the solution of compounds. The changes in the absorption spectra of compounds **7–10** are shown in Figure S5 (see Supplementary Materials). It can be seen that increasing the concentration of DNA in the solution leads to a non-significant decrease in absorbance. Changes in the absorbance of compounds **7–10** were not significant: about 5% taking into account dilution during titration. Such changes may characterize the relatively weak binding of the studied compounds with DNA under experimental conditions. The largest spectral changes were observed for conjugate **8** (Figure 5).

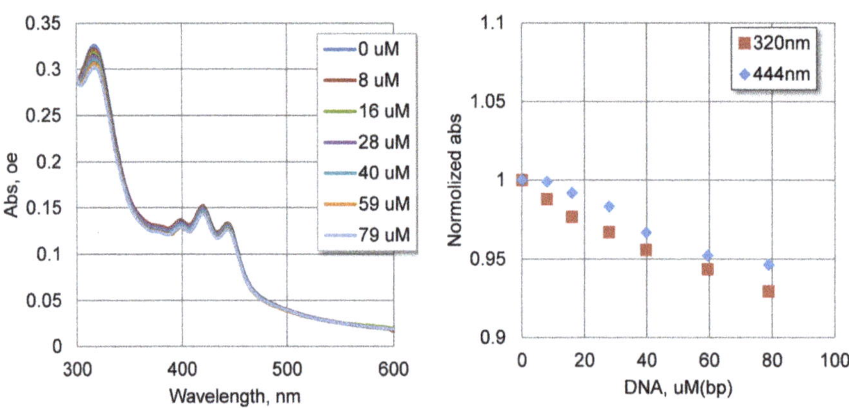

Figure 5. Changes of absorbance spectra upon DNA interaction for conjugate **8**. We used a 5 mkM compound in 10 mM potassium phosphate buffer pH = 8.0 and a calf thymus DNA in the range 0–80 µM (b.p.).

2.5. Antiproliferative Activity of Boronated Acridines 7–10

The antiproliferative activity of compounds **7–10** against cancer HCT116, MCF7, A549, and WI38 nonmalignant lung fibroblast cell lines was evaluated by means of a standard MTT colorimetric assay after 72 h of incubation (Table 1). All four compounds were found to be nontoxic against the lung cancer A549 cell line. However, against other cancer cells and nonmalignant cells, compounds showed activity in the mid-micromolar range that complicates use as new agents for BNCT.

Table 1. Antiproliferative activity of conjugates **7–10** and cisplatin against human cancer cells.

Cell Line	IC$_{50}$, µM				
	7	8	9	10	Cisplatin [55]
WI38	21 ± 4	21 ± 5	33 ± 3	34 ± 8	8 ± 3
A549	>200	>200	>200	>200	13 ± 3
HCT116	49 ± 20	61 ± 25	43 ± 18	41 ± 1	13 ± 4
MCF7	15 ± 3	25 ± 6	28 ± 3	13 ± 3	30 ± 9

3. Materials and Methods

3.1. General Methods

The acetylenic derivatives of cobalt bis(dicarbollide) [8-HC≡CCH$_2$O(CH$_2$CH$_2$O)$_2$-3,3'-Co(1,2-C$_2$B$_9$H$_{10}$)(1',2'-C$_2$B$_9$H$_{11}$)]K (**3**) [40], [8-HC≡CCH$_2$CH$_2$O(CH$_2$CH$_2$O)$_2$-3,3'-Co(1,2-C$_2$B$_9$H$_{10}$)(1',2'-C$_2$B$_9$H$_{11}$)]K (**4**) [40], [8-HC≡CCH$_2$O(CH$_2$)$_5$O-3,3'-Co(1,2-C$_2$B$_9$H$_{10}$)(1',2'-C$_2$B$_9$H$_{11}$)]Na (**5**) [54], [8-HC≡CCH$_2$CH$_2$O(CH$_2$)$_5$O-3,3'-Co(1,2-C$_2$B$_9$H$_{10}$)(1',2'-C$_2$B$_9$H$_{11}$)]Na (**6**) [54], and N$_3$CH$_2$CH$_2$NH$_2$ × HCl [56] were prepared according to the literature. 9-Methoxyacridine (Chemieliva Pharmaceutical Co., Ltd., Chongqing, China), diisopropylethylamine (Carl Roth GmbH, Karlsruhe, Germany), and CuI (PANREAC QUIMICA SA, Barcelona, Spain) were used without further purification. Ethanol, CH$_3$CN, CH$_2$Cl$_2$, and NaN$_3$ were commercially analytical grade reagents. The reaction progress was monitored via thin-layer chromatography (Merck F245 silica gel on aluminum plates). Acros Organics silica gel (0.060–0.200 mm) was used for column chromatography. The NMR spectra at ^{1}H (400.1 MHz), ^{11}B (128.4 MHz), and ^{13}C (100.0 MHz) were recorded with a Varian Inova 400 spectrometer (Varian, Palo Alto, CA, USA). The residual signal of the NMR solvent relative to Me$_4$Si was taken as the internal reference for ^{1}H- and ^{13}C-NMR spectra. ^{11}B-NMR spectra were referenced using BF$_3$·Et$_2$O as an external standard. Infrared spectra were recorded on a Spectra SF 2000 (OKB SPECTRUM, Saint-Petersburg, Russia) instrument. High-resolution mass spectra (HRMS) were measured on a mictOTOF II (Bruker Daltonic, Bremen, Germany) instrument using electrospray ionization (ESI). The measurements were performed in a negative ion mode (interface capillary voltage 3000 V) and positive ion mode (interface capillary voltage 4500 V), mass range from m/z 50 to m/z 3000, external or internal calibration was carried out with ESI Tuning Mix, Agilent. A syringe injection was used for solutions in acetonitrile (flow rate 3 μL/min). Nitrogen was applied as a dry gas; the interface temperature was set at 180 °C. Elemental analyses were performed at the Laboratory of Microanalysis of the A.N. Nesmeyanov Institute of Organoelement Compounds.

3.1.1. Synthesis of N^9-Azidoacridine 2

We added 1 mL of NEt$_3$ to a suspension of 9-methoxyacridine **1** (0.5 g, 2.4 mmol) and 2-azidoethanamine hydrochloride (0.32 g, 2.6 mmol) in 20 mL of CH$_3$CN. The reaction mixture was stirred under reflux for 15 h and then cooled to room temperature. To the cooling reaction mixture, a few drops of HCl were added before pH = 3 and left in the air for 1 h. Then, the solution was evaporated, and residue was crystallized from EtOH. The product was filtered, washed with cold EtOH (5 mL), and air dried to give pale yellow crystals of **2** (0.54 g, yield 76%). ^{1}H NMR (400 MHz, D$_2$O) д 7.61 (d, 2H, 2 × CH$_{Ar}$, J = 9.0 Hz), 7.49 (t, 2H, 2 × CH$_{Ar}$), 7.11(t, 2H, 2 × CH$_{Ar}$), 6.96 (d, 2H, 2 × CH$_{Ar}$, J = 9.0 Hz), 3.72 (CH$_2$), 3.63 (CH$_2$) ppm. ^{13}C NMR (101 MHz, D$_2$O): 158.9 (C$_{Ar}$), 138.6 (C$_{Ar}$), 135.3 (CH$_{Ar}$), 124.0 (2 × CH$_{Ar}$), 118.0 (CH$_{Ar}$), 111.5 (C$_{Ar}$), 57.4 (C$_{Ar}$), 49.7 (NCH$_2$), 47.6 (NCH$_2$) ppm. IR (KBr, н, cm^{-1}): 2065 (N$_3$). HRMS (ESI) m/z for [C$_{15}$H$_{13}$N$_5$]$^+$ calcd 264.1244 [M]$^+$, found: 264.1246 [M]$^+$.

General Procedure for the Synthesis of the Conjugates of Cobalt Bis(Dicarbollide) with Acridine 7–10

A mixture of 9-azidoacridine **2** (1 eq.), the alkynyl derivatives of cobalt bis(dicarbollide) **3–6** (1 eq.), diisopropylethylamine (0.5–1 mL), and CuI (0.1 eq.) in 10–15 mL of EtOH was heated under reflux for 2 h. Then, the reaction mixture was cooled to room temperature, and inorganic precipitate was filtered. Then, organic solvent was removed in vacuo. The residue was extracted using EtOAc (100 mL) and washed with 1M HCl (4 × 50 mL) and brine (2 × 50 mL) and dried (Na$_2$SO$_4$). Then, the ethyl acetate was evaporated. The residue was dissolved in 5 mL of acetone. To the resulting solution, 1 g of CsCl in 100 mL of water was added. The crude product was purified on a silica column using CH$_2$Cl$_2$-CH$_3$CN (2/1) as an eluent to give the desired products **7–10**.

3.1.2. Synthesis of Conjugate **7**

Conjugate **7** was prepared from compound **2** (0.09 g, 0.30 mmol), the alkynyl derivative of cobalt bis(dicarbollide) **3** (0.14 g, 0.30 mmol), diisopropylethylamine (1 mL, 0.74 g, 5.73 mmol), and CuI (0.006 g, 0.03 mmol) in 15 mL of EtOH. The product was obtained as an orange solid of **4** (0.19 g, yield 73%). ^{1}H NMR (400 MHz, acetone-d$_6$) д 8.52 (d, 2H, 2 × CH_{Ar}, J = 9.0 Hz), 8.02 (s, 4H, 4 × CH_{Ar}), 7.83 (s, 1H, CHCN$_3$), 7.62 (m, 2H, 2 × CH_{Ar}), 5.08 (s, 2H, CH_2N), 4.90 (s, 2H, CH_2NH), 4.41 (s, 2H, OCH_2C), 4.19 (br. s, 2H, CH_{carb}), 4.11 (br. s, 2H, CH_{carb}), 3.78 (s, 2H, BOCH_2), 3.57 (s, 2H, CH_2O), 3.49 (s, 2H, CH_2O), 3.22 (s, 2H, CH_2O) ppm. ^{11}B NMR (128 MHz, acetone-d$_6$): 23.8 (1B, s), 5.2 (1B, d, J = 144 Hz), 0.3 (1B, d, J = 152), −2.3 (1B, d, J = 164 Hz), −4.5 (2B, d, J = 142 Hz), −7.2 (4B, d, J = 124 Hz), −9.0 (2B, d, J unsolved), −17.3 (2B, d, J = 164 Hz), −20.2 (2B, d, J = 150 Hz), −22.5 (1B, d, J = 150 Hz), −28.6 (1B, d, J = 142 Hz) ppm. ^{13}C NMR (101 MHz, acetone-d$_6$): 158.3 (C$_{Ar}$), 144.8 (C$_{Ar}$), 140.2 (CN$_3$CH), 134.9 (CH$_{Ar}$), 125.1 (CH$_{Ar}$), 123.9 (CH$_{Ar}$, CN$_3$CH), 119.9 (CH$_{Ar}$), 113.5 (C$_{Ar}$), 72.2 (OCH$_2$), 70.3 (OCH$_2$), 68.8 (OCH$_2$), 63.8 (OCH$_2$), 59.7 (OCH$_2$), 53.3 (CH$_{carb}$), 50.0 (NCH$_2$), 49.6 (NCH$_2$), 46.6 (CH$_{carb}$) ppm. IR (KBr, н, cm^{-1}): 2568 (BH), 1581 (triazole). Found: C 36.41, H 5.14, B 22.79, N 8.07; Calc. for C$_{26}$H$_{45}$B$_{18}$CoN$_5$O$_3$Cs C 36.22, H 5.26, B 22.57, N 8.12. HRMS (ESI) m/z for [C$_{26}$H$_{45}$B$_{18}$CoN$_5$O$_3$]$^-$ calcd 729.4658 [M]$^-$, found 729.4657 [M]$^-$.

3.1.3. Synthesis of Conjugate **8**

Conjugate **8** was prepared from compound **2** (0.062 g, 0.21 mmol), alkynyl derivative of cobalt bis(dicarbollide) **4** (0.10 g, 0.21 mmol), diisopropylethylamine (0.5 mL, 0.37 g, 2.86 mmol), and CuI (0.004 g, 0.02 mmol) in 10 mL of EtOH. The product was obtained as an orange solid of **4** (0.14 g, yield 75%). ^{1}H NMR (400 MHz, acetone-d$_6$) д 8.54 (d, 2H, 2 × CH_{Ar}, J = 9.0 Hz), 8.02 (m, 4H, 4 × CH_{Ar}), 7.87 (s, 1H, CHCN$_3$), 7.66 (m, 2H, 2 × CH_{Ar}), 5.08 (m, 2H, CH_2N), 4.90 (m, 2H, CH_2NH), 4.22 (br. s, 2H, CH_{carb}), 4.19 (br. s, 2H, CH_{carb}), 3.67 (m, 2H, OCH_2CH$_2$C), 3.48 (m, 8H, BOCH_2, 3 × CH_2O), 2.80 (m, 2H, CH_2O) ppm. ^{11}B NMR (128 MHz, acetone-d$_6$): 23.4 (1B, s), 4.4 (1B, d, J = 145 Hz), 0.6 (1*B*, d, J = 168), −2.4 (1*B*, d, J unsolved), −4.3 (2*B*, d, J = 156 Hz), −7.2 (2*B*, d, J = 130 Hz), −8.0 (4*B*, d, J = 124 Hz), −17.4 (2*B*, d, J unsolved), −20.4 (2*B*, d, J = 162 Hz), −21.9 (1*B*, d, J = 160 Hz), −28.7 (1*B*, d, J = 152 Hz) ppm. ^{13}C NMR (101 MHz, acetone-d$_6$): 158.6 (C$_{Ar}$), 145.6 (C$_{Ar}$), 139.9 (CN$_3$CH), 135.6 (CH$_{Ar}$), 125.2 (CH$_{Ar}$), 124.3 (CH$_{Ar}$), 123.4 (CN$_3$CH), 119.2 (CH$_{Ar}$), 112.9 (C$_{Ar}$), 71.9 (OCH$_2$), 70.2 (OCH$_2$), 69.8 (OCH$_2$), 69.4 (OCH$_2$), 68.6 (OCH$_2$), 53.7 (CH$_{carb}$), 49.5 (NCH$_2$), 48.8 (NCH$_2$), 46.5 (CH$_{carb}$), 29.5 (CH$_2$) ppm. IR (KBr, н, cm^{-1}): 2553 (BH), 1593 (triazole). Found: C 37.17, H 5.28, B 22.51, N 7.68; Calc. for C$_{27}$H$_{47}$B$_{18}$CoN$_5$O$_3$Cs C 37.01, H 5.41, B 22.21, N 7.99. HRMS (ESI) m/z for [C$_{27}$H$_{47}$B$_{18}$CoN$_5$O$_3$]$^-$ calcd 743.4815 [M]$^-$, found: 743.4818 [M]$^-$.

3.1.4. Synthesis of Conjugate **9**

Conjugate **9** was prepared from compound **2** (0.09 g, 0.30 mmol), the alkynyl derivative of cobalt bis(dicarbollide) **5** (0.14 g, 0.30 mmol), diisopropylethylamine (1 mL, 0.74 g, 5.73 mmol), and CuI (0.006 g, 0.03 mmol) in 15 mL of EtOH. The product was obtained as an orange solid of **4** (0.18 g, yield 70%). ^{1}H NMR (400 MHz, acetone-d$_6$) д 8.58 (d, 2H, 2 × CH_{Ar}, J = 9.0 Hz), 8.07 (m, 2H, 2 × CH_{Ar}), 7.99 (m, 3H, 2 × CH_{Ar}, CHCN$_3$), 7.67 (t, 2H, 2 × CH_{Ar}), 5.14 (m, 2H, CH_2N), 4.96 (m, 2H, CH_2NH), 4.46 (s, 2H, OCH_2C), 4.25 (br. s, 2H, CH_{carb}), 4.17 (br. s, 2H, CH_{carb}), 3.46 (t, 2H, BOCH_2), 3.33 (t, 2H, CH_2), 1.44 (dq, 4H, 2 × CH_2), 1.31 (m, 2H, CH_2) ppm. ^{11}B NMR (128 MHz, acetone-d$_6$): 23.4 (1B, s), 4.3 (1*B*, d, J = 144 Hz), 0.0 (1*B*, d, J = 152), −2.6 (1*B*, d, J unsolved), −4.6 (2*B*, d, J = 162 Hz), −7.4 (2*B*, d, J = 122 Hz), −8.2 (4*B*, d, J = 124 Hz), −17.4 (2*B*, d, J = 162 Hz), −20.3 (2*B*, d, J = 144 Hz), −22.6 (1*B*, d, J = 158 Hz), −28.4 (1*B*, d, J = 150 Hz) ppm. ^{13}C NMR (101 MHz, acetone-d$_6$): 158.8 (C$_{Ar}$), 145.6 (C$_{Ar}$), 139.8 (CN$_3$CH), 135.8 (CH$_{Ar}$), 125.3 (CH$_{Ar}$), 124.4 (CH$_{Ar}$), 124.0 (CN$_3$CH), 119.0 (CH$_{Ar}$), 112.9 (C$_{Ar}$), 70.0 (OCH$_2$), 68.8 (OCH$_2$), 63.8 (OCH$_2$), 53.9 (CH$_{carb}$), 49.3 (NCH$_2$), 48.9 (NCH$_2$), 46.4 (CH$_{carb}$), 31.6 (CH$_2$), 29.6 (CH$_2$), 22.7 (CH$_2$) ppm. IR (KBr, н, cm^{-1}): 2553 (BH), 1583 (triazole). Found: C 37.83, H 5.32, B 22.91,

N 8.33; Calc. for $C_{27}H_{47}B_{18}CoN_5O_2Cs$ C 37.70, H 5.51, B 22.62, N 8.14. HRMS (ESI) m/z for $[C_{27}H_{47}B_{18}CoN_5O_2]^-$ calcd 727.4866 $[M]^-$, found: 727.4868 $[M]^-$.

3.1.5. Synthesis of Conjugate **10**

Conjugate 10 was prepared from compound **2** (0.063 g, 0.21 mmol), the alkynyl derivative of cobalt bis(dicarbollide) **6** (0.10 g, 0.21 mmol), diisopropylethylamine (0.5 mL, 0.37 g, 2.86 mmol), and CuI (0.004 g, 0.02 mmol) in 10 mL of EtOH. The product was obtained as an orange solid of **4** (0.13 g, yield 71%). ^{1}H NMR (400 MHz, acetone-d$_6$): 8.56 (d, 2H, $2 \times CH_{Ar}$, J = 9.0 Hz), 8.08 (m, 2H, $2 \times CH_{Ar}$), 7.98 (d, 2H, $2 \times CH_{Ar}$, J = 9.1 Hz), 7.83 (s, 1H, -CHCN$_3$), 7.68 (m, 2H, $2 \times CH_{Ar}$), 5.10 (m, 2H, CH$_2$N), 4.91 (m, 2H, CH$_2$NH), 4.25 (br. s, 2H, CH$_{carb}$), 4.18 (br. s, 2H, CH$_{carb}$), 3.47 (m, 4H, BOCH$_2$, OCH$_2$CH$_2$C), 3.31 (m, 2H, CH$_2$), 2.82 (m, 4H, $2 \times CH_2$), 1.43 (m, 2H, CH$_2$), 1.32 (m, 2H, CH$_2$) ppm. ^{11}B NMR (128 MHz, acetone-d$_6$): 23.4 (1B, s), 4.3 (1B, d, J = 162 Hz), 0.4 (1B, d, J = 142), −2.6 (1B, d, J = 146 Hz), −4.6 (2B, d, J = 138 Hz), −7.4 (2B, d, J unsolved), −8.3 (4B, d, J = 134 Hz), −17.5 (2B, d, J = 164 Hz), −20.4 (2B, d, J = 148 Hz), −22.8 (1B, d, J unsolved), −28.6 (1B, d, J = 162 Hz) ppm. ^{13}C NMR (101 MHz, acetone-d$_6$): 158.7 (C$_{Ar}$), 145.6 (C$_{Ar}$), 139.9 (CN$_3$CH), 135.8 (CH$_{Ar}$), 125.3 (CH$_{Ar}$), 124.4 (CH$_{Ar}$), 123.2 (CN$_3$CH), 119.0 (CH$_{Ar}$), 112.9 (C$_{Ar}$), 70.5 (OCH$_2$), 69.1 (OCH$_2$), 68.7 (OCH$_2$), 53.9 (CH$_{carb}$), 49.4 (NCH$_2$), 48.7 (NCH$_2$), 46.4 (CH$_{carb}$), 31.6 (CH$_2$), 29.4 (CH$_2$), 26.3 (CH$_2$), 22.7 (CH$_2$) ppm. IR (KBr, н, cm^{-1}): 2547 (BH), 1583 (triazole). Found: C 38.35, H 5.76, B 22.48, N 8.13; Calc. for $C_{28}H_{49}B_{18}CoN_5O_2Cs$ C 38.47, H 5.65, B 22.26, N 8.01. HRMS (ESI) m/z for $[C_{28}H_{49}B_{18}CoN_5O_2]^-$ calcd 741.5023 $[M]^-$, found: 741.5028 $[M]^-$.

3.2. Absorbance Spectroscopy

Absorption spectra were recorded on a Jasco v550 (Japan) spectrophotometer in a thermostatted cuvette with 1 cm optical path at 25 °C. A solution of 5 mkM compounds in 10 mM potassium phosphate buffer pH = 8.0 and a calf thymus DNA (Sigma-Aldrich) concentration in the range 0–80 µM (b.p.) was used to obtain spectral changes in DNA interactions.

3.3. Cells and MTT Assay

The human HCT116 colorectal carcinoma, MCF7 breast adenocarcinoma, A549 non-small cell lung carcinoma, and WI38 nonmalignant lung fibroblast cell lines were obtained from the European collection of authenticated cell cultures (ECACC; Salisbury, UK). All cells were grown in DMEM medium (Gibco™, Cork, Ireland) supplemented with 10% fetal bovine serum (Gibco™, Cork, Brazil). The cells were cultured in an incubator at 37 °C in a humidified 5% CO$_2$ atmosphere and subcultured 2 times a week. The MTT (3-(4,5-dimethylthiazol-2-yl)-2,5-diphenyltetrazolium bromide) assay was performed as described earlier [46]. In brief, the cells were seeded in 96-well plates («*TPP*», Switzerland) at a 1×10^4 cells/well in 100 µL. After 24 h incubation at 37 °C, the cells were incubated with the tested compounds in concentrations from 0 to 200 µM.

3.4. Crystallographic Data

At 100K, crystals of **2** ($C_{15}H_{14}ClN_5$, M = 299.76) are orthorhombic, space group P2$_1$2$_1$2$_1$: a = 7.0854(3), b = 11.0430(4), c = 17.4107(7) E, V = 1362.28(9) E^3, Z = 4 (Z' = 1), d$_{calc}$ = 1.462 g·cm^{-3}, F(000) = 624. Intensities of 14,766 reflections were measured with a Bruker D8 Quest diffractometer equipped with the Photon III detector [λ(MoKα) = 0.71072E, ш-scans, 2θ < 61°], and 3991 independent reflections [R$_{int}$ = 0.0451] were used in further refinement. The structure was solved using the direct method and refined through the full-matrix least-squares technique against F^2 in the anisotropic–isotropic approximation. The hydrogen atoms were found from the difference Fourier synthesis of electron density. All the hydrogen atoms were refined in the isotropic approximation without constraints imposed on the positional parameters. For **1**, the refinement converged to wR2 = 0.0844 and GOF = 0.959 for all independent reflections (R1 = 0.0364 was calculated against F for 3185 observed reflections with I > 2σ(I)). All calculations were performed using the SHELX [57] and OLEX2 [58] program packages. CCDC 2,257,438 contains the supplementary crystallographic data for **1**. These data can be obtained free of charge via https://www.ccdc.cam.ac.uk/structures/ (accessed on 3 July 2023) (or from the CCDC, 12 Union Road, Cambridge, CB21EZ, UK; or deposit@ccdc.cam.ac.uk). The X-ray diffraction study was performed using the equipment of the JRC PMR IGIC RAS.

The DFT calculations for the isolated cation of **1** as well for two conformations (relaxed and distorted) of the unsubstituted acridinium cation were performed using the Gaussian09 program [59]. The electronic energy calculations were performed using the def2TZVP basis set [60,61] and the PBE0 functional [62,63] with Grimme's empirical dispersion correction [64] and Becke-Johnson damping [65]. The geometry optimization procedures were performed invoking the standard cutoff criteria and ultrafine grids. According to the normal mode calculations, both fully relaxed structures (the cation of **1** and the equilibrium unsubstituted acridinium cation) correspond to energy minima. The RDG and sign(λ_2)·c$^{\circledR}$ functions were calculated using the MultiWFN program [66], whereas the topological analysis of electron density function was performed in the AIMAll program [67].

4. Conclusions

The copper(I)-catalyzed 1,3-dipolar [3 + 2] cycloaddition reaction of cobalt bis(dicarbollide) alkynes derivatives with azidoacridine yield the corresponding products containing 18 atoms of boron per molecule. The novel conjugates demonstrated antiproliferative activity against two tumor and one non-tumor human cell lines. DNA interaction studies using absorbance spectroscopy showed the weak binding of the obtained compounds with DNA. Among all other compounds, the acridine conjugate obtained from the 1,4-dioxane derivative of cobalt bis(dicarbollide) and propargyl alcohol showed the best result (the largest spectral changes were observed for this compound). It should be noted that the preliminary results indicated some potential, suggesting binding to DNA. Thus, the toxic boronated acridines presented in this work contain novel, unexplored structural features whose relevance in the field deserves investigation in order to achieve still-better non-toxic BNCT agents, capable of binding to DNA. Further efforts are warranted to explore the effect of other kinds of boronated substituents on the 9-aminoacridine scaffold on the activity and toxicity of boronated acridines. As compounds with a possibly more affinitive interaction with DNA, the introduction of active groups carrying a positive charge could be suggested, which would increase water solubility and make the acridine core more affinitive to DNA, possibly enhancing binding to DNA as a target.

Supplementary Materials: The following supporting information can be downloaded at: https://www.mdpi.com/article/10.3390/molecules28186636/s1, Figure S1 with main crystallographic data for **2**. Figure S2 with the area of substituent in the isolated optimized cation from **2**. Figure S3 with changes of absorbance spectra upon DNA interaction for conjugates **7,9,10**. Figures S4–S8: ESI-HRMS spectra of compounds **2, 7–10**, Figures S9–S13: IR spectra of compounds **2, 7–10**, Figures S14–S31: ^{1}H, ^{11}B{^{1}H}, ^{11}B and ^{13}C{^{1}H} spectra of compounds **2, 7–10**.

Author Contributions: Conceptualization, A.A.D. and I.B.S.; methodology, A.A.D. and I.V.A.; validation, N.V.D., A.A.A.; formal analysis, X-ray diffraction, I.V.A.; writing—original draft, A.A.D., I.V.A., A.A.N., and D.N.K.; review and editing, I.B.S., A.A.N. and V.I.B.; funding acquisition, A.A.D. All authors have read and agreed to the published version of the manuscript.

Funding: This work was supported by the Russian Science Foundation (RSF) No 22-73-00160.

Institutional Review Board Statement: Not applicable.

Informed Consent Statement: Not applicable.

Data Availability Statement: The data presented in this study are available in the Supplementary Materials.

Acknowledgments: The NMR spectra were obtained using equipment from the Center for Molecular Structure Studies at A.N. Nesmeyanov Institute of Organoelement Compounds, operating with financial support from the Ministry of Science and Higher Education of the Russian Federation. The theoretical calculations were supported by the Ministry of Science and Higher Education of the Russian Federation as part of the State Assignment of the Kurnakov Institute of General and Inorganic Chemistry of the Russian Academy of Sciences.

Conflicts of Interest: The authors declare no conflict of interest.

Sample Availability: Samples of compounds **7–10** are available from the authors.

References

1. Miyatake, S.I.; Kawabata, S.; Goto, H.; Narita, Y.; Suzuki, M.; Hirose, K.; Takai, Y.; Ono, K.; Ohnishi, T.; Tanaka, H.; et al. Accelerator-based BNCT in rescue treatment of patients with recurrent GBM: A multicenter phase II study. *J. Clin. Oncol.* **2020**, *38*, 2536. [CrossRef]
2. Kato, T.; Hirose, K.; Tanaka, H.; Mitsumoto, T.; Motoyanagi, T.; Arai, K.; Harada, T.; Takeuchi, A.; Kato, R.; Yajima, S.; et al. Design and construction of an accelerator-based boron neutron capture therapy (AB-BNCT) facility with multiple treatment rooms at the Southern Tohoku BNCT Research Center. *Appl. Radiat. Isot.* **2020**, *156*, 108961–108969. [CrossRef] [PubMed]
3. Barth, R.F.; Coderre, J.A.; Vicente, M.G.; Blue, T.E. Boron neutron capture therapy of cancer: Current status and future prospects. *Clin. Cancer Res.* **2005**, *11*, 3987–4002. [CrossRef] [PubMed]
4. Bregadze, V.I.; Sivaev, I.B. *Boron Science: New Technologies and Applications*; Hosmane, N.S., Ed.; CRC Press: Boca Raton, FL, USA, 2012; pp. 181–207. ISBN 9781439826621.
5. Barth, R.F.; Zhang, Z.; Liu, T. A realistic appraisal of boron neutron capture therapy as a cancer treatment modality. *Cancer Commun.* **2018**, *38*, 36–43. [CrossRef]
6. Barth, R.F.; Mi, P.; Yang, W. Boron delivery agents for neutron capture therapy of cancer. *Cancer Commun.* **2018**, *38*, 35–46. [CrossRef]
7. Sivaev, I.B.; Bregadze, V.I. Chemistry of cobalt bis(dicarbollides). A review. *Collect. Czech. Chem. Commun.* **1999**, *64*, 783–805. [CrossRef]
8. Dash, B.P.; Satapathy, R.; Swain, B.R.; Mahanta, C.S.; Jena, B.B.; Hosmane, N.S. Cobalt bis(dicarbollide) anion and its derivatives. *J. Organomet. Chem.* **2017**, *170*, 849–850. [CrossRef]
9. Fuentes, I.; Teixidor, F.; Viñas, C.; García-Mendiola, T.; Lorenzo, E.; Sato, S.; Nakamura, H.; Pita, M.; Marques, F. Metallacarboranes on the road to anticancer therapies: Cellular uptake, DNA interaction, and biological evaluation of cobaltabisdicarbollide [COSAN]⁻. *Chem. Eur. J.* **2018**, *24*, 17239–17254. [CrossRef]
10. Tarrés, M.; Canetta, E.; Paul, E.; Forbes, J.; Azzouni, K.; Viñas, C.; Teixidor, F.; Harwood, A.J. Biological interaction of living cells with COSAN-based synthetic vesicles. *Sci. Rep.* **2015**, *5*, 7804–7811. [CrossRef]
11. Spryshkova, R.A.; Karaseva, L.I.; Brattsev, V.A.; Serebriakov, N.G. Toxicity of functional derivatives of polyhedral carboranes. *Med. Radiol.* **1981**, *26*, 62–64.
12. Rokitskaya, T.I.; Kosenko, I.D.; Sivaev, I.B.; Antonenko, Y.N.; Bregadze, V.I. Fast flip-flop of halogenated cobalt bis(dicarbollide) anion in a lipid bilayer membrane. *Phys. Chem. Chem. Phys.* **2017**, *19*, 25112–25128. [CrossRef] [PubMed]
13. Verdiá-Báguena, C.; Alcaraz, A.; Aguilella, V.M.; Cioran, A.M.; Tachikawa, S.; Nakamura, H.; Teixidor, F.; Viñas, C. Amphiphilic COSAN and I₂-COSAN crossing synthetic lipid membranes: Planar bilayers and liposomes. *Chem. Commun.* **2014**, *50*, 6700–6703. [CrossRef] [PubMed]
14. Assaf, K.I.; Begaj, B.; Frank, A.; Nilam, M.; Mougharbel, A.S.; Kortz, U.; Nekvinda, J.; Grüner, B.; Gabel, D.; Nau, W.M. High-affinity binding of metallacarborane cobalt bis(dicarbollide) anions to cyclodextrins and application to membrane translocation. *J. Org. Chem.* **2019**, *84*, 11790–11798. [CrossRef] [PubMed]
15. Hartman, T.; Carlsson, J. Radiation dose heterogeneity in receptor and antigen mediated boron neutron capture therapy. *Radiother. Oncol.* **1994**, *31*, 61–75. [CrossRef] [PubMed]
16. Gabel, D.; Foster, S.; Fairchild, R.G. The Monte Carlo simulation of the biological effect of the ^{10}B(n,α)^{7}Li reaction in cells and tissue and its implication for Boron Neutron Capture Therapy. *Radiat. Res.* **1987**, *111*, 14–25. [CrossRef]
17. Crossley, E.L.; Ziolkowski, E.J.; Coderre, J.A.; Rendina, L.M. Boronated DNA-binding compounds as potential agents for Boron Neutron Capture Therapy. *Mini-Rev. Med. Chem.* **2007**, *7*, 303–313. [CrossRef]
18. Tjarks, W.; Ghaneolhosseini, H.; Henssen, C.L.; Malmquist, J.; Sjöberg, S. Synthesis of *para*- and *nido*-carboranyl phenanthridinium compounds for neutron capture therapy. *Tetrahedron Lett.* **1996**, *37*, 6905–6908. [CrossRef]
19. Ghaneolhosseini, H.; Tjarks, W.; Sjöberg, S. Synthesis of boronated phenanthridinium derivatives for potential use in boron neutron capture therapy (BNCT). *Tetrahedron* **1997**, *53*, 17519–17526. [CrossRef]
20. Tjarks, W.; Malmquist, J.; Gedda, L.; Sjöberg, S.; Carlsson, J. Synthesis and initial biological evaluation of carborane-containing phenanthridinium derivatives. In *Cancer Neutron Capture Therapy*; Mishima, Y., Ed.; Springer: New York, NY, USA, 1996; pp. 121–126.
21. Ghaneolhosseini, H.; Sjöberg, S. Synthesis of a boronated naphthalimide for potential use in boron neutron capture therapy (BNCT). *Acta Chem. Scand.* **1999**, *53*, 298–300. [CrossRef]

22. Nekvinda, J.; Różycka, D.; Rykowski, S.; Wyszko, E.; Fedoruk-Wyszomirska, A.; Gurda, D.; Orlicka-Płocka, M.; Giel-Pietraszuk, M.; Kiliszek, A.; Rypniewski, W.; et al. Synthesis of naphthalimide-carborane and metallacarborane conjugates: Anticancer activity, DNA binding ability. *Bioorg. Chem.* **2020**, *94*, 103432. [CrossRef]
23. Rykowski, S.; Gurda-Woźna, D.; Orlicka-Płocka, M.; Fedoruk-Wyszomirska, A.; Giel-Pietraszuk, M.; Wyszko, E.; Kowalczyk, A.; Stączek, P.; Bak, A.; Kiliszek, A.; et al. Design, synthesis, and evaluation of novel 3-carboranyl-1,8-naphthalimide derivatives as potential anticancer agents. *Int. J. Mol. Sci.* **2021**, *22*, 2772. [CrossRef] [PubMed]
24. Laskova, J.; Kosenko, I.; Serdyukov, A.; Sivaev, I.; Bregadze, V.I. Synthesis of naphthalimide derivatives of *closo*-dodecaborate and *nido*-carborane. *J. Organomet. Chem.* **2021**, *959*, 122186. [CrossRef]
25. Bogucka-Kocka, A.; Kołodziej, P.; Makuch-Kocka, A.; Różycka, D.; Rykowski, S.; Nekvinda, J.; Grüner, B.; Olejniczak, A.B. Nematicidal activity of naphthalimide–boron cluster conjugates. *Chem. Commun.* **2022**, *58*, 2528–2531. [CrossRef] [PubMed]
26. Rykowski, S.; Gurda-Woźna, D.; Orlicka-Płocka, M.; Fedoruk-Wyszomirska, A.; Giel-Pietraszuk, M.; Wyszko, E.; Kowalczyk, A.; Stączek, P.; Biniek-Antosiak, K.; Rypniewski, W.; et al. Design of DNA intercalators based on 4-carboranyl-1,8-naphthalimides: Investigation of their DNA-binding ability and anticancer activity. *Int. J. Mol. Sci.* **2022**, *23*, 4598. [CrossRef]
27. Rykowski, S.; Gurda-Wozna, D.; Fedoruk-Wyszomirska, A.; Orlicka-Płocka, M.; Kowalczyk, A.; Staczek, P.; Denel-Bobrowska, M.; Biniek-Antosiak, K.; Rypniewski, W.; Wyszko, E.; et al. Carboranyl-1,8-naphthalimide intercalators induce lysosomal membrane permeabilization and ferroptosis in cancer cell lines. *J. Enzyme Inhib. Med. Chem.* **2023**, *38*, 2171028. [CrossRef]
28. Soloway, A.H. Boron compounds in cancer therapy. In *Progress in Boron Chemistry*; McCloskey, A.L., Steinberg, H., Eds.; Pergamon Press: New York, NY, USA, 1964; Volume 1, pp. 203–234.
29. Davis, M.A.; Soloway, A.H. Carboranes. III. Boron-containing acridines. *J. Med. Chem.* **1967**, *10*, 730–732. [CrossRef]
30. Ghaneolhosseini, H.; Tjarks, W.; Sjoberg, S. Synthesis of novel boronated acridines- and spermidines as possible agents for BNCT. *Tetrahedron* **1998**, *54*, 3877–3884. [CrossRef]
31. Różycka, D.; Kowalczyk, A.; Bobrowska, M.D.; Kuźmycz, O.; Gapińska, M.; Stączek, P.; Olejniczak, A.B. Acridine/acridone–carborane conjugates as strong DNA-binding agents with anticancer potential. *ChemMedChem* **2023**, *18*, e202200666. [CrossRef]
32. Cebula, J.; Fink, K.; Goldeman, W.; Szermer-Olearnik, B.; Nasulewicz-Goldeman, A.; Psurski, M.; Cuprych, M.; Kędziora, A.; Dudek, B.; Bugla-Płoskońska, G.; et al. Structural patterns enhancing the antibacterial activity of metallacarborane-based antibiotics. *ChemRxiv* **2023**. [CrossRef]
33. Howell, L.A.; Gulam, R.; Mueller, A.; O'Connell, M.A.; Searcey, M. Design and synthesis of threading intercalators to target DNA. *Bioorg. Med. Chem. Lett.* **2010**, *20*, 6956–6959. [CrossRef]
34. Howell, L.A.; Bowater, R.A.; O'Connell, M.A.; Reszka, A.P.; Neidle, S.; Searcey, M. Synthesis of small molecules targeting multiple DNA structures using click chemistry. *ChemMedChem* **2012**, *7*, 792–804. [CrossRef] [PubMed]
35. Denny, W.A.; Cain, B.F.; Atwell, G.J.; Hansch, C.; Panthananickal, A.; Leo, A.J. Potential antitumor agents. Quantitative relationships between experimental antitumor activity, toxicity, and structure for the general class of 9-anilinoacridine antitumor agents. *J. Med. Chem.* **1982**, *25*, 276–315. [CrossRef] [PubMed]
36. Tonelli, M.; Vettoretti, G.; Tasso, B.; Novelli, F.; Boido, V.; Sparatore, F.; Busonera, B.; Ouhtit, A.; Farci, P.; Blois, S.; et al. Acridine derivatives as anti-BVDV agents. *Antiv. Res.* **2011**, *91*, 133–141. [CrossRef] [PubMed]
37. Rostovtsev, V.V.; Green, L.G.; Fokin, V.V.; Sharpless, K.B. A stepwise Huisgen cycloaddition process: Copper(I)-catalyzed regioselective ligation of azides and terminal alkynes. *Angew. Chem. Int. Ed.* **2002**, *41*, 2596–2599. [CrossRef]
38. Rani, A.; Singh, G.; Singh, A.; Maqbool, U.; Kaur, G.; Singh, J. CuAAC-ensembled 1,2,3-triazole-linked isosteres as pharmacophores in drug discovery: Review. *RSC Adv.* **2020**, *10*, 5610–5635. [CrossRef] [PubMed]
39. Smyshliaeva, L.A.; Varaksin, M.V.; Fomina, E.I.; Joy, M.N.; Bakulev, V.A.; Charushin, V.N.; Chupakhin, O.N. Cu(I)-catalyzed cycloaddition of vinylacetylene *ortho*-carborane and arylazides in the design of 1,2,3-triazolyl-modified vinylcarborane fluorophores. *Organometalics* **2020**, *39*, 3679–3688. [CrossRef]
40. Bregadze, V.I.; Semioshkin, A.A.; Las´kova, J.N.; Berzina, M.Y.; Lobanova, I.A.; Sivaev, I.B.; Grin, M.A.; Titeev, R.A.; Brittal, D.I.; Ulybina, O.V.; et al. Novel types of boronated chlorine e_6 conjugates via «click chemistry». *Appl. Organometal. Chem.* **2009**, *23*, 370–374. [CrossRef]
41. Wojtczak, B.A.; Andrysiak, A.; Grüner, B.; Leśnikowski, Z.J. "Chemical ligation": A versatile method for nucleoside modification with boron clusters. *Chem. Eur. J.* **2008**, *14*, 10675–10682. [CrossRef]
42. Druzina, A.A.; Shmalko, A.V.; Sivaev, I.B.; Bregadze, V.I. Cyclic oxonium derivatives of cobalt and iron bis (dicarbollide)s and their use in organic synthesis. *Russ. Chem. Rev.* **2021**, *90*, 785–830. [CrossRef]
43. Groom, C.R.; Bruno, I.J.; Lightfoot, M.P.; Ward, S.C. The Cambridge Structural Database. *Acta Cryst. B* **2016**, *72*, 171–179. [CrossRef]
44. Aghabozorg, H.; Ahmadvand, S.; Mirzaei, M.; Khavasi, H.R. Bis(9-aminoacridinium) bis(pyridine-2,6-dicarboxylato) cuprate(II) trihydrate. *Acta Cryst. E* **2010**, *66*, m1318–m1319. [CrossRef] [PubMed]
45. Eshtiagh-Hosseini, H.; Mirzaei, M.; Eydizadeh, E.; Yousefi, Z.; Molčanov, K. Bis(9-aminoacridinium) bis(pyridine-2,6-dicarboxylato-$\kappa^3 O^2,N,O^6$)manganate(II) trihydrate. *Acta Cryst. E* **2011**, *67*, m1411–m1412. [CrossRef] [PubMed]
46. Dhanabalan, N.; Thanigaimani, K.; Arshad, S.; Razak, I.A.; Santhanaraj, K.J. 9-Aminoacridin-10-ium 4-aminobenzoate dihydrate. *Acta Cryst. E* **2014**, *70*, o657–o658. [CrossRef] [PubMed]
47. Mirzaei, M.; Eshtiagh-Hosseini, H.; Eydizadeh, E.; Yousefi, Z.; Molčanov, K. 9-Aminoacridinium bis(pyridine-2,6-dicarboxylato-$\kappa^3 O^2,N,O^6$)ferrate(III) tetrahydrate. *Acta Cryst. E* **2012**, *68*, m761–m762. [CrossRef] [PubMed]

48. Mirzaei, M.; Eshtiagh-Hosseini, H.; Eydizadeh, E.; Yousefi, Z.; Molčanov, K. Bis(9-aminoacridinium) bis(pyridine-2,6-dicarboxylato)zincate(II) trihydrate. *Acta Cryst. E* **2012**, *68*, m355–m356. [CrossRef]
49. Johnson, E.R.; Keinan, S.; Mori-Sanchez, P.; Contreras-Garcia, J.; Cohen, A.J.; Yang, W. Revealing noncovalent interactions. *J. Am. Chem. Soc.* **2010**, *132*, 6498–6506. [CrossRef]
50. Pendas, A.M.; Francisco, E.; Blanco, M.A.; Gatti, C. Bond paths as privileged exchange channels. *Chem. Eur. J.* **2007**, *13*, 9362–9371. [CrossRef]
51. Anisimov, A.A.; Ananyev, I.V. Interatomic exchange-correlation interaction energy from a measure of quantum theory of atoms in molecules topological bonding: A diatomic case. *J. Comput. Chem.* **2020**, *41*, 2213–2222. [CrossRef]
52. Romanova, A.; Lyssenko, K.; Ananyev, I. Estimations of energy of noncovalent bonding from integrals over interatomic zero-flux surfaces: Correlation trends and beyond. *J. Comput. Chem.* **2018**, *39*, 1607–1616. [CrossRef]
53. Mishima, Y.; Ichihashi, M.; Hatta, S.; Honda, C.; Yamamura, K.; Nakagawa, T. New thermal neutron capture therapy for malignant melanoma: Melanogenesis-seeking ^{10}B molecule-melanoma cell interaction from in vitro to first clinical trial. *Pigm. Cell Res.* **1989**, *2*, 226–234. [CrossRef]
54. Kosenko, I.D.; Lobanova, I.A.; Chekulaeva, L.A.; Godovikov, I.A.; Bregadze, V.I. Synthesis of 1,4-disubstituted 1,2,3-triazoles based on cobalt bis (1,2-dicarbollide). *Russ. Chem. Bull.* **2013**, *62*, 497–503. [CrossRef]
55. Laskova, J.; Serdyukov, A.; Kosenko, I.; Ananyev, I.; Titova, E.; Druzina, A.; Sivaev, I.; Antonets, A.A.; Nazarov, A.A.; Bregadze, V.I. New azido coumarins as potential agents for fluorescent labeling and their "click" chemistry reactions for the conjugation with *closo*-dodecaborate anion. *Molecules* **2022**, *27*, 8575–8589. [CrossRef] [PubMed]
56. Zhu, Y.; Liu, X.; Zhang, Y.; Wang, Z.; Lasanajak, Y.; Song, X. Anthranilic acid as a versatile fluorescent tag and linker for functional glycomics. *Bioconjugate Chem.* **2018**, *29*, 3847–3855. [CrossRef]
57. Sheldrick, G.M. A short history of SHELX. *Acta. Cryst. A* **2008**, *64*, 112–122. [CrossRef] [PubMed]
58. Dolomanov, O.V.; Bourhis, L.J.; Gildea, R.J.; Howard, J.A.K.; PuschmannJ, H. OLEX2: A complete structure solution, refinement and analysis program. *Appl. Cryst.* **2009**, *42*, 339–341. [CrossRef]
59. Frisch, M.J.; Trucks, G.W.; Schlegel, H.B.; Scuseria, G.E.; Robb, M.A.; Cheeseman, J.R.; Scalmani, G.; Barone, V.; Petersson, G.A.; Nakatsuji, H.; et al. (Eds.) *Gaussian 09, Revision D.01*; Gaussian, Inc.: Wallingford, CT, USA, 2016.
60. Weigend, F.; Ahlrichs, R. Balanced basis sets of split valence, triple zeta valence and quadruple zeta valence quality for H to Rn: Design and assessment of accuracy. *Phys. Chem. Chem. Phys.* **2005**, *7*, 3297–3305. [CrossRef]
61. Weigend, F. Accurate Coulomb-fitting basis sets for H to Rn. *Phys. Chem. Chem. Phys.* **2006**, *8*, 1057–1065. [CrossRef]
62. Perdew, J.P.; Ernzerhof, M.; Burke, K. Rationale for mixing exact exchange with density functional approximations. *J. Chem. Phys.* **1996**, *105*, 9982–9985. [CrossRef]
63. Adamo, C.; Vincenzo, B. Toward reliable density functional methods without adjustable parameters: The PBE0 model. *J. Chem. Phys.* **1999**, *110*, 6158–6170. [CrossRef]
64. Grimme, S.; Antony, J.; Ehrlich, S.; Krieg, H. A consistent and accurate ab initio parametrization of density functional dispersion correction (DFT-D) for the 94 elements H-Pu. *J. Chem. Phys.* **2010**, *132*, 154104–154124. [CrossRef]
65. Grimme, S.; Antony, J.; Ehrlich, S.; Krieg, H. Effect of the damping function in dispersion corrected density functional theory. *J. Chem. Phys.* **2011**, *132*, 1456–1465. [CrossRef] [PubMed]
66. Lu, T.; Chen, F. Multiwfn: A multifunctional wavefunction analyzer. *J. Comput. Chem.* **2012**, *33*, 580–592. [CrossRef] [PubMed]
67. Keith, T.A. (Ed.) *AIMAll (Version 19.10.12)*; TK Gristmill Software: Overland Park, KS, USA, 2019.

Article

The Synthesis, Characterization, and Fluxional Behavior of a Hydridorhodatetraborane

Fatou Diaw-Ndiaye, Pablo J. Sanz Miguel, Ricardo Rodríguez and Ramón Macías *

Departamento de Química Inorgánica, Instituto de Síntesis Química y Catálisis Homogénea (ISQCH), Universidad de Zaragoza-CSIC, 50009 Zaragoza, Spain
* Correspondence: rmacias@unizar.es

Abstract: The octahydridotriborate anion plays a crucial role in the field of polyhedral boron chemistry, facilitating the synthesis of higher boranes and the preparation of diverse transition metal complexes. Among the stable forms of this anion, CsB_3H_8 (or $(n\text{-}C_4H_9)_4N)[B_3H_8]$ have been identified. These salts serve as valuable precursors for the synthesis of metallaboranes, wherein the triborate anion acts as a ligand coordinating to the metal center. In this study, we have successfully synthesized a novel rhodatetraborane dihydride, $[Rh(\eta^2\text{-}B_3H_8)(H)_2(PPh_3)_2]$ (**1**), which represents a Rh(III) complex featuring a bidentate chelate ligand fasormed by $B_3H_8{}^-$. Extensive characterization of this rhodatetraborane complex has been performed using NMR spectroscopy in solution and X-ray diffraction analysis in the solid state. Notably, the complex exhibits intriguing fluxional behavior, which has been investigated using NMR techniques. Moreover, we have explored the reactivity of complex **1** towards pyridine (py) and dimethylphenylphosphine (PMe_2Ph). Our findings highlight the labile nature of this four-vertex rhodatetraborane as it undergoes disassembly upon attack from the corresponding Lewis base, resulting in the formation of borane adducts, LBH_3, where L = py, PMe_2Ph. Furthermore, in these reactions, we report the characterization of new cationic hydride complexes, such as $[Rh(H)_2(PPh_3)_2\,(py)]^+$ (**2**) and $[Rh(H)_2(PMe_2Ph)_4]^+$. Notably, the latter complex has been characterized as the octahydridotriborate salt $[Rh(H)_2(PMe_2Ph)_4][B_3H_8]$ (**3**), which extends the scope of rhodatetraborane derivatives.

Keywords: boranes; metallaboranes; metal hydrides; fluxionality; solid-state structure

Citation: Diaw-Ndiaye, F.; Sanz Miguel, P.J.; Rodríguez, R.; Macías, R. The Synthesis, Characterization, and Fluxional Behavior of a Hydridorhodatetraborane. *Molecules* **2023**, *28*, 6462. https://doi.org/10.3390/molecules28186462

Academic Editors: Michael A. Beckett and Igor B. Sivaev

Received: 25 July 2023
Revised: 30 August 2023
Accepted: 31 August 2023
Published: 6 September 2023

1. Introduction

The synthesis of $[arachno\text{-}B_3H_8]^-$ has gathered significant attention in recent years due to its pivotal role in the preparation of higher boranes, such as $closo\text{-}B_{12}H_{12}{}^{2-}$, and a wide array of transition metal complexes [1–4].

Many salts of this anion have been prepared using B_2H_6, but concerns regarding the toxicity and flammability of diborane have prompted the exploration of alternative synthetic routes. Among the various methods, the reaction of $Na[BH_4]$ with I_2 stands out [2]; however, the presence of iodide anions poses challenges in obtaining the desired product. Recent investigations have focused on replacing iodine with various metal halides as oxidants, offering improved efficiency and selectivity in the synthesis of $[arachno\text{-}B_3H_8]^-$.

The crystal structure of $[(H_3N)_2BH_2][B_3H_8]$ has been determined by Peters and Nordman, shedding light on the structural composition of the octahydridotriborate anion. The anion consists of a triangular arrangement of boron atoms, with two bridging and six terminal hydrogens [3].

The inherent stability and accessibility of the $arachno\text{-}B_3H_8{}^-$ anion have paved the way for extensive investigations into its reactivity with various metal complexes [4]. The resulting $arachno$-2-metallaboranes can be synthesized through ligand substitution reactions, offering a general route for the incorporation of transition metals into the cluster framework:

$$L_nMX + B_3H_8{}^- \rightarrow [L_{n-1}MB_3H_8] + L + X^- \quad (\text{L = neutral ligand; X = halide})$$

The reactivity of *arachno*-$B_3H_8{}^-$ has been explored with a wide range of transition metals, spanning across the periodic table, including Ti, Cr, Mo, W, Mn, Re, Fe, Ru, Os, Ir [5], Cu, Ag and Zn [1,3,4].

The reaction between a transition metal complex and the octahydridotriborate anion is, *a priori*, the most direct route to the synthesis of *arachno*-2-metallatetraboranes. However, some of the reported metallatetraboranes were prepared from reactions of either monocyclopentadienyl metal chlorides or hydride-ligated complexes of transition metals from groups 5–9 with monoboranes ($LiBH_4$ or BH_3THF) [6–8]. This synthetic procedure was developed mainly by Fehlener and co-workers at Notre Dame University (Notre Dam, IN, USA) [9]; and Gosh and co-workers have been using it for a good number of years, at the Indian Institute of Technology Madras (Chennai, India), in the pursuit of new metallaboranes [10].

Alternatively, reactions between metal complexes and larger boranes, such as pentaborane, were also a route to tetraboranes, via cluster dismantling processes [11].

To expand the scope of transition element *arachno*-metallaboranes and explore novel structures and dynamic processes, our study focused on investigating the reactivity between Wilkinson's catalyst, $[RhCl(PPh_3)_3]$, and the octahydridotriborate anion, $[B_3H_8]^-$. This investigation resulted in the successful synthesis of dihydridorhodatetraborane, $[Rh(\eta^2\text{-}B_3H_8)(H)_2(PPh_3)_2]$ (**1**). The compound was comprehensively characterized using NMR spectroscopy and X-ray diffraction analysis. Notably, the newly synthesized metallatetraborane exhibited a chemical non-rigidity, which was studied using NMR spectroscopy at variable temperatures. In addition, we have carried out an exploratory study of the reactivity of **1** with Lewis bases, resulting in the characterization of hydride metal complexes.

2. Results and Discussion

2.1. Synthesis of $[Rh(\eta^2\text{-}B_3H_8)(H)_2(PPh_3)_2]$ (**1**)

The reaction of the Wilkinson's catalyst with the cesium salt CsB_3H_8 in ethanol leads to the formation of dihydridorhodatetraborane (**1**) (Scheme 1).

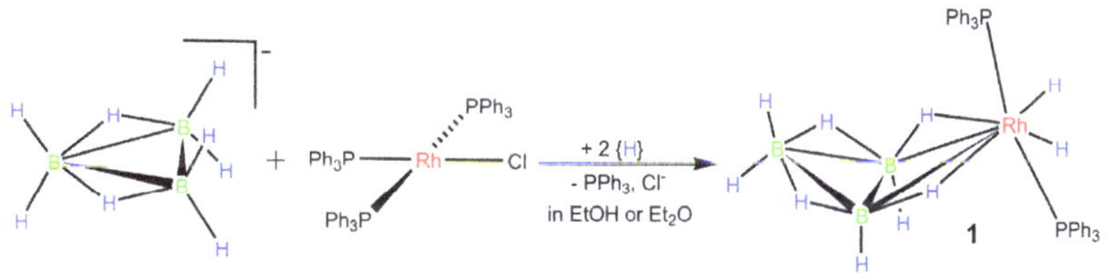

Scheme 1. Reaction between the Wilkinson's catalyst and the octahydridotriborate anion to give $[Rh(\eta^2\text{-}B_3H_8)(H)_2(PPh_3)_2]$ (**1**).

Due to the limited solubility of the cesium salt in ethanol and the insolubility of the rhodium complex (Wilkinson's catalyst) in the same solvent, the reaction proceeds in a heterogeneous solid–liquid phase. The reaction mixture initially forms a brick-red suspension, which transforms into a red-orange product, corresponding to the formation of dihydridorhodatetraborane (**1**). The product is then collected by filtration using a sintered disc filter funnel.

As an alternative approach, we conducted the reaction using the *tris*(dioxane) solvate $NaB_3H_8\cdot3(C_4H_8O_2)$ as the starting material in diethyl ether, which also exhibits limited solubility for Wilkinson's catalyst. Similar to the ethanol system, this synthesis is characterized by a heterogeneous reaction. The resulting yellow product, identified as hydridorhodathiaborane, is easily filtered under ambient conditions, yielding **1**.

2.2. X-ray Diffraction Analysis

The Cambridge Crystallographic Data Centre (CCDC) provides X-ray diffraction analyses for fifteen *arachno*-metallaboranes, incorporating {ML$_n$}-fragments of Nb [7], Cr [12], Mo [10], W [6,10], Mn [13], Re [8,13], Ru [11,14,15], Os [16] and Cu [17]. However, the availability of comparative structural data across the periodic table, for this particular class of four-vertex *arachno*-metallaboranes, is limited. It is important to emphasize that the crystal structure of compound **1** represents the first example of a Group 9 *arachno*-2-metallaborane characterized by X-ray diffraction analysis.

Single crystals of the compound were obtained by diffusing hexane into a solution of **1** in CH_2Cl_2. Figure 1 depicts an ORTEP-type drawing, illustrating selected interatomic distances and angles. The rhodium center in the compound exhibits an octahedral coordination sphere, where the $B_3H_8{}^-$ moiety acts as a bidentate η^2-ligand through two B–H–Rh bridge bonds. These bridge bonds are located *trans* to the *exo*-polyhedral hydride ligands. Completing the coordination number 6 around the metal, two Ph$_3$P ligands are mutually *trans*. Consequently, the molecule can be classified as an eighteen-electron, six-coordinate, octahedral d^6 rhodium(III) complex.

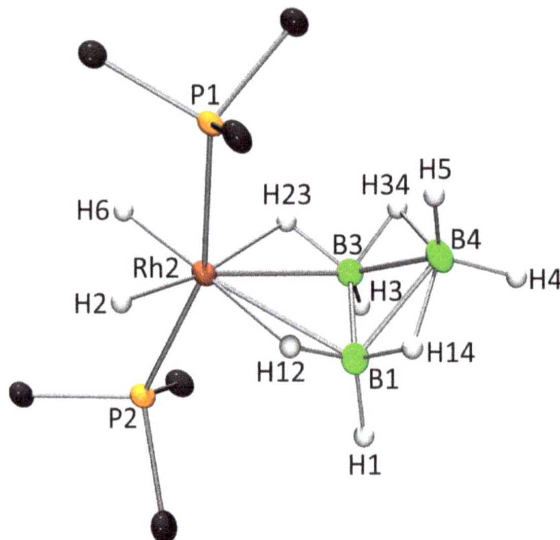

Figure 1. ORTEP-type of drawing for [Rh(η^2-B$_3$H$_8$)(H)$_2$(PPh$_3$)$_2$] (**1**), showing the cluster numbering system employed, with 50% thermal ellipsoids for non-hydrogen atoms. The phenyl rings (except for the ipso carbon atoms) are omitted to aid clarity. Selected interatomic distances (Å) and angles (°) with esds in parenthesis: Rh2–P1 2.3103(11), Rh2–P2 2.2911(11), Rh2–H 1.58(2) (average value of the two hydride ligands), Rh2–B1 2.428(5), Rh1–B3 2.420(5), B1–B3 1.764(8), B1–B4 1.790(8), B3–B4 1.805(8), P1–Rh2–P2 158.68(4), B1–Rh2–P1 103.27(13), B1–Rh2–P2 95.72(13), B3–Rh2–P1 106.40(13), B3–Rh2–P2 94.10(13), H–Rh2–P1 81.1(18) (average value of the two hydride ligands), B1–Rh2–B3 42.67(18), B3–B1–Rh2 68.4(2), B1–B3–Rh2 68.9(2), B3–B1–B4 61.0(3), B1–B3–B4 60.2(3), B1–B4–B3 58.8(3). The dihedral angle between planes {B1B3B4} and {B1B3Rh2} is 121.91(3).

Alternatively, compound **1** can be described as a four-vertex *arachno*-cluster, which can be related to the parent *arachno*-B$_4$H$_{10}$ by replacing a BH$_2$ 'wing-tip' with the d^6-{Rh(H)$_2$(PPh$_3$)$_2$} fragment. This description follows the architectural patterns proposed by Williams [18], where the four-vertex butterfly type *arachno*-cluster is derived from an octahedron by removing two adjacent vertices. According to the polyhedral skeletal electron pair theory (PSEPT) [19,20], these clusters are expected to have seven skeletal electron pairs (n + 3; where n is the number of vertices of the polyhedral cluster). Applying the PSEPT electron-counting rules, the {Rh(H)$_2$(PPh$_3$)$_2$} group in compound **1** can be considered as a vertex contributing three electrons to the cluster framework bonding [9 e$^-$(Rh) + 4 e$^-$(2PPh$_3$) + 2 e$^-$(2H) − 12 e$^-$ = 3 e$^-$], resembling the BH$_2$ 'wing-tip' in *arachno*-B$_4$H$_{10}$.

The distances between Rh2 and P1, as well as Rh2 and P2, are determined to be 2.3103(11) Å and 2.2911(11) Å, respectively. These bond lengths are significantly shorter compared to the Ru–P lengths observed in the *arachno*-2-ruthenatetraborane, [Ru(η^2-B$_3$H$_8$)(H)$_2$(PPh$_3$)$_2$], which also features two mutually *trans* PPh$_3$ ligands. In the ruthenate-traborane, the Ru–P bond lengths are measured to be 2.373(1) Å and 2.364(1) Å. However, when the hydrotris(pyrazol-1-yl)borate-ligated ruthenatetraborane, [Ru(η^2-B$_3$H$_8$)(PPh$_3$){K^3-HB(pz)$_3$}], is considered, the Ru–P bond distance is slightly shorter at 2.317(1) Å. Analysis from the Cambridge Crystallographic Data Centre (CCDC) reveals that the mean bond length for Rh–PPh$_3$ is 2.318 Å, while the mean bond distance for Ru–PPh$_3$ is slightly longer at 2.350 Å. Based on these findings, it can be concluded that the M–P bond distances fall within the crystallographic data, indicating that the Ru–PPh$_3$ bonds are on average longer than the corresponding Rh–PPh$_3$ lengths.

It is noteworthy to highlight that the {Ru(CO)(H)(PPh$_3$)$_2$} and {Ru(PPh$_3$){K^3-HB(pz)$_3$} fragments present in the ruthenaboranes are isolobal and isoelectronic with the {Rh(H)$_2$(PPh$_3$)$_2$} group observed in compound **1**. These fragments contribute three electrons to the cluster framework, thus fulfilling the expected seven skeletal electron pairs (seps) characteristic of a four-vertex *arachno*-cluster.

In the crystal structure of compound **1**, the rhodatetraborane clusters exhibit a packing arrangement along the crystallographic axis *a*. These clusters form ribbons that associate in pairs through sextuple phenyl embrace (SPE) interactions. The separation between the P atoms (P···P) and the collinearity of Rh–P···P–Rh are measured at 7.9 Å and 165°, respectively, falling within the reported range for this attractive edge-to-face interaction (Figure 2). The SPE interaction arises from intermolecular edge-to-face C–H···π attractive forces facilitated by the presence of phenyl rings [21,22].

All metal octahydridotriboranes stored in the CCDC exhibit a notable similarity, with mean distance values of 1.745 Å, 1.797 Å and 1.794 Å for the B1–B3, B1–B4 and B3–B4 linkages, respectively. Among these linkages, the B1–B3 edge involved in the M–B–B interaction (M = Nb, Cr, Mo, W, Mn, Re, Ru, Os, Cu) shows the shortest mean value. However, the length of the B1–B3 edge varies within the range of 1.707 Å to 1.833 Å, which is larger than the range of 1.772 Å to 1.817 Å observed for the other two B–B connections involving B–B–B bonds. This structural feature is expected due to the variation in the metal center across the metallatetraborane series. The M–B–B interaction, facilitated by two bridging hydrogen atoms, is expected to influence the B1–B3 distance, leading to significant differences among the compounds.

In the parent *arachno*-B$_4$H$_{10}$ cluster, the 'hinge' B1–B3 edge exhibits a mean value of 1.722 Å, which is the shortest among the five B–B bond distances present in this four-vertex *arachno*-cluster. The other B–B distances in *arachno*-B$_4$H$_{10}$ range between 1.844 Å and 1.847 Å, based on the average values derived from the five structures available in the CCDC. The butterfly dihedral angle, characterizing the molecular structure of B$_4$H$_{10}$, is measured to be 118.4 ± 0.4°. In comparison, the dihedral angle for the metal compounds falls within the range of 124.5 ± 5.2°. The smallest dihedral angle observed among the metallatetraboranes is 119.3°, which corresponds to the octahydridotriborato-*bis*(triphenylphosphine)copper(I) compound [17]. This compound features a {(PPh$_3$)$_2$Cu} vertex with both *endo*- and *exo*-triphenyl ligands, closely resembling the {BH$_2$} vertex in *arachno*-tetraborane(10). The coordination of the {(PPh$_3$)$_2$Cu} and {BH$_2$} vertices, bound to the {{η^2-B$_3$H$_8$} fragment, exhibits a tetrahedral geometry. Consequently, significant structural similarities between these two molecules are expected, as evidenced by the similarity in their dihedral angles.

Among the tetrametallaboranes determined through crystallography and deposited in the CCDC, different metal fragments act as vertices, formally replacing the {BH$_2$} vertex in *arachno*-B$_4$H$_{10}$. As a result, notable differences in the dihedral angles are observed, depending on the nature of the *endo-* and *exo-*ligands. For instance, the [Ru(η^2-B$_3$H$_8$)(Cl)(η^6-(CH$_3$)$_6$C$_6$))] cluster features a *pseudo*-octahedral ruthenium(II) center, bonded to *exo*-hexamethylbenzene and *endo*-chloro ligands. This metal fragment leads to a relatively flat structure with a dihedral angle of 129.7°. Another tetrametallaborane, [Nb(η^2-B$_3$H$_8$)(η^5-(C$_5$H$_5$)$_2$], exhibits a dihedral angle of 125.7°. In this case, the coordination number around the niobium(III) center is eight, with six positions occupied by the cyclopentadienyl ligands, each acting as a tridentate ligand, and the remaining two positions occupied by the bidentate octahydridotriborate anion.

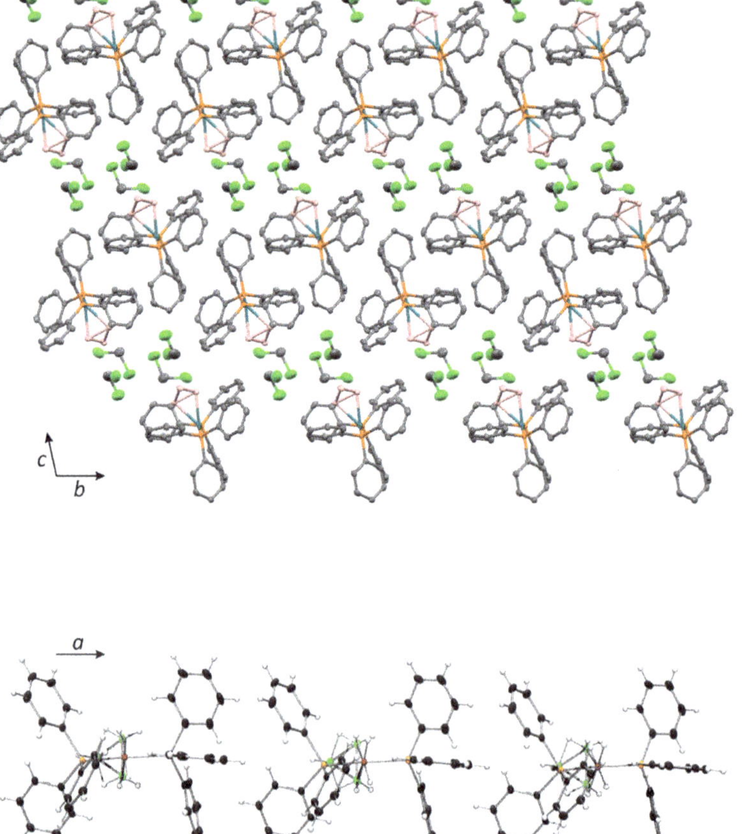

Figure 2. Packing of individual molecules of **1** that form ribbons along the *a* axis (above); detail of edge-to-face phenyl interactions between the molecules within the ribbons (below).

2.3. NMR Characterization and Comparison

The assignments provided in Table 1 for the resonances are reasonably determined based on their relative intensities and by comparing them with the resonances observed in the previously reported hydridoiridatetraborane, [Ir(η^2-B$_3$H$_8$)(H)$_2$(PPh$_3$)$_2$] [5]. To further support these assignments, DFT calculations were performed on a model compound, [Rh(η^2-B$_3$H$_8$)(H)$_2$(PH$_3$)$_2$]. The calculated data confirmed the resonance assignments and provided additional insight into the molecular structure and behavior of the compound.

Table 1. ^{11}B, ^{1}H and ^{31}P NMR data for compound [Rh(η^2-B$_3$H$_8$)(H)$_2$(PPh$_3$)$_2$] (compound **1**), compared to the corresponding DFT/GIAO ^{11}B-nuclear shielding data, calculated for the PH$_3$ model, [Rh(η^2-B$_3$H$_8$)(H)$_2$(PH$_3$)$_2$] [in brackets].

(a) Cluster Data:

Assignment [1]	δ (^{11}B) [2]	Assignment [1]	δ (^{1}H) [3]
B4	−1.0 [−3.8]	*exo*-H4	+2.54 [+2.74]
B1,3	−38.9 [−41.0]	*endo*-H5	+1.82 [+2.33]
		exo-H1, *exo*-H3	−0.11 [+1.02]
		H1,4; H3,4 (B–*H*–B)	−1.11 [−0.65]
		H1,2; H3,2 (Rh–*H*–B)	−7.07 [−5.10]
		H2, H6 (Rh–*H*)	−11.79 [4] [−6.37]

(b) Phosphorous-31 Data:

Assignment	δ (^{31}P) [3]	$^{1}J(^{103}$Rh–^{31}P)/Hz	$^{2}J(^{31}$P1–^{31}P2)/Hz
P1	39.9	111	
P2	44.7	111	367

[1] Based on the symmetry of **1**, ^{1}H–{^{11}B} selective experiments and DFT calculations. [2] CD$_2$Cl$_2$ solution at 298 K. [3] CD$_2$Cl$_2$ solution at 223 K. [4] ^{1}H–{^{11}B(off)}: broad, apparent quintet, J = 15.3 Hz; ^{1}H–{^{31}P}: broad, apparent triplet, which corresponds to a second order spectrum, analyzed to give $^{1}J(^{103}$Rh-^{1}H1,2) = $^{1}J(^{103}$Rh-^{1}H2,3) = 18 Hz, $^{1}J(^{103}$Rh-^{1}H2) = $^{1}J(^{103}$Rh-^{1}H6) = 16 Hz, $^{2}J(^{1}$H1,2-^{1}H2) = 2 Hz, $^{2}J(^{1}$H1,2-^{1}H6) = 12 Hz.

The NMR data obtained for compound **1** are fully in accord with the solid-state structure determined by X-ray diffraction analysis. The room temperature ^{11}B NMR spectrum shows two distinct resonances at δ(^{11}B) −1.0 and −38.9 ppm, with a relative intensity ratio of 1:2. These resonances can be attributed to the B4 and B1,3 vertices, respectively. Interestingly, the uncoupled ^{11}B spectrum does not display the expected ^{1}J(^{11}B-^{1}H) coupling constants (refer to Figure S1). This observation can be rationalized considering the chemical non-rigidity of the *endo-* and bridging-hydrogen atoms present in the {(η^2-B$_3$H$_8$)} ligand (*vide infra*).

Figure 3 illustrates a stick representation of the chemical shifts and relative intensities in the ^{11}B spectra for a series of isostructural and isoelectronic *arachno*-metallatetraboranes similar to compound **1**. These four-vertex clusters exhibit highly similar overall ^{11}B shielding patterns. The resonances corresponding to the metal-bound B1–B3 positions are grouped within a narrow region of δ(^{11}B) from −36.0 to −41.0 ppm. On the other hand, the signals associated with the "wing-tip" B4 position are observed between δ(^{11}B) −1.0 and +6.0 ppm.

The observed marked similarity in the ^{11}B NMR resonances is somewhat surprising, considering that the metal fragments change from early to late-transition elements, each bearing different ligands such as CO, PPh$_3$, dppe, C$_5$H$_5$, C$_5$Me$_5$ and hydrides, and the fact that the spectra were recorded in different solvents such as toluene-d^8, CD$_2$Cl$_2$ and CDCl$_3$.

This finding suggests that the fundamental nature of the metal-to-{η^2-B$_3$H$_8$} fragment interaction is maintained throughout this series of compounds. According to the electron-counting rules [19,20] the {Nb(η^5-C$_5$H$_5$)$_2$}- [7], {W(PMe$_3$)$_3$(H)$_3$}- [6], {Re(CO)$_4$}- [5], {Mn(CO)$_2$(dppe)}- [13], {Os(CO)(PPh$_3$)$_2$(H)}- [5], {Ru(CO)(PPh$_3$)$_2$(H)}-, {Fe(η^5-C$_5$Me$_5$)(CO)}-, {Ir(H)$_2$(PPh$_3$)$_2$} [5] and {Rh(H)$_2$(PPh$_3$)$_2$}-vertices contribute three electrons to the cluster framework. This electron count yields seven seps, as expected for four-vertex arachno-2-metallatetraboranes. The consistency in electron counting and the resultant ^{11}B resonances further support the notion that the interaction between the metal fragment and the {η^2-B$_3$H$_8$} ligand is maintained across this series of compounds.

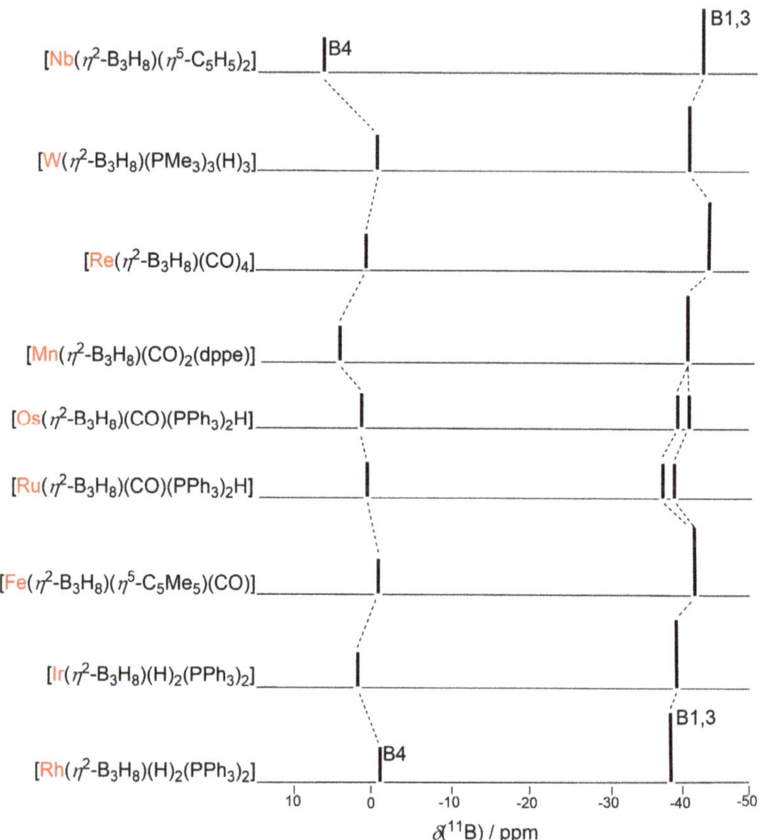

Figure 3. Stick representation of the ^{11}B-{^{1}H} NMR spectra of a series of *arachno*-metallaboranes. The NMR data were measured in different solvents such as toluene-d^8, CD$_2$Cl$_2$ and CDCl$_3$.

At 298 K, the ^{1}H-{^{11}B} NMR spectrum of compound **1** exhibits three signals at δ(^{1}H) +2.66 ppm, −6.96 ppm and −11.95 ppm, with a relative intensity ratio of 1:2:2. These spectroscopic data do not match the expected pattern based on the molecular structure of the dihydrorhodatetraborane. According to the C$_s$ point group symmetry, we would anticipate five proton resonances with a relative intensity ratio of 1:1:2:2:2, along with aromatic Ph signals (30H). However, when the ^{1}H-{^{11}B} spectrum is measured at 223 K, the expected pattern is observed. Peaks appear at δ(^{1}H) +2.54 (1H), +1.82 (1H), −0.11 (2H), −1.11 (2H), −7.07 (2H) and −11.79 ppm.

The lowest frequency signal at −11.79 ppm, corresponding to the Rh–H hydride ligands, does not broaden in the proton-coupled spectrum. This hydride resonance exhibits the characteristic pattern of a broad quintet, which appears as an apparent broad triplet in the ^{1}H-{^{31}P} spectrum (refer to Figure S2). However, the chemically equivalent hydride ligands, H2 and H6 in Figure 1, couple unequally to the H1,2 and H2,3 nuclei, resulting in magnetic non-equivalence (Figure S3). Consequently, the ^{1}H-{^{31}P} spectrum for the Rh–H ligands (H2, H6 in Figure 1) displays second-order behavior (Figure S3).

In the dihydridoiridatetraborane analogue, [Ir(η^2-B$_3$H$_8$)(H)$_2$(PPh$_3$)$_2$], the hydride signal appears at δ(^{1}H) −13.30 p.p.m, appearing as a triplet of doublets due to cisoid coupling to two ^{31}P nuclei with very similar coupling constants. Additionally, a small transoid coupling, ^{2}J(^{1}H$_{bridge}$-^{1}H$_t$) = 7.0 Hz, is observed. In compound **1**, there are also two ^{31}P nuclei with similar coupling constants. However, the proton pattern of the hydride nuclei shows second-order effects, as discussed above. Interestingly, the calculated transoid $^2J(^1H1,2-^1H6)$ coupling constant in compound **1** is significantly larger compared to that observed in the dihydridoiridatetraborane analogue [5].

The two ^{31}P nuclei in compound **1** form an AB-spin system with strong coupling, which is evident from the presence of a "roof effect" in the $^{31}P-\{^1H\}$ spectrum at 202 MHz (refer to Figure S4). In a strong coupling regime, the separation of the two central states is determined by the formula $C = [((\delta\upsilon)^2 + J^2]^{\frac{1}{2}}$, where $\delta\upsilon$ represents the difference in resonance frequencies of the two spins and J is the scalar coupling constant [23]. The large $^2J(^{31}P1-^{103}Rh-^{31}P2)$ of 367 Hz indicates a mutually trans-disposition of the two phosphorus atoms, confirming the molecular structure determined by X-ray diffraction analysis (Figure 1).

The proton signals at $\delta(^1H)$ +2.54 (1H), +1.82 (1H), −0.11 (2H), −1.11 (2H) and −7.07 (2H) exhibit significant broadening in the ^1H spectrum compared to the $^1H-\{^{11}B\}$ spectrum. This indicates that these resonances correspond to ^1H nuclei directly bound to boron atoms. The molecular structure of compound **1**, along with the $^1H-\{^{11}B\}$ selective experiments and the observed broadening patterns, has facilitated the complete assignment of the proton resonances to their respective positions within the structure of **1**.

In Figure 4, it is observed that the ^1H resonances assigned to the B1,3 *exo*-hydrogen atoms are grouped together between −0.51 and +1.30 ppm, forming a "low-frequency" cluster. On the other hand, the B4–H_{exo} signals are grouped between +1.83 and +4.81 ppm, forming a "high-frequency" cluster. Within this high-frequency group, the B4 *exo*-hydrogen resonance experiences significant deshielding when the metal atom is Nb, Re, Os, or Ir. This results in a large chemical shift difference for this particular resonance between $[Rh(\eta^2-B_3H_8)(H)_2(PPh_3)_2]$ (**1**) and $[Ir(\eta^2-B_3H_8)(H)_2(PPh_3)_2]$, as well as between $[Ru(\eta^2-B_3H_8)(CO)(PPh_3)_2H]$ and $[Os(\eta^2-B_3H_8)(CO)(PPh_3)_2H]$. If we consider that the metal center is located antipodal to the B4 vertex through an axis connecting M2 and B4, this effect can be attributed to the change from a second-row transition metal center to a third-row transition metal center.

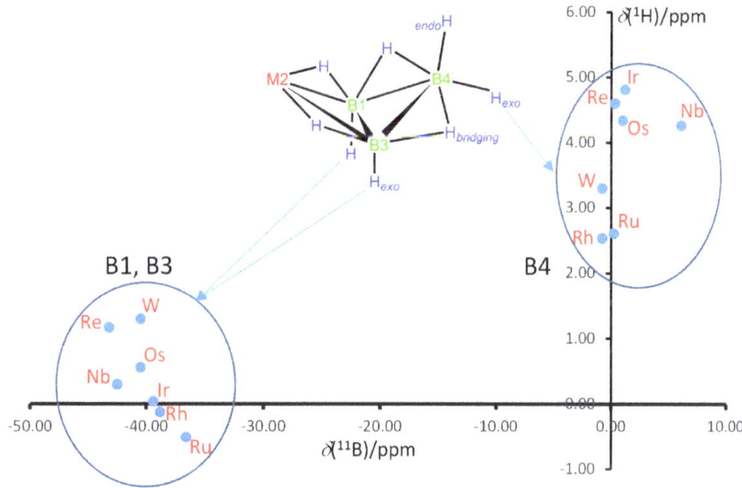

Figure 4. Plot of $\delta(^{11}B)$ versus δ (^1H) for directly bound [BH(terminal)] units in the following *arachno*-metallatetraboranes: $[Nb(\eta^2-B_3H_8)(\eta^5-C_5H_5)_2]$, $[W(\eta^2-B_3H_8)(PMe_3)_3(H)_3]$, $[Re(\eta^2-B_3H_8)(CO)_4]$, $[Os(\eta^2-B_3H_8)(CO)(PPh_3)_2H]$, $[Ru(\eta^2-B_3H_8)(CO)(PPh_3)_2H]$, $[Ir(\eta^2-B_3H_8)(H)_2(PPh_3)_2]$ and $[Rh(\eta^2-B_3H_8)(H)_2(PPh_3)_2]$ (**1**).

An interesting observation was made regarding the anomalous low proton shielding of *exo*-terminal protons that are positioned antipodal to third-row metal centers in twelve-vertex *closo*-metallaheteroborane systems. This phenomenon has been recognized as a diagnostic characteristic of this structural feature [24–30]. Similarly, we can utilize the strong deshielding of B4-H_{exo} protons as a diagnostic indicator for the presence of third-row transition metal centers in four-vertex *arachno*-2-metallatetraboranes. This provides valuable insights into the structural composition of these compounds.

2.4. Fluxional Behavior

In order to investigate the chemical non-rigidity and fluxional behavior of compound **1**, a variable temperature (VT) NMR study was conducted in CD_2Cl_2. Figure 5 illustrates the changes observed in the ^1H-$\{^{11}B\}$ NMR spectrum as the temperature was varied. The proton signals corresponding to B4-H_{endo}, B1,3-H_{exo}, and B4-$H_{bridging}$-B1/B3 hydrogen atoms gradually broaden and eventually disappear, indicating an intramolecular proton exchange process in compound **1**. Notably, this process does not involve the B4-H_{exo} and Rh-H-B1/B3 bridging hydrogen atoms nor the Rh-H hydride ligands.

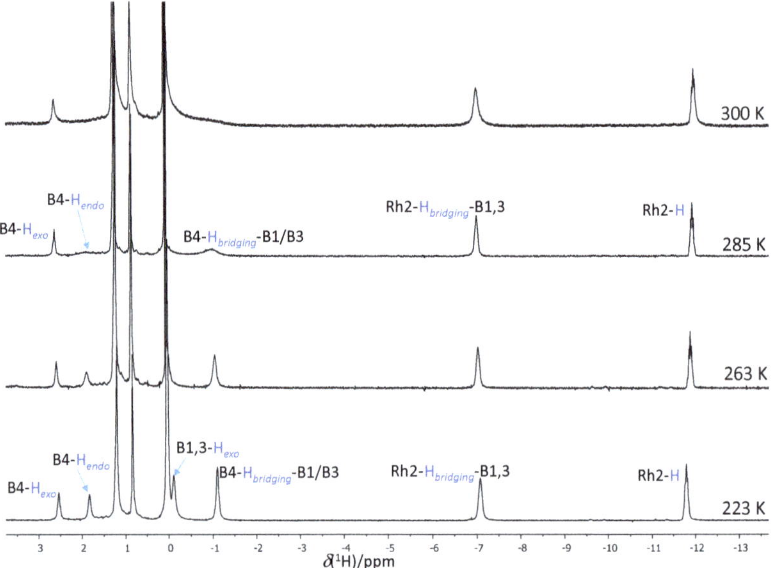

Figure 5. ^1H-$\{^{11}B\}$ NMR spectra, in CD_2Cl_2, at different temperatures, which demonstrate an intramolecular fluxional process for compound **1**.

The coalescence temperature, determined as the point at which the ^1H signals merge, was estimated to be 300 K. Using this information, the activation energy ($\Delta G^{\ddagger}$) for the asymmetric population system was calculated to be 10 kcal/mol (see Supporting Information for the analysis) [31].

To investigate the possible exchange of B4–H_{exo}, the Rh–$H_{briding}$–B1,3 and the Rh–H hydrogen atoms at higher temperatures, NMR spectra of compound **1** were measured at +67 °C in deuterated 1,1,2,2-tetrachloroethane. The ^1H-$\{^{11}B\}$ spectrum revealed the formation of a new hydridorhodatetraborane, exhibiting proton resonances at δ_H −6.78 and −11.67 ppm, which were assigned to Rh-H-B and Rh-H hydrogen atoms, respectively (Figure S5).

A further increase in the temperature to +97 °C resulted in the decomposition of both compound **1** and the new hydridorhodaborane. The products of this decomposition included borane triphenylphosphine (Ph_3P–BH_3) and hydride-ligated complexes, as evidenced by the presence of several doublets in the ^{31}P-$\{^1H\}$ spectrum (Figure S6). Additionally, the ^{11}B NMR spectrum showed peaks between δ_B +2.5 and +10.0 ppm, which did not exhibit $^1J(^{11}B–^1H)$ coupling, suggesting the formation of species containing O-B bonds (Figure S7).

The intramolecular hydrogen atom exchange observed in compound **1** shares similarities with the reported behavior of octahydridotriborate complexes such as [Mn(η^2-$B_3H_7Br)(CO)_4$] and [Ru(η^2-$B_3H_8)(CO)(H)(PPh_3)_2$] [32], where the M2–H–B1/B3 bridging atoms also remain static. In these cases, the fluxional process occurs with an activation energy, $\Delta G^\ddagger$, of approximately 12.2 kcal/mol at +23 °C for the manganesaborane. Interestingly, analogous compounds of third-row transition elements, such as [Os(η^2-$B_3H_8)(CO)(H)(PPh_3)_2$] and [Ir(η^2-B_3H_8) $(H)_2(PPh_3)_2$], which are CO-ligated ruthenaborane and compound **1** analogues, respectively, do not exhibit fluxional behavior.

Several mechanisms have been proposed to explain hydrogen exchange in four-vertex *arachno*-2-metallatetraboranes. In the case of covalent metal-octahydridotriborate $Be(B_3H_8)_2$, for instance, a rearrangement involving a Be-to-B_3H_8 bond change from η^2 to η^1, facilitated by Be–H–B bonds, followed by hydrogen atom exchange around the two BH_3 units of the Be-$\{\eta^1$-$B_3H_8\}$ fragment, has been suggested. This mechanism ultimately leads to complete proton and boron exchange at high temperatures [33]. However, this mechanism cannot be applied to explain the observed exchange in compound **1**, as the Rh–H–B hydrogen atoms do not participate in the dynamic process.

Similar fluxional behavior has been observed in $L_nCuB_3H_8$ species, where low-energy exchange of hydrogen and boron atoms occurs [34]. This behavior is reminiscent of the "free" $B_3H_8^-$ anion, for which the energy barrier for complete scrambling of hydrogen and boron atoms has been calculated as 5.2 kcal/mol [35,36]. The fluxional process in copper-octahydridotriborate complexes involves a *pseudo*-rotatory motion of the $\{L_nCu\}$ fragment around the $\{B_3H_8\}$ ligand, supported by Cu–H–B bonds of different hapticity. Additionally, $(CH_3)_2GaB_3H_8$ and $(CH_3)_2AlB_3H_8$ have been found to exhibit fluxional behavior in solution, and the mechanism explaining the intramolecular exchange of hydrogen and boron atoms also involves metal-to-octahydrotriborane hapticity [37].

The fluxional process observed in the complex [Mn(η^2-$B_3H_7Br)(CO)_4$], where the Mn–H–B hydrogen atoms are not involved in the exchange, was proposed to involve a rotation around the B4–Br_{exo} bond coupled with rotation about either B–H bond in the metal-boron bridge. Similarly, in the case of compound **1**, we can propose a concerted rotation of the B4–H_{endo}, B1,3–H_{exo} and B4–H–B1,3 bridging hydrogen atoms around the B4–H_{exo} bond as a mechanism to explain the observed fluxional exchange (Scheme 2). This rotational motion would allow for the dynamic rearrangement of hydrogen atoms without involving the Rh–H–B or Rh–H–B1,3 bonds.

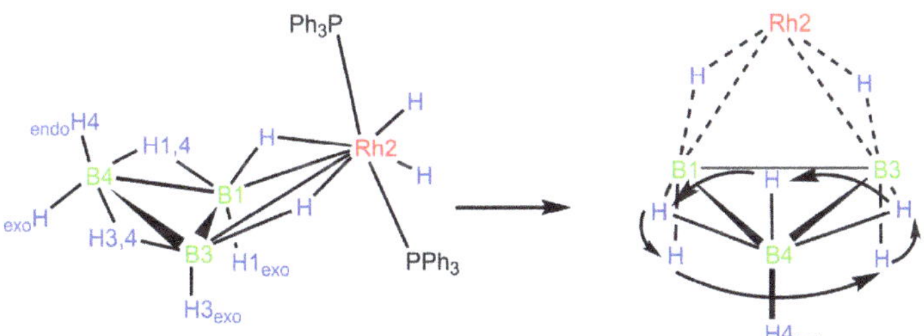

Scheme 2. A proposed mechanism of intramolecular hydrogen exchange in **1**.

2.5. Reactions of 1 with Lewis Bases

We conducted preliminary and exploratory studies on the reactivity of *arachno*-2-rhodatetraborane (**1**) with pyridine (py) and dimethylphenylphosphine (PMe_2Ph). The reactions were performed on a small scale in NMR tubes, and the results presented and discussed in this section should be considered as initial findings.

The ^{31}P-$\{^1H\}$ NMR spectrum of the reaction mixture, obtained by adding pyridine to a CD_2Cl_2 solution of **1** in a 5 mm NMR tube at 233 K, reveals a doublet at $\delta(^{31}P)$ +47.2 ppm, along with the resonances of the starting rhodatetraborane (Figure S8). In the ^{1}H-$\{^{31}$P$\}$ NMR spectrum at 223 K, two new signals appear at $\delta(^1$H) -16.89 and -17.96, with a 1:1 relative intensity ratio, exhibiting the patterns of a *pseudo*-triplet and a doublet of doublets (dd), respectively (Figure S9). In the ^{1}H-$\{^{11}$B$\}$ NMR spectrum, the apparent triplet transforms into an apparent quintet, and the dd becomes a triplet of doublets (Figure S10). The two-dimensional ^{1}H–^{31}P-HMBC spectrum reveals clear cross peaks between the ^{31}P doublet and the two hydride signals, and the ^{1}H-^{1}H correlation further confirms the coupling between both hydrides (Figures S11 and S12).

In the ^{11}B NMR spectrum, a broad peak of low intensity is observed at $\delta(^{11}$B) $+19.3$ ppm, accompanied by smaller intensity peaks between $+2$ and -5 ppm. The highest intensity signals correspond to a quartet at $\delta(^{11}$B) -12.1 p.p.m. and a multiplet at -37.8 ppm. The latter signal transforms into a doublet under ^{1}H decoupling (Figure S13). The main ^{11}B resonances can be confidently assigned to the pyridine and phosphine adducts, py-BH$_3$ and PPh$_3$-BH$_3$. However, the assignment of the lower intensity triplet at -9.4 ppm (close to the quartet of the pyridine borane) remains uncertain.

Based on the observed NMR data, it is proposed that the reaction of compound **1** with pyridine results in the formation of borane adducts and a new cationic rhodium(III) complex, [Rh(H)$_2$(PPh$_3$)$_2$(py)$_2$]$^+$ (compound **2**), which exhibits an octahedral structure (Scheme 3). However, the formation of anionic species, in particular the borate anions, is not clearly observed in the NMR spectra. The low-intensity signals observed in the ^{11}B NMR spectrum, some of which do not show $^1J(^1$H-^{11}B) coupling and others that appear as triplets, could potentially correspond to borate anions. Further characterization is required to determine the exact nature of the anionic species formed in the reaction with pyridine.

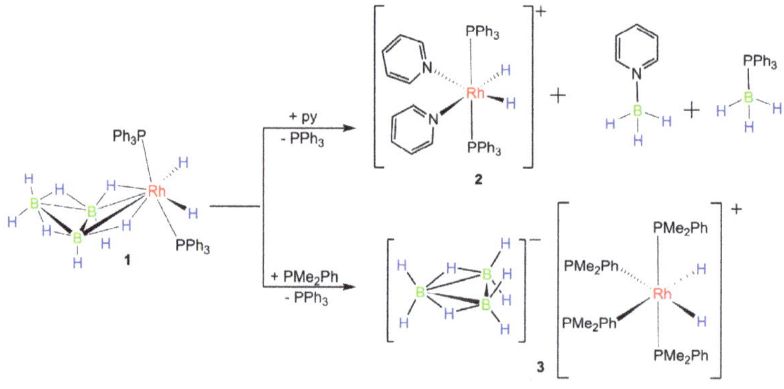

Scheme 3. Reactions of **1** with the ligands py and PMe$_2$Ph.

Upon addition of PMe$_2$Ph to a CD$_2$Cl$_2$ solution of compound **1**, the ^{31}P-$\{^1$H$\}$ NMR spectrum shows the appearance of two new doublets of triplets at $\delta(^{31}$P) $+0.3$ and -10.6 ppm. The spectrum at 233 K also reveals signals corresponding to free PMe$_2$Ph and PPh$_3$ at $\delta(^{31}$P) -45.6 and -7.3 ppm, respectively (Figure S15). In the ^{1}H-$\{^{11}$B$\}$ spectrum, a new hydride resonance is observed at $\delta(^1$H) -10.15 ppm, exhibiting the pattern of a doublet of *pseudo*-quartets. This hydride resonance appears as a simple doublet in the ^{1}H–$\{^{31}$P$\}$ spectrum (Figures S16 and S17). These observations strongly suggest the formation of the octahedral rhodium(III) cationic complex [Rh(H)$_2$(PMe$_2$Ph)$_4$]$^+$, in which the hydride ligands occupy *cis* positions to each other (Scheme 3).

The ^{11}B NMR spectrum exhibits a septet at $\delta(^{11}$B) -30.5 ppm, which can be assigned to the free octadecahydridoborate anion, B$_3$H$_8{}^-$. There is also a multiplet at -37.7 ppm, which becomes a doublet upon ^{1}H decoupling, corresponding to PhMe$_2$P–BH$_3$. Additionally, the spectrum shows signals of low intensity at -40.6, -44.7 and -45.4 ppm, as well as a broad peak of higher intensity at 16.3 ppm (Figure S18). These signals may correspond to uncharacterized metallaborane species present at low concentrations.

Overall, the data described for the reaction between **1** and PMe_2Ph strongly suggest the formation of the salt $[Rh(H)_2(PMe_2Ph)_4][B_3H_8]$ (**3**).

3. Conclusions

The reaction in ethanol of $Cs[B_3H_8]$ with the Wilkinson's catalyst provides a convenient method for the preparation of the *arachno*-2-rhodatetraborane, **1**. This reaction involves the oxidative addition of two hydrogen atoms to the rhodium(I) center to form a $\{Rh(III)(H)_2(PPh_3)_2\}^+$ cationic fragment that binds the $[B_3H_8]^-$ anionic ligand. The origin of the two additional hydrogen atoms is unclear and we can envision that some of the octahydridotriborate anion could donate them. Alternatively, the presence of ethanol could potentially act as a hydrogen transfer agent, facilitating the addition of hydrogen atoms to the rhodium center.

In the crystal structure, the sextuple phenyl embrace is an important driving force leading to the formation of ribbons in the lattice.

The fluxional behavior observed in compound **1** is similar to the non-rigid behavior found in other *arachno*-2-metallatetraboranes; based on the literature, we have proposed a probable mechanism of H atom exchange that involves the *endo*-H5 hydrogen atom, the *exo*-H1 and *exo*-H3 as well as the B1-H1,4-B4 and B3-H3,4-B4 bridging hydrogen atoms. It has been found that **1** is thermally unstable, decomposing at temperatures between +67 and +97 °C; this behavior suggests that the rhodatetraborane may exhibit a rich reaction chemistry *versus* different reagents.

We have explored this hypothesis in reactions of **1** with the Lewis bases, dimethylphenylphosphine, PMe_2Ph and pyridine. In these reactions, we have found that the $\{\eta^2\text{-}B_3H_8\}$ anionic ligand is labile, and it is cleaved by PMe_2Ph to form the salt $[Rh(H)_2(PMe_2Ph)_4][B_3H_8]$ (**3**). Alternatively, the reaction with pyridine demonstrates that dismantling of the $\eta^2\text{-}B_3H_8^-$ ligand can also lead to the formation of pyridine and triphenylphosphine adducts, $L–BH_3$, and to cationic complexes such as $[Rh(H)_2(PPh_3)_2(py)_2]^+$ (**2**).

The observed fluxional behavior and thermal instability highlight the versatility and potential reactivity of the rhodatetraborane compound **1**, making it an interesting candidate for further exploration in various chemical reactions and applications.

4. Materials and Methods

4.1. General

Reactions were carried out under an argon atmosphere using standard Schlenk line techniques. Solvents were obtained from a Solvent Purification System from Innovative Technology Inc. $NaB_3H_8 \cdot 3(C_4H_8O_2)$ was purchased from Katchem spol. s r. o., and used as received. The deuterated solvent CD_2Cl_2 was deaerated, following freeze–pump–thaw methods, and dried over 3 Å molecular sieves.

Infrared spectra were recorded on a Perkin-Elmer 100 spectrometer, using a Universal ATR Sampling Accessory. Solution NMR spectra were recorded on Bruker Avance AV 300-MHz, AV 400-MHz and AV 500-MHz spectrometers, using ^{11}B, $^{11}B\text{-}\{^1H\}$, ^1H, $^1H\text{-}\{^{11}B\}$, $^1H\text{-}\{^{11}B(\text{selective})\}$, $^1H\text{-}^{31}P$-HMBC and $^1H\text{-}^1H$-COSY techniques. The ^1H NMR chemical shifts were measured relative to the partially deuterated solvent peaks but are reported in ppm relative to tetramethylsilane. ^{11}B chemical shifts are quoted relative to $[BF_3 \cdot OEt_2]$.

4.2. Crystal Structure Determination

X-ray diffraction data were collected on an APEX DUO Bruker diffractometer, using graphite-monochromated Mo Kα radiation (λ = 0.71073 Å). Diffracted intensities were integrated [38] and corrected for absorption effects using the multi-scan method [39,40]. Both programs are included in the APEX4 package. All the structures were solved by direct methods with SHELXS [41] and refined by full-matrix least squares on F2 with SHELXL [42]. Hydrogen atoms were located from difference Fourier maps and refined isotropically.

Single crystals of **1** suitable for X-ray analysis were grown in a 5 mm NMR tube in a fridge at 4 °C by slow diffusion of hexane into a CH_2Cl_2 solution of the salt.

Structural data for $[Rh(\eta^2\text{-}B_3H_8)(H)_2(PPh_3)_2]\cdot 2CH_2Cl_2$ (**1**$\cdot 2CH_2Cl_2$, 100 K): Mr = 839.81, colorless prism, triclinic $P-1$, a = 12.3834(12) Å, b = 12.9413(12) Å, c = 13.9452(13) Å, α = 76.211(2)°, β = 85.014(2)°, γ = 66.7050(10)°, V = 1993.4(3) Å^3, Z = 2, T = 100(2) K, Dcalcd = 1.399 g cm^{-3}, μ = 0.803 mm^{-1}, absorption correction factors min. 0.824 max. 0.924. 32,874 reflections, 8938 unique (R_{int} = 0.0659), 6581 observed, R_1 = 0.0558 [$I > 2\sigma(I)$], wR$_2(F^2)$ = 0.1562 (all data), GOF = 1.060. CCDC 2281251.

4.3. Mass Spectrometry

The mass spectrum for compound **1** was measured on a Thermo-Finnigan LCQ-Fleet Ion Trap instrument using electrospray ionization (ESI) with samples dissolved in acetonitrile (approximately 100 ng mL^{-1}) and introduced to the ion source by infusion at a rate of 6 μL min^{-1}: source voltage 3.2 kV, tube lens voltage −90.7 V, capillary voltage −32.0 V, capillary temperature 360 °C, drying gas flow 7 L min^{-1}.

4.4. Computational Details

The calculations were performed using the Gaussian 09 package [43]. The structure of the model molecule, $[Rh(B_3H_8)(H)_2(PH_3)_2]$, was initially optimized using standard methods with the B3LYP/6-31+G(d) methodology and basis sets. The final optimization, including frequency analyses to confirm the true minima, together with GIAO nuclear-shielding calculations, was performed using B3LYP methodology with the 6-31++G(d) basis-set. GIAO nuclear shielding calculations were performed on the final optimized geometry, and computed ^{11}B shielding values were related to chemical shifts by comparison with the computed value for B_2H_6, which was taken to be $\delta(^{11}B)$ +16.6 ppm relative to the $BF_3(OEt_2)$ = 0.0 ppm standard.

4.5. Preparation of $[Rh(\eta^2\text{-}B_3H_8)(H)_2(PPh_3)_2]$ (**1**)

Method A: White powdery CsB_3H_8 [44] (0.0808 g, 0.470 mmol) was added to 10 mL of ethanol in a Schlenk tube (the ethanol was previously degassed with argon for five minutes). The tube was slightly heated to facilitate the formation of a solution, upon which $[RhCl(PPh_3)_3]$ (0.4311 g; 0.470 mmol) was added to form a red-orange suspension. The reaction mixture was stirred at room temperature for three hours to give a yellow solid in suspension. The product was filtered through a frit, in air, to yield a yellow-mustard solid and an orange filtrate. The solid was collected in a Schlenk tube, dissolved in dichloromethane and filtered, under argon, through a silica gel layer. The resulting yellow solid was crystallized from CH_2Cl_2/Hexane (1:2). This final product was studied using NMR spectroscopy, demonstrating that its composition corresponded to the hydridrorhodatetraborane **1**. The total yield after drying under vacuum for several hours was 0.26 mg (0.387 mmol, 82.34%).

^{1}H-{^{11}B} (500 MHz, CD_2Cl_2, 223 K): δ + 7.66 to +7.32 ppm (m, aromatic signals, C_6H_5, 30H). IR (ATR): ν_{max}/cm^{-1} 3054–2962 (w, C-H), 2508, 2452, 2378 (s, BH), 2451 (s, BH), 2418 (s, BH), 1585, 1568, 1480 (C=C aromatics). HR-MS(ESI): m/z calcd exact mass for $C_{36}H_{40}B_3P_2Rh$, [M]+, 670.1939; this anticipated parent ion is clearly absent. Instead, the spectrum exhibits high intensity peaks at 627.0867, 628.0884 and 629.0901 u, with an isotopic pattern that matches well that calculated for the ion $[C_{36}H_{30}P_2Rh]^+$, $[M - (B_3H_8 + H_2)]^+$. This ion corresponds to the $\{Rh(PPh_3)_2\}$ fragment, demonstrating that the rhodatetraborane **1** undergoes facile cleavage upon ionization (Figure S19).

Method B: White powdery $NaB_3H_8\cdot 3(C_4H_8O_2)$ (0.1261 g, 0.385 mmol) was dissolved in 10 mL of dry ether, in a Schlenk tube, which was immersed in an isopropanol bath at −30 °C. Subsequently, the Wilkinson's catalyst was added (0.3559 g, 0.385 mmol), under a flow of argon, to the sodium octahydridotriborate dioxane solution. The resulting brick-red suspension was stirred at room temperature under an atmosphere of argon. After one hour of stirring the temperature was increased to +5 °C and the color of the suspension became orange-red. The reaction was maintained for another three hours to give a brown-yellow suspension, immersed in the isopropanol bath at +10 °C. We allowed the solid to settle down and decanted the supernatant with a pipette, under a flow of argon. The decanted liquid was dried under vacuum to give an orange-yellow solid, whereas the sediment formed a beige solid, after drying. The NMR spectra of the former fraction (the decanted liquid) showed the presence of $O=PPh_3$ and Ph_3P-BH_3, as major and minor components, respectively. The ether-insoluble product corresponded to the four-vertex rhodatetraborane. This method afforded 20 mg of **1** (8%).

4.6. Reactions of [Rh(η^2-B$_3$H$_8$)(H)$_2$(PPh$_3$)$_2$] (**1**) with Lewis Bases

*Reaction of [Rh(η^2-B$_3$H$_8$)(H)$_2$(PPh$_3$)$_2$] (**1**) with py.* 10.2 mg of 1 (1.50×10^{-2} mmol) treated with 1.20 µg of pyridine (1.50×10^{-2} mmol), in a Schlenk tube immersed in a bath of isopropanol at $-30\ ^\circ$C. The resulting yellow solution was stirred, under an atmosphere of argon for 5 h; during this time, the temperature was raised to $+10\ ^\circ$C. The reaction was stirred for another 20 min at room temperature. The solvent was evaporated under vacuum to give an orange solid, which was dissolved in deuterated dichloromethane and studied using NMR spectroscopy. ^{31}P-$\{^1H\}$ (162 MHz, 233 K): δ +47.2 ppm [$^1J(^{31}P-^{103}Rh)$ = 118 Hz], together with the signal of $O=PPh_3$. ^1H-$\{^{31}P\}$ (400 MHz, 233 K): δ +8.59 (d, J = 5.0 Hz, *ortho*-NC$_5$H$_5$, 2H), +8.59 (d, J = 5.0 Hz, *o*-NC$_5$H$_5$, 2H), +8.34 (br. s, *o*-NC$_5$H$_5$, 2H), +7.95 (t., *p*-NC$_5$H$_5$, 2H), +7.95 (t., *p*-NC$_5$H$_5$, 2H), +6.54 (br. s, *m*-NC$_5$H$_5$, 2H), between +7.73 and +6.80 (m, C$_6$H$_5$-rings and NC$_5$H$_5$), -16.89 (t, $^1J(^{103}Rh$-$^1H2,6)$ + $^2J(^1H2$-$^1H2,3)$ = $^2J(^1H6$-$^1H1,2)$ = 12.9 Hz, 2H) and -17.96 (dd, $^1J(^{103}Rh$-$^1H2,6)$ = 23.4, $^2J(^1H2$-$^1H2,3)$ = $^2J(^1H6$-$^1H1,2)$ = 10.9 Hz, 2H). ^1H-$\{^{11}B\}$ (400 MHz, 233 K): δ -16.89 (app. quintet, $^1J(^{103}Rh$-$^1H2,6)$ + $^2J(^{31}P$-$^{31}P)$ + $^2J(^1H2$-$^1H2,3)$ = 12.9 Hz, 2H) and -17.96 (td, J = 25.5, 13.1 Hz, Rh-H2). δ ^{11}B (400 MHz, 298 K): δ +18.8 (br.), +1.47 (s), -1.5 (t), -9.31 (t, 95 Hz), -12.0 (q, $^1J(^{11}B$-$^1H)$ = 98 Hz, py-BH$_3$), -37.8 (dq, 62 Hz, Ph$_3$PBH$_3$), -27.4 (t, 98 Hz).

*In situ characterization of [Rh(H)$_2$(PMe$_2$Ph)$_4$][B$_3$H$_8$] (**3**).* 12.6 mg of 1 (1.88×10^{-2} mmol), dissolved in CD$_2$Cl$_2$, in a NMR tube, which was immersed in an isopropanol bath at $-30\ ^\circ$C, and 2.6 mg (2.7 µL) of PMe$_2$Ph (1.88×10^{-2} mmol) was added into the NMR tube, under a flow of argon. The reaction was studied using NMR spectroscopy, starting at 233 K and then heating the sample to room temperature. ^{31}P-$\{^1H\}$ (162 MHz, 233 K): δ +26.7 ppm [s, $O=PPh_3$], +19.7 (very br., PhMe$_2$P-BH$_3$), +0.3 [dt, $^1J(^{103}Rh-^{31}P)$ = 97 Hz, $^2J(^{31}P1-^{31}P2)$ = 24 Hz], -7.2 (s, PPh$_3$), -10.6 p.pm [dt, $^1J(^{103}Rh-^{31}P)$ = 86 Hz, $^2J(^{31}P1-^{31}P2)$ = 24 Hz], together with the signals of 1 (Table 1). ^1H-$\{^{31}P\}$ (400 MHz, 233 K): δ +7.84 to 6.94 (m, aromatics, C$_6$H$_5$), +1.56 (s, CH$_3$), +1.46 (s, CH$_3$), -10.25 (d, $^1J(^{103}Rh$-$^1H)$ = 13.8 Hz. ^1H-$\{^{11}B\}$ (400 MHz, 233 K): +1.2 (d, $^2J(^{31}P-^1H)$, PhMe$_2$-BH_3), +0.23(s, B$_3$H$_8^-$), -10.25 (d of *pseudo*-quintets, second order, $^2J(^{11}P-^1H_{trans})$ = 147.9 Hz, $^1J(^{103}Rh$-$^1H)$ = 14.0 Hz, $^2J(^{11}P$-$^1H_{cis})$ = 17.2 Hz, 2H) and -17.96 (dd, $^1J(^{103}Rh$-$^1H2,6)$ = 23.4, $^2J(^1H2$-$^1H2,3)$ = $^2J(^1H6$-$^1H1,2)$ = 10.9 Hz, 2H) ppm. ^{11}B (400 MHz, 298 K): δ -16.4 (br. s), -30.5 (sept, B$_3$H$_8^-$), -37.7 (quartet of d, $^1J(^{11}B$-$^{31}P)$ = 59 Hz, $^1J(^{11}B$-$^1H)$ = 102 Hz, PhMe$_2$P-BH$_3$), -45.2 (br. s).

Supplementary Materials: The following supporting information can be downloaded at https://www.mdpi.com/article/10.3390/molecules28186462/s1, Figures S1–S19: ^{11}B, ^1H, ^{31}P, ^1H-^{31}P-HMBC, ^1H-^1H-COSY NMR spectra for **1**, and for the reaction mixtures with pyridine and dimethylphenylphosphine; Table S1: calculated Cartesian coordinates for PH$_3$-ligated model compound [Rh(η^2-B$_3$H$_8$(H)$_2$(PH$_3$)$_2$].

Author Contributions: F.D.-N. synthesis, spectroscopic characterization, single crystal growth, preliminary analysis and preparation of the experimental data. P.J.S.M. and R.M. conceptualization, formal analysis and supervision. R.R. project administration and funding acquisition. Writing—original draft preparation, review and editing. R.M. Visualization, proofreading. Additional conceptualization and discussion of the work, P.J.S.M. and R.R. All authors have read and agreed to the published version of the manuscript.

Funding: PID2021-122406NB-I00 funded by MCIN/AEI/10.13039/501100011033 and by "ERDF A way of making Europe"; and Grupo de Referencia: Catálisis Homogénea Enantioselectiva, ChiralCat E05-23R, funded by Gobierno de Aragón.

Informed Consent Statement: Not applicable.

Data Availability Statement: Not applicable.

Acknowledgments: We want to acknowledge the use *of Servicio General de Apoyo a la Investigación-SAI*, University of Zaragoza. To an outstanding and creative scientist, John D. Kennedy, Emeritus Professor at University of Leeds, on his 80th birthday. He has been a superb mentor to a good number of scientists. We worked for long hours in "the 400 MHz", we waited for the new results and we discussed exciting experiments at "The Pack Horse". Those days have "dyed our souls" and will never leave us. Thank you *hombre roto de Leeds*.

Conflicts of Interest: The authors declare no conflict of interest.

Sample Availability: Not applicable.

References

1. Barton, L.; Strivastava, D.K. *Comprehensive Organometallic Chemistry II*; Abel, E.W., Stone, F.G.A., Wilkinson, G., Eds.; Pergamon: New York, NY, USA, 1995; Volume 1, pp. 275–372.
2. Cotton, F.A.; Wilkinson, G.; Murillo, C.A.; Bochmann, M. *Advanced Inorganic Chemistry*, 6th ed.; Wiley: New York, NY, USA, 1999.
3. Kennedy, J.D. The Polyhedral Metalloboranes. 1. Metalloborane Clusters with 7 Vertices And Fewer. In *Progress in Inorganic Chemistry*; John Wiley & Sons: New York, NY, USA, 1984; Volume 32, pp. 519–679.
4. Weller, A.S. d- and f-Block Metallaboranes. In *Comprehensive Organometallic Chemistry III*; Crabtree, R.H., Mingos, D.M.P., Eds.; Elsevier: Oxford, UK, 2007; Volume 3, pp. 133–174.
5. Bould, J.; Greenwood, N.N.; Kennedy, J.D. The first osmaboranes and a new iridatetraborane. *J. Organomet. Chem.* **1983**, *249*, 11–21. [CrossRef]
6. Grebenik, P.D.; Leach, J.B.; Green, M.L.H.; Walker, N.M. Transition metal mediated homologation of $BH_3 \cdot THF$: Synthesis and crystal structure of $[WH_3(PMe_3)_3B_3H_8]$. *J. Organomet. Chem.* **1988**, *345*, C31–C34. [CrossRef]
7. Grebenik, P.D.; Leach, J.B.; Pounds, J.M.; Green, M.L.H.; Mountford, P. Niobium metallaboranes: A novel metallaborane analogue of pentaborane(11). *J. Organomet. Chem.* **1990**, *382*, C1–C5. [CrossRef]
8. Ghosh, S.; Beatty, A.M.; Fehlner, T.P. The Reaction of Cp^*ReH_6, $Cp^* = C_5Me_5$, with Monoborane to Yield a Novel Rhenaborane. Synthesis and Characterization of arachno-$Cp^*ReH_3B_3H_8$. *Collect. Czech. Chem. Commun.* **2002**, *67*, 808–812. [CrossRef]
9. Fehlner, T.P. Systematic Metallaborane Chemistry. *Organometallics* **2000**, *19*, 2643–2651. [CrossRef]
10. Geetharani, K.; Kumar Bose, S.; Pramanik, G.; Kumar Saha, T.; Ramkumar, V.; Ghosh, S. An Efficient Route to Group 6 and 8 Metallaborane Compounds: Synthesis of arachno-$[Cp^*Fe(CO)B_3H_8]$ and closo-$[(Cp^*M)_2B_5H_9]$ (M = Mo, W). *Eur. J. Inorg. Chem.* **2009**, *2009*, 1483–1487. [CrossRef]
11. Green, M.L.H.; Leach, J.B.; Kelland, M.A. Synthesis and Interconversion of Some Small Ruthenaboranes: Reaction of a Ruthenium Borohydride with Pentaborane(9) to Form Larger Ruthenaboranes. *Organometallics* **2007**, *26*, 4031–4037. [CrossRef]
12. Guggenberger, L.J. Crystal structure of the tetramethylammonium salt of the octahydrotriborotetracarbonylchromium anion, $(CO)_4CrB_3H_8$. *Inorg. Chem.* **1970**, *9*, 367–373. [CrossRef]
13. Beckett, M.A.; Brassington, D.S.; Coles, S.J.; Gelbrich, T.; Hursthouse, M.B. Synthesis and characterisation of a series of Group 7 metal 2,2,2,2-dicarbonylbis(triorganophosphine)-arachno-2-metallatetraboranes, $[M(CO)_2L_2(B_3H_8)]$ (M=Re, Mn); crystal and molecular structures of $[Re(CO)_2(dppf)(B_3H_8)]$ and $[Mn(CO)_2(dppe)(B_3H_8)]$. *Polyhedron* **2003**, *22*, 1627–1632. [CrossRef]
14. Burns, I.D.; Hill, A.F.; Williams, D.J. Ruthenatetraboranes: Molecular Structure of $[Ru(B_3H_8)(PPh_3)\{\kappa^3-HB(pz)_3\}]$ (pz = Pyrazol-1-yl). *Inorg. Chem.* **1996**, *35*, 2685–2687. [CrossRef]
15. Bown, M.; Ingham, S.L.; Norris, G.E.; Waters, J.M. exo-2-([eta]6-Hexamethylbenzene)-endo-2-chloro-2-ruthena-arachno-tetraborane(8). *Acta Crystallogr. Sect. C* **1995**, *51*, 1503–1505. [CrossRef]
16. Bould, J.; Rath, N.P.; Barton, L. $[(CO)H(PPh_3)_2$-arachno-$OsB_3H_8]$. *Acta Crystallogr. Sect. C* **1996**, *52*, 1388–1390. [CrossRef]
17. Lippard, S.J.; Melmed, K.M. Transition metal borohydride complexes. III. Structure of octahydrotriboratobis(triphenylphosphine)-copper(I). *Inorg. Chem.* **1969**, *8*, 2755–2762. [CrossRef]
18. Williams, R.E. Coordination number pattern recognition theory of carborane structures. *Adv. Inorg. Chem. Radiochem.* **1976**, *18*, 67–142.
19. Olah, G.A.; Wade, K.; Williams, R.E. *Electron Deficient Boron and Carbon Clusters*; John Wiley & Sons: New York, NY, USA, 1991.
20. Wade, K. Structural and bonding patterns in cluster chemistry. *Adv. Inorg. Chem. Radiochem.* **1976**, *18*, 1–66.
21. Dance, I.; Scudder, M. The sextuple phenyl embrace, a ubiquitous concerted supramolecular motif. *J. Chem. Soc. Chem. Commun.* **1995**, *10*, 1039–1040. [CrossRef]
22. Dance, I.; Scudder, M. Molecules embracing in crystals. *CrystEngComm* **2009**, *11*, 2233–2247. [CrossRef]
23. Hore, P.J. *Nuclear Magnetic Resonance*; Oxford University Press: Oxford, UK, 2011.
24. Dhubhghaill, O.N.; Spalding, T.R.; Ferguson, G.; Kaitner, B.; Fontaine, X.L.R.; Kennedy, J.D.; Reed, D. Metallaheteroborane Chemistry. 4. The Synthesis of closo-$[2,2-(PPh_3)_2$-2-H-1,2-$XMB_{10}H_{10}]$ (X = Se Or Te, M = Rh Or Ir) Compounds, their Characterization by Nuclear Magnetic-Resonance Techniques, and the crystal and molecular structure of the X = Te, M = Rh complex. *J. Chem. Soc. Dalton Trans.* **1988**, *11*, 2739–2745.

25. Ferguson, G.; Kennedy, J.D.; Fontaine, X.L.R.; Faridoon; Spalding,, T.R. Metallaheteroborane chemistry.3. Synthesis of [2,2-(PEt$_3$)$_2$-1,2-TePtB$_{10}$H$_{10}$], [2,2-(PBu$^n_3$)2-1,2-TePTtB$_{10}$H$_{10}$], [2,2-(PMe$_2$Ph)$_2$-1,2-TePtB$_{10}$H$_{10}$], their characterization by nuclear magnetic-resonance spectroscopy, and the crystal and molecular-structure of [2,2-(PEt$_3$)$_2$-1,2-TePtB$_{10}$H$_{10}$]. *J. Chem. Soc. Dalton Trans.* **1988**, *10*, 2555–2564.

26. O'Connell, D.; Patterson, J.C.; Spalding, T.R.; Ferguson, G.; Gallagher, J.F.; Li, Y.; Kennedy, J.D.; Macías, R.; Thornton-Pett, M.; Holub, J. Conformational polymorphism and fluxional behaviour of M(PR$_3$)$_2$ units in *closo*-twelve-atom metallaheteroboranes with MX$_2$B$_9$ (X = C or As) and MZB$_{10}$ cages (Z = S, Se or Te). *J. Chem. Soc. Dalton Trans.* **1996**, *15*, 3323–3333. [CrossRef]

27. McGrath, M.; Spalding, T.R.; Fontaine, X.L.R.; Kennedy, J.D.; Thornton-Pett, M.J. Metallaheteroborane chemistry. Part 9. Syntheses and spectroscopy of platinum and palladium phosphine complexes containing η^5-{As$_2$B$_9$}-based cluster ligands. Crystal structures of [3,3-L$_2$-*closo*-3,1,2-PtAs$_2$B$_9$H$_9$] (L = PPh$_3$ or PMe$_2$Ph) and [3-Cl-3,8-(PPh$_3$)$_2$-*closo*-3,1,2-PdAs$_2$B$_9$H$_8$]. *J. Chem. Soc. Dalton Trans.* **1991**, 3223–3233.

28. Fontaine, X.L.R.; Kennedy, J.D.; McGrath, M.; Spalding, T.R. Metallheteroborane Chemistry. 8. NMR study of some arsena- and stibaboranes and of the rhodadiarsenaboranes [3,3-(PPh$_3$)$_2$-3-(H)-*closo*-3,1,2-RhAs$_2$B$_9$H$_9$] and [3-(η^5-C$_5$Me$_5$)-*closo*-3,1,2-RhAs$_2$B$_9$H$_9$]. *Magn. Reson. Chem.* **1991**, *29*, 711–720. [CrossRef]

29. Fontaine, X.L.R.; Greenwood, N.N.; Kennedy, J.D.; Nestor, K.; Thorntonpett, M.; Hermanek, S.; Jelinek, T.; Stibr, B. Polyhedral Metallacarbaborane Chemistry—Preparation, Molecular-Structure, And Nuclear Magnetic-Resonance Investigation of [3-(η^5-C$_5$Me$_5$)-*closo*-3,1,2-M$_2$B$_9$H$_{11}$] (M = Rh Or Ir). *J. Chem. Soc. Dalton Trans.* **1990**, *2*, 681–689. [CrossRef]

30. Ferguson, G.; Gallagher, J.F.; McGrath, M.; Sheehan, J.P.; Spalding, T.R.; Kennedy, J.D. Metallaheteroborane chemistry. Part 11. Selective syntheses of the palladium heteroborane complexes [2,2-(PR$_3$)$_2$-*closo*-2,1-PdEB$_{10}$H$_{10}$] (R$_3$= Ph$_3$, MePh$_2$ or Me$_2$Ph; E = Se or Te) and [2-X-2-(PPh$_3$)-*closo*-2,1-PdTeB$_{10}$H$_9$(PPh$_3$)] (X = Cl, Br, I, CN, SCN or O$_2$CMe). *J. Chem. Soc. Dalton Trans.* **1993**, 27–34. [CrossRef]

31. Shanan-Atidi, H.; Bar-Eli, K. A convenient method for obtaining free energies of activation by the coalescence temperature of an unequal doublet. *J. Phys. Chem.* **1970**, *74*, 961–963. [CrossRef]

32. Chen, M.W.; Calabrese, J.C.; Gaines, D.F.; Hillenbrand, D.F. Low-temperature crystal and molecular structure of tetracarbonyl[2-bromoheptahydrotriborato(1-)]manganese, (CO)4MnB3H7Br, and a proton NMR study of the kinetics of its intramolecular hydrogen exchange in solution. *J. Am. Chem. Soc.* **1980**, *102*, 4928–4933. [CrossRef]

33. Gaines, D.F.; Walsh, J.L.; Morris, J.H.; Hillenbrand, D.F. The chemistry of beryllaboranes. Characterization and reactions of beryllium bis(tetrahydroborate), Be(BH4)2, and beryllium bis(octahydrotriborate), Be(B$_3$H$_8$)$_2$. *Inorg. Chem.* **1978**, *17*, 1516–1522. [CrossRef]

34. Bushweller, C.H.; Beall, H.; Dewkett, W.J. Stereodynamics of L$_2$CuB$_3$H$_8$. Rate of triborohydride ion (B$_3$H$_8^-$) rearrangement as a function of L. *Inorg. Chem.* **1976**, *15*, 1739–1740. [CrossRef]

35. Beall, H.; Bushweller, C.H. Dynamical processes in boranes, borane complexes, carboranes, and related compounds. *Chem. Rev.* **1973**, *73*, 465–486. [CrossRef]

36. Serrar, C.; Es-sofi, A.; Boutalib, A.; Ouassas, A.; Jarid, A.; Nebot-Gil, I.; Tomás, F. Theoretical Study of the Structural and Fluxional Behavior of Copper(I)-Octahydrotriborate Complex. *J. Phys. Chem. A* **2001**, *105*, 9776–9780. [CrossRef]

37. Borlin, J.; Gaines, D.F. Internal exchange in new Group III metalloborane derivatives, dimethylaluminum ((CH$_3$)$_2$AlB$_3$H$_8$), and dimethylgallium triborane(8) (CH$_3$)$_2$GaB$_3$H$_8$). *J. Am. Chem. Soc.* **1972**, *94*, 1367–1369. [CrossRef]

38. Bruker-AXS SAINT. *Area-Detector Integration Software, version 6.01*; Bruker-AXS SAINT: Madison, WI, USA, 2001.

39. Bruker-AXS SADABS. *Area Detector Absorption Program*; Bruker-AXS SADABS: Madison, WI, USA, 1996.

40. Krause, L.; Herbst-Irmer, R.; Sheldrick, G.M.; Stalke, D. Comparison of silver and molybdenum microfocus X-ray sources for single-crystal structure determination. *J. Appl. Crystallogr.* **2015**, *48*, 3–10. [CrossRef] [PubMed]

41. Sheldrick, G.M. A short history of SHELX. *Acta Crystallogr. Sect. A Found. Crystallogr.* **2008**, *64*, 112–122. [CrossRef] [PubMed]

42. Sheldrick, G. Crystal structure refinement with SHELXL. *Acta Crystallogr. Sect. C* **2015**, *71*, 3–8. [CrossRef] [PubMed]

43. Frisch, M.J.; Trucks, G.W.; Schlegel, H.B.; Scuseria, G.E.; Robb, M.A.; Cheeseman, J.R.; Scalmani, G.; Barone, V.; Mennucci, B.; Petersson, G.A.; et al. *Gaussian 09, Revision A.02*; Gaussian, Inc.: Wallingford, CT, USA, 2009.

44. Chen, X.-M.; Ma, N.; Liu, X.-R.; Wei, C.; Cui, C.-C.; Cao, B.-L.; Guo, Y.; Wang, L.-S.; Gu, Q.C. Facile Synthesis of Unsolvated Alkali Metal Octahydrotriborate Salts MB$_3$H$_8$ (M = K, Rb, and Cs), Mechanisms of Formation, and the Crystal Structure of KB$_3$H$_8$. *Angew. Chem.* **2019**, *131*, 2746–2750. [CrossRef]

molecules

Article

Synthesis, Reactivity and Coordination Chemistry of Group 9 PBP Boryl Pincer Complexes: [(PBP)M(PMe$_3$)$_n$] (M = Co, Rh, Ir; n = 1, 2)

Philipp M. Rutz [1], Jörg Grunenberg [2] and Christian Kleeberg [1,*]

[1] Institute of Inorganic and Analytical Chemistry, Technische Universität Braunschweig, Hagenring 30, 38106 Braunschweig, Germany
[2] Institute of Organic Chemistry, Technische Universität Braunschweig, Hagenring 30, 38106 Braunschweig, Germany
* Correspondence: ch.kleeberg@tu-braunschweig.de; Tel.: +49-531-391-5392

Abstract: The unsymmetrical diborane(4) derivative [(d(CH$_2$P(iPr)$_2$)abB)–Bpin] (**1**) proved to be a versatile PBP boryl pincer ligand precursor for Co(I) (**2a**, **4a**), Rh(I) (**2–3b**) and Ir(I/III) (**2–3c**, **5–6c**) complexes, in particular of the types [(d(CH$_2$P(iPr)$_2$)abB)M(PMe$_3$)$_2$] (**2a–c**) and [(d(CH$_2$P(iPr)$_2$)abB)M–PMe$_3$] (**2b–c**). Whilst similar complexes have been obtained before, for the first time, the coordination chemistry of a homologous series of PBP pincer complexes, in particular the interconversion of the five- and four-coordinate complexes **2a–c/3a–c**, was studied in detail. For Co, instead of the mono phosphine complex **2a**, the dinitrogen complex [(d(CH$_2$P(iPr)$_2$)abB)Co(N$_2$)(PMe$_3$)] (**4a**) is formed spontaneously upon PMe$_3$ abstraction from **2a** in the presence of N$_2$. All complexes were comprehensively characterised spectroscopically in solution via multinuclear (VT-)NMR spectroscopy and structurally in the solid state through single-crystal X-ray diffraction. The unique properties of the PBP ligand with respect to its coordination chemical properties are addressed.

Keywords: boron; diborane(4); boryl complex; PBP pincer ligand

Citation: Rutz, P.M.; Grunenberg, J.; Kleeberg, C. Synthesis, Reactivity and Coordination Chemistry of Group 9 PBP Boryl Pincer Complexes: [(PBP)M(PMe$_3$)$_n$] (M = Co, Rh, Ir; n = 1, 2). *Molecules* **2023**, *28*, 6191. https://doi.org/10.3390/molecules28176191

Academic Editors: Michael A. Beckett and Igor B. Sivaev

Received: 29 June 2023
Revised: 17 August 2023
Accepted: 18 August 2023
Published: 22 August 2023

1. Introduction

Since their first report in 2009 by Nozaki, Yamashita and a co-worker, PBP pincer ligands with a diaminoboryl framework have been explored with respect to their coordination chemistry with various transition metals, in particular cobalt, rhodium and iridium, as well as with respect to potential applications in different catalytic and stoichiometric processes [1–7]. Whilst the majority of boryl pincer complexes are of this PBP diaminoboryl type, a number of boryl pincer complexes with other ligands frameworks, often quite unique ones, have also been reported [8–14]. Transition metal PBP diaminoboryl pincer complexes are fundamentally accessible through the oxidative addition of a hydridoborane ligand precursor, possibly followed by further modifications, a route already developed by Nozaki and Yamashita in their seminal work [1–7]. To overcome the inherent obstacles by this 'B–H oxidative addition route', we recently developed an unsymmetrical diborane(4), pinB–B(d(CH$_2$P(iPr)$_2$)ab) (**1**) (pin = (OCMe$_2$)$_2$, d(CH$_2$P(iPr)$_2$)ab = 1,2-(N(CH$_2$P(iPr)$_2$))$_2$C$_6$H$_4$), as a versatile PBP ligand precursor. This precursor provides direct access to PBP complexes through σ bond metathesis, as exemplified with the copper boryl complex [(d(CH$_2$P(iPr)$_2$)abB)Cu]$_2$, and, alternatively, oxidative addition, as exemplified with the platinum *bis*-boryl complexes *cis*-[(d(CH$_2$P(iPr)$_2$)abB)(iPr$_3$P)Pt–Bpin] and trans-[(d(CH$_2$P(iPr)$_2$)abB)Pt–Bpin] (Scheme 1) [15].

In the present work, we endeavoured to explore the use of pinB–B(d(CH$_2$P(iPr)$_2$)ab) (**1**) as a precursor for a series of group 9 PBP boryl pincer complexes and study their fundamental coordination chemistry. To facilitate the access to a range of PBP boryl pincer complexes, we chose three easily available group 9 metal complexes [(Me$_3$P)$_4$Co–Me], [(Me$_3$P)$_3$Rh–Cl] and [(cod)Ir–Cl]$_2$ as precursors [16–18].

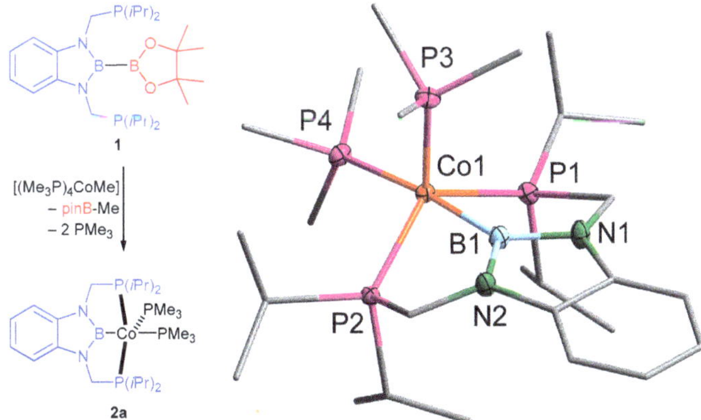

Scheme 1. Formation of PBP pincer boryl complexes from a diborane(4) precursor [15].

2. Results

2.1. Cobalt Complexes

The reaction of **1** with [(Me$_3$P)$_4$Co–Me] results in the mono boryl complex [(d(CH$_2$P(iPr)$_2$)abB)Co–(PMe$_3$)$_2$] (**2a**) (Scheme 2), presumably via an oxidative addition/ reductive elimination pathway [19–21]. The reaction delivers **2a** after 24 h at 50 °C as dark orange crystals in a 66% isolated yield. A single crystal X-ray diffraction study on **2a** revealed a five coordinate 18-valence electron Co(I) complex (Scheme 2). The complex **2a** crystallises in an achiral non-centrosymmetric space group of the type *Pca*2$_1$ with four molecules in the unit cell (Z = 4, Z′ = 1) (Supplementary Materials) [22].

Scheme 2. Formation of PBP cobalt boryl complexes **2a** (**left**) and its molecular structure (**right**). Selected distances [Å] and angles [°]: Co1–B1 1.936(2), Co1–P1 2.2063(4), Co1–P2 2.1859(3), Co1–P3 2.1643(4), Co1–P4 2.1968(4), P1–Co1–P2 125.55(2), P2–Co1–P3 112.03(1), P1–Co1–P3 110.56(2), B1–Co1–P4 173.85(5), B1–Co1–P3 83.24(5), Co1–[P1,P2,P3] 0.4372(3).

The coordination environment at the cobalt atom in **2a** is best described as distorted trigonal bipyramidal with the boryl ligand and one PMe$_3$ ligand in the apical positions, and the angle between these positions deviates by 7° from linearity. Moreover, the strong σ donor properties of the boryl ligand result in an elongation of the Co1–P4 distance of the apical PMe$_3$ ligand, compared to distance Co1–P3 of the equatorial PMe$_3$ ligand by 0.03 Å.

The equatorial positions are occupied by the two pincer phosphine donors and a second PMe$_3$ ligand, resulting in a sum of angles in the equatorial plane [P1,P2,P3] of 348.14°, whereby the angle P1–Co1–P2, involving the two pincer phosphorus atoms, is slightly larger than the other angles. For the deviation of the sum of angles, from 360° accounts for the significant displacement of Co1 from the [P1,P2,P3] plane by 0.4372(3) Å towards the P4 atom. This distortion of the trigonal bipyramidal coordination environment at the cobalt atom is due to the restraints imposed by the five-ring chelates in **2a**. Whilst the solid-state molecular structure of **2a** does not exhibit any crystallographic symmetry, it is virtually C_s symmetric, with a mirror plane through the atoms [B1,P3,P4,Co1] (Figure S37).

An analogous unrestraint mono boryl complex [(PMe$_3$)$_4$Co–Bcat] (cat = 1,2-O$_2$C$_6$H$_4$) exhibits a slightly longer B–Co distance of 1.9545(4) Å and a slightly shorter trans-B P–Co distance of 2.1897(1) Å, together with a less pronounced displacement of the cobalt atom from the equatorial ligand plane [21]. The equatorial Co–P distance in [(PMe$_3$)$_4$Co–Bcat], however, is more equally distributed around 2.17 Å. The closely related square planar PBP complex [(d(CH$_2$P(tBu)$_2$)abB)Co–N$_2$], reported to undergo reversible H$_2$ activation by Peters and a co-worker, exhibits similar Co–B and Co–P distances of 1.946(1) Å and 2.1884(4)/2.1901(3) Å and also significant deviation of the P–Co–P angle from linearity of 156.26(1)° as a result of the five-ring chelation [3].

The ^{31}P{^{1}H} NMR spectrum of **2a** comprises three distinct signals, two signals of the two distinct PMe$_3$ ligands, one in the apical—trans boryl—position around 0 ppm and the second around –16 ppm for the equatorial PMe$_3$ ligand. The third signal around 83 ppm is assigned to the two equivalent PBP pincer ligand P(iPr)$_2$ groups (Figures 1(top) and S4). Whilst these signals do not exhibit any fine structure at room temperature (Figure S4), at lower temperatures, the signals split in a doublet of doublets at 82.9 ppm (–69 °C) for the P(iPr)$_2$ groups, an apparent broadened quartet at 0.7 ppm (–69 °C) for the apical PMe$_3$ ligand and a triplet of doublets at –16.2 ppm (–69 °C) for the equatorial PMe$_3$ ligand (Figures 1(top) and S4). This is in agreement with the mutual couplings within an A$_2$MN spin system. This agrees with a conformation of the complex in solution similar to the one found in the solid state.

However, the temperature-dependent broadening is indicative of dynamic processes present in solution. A ^{1}H-^{1}H NOESY NMR spectrum at room temperature gives a fitting picture. Distinct NOE contacts between the PMe$_3$ signals and the methine C*H*Me$_2$ signals allow for the assignment of the PMe$_3$ ligands to the apical and equatorial positions, respectively. Exchange signals are observed between the two PMe$_3$ ligands, but also between pairs of methyl groups of the two distinct isopropyl moieties and the methine protons of these groups (Figure 1 (bottom)). This is fundamentally in agreement with two possible exchange mechanisms: (i) via the dissociation of a PMe$_3$ ligand with a transient four coordinate 16-electron complex [(d(CH$_2$P(iPr)$_2$)abB)Co–PMe$_3$] and the re-association of a PMe$_3$ ligand; (ii) a concerted mechanism exchanging the PMe$_3$ ligand positions via a (distorted) square pyramidal intermediate is feasible. Note also that the NMR data do not suggest any appreciable dissociation of PMe$_3$ from **2a**, contrary to the heavier rhodium homolog **2b** (vide infra).

The reaction of **2a** with an equimolar amount of BAr$_3$ as a Lewis acid should lead to abstraction of a PMe$_3$ ligand and, after reorganisation, to the complex [(d(CH$_2$P(iPr)$_2$)abB)Co–PMe$_3$] (**3a**). However, whilst one PMe$_3$ ligand can indeed be abstracted by BPh$_3$, the complex **3a** is not isolated. Instead, in a dinitrogen atmosphere, its dinitrogen adduct [(d(CH$_2$P(iPr)$_2$)abB)Co–(N$_2$)(PMe$_3$)] (**4a**) crystallises in minute amounts after several days at –40° C (Scheme 3).

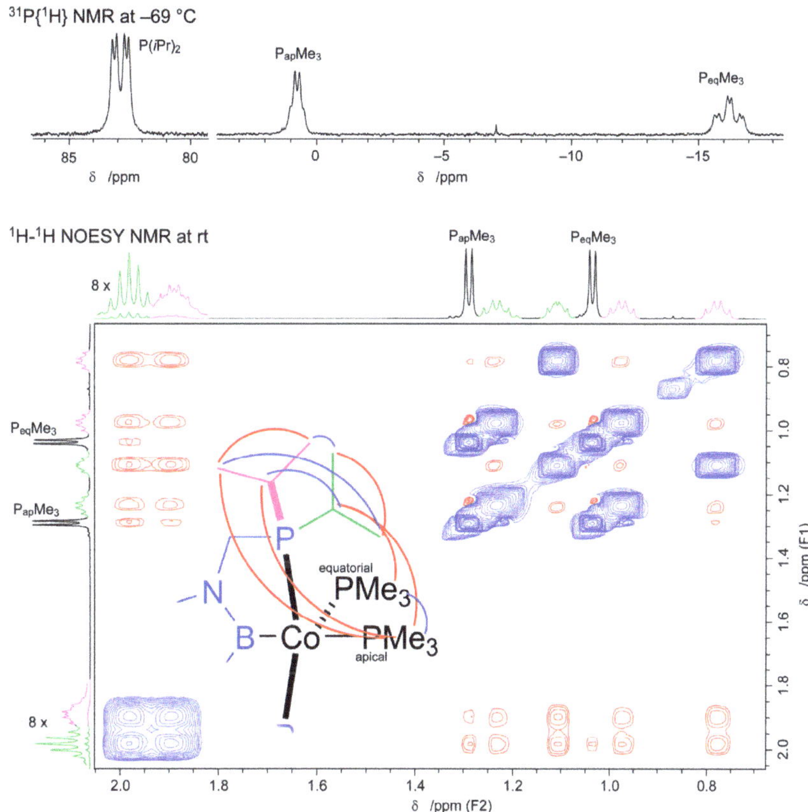

Figure 1. ^{31}P{^{1}H} NMR spectrum of **2a** at −69 °C (**top**), and a section of the ^{1}H-^{1}H NOESY NMR spectrum of **2a** (**bottom**), selected exchange (blue) and NOE (red) correlations are depicted (PhMe-d$_8$, 400.4/162.1 MHz, rt).

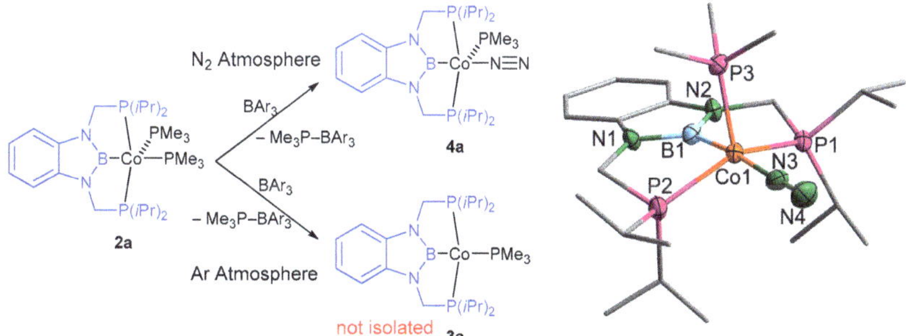

Scheme 3. Reaction of **2a** with BAr$_3$ (Ar = Ph, C$_6$F$_5$) and molecular structures of [(d(CH$_2$P(iPr)$_2$)abB)Rh–(N$_2$)(PMe$_3$)] (**4a**). Selected distances [Å] and angles [°]: Co1–B1 1.942(6), Co1–P1 2.1949(15), Co1–P2 2.2175(16), Co1–P3 2.1798(16), Co1–N3 1.816(5), N3–N4 1.118(7), P1–Co1–P2 134.51(7), P2–Co1–P3 110.78(6), P1–Co1–P3 104.85(6), B1–Co1–P3 89.5(2), B1–Co1–N3 172.2(2), Co1–[P1,P2,P3] 0.3915(9).

The N$_2$ complex **4a** crystallises in a space group of the type $P2_1/c$ with two independent molecules in the asymmetric unit (Z = 8, Z' = 2) (Supplementary Materials) [22].

Both molecules exhibit only a marginal geometric difference, and only one is discussed exemplarily (Figure S40).

As for **2a**, the coordination geometry of **4a** is best described as distorted trigonal bipyramidal with the boryl ligand and the N_2 ligand in the apical positions. The B–Co distance remains virtually unchanged by this substitution of the trans boryl ligand and is also identical to the distance found in the closely related four-coordinate PBP complex [(d(CH$_2$P(*t*Bu)$_2$)abB)Co–N$_2$] reported by Peters et al. of 1.946(1) Å [3]. This indicates again that the B–Co distance is largely determined by the geometrical restraints of the five-ring chelates (vide infra). The close to linear B–Co–N, the angle deviates only by less than 7° from the value found in **2a** and in Peter's N_2 complex [3]. The equatorial ligands experience more substantial changes, although their sum of angles around the cobalt atom increases only slightly by 2° to 350.14°, and consistently, the deviation of the cobalt atom from the [P1,P2,P3] plane decreases by 0.05 Å. The angle between the pincer P atoms deviates significantly by 9°; hence, **4a** is more distorted from an ideal trigonal bipyramidal geometry towards a square pyramidal arrangement than **2a**. However, the reduced steric demand of the ligand in the apical position trans to the boryl ligand in **4a** as compared to **2a** leads to a relaxation of the B1–Co1–P3 angle by about 7°. The N–N distance in the N_2 ligand in **4a** compares well with the distance of 1.119(2) Å found in Peter's N_2 complex; the N–Co distance, however, is in **4a** slightly—by 0.035 Å—enlarged [3].

As we failed to isolate **4a** in any appreciable amounts, we resorted to its spectroscopic in situ characterisation (Figure 2). Performing the reaction of **2a** with B(C$_6$F$_5$)$_3$ in toluene and monitoring this reaction via IR spectroscopy gives clear evidence of the immediate formation of an N_2 complex, based on the appearance of a strong IR band at 2061 cm^{-1} if the reaction is conducted under an N_2 atmosphere, whereas only a minute signal is observed under an argon atmosphere, presumably due to the presence of adventitious N_2 (Scheme 3). This compares well to the N≡N stretching frequency of 2013 cm^{-1} reported for the related complex [(d(CH$_2$P(*t*Bu)$_2$)abB)Co–N$_2$] (vide supra) by Peters et al. [3].

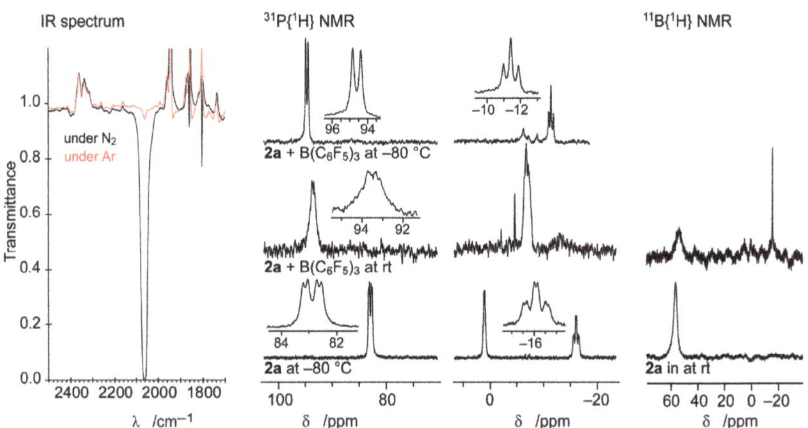

Figure 2. IR (in PhMe), ^{31}P{^{1}H} NMR and ^{11}B{^{1}H} spectra of **2a** + B(C$_6$F$_5$)$_3$ at rt and –80 °C and of isolated **2a** (PhMe-d$_8$, 162.1/96.3 MHz, rt).

Following the reaction of **2a** with B(C$_6$F$_5$)$_3$ under an N_2 atmosphere via ^{31}P and ^{11}11B NMR spectroscopy (Figure 2) gives a consistent picture: upon addition of the Lewis acid, the chemical shifts change from those of **2a** (Figure 2). Whilst the ^{11}B{^{1}H} NMR signal shifts only by about 3 ppm, it gives evidence for the presence of the PBP boryl ligand. The changes in the ^{31}P{^{1}H} NMR spectrum are more substantial. The two ^{31}P NMR signals of the PMe$_3$ ligands in **2a** change to a broad signal at −13 ppm and a second comparably narrow signal at −7 ppm without an appreciable fine structure. Upon cooling, however, the latter signal broadens, and its intensity reduces, whist the former signal changes into a well-developed triplet (-11.4 ppm, $^2J_{PP}$ = 73 Hz) at $-80\,°C$ (Figure 2). The latter triplet corresponds to the doublet at higher chemical shifts (94.6 ppm, $^2J_{PP}$ = 73 Hz). This is readily explained by the abstraction of one PMe$_3$ ligand to give the Lewis acid base adduct Me$_3$P–B(C$_6$F$_5$)$_3$ (δ_{31P} = -6.1 ppm, δ_{11B} = -14.7 ppm in CD$_2$Cl$_2$) and a PBP pincer cobalt complex bearing only one additional PMe$_3$ ligand Me$_3$P–B(C$_6$F$_5$)$_3$ is only sparingly soluble and to a large extend removed prior to the measurement. The remaining dissolved adduct, however, precipitates upon cooling resulting in a reduced ^{31}P NMR signal at lower temperatures. The chemical shift of -11.4 ppm and the P–P coupling constant of around 80 Hz suggest that this PMe$_3$ ligand occupies an equatorial position in a trigonal bipyramidal complex, as it resembles the chemical shift, but in particular, the higher P$_{eq}$–P$_{PBP}$ coupling constant found in **2a**. In other words, the complex that is quantitatively formed is not the four-coordinate complex **3a** but a five coordinate complex with a single PMe$_3$ ligand in an equatorial position—the nitrogen complex **4a**.

Gas-phase DFT computations on the thermodynamics of the complexes **3a** and **4a** and their heavier homologues (vide infra) as central atoms indeed show that for cobalt as the central atom, the formation of a five-coordinate N$_2$ complex of the type [(d(CH$_2$P(iPr)$_2$)abB)M–(N$_2$)(PMe$_3$)] is strongly favoured over the four coordinate complex [(d(CH$_2$P(iPr)$_2$)abB)M–(PMe$_3$)] by ΔG_{298} = -22.4 kJ mol^{-1} (ΔE_0 = -70.6 kJ mol^{-1}), despite the entropic penalty occurring from the coordination of gaseous N$_2$. However, for the rhodium and iridium analogue, the coordination of an N$_2$ ligand to the latter four-coordinate complex is—in agreement with our observations (vide infra)—disfavoured by ΔG_{298} = 51.3 kJ mol^{-1} (ΔE_0 = 8.2 kJmol^{-1}) for rhodium and ΔG_{298} = 51.4 kJ mol^{-1} (ΔE_0 = 8.4 kJ mol^{-1}) for iridium (Supplementary Materials) [22]. The computed N≡N stretching frequency in **4a** of 2170 cm^{-1} is by about 100 cm^{-1} off the experimental values, but within the expected range considering the harmonic nature of the computation and other approximations [22].

Due to an initial computation of the force constant between Co and N$_2$, the bonding in **4a** is quite strong (Co–N: 2.33 N cm^{-1}), whilst the trans-B Co–P bond in **2a** shows the expected kinetic lability (Co–P: 1.36 N cm^{-1}) of a spectator ligand. More importantly, the electronic coupling in **4a** between the N–N bond and the Co–N coordination is pronounced (Co–N/N–N coupling force constant: -0.02 cm N^{-1}) and synergistic (negative sign), pointing to an effective back donation [23]. And indeed, the experimental N$_2$ IR wavenumber of 2061 cm^{-1} is in line with a modest activation relative to free N$_2$ (~2330 cm^{-1}). Finally, the Co–B bond trans to the N$_2$ ligand seems to be very strong (Co–B: 2.28 N cm^{-1}), reducing the flexibility to access different coordination geometries [24].

2.2. Rhodium Complexes

The reaction of **1** with [(Me$_3$P)$_3$Rh–Cl] in the presence of KOtBu leads to the formation of a rhodium(I) boryl complex (Scheme 4). It may be speculated that the reaction proceeds via an intermediate rhodium alkoxido complex as discussed for the formation of the related complex [(dmabB)Rh(PMe$_3$)$_3$] (dmab = 1,2-(NMe)$_2$C$_6$H$_4$) [20]. However, the reaction of **1** with [(Me$_3$P)$_3$Rh–Cl] in the presence of KOtBu leads to the formation of an equilibrium mixture of the square planar complex [(d(CH$_2$P(iPr)$_2$)abB)Rh–PMe$_3$] (**3b**) and the five coordinate complex [(d(CH$_2$P(iPr)$_2$)abB)Rh(PMe$_3$)$_2$] (**2b**) (Scheme 4), whilst in the absence of KOtBu, no reaction is observed (Supplementary Materials) [22] (Figures S12 and S13). After recrystallisation from diethyl ether, the four coordinate complex **3b** is obtained as bright orange crystals at a 70% yield, whereas crystallisation from n-pentane in the presence of an excess PMe$_3$ leads to the isolation of the five-coordinate complex **2b** as crystalline material at a 43% yield. The spontaneous dissociation of one PMe$_3$ ligand from **2b** to give **3b** is not contradicting gas-phase DFT computational data (Table S10), suggesting an endothermic (15 kJ mol^{-1}) dissociation from **2b** to **3b** + PMe$_3$, but overall, an entropy driven exergonic process (-47 kJ mol^{-1}) (Supplementary Materials) [22].

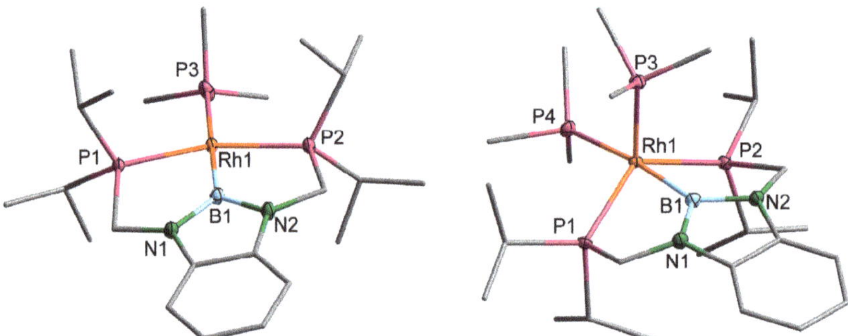

Scheme 4. Formation of PBP rhodium boryl complexes **3b** and **2b**.

Both complexes **2b** and **3b** crystallise in monoclinic space groups of the type $P2_1/n$ and $P2_1/c$, respectively, and contain one complex molecule in the asymmetric unit (Z = 4, Z′ = 1) (Supplementary Materials) [22]. The molecular structure of complex **2b** is analogous to that of the cobalt homologue **2a** (Figure S38). The rhodium ion is distorted trigonal bipyramidaly coordinated by the boryl pincer ligand and one PMe₃ ligand in the apical positions (Figure 3, right). The sum of angles in the equatorial plane [P1,P2,P3] comprising the pincer phosphorus atoms and one PMe₃ ligand is with 347° only insignificantly smaller than in **2a**, whereby the angle P1–Co1–P2, involving the pincer phosphorus atoms, is by about 2° larger than in **2a**. The displacement of Rh1 from the [P1,P2,P3] plane is by 0.05 Å larger than in **2a**, an effect of the increased radius of the rhodium ion within the restraining pincer coordination environment.

Figure 3. Molecular structures of the complexes [(d(CH₂P(iPr)₂)abB)Rh–PMe₃] (**3b**) (**left**) and [(d(CH₂P(iPr)₂)abB)Rh(PMe₃)₂] (**2b**) (**right**). Selected distances [Å] and angles [°], **3b**: Rh1–B1 2.0221(5), Rh1–P1 2.2658(1), Rh1–P2 2.2794(1), Rh1–P3 2.3555(1), P1–Rh1–P2 152.622(5), B1–Rh1–P3 177.02(2), Rh1–[P1,P2,P3,B1] 0.0264(3); **2b**: Rh1–B1 2.0256(7), Rh1–P1 2.3262(2), Rh1–P2 2.3364(2), Rh1–P3 2.3167(2), Rh1–P4 2.3705(5), P1–Rh1–P2 127.894(6), P2–Rh1–P3 108.475(7), P1–Rh1–P3 110.722(7), B1–Rh1–P4 172.54(2), Rh1–[P1,P2,P3] 0.4833(3).

Complex **3b** is best described as a distorted square planar complex with a nearly linear B1–Rh1–P3 angle and a significantly (by 27°), from linearity, deviating P1–Rh1–P2 angle. However, this angle is significantly closer to linearity than the respective angle in the five-coordinate complex **2b** (Figure 3, left). The change in the Rh···P/B distances between **2a** and **3b** is comparably small, despite the change in the coordination number. Most pronounced is a decrease in the pincer phosphorus atoms to rhodium distances in comparison to **2b** by about 0.06 Å, which may be attributed to the less strained ligand conformation in the more planar **3b**.

The equilibrium between **2b** and **3b**, as a fundamental aspect of their coordination chemistry, was further studied via NMR spectroscopy. NMR titration of **3b** with increasing amounts of PMe₃ shows a highly dynamic behaviour in the ³¹P{¹H} NMR spectra at room temperature (Figures 4 and S17–S19). Only one set of signals of the PBP ligand and the trans-B PMe₃ ligand is observed, respectively. Whilst the ³¹P NMR signal of the PBP ligand changes appreciably from 84 ppm to 75 ppm with increasing amounts of PMe₃ added, the signal of the trans-B PMe₃ ligand, in **2b**, is only marginally influenced (−27.3 to −26.4 ppm).

An additional signal is observed shifting from -37 ppm at low amounts of PMe$_3$ to -62 ppm after the addition of an excess of PMe$_3$. This is readily explained by a rapid exchange among **3b**, **2b** and free PMe$_3$ on the NMR time scale and consequently, the observation of an averaged chemical shift of the exchanging PMe$_3$ moieties throughout this process. In agreement with that, the spectrum observed for isolated **2b** is very virtually identical to the spectrum of **3b** after the addition of an equimolar amount of PMe$_3$.

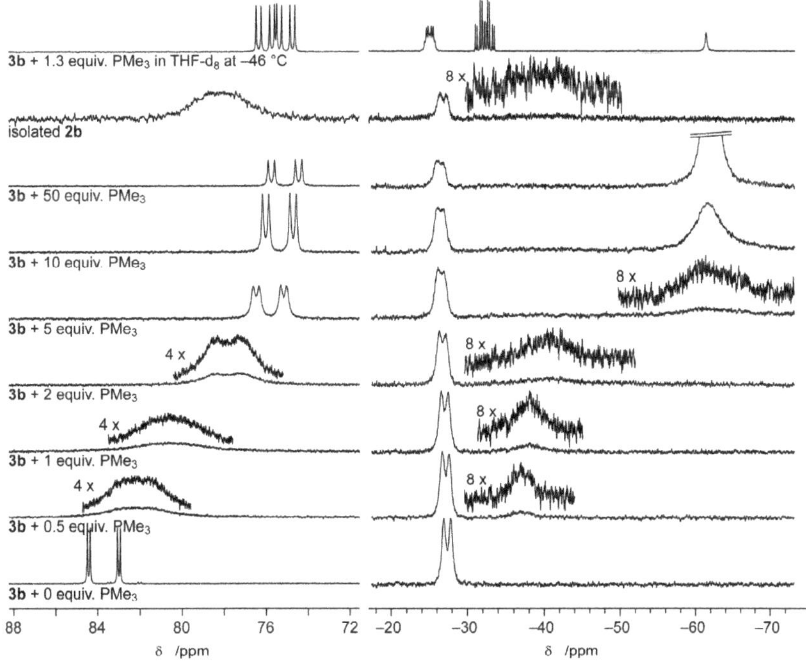

Figure 4. In situ ^{31}P{^{1}H} NMR spectra of the reaction of **3b** with different amounts of PMe$_3$ (121.6 MHz, C$_6$D$_6$, rt), isolated **2b** and **3b** with 1.3 equiv. PMe$_3$ at -46 °C (162.1 MHz, THF-d$_8$).

At low temperatures, however, the exchange among **3b**, **2b** and free PMe$_3$ becomes slow on the NMR timescale, and well-resolved signals for **2b** and free PMe$_3$ are observed (Figures 4, S14 and S15). The ^{31}P{^{1}H} NMR spectrum of **2b** itself at -46 °C comprises three signals (A, M and N) of an A$_2$MNX spin system with the expected ^{31}P-^{31}P and ^{31}P-^{103}Rh couplings (Figure S16, Table S3). Following the reaction of **3b** with different amounts of PMe$_3$ via UV-Vis spectroscopy corroborates the rapid equilibrium between **3b** and **2b** being rather on the side of **3b** and free PMe$_3$ (Figures S20 and S21).

In conclusion, it may be stated that the five-coordinate trigonal bipyramidal complex **2b**, in contrast to the Co analogue, easily dissociates one PMe$_3$ ligand to give the distorted square planar complex **3b**. The virtual indifference in the ^{31}P NMR chemical shift (and line shape) of the apparently not-exchanging trans-B PMe$_3$ ligand around 27 ppm suggests that this exchange does not affect this ligand but involves only the equatorial PMe$_3$ ligand.

2.3. Iridium Complexes

Whilst for the formation of the cobalt and rhodium PBP pincer complexes **2a** and **2b**/**3b**, it may be arguable whether activation of the diborane precursor **1** proceeds via a σ bond metathesis or an oxidative addition/reductive elimination pathway, the reaction of **1** with the iridium(I) complex [Ir(cod)Cl]$_2$ (cod = 1,5-cyclooctadien) to give the *bis*-boryl complex [(d(CH$_2$P(*i*Pr)$_2$)abB)Ir(Bpin)(Cl)] (**5c**) is obviously an oxidative addition reaction

(Scheme 5). This five-coordinate complex reacts with excess PMe$_3$ to give the six-coordinate complex [(d(CH$_2$P(iPr)$_2$)abB)Ir(Bpin)(PMe$_3$)(Cl)] (**6c**).

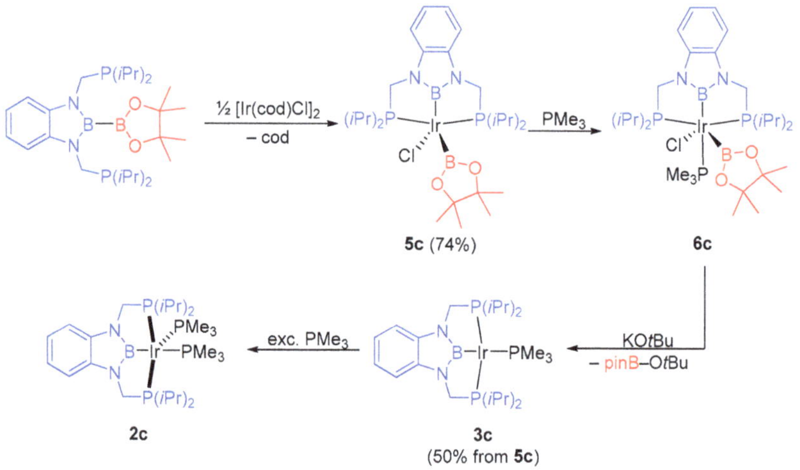

Scheme 5. Consecutive formation of the PBP iridium boryl compels **5c**, **6c**, **3c** and **2c**.

Both complexes **5c** and **6c** crystallise in monoclinic space groups of the type $P2_1/c$. The solid-state structure of **5c** contains one complex molecule in the asymmetric unit (Z = 4, Z′ = 1), whereas **6c** comprises two independent molecules in the asymmetric unit (Z = 8, Z′ = 2). The Bpin moiety in **5c** shows some positional disorder that is neglected in the further discussion; for **6c**, however, one of the independent molecules shows severe disorder and is not considered for further geometrical analysis (Supplementary Materials) [22].

The trigonal bipyramidal geometry of **5c** may be considered typical for a five-coordinate Ir *bis*-boryl complex with phosphine ligands (Figure 5). All five structurally characterised complexes of this type adopt a trigonal bipyramidal geometry with the two phosphine ligands in the axial positions (P–Ir–P angle 157–172°, for PXP pincer ligands P–Ir–P angle 157–162°) and small B···B distances and B–Ir–B angles in the ranges of 2.22–2.41 Å and 65.8–76.7°, respectively [25–29].

In **6c**, the PMe$_3$ ligand adopts a position trans to the PBP pincer boryl ligand, whilst the chlorido ligand occupies a position trans to the Bpin ligand (Figure 5). As a result, **6c** may best be described as a strongly distorted octahedral complex with the Bpin and chlorido ligand in the axial positions. Structurally, the extension of the coordination sphere to the distorted octahedral complex **6c** is accompanied by some ligand reorganisation. The P1–Ir–P2 angle reduces upon coordination by about 3° to deviate more from linearity, whereas the B–Ir–B angle deviates in **6c** by about 6° less from 90° than in **5c** (in accordance with the B···B distance increasing from **5c** to **6c** by 0.25 Å). The d(CH$_2$P(iPr)$_2$)abB ligand backbone in **6c** (mean plane [B1,N1,N2,C$_6$H$_4$]) includes an angle of 24.8(8)° with the equatorial plane of the complex (mean plane [P1,P2,P3,B1,Ir1]), 20° more than in the five-coordinate **5c**. This is a result of the increased steric encumbrance induced by the extension of the coordination sphere in **6c**. The B–Ir distance increases slightly upon PMe$_3$ coordination in **6c** because of the presence of trans ligands. This is more significant for B1, which is trans to the stronger trans influencing ligand PMe$_3$ as opposed to the chlorido ligand for B2. The Cl–Ir distance increases accordingly, whereas the P1/P2–Ir1 distances remain virtually unaffected. Again, because of the strong trans influence of the boryl ligand, the P–Ir distance of the PMe$_3$ ligand is longer than those of the pincer phosphine atoms by about 0.06 Å [30].

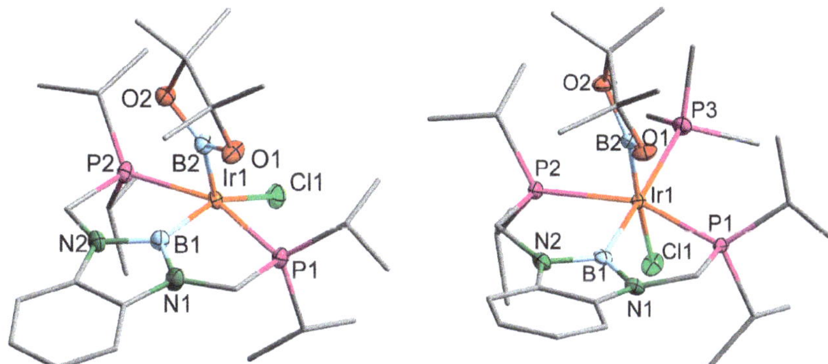

Figure 5. Molecular structures of the complexes [(d(CH$_2$P(iPr)$_2$)abB)Ir(Bpin)(Cl)] (**5c**) (**left**, disorder omitted for clarity) and [(d(CH$_2$P(iPr)$_2$)abB)Ir(Bpin)(Cl)(PMe$_3$)] (**6c**) (**right**, one of two independent molecules shown) (Supplementary Materials) [22]. Selected distances [Å] and angles [°], **5c**: Ir1–B1 1.986(2), Ir1–B2 2.012(2), Ir1–P1 2.3354(4), Ir1–P2 2.3306(4), Ir1–Cl1 2.4144(4), P1–Ir1–P2 156.45(2), B1–Ir1–B2 72.16(8), B1–Ir1–Cl1 153.23(6), B2–Ir1–Cl1 134.49(6), ∠[P1,P2,B1,Ir1][B1,N1,N2,C$_6$H$_4$] 4.3(2), B1–B2 2.354(3); **6c**: Ir1–B1 2.052(4), Ir1–B2 2.050(4), Ir1–P1 2.3391(8), Ir1–P2 2.3627(9), Ir1–P3 2.4155(9), Ir1–Cl1 2.5667(9), P1–Ir1–P2 153.28(3), B1–Ir1–B2 78.9(1), B1–Ir1–Cl1 107.6(1), B1–Ir1–P3 172.6(1), B2–Rh1–Cl1 173.4(1), B2–Rh1–P3 94.0(1), ∠[P1,P2,P3,B1,Ir1][B1,N1,N2,C$_6$H$_4$] 24.78(8), B1–B2 2.605(2).

The solution-state ^{1}H, ^{31}P and ^{13}C NMR spectroscopic data for **5c** and **6c** fulfil the expectations and can readily be explained by the solid-state structures. Surprising, however, are the ^{11}B NMR chemical shifts. For both complexes, two very distinct, somewhat broadened singlets at chemical shifts of 39.7 ppm ($\Delta w_{\frac{1}{2}}$ = 340 Hz) and 19.9 ppm ($\Delta w_{\frac{1}{2}}$ = 330 Hz) for **5c** and of 48.8 ppm ($\Delta w_{\frac{1}{2}}$ = 460 Hz) and 26.6 ppm ($\Delta w_{\frac{1}{2}}$ = 450 Hz) for **6c** are observed in THF-d$_8$ at room temperature. This chemical shift range is somewhat different from the ^{11}B NMR data for the reported Ir(III) boryl in a range of 29–35 ppm for five-coordinate and of 30–43 ppm for six-coordinate complexes, respectively [1,2,7,25,26,28,31].

Whilst complex **6c** is stable under inert conditions, it reacts readily with an equimolar amount of KOtBu to give the Ir(I) PBP pincer complex **3c** (Figure 6, Scheme 5). Monitoring this reaction via in situ NMR spectroscopy (Figures 6 and S34–S36) shows an essentially clean conversion to **3c**, as indicated by its characteristic signals around 80 ppm (doublet, $J_{P–P}$ = 5 Hz) for the pincer phosphorus atoms and a broadened singlet for the PMe$_3$ ligand around –20 ppm (Figure S35 Supplementary Materials) [22]. Only minor amounts of a so far unidentified side product with a ^{31}P NMR signal at 45 ppm (II) are observed. However, upon closer evaluation, two transient species are observed during this reaction. At one hand side, the five-coordinate complex **2c** (vide infra) is formed in small amounts in the beginning but is later on fully consumed (Figure 6). On the other side, a species with a ^{31}P{^{1}H} NMR singlet signal at 64.5 ppm (I) is observed. In agreement with this, the ^{11}B NMR data suggest the presence of a transient boryl intermediate at 40 ppm, whereas **3c** itself exhibits a moderately broad ^{11}B{^{1}H} NMR signal around 56 ppm (Figure S36 Supplementary Materials) [22,32]. It may be assumed that the conversion of **6c** to **3c** proceeds via the initial coordination of a OtBu ligand followed by (possibly after some reorganisation) the reductive elimination to an Ir(I) PBP complex, **3c** or a closely related species. The intermediate presence of **2c** may be explained by the intermediate liberation of PMe$_3$ and its transient addition to **3c** during this process. An in situ ^{31}P{^{1}H} NMR spectrum of a mixture of **3c** and excess PMe$_3$ corroborates the facile formation of **2c** (Figure 6, top). Moreover, it must be emphasised that the system **2c**/**3c**/PMe$_3$ exhibits much less dynamic behaviour than the homologous rhodium system **2b**/**3b**/PMe$_3$ (vide supra). Contrary to the latter, even in the presence of excess PMe$_3$ at room temperature, a well-resolved, ^{31}P{^{1}H} NMR spectrum (A$_2$MN spin system) with a narrow linewidth is observed, indicating only comparably slow exchange of a coordinated PMe$_3$ ligand with free PMe$_3$. Contrary to **2b**, distinct signals for both PMe$_3$ ligands are observable for **2c** at room temperature in the presence of free PMe$_3$ (Figure 6, top). One of these signals (around -70 ppm), however, sharpens upon only moderate cooling to an apparent quartet (Figure S27 Supplementary Materials) [22]. In agreement with that, in situ UV-Vis spectroscopic data of **3c** in the presence of different amounts of PMe$_3$ indicate a rapid equilibration, rather on the side of **2c** (Figures S30 and S31). The ^{11}B NMR shift of **2c** of around 55 ppm is virtually unaffected by the change in the coordination number.

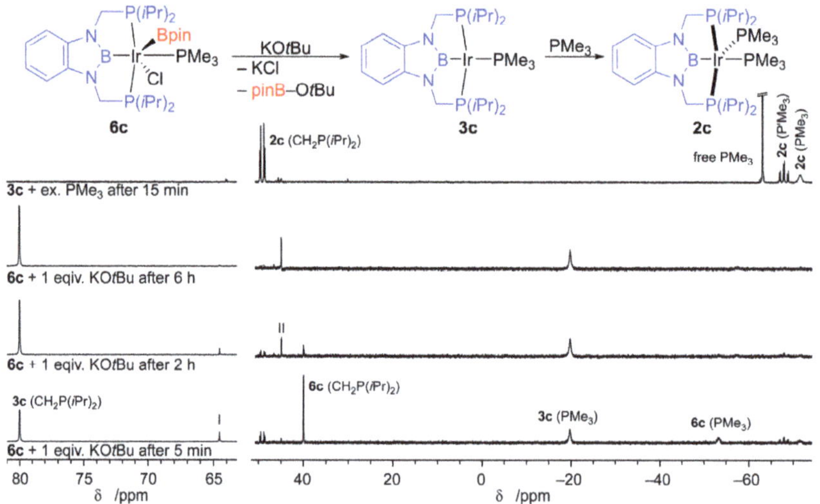

Figure 6. In situ ^{31}P{^{1}H} NMR spectra of the reaction of **6c** with KOtBu (121.6 MHz, THF-d$_8$, rt).

In conclusion, it may be stated that the five-coordinate trigonal bipyramidal complex **2c**, similarly to the cobalt analogue **2a** and opposed to the rhodium homologue, shows only little dynamic behaviour in solution and does not readily dissociate a PMe$_3$ ligand to give the distorted square planar complex **3c**. However, gas-phase DFT computational data suggest similar thermodynamic data for the dissociation of PMe$_3$ from **2c** ($\Delta E_0 = 16$ kJ mol^{-1}, $\Delta G_{298} = -48$ kJ mol^{-1}) as for the rhodium analogue **2b** (Supplementary Materials) [22].

The complexes **2c** and **3c** crystallise isostructurally with the homologous rhodium complexes in monoclinic space groups of the types $P2_1/n$ and $P2_1/c$, respectively ($Z = 4$, $Z' = 1$) (Supplementary Materials) [22]. As a consequence, the molecular structure of **2c** (Figure 7, right) differs only marginally form the structure of the lighter homologue **2b** and from the cobalt homologue **2a** (Figure S38).

The sum of angles in the equatorial plane [P1,P2,P3] of the distorted trigonal bipyramidal complex **2c** is, with 347.76°, only insignificantly different from that in **2a** and **2b**. The angle P1–Ir1–P2, involving the pincer phosphorus atoms, is larger than that in **2a** by about 2° and, hence, virtually identical to that in **2b**. The displacement of Ir1 from the [P1,P2,P3] plane is in the middle between the values for two lighter homologues, by 0.03 Å larger than in **2a** and by 0.02 Å smaller than in **2b**. Generally, the M–P distances, however, increase from **2a** to **2b** and **2c** by about 0.12 Å, most significantly between the cobalt and the rhodium complex.

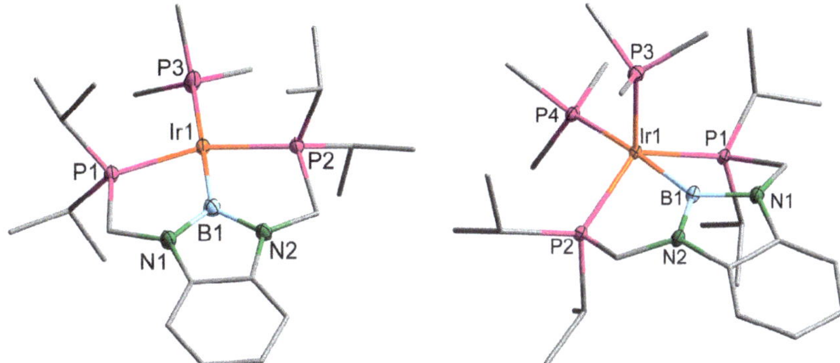

Figure 7. Molecular structures of the complexes [(d(CH$_2$P(*i*Pr)$_2$)abB)Ir–PMe$_3$] (**3c**) and (**left**) [(d(CH$_2$P(*i*Pr)$_2$)abB)Ir(PMe$_3$)$_2$] (**2c**) (**right**). Selected distances [Å] and angles [°], **3c**: Ir1–B1 2.034(2), Ir1–P1 2.2764(4), Ir1–P2 2.2662(2), Ir1–P3 2.3355(5), P1–Ir1–P2 152.95(2), B1–Ir1–P3 176.73(6), Ir1–[P1,P2,P3,B1] 0.0386(3); **2c**: Ir1–B1 2.055(3), Ir1–P1 2.3113(6), Ir1–P2 2.3165(6), Ir1–P3 2.2911(6), Ir1–P4 2.3521(6), P1–Ir1–P2 127.71(2), P2–Ir1–P3 108.84(8), P1–Ir1–P3 111.21(2), B1–Ir1–P4 172.02(8), Ir1–[P1,P2,P3] 0.4666(3).

Overall, the PBP pincer ligand shows, within the series **2a**, **2b 2c**, a high ability to coordinate different metal ions. The high flexibility of this ligand is also illustrated by a comparison of the five-coordinate complexes **5c** and **2c**. For both complexes, a trigonal bipyramidal geometry is observed; however, whilst in **2c**, the phosphorus atoms of the PBP pincer ligand occupy two equatorial positions and the boryl moiety is bound in an axial position, in **5c**, two phosphorus atoms coordinate in the two axial positions and the boron atom in an equatorial position. This is illustrated by P–M–P angles included by the pincer phosphorus atoms decreasing by 30° from **5c** to **2c**.

The solid-state structure of the distorted square planar complex **3c** is again very similar to that of its rhodium homologue **3b** (Figure S39) with a nearly linear B1–Ir1–P3 angle and a P1–Ir1–P2 angle of 152.95(2)° deviating significantly from linearity. Noteworthy is the Ir1–B1 distance in **3c** that is slightly (0.01 Å) longer, whereas the pincer P–M distances are identical, and the trans-B P–M distance is slightly shorter (0.02 Å) than the respective distance in the rhodium homologue **3b**.

3. Discussion

A series of either group 9 PBP diaminoboryl pincer complexes was synthesised using the unsymmetrical diborane(4) **1** as a PBP pincer precursor and fully characterised. In an extension of our earlier work [15], this exemplifies again the versatility of this compound as a PBP pincer ligand precursor. The Co^I and Rh^I complexes [(d(CH$_2$P(*i*Pr)$_2$)abB)Co–(PMe$_3$)$_2$] (**2a**) and [(d(CH$_2$P(*i*Pr)$_2$)abB)Rh–(PMe$_3$)$_n$] (**2b** (n = 2), **3b** (n = 1)), respectively, were obtained in a one-step reaction from the respective Co^I and Rh^I precursors (Schemes 2 and 4). Whilst an oxidative addition/reductive elimination pathway is, for both reactions, feasible, in the rhodium case, a σ bond metathesis pathway may be feasible, considering results based on a related non-pincer ligand [20]. The heavier Ir^I homologue, however, was obtained via the isolated intermediate Ir^{III} complexes [(d(CH$_2$P(*i*Pr)$_2$)abB)Ir(Bpin)(Cl)] (**5c**) and [(d(CH$_2$P(*i*Pr)$_2$)abB)Ir(Bpin)(PMe$_3$)(Cl)] (**6c**). Complex **5c** is formed upon an oxidative addition reaction of the diborane(4) **1** with [Ir(cod)Cl]$_2$ (cod = 1,5-cyclooctadien) and subsequently reacts via PMe$_3$ addition to **6c**. The coordination chemistry of the resulting homologous complexes [(d(CH$_2$P(*i*Pr)$_2$)abB)M(PMe$_3$)$_2$] (**2a–c**) and [(d(CH$_2$P(*i*Pr)$_2$)abB)M–PMe$_3$] (**3b,c**) was studied structurally in the solid state, as well as spectroscopically in solution. However, for M = Rh and Ir, both complexes are structurally very similar but differ in the dynamic behaviour and the relative accessibility of the four (**3b,c**) vs. the five (**2b,c**) coordinated complexes. For Co only the five-coordinate complex **2a** is accessible, whereas Lewis acid-promoted PMe$_3$ abstraction under a dinitrogen atmosphere leads to the formation of the surprisingly stable N$_2$ complex [(d(CH$_2$P(*i*Pr)$_2$)abB)Co–(N$_2$)(PMe$_3$)] (**4a**).

Having, with the unsymmetrical diborane(4) [(d(CH$_2$P(*i*Pr)$_2$)abB)–Bpin] (**1**), a well accessible and versatile PBP ligand precursor that is capable of oxidative addition (Pt^{II}, Co^I, Rh^I (possibly), Ir^I) and σ bond metathesis (Cu^I and possibly Rh^I) reactions [15,20] will stimulate the further development of PBP pincer ligands. In conclusion, PBP diaminoboryl pincer ligands are a ligand class with remarkable ligand properties with respect to their high σ donor strength and weak π acceptor properties—leading to a strong trans effect and influence [30]—that provide stability for the inherently reactive B–M bond due to their pincer framework. Furthermore, PBP pincer ligands are tuneable based on the backbone and P atoms substituents, making them interesting for a broad range of applications from catalysis to the stabilisation of reactive intermediates.

4. Materials and Methods

4.1. General Considerations

pinB–B(d(CH$_2$P(*i*Pr)$_2$)ab) (**1**), [(Me$_3$P)$_4$CoMe], [(Me$_3$P)$_3$RhCl] and [(cod)IrCl]$_2$ were prepared according to literature procedures [15–18,33]. All other compounds were commercially available and were used as received; their purity and identity were checked using appropriate spectroscopic methods. Unless otherwise noted, all solvents were dried using an MBraun solvent purification system, deoxygenated using the freeze-pump-thaw method and stored under purified nitrogen. Unless noted otherwise, all manipulations were performed using standard Schlenk techniques under an atmosphere of purified nitrogen or in a nitrogen-filled glove box (MBraun). NMR spectra were recorded on Bruker Avance II 300, Avance III HD 300 and Avance III 400 spectrometers. NMR tubes equipped with screw caps (WILMAD) were used, and the solvents were dried over potassium/benzophenone and degassed. Chemical shifts (δ) are given in ppm, using the (residual) resonance signal of the solvents for calibration (C$_6$D$_6$: ^{1}H NMR: 7.16 ppm, ^{13}C NMR: 128.06 ppm; PhMe-d$_8$: ^{1}H NMR: 2.08 ppm, ^{13}C NMR: 20.43 ppm; THF-d$_8$: ^{1}H NMR: 1.72 ppm, ^{13}C NMR: 25.31 ppm) [34]. ^{11}B and ^{31}P NMR chemical shifts are reported relative to pseudo external BF$_3$·Et$_2$O and 85% H$_3$PO$_{4\text{(aq)}}$, respectively. ^{13}C{^{1}H}, ^{11}B{^{1}H} and ^{31}P{^{1}H} NMR spectra were recorded employing composite pulse ^{1}H decoupling. ^{11}B NMR spectra were processed applying a back linear prediction, in order to suppress the broad background signal due to the boron in the NMR tube and instrument. A Lorentz-type window function (LB = 10 Hz) was used, and the spectra were carefully evaluated to ensure that no genuinely broad signals of the sample were suppressed. Simulations were conducted with the TOPSPIN/DAISY program package (Bruker). Melting points were determined in flame-sealed capillaries under nitrogen using a Büchi 535 apparatus and are not corrected. Elemental analyses were performed at the Institut für Anorganische und Analytische Chemie of the Technische Universität Braunschweig using an Elementar vario MICRO cube instrument. A Bruker Vertex 70 spectrometer was used for recording IR spectra. The IR spectra were recorded in PhMe solutions in a cuvette of an approximately 1 mm optical path length equipped with NaCl windows.

X-ray Structure Determination. The single crystals were transferred into inert perfluoroether oil inside a nitrogen-filled glovebox and, outside the glovebox, rapidly mounted on top of a CryoLoop (Hampton Research) and placed on the diffractometer in the cold nitrogen gas stream of a Cryostream 800 cooling system (Oxford Cryosystems) [35]. The data were collected on a Rigaku Oxford Diffraction Synergy-S instrument using either mirror-focused MoKα or CuKα radiation (Rigaku PhotonJet microfocus sources). The reflections were indexed and integrated, and appropriate absorption corrections were applied as implemented in the CrysAlisPro software package [36]. The structures were solved employing the program SHELXT and refined anisotropically for all non-hydrogen atoms via full-matrix least squares based on all F^2 values using SHELXL software [37–39]. Generally, hydrogen atoms were refined employing a riding model; methyl groups were treated as rigid bodies and were allowed to rotate about the E–CH$_3$ bond. During refinement and analysis of the crystallographic data, the programs OLEX^2, PLATON, Mercury and Diamond were used [40–43]. Unless noted otherwise non-C,H atoms are depicted as ellipsoids at the 50% probability level, whereas the carbon atom framework is depicted as a stick model (grey), and hydrogen atoms are omitted for clarity. Adapted numbering schemes may be used to improve the readability. Further crystallographic details can be found in the Supplementary Materials available.

4.2. Experimental Procedures and Analysis Data

4.2.1. [(d($CH_2P(iPr)_2$)abB)Co(PMe_3)$_2$] (2a)

In a Schlenk-flask, d($CH_2P(iPr)_2$)abB–Bpin (**1**) (100 mg, 0.198 mmol, 1 equiv.) and [($Me_3P)_4$CoMe] (75 mg, 0.198 mmol, 1 equiv.) were dissolved in toluene (50 mL) and stirred for 24 h at 50 °C whilst a reduced pressure was applied for about 50% of the time (the pressure was normalised overnight). The solvent was completely removed in vacuo and the brown residue was dissolved in *n*-pentane and recrystallised at −40 °C. The resulting dark orange crystals were washed with cold *n*-pentane (1 mL) and dried in vacuo (77 mg, 0.131 mmol, 66%).

^{1}H NMR (PhMe-d$_8$, 400.4 MHz, rt) δ = 6.96–6.91 (m, 2 H, 3-HC_{Ar}), 6.74–6.79 (m, 2 H, 2-HC_{Ar}), 3.50 (d, $^2J_{H\text{-}H}$ = 11.0 Hz, 2 H, CHH'), 3.47 (d, $^2J_{H\text{-}H}$ = 11.0 Hz, $^2J_{H\text{-}P}$ = 4 Hz, 2 H, CHH'), 1.98 (app. sept., $^3J_{H\text{-}H}$ = 7.6 Hz, $^3J_{H\text{-}H}$ = 7.0 Hz, 2 H, CH(CH$_3$)$_2$), 1.89 (m, $^3J_{H\text{-}H}$ = 7.0 Hz, $^3J_{H\text{-}H}$ = 7.4 Hz, $J_{H\text{-}P}$ = 2.5, 4.0 Hz, 2 H, C'H(CH$_3$)$_2$), 1.29 (d, $^2J_{H\text{-}P}$ = 5.0 Hz, 9 H, P$_{ap}$(CH_3)$_3$), 1.23 (app. q, $^3J_{H\text{-}H}$ = 7.6 Hz, $J_{H\text{-}P}$ = 6.8, 6.0 Hz, 6 H, CH(CH_3)(C'H$_3$)), 1.11 (m, $^3J_{H\text{-}H}$ = 7.0 Hz, $J_{H\text{-}P}$ = 5.7, 3.6 Hz, 6 H, CH(CH_3)(C'H$_3$)), 1.04 (d, $^2J_{H\text{-}P}$ = 4.8 Hz, 9 H, P$_{eq}$(CH_3)$_3$), 0.98 (app. q, $^3J_{H\text{-}H}$ = 7.0 Hz, $J_{H\text{-}P}$ = 6.9, 6.5 Hz, 6 H, C'H(CH_3)(C'H$_3$)), 0.78 (app. q, $^3J_{H\text{-}H}$ = 7.4 Hz, $J_{H\text{-}P}$ = 6.5, 5.7 Hz, 6 H, C'H(CH_3)(C'H$_3$)). **^{13}C{^{1}H} NMR (PhMe-d$_8$, 100.7 MHz, rt)** δ = 140.6 (app. t, $J_{C\text{-}P}$ = 6 Hz, 1-C_{Ar}), 117.2 (s, 3-HC_{Ar}), 106.5 (s, 2-HC_{Ar}), 44.5 (m, CHH'), 32.0 (app. dt, $J_{C\text{-}P}$ = 19, 3 Hz, CH(CH$_3$)$_2$), 29.1 (app. t, $J_{C\text{-}P}$ = 5 Hz, C'H(CH$_3$)$_2$), 27.8 (m, P$_{ap}$(CH_3)$_3$), 25.7 (app. dq, $J_{C\text{-}P}$ = 15, 4 Hz, P$_{eq}$(CH_3)$_3$), 21.9 (s, CH(CH_3)(C'H$_3$)), 20.3 (s, CH(CH_3)(C'H$_3$)), 19.4 (s, C'H(CH_3)(C'H$_3$)), 18.9 (app. t, $J_{C\text{-}P}$ = 3 Hz, C'H(CH_3)(C'H$_3$)). **^{31}P{^{1}H} NMR (PhMe-d$_8$, 162.1 MHz, rt)** δ = 83.0 (br. s, $\Delta w_{\frac{1}{2}}$ = 217 Hz, $CH_2P(iPr)_2$), −1.2 (br. s, $\Delta w_{\frac{1}{2}}$ = 132 Hz, P(CH$_3$)$_3$), −17.4 (br. s, $\Delta w_{\frac{1}{2}}$ = 245 Hz, P(CH$_3$)$_3$). **^{11}B{^{1}H} NMR (PhMe-d$_8$, 128.5 MHz, rt)** δ 57.0 (br. s, $\Delta w_{\frac{1}{2}}$ = 360 Hz). **^{1}H NMR (PhMe-d$_8$, 400.4 MHz, −69 °C)** δ = 7.20–7.14 (m, 2 H, C_{Ar}), 6.95–6.89 (m, 2 H, HC_{Ar}), 3.51–3.33 (m, 4 H, CHH'), 1.92 (br. s, 2 H, CH(CH$_3$)$_2$), 1.80 (br. app. sept., $J_{H\text{-}H}$ = 7 Hz, 2 H, C'H(CH$_3$)$_2$), 1.22 (d, $^2J_{H\text{-}P}$ = 5 Hz, 9 H, P(CH_3)$_3$), 1.20 (br. s, 6 H, CH(CH_3)(C'H$_3$)), 1.06 (d, $^2J_{H\text{-}P}$ = 5 Hz, 9 H, P(CH_3)$_3$), 1.03 (br. s, 6 H, CH(CH_3)(C'H$_3$)), 0.99–0.90 (m, 6 H, C'H(CH_3)(C'H$_3$)), 0.75 (br. s, 6 H, C'H(CH_3)(C'H$_3$)). **^{31}P{^{1}H} NMR (PhMe-d$_8$, 162.1 MHz, −69 °C)** δ = 82.9 (dd, $^2J_{P\text{-}P}$ = 80, 30 Hz, $CH_2P(iPr)_2$), 0.7 (app. q, $^2J_{P\text{-}P}$ = 30, 28 Hz, P$_{ap}$(CH$_3$)$_3$), −16.2 (td, q, $^2J_{P\text{-}P}$ = 80, 28 Hz, P$_{eq}$(CH$_3$)$_3$). **^{1}H NMR (C$_6$D$_6$, 300.1 MHz, rt)** δ = 7.10–7.03 (m, 2 H, HC_{Ar}), 6.92–6.85 (m, 2 H, HC_{Ar}), 3.50 (m, 4 H, CHH'), 2.06–1.85 (m, 4 H, CH(CH$_3$)$_2$), 1.29 (d, $^2J_{H\text{-}P}$ = 5.0 Hz, 9 H, P(CH_3)$_3$), 1.28–1.19 (m, 6 H, CH(CH_3)(C'H$_3$)), 1.14–1.07 (m, 6 H, CH(CH_3)(C'H$_3$)), 1.07 (d, $^2J_{H\text{-}P}$ = 4.8 Hz, 9 H, P(CH_3)$_3$), 1.03–0.94 (m, 6 H, C'H(CH_3)(C'H$_3$)), 0.87–0.76 (m, 6 H, C'H(CH_3)(C'H$_3$)). **^{11}B{^{1}H} NMR (C$_6$D$_6$, 96.3 MHz, rt)** δ 57.2 (br. s, $\Delta w_{\frac{1}{2}}$ = 460 Hz). **^{31}P{^{1}H} NMR (C$_6$D$_6$, 121.5 MHz, rt)** δ 82.9 (br. s, $\Delta w_{\frac{1}{2}}$ = 200 Hz, $CH_2P(iPr)_2$), −1.5 (br. s, $\Delta w_{\frac{1}{2}}$ = 135 Hz, P(CH$_3$)$_3$), −17.5 (br. s, $\Delta w_{\frac{1}{2}}$ = 240 Hz, P'(CH$_3$)$_3$). **Anal. Calcd. for C$_{26}$H$_{54}$BCoN$_2$P$_4$ (2a):** C, 53.08; H, 9.25; N, 4.76. Found: C, 52.84; H, 9.27; N, 5.13. **m.p.:** 160–163 °C.

4.2.2. [(d($CH_2P(iPr)_2$)abB)Co(N_2)(PMe_3)] (4a)

Single crystals of **4a**: In a nitrogen-filled glovebox, **2a** (10 mg, 17 µmol, 1 equiv.) and triphenylborane (4.1 mg, 17 µmol, 1 equiv.) were dissolved in C$_6$D$_6$ (0.7 mL). After 3 d at room temperature, the solution was layered with *n*-pentane. Colourless crystals of Me$_3$P–BPh$_3$ separated. The supernatant solution was decanted, and the solvent was removed in vacuo. The residue was dissolved in toluene (0.5 mL), and the solution was layered with *n*-pentane and cooled to −40 °C. Colourless crystals formed overnight, from which the supernatant solution was decanted and cooled again to $-40\,^\circ$C. A few orange single crystals of **4a** suitable for x-ray diffraction were obtained from this solution. In situ IR characterisation of **4a** was as follows: in a nitrogen-filled glovebox, **2a** (10 mg, 17 µmol, 1 equiv.) and tris(pentafluorophenyl)borane (8.7 mg, 17 µmol, 1 equiv.) were dissolved in toluene (0.4 mL) and transferred into an IR cuvette. An IR spectrum of this solution was recorded. The reaction under an Ar atmosphere was conducted analogously in an Ar-filled glovebox. In situ NMR characterisation of **4a** was performed as follows: in a nitrogen-filled glovebox, **2a** (16.1 mg, 27 µmol, 1 equiv.) and tris(pentafluorophenyl)borane (14 mg, 27 µmol, 1 equiv.) were dissolved in toluene-d_8 and filtered through a small pad of celite. NMR spectra of this solution were recorded.

^{1}H NMR (PhMe-d_8, 400.4 MHz, rt) δ = 6.88 (br. s, 2 H, HC_{Ar}), 6.66 (br. s, 2 H, HC_{Ar}), 3.50 (br. s, 2 H, CH_2), 3.34 (br. s, 2 H, CH_2), 2.11 (overlapping with the residual solvent signal, $CH(CH_3)_2$), 1.48–0.65 (P($CH_3)_3$) and $CH(CH_3)_2$). **^{31}P$\{^1$H$\}$ NMR (PhMe-d_8, 162.1 MHz, rt)** δ = 93.5 (br. s, $\Delta w_{\frac{1}{2}}$ = 211 Hz, $CH_2P(iPr)_2$), -13.4 (br. s, Co–$P(CH_3)_3$). **^{11}B$\{^1$H$\}$ NMR (PhMe-d_8, 128.5 MHz, rt)** δ 54.3 (br. s, $\Delta w_{\frac{1}{2}}$ = 630 Hz). **^{1}H NMR (PhMe-d_8, 400.4 MHz, $-69\,^\circ$C)** δ = 6.77 (br. s, 2 H, HC_{Ar}), 3.35 (br. s, 2 H, CH_2), 3.13 (br. d, 2 H, CH_2), 1.95 (br. s, 4 H, $CH(CH_3)_2$), 1.40 (br. s, 6 H, $CH(CH_3)_2$), 1.19 (br. s, 6 H, $CH(CH_3)_2$), 1.04 (br. s, 15 H, $CH(CH_3)_2$) and P($CH_3)_3$), 0.80 (br. s, 6 H, $CH(CH_3)_2$). **^{31}P$\{^1$H$\}$ NMR (PhMe-d_8, 162.1 MHz, $-80\,^\circ$C)** δ = 94.6 (d, $^2J_{P\text{-}P}$ = 74 Hz, $CH_2P(iPr)_2$), -11.4 (t, $^2J_{P\text{-}P}$ = 74 Hz, Co–$P(CH_3)_3$).

4.2.3. [(d(CH$_2$P(iPr)$_2$)abB)Rh(PMe$_3$)$_2$] (2b)

The reaction was performed as described for **3b** on a 55 µmol scale (vide infra). After filtration, an excess of PMe$_3$ (30 µL, 22 mg, 0.3 mmol, 5.5 equiv.) was added, and the resulting yellow solution was cooled to $-40\,^\circ$C. After 48 h, bright yellow crystals suitable for X-ray crystallography had formed. The supernatant solution was decanted, and the crystals were dried in vacuo (15 mg, 24 µmol, 43%). NMR spectra of the isolated material show an equilibrium among **2b**, **3b** and free PMe$_3$ (Figures S17–S19). NMR spectra of **2b** were recorded from a solution of **3b** (15 mg, 28 µmol) in THF-d_8 (0.7 mL) after the addition of PMe$_3$ (3.7 µL, 2.7 mg, 37 µmol, 1.3 equiv.).

^{1}H NMR (THF-d_8, 400.4 MHz, -46° C) δ 6.58–6.47 (m, 4 H, 2,3-HC_{Ar}), 3.61–3.42 (m, 4 H, CH_2), 2.12 (app. sept., J = 6.9 Hz, 2 H, $CH(CH_3)_2$), 1.64 (br. s, 2 H, $\Delta w_{\frac{1}{2}}$ = 25 Hz, $CH(CH_3)_2$), 1.43 (d, $^2J_{H\text{-}P}$ = 4.8 Hz, 9 H, P($CH_3)_3$), 1.34–1.21 (m, 12 H, $CH(CH_3)_2$), 1.11 (d, $^2J_{H\text{-}P}$ = 4.8 Hz, 9 H, P′($CH_3)_3$), 1.03 (app. q, J = 5.6 Hz, 6 H, $CH(CH_3)_2$), 0.94 (d, $^2J_{H\text{-}P}$ = 2 Hz, 2.9 H, free P($CH_3)_3$), 0.7 (app. q, J = 7.1 Hz, 6 H, $CH(CH_3)_2$). **^{11}B$\{^1$H$\}$ NMR (THF-d_8, 128.5 MHz, rt)** δ 55.4 (s, $\Delta w_{\frac{1}{2}}$ = 365 Hz). **^{31}P$\{^1$H$\}$ NMR (THF-d_8, 162.1 MHz, rt)** δ 76.6 (br. d, $^1J_{P\text{-}Rh}$ = 153 Hz, $\Delta w_{\frac{1}{2}}$ = 150 Hz, $CH_2P(iPr)_2$), -26.4 (br. d, $^1J_{P\text{-}Rh}$ = 97 Hz, $\Delta w_{\frac{1}{2}}$ = 105 Hz, P($CH_3)_3$), -37 (br. s, $\Delta w_{\frac{1}{2}}$ = 1000 Hz, P($CH_3)_3$), -54 (br. s, $\Delta w_{\frac{1}{2}}$ = 1600 Hz, free P($CH_3)_3$). **^{31}P$\{^1$H$\}$ NMR (THF-d_8, 162.1 MHz, $-46\,^\circ$C)** δ 75.5 (ddd, $^1J_{P\text{-}Rh}$ = 157 Hz, $^2J_{P\text{-}P}$ = 38, 103 Hz, $CH_2P(iPr)_2$), -25.1 (app. dq, $^1J_{P\text{-}Rh}$ = 105 Hz, $^2J_{P\text{-}P}$ = 43, 38 Hz, $P_{ap}(CH_3)_3$), -32.3 (dtd, $^1J_{P\text{-}Rh}$ = 157 Hz, $^2J_{P\text{-}P}$ = 103, 43 Hz, $P_{eq}(CH_3)_3$). **Anal. Calcd. for C$_{26}$H$_{54}$BN$_2$P$_4$Rh (2b):** C, 49.39; H, 8.61; N, 4.30. Found: C, 48.91; H, 8.56; N, 4.47.

4.2.4. [(d(CH$_2$P(iPr)$_2$)abB)Rh(PMe$_3$)] (3b)

In a nitrogen-filled glovebox, d(CH$_2$P(iPr)$_2$)abB–Bpin (**1**) (41 mg, 82 µmol, 1 equiv.) and [Rh(PMe$_3$)$_3$Cl] (30 mg, 82 µmol, 1 equiv.) were combined and dissolved in toluene (10 mL). A solution of KOtBu (9 mg, 82 µmol, 1 equiv.) in THF (2 mL) was added, and the bright orange solution was stirred for 5 min at room temperature. The solvent was removed in vacuo. The residue was extracted with n-pentane (2 × 3.5 mL) and filtered through a pad of celite. The solvent was removed in vacuo. The orange residue was recrystallised from diethyl ether (3 mL) at $-40\,^\circ$C to give bright orange crystals of [(d(CH$_2$P(iPr)$_2$)abB)Rh(PMe$_3$)] (**3b**) (30 mg, 56 µmol, 70%).

^{1}H NMR (THF-d_8, 400.4 MHz, rt) δ 6.72–6.67 (m, 2 H, HC_{Ar}), 6.67–6.62 (m, 2 H, HC_{Ar}), 3.64 (app. t, J = 2, 2 Hz, 4 H, NCH$_2$P), 2.18 (app. sept., J = 7, 6, 1.5, 1.5 Hz, 4 H, $CH(CH_3)_2$),

1.39 (dd, $^2J_{\text{H-P}}$ = 4.3, $^3J_{\text{H-Rh}}$ = 0.6 Hz, 9 H, P(CH_3)$_3$), 1.19 (app. q, J = 7.0, 7.4, 7.4 Hz, 12 H, CH(CH_3)$_2$), 1.09 (app. q, J = 6.0, 6.3, 6.3 Hz, 12 H, CH(CH_3)$_2$). **^{13}C{^{1}H} NMR (THF-d$_8$, 101.7 MHz, rt)** δ 140.7 (app. td, $J_{C\text{-}P}$ = 9 Hz, $J_{C\text{-}Rh}$ = 1.5 Hz, 1-C_{Ar}), 117.3 (s, HC_{Ar}), 104.9 (s, HC_{Ar}), 43.9 (m, CH_2), 28.6 (app. t, $J_{C\text{-}P}$ = 8.5 Hz, CH(CH_3)($C'H_3$)), 23.0 (app. dtd, $J_{C\text{-}P}$ = 13, 3 Hz, $^2J_{C\text{-}Rh}$ = 1 Hz, P(CH_3)$_3$), 20.8 (app. t, $J_{C\text{-}P}$ = 5 Hz, CH(CH_3)($C'H_3$)), 20.8 (br. s, CH(CH_3)($C'H_3$)). **^{11}B{^{1}H} NMR (THF-d$_8$, 128.5 MHz, rt)** δ 52.4 (s, $\Delta w_{\frac{1}{2}}$ = 400 Hz). **^{31}P{^{1}H} NMR (THF-d$_8$, 162.1 MHz, rt)** δ 84.1 (dd, $^1J_{P\text{-}Rh}$ = 173 Hz, $^2J_{P\text{-}P}$ = 17 Hz, CH_2P(*i*Pr)$_2$), -27.1 (br. d, $^1J_{P\text{-}Rh}$ = 111 Hz, $\Delta w_{\frac{1}{2}}$ = 90 Hz, P(CH_3)$_3$). **^{31}P{^{1}H} NMR (THF-d$_8$, 162.1 MHz, -102 °C)** δ 83.8 (dd, $^1J_{P\text{-}Rh}$ = 171 Hz, $^2J_{P\text{-}P}$ = 17 Hz, CH_2P(*i*Pr)$_2$), -25.4 (dt, $^1J_{P\text{-}Rh}$ = 113 Hz, $^2J_{P\text{-}P}$ = 17 Hz, P(CH_3)$_3$). **Anal. Calcd. for C$_{23}$H$_{45}$BN$_2$P$_3$Rh (3b):** C, 49.66; H, 8.15; N, 5.04. Found: C, 49.43; H, 8.11; N, 5.38. **m.p.:** 199–200 °C.

4.2.5. [(d(CH$_2$P(*i*Pr)$_2$)abB)Ir(PMe$_3$)$_2$] (2c)

In a nitrogen-filled glovebox, **3c** (30 mg, 46 µmol, 1 equiv.) was dissolved in THF (5 mL), and PMe$_3$ (23.6 µL, 17.7 mg, 0.23 mmol, 5 equiv.) was added. The solvent was removed under in vacuo conditions. The light yellow residue was recrystallised from diethyl ether (2 mL) at -40 °C to give light yellow crystals of [(d(CH$_2$P(*i*Pr)$_2$)abB)Ir(PMe$_3$)$_2$] (2c) (7 mg, 9.7 µmol, 21%).

^{1}H NMR (PhMe-d$_8$, 400.4 MHz, rt) δ 6.95–6.89 (m, 2 H, 3-HC_{Ar}), 6.83–6.77 (m, 2 H, 2-HC_{Ar}), 3.53–3.39 (m, 4 H, CH_2), 1.96 (app. br. sept., J = 7 Hz, 2 H, CH(CH_3)$_2$), 1.63 (br. s, 2 H, $\Delta w_{\frac{1}{2}}$ = 25 Hz, C'H(CH$_3$)$_2$), 1.51 (d, $^2J_{\text{H-P}}$ = 5.9 Hz, 9 H, P(CH_3)$_3$), 1.24 (d, $^2J_{\text{H-P}}$ = 6.3 Hz, 9 H, P'(CH_3)$_3$), 1.20–1.08 (m, 12 H, CH(CH_3)$_2$), 0.89 (app. q, J = 6.9 Hz, 6 H, C'H(CH_3)($C'H_3$)), 0.64 (app. q, J = 7.1 Hz, 6 H, C'H(CH_3)($C'H_3$)). **^{11}B{^{1}H} NMR (PhMe-d$_8$, 128.5 MHz, rt)** δ 55.2 (s, $\Delta w_{\frac{1}{2}}$ = 475 Hz). **^{31}P{^{1}H} NMR (PhMe-d$_8$, 162.1 MHz, -35 °C)** δ 50.7 (dd, $^2J_{P\text{-}P}$ = 27, 111 Hz, CH_2P(*i*Pr)$_2$), -65.7 (td, $^2J_{P\text{-}P}$ = 27, 111 Hz, P'(CH_3)$_3$), -69.9 (app br. q, $^2J_{P\text{-}P}$ = 27, 27 Hz, P(CH_3)$_3$). **^{31}P{^{1}H} NMR (PhMe-d$_8$, 162.1 MHz, rt)** δ 50.7 (dd, $^2J_{P\text{-}P}$ = 27, 112 Hz, CH_2P(*i*Pr)$_2$), -65.6 (td, $^2J_{P\text{-}P}$ = 27, 112 Hz, P'(CH_3)$_3$), -69.9 (br. s, $\Delta w_{\frac{1}{2}}$ = 95 Hz, P(CH_3)$_3$). **^{31}P{^{1}H} NMR (THF-d$_8$, 121.5 MHz, rt)** δ 50.6 (dd, $^2J_{P\text{-}P}$ = 28, 111 Hz, CH_2P(*i*Pr)$_2$), -66.4 (td, $^2J_{P\text{-}P}$ = 27, 111 Hz, P'(CH_3)$_3$), -70.0 (br. s, $\Delta w_{\frac{1}{2}}$ = 100 Hz, P(CH_3)$_3$). **^{13}C{^{1}H} NMR (PhMe-d$_8$, 100.7 MHz, rt)** δ = 141.2 (app. t, $J_{C\text{-}P}$ = 5 Hz, 1-C_{Ar}), 117.3 (s, 3-HC_{Ar}), 107.5 (s, 2-HC_{Ar}), 47.7 (app. td, app. t, $J_{C\text{-}P}$ = 19, 10 Hz, CH_2), 30.3 (overlapping m, C'H(CH$_3$)$_2$ and P(CH_3)$_3$), 28.9 (app. t, $J_{C\text{-}P}$ = 11 Hz, CH(CH$_3$)$_2$), 27.9 (br. d, $J_{C\text{-}P}$ = 19 Hz, P'(CH_3)$_3$), 21.2 (s, CH(CH_3)$_2$), 19.8 (s, CH(CH_3)$_2$), 19.7 (s, C'H(CH_3)($C'H_3$)), 18.8 (s, C'H(CH_3)($C'H_3$)). **Anal. Calcd. for C$_{26}$H$_{54}$BN$_2$P$_4$Ir (2c):** C, 43.27; H, 7.54; N, 3.88. Found: C, 42.79; H, 7.27; N, 4.02.

4.2.6. [(d(CH$_2$P(*i*Pr)$_2$)abB)Ir(PMe$_3$)] (3c)

In a Schlenk-flask, **5c** (50 mg, 68 µmol, 1 equiv.) was dissolved in THF (5 mL). Trimethylphosphine (34.7 µL, 26 mg, 0.342 mmol, 5 equiv.) was added, and the solution was stirred for 5 min at room temperature. The solvent was removed in vacuo. The colourless residue was dissolved in THF (5 mL), and a solution of KO*t*Bu (7.6 mg, 68 µmol, 1 equiv.) in THF (1 mL) was added. The resulting red–green solution was stirred for 2 h at room temperature. The solvent was removed in vacuo, and the residue was extracted with *n*-pentane (2 × 5 mL). The extract was filtered through a pad of celite and stored at -40 °C. After 24 h, dark red crystals with a greenish hue had separated. The supernatant solution was decanted, and the residue was washed with cold *n*-pentane (2 × 1 mL) and dried in vacuo (22 mg, 34 µmol, 50%).

^{1}H NMR (PhMe-d$_8$, 400.4 MHz, rt) δ 7.08–7.03 (m, 2 H, 3-HC_{Ar}), 6.99–6.93 (m, 2 H, 2-HC_{Ar}), 3.63 (app. t, $J_{\text{H-P}}$ = 2.1, 2.1 Hz, 4 H, CH_2), 2.04 (app. sept. t, $^3J_{\text{H-H}}$ = 7.0, 7.0 Hz, $J_{\text{H-P}}$ = 2.2, 2.2 Hz, 4 H, CH(CH$_3$)$_2$), 1.42 (d, $^2J_{\text{H-P}}$ = 5.7 Hz, 9 H, PMe$_3$), 1.05 (app. q, $^3J_{\text{H-H}}$ = 7.0 Hz, $J_{\text{H-P}}$ = 7.9, 7.9 Hz, 12 H, CH(CH_3)($C'H_3$)), 0.97 (app. q, $^3J_{\text{H-H}}$ = 7.0 Hz, $J_{\text{H-P}}$ = 6.1, 6.1 Hz, 12 H, CH(CH_3)($C'H_3$)). **^{1}H NMR (THF-d$_8$, 300.1 MHz, rt)** δ 6.73–6.66 (m, 2 H, HC_{Ar}), 6.66–6.58 (m, 2 H, HC_{Ar}), 3.74 (app. t, $J_{\text{H-P}}$ = 2.0, 2.0 Hz, 4 H, CH_2), 2.32

(app. sept. t, $^3J_{H\text{-}H}$ = 7.0, 7.0 Hz, $J_{H\text{-}P}$ = 2.2, 2.2 Hz, 4 H, CH(CH$_3$)$_2$), 1.57 (d, $^2J_{H\text{-}P}$ = 5.7 Hz, 9 H, PMe$_3$), 1.17 (app. q, $^3J_{H\text{-}H}$ = 7.0 Hz, $J_{H\text{-}P}$ = 7.9, 7.9 Hz, 12 H, CH(CH_3)(C'H$_3$)), 1.11 (app. q, $^3J_{H\text{-}H}$ = 7.0 Hz, $J_{H\text{-}P}$ = 6.1, 6.1 Hz, 12 H, CH(CH$_3$)(C'H_3)). ^{13}C{^{1}H} NMR (PhMe-d$_8$, **100.7 MHz, rt)** δ 140.1 (app. t, $J_{C\text{-}P}$ = 8 Hz, C$_{Ar}$), 117.7 (s, 3-HC$_{Ar}$), 108.6 (s, 2-HC$_{Ar}$), 44.4 (app. td, $J_{C\text{-}P}$ = 21, 12 Hz, NCH$_2$P), 28.6 (app. t, $J_{C\text{-}P}$ = 12 Hz, CH(CH$_3$)$_2$), 24.8 (app. dt, $J_{C\text{-}P}$ = 24, 2 Hz, PMe$_3$), 20.2 (app. dt, $J_{C\text{-}P}$ = 4 Hz, CH(CH_3)(C'H$_3$)), 19.3 (br. s, CH(CH$_3$)(C'H_3)). ^{11}B{^{1}H} NMR (PhMe-d$_8$, 128.5 MHz, rt) δ 57.5 (s, $\Delta w_{\frac{1}{2}}$ = 430 Hz). ^{11}B{^{1}H} NMR (THF-d$_8$, **96.3 MHz, rt)** δ 55.3 (s, $\Delta w_{\frac{1}{2}}$ = 380 Hz). ^{31}P{^{1}H} NMR (PhMe-d$_8$, 162.1 MHz, rt) δ 81.5 (d, $J_{P\text{-}P}$ = 5 Hz, CH$_2$P(iPr)$_2$), $-$18.6 (br s, P(CH$_3$)$_3$). ^{31}P{^{1}H} NMR (THF-d$_8$, 121.5 MHz, rt) δ 80.0 (d, $J_{P\text{-}P}$ = 5 Hz, CH$_2$P(iPr)$_2$), $-$20.1 (br s, P(CH$_3$)$_3$). **Anal. Calcd. for C$_{23}$H$_{45}$BIrN$_2$P$_3$ (3c):** C, 42.79; H, 7.03; N, 4.34. Found: C, 43.22; H, 7.25; N, 4.59. **m.p.:** 204–206 °C.

4.2.7. [(d(CH$_2$P(iPr)$_2$)abB)IrCl(Bpin)] (5c)

In a Schlenk-flask, **1** (100 mg, 0.198 mmol, 1 equiv.) and [Ir(cod)Cl]$_2$ (66.5 mg, 99 μmol, 1 equiv. Ir) were combined in *n*-pentane (50 mL). The yellow suspension was stirred at room temperature overnight before all volatiles were removed in vacuo. The bright yellow residue was recrystallised from *n*-pentane (20 mL) at $-40\,^\circ$C to give **5c** as bright yellow crystals (107 mg, 0.146 mmol, 74%).

^{1}**H NMR (THF-d$_8$, 400.4 MHz, rt)** δ 6.77–6.71 (m, 2 H, 2-HC$_{Ar}$), 6.69–6.64 (m, 2 H, 3-HC$_{Ar}$), 3.87 (app. dt, $^2J_{H\text{-}H}$ = 11.6 Hz, $J_{H\text{-}P}$ = 2.0, 2.0 Hz, 2 H, CHH'), 3.77 (app. dt, $^2J_{H\text{-}H}$ = 11.3 Hz, $J_{H\text{-}P}$ = 3.0, 3.0 Hz, 2 H, CHH'), 3.02 (m, $^3J_{H\text{-}H}$ = 7.2, 7.1 Hz, $J_{H\text{-}P}$ = 3.4, 2.2 Hz, 2 H, CH(CH$_3$)$_2$), 2.99 (m, $^3J_{H\text{-}H}$ = 7.6, 7.3 Hz, $J_{H\text{-}P}$ = 4.8, 5.8 Hz, 2 H, C'H(CH$_3$)$_2$), 1.47 (m, $^3J_{H\text{-}H}$ = 7.6 Hz, $J_{H\text{-}P}$ = 7.9, 9.0 Hz, 6 H, C'H(CH_3)(C'H$_3$)), 1.45 (app. q, $^3J_{H\text{-}H}$ = 7.2 Hz, $J_{H\text{-}P}$ = 7.4, 7.4 Hz, 6 H, CH(CH_3)(C'H$_3$)), 1.39 (app. q, $^3J_{H\text{-}H}$ = 7.3 Hz, $J_{H\text{-}P}$ = 6.5, 6.5 Hz, 6 H, C'H(CH$_3$)(C'H_3)), 1.14 (app. q, $^3J_{H\text{-}H}$ = 7.1 Hz, $J_{H\text{-}P}$ = 7.0, 7.0 Hz, 6 H, CH(CH$_3$)(C'H_3)), 0.81 (s, 12 H, OC(CH$_3$)$_2$). ^{13}**C{^{1}H} NMR (THF-d$_8$, 100.7 MHz, rt)** δ 140.8 (app. t, $J_{C\text{-}P}$ = 7 Hz, C$_{Ar}$), 118.3 (s, 3-HC$_{Ar}$), 108.2 (s, 2-HC$_{Ar}$), 83.2 (s, O$\underline{C}$(CH$_3$)$_2$), 45.7 (app. t, $J_{C\text{-}P}$ = 22 Hz, NCH$_2$P), 29.8 (app. t, $J_{C\text{-}P}$ = 12 Hz, CH(CH$_3$)$_2$), 28.9 (app. t, $J_{C\text{-}P}$ = 12 Hz, C'H(CH$_3$)$_2$), 24.8 (s, OC(CH$_3$)$_2$), 20.4 (app. t, $J_{C\text{-}P}$ = 3 Hz, C'H(CH_3)(C'H$_3$)), 19.7 (s, C'H(CH$_3$)(C'H_3)), 18.7 (s, CH(CH_3)(C'H$_3$)), 18.2 (s, C'H(CH$_3$)(C'H_3)). ^{11}**B{^{1}H} NMR (THF-d$_8$, 128.5 MHz, rt)** δ 39.7 (s, $\Delta w_{\frac{1}{2}}$ = 340 Hz), 19.9 (s, $\Delta w_{\frac{1}{2}}$ = 330 Hz). ^{31}**P{^{1}H} NMR (THF-d$_8$, 162.1 MHz, rt)** δ 66.4 (s). **Anal. Calcd. for C$_{26}$H$_{48}$B$_2$N$_2$O$_2$P$_2$IrCl (5c):** C, 42.67; H, 6.61; N, 3.83. Found: C, 42.48; H, 6.41; N, 4.06. **m.p.:** 232–234 °C.

4.2.8. [(d(CH$_2$P(iPr)$_2$)abB)IrCl(Bpin)(PMe$_3$)] (6c)

In a nitrogen-filled glovebox, **5c** (15 mg, 20 μmol, 1 equiv.) was dissolved in *n*-pentane (5 mL), and trimethylphosphine (10.4 μL, 7.8 mg, 0.102 mmol, 5 equiv.) was added. The solvent was removed after 5 min at room temperature to give **6c** as a colourless solid. Single crystalline **6c** was obtained from the above mixture upon crystallisation at $-40\,^\circ$C (6 mg, 7 μmol, 37%).

^{1}**H NMR (THF-d$_8$, 400.4 MHz, rt)** δ 6.74–6.69 (m, 2 H, 2-HC$_{Ar}$), 6.67–6.62 (m, 2 H, 3-HC$_{Ar}$), 3.88 (app. dt, $^2J_{H\text{-}H}$ = 11.0 Hz, $J_{H\text{-}P}$ = 2.2, 2.2 Hz, 2 H, CHH'), 3.64 (app. dt, $^2J_{H\text{-}H}$ = 11.0 Hz, $J_{H\text{-}P}$ = 2.2, 2.2 Hz, 2 H, CHH'), 3.06 (m, $^3J_{H\text{-}H}$ = 7.1, 7.1 Hz, $J_{H\text{-}P}$ = 3.5, 3.5 Hz, 2 H, CH(CH$_3$)$_2$), 2.61 (m, $^3J_{H\text{-}H}$ = 7.2, 7.2 Hz, $J_{H\text{-}P}$ = 3.7, 3.7 Hz, 2 H, C'H(CH$_3$)$_2$), 1.68 (d, $^2J_{H\text{-}P}$ = 7.2 Hz, 9 H, PMe$_3$), 1.42 (app. q, $^3J_{H\text{-}H}$ = 7.1 Hz, $J_{H\text{-}P}$ = 7.0, 7.0 Hz, 6 H, CH(CH_3)(C'H$_3$)), 1.41 (app. q, $^3J_{H\text{-}H}$ = 7.1 Hz, $J_{H\text{-}P}$ = 7.0, 7.0 Hz, 6 H, CH(CH$_3$)(C'H_3)), 1.38 (app. q, $^3J_{H\text{-}H}$ = 7.2 Hz, $J_{H\text{-}P}$ = 7.0, 7.0 Hz, 6 H, C'H(CH_3)(C'H$_3$)), 1.31 (app. q, $^3J_{H\text{-}H}$ = 7.2 Hz, $J_{H\text{-}P}$ = 7.1, 7.1 Hz, 6 H, C'H(CH$_3$)(C'H_3)), 0.66 (s, 12 H, OC(CH$_3$)$_2$). ^{13}**C{^{1}H} NMR (THF-d$_8$, 100.7 MHz, rt)** δ 142.4 (app. td, $J_{C\text{-}P}$ = 8, 2 Hz, C$_{Ar}$), 117.9 (s, 3-HC$_{Ar}$), 108.5 (s, 2-HC$_{Ar}$), 81.7 (s, O$\underline{C}$(CH$_3$)$_2$), 46.3 (app. td, $J_{C\text{-}P}$ = 20, 7 Hz, NCH$_2$P), 31.1 (app. t, $J_{C\text{-}P}$ = 14 Hz, CH(CH$_3$)$_2$), 27.6 (app. td, $J_{C\text{-}P}$ = 11, 2 Hz, C'H(CH$_3$)$_2$), 25.7 (s, OC(CH$_3$)$_2$), 20.5 (d, $^2J_{C\text{-}P}$ = 24 Hz, PMe$_3$), 21.1 (s, C'H(CH$_3$)(C'H_3)), 20.0 (s, C'H(CH_3)(C'H$_3$)), 19.6 (s, CH(CH_3)(C'H$_3$)), 19.0 (s, CH(CH$_3$)(C'H_3)). ^{11}**B{^{1}H} NMR (THF-d$_8$, 128.5 MHz, rt)** δ 48.8 (s, $\Delta w_{\frac{1}{2}}$ = 460 Hz), 26.6 (s,

$\Delta w_{\frac{1}{2}} = 450$ Hz). ^{31}P{^{1}H} **NMR (THF-d$_8$, 162.1 MHz, rt)** δ 41.4 (d, $J_{P-P} = 12$ Hz, CH$_2$P(iPr)$_2$), -51.9 (br. s, $\Delta w_{\frac{1}{2}} = 70$ Hz, PMe$_3$). **Anal. Calcd. for C$_{29}$H$_{57}$B$_2$N$_2$O$_2$P$_3$IrCl (6c):** C, 43.11; H, 7.11; N, 3.47. Found: C, 42.83; H, 7.03; N, 3.58. **m.p.:** 182–184 °C (decomp).

Supplementary Materials: The following supporting information can be downloaded at: https://www.mdpi.com/article/10.3390/molecules28176191/s1, additional spectroscopic and experimental details, crystallographic and computational details [44–49]. Crystallographic data (including structure factors) for the structures reported in this paper have been deposited with the Cambridge Crystallographic Data Centre. CCDC 2269774–2269781 contains the supplementary crystallographic data for this paper. These data can be obtained free of charge from The Cambridge Crystallographic Data Centre via www.ccdc.cam.ac.uk/structures (accessed on 28 June 2023).

Author Contributions: Conceptualization, C.K.; Funding acquisition, C.K.; Investigation, P.M.R.; Methodology, J.G. and C.K.; Writing—original draft, P.M.R., J.G. and C.K.; Writing—review & editing, P.M.R., J.G. and C.K. All authors have read and agreed to the published version of the manuscript.

Funding: This research was funded by the Deutsche Forschungsgemeinschaft (DFG) (KL 2243/5-1).

Institutional Review Board Statement: Not applicable.

Informed Consent Statement: Not applicable.

Data Availability Statement: The data presented in this study are available in the article or Supplementary Materials.

Acknowledgments: The authors thank AllyChem Co., Ltd. for the generous gift of diboron reagents.

Conflicts of Interest: The authors declare no conflict of interest.

Sample Availability: Not available.

References

1. Segawa, Y.; Yamashita, M.; Nozaki, K. Syntheses of PBP Pincer Iridium Complexes: A Supporting Boryl Ligand. *J. Am. Chem. Soc.* **2009**, *131*, 9201–9203. [CrossRef] [PubMed]
2. Segawa, Y.; Yamashita, M.; Nozaki, K. Diphenylphosphino- or Dicyclohexylphosphino-Tethered Boryl Pincer Ligands: Syntheses of PBP Iridium(III) Complexes and Their Conversion to Iridium-Ethylene Complexes. *Organometallics* **2009**, *28*, 6234–6242. [CrossRef]
3. Lin, T.-P.; Peters, J.C. Boryl-Mediated Reversible H$_2$ Activation at Cobalt: Catalytic Hydrogenation, Dehydrogenation, and Transfer Hydrogenation. *J. Am. Chem. Soc.* **2013**, *135*, 15310–15313. [CrossRef]
4. van der Vlugt, J.I. Boryl-Based Pincer Systems: New Avenues in Boron Chemistry. *Angew. Chem. Int. Ed.* **2010**, *49*, 252–255. [CrossRef]
5. Hasegawa, M.; Segawa, Y.; Yamashita, M.; Nozaki, K. Isolation of a PBP-Pincer Rhodium Complex Stabilized by an Intermolecular C–H σ Coordination as the Fourth Ligand. *Angew. Chem. Int. Ed.* **2012**, *51*, 6956–6960. [CrossRef] [PubMed]
6. Lin, T.-P.; Peters, J.C. Boryl−Metal Bonds Facilitate Cobalt/Nickel-Catalyzed Olefin Hydrogenation. *J. Am. Chem. Soc.* **2014**, *136*, 13672–13683. [CrossRef]
7. Tanoue, K.; Yamashita, M. Synthesis of Pincer Iridium Complexes Bearing a Boron Atom and ^{i}Pr-Substituted Phosphorus Atoms: Application to Catalytic Transfer Dehydrogenation of Alkanes. *Organometallics* **2015**, *34*, 4011–4017. [CrossRef]
8. Vondung, L.; Frank, N.; Fritz, M.; Alig, L.; Langer, R. Phosphine-Stabilized Borylenes and Boryl Anions as Ligands? Redox Reactivity in Boron-Based Pincer Complexes. *Angew. Chem. Int. Ed.* **2016**, *55*, 14450–14454. [CrossRef]
9. Lai, Q.; Bhuvanesh, N.; Ozerov, O.V. Unexpected B/Al Transelementation within a Rh Pincer Complex. *J. Am. Chem. Soc.* **2020**, *142*, 20920–20923. [CrossRef]
10. Shih, W.-C.; Gu, W.; MacInnis, M.C.; Timpa, S.D.; Bhuvanesh, N.; Zhou, J.; Ozerov, O.V. Facile Insertion of Rh and Ir into a Boron−Phenyl Bond, Leading to Boryl/Bis(phosphine) PBP Pincer Complexes. *J. Am. Chem. Soc.* **2016**, *138*, 2086–2089. [CrossRef]
11. Shih, W.-C.; Ozerov, O.V. Synthesis and Characterization of PBP Pincer Iridium Complexes and Their Application in Alkane Transfer Dehydrogenation. *Organometallics* **2017**, *36*, 228–233. [CrossRef]
12. Spokoyny, A.M.; Reuter, M.G.; Stern, C.L.; Ratner, M.A.; Seideman, T.; Mirkin, C.A. Carborane-Based Pincers: Synthesis and Structure of SeBSe and SBS Pd(II) Complexes. *J. Am. Chem. Soc.* **2009**, *131*, 9482–9483. [CrossRef] [PubMed]
13. El-Zaria, M.E.; Arii, H.; Nakamura, H. m-Carborane-Based Chiral NBN Pincer-Metal Complexes: Synthesis, Structure, and Application in Asymmetric Catalysis. *Inorg. Chem.* **2011**, *50*, 4149–4161. [CrossRef]
14. Eleazer, B.J.; Smith, M.D.; Popov, A.A.; Peryshkov, D.V. (BB)-Carboryne Complex of Ruthenium: Synthesis by Double B−H Activation at a Single Metal Center. *J. Am. Chem. Soc.* **2016**, *138*, 10531–10538. [CrossRef]

15. Rutz, P.M.; Grunenberg, J.; Kleeberg, C. Unsymmetrical Diborane(4) as a Precursor to PBP Boryl Pincer Complexes: Synthesis and Cu(I) and Pt(II) PBP Complexes with Unusual Structural Features. *Organometallics* **2022**, *41*, 3044–3054. [CrossRef]
16. Klein, H.-F.; Karsch, H.H. Methylkobaltverbindungen mit nicht chelatisierenden Liganden, I. Methyltetrakis(trimethylphosphin)kobalt und seine Derivate. *Chem. Berichte* **1975**, *108*, 944–955. [CrossRef]
17. Jones, R.A.; Real, F.M.; Wilkinson, G.; Galas, A.M.R.; Hursthouse, M.B.; Malik, K.M.A. Synthesis of trimethylphosphine complexes of rhodium and ruthenium. X-Ray crystal structures of tetrakis(trimethylphosphine)rhodium(I) chloride and chlorotris(trimethylphosphine)rhodium(I). *J. Chem. Soc. Dalton Trans.* **1980**, 511–518. [CrossRef]
18. Choudhury, J.; Podder, S.; Roy, S. Cooperative Friedel−Crafts Catalysis in Heterobimetallic Regime: Alkylation of Aromatics by π-Activated Alcohols. *J. Am. Chem. Soc.* **2005**, *127*, 6162–6163. [CrossRef]
19. Adams, C.J.; Baber, R.A.; Batsanov, A.S.; Bramham, G.; Charmant, J.P.H.; Haddow, M.F.; Howard, J.A.K.; Lam, W.H.; Lin, Z.; Marder, T.B.; et al. Synthesis and reactivity of cobalt boryl complexes. *Dalton Trans.* **2006**, *volume*, 1370–1373. [CrossRef]
20. Borner, C.; Brandhorst, K.; Kleeberg, C. Selective B–B Bond Activation in an Unsymmetrical Diborane(4) by [(Me$_3$P)$_4$Rh–X] (X = Me, OtBu): A Switch of Mechanism? *Dalton Trans.* **2015**, *44*, 8600–8604. [CrossRef]
21. Drescher, W.; Schmitt-Monreal, D.; Jacob, C.R.; Kleeberg, C. [(Me$_3$P)$_3$Co(Bcat)$_3$]: Equilibrium Oxidative Addition of a B–B Bond and Interconversion between the fac-Tris-Boryl and the mer-Tris-Boryl Complex. *Organometallics* **2020**, *39*, 538–543. [CrossRef]
22. Assefa, M.K.; Devera, J.L.; Brathwaite, A.D.; Mosley, J.D.; Duncan, M.A. Vibrational scaling factors for transition metal carbonyls. *Chem. Phys. Lett.* **2015**, *640*, 175–179. [CrossRef]
23. Grunenberg, J. Ill-defined concepts in chemistry: Rigid force constants vs. compliance constants as bond strength descriptors for the triple bond in diboryne. *Chem. Sci.* **2015**, *6*, 4086–4088. [CrossRef] [PubMed]
24. Del Castillo, T.J.; Thompson, N.B.; Suess, D.L.M.; Ung, G.; Peters, J.C. Evaluating Molecular Cobalt Complexes for the Conversion of N$_2$ to NH$_3$. *Inorg. Chem.* **2015**, *54*, 9256–9262. [CrossRef]
25. Schubert, H.; Leis, W.; Mayer, H.A.; Wesemann, L. A bidentate boryl ligand: Syntheses of platinum and iridium complexes. *Chem. Commun.* **2014**, *50*, 2738–2740. [CrossRef]
26. Clegg, W.; Lawlor, F.J.; Marder, T.B.; Nguyen, P.; Norman, N.C.; Orpen, A.G.; Quayle, M.J.; Rice, C.R.; Robins, E.G.; Scott, A.J.; et al. Boron–boron bond oxidative addition to rhodium(I) and iridium(I) centres. *J. Chem. Soc. Dalton Trans.* **1998**, 301–309. [CrossRef]
27. Press, L.P.; Kosanovich, A.J.; McCulloch, B.J.; Ozerov, O.V. High-Turnover Aromatic C–H Borylation Catalyzed by POCOP-Type Pincer Complexes of Iridium. *J. Am. Chem. Soc.* **2016**, *138*, 9487–9497. [CrossRef]
28. Lee, C.I.; DeMott, J.C.; Pell, C.J.; Christopher, A.; Zhou, J.; Bhuvanesh, N.; Ozerov, O.V. Ligand survey results in identification of PNP pincer complexes of iridium as long-lived and chemoselective catalysts for dehydrogenative borylation of terminal alkynes. *Chem. Sci.* **2015**, *6*, 6572–6582. [CrossRef]
29. Foley, B.J.; Bhuvanesh, N.; Zhou, J.; Ozerov, O.V. Combined Experimental and Computational Studies of the Mechanism of Dehydrogenative Borylation of Terminal Alkynes Catalyzed by PNP Complexes of Iridium. *ACS Catal.* **2020**, *10*, 9824–9836. [CrossRef]
30. Zhu, J.; Lin, Z.; Marder, T.B. Trans Influence of Boryl Ligands and Comparison with C, Si, and Sn Ligands. *Inorg. Chem.* **2005**, *44*, 9384–9390. [CrossRef]
31. Chotana, G.A.; Vanchura, B.A., II; Tse, M.K.; Staples, R.J.; Maleczka, R.E., Jr.; Smith, M.R., III. Getting the sterics just right: A five-coordinate iridium trisboryl complex that reacts with C–H bonds at room temperature. *Chem. Commun.* **2009**, 5731–5733. [CrossRef] [PubMed]
32. Putz, H.; Brandenburg, K. *Cole Research Group. 11B NMR Chemical Shifts*; SDSU Department of Chemistry & Biochemistr: San Diego, CA, USA, 2015.
33. Price, R.T.; Andersen, R.A.; Muetterties, E.L. Arene C-H bond activation: Reaction of (Me$_3$P)$_3$Rh(Me) with toluene to give (Me$_3$P)$_3$Rh(Ar) where Ar is o-, m- and p-tolyl. *J. Organomet. Chem.* **1989**, *376*, 407–417. [CrossRef]
34. Fulmer, G.R.; Miller, A.J.M.; Sherden, N.H.; Gottlieb, H.E.; Nudelman, A.; Stoltz, B.M.; Bercaw, J.E.; Goldberg, K.I. NMR Chemical Shifts of Trace Impurities: Common Laboratory Solvents, Organics, and Gases in Deuterated Solvents Relevant to the Organometallic Chemist. *Organometallics* **2010**, *29*, 2176–2179. [CrossRef]
35. Stalke, D. Cryo crystal structure determination and application to intermediates. *Chem. Soc. Rev.* **1998**, *27*, 171–178. [CrossRef]
36. Agilent Technologies. *CrysalisPro*, Version 1.171.40.84–1.171.41.122; Agilent Technologies: Santa Clara, CA, USA, 2020–2021.
37. Sheldrick, G.M. SHELXT—Integrated space-group and crystal-structure determination. *Acta Cryst.* **2015**, *A71*, 3–8. [CrossRef]
38. Sheldrick, G.M. Crystal structure refinement with SHELXL. *Acta Cryst.* **2015**, *C71*, 3–8. [CrossRef]
39. Sheldrick, G.M. A short history of SHELX. *Acta Cryst.* **2008**, *A64*, 112–122. [CrossRef]
40. Dolomanov, O.V.; Bourhis, L.J.; Gildea, R.J.; Howard, J.A.K.; Puschmann, H. OLEX2: A complete structure solution, refinement and analysis program. *J. Appl. Cryst.* **2009**, *42*, 339–341. [CrossRef]
41. Spek, A.L. Structure validation in chemical crystallography. *Acta Cryst.* **2009**, *D65*, 148–155. [CrossRef]
42. Macrae, C.F.; Sovago, I.; Cottrell, S.J.; Galek, P.T.A.; McCabe, P.; Pidcock, E.; Platings, M.; Shields, G.P.; Stevens, J.S.; Towler, M.; et al. Mercury 4.0: From visualization to analysis, design and prediction. *J. Appl. Cryst.* **2020**, *53*, 226–235. [CrossRef]
43. Putz, H.; Brandenburg, K. *Diamond—Crystal and Molecular Structure Visualization, Crystal Impact*; GbR: Bonn, Germany, 2018.
44. Frisch, M.J.; Trucks, G.W.; Schlegel, H.B.; Scuseria, G.E.; Robb, M.A.; Cheeseman, J.R.; Scalmani, G.; Barone, V.; Mennucci, B.; Petersson, G.A.; et al. *Gaussian 09, Revision D.01*; Gaussian, Inc.: Wallingford, CT, USA, 2013.

45. Tao, J.; Perdew, J.P.; Staroverov, V.N.; Scuseria, G.E. Climbing the Density Functional Ladder: Nonempirical Meta-Generalized Gradient Approximation Designed for Molecules and Solids. *Phys. Rev. Lett.* **2003**, *91*, 146401-1–146401-4. [CrossRef] [PubMed]
46. Staroverov, V.N.; Scuseria, G.E.; Tao, J.; Perdew, J.P. Comparative assessment of a new nonempirical density functional: Molecules and hydrogen-bonded complexes. *J. Chem. Phys.* **2003**, *119*, 12129–12137. [CrossRef]
47. Weigend, F.; Ahlrichs, R. Balanced basis sets of split valence, triple zeta valence and quadruple zeta valence quality for H to Rn: Design and assessment of accuracy. *Phys. Chem. Chem. Phys.* **2005**, *7*, 3297–3305. [CrossRef]
48. Brandhorst, K.; Grunenberg, J. Efficient computation of compliance matrices in redundant internal coordinates from Cartesian Hessians for nonstationary points. *J. Chem. Phys.* **2010**, *132*, 184101-1–184101-7. [CrossRef]
49. Brandhorst, K.; Grunenberg, J. How strong is it? The interpretation of force and compliance constants as bond strength descriptors. *Chem. Soc. Rev.* **2008**, *37*, 1558–1567. [CrossRef] [PubMed]

Article

Synthesis and Structural Characterization of Copper Complexes Containing "R-Substituted" Bis-7-Azaindolyl Borate Ligands

Miriam Jackson [1], Simon D. Thomas [1], Graham J. Tizzard [2], Simon J. Coles [2] and Gareth R. Owen [1,*]

[1] Chemical and Environmental Sciences, Faculty of Computing, Engineering and Science, University of South Wales, Pontypridd CF37 4AT, UK

[2] UK National Crystallography Service, School of Chemistry, University of Southampton, Highfield, Southampton SO17 1BJ, UK

* Correspondence: gareth.owen@southwales.ac.uk; Tel.: +44-1443-65-4527

Abstract: The coordination chemistry of scorpionate ligands based on borates containing the 7-azaindole heterocycle is relatively unexplored. Thus, there is a requirement to further understand their coordination chemistry. This article outlines the synthesis and characterization of a family of complexes containing anionic flexible scorpionate ligands of the type $[(R)(bis-7\text{-}azaindolyl)borohydride]^-$ ($[^{\mathbf{R}}\mathbf{Bai}]^-$), where R = Me, Ph or naphthyl. The three ligands were coordinated to a series of copper(I) complexes containing a phosphine co-ligand to form the complexes, $[Cu(^{\mathbf{Me}}\mathbf{Bai})(PPh_3)]$ (**1**), $[Cu(^{\mathbf{Ph}}\mathbf{Bai})(PPh_3)]$ (**2**), $[Cu(^{\mathbf{Naphth}}\mathbf{Bai})(PPh_3)]$ (**3**), $[Cu(^{\mathbf{Me}}\mathbf{Bai})(PCy_3)]$ (**4**), $[Cu(^{\mathbf{Ph}}\mathbf{Bai})(PCy_3)]$ (**5**) and $[Cu(^{\mathbf{Naphth}}\mathbf{Bai})(PCy_3)]$ (**6**). Additional copper(II) complexes, namely, $[Cu(^{\mathbf{Me}}\mathbf{Bai})_2]$ (**7**) and $[Cu(^{\mathbf{Ph}}\mathbf{Bai})_2]$ (**8**), were obtained during attempts to obtain single crystals from complexes **4** and **2**, respectively. Complexes **7** and **8** were also prepared independently from $CuCl_2$ and two equivalents of the corresponding $Li[^{\mathbf{R}}\mathbf{Bai}]$ salt alongside an additional complex, namely, $[Cu(^{\mathbf{Naphth}}\mathbf{Bai})_2]$ (**9**). The copper(I) and copper(II) complexes were characterized using spectroscopic and analytical methods. Furthermore, a crystal structure was obtained for eight of the nine complexes. In all cases, the boron-based ligand was found to bind to the metal centers via a $\kappa^3\text{-}N,N,H$ coordination mode.

Keywords: scorpionate; copper; borohydride; ligand; nitrogen; coordination

Citation: Jackson, M.; Thomas, S.D.; Tizzard, G.J.; Coles, S.J.; Owen, G.R. Synthesis and Structural Characterization of Copper Complexes Containing "R-Substituted" Bis-7-Azaindolyl Borate Ligands. *Molecules* **2023**, *28*, 4825. https://doi.org/10.3390/molecules28124825

Academic Editor: Renhua Qiu

Received: 18 May 2023
Revised: 9 June 2023
Accepted: 13 June 2023
Published: 17 June 2023

1. Introduction

Research involving "scorpionate ligands" has been in substantial development since the late 1960s [1–5]. The original Trofimenko-type scorpionate ligands were based around polypyrazolylborate ligands; however, a plethora of derivative ligands have been developed over the years. The polypyrazolylborates generally bind to metal centers via the available nitrogen donors and sometimes the B–H units, where a range of coordination modes, such as $\kappa^3\text{-}N,N,N$, $\kappa^3\text{-}N,N,H$ and $\kappa^2\text{-}N,N$, are common. Over the years, polypyrazolylborates gained a reputation as being robust ligands that form polycyclic chelates (dependent on the specific coordination mode). For example, the $\kappa^3\text{-}N,N,N$ coordination mode, which is found in the vast majority of systems, results in three six-membered rings within the resulting complex (Figure 1, right). Traditionally, they are considered inert spectator ligands that do not typically become involved within transformations at the metal center. This perspective changed following the emergence of a new set of ligand systems where the pyrazolyl heterocycle was exchanged with a heterocycle that contains an additional atom between the site of attachment to the central boron atom and the donor atom that binds to the metal center. This led to a considerable change in the landscape of the research field involving scorpionate ligands. In 2008, we suggested that this new set of ligands be termed "flexible scorpionates" [6,7] since the consequences of this extra atom in each of the heterocycles change their properties and potential reactivity quite significantly [8–10]. A range of heterocycles has now been utilized to generate flexible scorpionate ligands based on sulfur [9–16], nitrogen [17–21], phosphine [22,23], oxygen [24,25] and other donors [26,27]. To the best of our knowledge, there are only a small number of complexes reported containing nitrogen-based flexible scorpionate ligands. These are primarily based on the 7-azaindole heterocycle, as outlined herein. The tris(7-azaindolyl)borate (**Tai**) was the first to be reported by Wang in 2005 (Figure 1) [17]. This ligand was found to bind to metal centers with a range of coordination modes, including κ^3-*N,N,N* [17], κ^3-*N,N,H* [6,17,28–32] and κ^2-*N,H* [33]. With the **Tai** ligand, by far the most common coordination mode was found to be κ^3-*N,N,H* rather than κ^3-*N,N,N* (Figure 1, left and middle). This is the opposite scenario from the Trofimenko-type trispyrazolylborates, where κ^3-*N,N,N* is more common (Figure 1, right). The reason for this is down to the ring sizes formed upon chelation. For **Tai**, the κ^3-*N,N,H* coordination mode provides two six-membered rings and one eight-membered ring rather than the formation of three eight-membered rings in the κ^3-*N,N,N* mode.

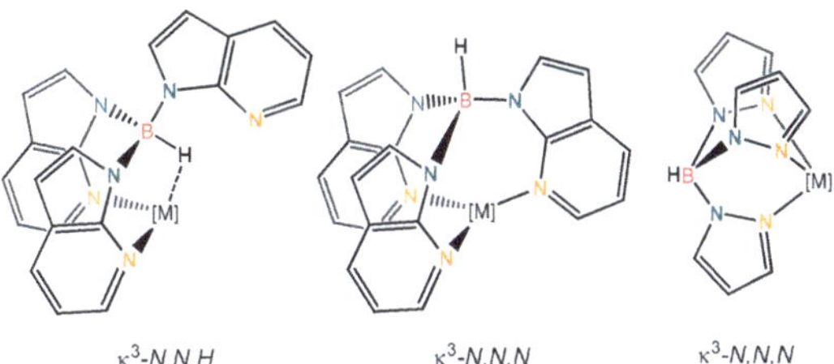

Figure 1. The κ^3-*N,N,H* coordination of tris(7-azaindolyl)borate to a metal center, highlighting two six-membered rings and one eight-membered ring formed upon chelation (**left**); the κ^3-*N,N,N* coordination of tris(7-azaindolyl)borate to a metal center, highlighting three eight-membered rings formed upon chelation (**middle**); and the κ^3-*N,N,N* coordination of trispyrazolylborate to a metal centre, highlighting the three six-membered rings formed upon chelation (**right**).

Building on the early investigations on the **Tai** ligand [17,33], our research group has carried out subsequent research highlighting some interesting transformations. For example, we were able to demonstrate reactivity involving the borohydride unit, revealing that the **Tai** ligand is able to form metallaboratrane-type complexes [34], where the hydride is removed from the ligand and the resulting borane ligand binds with either κ^4-*N,N,B,N* or κ^3-*N,N,B* coordination modes (Figure 2) [29,31,35,36]. We additionally prepared several complexes containing straightforward coordination of the **Tai** ligand [6,28–32]. Furthermore, we also prepared four derivative ligands in which one of the 7-azaindolyl substituents was replaced by an alkyl or aryl substituent and further investigated their coordination properties (Figure 3) [29,37,38]. The four ligands, namely, [Me]**Bai**, [Ph]**Bai**, [Naphth]**Bai** and [Mes]**Bai**, contain a central borohydride unit containing two 7-azaindolyl substituents and one additional R group (i.e., either methyl, phenyl, naphthyl or mesityl). In all cases, these ligands adopt the archetypical κ^3-*N,N,H* coordination mode (unless there is an activation of the B–H bond), where the R group points away from the transition metal center and the B–H---M interactions appear to be strong [28,32,38]. Despite this, even though the R group is positioned away from the metal centers upon coordination, it was found that the group influences the properties and subsequent reactivity of the resulting complexes quite significantly. For example, in a family of complexes of the type [RuH({[R]**Bai**}(PPh$_3$)$_2$] containing ligands with aromatic R groups, it was found that the naphthyl ring in [Naphth]**Bai** demonstrated hindered rotation about the B–C bond of the ligand [30]. Across several examples, we observed different product selectivities depending on the identity of the third "arm" of the ligand [29,36,37]. With limited knowledge about these ligands, there is a requirement to further explore the coordination properties of these derivative ligands. Therefore, we initiated an investigation to prepare a range of copper(I) phosphine and copper(II) complexes containing the [Me]**Bai**, [Ph]**Bai** and [Naphth]**Bai** ligands. This knowledge, particularly a deeper understanding of the nature of the interaction of the BH unit with the metal centers, will be invaluable in tuning the potential reactivity and exploring future applications. Herein, we outline the synthesis, characterization and structural characterization of a series of nine copper complexes containing these ligands.

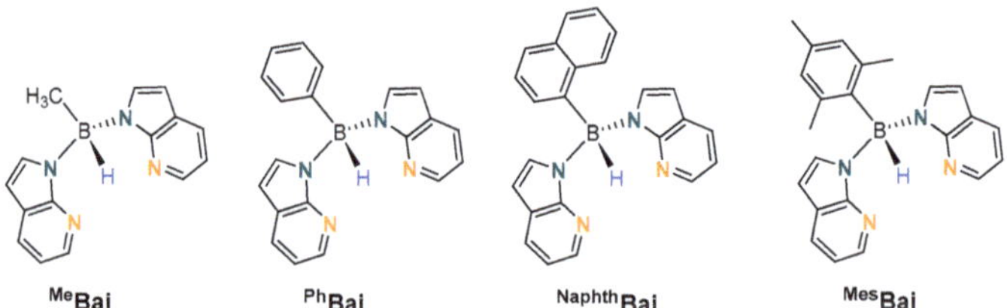

Figure 2. The κ^4-*N,N,B,N* and κ^3-*N,N,B* coordination modes formed by the removal of the hydride from the **Tai** ligand. In these examples, the hydride becomes incorporated into the organic co-ligand.

Figure 3. Derivative ligands, ^{Me}**Bai**, ^{Ph}**Bai**, ^{Naphth}**Bai** and ^{Mes}**Bai**.

2. Results and Discussion

2.1. Synthesis and Characterisation of Copper(I) Phosphine Complexes

The synthesis of copper(I) phosphine complexes containing scorpionate ligands is well known. They are readily synthesized via standard literature protocols [14,39]. Despite this, there is only one example of a copper complex containing a 7-azaindolylborate-based ligand known [17]. Therefore a set of copper complexes with the general formula [Cu(^{R}**Bai**)(PR'$_3$)] (where R = Me, Ph or Naphth and R' = Ph or Cy) were prepared. These were synthesized via a direct reaction of stoichiometric quantities of copper(I) chloride and the corresponding ligand precursor Li[^{R}**Bai**] in the presence of one equivalent of either triphenylphosphine or tricyclohexylphosphine (Scheme 1). The reactions were performed in methanol solvent from which the products precipitated out as white/off-white solids. These solids were washed further with acetonitrile and subsequently dried to provide the products, namely, complexes **1–6**, in yields ranging between 72% and 92% (see the Materials and Experimental Methods section for details; the ^{1}H, ^{13}C{^{1}H}, ^{31}P{^{1}H} and ^{11}B{^{1}H} NMR spectra and mass spectrometry data for these complexes are provided in the Supplementary Materials).

R/R' = Me/Ph (**1**); = Ph/Ph (**2**); = Naphth/Ph (**3**);
= Me/Cy (**4**); = Ph/Cy (**5**); = Naphth/Cy (**6**)

Scheme 1. Synthesis of [Cu{κ^3-*N,N,H*-^{R}**Bai**}(PR$_3$)] complexes **1–6**.

The solid products were found to be air and moisture stable. They demonstrated solubility in chlorinated solvents and partial solubility in aromatic solvents, acetonitrile and diethyl ether. It should be noted that they were found to decompose in tetrahydrofuran solutions. They were characterized by multinuclear NMR spectroscopy, IR spectroscopy and mass spectrometry. Selected characterization data for complexes **1–6** are presented in Table 1, along with data for the corresponding lithium ligand salt precursors for comparison. First, the $^{31}P\{^1H\}$ NMR spectra for complexes **1–6** in C_6D_6 revealed one single resonance with a moderate downfield shift with respect to their respective ligand precursors, namely, PPh_3 (-5.26 ppm) and PCy_3 (9.86 ppm). The phosphorus signals in the spectra for complexes **1**, **2** and **3** were located at 1.41 ppm, 1.17 ppm and 1.64 ppm, respectively, whilst the corresponding signals in **4**, **5** and **6** were found at 17.74 ppm, 18.08 ppm and 18.63 ppm, respectively. These values are in agreement with similar related scorpionate copper(I) phosphine complexes [14,17,39]. The ^{11}B NMR spectra for complexes **1–6** also revealed a single broad resonance (with half-height widths ranging between 180 Hz and 213 Hz), with chemical shifts in the region between -6.23 ppm and -7.65 ppm. The ^{11}B NMR signals would be expected to present as doublet signals due to the presence of the BH unit in the ligand (i.e., $^1J_{BH}$ coupling). The peak widths, however, were too large to directly observe any coupling. Nevertheless, tentative evidence for the coupling was obtained in the corresponding $^{11}B\{^1H\}$ NMR spectra, where the half-height widths of the signals were reduced to between 129 Hz and 121 Hz. It should be noted that these chemical shifts were only slightly shifted (ca. 1 ppm downfield) with respect to their corresponding ligand salt precursors. In particular, the ^{11}B NMR shifts for the $Cu(^{Naphth}Bai)(PR'_3)$ complexes were almost the same as the shifts for $Li[^{Naphth}Bai]$. These rather small downfield shifts upon coordination to copper centers are typical for copper complexes of this type, and thus, are expected [14,17,39]. Whilst this would seem to indicate that either there is no interaction or only a weak interaction between the BH units and the copper centers, the corresponding ^1H NMR and IR spectroscopic evidence and structural data outlined below are contrary to this (vida infra).

Table 1. Selected NMR (ppm; Hz) and IR (cm^{-1}) spectroscopic data for [^{R}**Bai**] ligand salt precursors and their corresponding copper complexes. The solvent used for the NMR studies was C_6D_6.

Ligand/Complex	$^{31}P\{^1H\}$ NMR	$^{11}B\{^1H\}$ NMR	h.h.w in $^{11}B/^{11}B\{^1H\}$	$^1H\{^{11}B\}$ NMR B-H	IR (B-H) Powder Film
Li[^{Me}Bai]	-	-8.14	-	4.66	2396, 2272
Li[^{Ph}Bai]	-	-6.80	-	5.47	2394, 2264
Li[^{Naphth}Bai]	-	-6.87	-	6.05	2429, 2272
[Cu(^{Me}Bai)(PPh$_3$)] (**1**)	1.41	-7.27	186/129	5.63	2143, 2089
[Cu(^{Ph}Bai)(PPh$_3$)] (**2**)	1.17	-6.24	184/121	6.58	2120, 2062
[Cu(^{Naphth}Bai)(PPh$_3$)] (**3**)	1.64	-6.23	180/133	6.97	2125, 2082
[Cu(^{Me}Bai)(PCy$_3$)] (**4**)	17.74	-7.65	181/123	5.49	2141, 2087
[Cu(^{Ph}Bai)(PCy$_3$)] (**5**)	18.08	-6.37	175/129	6.36	2169, 2076
[Cu(^{Naphth}Bai)(PCy$_3$)] (**6**)	18.63	-6.96	213/123	6.72	2189, 2090

For our initial analysis, we performed the NMR spectroscopic experiments in $CDCl_3$. However, the resulting spectra, particularly the ^1H and $^1H\{^{11}B\}$ NMR spectra, were poorly resolved. We found that the spectra were moderately improved and better resolved in C_6D_6. In all cases, the chemical shift patterns in both the ^1H and $^{13}C\{^1H\}$ spectra, and integrations within the ^1H NMR spectra, of the complexes were consistent with the presence of two 7-azaindolyl units, one phosphine ligand and one R group. This confirmed that all products were of the formula [Cu(^{R}**Bai**)(PR'$_3$)]. The $^1H\{^{11}B\}$ NMR spectra were utilized to obtain the chemical shift of hydrogen nuclei within the B–H---Cu unit. This is also included in Table 1. An examination of the aromatic region of the ^1H NMR spectra for complexes **1–6** demonstrated that the two azaindolyl rings within the ligand were chemically equivalent on the NMR timescale. For the bulkier naphthyl unit in the ^{Naphth}**Bai** complexes, hindered rotation might have been present, as was observed in a previous example [30]. This was found not to be the case in the copper complexes **3** and **6**. The *BH* resonances within these complexes are also of interest. These were located in the δ 5.49–6.97 region ppm for all complexes. These appeared as extremely broad signals in the standard ^{1}H NMR spectra. They did appear as broadened singlets in the corresponding ^{1}H{^{11}B} NMR spectra for complexes **2**, **3**, **5** and **6**. In the cases of **1** and **4**, quartet signals were expected. These signals within the spectra did show some apparent coupling; however, this was not resolved. In comparison with the ligand salt precursors, these signals were shifted downfield by approximately 1 ppm in each case upon coordination with the copper centers. Finally, the proton signals corresponding to the BCH$_3$ unit were observed within the ^{1}H NMR spectra for **1** and **4**. These appeared as more resolved doublets in the corresponding ^{1}H{^{11}B} NMR spectra at 1.15 ppm ($^3J_{HH}$ = 4.2 Hz) and 1.17 ppm ($^3J_{HH}$ = 4.4 Hz), respectively.

The B–H---Cu units in the six complexes were also explored using IR spectroscopy (see Table 1). In each case, powder film samples gave two characteristic bands between 2062 cm^{-1} and 2090 cm^{-1} for the first band and 2120 cm^{-1} and 2189 cm^{-1} for the second band. A large reduction in the stretching frequencies with respect to the lithium salt precursors confirmed a significant interaction of the B–H units with the copper centers, and thus, suggested the κ^3-*N,N,H* coordination mode for the ^{R}**Bai** ligands. The compounds were also analyzed using mass spectrometry. The molecular ion peaks were found in most cases with two exceptions, where expected fragmentation had occurred under the mass spectrometry conditions. Interestingly, complex **4** exhibited luminescence under UV light in the solid state at room temperature. This is the only complex within this investigation to exhibit this behaviour.

2.2. Structural Characterization of Copper(I) Phosphine Complexes

Single crystals of all six complexes were obtained, allowing for a detailed comparison across the three different scorpionate ligands. The first three structures analyzed were those containing the triphenylphosphine co-ligand. Single crystals of **1** were obtained via the slow evaporation of a methanolic solution, whilst single crystals of **2** and **3** were grown via the slow evaporation of acetonitrile from their saturated solutions. Complex **1** crystallized as colorless prism-shaped crystals, **2** as colorless rod-shaped crystals and **3** as colorless needle-shaped crystals. The structure for **3** contained a molecule of acetonitrile solvent of crystallization within the asymmetric unit. The molecular structures of these complexes are shown in Figure 4. Selected bond distances and angles for these complexes are shown in Table 2. Details on data collection and crystallographic parameters for all structures are provided in the Supplementary Materials.

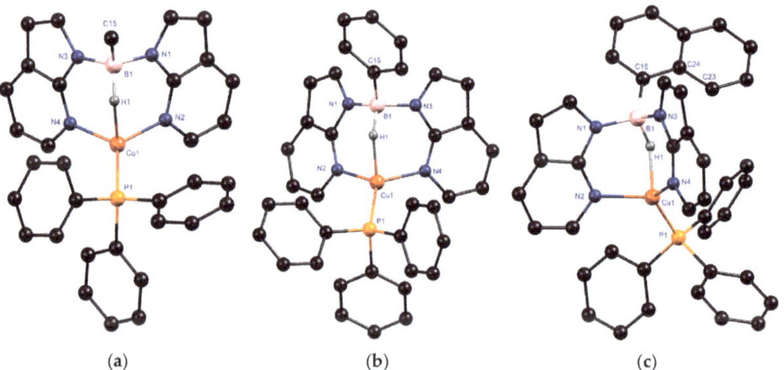

(a)   (b)   (c)

Figure 4. Molecular structures of [Cu{κ^3-*N,N,H*-R(H)B(7-azaindolyl)$_2$}(PPh$_3$)] (R = Me, **1** (**a**); Ph, **2** (**b**); naphthyl, **3** (**c**)). Ball and stick representations. Hydrogen atoms, with the exception of those attached to the boron centers, are omitted for clarity.

Table 2. Selected distances (Å) and angles (°) for **1**–**3** along with a comparison with [Cu(**Tai**)(PPh$_3$)].

Distance (Å)/Angle (°)	1	2	3 • (MeCN)	Cu(Tai)(PPh$_3$) [a]
Cu(1)–N(2)/Cu(1)–N(4)	2.0228(5)/2.0174(5)	2.0259(5)/2.0206(6)	2.0350(8)/2.0477(9)	2.008(3)/2.009(3)
B(1)–N(1)/B(1)–N(3)	1.5507(9)/1.5652(9)	1.5522(9)/1.5598(10)	1.5561(14)/1.5567(14)	1.558(5)/1.547(4)/*1.550(4)* [b]
B(1)–C(15)	1.6106(9)	1.6127(11)	1.6181(16)	-
Cu(1)–P(1)	2.18341(17)	2.1825(2)	2.1899(3)	2.1844(12)
Cu(1)⋯B(1)	2.7757(7)	2.7741(8)	2.7860(14)	2.808(3)
B(1)–H(1)	1.230(8)	1.268(9)	1.282(12)	1.20(3)
Cu(1)⋯H(1)	1.847(8)	1.813(9)	1.803(11)	1.81(3)
N(2)–Cu(1)–P(1)	125.192(16)	118.485(15)	118.79(3)	125.31(8)
N(4)–Cu(1)–P(1)	123.969(17)	121.888(19)	124.43(3)	126.73(9)
N(4)–Cu(1)–N(2)	107.73(2)	108.70(2)	107.27(3)	105.87(12)
H(1)–Cu(1)–P(1)	104.7(2)	119.5(3)	118.3(3)	107(1)
H(1)–Cu(1)–N(2)	90.7(2)	88.9(2)	88.1(3)	88(1)
H(1)–Cu(1)–N(4)	90.6(2)	92.0(3)	91.1(3)	86(1)
Σ of angles of non-hydrogen substituents at copper	356.89(3)	349.07(3)	350.49(5)	357.91(17)
N(1)–B(1)–N(3)	109.77(5)	109.42(6)	106.89(8)	111.7(3)
C(15)–B(1)–N(1)	110.77(5)	114.67(6)	114.94(9)	*113.7(2)* [b, c]
C(15)–B(1)–N(3)	112.35(5)	110.53(5)	111.63(9)	*111.8(3)* [b, c]
Σ of angles of non-hydrogen substituents at boron	332.89(9)	334.62(9)	333.46(15)	337.2(5)
B–H⋯Cu(1)	127.7(5)	127.6(6)	128.3(7)	136(2)
Cu(1)–N–N–B(1) [d]	−0.75(4)/−2.07(4)	−4.72(4)/−1.55(4)	7.07(6)/12.13(6)	−5.7(2)/−7.8(2)
Position of the Ar group at B [e]	-	64.48(8)	57.66(14)	178.6(2) [b, c]

[a] The corresponding values for this structure were obtained from the published article or by measurements obtained from the published cif file using Mercury software; [b] the values in italics are those that corresponded to the uncoordinated 7-azaindolyl "arm" in the **Tai** ligand; [c] the values corresponded to a nitrogen atom in the third 7-azaindolyl unit in **Tai** rather than the corresponding carbon atoms in ^{R}**Bai**; [d] the two torsion angles involving the relative positions of the copper and boron atoms about the two nitrogen atoms of the coordinated 7-azaindolyl rings; [e] defined by the modulus of the smallest torsion angle M--B–C$_{ipso}$–C$_{ortho}$ for ^{R}**Bai** or M--B–N--N for **Tai**.

The structures of **1**–**3** confirmed the coordination of one triphenylphosphine ligand and one [^{R}**Bai**]$^-$ ligand to the copper centers. In all three complexes, the scorpionate ligand was coordinated via the two pyridyl nitrogen donors on the azaindolyl rings and the borohydride unit, thus adopting a κ^3-*N,N,H* coordination motif. The Cu(1)⋯H(1) distances for the three complexes were found to be 1.847(8) Å (for **1**), 1.813(9) Å (for **2**) and 1.803(11) Å (for **3**). These are typical for complexes of this type (c.f. the sum of the covalent radii of Cu and H = 1.63 Å [40]). The Cu---H interaction decreased on changing from Me to Ph to Naphth. The corresponding B(1)–H(1) distances for **1**–**3** were 1.230(8) Å, 1.268(9) Å and 1.282(12) Å, respectively. The Cu(1)⋯H(1)–B(1) interaction impacted the overall geometry of the copper centers. The geometry at the metal centers was somewhat distorted between tetrahedral and trigonal pyramidal. The sums of the angles defining the trigonal plane, involving a copper atom, two nitrogen atoms and phosphine, were 356.89(3)° for **1**, 349.07(3)° for **2** and 350.49(9)° for **3**. The idealized bond angles for trigonal pyramidal are 120° and 90°, whilst for trigonal pyramidal, it is 109.5°. Ignoring those involving the hydrogen atom (which would sit on the axial site of the former geometry), complexes **1**, **2** and **3** had angles ranging between 107.73(2) and 125.192(16), 108.70(2) and 121.888(19) and 107.27(3) and 124.43(3), respectively. The N(2)–Cu–N(4) angles that arose from the chelation of the ^{R}**Bai** ligand provided the smallest of the aforementioned angles. The only other known copper complex containing a 7-azaindolyl-based scorpionate ligand is [Cu(**Tai**)(PPh$_3$)] [17]. This adopts the same coordination mode (i.e., κ^3-*N,N,H*) as found in the copper complexes reported herein despite the fact that the **Tai** ligand contains three 7-azaindolyl "arms". In [Cu(**Tai**)(PPh$_3$)], the pyridyl unit of the uncoordinated 7-azaindolyl points directly away in the opposite direction from the Cu$\cdots$H–B vector, where the Cu--B–N---N torsion angle is 178.6(2)$^\circ$. In the case of complexes **2** and **3**, the aryl groups were orientated at quite different positions with respect to the Cu$\cdots$H–B vector. In the case of **2**, the smallest M--B–C$_{ipso}$–C$_{ortho}$ torsion angle was 64.48(8)$^\circ$, whilst in **3**, the corresponding angle was 57.66(14)$^\circ$. These showed that the aryl groups in the ^{Ph}**Bai** and ^{Naphth}**Bai** complexes had twisted orientations relative to the Cu$\cdots$H–B unit, where the aryl group avoided steric clashes with the C–H bonds of the azaindolyl rings. Interestingly, the packing within the structure for **1** showed π-π stacking involving one of two azaindolyl "arms" of the ligand on one complex with the same on the adjacent complex. The interaction involved the pyridyl ring units, where the distance between the two centroids defined by the six atoms within each of the pyridyl rings was 3.62841(3) Å. There were no apparent π-π stacking interactions in the other two complexes containing the ^{Ph}**Bai** and ^{Naphth}**Bai** ligands. The reason behind this was most likely due to the smaller volume of space the methyl group occupied, allowing for the azaindolyl units to become closer to each other.

For the most part, the bonding features involving the "Cu{BH(7-azaindolyl)$_2$(PPh$_3$)]" core, which is common to all of the aforementioned complexes, were similar to each other. Notable exceptions were the N–B–N bond angles involving the two coordinated 7-azaindolyl units, which were 109.77(5)$^\circ$ (for **1**), 109.42(6)$^\circ$ (for **2**), 106.89(8)$^\circ$ (for **3**) and 111.7(3)$^\circ$ (for [Cu(**Tai**)(PPh$_3$)]). This suggested some influence of the third "arm" of the ligand on the coordination of the ligand within the complexes.

The crystal structures of three complexes containing the tricyclohexylphosphine co-ligand were analyzed next (Figure 5). Single crystals of [Cu(^{Me}**Bai**)(PCy$_3$)] (**4**) were obtained via concentration of a saturated methanol solution, whilst crystals of [Cu(^{Ph}**Bai**)(PCy$_3$)] (**5**) and [Cu(^{Naphth}**Bai**)(PCy$_3$)] (**6**) were grown via slow evaporation from saturated acetonitrile solutions. All three complexes crystallized as colorless block-shaped crystals. The calculated structure for **6** was found to contain one molecule of acetonitrile solvent within the asymmetric unit. Selected bond distances and angles for these complexes are shown in Table 3.

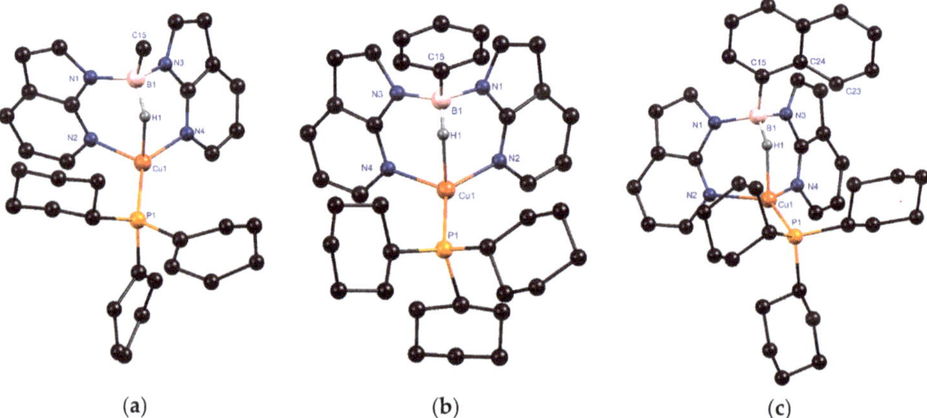

(a) (b) (c)

Figure 5. Molecular structures of [Cu{κ^3-*N,N,H*-R(H)B(7-azaindolyl)$_2$}(PCy$_3$)] (R = Me, **4** (**a**); Ph, **5** (**b**); naphthyl, **6** (**c**)). Ball and stick representations. Hydrogen atoms, with the exception of the one attached to each of the boron centers, are omitted for clarity.

Complexes **4**, **5** and **6** share many similarities with the corresponding triphenylphosphine complexes discussed above. Most bonding parameters were analogous to those of complexes **1–3**. In particular, the crystal structures also confirmed the κ^3–*N,N,H* coordination modes of the ^{R}**Bai** ligands within these complexes. As a general trend, the Cu(1)–P(1) distances in these three complexes were slightly longer than the corresponding triphenylphosphine complexes. This was due to the greater steric bulk of the tricyclohexylphosphine co-ligand. The positioning of the aryl rings with respect to the Cu$\cdots$H–B vector in **5** and **6** was larger than those found in **2** and **3**, where minor differences can be attributed to slightly increased N(1)–B(1)–N(3) angles within these PCy$_3$ complexes. The smallest M--B–C$_{ipso}$–C$_{ortho}$ torsion angles in **5** and **6** were 87.5(4)° and 70.00(6)°. As with the ^{Me}**Bai** complex **1**, there was also π-π stacking in complex **4**. This again involved one of two azaindolyl "arms" on one complex with the same on the adjacent complex. In this case, however, the interaction involved the indolyl units where the distance between the two centroids defined by the five atoms within each of the indolyl rings was 3.72074(12) Å. As with **2** and **3**, there were no apparent π-π stacking interactions in the structures for **5** and **6**.

Table 3. Selected distances (Å) and angles (°) for **4–6**.

Distance (Å)/Angle (°)	4	5 [a]	6 • (MeCN)
Cu(1)–N(2)/Cu(1)–N(4)	2.0740(16)/1.9955(16)	2.034(3)/2.032(3)	2.0384(4)/2.0515(4)
B(1)–N(1)/B(1)–N(3)	1.560(3)/1.565(3)	1.555(5)/1.552(5)	1.5646(7)/1.5545(7)
B(1)–C(15)	1.603(3)	1.620(6)	1.6161(7)
Cu(1)–P(1)	2.1802(5)	[2.205(3)]/[2.208(3)] [a]	2.19484(15)
Cu(1)$\cdots$B(1)	2.763(2)	2.776(4)	2.7995(7)
B(1)–H(1)	1.23(2)	1.05(6)	1.243(7)
Cu(1)$\cdots$H(1)B(1)	1.85(2)	1.92(6)	1.847(6)
N(2)–Cu(1)–P(1)	118.87(5)	[118.17(11)]/[129.61(11)] [a]	126.875(13)
N(4)–Cu(1)–P(1)	137.38(5)	[129.65(11)]/[118.19(11)] [a]	123.195(13)
N(4)–Cu(1)–N(2)	99.77(6)	109.95(12)	104.488(17)
H(1)–Cu(1)–P(1)	105.1(7)	[109.1(18)]/[109.1(18)] [a]	112.0(2)
H(1)–Cu(1)–N(2)	93.4(6)	86.9(17)	88.0(2)
H(1)–Cu(1)–N(4)	88.7(7)	86.7(17)	90.1(2)
Σ of angles of non-hydrogen substituents at copper	356.02(9)	[357.7(2)]/[357.7(2)] [a]	354.56(3)
N(1)–B(1)–N(3)	110.42(16)	110.6(3)	107.81(4)
C(15)–B(1)–N(1)	111.14(15)	112.0(3)	113.59(4)
C(15)–B(1)–N(3)	111.19(17)	112.1(3)	112.34(4)
Σ of angles of non-hydrogen substituents at boron	332.8(3)	334.7(5)	333.74(7)
B–H$\cdots$Cu(1)	127.0(16)	136(5)	128.8(5)
Cu(1)–N–N–B(1) [b]	12.24(11)/20.97(11)	1.7(2)/−1.7(2)	−7.41(3)/−8.30(3)
Position of the Ar group at B [c]	-	87.5(4)	70.00(6)

[a] The structure for complex **5** showed disorder with the PCy$_3$ ligand over two positions (ca. 50:50), and thus, there are two values for all measurements involving P(1); [b] the two torsion angles that involved the relative positions of the copper and boron atoms about the two nitrogen atoms of the coordinated 7-azaindolyl rings; [c] defined by the modulus of the smallest torsion angle M--B–C$_{ipso}$–C$_{ortho}$.

2.3. Synthesis and Crystallization of Copper(II) Complexes

During our investigations and attempts to obtain crystal structures for complexes **2** and **4** above, we also isolated single crystals, which were solved as [Cu(^{R}**Bai**)$_2$] (where R = Me (**7**) and Ph (**8**)). The analysis of their structures is outlined below. These two complexes, along with the additional complex [Cu(^{Naphth}**Bai**)$_2$] (**9**), were prepared independently via a reaction of copper(II) chloride with two equivalents of the corresponding Li[^{R}**Bai**] ligand precursor in methanol at room temperature (Scheme 2). The three complexes were obtained in high yields between 75 and 91%. The IR spectra for the three paramagnetic compounds each showed single bands at 2226 cm^{-1} (for **7**), 2269 cm^{-1} (for **8**) and 2221 cm^{-1} (for **9**). These corresponded to the coordinated B–H stretching frequency, confirming the κ^3-N,N,H coordination mode of the two ligands in each complex. In these cases, the reduction of the stretching frequencies within these bis-ligand complexes compared with the ligand precursors was less pronounced, suggesting that the Cu$\cdots$H–B interactions were weaker. This may be related to the increased coordination number of the copper centers in **7–9** in comparison with the tetracoordinated copper centers in complexes **1–6**. Mass spectrometry (ES$^+$) confirmed the molecular composition of the complexes with the following molecular ion peaks: 586 a.m.u. for [Cu(^{Me}**Bai**)$_2$]$^+$, 710 a.m.u. for [Cu(^{Ph}**Bai**)$_2$]$^+$ and 809 a.m.u. for [Cu(^{Naphth}**Bai**)$_2$]$^+$.

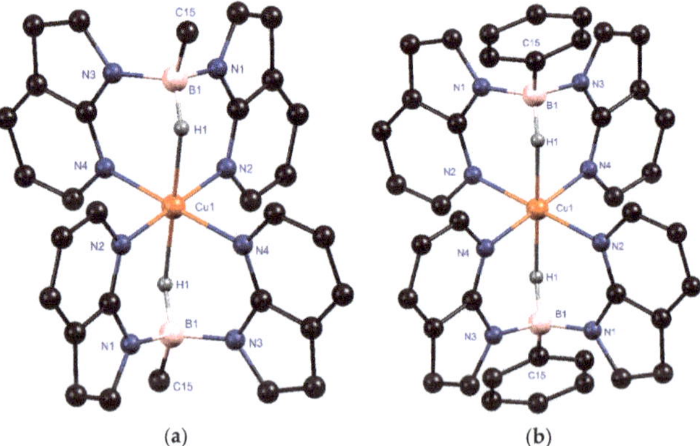

Scheme 2. Synthesis of [Cu{κ^3-N,N,H-^{R}**Bai**}$_2$] complexes **7–9**, either directly from CuCl$_2$ and two equivalents of pro-ligand or obtained via recrystallized attempts from **2** and **4**.

The formation of the mononuclear bis-ligand complexes [Cu(^{Me}**Bai**)$_2$] (**7**) and [Cu(^{Ph}**Bai**)$_2$] (**8**) was confirmed via X-ray crystallography (Figure 6). Selected bond distances and angles for these two complexes are shown in Table 4. Single crystals of **7** were obtained via concentration of a methanol solution containing complex **4**. Crystals of **8** were obtained by layering a concentrated dichloromethane solution of complex **2** with hexane. Complex **7** crystallized as brown prisms, containing one complex within the asymmetric unit, whilst complex **8** crystallized as light brown blocks, with two independent complexes and a disordered molecule of dichloromethane solvent within the unit cell. In the cases of both **7** and **8**, the asymmetric unit consisted of half of each complex (i.e., [Cu(^{R}**Bai**)]).

(a)   (b)

Figure 6. Molecular structures of [Cu{κ^3-N,N,H-R(H)B(7-azaindolyl)$_2$}$_2$] (R = Me, **7** (**a**); Ph, **8** (**b**)). Ball and stick representations. Hydrogen atoms, with the exception of those attached to the boron centers, are omitted for clarity.

Table 4. Selected distances (Å) and angles (°) for **7** and **8**.

Distance (Å)/Angle (°)	7	8 • (CH$_2$Cl$_2$) [a]
Cu(1)–N(2)/Cu(1)–N(4)	2.0161(19)/2.0316(16)	[2.0455(13)/2.0235(12)]/[2.0358(12)/2.0159(12)]
B(1)–N(1)/B(1)–N(3)	1.560(3)/1.553(3)	[1.547(2)/1.554(2)]/[1.5478(19)/1.554(2)]
B(1)–C(15)	1.606(3)	[1.599(2)]/[1.603(2)]
Cu(1)⋯B(1)	2.944(2)	[2.9384(16)]/[2.9281(15)]
B(1)–H(1)	1.18(2)	[1.16(2)]/[1.14(2)]
Cu(1)⋯H(1)B(1)	2.08(2)	[2.00(2)]/[2.03(2)]
N(2)–Cu(1)–N(4)	89.77(6)	[92.94(5)]/[91.10(5)]
N(2)′–Cu(1)–N(4)	90.23(6)	[87.06(5)]/[88.90(5)]]
H(1)–Cu(1)–N(2)	87.6(6)	[85.0(6)]/[85.1(6)]
H(1)–Cu(1)–N(4)	88.1(6)	[85.3(6)]/[86.3(6)]
N(1)–B(1)–N(3)	107.99(16)	[108.88(12)]/[108.58(11)]
C(15)–B(1)–N(1)	111.40(17)	[112.15(12)]/[114.06(11)]
C(15)–B(1)–N(3)	111.98(16)	[115.15(12)]/[112.51(11)]
$\sum$ of angles of non-hydrogen substituents at boron	331.4(3)	[336.2(2)]/[335.15(19)]
B–H⋯Cu(1)	126.5(16)	[135.2(15)]/[133.7(15)]
Cu(1)–N–N–B(1) [b]	0.31(11)/−16.58(12)	[5.73(9)/−7.65(9)]/[−9.25(8)/6.28(8)]
Position of the Ar group at B [c]	-	[77.96(15)]/[78.03(15)]

[a] There were two independent complexes of equal occupancy within this structure, and thus, two sets of values are provided in the table. The values for each complex are separated by square parenthesis. [b] The two torsion angles that involved the relative positions of the copper and boron atoms about the two nitrogen atoms of the coordinated 7-azaindolyl rings. [c] Defined by the modulus of the smallest torsion angle M--B–C$_{ipso}$–C$_{ortho}$.

The crystal structures confirmed the coordination of two ^{R}**Bai** ligands, both with κ^3-*N,N,H* coordination modes, as expected. This resulted in octahedral copper(II) centers, where the BH units were positioned *trans* to each other and all four pyridyl nitrogen donors of the two scorpionate ligands lay on a crystallographic plane. The N(2)–Cu–N(4) angles that arose from the chelation of the ^{R}**Bai** ligands were significantly smaller in these complexes compared with the copper(I) phosphine complexes outlined above. This was to accommodate the change in geometry at the copper centers. Interestingly, the intra-ligand N–Cu–N angle in **7** was smaller than the inter-ligand N–Cu–N angle, c.f. 89.77(6)° with 90.23(6)°. The opposite was found in the case of **8**, which had angles of 92.94(5)° and 91.10(5)° for the intra-ligand angles and 87.06(5)° and 88.90(5)° for the inter-ligand angles. There appeared to be great flexibility in the κ^3-*N,N,H* coordination and chelation of the ^{R}**Bai** ligands. The N–Cu–N angle may have had an impact on the positioning of the BH units within the complexes, reducing their interaction with the copper centers. This is of interest since the copper centers in **7** and **8** are in a higher oxidation state [c.f. Cu(II) vs. Cu(I)] and higher coordination number (c.f. six-coordinate vs. four-coordinate). The Cu(1)---B(1) distances in complexes **1** – **6** above ranged between 2.763(2) Å and 2.7995(7) Å. The boron centers were further away in the bis-ligand complexes. In these cases, the Cu(1)---B(1) distances ranged between 2.9281(15) Å and 2.944(2) Å. The corresponding Cu(1)⋯H(1) distances in **7** and **8** ranged between 2.00(2) Å and 2.08(2) Å. Finally, as with the two structures containing the ^{Me}**Bai** above, there was π-π stacking in complex **7**. This involved one pyridyl component within one complex with an indolyl component in an adjacent complex. The distance between the two respective centroids was 3.82028(11) Å. The π-π stacking appeared to be specifically related to the steric properties of this ligand in particular.

3. Materials and Experimental Methods

The syntheses of the complexes were carried out using standard Schlenk techniques. Solvents were sourced as extra dry from *"Acros Organics"* and were stored over either 4 Å or 3 Å molecular sieves. The NMR solvents, $CDCl_3$ and C_6D_6 were stored in a Young's ampule over 4 Å molecular sieves under a N_2 atmosphere and were degassed through freeze–thaw cycles prior to use. Reagents were used as purchased from commercial sources. The ligand salts Li[**^{R}Bai**] (where R = Me, Ph or Naphth) [29,37,38] were synthesized according to standard literature procedures. NMR spectroscopy experiments were conducted on a Bruker 400 MHz AscendTM 400 spectrometer. All spectra were referenced internally to the residual protic solvent (^{1}H) or the signals of the solvent (^{13}C). Proton (^{1}H) and carbon (^{13}C) assignments were further supported using HSQC, HMBC and COSY NMR experiments. In these cases, the apparent coupling constant is provided. Infrared spectra were recorded on a PerkinElmer Spectrum Two ATR FT-IR spectrometer as powder films. Elemental analysis was performed at London Metropolitan University using their elemental analysis service. Mass spectra were recorded at Cardiff University analytical services. Crystallographic details on data collection and parameters are outlined in the Supplementary Materials.

3.1. Synthesis of [Cu{κ³-N,N,H-^{Me}Bai}(PPh₃)] (1)

A clean dry Schlenk flask was charged with CuCl (50 mg, 0.51 mmol), ligand precursor [Li(^{Me}**Bai**)] (133 mg, 0.51 mmol) and PPh₃ (262 mg, 1 mmol). Methanol (20 mL) was subsequently added, and the reaction mixture was stirred for 2 h at room temperature. The solvent was then removed via filtration to isolate the precipitate [Cu(^{Me}**Bai**)(PPh₃)] as a white solid, which was washed with acetonitrile and dried (247 mg, 0.42 mmol, 83%). NMR δ ppm: ^{1}H (C_6D_6, 400 MHz), 7.94 (d, $^3J_{HH}$ = 5.02 Hz, 2H, ^{Aza}CH), 7.85 (d, $^3J_{HH}$ = 3.2 Hz, 2H, ^{Aza}CH), 7.53–7.59 (overlapping m, 8H, P(C_6H_5) + ^{Aza}CH), 6.92–7.00 (m, 9H, P(C_6H_5)), 6.51 (dd, $^3J_{HH}$ = 5.95 Hz, 2H, ^{Aza}CH), 6.43 (d, $^3J_{HH}$ = 3.24 Hz, 2H, ^{Aza}CH), 5.63 (d, $^3J_{HH}$ = 2.74 Hz, H, BH), 1.15 (broad doublet, 3H, BCH₃, apparent J_{HH} = 3.77 Hz, this signal became more resolved in the corresponding ^{1}H{^{11}B} experiment, $^3J_{HH}$ = 4.24 Hz), ^{13}C{^{1}H} (C_6D_6, 100 MHz), 151.3 (^{Aza}C), 140.8 (^{Aza}CH), 134.4 (d, $^1J_{CP}$ = 31.1 Hz, P^{ipso}(C_6H_5)), 134.3 (d, $^2J_{CP}$ = 15.4 Hz, P^{ortho}(C_6H_5)), 132.1 (^{Aza}CH), 130.2 (d, $^4J_{CP}$ = 1.4 Hz, P^{para}(C_6H_5)), 129.2 (d, $^3J_{CP}$ = 9.5 Hz, P^{meta}(C_6H_5)), 128.9 (^{Aza}CH), 124.8 (^{Aza}C), 114.0 (^{Aza}CH), 100.6 (^{Aza}CH), 1.5 (BCH₃). ^{31}P{^{1}H} NMR (δ, C_6D_6): 1.41. ^{11}B NMR (δ, C_6D_6): −7.27 (s, h.h.w. = 186 Hz). ^{11}B{^{1}H} NMR (δ, C_6D_6): −7.27 (s, h.h.w. = 129 Hz). IR (cm^{-1}, powder film): 2089, 2143. MS ES+ (*m/z*): 587 [Cu(^{Me}**Bai**)PPh₃]$^+$.

3.2. Synthesis of [Cu{κ³-N,N,H-^{Ph}Bai}(PPh₃)] (2)

A clean dry Schlenk flask was charged with CuCl (100 mg, 1.00 mmol), ligand precursor [**Li(^{Ph}Bai)**] (330 mg, 1.00 mmol) and PPh₃ (262 mg, 1.00 mmol). Methanol (20 mL) was subsequently added and the reaction mixture was stirred for 2 h at room temperature. The solvent was then removed via filtration and the precipitate [Cu(^{Ph}**Bai**)(PPh₃)] was washed with acetonitrile (2 × 10 mL) to give the product as a white solid (598 mg, 0.92 mmol, 92%). NMR δ ppm: ^{1}H (C_6D_6, 400 MHz), 7.93 (d, $^3J_{HH}$ = 4.81 Hz, 2H, ^{Aza}CH), 7.84 (3H, overlapping ^{Aza}CH + (C_6H_5)), 7.52 (m, 8H, P(C_6H_5) + (C_6H_5)), 7.38 (t, $^3J_{HH}$ = 7.52 Hz, 2H, ^{Aza}CH), 7.30 (t, $^3J_{HH}$ = 7.89 Hz, 2H, (C_6H_5)), 6.88–6.96 (m, 9H, P(C_6H_5)), 6.58 (unresolved, H, BH), 6.53 (dd, $^3J_{HH}$ = 2.5 Hz, $^3J_{HH}$ = 5.17 Hz, 2H, ^{Aza}CH), 6.42 (d, $^3J_{HH}$ = 3.27 Hz, 2H, ^{Aza}CH). ^{13}C {^{1}H} (C_6D_6, 100 MHz), 151.6 (^{Aza}C), 140.3 (^{Aza}CH), 134.0 (^{Aza}CH), 133.7 (d, $^2J_{CP}$ = 15.4 Hz, P^{ortho}(C_6H_5)), 133.14 (^{Aza}CH), 129.6 (d, $^4J_{CP}$ = 1.2 Hz, P^{para}(C_6H_5)), 128.6 (d, $^3J_{CP}$ = 10.0 Hz, P^{meta}(C_6H_5)), 127.7 (d (overlapped with C_6D_6), $^1J_{CP}$ = 23.9 Hz, P^{ipso}(C_6H_5)), 127.0 (C_6H_5), 125.3 (C_6H_5), 124.3 (^{Aza}C), 113.8 (^{Aza}CH), 100.4 (^{Aza}CH), 1.04 (BCH). ^{31}P{^{1}H} NMR (δ, C_6D_6): 1.17. ^{11}B NMR (δ, C_6D_6): −6.24 (s, h.h.w. = 184 Hz). ^{11}B{^{1}H} NMR (δ, C_6D_6): −6.24 (s, h.h.w. = 121 Hz). IR (cm^{-1}, powder film): 2120, 2062. MS ES$^+$ (*m/z*): 649 [Cu(^{Ph}**Bai**)PPh₃]$^+$.

3.3. Synthesis of Synthesis of [Cu{κ³-N,N,H-^{Naphth}**Bai**}(PPh₃)] (3)

A clean dry Schlenk flask was charged with CuCl (44 mg, 0.45 mmol), ligand precursor [Li(^{Naphth}**Bai**)] (200 mg, 0.45 mmol) and PPh₃ (112 mg, 0.45 mmol). Methanol (10 mL) was subsequently added, and the reaction mixture was stirred for 2 h at room temperature. The solvent was then removed via filtration. The precipitate was washed with MeCN (2 × 5 mL) and dried to give [Cu(^{Naphth}**Bai**)(PPh₃)] as an off-white solid (243 mg, 0.34 mmol, 79%). NMR δ ppm: ^{1}H (C₆D₆, 400 MHz), 8.20 (d, ^{3}J$_{HH}$ = 6.49 Hz, 1H, ^{Naphth}CH), 8.10 (d, ^{3}J$_{HH}$ = 8.34 Hz, 1H, ^{Naphth}CH), 7.97 (d, ^{3}J$_{HH}$ = 4.63 Hz, 2H, ^{Aza}CH), 7.81 (d, ^{3}J$_{HH}$ = 8.34 Hz, 2H, ^{Aza}CH), 7.67 (s, 2H, ^{Aza}CH), 7.51–7.58 (m, 8H, P(C₆H₅) + ^{Naphth}CH), 7.49 (t, ^{3}J$_{HH}$ = 7.32 Hz, 1H, ^{Naphth}CH), 7.22 (t, ^{3}J$_{HH}$ = 7.42 Hz, 1H, ^{Naphth}CH), 7.15 (s, 1H, ^{Naphth}CH), 6.97 (unresolved, H, BH), 6.81–6.97 (m, 9H, P(C₆H₅) + ^{Naphth}CH), 6.54 (dd, ^{3}J$_{HH}$ = 2.55 Hz, ^{3}J$_{HH}$ = 5.1 Hz, 2H, ^{Aza}CH), 6.35 (s, 2H, ^{Aza}CH). ^{13}C {^{1}H} (C₆D₆, 100 MHz), 151.6 (^{Aza}C), 143.1 (^{Naphth}C), 140.4 (^{Aza}CH), 137.7 (^{Naphth}C), 134.4 (^{Aza}C), 133.6 (d, ^{2}J$_{CP}$ = 15.4 Hz, P^{ortho}(C₆H₅)), 133.6 (d, ^{1}J$_{CP}$ = 27.5 Hz, P^{ipso}(C₆H₅)), 132.0 (^{Naphth}CH), 129.6 (^{Naphth}CH), 128.7 (d, ^{2}J$_{CP}$ = 10.6 Hz, P^{meta}(C₆H₅)), 128.2 (^{Naphth}CH), 128.6 (d, ^{2}J$_{CP}$ = 5.7 Hz, P^{para}(C₆H₅)), 127.0 (^{Naphth}CH), 125.3 (^{Aza}CH), 124.4 (^{Naphth}C), 124.4 (^{Naphth}C), 113.9 (^{Aza}CH), 100.5 (^{Aza}CH), 1.08 (BCH). ^{31}P{^{1}H} NMR (δ, C₆D₆): 1.64. ^{11}B NMR (δ, C₆D₆): −6.23 (s, h.h.w. = 180 Hz). ^{11}B{^{1}H} NMR (δ, C₆D₆): −6.23 (s, h.h.w. = 133 Hz). IR (cm^{-1}, powder film): 2125, 2082. MS ES$^+$ (*m/z*): 699 [Cu(^{Naphth}**Bai**)PPh₃]$^+$, 687 [Cu(Naphth)(7-azaindolyl)₂PPh₃)].

3.4. Synthesis of [Cu{κ³-N,N,H-^{Me}**Bai**}(PCy₃)] (4)

A clean dry Schlenk flask was charged with CuCl (100 mg, 1.00 mmol), ligand precursor [Li(^{Me}**Bai**)] (271 mg, 1.00 mmol) and PCy₃ (283 mg, 1.00 mmol). Methanol (20 mL) was subsequently added, and the reaction mixture was stirred for 2 h at room temperature. The solvent was then removed via filtration to give the precipitate [Cu(^{Me}**Bai**)(PCy₃)] as a white solid, which was washed with acetonitrile and dried (436 mg, 0.72 mmol, 72%). NMR δ ppm: ^{1}H (C₆D₆, 400 MHz), 8.27 (dd, ^{3}J$_{HH}$ = 3.61 Hz, ^{3}J$_{HH}$ = 1.40 Hz, 2H, ^{Aza}CH), 7.83 (dd, ^{3}J$_{HH}$ = 6.12 Hz, ^{3}J$_{HH}$ = 1.53 Hz, 2H, ^{Aza}CH), 7.59 (d, ^{3}J$_{HH}$ = 3.25 Hz, 2H, ^{Aza}CH), 6.69 (dd, ^{3}J$_{HH}$ = 2.60 Hz, ^{3}J$_{HH}$ = 5.04 Hz, 2H, ^{Aza}CH), 6.41 (d, ^{3}J$_{HH}$ = 3.16 Hz, 2H, ^{Aza}CH), 5.49 (unresolved, H, BH), 1.19–1.92 (m, 30H, P(C₆H₁₁)), 1.17 (broad doublet, 3H, BCH₃, apparent J$_{HH}$ = 4.31 Hz, this signal became more resolved in the corresponding ^{1}H{^{11}B} experiment, ^{3}J$_{HH}$ = 4.41 Hz), ^{13}C {^{1}H} (CDCl₃, 100 MHz), 150.3 (^{Aza}CH), 140.2 (^{Aza}CH), 131.1 (^{Aza}CH), 127.9 (^{Aza}CH), 123.7 (^{Aza}CH), 113.4 (^{Aza}CH), 99.0 (^{Aza}CH), 32.7 (d, ^{3}J$_{CP}$ = 15.37 Hz, P(C₆H₁₁)), 31.06 (d, ^{3}J$_{CP}$ = 2.3 Hz, P(C₆H₁₁)), 30.36 (d, ^{3}J$_{CP}$ = 3.98 Hz, P(C₆H₁₁)), 27.6 (d, ^{3}J$_{CP}$ = 10.70 Hz, P(C₆H₁₁)), 26.3 (BCH₃). ^{31}P{^{1}H} NMR (δ, C₆D₆): 17.74. ^{11}B NMR (δ, C₆D₆): −7.65 (s, h.h.w. = 181 Hz). ^{11}B{^{1}H} NMR (δ, C₆D₆): −7.65 (s, h.h.w. = 123 Hz). IR (cm^{-1}, powder film): 2141, 2087. MS ES$^+$ (*m/z*): the main peak here was 623.4, which corresponded to [Cu(PCy₃)₂]$^+$.

3.5. Synthesis of [Cu{κ³-N,N,H-^{Ph}**Bai**}(PCy₃)] (5)

A clean dry Schlenk flask was charged with CuCl (100 mg, 1.00 mmol), ligand precursor [Li(^{Ph}**Bai**)] (330 mg, 1.00 mmol) and PCy₃ (280 mg, 1.00 mmol). Methanol (20 mL) was subsequently added, and the reaction mixture was stirred for 2 h at room temperature. The solvent was then removed via filtration to give the precipitate [Cu(^{Ph}**Bai**)(PCy₃)] as a white solid (603 mg, 0.90 mmol, 90%). NMR δ ppm: ^{1}H (C₆D₆, 400 MHz), 8.28 (d, ^{3}J$_{HH}$ = 5.08 Hz, 2H, ^{Aza}CH), 7.90 (d, ^{3}J$_{HH}$ = 6.84 Hz, 2H, C₆H₅), 7.83 (d, ^{3}J$_{HH}$ = 3.35 Hz, 2H, ^{Aza}CH), 7.58 (d, ^{3}J$_{HH}$ = 7.71 Hz, 2H, C₆H₅), 7.45 (t, ^{3}J$_{HH}$ = 7.49 Hz, 2H, ^{Aza}CH), 7.31 (t, ^{3}J$_{HH}$ = 7.33 Hz, H, C₆H₅), 6.70 (dd, ^{3}J$_{HH}$ = 5.30 Hz, 2H, ^{Aza}CH), 6.40 (d, ^{3}J$_{HH}$ = 3.18 Hz, 2H, ^{Aza}CH), 6.36 (unresolved, H, BH), 1.28–1.86 (m, 18H, P(C₆H₁₁)), 0.91–1.12 (m, 12H, P(C₆H₁₁)), ^{13}C {^{1}H} (C₆D₆, 100 MHz), 150.7 (^{Aza}C), 139.1 (^{Aza}CH), 132.9 (C₆H₅), 132.3 (^{Aza}CH), 127.6 (^{Aza}CH), 126.9 (C₆H₅), 126.7 (C₆H₅), 126.5 (C₆H₅), 124.4 (C₆H₅), 123.6 (^{Aza}CH), 112.8 (^{Aza}CH), 99.5 (^{Aza}CH), 31.8 (d, ^{2}J$_{CP}$ = 15.1 Hz, P(C₆H₁₁)), 29.6 (s, P(C₆H₁₁)), 29.4 (d, ^{4}J$_{CP}$ = 4.1 Hz, P(C₆H₁₁)), 26.5 (d, ^{3}J$_{CP}$ = 10.7 Hz, P(C₆H₁₁)), 0.21 (BCH). ^{31}P{^{1}H}

NMR (δ, C_6D_6): 18.08. ^{11}B NMR (δ, C_6D_6): -6.37 (s, h.h.w. = 175 Hz). ^{11}B$\{^{1}$H$\}$ NMR (δ, C_6D_6): -6.37 (s, h.h.w. = 129 Hz). IR (cm^{-1}, powder film): 2169, 2076. Elemental analysis calc. for $C_{38}H_{49}BCuN_4P \bullet MeOH$ (%): C 66.90, H 7.77, N 8.00; found: C 66.30, H 7.16, N 8.59. MS ES$^+$ (m/z): 667 [Cu(^{Ph}**Bai**)PCy$_3$]$^+$, 710.2 [Cu(^{Ph}**Bai**)$_2$]$^+$.

3.6. Synthesis of [Cu{κ^3-N,N,H-^{Naphth}**Bai**}(PCy$_3$)] (6)

A clean dry Schlenk flask was charged with CuCl (44 mg, 0.45 mmol), ligand precursor [Li(^{Naphth}**Bai**)] (200 mg, 0.45 mmol) and PCy$_3$ (120 mg, 0.45 mmol). Methanol (10 mL) was subsequently added and the reaction mixture was stirred for 2 h at room temperature. The solvent was then removed via filtration. The precipitate was washed with MeCN (2 × 5 mL) to give [Cu(^{Naphth}**Bai**)(PCy$_3$)] as an off-white solid (243 mg, 0.34 mmol, 79%). NMR δ ppm: ^{1}H (C_6D_6, 400 MHz), 8.30 (d, $^3J_{HH}$ = 5.03 Hz, 2H, ^{Aza}CH), 8.20 (d, $^3J_{HH}$ = 6.82 Hz, 1H, Naphth-CH), 8.08 (d, $^3J_{HH}$ = 8.60 Hz, 1H, Naphth-CH), 7.82 (t, $^3J_{HH}$ = 8.04 Hz, 2H, Naphth-CH), 7.65 (d, $^3J_{HH}$ = 3.20 Hz, 2H, ^{Aza}CH), two overlapping peaks: 7.56 (d, $^3J_{HH}$ = 7.69 Hz, 2H, ^{Aza}CH), 7.55 (t, $^3J_{HH}$ = 7.79 Hz, 1H, Naphth-CH), 7.29 (t, $^3J_{HH}$ = 7.76 Hz, 1H, Naphth-CH), 6.72 (unresolved, 1H, BH), 6.70 (dd, $^3J_{HH}$ = 5.09 Hz, 2H, ^{Aza}CH), 6.34 (d, $^3J_{HH}$ = 3.31 Hz, 2H, ^{Aza}CH), 1.71–1.85 (m, 8H, P(C$_6$H$_{11}$)), 1.27–1.44 (m, 14H, P(C$_6$H$_{11}$)), 0.81–0.98 (m, 8H, P(C$_6$H$_{11}$)), remaining naphthalene proton peaks lie under the solvent peak at 7.26 ppm. ^{13}C$\{^{1}$H$\}$ (C_6D_6, 100 MHz), 150.9 (^{Aza}C), 140.1 (^{Aza}CH), 137.8 (^{Naphth}C), 134.5 (^{Naphth}C), 133.5 (^{Aza}CH), 131.8 (^{Naphth}CH), 129.1 (^{Naphth}CH), 128.6 (^{Naphth}CH), 128.5 (^{Aza}CH), 128.1 (^{Naphth}CH), 126.8 (^{Naphth}CH), 125.3 (^{Naphth}CH), 124.3 (^{Aza}C), 124.2 (^{Naphth}CH), 124.1 (^{Naphth}CH), 113.7 (^{Aza}CH), 100.5 (^{Aza}CH), 32.5 (d, $^3J_{CP}$ = 17.08 Hz, P(C$_6$H$_{11}$)), 30.3 (d, $^3J_{CP}$ = 3.99 Hz, P(C$_6$H$_{11}$)), 29.7 (s, P(C$_6$H$_{11}$)), 27.2 (d, $^3J_{CP}$ = 10.62 Hz, P(C$_6$H$_{11}$)), 1.1 (s, BCH$_3$). ^{31}P$\{^{1}$H$\}$ NMR (δ, C_6D_6): 18.63. ^{11}B NMR (δ, C_6D_6): -6.96 (s, h.h.w. = 213 Hz). ^{11}B$\{^{1}$H$\}$ NMR (δ, C_6D_6): -6.96 (s, h.h.w. = 123 Hz). IR (cm^{-1}, powder film): 2189, 2090. MS ES$^+$ (m/z): 717 [Cu(^{Naphth}**Bai**)PCy$_3$]$^+$.

3.7. Synthesis of [Cu{κ^3-N,N,H-^{Me}**Bai**}$_2$] (7)

Copper(II) chloride (19 mg, 0.14 mmol) and [Li(^{Me}**Bai**)] (100 mg, 0.29 mmol) were placed into a Schlenk flask, along with methanol (10 mL) under an inert atmosphere. The suspension was stirred for 2 h at room temperature. Subsequently, the supernatant was filtered off and the precipitate was washed with MeOH (10 mL) and MeCN (10 mL) to give a dark yellow solid (91%, 73 mg, 0.13 mmol). IR (cm^{-1}, powder film): 2226. MS ES$^+$ (m/z): 586.2 [Cu(^{Me}**Bai**)$_2$]$^+$.

3.8. Synthesis of [Cu{κ^3-N,N,H-^{Ph}**Bai**}$_2$] (8)

Copper(II) chloride (16 mg, 0.12 mmol) and [Li(^{Ph}**Bai**)] (100 mg, 0.24 mmol) were put into a Schlenk flask, and methanol (10 mL) was added under an inert atmosphere. The suspension was stirred for 2h at room temperature. Subsequently, the supernatant was filtered off and the precipitate was washed with MeOH (10 mL) and MeCN (10 mL) to give a dark yellow solid (80%, 68 mg, 0.1 mmol). IR (cm^{-1}, powder film): 2269. Elemental analysis calc. for $C_{40}H_{32}B_2CuN_8 \bullet \frac{1}{2}$ MeOH (%): C 67.01, H 4.72, N 15.44; found: C 66.80, H 4.37, N 15.13. MS ES$^+$ (m/z): 710.2 [Cu(^{Ph}**Bai**)$_2$]$^+$, 772.2 [Cu$_2$(^{Ph}**Bai**)$_2$ − H]$^+$.

3.9. Synthesis of [Cu{κ^3-N,N,H-^{Naphth}**Bai**}$_2$] (9)

Copper(II) chloride (15 mg, 0.11 mmol) and [Li(^{Naphth}**Bai**)] (100 mg, 0.22 mmol) were put into a Schlenk flask, and methanol (10 mL) was added under an inert atmosphere. The suspension was stirred for 2 h at room temperature. Subsequently, the supernatant was filtered off and the precipitate was washed with MeOH (10 mL) and MeCN (10 mL) to give a dark yellow solid (75 %, 67 mg, 0.08 mmol). IR (cm^{-1}, powder film): 2221. Elemental analysis calc. for $C_{48}H_{36}B_2CuN_8 \bullet MeOH$ (%): C 69.89, H 4.79, N 13.31; found: C 70.14, H 4.39, N 13.38. MS ES$^+$ (m/z): 809.2 [Cu(^{Naphth}**Bai**)$_2$]$^+$.

4. Conclusions

The synthesis and characterization of a series of copper(I) and copper(II) complexes containing the novel [^{R}**Bai**]$^-$ scorpionate ligands are reported herein. Eight of the nine new complexes were structurally characterized via X-ray crystallography. In all cases, a κ^3-*N,N,H* coordination mode, where both 7-azaindolyl "arms" were coordinated, along with the BH unit, was observed. The crystal structures confirmed significant B–H---Cu interactions. This was supported by solid-state infrared spectroscopy, where the structures were maintained, to some degree, in solution, as evidenced by multinuclear spectroscopy data. Furthermore, the structural characterization highlights a high degree of flexibility in terms of the size of the N–M–N angles and the approach of the B–H unit within the ^{R}**Bai** ligands upon coordination.

These new copper(I) and copper(II) complexes add to the family of complexes containing flexible 7-azaindole-based scorpionate ligands, which, despite being first reported in 2005, are still underdeveloped. This investigation consolidated the fact that the κ^3-*N,N,H* coordination mode is the preferred mode of binding. The crystal structures demonstrated flexibility in the N–M–N chelating angles. This had an impact on the positioning of the B–H---M interaction. The potential of the azaindolyl-based "flexible scorpionates" in terms of reactivity and applications has yet to be fully realized. The reactivity of the BH unit with the metal center is likely to be instrumental in defining this reactivity. These aspects are currently under investigation.

Supplementary Materials: The following supporting information can be downloaded at: https://www.mdpi.com/article/10.3390/molecules28124825/s1, which contains details on the crystallographic information parameters (Tables S1–S3), as well as crystallographic collection and refinement details and selected NMR spectra for complexes **1–6** (Figures S1.1–S6.4).

Author Contributions: M.J. and S.D.T. performed the experiments and drafted the results in the form of reports; G.J.T. and S.J.C. carried out the crystallography work; G.R.O. wrote the manuscript and directed the project. All authors have read and agreed to the published version of the manuscript.

Funding: The authors would like to thank the Welsh Government for funding this work. The project was partly funded by a Sêr Cymru II Capacity Builder Accelerator Award (80761-USW208) and partly funded by a KESS2 PhD studentship (M.J.). The Sêr Cymru II and KESS schemes were funded by the European Regional Development and European Social Funds. We are also grateful to Tata Steel UK who co-funded the PhD studentship.

Conflicts of Interest: The authors declare no conflict of interest.

Sample Availability: Not applicable.

References

1. Trofimenko, S. Boron-Pyrazole Chemistry. *J. Am. Chem. Soc.* **1966**, *88*, 1842–1844. [CrossRef]
2. Trofimenko, S. *Scorpionates: The Coordination of Poly(pyrazolyl)borate Ligands*; Imperial College Press: London, UK, 1999.
3. Pettinari, C. *Scorpionates II: Chelating Borate Ligands*; Imperial College Press: London, UK, 2008. [CrossRef]
4. Pettinari, C. Scorpionate Compounds. *Eur. J. Inorg. Chem.* **2016**, *2016*, 2209–2211. [CrossRef]
5. Muñoz-Molina, J.M.; Belderrain, T.R.; Pérez, P.J. Trispyrazolylborate coinage metals complexes: Structural features and catalytic transformations. *Coord. Chem. Rev.* **2019**, *390*, 171–189. [CrossRef]
6. Tsoureas, N.; Owen, G.R.; Hamilton, A.; Orpen, A.G. Flexible scorpionates for transfer hydrogenation: The first example of their catalytic application. *Dalton Trans.* **2008**, 6039–6044. [CrossRef]
7. Naktode, K.; Reddy, T.D.N.; Nayek, H.P.; Mallik, B.S.; Panda, T.K. Heavier group 2 metal complexes with a flexible scorpionate ligand based on 2-mercaptopyridine. *RSC Adv.* **2015**, *5*, 51413–51420. [CrossRef]
8. Owen, G.R. Hydrogen atom storage upon Z-class borane ligand functions: An alternative approach to ligand cooperation. *Chem. Soc. Rev.* **2012**, *41*, 3535–3546. [CrossRef]
9. Spicer, M.D.; Reglinski, J. Soft Scorpionate Ligands Based on Imidazole-2-thione Donors. *Eur. J. Inorg. Chem.* **2009**, *2009*, 1553–1574. [CrossRef]
10. Neshat, A.; Shahsavari, H.R.; Mastrorilli, P.; Todisco, S.; Haghighi, M.G.; Notash, B. A Borane Platinum Complex Undergoing Reversible Hydride Migration in Solution. *Inorg. Chem.* **2018**, *57*, 1398–1407. [CrossRef]
11. Nuss, G.; Ozwirk, A.; Harum, B.N.; Saischek, G.; Belaj, F.; Mösch-Zanetti, N.C. Copper Complexes with a Hybrid Scorpionate Ligand Containing Pyridazine-3-thione. *Eur. J. Inorg. Chem.* **2012**, *2012*, 4701–4707. [CrossRef]

12. Kimblin, C.; Bridgewater, B.M.; Hascall, T.; Parkin, G. The synthesis and structural characterization of bis(mercaptoimidazolyl) hydroborato complexes of lithium, thallium and zinc. *J. Chem. Soc. Dalton Trans.* **2000**, 891–897. [CrossRef]

13. Garner, M.; Reglinski, J.; Cassidy, I.; Spicer, M.D.; Kennedy, A.R. Hydrotris(methimazolyl)borate, a soft analogue of hydrotris(pyrazolyl)borate. Preparation and crystal structure of a novel zinc complex. *Chem. Commun.* **1996**, 1975–1976. [CrossRef]

14. Dodds, C.A.; Garner, M.; Reglinski, J.; Spicer, M.D. Coinage Metal Complexes of a Boron-Substituted Soft Scorpionate Ligand. *Inorg. Chem.* **2006**, *45*, 2733–2741. [CrossRef] [PubMed]

15. Nuss, G.; Saischek, G.; Harum, B.N.; Volpe, M.; Belaj, F.; Mösch-Zanetti, N.C. Pyridazine Based Scorpionate Ligand in a Copper Boratrane Compound. *Inorg. Chem.* **2011**, *50*, 12632–12640. [CrossRef] [PubMed]

16. Lam, T.L.; Tso, K.C.-H.; Cao, B.; Yang, C.; Chen, D.; Chang, X.-Y.; Huang, J.-S.; Che, C.-M. Tripodal S-Ligand Complexes of Copper(I) as Catalysts for Alkene Aziridination, Sulfide Sulfimidation, and C–H Amination. *Inorg. Chem.* **2017**, *56*, 4253–4257. [CrossRef] [PubMed]

17. Song, D.; Jia, W.L.; Wu, G.; Wang, S. Cu(i) and Zn(ii) complexes of 7-azaindole-containing scorpionates: Structures, luminescence and fluxionality. *Dalton Trans.* **2005**, 433–438. [CrossRef]

18. Saha, K.; Ramalakshmi, R.; Gomosta, S.; Pathak, K.; Dorcet, V.; Roisnel, T.; Halet, J.-F.; Ghosh, S. Design, Synthesis, and Chemistry of Bis(σ)borate and Agostic Complexes of Group 7 Metals. *Chem. A Eur. J.* **2017**, *23*, 9812–9820. [CrossRef]

19. Groshens, T.J. Synthesis, characterization, and coordination chemistry of the dihydrobis(5-aminotetrazol-1-yl)borate anion. *J. Coord. Chem.* **2010**, *63*, 1882–1892. [CrossRef]

20. Bailey, P.J.; Bell, N.L.; Gim, L.L.; Yucheng, T.; Funnell, N.; White, F.; Parsons, S. "Twisted" scorpionates: Synthesis of a tris(2-pyridonyl)borate (Thp) ligand; lessons in the requirements for successful B(L2D)3 type ligands. *Chem. Commun.* **2011**, *47*, 11659–11661. [CrossRef]

21. Schinabeck, A.; Rau, N.; Klein, M.; Sundermeyer, J.; Yersin, H. Deep blue emitting Cu(i) tripod complexes. Design of high quantum yield materials showing TADF-assisted phosphorescence. *Dalton Trans.* **2018**, *47*, 17067–17076. [CrossRef]

22. Grätz, M.; Bäcker, A.; Vondung, L.; Maser, L.; Reincke, A.; Langer, R. Donor ligands based on tricoordinate boron formed by B–H-activation of bis(phosphine)boronium salts. *Chem. Commun.* **2017**, *53*, 7230–7233. [CrossRef]

23. MacMillan, S.N.; Harman, W.H.; Peters, J.C. Facile Si–H bond activation and hydrosilylation catalysis mediated by a nickel–borane complex. *Chem. Sci.* **2014**, *5*, 590–597. [CrossRef]

24. Al-Harbi, A.; Rong, Y.; Parkin, G. Synthesis and Structural Characterization of Bis(2-oxoimidazolyl)hydroborato Complexes: A New Class of Bidentate Oxygen-Donor Ligands. *Inorg. Chem.* **2013**, *52*, 10226–10228. [CrossRef] [PubMed]

25. Al-Harbi, A.; Kriegel, B.; Gulati, S.; Hammond, M.J.; Parkin, G. Bis- and Tris(2-oxobenzimidazolyl)hydroborato Complexes of Sodium and Thallium: New Classes of Bidentate and Tridentate Oxygen Donor Ligands. *Inorg. Chem.* **2017**, *56*, 15271–15284. [CrossRef] [PubMed]

26. Landry, V.K.; Buccella, D.; Pang, K.; Parkin, G. Bis- and tris(2-seleno-1-methylimidazolyl)hydroborato complexes, {[Bse^{Me}]ZnX}$_2$(X = Cl, I), [Bse^{Me}]$_2$Zn and [Tse^{Me}]Re(CO)$_3$: Structural evidence that the [Bse^{Me}] ligand is not merely a "heavier" version of the sulfur counterpart, [Bm^{Me}]. *Dalton Trans.* **2007**, 866–870. [CrossRef]

27. Holler, S.; Tüchler, M.; Roschger, M.C.; Belaj, F.; Veiros, L.F.; Kirchner, K.; Mösch-Zanetti, N.C. Three-Fold-Symmetric Selenium-Donor Metallaboratranes of Cobalt and Nickel. *Inorg. Chem.* **2017**, *56*, 12670–12673. [CrossRef]

28. Tsoureas, N.; Nunn, J.; Bevis, T.; Haddow, M.F.; Hamilton, A.; Owen, G.R. Strong agostic-type interactions in ruthenium benzylidene complexes containing 7-azaindole based scorpionate ligands. *Dalton Trans.* **2011**, *40*, 951–958. [CrossRef]

29. Tsoureas, N.; Bevis, T.; Butts, C.P.; Hamilton, A.; Owen, G.R. Further Exploring the "Sting of the Scorpion": Hydride Migration and Subsequent Rearrangement of Norbornadiene to Nortricyclyl on Rhodium(I). *Organometallics* **2009**, *28*, 5222–5232. [CrossRef]

30. Tsoureas, N.; Hope, R.F.; Haddow, M.F.; Owen, G.R. Important Steric Effects Resulting from the Additional Substituent at Boron within Scorpionate Complexes Containing κ^3-*NNH* Coordination Modes. *Eur. J. Inorg. Chem.* **2011**, *2011*, 5233–5241. [CrossRef]

31. Tsoureas, N.; Kuo, Y.-Y.; Haddow, M.F.; Owen, G.R. Double addition of H$_2$ to transition metal–borane complexes: A 'hydride shuttle' process between boron and transition metal centres. *Chem. Commun.* **2011**, *47*, 484–486. [CrossRef]

32. Da Costa, R.C.; Rawe, B.W.; Iannetelli, A.; Tizzard, G.J.; Coles, S.J.; Guwy, A.J.; Owen, G.R. Stopping Hydrogen Migration in Its Tracks: The First Successful Synthesis of Group Ten Scorpionate Complexes Based on Azaindole Scaffolds. *Inorg. Chem.* **2019**, *58*, 359–367. [CrossRef]

33. Saito, T.; Kuwata, S.; Ikariya, T. Synthesis and Reactivity of Tris(7-azaindolyl)boratoruthenium Complex. Comparison with Poly(methimazolyl)borate Analogue. *Chem. Lett.* **2006**, *35*, 1224–1225. [CrossRef]

34. Hill, A.F.; Owen, G.R.; White, A.J.P.; Williams, D.J. The sting of the scorpion: A metallaboratrane. *Angew. Chem. Int. Ed.* **1999**, *38*, 2759–2761. [CrossRef]

35. Tsoureas, N.; Haddow, M.F.; Hamilton, A.; Owen, G.R. A new family of metallaboratrane complexes based on 7-azaindole: B–H activation mediated by carbon monoxide. *Chem. Commun.* **2009**, 2538–2540. [CrossRef]

36. Tsoureas, N.; Hamilton, A.; Haddow, M.F.; Harvey, J.N.; Orpen, A.G.; Owen, G.R. Insight into the Hydrogen Migration Processes Involved in the Formation of Metal–Borane Complexes: Importance of the Third Arm of the Scorpionate Ligand. *Organometallics* **2013**, *32*, 2840–2856. [CrossRef]

37. Da Costa, R.C.; Rawe, B.W.; Tsoureas, N.; Haddow, M.F.; Sparkes, H.A.; Tizzard, G.J.; Coles, S.J.; Owen, G.R. Preparation and reactivity of rhodium and iridium complexes containing a methylborohydride based unit supported by two 7-azaindolyl heterocycles. *Dalton Trans.* **2018**, *47*, 11047–11057. [CrossRef] [PubMed]

38. Owen, G.R.; Tsoureas, N.; Hope, R.F.; Kuo, Y.-Y.; Haddow, M.F. Synthesis and characterisation of group nine transition metal complexes containing new mesityl and naphthyl based azaindole scorpionate ligands. *Dalton Trans.* **2011**, *40*, 5906–5915. [CrossRef]
39. Lobbia, G.G.; Pettinari, C.; Santini, C.; Somers, N.; Skelton, B.; White, A.H. Synthesis, reactivity and solid-state structural studies of new phosphino copper(I) derivatives of hydrotris(3-methyl-2-thioxo-1-imidazolyl)borate. *Inorg. Chim. Acta* **2001**, *319*, 15–22. [CrossRef]
40. Cordero, B.; Gómez, V.; Platero-Prats, A.E.; Revés, M.; Echeverría, J.; Cremades, E.; Barragán, F.; Alvarez, S. Covalent radii revisited. *Dalton Trans.* **2008**, 2832–2838. [CrossRef]

Article

Synthesis and In Vitro Biological Evaluation of *p*-Carborane-Based Di-*tert*-butylphenol Analogs

Sebastian Braun [1], Sanja Jelača [2], Markus Laube [3], Sven George [4], Bettina Hofmann [4], Peter Lönnecke [1], Dieter Steinhilber [4], Jens Pietzsch [3,5], Sanja Mijatović [2], Danijela Maksimović-Ivanić [2] and Evamarie Hey-Hawkins [1,*]

[1] Institut für Anorganische Chemie, Universität Leipzig, Johannisallee 29, 04103 Leipzig, Germany; braun1993@gmx.net (S.B.); loenneck@rz.uni-leipzig.de (P.L.)

[2] Department of Immunology, Institute for Biological Research "Siniša Stanković", National Institute of Republic of Serbia, University of Belgrade, Bul. Despota Stefana 142, 11060 Belgrade, Serbia; sanja.jelaca@ibiss.bg.ac.rs (S.J.); sanjamama@ibiss.bg.ac.rs (S.M.); nelamax@ibiss.bg.ac.rs (D.M.-I.)

[3] Department of Radiopharmaceutical and Chemical Biology, Institute of Radiopharmaceutical Cancer Research, Helmholtz-Zentrum Dresden-Rossendorf, Bautzner Landstrasse 400, 01328 Dresden, Germany; m.laube@hzdr.de (M.L.); j.pietzsch@hzdr.de (J.P.)

[4] Institute of Pharmaceutical Chemistry, University of Frankfurt, Max-von-Laue-Straße 9, 60438 Frankfurt, Germany; s.george@em.uni-frankfurt.de (S.G.); hofmann@pharmchem.uni-frankfurt.de (B.H.); steinhilber@em.uni-frankfurt.de (D.S.)

[5] Faculty of Chemistry and Food Chemistry, Technische Universität Dresden, School of Science, Mommsenstrasse 4, 01062 Dresden, Germany

* Correspondence: hey@uni-leipzig.de

Citation: Braun, S.; Jelača, S.; Laube, M.; George, S.; Hofmann, B.; Lönnecke, P.; Steinhilber, D.; Pietzsch, J.; Mijatović, S.; Maksimović-Ivanić, D.; et al. Synthesis and In Vitro Biological Evaluation of *p*-Carborane-Based Di-*tert*-butylphenol Analogs. *Molecules* **2023**, *28*, 4547. https://doi.org/10.3390/molecules28114547

Academic Editors: Michael A. Beckett and Igor B. Sivaev

Received: 16 May 2023
Revised: 29 May 2023
Accepted: 31 May 2023
Published: 4 June 2023

Abstract: Targeting inflammatory mediators and related signaling pathways may offer a rational strategy for the treatment of cancer. The incorporation of metabolically stable, sterically demanding, and hydrophobic carboranes in dual cycloxygenase-2 (COX-2)/5-lipoxygenase (5-LO) inhibitors that are key enzymes in the biosynthesis of eicosanoids is a promising approach. The di-*tert*-butylphenol derivatives **R-830**, **S-2474**, **KME-4**, and **E-5110** represent potent dual COX-2/5-LO inhibitors. The incorporation of *p*-carborane and further substitution of the *p*-position resulted in four carborane-based di-*tert*-butylphenol analogs that showed no or weak COX inhibition but high 5-LO inhibitory activities in vitro. Cell viability studies on five human cancer cell lines revealed that the *p*-carborane analogs **R-830-Cb**, **S-2474-Cb**, **KME-4-Cb**, and **E-5110-Cb** exhibited lower anticancer activity compared to the related di-*tert*-butylphenols. Interestingly, **R-830-Cb** did not affect the viability of primary cells and suppressed HCT116 cell proliferation more potently than its carbon-based **R-830** counterpart. Considering all the advantages of boron cluster incorporation for enhancement of drug biostability, selectivity, and availability of drugs, **R-830-Cb** can be tested in further mechanistic and in vivo studies.

Keywords: carboranes; inflammation; cyclooxygenases; lipoxygenases; cancer

1. Introduction

Besides genetic mutations, epigenetic changes, diet, and lifestyle, chronic inflammatory processes are a significant factor in carcinogenesis [1,2]. The process of cancer formation is a stepwise progress which can be split into three phases, namely initiation, promotion, and progression. Cancer progression relies on the control of tumor microenvironment which is essential to provoke further tumor invasiveness and metastasis [3–5]. Hence, there is need for a balance between immunosuppressive and immunostimulatory mediators in the tumor- and inflammatory microenvironment [2,6–8]. Inflammatory mediators, such as cytokines, chemokines, and eicosanoids, can stimulate tumor cell proliferation and cancer progression by disrupting homeostatic mechanisms [3,9]. Eicosanoids, in particular prostaglandins (PGs) and leukotrienes (LTs), are hormone-like lipids that play a pivotal part in mediating the crosstalk between cells in the tumor microenvironment [2]. Targeting these inflammatory mediators by interfering with related signaling pathways may offer a rational strategy for the treatment of cancer. Pro-inflammatory PGs and LTs are mainly formed from arachidonic acid (AA) via cyclooxygenase-2 (COX-2) and 5-lipoxygenase (5-LO) signaling pathways [10–12]. Cyclooxygenases (COX) catalyze the rate-determining step in the biosynthesis of PGs [13–15]. The constitutively expressed COX-1 is involved in physiological "housekeeping" functions, such as the homeostasis of the gastrointestinal (GI) tract [16,17]. COX-2 is highly inducible by pro-inflammatory stimuli through macrophage activation or tumor promoters [12,18]. It can be found in macrophages, leukocytes, and fibroblasts, and COX-2 dependent production of pro-inflammatory PGs is upregulated during inflammatory diseases, such as rheumatoid arthritis, cardiovascular diseases, and asthma [10,12,15,19]. Furthermore, COX-2, in particular the downstream product prostaglandin E_2 (PGE$_2$), is able to promote carcinogenesis and progression in different cancer types, such as skin, breast, colon, and lung cancer, that are associated with poor prognosis [2,15,20–22]. As COX-1 and COX-2 share nearly homologous amino acid sequence identity, selective COX-2 inhibition remains challenging [17,19]. COX-2 is physiologically repressed by the active form of vitamin D and glucocorticoids [23,24]. Artificial COX inhibitors are categorized into two groups, namely non-steroidal anti-inflammatory drugs (NSAIDs) and COXIBs. As a result, of two crucial changes in the amino acid sequence, the COX-2 active site is ca. 17% larger and has a slightly different form than that of COX-1 [25]. Thus, size enlargement of inhibitors has been shown to enable COX-2 selective inhibition [19,20]. While non-selective COX inhibition by NSAIDs leads to the reduction of gastroprotective PGs, COX-2 selective inhibition suffers from adverse cardiovascular effects caused by dysbalanced prostaglandin I_2 (PGI$_2$) and thromboxane A_2 (TXA$_2$) levels [2,25–28].

5-LO is a non-heme iron-containing oxidoreductase that belongs to a heterogeneous family of lipid peroxidizing enzymes and catalyzes the conversion of AA into LTs with the help of 5-lipoxygenase-activating protein (FLAP) [10,29–31]. These important lipid mediators are involved in host defense, inflammatory processes, and cellular signaling [2,32]. 5-LO is highly inducible and can be stimulated by growth factors and pro-inflammatory stimuli [20,28,33]. Indeed, 5-LO upregulation and LTs overproduction are related to hypersensitivity reactions, inflammatory diseases, and allergic disorders, such as arthritis or psoriasis. It has also been observed in different types of epithelial cancers, such as lung, colon, and prostate cancer [9,12,33–36]. 5-LO can be downregulated by four different types of inhibitors, namely redox (non-competitive), iron-chelating, non-redox (competitive), and allosteric inhibitors [37–39]. 5-LO inhibitors, in general, suffer from having a short half-life and exhibiting hepatic toxicity [10,32].

PGs and LTs have complementary effects on the pathogenesis of cancer and tumor progression, and thus dual COX-2/5-LO inhibition may represent a promising strategy to treat cancer and to prevent the pathogenesis of cancer [33,34,40]. Dual COX-2/5-LO inhibitors block the release of both PGs and LTs from the corresponding COX-2 and 5-LO signaling pathways [10,17]. Dual COX-2/5-LO inhibition may prevent the upregulation of the respective opposed signaling pathway by blocking both targets. Consequently, balanced COX-2/5-LO inhibition may lower the risk for the appearance of severe adverse effects, such as GI injury and hypersensitive reactions [21,38,41–44].

Di-*tert*-butylphenols represent the predominant class of dual COX-2/5-LO inhibitors and share a conserved 2,6-di-*tert*-butylphenol motif substituted at the *para (p)*-position (Figure 1). Di-*tert*-butylphenols show a superior therapeutic index, i.e., the ratio of anti-inflammatory effect to GI safety profile, compared to that of classical NSAIDs [10,33,34]. Several di-*tert*-butylphenols have been synthesized and pharmacologically evaluated. **R-830** (3,5-di-*tert*-butyl-4-(2'-thienoyl)-phenol) produces anti-edemic, analgesic, antipyretic, and anti-erythemic effects in several animal models [45]. It further reduces PGE$_2$ production by stimulated rat synovial cells in the nanomolar range and significantly inhibited leukotriene B_4 (LTB$_4$) production by human neutrophils at micromolar concentrations [45,46]. The dual COX-2/5-LO inhibitor **KME-4** (α-3,5-di-*tert*-butyl-4-hydroxybenzylidene)-γ-butyrolactone) inhibits both guinea pig peritoneal polymorphonuclear leukocytes (PMNL) and rabbit platelet COX at low micromolar concentrations [47,48]. Moreover, in vivo experiments revealed that **KME-4** possess anti-arthritic activity in a dose-dependent manner, while showing lower ulcerogenic potential compared to other NSAIDs, such as indomethacin or naproxen [49,50].

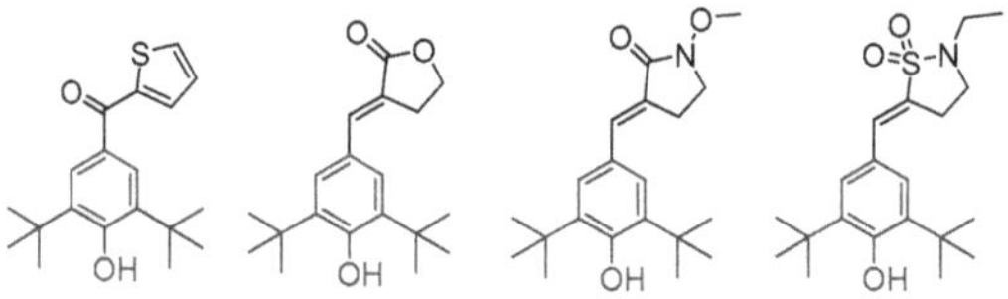

Figure 1. Chemical structures of selected di-*tert*-butylphenols showing the conserved 2,6-di-*tert*-butylphenol motif (blue).

The pyrrolidone derivative **E-5110** (*N*-methoxy-3-(3,5-di-*tert*-butyl-4-hydroxybenzylidene)-2-pyrrolidone) inhibits both PGE_2 production by cultured rat synovial cells at nanomolar concentrations and LTB_4 generation by human leukocytes stimulated in the micromolar range [51–54]. Further in vivo studies revealed that **E-5110** exhibits analgesic activity similar to indomethacin with lower ulcerogenic effect on rat gastric mucosa than indomethacin and piroxicam [54,55]. Both in vitro and in vivo toxicokinetic studies on the metabolism of **E-5110** demonstrated induction of cytochrome P_{450} 3A (CYP3A) activity, which is therefore expected to mainly contribute to its metabolism as well as its bioavailability [56]. **S-2474** ((*E*)-5-(3,5-di-*tert*-butyl-4-hydroxybenzylidene)-2-ethylisothiazo-lidine-1,1-dioxide) is a selective inhibitor of COX-2 and 5-lipoxygenase in vitro with half-maximal inhibitory concentration (IC_{50}) values in the low micromolar and nanomolar range, respectively [57]. Furthermore, it shows in vivo anti-inflammatory activity in rats with low ulcerogenic activity due to its COX-2 selectivity [58].

A promising approach to increase metabolic stability of carbon-based drugs is the incorporation of hydrophobic boron clusters, such as the 12-vertex dicarba-*closo*-dodecaborane (carborane), as bioisosters of substituted or unsubstituted phenyl rings [59–61]. Carboranes are icosahedral boron clusters in which at least one BH^- vertex in $(B_{12}H_{12})^{2-}$ is replaced by an isolobal CH group [61–63]. As their distorted icosahedral shape reflects the shape of a rotating phenyl ring, they can be considered benzene analogs [63–66]. In analogy to the nomenclature of the substitution of a benzene ring, three different regioisomers, namely 1,2-(*ortho*), 1,7-(*meta*), or 1,12-carborane (*para*), are reported [64,67,68]. Due to a slightly larger van-der-Waals diameter compared to a phenyl ring (carborane: 5.25 Å, phenyl ring: 4.72 Å) [59], carboranes are frequently used as phenyl mimetics for biologically active compounds [61,62,65,69]. Multiple non-covalent interactions, like dihydrogen bond formation, may increase the affinity to biological targets, including enzymes or receptor proteins [61,62,65]. Due to their remarkable properties, carboranes offer a wide range of possibilities for drug design in medicinal chemistry [61,70].

Our group has intensively investigated the introduction of carborane moieties into NSAIDs, COXIBs, and 5-LO inhibitors [71–77]. We have recently reported the first carborane-based dual COX-2/5-LO inhibitors that are derived from RWJ-63556 showing excellent inhibitory potential toward COX-2 and 5-LO, accompanied by high anticancer activity on the A375 melanoma cell line [78].

Herein, we report the synthesis of *p*-carborane-containing analogs of the di-*tert*-butylphenol derivatives **R-830**, **KME-4**, **E-5110**, and **S-2474**, their inhibitory potential toward COX-1, COX-2, and 5-LO, as well as their cytotoxicity on five human cancer cell lines.

2. Results and Discussion

2.1. Design and Synthesis of Di-tert-butylphenol Analogs

We recently reported the preparation and biological evaluation of eight meta (*m*)- and *p*-carborane analogs of the di-*tert*-butylphenol derivative tebufelone [79]. There, the incorporation of carborane moieties led to loss of COX inhibition, but retained 5-LO inhibitory potential, and to an overall enhanced anticancer activity. Inspired by these results, in particular by the *p*-carborane-containing analogs, we have now extended the investigation on the bioisosteric replacement of the bulky di-*tert*-butylphenyl motif with a *p*-carborane moiety to the di-*tert*-butylphenols **R-830**, **KME-4**, **E-5110**, and **S-2474**. These compounds are potent dual COX-2/5-LO inhibitors sharing a conserved 2,6-di-*tert*-butylphenol motif, but they differ in the *p*-position bearing different pharmacophores, such as thiophene, γ-sulfonyl lactam, lactam, and γ-lactones (Scheme 1). Katsumi and co-workers demonstrated that the structural combination of at least one *tert*-butyl group at the *ortho*-position and an oxygen atom at the *p*-position is essential for anti-inflammatory activity [51,53]. Accordingly, four *p*-carborane-containing di-*tert*-butylphenol analogs were synthesized by replacing the 2,6-di-*tert*-butylphenol motif with a 1-hydroxy-*p*-carborane moiety while keeping the pharmacophoric group (Scheme 1). 1-Hydroxy-*p*-carborane was prepared according to a published procedure [80] and converted quantitatively into the corresponding *tert*-butyldimethylsilyl (TBDMS) ether by a conventional silylation reaction with excess of *tert*-butyldimethylsilyl chloride (TBDMSCl) in the presence of triethylamine (NEt$_3$) catalyzed by 4-dimethylaminopyridine (DMAP). The TBDMS-protected *p*-carborane-containing **R-830** analog **1** was obtained by deprotonation and lithiation of the TBDMS-protected hydroxy-*p*-carborane with *n*BuLi followed by reaction with excess of the corresponding 2-thiophenecarbonyl chloride in good yield (64%). **1** was subsequently deprotected with tetra-*n*-butylammonium fluoride (TBAF) to obtain the desired analog **R-830-Cb** in almost quantitative yield (90%, Scheme 1, (I)). The key intermediate 1-(*tert*-butyl-dimethylsiloxy)-12-formyl-1,12-dicarba-*closo*-dodecaborane(12) (*p*-carboranylaldehyde **2**) was obtained by deprotonation and lithiation of the TBDMS-protected 1-hydroxy-*p*-carborane with *n*BuLi followed by addition of excess methyl formate in almost quantitative yield (86%). The *p*-carborane analog **S-2474-Cb** was prepared in a four-step sequence starting from 2-ethylisothiazolidine-1,1-dioxide and *p*-carboranylaldehyde **2** (Scheme 1, (II)). Selective α-proton abstraction of the γ-sulfonyl lactam precursor by *n*BuLi under kinetic control followed by addition of the *p*-carboranylaldehyde **2** led to formation of the intermediate alcohol **3** in moderate yield (53%). Compound **3** was further converted to the corresponding tosylate **4** by deprotonation of the hydroxyl group with sodium hydride (NaH) and subsequent reaction with excess *p*-toluenesulfonyl chloride (87% yield) to introduce an excellent leaving group. Reaction with NaH, a strong non-nucleophilic base, gave the TBDMS-protected *p*-carborane analog of **S-2474 5** in good yield (64%). Deprotection of the alcohol **5** with TBAF gave the *p*-carborane analog **S-2474-Cb** in excellent yield (96%, Scheme 1). The *p*-carborane-containing analogs **KME-4-Cb** and **E-5110-Cb** were prepared in two steps starting from *p*-carboranylaldehyde **2** and the corresponding Wittig reagents 3-(triphenylphosphoranylidene)-γ-butyrolactone and (1-methoxy-2-oxo-3-pyrrolidinyl)triphenyl-phosphonium bromide, respectively (Scheme 1, (III) and (IV)). The Wittig reagents were prepared according to literature procedures und represent key intermediates in the synthesis of both target compounds **KME-4-Cb** [81] and **E-5110-Cb** [54]. The TBDMS-protected analogs **6** and **7** were obtained by reaction of the *p*-carboranylaldehyde **2** and the corresponding Wittig reagents with or without the presence of NEt$_3$ in good to excellent yields (50–84%). Finally, the isolated TBDMS-protected compounds **6** and **7** were quantitatively converted to the final products **KME-4-Cb** (99%) and **E-5110-Cb** (93%) by TBAF-mediated deprotection of the hydroxyl groups.

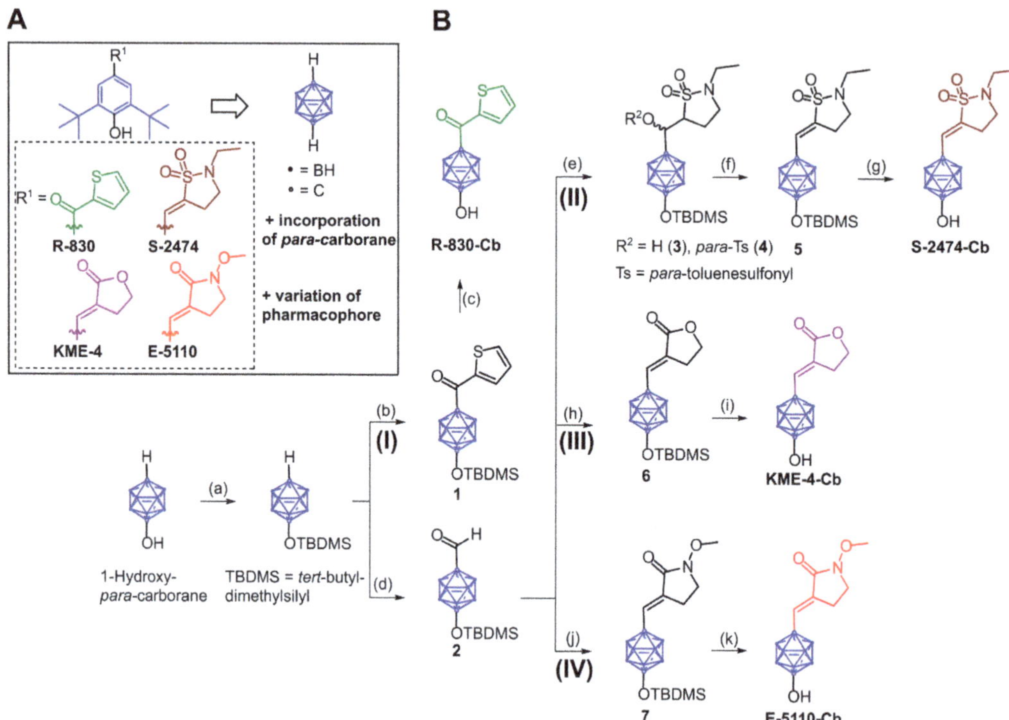

Scheme 1. (**A**) Synthesis of carborane-based di-*tert*-butylphenol analogs **R-830-Cb**, **S-2474-Cb**, **KME-4-Cb**, and **E-5110-Cb**. (**B**) Reagents and conditions: (**I**) Synthetic entry towards *p*-substituted di-*tert*-butylphenol analog **R-830-Cb**. (a) 1. TBDMSCl, NEt₃ and 4-dimethylaminopyridine (DMAP), DCM, 0 °C → RT, 48 h; 2. HCl_(aq), RT (97% yield); (b) 1. *n*BuLi, Et₂O, 0 °C → RT, 2 h; 2. 2-thiophenecarbonyl chloride, 0 °C → RT, 24 h; 3. HCl_(aq), RT (64% yield); (c) 1. TBAF, THF, 0 °C, 20–30 min; 2. H₂O, RT (90% yield); Key intermediate: (d) 1. *n*BuLi, Et₂O, 0 °C → RT, 2 h; 2. methyl formate, 0 °C → RT, 24 h; 3. HCl_(aq), RT (86% yield); (**II**) Synthetic entry towards *p*-substituted di-*tert*-butylphenol analog **S-2474-Cb**. (e) 1. 2-ethylisothiazolidine-1,1-dioxide, THF, −78 °C, 30 min; 2. *n*BuLi, THF, −78 °C, 1.5 h → RT; 3. HCl_(aq), RT (53% yield); 4. NaH, THF, RT, 2 h; 5. *p*-toluenesulfonyl chloride, 0 °C → RT, 24 h; 6. HCl_(aq), RT (87% yield); (f) 1. NaH, THF, RT, 2 h; 2. H₂O, RT (64% yield); (g) 1. TBAF, THF, 0 °C, 20–30 min; 2. H₂O, RT (96% yield); (**III**) Synthetic entry towards *p*-substituted di-*tert*-butylphenol analog **KME-4-Cb**; (h) 1. 3-(triphenylphosphoranylidene)-γ-butyrolactone, THF/DMF, 120 °C, 48 h; 2. H₂O, RT (84% yield); (i) 1. TBAF, THF, 0 °C, 25 min; 2. H₂O, RT (99% yield); (**IV**) Synthetic entry towards *p*-substituted di-*tert*-butylphenol analog **E-5110-Cb**; (j) 1. (1-methoxy-2-oxo-3-pyrrolidinyl)triphenyl-phosphonium bromide, NEt₃, THF, RT, 45 min; 2. DMF, 75 °C, 24 h; 3. H₂O, RT (50% yield); (k) 1. TBAF, THF, 0 °C, 30 min; 2. H₂O, RT (93% yield).

All compounds were fully characterized by NMR and IR spectroscopy, mass spectrometry and elemental analysis. Three-dimensional structures of intermediates **1**, **5**, **6**, and **7**, as well as of *p*-carborane-containing analogs **R-830-Cb**, **S-2474-Cb**, **KME-4-Cb**, and **E-5110-Cb**, were determined by single-crystal X-ray crystallography (Figure 2; for further details, see Supplementary Materials, Tables S3–S6).

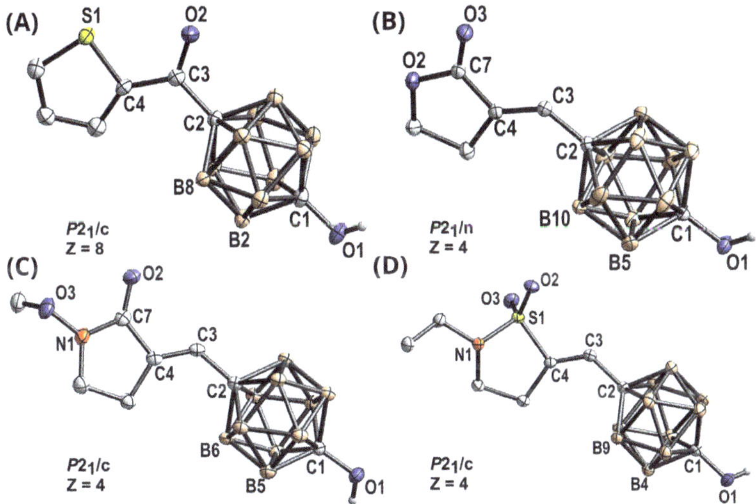

Figure 2. Molecular structure of compounds: (**A**) **R-830-Cb**, (**B**) **KME-4-Cb**, (**C**) **E-5110-Cb**, and (**D**) **S-2474-Cb**. Hydrogen atoms, except for O-H bonds, are omitted for clarity. Thermal ellipsoids are drawn at 50% probability level (beige = B, grey = C, blue = O, orange = N, yellow = S). Selected bond lengths, distances and bond angles are given in the Supplementary Materials, Tables S3–S6.

Solubility and chemical stability in organic solvents, such as aqueous dimethyl sulfoxide (DMSO), are crucial for biological investigations. Therefore, the stability of the *p*-carborane-containing compounds **R-830-Cb**, **S-2474-Cb**, **KME-4-Cb**, and **E-5110-Cb** in aqueous DMSO-d_6 at room temperature in air was studied by ^{1}H- and ^{11}B{^{1}H}-NMR spectroscopy over four weeks and confirmed that all compounds are stable (see Supplementary Materials, Figure S24).

2.2. Determination of Lipophilicity (logD) by HPLC

The lipophilicity was determined as $logD_{7.4,HPLC}$ value by a high-performance liquid chromatography (HPLC) method originally described by Donovan and Pescatore (Table 1) [82]. **R-830**, **KME-4**, **S-2474** and **E-5110** have log*D* values in the range of 3.35–3.91. Interestingly, the introduction of the lipophilic carborane cluster in **R-830-Cb**, **KME-4-Cb**, **S-2474-Cb**, and **E-5110-Cb** instead of the bulky but also lipophilic di-*tert*-butylphenyl motif resulted in a marked decrease in lipophilicity and log*D* values in the range of 1.50–2.37.

Table 1. Results of COX assay and $logD_{7.4, HPLC}$ determination.

Compound	% Inhibition at 100 µM [A,B]		$logD_{7.4, HPLC}$
	COX-1	COX-2	
R-830	93	85	3.91
R-830-Cb	18	6	2.37
KME-4	64	97	3.79
KME-4-Cb	26	n.i.	1.87
E-5110	66	105	3.35
E-5110-Cb	n.i.	n.i.	1.50
S-2474	38	86	3.75
S-2474-Cb	n.i.	n.i.	1.88

[A] n.i. = no inhibition (%inhibition below 5%); [B] Celecoxib served as reference for COX-2: pIC_{50} ($pIC_{50} = -\log_{10}$ (IC_{50} [M])) was found to be 7.05 ± 0.08 (mean ± SD, $n = 5$, resembles IC_{50} of 89 nM); SC-560 served as reference for COX-1: pIC_{50} was found to be 8.07 ± 0.63 (mean ± SD, $n = 3$, resembles IC_{50} of 8.4 nM).

2.3. Evaluation of Inhibitory Potential toward COX

All compounds were tested in vitro at a concentration of 100 µM for their inhibitory potential toward ovine COX-1 and human recombinant COX-2 using the COX Fluorescent Inhibitor Screening Assay Kit (Cayman Chemical Company) employing the selective COX-2 inhibitor celecoxib as well as the COX-1 inhibitor SC-560 as references (Table 1). While all di-*tert*-butylphenol derivatives showed the expected inhibition of COX-2 and COX-1 as reported in the literature [46–49,57], no or only weak inhibition (<26%) of both isoforms was observed for the respective *p*-carborane analogs (Table 1).

This finding is in line with previous reports claiming that the presence of at least one *tert*-butyl and a hydroxyl group is essential for anti-inflammatory activity [51,53] which could not be compensated by the hydroxy-substituted *p*-carborane cluster.

2.4. Evaluation of Inhibitory Potential toward 5-LO

All compounds were tested for their 5-LO inhibitory activity in an intact cell assay (polymorphonuclear leukocytes, PMNL) by using the selective 5-LO inhibitor BWA4C as control (Table 2). The four carborane-based analogs inhibited 5-LO product formation, with IC_{50} values in the nanomolar range, comparable to those of the respective reference compounds. These results confirm that the substitution of the di-*tert*-butylphenyl motif by *p*-carborane is well tolerated and leads to strong inhibitors of 5-LO product formation.

Table 2. IC_{50} values for inhibition of 5-LO product formation in intact PMNL for the carborane derivatives **R-830-Cb**, **KME-4-Cb**, **E-5110-Cb**, **S-2474-Cb**, and their respective reference compounds **R-830**, **KME-4**, **E-5110**, and **S-2474** (see Supplementary Materials, Figures S25 and S26 and (for 95% CIs) Table S2).

Compound	IC_{50} [µM] [(A,B)]	Compound	IC_{50} [µM]
R-830-Cb	0.65	**R-830**	0.26
KME-4-Cb	0.07	**KME-4**	0.15
E-5110-Cb	0.22	**E-5110**	0.12
S-2474-Cb	<0.03	**S-2474**	<0.03

[(A)] The selective 5-LO inhibitor BWA4C (0.1 µM) was used as control and inhibited 5-LO product formation by 92.5% $\pm$ 0.9; [(B)] Data are presented as the mean of at least three independent experiments with 95% CIs (CIs = confidence intervals).

2.5. In Vitro Determination of Cell Viability

Five human cell lines (A375 melanoma, A549 lung adenocarcinoma, HCT116 and HT29 colorectal carcinoma, MDA-MB-231 triple-negative breast adenocarcinoma) were used to study the antitumor effects of the di-*tert*-butylphenols and their respective carborane analogs. All selected cell lines were derived from inflammation-associated tumors [83,84]. Culturing cells in the presence of **R-830**, **KME-4**, **E-5110**, and **S-2474** resulted in a dose-dependent decrease in cell viability determined after 72 h (Table 3, Supplementary Materials, Figures S27 and S28).

Incorporation of a *p*-carborane moiety resulted in a significant decrease in cytotoxic potential for all tested compounds, except **R-830-Cb**, which showed antitumor potential in a similar micromolar range as the parent compound. In general, values obtained with the 3-(4,5-dimethylthiazol-2-yl)-2,5-diphenyltetrazolium bromide (MTT) assay were significantly lower than those obtained with the crystal violet (CV) assay, as expected, because dual COX-2/5-LO inhibitors affect the redox status of cells [85]. Since HCT116 cells were highly sensitive to the carborane analog **R-830-Cb** and the di-*tert*-butylphenols, the antitumor effects of the compounds might be independent of the inhibition of COX-2 and 5-LO [86,87]. Accordingly, **R-830-Cb** was selected for further in-detail investigation of the mechanism of action. Treatment of primary peritoneal exudate cells isolated from healthy mice with **R-830** or its carborane counterpart **R-830-Cb** showed that the latter one was completely non-toxic in the applied dose range, whereas **R-830** diminished the viability of primary cells at 100 µM (selectivity index (SI) $\approx$ 3, Supplementary Materials, Figure S29). These results suggest that the introduction of *p*-carborane preserves the cytotoxic potential for the malignant phenotype without affecting normal cells in the tested dose range.

Table 3. IC$_{50}$ values [μM] of di-*tert*-butylphenol derivatives and their carborane analogs for specific cancer cell lines. Data are presented as mean ± SD of three independent experiments for 3-(4,5-dimethylthiazol-2-yl)-2,5-diphenyltetrazolium bromide (MTT) and crystal violet (CV) tests. * Carbon-based reference compounds.

Compounds	Assays	A375	A549	HCT116	HT-29	MDA-MB-231
R-830 *	MTT	16.7 ± 3.1	19.0 ± 0.7	20.6 ± 0.4	19.4 ± 0.4	33.3 ± 1.2
	CV	29.4 ± 2.3	32.6 ± 1.3	30.8 ± 3.2	24.9 ± 1.3	34.0 ± 1.6
R-830-Cb	MTT	35.9 ± 0.8	66.5 ± 1.7	40.9 ± 3.1	65.6 ± 5.3	46.5 ± 0.1
	CV	47.2 ± 0.7	77.3 ± 1.7	49.1 ± 6	73.3 ± 3.4	55.6 ± 0.7
KME-4 *	MTT	13.8 ± 0.6	21.9 ± 0.7	18.4 ± 4.6	18.5 ± 1.5	80.9 ± 7.3
	CV	36.4 ± 1.3	36.3 ± 2.8	36.2 ± 1.9	36.2 ± 1.9	88.6 ± 2.4
KME-4-Cb	MTT	83.3 ± 2.3	119.7 ± 7.0	88.5 ± 4.3	139.3 ± 5.4	119.7 ± 3.9
	CV	155.2 ± 5.8	157.8 ± 3.0	127.0 ± 8.0	164.7 ± 8.5	132.7 ± 4.1
E-5110 *	MTT	12.9 ± 1.1	22.1 ± 1.6	25.1 ± 1.1	25.6 ± 2.6	34.2 ± 1.2
	CV	33.5 ± 1.7	41.1 ± 4.3	39.6 ± 2.8	39.8 ± 2.1	47.3 ± 1.7
E-5110-Cb	MTT	69.3 ± 1.7	176.6 ± 14.5	143.9 ± 10.5	119.9 ± 9.7	200 ± 0.0
	CV	103.2 ± 10.9	197.7 ± 5.7	193.1 ± 3.0	155.4 ± 6.0	>200
S-2474 *	MTT	13.2 ± 2.7	16.2 ± 1.2	17.5 ± 1.0	16.1 ± 1.1	34.8 ± 1.7
	CV	19.2 ± 0.7	17.9 ± 1.2	23.7 ± 1.9	23.8 ± 1.7	43.5 ± 1.8
S-2474-Cb	MTT	106.0 ± 6.3	121.8 ± 10.4	108.1 ± 11.5	81.8 ± 6.1	>200
	CV	135.5 ± 6.3	149.7 ± 4.3	133.4 ± 7.1	137.5 ± 7.8	>200

Analysis of apoptotic cell death by flow cytometry and fluorescent microscopy showed that the mechanism of antitumor action of **R-830** was preserved and even enhanced by the incorporation of *p*-carborane (Figure 3A,B). Treatment with **R-830-Cb** not only resulted in lower cell density, but also revealed the presence of numerous cells with abnormally shaped nuclei and condensed chromatin (Figure 3B).

Apostat staining confirmed that apoptosis was in correlation with the activation of caspases (Figure 3C). At the same time, and in parallel with the presence of apoptotic cells, cell division was more severely affected by **R-830-Cb** (Figure 3D). The decreased production of reactive oxygen species explains the observed effects of the applied treatments, namely inhibition of proliferation and induction of apoptosis (Figure 3E). It is widely accepted that drug-mediated hyperproduction of reactive oxygen species that are not buffered by cellular anti-oxidative protection is responsible for cell death induction [88]. However, the mechanisms of action of some therapeutics rely on the fact that tumor cells, especially colon carcinomas, produce a significantly higher amount of reactive oxygen species necessary for their metabolic and proliferative demands [89]. The described feature is acquired during disease progression, and it is tightly connected with abnormal blood supply and a hypoxic microenvironment [90]. Therefore, elimination of these molecules results in a rapid decrease in cell viability. Importantly, the intense autophagy detected by acridine orange (AO) supravital staining of cells exposed to **R-830-Cb**, revealed cytoprotective character for this compound, but not for **R-830** (Figure 3F,G).

Ultimately, simultaneous treatment of cells with **R-830-Cb** and specific autophagy inhibitors, chloroquine or 3-methyladenine (3-MA), showed that suppression of the autophagy process caused a drug-induced decrease in cell viability (Figure 3G). This means that the cytotoxic activity of the tested drug is enhanced by the inhibition of autophagy.

Considering the disadvantages of 2D cell cultures, since the cells are in a monolayer without all the components of the tumor microenvironment, the real improvement and potential of **R-830-Cb** should be discussed in a more complex experimental setting [91].

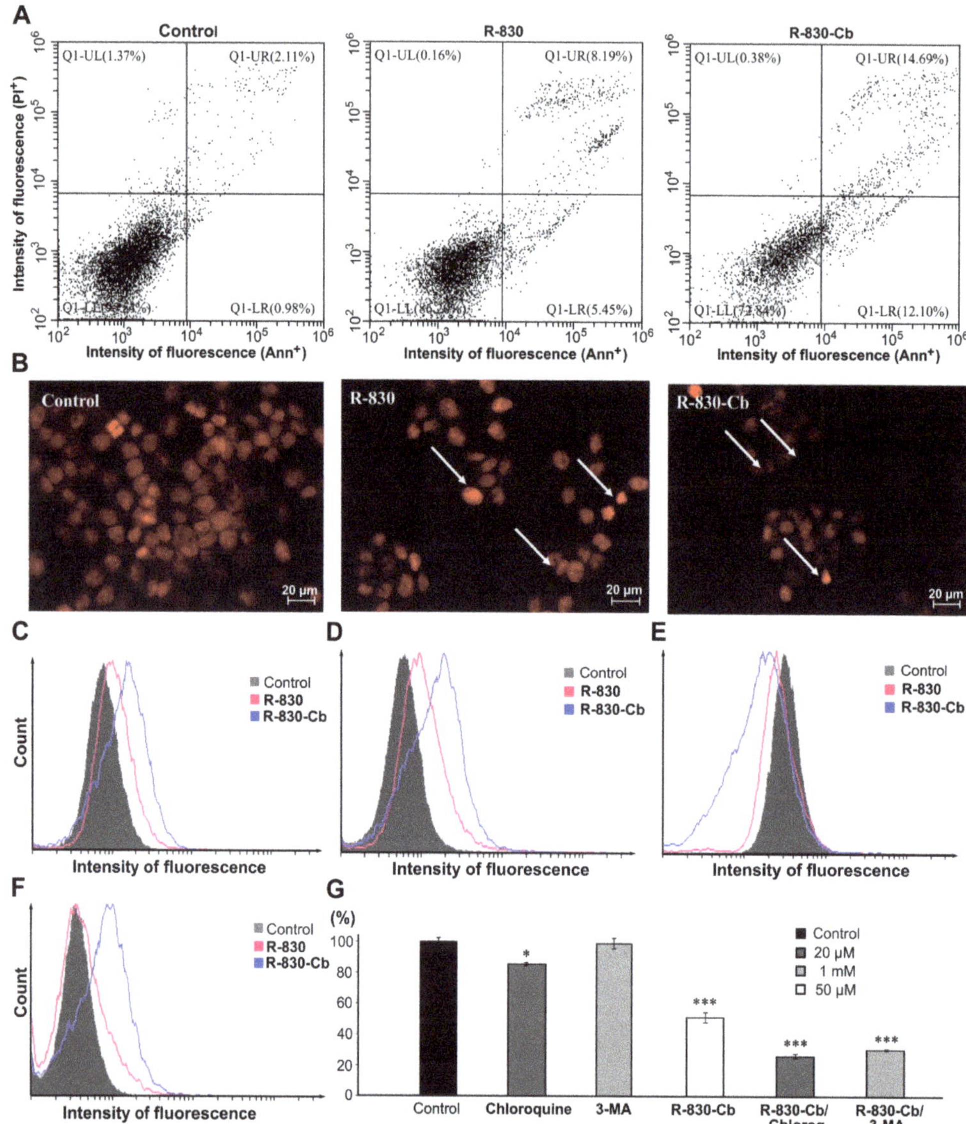

Figure 3. Mode of action of **R-830** and **R-830-Cb**. HCT116 cells were treated with an IC_{50} dose of **R-830** and **R-830-Cb** for 72 h and stained with (**A**) Annexin V (AnnV)/propidium iodide (PI), (**B**) PI (white arrows mark apoptotic nuclei), (**C**) Apostat, (**D**) Carboxyfluorescein diacetate succinimidyl ester (CFSE), (**E**) Dihydrorhodamine 123 (DHR 123), and (**F**) Acridine orange (AO). (**G**) HCT116 cells were co-treated with **R-830-Cb** and 3-methyladenine (3-MA) or chloroquine and viability was estimated by crystal violet (CV) assay. * $p < 0.05$; *** $p < 0.001$ in comparison to relevant control.

3. Experimental Section

3.1. Materials and Methods

All reactions were carried out under nitrogen using standard Schlenk techniques. All solvents were degassed, dried and purified with the solvent purification system SPS-800 by MBraun. Molecular sieves (3 and 4 Å) were activated at 300 °C in vacuum for a minimum of three hours for storage of dry solvents. Hexane has been used as a mixture of different isomers. For column chromatography, silica gel 60 with a particle size in the range of 0.035–0.070 mm (Acros) was used. A gradient of hexane and ethyl acetate (EtOAc) was usually used as the eluent system. Reactions were monitored with thin-layer chromatography using glass plates coated with silica gel (0.25 mm, silica gel 60, F_{254}, Merck). Carborane-containing areas were stained and identified with a 5% solution of palladium(II) chloride in methanol. All NMR spectra were recorded on a Bruker Avance DRX 400 spectrometer at the following measurement frequencies at 26 °C: ^{1}H-NMR 400.16 MHz, ^{11}B-NMR 128.38 MHz, ^{13}C-NMR 100.63 MHz. Tetramethylsilane (TMS) served as an internal standard for ^{1}H-NMR spectra, and deuterated chloroform ($CDCl_3$) (7.26 ppm) was used for referencing. Signal multiplicities were indicated by the following abbreviations and combinations: br = broad, s = singlet, d = doublet, t = triplet, q = quartet, dd = doublet of doublets, dt = doublet of triplets, td = triplet of doublets, ddd = doublet of doublets of doublets, dtd = doublet of triplets of doublets. The unit of the coupling constant J is given in hertz (Hz). The chemical shifts in the ^{13}C{^{1}H}-NMR spectra could be extracted from the spectra by ^{1}H broadband decoupling in the proton channel. Residual signals of $CDCl_3$, acetone-d_6, deuterated dimethyl sulfoxide (DMSO-d_6) at 77.2, 29.9, 39.5 ppm served as a reference. All ^{11}B{^{1}H} NMR spectra were also internally referenced to TMS with the aid of the Ξ scale [92]. The analysis of the data of the measurement was carried out using the software Mestrenova 14. ESI mass spectra were measured using a Bruker Daltonics Apex II FT-ICR spectrometer. All IR spectra were obtained using an ATR-IR spectrometer (Nicolet iS5, Thermo Fisher Scientific, Waltham, MA, USA) in the range of 400–4000 cm^{-1}. Elemental analyses were carried out with a Heraeus Vario El analyser. The melting points were determined in glass capillaries using a Gallenkamp MPD350·BM2.5 apparatus and are uncorrected. 1,12-Dicarba-*closo*-dodecaborane(12) (Katchem, Prague, Czech Republic), 2-ethylisothiazolidine-1,1-dioxide (BLD Pharmatech Ltd., Shanghai, China), 2-thiophenecarbonyl chloride (Alfa Aesar, Ward Hill, MA, USA) are commercially available. All chemicals were used without further purification. 2,6-Bis(1,1-dimethylethyl)-4-[(*E*)-(2-ethyl-1,1-dioxido-5-isothiazolidinylidene)methyl]phenol (**S-2474**) [58], 3-([3,5-bis(1,1-dimethylethyl)-4-hydroxyphenyl]methylene)dihydro-2(3*H*)-furanone (**KME-4**) [53], 3-([3,5-bis(1,1-dimethylethyl)-4-hydroxyphenyl]methylene)-1-methoxy-2-pyrrolid inone (**E-5110**) [54], 2,6-di-*tert*-butyl-4-(2'-thenoyl)phenol (**R-830**) [93], (1-methoxy-2-oxo-3-pyrroli dinyl)triphenyl-phosphonium bromide [54], 1-(*tert*-butyl-dimethylsiloxy)-1,12-dicarba-*closo*-dodecaborane(12) [94], and 3-(triphenylphosphoranylidene)-γ-butyrolactone [81] were prepared according to the literature.

3.2. Synthetic Procedures

3.2.1. 1-(*tert*-Butyl-dimethylsiloxy)-12-(thiophen-2'-carbonyl)-1,12-dicarba-*closo*-dodecaborane(12) (**1**)

1-(*tert*-Butyl-dimethylsiloxy)-1,12-dicarba-*closo*-dodecaborane(12) (1.00 eq., 0.83 g, 3.02 mmol) was dissolved in 25 mL dry diethyl ether. At 0 °C, *n*BuLi (1.05 eq., 1.43 M in hexane, 2.32 mL, 3.32 mmol) was added dropwise. Afterwards, the solution was stirred at room temperature for two hours. At 0 °C, a solution of 2-thiophenecarbonyl chloride in 10 mL dry diethyl ether (1.50 eq., 0.48 mL, 4.53 mmol) was added dropwise. The solution was stirred at room temperature for 24 h, then 20 mL of a 1 M HCl was added and the phases were separated. The aqueous phase was extracted three times with 25 mL diethyl ether, the combined organic phases were washed twice with 25 mL brine and then dried over magnesium sulfate. The solvent was removed under reduced pressure and the crude product was purified by column chromatography (SiO_2, hexane → hexane/EtOAc, 30:1, *v/v*). Colorless solid, yield 64% (0.74 g, 1.92 mmol); m.p.: 102–103 °C; TLC (hexane/EtOAc, 30:1, *v/v*): R_f = 0.51. ^{1}H-NMR (400.16 MHz, $CDCl_3$): δ [ppm] = 0.10 (s, 6H, 8-**CH$_3$**), 0.76 (s, 9H, 10-**CH$_3$**), 1.60–3.52 (br m, 10H, B**H**), 7.06 (dd, $^3J_{HH}$ = 5.0; 4.0 Hz, 2H, 2-**CH**), 7.64 (dd, $^3J_{HH}$ = 5.0, 1.1 Hz, 1H, 1-**CH**), 7.83 (dd, $^3J_{HH}$ = 4.0, 1.1 Hz, 1H, 3-**CH**); ^{13}C{^{1}H}-NMR (100.63 MHz, $CDCl_3$): δ [ppm] =−4.18 (8-**CH$_3$**), 17.8 (9-**C$_q$**), 25.1 (10-**CH$_3$**), 71.5 (6-**C$_q$**), 113.4 (7-**C$_q$**), 128.2 (2-**C$_q$**), 135.8 (1-**C$_q$**), 135.9 (3-**C$_q$**), 140.6 (4-**C$_q$**), 179.8 (5-**C$_q$**); ^{11}B{^{1}H}-NMR (128.37 MHz, $CDCl_3$): δ [ppm] = −15.4 (s, 5**B**), −12.8 (s, 5**B**); FTIR (ATR): $\tilde{v}$ [cm^{-1}] = 3126 (m, v_{C-H}), 2959 (m, v_{C-H}), 2929 (m, v_{C-H}), 2857 (m, v_{C-H}), 2620 (s, v_{B-H}), 2599 (s, v_{B-H}),

1644 (s, $\nu_{C=O}$), 1515 (m, $\nu_{c=c}$), 1470 (m, δ_{C-H}), 1462 (m, δ_{C-H}), 1409 (s, δ_{C-H}), 1355 (m, δ_{C-H}), 1247 (s, $\nu_{C=S}$), 1115 (s), 1064 (m), 1021 (s), 947 (s), 866 (w), 839 (m), 803 (m), 781 (s), 755 (m), 725 (s, ν_{B-B}), 688 (s), 673 (s), 656 (s), 603 (m), 574 (m), 493 (w), 466 (m); HR-ESI-MS (negative mode, CH_3CN/CH_2Cl_2): m/z [M–SiMe$_2$tBu]$^-$, calcd. for $C_7H_{13}B_{10}O_2S$: 271.1572, found: 271.1579; elemental analysis: calcd. (%) for $C_{13}H_{28}B_{10}O_2SSi$: C 40.60, H 7.34, found: C 40.75, H 7.14.

3.2.2. 1-Hydroxy-12-(thiophen-2′-carbonyl)-1,12-dicarba-*closo*-dodecaborane(12) (**R-830-Cb**)

TBDMS (*tert*-butyl-dimethylsilyl)-protected 1-hydroxy-1,12-dicarba-*closo*-dodecaborane derivative (**1**) (1.00 eq., 0.50 g, 1.11 mmol) was dissolved in 4.00 mL dry tetrahydrofuran (THF). At 0 °C, a tetra-*n*-butylammonium fluoride (TBAF) solution (1.05 eq., 1.00 M in THF, 1.36 mL, 1.36 mmol) was added dropwise. The reaction mixture was stirred for 20 min at 0 °C. The reaction was then stopped by adding 20 mL distilled water. After dilution with 50 mL EtOAc, the phases were separated. The aqueous phase was extracted three times with 25 mL EtOAc. The combined organic phases were washed with 25 mL brine and dried over magnesium sulfate. The volatiles were removed under reduced pressure and the crude product was purified by column chromatography (SiO_2, hexane/EtOAc, 2:1, v/v). Colorless solid, yield 90% (0.32 g, 1.18 mmol); m.p.: 131–132 °C; TLC (hexane/EtOAc, 2:1, v/v): R_f = 0.56. ^{1}H-NMR (400.16 MHz, CDCl$_3$): δ [ppm] = 1.56–3.43 (br m, 10H, B**H**), 3.56 (br s, 1H, 7-O**H**), 7.08 (t, $^3J_{HH}$ = 4.5 Hz, 1H, 2-C**H**), 7.66 (d, $^3J_{HH}$ = 5.0, 1.1 Hz, 1H, 1-C**H**), 7.83 (d, $^3J_{HH}$ = 4.0 Hz, 1H, 3-C**H**); ^{13}C{^{1}H}-NMR (100.63 MHz, CDCl$_3$): δ [ppm] = 72.1 (6-**C**$_q$), 110.5 (7-**C**$_q$), 128.4 (2-**C**$_q$), 136.1 (1-**C**$_q$), 136.4 (3-**C**$_q$), 140.3 (4-**C**$_q$), 179.6 (5-**C**$_q$); ^{11}B{^{1}H}-NMR (128.37 MHz, CDCl$_3$): δ [ppm] = −15.1 (s, 5B), −13.1 (s, 5B); FTIR (ATR): $\tilde{\nu}$ [cm^{-1}] = 3328 (b, ν_{O-H}), 2615 (s, ν_{B-H}), 2601 (s, ν_{B-H}), 1646 (s, $\nu_{C=O}$), 1629 (s, $\nu_{C=O}$), 1510 (m, $\nu_{C=C}$), 1449 (w, δ_{C-H}), 1403 (s, δ_{C-H}), 1355 (s, δ_{C-H}), 1345 (s, δ_{C-H}), 1261 (s, $\nu_{C=S}$), 1215 (s),1192 (s, ν_{C-O}), 1115 (s), 1063 (m), 1035 (w), 1010 (s), 935 (m), 865 (m), 741 (s), 722 (s, ν_{B-B}), 691 (s), 665 (s), 573 (w), 556 (w), 447 (m); HR-ESI-MS (negative mode, MeOH): m/z [M−H]$^-$ calcd. for $C_7H_{13}B_{10}O_2S$: 271.1572, found: 271.1570; elemental analysis: calcd. (%) for $C_7H_{14}B_{10}O_2S$: C 31.19, H 5.22, found: C 31.04, H 5.21.

3.2.3. 1-(*tert*-Butyl-dimethylsiloxy)-12-formyl-1,12-dicarba-*closo*-dodecaborane(12) (**2**)

1-(*tert*-Butyl-dimethylsiloxy)-1,12-dicarba-*closo*-dodecaborane(12) (1.00 eq., 2.14 g, 7.79 mmol) was dissolved in 100 mL dry diethyl ether. At 0 °C, *n*BuLi (1.05 eq., 1.43 M in hexane, 5.72 mL, 8.18 mmol) was added dropwise. Afterwards, the solution was stirred at room temperature for two hours. At −78 °C, methyl formate (3.50 eq., 2.20 mL, 27.3 mmol) was added dropwise. The solution was first stirred for one hour at −78 °C, then warmed to room temperature and stirred for 24 h. After addition of 30 mL 1 M HCl, the phases were separated. The aqueous phase was extracted three times with 50 mL diethyl ether, and the combined organic phases were washed twice with 25 mL brine and then dried over magnesium sulfate. The solvent was removed under reduced pressure and the crude product was purified by column chromatography (SiO_2, hexane → hexane/EtOAc, 20:1, v/v). Colorless solid, yield 86% (2.02 g, 6.68 mmol); m.p.: 40–41 °C; TLC (hexane): R_f = 0.57. ^{1}H-NMR (400.16 MHz, CDCl$_3$): δ [ppm] = 0.09 (s, 6H, 4-C**H**$_3$), 0.75 (s, 9H, 6-C**H**$_3$), 1.30–3.36 (br m, 10H, B**H**), 8.96 (s, 1H, 1-C**H**); ^{13}C{^{1}H}-NMR (100.63 MHz, CDCl$_3$): δ [ppm] = −4.20 (4-**C**H$_3$), 17.7 (5-**C**$_q$), 25.1 (6-**C**H$_3$), 70.7 (2-**C**$_q$), 113.1 (3-**C**$_q$), 187.2 (1-**C**$_q$); ^{11}B{^{1}H}-NMR (128.37 MHz, CDCl$_3$): δ [ppm] = −16.9 (s, 5B), −12.5 (s, 5B); FTIR (ATR): $\tilde{\nu}$ [cm^{-1}] = 2955 (m, ν_{C-H}), 2930 (m, ν_{C-H}), 2859 (m, ν_{C-H}), 2609 (s, ν_{B-H}), 1970 (w), 1735 (s, $\nu_{C=O}$), 1471 (m, δ_{C-H}), 1463 (w, δ_{C-H}),1390 (w, δ_{C-H}), 1362 (w, δ_{C-H}), 1252 (s, ν_{C-O}), 1168 (s, ν_{C-O}), 1031 (s, ν_{C-O}), 1004 (w), 962 (m), 781 (s), 730 (m, ν_{B-B}), 669 (m), 642 (m), 575 (w), 557 (m); HR-ESI-MS (negative mode, CH_3CN): m/z [M+Cl]$^-$, calcd. for $C_9H_{26}B_{10}ClO_2Si$: 339.2326, found: 339.2302; elemental analysis: calcd. (%) for $C_9H_{26}B_{10}O_2Si$: C 35.74, H 8.66, found: C 35.70, H 8.67.

3.2.4. 1-(*tert*-Butyl-dimethylsiloxy)-12-(hydroxymethyl-[2′-ethylisothiazolidine-1′,1′-dioxide])-1,12-dicarba-*closo*-dodecaborane(12) (**3**)

2-Ethylisothiazolidine-1,1-dioxide (1.10 eq., 1.08 g, 7.27 mmol) was dissolved in 15 mL dry THF. At $-78\,^{\circ}$C, *n*BuLi (1.15 eq., 1.36 M in hexane, 5.61 mL, 7.63 mmol) was added dropwise. Afterwards, the solution was stirred at $-78\,^{\circ}$C for 30 min. At $-78\,^{\circ}$C, aldehyde **2** (1.00 eq., 2.00 g, 6.61 mmol) was added dropwise. The solution was first stirred for 1.5 h at $-78\,^{\circ}$C, then it was warmed to room temperature and stopped by adding 15 mL 1 M HCl. The phases were separated, and the aqueous phase was extracted three times with 75 mL diethyl ether. The combined organic phases were washed twice with 25 mL brine and then dried over magnesium sulfate. The solvent was removed under reduced pressure and the crude product was purified by column chromatography (SiO$_2$, hexane/EtOAc 3:1 → 2:1, *v/v*). Colorless solid, yield 53% (1.58 g, 3.49 mmol); m.p.: 149–150 $^{\circ}$C; TLC (hexane/EtOAc 3:1, *v/v*): R_f = 0.55. ^{1}H-NMR (400.16 MHz, CDCl$_3$): δ [ppm] = 0.09 (s, 6H, 9-CH$_3$), 0.74 (s, 9H, 11-CH$_3$), 1.29–3.33 (br m, 10H, BH), 1.18 (t, $^{3}J_{HH}$ = 7.2 Hz, 2H, 1-CH$_3$), 2.22 (m, 1H, 4-CH$_2$), 2.37 (m, 1H, 4-CH$_2$), 2.97 (d, $^{3}J_{HH}$ = 3.9 Hz, 2H, 3-CH$_2$), 3.08 (m, 4H, 2/3-CH$_2$), 4.33 (d, $^{3}J_{HH}$ = 3.6 Hz, 1H, 6-CH); ^{13}C{^{1}H}-NMR (100.63 MHz, CDCl$_3$): δ [ppm] = -4.3 (9-CH$_3$), 13.2 (1-CH$_3$), 17.7 (10-C$_q$), 18.1 (4-CH$_2$), 39.4 (5-CH), 44.4 (2-CH$_2$), 60.8 (3-CH$_2$), 68.1 (6-C-OH), 77.3 (7-C$_q$), 111.6 (8-C$_q$); ^{11}B{^{1}H}-NMR (128.37 MHz, CDCl$_3$): δ [ppm] = -16.0 (s, 5B), -12.8 (s, 5B); FTIR (ATR): $\tilde{\nu}$ [cm^{-1}] = 3447 (b, ν_{O-H}), 2958 (m, ν_{C-H}), 2929 (m, ν_{C-H}), 2858 (m, ν_{C-H}), 2598 (s, ν_{B-H}), 1466 (m, δ_{C-H}), 1409 (w, δ_{C-H}), 1361 (w, δ_{C-H}), 1329 (w, δ_{C-H}), 1284 (s, $\nu_{S=O}$), 1266 (m), 1216 (s), 1183 (w, ν_{C-O}), 1129 (s, ν_{C-O}), 1109 (s, ν_{C-O}), 1078 (w), 1032 (s), 1004 (w), 989 (m), 953 (w), 940 (m), 887 (w), 837 (w), 783 (s), 741 (s, ν_{B-B}), 707 (w), 672 (m), 656 (w), 640 (w), 613 (w), 579 (m), 503 (w), 482 (w), 452 (w); HR-ESI-MS (positive mode, MeOH): *m/z* [M + H]$^+$, calcd. for C$_{14}$H$_{38}$B$_{10}$NO$_4$SSi: 454.3215, found: 454.3234; elemental analysis: calcd. (%) for C$_{14}$H$_{37}$B$_{10}$NO$_4$SSi: C 37.23, H 8.26, N 3.10, found: C 37.90, H 8.25, N 3.89.

3.2.5. 1-(*tert*-Butyl-dimethylsiloxy)-12-(*p*-toluenesulfonylmethyl-[2′-ethylisothiazolidine-1′,1′-dioxide])-1,12-dicarba-*closo*-dodecaborane(12) (**4**)

1-(*tert*-Butyl-dimethylsiloxy)-12-(hydroxymethyl-[2′-ethylisothiazolidine-1′,1′-dioxide])-1,12-dicarba-*closo*-dodecaborane(12) (**3**) (1.00 eq., 1.00 g, 2.21 mmol) was dissolved in 7.00 mL dry THF. At 0 $^{\circ}$C, NaH (1.05 eq., 60% dispersion in mineral oil, 93.1 mg, 0.69 mmol) was added. The reaction mixture was stirred for two hours at room temperature. At 0 $^{\circ}$C, *p*-toluenesulfonyl chloride (2.00 eq., 0.84 g, 4.42 mmol) was added and the solution was stirred for 24 h at room temperature. The reaction was then stopped by adding 30 mL of a 1 M HCl. After dilution with 75 mL diethyl ether, the phases were separated. The aqueous phase was extracted three times with 25 mL diethyl ether. The combined organic phases were washed with 25 mL brine and dried over magnesium sulfate. The volatiles were removed under reduced pressure and the crude product was purified by column chromatography (SiO$_2$, hexane/EtOAc, 4:1 → 3:1, *v/v*). Colorless solid, yield 87% (1.17 g, 1.93 mmol); m.p.: 181–182 $^{\circ}$C; TLC (hexane/EtOAc, 4:1, *v/v*): R_f = 0.54. ^{1}H-NMR (400.16 MHz, CDCl$_3$): δ [ppm] = 0.05 (s, 6H, 13-CH$_3$), 0.72 (s, 9H, 15-CH$_3$), 1.09–3.24 (br m, 10H, BH), 1.00 (t, $^{3}J_{HH}$ = 7.3 Hz, 2H, 1-CH$_3$), 2.26 (dtd, $^{3}J_{HH}$ = 13.1, 7.0, 6.2 Hz, $^{4}J_{HH}$ = 3.2 Hz, 2H, 4-CH$_2$), 2.44 (s, 3H, 10-CH$_3$), 2.71 (m, 1H, 3-CH$_2$), 2.82 (m, 1H, 3-CH$_2$), 2.92 (m, 2H, 2-CH$_2$), 3.21 (ddd, $^{3}J_{HH}$ = 8.6, 6.4 Hz, $^{4}J_{HH}$ = 2.3 Hz, 1H, 5-CH), 5.29 (d, $^{3}J_{HH}$ = 2.3 Hz, 1H, 6-CH), 7.31 (d, $^{3}J_{HH}$ = 8.1 Hz, 1H, 8-CH), 7.77 (d, $^{3}J_{HH}$ = 8.2 Hz, 1H, 9-CH); ^{13}C{^{1}H}-NMR (100.63 MHz, CDCl$_3$): δ [ppm] = -4.3 (13-CH$_3$), 12.7 (1-CH$_3$), 17.7 (14-C$_q$), 20.5 (4-CH$_2$), 21.8 (10-CH$_3$), 25.1 (15-CH$_3$), 38.9 (3-CH$_2$), 43.5 (2-CH$_2$), 60.6 (5-CH), 70.1 (11-C$_q$), 77.3 (6-CH), 112.6 (12-C$_q$), 128.5 (9-CH), 129.4 (8-CH), 134.2 (10-CH$_3$), 145.0 (7-C$_q$); ^{11}B{^{1}H}-NMR (128.37 MHz, CDCl$_3$): δ [ppm] = -15.8 (s, 5B), -12.6 (s, 5B); FTIR (ATR): $\tilde{\nu}$ [cm^{-1}] = 2956 (m, ν_{C-H}), 2930 (m, ν_{C-H}), 2857 (m, ν_{C-H}), 2619 (s, ν_{B-H}), 2592 (s, ν_{B-H}), 1595 (w), 1463 (w, δ_{C-H}), 1450 (w, δ_{C-H}), 1372 (m, δ_{C-H}), 1341 (w, δ_{C-H}), 1293 (m, δ_{C-H}), 1254 (s, $\nu_{S=O}$), 1189 (m, ν_{C-O}),1172 (s, ν_{C-O}), 1132 (s), 1109 (s), 1095 (w), 1076 (w), 1034 (w), 1019 (w), 1005 (w), 972 (s), 894 (m), 870 (w), 838 (s), 810 (s), 783 (s), 751 (s, ν_{B-B}), 720 (w), 670 (s), 639 (w), 631 (w), 577 (m), 559 (m), 544 (s), 498 (w), 455 (w),

442 (w), 435 (w), 429 (w); HR-ESI-MS (positive mode, MeOH): *m/z* [M + H]$^+$, calcd. for C$_{21}$H$_{44}$B$_{10}$NO$_6$S$_2$Si: 608.3304, found: 608.3350; elemental analysis: calcd. (%) for C$_{21}$H$_{43}$B$_{10}$NO$_6$S$_2$Si: C 41.63, H 7.51, N 2.31 found: C 42.18, H 7.28, N 2.22.

3.2.6. (3*E*)-1-(*tert*-Butyl-dimethylsiloxy)-12-(methylene-[2′-ethylisothiazolidine-1′,1′-dioxide])-1,12-dicarba-*closo*-dodecaborane(12) (5)

1-(*tert*-Butyl-dimethylsiloxy*)*-12-(*p*-toluenesulfonylmethyl-[2′-ethylisothiazolidine-1′,1′-dioxide])-1,12-dicarba-*closo*-dodecaborane(12) (**4**) (1.00 eq., 0.46 g, 0.75 mmol) was dissolved in 3.80 mL dry THF. At 0 °C, NaH (1.00 eq., 60% dispersion in mineral oil, 30.4 mg, 0.75 mmol) was added. The reaction mixture was stirred for two hours at room temperature. The reaction was then stopped by adding 25 mL distilled water. After dilution with 25 mL EtOAc, the phases were separated. The aqueous phase was extracted three times with 25 mL EtOAc. The combined organic phases were washed with 25 mL brine and dried over magnesium sulfate. The volatiles were removed under reduced pressure and the crude product was purified by column chromatography (SiO$_2$, hexane/EtOAc, 4:1 → 0:1, *v*/*v*). Colorless solid, yield 64% (0.21 g, 0.48 mmol); m.p.: 147–148 °C; TLC (hexane/EtOAc, 4:1, *v*/*v*): R_f = 0.50. ^{1}H-NMR (400.16 MHz, CDCl$_3$): δ [ppm] = 0.08 (s, 6H, 9-**CH**$_3$), 0.75 (s, 9H, 11-**CH**$_3$), 1.14–3.34 (br m, 10H, B**H**), 1.23 (t, $^3J_{HH}$ = 7.2 Hz, 2H, 1-**CH**$_3$), 2.89 (td, $^3J_{HH}$ = 6.6 Hz, $^4J_{HH}$ = 2.9 Hz, 2H, 4-**CH**$_2$), 3.08 (q, $^3J_{HH}$ = 7.3 Hz, 2H, 2-**CH**$_2$), 3.14 ($^3J_{HH}$ = 6.6 Hz, 2H, 3-**CH**$_2$), 6.09 (t, $^4J_{HH}$ = 2.8 Hz, 1H, 6-**CH**); ^{13}C{^{1}H}-NMR (100.63 MHz, CDCl$_3$): δ [ppm] = −4.4 (9-**CH**$_3$), 12.7 (1-**CH**$_3$), 17.6 (10-**C**$_q$), 21.9 (4-**CH**$_2$), 24.9 (11-**CH**$_3$), 39.1 (2-**CH**$_2$), 42.8 (3-**CH**$_2$), 63.9 (7-**C**$_q$), 112.6 (12-**C**$_q$), 111.8 (8-**C**$_q$), 126.9 (6-**CH**), 140.5 (5-**C**$_q$); ^{11}B{^{1}H}-NMR (128.37 MHz, CDCl$_3$): δ [ppm] = −15.0 (s, 5B), −12.4 (s, 5B); FTIR (ATR): $\tilde{v}$ [cm^{-1}] = 2948 (m, v_{C-H}), 2930 (m, v_{C-H}), 2859 (w, v_{C-H}), 2619 (s, v_{B-H}), 1667 (w, $v_{C=C}$), 1471 (w, δ_{C-H}), 1461 (m, δ_{C-H}), 1422 (w, δ_{C-H}), 1390 (m, δ_{C-H}), 1361 (w, δ_{C-H}), 1309 (m, δ_{C-H}), 1297 (s, $v_{S=O}$), 1262 (s, $v_{S=O}$), 1250 (s, $v_{S=O}$), 1208 (m, v_{C-O}),1189 (m, v_{C-O}), 1172 (m), 1156 (m), 1136 (w), 1102 (w), 1073 (m), 1037 (s), 1023 (m), 1005 (w), 972 (w), 939 (m), 837 (s), 781 (s), 750 (w), 729 (s, v_{B-B}), 698 (m), 670 (m), 643 (w), 596 (m), 572 (m), 561 (m), 533 (s), 509 (s), 466 (m), 448 (m), 441 (w), 435 (w); HR-ESI-MS (positive mode, MeOH): *m/z* [M + H]$^+$, calcd. for C$_{14}$H$_{36}$B$_{10}$NO$_3$SSi: 436.3110, found: 436.3120; elemental analysis: calcd. (%) for C$_{14}$H$_{35}$B$_{10}$NO$_3$SSi: C 38.77, H 8.13, N 3.23 found: C 39.49, H 8.27, N 2.98.

3.2.7. (3*E*)-1-Hydroxy-12-(methylene-[2′-ethylisothiazolidine-1′,1′-dioxide])-1,12-dicarba-*closo*-dodecaborane(12) (S-2474-Cb)

(3*E*)-1-(*tert*-Butyl-dimethylsiloxy)-12-(methylene-[2′-ethylisothiazolidine-1′,1′-dioxide])-1,12-dicarba-*closo*-dodecaborane(12) (**5**) (1.00 eq., 100 mg, 0.23 mmol) was dissolved in 0.80 mL dry THF. At 0 °C, a TBAF solution (1.05 eq., 1.00 M in THF, 0.24 mL, 0.24 mmol) was added dropwise. The reaction mixture was stirred for 20 min at 0 °C. The reaction was then stopped by adding 10 mL distilled water. After dilution with 25 mL EtOAc the phases were separated. The aqueous phase was extracted three times with 25 mL EtOAc. The combined organic phases were washed twice with 20 mL brine and dried over magnesium sulfate. The volatiles were removed under reduced pressure and the crude product was purified by column chromatography (SiO$_2$, hexane/EtOAc, 1:1 → 0:1, *v*/*v*). Colorless solid, yield 96% (70.0 mg, 0.22 mmol); m.p.: 220–222 °C; TLC (hexane/EtOAc, 2:1, *v*/*v*): R_f = 0.46. ^{1}H-NMR (400.16 MHz, acetone-d$_6$): δ [ppm] = 1.48–3.30 (br m, 10H, B**H**), 1.17 (t, $^3J_{HH}$ = 7.3 Hz, 2H, 1-**CH**$_3$), 2.95 (dt, $^3J_{HH}$ = 6.5 Hz, $^4J_{HH}$ = 3.3 Hz, 2H, 4-**CH**$_2$), 3.00 (q, $^3J_{HH}$ = 7.3 Hz, 2H, 2-**CH**$_2$), 3.20 (t, $^3J_{HH}$ = 6.5 Hz, 2H, 3-**CH**$_2$), 5.96 (t, $^4J_{HH}$ 2.9 Hz, 1H, 6-**CH**); ^{13}C{^{1}H}-NMR (100.63 MHz, acetone-d$_6$): δ [ppm] = 13.2 (1-**CH**$_3$), 22.9 (4-**CH**$_2$), 40.0 (2-**CH**$_2$), 43.8 (3-**CH**$_2$), 65.0 (7-**C**$_q$), 111.8 (8-**C**$_q$), 126.0 (6-**CH**), 143.6 (5-**C**$_q$); ^{11}B{^{1}H}-NMR (128.37 MHz, acetone-d$_6$): δ [ppm] = −14.7 (s, 5B), −12.7 (s, 5B); FTIR (ATR): $\tilde{v}$ [cm^{-1}] = 3284 (b, v_{OH}), 2987 (m, v_{C-H}), 2933 (m, v_{C-H}), 2880 (m, v_{C-H}), 2611 (s, v_{B-H}), 2595 (s, v_{B-H}), 1665 (m, $v_{C=C}$), 1466 (w, δ_{C-H}), 1454 (m, δ_{C-H}), 1423 (w, δ_{C-H}), 1389 (m, δ_{C-H}), 1352 (w, δ_{C-H}), 1278 (s, v_{C-O}), 1235 (w, v_{C-O}), 1151 (s, $v_{S=O}$), 1132 (s, $v_{S=O}$), 1098 (s), 1023 (s), 959 (m), 935 (w), 897 (w), 832 (m), 795 (m), 728 (s, v_{B-B}), 700 (w), 602 (m), 570 (w), 546 (s), 507 (s), 487 (m), 457 (m); HR-ESI-MS (positive mode, MeOH): m/z [M + H]$^+$, calcd. for $C_8H_{22}B_{10}NO_2S$: 322.2245, found: 322.2255; elemental analysis: calcd. (%) for $C_8H_{21}B_{10}NO_2S$: C 30.08, H 6.63, N 4.39, found: C 30.05, H 6.53, N 4.42.

3.2.8. (3*E*)-1-(*tert*-Butyldimethylsiloxy)-12-(methylene-[dihydrofurane-2′(3*H*)-one])-1,12-dicarba-*closo*-dodecaborane(12) (**6**)

Aldehyde **2** (1.00 eq., 0.70 g, 2.31 mmol) and 3-(triphenylphosphoranylidene)-γ-butyrolactone(1.50 eq., 1.20 g, 3.47 mmol) were dissolved in 11 mL dry THF and 1.40 mL dry (*N,N*)-dimethylformamide (DMF). The reaction mixture was heated under reflux for 48 h. After cooling to room temperature, the solvent was removed under reduced pressure. The residue was then dissolved in 100 mL EtOAc and 75 mL distilled water. The phases were separated and the aqueous phase was extracted three times with 50 mL EtOAc. The combined organic phases were washed twice with 50 mL brine and then dried over magnesium sulfate. The solvent was removed under reduced pressure and the crude product was purified by column chromatography (SiO$_2$, hexane/EtOAc, 5:1 → 3:1, v/v). Colorless solid, yield 84% (0.72 g, 1.94 mmol); m.p.: 149–151 °C; TLC (hexane/EtOAc, 5:1, v/v): R_f = 0.62. ^{1}H-NMR (400.16 MHz, CDCl$_3$): δ [ppm] = 0.08 (s, 6H, 8-**CH$_3$**), 0.74 (s, 9H, 10-**CH$_3$**), 1.34–3.31 (br m, 10H, **BH**), 2.95 (td, $^3J_{HH}$ = 7.2 Hz; $^4J_{HH}$ = 3.0 Hz, 2H, 3-**CH$_2$**), 4.30 (t, $^3J_{HH}$ = 7.2 Hz, 2H, 1-**CH$_2$**), 6.38 (t, $^4J_{HH}$ = 3.1 Hz, 1H, 5-**CH**); ^{13}C{^{1}H}-NMR (100.63 MHz, CDCl$_3$): δ [ppm] = −4.2 (8-**CH$_3$**), 17.7 (9-**C$_q$**), 25.1 (3-**CH$_3$**), 64.7 (6-**C$_q$**), 65.5 (1-**CH$_2$**), 111.8 (7-**C$_q$**), 128.6 (4-**C$_q$**), 134.0 (5-**CH**), 170.8 (2-**C$_q$**); ^{11}B{^{1}H}-NMR (128.37 MHz, CDCl$_3$): δ [ppm] = −15.1 (s, 5**B**), −12.5 (s, 5**B**); FTIR (ATR): $\tilde{v}$ [cm^{-1}] = 2925 (m, v_{C-H}), 2885 (w, v_{C-H}), 2856 (m, v_{C-H}), 2608 (s, v_{B-H}), 1725 (s, $v_{C=O}$), 1676 (m, $v_{C=C}$), 1470 (m, δ_{C-H}), 1428 (w, δ_{C-H}), 1380 (m, δ_{C-H}), 1360 (m, δ_{C-H}), 1350 (m, δ_{C-H}), 1261 (s, v_{C-O-C}), 1261 (s, v_{C-O-C}), 1250 (s, v_{C-O-C}), 1207 (s), 1072 (m), 1029 (s), 965 (m), 939 (w), 836 (s), 779 (s), 781 (s), 736 (m, v_{B-B}), 695 (m), 646 (m), 628 (w), 574 (m), 517 (m), 461 (m); HR-ESI-MS (positive mode, CH$_3$CN): m/z [M + NH$_4$]$^+$, calcd. for $C_{13}H_{34}B_{10}NO_3Si$: 390.3233, found: 390.3228; elemental analysis: calcd. (%) for $C_{13}H_{30}B_{10}O_3Si$: C 42.14, H 8.16, found: C 42.22, H 8.04.

3.2.9. (3*E*)-1-Hydroxy-12-(methylene-[dihydrofurane-2′(3*H*)-one])-1,12-dicarba-*closo*-dodecaborane(12) (**KME-4-Cb**)

(3*E*)-1-(*tert*-Butyl-dimethylsiloxy)-12-(methylene-[dihydrofuran-2′(3*H*)-one])-1,12-dicarba-*closo*-dodecaborane(12) (**6**) (1.00 eq., 0.29 g, 0.78 mmol) was dissolved in 2.50 mL dry THF. At 0 °C, a TBAF solution (1.05 eq., 1.00 M in THF, 0.82 mL, 0.82 mmol) was added dropwise. The reaction mixture was stirred for 25 min at 0 °C. The reaction was then stopped by adding 20 mL distilled water. After dilution with 25 mL EtOAc the phases were separated. The aqueous phase was extracted three times with 25 mL EtOAc. The combined organic phases were washed with 20 mL brine and dried over magnesium sulfate. The volatiles were removed under reduced pressure and the crude product was purified by column chromatography (SiO$_2$, hexane/EtOAc, 1:1, v/v). Colorless solid, yield 99% (0.19 g, 0.77 mmol); m.p.: 210–212 °C; TLC (hexane/EtOAc, 1:1, v/v): R_f = 0.62. ^{1}H-NMR (400.16 MHz, acetone-d$_6$): δ [ppm] = 1.47–3.27 (br m, 10H, **BH**), 3.03 (td, $^3J_{HH}$ = 7.2 Hz, $^4J_{HH}$ = 3.1 Hz, 2H, 3-**CH$_2$**), 4.35 (t, $^3J_{HH}$ = 7.1 Hz, 2H, 1-**CH$_2$**), 6.20 (t, $^4J_{HH}$ = 3.1 Hz, 1H, 5-**CH**); ^{13}C{^{1}H}-NMR (100.63 MHz, acetone-d$_6$): δ [ppm] = 26.0 (3-**CH$_3$**), 66.0 (6-**C$_q$**), 66.3 (1-**CH$_2$**), 111.5 (7-**C$_q$**), 131.5 (4-**C$_q$**), 132.1 (5-**CH**), 170.9 (2-**C$_q$**); ^{11}B{^{1}H}-NMR (128.37 MHz, acetone-d$_6$): δ [ppm] = −14.8 (s, 5**B**), −12.6 (s, 5**B**); FTIR (ATR): $\tilde{v}$ [cm^{-1}] = 3157 (b, v_{O-H}), 2977 (w, v_{C-H}), 2920 (w, v_{C-H}), 2603 (s, v_{B-H}), 1727 (s, $v_{C=O}$), 1672 (m, $v_{C=C}$), 1577 (w, $v_{C=C}$), 1480 (m, δ_{C-H}), 1452 (w, δ_{C-H}), 1427 (m, $v_{C=C}$), 1387 (s), 1358 (s, δ_{C-H}),1282 (w, v_{C-O}), 1209 (s, v_{C-O-C}), 1088 (m), 1032 (s), 976 (m), 866 (w), 776 (w), 754 (w), 732 (w), 715 (w), 693 (m), 634 (m), 588 (w), 572 (w), 524 (w), 483 (w); HR-ESI-MS (positive mode, CH$_3$CN): m/z [M + H]$^+$, calcd. for $C_7H_{17}B_{10}O_3$: 257.2102, found: 257.2184; elemental analysis: calcd. (%) for $C_7H_{16}B_{10}O_3$: C 32.80, H 6.29, found: C 32.19, H 6.20.

3.2.10. (3*E*)-1-(*tert*-Butyl-dimethylsiloxy)-12-(methylene-[1′-methoxypyrrolidine-2′-one])-1,12-dicarba-*closo*-dodecaborane(12) (**7**)

(1-Methoxy-2-oxo-3-pyrrolidinyl)triphenyl-phosphonium bromide (1.50 eq., 0.95 g, 2.08 mmol) and NEt$_3$ (1.50 eq., 0.29 mL, 2.08 mmol) were dissolved in 7.00 mL dry THF. The solution was stirred for 45 min at room temperature. Afterwards, a solution of aldehyde **2** (1.00 eq., 0.38 g, 1.38 mmol) in 2.00 mL dry DMF was added dropwise. The reaction mixture was stirred for 24 h at 75 °C. After cooling to room temperature, the solvent was removed under reduced pressure. The residue was then dissolved in 50 mL EtOAc and 30 mL distilled water. The phases were separated, and the aqueous phase was extracted three times with 25 mL EtOAc. The combined organic phases were washed twice with 30 mL brine and then dried over magnesium sulfate. The solvent was removed under reduced pressure and the crude product was purified by column chromatography (SiO$_2$, hexane/EtOAc, 1:1 → 0:1, *v/v*). Colorless solid, yield 50% (0.28 g, 0.69 mmol); m.p.: 168–170 °C; TLC (hexane/EtOAc, 1:1, *v/v*): R_f = 0.60. ^{1}H-NMR (400.16 MHz, CDCl$_3$): δ [ppm] = 0.07 (s, 6H, 9-**CH$_3$**), 0.74 (s, 9H, 11-**CH$_3$**), 1.42–3.30 (br m, 10H, **BH**), 2.79 (td, $^3J_{HH}$ = 6.4 Hz, $^4J_{HH}$ = 2.9 Hz, 2H, 4-**CH$_2$**), 3.50 (t, $^3J_{HH}$ = 6.4 Hz, 2H, 3-**CH$_2$**), 3,78 (s, 3H, 1-**CH$_3$**), 6.19 (t, $^4J_{HH}$ = 2,8 Hz, 1H, 6-**CH**); ^{13}C{^{1}H}-NMR (100.63 MHz, CDCl$_3$): δ [ppm] = −4.2 (9-**CH$_3$**), 17.7 (10-**C$_q$**), 21.1 (4-**CH$_2$**), 25.1 (11-**CH$_3$**), 42.8 (3-**CH$_2$**), 62.2 (1-**CH$_3$**), 65.2 (7-**C$_q$**), 111.3 (8-**C$_q$**), 128.5 (6-**C$_q$**), 131.3 (5-**C$_q$**), 163.7 (2-**C$_q$**); ^{11}B{^{1}H}-NMR (128.37 MHz, CDCl$_3$): δ [ppm] = −15.0 (s, 5**B**), −12.5 (s, 5**B**); FTIR (ATR): $\tilde{v}$ [cm^{-1}] = 2953 (w, v_{C-H}), 2932 (m, v_{C-H}), 2896 (w, v_{C-H}), 2858 (w, v_{C-H}), 2603 (s, v_{B-H}), 1994 (w), 1708 (s, $v_{C=O}$), 1672 (m, $v_{C=C}$), 1472 (m, $v_{C=C}$), 1461 (w, δ_{C-H}), 1428 (w, δ_{C-H}), 1399 (w, δ_{C-H}), 1362 (w, δ_{C-H}), 1344 (m, δ_{C-H}), 1249 (s, v_{C-O-C}), 1119 (w), 1038 (s), 1003 (w), 970 (m), 926 (m), 837 (s), 779 (s), 731 (m, v_{B-B}), 694 (m), 666 (w), 628 (w), 603 (m), 521 (w), 482 (m), 466 (m), 449 (m), 431 (m), 461 (m); HR-ESI-MS (positive mode, CH$_3$CN): *m/z* [M + H]$^+$, calcd. for C$_{14}$H$_{34}$B$_{10}$NO$_3$Si: 402.3233, found: 402.3248; elemental analysis: calcd. (%) for C$_{14}$H$_{33}$B$_{10}$NO$_3$Si: C 42.08, H 8.32, N 3.51, found: C 42.39, H 8.07, N 3.53.

3.2.11. (3*E*)-1-Hydroxy-12-(methylene-[1′-methoxypyrrolidine-2′-one])-1,12-dicarba-*closo*-dodecaborane(12) (**E-5110-Cb**)

(3*E*)-1-(*tert*-Butyl-dimethylsiloxy)-12-(methylene-[1′-methoxypyrrolidine-2′-one])-1,12-dicarba-*closo*-dodecaborane(12) (**7**) (1.00 eq., 0.18 g, 0.45 mmol) was dissolved in 1.50 mL dry THF. At 0 °C, a TBAF solution (1.05 eq., 1.00 M in THF, 0.47 mL, 0.47 mmol) was added dropwise. The reaction mixture was stirred for 30 min at 0 °C. The reaction was then stopped by adding 20 mL distilled water. After dilution with 25 mL EtOAc the phases were separated. The aqueous phase was extracted three times with 25 mL EtOAc. The combined organic phases were washed twice with 20 mL brine and dried over magnesium sulfate. The volatiles were removed under reduced pressure and the crude product was purified by column chromatography (SiO$_2$, hexane/EtOAc, 2:1, *v/v*). Colorless solid, yield 93% (0.12 g, 0.42 mmol); m.p.: 222–224 °C; TLC (hexane/EtOAc, 2:1, *v/v*): R_f = 0.52. ^{1}H-NMR (400.16 MHz, acetone-d$_6$): δ [ppm] = 1.48–3.28 (br m, 10H, **BH**), 2.84 (ddd, $^3J_{HH}$ = 6.8, 5.9 Hz, $^4J_{HH}$ = 2.9 Hz, 2H, 4-**CH$_2$**), 3.58 (dd, $^3J_{HH}$ = 6.7, 5.9 Hz, 2H, 3-**CH$_2$**), 3.73 (s, 3H, 1-**CH$_3$**), 5.99 (t, $^4J_{HH}$ = 2.9 Hz, 1H, 6-**CH**); ^{13}C{^{1}H}-NMR (100.63 MHz, acetone-d$_6$): δ [ppm] = 21.9 (4-**CH$_2$**), 43.3 (3-**CH$_2$**), 62.1 (1-**CH$_3$**), 66.4 (7-**C$_q$**), 111.1 (8-**C$_q$**), 126.8 (6-**C$_q$**), 134.5 (5-**C$_q$**), 163.2 (2-**C$_q$**); ^{11}B{^{1}H}-NMR (128.37 MHz, acetone-d$_6$): δ [ppm] = −14.7 (s, 5**B**), −12.6 (s, 5**B**); FTIR (ATR): $\tilde{v}$ [cm^{-1}] = 3165 (b, v_{O-H}), 2941 (w, v_{C-H}), 2603 (s, v_{B-H}), 1686 (s, $v_{C=O}$), 1662 (s, $v_{C=O}$), 1487 (w, $v_{C=C}$), 1458 (w, δ_{C-H}), 1439 (m, δ_{C-H}), 1393 (w, δ_{C-H}), 1355 (w, δ_{C-H}), 1286 (m), 1215 (s), 1126 (m), 1036 (s), 1003 (m), 955 (m), 923 (w), 894 (w), 840 (w), 749 (w, v_{B-B}), 692 (s), 618 (s), 571 (w), 522 (w), 489 (w); HR-ESI-MS (positive mode, MeOH): *m/z* [M + H]$^+$, calcd. for C$_8$H$_{20}$B$_{10}$NO$_3$: 288.2368, found: 288.2388; elemental analysis: calcd. (%) for C$_8$H$_{19}$B$_{10}$NO$_3$: C 33.67, H 6.71, N 4.91, found: C 33.22, H 6.68, N 4.80.

3.3. HPLC Measurements

3.3.1. Analysis of Purity Ultra Performance Liquid Chromatography (UPLC)

Samples were monitored at 254 nm using the following UPLC system: column Aquity UPLC® BEH C18 column (waters, 150 × 2.1 mm, 1.7 μM, 130 Å), UPLC (waters, Milford, MA, USA): binary solvent manager UPB, sample manager UPA, column manager UPM, and diode array detector PDA, γ detector Gabi Star (Raytest), flow rate 0.4 mL/min, eluent: (A) 0.1% trifluoroacetic acid in H_2O, (B) MeCN; gradient: $t_{0\,min}$ 95/5—$t_{0.3\,min}$ 95/5—$t_{5.3\,min}$ 5/95—$t_{6.5\,min}$ 5/95—$t_{6.8\,min}$ 95/5—$t_{10\,min}$ 95/5). All compounds were found to have a purity of >97% (see Supplementary Materials, Figures S21–S23).

3.3.2. Determination of Lipophilicity (logD) by HPLC

The $logD_{7.4,\,HPLC}$ value was determined as previously reported by us [95] utilizing an HPLC method originally described by Donovan and Pescatore [82]. The following HPLC system was used: Agilent 1100 HPLC (binary pump G1312A, autosampler G1313A, column oven G1316A, degasser G1322A, UV detector G1314A, γ detector Gabi Star (Raytest); column ODP-50 4B (Shodex Asahipak 50 × 4.6 mm) with C18 pre-column; eluent: MeOH/phosphate buffer (10 mM, pH 7.4), gradient t_0 min 30/70—$t_{25\,min}$ 95/5—$t_{27\,min}$ 95/5—$t_{28\,min}$ 30/70—$t_{40\,min}$ 30/70, flow rate = 0.6 mL/min. Oxycarboxin (t_R 9.02 min, $logD_{7.4}$ 1.13) and triphenylene (t_R 29.47 min, $logD_{7.4}$ 5.49) served as references. Toluene was used as control and $logD_{7.4}$ was found to be 2.72 (literature $logD_{7.4}$ 2.72) [81].

3.4. COX Inhibition Studies

COX inhibition activity against ovine COX-1 and human COX-2 was determined using the fluorescence-based COX assay COX Fluorescent Inhibitor Screening Assay Kit (Cayman Chemical Company, Ann Arbor, MI, USA) according to the manufacturer's instructions as previously reported by us. All compounds were screened at a concentration of 100 μM in duplicate [76].

3.5. 5-LO Inhibitory Studies

For the determination of 5-LO inhibitory activities in intact cells, freshly isolated polymorphonuclear leukocytes (PMNL) ($5 × 10^6$) were re-suspended in phosphate-buffered saline (PBS) (pH 7.4) containing 1 mg/mL glucose and 1 μM $CaCl_2$. After preincubation with **R-830**, **S-2474**, **KME-4**, and **E-5110** in DMSO for 15 min at 37 °C, 5-LO product formation was stimulated by addition of calcium ionophore A23187 (2.5 μM in MeOH) and exogenous arachidonic acid (20 μM in EtOH). After 10 min at 37 °C, the reaction was stopped by addition of ice-cold methanol (1 mL). HCl (30 μL, 1 M), prostaglandin B_1 (200 ng) and PBS (500 μL) were added and the formed metabolites were extracted and analyzed by HPLC as described previously [96]. 5-LO product formation was determined as the number of 5-LO products produced (nanograms) per 10^6 cells, which includes leukotriene B_4 (LTB_4), its all-*trans* isomers and 5-hydroperoxyicosatetraenoic acid (5-H(p)ETE). Cysteinyl LTs C_4, D_4, and E_4 as well as oxidation products of LTB_4 were not detected. Data were normalized to vehicle control (DMSO) and IC_{50} values and 95% confidence intervals (CIs) of at least three independent measurements were calculated (Table 2, see Supplementary Materials, Table S2, Figures S25 and S26). The selective 5-LO inhibitor BWA4C (0.1 μM) was used as control and inhibited 5-LO product formation by 92.5% ± 0.9.

3.6. Cell Viability Studies—MTT and CV Assays

All cell lines were seeded overnight and treated with **R-830**, **R-830-Cb**, **KME-4**, **KME-4-Cb**, **E-5110**, **E-5110-Cb**, **S-2474**, and **S-2474-Cb** for 72 h. After the incubation period, the supernatant was discarded from the wells, and the cells were washed two times with 200 μL of PBS. Next, MTT solution at a final concentration of 0.5 mg/mL was added and incubated at 37 °C until purple formazan crystals were formed (30 min to one hour). After incubation, the dye was discarded. In order to dissolve the formed formazan crystals DMSO was added, and the absorbance was measured at λ_{max} = 540 nm, with the reference/background wavelength of 670 nm. All results are expressed as a percentage of the control value which was arbitrary set to 100%. After 72 h treatment, cells were washed with 200 µL of PBS and fixed with 4% paraformaldehyde (PFA) for 15 min at room temperature (RT). Next, cells were stained with 1% CV solution. After 20 min, cells were washed in tap water and dried on air. Dye was dissolved in 33% acetic acid and the absorbance was measured at $\lambda_{max} = 540$ nm, with the reference/background wavelength of 670 nm. Results were expressed as a percentage of the control value which was arbitrary set to 100%.

3.7. Statistical Analyses

All experiments were repeated at least three independent times and data presented represents the means $\pm$ SD of replicates. To evaluate the significance between groups Student's *t*-test was used and two-sided *p* values of less than 0.05 were considered statistically significant.

Further materials, methods, and procedures for physicochemical characterization and biological evaluation: see Supplementary Materials (i.e., Supplementary Materials) for further details.

4. Conclusions

The substitution of the bulky di-*tert*-butylphenyl moiety in the potent dual COX-2/5-LO inhibitors **R-830**, **KME-4**, **E-5110**, and **S-2474** by *p*-carborane gave rise to the carborane-based analogs **R-830-Cb**, **KME-4-Cb**, **E-5110-Cb**, and **S-2474-Cb**. Replacing the bulky but also hydrophobic di-*tert*-butylphenyl moiety with the hydrophobic boron cluster resulted in a significant decrease in lipophilicity, suggesting that intermolecular interactions, such as hydrogen bonds or dihydrogen bonds, are contributing to the compounds' lipophilicity [97]. The carborane analogs show no or only weak COX inhibition, as anticipated based on previous reports about the anti-inflammatory activity of related mono- or di-*tert*-butylphenols.

However, in vitro studies on intact cells revealed that all carborane-based analogs are strong inhibitors of 5-LO product formation with IC_{50} values in the nanomolar range, comparable to the respective organic counterparts, indicating that the bioisosteric replacement of the di-*tert*-butylphenyl motif by *p*-carborane is well tolerated.

Furthermore, the introduction of *p*-carborane generally decreased the cytotoxic potential of dual inhibitors, except in the case of **R-830**. Interestingly, the *p*-carborane analog **R-830-Cb** was non-toxic to primary cells but effective against tested cell lines derived from various types of inflammation-related tumors. Its cytotoxic potential was mediated by potent inhibition of ROS, resulting in inhibited proliferation and caspase-dependent apoptosis.

Thus, carborane-based analog **R-830-Cb** is a promising candidate for further assessment and detailed mechanistic as well as in vivo studies.

Supplementary Materials: The following Supplementary Materials can be downloaded at https://www.mdpi.com/article/10.3390/molecules28114547/s1. Section 1: General methods, materials, and procedures; Section 2: Chemical structures of intermediate compounds and selected spectra of analogs **R-830-Cb**, **KME-4-Cb**, **E-5110-Cb**, and **S-2474-Cb**; Section 3: Determination of purity by HPLC measurements; Section 4: NMR spectroscopic stability studies; Section 5: Determination of lipophilicity (log*D*) by HPLC; Section 6: Biological evaluation; Section 7: Single-crystal X-ray diffraction data for compounds **1**, **5**, **6**, **7**, **R-830-Cb**, **S-2474-Cb**, **KME-4-Cb**, and **E-5110-Cb**. Additional references are cited within the Supplementary Materials [98–102].

Author Contributions: Conceptualization, S.B. and E.H.-H.; methodology, S.B., S.J., M.L., B.H., P.L., S.M., D.M.-I. and E.H.-H.; software, S.B., S.J., M.L., B.H., P.L., S.M., D.M.-I. and E.H.-H.; validation, S.B., S.J., M.L., S.G., B.H., P.L., D.S., J.P., S.M., D.M.-I. and E.H.-H.; formal analysis, S.B., M.L., B.H., J.P., D.M.-I. and E.H.-H.; investigation, S.B., S.J., M.L., S.G., B.H., P.L., S.M., D.M.-I. and E.H.-H.; resources, D.S., J.P., S.M., D.M.-I. and E.H.-H.; data curation, S.B., S.J., M.L., B.H., P.L., S.M., D.M.-I. and E.H.-H.; writing—original draft preparation, S.B.; writing—review and editing, S.B., S.J., M.L., B.H., P.L., J.P., S.M., D.M.-I. and E.H.-H.; visualization, S.B., S.J., M.L., B.H., P.L., J.P., S.M., D.M.-I. and E.H.-H.; supervision, E.H.-H.; project administration, E.H.-H.; funding acquisition, E.H.-H. All authors have read and agreed to the published version of the manuscript.

Funding: This research was funded by the European Social Fund (S.B., E.H.-H.), the DFG (PI 304/7-1 (M.L. and J.P.), He 1376/54-1 (E.H.-H.), SFB 1039 (D.S.), Transregio 314061271-CRC/TRR 205/1-2, B10 (J.P.)), the Graduate School Leipzig School of Natural Sciences—Building with Molecules and Nano-objects (BuildMoNa) (S.B., E.H.-H.), the Ministry of Science, Technological Development, and Innovation of the Republic of Serbia, grant number 451-03-47/2023-01/200007 (D.M.-I., S.M., S.J.). The partial support by the LiSyM Cancer phase I joint collaborative project DEEP-HCC (Federal Ministry of Education and Research; No. 031L0258B) is also acknowledged.

Institutional Review Board Statement: Animal handling and all study protocols were in agreement with the European Community guidelines (EEC Directive of 1986; 86/609/EEC) and the local guidelines. Experimental protocols were approved by the local Institutional Animal Care and Use Committee (IACUC) and by the national licensing committee at the Department of Animal Welfare, Veterinary Directorate, Ministry of Agriculture, Forestry and Water Management of the Republic of Serbia (Permission No. 323-07-02147/2023-05).

Informed Consent Statement: Not applicable.

Data Availability Statement: The data presented in this study are available in the article or Supplementary Materials.

Conflicts of Interest: The authors declare no conflict of interest.

Abbreviations

Abbreviation	Meaning
A375	Human melanoma cell line
A549	Human lung carcinoma cell line
AA	Arachidonic acid
AnnV	Annexin V
AO	Acridine orange
Apostat	FITC-conjugated pan-caspase inhibitor
ATR-IR	Attenuated total reflection infrared spectroscopy
b/br	Broad
BWA4C	N-[(E)-3-(3-Phenoxyphenyl)prop-2-enyl]acetohydroxamic acid
$CDCl_3$	Deuterated chloroform
CFSE	Carboxyfluorescein N-succinimidyl ester
CIs	Confidence intervals
COX	Cyclooxygenase
COXIBs	COX-2 specific inhibitors
CV	Crystal violet
CYP	Cytochrome P_{450}
d	Doublet
dd	Doublet of doublets
ddd	Doublet of doublets of doublets
dtd	Doublet of triplets of doublets
DHR 123	Dihydrorhodamine 123
DMAP	4-Dimethylaminopyridine
FLAP	5-Lipoxygenase-activating protein
FT-ICR	Fourier-transform ion cyclotron resonance mass spectrometry
GI	Gastrointestinal
HCT116	Human colorectal carcinoma cell line
5-H(p)ETE	5-Hydroperoxyeicosatetraenoic acid
HPLC	High-performance liquid chromatography
HR-ESI-MS	High-resolution electrospray ionization mass spectrometry
HT29	Human colorectal adenocarcinoma cell line
IC_{50}	Half-maximal inhibitory concentration

log$D$	Partition coefficient
LO	Lipoxygenase
LTs	Leukotrienes
LTB$_4$	Leukotriene B$_4$
m	Medium (IR), multiplet (NMR), *meta*
M	Molarity
3-MA	3-Methyladenine
MDA-MB-231	M. D. Anderson-Metastatic breast-231 triple negative adenocarcinoma cell line
m.p.	Melting point
MTT	3-(4,5-Dimethylthiazol-2-yl)-2,5-diphenyl-2*H*-tetrazolium bromide
*n*BuLi	*n*-Butyllithium
NEt$_3$	Triethylamine
n.i.	no inhibition
NMR	Nuclear magnetic resonance
NSAIDs	Non-steroidal anti-inflammatory drugs
p	*para*
PBS	Phosphate-buffered saline
PFA	Paraformaldehyde
PGs	Prostaglandins
PGE$_2$	Prostaglandin E$_2$
PGI$_2$	Prostaglandin I$_2$
PI	Propidium iodide
PMNL	Polymorphonuclear leukocytes
ppm	Parts per million
p-Ts	*p*-Toluenesulfonyl
q	Quartet
R_F	Retention factor
RT	Room temperature
s	Strong (IR), singlet (NMR)
SC-560	5-(4-Chlorophenyl)-1-(4-methoxyphenyl)-3-trifluoromethyl pyrazole
SD	Standard deviation
SI	Selectivity index
t	Triplet
td	Triplet of doublets
TBAF	Tetra-*n*-butylammonium fluoride
TBDMSCl	*Tert*-butyldimethylsilyl chloride
TLC	Thin-layer chromatography
TMS	Tetramethylsilane
t_R	Retention time
TXA$_2$	Thromboxane A$_2$
UPLC	Ultra-performance liquid chromatography
w	Weak

References

1. Ting, A.H.; McGarvey, K.M.; Baylin, S.B. The cancer epigenome-components and functional correlates. *Genes Dev.* **2006**, *20*, 3215–3231. [CrossRef] [PubMed]
2. Wang, D.; Dubois, R.N. Eicosanoids and cancer. *Nat. Rev. Cancer* **2010**, *10*, 181–193. [CrossRef] [PubMed]
3. Sanz-Motilva, V.; Martorell-Calatayud, A.; Nagore, E. Non-steroidal Anti-Inflammatory Drugs and Melanoma. *Curr. Pharm. Des.* **2012**, *18*, 3966–3978. [CrossRef]
4. Kinzler, K.W.; Vogelstein, B. Lessons from HEREDITARY colorectal cancer. *Cell* **1996**, *87*, 159–170. [CrossRef]
5. Prendergast, G.C.; Metz, R.; Muller, A. Towards a genetic definition of cancer-associated inflammation: Role of the IDO pathway. *Am. J. Clin. Pathol.* **2010**, *176*, 2082–2087. [CrossRef] [PubMed]
6. Hanahan, D.; Weinberg, R.A. The hallmarks of cancer. *Cell* **2000**, *100*, 57–70. [CrossRef] [PubMed]
7. Arneth, B. Tumor Microenvironment. *Medicina* **2019**, *56*, 15. [CrossRef] [PubMed]
8. Johnson, A.M.; Kleczko, E.K.; Nemenoff, R. Eicosanoids in Cancer: New Roles in Immunoregulation. *Front. Pharmacol.* **2020**, *11*, 595498. [CrossRef]
9. Moore, G.Y.; Pidgeon, G. Cross-Talk between Cancer Cells and the Tumour Microenvironment: The Role of the 5-Lipoxygenase Pathway. *Int. J. Mol. Sci.* **2017**, *18*, 236. [CrossRef]

10. Charlier, C.; Michaux, C. Dual inhibition of cyclooxygenase-2 (COX-2) and 5-lipoxygenase (5-LOX) as a new strategy to provide safer non-steroidal anti-inflammatory drugs. *Eur. J. Med. Chem.* **2003**, *38*, 645–659. [CrossRef]

11. Jacob, P.J.; Manju, S.L.; Ethiraj, K.R.; Elias, G. Safer anti-inflammatory therapy through dual COX-2/5-LOX inhibitors: A structure-based approach. *Eur. J. Pharm. Sci.* **2018**, *121*, 356–381.

12. Brücher, B.L.D.M.; Jamall, I.S. Eicosanoids in carcinogenesis. *4open* **2019**, *2*, 9. [CrossRef]

13. Sharma, V.; Bhatia, P.; Alam, O.; Naim, M.J.; Nawaz, F.; Sheikh, A.A.; Jha, M. Recent advancement in the discovery and development of COX-2 inhibitors: Insight into biological activities and SAR studies (2008–2019). *Bioorg. Chem.* **2019**, *89*, 103007. [CrossRef] [PubMed]

14. Rabbani, M.; Ismail, S.M.; Zarghi, A. Selective COX-2 inhibitors as anticancer agents: A patent review (2014–2018). *Expert Opin. Ther. Pat.* **2019**, *29*, 407–427. [CrossRef]

15. Mohsin, N.-A.; Irfan, M. Selective cyclooxygenase-2 inhibitors: A review of recent chemical scaffolds with promising anti-inflammatory and COX-2 inhibitory activities. *Med. Chem. Res.* **2020**, *29*, 809–830. [CrossRef]

16. Roos, J.; Grösch, S.; Werz, O.; Schröder, P.; Ziegler, S.; Fulda, S.; Paulus, P.; Urbschat, A.; Kühn, B.; Maucher, I.; et al. Regulation of tumorigenic Wnt signaling by cyclooxygenase-2, 5-lipoxygenase and their pharmacological inhibitors: A basis for novel drugs targeting cancer cells? *Pharmacol. Ther.* **2016**, *157*, 43–64. [CrossRef]

17. Meshram, M.A.; Bhise, U.O.; Makhal, P.N.; Kaki, V. Synthetically-tailored and nature-derived dual COX-2/5-LOX inhibitors: Structural aspects and SAR. *Eur. J. Med. Chem.* **2021**, *225*, 113804. [CrossRef]

18. Arias-Negrete, S.; Keller, K.; Chadee, K. Proinflammatory cytokines regulate cyclooxygenase-2 mRNA expression in human macrophages. *Biochem. Biophys. Res. Commun.* **1995**, *208*, 582–589. [CrossRef]

19. Vecchio, A.J.; Simmons, D.M.; Malkowski, M. Structural basis of fatty acid substrate binding to cyclooxygenase-2. *J. Biol. Chem.* **2010**, *285*, 22152–22163. [CrossRef]

20. Zarghi, A.; Arfaei, S. Selective COX-2 Inhibitors: A Review of Their Structure-Activity Relationships. *Iran. J. Pharm. Sci.* **2011**, *10*, 655.

21. Goossens, L.; Pommery, N.; Hénichart, J. COX-2/5-LOX dual acting anti-inflammatory drugs in cancer chemotherapy. *Curr. Top. Med. Chem.* **2007**, *7*, 283–296. [CrossRef] [PubMed]

22. Haase-Kohn, C.; Laube, M.; Donat, C.K.; Belter, B.; Pietzsch, J. CRISPR/Cas9 Mediated Knockout of Cyclooxygenase-2 Gene Inhibits Invasiveness in A2058 Melanoma Cells. *Cells* **2022**, *11*, 749. [CrossRef] [PubMed]

23. Mitchell, J.A.; Belvisi, M.G.; Akarasereenont, P.; Robbins, R.A.; Kwon, O.J.; Croxtall, J.; Barnes, P.J.; Vane, J. Induction of cyclo-oxygenase-2 by cytokines in human pulmonary epithelial cells: Regulation by dexamethasone. *Br. J. Pharmacol.* **1994**, *113*, 1008–1014. [CrossRef]

24. Moreno, J.; Krishnan, A.V.; Swami, S.; Nonn, L.; Peehl, D.M.; Feldman, D. Regulation of prostaglandin metabolism by calcitriol attenuates growth stimulation in prostate cancer cells. *Cancer Res.* **2005**, *65*, 7917–7925. [CrossRef] [PubMed]

25. Ahmadi, M.; Bekeschus, S.; Weltmann, K.-D.; Woedtke, T.; Wende, K. Non-steroidal anti-inflammatory drugs: Recent advances in the use of synthetic COX-2 inhibitors. *RSC Med. Chem.* **2022**, *13*, 471–496. [CrossRef]

26. Orlando, B.J.; Lucido, M.J.; Malkowski, M. The structure of ibuprofen bound to cyclooxygenase-2. *J. Struct. Biol.* **2015**, *189*, 62–66. [CrossRef]

27. Lucido, M.J.; Orlando, B.J.; Vecchio, A.J.; Malkowsk, M. Crystal Structure of Aspirin-Acetylated Human Cyclooxygenase-2: Insight into the Formation of Products with Reversed Stereochemistry. *Biochemistry* **2016**, *55*, 1226–1238. [CrossRef]

28. Haeggström, J.Z.; Funk, C. Lipoxygenase and leukotriene pathways: Biochemistry, biology, and roles in disease. *Chem. Rev.* **2011**, *111*, 5866–5898. [CrossRef]

29. Reddy, K.K.; Rajan, V.K.V.; Gupta, A.; Aparoy, P.; Reddanna, P. Exploration of binding site pattern in arachidonic acid metabolizing enzymes, Cyclooxygenases and Lipoxygenases. *BMC Res. Notes* **2015**, *8*, 152. [CrossRef]

30. Reddy, S.; Zhang, S. Polypharmacology: Drug discovery for the future. *Expert Rev. Clin. Pharmacol.* **2013**, *6*, 41–47. [CrossRef]

31. Pidgeon, G.; Cathcart, M.-C. *The Role of Cyclooxygenases and Lipoxygenases in the Regulation of Tumor Angiogensis*; Taylor & Francis: New York, NY, USA, 2013; Chapter 7.

32. Sinha, S.; Doble, M.; Manju, S. 5-Lipoxygenase as a drug target: A review on trends in inhibitors structural design, SAR and mechanism based approach. *Bioorg. Med. Chem.* **2019**, *27*, 3745–3759. [CrossRef]

33. Orafaie, A.; Mousavian, M.; Orafai, H.; Sadeghian, H. An overview of lipoxygenase inhibitors with approach of in vivo studies. *Prostaglandins Other Lipid Mediat.* **2020**, *148*, 106411. [CrossRef] [PubMed]

34. de Leval, X.; Julemont, F.; Delarge, J.; Pirotte, B.; Dogne, J.-M. New trends in dual 5-LOX/COX inhibition. *Curr. Med. Chem.* **2002**, *9*, 941–962. [CrossRef] [PubMed]

35. Julémont, F.; Dogné, J.-M.; Pirotte, B.; de Leval, X. Recent development in the field of dual COX / 5-LOX inhibitors. *Mini-Rev. Med. Chem.* **2004**, *4*, 633–638. [CrossRef]

36. Le Filliatre, G.; Sayah, S.; Latournerie, V.; Renaud, J.F.; Finet, M.; Hanf, R. Cyclooxygenase and lipoxygenase pathways in mast cell dependent-neurogenic inflammation induced by electrical stimulation of the rat saphenous nerve. *Br. J. Pharmacol.* **2001**, *132*, 1581–1589. [CrossRef]

37. He, C.; Wu, Y.; Lai, Y.; Cai, Z.; Liu, Y.; Lai, L. Dynamic eicosanoid responses upon different inhibitor and combination treatments on the arachidonic acid metabolic network. *Mol. BioSystems* **2012**, *8*, 1585–1594. [CrossRef] [PubMed]

38. Busse, W.W. Leukotrienes and Inflammation. *Am. J. Respir. Crit. Care. Med.* **1996**, *157*, 210–213. [CrossRef] [PubMed]

39. Mahboubi-Rabbani, M.; Zarghi, A. Lipoxygenase Inhibitors as Cancer Chemopreventives: Discovery, Recent Developments and Future Perspectives. *Curr. Med. Chem.* **2021**, *28*, 1143–1175. [CrossRef] [PubMed]

40. Davis, R.; Brogden, R.N. Nimesulide. *Drugs* **1994**, *48*, 431–454. [CrossRef]

41. Khanapure, S.P.; Garvey, D.S.; Janero, D.R.; Letts, L.G. Eicosanoids in Inflammation: Biosynthesis, Pharmacology, and Therapeutic Frontiers. *Curr. Top. Med. Chem.* **2007**, *7*, 311–340. [CrossRef]

42. Braeckmann, R.A.; Granneman, G.R.; Locke, C.S.; Machinist, J.M.; Cavanaugh, J.H.; Awni, W. The Pharmacokinetics of Zileuton in Healthy Young and Elderly Volunteers. *Clin. Pharmacokinet.* **1995**, *29*, 42–48. [CrossRef] [PubMed]

43. Paredes, Y.; Massicotte, F.; Pelletier, J.-P.; Martel-Pelletier, J.; Laufer, S.; Lajeunesse, D. Study of the role of leukotriene B4 in abnormal function of human subchondral osteoarthritis osteoblasts: Effects of cyclooxygenase and/or 5-lipoxygenase inhibition. *Arthritis Rheumatol.* **2002**, *46*, 1804–1812. [CrossRef] [PubMed]

44. Martel-Pelletier, J.; Lajeunesse, D.; Reboul, P.; Pelletier, J.-P. Therapeutic role of dual inhibitors of 5-LOX and COX, selective and non-selective non-steroidal anti-inflammatory drugs. *Ann. Rheum. Dis.* **2003**, *62*, 501–509. [CrossRef] [PubMed]

45. Jovanovic, D.V.; Fernandes, J.C.; Martel-Pelletier, J.; Jolicoeur, F.-C.; Reboul, P.; Laufer, S.; Tries, S.; Pelletier, J.-P. In vivo dual inhibition of cyclooxygenase and lipoxygenase by ML-3000 reduces the progression of experimental osteoarthritis: Suppression of collagenase 1 and interleukin-1? synthesis. *Arthritis Rheum.* **2001**, *44*, 2320–2330. [CrossRef]

46. Ruiz, J. QSAR study of dual cyclooxygenase and 5-lipoxygenase inhibitors 2,6-di-tert-butylphenol derivatives. *Bioorg. Med. Chem.* **2003**, *11*, 4207–4216. [CrossRef]

47. Moore, G.G.; Swingle, K.F. 6-Di-tert-butyl-4-(2′-thenoyl)phenol(R-830): A novel nonsteroidal anti-inflammatory agent with antioxidant properties. *Agents Actions* **1982**, *12*, 674–683. [CrossRef]

48. Blackham, A.; Norris, A.A.; Woods, F. Models for evaluating the anti-inflammatory effects of inhibitors of arachidonic acid metabolism. *J. Pharm. Pharmacol.* **1985**, *37*, 787–793. [CrossRef]

49. Hidaka, T.; Hosoe, K.; Ariki, Y.; Takeo, K.; Yamashita, T.; Katsumi, I.; Kondo, H.; Yamashita, K.; Watanabe, K. Pharmacological properties of a new anti-inflammatory compound, alpha-(3,5-di-tert-butyl-4-hydroxybenzylidene)-gamma-butyrolacto ne (KME-4), and its inhibitory effects on prostaglandin synthetase and 5-lipoxygenase. *Jpn. J. Clin. Pharmacol. Ther.* **1984**, *36*, 77–85.

50. Hidaka, T.; Takeo, K.; Hosoe, K.; Katsumi, I.; Yamashita, T.; Watanabe, K. Inhibition of polymorphonuclear leukocyte 5-lipoxygenase and platelet cyclooxygenase by alpha-(3,5-di-tert-butyl-4-hydroxybenzylidene)-gamma-butyrolacto ne (KME-4), a new anti-inflammatory drug. *Jpn. J. Clin. Pharmacol. Ther.* **1985**, *38*, 267–272.

51. Hidaka, T.; Hosoe, K.; Katsumi, I.; Yamashita, T.; Watanabe, K. The effect of alpha-(3,5-di-t-butyl-4-hydroxybenzylidene)-gamma-butyrolactone (KME-4), a new anti-inflammatory drug, on leucocyte migration in rat carrageenan pleurisy. *J Pharm Pharmacol.* **1986**, *38*, 244–247. [CrossRef]

52. Hidaka, T.; Hosoe, K.; Yamashita, T.; Watanabe, K. Effect of alpha-(3,5-di-tert-butyl-4-hydroxybenzylidene)-gamma-butyrolactone (KME-4), a new anti-inflammatory drug, on the established adjuvant arthritis in rats. *Jpn. J. Clin. Pharmacol. Ther.* **1986**, *42*, 181–187.

53. Katsumi, I.; Kondo, H.; Yamashita, K.; Hidaka, T.; Hosoe, K.; Yamashita, T.; Watanabe, K. Studies on styrene derivatives. I. Synthesis and antiinflammatory activities of alpha-benzylidene-gamma-butyrolactone derivatives. *Chem. Pharm. Bull.* **1986**, *34*, 121–129. [CrossRef] [PubMed]

54. Ikuta, H.; Shirota, H.; Kobayashi, S.; Yamagishi, Y.; Yamada, K.; Yamatsu, I.; Katayama, K. Synthesis and antiinflammatory activities of 3-(3,5-di-tert-butyl-4-hydroxybenzylidene)pyrrolidin-2-ones. *J. Med. Chem.* **1987**, *30*, 1995–1998. [CrossRef] [PubMed]

55. Shirota, H.; Goto, M.; Hashida, R.; Yamatsu, I.; Katayama, K. Inhibitory effects of E-5110 on interleukin-1 generation from human monocytes. *Agents Actions* **1989**, *27*, 322–324. [CrossRef]

56. Shirota, H.; Katayama, K.; Ono, H.; Chiba, K.; Kobayashi, S.; Terato, K.; Ikuta, H.; Yamatsu, I. Pharmacological properties of N-methoxy-3-(3,5-ditert-butyl-4-hydroxybenzylidene)-2-pyrrolidone (E-5110), a novel nonsteroidal antiinflammatory agent. *Agents Actions* **1987**, *21*, 250–252. [CrossRef]

57. Katayama, K.; Shirota, H.; Kobayashi, S.; Terato, K.; Ikuta, H.; Yamatsu, I. In vitro effect of N-methoxy-3-(3,5-ditert-butyl-4-hydroxybenzylidene)-2-pyrrolidon e (E-5110), a novel nonsteroidal antiinflammatory agent, on generation of some inflammatory mediators. *Agents Actions* **1987**, *21*, 269–271. [CrossRef]

58. Inagaki, M.; Tsuri, T.; Jyoyama, H.; Ono, T.; Yamada, K.; Kobayashi, M.; Hori, Y.; Arimura, A.; Yasui, K.; Ohno, K.; et al. Novel antiarthritic agents with 1,2-isothiazolidine-1,1-dioxide (gamma-sultam) skeleton: Cytokine suppressive dual inhibitors of cyclooxygenase-2 and 5-lipoxygenase. *J. Med. Chem.* **2000**, *43*, 2040–2048. [CrossRef]

59. Scholz, M.; Hey-Hawkins, E. Carboranes as pharmacophores: Properties, synthesis, and application strategies. *Chem. Rev.* **2011**, *111*, 7035–7062. [CrossRef]

60. Stockmann, P.; Gozzi, M.; Kuhnert, R.; Sárosi, M.B.; Hey-Hawkins, E. New keys for old locks: Carborane-containing drugs as platforms for mechanism-based therapies. *Chem. Soc. Rev.* **2019**, *48*, 3497–3512. [CrossRef]

61. Marfavi, A.; Kavianpour, P.; Rendina, L. Carboranes in drug discovery, chemical biology and molecular imaging. *Nat. Rev. Chem.* **2022**, *6*, 486–504. [CrossRef]

62. Messner, K.; Vuong, B.; Tranmer, G. The Boron Advantage: The Evolution and Diversification of Boron's Applications in Medicinal Chemistry. *Pharmaceuticals* **2022**, *15*, 264. [CrossRef] [PubMed]

63. Hoffmann, R. Building Bridges between Inorganic and Organic Chemistry (Nobel Lecture). *Angew. Chem. Int. Ed.* **1982**, *21*, 711–724. [CrossRef]

64. Issa, F.; Kassiou, M.; Rendina, L. Boron in drug discovery: Carboranes as unique pharmacophores in biologically active compounds. *Chem. Rev.* **2011**, *111*, 5701–5722. [CrossRef] [PubMed]

65. Chen, Y.; Du, F.; Tang, L.; Xu, J.; Zhao, Y.; Wu, X.; Li, M.; Shen, J.; Wen, Q.; Cho, C.H.; et al. Carboranes as unique pharmacophores in antitumor medicinal chemistry. *Mol. Ther. Oncolytics* **2022**, *24*, 400–416. [CrossRef]

66. Ali, F.; Hosmane, N.; Zhu, Y. Boron Chemistry for Medical Applications. *Molecules* **2020**, *25*, 828. [CrossRef]

67. Bregadze, V.I. Dicarba-closo-dodecaboranes C2B10H12 and their derivatives. *Chem. Rev.* **1992**, *92*, 209–223. [CrossRef]

68. Grimes, R.N. *Carboranes*, 2nd ed.; Elsevier Science & Technology: Saint Louis, MO, USA, 2011.

69. Zargham, E.O.; Mason, C.A.; Lee, M. The Use of Carboranes in Cancer Drug Development. *Int. J. Cancer Clin. Res.* **2019**, *6*, 110.

70. Das, B.C.; Nandwana, N.K.; Das, S.; Nandwana, V.; Shareef, M.A.; Das, Y.; Saito, M.; Weiss, L.M.; Almaguel, F.; Hosmane, N.S.; et al. Boron Chemicals in Drug Discovery and Development: Synthesis and Medicinal Perspective. *Molecules* **2022**, *27*, 2615. [CrossRef]

71. Neumann, W.; Xu, S.; Sárosi, M.B.; Scholz, M.S.; Crews, B.C.; Ghebreselasie, K.; Banerjee, S.; Marnett, L.J.; Hey-Hawkins, E. nido-Dicarbaborate Induces Potent and Selective Inhibition of Cyclooxygenase-2. *ChemMedChem* **2016**, *11*, 175–178. [CrossRef]

72. Buzharevski, A.; Paskas, S.; Laube, M.; Lönnecke, P.; Neumann, W.; Murganic, B.; Mijatović, S.; Maksimović-Ivanić, D.; Pietzsch, J.; Hey-Hawkins, E. Carboranyl Analogues of Ketoprofen with Cytostatic Activity against Human Melanoma and Colon Cancer Cell Lines. *ACS Omega* **2019**, *4*, 8824–8833. [CrossRef]

73. Buzharevski, A.; Paskas, S.; Sárosi, M.-B.; Laube, M.; Lönnecke, P.; Neumann, W.; Mijatović, S.; Maksimović-Ivanić, D.; Pietzsch, J.; Hey-Hawkins, E. Carboranyl Analogues of Celecoxib with Potent Cytostatic Activity against Human Melanoma and Colon Cancer Cell Lines. *ChemMedChem* **2019**, *14*, 315–321. [CrossRef]

74. Kuhnert, R.; Sárosi, M.-B.; George, S.; Lönnecke, P.; Hofmann, B.; Steinhilber, D.; Steinmann, S.; Schneider-Stock, R.; Murganić, B.; Mijatović, S.; et al. Carborane-Based Analogues of 5-Lipoxygenase Inhibitors Co-inhibit Heat Shock Protein 90 in HCT116 Cells. *ChemMedChem* **2019**, *14*, 255–261. [CrossRef]

75. Buzharevski, A.; Paskaš, S.; Sárosi, M.-B.; Laube, M.; Lönnecke, P.; Neumann, W.; Murganić, B.; Mijatović, S.; Maksimović-Ivanić, D.; Pietzsch, J.; et al. Carboranyl Derivatives of Rofecoxib with Cytostatic Activity against Human Melanoma and Colon Cancer Cells. *Sci. Rep.* **2020**, *10*, 4827. [CrossRef] [PubMed]

76. Useini, L.; Mojić, M.; Laube, M.; Lönnecke, P.; Dahme, J.; Sárosi, M.B.; Mijatović, S.; Maksimović-Ivanić, D.; Pietzsch, J.; Hey-Hawkins, E. Carboranyl Analogues of Mefenamic Acid and Their Biological Evaluation. *ACS Omega* **2022**, *7*, 24282–24291. [CrossRef] [PubMed]

77. Useini, L.; Mojić, M.; Laube, M.; Lönnecke, P.; Mijatović, S.; Maksimović-Ivanić, D.; Pietzsch, J.; Hey-Hawkins, E. Carborane Analogues of Fenoprofen Exhibit Improved Antitumor Activity. *ChemMedChem* **2022**, *18*, e202200583. [CrossRef] [PubMed]

78. Braun, S.; Paskaš, S.; Laube, M.; George, S.; Hofmann, B.; Lönnecke, P.; Steinhilber, D.; Pietzsch, J.; Mijatović, S.; Maksimović-Ivanić, D.; et al. In Vitro Cytostatic Effect on Tumor Cells by Carborane-Based Dual Cyclooxygenase-2 and 5-Lipoxygenase Inhibitors. *Adv. Ther.* **2023**, *6*, 2200252. [CrossRef]

79. Braun, S.; Paskaš, S.; Laube, M.; George, S.; Hofmann, B.; Lönnecke, P.; Steinhilber, D.; Pietzsch, J.; Mijatović, S.; Maksimović-Ivanić, D.; et al. Carborane-based Tebufelone Analogs and their Biological Evaluation In Vitro. *ChemMedChem* **2023**, *18*, e202300206. [CrossRef]

80. Ohta, K.; Goto, T.; Yamazaki, H.; Pichierri, F.; Endo, Y. Facile and efficient synthesis of C-hydroxycarboranes and C,C'-dihydroxycarboranes. *Inorg. Chem.* **2007**, *46*, 3966–3970. [CrossRef]

81. Fournier, J.; Arseniyadis, S.; Cossy, J. A modular and scalable one-pot synthesis of polysubstituted furans. *Angew. Chem. Int.* **2012**, *51*, 7562–7566. [CrossRef]

82. Donovan, S.F.; Pescatore, M. Method for measuring the logarithm of the octanol–water partition coefficient by using short octadecyl–poly(vinyl alcohol) high-performance liquid chromatography columns. *J. Chromatogr. A* **2002**, *952*, 47–61. [CrossRef]

83. Schmitt, M.; Greten, F. The inflammatory pathogenesis of colorectal cancer. *Nat. Rev. Immunol.* **2021**, *21*, 653–667. [CrossRef] [PubMed]

84. Zhao, H.; Wu, L.; Yan, G.; Chen, Y.; Zhou, M.; Wu, Y.; Li, Y. Inflammation and tumor progression: Signaling pathways and targeted intervention. *Signal Transduct. Target. Ther.* **2021**, *6*, 263. [CrossRef]

85. Cho, K.-J.; Seo, J.-M.; Kim, J.-H. Bioactive lipoxygenase metabolites stimulation of NADPH oxidases and reactive oxygen species. *Mol. Cells* **2011**, *32*, 1–5. [CrossRef] [PubMed]

86. Weisser, H.; Göbel, T.; Krishnathas, G.M.; Kreiß, M.; Angioni, C.; Sürün, D.; Thomas, D.; Schmid, T.; Häfner, A.-K.; Kahnt, A. Knock-out of 5-lipoxygenase in overexpressing tumor cells-consequences on gene expression and cellular function. *Cancer Gene Ther.* **2023**, *30*, 108–123. [CrossRef]

87. Agarwal, B.; Swaroop, P.; Protiva, P.; Raj, S.V.; Shirin, H.; Holt, P. COX-2 is needed but not sufficient for apoptosis induced by COX-2 selective inhibitors in colon cancer cells. *Apoptosis* **2003**, *8*, 649–654. [CrossRef] [PubMed]

88. Redza-Dutordoir, M.; Averill-Bates, D. Activation of apoptosis signalling pathways by reactive oxygen species. *Biochim. Biophys. Acta* **2016**, *1863*, 2977–2992. [CrossRef]

89. Nakamura, H.; Takada, K. Reactive oxygen species in cancer: Current findings and future directions. *Cancer Sci.* **2021**, *112*, 3945–3952. [CrossRef]

90. Weinberg, F.; Ramnath, N.; Nagrath, D. Reactive Oxygen Species in the Tumor Microenvironment: An Overview. *Cancers* **2019**, *11*, 1191. [CrossRef]

91. Kapałczyńska, M.; Kolenda, T.; Przybyła, W.; Zajączkowska, M.; Teresiak, A.; Filas, V.; Ibbs, M.; Bliźniak, R.; Łuczewski, Ł.; Lamperska, K. 2D and 3D cell cultures—A comparison of different types of cancer cell cultures. *Arch. Med. Sci.* **2018**, *14*, 910–919. [CrossRef]

92. Becker, E.D.; de Menezes, S.M.C.; Goodfellow, R.; Granger, P. NMR nomenclature. Nuclear spin properties and conventions for chemical shifts(IUPAC Recommendations 2001). *Pure Appl. Chem.* **2001**, *73*, 1795–1818.

93. Liu, G.; Xu, B. Hydrogen bond donor solvents enabled metal and halogen-free Friedel–Crafts acylations with virtually no waste stream. *Tetrahedron Lett.* **2018**, *59*, 869–872. [CrossRef]

94. Goto, T.; Ohta, K.; Suzuki, T.; Ohta, S.; Endo, Y. Design and synthesis of novel androgen receptor antagonists with sterically bulky icosahedral carboranes. *Bioorg. Med. Chem.* **2005**, *13*, 6414–6424. [CrossRef] [PubMed]

95. Wodtke, R.; Wodtke, J.; Hauser, S.; Laube, M.; Bauer, D.; Rothe, R.; Neuber, C.; Pietsch, M.; Kopka, K.; Pietzsch, J.; et al. Development of an 18F-Labeled Irreversible Inhibitor of Transglutaminase 2 as Radiometric Tool for Quantitative Expression Profiling in Cells and Tissues. *J. Med. Chem.* **2021**, *64*, 3462–3478. [CrossRef] [PubMed]

96. Brungs, M.; Rådmark, O.; Samuelsson, B.; Steinhilber, D. Sequential induction of 5-lipoxygenase gene expression and activity in Mono Mac 6 cells by transforming growth factor beta and 1,25-dihydroxyvitamin D3. *Proc. Natl. Acad. Sci. USA* **1995**, *92*, 107–111. [CrossRef]

97. Kempińska, D.; Chmiel, T.; Kot-Wasik, A.; Mróz, A.; Mazerska, Z.; Namieśnik, J. State of the art and prospects of methods for determination of lipophilicity of chemical compounds. *Trends Anal. Chem.* **2019**, *113*, 54–73. [CrossRef]

98. Rigaku Oxford Diffraction. *CrysAlisPro Software System*; Rigaku Corporation: Wroclaw, Poland, 2006.

99. SCALE3 ABSPACK. *Empirical Absorption Correction Using Spherical Harmonics*; Oxford Diffraction: Abingdon, UK, 2006.

100. Sheldrick, G.M. SHELXT–Integrated space-group and crystal-structure determination. *Acta Cryst. A* **2015**, *71*, 3–8. [CrossRef] [PubMed]

101. Sheldrick, G.M. Crystal structure refinement with SHELXL. *Acta Cryst. C* **2015**, *71*, 3–8. [CrossRef]

102. Putz, H.; Brandenburg, K. *DIAMOND: Crystal and Molecular Structure Visualization*; Crystal Impact GbR: Bonn, Germany, 2022.

Article

Dehydroborylation of Terminal Alkynes Using Lithium Aminoborohydrides

P. Veeraraghavan Ramachandran * and Henry J. Hamann

Herbert C. Brown Center for Borane Research, Department of Chemistry, Purdue University, 560 Oval Drive, West Lafayette, IN 47907, USA
* Correspondence: chandran@purdue.edu

Abstract: Dehydrogenative borylation of terminal alkynes has recently emerged as an atom-economical one-step alternative to traditional alkyne borylation methodologies. Using lithium aminoborohydrides, formed in situ from the corresponding amine-boranes and *n*-butyllithium, a variety of aromatic and aliphatic terminal alkyne substrates were successfully borylated in high yield. The potential to form mono-, di-, and tri-*B*-alkynylated products has been shown, though the mono-product is primarily generated using the presented condition. The reaction has been demonstrated at large (up to 50 mmol) scale, and the products are stable to column chromatography as well as acidic and basic aqueous conditions. Alternately, the dehydroborylation can be achieved by treating alkynyllithiums with amine-boranes. In that respect, aldehydes can act as starting materials by conversion to the 1,1-dibromoolefin and in situ rearrangement to the lithium acetylide.

Keywords: amine-borane; dehydrogenative borylation; lithium aminoborohydrides; alkynylborane-amines; 1,1-dibromoolefin

1. Introduction

Alkynylboron compounds are valuable intermediates in organic synthesis and can be used as building blocks in a wide range of transformations [1], including coupling reactions [2]. Traditional synthetic routes to these compounds (Scheme 1(1)) involve conversion to an alkynylmetal, exchange with a boron source, and treatment with a dry Brønsted acid [3]. A more atom-economical approach to this transformation is a one-step, direct coupling between the terminal alkyne and the boron source, producing dihydrogen gas as a byproduct. Efforts towards this one-step, dehydrogenative borylation have been frustrated by the tendency of alkynes to undergo hydroboration rather than boronation upon reaction with dioxaborolane and diboron reagents [4,5]. A transition metal-catalyzed dehydrogenative borylation was first reported in 2013, utilizing an iridium SiNN pincer complex (Scheme 1(2)) [6]. Since that initial report, a variety of transition metals (Ag [7], Pd [8], Zn [9,10], Cu [11], Fe [12]) and ligand systems have been reported to accomplish the dehydrocoupling (Scheme 1(2)). However, many of these methodologies use expensive metal catalysts or intricate ligands.

Recently, several methods utilizing main group elements to catalyze the dehydrogenative borylation of terminal alkynes have been developed. Pucheault and coworkers employed magnesium halide (Grignard) catalysts to produce the corresponding acetylides from terminal alkynes [13]. Reaction with diisopropylaminoborane and tandem deprotonation by alkyne regenerated the magnesium acetylide and provided the dehydrocoupled diisopropylaminoalkynylborane product. Grignard promoted in situ dehydrogenation of diisopropylamine-borane to form diisopropylaminoborane was found to be an equally effective starting point for the dehydrocoupling (Scheme 1(3)).

An aluminum-catalyzed borylation of terminal alkynes reported by Thomas and coworkers utilizes the intramolecular alane–amine complex formed from the reaction of 2-lithio-*N*,*N*-dimethylaniline with Me$_2$AlCl [14]. The tethered Lewis pair catalyzes the

Citation: Ramachandran, P.V.; Hamann, H.J. Dehydroborylation of Terminal Alkynes Using Lithium Aminoborohydrides. *Molecules* **2023**, *28*, 3433. https://doi.org/10.3390/molecules28083433

Academic Editors: Michael A. Beckett and Igor B. Sivaev

Received: 22 March 2023
Revised: 10 April 2023
Accepted: 10 April 2023
Published: 13 April 2023

formation of an alkynyl aluminum intermediate, which yields the desired alkynylboronate upon reaction with pinacol borane (Scheme 1(3)). While the desired boronate is provided, the highly reactive nature of aluminum compounds makes their handling difficult.

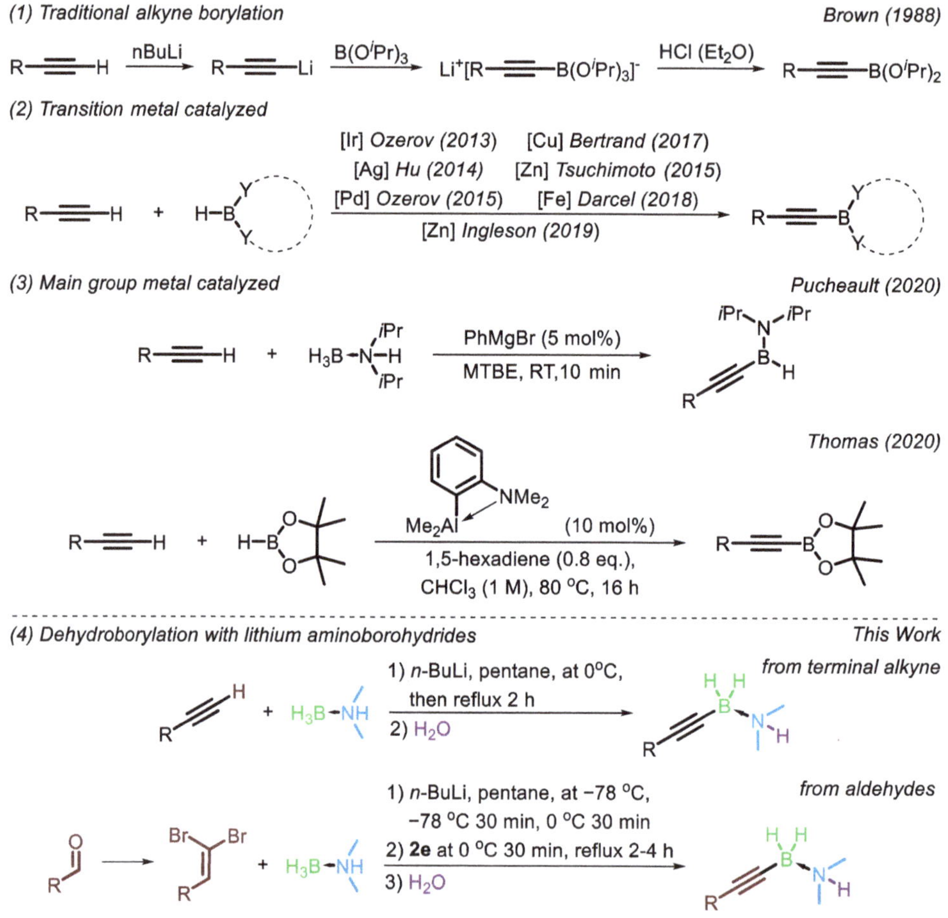

Scheme 1. Past and present alkyne borylation methodologies; Brown [3], Ozerov [6], Hu [7], Ozerov [8], Bertrand [11], Tsuchimoto [9], Darcel [12], Ingleson [10], Pucheault [13], Thomas [14].

Hydrometallation was a concern for both the magnesium and aluminum catalyzed dehydrocoupling reactions. It was found in each case that by producing an alkynylmetal intermediate, hydrometallation could be circumvented. As part of our work exploring the synthetic utility of amine-boranes [15,16] we have recently discovered a direct C-H dehydrogenative borylation of terminal alkynes. Lithium aminoborohydrides (LABs) generated from air- and moisture-stable amine-borane precursors have been found to be effective reagents for the dehydroborylation (Scheme 1(4)).

2. Results and Discussion

LAB reagents are generated from the reaction of an amine-borane with n-butyllithium (n-BuLi) and are best known for their powerful and selective reducing properties, similar to lithium aluminum hydride [17]. When prepared with a slight deficit of n-BuLi, LAB reagents are air-stable solids. Initially, we were investigating LAB reagents as activated derivatives of amine-boranes for the hydroboration of alkynes, as most amine-boranes themselves do not hydroborate under ambient conditions [18,19]. A reaction of phenylacetylene (**1a**) and lithium piperidinoborohydride prepared from piperidine-borane (**2g**) was found to provide a series of alkynylated amine-boranes, not the hydroboration product (Scheme 2).

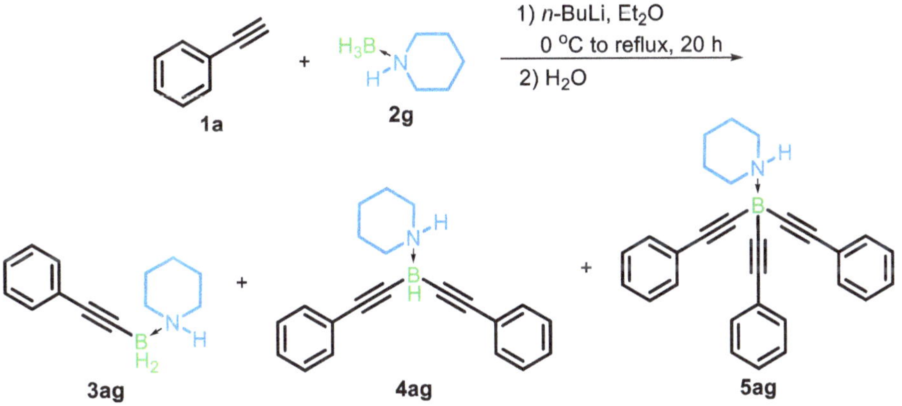

Scheme 2. Initially obtained mono-, di- and trialkynylated amine-boranes.

The formation of the mono- (**3ag**), di- (**4ag**), and tri-*B*-alkynylated (**5ag**) amine-borane was confirmed by ^{11}B NMR and HRMS (see Supporting Information). This preliminary experiment demonstrated the potential of LABs to undergo dehydrocoupling with up to 3 terminal alkynes. With this result, we set about standardizing the reaction conditions, starting with the order of addition. Using 4-methoxyphenylacetylene (**1b**) and dimethylamine-borane (DMAB) (**2e**), three experiments were performed in diethyl ether (Et$_2$O): (1) the LAB reagent was formed using **2e** and *n*-BuLi, followed by addition of **1b**, (2) the lithium acetylide was formed using **1b** and *n*-BuLi followed by addition of **2e**, and (3) *n*-BuLi was added to a mixture of **1b** and **2e**. In each case the formation of lithium *N,N*-dimethylaminoborohydride was detected by ^{11}B NMR (δ −14.88 (q, *J* = 83.4 Hz)), and the total conversion of **1b** to one of the alkynylborane products after 24 h at room temperature was similar for each equivalence examined (see Supporting Information). For operational simplicity, method 3 was used going forward to further examine the reagent equivalence, where it was found that utilizing **1b** and **2e** in a ratio of 1 to 2, with 1.865 eq. of *n*-BuLi provided nearly quantitative conversion of the alkyne, with a favorable 85:15 ratio of mono-*B*-substituted (**3be**) product to the di-*B*-substituted (**4be**) product (see Supporting Information). Using this stoichiometry, none of the tri-*B*-substituted (**5be**) product was detected. It is proposed that the excess of amine-borane helps to limit the formation of **4** and **5**, while the modest deficiency of *n*-BuLi prevents the concurrent formation of the lithium acetylide and LAB reagent.

Shortening the reaction duration increased the proportion of mono-B-substituted (**3be**) product at the expense of overall alkyne conversion, indicating a competition between the lithium alkynylaminoborohydride intermediate and the still-present LAB reagent for the remaining alkyne substrate. The formation of di-B-substituted product was minimized during the solvent study. Dictated by the reactivity of the LAB reagent, a series of primarily ethereal and hydrocarbon solvents were examined (Table 1 entries 1–7). When hydrocarbon solvents were employed, the intermediate product was visibly less soluble and the formation of the **4be** product was almost entirely suppressed. This led to the production of **3be** in both toluene and pentane with a total alkyne conversion of 79% and 92%, respectively. Fine-tuning of the reaction conditions (see Supporting Information) revealed that in refluxing pentane, the reaction was complete within 2 h, with a total alkyne conversion of 97% and a 99:1 ratio of **3be** to **4be** (Table 1 Entry 8).

Table 1. Optimization of the reaction solvent and amine-borane [a].

Entry	Condition	Solvent	Amine-Borane	Products	Ratio [b]	Conversion [c]
1	rt, 24 h	Et$_2$O	2e	3ae/4ae	85:15	95%
2	rt, 24 h	THF	2e	3ae/4ae	73:27	51%
3	rt, 24 h	DME	2e	3ae/4ae	80:20	30%
4	rt, 24 h	DCM	2e	3ae/4ae	87:13	80%
5	rt, 24 h	toluene	2e	3ae/4ae	100:0	79%
6	rt, 24 h	pentane	2e	3ae/4ae	99:1	92%
7	rt, 24 h	Et$_3$N	2e	3ae/4ae	85:15	80%
8	reflux, 2 h	pentane	2e	3ae/4ae	99:1	97%
9	reflux, 2 h	pentane	2a	3aa/4aa	-	trace
10	reflux, 2 h	pentane	2b	3ab/4ab	-	trace
11	reflux, 2 h	pentane	2c	3ac/4ac	99:1	16%
12	reflux, 2 h	pentane	2d	3ad/4ad	92:8	82%
13	reflux, 2 h	pentane	2f	3af/4af	99:1	66%
14	reflux, 2 h	pentane	2g	3ag/4ag	99:1	83%
15	reflux, 2 h	pentane	2h	3ah/4ah	-	trace
16	reflux, 2 h	pentane	2i	3ai/4ai	93:7	14%
17	reflux, 2 h	pentane	2j	3aj/4aj	-	0%

[a] Reactions were carried out using optimized stoichiometry with 4-methoxyphenylacetylene. [b] Ratio of mono- to di-B-substituted product. [c] Alkyne conversion includes both mono- and di-B-substituted products.

Following the determination of the optimal reaction solvent and conditions, an investigation of the amine-borane reagent (**2a–2j**) was undertaken. The amine-boranes examined were prepared using either a salt metathesis [20] (**2a** and **2e**) or a sodium bicarbonate-mediated [21] (**2b–2d**, **2f–2j**) reaction. Fortuitously, **2e** used in the earlier investigations showed the greatest total alkyne conversion. Unhindered primary amine-boranes **2a** and **2b** showed only trace conversion, while bulkier **2c** and **2d** gave 16% and 82% conversion, respectively. Similar results were obtained for the other secondary amine-boranes examined. Diethylamine- (**2f**) and piperidine-borane (**2g**) gave 66% and 83% conversion respectively, but morpholine-borane (**2h**) gave only trace product. The highly hindered diisopropylamine-borane (**2i**) gave only 14% conversion, in line with what was observed previously for a similar reaction using **2i** [13]. The complete absence of product when using triethylamine-borane (**2j**) bolsters the proposed importance of the formation of the LAB reagent as the active intermediate in the reaction.

The LAB intermediate was additionally observed during an experiment tracking the reaction of **1b** and **2e** with ^{11}B NMR spectroscopy (Figure 1). The initial quartet at δ −14.76 ppm consistent with **2e** (Figure 1a), was determined to have converted to the corresponding LAB reagent based on the coupling constant ($J = 96.69$ Hz vs. 83.26 Hz) (Figure 1b). Reaction of the LAB reagent with **1b** then produced a lithium dimethylaminoalkynylborohydride, represented by a triplet near δ −16 ppm (Figure 1c). Following the water quench, the product **3be** was detected as a triplet near δ −15 ppm (Figure 1d).

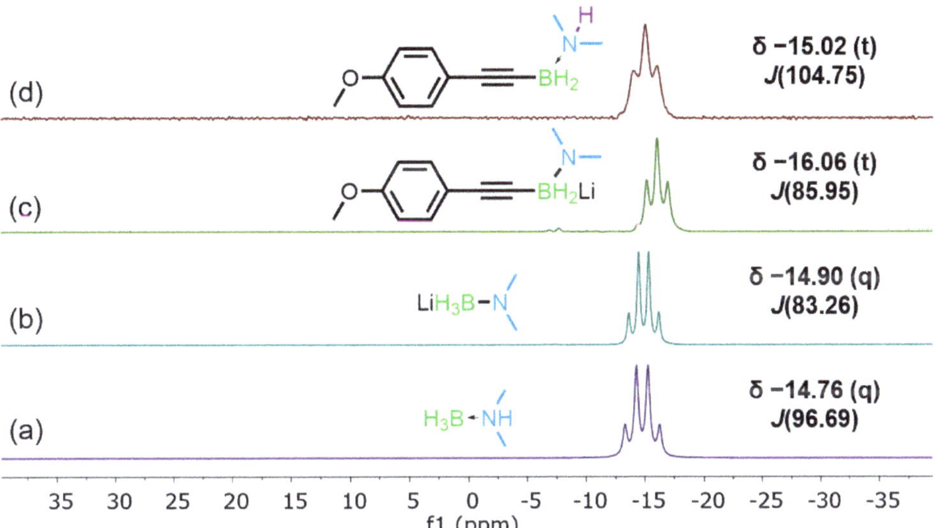

Figure 1. ^{11}B NMR peaks observed during reaction monitoring experiment; (**a**) dimethylamine-borane starting material, (**b**) lithium dimethylaminoborohydride after *n*-BuLi addition, (**c**) lithium dimethylaminoalkynylborohydride following dehydroborylation, (**d**) final product obtained after water quench.

Using the optimized conditions for **2e**, the substrate scope was examined (Scheme 3). Aromatic terminal alkynes readily underwent the dehydrocoupling, including those with electron donating groups (**1b**), electron withdrawing groups (**1c** and **1d**), and those with substitutions at the ortho (**1e**), meta (**1f**), and para (**1g**) positions. Aromatic homologues (**1h**), as well as straight chain (**1i** and **1j**), branched (**1k**), and cyclic (**1l**) aliphatic terminals alkynes, were also amenable to borylation. The protocol was also applied to functionalized alkynes, including trimethylsilyl- (**1m**), and trifluromethylacetylene (**1n**). Selectivity for borylation at only the terminal alkyne position was demonstrated by the successful reactions of substrates with an internal alkene (**1o**) and internal alkyne (**1p**). The reactions of aryl alkynes were completed within 2 h, while alkyl alkyne substrate required up to 6 h for completion. The mono-*B*-substituted to di-*B*-substituted product ratio was greater than 80:20 for all substrates, though most substrates provided a ratio of 95:5 or greater. Reactions with substrates containing reducible groups or those that could be easily deprotonated were not successful, including esters, nitriles, alcohols, amines, and nitro groups.

Following examination of the scope of usable terminal alkyne substrates, a series of gram-scale reactions were performed using 4-methoxyphenylacetylene (**1b**) and **2e**. Reactions at 10 mmol and 50 mmol scale with respect to **1b** each gave results in line with the 1 mmol reaction. Within 2–2.5 h, yields of 99% and 91% were obtained for the 10 mmol and 50 mmol scale reactions, respectively (Scheme 4). To circumvent the large-scale chromatographic separation of the product **3be** from the excess remaining **2e**, an aqueous workup procedure was used. It was found that **2e** would decompose into water-soluble products in the presence of dilute (1 M) HCl, while **3be** would remain intact, allowing it to be extracted using ethyl acetate. Basic aqueous solutions (1 M NaOH) were also tolerated by **3be**.

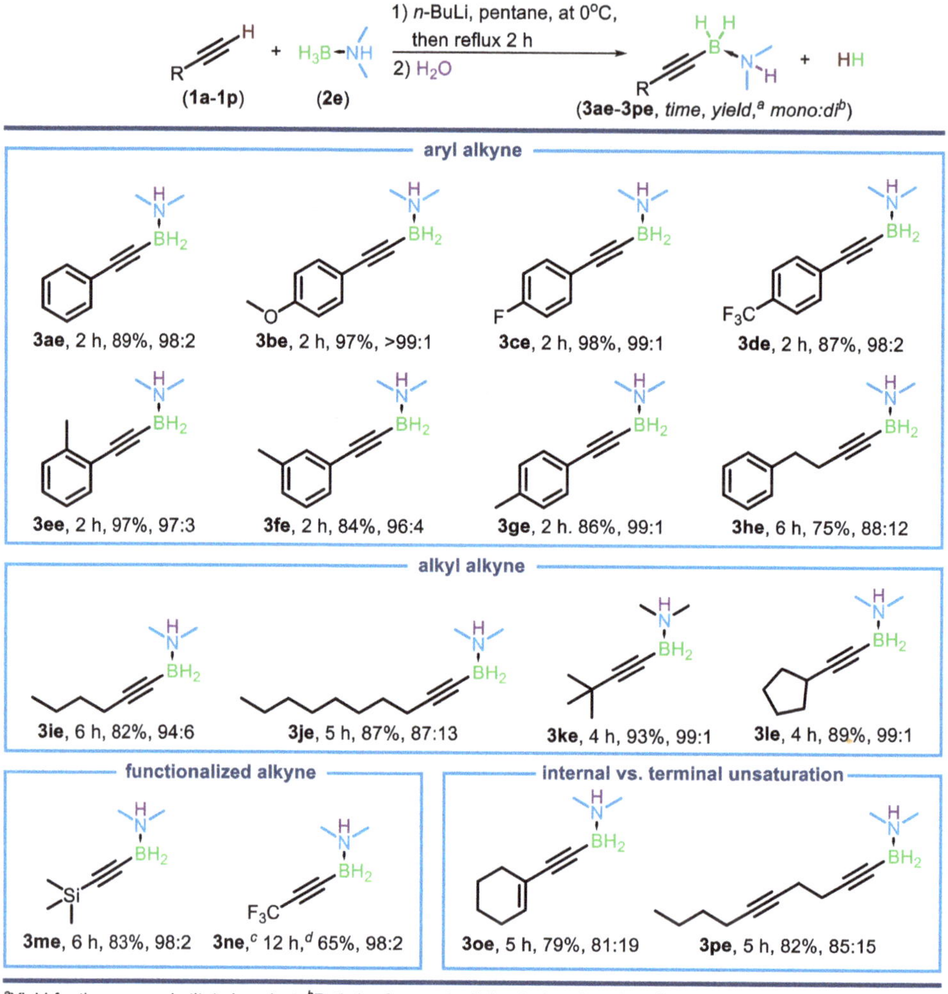

^{a}Yield for the mono-substituted product. ^{b}Ratio is of mono- to di-B-substituted product. ^{c}Dry ice/acetone condenser used due to substrate boiling point. ^{d}Reaction could not be monitored due to substrate boiling point.

Scheme 3. Scope of terminal alkyne substrates.

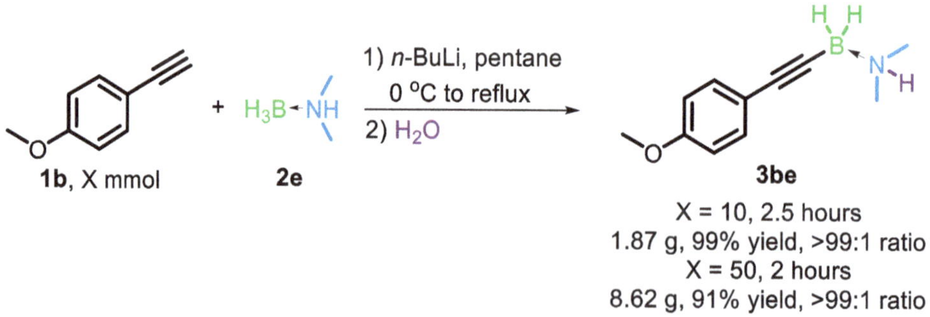

Scheme 4. Gram-scale experiments for the synthesis of **3be**.

To expand the scope of available terminal alkyne substrates, a modified method was envisioned. Terminal alkynes can be produced from aldehydes by reaction with a triphenylphosphine-dibromomethylene ylide formed from carbon tetrabromide and triphenylphosphine to yield a 1,1-dibromoolefin. Reaction of the dibromoolefin intermediate with *n*-BuLi yields the terminal alkyne, with the overall process known as the Corey–Fuchs reaction [22]. By isolating the 1,1-dibromoolefin intermediate and using that as the starting material, the Fritsch–Buttenberg–Wiechell rearrangement [23–25] would be performed in situ, and the lithium acetylide generated would go on to provide the borylated alkyne upon reaction with **2e**. To examine this potential alternative route to the alkynylborane-amine products, **3be** was selected as a target to have a direct comparison between the results of two methods. Using 4-methoxybenzaldehyde, 1-(2,2-dibromovinyl)-4-methoxybenzene (**6a**) was prepared using a reported procedure [22]. An initial reaction maintaining a **6a** to **2e** ratio of 1 to 2, while increasing the *n*-BuLi equiv. to 3.865, provided only 10% of the desired **7ae**. This low yield is attributed to the presence of excess *n*-BuLi, allowing both LAB reagent and lithium acetylide to be present in the reaction. Refinement of the reaction stoichiometry (see Supporting Information) led to an optimal **6a** to **2e** to *n*-BuLi ratio of 1:2:2.3. These conditions afforded **7ae** in 89% yield with a mono- to diborylated product ratio $\geq$99:$\leq$1. Following this success, a series of 1,1-dibromoolefins (**6b–6d**) were prepared from 4-(methylthio)benzaldehyde, thiophene-2-carbaldehyde, and isovaleraldehyde. Substrates **6b–6d** were then subjected to the optimized conditions, and provided the corresponding alkynylated borane-amines **7be**, **7ce**, and **7de** in 87%, 99%, and 89%, respectively, demonstrating the compatibility of aryl, heteroaromatic, and alkyl 1,1-dibromoolefins with the devised protocol (Scheme 5).

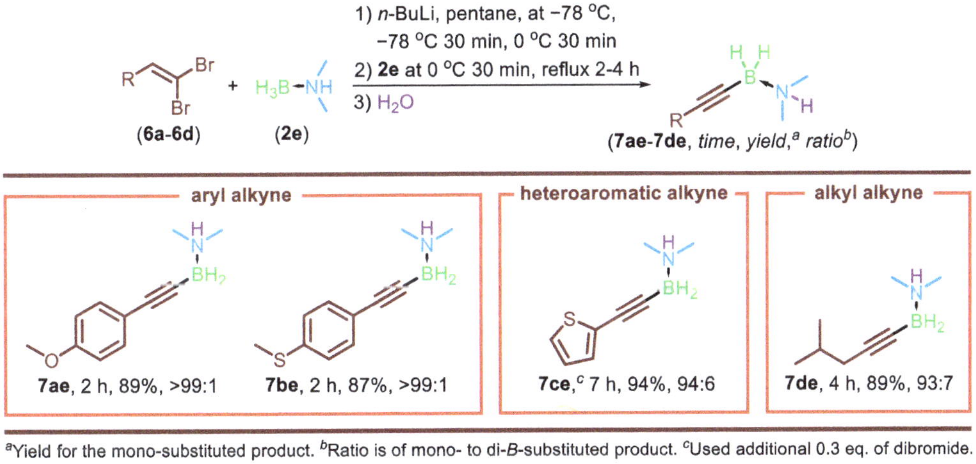

^{a}Yield for the mono-substituted product. ^{b}Ratio is of mono- to di-*B*-substituted product. ^{c}Used additional 0.3 eq. of dibromide.

Scheme 5. Alkynylborane-amines produced from 1,1-dibromoolefins.

3. Materials and Methods

3.1. General Information

All reagents and starting materials were purchased from Sigma-Aldrich (St. Louis, MO, USA), Oakwood (Estill, SC, USA) or Fisher Scientific (Waltham, MA, USA). Amines, ammonium salts, sodium borohydride, sodium bicarbonate, and reagent-grade tetrahydrofuran (75 to 400 ppm BHT) were used as received for preparation of amine-boranes. For the preparation of alkynylborane-amines, solid alkynes and *n*-butyllithium (1.6 M in hexane) were used as received and liquid alkynes were distilled over lithium aluminum hydride under nitrogen. Anhydrous pentane was prepared by distillation from calcium hydride and stored under nitrogen atmosphere. For the preparation of dibromide substrates, the aldehydes, carbon tetrabromide, triphenylphosphine, and reagent-grade dichloromethane were used as received. Additional solvents used for optimization reactions were distilled from sodium/benzophenone (diethyl ether, tetrahydrofuran, dimethoxyethane), calcium hydride (dichloromethane, toluene), or sodium hydroxide (triethylamine) and stored under nitrogen. Thin-layer chromatography (TLC, Silver Spring, MD, USA) was performed on F60 silica gel plates purchased from Macherey-Nagel (Allentown, PA, USA) and visualized using iodine on silica or UV light. Column chromatography was performed using 60 M Kieselgel silica gel. The identities of the products were confirmed by nuclear magnetic resonance (NMR) spectroscopy and measured in δ values in parts per million (ppm). Spectra of products were recorded from a Bruker (Billerica, MA, USA) 400 MHz, Varian (Palo Alto, CA, USA) INOVA 300 MHz, or Varian (Palo Alto, CA, USA) MERCURY 300 MHz NMR spectrometer. The ^{1}H NMR (300 MHz or 400 MHz) spectra were recorded at ambient temperature and calibrated against the residual solvent peak of $CDCl_3$ (δ = 7.26 ppm) as an internal standard. Coupling constants (J) are given in hertz (Hz), and signal multiplicities are described of NMR data as s = singlet, d = doublet, dd = doublet of doublets, dt = doublet of triplets, ddd = doublet of doublet of doublets, ddt = doublet of doublet of triplets, dqd = doublet of quartet of doublets, dqt = doublet of quartet of triplets, dtd = doublet of triplet of doublets, dddd = doublet of doublet of doublet of doublets, t = triplet, td = triplet of doublets, tt = triplet of triplets, tdd = triplet of doublet of doublets, tdt = triplet of doublet of triplets, q= quartet, p= pentet, h = hextet, m = multiplet, and br = broad. The ^{13}C NMR (75 MHz or 101 MHz) spectra were recorded at ambient temperature and calibrated using $CDCl_3$ (δ = 77.0 ppm) as an internal standard. ^{11}B NMR (96 MHz) spectra were recorded at ambient temperature and chemical shifts are reported relative to the external standard, BF_3:OEt_2 (δ = 0 ppm). ^{19}F NMR (282 MHz) spectra were recorded at ambient temperature and chemical shifts are reported relative to the external standard—$CFCl_3$ (δ = 0 ppm).

3.2. Experimental

3.2.1. General Procedure for the Salt Metathesis Synthesis of Amines-Boranes

Following a previously reported procedure [26], sodium borohydride (0.76 g, 20 mmol) and the desired ammonium salt (20 mmol) were weighed to a 100 mL dry round bottom flask, containing a magnetic stir-bar. Then, at rt, reagent-grade tetrahydrofuran (20.0 mL) was added, and the mixture was stirred. Reaction progress was monitored using ^{11}B NMR spectroscopy. (Note: A drop of DMSO is added to the reaction aliquot prior to running the ^{11}B NMR experiment). Upon completion of the reaction, as determined by ^{11}B NMR, the reaction contents were filtered through celite and sodium sulfate and the solid residue washed with additional THF. Removal of the solvent from the filtrate using rotary evaporation yielded the corresponding amine-boranes (**2a**, **2e**). Residual solvent was removed by placing under high vacuum for ~12 h.

3.2.2. General Procedure for the Bicarbonate-Mediated Synthesis of Amines-Boranes

Following a previously reported procedure [21], sodium borohydride (1.51 g, 2 eq., 40 mmol) and powdered sodium bicarbonate (6.72 g, 4 eq., 80 mmol) were weighed to a 100 mL dry round bottom flask, containing a magnetic stir-bar. The desired amine (1 eq., 20 mmol) was added to the flask via syringe, followed by addition of reagent-grade tetrahydrofuran (20 mL) at rt. Under vigorous stirring, water (0.36 mL, 4 eq., 80 mmol) was added dropwise to prevent excessive frothing. Reaction progress was monitored using ^{11}B NMR spectroscopy. (Note: A drop of DMSO is added to the reaction aliquot prior to running the ^{11}B NMR experiment). Upon completion of the reaction, as determined by ^{11}B NMR, the reaction contents were filtered through celite and sodium sulfate and the solid residue washed with additional THF. Removal of the solvent from the filtrate using rotary evaporation yielded the corresponding amine-boranes (**2b–2d**, **2f–2j**). Residual solvent was removed by placing under high vacuum for ~12 h.

3.2.3. Characterization of Amines-Boranes

*Methylamine-borane (**2a**)*: The compound was prepared as described in the salt metathesis procedure and obtained as a white solid (mass = 0.61 g, 68% yield). ^{1}H NMR (300 MHz, CDCl$_3$) δ 3.78 (s, 2H), 2.55 (t, *J* = 6.3 Hz, 3H), 1.49 (dd, *J* = 189.1, 92.0 Hz, 3H). ^{13}C NMR (75 MHz, CDCl$_3$) δ 34.6. ^{11}B NMR (96 MHz, CDCl$_3$) δ −18.78 (q, *J* = 95.1 Hz). Compound characterization is in agreement with previous reports for this compound [15].

*Propylamine-borane (**2b**)*: The compound was prepared as described in the bicarbonate-mediated procedure and obtained as a white solid (mass = 1.38 g, 95% yield). ^{1}H NMR (300 MHz, CDCl$_3$) δ 3.81 (s, 2H), 2.75 (p, *J* = 7.2 Hz, 2H), 1.63 (h, *J* = 7.4 Hz, 2H), 0.93 (t, *J* = 7.4 Hz, 3H). ^{13}C NMR (75 MHz, CDCl$_3$) δ 50.4, 22.4, 11.0. ^{11}B NMR (96 MHz, CDCl$_3$) δ −19.84 (q, *J* = 92.0 Hz). Compound characterization is in agreement with previous reports for this compound [21].

*Cyclohexylamine-borane (**2c**)*: The compound was prepared as described in the bicarbonate-mediated procedure and obtained as a white solid (mass = 2.19 g, 97% yield). ^{1}H NMR (300 MHz, CDCl$_3$) δ 3.64 (s, 2H), 2.68 (dqt, *J* = 14.1, 7.6, 4.0 Hz, 1H), 2.13 (d, *J* = 10.4 Hz, 2H), 1.80–1.70 (m, 2H), 1.67–1.56 (m, 1H), 1.37–1.07 (m, 5H). ^{13}C NMR (75 MHz, CDCl$_3$) δ 57.0, 32.3, 25.3, 24.5. ^{11}B NMR (96 MHz, CDCl$_3$) δ −0.71−−41.09 (m). Compound characterization is in agreement with previous reports for this compound [21].

*t-Butylamine-borane (**2d**)*: The compound was prepared as described in the bicarbonate-mediated procedure and obtained as a white solid (mass = 1.65 g, 95% yield). ^{1}H NMR (300 MHz, CDCl$_3$) δ 3.76 (s, 2H), 1.27 (s, 9H). ^{13}C NMR (75 MHz, CDCl$_3$) δ 53.1, 28.0. ^{11}B NMR (96 MHz, CDCl$_3$) δ −23.25 (q, *J* = 96.7 Hz). Compound characterization is in agreement with previous reports for this compound [21].

*Dimethylamine-borane (**2e**)*: The compound was prepared as described in the salt metathesis procedure and obtained as a white solid (mass = 1.07 g, 91% yield). ^{1}H NMR (300 MHz, CDCl$_3$) δ 4.31 (s, 1H), 2.46 (d, *J* = 5.8 Hz, 6H), 1.43 (dd, *J* = 188.2, 91.9 Hz, 3H). ^{13}C NMR (75 MHz, CDCl$_3$) δ 44.2. ^{11}B NMR (96 MHz, CDCl$_3$) δ −14.76 (q, *J* = 96.7 Hz). Compound characterization is in agreement with previous reports for this compound [15].

*Diethylamine-borane (**2f**)*: The compound was prepared as described in the bicarbonate-mediated procedure and obtained as a white solid (mass = 1.65 g, 95% yield). ^{1}H NMR (300 MHz, CDCl$_3$) δ 3.20 (s, 1H), 2.83 (dqt, *J* = 9.7, 7.3, 3.8 Hz, 4H), 1.25 (t, *J* = 7.3 Hz, 6H). ^{13}C NMR (75 MHz, CDCl$_3$) δ 48.5, 11.3. ^{11}B NMR (96 MHz, CDCl$_3$) δ −17.07 (q, *J* = 96.7 Hz). Compound characterization is in agreement with previous reports for this compound [21].

*Piperidine-borane (**2g**)*: The compound was prepared as described in the bicarbonate-mediated procedure and obtained as a white solid (mass = 1.96 g, 99% yield). ^{1}H NMR (300 MHz, CDCl$_3$) δ 3.75 (s, 1H), 3.22 (d, *J* = 13.4 Hz, 2H), 2.59–2.36 (m, 2H), 1.75 (d, *J* = 10.1 Hz, 3H), 1.51 (ddt, *J* = 27.9, 14.2, 3.6 Hz, 2H), 1.32 (tdd, *J* = 16.3, 8.7, 4.6 Hz, 1H). ^{13}C NMR (75 MHz, CDCl$_3$) δ 53.4, 25.4, 22.6. ^{11}B NMR (96 MHz, CDCl$_3$) δ −15.55 (q, *J* = 96.7 Hz). Compound characterization is in agreement with previous reports for this compound [21].

*Morpholine-borane (**2h**)*: The compound was prepared as described in the bicarbonate-mediated procedure and obtained as a white solid (mass = 1.98 g, 98% yield). ^{1}H NMR (300 MHz, CDCl$_3$) δ 4.40 (s, 1H), 3.91 (dd, *J* = 12.6, 3.3 Hz, 2H), 3.55 (td, *J* = 12.3, 2.2 Hz, 2H), 3.05 (d, *J* = 13.7 Hz, 2H), 2.86–2.64 (m, 2H), 2.35–0.65 (m, 3H). ^{13}C NMR (75 MHz, CDCl$_3$) δ 65.6, 51.8. ^{11}B NMR (96 MHz, CDCl$_3$) δ −15.47 (q, *J* = 98.0, 96.7 Hz). Compound characterization is in agreement with previous reports for this compound [21].

*Diisopropylamine-borane (**2i**)*: The compound was prepared as described in the bicarbonate-mediated procedure and obtained as a white solid (mass = 2.12 g, 92% yield). ^{1}H NMR (300 MHz, CDCl$_3$) δ 3.18 (s, 1H), 3.10 (dqd, *J* = 13.1, 6.5, 3.5 Hz, 2H), 1.15 (t, *J* = 6.3 Hz,

12H). ^{13}C NMR (75 MHz, CDCl$_3$) δ 51.6, 20.6, 18.6. ^{11}B NMR (96 MHz, CDCl$_3$) δ −21.81 (q, J = 96.7 Hz). Compound characterization is in agreement with previous reports for this compound [27].

Triethylamine-borane (2j): The compound was prepared as described in the bicarbonate-mediated procedure and obtained as a white solid (mass = 2.16 g, 94% yield). ^{1}H NMR (300 MHz, CDCl$_3$) δ 2.61 (q, J = 7.3 Hz, 6H), 1.02 (t, J = 7.3 Hz, 9H). ^{13}C NMR (75 MHz, CDCl$_3$) δ 51.9, 8.1. ^{11}B NMR (96 MHz, CDCl$_3$) δ −13.81 (q, J = 99.4, 98.0 Hz). Compound characterization is in agreement with previous reports for this compound [21].

3.2.4. General Procedure for the 1 mmol Scale Synthesis of Alkynylborane-Amines

In an oven-dried, 25 mL, round-bottom flask with a side arm and containing a stir bar, the dimethylamine-borane (0.118 g, 0.002 mole, 2 eq.) is weighed along with the alkyne (0.001 mole, 1 eq.) if the alkyne substrate is a solid. A reflux condenser is affix to the flask and the whole apparatus is thoroughly sealed and flushed with nitrogen. Freshly distilled pentane (1 mL) is then charged into the flask through the side arm via syringe. If the alkyne substrate is a liquid, it is added at this time. The reaction mixture is stirred and brought to 0 °C using an ice bath. Then, at 0 °C, *n*-butyllithium (0.001865 mole, 1.865 eq.) is added dropwise as a 1.6 M solution in hexanes through the side arm via syringe. The reaction mixture is stirred for 5 min at 0 °C, and additional pentane (1 mL) is added to help rinse any reaction components from the walls of the flask. The mixture is brought to reflux and monitored by TLC or by ^{1}H NMR (with a drop of water added to quench the aliquot). The reaction is typically complete in just 2 h for aromatic alkyne substrates and 6 h for aliphatic alkyne substrates. Reaction completion is judged by the absence of the acetylenic proton in the ^{1}H NMR spectrum. Once completed, the reaction mixture is allowed to cool to room temperature, quenched with water (2 mL), and stirred for 30 min. The organic phase is separated, and the aqueous phase is extracted with diethyl ether (3 × 15 mL). All organic layers are combined, dried with sodium sulfate, filtered through cotton, and condensed by rotary evaporation. The crude mixture is then subjected to column chromatography using a 7:2:1, dichloromethane:hexane:diethyl ether solvent system, yielding the corresponding alkynylborane-amines (**3ae–3pe**). Residual solvent was removed by placing under high vacuum for ~12 h. As an alternative to column chromatography, the excess dimethylamine-borane can be removed by addition of 1 M HCl (5 mL) following the water quench and stirring for 30 min at rt prior to performing the organic extraction.

3.2.5. Characterization of Alkynylborane-Amines from Terminal Alkynes

(Phenylethynyl)borane-dimethylamine (3ae): The compound was prepared as described in the 1 mmol scale procedure and obtained in 2 h as a white solid (mass = 142 mg, 89% yield, 98:2 mono:di ratio); melting point: 74–76 °C (Meltemp). ^{1}H NMR (300 MHz, CDCl$_3$) δ 7.40 (dt, J = 7.5, 1.7 Hz, 2H), 7.35–7.11 (m, 3H), 4.38 (s, 1H), 2.53 (d, J = 5.7 Hz, 6H). ^{13}C NMR (75 MHz, CDCl$_3$) δ 131.2, 127.0, 125.2, 100.0, 42.6. ^{11}B NMR (96 MHz, CDCl$_3$) δ −15.33 (t, J = 100.4 Hz).

((4-Methoxyphenyl)ethynyl)borane-dimethylamine (3be): The compound was prepared as described in the 1 mmol scale procedure and obtained in 2 h as a white solid (mass = 183 mg, 97% yield, >99:1 ratio); melting point: 118–120 °C (Meltemp). ^{1}H NMR (300 MHz, CDCl$_3$) δ 7.35 (d, J = 8.7 Hz, 2H), 6.78 (d, J = 8.8 Hz, 2H), 4.46 (s, 1H), 3.76 (s, 3H), 2.56 (d, J = 5.8 Hz, 6H). ^{13}C NMR (75 MHz, CDCl$_3$) δ 158.4, 132.6, 117.5, 113.7, 99.7, 55.3, 42.5. ^{11}B NMR (96 MHz, CDCl$_3$) δ −15.14 (t, J = 93.3 Hz).

((4-Fluorophenyl)ethynyl)borane-dimethylamine (3ce): The compound was prepared as described in the 1 mmol scale procedure and obtained in 2 h as a white solid (mass = 174 mg, 98% yield, 99:1 ratio); melting point: 102–104 °C (Meltemp). ^{1}H NMR (300 MHz, CDCl$_3$) δ 7.45–7.33 (m, 2H), 7.00–6.89 (m, 2H), 2.67 (d, J = 5.8 Hz, 6H). ^{13}C NMR (75 MHz, CDCl$_3$) δ 163.1, 159.9, 132.9 (d, J = 7.9 Hz), 121.3 (d, J = 3.0 Hz), 115.9–114.6 (m), 98.9,

42.6. ^{11}B NMR (96 MHz, CDCl$_3$) δ −15.17 (t, *J* = 99.9 Hz). ^{19}F NMR (282 MHz, CDCl$_3$) δ −114.72–114.96 (m).

((4-Ttrifluoromethyl)phenyl)ethynyl)borane-dimethylamine (**3de**): The compound was prepared as described in the 1 mmol scale procedure and obtained in 2 h as a white solid (mass = 198 mg, 87% yield, 98:2 ratio); white solid, melting point: 80–82 °C (Meltemp). ^{1}H NMR (300 MHz, CDCl$_3$) δ 7.49 (s, 4H), 2.69 (d, *J* = 5.8 Hz, 6H). ^{13}C NMR (75 MHz, CDCl$_3$) δ 131.3, 129.1, 128.7, 128.2, 125.8, 125.0, 124.9, 122.2, 98.9, 42.6. ^{11}B NMR (96 MHz, CDCl$_3$) δ −15.23 (t, *J* = 100.6 Hz). ^{19}F NMR (282 MHz, CDCl$_3$) δ −64.14 (s).

((2-Methylphenyl)ethynyl)borane-dimethylamine (**3ee**): The compound was prepared as described in the 1 mmol scale procedure and obtained in 2 h as a clear, colorless liquid (mass = 168 mg, 97% yield, 97:3 ratio); ^{1}H NMR (300 MHz, CDCl$_3$) δ 7.39 (d, *J* = 7.4 Hz, 1H), 7.13 (dddd, *J* = 14.8, 9.1, 7.2, 2.3 Hz, 3H), 4.28 (s, 1H), 2.58 (d, *J* = 5.8 Hz, 6H), 2.45 (s, 3H). ^{13}C NMR (75 MHz, CDCl$_3$) δ 139.3, 131.5, 129.2, 126.9, 125.3, 125.0, 42.5, 21.2. ^{11}B NMR (96 MHz, CDCl$_3$) δ −15.17 (t, *J* = 99.6 Hz).

((3-Methylphenyl)ethynyl)borane-dimethylamine (**3fe**): The compound was prepared as described in the 1 mmol scale procedure and obtained in 2 h as a clear, colorless liquid (mass = 146 mg, 84% yield, 96:4 ratio); ^{1}H NMR (300 MHz, CDCl$_3$) δ 7.29–7.18 (m, 2H), 7.13 (t, *J* = 7.5 Hz, 1H), 7.02 (d, *J* = 7.6 Hz, 1H), 4.39 (s, 1H), 2.55 (d, *J* = 5.8 Hz, 6H), 2.28 (s, 3H). ^{13}C NMR (75 MHz, CDCl$_3$) δ 131.9, 128.3, 128.0, 127.9, 125.0, 100.2, 42.6, 21.4. ^{11}B NMR (96 MHz, CDCl$_3$) δ −15.40 (t, *J* = 100.0 Hz).

((4-Methylphenyl)ethyny)borane-dimethylamine (**3ge**): The compound was prepared as described in the 1 mmol scale procedure and obtained in 2 h as a white solid (mass = 149 mg, 86% yield, 99:1 ratio); melting point: 75–77 °C (Meltemp). ^{1}H NMR (300 MHz, CDCl$_3$) δ 7.30 (d, *J* = 7.9 Hz, 2H), 7.04 (d, *J* = 7.8 Hz, 2H), 4.44 (s, 1H), 2.52 (d, *J* = 5.8 Hz, 6H), 2.29 (s, 3H). ^{13}C NMR (75 MHz, CDCl$_3$) δ 136.8, 131.1, 128.9, 122.2, 100.1, 42.5, 21.5. ^{11}B NMR (96 MHz, CDCl$_3$) δ −15.10 (t, *J* = 99.7 Hz).

(4-Phenylbut-1-yn-1-yl)borane-dimethylamine (**3he**): The compound was prepared as described in the 1 mmol scale procedure and obtained in 6 h as a white solid (mass = 140 mg, 75% yield, 88:12 ratio); melting point: 69–72 °C (Meltemp). ^{1}H NMR (300 MHz, CDCl$_3$) δ 7.36–7.08 (m, 5H), 4.14 (s, 1H), 2.83 (t, *J* = 7.6 Hz, 2H), 2.52 (t, *J* = 7.8 Hz, 2H), 2.44 (d, *J* = 5.7 Hz, 6H). ^{13}C NMR (75 MHz, CDCl$_3$) δ 141.3, 128.4, 128.2, 42.2, 36.1, 22.3. ^{11}B NMR (96 MHz, CDCl$_3$) δ −15.35 (t, *J* = 99.7 Hz).

(Hexynyl)borane-dimethylamine (**3ie**): The compound was prepared as described in the 1 mmol scale procedure and obtained in 6 h as a clear, colorless liquid (mass = 114 mg, 82% yield, 94:6 ratio); ^{1}H NMR (300 MHz, CDCl$_3$) δ 2.61 (d, *J* = 5.8 Hz, 6H), 2.37–2.06 (m, 2H), 1.63–1.28 (m, 4H), 1.02–0.72 (m, 3H). ^{13}C NMR (75 MHz, CDCl$_3$) δ 100.7, 42.2, 31.9, 22.2, 19.8, 13.8. ^{11}B NMR (96 MHz, CDCl$_3$) δ −15.51 (t, *J* = 98.8 Hz).

(Decynyl)borane-dimethylamine (**3je**): The compound was prepared as described in the 1 mmol scale procedure and obtained in 5 h as a clear, colorless liquid (mass = 170 mg, 87% yield, 87:13 ratio); ^{1}H NMR (300 MHz, CDCl$_3$) δ 4.40 (s, 1H), 2.59 (d, *J* = 5.8 Hz, 6H), 2.21 (t, *J* = 7.2 Hz, 2H), 1.51 (p, *J* = 6.9 Hz, 2H), 1.44–1.34 (m, 2H), 1.34–1.17 (m, 8H), 0.87 (t, *J* = 6.4 Hz, 3H). ^{13}C NMR (75 MHz, CDCl$_3$) δ 100.7, 42.2, 31.9, 29.8, 29.3, 29.3, 29.2, 22.7, 20.1, 14.2. ^{11}B NMR (96 MHz, CDCl$_3$) δ −15.50 (t, *J* = 100.8 Hz).

(3,3-Dimethylbut-1-yn-1-yl)borane-dimethylamine (**3ke**): The compound was prepared as described in the 1 mmol scale procedure and obtained in 4 h as a clear, white solid (mass = 1370 mg, 93% yield, 99:1 ratio);White solid, Melting point: 118–120 °C (Meltemp). ^{1}H NMR (300 MHz, CDCl$_3$) δ 4.03 (s, 1H), 2.61 (d, *J* = 5.9 Hz, 6H), 1.23 (s, 9H). ^{13}C NMR (75 MHz, CDCl$_3$) δ 109.7, 42.2, 31.9, 28.1. ^{11}B NMR (96 MHz, CDCl$_3$) δ −15.13 (t, *J* = 98.4 Hz).

((Cyclopentyl)ethynyl)borane-dimethylamine (**3le**): The compound was prepared as described in the 1 mmol scale procedure and obtained in 4 h as a clear, white solid (mass = 135 mg, 89% yield, 99:1 ratio); melting point: 67–70 °C (Meltemp). ^{1}H NMR (300 MHz, CDCl$_3$) δ 4.36 (s, 1H), 2.60 (d, J = 5.7 Hz, 6H), 1.99–1.81 (m, 2H), 1.81–1.64 (m, 2H), 1.64–1.43 (m, 4H). ^{13}C NMR (75 MHz, CDCl$_3$) δ 42.2, 34.6, 31.5, 25.0. ^{11}B NMR (96 MHz, CDCl$_3$) δ −15.47 (t, J = 98.6 Hz).

((Trimethylsilyl)ethynyl)borane-dimethylamine (**3me**): The compound was prepared as described in the 1 mmol scale procedure and obtained in 6 h as a clear, white solid (mass = 129 mg, 83% yield, 98:2 ratio); melting point: 78–80 °C (Meltemp) ^{1}H NMR (300 MHz, CDCl$_3$) δ 2.64 (d, J = 5.8 Hz, 6H), 0.14 (s, 9H). ^{13}C NMR (75 MHz, CDCl$_3$) δ 42.7, 1.1. ^{11}B NMR (96 MHz, CDCl$_3$) δ −15.59 (t, J = 99.9 Hz).

(3,3,3-Trifluoroprop-1-yn-1-yl)borane-dimethylamine (**3ne**): The compound was prepared as described in the 1 mmol scale procedure and obtained in 12 h as a clear, pale yellow liquid (mass = 98 mg, 65% yield, 98:2 ratio); ^{1}H NMR (300 MHz, CDCl$_3$) δ 3.82 (s, 1H), 2.63 (d, J = 5.8 Hz, 6H), 1.90 (dd, J = 191.7, 97.4 Hz, 2H). ^{13}C NMR (75 MHz, CDCl$_3$) δ 113.9 (q, J = 254.5 Hz), 42.7. ^{11}B NMR (96 MHz, CDCl$_3$) δ −16.30 (t, J = 102.1 Hz). ^{19}F NMR (282 MHz, CDCl$_3$) δ −50.14.

((Cyclohex-1-en-1-yl)ethynyl)borane-dimethylamine (**3oe**): The compound was prepared as described in the 1 mmol scale procedure and obtained in 5 h as a white solid (mass = 129 mg, 79% yield, 81:19 ratio); melting point: 80–82 °C (Meltemp). ^{1}H NMR (300 MHz, CDCl$_3$) δ 5.98 (tt, J = 3.7, 1.6 Hz, 1H), 4.34 (s, 1H), 2.60 (d, J = 5.9 Hz, 6H), 2.13 (tt, J = 5.7, 2.3 Hz, 2H), 2.05 (h, J = 3.1 Hz, 2H), 1.58 (tdt, J = 10.4, 5.6, 2.7 Hz, 4H). ^{13}C NMR (75 MHz, CDCl$_3$) δ 131.8, 122.0, 102.1, 42.4, 30.1, 25.6, 22.6, 21.8. ^{11}B NMR (96 MHz, CDCl$_3$) δ −15.20 (t, J = 99.7 Hz).

(Deca-1,5-diyn-1-yl)borane-dimethylamine (**3pe**): The compound was prepared as described in the 1 mmol scale procedure and obtained in 5 h as a clear, colorless liquid (mass = 157 mg, 82% yield, 85:15 ratio); ^{1}H NMR (300 MHz, CDCl$_3$) δ 4.11 (s, 1H), 2.61 (d, J = 5.8 Hz, 6H), 2.50–2.30 (m, 4H), 2.14 (tt, J = 6.9, 2.0 Hz, 2H), 1.56–1.31 (m, 4H), 0.90 (t, J = 7.1 Hz, 2H). ^{13}C NMR (75 MHz, CDCl$_3$) δ 99.0, 80.8, 79.0, 42.2, 31.2, 22.0, 20.7, 20.0, 18.5, 13.7. ^{11}B NMR (96 MHz, CDCl$_3$) δ −15.39 (t, J = 100.2 Hz).

3.2.6. General Procedure for the Synthesis of Dibromide Substrates

Into a 250 mL round bottom flask containing a stir bar was weighed CBr$_4$ (6.63 g, 20 mmol). The flask was purged using N$_2$, then CH$_2$Cl$_2$ (30 mL) was added, and the mixture was stirred to dissolve the solid. The mixture was brough to 0 °C using an ice bath then PPh$_3$ (10.5 g, 40 mmol) was added, and the mixture was stirred 1 h at 0 °C. The aldehyde (10 mmol) was then added, and the mixture was stirred for an additional 1 h at 0 °C. Then, at 0 °C, the reaction was quenched with H$_2$O (10 mL). The organics were separated, then washed with brine (10 mL), dried using MgSO$_4$ and concentrated in a 1 L flask using rotary evaporation. Hexane (250 mL) was added to the residue and the supernatant layer was collected. The remaining residue was redissolved in CH$_2$Cl$_2$ (50 mL) and concentrated. Hexane (250 mL) was added to the residue and the supernatant layer was collected. This process was repeated once more. The combined supernatant layers were passed through silica gel (100 g) and the eluent was concentrated to yield the dibromide product.

3.2.7. Characterization of Dibromide Substrates

1-(2,2-Dibromovinyl)-4-methoxybenzene (**6a**): The compound was prepared as described in the dibromide synthesis procedure and obtained as a white solid (mass = 1.87 g, 64% yield). ^{1}H NMR (300 MHz, CDCl$_3$) δ 7.56–7.47 (m, 2H), 7.41 (s, 1H), 6.94–6.85 (m, 2H), 3.82 (s, 3H). ^{13}C NMR (75 MHz, CDCl$_3$) δ 159.5, 136.1, 129.8, 127.7, 113.7, 87.2, 55.3. Compound characterization is in agreement with previous reports for this compound [28].

(4-(2,2-Dibromovinyl)phenyl)(methyl)sulfane (**6b**): The compound was prepared as described in the dibromide synthesis procedure and obtained as a pale-yellow solid (mass = 1.42 g, 46% yield). ^{1}H NMR (300 MHz, CDCl$_3$) δ 7.50–7.45 (m, 2H), 7.42 (s, 1H), 7.25–7.18 (m, 2H), 2.49 (s, 3H). ^{13}C NMR (75 MHz, CDCl$_3$) δ 139.5, 136.1, 131.6, 128.7, 125.7, 88.9, 15.4. Compound characterization is in agreement with previous reports for this compound [29].

2-(2,2-Dibromovinyl)thiophene (**6c**): The compound was prepared as described in the dibromide synthesis procedure and obtained as a white solid (mass = 2.04 g, 76% yield). ^{1}H NMR (300 MHz, CDCl$_3$) δ 7.66 (d, *J* = 0.7 Hz, 1H), 7.39 (ddt, *J* = 5.1, 1.3, 0.6 Hz, 1H), 7.25 (ddt, *J* = 3.6, 1.2, 0.6 Hz, 1H), 7.04 (ddd, *J* = 5.2, 3.7, 0.5 Hz, 1H). ^{13}C NMR (75 MHz, CDCl$_3$) δ 137.9, 130.7, 129.9, 127.0, 126.4, 86.9. Compound characterization is in agreement with previous reports for this compound [28].

1,1-Dibromo-4-methylpent-1-ene (**6d**): The compound was prepared as described in the dibromide synthesis procedure and obtained as a pale-yellow liquid (mass = 2.01 g, 83% yield). ^{1}H NMR (400 MHz, CDCl$_3$) δ 6.40 (td, *J* = 7.3, 2.0 Hz, 1H), 1.99 (td, *J* = 7.1, 2.0 Hz, 2H), 1.75 (dtd, *J* = 13.4, 6.7, 2.0 Hz, 1H), 0.93 (dd, *J* = 6.7, 2.1 Hz, 6H). ^{13}C NMR (101 MHz, CDCl$_3$) δ 137.7, 88.8, 41.7, 27.7, 22.1. Compound characterization is in agreement with previous reports for this compound [30].

3.2.8. General Procedure for the Synthesis of Alkynylborane-Amines from Dibromide Substrates

In an oven-dried, 25 mL, round-bottom flask with a side arm and containing a stir bar, the dibromide (0.001 mole, 1 eq.) is weighed. A reflux condenser is affixed to the flask and the whole apparatus is thoroughly sealed and flushed with nitrogen. The flask was sealed and purged with nitrogen. Freshly distilled pentane (4 mL) is then charged into the flask through the side arm via syringe. The mixture is brought to −78 °C using a dry ice/acetone bath, and stirred to cool. Then, *n*-BuLi (2.3 eq., 2.3 mmol, 1.5 mL) is added dropwise at −78 °C. The reaction is stirred for 30 min at −78 °C then then brought to 0 °C using an ice bath and stirred for an additional 30 min @ 0 °C. Dimethylamine-borane (0.118 g, 0.002 mole, 2 eq.) is added to the flask @ 0 °C and stirred for an additional 30 min at rt. Then, 2 mL of pentane is added to wash the walls of the flask and the solution brought to reflux and monitored by TLC or by ^{1}H NMR (with a drop of water added to quench the aliquot). Reaction completion is judged by the absence of the acetylenic proton in the by ^{1}H NMR spectrum. Once completed, the reaction mixture is allowed to cool to room temperature, quenched with water (2 mL), and stirred for 30 min at rt, followed by 1 M HCl (5 mL) and stirred for an additional 30 min at rt. The organic layer is separated and the water extracted with ethyl acetate 3 times. The combined extracts are dried with sodium sulfate and filtered through cotton, then condensed using rotary evaporation, yielding the corresponding alkynylborane-amines (**7ae–7de**). Residual solvent is removed by placing under high vacuum for ~12 h.

3.2.9. Characterization of Alkynylborane-Amines from Dibromide Substrates

((4-Methoxyphenyl)ethynyl)borane-dimethylamine (**7ae**): The compound was prepared as described in the alkynylborane-amine from dibromide procedure and obtained in 2 hours as a white solid (mass = 168 mg, 89% yield, >99:1 ratio); melting point: 118–120 °C (Meltemp). ^{1}H NMR (300 MHz, CDCl$_3$) δ 7.39–7.32 (m, 2H), 6.82–6.75 (m, 2H), 3.94 (s, 2H), 3.79 (s, 3H), 2.66 (d, *J* = 5.8 Hz, 6H). ^{13}C NMR (75 MHz, CDCl$_3$) δ 158.20, 132.33, 117.24, 113.46, 99.50, 55.07, 42.30. ^{11}B NMR (96 MHz, CDCl$_3$) δ −15.14 (t, *J* = 89.1 Hz).

((4-(Methylthio)phenyl)ethynyl)borane-dimethylamine (**7be**): The compound was prepared as described in the alkynylborane-amine from dibromide procedure and obtained in 2 h as a white solid (mass = 178 mg, 87% yield, >99:1 ratio); melting point: 118–121 °C (Meltemp). ^{1}H NMR (300 MHz, CDCl$_3$) δ 7.40–7.28 (m, 2H), 7.18–7.05 (m, 2H), 3.79 (s, 1H), 2.68 (d,

J = 5.8 Hz, 6H), 2.47 (s, 3H). ^{13}C NMR (75 MHz, CDCl$_3$) δ 137.2, 131.6, 125.9, 121.9, 42.6, 15.7. ^{11}B NMR (96 MHz, CDCl$_3$) δ −15.10 (t, J = 98.0 Hz).

(Thiophen-2-ylethynyl)borane-dimethylamine (**7ce**): The compound was prepared as described in the alkynylborane-amine from dibromide procedure and obtained in 7 hours as a pale-yellow solid (mass = 155 mg, 94% yield, 94:6 ratio); melting point: 139–142 °C (Meltemp). ^{1}H NMR (300 MHz, CDCl$_3$) δ 7.14–7.07 (m, 2H), 6.91 (dd, J = 5.2, 3.6 Hz, 1H), 3.79 (s, 1H), 2.67 (d, J = 5.8 Hz, 6H). ^{13}C NMR (101 MHz, CDCl$_3$) δ 130.2, 126.6, 125.7, 125.1, 42.5. ^{11}B NMR (96 MHz, CDCl$_3$) δ −15.13 (t, J = 102.1 Hz).

(4-Methylpent-1-yn-1-yl)borane-dimethylamine (**7de**): The compound was prepared as described in the alkynylborane-amine from dibromide procedure and obtained in 4 hours as a clear, colorless liquid (mass = 123 mg, 89% yield, 93:7 ratio); ^{1}H NMR (300 MHz, CDCl$_3$) δ 3.82 (s, 1H), 2.62 (d, J = 5.8 Hz, 6H), 2.11 (dt, J = 6.7, 1.9 Hz, 2H), 1.85–1.70 (m, 1H), 0.97 (d, J = 6.6 Hz, 6H). ^{13}C NMR (101 MHz, CDCl$_3$) δ 42.2, 29.3, 28.6, 22.1. ^{11}B NMR (96 MHz, CDCl$_3$) δ −15.26 (t, J = 99.4 Hz).

4. Conclusions

Dehydrogenative borylation utilizing air- and moisture-stable amine-boranes has been presented. Terminal alkynes, as well as 1,1-dibromoolefins derived from aldehydes, act as suitable substrates for the reaction with lithium aminoborohydrides generated in situ from the amine-borane and *n*-butyllithium or treating an amine-borane with an alkynyllithium. A wide variety of substrates are amenable to this methodology, including aromatic, heteroaromatic, and aliphatic substrates, with the primary factor limiting the substrate scope being the reducing capabilities of the active lithium aminoborohydride reagent. The *B*-alkynylated amine-borane products are obtained in high yield and are stable to column chromatography as well as acidic and basic aqueous conditions. Though the mono-*B*-alkynylated product is primarily generated using the presented conditions, the potential to form mono-, di-, and tri-B-alkynylated products has been demonstrated. The reaction has additionally been demonstrated at large (up to 50 mmol) scale. Work is currently underway investigating the applications and transformation of these alkynylated amine-borane compounds.

Supplementary Materials: The following Supporting Information can be downloaded at: https://www.mdpi.com/article/10.3390/molecules28083433/s1, Characterization and NMR spectra of initial mono-, di-, and tri-B-alkynylated borane-amines, reaction optimization data, and NMR spectra of amine-boranes, alkynylborane-amine from terminal alkynes, dibromide substrates, and alkynylborane-amine from dibromides.

Author Contributions: Conceptualization, supervision, resources, project administration, and funding acquisition, P.V.R. Both authors have contributed to methodology, validation, investigation, data curation, writing—original draft preparation, and review and editing; visualization, H.J.H. All authors have read and agreed to the published version of the manuscript.

Funding: This research was funded by the Herbert C. Brown Center for Borane Research of Purdue University.

Institutional Review Board Statement: Not applicable.

Informed Consent Statement: Not applicable.

Data Availability Statement: All of the ^{1}H, ^{11}B, ^{13}C, and ^{19}F NMR spectra are available in the Supporting Information.

Acknowledgments: Initial experiments by and helpful discussions with Ameya S. Kulkarni are gratefully acknowledged.

Conflicts of Interest: The authors declare no conflict of interest.

Sample Availability: Samples of the compounds are not available from the authors.

References

1. Jiao, J.; Nishihara, Y. Alkynylboron compounds in organic synthesis. *J. Organomet. Chem.* **2012**, *721–722*, 3–16. [CrossRef]
2. Soderquist, J.A.; Matos, K.; Rane, A.; Ramos, J. Alkynylboranes in the Suzuki-Miyaura coupling. *Tetrahedron Lett.* **1995**, *36*, 2401–2402. [CrossRef]
3. Brown, H.C.; Bhat, N.G.; Srebnik, M. A Simple, General Synthesis of 1-Alkynyldiisopropoxyboranes. *Tetrahedron Lett.* **1988**, *29*, 2631–2634. [CrossRef]
4. Saptal, V.B.; Wang, R.; Park, S. Recent advances in transition metal-free catalytic hydroelementation (E = B, Si, Ge, and Sn) of alkynes. *RSC Adv.* **2020**, *10*, 43539–43565. [CrossRef]
5. Geier, S.J.; Vogels, C.M.; Melanson, J.A.; Westcott, S.A. The transition metal-catalysed hydroboration reaction. *Chem. Soc. Rev.* **2022**, *51*, 8877–8922. [CrossRef]
6. Lee, C.-I.; Zhou, J.; Ozerov, O.V. Catalytic Dehydrogenative Borylation of Terminal Alkynes by a SiNN Pincer Complex of Iridium. *J. Am. Chem. Soc.* **2013**, *135*, 3560–3566. [CrossRef]
7. Hu, J.-R.; Liu, L.-H.; Hu, X.; Ye, H.-D. Ag(I)-catalyzed C–H borylation of terminal alkynes. *Tetrahedron* **2014**, *70*, 5815–5819. [CrossRef]
8. Pell, C.J.; Ozerov, O.V. Catalytic dehydrogenative borylation of terminal alkynes by POCOP-supported palladium complexes. *Inorg. Chem. Front.* **2015**, *2*, 720–724. [CrossRef]
9. Tsuchimoto, T.; Utsugi, H.; Sugiura, T.; Horio, S. Alkynylboranes: A Practical Approach by Zinc-Catalyzed Dehydrogenative Coupling of Terminal Alkynes with 1,8-Naphthalenediaminatoborane. *Adv. Synth. Catal.* **2015**, *357*, 77–82. [CrossRef]
10. Procter, R.J.; Uzelac, M.; Cid, J.; Rushworth, P.J.; Ingleson, M.J. Low-Coordinate NHC–Zinc Hydride Complexes Catalyze Alkyne C–H Borylation and Hydroboration Using Pinacolborane. *ACS Catal.* **2019**, *9*, 5760–5771. [CrossRef]
11. Romero, E.A.; Jazzar, R.; Bertrand, G. Copper-catalyzed dehydrogenative borylation of terminal alkynes with pinacolborane. *Chem. Sci.* **2017**, *8*, 165–168. [CrossRef] [PubMed]
12. Wei, D.; Carboni, B.; Sortais, J.-B.; Darcel, C. Iron-Catalyzed Dehydrogenative Borylation of Terminal Alkynes. *Adv. Synth. Catal.* **2018**, *360*, 3649–3654. [CrossRef]
13. Birepinte, M.; Liautard, V.; Chabaud, L.; Pucheault, M. Magnesium-Catalyzed Tandem Dehydrogenation-Dehydrocoupling: An Atom Economical Access to Alkynylboranes. *Chem. Eur. J.* **2020**, *26*, 3236–3240. [CrossRef] [PubMed]
14. Willcox, D.R.; De Rosa, D.M.; Howley, J.; Levy, A.; Steven, A.; Nichol, G.S.; Morrison, C.A.; Cowley, M.J.; Thomas, S.P. Aluminium-Catalyzed C(sp)−H Borylation of Alkynes. *Angew. Chem. Int. Ed.* **2021**, *60*, 20672–20677. [CrossRef] [PubMed]
15. Ramachandran, P.V.; Hamann, H.J.; Choudhary, S. Amine-boranes as Dual-Purpose Reagents for Direct Amidation of Carboxylic Acids. *Org. Lett.* **2020**, *22*, 8593–8597. [CrossRef] [PubMed]
16. Ramachandran, P.V.; Alawaed, A.A.; Hamann, H.J. A Safer Reduction of Carboxylic Acids with Titanium Catalysis. *Org. Lett.* **2022**, *24*, 8481–8486. [CrossRef] [PubMed]
17. Pasumansky, L.; Goralski, C.T.; Singaram, B. Lithium Aminoborohydrides: Powerful, Selective, Air-Stable Reducing Agents. *Org. Process Res. Dev.* **2006**, *10*, 959–970. [CrossRef]
18. Ramachandran, P.V.; Drolet, M.P.; Kulkarni, A.S. A non-dissociative open-flask hydroboration with ammonia borane: Ready synthesis of ammonia-trialkylboranes and aminodialkylboranes. *Chem. Commun.* **2016**, *52*, 11897–11900. [CrossRef]
19. Ramachandran, P.V.; Drolet, M.P. Direct, high-yielding, one-step synthesis of vic-diols from aryl alkynes. *Tetrahedron Lett.* **2018**, *59*, 967–970. [CrossRef]
20. Taylor, M.D.; Grant, L.R.; Sands, C.A. A convenient preparation of pyridine-borane. *J. Am. Chem. Soc.* **1955**, *77*, 1506–1507. [CrossRef]
21. Ramachandran, P.V.; Kulkarni, A.S.; Zhao, Y.; Mei, J.G. Amine-boranes bearing borane-incompatible functionalities: Application to selective amine protection and surface functionalization. *Chem. Commun.* **2016**, *52*, 11885–11888. [CrossRef] [PubMed]
22. Corey, E.J.; Fuchs, P.L. A synthetic method for formyl→ethynyl conversion (RCHO→RC≡CH or RC≡CR'). *Tetrahedron Lett.* **1972**, *13*, 3769–3772. [CrossRef]
23. Buttenberg, W.P. Condensation des Dichloracetals mit Phenol und Toluol. *Justus Liebigs Ann. Chem.* **1894**, *279*, 324–337. [CrossRef]
24. Fritsch, P., IV. Ueber die Darstellung von Diphenylacetaldehyd und eine neue Synthese von Tolanderivaten. *Justus Liebigs Ann. Chem.* **1894**, *279*, 319–323. [CrossRef]
25. Wiechell, H. Condensation des Dichloracetals mit Anisol und Phenetol. *Justus Liebigs Ann. Chem.* **1894**, *279*, 337–344. [CrossRef]
26. Noth, H.; Beyer, H. Beiträge zur Chemie des Bors, I. Darstellung und Eigenschaften der Alkylamin-borane, $R_{3-n}H_nN.BH_3$. *Chem. Ber. Recl.* **1960**, *93*, 928–938. [CrossRef]
27. Ramachandran, P.V.; Hamann, H.J.; Lin, R. Activation of sodium borohydride via carbonyl reduction for the synthesis of amine- and phosphine-boranes. *Dalton Trans.* **2021**, *50*, 16770–16774. [CrossRef] [PubMed]
28. Rao, M.L.N.; Jadhav, D.N.; Dasgupta, P. Pd-Catalyzed Domino Synthesis of Internal Alkynes Using Triarylbismuths as Multicoupling Organometallic Nucleophiles. *Org. Lett.* **2010**, *12*, 2048–2051. [CrossRef]

29. Lehane, K.N.; Moynihan, E.J.A.; Brondel, N.; Lawrence, S.E.; Maguire, A.R. Impact of sulfur substituents on the C–H$\cdots$O interaction of terminal alkynes in crystal engineering. *CrystEngComm* **2007**, *9*, 1041–1050. [CrossRef]
30. Fenneteau, J.; Vallerotto, S.; Ferrié, L.; Figadère, B. Liebeskind–Srogl cross-coupling on γ-carboxyl-γ-butyrolactone derivatives: Application to the side chain of amphidinolides C and F. *Tetrahedron Lett.* **2015**, *56*, 3758–3761. [CrossRef]

Article

A Window into the Workings of *anti*-$B_{18}H_{22}$ Luminescence—Blue-Fluorescent Isomeric Pair 3,3′-Cl_2-$B_{18}H_{20}$ and 3,4′-Cl_2-$B_{18}H_{20}$ (and Others) [†]

Marcel Ehn [1], Dmytro Bavol [1], Jonathan Bould [1], Vojtěch Strnad [1,2], Miroslava Litecká [1], Kamil Lang [1], Kaplan Kirakci [1], William Clegg [3], Paul G. Waddell [3] and Michael G. S. Londesborough [1,*]

[1]   Institute of Inorganic Chemistry of the Czech Academy of Sciences, 250 68 Řež, Czech Republic; ehn@email.cz (M.E.); bavol@iic.cas.cz (D.B.); jbould@gmail.com (J.B.); strnvoj@seznam.cz (V.S.); litecka@iic.cas.cz (M.L.); lang@iic.cas.cz (K.L.); kaplan@iic.cas.cz (K.K.)

[2]   Department of Physical Chemistry, University of Chemistry and Technology in Prague, Technická 5, Dejvice, 166 28 Prague, Czech Republic

[3]   Chemistry, School of Natural and Environmental Sciences, Newcastle University, Newcastle upon Tyne NE1 7RU, UK; bill.clegg@newcastle.ac.uk (W.C.); paul.waddell@newcastle.ac.uk (P.G.W.)

*   Correspondence: michaell@iic.cas.cz

[†]   Dedicated on the occasion of his 80th birthday to Prof. John D. Kennedy, who contributed so thoroughly to the development of macropolyhedral borane chemistry and who is a friend to many of us.

Abstract: The action of $AlCl_3$ on room-temperature tetrachloromethane solutions of *anti*-$B_{18}H_{22}$ (**1**) results in a mixture of fluorescent isomers, 3,3′-Cl_2-$B_{18}H_{20}$ (**2**) and 3,4′-Cl_2-$B_{18}H_{20}$ (**3**), together isolated in a 76% yield. Compounds **2** and **3** are capable of the stable emission of blue light under UV-excitation. In addition, small amounts of other dichlorinated isomers, 4,4′-Cl_2-$B_{18}H_{20}$ (**4**), 3,1′-Cl_2-$B_{18}H_{20}$ (**5**), and 7,3′-Cl_2-$B_{18}H_{20}$ (**6**) were isolated, along with blue-fluorescent monochlorinated derivatives, 3-Cl-$B_{18}H_{21}$ (**7**) and 4-Cl-$B_{18}H_{21}$ (**8**), and trichlorinated species 3,4,3′-Cl_3-$B_{18}H_{19}$ (**9**) and 3,4,4′-Cl_3-$B_{18}H_{19}$ (**10**). The molecular structures of these new chlorinated derivatives of octadecaborane are delineated, and the photophysics of some of these species are discussed in the context of the influence that chlorination bears on the luminescence of *anti*-$B_{18}H_{22}$. In particular, this study produces important information on the effect that the cluster position of these substitutions has on luminescence quantum yields and excited-state lifetimes.

Keywords: *anti*-$B_{18}H_{22}$; luminescence; fluorescence; cluster boron hydrides; chlorination; halogenation; substitution; excited-state lifetime; quantum yield

Citation: Ehn, M.; Bavol, D.; Bould, J.; Strnad, V.; Litecká, M.; Lang, K.; Kirakci, K.; Clegg, W.; Waddell, P.G.; Londesborough, M.G.S. A Window into the Workings of *anti*-$B_{18}H_{22}$ Luminescence—Blue-Fluorescent Isomeric Pair 3,3′-Cl_2-$B_{18}H_{20}$ and 3,4′-Cl_2-$B_{18}H_{20}$ (and Others). *Molecules* **2023**, *28*, 4505. https://doi.org/10.3390/molecules28114505

Academic Editors: Michael A. Beckett and Igor B. Sivaev

Received: 3 May 2023
Revised: 30 May 2023
Accepted: 30 May 2023
Published: 1 June 2023

1. Introduction

With the obvious exception of carbon, it is boron that amongst all the elements boasts the greatest number and diversity of hydride compounds [1]. These species produce polyhedral cluster assemblies with an ostensible 12-vertex icosahedral size limit. This size constraint, however, can be surpassed by the intimate fusion of two or more clusters to form 'macropolyhedral' species, in which constituent subclusters conjoin via shared polyhedral edges or faces [2]. We are interested in how the chemical modification of such macropolyhedral boron hydride clusters produces changes in their physical properties, and in particular, with regard to the interaction of these molecules with laser light. At very high excitation energies, our investigations hold relevance to the generation of boron–proton plasmas, and, at lower excitation energies, emission from electronically excited states that form the basis for an important class of light sources. Here, there is a renewed interest in the binary boron hydride cluster compound *anti*-$B_{18}H_{22}$ **1** due to the discovery of its utility as a new class of laser material [3]. It is also of particular interest as it fluoresces in the elusive blue region of the spectrum, with a quantum yield approaching unity [4];

it is highly photostable [3], and it is readily soluble in organic polymer matrices [3,5]. These are all factors of general practical benefit regarding the fabrication of optical devices. The photophysics of **1** may be modified by replacing some of its cluster terminal hydrogen atoms with substituents or ligands such as –SH [6], bromine [7,8], iodine [9], alkyls [10], and pyridine [11,12] to give molecules capable of, for example, environment-sensitive thermochromic luminescence [11], single-molecule multiple emissions [7], and the photosensitisation of oxygen [9]. Recently, metal-catalysed and metal-free nucleophilic substitution of the iodine group in 7-I-$B_{18}H_{21}$ [9,13] has been demonstrated [14], leading to a series of luminescent B-N, B-O, and B-S substituted octadecaborane derivatives [14]. In addition, a monochlorinated derivative of **1**, the blue fluorescent 7-Cl-$B_{18}H_{21}$, has been reported, featuring a quantum yield of luminescence of 0.8 [15]. Collectively, these entries to the literature are the beginnings of a systematic survey of the functionalisation of **1**, from which there are starting to emerge several broader patterns linking the photophysical behaviour of these molecules to their chemical structure. Accompanying these trends are tentative explanations based on, for example, the principles of 'Total Cluster Volume' [10], cluster electron density mapping [10], and the cluster-position of substitution [14]. Here, we report an extension to this growing portfolio of compounds by the synthesis of several chlorinated derivatives of **1**, the highest yielding examples of which are blue-fluorescent isomers 3,3'-Cl_2-$B_{18}H_{20}$ (**2**) and 3,4'-Cl_2-$B_{18}H_{20}$ (**3**). Compounds **2** and **3** are the first examples of halogenation of **1** at the B3 or B3' positions, and thus afford further opportunity to deepen the understanding on what are the chemical factors influencing luminescence from *anti*-$B_{18}H_{22}$ and its derivatives.

2. Results and Discussion

Synthesis and structural characterization. Addition of aluminium chloride to room-temperature tetrachloromethane solutions of *anti*-$B_{18}H_{22}$ **1** results primarily in the formation of a mixture of two of its dichlorinated derivatives, isomers 3,3'-Cl_2-$B_{18}H_{20}$ (**2**) and 3,4'-Cl_2-$B_{18}H_{20}$ (**3**) (Figure 1), which, after work-up and purification via sublimation, were obtained in an non-optimised yield of 76%. The mass spectrum of the crude product mixture (Figure S1) indicates the additional presence of lesser amounts of mono- and trichlorinated derivatives of **1**. The generally smaller quantities of these species precluded their full examination, but we were able to collect good NMR, UV-vis, and crystallographic data on some of these minor products that we present later in this manuscript. Temperature and duration of reaction seem to bear an influence on both the rate at which the chlorinated derivatives are formed as well as the ratio in which they are produced. Thus, typically, a room temperature reaction requires a period of 4–5 days for the complete consumption of **1**. The same synthesis carried out at 55 °C gives a full conversion to products in only 90 min, whereas at an ice-bath temperature of 2 °C a period of 30 days is required. The product mix ratios (Table 1) were quantitatively determined by analytical HPLC and an analysis of the ^{11}B NMR spectra of the reaction mixtures and, in particular, a comparison of the integrations of the singlets at $\delta(^{11}B)$ +23.1 ppm for the substituted B3 and/or B3' position(s), and at −24.3 ppm for the substituted B4 and/or B4' position(s) (Figure S2). These ratios show that at higher temperatures, substitution at the B4 and/or B4' position(s) becomes more probable, and the formation of compound **3** and trichlorinated derivatives **9** and **10** are therefore more likely. Lower temperatures or shorter reaction times boost the amount of monochlorinated derivatives **7** and **8** in the product mix (primarily **7**), and longer reaction times, over a certain temperature, support trichlorination and the formation of **9** and **10** (primarily **9**).

HPLC, fractional sublimation, and crystallization were used to separate the various products of chlorination of **1**. Our discussion of these new species begins with the two main products, dichlorinated **2** and **3**.

The molecular structures of **2** and **3**, as determined by single-crystal X-ray diffraction studies (SCXRD), are shown in Figure 1. Both isomers preserve the classical octadecaborane polyhedral architecture of **1**, with no perturbation of significance in any of the B-B connectivities. Equally, there is very little difference between B-Cl bond lengths for the different substituent positions that all measure approximately 1.8 Å, which is a similar value to B-Cl bonds in other known macropolyhedral borane systems [16,17]. Isomers **2** and **3** provide identical mass spectra with m/z 284.32, due to their $[B_{18}H_{19}Cl_2]^-$ ions (Figure S5), but may be distinguished by their different absorption spectra and retention times when passed through an HPLC system prior to injection into the mass spectrometric device (Figure S6). The relative positions of substitution in isomers **2** and **3** result in the preservation of C_i symmetry in **2** and asymmetry in **3**. These differences are clearly seen in the NMR data for the compounds (see Figure S7 vs. Figure S8 for direct comparison, and Figure 2 for wider comparison with all chlorinated derivatives of **1** described in this manuscript).

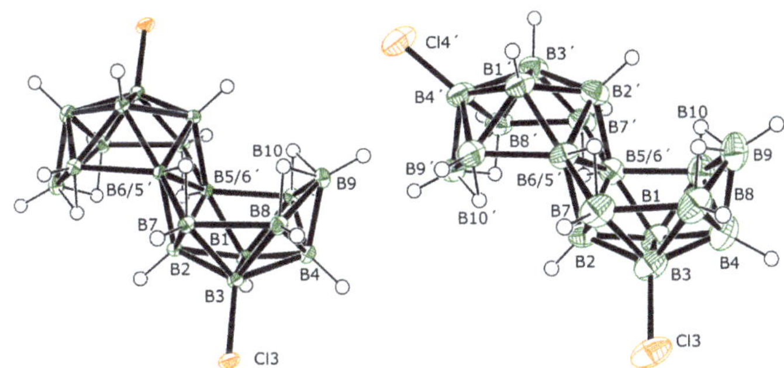

Figure 1. The crystallographically determined molecular structures of $3,3'$-Cl_2-$B_{18}H_{20}$ (**2**) and $3,4'$-Cl_2-$B_{18}H_{20}$ (**3**), drawn with 50% probability ellipsoids for non-H atoms. For a list of interatomic distances see Tables S3 and S4.

Table 1. The ratio of chlorinated derivatives of **1** formed using different reaction conditions. Percentages were calculated by the integration of peaks separated on HPLC (Apex retention times: 3.51 min for **9** and **10**; 7.18 min for **2**; 8.68 min for **3**; and 15.46 min for **7** and **8**). For more details on HPLC conditions, see Section 3.

React. Conditions	Monochlorinated Deriv. (7, 8)	$3,3'$-Cl_2-$B_{18}H_{20}$ (2)	$3,4'$-Cl_2-$B_{18}H_{20}$ (3)	Other Dichlorinated Deriv. (4 to 6)	Trichlorinated Deriv. (9, 10)
+2 °C/30 days	43%	20%	34%	<1%	3%
RTP/5 days	2%	41%	35%	<1%	21%
+55 °C/90 min	13%	29%	50%	<1%	8%

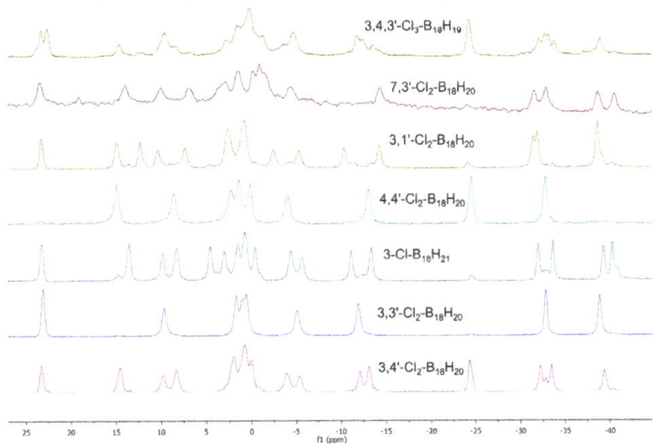

Figure 2. Stacked ^{11}B-$\{^1H\}$ NMR spectra of all chlorinated derivatives of *anti*-$B_{18}H_{22}$ (**1**).

Table 2 shows a listing of the measured and calculated NMR data for compound **2** that are entirely consistent with its C_i symmetric molecular structure shown in Figure 1. The excellent correlation between the measured data and those calculated for the molecular configuration shown in Figure 1 provide further weight to this structure being representative of the bulk sample of **2**. The degree of perturbation from the $\delta(^{11}B)$ chemical shifts for parent compound **1** is small for **2**, with only the chlorine-substituted $B(3,3')$ chemical shift changing significantly (Figure 3). The chlorine atoms deshield the $B(3,3')$ nuclei, shifting the resonance downfield by about 10 ppm.

Table 2. Measured proton and boron-11 NMR data for $3,3'\text{-}Cl_2\text{-}B_{18}H_{20}$ (**2**) at 291 K in $CDCl_3$ solution together with calculated values for comparison. (^{11}B and $^{11}B\text{-}\{^1H\}$ NMR spectra are shown in Figure S7).

Assign.	$\delta(^{11}B)$/ppm		$\delta(^1H)$/ppm
	Meas.	Calc.	Meas.
3,3'	+23.4	+26.1	--- [a]
10,10'	+9.8	+10.2	+4.06
5,6	+1.8	+1.9	--- [b]
1,1'	+0.7	−0.1	+3.28
9,9'	+0.2	−0.2	+3.45
8,8'	−5.1	−4.8	+3.28
7,7'	−11.9	−11.2	+3.20
2,2'	−32.8	−33.0	−0.04
4,4'	−38.9	−38.3	+0.59
9,10		−1.28	−0.53
6,7		−1.28	−0.53
8,9		−3.05	−2.41

[a] Chlorine substituent, [b] Site of conjunction.

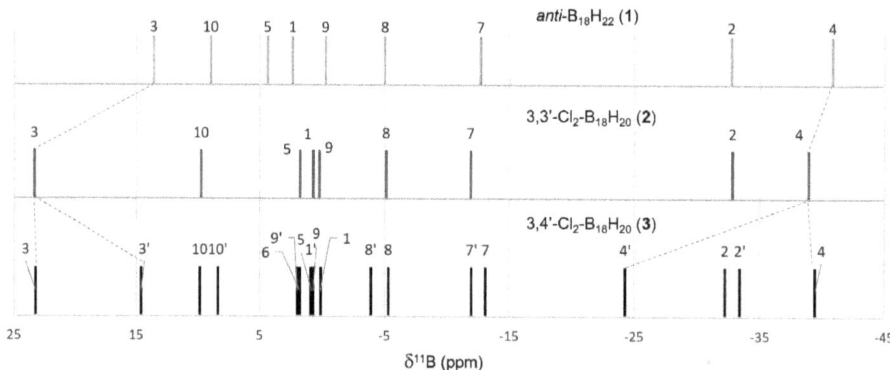

Figure 3. A comparison of $^{11}B\text{-}\{^1H\}$ NMR chemical shifts for $anti\text{-}B_{18}H_{22}$ (**1**), $3,3'\text{-}Cl_2\text{-}B_{18}H_{20}$ (**2**) and $3,4'\text{-}Cl_2\text{-}B_{18}H_{20}$ (**3**). Numbering for symmetrical **1** and **2** are simplified to single numbers, i.e., $3 = (3,3')$ or $10 = (10,10')$, etc.

Table 3 lists the measured and calculated NMR data for compound **3**. The asymmetry of **3** is immediately apparent in its ^{11}B NMR spectrum, which displays 18 peaks. These peaks may be conveniently split into two sets of nine resonances for each of the molecule's two subclusters: $\{3\text{-}Cl\text{-}B_9\}$ and $\{4'\text{-}Cl\text{-}B_9\}$. The nine peaks for the $\{3\text{-}Cl\text{-}B_9\}$-subcluster essentially overlap with those for the symmetrical compound **2**. The $\{4'\text{-}Cl\text{-}B_9\}$-subcluster, on the contrary, displays change; the $B3'\text{-}H$ peak shifts upfield by about 9 ppm to a position similar to that in parent compound **1**, and there is a roughly 16 ppm downfield shift of the $B4'\text{-}Cl$ peak (see dotted lines in Figure 3).

Table 3. Measured proton and boron-11 NMR data for 3,4′-Cl$_2$-B$_{18}$H$_{20}$ (**3**) at 291 K in CDCl$_3$ solution together with calculated values for comparison. (^{11}B and ^{11}B-{^{1}H} NMR spectra are shown in Figure S8).

Assign.	δ(^{11}B)/ppm		δ(^{1}H)/ppm
	Meas.	Calc.	Meas.
3	+23.3	+25.8	--- [a]
3′	+14.6	+14.3	+4.12
10	+9.9	+9.0	+4.08
10′	+8.4	+7.5	+4.41
9′	+2.0	+1.0	+3.41
6	+1.8	−0.2	--- [b]
1′	+0.9	−1.0	+3.41
5	+0.8	−1.2	--- [b]
9	+0.7	−1.3	+3.31
1	+0.1	−2.6	+3.41
8′	−3.9	−6.1	+3.43
8	−5.3	−7.7	+3.30
7′	−12.0	−14.2	+3.04
7	−13.1	−15.3	+3.20
4′	−24.3	−24.1	--- [a]
2	−32.2	−34.8	−0.01
2′	−33.4	−36.2	+0.02
4	−39.4	−42.2	+0.64
9′,10′			−0.07
9,10			−0.47
5,7′			−0.50
6,7			−1.07
8′,9′			−2.26
8,9			−2.44

[a] Chlorine substituent, [b] Site of conjunction.

Overall, the ^{11}B NMR properties of compounds **2** and **3** suggest them to be the same two dichlorinated *anti*-B$_{18}$H$_{20}$Cl$_2$ species mentioned, but structurally undefined and uncharacterised, in a previous report describing the chlorination of **1** with SO$_2$Cl$_2$ [15].

The charge distribution in *anti*-B$_{18}$H$_{22}$ is such that the basal boron atoms B(1), B(2), B(3) and, in particular, B(4) (see Figure 4, and Figure 1 for numbering) possess the greatest share of the available electron density. This distinction is apparent in the products of electrophilic substitution of **1** recorded in the literature so far. Thus, it is the case for recently published brominated and iodinated derivatives of **1** that, under similar reaction conditions to those used here, bromination results mainly in the monohalogenated product 4-Br-B$_{18}$H$_{21}$ (although a partially characterised dibrominated derivate 4,4′-Br$_2$-B$_{18}$H$_{20}$ was also observed in mass spectrometric analysis and NMR) [7], and iodination to high yields of the dihalogenated species 4,4′-I$_2$-B$_{18}$H$_{20}$. (n.b. the reaction of the dianionic form of **1** with iodine in alcohol solutions gives the monosubstituted 7-I-B$_{18}$H$_{21}$) [9]. It is therefore apparent from this work that the chlorination of **1** is far more versatile, and the substitution of its octadecaborane cluster is more multi-directed than is the case for bromination or iodination, in particular with regard to the activation of the B3 cluster position towards substitution. The relatively lower energy of the 3,3′-Cl$_2$-B$_{18}$H$_{20}$ isomer **2**, calculated at 2.3 kcal/mol, would, all other things being equal, lead to a Boltzmann distribution of 98% for **2** and 2% for isomer **3**. However, the considerably more negative charge associated with the B(4 and 4′) cluster atoms (see Figure 4) presumably competes with the differences in zero-point total energies, especially at higher temperatures, leading to the observed ratios of **2**:**3** formed during synthesis at 55 °C.

Above and beyond the two main products from this reaction, compounds **2** and **3**, an effort was made to isolate and characterise the lower-yielding side products. Laborious work involving HPLC separation and identification using mass spectrometry and UV absorption spectroscopy resulted in the imperfect isolation of an additional six chlorinated derivatives of **1**. Of these six compounds, three are further isomers of dichlorinated derivatives of **1**, specifically $4,4'\text{-}Cl_2\text{-}B_{18}H_{20}$ (**4**), $3,1'\text{-}Cl_2\text{-}B_{18}H_{20}$ (**5**), and $7,3'\text{-}Cl_2\text{-}B_{18}H_{20}$ (**6**). The molecular structures of each of these isomers were established by SCXRD analyses, the results of which are shown in Figure 5.

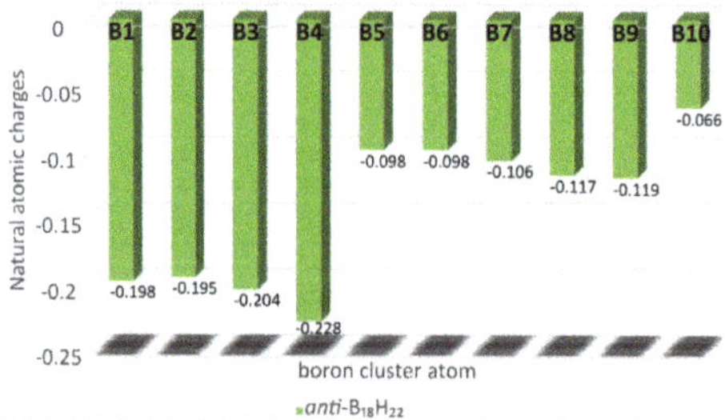

Figure 4. Natural atomic charges for boron positions in $anti\text{-}B_{18}H_{22}$ (**1**). Numbering for symmetrically equivalent atoms are simplified to single numbers, i.e., 3 = (3,3′) or 10 = (10,10′), etc.

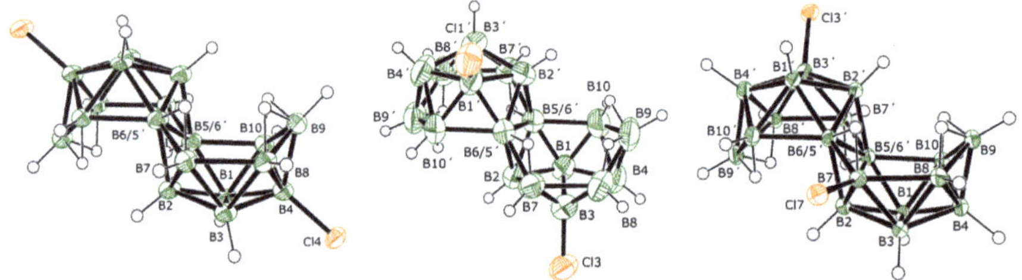

Figure 5. The crystallographically determined molecular structures of $4,4'\text{-}Cl_2\text{-}B_{18}H_{20}$ (**4**), $3,1'\text{-}Cl_2\text{-}B_{18}H_{20}$ (**5**), and $7,3'\text{-}Cl_2\text{-}B_{18}H_{20}$ (**6**), drawn with 50% probability ellipsoids for non-H atoms. For a list of interatomic distances see Tables S9–S11.

These molecular structures were supported by mass spectrometry and, in the cases of **4** and **5**, a comparison of measured NMR data with calculated NMR chemical shifts for specific optimised molecular structures (Tables 4 and 5). Unfortunately, only ^{11}B NMR data were collected for compound **6**, which was isolated as a single crystal in one of many fractional sublimation procedures used to purify other dichlorinated species. These measured data nevertheless match well with those calculated for the ^{11}B chemical shifts (Table 6) of a molecular geometry for **6** shown in Figure 5.

Table 4. Measured proton and boron-11 NMR data for $4,4'\text{-}Cl_2\text{-}B_{18}H_{20}$ (**4**) at 291 K in CDCl$_3$ solution together with calculated values for comparison. (^{11}B and ^{11}B-{^{1}H} NMR spectra are shown in Figure S12).

Assign.	$\delta(^{11}\text{B})$/ppm		$\delta(^1\text{H})$/ppm
	Meas.	Calc.	Meas.
3,3′	+14.6	+15.1	+4.13
10,10′	+8.4	+7.5	+4.42
5,6	+2.1	+0.2	--- a
9,9′	+1.3	−1.0	+3.78
1,1′	+0.0	−1.1	+3.30
8,8′	−4.3	−6.5	+3.44
7,7′	−13.2	−14.9	+3.03
4,4′	−24.6	−23.6	--- b
2,2′	−35.9	−35.9	+3.03
9,10			+0.03
6,7			−0.96
8,9			−2.18

^{a} Site of conjunction, ^{b} Chlorine substituent.

Table 5. Measured proton and boron-11 NMR data for 3,1'-Cl$_2$-B$_{18}$H$_{20}$ (**5**) at 291 K in CDCl$_3$ solution together with calculated values for comparison. (^{11}B and ^{11}B-{^{1}H} NMR spectra are shown in Figure S13).

| Assign. | δ(^{11}B)/ppm | | δ(^{1}H)/ppm |
	Meas.	Calc.	Meas.
3	+23.3	+26.1	--- [a]
3'	+14.9	+15.2	+4.31
1	+12.3	+13.7	--- [a]
10'	+10.3	+9.0	+4.10
10	+7.3	+6.3	+4.36
5	+2.6	+1.9	--- [b]
1'	+2.5	+1.2	+3.41
6	+0.7	−0.7	--- [b]
9	+0.7	−2.7	+3.47
9'	+0.7	−2.9	+3.39
8	−2.4	−4.8	+3.09
8'	−5.3	−7.6	+3.34
7	−10.3	−12.7	+3.09
7'	−14.3	−15.9	+3.27
2	−31.4	−34.6	+0.33
2'	−31.9	−34.8	+0.33
4	−38.6	−42.1	+0.68
4'	−38.8	−41.2	+0.68
9',10'			−0.10
9,10			−0.19
5,7'			−0.43
6,7			−0.90
8',9'			−2.44
8,9			−2.80

[a] Chlorine substituent, [b] Site of conjunction.

Table 6. Measured boron-11 NMR data for 7,3'-Cl$_2$-B$_{18}$H$_{20}$ (**6**) at 291 K in CDCl$_3$ solution together with calculated values for comparison. (^{11}B and ^{11}B-{^{1}H} NMR spectra are shown in Figure S14).

| Assign. | δ(^{11}B)/ppm | |
	Meas.	Calc.
3' [a]	+23.4	+23.6
3	+14.0	+11.8
10'	+10.0	+10.3
10	+6.8	+7.0
5 [b]	+3.7	+3.1
7 [a]	+3.1	+3.6
9	+2.9	+3.6
1'	+1.4	+1.5
9'	+1.4	+1.8
1	−0.2	−1.5
6 [b]	−1.0	−1.7
8	−1.5	−1.7
8'	−4.4	−4.9
7'	−14.3	−13.2
2'	−31.5	−30.5
2	−32.8	−32.0
4'	−38.7	−37.0
4	−40.5	−39.1

[a] Chlorine substituent, [b] Site of conjunction.

Two mono-chlorinated derivatives of **1**, 3-Cl-B$_{18}$H$_{21}$ (**7**) and 4-Cl-B$_{18}$H$_{21}$ (**8**), were also observed. Although the isolation of these monochlorinated derivatives from other, multichlorinated, compounds was facile, the complete separation of **7** from **8** proved too difficult. However, SCXRD studies on the solid solutions of these compounds resulted in the molecular structures shown in Figure 6. Again, the octadecaborane cluster structure of **1** is unperturbed; however, the B-Cl distances appear to be about 0.1 Å shorter than those lengths for the dichlorinated species described here, although this may be primarily due to the disorder present in the solid solution.

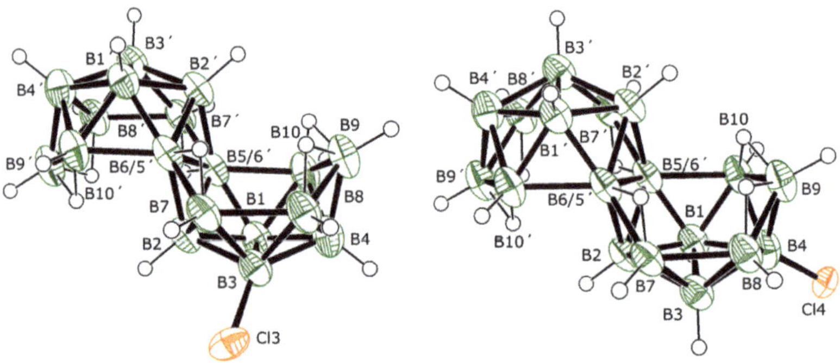

Figure 6. The crystallographically determined molecular structures of 3-Cl-B$_{18}$H$_{21}$ (**7**), and 4-Cl-B$_{18}$H$_{21}$ (**8**), drawn with 50% probability ellipsoids for non-H atoms. For a list of interatomic distances see Table S15. The apparent angular distortion of the 3-Cl substituent is an artefact of disorder, which could be resolved only for the Cl atoms, so the experimentally determined boron atom positions are an average weighted towards the 4-Cl isomer.

Analytical chromatography of the multi-chlorinated mixture led to the separation of compounds **7** and **8** from the other multi-chlorinated species. Initial separations generally gave a mixture of **7** and **8** with a mass spectrum of *m/z* 250.32, due to their [B$_{18}$H$_{20}$Cl]$^-$ ions (Figure S16). NMR of these mixtures indicated the presence of about 85% of 3-Cl-B$_{18}$H$_{21}$ (**7**) to 15% of 4-Cl-B$_{18}$H$_{21}$ (**8**), based on an integration of the B3-Cl singlet peak (located at δ(^{11}B) +23.2 ppm) for the former and the B4-Cl singlet peak (located at -24.5 ppm) for the latter compound. Crystals (of the three samples measured by SCXRD) of this sample that were obtained via sublimation show, however, a reversal of this ratio (approximately 75% of **8** to 25% of **7**, with small variations in different crystals), suggesting that **8** is the more volatile of the two isomers. Indeed, this physical difference enabled the effective separation of **8** from **7** via careful sublimation and the isolation of the 3-Cl derivative in good purity for NMR and UV-vis studies. There is a very good correlation between measured NMR data for **7** and chemical shifts calculated for the monochlorinated 3-Cl-B$_{18}$H$_{21}$ structure (Table 7).

Table 7. Measured proton and boron-11 NMR data for 3-Cl-B$_{18}$H$_{21}$ (**7**) at 291 K in CDCl$_3$ solution together with calculated values for comparison. (^{11}B and ^{11}B-{^{1}H} NMR spectra are shown in Figure S17).

Assign.	δ(^{11}B)/ppm		δ(^{1}H)/ppm
	Meas.	Calc.	Meas.
3	+23.2	+26.1	--- [a]
3$'$	+13.5	+13.2	+3.89
10	+9.8	+9.0	+4.09
10$'$	+8.3	+7.6	+4.08
6	+4.5	+4.0	--- [b]
1	+3.0	+0.2	+3.42
5	+1.5	+0.1	--- [b]
9$'$	+0.7	-1.8	+3.36
1$'$	+0.6	-1.8	+3.37

Table 7. *Cont.*

Assign.	$\delta(^{11}B)$/ppm		$\delta(^{1}H)$/ppm
	Meas.	**Calc.**	**Meas.**
9	-0.4	-3.0	$+3.10$
$8'$	-4.4	-6.5	$+3.06$
8	-5.6	-8.0	$+3.29$
$7'$	-11.1	-13.2	$+0.09$
7	-13.3	-15.2	
2	-31.9	-34.6	
$2'$	-33.6	-36.3	
4	-39.3	-42.1	
$4'$	-40.2	-43.5	

[a] Chlorine substituent, [b] Site of conjunction.

Finally, an attempt was made to determine the location of the chlorine substituents on the very small amount of isolated trichlorinated species using a comparison of the collected experimental ^{11}B NMR spectroscopic data with the computed chemical shifts for a series of seven trichlorinated isomers with the lowest zero-point corrected total energies, as shown in Table 8.

Table 8. Calculated zero-point corrected energies for the seven most stable Cl_3-$B_{18}H_{19}$ isomers relative to $1,3,3'$-Cl_3-$B_{18}H_{19}$ as the lowest energy isomer. Associated colour code indicates the extent to which zero-point corrected energies for individual isomers increase—from green (lowest energy) to red (highest energy).

Isomer	kcal/mol
$1,3,3'$-Cl_3-$B_{18}H_{19}$	0.0
$1,3,1'$-Cl_3-$B_{18}H_{19}$	0.3
$3,4,3'$-Cl_3-$B_{18}H_{19}$	1.1
$1,4,3'$-Cl_3-$B_{18}H_{19}$	1.4
$1,3',4'$-Cl_3-$B_{18}H_{19}$	1.5
$1,4,1'$-Cl_3-$B_{18}H_{19}$	1.6
$3,4,4'$-Cl_3-$B_{18}H_{19}$	2.5

Figure S18 shows the measured ^{11}B NMR spectrum of the HPLC-separated fraction with a single mass spectrometric peak at m/z 319.22 (Figure S19), which computes precisely to the expected mass of $[Cl_3$-$B_{18}H_{18}]^-$. This spectrum appears to comprise two molecular species present in an approximate 2:1 ratio. Figures S20–S33 and Tables S20–S33 are the calculated spectra for all of the seven lowest energy trichlorinated isomers listed in Table 8. A comparison of these measured and calculated data suggest that the main component is very likely $3,4,3'$-Cl_3-$B_{18}H_{19}$ (**9**) and the minor component most probably $3,4,4'$-Cl_3-$B_{18}H_{19}$ (**10**). However, in absence of any reliable crystallographic data, these are tentative assignments. The measured and calculated NMR data for $3,4,3'$-Cl_3-$B_{18}H_{19}$, compound **9**, are shown in Table 9.

Table 9. Measured boron-11 NMR data for $3,4,3'$-Cl_3-$B_{18}H_{19}$ (**9**) at 291 K in $CDCl_3$ solution together with calculated values for comparison. (^{11}B and ^{11}B-$\{^{1}H\}$ NMR spectra are shown in Figure S18).

Assign.	$\Delta(^{11}B)$/ppm	
	Meas.	**Calc.**
$3'$ [a]	$+22.1$	$+19.4$
3 [a]	$+21.3$	$+18.6$
$10'$	$+8.5$	$+5.9$
10	$+8.5$	$+5.5$
1	$+0.6$	-3.1
$9'$	$+0.1$	-3.6
5 [b]	-0.1	-4.1

Table 9. *Cont.*

Assign.	$\Delta(^{11}B)$/ppm	
	Meas.	Calc.
9	−0.1	−4.3
1′	−0.1	−4.6
6 [b]	−0.1	−4.8
8	−5.9	−9.4
8′	−5.9	−9.5.7
7′	−12.8	−15.0
7	−13.5	−15.6
4 [a]	−25.4	−25.4
2	−33.8	−35.3
2′	−34.2	−35.7
4	−40.1	−41.5

[a] Chlorine substituent, [b] Site of conjunction.

A detailed comparison of the molecular geometry of these various chlorinated clusters, together with the parent *anti*-$B_{18}H_{22}$ (**1**) [18], shows that the B–B (and B–Cl) bond lengths are rather insensitive to the substituent positions. Because structural disorder has an averaging effect on parts of the molecules in which disorder is not resolved and twinning adversely affects precision, such an analysis can be carried out reliably only on ordered structures, which here means one form of compound **2** (not the forms in the Supporting Information), and compounds **3**, **4**, and **6**; two of these (**2** and **4**) have more than one crystallographically independent molecule, all of which can be included in the comparison. A full list of bond lengths and their differences from the corresponding bonds in compound **1** are given in Table S38. These differences between the substituted and parent clusters range from −0.045 Å to +0.034 Å, while the total range of all B–B bond lengths is 0.294 Å (from 1.706 to 2.002 Å, with individual crystallographic standard uncertainties between about 0.002 and 0.02 Å); they follow no clear pattern relative to the positions of substitution. We therefore conclude that the cluster bonding is essentially unaffected by chlorination.

Photophysical properties of some of the chlorinated derivatives of 1. As mentioned in the Introduction, compound **1** [3–5] and its derivatives [6–12,14–16] display various modes of useful luminescence. As such, we were keen to understand the photophysics of the new chlorinated derivatives of **1** reported here. Table 10 and Figure 7 show the main photophysical data recorded for new compounds **2**, **4**, and **7**. Absorption and excitation spectra for these species may be found in the SI (Figures S34–S36). Overall, there is little difference in the absorption and emission maxima for these derivatives, which all absorb in the UV-A region (absorption maxima in the 324–344 nm range) and emit in the blue region (emission maxima in the 406–435 nm range). Thus, chlorination seems to have only a small influence over the relative energies of the contributing HOMO and LUMO systems. However, from these data a trend is apparent that increasing the number of chlorine substituents both shortens the lifetimes of the fluorescent excited states (τ_L) and decreases the quantum yield of fluorescence (Φ_L). So, for example, mono-chlorinated derivative **7** has about half of the fluorescent lifetime of parent compound **1** and approximately half of its Φ_L, and dichlorinated **2** has just over a tenth of the fluorescent lifetime of **1** and roughly a twelfth of its Φ_L.

Table 10. Photophysical properties of the chlorinated derivatives of *anti*-$B_{18}H_{22}$ in air-saturated hexane at room temperature [a].

	λ_{abs}/nm	λ_L/nm	τ_L/ns	Φ_L (λ_{exc}/nm)
anti-$B_{18}H_{22}$ (**1**) [b]	335	406	11.2	0.97 (335 nm)
7-Cl-$B_{18}H_{21}$ (**11**) [c]	344	418	8.3	0.80 (340 nm)
3-Cl-$B_{18}H_{21}$ (**7**)	327	407	6.1	0.52 (320 nm)
4,4′-Cl_2-$B_{18}H_{20}$ (**4**)	344	435	1.5	0.17 (340 nm)
3,3′-Cl_2-$B_{18}H_{20}$ (**2**)	324	410	1.2	0.07 (320 nm)

[a] λ_{abs}—absorption maximum; λ_L—emission maximum (λ_{exc} = 320 nm); τ_L—fluorescence lifetime in air-saturated hexane (λ_{exc} = 340 nm, λ_{em} = 450 nm,); Φ_L—fluorescence quantum yield in air-saturated hexane (experimental error of Φ_L is ±0.01). [b] data taken from reference [4]. [c] data taken from reference [13].

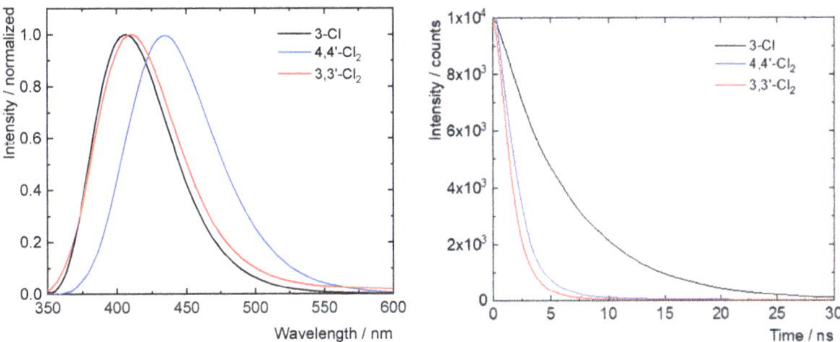

Figure 7. Normalised fluorescence spectra (**left**) and fluorescence decay spectra (**right**) for compounds **2**, **4** and **7**.

A recent article by Spokoyny et al. [15] proposes that the location of substituents on the octadecaborane cluster plays a crucial role in the efficiency of emission from the molecule (its quantum yield of luminescence). In their paper [15], a comparison was made between $7\text{-I-B}_{18}H_{21}$ and $4\text{-I-B}_{18}H_{21}$ isomers, with the former displaying a two-fold greater quantum yield of phosphorescence. The mono-chlorinated species $7\text{-Cl-B}_{18}H_{21}$ (**11**) is also reported, the photophysical properties of which are also shown here in Table 10. Comparison of the data for **7** and **10** reveals that chlorination of the basal B3 position in **7** leads to a roughly 25% shorter fluorescent lifetime and 25% lower quantum yield than in **11**, where chlorination is at the open-face 'gunwale' B7 position. A further interesting comparison is between the photophysics of compounds **2** and **4**: although both species have a similarly short excited-state lifetime, compound **4** has more than twice the quantum yield of fluorescence than **2**. This suggests that chlorination (and by tentative extrapolation, substitution) of **1** at the B4/4$'$ positions reduces the efficiency of fluorescence less than when at the B3/3$'$ cluster sites.

3. Experimental

3.1. Materials

Anti-$B_{18}H_{22}$ was made using the oxidative fusion of the nonaborane anion made from *nido*-$B_{10}H_{14}$, as described by Gaines et al. [19]. *nido*-$B_{10}H_{14}$ purchased from Katchem s.r.o. (Kralupy, Czech Republic) and $AlCl_3$ (Sigma Aldrich, Merck Life Science spol. S r.o. Prague, Czech Republic) were both sublimed before use; CCl_4, *n*-hexane, and CH_2Cl_2 (all Lach-ner, Prague, Czech Republic) were used without purification. Silica gel for chromatography was 60–200 μm (Lach-ner, Prague, Czech Republic). The authors recommend that, prior to any use of *nido*-$B_{10}H_{14}$, a consultation of the information on its toxicity and required safety precautions are made (see https://cameochemicals.noaa.gov/chemical/503#es and https://inchem.org/documents/icsc/icsc/eics0712.htm, accessed on 2 May 2023).

3.2. Chlorinated Derivatives of Anti-$B_{18}H_{22}$ (1)

A 100 mL RB-flask fitted with a Teflon stop-cock was charged with *anti*-$B_{18}H_{22}$ (360 mg, 1.7 mmol) and freshly sublimed $AlCl_3$ (2.7 g, 20 mmol), to which was added 20 mL of CCl_4 under a stream of argon gas. The vessel was then closed, and the contents stirred. The reaction was monitored by NMR spectroscopy at regular intervals. On the complete consumption of the *anti*-$B_{18}H_{22}$ starting material (4–5 days at RTP or approx. 90 min at 55 °C), the reaction vessel was placed in an ice-bath, allowed to cool, and then cold distilled water (25 mL) that had been acidified with a small volume of $HCl_{(aq)}$ (3 mL) was added slowly whilst stirring so as to destroy any remaining $AlCl_3$. The two-layer liquid was then filtered, and the water layer separated and extracted three times with 15 mL measures of dichloromethane. The combined CCl_4 and CH_2Cl_2 solutions were then reduced to dryness on a rotary evaporator and the remaining solids sublimed at 120 °C.

Subsequent crystallisation from a minimum amount of hot hexanes gave $3,3'$-Cl_2-$B_{18}H_{20}$ (**2**) and $3,4'$-Cl_2-$B_{18}H_{20}$ (**3**) jointly in a 76% yield (373 mg, 1.31 mmol). The remaining minor components were isolated, as described in the manuscript, using HPLC techniques.

3.3. General Methods

Preparative column chromatography was carried out using a 4 cm diameter column of silica gel G (60–200 μm Lach-ner). **NMR** spectra were recorded on a JEOL 600 MHz (14.1 T) spectrometer. Structural assignments were made using ^{11}B, ^{11}B-$\{^1H\}$, ^1H, ^1H-$\{^{11}B(\text{broadband})\}$, ^1H-$\{^{11}B(\text{selective})\}$, COSY, and HMQC (Heteronuclear Multiple-Quantum Correlation) techniques. **Mass Spectrometry** measurements were performed on a Thermo-Finnigan LCQ-Fleet Ion Trap instrument using electrospray ionization (ESI) with the detection of negative ions. For ESI, samples dissolved in acetonitrile (concentrations approximately 100 ng·mL^{-1}) were introduced to the ion source by infusion at 5 μL·min^{-1} with a source voltage of 4.5 kV, a tube lens voltage of -90.8 V, a capillary voltage of -20.3 V, a capillary temperature of 275 °C, and a nebulizing sheath gas and auxiliary gas flow of 1.8 L·min^{-1} and 5.9 L·min^{-1}, respectively. Molecular ions $[M]^-$ were detected for all univalent anions as base peaks in the spectra. The experimental and calculated isotopic distribution patterns were fully agreed upon for all isolated compounds. The isotopic distribution in the boron plot of all peaks was in complete agreement with the calculated spectral pattern. The data are presented for the most abundant mass in the boron distribution plot (100%) and for the peak on the right side of the boron plot corresponding to the m/z value. **Analytical chromatography** was carried out on a Thermo Finnigan Surveyor HPLC system equipped with a Photo Diode Array detector. Chromatographic conditions: column LiChroCART RP-select B 60 Å (5 μm, 250 × 3.00 mm *I.D.*). Mobile phase preparation: 9 mL of 0.5 M stock solution of propylamine acetate (PAA) was diluted by water (HPLC grade) to 430 mL (pH was adjusted to 4.5). This buffer was mixed with 570 mL of ACN (HPLC gradient grade). Flow rate 0.4 mL·min^{-1}, detection PDA (190–800 nm). Samples with a concentration of approximately 1 mg·mL^{-1} in the mobile phase were injected in a volume of 3 μL. *Semi-***preparative chromatography** was realized on LaPrep Sigma, VWR system equipped with multiple-wavelength detector and a fully automated fraction collector Foxy R2, Teledyne ISCO. Chromatographic conditions: column TESSEK Separon SGX C18 (7 μm, 250 × 25.00 mm *I.D.*). Mobile phase preparation: 12 mL of 1.0 M stock solution of propylamine acetate (PAA) was diluted by water (HPLC grade) to 300 mL (pH was adjusted to 4.2). This buffer was mixed with 700 mL of ACN (HPLC gradient grade). Flow rate 8.0 mL·min^{-1} and fixed wavelengths: 225, 335, and 355 nm. Samples with a concentration of approximately 10 mg·mL^{-1} in the mobile phase were injected in a volume of 2 mL. **UV-vis absorption and fluorescence** properties were recorded on a Perkin Elmer Lambda 35 spectrometer and FLS1000 (Edinburgh Instrument, Livingston, UK) on the air-saturated solutions of the respective compounds. Fluorescence lifetime experiments were performed upon excitation at 340 nm (EPLED-340, 340 ± 10 nm) and decay curves were fitted to exponential functions by the iterative reconvolution procedure of the Fluoracle software (v. 2.13.2, Edinburgh Instrument, Livingston, UK). Fluorescence quantum yields were measured using a Quantaurus QY C11347-1 spectrometer (Hamamatsu, Shizuoka, Japan).

3.4. X-ray Diffraction Analyses

Single crystals of all compounds were grown either by slow evaporation of concentrated solutions of the compounds in *n*-hexane in a glass vial or from the sublimation of small amounts sealed in an evacuated glass NMR tube and placed with the bottom third of the tube in a sand-bath maintained at ca. 90 °C. Single-crystal diffraction data were measured on Rigaku Oxford Diffraction XtaLAB Synergy diffractometers in Prague and Newcastle. Structures were solved with SHELXT [20] and refined with SHELXL [21], with the aid of the OLEX2 interface [22]. Table 11 gives the main crystal and refinement data for compounds **2–8**; corresponding information for polymorphic forms is given in Table S37. In the various structures the asymmetric unit contains one-half, one, or more than one molecule, with some molecules lying on inversion centres. In some cases, this means an unsymmetrical molecule is disordered; lower-symmetry models were explored and rejected. For the co-crystallised compounds **7** and **8**, disorder of the 3 and 4 positions occurs together with disorder of the molecule over an inversion centre; three separate crystals from two sublimation experiments were examined and gave consistent results with only minor variations of the composition (only one result is given here). Crystals of **4** and **2a** (one of the polymorphs of **2**) were twinned, adversely affecting the precision of the structure of **4** (for which the resolved two twin components are probably an approximation to multiple twinning). Some structures were found to contain minor amounts of other isomers, involving further disorder; two of the four forms of the 3,3' compound **2** (**2b** and **2c**) contain small amounts of the 3,4' isomer, while there is some minor additional 4 and/or 4' Cl substitution identified in the structure of the 3,1' isomer **5**. In most cases, particularly for fully ordered structures, H atoms could be located in difference maps and refined freely; for less well-determined H atoms, restraints were applied to maintain approximately the same geometry as in the ordered structures.

Table 11. Crystal and refinement data for compounds **2–8**.

	2	3	4	5	6	7/8
Chemical formula	$B_{18}H_{20}Cl_2$	$B_{18}H_{20}Cl_2$	$B_{18}H_{20}Cl_2$	$B_{18}H_{20}Cl_2$	$B_{18}H_{20}Cl_2$	$B_{18}H_{21}Cl$
M_r	285.6	285.6	285.6	285.6	285.6	251.2
Crystal system	monoclinic	orthorhombic	triclinic	monoclinic	monoclinic	monoclinic
Space group	$P2_1/c$	$Aba2$	$P\bar{1}$	$P2_1$	$C2/c$	$P2_1/c$
a (Å)	10.98586(12)	18.9603(5)	11.6342(9)	7.2776(2)	14.1953(3)	7.2553(3)
b (Å)	12.76094(15)	14.4383(3)	11.6732(9)	10.6035(2)	11.8073(2)	10.8304(4)
c (Å)	11.21906(13)	12.0489(2)	11.6840(8)	10.4842(3)	20.3684(4)	10.0028(3)
α (°)			89.695(6)			
β (°)	94.9949(10)		86.618(6)	100.525(4)	112.580(2)	106.432(3)
γ (Å)			84.082(6)			
V (Å^3)	1566.83(3)	3298.44(12)	1575.6(2)	795.43(4)	3152.21(11)	753.89(5)
Z	4	8	4	2	8	2
Crystal size (mm^3)	0.21 × 0.17 × 0.12	0.10 × 0.06 × 0.01	0.09 × 0.08 × 0.02	0.14 × 0.10 × 0.06	0.12 × 0.07 × 0.05	0.09 × 0.06 × 0.02
T (K)	100	150	100	100	100	150
Radiation, λ (Å)	Cu$K\alpha$, 1.54184	Cu$K\alpha$, 1.54184	Cu$K\alpha$, 1.54184	Mo$K\alpha$, 0.71073	Cu$K\alpha$, 1.54184	Cu$K\alpha$, 1.54184
Reflections measured	20,387	15,233	14,596	11,545	10,723	7108
Unique reflections	3301	2229	6177	3144	3315	1504
R_{int}	0.0432	0.0347	0.1203	0.0301	0.0248	0.0272
Parameters, restraints	261, 0	260, 2	482, 772	282, 79	261, 0	139, 1
R (F, $F^2 > 2\sigma$)	0.0276	0.0362	0.1626	0.0721	0.0340	0.0597
R_w (F^2, all data)	0.0719	0.1027	0.4501	0.2246	0.0977	0.1568
Goodness of fit (F^2)	1.121	1.046	1.065	1.095	1.069	1.114
Flack parameter				0.02(5)		
Max, min $\Delta\rho$ (e Å^{-3})	0.31, −0.24	0.42, −0.16	1.86, −0.69	0.33, −0.40	0.61, −0.26	0.21, −0.22
CCDC deposition number	2256510	2256511	2256512	2256513	2256514	2256515

4. Conclusions

As compared to the halogenation of **1** with bromine and iodine, its chlorination using $AlCl_3$ in CCl_4 produces a larger number of substituted derivatives with halogenation occurring in a greater variety of positions on the 18-vertex boron cluster. The specific sites of chlorination in the two main products, dichlorinated isomers **2** and **3**, seem to be driven by the competing factors of charge distribution in **1** and the relative zero-point corrected total energies of the products themselves. Regarding the photophysics of these compounds, chlorination of **1** appears to reduce the excited-state lifetime of fluoresence as well as its quantum yield of fluorescence. These reductions are not linear, with dichlorination resulting in disproportionately greater reduction compared to monochlorination. Although the current sample size is still small, the data also suggest that substitution at the B4 position leads to a greater bathochromic shift in the emission wavelength compared to B3-substituted species, but it dampens the fluorescence less. Furthermore, although substitution at the 'gunwale' open-face B7 position is less stable in comparison to substitution at the basal B3 or B4 positions, it does lead to superior quantum yields of fluorescence.

In closing, it is evident that, whereas the profound reduction in the lifetime of excited states of these molecules caused by chlorination is detrimental to the system's fluorescence quantum yield, it would presumably reduce the probability of excited-state absorption (ESA) of the pump and the stimulated emission energy that has been shown to be the major mitigating factor to the better laser performance of compound **1** [23]. It is therefore conceivable that the combination of chlorine substituents (to reduce the excited-state lifetime) with alkyl substituents (to boost quantum yield) [10] could produce a B_{18}-derivative molecular system with good emission properties and fewer ESA complications, and hence an interesting laser performance.

Supplementary Materials: The following supporting information can be downloaded at: https://www.mdpi.com/article/10.3390/molecules28114505/s1, Relevant NMR spectra and mass spectra for new compounds **2–9**; Absorption and fluorescence excitation spectra of relevant compounds; Interatomic distances for compounds **2–8**; Crystal data for polymorphic forms of compound **2**.

Author Contributions: M.G.S.L. (conceptualization of work, synthesis, data interpretation, corresponding author); M.E. (synthesis and NMR spectroscopy); D.B. (mass spectrometry, preparative and analytical chromatography separation); M.E., V.S. and J.B. (NMR spectroscopy and calculational study); M.L., W.C. and P.G.W. (single-crystal X-ray diffraction analyses); K.L. and K.K. (photophysics). All authors have read and agreed to the published version of the manuscript.

Funding: This work was funded by the Czech Science Foundation grant number 23-07563S. For crystallographic instrumentation in Prague, we also thank the Research Infrastructure NanoEnviCz project, supported by the Ministry of Education, Youth and Sports of the Czech Republic, Project no. LM2018124. Crystallographic instrumentation in Newcastle was funded by EPSRC (UK), Project no. EP/W021129/1, and by Newcastle University (no Project number). Newcastle University and EPSRC (UK): equipment funding.

Data Availability Statement: DFT calculations for the chlorinated boranes described here were performed using the Gaussian 09 package [24] at the B3LYP/6-311++G(d,p) level for all atoms. Frequency analyses were carried out to confirm the validity of minima. GIAO NMR nuclear shielding predictions were performed at the appropriate level on the final optimised geometries and the calculated boron nuclear shielding values were related to chemical shifts $\delta(^{11}B)$ by comparison with the computed value for B_2H_6 which was taken to be $\delta(^{11}B)$ +16.6 ppm relative to the $F_3B.OEt_2$ at 0.0 ppm. Zero-point corrected energy calculations were performed using the Gaussian 16W package source at the B3LYP/cc-pVDZ level for all atoms.

Conflicts of Interest: The authors declare no conflict of interest.

Sample Availability: Samples of the main product compounds from the chlorination of **1** are available from the authors.

Abbreviations

SCXRD—single-crystal X-ray diffraction; NMR—Nuclear magnetic resonance; UV—ultra-violet; HPLC—high performance liquid chromatography; HOMO—Highest occupied molecular orbital; LUMO—Lowest unoccupied molecular orbital; RTP—room temperature and pressure.

References

1. Beall, H.; Gaines, D.F. Boron Hydrides. In *Encyclopedia of Physical Science and Technology*, 3rd ed.; Academic Press: Cambridge, MA, USA, 2003; pp. 301–316.
2. Shea, S.L.; Bould, J.; Londesborough, M.G.S.; Perera, S.D.; Franken, A.; Ormsby, D.L.; Jelínek, T.; Štíbr, B.; Holub, J.; Kilner, C.A.; et al. Polyhedral boron-containing cluster chemistry: Aspects of architecture beyond the icosahedron. *Pure Appl. Chem.* **2003**, *75*, 1239–1248. [CrossRef]
3. Cerdán, L.; Braborec, J.; Garcia-Moreno, I.; Costela, A.; Londesborough, M.G.S. A Borane Laser. *Nat. Commun.* **2015**, *6*, 5958. [CrossRef]
4. Londesborough, M.G.S.; Hnyk, D.; Bould, J.; Serrano-Andrés, L.; Sauri, V.; Oliva, J.M.; Kubat, P.; Polivka, T.; Lang, K. Distinct Photophysics of the Isomers of $B_{18}H_{22}$ Explained. *Inorg. Chem.* **2012**, *51*, 1471–1479. [CrossRef]

5. Ševčík, J.; Urbánek, P.; Hanulíková, B.; Čapková, T.; Urbánek, M.; Antoš, J.; Londesborough, M.G.S.; Bould, J.; Ghasemi, B.; Petřkovský, L.; et al. The Photostability of Novel Boron Hydride Blue Emitters in Solution and Polystyrene Matrix. *Materials* **2021**, *14*, 589. [CrossRef]

6. Sauri, V.; Oliva, J.M.; Hnyk, D.; Bould, J.; Braborec, J.; Merchan, M.; Kubat, P.; Císařová, I.; Lang, K.; Londesborough, M.G.S. Tuning the Photophysical Properties of *anti*-$B_{18}H_{22}$: Efficient Intersystem Crossing between Excited Singlet and Triplet States in New 4,4′-(HS)$_2$-*anti*-$B_{18}H_{20}$. *Inorg. Chem.* **2013**, *52*, 9266–9274. [CrossRef]

7. Anderson, K.P.; Waddington, M.A.; Balaich, G.J.; Stauber, J.M.; Bernier, N.A.; Caram, J.R.; Djurovich, P.I.; Spokoyny, A. A Molecular Boron Cluster-Based Chromophore with Dual Emission. *Dalton Trans.* **2020**, *49*, 16245–16251. [CrossRef] [PubMed]

8. Anderson, K.P.; Hua, A.S.; Plumley, J.B.; Ready, A.D.; Rheingold, A.L.; Peng, T.L.; Djurovich, P.I.; Kerestes, C.; Snyder, N.A.; Andrews, A.; et al. Benchmarking the dynamic luminescence properties and UV stability of $B_{18}H_{22}$-based materials. *Dalton Trans.* **2022**, *51*, 9223–9228. [CrossRef] [PubMed]

9. Londesborough, M.G.S.; Dolanský, J.; Bould, J.; Braborec, J.; Kirakci, K.; Lang, K.; Císařová, I.; Kubát, P.; Roca-Sanjuán, D.; Francés-Monerris, A.; et al. Effect of Iodination on the Photophysics of the Laser Borane *anti*-$B_{18}H_{22}$: Generation of Efficient Photosensitizers of Oxygen. *Inorg. Chem.* **2019**, *58*, 10248–10259. [CrossRef]

10. Bould, J.; Lang, K.; Kirakci, K.; Cerdán, L.; Roca-Sanjuán, D.; Francés-Monerris, A.; Clegg, W.; Waddell, P.G.; Fuciman, M.; Polívka, T.; et al. A Series of Ultra-Efficient Blue Borane Fluorophores. *Inorg. Chem.* **2020**, *59*, 17058–17070. [CrossRef]

11. Londesborough, M.G.S.; Dolanský, J.; Cerdán, L.; Lang, K.; Jelínek, T.; Oliva, J.M.; Hnyk, D.; Roca-Sanjuán, D.; Francés-Monerris, A.; Martinčík, J.; et al. Thermochromic Fluorescence from $B_{18}H_{20}(NC_5H_5)_2$: An Inorganic–Organic Composite Luminescent Compound with an Unusual Molecular Geometry. *Adv. Opt. Mater.* **2017**, *5*, 1600694. [CrossRef]

12. Londesborough, M.G.S.; Dolanský, J.; Jelínek, T.; Kennedy, J.D.; Císařová, I.; Kennedy, R.D.; Roca-Sanjuán, D.; Francés-Monerris, A.; Lang, K.; Clegg, W. Substitution of the laser borane *anti*-$B_{18}H_{22}$ with pyridine: A structural and photophysical study of some unusually structured macropolyhedral boron hydrides. *Dalton Trans.* **2018**, *47*, 1709–1725. [CrossRef] [PubMed]

13. Olsen, F.P.; Vasavada, R.C.; Hawthorne, M.F. The chemistry of *n*-$B_{18}H_{22}$ and *i*-$B_{18}H_{22}$. *J. Am. Chem. Soc.* **1968**, *90*, 3946–3951. [CrossRef]

14. Anderson, K.P.; Djurovich, P.I.; Rubio, V.P.; Liang, A.; Spokoyny, A.M. Metal-Catalyzed and Metal-Free Nucleophilic Substitution of 7-I-$B_{18}H_{21}$. *Inorg. Chem.* **2022**, *61*, 15051–15057. [CrossRef] [PubMed]

15. Anderson, K.P.; Rheingold, A.L.; Djurovich, P.I.; Soman, O.; Spokoyny, A.M. Synthesis and luminescence of monohalogenated $B_{18}H_{22}$ clusters. *Polyhedron* **2022**, *227*, 116099. [CrossRef]

16. Londesborough, M.G.S.; Lang, K.; Clegg, W.; Waddell, P.G.; Bould, J. Swollen Polyhedral Volume of the *anti*-$B_{18}H_{22}$ Cluster via Extensive Methylation: Anti-$B_{18}H_8Cl_2Me_{12}$. *Inorg. Chem.* **2020**, *59*, 2651–2654. [CrossRef]

17. Londesborough, M.G.S.; Macias, R.; Kennedy, J.D.; Clegg, W.; Bould, J. Macropolyhedral Nickelaboranes from the Metal-Assisted Fusion of KB_9H_{14}. *Inorg. Chem.* **2019**, *58*, 13258–13267. [CrossRef]

18. Hamilton, E.J.M.; Kultyshev, R.G.; Du, B.; Meyers, E.A.; Liu, S.; Hadad, C.M.; Shore, S.G. A Stacking Interaction between a Bridging Hydrogen Atom and Aromatic π Density in the *n*-$B_{18}H_{22}$–Benzene System. *Chem. Eur. J.* **2006**, *12*, 2571–2578. [CrossRef]

19. Gaines, D.F.; Nelson, C.K.; Steehler, G.A. Preparation of *n*-Octadecaborane(22), *n*-$B_{18}H_{22}$, by Oxidative Fusion of Dodecahydrononaborane(l-) Clusters. *JACS* **1984**, *106*, 7266–7267. [CrossRef]

20. Sheldrick, G.M. SHELXT—Integrated space-group and crystal-structure determination. *Acta Crystallogr. Sect. A* **2015**, *71*, 3–8. [CrossRef]

21. Sheldrick, G.M. Crystal structure refinement with SHELXL. *Acta Crystallogr. Sect. C* **2015**, *71*, 3–8. [CrossRef]

22. Dolomanov, O.V.; Bourhis, L.J.; Gildea, R.J.; Howard, J.A.K.; Puschmann, H. OLEX2: A complete structure solution, refinement and analysis program. *J. Appl. Crystallogr.* **2009**, *42*, 339–341. [CrossRef]

23. Cerdán, L.; Francés-Monerris, A.; Roca-Sanjuán, D.; Bould, J.; Dolanský, J.; Fuciman, M.; Londesborough, M.G.S. Unveiling the Role of Upper Excited States in the Photochemistry and Laser Performance of *anti*-$B_{18}H_{22}$. *J. Mater. Chem. C* **2020**, *8*, 12806–12818. [CrossRef]

24. Frisch, M.J.; Trucks, G.W.; Schlegel, H.B.; Scuseria, G.E.; Robb, M.A.; Cheeseman, J.R.; Scalmani, G.; Barone, V.; Petersson, G.A.; Nakatsuji, H.; et al. Gaussian, Versions 09 and 16. Gaussian, Inc.: Wallingford, CT, USA, 2013 and 2016. Available online: https://gaussian.com/ (accessed on 2 May 2023).

Article

Two New Aluminoborates with 3D Porous-Layered Frameworks

Chen Wang, Juan Chen, Chong-An Chen, Zhen-Wen Wang and Guo-Yu Yang *

MOE Key Laboratory of Cluster Science, School of Chemistry and Chemical Engineering, Beijing Institute of Technology, Beijing 100081, China; cw@bit.edu.cn (C.W.); chenjuan@bit.edu.cn (J.C.); cca@bit.edu.cn (C.-A.C.); wzw@bit.edu.cn (Z.-W.W.)
* Correspondence: ygy@bit.edu.cn

Abstract: Two new aluminoborates, $NaKCs[AlB_7O_{13}(OH)]\cdot H_2O$ (**1**) and $K_4Na_5[AlB_7O_{13}(OH)]_3\cdot 5H_2O$ (**2**), have been hydro(solvo)thermally made with mixed alkali metal cationic templates. Both **1** and **2** crystallize in the monoclinic space group $P2_1/n$ and contain similar units of $[B_7O_{13}(OH)]^{6-}$ cluster and AlO_4 tetrahedra. The $[B_7O_{13}(OH)]^{6-}$ cluster is composed of three classical B_3O_3 rings by vertex sharing, of which two of them connect with AlO_4 tetrahedra to constitute monolayers, and one provides an O atom as a bridging unit to link two oppositely orientated monolayers by Al-O bonds to form 3D porous-layered frameworks with 8-MR channels. UV-Vis diffuse reflectance spectra indicate that both **1** and **2** exhibit short deep-UV cutoff edges below 190 nm, revealing that they have potential applications in deep-UV regions.

Keywords: aluminoborates; oxoboron cluster; 3D porous layers; solvothermal syntheses

1. Introduction

Crystalline borates send out an enchanting charm in the sciences because of their multifarious structures and widespread applications in microporous and nonlinear optical (NLO) materials [1–5]. In 1975, the NLO properties of $KB_5O_8\cdot 4H_2O$ [6] were studied by C F. Dewey et al. for the first time, pointing out the new research direction for the structure and properties of borates. Subsequently, high-temperature solid-state reactions and the boric flux method became the main methods of synthesizing borates [7,8]. Until 2004, Yang's group applied the hydro(solvo)thermal method for the borate system and gradually introduced inorganic cations, organic amines, transition metal complexes, or chiral metal complexes as structure directing agents (SDAs) [9]. It is significant that the SDAs play an important role in the formation of structure by host–guest symmetry and charge matching [10], which effectively regulate the inorganic skeleton and successfully acquire abundant borates with novel open frameworks. In recent years, researchers have paid more attention to the alkali and alkaline earth metal borates [11] because of their better chemical stabilities, higher transmittances, greater damage thresholds, and almost no absorption properties of ultraviolet (UV) light [12], such as the well-known NLO materials: β-BaB_2O_4 (BBO) [13], $CsLiB_6O_{10}$ (CLBO) [14], and LiB_3O_5 (LBO) [15]. These research achievements have enormously inspired scientists' enthusiasm and curiosity for pursuing newer borates.

In terms of structure, boron atoms typically adopt three or four coordination geometries with oxygen atoms to form BO_3 triangles or BO_4 tetrahedra. The combination of these two units via corner- or edge-sharing generates various oxoboron clusters, which can further polymerize through H-bonds and covalent bonds to constitute 1D chains, 2D layers, and 3D frameworks [16–20]. Moreover, in order to expand the structural diversity of borates, Al^{3+} was introduced into the borates' framework [21]. It is worth noting that Al is in the same group as boron but has more plentiful coordination modes, such as the AlO_4 tetrahedron, AlO_5 tetragonal pyramid, and AlO_6 octahedron [22–24]. The developments of aluminoborates (ABOs) were slow since Al was firstly led into the borates system by

Citation: Wang, C.; Chen, J.; Chen, C.-A.; Wang, Z.-W.; Yang, G.-Y. Two New Aluminoborates with 3D Porous-Layered Frameworks. *Molecules* **2023**, *28*, 4387. https://doi.org/10.3390/molecules28114387

Academic Editors: Michael A. Beckett and Igor B. Sivaev

Received: 30 April 2023
Revised: 20 May 2023
Accepted: 22 May 2023
Published: 27 May 2023

Lehmann and Teske in 1973 [25], mainly because the single crystal structures of limited ABOs were difficult to be determined [26]. However, this difficulty was solved in 2009. Our group put forward the use of Al(i-PrO)$_3$ to replace the traditional inorganic Al sources under hydro(solvo)thermal conditions [27]. Except for its better solubility in organic solvents, the synergism between chiral AlO$_4$ tetrahedra formed in hydrolysis progress and acentric structures [28]. By guiding with different types of SDAs, numerous ABOs have been reported [29–32].

Herein, two novel 3D porous layered ABOs, NaKCs[AlB$_7$O$_{13}$(OH)]·H$_2$O (**1**) and K$_4$Na$_5$[AlB$_7$O$_{13}$(OH)]$_3$·5H$_2$O (**2**), were solvothermally made. The 3D porous-layered frameworks of **1** and **2** were both built by the alternation of [B$_7$O$_{13}$(OH)]$^{6-}$ clusters and AlO$_4$ units. The [B$_7$O$_{13}$(OH)]$^{6-}$ cluster was composed of three classical B$_3$O$_3$ rings, of which two of them constructed monolayers with AlO$_4$ tetrahedra, and adjacent layers were connected through bridging O atoms provided by another B$_3$O$_3$ ring. The evident difference between **1** and **2** was that the asymmetric unit of **2** contained three crystallographically independent [AlB$_7$O$_{13}$(OH)]$^{3-}$, and they were linked, in turn, along the *b*-axis. Different cationic diameters also resulted in the diverse curvature of the porous layers. The structure, comparison, and synthesis of the above two compounds will be discussed in detail in the following sections.

2. Results and Discussion

2.1. Synthesis Procedure

Compounds **1** and **2** both adopted two kinds of boron sources as reactants, and the products could not be obtained without any one, which were confirmed through experiments. Furthermore, **1** used H$_3$BO$_3$ and Na$_2$B$_4$O$_7$·10H$_2$O, while **2** used H$_3$BO$_3$ and NaBO$_2$·4H$_2$O. Wherein, H$_3$BO$_3$, through self-polymerization, could build various oxoboron clusters. Meanwhile, polyborate (Na$_2$B$_4$O$_7$·10H$_2$O, NaBO$_2$·4H$_2$O) could not only enhance the pH of whole system but also may further recombine new oxoboron clusters through the degradation of polyanions. In addition, the reaction temperature and pH were also major factors affecting the reaction, and higher temperatures and pH levels were more conducive to improving the polymerization of the oxoboron clusters (Temperature: 210 °C for **1** and 230 °C for **2**; pH: 7 for **1** and 9 for **2**). The possible chemical equations during the reaction are given below, respectively.

$$3H_3BO_3 + [B_4O_7]^{2-} + 4OH^- + Al^{3+} = [AlB_7O_{13}(OH)]^{3-} + 6H_2O \tag{1}$$

$$H_3BO_3 + 13[BO_2]^- + 2Al^{3+} = 2[AlB_7O_{13}(OH)]^{3-} + OH^- \tag{2}$$

2.2. Structure of **1**

Single crystal X-ray analyses display that **1** crystalizes in the monoclinic space group $P2_1/n$. The asymmetric unit of **1** contains one [B$_7$O$_{13}$(OH)]$^{6-}$ cluster, one Na, one K, one Cs, and one water molecule (Figure 1a). The [B$_7$O$_{13}$(OH)]$^{6-}$ cluster consists of three familiar B$_3$O$_3$ rings, in which five BO$_3$ triangle units and two BO$_4$ tetrahedral groups are connected by vertex sharing. The B-O distances in the range of 1.329 Å to 1.402 Å, and the O-B-O bond angles lie in the range of 114.9°–122.7° for the BO$_3$ triangles. Meanwhile, the B-O distances range from 1.427 Å to 1.511 Å, and the O-B-O bond angles vary from 106.8° to 148.9° for BO$_4$ tetrahedra.

Each [B$_7$O$_{13}$(OH)]$^{6-}$ cluster was further connected to four AlO$_4$ tetrahedral via O1, O3, O7, and O14 atoms, and vice versa (Figure 1b). The alternating connection of oxoboron clusters and AlO$_4$ tetrahedra formed a 2D fluctuating monolayer with 13-MR windows, which were arranged in two orientations (Figure 1c). The windows consisted of Al-B2-B3-B5-B7-Al-B2-B1-Al-B7-B5-B3-B1(Figure S1a). Two monolayers exhibited axial symmetry along the *b*-axis and were linked by Al-O bonds, constituting a 3D double-layer structure with two unequal types of 8-MR channels, which possessed the same components but had different shapes. These two channels were delineated by Al-B1-B3-B4-Al-B2-B3-B4 (Figure S1b) and placed in an -ABAB- sequence in the *ac* plane (Figure 1d,e).

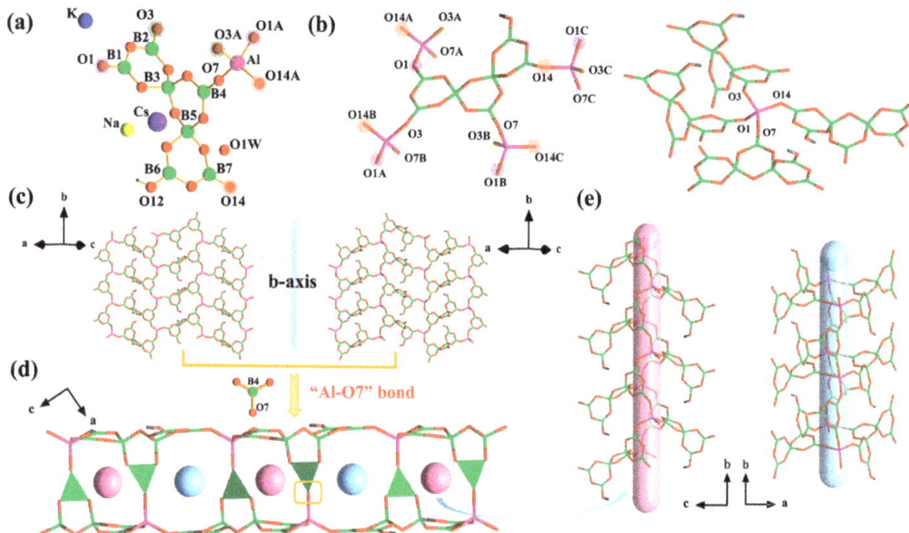

Figure 1. (**a**) Asymmetric unit of **1**; (**b**) Coordination environment of B$_7$ cluster and AlO$_4$ tetrahedra in **1**; (**c**) Two orientated monolayers in **1**; (**d**) The 3D porous-layered framework with two channels in **1**; (**e**) View of the two types of the 8-MR channels.

As for the coordination of metal cations, K was surrounded by eight O atoms, Na was coordinated by seven O atoms, and Cs was bonded by eight O atoms (Figure S2). It should be noted that these metal atoms shared oxygen atoms to lead a three-dimensional metal–oxygen net similar to a ladder (Figure 2a). Specifically, Na atoms were located between adjacent porous layers, K atoms were situated on the wall of 8-MR channels, and Cs and O1W atoms were seated in the neighboring channels, respectively. In general, the porous layers stacked along the *ac* plane in an -AAA- sequence, composing the complete dense 3D network with cations (Figure 2b).

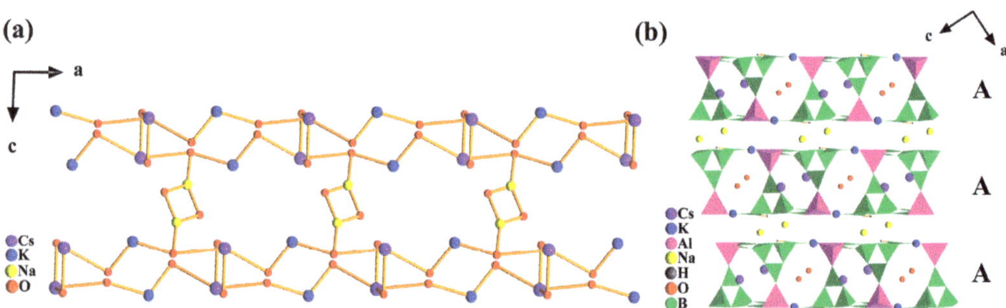

Figure 2. (**a**) The metal–oxygen net in **1**; (**b**) The complete dense 3D network in **1**.

2.3. Structure of **2**

Notably, **2** crystallized in the monoclinic space group $P2_1/n$, and its asymmetric unit contained three crystallographically independent $[AlB_7O_{13}(OH)]^{3-}$, four K atoms, five Na atoms, and five water molecules (Figure 3a). The $[B_7O_{13}(OH)]^{6-}$ cluster was the same as it was in **1**, and the distances of the B-O bond varied from 1.339 Å to 1.521 Å.

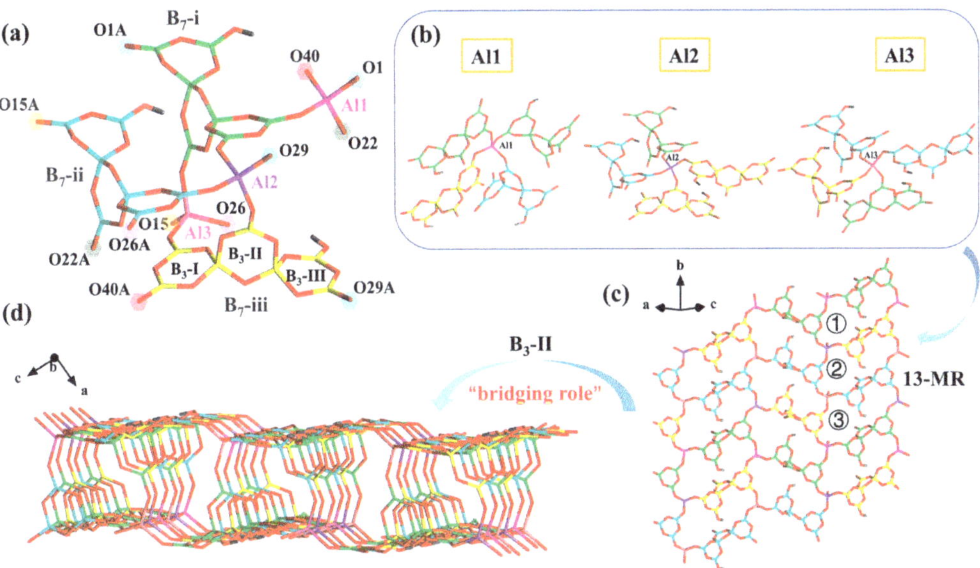

Figure 3. (**a**) Asymmetric unit of **2**; (**b**) Coordination environment of three different AlO_4 tetrahedra in **2**; (**c**) The monolayer with three kinds of 13-MR rings in **2**; (**d**) The 3D framework in **2**.

Each AlO_4 tetrahedra connected with four neighboring {B₇} clusters of three different types. The Al1 was bonded with two B₇-i, one B₇-ii, and one B₇-iii. The Al2 was linked with two B₇-iii, one B₇-I, and one B₇-ii, whereas Al3 was joined to two B₇-ii, one B₇-I, and one B₇-iii, separately (Figure 3b). These three $[AlB_7O_{13}(OH)]^{3-}$ clusters were connected, in turn, along the *b*-axis, constituting a monolayer with three kinds of 13-member rings (Figure 3c). The AlO_4 tetrahedra and B_3O_3-II (B₃-II) rings were interconnected by sharing O atoms, linking two adjacent single layers with opposite orientations into a 3D porous structure, and the porous layers were stacked in an -AAA- sequence in the *ac* plane (Figure 4b). From this point of view, the B_3O_3-I (B₃-I) and B_3O_3-III (B₃-III) connected with the AlO_4 tetrahedra to constitute monolayers, while the B₃-II as a bridging unit linked two oppositely orientated layers to the 3D frameworks (Figure 3d).

Two kinds of channels existed, and each B_3O_3 ring played a different role (Figure 4a). Channel A was made of two AlO_4 tetrahedra, two B₃-I, and B₃-II rings. The B₃-I was responsible for bonding adjacent AlO_4 tetrahedra in order to extend along the *b*-axis, while B₃-II played an effect on linking the B₃-I and AlO_4 tetrahedra to form a closed window. The B₃-III could be seen as a decoration hanging on the channel wall. However, the situation of channel B was diverse. It consisted of two AlO_4 tetrahedra, two B₃-II, and B₃-III rings. The B₃-II and B₃-III only played a part in a closed window, while another B₃-I linkedup the neighboring windows. From this perspective, channel B was composed of parallel windows, as channel A's were linked end to end. In view of this, channel A could be seen as a "sine wave" model, while channel B could be regarded as a "parallel wave" model.

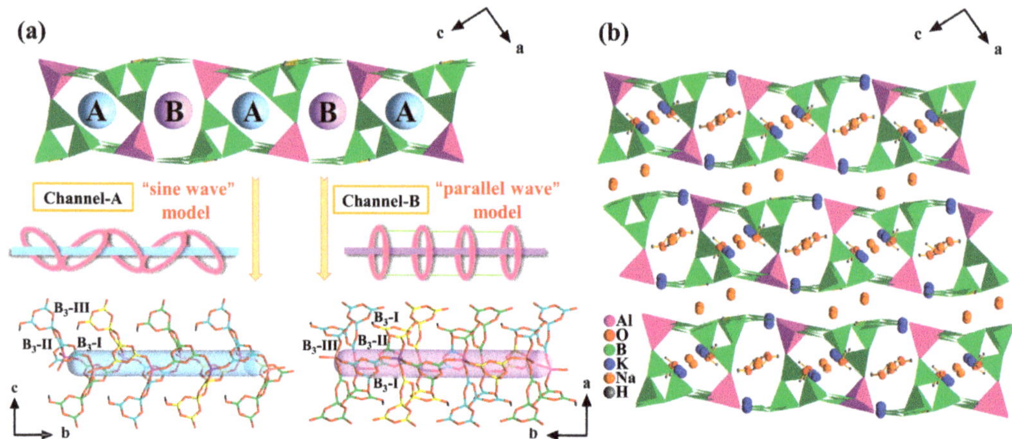

Figure 4. (**a**) View of two different types of channels in **2**; (**b**) The 3D porous-layered framework in **2**.

As for the metal cations, an Na atom was coordinated with seven O atoms, and the K atom was surrounded by five O atoms (Figure S3). It is worth noting that Na1 and the water molecules were filled in each channel, and Na2 was located in the interval between contiguous porous layers. Likewise, the K atoms had two locations. K1 was situated in channel A, and K2 was seated on the wall of channel B (Figure 4b). The metal–oxygen chain extended along *a*-axis and combined with the B-O network and the AlO_4 tetrahedra, enhancing the stability of compound **2** (Figure S4).

2.4. Structure Comparison

To discuss in detail, compounds **1** and **2** exhibited a few similarities as well as distinctions. On the one hand, there were the same fundamental building blocks (FBBs) of both two, namely, $[AlB_7O_{13}(OH)]^{3-}$, constituting the similar 3D porous-layered frameworks. On the other hand, the asymmetric unit of **2** contained three crystallographically independent $[AlB_7O_{13}(OH)]^{3-}$, and they were connected, in turn, along the *b*-axis, being consistent with the cell parallel of **2**, being three times longer than that of **1**. Meanwhile, the cations were dissimilar to induce the various distortion of porous layers due to the different ionic radius. Furthermore, **1** was the approximately parallel layer, and **2** was the fluctuant layer, showing the distinct shapes for channels, whereas the aperture of **1** was even larger. Moreover, it was significant that the cations were in different positions of the two compounds: the K^+ in **2** replaced the Cs^+ in **1**, and a part of Na^+ was filled in the channels of **2**, whereas they were only located in the interlayers in **1**. Additionally, there were more water molecules in **2**, situated in each channel, and the abundant hydrogen bonds made the whole structure more stable.

To date, there have been limited 3D porous-layered ABOs reported on, such as $[H_3O]K_{3.52}Na_{3.48}\{Al_2[B_7O_{13}(OH)][B_5O_{10}][B_3O_5]\}[CO_3]$ [33] (**3**), $K_2[Al_2B_7O_{14}(OH)(en)_{0.5}]$ ·H_2O [34] (**4**), and $Ba_3Al_2[B_3O_6(OH)]_2[B_4O_7(OH)_2]$ [35] (**5**), with their respective characteristics. Firstly, the kind of window related to oxoboron clusters participated in the consistency of monolayers (Figure S5). There was one type of window in **1** and **5** because the monolayer was formed by single oxoboron cluster, whereas three types of windows in **2** and **3** existed, owing to three oxoboron clusters that all made contributions to the monolayers, the same state for **4**. Secondly, the aperture of the window was influenced by the oxoboron clusters' sizes. There were larger windows in **1**, as its FBBs were composed of seven $BO_{3/4}$ units. The same 13-MR window also occurred in **3**, but the $[B_5O_{10}]^{5-}$ and $[B_3O_7]^{5-}$ clusters were not enough to support such a large ring. Thus, a part of the 13-MRs were split into 8-MR and 10-MR. Thirdly, the bridging unit of the porous layers was different (Figure S6). The oxoboron clusters originating from monolayers providing the bridging units in compounds

1–3, and the AlO_4 tetrahedra were effective of this in **4**. However, in compound **5**, the individual $[B_4O_7(OH)_2]^{4-}$ cluster only played a part in connecting the adjacent monolayers. In terms of structure, there were unprotonated B_3O_3 rings perpendicular to the monolayers in **1–3**, which made their own could act as bridging units. However, in **4** and **5**, the terminal oxygens, extending outward, were all protonated. Thus, only other units could act as bridging units in these frameworks. Fourthly, the warping degree of the porous layers was diverse. The frameworks of **1**, **4**, and **5** were approximately parallel layers, possibly because the larger cationic radius made an effect on supporting the channels in Cs^+ and Ba^{2+}, while ethylenediamine molecules played this role in **4**. However, there was K^+ or Na^+ in **2** and **3**, making them show the fluctuant layers.

2.5. Powder XRD Patterns

The experimental PXRD patterns of **1** and **2** were consistent with the single crystal data's simulated patterns, which illustrated that the samples were phase pure. The disagreement of the diffraction peak intensities between the experimental and simulated patterns were caused by the variations in the crystal orientations of the samples (Figure S7).

2.6. IR Spectra

It was homologous for **1** and **2** that the absorption bands and peaks were within 4000–500 cm^{-1} in the infrared spectra. Thus, only **1** was described in detail. The absorption peaks at 3440 cm^{-1} were the stretching vibrations of the -OH groups, while the peaks at 1624 cm^{-1} were the vibrations of H-O-H. The absorption bands ranging from 1445 to 1213 cm^{-1} were in accord with the asymmetric stretching of B-O in BO_3 units, and the bands from 1095 to 990 cm^{-1} were attributed to the asymmetric stretching of the BO_4 units. The peaks at 905 and 850 cm^{-1} were assigned to the symmetric stretching of BO_3 and BO_4, individually. The bands from 728 to 675 cm^{-1} belonged to the bending vibrations of these units. Moreover, the peaks in the range of 787 to 768 cm^{-1} corresponded with the stretching vibrations of the AlO_4 groups (Figure S8).

2.7. UV-Vis Absorption Spectra

As shown in Figure 5, the UV-Vis diffuse reflectance spectra that has been tested ranged from 190 to 800 nm. The Kubelka–Munk function $F(R) = (1 - R)^2/2R = \alpha/S$ was used to calculate the absorption date (α/S), where R was the reflectance, α was the absorption coefficient, and S was the scattering coefficient. The band gaps of **1** and **2** were 6.11 eV and 5.30 eV, indicating that they were wide-band semiconductors. The UV cut-off edges of both **1** and **2** were below 190 nm, revealing that they had potential applications in ultraviolet regions.

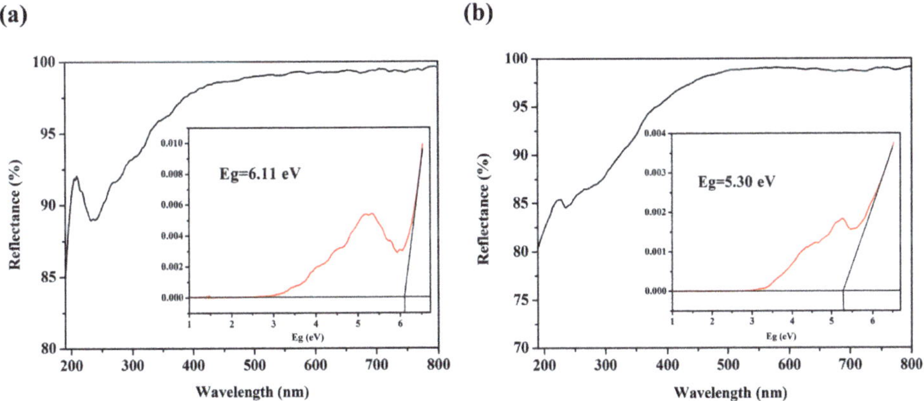

(a)   **(b)**

Figure 5. UV-Vis absorption spectra of **1** (**a**) and **2** (**b**).

2.8. Thermal Analysis

The thermal properties of compounds **1** and **2** were measured under the air atmosphere with a heating rate of 10°/min from 25 to 1000 °C. The 5.14% (Cal: 5.01%) weight losses from 125 °C to 463 °C in **1** and 8.97% (Cal: 8.70%) in the range of 102 °C to 441 °C in **2** were due to the removal of water molecules and the dehydration of -OH groups (Figure S9).

3. Materials and Methods

3.1. General Procedure

All chemical reagents were commercially available and used without further purification. Powder X-ray diffraction (PXRD) patterns were collected on a Bruker D8 Advance X-ray diffractometer with Cu Kα radiation ($\lambda = 1.54056$ Å) in the angular range of 2θ scanning from 5–50° at room temperature. Infrared (IR) spectra were tested on a Nicolet iS10 instrument with wavenumbers ranging from 4000 to 40 cm^{-1}. UV-Vis diffuse reflectance spectra were recorded in the range of 190–800 nm on a Shimadzu UV-3600 spectrometer. Thermogravimetric analyses were performed on a Mettler Toledo TGA/DSC 1100 analyzer from 25 to 1000 °C, with a heating rate of 10 °C h^{-1}, under an air atmosphere.

3.2. Syntheses

3.2.1. Syntheses of **1**

A mixture of H_3BO_3 (0.123 g, 2.0 mmol), $K_2B_4O_7\cdot4H_2O$ (0.159 g, 0.5 mmol), $Na_2B_4O_7\cdot10H_2O$ (0.193 g, 0.5 mmol), Cs_2CO_3 (0.187 g, 0.5 mmol), and $Al(i\text{-}PrO)_3$ (0.206 g, 1.0 mmol) was added into a mixed solution of 3 mL ethanol and 2 mL distilled water. After continuous stirring for 2 h at room temperature, the resulting solution was sealed in a 25 mL Teflon-lined stainless-steel autoclave. Subsequently, it was heated in an oven at 210 °C for 5 days under an autogenous pressure. The colorless lamellar crystals were obtained after cooling down to room temperature and being washed with distilled water (Figure S10).

3.2.2. Syntheses of **2**

A mixture of H_3BO_3 (0.362 g, 6.0 mmol), $NaBO_2\cdot4H_2O$ (0.288 g, 2.0 mmol), K_2CO_3 (0.063 g, 0.5 mmol), and $Al(i\text{-}PrO)_3$ (0.211 g, 1.0 mmol) was added into a mixed solution of 4 mL ethanol and 1 mL distilled water with constant stirring for 1h. Then, it was sealed in a 25 mL Teflon-lined stainless-steel autoclave and heated at 230 °C for 5 days. The colorless block crystals were obtained under the same procedures as **1**.

3.3. X-ray Crystallography

The single crystal X-ray diffraction data of **1** and **2** were tested and collected on a Gemini A Ultra CCD diffractometer with graphite monochromated Mo Kα ($\lambda = 0.71073$ Å) radiation in the ω scanning mode at room temperature. The structures were solved by direct methods and refined on F^2 by the full-matrix least-squares method with the SHELX-2014 program package [36]. All non-hydrogen atoms in the compounds were refined with anisotropic displacement parameters. The hydrogen atoms were placed by geometrical calculations and fixed through structural refinement. Crystallographic data were deposited with the Cambridge Crystallographic Data Centre: CCDC 2256812 for **1** and CCDC 2256819 for **2**. Detailed crystallographic data of two compounds are listed in Table 1.

Table 1. Crystallographic data and structural refinements for **1**, **2**.

	1	2
Formula	$NaKCsAlB_7O_{15}H_3$	$K_4Na_5Al_3B_{21}O_{47}H_{13}$
Molecular weight	540.66	1346.41
Crystal system	Monoclinic	Monoclinic
Space group	$P2_1/n$	$P2_1/n$
a/Å	11.3647 (16)	11.6261 (3)
b/Å	6.9730 (8)	20.9721 (6)

Table 1. *Cont.*

	1	2
$c/\text{Å}$	17.6729 (19)	16.7820 (5)
$\alpha/°$	90	90
$\beta/°$	91.880 (10)	93.646 (2)
$\gamma/°$	90	90
$V/\text{Å}^3$	1399.8 (3)	4083.6 (2)
Z	4	4
$Dc/\text{g cm}^{-3}$	2.556	2.183
μ/mm^{-1}	3.118	0.699
$F(000)$	1016	2648
Goodness-of-fit on F^2	1.069	1.079
R indices $[I > 2\sigma(I)]$ [1]	0.0484 (0.1046)	0.0471 (0.1519)
R indices (all data)	0.0751 (0.1216)	0.0526 (0.1563)

[1] $R_1 = \Sigma||F_0| - |F_c||/\Sigma|F_0|$. $wR_2 = \{\Sigma w[(F_0)^2 - (F_c)^2]^2/\Sigma w[(F_0)^2]^2\}^{1/2}$.

4. Conclusions

In summary, two new aluminoborates with mixed alkali metal cations were successfully obtained under hydrothermal conditions. Both **1** and **2** included the same fundamental building units, $[B_7O_{13}(OH)]^{6-}$ clusters, and AlO_4 tetrahedra, and the alternation of them made four connected networks with 8-MR channels and 13-MR windows along the *b*-axis, constituting the 3D porous-layered frameworks. The UV-Vis diffuse reflectance spectra indicated that both **1** and **2** exhibited the short deep-UV cutoff edges below 190 nm, and the bandgaps of them were 6.11 and 5.30 eV, revealing that they had potential applications in deep-UV regions. The successfully synthesis of the two above novel structures expanded the possibilities of ABOs structures and revealed the effect of metal cations on constructing frameworks. In the future, we will continue to explore the synthesis of distinctive ABOs with various alkali and alkaline earth metals.

Supplementary Materials: The following supporting information can be downloaded at: https://www.mdpi.com/article/10.3390/molecules28114387/s1, Figure S1: Composition of 13-MR (a) and 8-MR (b) windows in **1**; Figure S2: The coordination of metal cations in **1**; Figure S3: The coordination of metal cations in **2**; Figure S4: Metal-oxygen chain in **2**; Figure S5: The 2D monolayers in **1** (a), **3** (b), **4** (c), and **5** (d), respectively; Figure S6: The porous-layered structures in **3** (a), **4** (b), and **5** (c), respectively; Figure S7: PXRD of **1** (a) and **2** (b); Figure S8: IR spectra of **1** (a) and **2** (b); Figure S9: TG-DSC curves of **1** (a) and **2** (b); Figure S10: The morphology of compounds **1** and **2**, respectively.

Author Contributions: Conceptualization, C.W.; methodology, C.W. and J.C.; data curation, C.W., J.C., C.-A.C. and Z.-W.W.; writing—original draft preparation, C.W.; writing—review and editing, G.-Y.Y. All authors have read and agreed to the published version of the manuscript.

Funding: This research was funded by the National Natural Science Foundation of China, grant numbers 21831001, 21571016, 91122028, and 20725101.

Institutional Review Board Statement: Not applicable.

Informed Consent Statement: Not applicable.

Data Availability Statement: The raw data supporting the conclusions of this article will be made available by the authors without undue reservation.

Conflicts of Interest: The authors declare no conflict of interest.

Sample Availability: Not applicable.

References

1. Lin, Z.E.; Yang, G.Y. Oxo Boron Clusters and Their Open Frameworks. *Eur. J. Inorg. Chem.* **2011**, *2011*, 3857–3867. [CrossRef]
2. Wang, S.; Alekseev, E.V.; Depmeier, W.; Albrecht-Schmitt, T.E. Recent progress in actinide borate chemistry. *Chem. Commun.* **2011**, *47*, 10874–10885.

3. Li, J.H.; Jiang, X.F.; Wei, Q.; Xue, Z.Z.; Wang, G.M.; Yang, G.Y. Dual-Ligand-Oriented Design of Noncentrosymmetric Complexes with Nonlinear-Optical Activity. *Inorg. Chem.* **2022**, *61*, 16509–16514. [CrossRef]

4. Qiu, Q.M.; Yang, G.Y. From $[B_6O_{13}]^{8-}$ to $[GaB_5O_{13}]^{8-}$ to $[Ga\{B_5O_9(OH)\}\{BO(OH)_2\}]^{2-}$: Synthesis, structure and nonlinear optical properties of new metal borates. *CrystEngComm* **2021**, *23*, 5200–5207. [CrossRef]

5. Li, X.Y.; Wei, Q.; Hu, C.L.; Pan, J.; Li, B.X.; Xue, Z.Z.; Li, X.Y.; Li, J.H.; Mao, J.G.; Wang, G.M. Achieving Large Second Harmonic Generation Effects via Optimal Planar Alignment of Triangular Units. *Adv. Mater.* **2022**, *33*, 2210718. [CrossRef]

6. Dewey, C.F.; Cook, W.R.; Hodgson, R.T.; Wynne, J.J. Frequency doubling in $KB_5O_8 \cdot 4H_2O$ and $NH_4B_5O_8 \cdot 4H_2O$ to 217.3 nm. *Appl. Phys. Lett.* **1975**, *26*, 714–716. [CrossRef]

7. Wu, H.P.; Pan, S.L.; Poeppelmeier, K.R.; Li, H.Y.; Jia, D.Z.; Chen, Z.H.; Fan, X.Y.; Yang, Y.; Rondinelli, J.M.; Luo, H.S. $K_3B_6O_{10}Cl$: A new structure analogous to perovskite with a large second harmonic generation response and deep UV absorption edge. *J. Am. Chem. Soc.* **2011**, *133*, 7786–7790. [CrossRef] [PubMed]

8. Chen, C.A.; Yang, G.Y. Syntheses, structures and optical properties of two B_3O_7 cluster-based borates. *CrystEngComm* **2022**, *24*, 1203–1210. [CrossRef]

9. Awarded, A.H.; Honored, C.B.; Gellman, S.H. Organic Materials/ Asymmetric Catalysis/ Peptide Chemistry. *Angew. Chem. Int. Ed.* **2007**, *46*, 21.

10. Pan, C.Y.; Liu, G.Z.; Zheng, S.T.; Yang, G.Y. $GeB_4O_9 \cdot H_2en$: An organically templated borogermanate with large 12-ring channels built by B_4O_9 polyanions and GeO_4 units: Host-guest symmetry and charge matching in triangular-tetrahedral frameworks. *Chem. Eur. J.* **2008**, *14*, 5057–5063.

11. Wang, K.; Li, X.F.; He, C.; Li, J.H.; An, X.T.; Wei, L.; Wei, Q.; Wang, G.M. $NaSb_3O_2(SO_4)_3 \cdot H_2O$: A New Alkali-Metal Antimony(III) Sulfate with a Unique $Sb_6O_{20}H_4$ Unit and Moderate Birefringence. *Cryst. Growth Des.* **2021**, *22*, 478–484. [CrossRef]

12. Qiu, Q.M.; Yang, G.Y. Three mixed-alkaline-metal borates with $\{Li@B_{12}O_x(OH)_{24-x}\}$ (x = 18, 22) clusters: From isolated oxoboron cluster to unusual layer. *CrystEngComm* **2021**, *23*, 6518–6525. [CrossRef]

13. Chen, C.T.; Wu, B.C.; Jiang, A.D.; You, G.M. A New-Type Ultraviolet SHG Crystal-Beta-BaB_2O_4. *Sci. Sin. Ser. B.* **1985**, *28*, 235–243.

14. Chen, C.T.; Wu, Y.C.; Jiang, A.D.; Wu, B.C.; You, G.M.; Li, R.K.; Lin, S.J. New nonlinear-optical crystal: LiB_3O_5. *J. Opt. Soc. Am. B* **1989**, *6*, 616–621. [CrossRef]

15. Mori, I.K.Y.; Nakajima, S.; Sasaki, T.; Nakai, S. New nonlinear optical crystal: Cesium lithium borate. *Appl. Phys. Lett.* **1995**, *67*, 1818–1820. [CrossRef]

16. Zhang, T.J.; Pan, R.; He, H.; Yang, B.F.; Yang, G.Y. Solvothermal Synthesis and Structure of Two New Borates Containing $[B_7O_9(OH)_5]^{2-}$ and $[B_{12}O_{18}(OH)_6]^{6-}$ Clusters. *J. Cluster Sci.* **2015**, *27*, 625–633. [CrossRef]

17. Chen, J.; Wang, J.J.; Chen, C.A.; Yang, G.Y. Two New Borates Built by Different Types of $\{B_9\}$ Cluster Units. *Chem. Res. Chin. Univ.* **2022**, *38*, 744–749. [CrossRef]

18. Wu, H.Q.; He, H.; Yang, B.F.; Yang, G.Y. A new mixed metal borate of $BaPb[B_5O_9(OH)] \cdot H_2O$ with acentric structure. *Inorg. Chem. Commun.* **2013**, *37*, 77–79. [CrossRef]

19. Liu, W.F.; Su, Z.M.; Jia, Z.Y.; Yang, G.Y. Syntheses, Structures and Characterizations of Two New Polyborates Containing Heptaborate Sub-clusters. *J. Cluster Sci.* **2019**, *30*, 1139–1144.

20. Irina, V.K.; Varvara, V.A.; Grigorii, A.B.; Evgeniy, A.S.; Yulia, V.I.; Sergey, P.G. A New Approach for the Synthesis of Powder Zinc Oxide and Zinc Borates with Desired Properties. *Inorganics* **2022**, *10*, 212.

21. Chen, C.A.; Liu, W.F.; Yang, G.Y. Two layered galloborates from centric to acentric structures: Syntheses and NLO properties. *Chem. Commun.* **2022**, *58*, 8718–8721. [CrossRef]

22. Liu, Y.; Pan, R.; Cheng, J.W.; He, H.; Yang, B.F.; Zhang, Q.; Yang, G.Y. A Series of Aluminoborates Templated or Supported by Zinc-Amine Complexes. *Chem. Eur. J.* **2015**, *21*, 15732–15739. [CrossRef] [PubMed]

23. Qin, D.; Zhang, T.J.; Ma, C.B.; Yang, G.Y. Two novel 3D borates: Porous-layer and layer-pillar frameworks. *Dalton Trans.* **2020**, *49*, 3824–3829. [CrossRef]

24. Wei, Q.; Sun, S.J.; Zhang, J.; Yang, G.Y. Extending Unique 1D Borate Chains to 3D Frameworks by Introducing Metallic Nodes. *Chem. Eur. J.* **2017**, *23*, 7614–7620. [PubMed]

25. Lehmann, H.A.; Teske, K. Uber einige neue Borate des Aluminiums. *Z. Anorg. Allg. Chem.* **1973**, *400*, 169–175. [CrossRef]

26. Ju, J.; Lin, J.; Li, G.; Yang, T.; Li, H.; Liao, F.; Loong, C.K.; You, L. Aluminoborate-based molecular sieves with 18-octahedral-atom tunnels. *Angew. Chem. Int. Ed.* **2003**, *42*, 5607–5610. [CrossRef] [PubMed]

27. Rong, C.; Yu, Z.W.; Wang, Q.; Zheng, S.T.; Pan, C.Y.; Deng, F.; Yang, G.Y. Aluminoborates with Open Frameworks: Syntheses, Structures, and Properties. *Inorg. Chem.* **2009**, *48*, 3650–3659. [CrossRef]

28. Wei, L.; Wei, Q.; Lin, Z.E.; Meng, Q.; He, H.; Yang, B.F.; Yang, G.Y. A 3D aluminoborate open framework interpenetrated by 2D zinc-amine coordination-polymer networks in its 11-ring channels. *Angew. Chem. Int. Ed.* **2014**, *53*, 7188–7191. [CrossRef]

29. Dong, Y.Z.; Chen, C.A.; Chen, J.; Cheng, J.W.; Li, J.H.; Yang, G.Y. Two porous-layered borates built by $B_7O_{13}(OH)$ clusters and AlO_4/GaO_4 tetrahedra. *CrystEngComm* **2022**, *24*, 8027–8033. [CrossRef]

30. Li, X.Y.; Li, J.H.; Cheng, J.W.; Yang, G.Y. Two Acentric Aluminoborates Incorporated d(10) Cations: Syntheses, Structures, and Nonlinear Optical Properties. *Inorg. Chem.* **2023**, *62*, 1264–1271. [CrossRef] [PubMed]

31. Qin, D.; Pei, H.L.; Chen, C.A.; Yang, G.Y. Two Chiral Metal Complex Templated Aluminoborates Constructed from Three Types of Oxoboron Clusters. *Inorg. Chem.* **2021**, *60*, 6576–6584. [CrossRef] [PubMed]

32. Wu, L.Z.; Cheng, L.; Shen, J.N.; Yang, G.Y. From discrete borate cluster to three-dimensional open framework. *CrystEngComm* **2013**, *15*, 4483–4488. [CrossRef]
33. Chen, C.A.; Pan, R.; Yang, G.Y. Syntheses and structures of a new 2D layered borate and a novel 3D porous-layered aluminoborate. *Dalton Trans.* **2020**, *49*, 3750–3757. [CrossRef] [PubMed]
34. Chen, C.A.; Pan, R.; Zhang, T.J.; Li, X.Y.; Yang, G.Y. Three Inorganic-Organic Hybrid Gallo-/Alumino-Borates with Porous-Layered Structures Containing [MB$_4$O$_{10}$(OH)] (M = Al/Ga) Cluster Units. *Inorg. Chem.* **2020**, *59*, 18366–18373. [CrossRef]
35. Cheng, L.; Wei, Q.; Wu, H.Q.; Zhou, L.J.; Yang, G.Y. Ba$_3$M$_2$[B$_3$O$_6$(OH)]$_2$[B$_4$O$_7$(OH)$_2$] (M = Al, Ga): Two novel UV nonlinear optical metal borates containing two types of oxoboron clusters. *Chem. Eur. J.* **2013**, *19*, 17662–17667. [CrossRef]
36. Sheldrick, G.M. A short history of SHELX. *Acta Crystallogr. Sect. A Found. Crystallogr.* **2008**, *64*, 112–122. [CrossRef]

 *molecules*

Article

The Nature of the (Oligo/Hetero)Arene Linker Connecting Two Triarylborane Cations Controls Fluorimetric and Circular Dichroism Sensing of Various ds-DNAs and ds-RNAs

Lidija-Marija Tumir [1], Dijana Pavlović Saftić [1], Ivo Crnolatac [1], Željka Ban [1], Matea Maslać [1], Stefanie Griesbeck [2], Todd B. Marder [2,*] and Ivo Piantanida [1,*]

[1]　Division of Organic Chemistry and Biochemistry, Ruđer Bošković Institute, 10000 Zagreb, Croatia; tumir@irb.hr (L.-M.T.); dijana.saftic@irb.hr (D.P.S.); ivo.crnolatac@irb.hr (I.C.); zeljka.ban@irb.hr (Ž.B.); matea.maslac@gmail.com (M.M.)

[2]　Institut für Anorganische Chemie and Institute for Sustainable Chemistry & Catalysis with Boron, Julius-Maximilians-Universität Würzburg, 97074 Würzburg, Germany; stefanie.griesbeck@gmx.de

*　Correspondence: todd.marder@uni-wuerzburg.de (T.B.M.); pianta@irb.hr (I.P.); Tel.: +385-1-457-1326 (I.P.)

Abstract: A series of tetracationic bis-triarylborane dyes, differing in the aromatic linker connecting two dicationic triarylborane moieties, showed very high submicromolar affinities toward ds-DNA and ds-RNA. The linker strongly influenced the emissive properties of triarylborane cations and controlled the fluorimetric response of dyes. The fluorene-analog shows the most selective fluorescence response between AT-DNA, GC-DNA, and AU-RNA, the pyrene-analog's emission is non-selectively enhanced by all DNA/RNA, and the dithienyl-diketopyrrolopyrrole analog's emission is strongly quenched upon DNA/RNA binding. The emission properties of the biphenyl-analog were not applicable, but the compound showed specific induced circular dichroism (ICD) signals only for AT-sequence-containing ds-DNAs, whereas the pyrene-analog ICD signals were specific for AT-DNA with respect to GC-DNA, and also recognized AU-RNA by giving a different ICD pattern from that observed upon interaction with AT-DNA. The fluorene- and dithienyl-diketopyrrolopyrrole analogs were ICD-signal silent. Thus, fine-tuning of the aromatic linker properties connecting two triarylborane dications can be used for the dual sensing (fluorimetric and CD) of various ds-DNA/RNA secondary structures, depending on the steric properties of the DNA/RNA grooves.

Keywords: triarylborane; fluorescent probe; circular dichroism; DNA recognition; RNA recognition

Citation: Tumir, L.-M.; Pavlović Saftić, D.; Crnolatac, I.; Ban, Ž.; Maslać, M.; Griesbeck, S.; Marder, T.B.; Piantanida, I. The Nature of the (Oligo/Hetero)Arene Linker Connecting Two Triarylborane Cations Controls Fluorimetric and Circular Dichroism Sensing of Various ds-DNAs and ds-RNAs. *Molecules* **2023**, *28*, 4348. https://doi.org/10.3390/molecules28114348

Academic Editors: Michael A. Beckett and Igor B. Sivaev

Received: 4 May 2023
Revised: 23 May 2023
Accepted: 24 May 2023
Published: 25 May 2023

1. Introduction

DNA and RNA are biomolecules that are essential for life, including the growth, operation, and reproduction of living organisms. These biomolecules are thus promising pharmaceutical targets for small molecules, which can bind to them, influence their biological properties, and signal binding events by a specific response. An important task is the design of small molecular sensors which are able to non-covalently bind selectively and recognize certain structures, i.e., discriminate among proteins, DNA and RNA polynucleotides, certain sequences (e.g., AT or GC), specific structural motifs, etc. [1]. Different spectroscopic signals from a small molecule probe upon binding to different biomolecules and/or different structural motifs is a very desirable characteristic, as it allows the observation of different processes and organelles within a cell.

Bis-triarylborane derivatives proved to be suitable fluorophores [2–4] for cell imaging [5–9], whereby the linker connecting the strongly π-accepting triarylborane units affects the photophysical properties of the fluorophores, type, and number of non-covalent interactions of the dye with biomolecules and, consequently, its affinity and selective recognition of specific structures.

We recently demonstrated that the linker connecting two triarylborane moieties in tetracationic bis-triarylboranes has a profound impact on its fluorescence response. For example, the bis-thiophene analog (**1, Scheme 1**) showed a high emission increase, which was selective between DNA/RNA and protein [10], while the emission of its analog with 1,3-butadiyne linker (introduced as Raman-probe) [11,12] was strongly quenched by any DNA/RNA/protein. The opposite emission responses clearly demonstrated the impact of the linker on the emission properties of this type of bis-triarylborane fluorophore. Thus, to test the impact of the linker in more detail, we combined Raman-responsive chromophores with fluorophores within the same linker, which additionally demonstrated the impact of the linker on the emission response to DNA/RNA binding by introducing an aggregation-induced emission response [13]. Bis(triarylborane) tetracations linked by diethynylanthracene- and diethynyl-bis-thiophene bridges showed increased antiviral activity, a very strong and quick rise in cytotoxicity, and a very significant increase in reactive oxygen species (ROS) activity upon visible light (400–700 nm) irradiation [14].

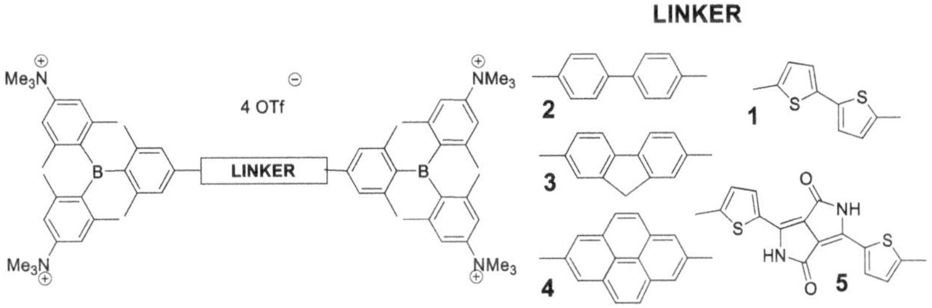

Scheme 1. Schematic presentation of compounds **2–5** and, for reference, the previously studied analog **1** [10].

In a previous paper [15], we demonstrated that the emission color of compounds **2–5** could be tuned from blue to pink by changing the linker, nicely covering the complete range of the visible spectrum. In addition, experiments on cell lines showed that all compounds very efficiently enter living cells, accumulating mostly in lysosomes, and having a negligible impact on cell proliferation, thus behaving as promising new fluorescent probes. The most intriguing compound, **5**, emits in the red region and has a large two-photon absorption cross-section. However, such promising results in intracellular sensing led to the question of what the target of our small molecules is not a trivial question, as till now, we demonstrated that close analogs (**1**) bind with similar affinity to DNA, RNA, and proteins, but, fortunately, due to its different emission properties, we attributed a protein-like target in cells rather than DNA/RNA.

To study in detail the relation between linker properties and DNA/RNA/protein binding interactions, we decided to vary steric and structural properties systematically of the linear bis-aryl-linker (Scheme 1), starting from biphenyl (**2**), characterized by high rotational freedom around the C-C single bond, then introducing rotationally "locked" fluorene (**3**) and, finally, the larger, planar pyrene (**4**) moiety. In addition, also we studied an extended version of bis-thiophene analog (**1**) by inserting a diketopyrrolopyrrole unit between two thiophenes (**5**) to see whether elongation, the high rotational flexibility and possibility of additional H-bonding of the linker within DNA/RNA grooves would have an impact on binding and sensing of various DNAs/RNAs.

2. Results and Discussion

Compounds **2–5** were prepared as described previously, and their spectrophotometric properties were previously characterized in various solvents (Supplementary Information, Table S2) [10,15]. For the purpose of the current study, UV/vis and fluorimetric spectra of **2–5** were collected at pH 7, in sodium cacodylate buffer ($I = 0.05$ M), in general, all spectral properties agreed well with spectra in our previous studies (Figure 1, Supplementary Information, Figures S1–S8).

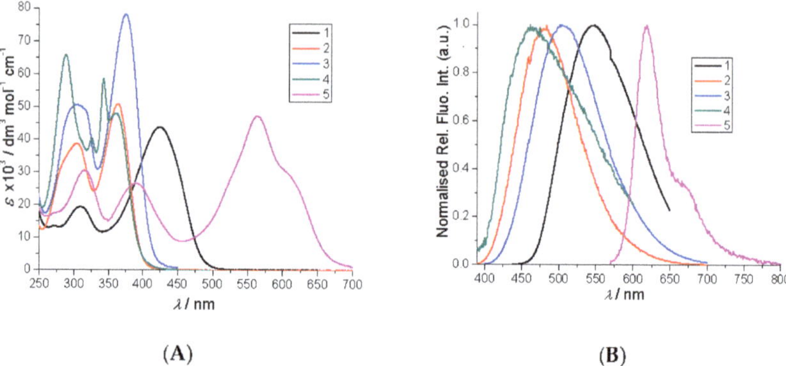

Figure 1. UV/vis (**A**) and fluorescence spectra ((**B**) λ_{exc} at abs. maxima; normalized to the emission maximum) of **1–5**. Done at pH = 7.0, buffer sodium cacodylate, I = 0.05 M.

More detailed analysis revealed that basic linker-modification for biphenyl (**2**)—fluorene (**3**)—pyrene (**4**) did not have a significant impact on the light absorption properties of the compounds (Figure 1A), in contrast to the introduction of the bis-thiophene linker (**1**), which red-shifted the absorbance ca. 70 nm and, in particular, the dithienyl-diketopyrrolopyrrole (**5**) induced a bathochromic shift of over 200 nm. However, the emission properties were more sensitive to the linker (Figure 1B), with pyrene analog **4** showing the smallest Stokes shift and emitting at the shortest wavelength.

All compounds showed stable fluorescence spectra over a long period, with the exception of biphenyl-analog **2**, although UV/Vis spectrum at the corresponding conditions was unchanged. For that reason, we used UV/Vis spectroscopy to study the interactions of **23** with DNA/RNA.

2.1. Interactions with ds-DNA and ds-RNA

2.1.1. Thermal Denaturation of ds-Polynucleotides

Thermally induced dissociation of the ds-polynucleotides occurs at a well-defined temperature (T_m value), thus being used for the characterization of various ds-DNA or ds-RNA-related processes. For example, non-covalent binding of small molecules to ds-polynucleotides usually increases the thermal stability of the ds-helices, and this increase (ΔT_m value) can be correlated with the various binding modes [16] and corroborated with other independent methods.

We tested the impact of **2–5** on the thermal stability of various ds-DNA and ds-RNA (Supplementary Information, Figures S24–S26), and the results are summarised in Table 1. Most compounds similarly stabilized all ds-DNA/RNA, the only exception being bulky linker-analog **5**, for which stabilization of DNA/RNA was significantly weaker, likely due to the hindered insertion into DNA/RNA grooves caused by the more sterically demanding linker.

Table 1. The [a] ΔT_m values (°C) of studied ds-polynucleotides upon addition of ratio [b] r = 0.1 of **1–5** at pH = 7.0 (sodium cacodylate buffer, I = 0.05 M).

	ct-DNA	p(dAdT)$_2$	poly rA—poly rU
[c] **1**	7.3	10.0	9.5
2	4.5	6.7	12.1
3	10.0	8.2	9.3
4	8.0	9.6	7.4
5	3.4	-	0.6

[a] Error in ΔT_m: ±0.5 °C; [b] r = [compound]/[polynucleotide], [c] Previously published results [10].

2.1.2. Spectrophotometric Titrations of **2–5** with ds-DNA/RNA

The addition of DNA/RNA caused significant changes in the UV/Vis spectra of all compounds, characterized by hypochromic and bathochromic effects (Figure 2 and Supplementary Information, Figures S9–S12). However, in most cases, changes in the UV/Vis spectra were observed up to the point of saturation of the binding sites on ds-DNA/RNA (ratio $r_{[dye]/[polynucleotide]} \sim 0.2$), suggesting very high affinities. Such changes do not allow accurate processing of the titration data by non-linear fitting procedures adapted from the Scatchard equation [17,18], which are used for the calculation of binding constants. The exception was biphenyl-derivative **2**, allowing an estimation of the binding constants (Figure 2a, Table 2).

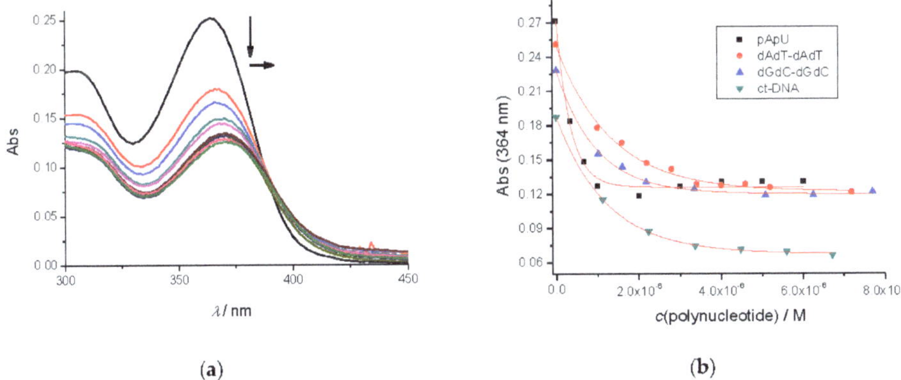

Figure 2. (**a**) The UV/vis titration of **2** ($c = 1 \times 10^{-6}$ M) with poly(dAdT)$_2$; (**b**) dependence of absorption at λ_{max}= 364 nm on c(polynucleotide), the red line is non-linear least square fitting of experimental data to Scatchard eq. (McGhee, von Hippel formalism) [17,18]. Done at pH 7, sodium cacodylate buffer, I = 0.05 M.

Table 2. Binding constants (logKs) and spectroscopic properties of **1–5** with polynucleotides calculated by processing fluorimetric [a] or [b] UV/vis titrations.

	1 [10]	2	3	4	5
ct-DNA	7.0	[b] 6.1	6.9	7.1	7.7
p(dAdT)$_2$	7.9	[b] 6.1	6.9	8.7	7.7
p(dGdC)$_2$	7.6	[b] 6.0	7.4	8.0	7.9
prAprU	[b]7	[b] 7.5	[c] 7.9	7.4	7.7

[a] Calculated using the Scatchard eq. [17,18] for a fixed n = 0.2; colors denote INCREASE or QUENCHING of fluorescence emission at an excess of DNA/RNA over dye. [b] UV/Vis titrations. [c] Strong hypsochromic shift of the emission maximum caused quenching at starting λ_{max} and an increase at the new λ_{max}.

To circumvent this problem, we took advantage of the strong fluorescence of **3–5**, enabling fluorimetric titrations at several orders of magnitude lower concentrations of the dyes, thus allowing the collection of at least part of the titration data at an excess of DNA/RNA over dye. Detailed comparison of fluorimetric titrations (Figures 3–5; Table 2, Supplementary Information, Figures S13–S23) revealed that the emission response was strongly affected by the linker, as well as by the secondary structure of the ds-polynucleotide.

Particularly intriguing was the fluorimetric response of fluorene analog **3**, showing emission quenching only for GC-DNA (Figure 3b), whereas, for AT-containing DNAs, strong aggregation-induced quenching was followed by an emission increase at r < 0.2 (Figure 3a). Such selectivity in response can be correlated to the steric properties of the DNA minor groove (Supplementary Information Table S1), whereby the AT-DNA groove is very efficient for small molecule binding (thus, all groove binders bind to AT-DNA [19]) and also supports aggregation at an excess of a dye over DNA [20] (r > 0.2, emission quenching), whereas sterically hindered GC-DNA does not support aggregation but only loose binding of single molecules—thus allowing a single binding mode. As among ds-DNA bases, guanine is the easiest nucleobase to oxidize [21]; it commonly induces pronounced emission quenching [22], at variance to AT-sequences which can induce emission increase.

An even more intriguing response of **3** was observed for the AU-RNA exclusively (Figure 3c). Thus, a strong hypsochromic shift (−25 nm) of the emission maximum was observed, whereas the relative intensity of emission changed only negligibly. The selective response of **3** could be attributed to the ds-RNA secondary structure (see Supplementary Information Table S1 for structural details), characterized by a very deep and narrow major groove, which can accommodate small molecules differently from the minor groove of DNAs, consequently resulting in the finely tuned fluorimetric response of **3**.

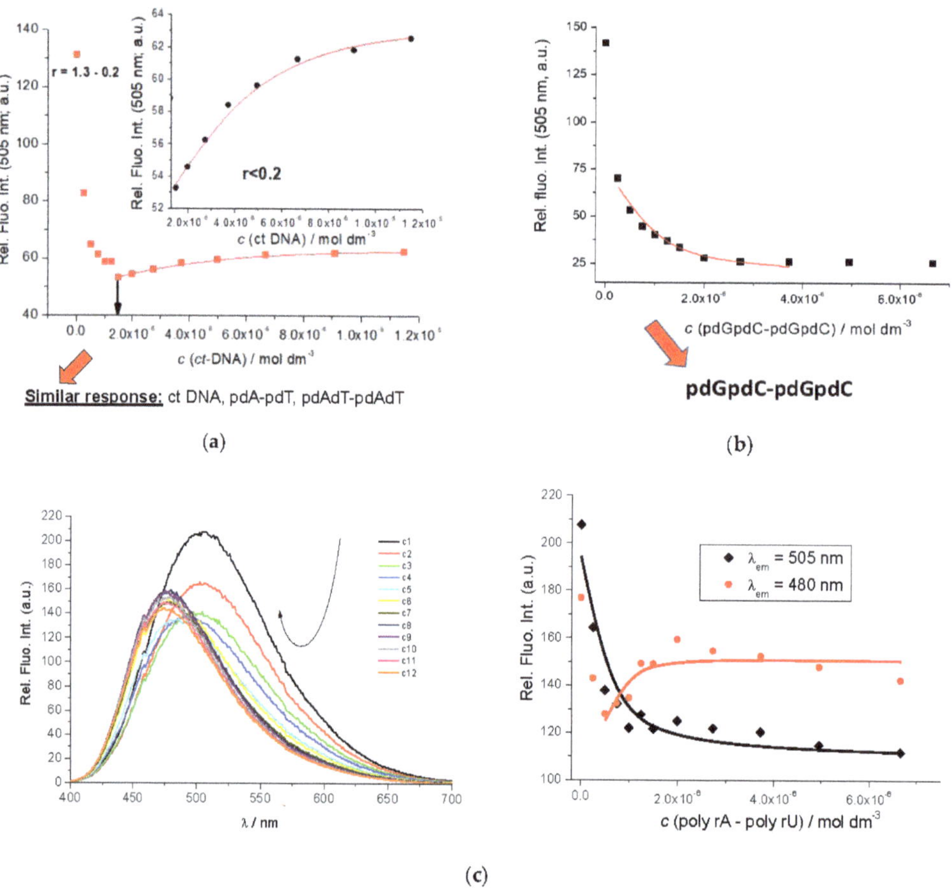

Figure 3. Fluorimetric titration of **3**, λ_{exc} = 375 nm, λ_{em} = 505 nm, c = 2 × 10^{-7} M with (**a**) ct-DNA, AT-DNAs; (**b**) poly dGdC-poly dGdC; (**c**) poly rA—poly rU; Experimental (■ or ■) and calculated (— or —) fluorescence intensities of **3** upon addition of RNA. Done at pH = 7.0, Na cacodylate buffer, I = 0.05 M.

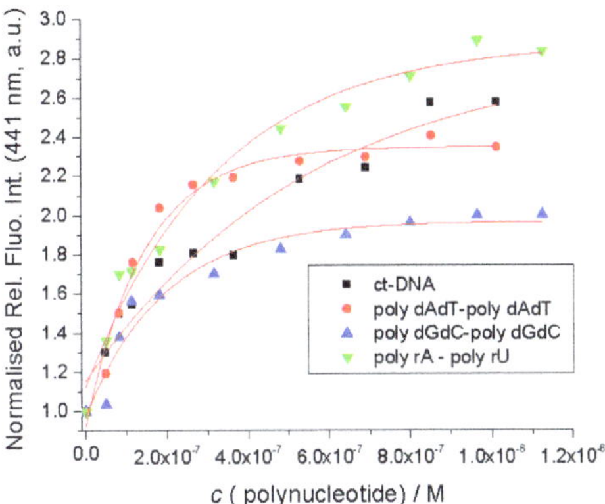

Figure 4. Fluorimetric titration of **4**, $\lambda_{exc} = 345$ nm, $c = 5.0 \times 10^{-8}$ M with polynucleotides; experimental (●, ▲, ■, ▲) and fit the Scatchard eq. (—) fluorescence intensities at $\lambda_{em} = 441$ nm. Done at pH = 7.0, Na cacodylate buffer, $I = 0.05$ M.

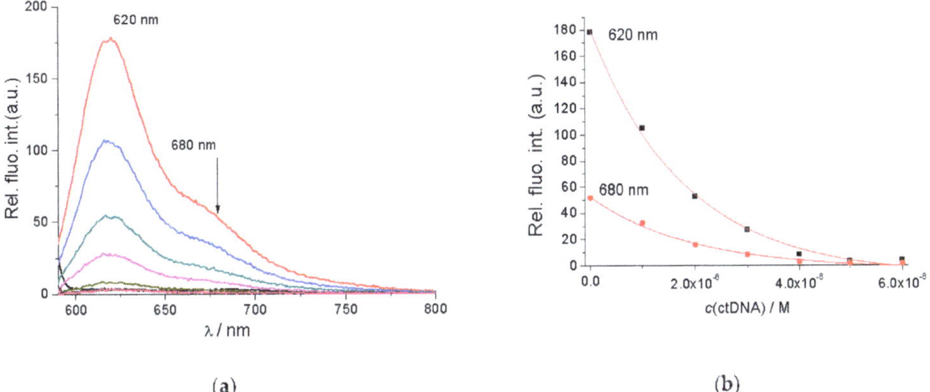

(a)   (b)

Figure 5. (a) Changes in the fluorescence spectrum of **5** ($c = 5.0 \times 10^{-8}$ M, $\lambda_{exc} = 562$ nm) upon titration with **ctDNA**; (b) Dependence of **5** emissions at $\lambda_{max} = 620$ nm and $\lambda = 680$ nm experimental (■, ●) and fit the Scatchard eq. (—) fluorescence intensities on c(ctDNA), at pH 7, sodium cacodylate buffer, $I = 0.05$ M.

In comparison to fluorene (**3**), further increase of rigidity and the aromatic surface in pyrene-analog (**4**) resulted in a complete loss of selectivity; its emission change was exclusively enhanced by any ds-DNA/RNA studied (Figure 4).

Intriguingly, whereas previously studied very flexible bis-thiophene-linker (**1**) showed only emission increase for any DNA/RNA added [10], the insertion of diketopyrrolopyrrole between two thiophenes (**5**) completely reversed the response, showing exclusively fluorescence quenching for any of ds-DNA/RNA (Figure 5).

Correlation of the emission responses to linker properties of **1**, **3–5** revealed that neither length nor rigidity/planarity of the linker controls the emission of the complexes formed. In fact, it seems that the decisive role involves a fine interplay between the binding site (DNA/RNA groove size and shape differing for each DNA/RNA studied) and the electronic properties of the compete chromophore system (including communication between the triarylborane and the linker).

Processing of fluorimetric titration data by means of the Scatchard equation (McGhee, vonHippel formalism) [17,18] yielded binding constants (Table 2), whereby somewhat lower values obtained for **2** can be attributed to the less sensitive titration method (UV/vis). The other compounds showed mostly rather similar binding affinities to all ds-DNAs/RNAs studied (logKs = 7–8), pointing out that the different emission responses do not reflect differences in binding interactions.

2.1.3. Circular Dichroism

So far, we have studied the non-covalent interactions at 25 °C by monitoring the spectroscopic properties of compounds **2–5** upon the addition of the polynucleotides. In order to obtain insight into the changes in polynucleotide properties induced by the small molecule binding, we chose CD spectroscopy as a highly sensitive method for the examination of conformational changes in the secondary structure of polynucleotides [23]. In addition, **2–5**, as achiral small molecules, can still generate an induced CD spectrum (ICD) upon binding to polynucleotides, which could give useful information about the modes of interaction [24,25].

The fluorene analog **3** did not show any induced (I)CD bands but only caused a decrease in the CD bands of DNA/RNA < 300 nm (Supplementary Information, Figures S29 and S30). Compound **5** showed only a very weak negative induced (I)CD band at 315 nm for GC-DNA, and for other DNAs/RNAs, the addition of **5** only caused a decrease in the CD bands of DNA/RNA < 300 nm (Supplementary Information, Figures S32 and S33). Such a response is in accord with previously studied compound **1**, as well as its analog with the 1,3-butadiyne linker [11], attributed to the binding of a dye to DNA/RNA grooves (thus only slightly affecting the helicity of DNA/RNA), whereas the absence of ICD bands was a consequence of the unfavorable orientation of the dye transition dipole moments with respect to the DNA/RNA chiral axis. Namely, if the dominant transition dipole moment of the dye is coplanar with the base pairs, it will yield a weak negative ICD, whereas if it is at a ca. 45° angle to the DNA chiral axis, it will give a strong positive ICD [24,25]; however, if is at an angle in-between, the positive and negative components could cancel each other out.

However, for several DNAs/RNAs, the biphenyl analog **2** and pyrene analog **4** showed moderate or even strong ICD bands > 300 nm. For example, **2** showed ICD bands only for AT-containing DNAs but no signal > 300 nm for AU-RNA and GC-DNA (Figure 6a). Such a selective response can be attributed to the well-defined minor groove of AT-sequences (see Supplementary Information Table S1), in which rod-like bis-triarylboranes fit nicely [10,11], and, in the case of **2**, the biphenyl-linker orients uniformly in respect to the AT-DNA helical axis, thus producing ICD bands [24,25].

For pyrene analog **4**, the selectivity of the ICD band > 300 nm response is even more pronounced. The strong positive ICD bands at >350 nm obtained for AU-RNA and AT-DNA are similar (Figure 6b), suggesting that the pyrene-linker is uniformly oriented in the grooves of narrow width at an angle of ca. 45° with respect to the polynucleotide chiral axis [24,25] (Supplementary Information Table S1; see RNA major groove and AT-DNA minor groove), whereas the amino group of guanine in GC-DNA sterically hinders insertion of pyrene and, thus, completely abolished any ICD band > 300 nm (Supplementary Information Figure S31). Furthermore, in the 230–300 nm range (where intrinsic CD bands of DNA/RNA appear), there is a clear difference between the **4**/AU-RNA and **4**/AT-DNA CD spectra, which can be mostly attributed to strong differences in secondary structure between the A-helix of RNA and B-helix of DNA, as well as a resulting different orientation of bound **4** within the corresponding binding sites.

Comparison of all CD data reveals that fine-tuning of linker-chromophore transition dipole moments with respect to the DNA/RNA chiral axis can control selective ICD response to a particular type of DNA or RNA binding site.

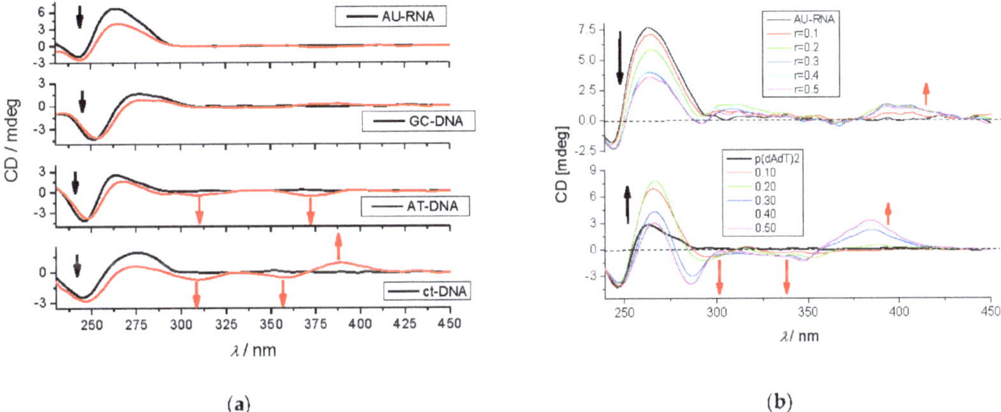

Figure 6. (a) CD titration of polynucleotides ($c = 2 \times 10^{-5}$ M) with **2** at molar ratio $r_{[\text{compound}]/[\text{polynucleotide}]} = 0.3$. Red arrows denote ICD bands > 300 nm, and black arrows denote the decrease of CD bands < 300 nm belonging to the DNA/RNA helix; (b) CD titration of AT-DNA and AU-RNA ($c = 2 \times 10^{-5}$ M) with **4** at molar ratios $r_{[\text{compound}]/[\text{polynucleotide}]} = 0$, 0.1, 0.2, 0.3, 0.4, and 0.5. Red arrows denote ICD bands > 300 nm, and black arrows denote a decrease of CD bands < 300 nm belonging to the DNA/RNA helix. The addition of **4** to GC-DNA or ct-DNA did not yield any ICD bands. Done at pH 7.0, buffer sodium cacodylate, $I = 0.05$ M.

3. Materials and Methods

3.1. General Procedures

UV-Vis absorption spectra of compounds, UV-Vis titrations, and thermal melting experiments were measured on a Varian Cary 100 Bio spectrometer. Fluorescence spectra were recorded on a Varian Cary Eclipse fluorimeter. CD spectra were recorded on the JASCO J815 spectrophotometer. UV-Vis, fluorescence, and CD spectra were recorded using 1 cm path quartz cuvettes. Polynucleotides were purchased as noted: calf thymus (*ct*)-DNA, poly dAdT—poly dAdT, poly dGdC—poly dGdC, and poly rA—poly rU (Sigma) and dissolved in sodium cacodylate buffer, $I = 0.05$ M, pH = 7.0. The calf thymus (*ct*-) DNA was additionally sonicated and filtered through a 0.45 µm filter [26]. Polynucleotide concentration was determined spectroscopically and expressed as the concentration of phosphates [27]. Stock solutions of compounds **2–5** were prepared by dissolving the compounds in H_2O or DMSO; total DMSO content was below 1% in UV-Vis and below 0.1% in fluorimetric measurements. All measurements were performed in sodium cacodylate buffer, $I = 0.05$ M, pH = 7.0. Concentrations of **2–5** below 2×10^{-5} M were used for UV-Vis absorbance measurements to avoid intermolecular association.

3.2. UV/Vis, CD, and Fluorescence Titrations

UV-Vis and fluorimetric titrations were performed by adding portions of polynucleotide solution into the solution of the compound studied. After mixing polynucleotides with the compounds, it was observed that equilibrium was reached in less than 120 s. UV-Vis titrations were measured in the wavelength range of 250–450 nm. In fluorimetric titrations, the concentrations of **2–5** were 5×10^{-8}–2×10^{-7} M. The excitation wavelength of $\lambda_{\text{exc}} > 350$ nm was used to avoid absorption of excitation light caused by increasing absorbance of the polynucleotide. Titration data obtained for ds-DNA and ds-RNA were fitted by the non-linear least square method (using Origin 7.0 software) to the McGhee, von-Hippel formalism of Scatchard equation [17,18], whereby simultaneously binding constant (logKs) and Scatchard ratio n$_{[\text{bound molecule/DNA}]}$ were calculated. Calculations mostly gave values of ratio n = 0.2 ± 0.05, but for easier comparison, all *Ks* values were re-calculated for a fixed n = 0.2. Values for *Ks* have satisfactory correlation coefficients (>0.98).

CD experiments were performed by adding portions of stock solutions of compound **2–5** into a solution of the polynucleotides (c $\approx$ 1–2 $\times$ 10^{-5} M). Compounds **2–5** are achiral and, therefore, do not possess intrinsic CD spectra. CD spectra were recorded with a scan speed of 200 nm/min. Buffer background was subtracted from each spectrum, while each spectrum was the result of three accumulations.

3.3. Thermal Denaturation Experiments

Thermal denaturation curves for ds-DNA, ds-RNA, and their complexes with compounds **2–5** were determined by following the absorption change at 260 nm as a function of temperature [16]. The absorbance scale was normalized. T_m values are the midpoints of the transition curves determined from the maximum of the first derivative and checked graphically by the tangent method [28]. The ΔT_m values were calculated by subtracting Tm of the free nucleic acid from the T_m of the complex. Every ΔT_m value reported herein was the average of at least two measurements. The error in ΔT_m is $\pm 0.5\ ^\circ$C.

4. Conclusions

We studied how DNA and RNA interacted with tetracationic quadrupolar chromophores, which include three-coordinate triarylboron π-acceptors that were connected by various π-linkers: biphenyl (**2**), fluorine (**3**), pyrene (**4**), and dithienyl-diketopyrrolopyrrole (**5**). Our primary objective was to investigate the effect of the different linkers joining two triarylborane units on their binding to DNA/RNA. It was previously found that the compounds enter the HeLa cells and localize at the lysosomes without any effect on cell viability [15]. However, regarding their use in bioimaging, only compounds **1**, **3**, and **5** absorb in the visible range (>390 nm), optimal for bioimaging, whereas others require either UV excitation or two-photon absorption in the NIR.

All compounds very strongly bind to all ds-DNAs or ds-RNAs studied, thus again stressing the general applicability of our bis-triarylborane dications as newly developed dyes for ds-DNA/RNA sensing. The binding sites of the dyes are the ds-DNA minor groove or ds-RNA major groove.

The relation between the fluorimetric response of the dyes upon DNA/RNA binding and the structure of the dye or the secondary structure of the DNA/RNA was quite complex. The electronic and steric characteristics of the chromophore systems (triarylborane moieties and the π-linkers between them) and binding site (DNA/RNA groove size and shape varying for each DNA/RNA investigated, Table S1) influenced each other, so their interactions yield redistribution of the chromophore dipole moments and fine conformation tuning, that finally yield the particular fluorescence responses.

The results presented herein show that large and rigid aromatic linkers (**4**, **5**) give similar emission response to all DNAs/RNAs studied, whereas somewhat smaller linker, like fluorene **3**, has some adaptability to the particular structure of the binding site of different ds-DNA or ds-RNA, which results in fluorimetric recognition between AT-DNAs, GC-DNA, and AU-RNA.

Circular dichroism (CD) spectroscopy, as a structurally sensitive method, demonstrated that all dyes studied cause a minor decrease in DNA/RNA helicity, in line with a groove binding site. However, induced (I)CD bands were observed in the 300–400 nm range only for some dyes, with no apparent relation to the linker rigidity or steric volume. For instance, **3** (fluorene linker) and **5** (dithienyl-diketopyrrolopyrrole linker) yielded only negligible induced CD signals, suggesting the unfavorable orientation of the dye transition dipole moments with respect to the DNA/RNA chiral axis. Oppositely, the addition of **2** (biphenyl linker) yielded pronounced ICD signals specifically upon binding to DNA containing AT-sequences as the consequence of a tight fit in the AT-DNA minor groove. Compound **4**, with a much larger, planar aromatic surface linker (pyrene), also gave a specific ICD band for AT-DNA but not for GC-DNA. However, at variance to the flexible biphenyl-**2**, pyrene analog **4** also yielded a strong ICD band for AU-RNA, pointing out a tight and well-oriented fit within the ds-RNA major groove.

This study shows that systematic variation of the properties of linkers between tri-arylborane π-acceptors (flexibility vs. rigidity, aromatic surface size, and participation of nitrogen/sulfur heteroatoms) can have a strong impact on the fluorimetric and circular dichroism recognition of various ds-DNA/RNA secondary structures, depending on the properties of the DNA/RNA grooves. Obtained results are in line with complementary advances in the field of strongly sensitive fluorimetric probes for biomacromolecules [29–31].

Supplementary Materials: The following supporting information can be downloaded at: https://www.mdpi.com/article/10.3390/molecules28114348/s1. References [15,32,33] are cited in the Supplementary Materials.

Author Contributions: Investigation, L.-M.T., I.C., D.P.S., S.G., Ž.B. and M.M.; methodology, data curation, L.-M.T.; conceptualization, validation, writing—review and editing, T.B.M.; conceptualization, supervision, funding acquisition, writing—original draft preparation, L.-M.T. and I.P. All authors have read and agreed to the published version of the manuscript.

Funding: The financial support of the Croatian Science Foundation projects IP-2018-01-5475 and the Julius-Maximilians-Universität Würzburg are gratefully acknowledged.

Institutional Review Board Statement: Not applicable.

Informed Consent Statement: Not applicable.

Data Availability Statement: Not applicable.

Conflicts of Interest: The authors declare no conflict of interest.

Sample Availability: Stock solutions of samples are available from the authors.

References

1. Radić Stojković, M.; Škugor, M.; Tomić, S.; Grabar, M.; Smrečki, V.; Dudek, Ł.; Grolik, J.; Eilmes, J.; Piantanida, I. A short, rigid linker between pyrene and guanidiniocarbonyl-pyrrole induced a new set of spectroscopic responses to the ds-DNA secondary structure. *Org. Biomol. Chem.* **2013**, *11*, 4077–4085. [CrossRef] [PubMed]
2. Ji, L.; Griesbeck, S.; Marder, T.B. Recent developments in and perspectives on three-coordinate boron materials: A bright future. *Chem. Sci.* **2017**, *8*, 846–863. [CrossRef] [PubMed]
3. Entwistle, C.D.; Marder, T.B. Boron Chemistry Lights the Way: Optical Properties of Molecular and Polymeric Systems. *Angew. Chem. Int. Ed. Engl.* **2002**, *41*, 2927–2931. [CrossRef]
4. Entwistle, C.D.; Marder, T.B. Applications of Three-Coordinate Organoboron Compounds and Polymers in Optoelectronics, Special Issue on Organic Electronics. *Chem. Mater.* **2004**, *16*, 4574–4585. [CrossRef]
5. Griesbeck, S.; Zhang, Z.; Gutmann, M.; Lühmann, T.; Edkins, R.M.; Clermont, G.; Lazar, A.N.; Haehnel, M.; Edkins, K.; Eichhorn, A.; et al. Water-Soluble Triarylborane Chromophores for One- and Two-Photon Excited Fluorescence Imaging of Mitochondria in Cells. *Chem. Eur. J.* **2016**, *22*, 14701–14706. [CrossRef]
6. Berger, S.M.; Marder, T.B. Applications of Triarylborane Materials in Cell Imaging and Sensing of Bio-relevant Molecules such as DNA, RNA, and Proteins. *Mater. Horiz.* **2022**, *9*, 112–120.
7. Berger, S.M.; Rühe, J.; Schwarzmann, J.; Phillipps, A.; Richard, A.-K.; Ferger, M.; Krummenacher, I.; Tumir, L.-M.; Ban, Ž.; Crnolatac, I.; et al. Bithiophene-Cored, *mono-*, *bis-*, and *tris-*(Trimethylammonium)-Substituted, *bis*-Triarylborane Chromophores: Effect of the Number and Position of Charges on Cell Imaging and DNA/RNA Sensing. *Chem. Eur. J.* **2021**, *27*, 14057–14072. [CrossRef]
8. Griesbeck, S.; Ferger, M.; Czernetzi, C.; Wang, C.; Bertermann, R.; Friedrich, A.; Haehnel, M.; Sieh, D.; Taki, M.; Yamaguchi, S.; et al. Optimization of Aqueous Stability versus π-Conjugation in Tetracationic Bis(triarylborane) Chromophores: Applications in Live-Cell Imaging. *Chem. Eur. J.* **2019**, *25*, 7679–7688. [CrossRef]
9. Griesbeck, S.; Michail, E.; Rauch, F.; Ogasawara, H.; Wang, C.; Sato, Y.; Edkins, R.; Zhang, Z.; Taki, M.; Lambert, C.; et al. The Effect of Branching on One- and Two-Photon Absorption, Cell Viability and Localization of Cationic Triarylborane Chromophores with Dipolar vs. Octupolar Charge Distributions for Cellular Imaging. *Chem. Eur. J.* **2019**, *25*, 13164–13175. [CrossRef]
10. Ban, Ž.; Griesbeck, S.; Tomić, S.; Nitsch, J.; Marder, T.B.; Piantanida, I. A Quadrupolar Bis-Triarylborane Chromophore as a Fluorimetric and Chirooptic Probe for Simultaneous and Selective Sensing of DNA, RNA and Proteins. *Chem. Eur. J.* **2020**, *26*, 2195–2203. [CrossRef]
11. Amini, H.; Ban, Ž.; Ferger, M.; Lorenzen, S.; Rauch, F.; Friedrich, A.; Crnolatac, I.; Kendel, A.; Miljanić, S.; Piantanida, I.; et al. Tetracationic Bis-Triarylborane 1,3-Butadiyne as a Combined Fluorimetric and Raman Probe for Simultaneous and Selective Sensing of Various DNA, RNA, and Proteins. *Chem. Eur. J.* **2020**, *26*, 6017–6028. [CrossRef] [PubMed]

12. Ferger, M.; Roger, C.; Köster, E.; Rauch, F.; Lorenzen, S.; Krummenacher, I.; Friedrich, A.; Košćak, M.; Nestić, D.; Braunschweig, H.; et al. Electron-Rich EDOT Linkers in Tetracationic bis-Triarylborane Chromophores: Influence on Water-Stability, Biomacromolecule Sensing, and Photoinduced Cytotoxicity. *Chem. Eur. J.* **2022**, *28*, e202201130. [CrossRef] [PubMed]
13. Ferger, M.; Ban, Ž.; Krošl, I.; Tomić, S.; Dietrich, L.; Lorenzen, S.; Rauch, F.; Sieh, D.; Friedrich, A.; Griesbeck, S.; et al. Bis(phenylethynyl)arene Linkers in Tetracationic Bis-triarylborane Chromophores Control Fluorimetric and Raman Sensing of Various DNAs and RNAs. *Chem. Eur. J.* **2021**, *27*, 5142–5159. [CrossRef] [PubMed]
14. Božinović, K.; Nestić, D.; Lambert, C.; Michail, E.; Majhen, D.; Ferger, M.; Marder, T.B.; Košćak, M.; Piantanida, I. Diethynylarene-linked bis(triarylborane)cations as theranostic agents for tumor cell and virus-targeted photodynamic therapy. *J. Photochem. Photobiol. B Biol.* **2022**, *234*, 112523. [CrossRef]
15. Griesbeck, S.; Michail, E.; Wang, C.; Ogasawara, H.; Lorenzen, S.; Gerstner, L.; Zang, T.; Nitsch, J.; Sato, Y.; Bertermann, R.; et al. Tuning the π-bridge of quadrupolar triarylborane chromophores for one- and two-photon excited fluorescence imaging of lysosomes in live cells. *Chem. Sci.* **2019**, *10*, 5405–5422. [CrossRef]
16. Mergny, J.-L.; Lacroix, L. Analysis of Thermal Melting Curves. *Oligonucleotides* **2003**, *13*, 515–537. [CrossRef]
17. Scatchard, G. The Attractions of Proteins for Small Molecules and Ions. *Ann. N.Y. Acad. Sci.* **1949**, *51*, 660–672. [CrossRef]
18. McGhee, J.D.; von Hippel, P.H. Theoretical aspects of DNA-protein interactions: Co-operative and non-co-operative binding of large ligands to a one-dimensional homogeneous lattice. *J. Mol. Biol.* **1974**, *86*, 469–489. [CrossRef]
19. Demeunynck, M.; Demeunynck, M.; Bailly, C.; Wilson, W.D. *Small Molecule DNA and RNA Binders: From Synthesis to Nucleic Acid Complexes*; Demeunynck, M., Bailly, C., Wilson, W.D., Eds.; Wiley-VCH: Weinheim, Germany, 2003.
20. Tumir, L.M.; Crnolatac, I.; Deligeorgiev, T.; Vasilev, A.; Kaloyanova, S.; Branilovic, M.G.; Tomic, S.; Piantanida, I. Kinetic Differentiation between Homo- and Alternating AT DNA by Sterically Restricted Phosphonium Dyes. *Chem. Eur. J.* **2012**, *18*, 3859–3864. [CrossRef]
21. Steenken, S.; Jovanovic, S.V. How Easily Oxidizable Is DNA? One-Electron Reduction Potentials of Adenosine and Guanosine Radicals in Aqueous Solution. *J. Am. Chem. Soc.* **1997**, *119*, 617–618. [CrossRef]
22. Piantanida, I.; Palm, B.S.; Žinić, M.; Schneider, H.J. A new 4,9-diazapyrenium intercalator for single- and double-stranded nucleic acids: Distinct differences from related diazapyrenium compounds and ethidium bromide. *J. Chem. Soc. Perk. Trans.* **2001**, *2*, 1808–1816. [CrossRef]
23. Rodger, A.; Nordén, B. *Circular Dichroism and Linear Dichroism*; Oxford University Press: New York, NY, USA, 1997; Chapter 2.
24. Eriksson, M.; Nordén, B. Linear and circular dichroism of drug-nucleic acid complexes. In *Methods in Enzymology*; Academic Press: Cambridge, MA, USA, 2001; Volume 340, pp. 68–98.
25. Šmidlehner, T.; Piantanida, I.; Pescitelli, G. Polarization spectroscopy methods in the determination of interactions of small molecules with nucleic acids—Tutorial. *Beil. J. Org. Chem.* **2018**, *14*, 84–105. [CrossRef]
26. Chaires, J.B.; Dattagupta, N.; Crothers, D.M. Studies on interaction of anthracycline antibiotics and deoxyribonucleic acid: Equilibrium binding studies on interaction of daunomycin with deoxyribonucleic acid. *Biochemistry* **1982**, *21*, 3933–3940. [CrossRef] [PubMed]
27. Chalikian, T.V.; Völker, J.; Plum, G.E.; Breslauer, K.J. A more unified picture for the thermodynamics of nucleic acid duplex melting: A characterization by calorimetric and volumetric techniques. *Proc. Natl. Acad. Sci. USA* **1999**, *96*, 7853–7858. [CrossRef] [PubMed]
28. Tumir, L.M.; Piantanida, I.; Juranovic, I.; Meic, Z.; Tomic, S.; Zinic, M. Recognition of homo-polynucleotides containing adenine by a phenanthridinium bis-uracil conjugate in aqueous media. *Chem. Commun.* **2005**, *20*, 2561–2563. [CrossRef] [PubMed]
29. Deng, T.; Qi, Z.W.; Wu, Y.L.; Zhao, J.; Wang, L.; Peng, D.F.; Zhang, Y.; Huang, X.A.; Liu, F. A photosensitizing perylenediimide dye lights up cell nucleolus through visible light-mediated intracellular translocation. *Dye. Pigment.* **2021**, *196*, 109722. [CrossRef]
30. Cesaretti, A.; Calzoni, E.; Montegiove, N.; Bianconi, T.; Alebardi, M.; La Serra, M.A.; Consiglio, G.; Fortuna, C.G.; Elisei, F.; Spalletti, A. Lighting-Up the Far-Red Fluorescence of RNA-Selective Dyes by Switching from Ortho to Para Position. *Int. J. Mol. Sci.* **2023**, *24*, 4812. [CrossRef]
31. Deng, K.L.; Wang, L.; Xia, Q.; Liu, R.Y.; Qu, J.Q. A nucleic acid-specific fluorescent probe for nucleolus imaging in living cells. *Talanta* **2019**, *192*, 212–219. [CrossRef]
32. Saenger, W. *Principles of Nucleic Acid Structure*; Springer: New York, NY, USA, 1983; p. 226.
33. Cantor, C.R.; Schimmel, P.R. *Biophysical Chemistry Part III: The Behavior of Biological Macromolecules*; W.H. Freeman and Company: San Francisco, CA, USA, 1980; pp. 1109–1181.

Article

Metallacarborane Synthons for Molecular Construction—Oligofunctionalization of Cobalt Bis(1,2-dicarbollide) on Boron and Carbon Atoms with Extendable Ligands

Krzysztof Śmiałkowski [1,2], Carla Sardo [1,3] and Zbigniew J. Leśnikowski [1,*]

[1] Laboratory of Medicinal Chemistry, Institute of Medical Biology Polish Academy of Sciences, Lodowa 106, 93-232 Lodz, Poland; ksmialkowski@cbm.pan.pl (K.Ś.); csardo@unisa.it (C.S.)

[2] Lodz Institutes of the Polish Academy of Science, The Bio-Med-Chem Doctoral School, University of Lodz, 90-237 Lodz, Poland

[3] Department of Pharmacy, University of Salerno, Via Giovanni Paolo II, 132, 84084 Fisciano, SA, Italy

[*] Correspondence: zlesnikowski@cbm.pan.pl; Tel.: +48-42-209-33-80

Abstract: The exploitation of metallacarboranes' potential in various fields of research and practical applications requires the availability of convenient and versatile methods for their functionalization with various functional moieties and/or linkers of different types and lengths. Herein, we report a study on cobalt bis(1,2-dicarbollide) functionalization at 8,8′-boron atoms with different hetero-bifunctional moieties possessing a protected hydroxyl function allowing further modification after deprotection. Moreover, an approach to the synthesis of three and four functionalized metallacarboranes, at boron and carbon atoms simultaneously via additional functionalization at carbon to obtain derivatives carrying three or four rationally oriented and distinct reactive surfaces, is described.

Keywords: metallacarboranes; cobalt bis(1,2-dicarbollide); oligofunctionalization; alkylation; stereochemistry

Citation: Śmiałkowski, K.; Sardo, C.; Leśnikowski, Z.J. Metallacarborane Synthons for Molecular Construction—Oligofunctionalization of Cobalt Bis(1,2-dicarbollide) on Boron and Carbon Atoms with Extendable Ligands. *Molecules* **2023**, *28*, 4118. https://doi.org/10.3390/molecules28104118

Academic Editors: Michael A. Beckett and Igor B. Sivaev

Received: 13 April 2023
Revised: 4 May 2023
Accepted: 5 May 2023
Published: 16 May 2023

1. Introduction

Boron clusters, a class of polyhedral caged compounds, are playing an increasingly prominent role in the development of a broad range of technologies in fields such as material science [1,2] and medicinal chemistry [3–5]. They also find applications as labels for nucleic acid fragments [6–8]. Conjugates of DNA oligonucleotides and functionalized boron clusters have recently been proposed as building blocks for the construction of new nanomaterials for biomedical applications [9,10].

The ability of a boron cluster's type of dicarbaborate anion (*nido*-7,8-$C_2B_9H_{11}$) to coordinate a wide spectrum of metal ions, such as Fe, Co, Cr, Ta, Mo, W, V, and Nb, and form metal complexes, namely, metallacarboranes, extends the range of its prospective applications [11–14]. Among them, the most widely used metallocarborane is cobalt bis(1,2-dicarbolide), which is a sandwich of two $[C_2B_9H_{11}]^{2-}$ (dicarbollide) units with a cobalt ion in the center of the complex structure.

The broad technological potential of metallacarboranes requires access to a diverse array of functionalities (reactive functional groups, alkyl chains, spacers or polymers of various lengths, pharmacologically active species, etc.). Furthermore, it is necessary for more than one of these kinds of functionalities to coexist in the same metallacarborane structure to produce hetero-functionalized, tailor-made metallacarborane derivatives. The possibility to arrange these functionalities in a specific spatial orientation with respect to the topology of the 3D cluster core facilitates the achievement of the desired covalent modification, providing atomic-level precision and allowing for the control of size and surface composition. Finally, a highly relevant aspect regarding molecular design is the possibility of distancing the metallacarborane cage from the introduced functional group and the prospect of their further modification and/or extension.

The successful use of functionalized 1,2-dicarbadodecaborane as a core unit in the synthesis of building blocks containing a boron cluster and DNA for the construction of functional nanoparticles carrying therapeutic nucleic acids [9,10] prompted us to extend this technology towards metallacarboranes. Due to the different shapes of metallacarboranes, it may be possible to obtain nanoparticles with a different topology than that of 1,2-dicarbadodecaborane, which, in turn, can affect their biological properties. The ability of boron clusters to function as membrane carriers for a broad range of cargo molecules, thereby facilitating therapeutic nucleic acid cellular uptake, constitutes an additional potential advantage of metallacarborane-containing DNA nanoparticles [15–18].

2. Results

Herein, the results of a study on the oligofunctionalization of cobalt bis(1,2-dicarbollide) at boron atoms 8 and 8′ and 1,1′ and/or 2,2′ at carbon atoms with extendable ligands are described. Ligands such as hydroxyalkyls, which possess a hydroxy function separated from the metallacarborane core by spacers, were used. They offer the possibility of maximizing the distance of functional moieties from each other and the cluster core to avoid the influence of metallacarborane (known as the "metallacarborane effect") [19] on the chemical properties of these moieties. Analogous (but unprotected and substituted at cage carbon atoms only) alkylhydroxy derivatives of cobalt bis(1,2-dicarbollide) were described by Grüner and colleagues [20,21].

Moreover, derivatives were designed to test the applicability of two types of protections commonly employed in the chemical synthesis of DNA hydroxyl-protecting groups, namely, trityl and alkylsilyl protections, to allow for chemo-selection in a subsequent chemical manipulation. These results are complemented by studies on the further functionalization of the metallocarborane substituted on the 8,8′ boron atoms via substitution on the 1,1′ and 2,2′ carbon atoms of the carborane ligands.

2.1. Functionalization at 8 and 8′ Boron Atoms via Direct Alkylation of Hydroxy Groups in 8,8′-Dihydroxy-bis(1,2-dicarbollide)-3-cobalt(1-)ate (2)

The functionalization procedure began with converting cobalt bis(1,2-dicarbollide) (**1**) into easily available 8,8′-dihydroxy-bis(1,2-dicarbollido)-3-cobalt(1-)ate (**2**) in a reaction with 80% aqueous sulfuric acid [22]. The substitution reaction proceeded selectively at the 8 and 8′ boron atoms, which are the ones with maximum electron density [23].

As the alkylating agents used for compound **2** were 4-(trityloxy)butyl-4-methylbenzenesulfonate (**3**) [24] or (3-bromopropoxy)-*tert*-butyldimethylsilane (**4**), which differ in terms of leaving groups (tosyl or bromine) and hydroxyl group protection (trityl, -CPh$_3$ or *tert*-butyldimethylsilyl, TBDMS), NaH was used as a base to activate the hydroxyl groups in **2** (Scheme 1).

For the reactions involving both tosylate **3** and bromide **4**, even if an excessive amount of NaH was used, the complete alkylation of both hydroxyl groups was not achieved after 24 h, yielding a mixture of mono-substitution products **5** or **6** (minor products) and bis-substitution products **7** or **8** (major products). Both mono- and bis-alkylated products are easily separable via column chromatography conducted on silica gel using a gradient of methanol or acetonitrile in chloroform as the eluting solvent system.

The final yield of trityl-protected **7** after purification was found to vary considerably, ranging from 30 to 60%, although the yield of conversion detected in the crude reaction mixture was high, with a bis-functionalized derivative as the major product formed. This variability in the recovery of the tritylated products **5** and **7** can be ascribed to the relative instability of the trityl protection in **5** and **7**. A reason for this instability could be the known Lewis acidity of the boron cluster cage [25] and the acid lability of trityl protection. Interestingly, it seems that one of the factors influencing this effect may be the distance of the trityl protection from the metallocarborane core because in compounds **15, 16,** and **21,** which contain longer linkers, the instability of the trityl groups does not appear to be a problem.

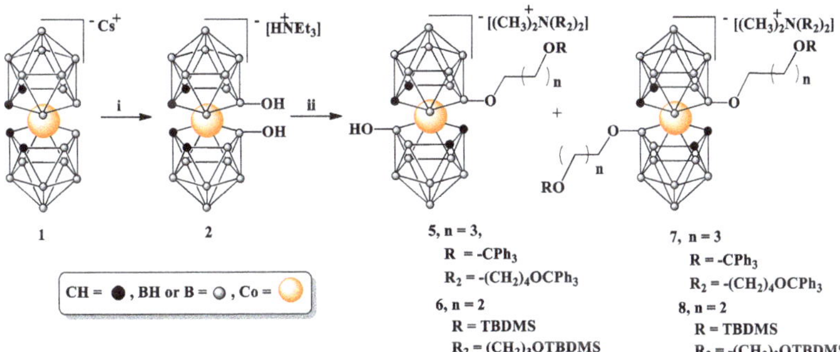

Scheme 1. Alkylation of 8,8′-dihydroxy-bis(1,2-dicarbollide)-3-cobalt(1-)ate (**2**) with Ts(CH$_2$)$_4$OCPh$_3$ (**3**) or Br(CH$_2$)$_3$OTBDMS (**4**) as electrophiles. i. H$_2$SO$_{4\ 80\%}$ 140 °C; ii. NaH, **3** or **4**, 80 °C or 60 °C, and DMF; TBDMS = -Si(CH$_3$)$_2$C(CH$_3$)$_3$.

Compounds **6** and **8** containing silyl protections are reasonably stable anions that can be purified and separated via column chromatography on silica gel. TBDMS protecting groups were quantitatively removed from terminal hydroxyls via treatment with TBAF in THF at room temperature, as demonstrated for **8** (Scheme 2). This treatment led to a counterion exchange in **9**, which was demonstrated using NMR analysis.

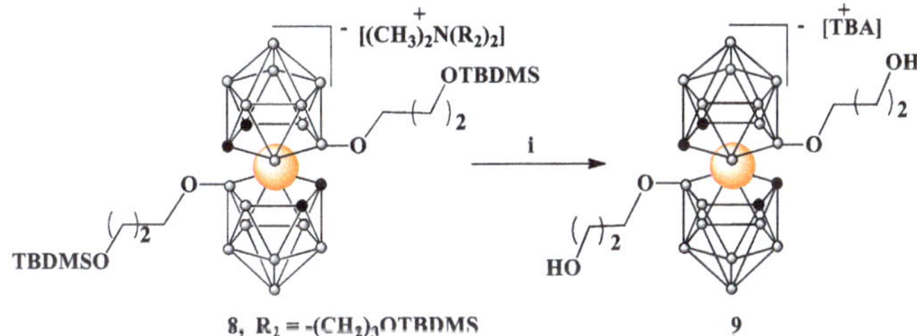

Scheme 2. Deprotection of TBDMS protected **8**. i. TBAF in THF. TBDMS = -Si(CH$_3$)$_2$C(CH$_3$)$_3$, TBA = N[(CH$_2$)$_3$CH$_3$]$_4$.

As indicated via ^{1}H NMR (Supplementary Information, Figures S2, S7 and S12), alkylated compounds **5–8** were isolated in the form of tetraalkylammonium (CH$_3$)$_2$NR$_2$ salt, where R$_2$ is -(CH$_2$)$_4$OCPh$_3$ or -(CH$_2$)$_3$OSi(CH$_3$)$_2$C(CH$_3$)$_3$ (Scheme 1). The unexpected structure of the tetraalkylammonium counterions is most probably due to the in situ formation of sodium dimethylamine as a result of a side reaction involving the reduction of DMF, used as a solvent by NaH, and its reaction with the alkylating agent [26]. The modest nucleophilicity of the hydroxyls on one side and the erosion of reagents because of this competing reaction with solvent on the other side account for the use of an excess of reagents to achieve good yields. These undesirable properties of the pair of NaH and DMF are also responsible for the incomplete alkylation of both hydroxyl groups in **2** mentioned above under the conditions employed, despite the other advantages of DMF solvent.

Next, we attempted to functionalize the carbon atoms of the bisubstituted on the boron atoms compound **8** (Scheme 3). The reaction was carried out in anhydrous DME at a temperature ranging from −70 °C to room temperature according to a typical procedure. First, to a solution of **8** in DME, nBuLi in hexane and then ethylene oxide solution in THF were added. After 24 h, the reaction was quenched, which, after a standard workup, yielded the mono-substitution product **10** formed as the minor product and the bis-substitution product **11** formed as the major one. Although moderate amounts of the target compounds can be obtained using this approach, the synthetic yields are generally not high.

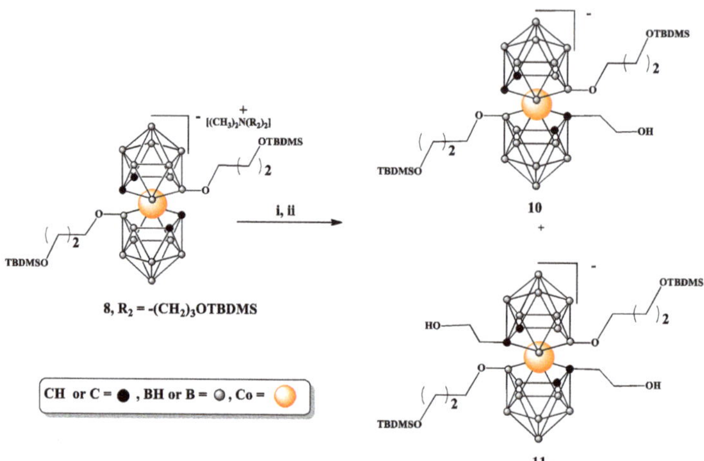

Scheme 3. Substitution on carbon atoms in 8,8′-alkoxy-functionalized metallacarborane **8**. i. nBuLi, DME; ii. ethylene oxide. TBDMS = -Si(CH$_3$)$_2$C(CH$_3$)$_3$.

We hypothesize that the reasons for the limited susceptibility of the type **8** compounds to functionalization on carbon atoms may include the electron-donating effect of boron cluster ligands; steric hindrance due to the presence of large substituents at the 8 and 8′ positions, thereby hampering the electrophile's access to the activated carbon atoms; the formation of intramolecular hydrogen bonds between the oxygen atoms of the substituents at the 8,8′ position and acidic carborane C-H groups (Figure 1) analogously to C-H···X-B hydrogen bonds [27,28]; and rotations of 1,2-dicarbollide ligands around an axis, thereby decreasing the susceptibility of C-H groups to activation and alkylation. Consequently, we decided to test a different approach based on derivatives of cobalt bis(1,2-dicarbollide) with arrested rotation and containing a phosphorothioate bridging moiety, namely, **13**.

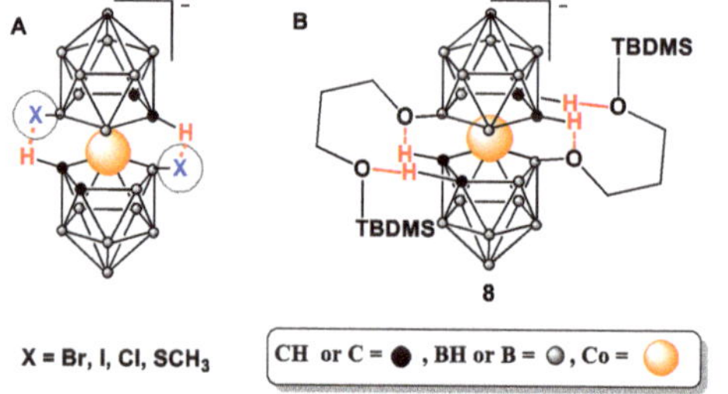

Figure 1. Intramolecular hydrogen bonds in halogenated cobalt bis(1,2-dicarbollide) (**A**) and hypothetical intramolecular interactions in 8,8′-bis-alkylated 8 (**B**).

*2.2. Functionalization at Boron Atoms 8 and 8′ via S-Alkylation of 8,8′-O,O-[Cobalt bis(1,2-dicarbollide)]phosphorothioate (**13**)*

2.2.1. Synthesis of 8,8′-Bridged 8,8′-O,O-[Cobalt bis(1,2-dicarbollide)]phosphorothioate (**13**)

The target compound **13** was obtained through a simple, two-step procedure (Scheme 4). In the first step, the 8,8′-dihydroxy-derivative **2** was converted into 8,8′-bridged 8,8′-O,O-[cobalt bis(1,2-dicarbollide)] H-phosphonate acid ester (**12**) via a reaction with tris(1H-imidazol-1-yl)phosphine [29] and the in situ hydrolysis of the resultant imidazolide.

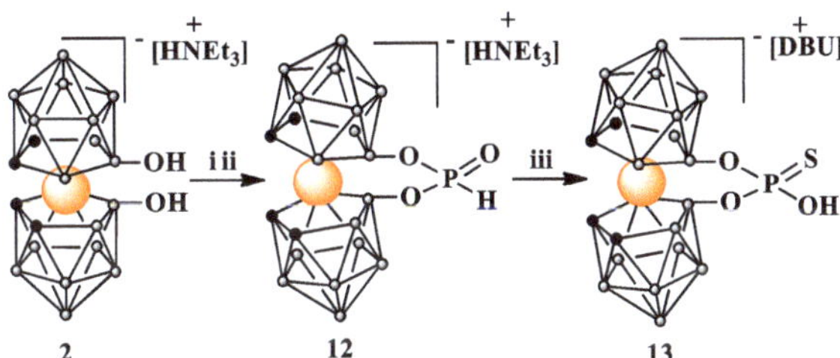

Scheme 4. Synthesis of H-phosphonate (**12**) and thiophosphate (**13**) esters of 8,8′-dihydroxy-bis(1,2-dicarbollido)-3-cobalt(1-)ate (**2**); i. PCl₃, imidazole, Et₃N in THF; ii. H₂O, iii. S₈, DBU in MeOH. DBU = 1,8-diazabicyclo(5.4.0)undec-7-en.

In the second step, H-phosphonate acid ester **12** was dissolved in anhydrous methanol; then, elemental sulfur S_8 was added. A strong organic base, 1,8-diazabicyclo(5.4.0)undec-7-en (DBU), was added to the resultant suspension, and the reaction was left for 96 h at room temperature with stirring. After the evaporation of the methanol, the residue containing the crude product was purified via silica gel column chromatography using a gradient of acetonitrile in chloroform as the eluting solvent system.

2.2.2. S-Alkylation of 8,8′-Bridged 8,8′-O,O-[Cobalt bis(1,2-dicarbollide)]phosphorothioate (**13**) with Linear and Branched Alkylating Agents

Sulfur is a larger atom than oxygen, rendering its electrons more polarizable and the atom itself more nucleophilic. The alkylation of the sulfur atom of phosphorothioates is a viable method for the synthesis of their S-alkylated derivatives. This method takes advantage of the excellent nucleophilic properties of sulfur and is commonly used in organophosphorus chemistry [30].

Using these advantageous properties of phosphorothioates, we attached both linear **13** and **14** or branched **20** ligands containing hydroxyl functions protected by a trityl group to the metallocarborane derivative **13**. The alkylation reaction proceeded smoothly, and the yields of the isolated alkylated products **15**, **16**, and **21** were high (Scheme 5).

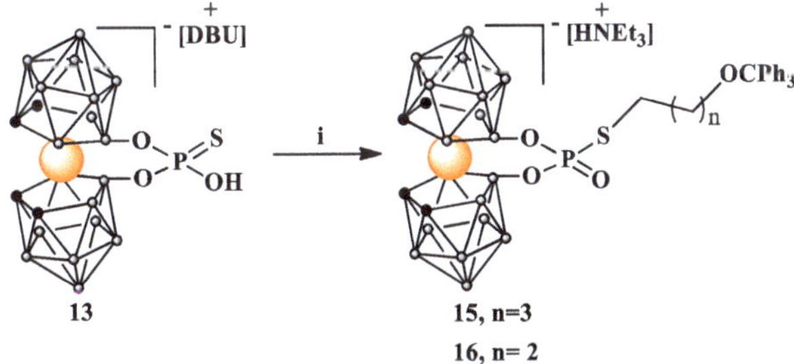

Scheme 5. Alkylation of 8,8′-O,O-[cobalt bis(1,2-dicarbollide)]phosphorothioate (**13**) with linear alkylating agent. i. Ts(CH₂)₄OCPh₃ (**3**) or Br(CH₂)₃OCPh₃ (**14**), acetone, NEt₃, 60 °C, and 24 h. DBU = 1,8-diazabicyclo(5.4.0)undec-7-en.

It is worth noting that derivatives of known 8,8′-bridged 8,8′-O,O-[cobalt bis(1,2-dicarbollide)]phosphate [22], a counterpart of **13** containing an 8,8′-O,O-phosphate bridge instead of a phosphorothioate one, have not yet been described. The expected low nucleophilicity of the phosphorus center resulting from the metallocarborane effect [19] and the lower nucleophilicity of oxygen compared to sulfur atom can be one of the reasons for this.

Consequently, a study on the functionalization of **13**′s carbon atoms was undertaken. For this purpose, a method for the synthesis of a branched alkylating agent (**20**), in this case, a glycerol derivative, was first developed. In the first step, bis(trityloxy)propan-2-ol

(**18**) was synthesized according to a method reported in the literature [31]. Next, **18** was reacted with 1,4-bis(*p*-toluenesulfonyloxy)butane (**19**), yielding a branched alkylating agent with an elongated linker (**20**) (Scheme 6).

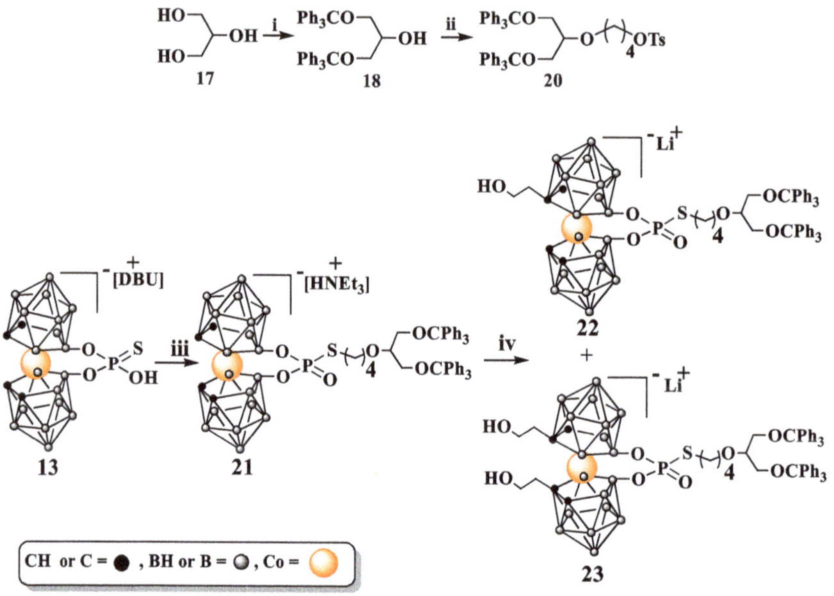

Scheme 6. Synthesis of branched alkylation agent 4-(1,3-bis(trityloxy)propan-2-yloxy)butyl-4-methylbenzenesulfonate (**20**), its use to alkylate 8,8′-O,O-[cobalt bis(1,2-dicarbollide)]phosphorothioate (**13**), and functionalization of **21** on carbon atoms: i. ClCPh₃ (trityl chloride), pyridine; ii. NaH, DMF, 1,4-bis-(4-methylbenzenesulfonate)butane (**19**); iii. Et₃N, **20** in a acetone/EtOAc; iv. nBuLi, DME; ethylene oxide; DBU = 1,8-diazabicyclo(5.4.0)undec-7-en.

The functionalization of **21** on carbon atoms was achieved via the activation of cage C-H groups with nBuLi followed by treatment with ethylene oxide. As expected, carrying out the synthesis of **22** and **23** required overcoming the low susceptibility of intermediate **21** to substitution on carbon atoms. However, mono- and disubstituted derivatives **22** and **23** were obtained with yields that enabled their full characterization and use for further chemical manipulations. The accessibility and high yields offered by the synthesis of intermediates **13** and **21** allow for the easy scaling-up of the synthesis of **22** and **23** (if needed).

3. Discussion

In the literature, the description of metallacarborane building blocks with hydroxyl or other functional groups separated from a cluster cage by an alkyl spacer seems limited. In this contribution, we report the development of methods for the functionalization of cobalt bis(1,2-dicarbollide) (**1**) with hetero-bifunctional derivatives of silyl- or trityl-protected alcohols attached directly to boron atoms and trityl-protected branched alcohols attached to metallacarborane through a phosphorothioate bridge. An approach to the synthesis of oligofunctionalized metallacarboranes via carbon deprotonation through nBuLi to obtain derivatives carrying substituents at both boron and carbon atoms is also described.

After a long period of discussion, the aromatic, three-dimensional (3D) nature of boron clusters is now widely accepted [32,33]. As in the case of aromatic organic molecules, the attachment of a substituent to one of the atoms of a 3D aromatic boron cluster system changes the distribution of the electron density of the entire molecule and affects the properties of other reactive centers. This makes the oligofunctionalization of metallacarboranes, especially when performed simultaneously on boron and carbon atoms, a particular challenge. Additionally, the complicated stereochemistry of substituted metallacarboranes owing to the presence of various types of chirality in the same molecule further compounds this challenge.

One of the practical manifestations of an effect of synchronized changes in electron density within the whole metallacarborane molecule on the properties of reactive centers is the inclination for the formation of disubstituted derivatives on boron atoms 8 and $8'$ and carbon atoms 1,2 and $1',2'$. This is illustrated by the preferential formation of disubstituted derivatives compared to monosubstituted ones in the case of **7** vs. **5**; **8** vs. **6**; **11** vs. **10**; and **23** vs. **22**. This is clearly due to the activation of a second nucleophilic center, such as B-OH or C-H groups, by a strong base after a previous substitution at the first center. However, this does not change the fact that the overall alkylation efficiency of the C-H groups in the derivatives of **1** already substituted on 8 and $8''$ boron atoms is low.

For example, the monoalkylation derivatives of 8,8′-dihydroxy-bis(1,2-dicarbollide)-3-cobalt(1-)ate (**2**) and bisalkylation derivatives were isolated in a ratio of 1:25 for **5** and **7** and 1:10 for **6** and **8**. The mono- and bis-substituted at carbon atoms derivatives of the bis-substituted at boron atoms of compound **8** were isolated in a 1:3 ratio for **10** and **11**. The same trend, though less pronounced, was observed for the boron and carbon functionalized derivatives **22** and **23** with arrested rotation, for which the ratio of mono-substitution on carbon bis-substitution was 1:1.2.

Another property significantly influencing the chemistry of oligofunctionalized metallacarboranes is their stereochemistry. The phenomenon of boron cluster chirality was recognized early [34–36]; however, this did not initially arouse much interest. The increasing use of boron clusters in the field of new materials, nanotechnology, and medical chemistry renders the stereochemistry of boron clusters increasingly important.

Strangely, though the most symmetric species in nature is the $B_{12}H_{12}^-$ ion, relatively minor changes in the boron cluster structure might render its basic framework dissymmetric enough to provoke chirality [37,38]. This is particularly evident in the case of oligofunctionalized metallacarboranes. The derivatives of type **22** and **23** are extreme examples containing various sources of chirality such as a chiral center at the phosphorus atom, axial and planar chirality due to bending of the metallacarborane molecule, and the existence of a number of isomers due to substitution at carbon atoms in addition to boron substitution.

A comparison of the ^{11}B-NMR spectra for compounds **13**, **15**, **16**, **21**, **22**, and **23** with arrested rotation as well as their ^{31}P-NMR spectra clearly shows changes in the number of isomers of the individual products depending on the type of substitution. Compounds **13**, **15**, **16**, and **21** show a singlet at δ of about 23 ppm attributed to 2B(8,8′) in the ^{11}B-NMR spectrum and a singlet at 48.62 ppm for **13** of about 16 ppm for compounds **15**, **16**, and **21** in the ^{31}P-NMR spectrum; these findings are consistent with the relative symmetry of the complexes.

On the contrary, the ^{11}B- and ^{31}P-NMR spectra of mono- and bisalkylated at carbon-atoms derivatives **22** and **23** show a dramatic change in symmetry due to the many possible combinations of substitutions on carbon atoms and the formation of a center of chirality on the phosphorus atom in monosubstituted **22** and in some isomers of bis-substituted **23**. In consequence, four signals at 25.37, 24.65, 23.69, and 22.93 ppm with an integral intensity ratio of 2:2:1:1 corresponding to B8,8′ for **22** and five signals at 25.35, 24.40, 23.64, 22.80, and 22.23 ppm with an approximate integral intensity ratio of 1:2:1.5:1:1 for **23** were observed. A similar effect was observed in the ^{31}P-NMR spectra, with seven signals at 15.10, 14.94, 14.57, 14.42, 14.11, 13.74, and 13.49 ppm for **22** and five signals at 14.99, 14.39, 14.16, 14.01, and 13.38 ppm for **23**, thus reflecting the asymmetry of these compounds and formations of isomers.

The sensitivity of standard chromatographic techniques is insufficient for distinguishing these isomers and only allows for the separation of mono- and disubstituted derivatives on carbon atoms, which have already been substituted on boron atoms in the case of **10** and

11 and **22** and **23**. A complete or partial separation of the isomers is probably attainable using HPLC; however, with regard to further practical applications of these derivatives, it is not necessary at the present stage.

4. Materials and Methods

4.1. Materials

All solvents were purchased in the highest available quality required for their application in this study. The reactions requiring anhydrous conditions were carried out under argon atmosphere using anhydrous solvents treated with activated molecular sieves for at least 24 h. Molecular sieves (4Å and 3Å) were purchased from Alfa Aesar (Karlsruhe, Germany) and heat-activated under vacuum before use. Triethylamine, trityl chloride, toluene-sulfonyl chloride, $NaHCO_3$, Na_2SO_4, 1,4-butanediol, P_2O_5, sodium hydride, n-butyllithium (1.6 M solution in hexane), (3-bromopropoxy)-*tert*-butyldimethylsilane, (**4**) and ethylene oxide (2.9–3.1 M solution in THF) were purchased from Sigma Aldrich (Steinheim, Germany). Cobalt bis(1,2-dicarbollide) was purchased from Katchem Ltd., Prague, Czech Republic.

4.2. Methods

Chromatography. Ultramate 3000 HPLC system (DIONEX, Sunnyvale, USA). equipped with a photodiode array detector (with fixed wavelengths of 210, 270, 310, and 330 nm) was used to determine the purity of products. The method consisted of a gradient elution from 20% to 90% aqueous acetonitrile using a Hypersil Gold (5μm particle size) reverse-phase column at 25 °C. HPLC data were acquired and processed using Chromeleon 6.8 software (DIONEX, Sunnyvale, CA, USA). Chromatography for purification of products was performed on a 230–400 mesh silica-gel Sigma Aldrich (Steinheim, Germany) filled glass column. Analytical TLC was performed on F254 silica gel plates purchased from Sigma Aldrich. Compounds were visualized using UV light (254 nm) and/or via staining with 0.05% *w/v* palladium chloride solution in MeOH/HCl.

NMR spectroscopy. ^1H, ^{11}B, $^{11}B\{^1H\}$, $^{13}C\{^1H\}$, ^{31}P, and $^{31}P\{^1H\}$ NMR spectra were recorded with a Bruker Advance III 600 MHz spectrometer.

UV-Vis spectrophotometry and **line-fitting analysis.** Measurements were performed using a Jasco V-750 UV spectrophotometer at room temperature in acetonitrile.

MS and **FT-IR.** MALDI-TOF MS spectra were recorded using a Voyager–Elite mass spectrometer (PerSeptive Biosystems) with 3-hydroxypicolinic acid (HPA) as the matrix. ESI MS mass spectra were recorded using Agilent 6546 LC/Q-TOF (Santa Clara, Kalifornia, United States). Negative ions were detected. Infrared absorption spectra were recorded using a Nicolet 6700 FT-IR spectrometer (Thermo Scientific) equipped with a Smart orbit diamond Attenuated Total Reflectance (ATR) accessory. Samples to be analyzed were placed on a diamond ATR element in a solid form or through casting from CH_2Cl_2 solution. Data were acquired and processed using Omnic 8.1 software (Thermo Scientific, Waltham, CA, USA).

8,8′-dihydroxy-bis(1,2-dicarbollido)-3-cobalt(1-)ate HNEt₃ (**2**) was synthesized according to the procedure reported by Plešek et al. [22].

4-Trityloxybutyl 4-methylbenzenesulfonate (**3**) was obtained as described in [39].

3-Bromo-1-trityloxypropane (**14**) was synthesized as described previously [10].

1,3-bis(trityloxy)propan-2-ol (**18**) was synthesized according to a procedure in the literature [31].

1,4-bis-(4-methylbenzenesulfonate)butane (**19**) was synthesized according to a modified procedure from the literature [40].

Synthesis of 3,3′-Co[[(8-O(CH₂)₄OCPh₃]-1,2-C₂B₉H₁₀](8′-OH-1,2-C₂B₉H₁₀) (CH₃)₂N[(CH₂)₄ OPh₃]₂ (**5**) *and 3,3′-Co[(8-O(CH2)₄OCPh₃-1,2-C₂B₉H₁₀)]₂ (CH₃)₂N[(CH₂)₄Oph₃]₂* (**7**). Compound **2** (50 mg, 0.10 mmol) was dissolved in 0.5 mL of anhydrous dimethoxyethane (DME) and added to 60% NaH in mineral oil dispersion (4.4 mg, 0.10 mmol NaH) under argon atmosphere. The reaction mixture was stirred for 2 h at room temperature. After this time has elapsed, the solvent was evaporated under reduced pressure, and the resultant solid residue was dissolved in 0.5 mL of anhydrous dimethylformamide (DMF) and added to a second aliquot of 60% NaH in mineral oil dispersion (26.2 mg, 0.65 mmol NaH). After sitting for 2 h at room temperature, the mixture was dropped into a solution of **3** (318.8 mg, 0.65 mmol) in 0.5 mL of anhydrous DMF; then, the reaction mixture was heated at 80 °C (oil bath temperature) for 24 h. The post-reaction mixture was concentrated; then, 3 mL of H_2O was added. The resulting precipitate was washed with several aliquots of H_2O. Then, the solid was dissolved in 5 mL of CH_2Cl_2, and the solution was dried with anhydrous Na_2SO_4, filtered, and concentrated. Repeated column chromatography on silica gel using a gradient of CH_3OH in CH_3Cl (0 to 20%) as an eluting solvent system yielded **5** and **7** as the first and second bands, respectively. Fractions containing compound **7** were collected, evaporated to dryness, and then crystallized from hexane, furnishing 96% pure product as determined by HPLC. Monosubstituted compound **5** was obtained as a red oil.

(**5**): Yield: 7 mg (5%); **TLC** ($CHCl_3$:MeOH 4:1); R_f: 0.53; **MALDI-MS** (m/z): 670.5 (calc. for $C_{27}B_{18}H_{44}O_3Co_1$: 670.17). Since excessively small quantities of this product were obtained, it was not further analyzed via NMR.

(**7**): Yield: 180 mg (62%); **TLC** ($CHCl_3$:MeOH 4:1); R_f: 0.64; **^{1}H NMR** (500 MHz, CD_3CN) δ: 7.43 (m, 24H, $\underline{H_{arom}}$), 7.32 (m, 24H, $\underline{H_{arom}}$), 7.25 (m, 12H, $\underline{H_{arom}}$), 4.18 (s, 4H, $\underline{CH}_{carborane}$), 3.31 (t, J = 6.2 Hz, 4H, BO$\underline{CH_2}$CH$_2$CH$_2$CH$_2$OTr), 3.07 (m, 8H, $\underline{CH_2}$OTr), 2.99 (t, J = 6.4 Hz, 4H, N$\underline{CH_2}$CH$_2$CH$_2$CH$_2$OTr), 2.86 (s, 6H, N($\underline{CH_3}$)$_2$), 1.68 (m, 4H, NCH$_2\underline{CH_2}$CH$_2$CH$_2$OTr), 1.60 (m, 8H, $\underline{CH_2}$CH$_2$OTr), and 1.48 (m, 4H, BOCH$_2\underline{CH_2}$CH$_2$CH$_2$OTr); **^{13}C{^{1}H} NMR** (126 MHz, CD_3CN) δ: 145.53 and 145.18 (12C, $aromatic_{trityl}$); 129.44 and 129.41 (24C, $aromatic_{trityl}$); 129.37 and 128.83 (24C, $aromatic_{trityl}$); 128.21 (12C, $aromatic_{trityl}$); 87.43 and 87.06 (4C, $\underline{C}(Ph)_3$), 69.43 (2C, BO$\underline{CH_2}$CH$_2$CH$_2$CH$_2$OTr), 64.30 (2C, BOCH$_2$CH$_2$CH$_2\underline{CH_2}$OTr), 63.30 (2C, NCH$_2$CH$_2$CH$_2\underline{CH_2}$OTr), 62.33 (2C, N$\underline{CH_2}$CH$_2$CH$_2$CH$_2$OTr), 51.56 (4C, $\underline{CH}_{carb}$), 30.39, 29.68 (4C, $\underline{CH_2}$CH$_2$OTr), 27.51, 27.13 (2C, BOCH$_2\underline{CH_2}$CH$_2$CH$_2$OTr), and 20.29 (NCH$_2\underline{CH_2}$CH$_2$CH$_2$OTr); **^{11}B{^{1}H} NMR** (160 MHz, CD_3CN) δ: 20.63 (s, 2B, B$^{8,8'}$), −3.56 (s, 2B, B$^{10,10'}$), −7.52 (s, 4B, B$^{4,4',7,7'}$), −9.03 (s, 4B, B$^{9,9',12,12'}$), −20.53 (s 4B, B$^{5,5',11,11'}$), and −28.38 (s, 2B, B$^{6,6'}$) **^{11}B NMR** (160 MHz, CD_3CN) δ: 20.66 (s, 2B, B$^{8,8'}$), −3.59 (d, 2B, B$^{10,10'}$), −8.22 (m, 8B, B$^{9,9',12,12',4,4',7,7'}$), −20.48 (d, 2B, B$^{5,5',11,11'}$), and −28.09 (d, 2B, B$^{6,6'}$); **FT-IR** (cm^{-1}): 3083.53, 3055.16, 3029.62 (ν C-H aromatic, (C-H$_{carb}$); 2927.18 (ν C-H$_{asym}$, CH_2); 2867.43 (ν C-H, CH_2O, ν C-H$_{sym}$, CH_2); 2543.23 (ν B-H); 1596.06, 1488.72, 1447.49 (ν C=C); 1152.23; 745.10 and 693.37 (δ C-H aromatic); 703.89 (ν B-B); **MALDI-MS** (m/z): found 984.2 (calc. for $C_{50}B_{18}I_{66}O_4Co_1$: 984.59).

Synthesis of 3,3′-Co[8-O(CH$_2$)$_3$OTBDMS]-1,2-C$_2$B$_9$H$_{10}$)(8′-OH-1,2-C$_2$B$_9$H$_{10}$) (CH$_3$)$_2$N[(CH$_2$)$_3$OTBDMS)]$_2$ (**6**) *and 3,3′-Co[8-O(CH$_2$)$_3$OTBDMS-1,2-C$_2$B$_9$H$_{10}$)]$_2$ (CH$_3$)$_2$N[(CH$_2$)$_3$OTBDMS)]$_2$* (**8**). Compound **2** (300 mg, 0.65 mmol) was dissolved in 3 mL of anhydrous DME under argon atmosphere and added to 60% NaH in mineral oil dispersion (26.2 mg, 0.65 mmol NaH). The reaction mixture was stirred for 2 h at room temperature. Subsequently, the solvent was evaporated, and the solid residue was dissolved in 3 mL of anhydrous DMF and added to a second portion of 60% NaH in mineral oil dispersion (157 mg, 3.93 mmol NaH). After sitting for 2 h at room temperature, the mixture was heated to 60 °C (oil bath temperature) and 911 µL of (3-bromopropoxy)(*tert*-butyl)dimethylsilane (**4**) (3.93 mmol) was added dropwise. After 22 h, an additional quantity of **4** (600 µL, 2.59 mmol) was added, and the mixture was stirred for next 24 h at 60 °C. A white solid was formed. The post-reaction mixture was then filtered through syringe filter (5 µm, PTFE, Carl Roth), and DMF was evaporated under reduced pressure. The solid residue was treated with 2 mL of $CHCl_3$ and filtered; then, the the filtrate was concentrated and loaded on a silica gel column for separation of products. First, 100% $CHCl_3$ and then 5% and 10% CH_3CN in $CHCl_3$ were used as eluting solvent systems. Compound **8** was isolated as first fraction and obtained in the form of orange crystals after solvent evaporation; its purity was above 95% as determined via HPLC. The second fraction containing **6** was obtained as red oil after solvent evaporation. Both products **6** and **8** were isolated as N,N-bis[(3-(*tert*-butyldimethylsilyloxypropyl)]-N,N-dimethyl ammonium salts [26].

(6): Yield: 60 mg (10%); **TLC** (CHCl$_3$:CH$_3$CN 3:1) **R$_f$:** 0.53; **^{1}H NMR** (600 MHz, CD$_3$CN) δ: 6.05 (s, 1H, $\underline{OH}$), 3.73 (t, J = 6.5 Hz, 2H, BOCH$_2$CH$_2$$\underline{CH_2}$OSi), 3.69 (m, 4H, N$\underline{CH_2}CH_2CH_2$OSi), 3.65 (m, 2H, BO$\underline{CH_2}CH_2CH_2$OSi), 3.56 (s, 2H, $\underline{CH_{carb}}$), 3.45 (s, 2H, $\underline{CH_{carb}}$), 3.27 (m, 4H, NCH$_2$CH$_2$$\underline{CH_2}$OSi), 2.97 (s, 6H, N($\underline{CH_3}$)$_2$), 1.88 (m, 4H, NCH$_2$$\underline{CH_2}CH_2$OSi), 1.74 (m, J = 6.4 Hz, 2H, BOCH$_2$$\underline{CH_2}CH_2$OSi), 0.90 (s, 9H, BOCH$_2CH_2CH_2$OSi(CH$_3$)$_2$C($\underline{CH_3}$)$_3$), 0.89 (s, 18H, NCH$_2CH_2CH_2$OSi(CH$_3$)$_2$C($\underline{CH_3}$)$_3$), 0.07 (s, 6H, BOCH$_2CH_2CH_2$OSi($\underline{CH_3}$)$_2$C(CH$_3$)$_3$), and 0.05 (s, 12H, NCH$_2CH_2CH_2$OSi($\underline{CH_3}$)$_2$C(CH$_3$)$_3$); ^{13}C{^{1}H} NMR (126 MHz, CD$_3$CN) δ: 67.05 (1C, BO$\underline{CH_2}$CH$_2$CH$_2$OSi), 62.56 (2C, NCH$_2$CH$_2$$\underline{CH_2}$OSi), 61.23 (1C, BOCH$_2CH_2$$\underline{CH_2}$OSi), 60.17 (2C, N$\underline{CH_2}CH_2CH_2$OSi), 52.03 (2C, N($\underline{CH_3}$)$_2$), 46.23 (2C, CH$_{carb}$), 45.51 (2C, $\underline{CH_{carb}}$), 35.92 (1C, BOCH$_2$$\underline{CH_2}CH_2$OSi), 26.45 (2C, NCH$_2$$\underline{CH_2}CH_2$OSi), 26.27 (6C, OSi(CH$_3$)$_2$C($\underline{CH_3}$)$_3$), 26.11 (3C, OSi(CH$_3$)$_2$C($\underline{CH_3}$)$_3$), 18.81 (1C, OSi(CH$_3$)$_2$$\underline{C}$(CH$_3$)$_3$), 18.72 (2C, OSi(CH$_3$)$_2$$\underline{C}$($\overline{CH_3}$)$_3$), −5.10 (2C, OSi($\underline{CH_3}$)$_2$C($\overline{CH_3}$)$_3$), and −5.40 (4C, OSi($\underline{CH_3}$)$_2$C(CH$_3$)$_3$); ^{11}B{^{1}H} NMR (193 MHz, CD$_3$CN) δ: 27.16 (s, 1B, B$^{8'}$), 25.10 (s, 1B, B^8), −5.02 to −9.09 (10B, overlapped B$^{9,9',10,10'12,12',4,4',7,7'}$), −20.10 to −20.69 (4B, B$^{5,5',11,11'}$), and −29.18 to −30.16 (2B, B$^{6,6'}$); ^{11}B NMR (160 MHz, CD$_3$CN) δ: 27.16 (s, 1B, B$^{8'}$), 25.09 (s, 1B, B^8), −3.22 to −9.40 (10B, overlapped B$^{9,9',10,10'12,12',4,4',7,7}$) −19.72 to −21.11 (4B, B$^{5,5',11,11'}$), and −28.81 to −30.49 (2B, B$^{6,6'}$); **FT-IR** (cm^{-1}): 3278.64, 2952.00, 2927.58, 2855.64, 2539.99, 1470.65, 1387.53, 1360.69, 1252.66, 1156.76, 1093.25, 1005.54, 972.16, 938.48, 880.24, 832.61, 774.98, 718.29, and 661.26; **ESI-MS** (m/z): found 530.39 and 472.35 [M-tButyl]$^-$ (calc. for C$_{13}$B$_{18}$H$_{43}$Si$_1$O$_3$Co$_1$: 529.09).

(8): Yield: 572 mg (80%); **TLC** (CHCl$_3$:CH$_3$CN 3:1) **R$_f$** 0.70; **^{1}H NMR** (500 MHz, CD$_3$CN) δ: 4.18 (s, 4H, $\underline{CH_{carb}}$), 3.69 (m, 4H, BOCH$_2$CH$_2$$\underline{CH_2}$OSi), 3.64 (t, J = 6.4 Hz, 4H, N$\underline{CH_2}$CH$_2$CH$_2$OSi), 3.38 (t, J = 6.0 Hz, 4H, and BO$\underline{CH_2}$CH$_2$CH$_2$OSi), 3.27 (m, 4H, NCH$_2$CH$_2$$\underline{CH_2}$OSi), 2.97 (s, 6H, N($\underline{CH_3})_2$), 1.88 (m, 4H, N$\underline{CH_2}CH_2CH_2$OSi), 1.58 (m, J = 6.2 Hz, 4H, BOCH$_2$$\underline{CH_2}CH_2$OSi), 0.90 (s, 18H, OSi(CH$_3$)$_2$C($\underline{CH_3})_3$), 0.88 (s, 18H, OSi(CH$_3$)$_2$C$(\underline{CH_3})_3$), 0.07 (s, 12H, OSi($\underline{CH_3}$)$_2$C(CH$_3$)$_3$), *and 0.04 (s, 12H, OSi($\underline{CH_3}$)$_2$C(CH$_3$)$_3$);* ^{13}C NMR *(126 MHz, CD$_3$CN)* δ: 66.18 (2C, BOCH$_2$CH$_2$$\underline{CH_2}$OSi), 62.54 (2C, NCH$_2CH_2$$\underline{CH_2}$OSi), 60.95 (2C, BO$\underline{CH_2}CH_2CH_2$OSi), 60.17 (2C, N$\underline{CH_2}CH_2CH_2$OSi), 52.06 (2C, N($\underline{CH_3})_2$), 51.60 (4C, $\underline{CH_{carb}}$), 35.97 (2C, BOCH$_2$$\underline{CH_2}CH_2$OSi), 26.46 (2C, NCH$_2$$\underline{CH_2}CH_2$OSi), 26.28 (6C, OSi($\overline{CH_3}$)$_2$C(CH$_3$)$_3$), 26.12 (6C, OSi(CH$_3$)$_2$C($\underline{CH_3}$)$_3$ 18.83 (2C, OSi($\overline{CH_3}$)$_2$$\underline{C}$(CH$_3$)$_3$), 18.72 (2C, OSi(CH$_3$)$_2$$\underline{C}$(CH$_3$)$_3$), −5.09 (4C, OSi($\underline{CH_3}$)$_2$C(CH$_3$)$_3$), and −5.39 (4C, OSi($\underline{CH_3}$)$_2$C(CH$_3$)$_3$); ^{11}B{^{1}H} NMR (160 MHz, CD$_3$CN) δ: 20.61 (s, 2B, B$^{8,8'}$), −3.61 (s, 2B, B$^{10,10'}$), −7.45 (s, 4B, B$^{4,4',7,7'}$), −9.03 (s, 4B, B$^{9,9',12,12'}$), −20.51 (s, 4B, B$^{5,5',11,11'}$), and −28.38 (s, 2B, B$^{6,6'}$); ^{11}B NMR (160 MHz, CD$_3$CN) δ: 20.76 (s, 2B, B$^{8,8'}$), −3.59 (d, 2B, B$^{10,10'}$), −7.18 to −9.37 (m, 8B, overlapped B$^{4,4',7,7',9,9',12,12'}$), −20.48 (d, 4B, B$^{5,5',11,11'}$), and −28.02 (d, 2B, B$^{6,6'}$); **FT-IR** (cm^{-1}): 3048.6 (C-H$_{carb}$); 2949.4 (ν C-H$_{asym}$, CH$_3$); 2927.4 (ν C-H$_{asym}$, CH$_2$), 2891.6 and 2884.2 (ν C-H$_{sym}$ CH$_3$); 2856.2 (ν C-H, CH$_2$O, ν C-H$_{sym}$, CH$_2$); 2605.2 and 2550.6 (ν B-H); 1470.8 (δ C-H$_{sym}$, CH$_2$) and 1436.0; 1386.2; 1359.5; 1251.2 (Si-CH$_3$) and 1161.9 (Si-O-C); 1093.1 (ν Si-O); 1019.3; 1006.3; 975.2; 955.3; 942.7; 881.; 831.8 and 772.4 (ν Si-C); 710.9; 661.0; **MALDI-MS** (m/z): found 700.5 (calc. for C$_{22}$B$_{18}$H$_{62}$CoO$_4$Si$_2$ 700.43).

Deprotection of compound **8** *to obtain [3,3'-Co(8-O(CH$_2$)$_3$OH-1,2-C$_2$B$_9$H$_{10}$)$_2$] TBA* **(9).** Compound **8** (20 mg, 0.018 mmol) was dissolved in 0.2 mL of tetrahydrofuran (THF). A total of 69 μL of terabutylammonium fluoride (TBAF) (1M solution in THF, 0.069 mmol) was added to the obtained solution, and the reaction mixture was stirred at room temperature overnight. Then, solvent was evaporated under reduced pressure, and the resulting oil was dissolved in CHCl$_3$ and applied to the silica gel column prepared in the same solvent. Elution in a gradient 1 to 20% CH$_3$CN in CHCl$_3$ provided 96% pure **9** (determined by HPLC) as TBA salt. **Yield:** 12.5 mg (95%); **TLC** (CHCl$_3$:CH$_3$CN 3:1) **R$_f$:** 0.47; **^{1}H NMR** (500 MHz, CD$_3$CN) δ: 4.17 (s, 4H, $\underline{CH_{carb}}$), 3.52 (t, J = 6.3 Hz, 4H, BOCH$_2$CH$_2$$\underline{CH_2}$OH), 3.43 (t, J = 6.0 Hz, 4H, BO$\underline{CH_2}$CH$_2$$\underline{CH_2}$OH), 3.10–3.03 (m, 8H, N$\underline{CH_2}CH_2CH_2CH_3$), 2.58 (s, 2H, $\underline{OH}$), 1.59 (m, J = 8.2, 3.7 Hz, 12H, BOCH$_2$$\underline{CH_2}CH_2OH, N\underline{CH_2}CH_2CH_3$), 1.41–1.28 (m, 8H, NCH$_2CH_2$$\underline{CH_2}CH_3$), and 0.96 (t, J = 7.4 Hz, 12H, NCH$_2$CH$_2$CH$_2$$\underline{CH_3}$); ^{13}C NMR (126 MHz, CD$_3$CN) δ: 67.47 (2C, BOCH$_2$CH$_2$$\underline{CH_2}$OH), 60.63 (2C, BO$\underline{CH_2}CH_2CH_2$OH), 59.23 (4C, N$\underline{CH_2}CH_2CH_2CH_3$), 51.39 (4C, $\underline{CH_{carborane}}$), 35.54 (2C, BOCH$_2$$\underline{CH_2}CH_2$OH), 24.21 (4C, NCH$_2$$\underline{CH_2}CH_2CH_3$), 20.25 (4C, NCH$_2CH_2$$\underline{CH_2}CH_3$), and 13.72 (4C, NCH$_2CH_2CH_2$$\underline{CH_3}$);

^{11}B{^{1}H} **NMR** (160 MHz, CD$_3$CN) δ: 20.95 (s, 2B, B$^{8,8'}$), −3.49 (s, 2B, B$^{10,10'}$), −7.71 (s, 4B B$^{4,4',7,7'}$), −8.85 (s, 4B, B$^{9,9',12,12'}$), −20.44 (s, 4B, B$^{5,5',11,11'}$), and −28.14 (s, 2B, B$^{6,6'}$); ^{11}**B NMR** (160 MHz, CD$_3$CN) δ: 20.97 (s, 2B, B$^{8,8'}$), −3.48 (d, 2B, B$^{10,10'}$), −7.18 to −9.26 (m, 8B, B$^{4,4'7,7'9,9',12,12'}$), −20.42 (d, 4B, B$^{5,5',11,11'}$), and −28.19 (d, 2B, B$^{6,6'}$); **FT-IR** (cm^{-1}): 3588.76 (ν O-H); 3450.79 (ν N^{+}-R); 3052.83 (C-H$_{carb}$); 2962.10 (ν C-H$_{asym}$, CH$_2$); 3932.36 and 2873.70 (ν C-H, CH$_2$O, ν C-H$_{sym}$, CH$_2$); 2529.19 (ν B-H); 1467.75 (ν N^{+}-R), 1380.63; 1161.25 (HO-C); 1105.54; 1062.05; 1007.15; 969.00; 945.10; 920.29; 876.90; 789.11; 735.69; 696.07; 664.83; **MALDI-MS** (*m*/*z*): found 472.8 (calc. for C$_{10}$B$_{18}$H$_{34}$O$_4$Co$_1$ 471.90).

Synthesis of 3,3′-Co{[8-O(CH$_2$)$_3$OTBDMS-1-(CH$_2$)$_2$OH]-1,2-C$_2$B$_9$H$_9$)}[8′-O(CH$_2$)$_3$OTBDMS-1′,2′-C$_2$B$_9$H$_{10}$)]$^{-}$ (10) and 3,3′-Co[(8-O(CH$_2$)$_3$OTBDMS-1-(CH$_2$)$_2$OH-1,2-C$_2$B$_9$H$_9$)]$_2^{-}$ (11). Compound **8** (50 mg, 0.04 mmol) was dried via co-evaporation with anhydrous benzene and then kept under vacuum, over P$_2$O$_5$ overnight. Then, it was dissolved in anhydrous DME (1 mL), and the solution was cooled in CO$_2$/isopropanol cooling bath. After 15 min, n-BuLi (43 µL, 1.6 M solution in hexane, 1.5 eq) was added, and the reaction mixture was stirred for 10 min. Afterwards, cooling bath was removed, and the mixture was stirred for next 10 min. Then, the reaction mixture was cooled again in cooling bath and another portion of n-BuLi (43 µL) was added. After 15 min, ethylene oxide (60 µL, 2.9–3.1 M solution in THF, 4.5 eq) was added, and the reaction was left overnight in cooling bath. Then, CH$_2$Cl$_2$ (3 mL) was added to the reaction mixture, the reaction was quenched via addition of water, and the organic solution was washed three times with 5 mL portions of water. Organic layer was separated and dried over MgSO$_4$; then, solvents were evaporated. Crude product was purified and mono- and bis-substituted products were separated through silica gel column chromatography using a gradient of MeOH in CH$_2$Cl$_2$ from 0 to 3% of MeOH.

(10): Yield: 4.7 mg (9%); **TLC** (MeOH:CH$_2$Cl$_2$ 1:12.5): **R$_f$:** 0.28; **ESI-MS** (*m*/*z*): found 744.55, (calc. for C$_{24}$B$_{18}$H$_{66}$O$_5$Si$_2$Co$_1$ 744.48). Since excessively small quantities of this product were obtained, it was not further analyzed via NMR.

(11): Yield: 15 mg (27%); **TLC** (MeOH:CH$_2$Cl$_2$ 1:12.5): **R$_f$:** 0.16; ^{1}**H NMR** (500 MHz, CD$_3$CN) δ: 4.32–4.08 (s, 2H, diastereoizomeric C*H$_{carborane}$*), 3.77 to 3.52 (m, 16H, overlapped NC*H$_2$*CH$_2$CH$_2$O, NCH$_2$CH$_2$*CH$_2$*O, BOC*H$_2$*CH$_2$CH$_2$O, BOCH$_2$CH$_2$*CH$_2$*O), 3.52 to 3.39 (m, 4H, HOC*H$_2$*CH$_2$C$_{carb}$), 3.39 to 3.31 (m, 4H, HOCH$_2$*CH$_2$*C$_{carb}$), 3.03 (s, 6H, N(*CH$_3$*)$_2$), 1.74 (m, 4H, NCH$_2$*CH$_2$*CH$_2$O), 1.65 (m, 4H, BOCH$_2$*CH$_2$*CH$_2$O), 0.90 (s, 18H, NCH$_2$CH$_2$CH$_2$OSi(CH$_3$)$_2$C(*CH$_3$*)$_3$), 0.88 (s, 18H, BOCH$_2$CH$_2$CH$_2$OSi(CH$_3$)$_2$C(*CH$_3$*)$_3$), 0.07 (s, 12H, BOCH$_2$CH$_2$CH$_2$OSi(*CH$_3$*)$_2$C(CH$_3$)$_3$), and 0.05 (s, 12H, NCH$_2$CH$_2$CH$_2$OSi(*CH$_3$*)$_2$C(CH$_3$)$_3$; ^{13}**C{^{1}H} NMR** (126 MHz, CD$_3$CN) δ: 67.39 (2C, BOC*H$_2$*CH$_2$CH$_2$O), 66.93 (1C, C*H$_{carborane}$*), 64.82 (1C, C*H$_{carborane}$*), 64.10 (2C, NCH$_2$CH$_2$*CH$_2$*OSi), 62.28 (2C, BOCH$_2$CH$_2$*CH$_2$*O), 61.00 (2C, HOC*H$_2$*CH$_2$C$_{carb}$), 57.17 (2C, NC*H$_2$*CH$_2$CH$_2$O), 56.24 (2C, N(*CH$_3$*)$_2$), 53.22 (2C, C$_{carborane}$), 45.17 (2C, HOCH$_2$*CH$_2$*C$_{carb}$), 36.89 (2C, NCH$_2$*CH$_2$*CH$_2$O), 27.28 (2C, BOCH$_2$*CH$_2$*CH$_2$O), 26.98 (6C, OSi(CH$_3$)$_2$C(*CH$_3$*)$_3$), 26.77 (6C, OSi(CH$_3$)$_2$C(*CH$_3$*)$_3$), 19.53 (2C, OSi(CH$_3$)$_2$*C*(CH$_3$)$_3$), 19.39 (2C, OSi(CH$_3$)$_2$*C*(CH$_3$)$_3$), −4.38 (4C, OSi(*CH$_3$*)$_2$C(CH$_3$)$_3$), and −4.73 (4C, OSi(*CH$_3$*)$_2$C(CH$_3$)$_3$). ^{11}**B{^{1}H} NMR** (120 MHz, CD$_3$CN) δ: 29.65, 25.25, 24.28, 23.33, 21.98 (in ratio: 3:1.5:1:1:10), 31.27 to 19.66 (m, overlapped diastereoizomeric B$^{8,8'}$), −2.63 to −13.75 (m, overlapped diastereoizomeric, B$^{10,10',9,9',12,12',4,4',7,7'}$), −14.09 to −21.65 (m, overlapped diastereoizomeric B$^{5,5',11,11'}$), and −22.11 to −29.38 (m, overlapped diastereoizomeric B$^{6,6'}$); ^{11}**B NMR** (120 MHz, CD$_3$CN) δ: 30.85–20.54 (m, overlapped diastereoizomeric B$^{8,8'}$), −2.16 to −13.03 (m, overlapped diastereoizomeric, B$^{10,10',9,9',12,12',4,4',7,7'}$), −13.91 to −21.45 (m, overlapped diastereoizomeric, B$^{5,5',11,11'}$), and −21.52 to −27.10 (m, overlapped diastereoizomeric, B$^{6,6'}$); **ESI-MS** (*m*/*z*): found 788.58 (calc. for C$_{26}$B$_{18}$H$_{70}$O$_6$Si$_2$Co$_1$ 788.53).

Synthesis of 8,8′-bridged [8,8′-O$_2$P(O)H-3,3′-Co(1,2-C$_2$B$_9$H$_{10}$)$_2$] HNEt$_3$ H-phosphonate (12). Imidazole (0.51 g, 7.5 mmol) was dissolved in minimum ammount of anhydrous acetonitrile. The solvent was evaporated under reduced pressure, and the procedure was repeated twice. After drying under vacuum for 1.5 h, imidazole was redissolved in 16 mL of anhydrous THF and the solution was cooled to −70 °C in a dry ice/isopropanol bath under argon atmosphere. PCl$_3$ (210 µL, 2.40 mmol) was added dropwise followed by Et$_3$N (1 mL, 7.17 mmol) mixed with 1 mL of anhydrous THF. The entire mixture was stirred at −70 °C for 15 min and then solution of 8,8′-dihydroxy-bis(1,2-dicarbollido)-3-cobalt(1-)ate HNEt$_3$ (**2**) (320 mg, 0.69 mmol) in 13 mL of THF was added dropwise. After further 30 min, the reaction mixture was removed from cooling bath and allowed to warm to room temperature. After another 1 h, the reaction was quenched with 30 mL of water and extracted four times with 40 mL of diethyl ether (Et$_2$O). The combined ether extracts were dried with MgSO$_4$ and the solvent was evaporated. The resultant solid crude product was dryed under vacuum and then purified by silica gel (230–400 mesh) column chromatography using CH$_3$CN:CHCl$_3$ 1:4 as eluting solvent system. **Yield:** 320 mg (91%); **TLC** (CH$_3$CN:CHCl$_3$ 1:2); **R$_f$:** 0.35; **^{1}H NMR** (500 MHz, CD$_3$CN) δ: 7.47, 6.07 (1H, *P-H*), 3.70 (s, 4H, *CH$_{carborane}$*), 3.10 (m, 6H, N*CH$_2$*CH$_3$), and 1.24 (t, 9H, NCH$_2$*CH$_3$*); **^{13}C{^{1}H} NMR** (125 MHz, CD$_3$CN) δ: 47.96 (4C, C$_{carb}$), 47.74 (3C, N*CH$_2$*CH$_3$), and 9.24 (3C, NCH$_2$*CH$_3$*); **^{11}B{^{1}H} NMR** (120 MHz, CD$_3$CN) δ: 23.02 (s, 2B, B$^{8,8′}$), −2.83 (s, 2B, B$^{10,10′}$), −5.70 (s, 4B, B$^{9,9′,12,12′}$), −7.94 (s, 2B, B$^{4,4′}$), −8.74 (s, 2B, B$^{7,7′}$), −18.89 (s, 4B, B$^{5,5′,11,11′}$), and −27.85 (s, 2B, B$^{6,6′}$); **^{11}B NMR** (125 MHz, CD$_3$CN) δ: 23.02 (s, 2B, B$^{8,8′}$), −2.83 (d, 2B, B$^{10,10′}$), −5.70 (d, 4B, B$^{9,9′,12,12′}$), −8.36 (t, 4B, B$^{4,4′,7,7′}$), −18.91 (d, 4B, B$^{5,5′,11,11′}$), and −27.84 (d, 2B, B$^{6,6′}$); **^{31}P{^{1}H} NMR** (202 MHz, CD$_3$CN) δ: −3.01 (s, P-H); **^{31}P NMR** (202 MHz, CD$_3$CN) δ: −3.00 (d, P-H); **ATR-IR (cm^{-1}):** 3621, 3029, 2993, 2544, 1609, 1474, 1446, 1393, 1218, 1152, 1137, 1094, 1025, 993, 981, 920, 903, 871, 849, 787, 743, 691, and 666. **UV-Vis** λ_{max} (nm): 297 and 445. **ESI-MS** (*m/z*): found 402.24 (calc. for C$_4$H$_{21}$O$_3$B$_{18}$P$_1$Co$_1$: 401.71).

Synthesis of 8,8′-bridged [8,8′-O$_2$P(O)SH-3,3′-Co(1,2-C$_2$B$_9$H$_{10}$)$_2$] HDBU phosphorothioate (**13**). H-phosphonate acid ester **12** (130 mg, 0.26 mmol) was dissolved in anhydrous MeOH (6.5 mL). The solution was added under argon atmosphere to S$_8$ (85 mg, 2.6 mmol). Then, 1,8-Diazabicyclo(5.4.0)undec-7-en (DBU) (160 µL, 1.05 mmol) was added, and the mixture was stirred for 96 h at room temperature. Subsequently, solvent was evaporated under reduced pressure. The crude product was dissolved in CH$_3$CN and then purified via silica gel column chromatography using a gradient of CH$_3$CN:CHCl$_3$ from 1:4 to 1:1 as eluting solvent system. Finally, product **13** was eluted from the column using 100% MeOH as eluent. **Yield:** 105 mg (70%); **TLC** (CH$_3$CN:CHCl$_3$ 2:1) **R$_f$:** 0.5; **^{1}H NMR** (500 MHz, CD$_3$CN) δ: 9.14 (s, 1H, *NH*), 3.58 (s, 4H, *CH$_{carborane}$*), 3.50 (m, 2H, NH*CH$_2$*CH$_2$), 3.44 (t, *J* = 5.9 Hz, 2H, CH$_2$N*CH$_2$*CH$_2$CH$_2$NH), 3.31 (s, 2H, *CH$_2$*NCH$_2$CH$_2$*CH$_2$*NH), 2.69 (dd, *J* = 6.6, 3.5 Hz, 2H, NHC*CH$_2$*), 1.97 (dd, *J* = 6.6 Hz, 2H, CH$_2$NCH$_2$*CH$_2$*CH$_2$NH), 1.72 (m, 4H, NHCCH$_2$*CH$_2$*CH$_2$), and 1.65 (dt, *J* = 14.9, 5.2 Hz, 2H, NHCCH$_2$CH$_2$CH$_2$*CH$_2$*); **^{13}C{^{1}H} NMR** (125 MHz, CD$_3$CN) δ: 166.94 (NH*C*CH$_2$), 54.98 (*CH$_2$*NCH$_2$CH$_2$CH$_2$NH), 49.26 (CH$_2$N*CH$_2$*CH$_2$CH$_2$NH), 46.76 (C$_{carb}$), 46.63 (C$_{carb}$), 39.04 (NH*CH$_2$*), 33.42 (NHC*CH$_2$*), 29.47 (NHCCH$_2$*CH$_2$*CH$_2$), 27.05 (NHCCH$_2$CH$_2$*CH$_2$*), 24.52 (NHCCH$_2$CH$_2$*CH$_2$*), and 19.96 (CH$_2$NCH$_2$*CH$_2$*CH$_2$NH); **^{11}B{^{1}H} NMR** (120 MHz, CD$_3$CN) δ: 23.46 (s, 2B, B$^{8,8′}$), −3.54 (s, 2B, B$^{10,10′}$), −5.72 (s, 4B, B$^{9,9′,12,12′}$), −8.73 (s, 4B, B$^{4,4′,7,7′}$), −19.49 (s, 4B, B$^{5,5′,11,11′}$), and −28.25 (s, 2B, B$^{6,6′}$); **^{11}B NMR** (120 MHz, CD$_3$CN) δ: 23.46 (s, 2B, B$^{8,8′}$), −3.50 (d, 2B, B$^{10,10′}$), −5.74 (d, 4B, B$^{9,9′,12,12′}$), −8.71 (d, 4B, B$^{4,4′,7,7′}$), −19.53 (d, 4B, B$^{5,5′,11,11′}$), and −28.34 (d, 2B, B$^{6,6′}$); **^{31}P{^{1}H}NMR** (202 MHz, CD$_3$CN) δ: 48.63 (s); **^{31}P NMR** (202 MHz, CD$_3$CN) δ: 48.62 (s); **ATR-IR (cm^{-1}):** 3383 (ν OH), 3223 (ν NH), 3091 (ν NH), 3026, 2926 (ν CH), 2856 (ν CH), 2799 (ν CH), 2545 (ν BH), 1725, 1640, 1607, 1465, 1444, 1363, 1321, 1292, 1205, 1157, 1103, 1076, 978, 936, 910, 887, 836, 747, and 690. **UV-Vis** λ_{max} (nm) 215, 296, and 450. **ESI-MS** (*m/z*): found: 434.20 (calc. for C$_4$H$_{21}$O$_3$B$_{18}$P$_1$S$_1$Co$_1$ 433.78).

Synthesis of 8,8′-bridged [8,8′-O$_2$P(O)S(CH$_2$)$_n$OCPh$_3$-3,3′-Co(1,2-C$_2$B$_9$H$_{10}$)$_2$] HNEt$_3$ S-alkylated phosphorothioates **15** *and* **16**. [8,8′-O$_2$P(O)SH-3,3′-Co(1,2-C$_2$B$_9$H$_{10}$)$_2$] HDBU (**13**) (13 mg, 0.02 mmol) was dissolved in acetone (0.520 mL); then, Et$_3$N (65 µL, 0.46 mmol) was added under stirring at room temperature. The mixture was heated to 60 °C in an oil bath; then, alkylating agent **3** or **14** (0.04 mmol, dissolved in 130 µL of CH$_2$Cl$_2$) was added. The reaction mixture was maintained overnight at 60 °C with stirring; then, it was cooled to room temperature, and solvents were evaporated under reduced pressure. The residue was dispersed in CH$_2$Cl$_2$ and then filtered and the solution was loaded on silica gel column prepared in CH$_2$Cl$_2$. Chromatography was performed using a gradient of MeOH in CH$_2$Cl$_2$ from 0 to 3% MeOH. (**15**): **Yield:** 17 mg (90%); **TLC** (CH$_3$CN:CHCl$_3$ 1:2): **R$_f$:** 0.71;

^{1}H NMR (500 MHz, CD$_3$CN): δ 7.41 (d, *J* = 7.5 Hz, 6H, *H$_{arom}$*), 7.31 (t, *J* = 7.6 Hz, 6H, *H$_{arom}$*), 7.23 (t, *J* = 7.2 Hz, 3H, *H$_{arom}$*), 3.67 (s, 4H, *CH$_{carborane}$*), 3.07 (q, *J* = 7.3 Hz, 6H, N*CH$_2$*CH$_3$), 3.00 (t, *J* = 6.0 Hz, 2H SCH$_2$CH$_2$CH$_2$*CH$_2$*OTr), 2.78 (dt, 2H, S*CH$_2$*CH$_2$CH$_2$CH$_2$OTr), 1.72 (m, *J* = 7.1 Hz, 2H SCH$_2$CH$_2$*CH$_2$*CH$_2$OTr), 1.64 (m, *J* = 6.9 Hz, 2H, SCH$_2$*CH$_2$*CH$_2$CH$_2$Otr), and 1.21 (t, *J* = 7.3 Hz, 9H, NCH$_2$*CH$_3$*); **^{13}C{^{1}H} NMR** (126 MHz, CD$_3$CN): δ 145.55 (3C, *aromatic$_{trityl}$*), 129.54 (6C, *aromatic$_{trityl}$*), 128.87 (6C, *aromatic$_{trityl}$*), 128.01 (3C, *aromatic$_{trityl}$*), 87.30 (1C, O*C(Ph)$_3$*), 63.90 (1C, SCH$_2$CH$_2$CH$_2$*CH$_2$*OTr), 47.76 (overlapped 3C, HN*CH$_2$*CH$_3$, 4C, *CH$_{carborane}$*), 31.20 (1C, S*CH$_2$*CH$_2$CH$_2$CH$_2$OTr), 29.75 (1C, SCH$_2$CH$_2$*CH$_2$*CH$_2$OTr), 28.84 (1C, SCH$_2$*CH$_2$*CH$_2$CH$_2$OTr), and 9.29 (1C, NCH$_2$*CH$_3$*); **^{11}B{^{1}H} NMR** (160 MHz, CD$_3$CN): δ 23.08 (s, 2B, B$^{8,8'}$), −2.86 (s, 2B, B$^{10,10'}$), −5.54 (s, 4B, B$^{9,9',12,12'}$), −8.36 (s, 4B, B$^{4,4',7,7'}$), −18.87 (s, 4B, B$^{5,5',11,11'}$), and −27.70 (s, 2B, B$^{6,6'}$); **^{11}B NMR** (160 MHz, CD$_3$CN): δ 23.06 (s, 2B, B$^{8,8'}$), −2.87 (d, 2B, B$^{10,10'}$), −5.57 (d, 4B, B$^{9,9',12,12'}$), −8.16 (d, 4B, B$^{4,4',7,7'}$), −18.92 (d, 4B, B$^{5,5',11,11'}$), and −27.63 (d, 2B, B$^{6,6'}$); **^{31}P NMR{^{1}H}** (202 MHz, CD$_3$CN): δ 16.72 (s), 11.37 (s); **^{31}P NMR** (202 MHz, CD$_3$CN): δ 16.72 (t) and 11.37 (t); **ATR-IR (cm^{-1})** 3032 (ν CH aromatic), 2987 (ν CH aliphatic), 2925 (ν CH aliphatic), 2851 (ν CH aliphatic), 2681, 2566 (ν BH), 1727, 1595, 1474, 1447, 1392, 1264, 1200, 1134, 1103, 1068, 1032, 1016, 980, 941, 917, 901, 849, 735 (aromatic CH bending), and 705 (aromatic CH bending). **UV-Vis** λ$_{max}$ (nm) 196, 299, and 440. **ESI-MS** (*m*/*z*): found: 748.37 (calc. for C$_{27}$B$_{18}$H$_{43}$O$_4$P$_1$S$_1$Co$_1$: 748.24).

(16): Yield: 16 mg (86%); **TLC** (CH$_3$CN:CHCl$_3$ 1:2) **Rf:** 0.69; **^{1}H NMR** (500 MHz, CD$_3$CN) δ: 7.43 (m, 6H, *H$_{arom}$*), 7.33 (m, 6H, *H$_{arom}$*), 7.25 (m, 3H, *H$_{arom}$*), 3.67 (s, 4H, *CH$_{carborane}$*), 3.10 (t, *J* = 6.0 Hz, 2H SCH$_2$CH$_2$*CH$_2$*OTr), 3.08 (q, *J* = 7.3 Hz, 6H, N*CH$_2$*CH$_3$), 2.97 (dt, *J* = 14.9, 7.3 Hz, 2H, PS*CH$_2$*CH$_2$CH$_2$OTr), 2.58 (s, 1H, *CH$_3$OH*), 1.27 (t, *J* = 5.4 Hz, 2H PSCH$_2$*CH$_2$*CH$_2$OTr), 1.22 (t, *J* = 7.3 Hz, 9H, NCH$_2$*CH$_3$*), and 1.18 (s, 3H, *CH$_3$OH*); **^{13}C{^{1}H} NMR** (126 MHz, CD$_3$CN) δ: 145.27 (3C, *aromatic$_{trityl}$*), 129.45 (6C, *aromatic$_{trityl}$*), 128.78 (6C, *aromatic$_{trityl}$*), 127.94 (3C, *aromatic$_{trityl}$*), 87.25 (O*C(Ph)$_3$*), 62.71 (PSCH$_2$CH$_2$*CH$_2$*OTr), 55.13 (4C, *CH$_{carb}$*), 47.64 (HN*CH$_2$*CH$_3$), 47.59, 32.14, 31.98 (d, *J* = 6.1 Hz), 30.29, 29.66, 28.41 (d, *J* = 3.8 Hz), and 9.17 (HNCH$_2$*CH$_3$*); **^{11}B{^{1}H} NMR** (160 MHz, CD$_3$CN): δ (ppm) 23.08 (s, 2B, B$^{8,8'}$), −2.86 (s, 2B, B$^{10,10'}$), −5.53 (s, 4B, B$^{9,9',12,12'}$), −8.36 (s, 4B, B$^{4,4',7,7'}$), −18.87 (s, 4B, B$^{5,5',11,11'}$), and −27.70 (s, 2B, B$^{6,6'}$); **^{11}B NMR** (160 MHz, CD$_3$CN): δ 23.07 (s, 2B, B$^{8,8'}$), −2.92 (d, 2B, B$^{10,10'}$), −5.60 (d, 4B, B$^{9,9',12,12'}$), −8.38 (d, 4B, B$^{4,4',7,7'}$), −18.98 (d, 4B, B$^{5,5',7,7'}$), and −27.88 (d, 2B, B$^{6,6'}$); **^{31}P{^{1}H} NMR** (202 MHz, CD$_3$CN) δ: 16.54 (s); **^{31}P NMR** (202 MHz, CD$_3$CN) δ: 16.54 (t); **ATR-IR (cm^{-1})** 3032 (ν CH aromatic), 2987 (ν CH aliphatic), 2925 (ν CH aliphatic), 2851 (ν CH aliphatic), 2684, 2567 (ν BH), 1699, 1596, 1474, 1447, 1392, 1265, 1200, 1135, 1104, 1066, 1032, 1016, 980, 941, 917, 902, 849, 735 (aromatic CH bending), and 705 (aromatic CH bending). **UV-Vis** λ$_{max}$ (nm) 194, 299, and 436. **ESI-MS** (*m*/*z*): 734.35 (calc. for C$_{26}$B$_{18}$H$_{41}$O$_4$P$_1$S$_1$Co$_1$ 734.17).

*Synthesis of 4-[1,3-bis(trityloxy)propan-2-yl-oxy]butyl-4-methylbenzenesulfonate (**20**).* The reaction was performed under argon atmosphere in anhydrous conditions. 1,3-Bis(trityloxy) propan-2-ol (**18**) (2.35 g, 4.07 mmol) was dissolved in 18 mL of anhydrous DMF; then, NaH$_{60\%}$ (195 mg, 4.87 mmol) was added. After stirring for 15 min, 1,4-bis(*p*-toluenesulfonyloxy) butane [40] (4.23 g, 10.63 mmol), dissolved in 18 mL of DMF, was added. The reaction mixture was stirred for another 2 h at room temperature and was then cooled in ice bath; subsequently, an excess of NaH was centrifuged. The supernatant was poured into a cooled 40 mL volume of phosphate buffer. The mixture was extracted with AcOEt (4x 100 mL). Organic extracts were combined, washed with H$_2$O, and dried over MgSO$_4$. Solvents were evaporated under reduced pressure. The crude product was purified via silica gel column chromatography using a gradient of AcOEt in hexane from 0% to 10% as eluting solvent system. **Yield:** 883 mg (27%); **TLC** (Hexane:AcOEt 2:1) **R$_f$:** 0.55; **^{1}H NMR** (600 MHz, CDCl$_3$) δ: 7.78 (d, 2H, *H$_{arom}$*), 7.41 (d, 12H, *H$_{arom}$*), 7.27 (m, 20H, *H$_{arom}$*), 4.04 (t, 2H, OCH$_2$CH$_2$CH$_2$*CH$_2$*OSO$_2$), 3.55 (p, 1H, TrOCH$_2$*CHO*CH$_2$OTr), 3.48 (t, 2H, O*CH$_2$*CH$_2$CH$_2$CH$_2$OSO$_2$), 3.23 (ddd, 4H, TrO*CH$_2$*CH), 2.43 (s, 3H, *CH$_{3tosyl}$*), 1.75 (m, 2H, OCH$_2$CH$_2$*CH$_2$*CH$_2$OSO$_2$), and 1.58 (m, 2H, OCH$_2$*CH$_2$*CH$_2$CH$_2$OSO$_2$); **^{13}C{^{1}H} NMR** (125 MHz, CD$_3$CN) δ: 144.62 (1C, *aromatic$_{tosyl}$*), 144.05 (6C, *aromatic$_{trityl}$*), 133.19 (1C, *aromatic$_{tosyl}$*), 129.81 (2C, *aromatic$_{tosyl}$*), 128.73 (12C,

$aromatic_{trityl}$), 127.88 (2C, $aromatic_{tosyl}$), 127.76 (12C, $aromatic_{trityl}$), 126.91 (6C, $aromatic_{trityl}$), 86.52 (2C, OC(Ph)$_3$), 78.59 (1C, TrOCH$_2$$CHOCH_2$OTr), 70.51 (1C, OCH$_2CH_2CH_2$$CH_2OSO_2$-), 69.47 (1C, O$CH_2CH_2CH_2CH_2OSO_2$-), 63.39 (2C, TrO$CH_2$CH), 26.10 (1C, OCH$_2CH_2$$CH_2CH_2OSO_2$), 25.86 (1C, OCH$_2$$CH_2CH_2CH_2OSO_2$), and 21.64 (1C, CH$_3$^{tosyl}); **ATR-IR (cm^{-1}):** 3054, 3018, 2946, 2929, 2869, 1978, 1732, 1596, 1488, 1447, 1352, 1304, 1218, 1187, 1172, 1122, 1094, 1065, 1029, 992, 950, 922, 841, 811, 768, 745, and 699; **UV-Vis** λ_{max} (nm): 198, 229, and 260; **ESI-MS** (*m/z*): found 825.32 [M + Na]$^-$, 841.29 [M + K]$^-$, (calc. for C$_{52}$H$_{50}$O$_6$S$_1$ 803.01).

Synthesis of 8,8′-bridged {8,8′-O$_2$P(O)S[(CH$_2$)$_4$OCH(CH$_2$OCPh$_3$)$_2$]-3,3′-Co(1,2-C$_2$B$_9$H$_{10}$)$_2$} HNEt$_3$ (21). [8,8′-O$_2$P(O)SH-3,3′-Co(1,2-C$_2$B$_9$H$_{10}$)$_2$] HDBU (**13**) (440 mg, 0.75 mmol) was dissolved in 20 mL of anhydrous acetone; then, anhydrous Et$_3$N (2.15 mL, 15.42 mmol) was added to the resultant solution under stirring at room temperature. The mixture was heated to 60 °C; then, 4-[1,3-bis(trityloxy)propan-2-yl-oxy]butyl-4-methylbenzenesulfonate (**20**) (901 mg, 1.12 mmol), which was dissolved in 20 mL of anhydrous AcOEt, was added dropwise. After stirring overnight at 60 °C, the mixture was cooled, and the solvents were evaporated under reduced pressure. The residue was dispersed in CH$_2$Cl$_2$, filtered, and the filtrate was loaded into the silica gel column prepared in CH$_2$Cl$_2$. Chromatography was performed using a gradient of CH$_3$OH in CH$_2$Cl$_2$ from 0 to 3% of CH$_3$OH. **Yield:** 560 mg (64%); **TLC** (CH$_3$CN:CHCl$_3$ 1:2), **R$_f$:** 0.60; **^{1}H NMR** (600 MHz, CD$_3$CN) δ: 7.41 (d, 12H, $H_{aromatic}$), 7.31 (m, 18H, $H_{aromatic}$), 3.70 (s, 4H, $CH_{carborane}$), 3.60 (p, 1H, TrOCH$_2$$CHOCH_2$OTr), 3.45 (t, 2H, PSCH$_2CH_2CH_2$$CH_2$O), 3.19 (ddd, 4H, CHO$CH_2$OTr), 3,10 (q, 6H, HN$CH_2CH_3$), 2.85 (dt, 2H, PS$CH_2CH_2CH_2CH_2$O), 1.71 (m, 2H, PSCH$_2CH_2$$CH_2CH_2$O), 1.61 (m, 2H, PSCH$_2$$CH_2CH_2CH_2$O), and 1.24 (t, 9H, NCH$_2$$CH_3$); **^{13}C{^{1}H} NMR** (125 MHz, CD$_3$CN) δ: 144.73 (6C, $aromatic_{trityl}$), 129.10 (12C, $aromatic_{trityl}$), 128.41 (12C, $aromatic_{trityl}$), 127.60 (6C, $aromatic_{trityl}$), 86.89 (2C, OC(Ph)$_3$), 78.47 (1C, TrOCH$_2$$CHOCH_2$OTr), 69.92 (1C, PSCH$_2CH_2CH_2$$CH_2$O), 63.66 (2C, CHO$CH_2$OTr), 47.28 (3C, HN$CH_2CH_3$), 47.18 (4C, $CH_{carborane}$), 31.89 (1C, PSCH_2CH$_2$CH$_2$CH$_2$O), 22.93 (1C, PSCH$_2$CH$_2$$CH_2CH_2$O), 13.96 (1C, PSCH$_2$$CH_2CH_2CH_2$O), and 8.80 (3C, HNCH$_2$$CH_3$); **^{11}B{^{1}H} NMR** (120 MHz, CD$_3$CN) δ: 23.02 (s, 2B, B$^{8,8'}$), −3.04 (s, 2B, B$^{10,10'}$), −5.67 (s, 4B, B$^{9,9',12,12'}$), −8.55 (s, 4B, B$^{4,4',7,7'}$), −19.07 (s, 4B, B$^{5,5',11,11'}$), and −27.99 (s, 2B, B$^{6,6'}$); **^{11}B NMR** (120 MHz, CD$_3$CN) δ: 23.02 (s, 2B, B$^{8,8'}$), −2.98 (d, 2B, B$^{10,10'}$), −5.64 (d, 4B, B$^{9,9',12,12'}$), −8.19 (d, 4B, B$^{4,4',7,7'}$), −19.05 (d, 4B, B$^{5,5',11,11'}$), and −27.99 (d, 2B, B$^{6,6'}$); **^{31}P{^{1}H}NMR** (202 MHz, CD$_3$CN) δ (ppm) 16.49 (s); **^{31}P NMR** (202 MHz, CD$_3$CN) δ: 16.48 (t); **ATR-IR (cm^{-1}):** 3031, 2929, 2870, 2564, 2564, 2161, 1978, 1644, 1595, 1489, 1447, 1322, 1202, 1134, 1096, 1031, 940, 899, 848, 763, 747, and 697; **UV-Vis** λ_{max} (nm): 196, 233, 299, and 453; **MS (ESI)** (*m/z*): found 1064.52 (calc. for C$_{49}$B$_{18}$H$_{63}$O$_6$P$_1$S$_1$Co$_1$ 1064.59).

Synthesis of 8,8′-bridged {8,8′-O$_2$P(O)S[(CH$_2$)$_4$OCH(CH$_2$OCPh$_3$)$_2$]-3,3′-Co[1-(CH$_2$)$_2$OH-1,2-C$_2$B$_9$H$_{10}$)](1′,2′-C$_2$B$_9$H$_{10}$)} HNEt$_3$ (22) and {8,8′-O$_2$P(O)S[(CH$_2$)$_4$OCH(CH$_2$OCPh$_3$)$_2$]-3,3′-Co[1-(CH$_2$)$_2$OH-1,2-C$_2$B$_9$H$_{10}$)] [1′-(CH$_2$)$_2$OH-1′,2′-C$_2$B$_9$H$_{10}$)]} HNEt$_3$ (23). Compound **21** (175 mg, 0.15 mmol) was dried via co-evaporation with anhydrous benzene and then kept under vacuum over P$_2$O$_5$ overnight. Then, it was dissolved in anhydrous DME (3 mL), and the solution was cooled in CO$_2$/isopropanol cooling bath. After 15 min, n-BuLi (140 μL, 1.6 M solution in hexane, 1.5 eq) was added, and the reaction mixture was stirred for 10 min. Afterwards, the cooling bath was removed, and the mixture was stirred for next 10 min. Then, the reaction mixture was cooled again in cooling bath and another portion of n-BuLi (140 μL) was added. After 15 min, ethylene oxide (200 μL, 2.9–3.1 M solution in THF)) was added, and the reaction was left overnight in cooling bath. Then, CH$_2$Cl$_2$ (5 mL) was added to the reaction mixture, the reaction was quenched via addition of water, and then the organic solution was washed three times with 5 mL portions of water. Organic layer was separated and dried over MgSO$_4$; then, solvents were evaporated. Crude product was purified, and mono- and bis-substituted products were separated via silica gel column chromatography using a gradient of MeOH in CH$_2$Cl$_2$ from 0 to 3% of MeOH. (**22**): **Yield:** 17 mg (10%); **TLC** (MeOH:CH$_2$Cl$_2$ 1:12.5): **R$_f$:** 0.13; **^{1}H NMR** (500 MHz, CD$_3$CN): δ (ppm) 7.41 (d, 12H, $H_{aromatic}$), 7.31 (m, 18H, $H_{aromatic}$), 3.90 (s, $CH_{carborane}$), 3.78 to 3.54 (m, overlapped, $CH_{carborane}$, HOCH_2CH$_2$C$_{carb}$,

TrOCH$_2$*CH*OCH$_2$OTr), 3.44 (t, 2H, PSCH$_2$CH$_2$CH$_2$*CH$_2$*O-), 3.19 (ddd, 4H, CHO*CH$_2$*OTr), 3.02 to 2.64 (m, overlapped, PS*CH$_2$*CH$_2$CH$_2$CH$_2$O-, HOCH$_2$*CH$_2$*C$_{carb}$), and 1.65 (m, 4H, overlapped, PSCH$_2$*CH$_2$CH$_2$*CH$_2$O); **^{11}B{^{1}H} NMR** (120 MHz, CD$_3$CN) δ (ppm) 25.37, 24.65, 23.69, 22.93 (in ratio 2:2:1:1) 26.6 to 21.37 (m, overlapped diastereoizomeric B$^{8,8'}$), 0.52 to −11.61 (m, overlapped diastereoizomeric, B$^{10,10',9,9',12,12',4,4',7,7'}$), −12.05 to −11.61 (d, overlapped diastereoizomeric B$^{5,5',11,11'}$), and −21.45 to −28.25 (s, overlapped diastereoizomeric B$^{6,6'}$); **^{11}B NMR** (120 MHz, CD$_3$CN) δ (ppm) 27.17 to 21.57 (m, overlapped diastereoizomeric B$^{8,8'}$), 2.85 to −11.77 (m, overlapped diastereoizomeric, B$^{10,10',9,9',12,12',4,4',7,7'}$), −11.82 to −21.00 (d, overlapped diastereoizomeric B$^{5,5',11,11'}$), and −21.10 to −29.92 (s, overlapped diastereoizomeric B$^{6,6'}$); **^{31}P{^{1}H}NMR** (202 MHz, CD$_3$CN) δ: 14.94, 14.58, 14.43, 14.12, and 13.49 (in ratio: 4:1:2:1.5:15); **^{31}P NMR** (202 MHz, CD$_3$CN) δ: 14.94 (t), 14.42 (t), 14.12 (t), 13.49 (t); **ATR-IR (cm^{-1}):** 3630, 3370, 3057, 3031, 2925, 2869, 2565, 2161, 1979, 1596, 1489, 1448, 1255, 1202, 1128, 1077, 1032, 985, 898, 871, 763, 746, and 699; **ESI-MS** (*m/z*): found: 1108.54 *m/z* (calc. for C$_{51}$B$_{18}$H$_{67}$O$_7$P$_1$S$_1$Co$_1$ 1108.64).

(23): **Yield:** 21 mg (12%); **TLC** (MeOH:CH$_2$Cl$_2$ 1:12.5): **R$_f$:** 0.27; **^{1}H NMR** (500 MHz, CD$_3$CN) δ: 7.40 (d, 12H, *H$_{aromatic}$*), 7.31 (m, 18H, *H$_{aromatic}$*), 3.84 to 3.51 (m, overlapped, *CH$_{carborane}$*, HO*CH$_2$*CH$_2$-, TrOCH$_2$*CH*OCH$_2$OTr,), 3.44 (t, 2H, PSCH$_2$CH$_2$CH$_2$*CH$_2$*O-), 3.18 (ddd, 4H, CHO*CH$_2$*OTr), 3.12 to 2.66 (m, overlapped, PS*CH$_2$*CH$_2$CH$_2$CH$_2$O, HOCH$_2$*CH$_2$*C$_{carborane'}$), and 1.65 to 1.55 (m, 4H, overllaped, PSCH$_2$*CH$_2$CH$_2$*CH$_2$O); **^{11}B{^{1}H} NMR** (120 MHz, CD$_3$CN) δ: 25.36, 24.41, 23.65, 22.80, 22.23 (in ratio 1:2:1.5:1:1), 26.59 to 20.45 (m, overlapped diastereoizomeric B$^{8,8'}$), 2.75 to −12.40 (m, overlapped diastereoizomeric, B$^{10,10',9,9',12,12',4,4',7,7'}$), and −12.34 to −24.64 (m overlapped diastereoizomeric B$^{5,5',11,11'6,6'}$); **^{11}B NMR** (120 MHz, CD$_3$CN) δ: 26.65 to 20.21 (m, overlapped diastereoizomeric B$^{8,8'}$), 2.32 to −12.88 (m, overlapped diastereoizomeric, B$^{10,10',9,9',12,12',4,4',7,7'}$), −12.84 to −25.84 (m overlapped diastereoizomeric B$^{5,5',11,11'6,6'}$); **^{31}P{^{1}H}NMR** (202 MHz, CD$_3$CN) δ: 15.00, 14.16, 14.02, 13.38 (in ratio 5:1:3:1); **^{31}P NMR** (202 MHz, CD$_3$CN) δ: 15.99 (t), 14.40 (s), 14.01 (t), and 13.38 (t); **ATR-IR (cm^{-1}):** 3566, 3357, 3056, 3027, 2920, 2889, 2857, 2565, 2166, 1596, 1489, 1448, 1291, 1255, 1201, 1120, 1078, 1032, 1001, 889, 871, 764, 746, and 699 **ESI-MS** (*m/z*): found: 1152.57 (calc. for C$_{53}$B$_{18}$H$_{71}$O$_8$P$_1$S$_1$Co 1152.69).

5. Conclusions

Although derivatives of metallocarboranes containing various simple substituents attached to the boron or carbon atoms of the complex carboranyl ligands are abundant, they do not usually allow for further chemical transformations. The adducts of some metallacarboranes and cyclic ethers are notable exceptions. In this work, we proposed methods for filling this gap, at least partially, by using extendable ligands.

The exploitation of icosahedral metallacarborane's immense potential in various fields of chemistry and technology requires the availability of convenient and versatile methods for their modification with various functional moieties and/or linkers of various types and lengths. Herein, we report a convenient approach to introducing extendible arms on 8,8'-dihydroxy cobalt bis(1,2-dicarbollide). This approach can be used to introduce different hetero-bifunctional electrophiles containing a protected hydroxyl function that allows for further modification.

Supplementary Materials: The following supporting information can be downloaded at: https://www.mdpi.com/article/10.3390/molecules28104118/s1. Supporting Information. Metallacarborane Synthons for Molecular Construction—Oligofunctionalization of Cobalt Bis(1,2-dicarbollide) on Boron and Carbon Atoms with Extendable Ligands. Figures S1–S91.

Author Contributions: Conceptualization, investigation, data curation, and writing—original draft preparation, K.Ś.; conceptualization, investigation, data curation, and writing—original draft preparation, C.S.; conceptualization, validation, formal analysis, investigation, data curation, writing—review and editing, supervision, project administration, and funding acquisition, Z.J.L. All authors have read and agreed to the published version of the manuscript.

Funding: This work was supported in part by the National Science Center, Poland, grant 2015/16/W/ST5/00413.

Institutional Review Board Statement: Not applicable.

Informed Consent Statement: Not applicable.

Data Availability Statement: Data is contained within the article or Supplementary Material.

Acknowledgments: The electrospray ionization (ESI) mass spectra for **10–13**, **15**, **16**, and **21–23**, were recorded using an Agilent 6546 LC/Q-TOF at the National Library of Chemical Compounds (NLCC) via a procedure established within the project POL-OPENSCREEN financed by the Ministry of Science and Higher Education (decision no. DIR/WK/2018/06 of 24 October 2018). The authors thank Dorota Borowiecka and Agata Kraj of NLCC for recording the mass spectra.

Conflicts of Interest: The authors declare no conflict of interest.

Sample Availability: Samples of the new compounds synthesized in this work are available from the authors.

References

1. Dash, B.P.; Satapathy, R.; Maguireband, J.A.; Hosmane, N.S. Polyhedral boron clusters in materials science. *New J. Chem.* **2011**, *35*, 1955–1972. [CrossRef]
2. Nuñez, R.; Romero, I.; Teixidor, F.; Viñas, C. Icosahedral boron clusters: A perfect tool for the enhancement of polymer features. *Chem. Rev.* **2016**, *45*, 5137–5434. [CrossRef] [PubMed]
3. Leśnikowski, Z.J. Challenges and opportunities for the application of boron clusters in drug design. *J. Med. Chem.* **2016**, *59*, 7738–7758. [CrossRef]
4. Messner, K.; Vuong, B.; Tranmer, G.K. The boron advantage: The evolution and diversification of boron's applications in medicinal chemistry. *Pharmaceuticals* **2022**, *15*, 264. [CrossRef] [PubMed]
5. Marfavi, A.; Kavianpour, P.; Rendina, L.M. Carboranes in drug discovery, chemical biology and molecular imaging. *Nat. Rev. Chem.* **2022**, *6*, 486–504. [CrossRef] [PubMed]
6. Olejniczak, A.B.; Sut, A.; Wróblewski, A.E.; Leśnikowski, Z.J. Infrared spectroscopy of nucleoside and DNA-oligonucleotide conjugates labeled with carborane or metallacarborane group. *Vibr. Spectrosc.* **2005**, *39*, 177–185. [CrossRef]
7. Olejniczak, A.B.; Lesnikowski, Z.J. Boron clusters as redox labels for nucleosides and nucleic acids. In *Handbook of Boron Science*; Hosmane, N., Eagling, R., Eds.; World Scientific: Hackensack, NJ, USA, 2018; Volume 4, pp. 1–13.
8. Kodr, D.; Yenice, C.P.; Simonova, A.; Saftić, D.P.; Pohl, R.; Sýkorová, V.; Ortiz, M.; Havran, L.; Fojta, M.; Lesnikowski, Z.J.; et al. Carborane—Or metallacarborane-linked nucleotides for redox labelling. Orthogonal multipotential coding of all four DNA bases for electrochemical analysis and sequencing. *J. Am. Chem. Soc.* **2021**, *143*, 7124–7134. [CrossRef]
9. Janczak, S.; Olejniczak, A.; Balabańska, S.; Chmielewski, M.K.; Lupu, M.; Viñas, C.; Lesnikowski, Z.J. The boron clusters as a platform for new materials: Synthesis of functionalized o-carborane ($C_2B_{10}H_{12}$) derivatives incorporating DNA fragments. *Chem. Eur. J.* **2015**, *21*, 15118–15122. [CrossRef]
10. Kaniowski, D.; Ebenryter-Olbinska, K.; Kulik, K.; Janczak, S.; Maciaszek, A.; Bednarska-Szczepaniak, K.; Nawrot, B.; Lesnikowski, Z.J. Boron clusters as a platform for new materials: Composites of nucleic acids and oligofunctionalized carboranes ($C_2B_{10}H_{12}$) and their assembly into functional nanoparticles. *Nanoscale* **2020**, *12*, 103–114. [CrossRef]
11. Sivaev, I.B. Ferrocene and transition metal bis(dicarbollides) as platform for design of rotatory molecular switches. *Molecules* **2017**, *22*, 2201. [CrossRef]
12. Dash, B.P.; Satapathy, R.; Swain, B.R.; Mahanta, C.S.; Jena, B.B.; Hosmane, N.S. Cobalt bis(dicarbollide) anion and its derivatives. *J. Oranomet. Chem.* **2017**, *849*, 170–194. [CrossRef]
13. Fink, K.; Boratyński, J.; Paprocka, M.; Goszczyński, T.M. Metallacarboranes as a tool for enhancing the activity of therapeutic peptides. *Ann. N. Y. Acad. Sci.* **2019**, *1457*, 128–141. [CrossRef] [PubMed]
14. Gozzi, M.; Schwarze, B.; Hey-Hawkins, E. Preparing (metalla)carboranes for nanomedicine. *ChemMedChem* **2021**, *16*, 1533–1565. [CrossRef] [PubMed]
15. Barba-bon, A.; Salluce, G.; Lostalé-Seijo, I.; Assaf, K.I.; Henning, A.; Montenegro, J.; Nau, W.M. Boron clusters as broadband membrane carriers. *Nature* **2022**, *603*, 637–642. [CrossRef] [PubMed]
16. Verdiá-Báguena, C.; Alcaraz, A.; Aguilella, V.M.; Cioran, A.M.; Tachikawa, S.; Nakamura, H.; Teixidor, F.; Viñas, C. Amphiphilic COSAN and I_2-COSAN crossing synthetic lipid membranes: Planar bilayers and liposomes. *Chem. Commun.* **2014**, *50*, 6700–6703. [CrossRef]
17. Tarrés, M.; Canetta, E.; Paul, E.; Forbes, J.; Azzouni, K.; Viñas, C.; Teixidor, F.; Harwood, A.J. Biological interaction with living cells of COSAN-based synthetic vesicles. *Sci. Rep.* **2015**, *5*, 7804. [CrossRef] [PubMed]
18. Avdeeva, V.V.; Garaev, T.M.; Malinina, E.A.; Zhizhin, K.Y.; Kuznetsov, N.T. Physiologically active compounds based on membranotropic cage carriers–derivatives of adamantane and polyhedral boron clusters. *Russ. J. Inorg. Chem.* **2022**, *67*, 28–47. [CrossRef]

19. Sardo, C.; Janczak, S.; Leśnikowski, Z.J. Unusual resistance of cobalt bis dicarbollide phosphate and phosphorothioate bridged esters towards alkaline hydrolysis: The "metallacarborane effect". *J. Organomet. Chem.* **2019**, *896*, 70–76. [CrossRef]
20. Grüner, B.; Švec, P.; Šícha, V.; Padělkova, Z. Direct and facile synthesis of carbon substituted alkylhydroxy derivatives of cobalt bis(1,2-dicarbollide), versatile building blocks for synthetic purposes. *Dalton Trans.* **2012**, *41*, 7498–7512. [CrossRef]
21. Nekvinda, J.; Švehla, J.; Císařová, I.; Grüner, B. Chemistry of cobalt bis(1,2-dicarbollide) ion; the synthesis of carbon substituted alkylamino derivatives from hydroxyalkyl derivatives via methylsulfonyl or p-toluenesulfonyl esters. *J. Organomet. Chem.* **2015**, *798*, 112–120. [CrossRef]
22. Plešek, J.; Grüner, B.; Báča, J.; Fusek, J. Syntheses of the B(8)-hydroxy- and B(8,8′)-dihydroxy-derivatives of the bis(1,2-dicarbollido)-3-cobalt(1-)ate ion by its reductive acetoxylation and hydroxylation: Molecular structure of $[8,8'-\mu-CH_3C(O)_2 < (1,2-C_2B_9 H_{10})_2-3-Co]_0$ zwitterion determined by X-ray diffraction analysis. *J. Organomet. Chem.* **2002**, *649*, 181–190.
23. Sivaev, I.; Bregadze, V.I. Chemistry of cobalt bis(dicarbollides). A review. *Collect. Czech. Chem. Commun.* **1999**, *64*, 783–805. [CrossRef]
24. Van den Berg, R.J.B.H.N.; Korevaar, C.G.N.; Herman, S.; Overkleeft, H.S.; van der Marel, G.A.; van Boom, J.H. Effective, high-yielding and stereospecific total synthesis of d-erythro-(2R,3S)-sphingosine from d-ribo-(2S,3S,4R)-phytosphingosine. *J. Org. Chem.* **2004**, *69*, 5699–5704. [CrossRef]
25. Sivaev, I.B.; Bregadze, V.I. Lewis acidity of boron compounds. *Coord. Chem. Rev.* **2014**, *270*, 75–88. [CrossRef]
26. Hesek, D.; Lee, M.; Noll, B.C.; Fisher, J.F.; Mobashery, S. Complications from dual roles of sodium hydride as a base and as a reducing Agent. *J. Org. Chem.* **2009**, *74*, 2567–2570. [CrossRef] [PubMed]
27. Planas, J.G.; Teixidor, F.; Viñas, C.; Light, M.E.; Hursthouse, M.B. Self-assembly of halogenated cobaltacarborane compounds: Boron-assisted C-H X-B hydrogen bonds? *Chem. Eur. J.* **2007**, *13*, 2493–2502. [CrossRef]
28. Stogniy, M.Y.; Suponitsky, K.Y.; Chizhov, A.O.; Sivaev, I.B.; Bregadze, V.I. Synthesis of 8-alkoxy and 8,8′-dialkoxy derivatives of cobaltbis(dicarbollide). *J. Organomet. Chem.* **2018**, *865*, 138–144. [CrossRef]
29. Stawinski, J.; Strömberg, R. Di- and oligonucleotide synthesis using H-phosphonate chemistry. In *Methods in Molecular Biology, Oligonucleotide Synthesis: Methods and Applications*; Herdewijn, P., Ed.; Humana Press Inc.: Totowa, NJ, USA, 2008; Volume 288, pp. 81–100.
30. Jones, D.J.; O'Leary, E.M.; O'Sullivan, T.P. Synthesis and application of phosphonothioates, phosphonodithioates, phosphorothioates, phosphinothioates and related compounds. *Tetrahedron Lett.* **2018**, *59*, 4279–4292. [CrossRef]
31. Hronowski, L.J.J.; Szarek, W.A.; Hay, G.W.; Krebs, A.; Depew, W.T. Synthesis and characterization of 1-O-β-lactosyl-(R,S)-glycerols and 1,3-di-O-β-lactosylglycerol. *Carbohydr. Res.* **1989**, *190*, 203–218. [CrossRef]
32. King, B.R. Three-dimensional aromaticity in polyhedral boranes and related molecules. *Chem. Rev.* **2001**, *101*, 1119–1152. [CrossRef]
33. Poater, J.; Viñas, C.; Bennour, I.; Escayola, S.; Solà, M.; Teixidor, F. Too persistent to give up: Aromaticity in boron clusters survives radical structural changes. *J. Am. Chem. Soc.* **2020**, *142*, 9396–9407. [CrossRef]
34. Fulcrand-El Kattan, G.; Lesnikowski, Z.J.; Yao, S.; Tanious, F.; Wilson, W.D.; Shinazi, R.F. Carboranyl oligonucleotides. 2. Synthesis and physicochemical properties of dodecathymidylate containing 5-(o-carboranyl-1-yl)-2′-O-deoxyuridine. *J. Am. Chem. Soc.* **1994**, *116*, 7494–7501.
35. Lesnikowski, Z.J.; Shi, J.; Schinazi, R.F. Nucleic acids and nucleosides containing carboranes. *J. Organomet. Chem.* **1999**, *581*, 156–169. [CrossRef]
36. Plešek, J. The age of chiral deltahedral borane derivatives. *Inorg. Chim. Acta* **1999**, *289*, 45–50. [CrossRef]
37. Grüner, B.; Plzàk, Z. High-performance liquid chromatographic separations of boron cluster compounds. *J. Chromatogr.* **1997**, *789*, 497–517. [CrossRef]
38. Horàkovà, H.; Grüner, B.; Vespalec, R. Emerging subject for chiral separation science: Cluster boron compounds. *Chirality* **2011**, *23*, 307–319. [CrossRef]
39. Van Swieten, P.F.; Rene Maehr, R.; van den Nieuwendijk, A.M.C.H.; Kessler, B.M.; Reich, M.; Wong, C.-S.; Kalbacher, H.; Leeuwenburgh, M.A.; Driessen, C.; van der Marel, G.A.; et al. Development of an isotope-coded activity-based probe for the quantitative profiling of cysteine proteases. *Bioorg. Chem. Med. Lett.* **2004**, *14*, 3131–3134. [CrossRef]
40. Martin, A.E.; Ford, T.M.; Bulkowski, J.E. Synthesis of selectively protected tri- and hexaamine macrocycles. *J. Org. Chem.* **1982**, *47*, 412–415. [CrossRef]

MDPI

Article

Synthesis and Structural Characterization of *p*-Carboranylamidine Derivatives [†]

Nicole Harmgarth [1], Phil Liebing [2], Volker Lorenz [1], Felix Engelhardt [1], Liane Hilfert [1], Sabine Busse [1], Rüdiger Goldhahn [3] and Frank T. Edelmann [3,*]

[1] Chemisches Institut, Otto-von-Guericke-Universität Magdeburg, Universitätsplatz 2, 39106 Magdeburg, Germany; volker.lorenz@ovgu.de (V.L.); liane.hilfert@ovgu.de (L.H.); sabine.busse@ovgu.de (S.B.)

[2] Institut für Anorganische und Analytische Chemie, Friedrich-Schiller-Universität Jena, Humboldtstr. 8, 07743 Jena, Germany; phil.liebing@uni-jena.de

[3] Institut für Physik, Otto-von-Guericke-Universität Magdeburg, Universitätsplatz 2, 39106 Magdeburg, Germany; ruediger.goldhahn@ovgu.de

* Correspondence: frank.edelmann@ovgu.de

[†] Dedicated to Professor John D. Kennedy on the occasion of his 80th birthday.

Abstract: In this contribution, the first amidinate and amidine derivatives of *p*-carborane are described. Double lithiation of *p*-carborane (**1**) with *n*-butyllithium followed by treatment with 1,3-diorganocarbodiimides, R–N=C=N–R (R = ^{i}Pr, Cy (= cyclohexyl)), in DME or THF afforded the new *p*-carboranylamidinate salts *p*-$C_2H_{10}B_{10}[C(N^iPr)_2Li(DME)]_2$ (**2**) and *p*-$C_2H_{10}B_{10}[C(NCy)_2Li(THF)_2]_2$ (**3**). Subsequent treatment of **2** and **3** with 2 equiv. of chlorotrimethylsilane (Me$_3$SiCl) provided the silylated neutral bis(amidine) derivatives *p*-$C_2H_{10}B_{10}[C\{^iPrN(SiMe_3)\}(=N^iPr)]_2$ (**4**) and *p*-$C_2H_{10}B_{10}[C\{CyN(SiMe_3)\}(=NCy)]_2$ (**5**). The new compounds **3** and **4** have been structurally characterized by single-crystal X-ray diffraction. The lithium carboranylamidinate **3** comprises a rare trigonal planar coordination geometry around the lithium ions.

Keywords: boron; carborane; *p*-carborane; lithiation; carboranylamidinate

Citation: Harmgarth, N.; Liebing, P.; Lorenz, V.; Engelhardt, F.; Hilfert, L.; Busse, S.; Goldhahn, R.; Edelmann, F.T. Synthesis and Structural Characterization of *p*-Carboranylamidine Derivatives. *Molecules* **2023**, *28*, 3837. https://doi.org/10.3390/molecules28093837

Academic Editors: Michael A. Beckett and Igor B. Sivaev

Received: 6 April 2023
Revised: 27 April 2023
Accepted: 27 April 2023
Published: 30 April 2023

1. Introduction

Icosahedral *closo*-carborane cage compounds of the general composition $C_2B_{10}H_{12}$ can be viewed as 3D molecular analogs of benzene [1]. They are of high scientific and technological interest due to a variety of practical applications, e.g., in materials science [2–13], homogeneous catalysis [14–21], and medicinal chemistry [22–29]. Moreover, carborane derivatives are widely employed in coordination chemistry and as building blocks in supramolecular, bio-inorganic, and organometallic chemistry [30–34]. Depending on the position of the two carbon atoms in the carborane cage, three isomers can be distinguished. As shown in Figure 1, these are *ortho*-carborane (*o*-carborane, 1,2-$C_2B_{10}H_{12}$), *meta*-carborane (*m*-carborane, 1,7-$C_2B_{10}H_{12}$), and *para*-carborane (*p*-carborane, 1,12-$C_2B_{10}H_{12}$) [35]. Most readily available among these is *o*-carborane, while the other two are made by thermal rearrangement of the *ortho*-isomer. This is why *p*-carborane is the most expensive precursor and its derivative chemistry is the least studied [35].

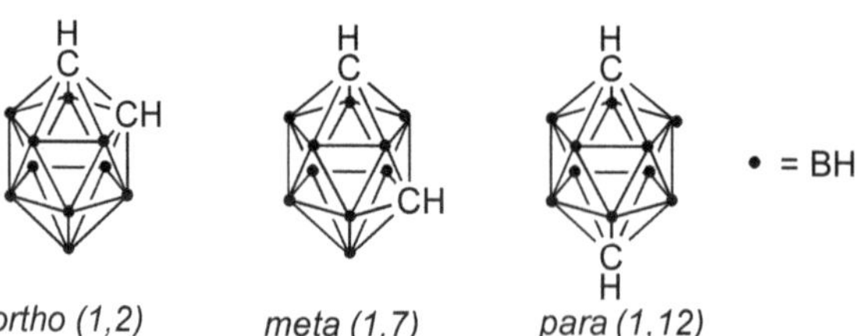

Figure 1. Three isomers of the icosahedral *closo*-carborane cage $C_2B_{10}H_{12}$.

In 2010, we reported a novel type of *o*-carborane-based *N*-chelating ligands which were named carboranylamidinates. These were obtained in the form of their lithium derivatives by in situ lithiation of *o*-carborane using *n*-butyllithium followed by treatment with one equivalent of 1,3-diorganocarbodiimides, R–N=C=N–R (R = ^{i}Pr, Cy (= cyclohexyl)). As illustrated in Scheme 1, the resulting carboranylamidinate anions were quite unique as they combined the highly versatile amidinate ligand system, [RC(NR′)$_2$] [36–40], with a σ-bond to the carborane cage. Subsequently, the lithium salts served as precursors for a variety of main-group and transition metal complexes comprising carboranylamidinate ligands [41–45]. In all these complexes, the carboranylamidinate ligands adopt the characteristic κC,κN-chelating coordination mode instead of the regular κN,κN′-chelating mode of metal-coordinated amidinate anions.

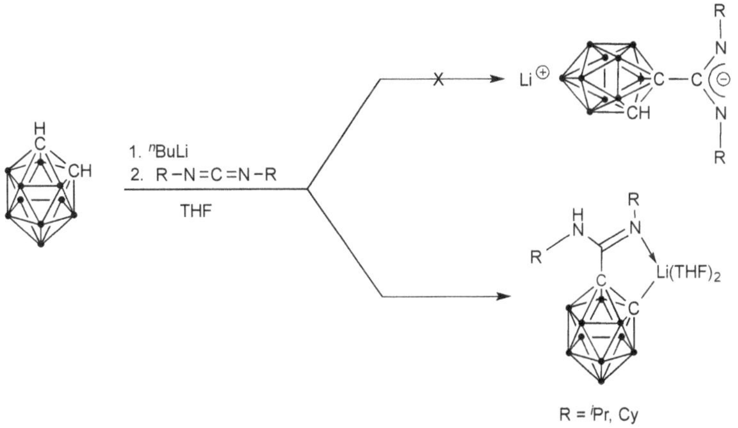

Scheme 1. Preparation of lithium carboranylamidinates derived from *o*-carborane.

Thus far, the formation of κC,κN-chelating carboranylamidinate anions has been limited to compounds derived from *o*-carborane. In 2014, we reported that similar reactions starting from *m*-carborane take a completely different course. As illustrated in Scheme 2, successive treatment of *m*-carborane with *n*-butyllithium and 1,3-di-*iso*-propylcarbodiimide did not lead to the formation of a related carboranylamidinate anion. Instead, an unprecedented deboronation reaction of the *m*-carborane took place, in which a BH group was detached from the carborane cage and incorporated into a *nido*-carborane-anellated diazadiborepine ring system. 1,3-dicyclohexylcarbodiimide reacted in a very similar manner but afforded a slightly modified seven-membered diazadiborepine ring system [46].

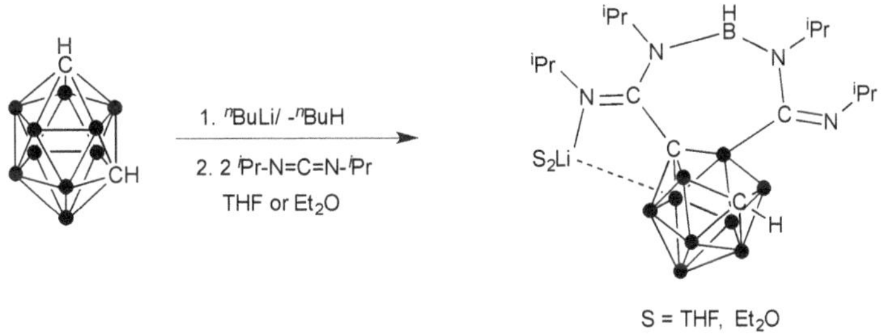

Scheme 2. Formation of polycyclic diazadiborepines from *m*-carborane.

Until now, the question remained of how the third isomer, *p*-carborane (**1**), would behave in the same reaction sequence of lithiation and carbodiimide addition. Here, we present the answer to this question.

2. Results and Discussion

2.1. Synthesis and Characterization

In the first set of experiments, THF solutions of *p*-carborane were metalated with 1 equiv. of *n*-butyllithium and then treated in situ with two different carbodiimides R–N=C=N–R (R = ^{i}Pr, Cy). Under these conditions, only small amounts (ca. 20% yield) of crystalline products could be isolated, which were difficult to separate from unreacted *p*-carborane (NMR control). This finding implied that the envisaged *mono*-amidinate derivatives shown in the upper equation in Scheme 3 were not formed as pure reaction products and that disubstitution was instead the preferred reaction pathway. This assumption was soon confirmed by adjusting the stoichiometric ratio of the reactants to 1:2:2 according to the second equation in Scheme 3. Under these conditions, the new compounds 2 (R = ^{i}Pr) and 3 (R = Cy) could be isolated as pure crystalline solids in significantly improved yields of 51% (**2**) and 46% (**3**), respectively. Both lithium amidinate salts are readily soluble in THF, DME, and diethyl ether. Crystallization from DME (**2**) and THF (**3**) afforded the nicely crystalline solvates depicted in Scheme 4.

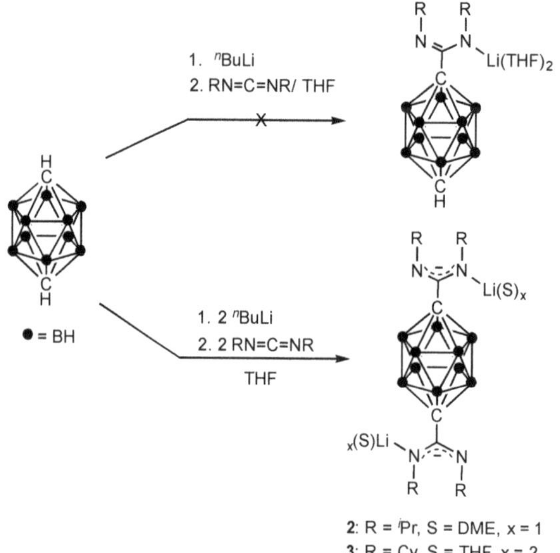

Scheme 3. Synthetic route to the title compounds **2** and **3**.

Scheme 4. Preparation of the silylated bis(amidine) derivatives **4** and **5**.

Both bis(anionic) title compounds **2** and **3** were fully characterized through the usual set of elemental analyses and spectroscopic methods. In the IR spectra, strong bands at 1523 cm^{-1} (**2**) and 1543 cm^{-1} (**3**) are typical for the stretching vibrations of the delocalized amidinate NCN units [36–40]. Medium strong bands in the range of 2590–2620 cm^{-1} could be assigned to the B–H stretching vibrations, while the ν_{as}(C-O-C) bands of the coordinated solvents appear around 1050 cm^{-1} as medium or strong bands. The ^{1}H and ^{13}C NMR spectra show the typical signals of the *iso*-propyl and cyclohexyl substituents, which do not need to be discussed here in detail (cf., Experimental Section and Supplementary Materials). In the ^{1}H NMR spectra, the B–H hydrogens give rise to broad multiplets extending over a range of ca. 1.5 ppm. The ^{13}C NMR chemical shifts of the carbon atoms of the NCN groups are 155.2 ppm (**2**) and 154.0 ppm (**3**), respectively. A ^{13}C resonance of the quaternary carbon atoms within the carborane cage could be detected only in the spectrum of **2** (δ 93.3 ppm). All cage boron atoms give rise to a single resonance around -14 ppm in the ^{11}B NMR spectra of both amidinate salts. Apparently, the centrosymmetric structure leads to very similar chemical shifts of the boron atoms so that the signals could not be further resolved. Finally, ^{7}Li NMR spectra displayed only one signal around 0.1 ppm. As expected for salt-like compounds, the mass spectra of **2** and **3** did not show the respective molecular ions but only fragment peaks of the unsolvated carboranylamidinate anions (cf., Experimental Section).

Remarkably, the formation of the lithium carboranylamidiate represents the first incidence of a "normal" reactivity of a lithiated carborane with carbodiimides. This means that 1,12-dilithiocarborane behaves toward carbodiimides like any other organolithium reagents and adds to the central carbon atom of the N=C=N moiety under the formation of regular amidinate anion of the type [RC(NR$'$)$_2$]$^-$ [36–40]. This finding reveals that all three C$_2$B$_{10}$H$_{12}$ isomers behave differently in their reactivity toward 1,3-diorganocarbodiimides.

As an initial reactivity study involving the lithium carboranylamidinate salts **2** and **3**, we investigated silylation reactions with chlorotrimethylsilane, Me$_3$SiCl, which should lead to the formation of neutral bis-silylated amidine derivatives, as illustrated in Scheme 4.

Both reactions were carried out in THF solutions at r.t. Work-up via extraction with toluene afforded the bis-silylated products **4** and **5** as colorless crystals in moderate yields (**4**: 54%, **5**: 43%). Both compounds dissolve freely in diethyl ether and toluene, and are moderately moisture-sensitive due to the presence of Si–N bonds. Besides an X-ray structural analysis of **4** (see next paragraph), all analytical and spectroscopic data were in excellent agreement with the formation of bis(silylated) *p*-carboranyl-bis(amidines). Highly characteristic in the IR spectra are the ν C=N bands at 1622 cm^{-1} (**4**) and 1627 cm^{-1} (**5**), respectively. These bands clearly indicate the transition from the delocalized amidinate NCN units in the salt-like amidinate precursors **2** and **3** (ν NCN 1523 cm^{-1} (**2**) and 1543 cm^{-1} (**3**)) to N–C=N moieties with localized carbon–nitrogen double and single bonds. Bands at 2590 cm^{-1} (**4**) and 2606 cm^{-1} (**5**) can be assigned to the B–H stretching vibrations, while typical Si–C stretch bands of the SiMe$_3$ groups appear at ν_{as} 726 cm^{-1} and 750 cm^{-1} as well as ν_s 651 cm^{-1} and 658 cm^{-1}, respectively. The ^{1}H, ^{13}C, and ^{29}Si NMR spectra of **4** and **5** all showed only one singlet resonance for the SiMe$_3$ groups. This is in agreement with the centrosymmetric molecular structure found in the X-ray structural analysis of **4** (see following paragraph). As was observed for the anionic precursors **2** and **3**, the ^{11}B NMR spectra of the bis-silylated derivatives also displayed only single resonances around -13.4 ppm.

2.2. Crystal and Molecular Structures

The title compounds **3** and **4** could be structurally characterized through X-ray diffraction studies. The molecular structures are depicted in Figures 2 and 3. Colorless, prism-shaped single-crystals of **3** were grown from concentrated solutions in THF at r.t., while compound **4** was obtained in the form of well-formed, colorless, block-like single-crystals upon slow crystallization from toluene at 4 °C. As illustrated in Figure 2, structure determination of compound **3** confirmed the presence of an anionic *p*-carboranylamidinate species formed by the addition of dilithiated *p*-carborane to the central C atom of the carbodiimide reagent. The overall molecular structure is centrosymmetric. With 1.953(3) Å, the Li-N2 bond length is typical for a coordinative bond. As in other typical lithium amidinates such as Li[MeC$_6$H$_4$C(NSiMe$_2$)$_2$](THF)$_2$ [36,47], the C–N distances in the amidinate NCN unit are quite similar (N(1)-C(2) 1.305(2), N(2)-C(2) 1.340(2)), indicating complete delocalization of the negative charge. However, the coordination of the lithium ion to the anionic amidinate moieties differs from the vast majority of other lithium amidinates in that it is not $\kappa N,\kappa N'$-chelating. Instead, the lithium ions are coordinated to only one nitrogen atom of the NCN moiety, resulting in a nearly trigonal planar coordination geometry around Li. There are only very few examples of similar monodentate amidinate coordination to lithium [48,49], and all of them result from steric crowding around the NCN unit, e.g., through very bulky terphenyl or triptycenyl substituents. Thus it can be assumed that steric congestion is also the reason for the rather unusual trigonal planar coordination of the lithium ions in compound **3**.

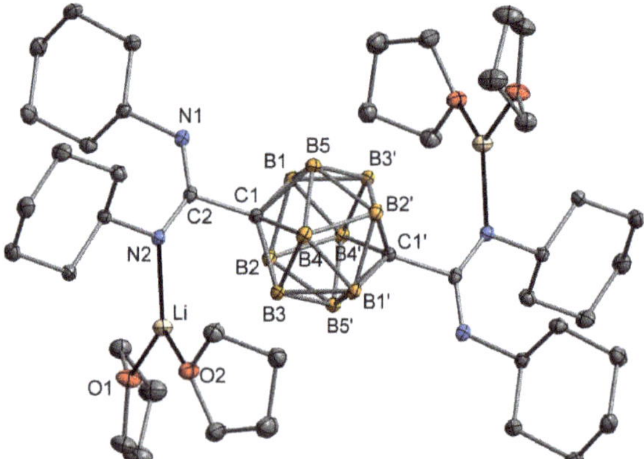

Figure 2. Molecular structure of **3** in the crystal. Displacement ellipsoids of the heavier atoms are drawn with 50% probability; H atoms are omitted for clarity. Selected bond lengths (Å) and angles (deg.): C(1)-C(2) 1.551(2), N(1)-C(2) 1.305(2), N(2)-C(2) 1.340(2), N(2)-Li 1.953(3), O(1)-Li 1.949(3), O(2)-Li 1.933(3), N(1)-C(2)-N(2) 136.9(1), N(1)-C(2)-C(1) 109.7(1), N(2)-C(2)-C(1) 113.4(1), C(2)-N(2)-Li 134.9(1). Symmetry code to generate equivalent atoms: '2–*x*, 1–*y*, 1–*z*.

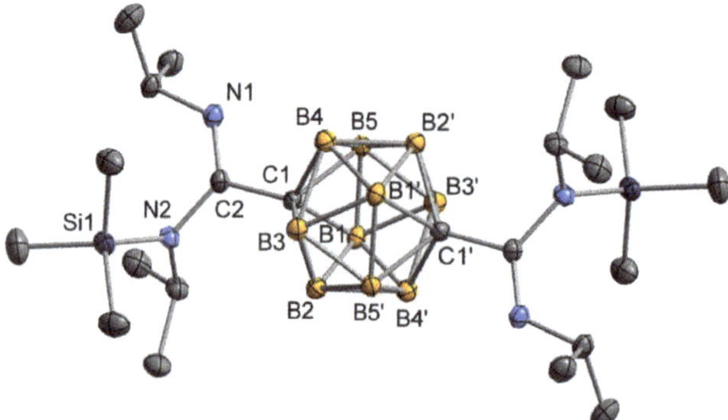

Figure 3. Molecular structure of **4** in the crystal. Displacement ellipsoids of the heavier atoms with 50% probability; H atoms are omitted for clarity. Selected bond lengths (Å) and angles (deg.): C(1)-C(2) 1.539(2), C(2)-N(1) 1.265(2), C(2)-N(2) 1.423(2), Si(1)-N(2) 1.751(1), N(1)-C(2)-N(2) 128.9(1), N(1)-C(2)-C(1) 113.8(1), N(2)-C(2)-C(1) 117.3(1), C(2)-N(2)-Si(1) 121.00(9). Symmetry code to generate equivalent atoms: '1–*x*,1–*y*, 1–*z*.

The neutral *p*-carboranyl-bis(amidine) derivative **4** crystallizes in the monoclinic space $P2_1/n$, and, like **3**, the molecule also shows crystallographically imposed centrosymmetry. The transition from the delocalized anionic NCN moieties in the amidinate salts **2** and **3** to a neutral amidine is clearly evidenced by the change in the C–N bond lengths. With distances of C(2)-N(1) 1.265(2) Å and C(2)-N(2) 1.423(2) Å, compound **4** clearly contains N–C=N units with isolated single and double bonds. In this respect, the molecular structure of **4** is closely related to the oxygen analogue *p*-carborane-1,12-dicarboxylic acid [50]. The N1-C2-N2 angle is 128.9(1)°, and the Si-N2 distance is 1.751(1) Å. Both values are in good agreement with those of the silylated *o*-carboranylamidine derivative o-$C_2B_{10}H_{10}[\kappa C,N$-C(^{i}PrNSiMe$_3$)(=N^{i}Pr)]SiMe$_3$ [51].

3. Experimental Section

3.1. General Procedures and Instrumentation

All reactions were carried out in oven-dried or flame-dried glassware under an inert atmosphere of dry argon employing standard Schlenk and glovebox (MBraun MBLab) techniques. The solvents *n*-pentane, toluene, DME, and THF were distilled from sodium/benzophenone under nitrogen atmosphere prior to use. *p*-carborane was obtained from Katchem spol. s.r.o., 278 01 Kralupy nad Vltavou, Czech Republic (https://katchem.cz/en). Other starting materials were purchased from Sigma-Aldrich and used without further purification. All NMR spectra (^{1}H, ^{13}C, ^{29}Si, ^{11}B, and ^{7}Li) were recorded in THF-d_8 solutions on a Bruker DPX 400 spectrometer. IR spectra were measured with a Bruker Vertex 70V spectrometer equipped with a diamond ATR unit between 4000 cm^{-1} and 50 cm^{-1}. Mass spectra were measured on a MAT 95 apparatus (EI, 70 eV). Microanalyses (C, H, N) were performed using a VARIO EL cube apparatus. Melting/decomposition points were determined using a Büchi Melting Point B-540.

3.2. Synthesis of Compound p-$C_2H_{10}B_{10}[C(N^iPr)_2Li(DME)]_2$ (2)

A total of 0.50 g (3.5 mmol) *p*-carborane, dissolved in THF (50 mL), was treated at r.t. with 2 equiv. of *n*-butyllithium (7.0 mmol, 4.40 mL of a 1.6 M solution in *n*-hexane). After stirring for 1 h, 0.88 g (7.0 mmol) of 1,3-di-*iso*-propylcarbodiimide was added via syringe, and stirring at r.t. was continued for 12 h. The resulting clear yellow solution was evaporated to dryness and the oily crude product was redissolved in a minimum volume of DME (ca. 10 mL). Product **2** was precipitated by the addition of *n*-pentane (ca. 50 mL) and isolated after drying under vacuum as a microcrystalline, pale yellow solid in 51% isolated yield (1.06 g). M.p. 215 °C (dec.). Elemental analysis calcd. for $C_{24}H_{58}B_{10}N_4O_4Li_2$ (M = 588.71 g mol^{-1}): C, 48.96%; H, 9.9%; N, 9.51%; found C, 48.91%; H, 9.77%; N, 9.64%. ^{1}H NMR (400 MHz, THF-d_8, 21 °C): δ = 3.53 (sept, 4 H, ^3J = 6.40 Hz, CH-^{i}Pr), 3.41, 3.25 (DME), 1.55–3.12 (m br, 10 H, BH), 0.78 (d, 24 H, ^3J = 6.00 Hz, CH$_3$-^{i}Pr) ppm. ^{13}C{^{1}H} NMR (100.6 MHz, THF-d_8, 23 °C): δ = 155.2 (NCN), 93.3 (C-NCN), 72.6, 58.9 (DME), 46.3 (CH-^{i}Pr), 25.6 (CH$_3$-^{i}Pr) ppm. ^{11}B{^{1}H} NMR (128.4 MHz, THF-d_8, 23 °C): δ = −14.1 ppm. ^{7}Li{^{1}H} NMR (155.5 MHz, THF-d_8, 23 °C): δ = 0.11 ppm. IR (ATR): ν_{max} 2955 m (ν_s CH$_3$), 2926 m (ν_{as} CH$_2$), 2860 w ($\nu_{as,s}$ CH$_3$/CH$_2$), 2597 m (ν BH), 1523 s (ν NCN), 1461 m ($\delta_{as,s}$ CH$_3$/CH$_2$), 1367 m (δ_s CH$_3$), 1357 m (δ_s CH$_3$), 1312 m, 1285 m, 1264 s, 1191 w, 1157 w, 1112 m, 1077 vs (ν_{as} C-O-C), 1021 m, 961 w, 901 w, 870 m, 801 w, 737 m, 679 w, 621 m, 599 m, 566 m, 506 m, 494 m, 460 w, 385 w, 374 w, 242 vs cm^{-1}. MS (EI, 70 eV): *m/z* (%) 396 (17) [((^{i}PrN)$_2$C)$_2$-$C_2H_{10}B_{10}$]$^+$, 353 (27) [((^{i}PrN)$_2$C)$_2$-$C_2H_{10}B_{10}$-^{i}Pr]$^+$, 270 (15) [(^{i}PrN)$_2$C-$C_2H_{10}B_{10}$]$^+$, 227 (49) [(^{i}PrN)NC-$C_2H_{10}B_{10}$]$^+$, 170 (25) [NC-$C_2H_{10}B_{10}$]$^+$, 143 (90) [$C_2H_9B_{10}$]$^+$, 58 (100) [^{i}PrN + H]$^+$.

3.3. Synthesis of Compound p-$C_2H_{10}B_{10}[C(NCy)_2Li(THF)_2]_2$ (3)

This compound was prepared in a similar manner as described for **2** but using 1.44 g (7.0 mmol) of *N,N′*-dicyclohexylcarbodiimide as a precursor. The resulting clear solution was concentrated to a total volume of ca. 30 mL which led to the formation of a white precipitate, which was then redissolved by brief heating. Colorless single crystals suitable for X-ray diffraction were obtained directly by storing the concentrated THF solution at r.t. for a few days. Yield: 1.37 g (46%). M.p. 238 °C (dec.). Elemental analysis calcd. for $C_{44}H_{86}B_{10}Li_2N_4O_4$ (M = 857.18 g mol^{-1}): C, 61.65%; H, 10.11%; N, 6.54%; found C, 61.59%; H, 10.01%; N, 6.67%. ^{1}H NMR (400 MHz, THF-d_8, 24 °C): δ = 3.20–3.29 (m br, 2 H, C*H*-Cy), 3.60 (THF), 2.99–3.09 (m br, 2 H, C*H*-Cy), 1.50–2.90 (m br, 10 H, B*H*), 1.76 (THF), 1.63–1.73 (m, 10 H, C*H$_2$*-Cy), 1.43–1.61 (m, 10 H, C*H$_2$*-Cy), 1.12–1.34 (m, 20 H, C*H$_2$*-Cy) ppm. ^{13}C{^{1}H} NMR (100.6 MHz, THF-d_8, 24 °C): δ = 154.2, 154.0 (NCN), 68.1 (THF), 56.4, 56.1, 54.6, 54.4 (CH-Cy), 36.5, 35.8, 35.0, 34.9, 34.8, 34.7, 30.6, 27.8, 27.7, 27.0, 26.9, 26.4 (CH$_2$-Cy), 26.3 (THF), 25.8, 25.7, 25.6, 25.4, 25.0, 24.9 (CH$_2$-Cy) ppm. ^{11}B{^{1}H} NMR (128.4 MHz, THF-d_8, 24 °C): δ = −14.0, −14.8, −15.8 ppm. ^{7}Li{^{1}H} NMR (MHz, THF-d_8, 24 °C): δ = 0.09 ppm. IR (KBr disk): ν_{max} 3419 w, 3222 w, 3091 w, 2979 m, 2927 vs (ν_{as} CH$_2$), 2852 s (ν_s CH$_2$), 2611 m (ν BH), 2119 w, 1959 w, 1657 m, 1543 s (ν NCN), 1510 m, 1463 m, 1449 m (δ_s CH$_2$), 1372 w, 1332 m, 1289 m, 1270 m, 1230 m, 1181 w, 1147 w, 1132 m, 1043 m (ν_{as} C-O-C), 970 w, 930 w, 888 m, 841 w, 833 w, 810 w, 794 w, 742 w, 699 w, 674 w, 589 w, 561 w, 511 w, 489 w, 477 w, 449 w, 437 w cm^{-1}. MS (EI, 70 eV): m/z (%) 557 (45) [((CyN)$_2$C)$_2$C$_2$H$_{10}$B$_{10}$ + H]$^+$, 474 (27) [((CyN)$_2$C)$_2$C$_2$H$_{10}$B$_{10}$ − C$_6$H$_{11}$ + H]$^+$, 392 (100) [((CyN)$_2$C)$_2$C$_2$H$_{10}$B$_{10}$ − 2C$_6$H$_{11}$ + 2H]$^+$, 268 (33) [(CyN)NCC$_2$H$_{10}$B$_{10}$ + H]$^+$, 143 (16) [C$_2$H$_9$B$_{10}$]$^+$.

3.4. Synthesis of Compound p-$C_2H_{10}B_{10}[C(^iPrN(SiMe_3)(=N^iPr)]_2$ (4)

A solution of compound **2** (3.5 g in 50 mL) THF was prepared as described above and treated in situ with 0.90 mL (7.0 mmol) of chlorotrimethylsilane. After stirring for 12 h at r.t., the solvent was removed in vacuum, and the orange-yellow residue was extracted with toluene (50 mL) and filtered in order to remove the by-product LiCl. The concentration of the filtrate to a total volume of ca. 20 mL followed by cooling to 4 °C for several days led to the formation of colorless, block-like single crystals which were suitable for X-ray diffraction. Yield: 1.03 g (54%). M.p. 265 °C. Elemental analysis calcd. for $C_{22}H_{54}B_{10}N_4Si_2$ (M = 538.98 g mol^{-1}): C, 49.03%; H, 10.10%; N, 10.39%; found C, 48.82%; H, 10.54%; N, 10.50%. ^{1}H NMR (400 MHz, THF-d_8, 21 °C): δ = 3.60 (sept, 2 H, ^3J = 6.00 Hz, CH-^{i}Pr), 3.22 (m, 2 H, CH-^{i}Pr), 1.70–3.10 (m br, 10 H, B*H*), 1.27 (d, 12 H, ^3J = 5.60 Hz, CH$_3$-^{i}Pr), 0.97 (d, 12 H, ^3J = 6.00 Hz, CH$_3$-^{i}Pr), 0.16 (s, 18 H, CH$_3$-SiMe$_3$) ppm. ^{13}C{^{1}H} NMR (100.6 MHz, THF-d_8, 22 °C): δ = 152.4 (NCN), 89.5 (C-NCN), 51.1, 50.6 (CH-^{i}Pr), 24.9, 23.6 (CH$_3$-^{i}Pr), 3.82 (CH$_3$-SiMe$_3$) ppm. ^{11}B{^{1}H} NMR (128.4 MHz, THF-d_8, 22 °C): δ = −13.4 ppm. ^{29}Si{^{1}H} NMR (79.5 MHz, THF-d_8, 21 °C): δ = −1.40 ppm. IR (ATR): ν_{max} 2987 w (ν_s CH$_3$), 2966 m, 2932 w, 2889 w (ν_{as} CH$_3$), 2641 w, 2623 m, 2590 m (ν BH), 1622 m (ν C=N), 1468 w (δ_{as} CH$_3$), 1450 w, 1405 w, 1375 m (δ_s CH$_3$), 1362 w (δ_s CH$_3$), 1319 m, 1255 m, 1207 s, 1159 m, 1139 m, 1117 m, 1092 m, 1042 w, 1008 m, 979 m, 930 w, 887 m, 858 m, 833 vs (ρ CH$_3$), 750 m, 726 m (ν_{as} SiC$_3$), 678 m, 651 m (ν_s SiC$_3$), 623 w, 607 w, 582 w, 548 w, 516 m, 471 m, 423 w, 379 w, 333 w, 260 m, 176 w, 140 w, 87 w, 68 w, 61 m cm^{-1}. MS (EI, 70 eV): m/z (%) 541 (3) [M]$^+$, 526 (5) [M − CH$_3$]$^+$, 498 (100) [M − 3CH$_3$]$^+$, 468 (29) [M − Si(CH$_3$)$_3$]$^+$, 342 (16) [(^{i}PrN)(^{i}PrNSiMe$_3$)CC$_2$H$_{10}$B$_{10}$]$^+$.

3.5. Synthesis of Compound p-$C_2H_{10}B_{10}[C(CyN(SiMe_3)(=NCy)]_2$ (5)

Compound **5** was prepared in the same manner as described above using 3.00 g (3.5 mmol) of **3** and 0.90 mL (7.0 mmol) chlorotrimethylsilane to afford colorless crystals after crystallization from a small amount of toluene at −32 °C. Yield: 1.05 g (43%). M.p. 159 °C. Elemental analysis calcd. for $C_{34}H_{72}B_{10}N_4Si_2$ (M = 701.25 g mol^{-1}): C, 59.23%; H, 10.35%; N, 7.99%; found C, 58.20%; H, 10.28%; N, 7.94%. ^{1}H NMR (400 MHz, THF-d_8, 38 °C): δ = 3.32–3.36 (m, 2 H, C*H*-Cy), 1.90–3.15 (m br, 10 H, B*H*), 2.73–2.79 (m, 2 H, C*H*-Cy), 1.04–1.86 (m, 24 H, C*H$_2$*-Cy), 0.15 (s, 18 H, CH$_3$-SiMe$_3$) ppm. ^{13}C{^{1}H} NMR (100.6 MHz, THF-d_8, 23 °C): δ = 152.0 (NCN), 89.6 (C-NCN), 60.1, 58.8 (CH-Cy), 35.8, 35.2, 35.1, 33.6, 27.6, 27.0, 26.8, 26.4, 24.6, 24.5 (CH$_2$-Cy), 4.0 (CH$_3$-SiMe$_3$) ppm. ^{11}B{^{1}H} NMR (128.4 MHz, THF-d_8, 23 °C): δ = −13.4 ppm. ^{29}Si{^{1}H} NMR (MHz, THF-d_8, 22 °C): δ = 0.20 ppm. IR (ATR): ν_{max} 2930 m (ν_{as} CH$_2$), 2852 m ($\nu_{as,s}$ CH$_3$/CH$_2$), 2644 s, 2606 m (ν BH), 2119 m,

1658 w, 1627 m (v C=N), 1495 w, 1450 m ($\delta_{as,s}$ CH$_3$/CH$_2$), 1407 w, 1386 w (δ_s CH$_3$), 1358 w, 1344 w, 1297 w, 1251 m, 1189 s, 1179 m, 1142 m, 1115 m, 1076 m, 1028 m, 996 m, 981 m, 956 m, 890 m, 857 m, 831 vs (ρ CH$_3$), 820 vs, 777 m, 750 m (v_{as} SiC$_3$), 730 m, 701 m, 673 m, 658 m (v_s SiC$_3$), 630 m, 564 w, 540 w, 520 w, 497 w, 474 m, 461 m, 409 m, 384 m, 341 m, 322 m, 291 m, 260 m, 207 w, 181 w, 162 w, 136 w, 117 w, 86 w, 69 w, 57 m cm^{-1}. MS (EI, 70 eV): m/z (%) 702 (3) [M]$^+$, 687 (3) [M − CH$_3$]$^+$, 629 (32) [M − Si(CH$_3$)$_3$]$^+$, 619 (71) [M − C$_6$H$_{11}$]$^+$, 604 (12) [M − C$_6$H$_{11}$ − CH$_3$]$^+$, 556 (2) [M − 2 Si(CH$_3$)$_3$]$^+$, 530 (67) [M − Si(CH$_3$)$_3$ − C$_6$H$_{11}$ − CH$_3$]$^+$, 279 (75) [CyN(Si(CH$_3$)$_3$)CNCy]$^+$, 83 (100) [C$_6$H$_{11}$]$^+$.

3.6. X-ray Crystallography

Single-crystal X-ray intensity data of **3** and **4** were collected on a STOE IPDS 2T diffractometer [52] equipped with a 34 cm image plate detector, using graphite-monochromated Mo-K$_\alpha$ radiation. The structure was solved by dual-space methods (SHELXT-2014/5) [53] and refined by full-matrix least-squares methods on F^2 using SHELXL-2017/1 [54]. Crystallographic data for the title compounds have been deposited at the CCDC, 12 Union Road, Cambridge CB21EZ, U.K. Copies of the data can be obtained free of charge via the depository numbers 2248726 (**3**) and 2248725 (**4**) (e-mail: deposit@ccdc.cam.ac.uk, http://www.ccdc.cam.ac.uk).

4. Conclusions and Future Outlook

In summary, we succeeded in the synthesis and full characterization of the first amidinate and amidine derivatives of *para*-carborane. Lithium carboranylamidinates based on *p*-carborane are readily accessible by the addition of in situ-prepared 1,12-dilithio-*p*-carborane to 1,3-diorganocarbodiimides, R–N=C=N–R (R = ^{i}Pr, Cy). This result showed that all three isomers of C$_2$B$_{10}$H$_{12}$ react in completely different manners with carbodiimides. An initial reactivity study involving treatment of **2** and **3** with 2 equiv. of Me$_3$SiCl revealed that neutral bis-silylated amidine derivatives are also easily prepared. It should be noted here that the oxygen analogue *p*-carborane-1,12-dicarboxylic acid has been successfully utilized as a linker in the design of carborane-based MOFs (=metal-organic frameworks) [55,56]. One could easily foresee that the amidinate and amidine derivatives of *p*-carborane reported here will play a similar fruitful role in MOF chemistry in the future.

Supplementary Materials: The following supporting information can be downloaded at: https://www.mdpi.com/article/10.3390/molecules28093837/s1: IR, NMR, and mass spectra for all title compounds as well as X-ray diffraction data for **3** and **4**.

Author Contributions: N.H. performed the experimental work; V.L. supervised the experimental work; P.L. and F.E. carried out the crystal structure determinations; L.H. measured the IR and NMR spectra; S.B. measured the mass spectra and performed the elemental analyses; F.T.E. conceived and supervised the experiments; R.G. provided the research infrastructure and read and edited the paper; F.T.E. wrote the manuscript. All authors have read and agreed to the published version of the manuscript.

Funding: This work was financially supported by the Otto-von-Guericke-Universität Magdeburg.

Institutional Review Board Statement: Not applicable.

Informed Consent Statement: Not applicable.

Data Availability Statement: Not applicable.

Conflicts of Interest: The authors declare no conflict of interest.

Sample Availability: Samples of the compounds are not available from the authors.

References

1. Poater, J.; Solà, M.; Viñas, C.; Teixidor, F. π Aromaticity and Three-Dimensional Aromaticity: Two sides of the Same Coin? *Angew. Chem. Int. Ed.* **2014**, *53*, 12191–12195. [CrossRef]
2. Brown, A.D.; Colquhoun, H.M.; Daniels, A.J.; MacBride, J.A.H.; Stephenson, I.R.; Wade, K. Polymers and ceramics based on icosahedral carboranes. Model studies of the formation and hydrolytic stability of aryl ether, ketone, amide and borane linkages between carborane units. *J. Mater. Chem.* **1992**, *2*, 793–804. [CrossRef]
3. Murphy, D.M.; Mingos, D.M.P.; Haggitt, J.L.; Poell, H.R.; Westcott, S.A.; Marder, T.B.; Taylor, N.J.; Kanis, D.R. Synthesis of icosahedral carboranes for second-harmonic generation. Part 2. *J. Mater. Chem.* **1993**, *3*, 139–148. [CrossRef]
4. Koshino, M.; Tanaka, T.; Solin, N.; Suenaga, K.; Isobe, H.; Nakamura, E. Imaging of Single Organic Molecules in Motion. *Science* **2007**, *316*, 853. [CrossRef]
5. Villagómez, C.J.; Sasaki, T.; Tour, J.M.; Grill, L. Bottom-up Assembly of Molecular Wagons on a Surface. *J. Am. Chem. Soc.* **2010**, *132*, 16848–16954. [CrossRef]
6. Dash, B.P.; Satapathy, R.; Gaillard, E.R.; Maguire, J.A.; Hosmane, N.S. Synthesis and Properties of Carborane-Appended C_3-Symmetrical Extended π Systems. *J. Am. Chem. Soc.* **2010**, *132*, 6578–6587. [CrossRef]
7. Bauduin, P.; Prevost, S.; Farràs, P.; Teixidor, F.; Diat, O.; Zemb, T. A Theta-Shaped Amphiphilic Cobaltabisdicarbollide Anion: Transition From Monolayer Vesicles to Micelles. *Angew. Chem. Int. Ed.* **2011**, *50*, 52998–55300.
8. Cioran, A.M.; Musteti, A.D.; Teixidor, F.; Krpetic, Ž.Č.; Prior, I.A.; He, Q.; Kiely, C.J.; Brust, M.; Viñas, C. Mercaptocarborane-Capped Gold Nanoparticles: Electron Pools and Ion Traps with Switchable Hydrophilicity. *J. Am. Chem. Soc.* **2012**, *134*, 212–221. [CrossRef]
9. Schwartz, J.J.; Mendoza, A.M.; Wattanatorn, N.; Zhao, Y.; Nguyen, V.T.; Spokoyny, A.M.; Mirkin, C.A.; Baše, T.; Weisse, P.S. Surface Dipole Control of Liquid Crystal Alignment. *J. Am. Chem. Soc.* **2016**, *138*, 5957–5967. [CrossRef]
10. Li, Z.; Núñes, R.; Light, M.E.; Ruiz, E.; Teixidor, F.; Viñas, C.; Ruiz-Molina, D.; Roscini, C.; Planas, J.G. Water-Stable Carborane-Based Eu^{3+}/Tb^{3+} Metal–Organic Frameworks for Tunable Time-Dependent Emission Color and Their Application in Anticounterfeiting Bar-Coding. *Chem. Mater.* **2022**, *34*, 4795–4808. [CrossRef]
11. Zhang, K.; Song, R.; Qi, J.; Zhang, Z.; Yu, C.; Li, K.; Zhang, Z.; Li, B. Colossal Barocaloric Effect in Carboranes as a Performance Tradeoff. *Adv. Funct. Mater.* **2022**, *32*, 2112622. [CrossRef]
12. Minnyaylo, E.O.; Kudryavtseva, A.I.; Zubova, V.Y.; Anisimov, A.A.; Zaitsev, A.V.; Ol'shevskaya, V.A.; Dolgushin, F.M.; Peregudov, A.S.; Muzafarov, A.M. Synthesis of mono- and polyfunctional organosilicon derivatives of polyhedral carboranes for the preparation of hybrid polymer materials. *New J. Chem.* **2022**, *46*, 11143–11148. [CrossRef]
13. Liu, K.; Zhang, J.; Shi, Q.; Ding, L.; Kiu, T.; Fang, Y. Precise Manipulation of Excited State Intramolecular Proton Transfer via Incorporating Charge Transfer toward High-Performance Film-Based Fluorescence Sensing. *J. Am. Chem. Soc.* **2023**, *145*, 7408–7415. [CrossRef]
14. Belmont, J.A.; Soto, J.; King III, R.E.; Donaldson, A.J.; Hewes, J.D.; Hawthorne, M.F. Metallacarboranes in catalysis. 8. I: Catalytic hydrogenolysis of alkenyl acetates. II: Catalytic alkene isomerization and hydrogenation revisited. *J. Am. Chem. Soc.* **1989**, *111*, 7475–7486. [CrossRef]
15. Teixidor, F.; Flores, M.A.; Viñas, C.; Kivekäs, R.; Sillanpää, R. [Rh(7-SPh-8-Me-7,8-$C_2B_9H_{10}$)(PPh$_3$)$_2$]: A New Rhodacarborane with Enhanced Activity in the Hydrogenation of 1-Alkenes. *Angew. Chem. Int. Ed.* **1996**, *35*, 2251–2253. [CrossRef]
16. Ferlekidis, A.; Goblet-Stachow, M.; Liégeois, J.F.; Pirotte, B.; Delarge, J.; Demonceau, A.; Fontaine, M.; Noels, A.F.; Chizhevsky, I.T.; Zinevich, T.V.; et al. Ligand effects in the hydrogenation of methacycline to doxycycline and epi-doxycycline catalysed by rhodium complexes molecular structure of the key catalyst [$closo$-3,3-($\eta^{2,3}$-$C_7H_7CH_2$)-3,1,2-RhC$_2$B$_9$H$_{11}$]. *J. Organomet. Chem.* **1997**, *536–537*, 405–412. [CrossRef]
17. Gozzi, M.; Schwarze, B.; Hey-Hawkins, E. Half- and Mixed-sandwich Metallacarboranes in Catalysis. In *Boron Chemistry in Organometallics, Catalysis, Materials and Medicine*; Hosmane, N.S., Eagling, R., Eds.; World Scientific: Singapore, 2018; pp. 27–80.
18. Fisher, S.P.; Tomich, A.W.; Lovera, S.O.; Kleinsasser, J.F.; Guo, J.; Asay, M.J.; Nelson, H.M.; Lavallo, V. Nonclassical Applications of *closo*-Carborane Anions: From Main Group Chemistry and Catalysis to Energy Storage. *Chem. Rev.* **2019**, *119*, 8262–8290. [CrossRef]
19. Gunther, S.O.; Lai, Q.; Senecal, T.; Huacuja, R.; Bremer, S.; Pearson, D.M.; DeMott, J.C.; Bhuvanesh, N.; Ozerov, O.V.; Klosin, J. Highly Efficient Carborane-Based Activators for Molecular Olefin Polymerization Catalysts. *ACS Catal.* **2021**, *11*, 3335–3342. [CrossRef]
20. Grishin, I.V.; Zimina, A.M.; Anufriev, S.A.; Knyazeva, N.A.; Piskunov, A.V.; Dolgushin, F.M.; Sivaev, I.B. Synthesis and Catalytic Properties of Novel Ruthenacarboranes Based on nido-[5-Me-7,8-$C_2B_9H_{10}$]$^{2-}$ and nido-[5,6-Me$_2$-7,8-$C_2B_9H_9$]$^{2-}$ Dicarbollide Ligands. *Catalysts* **2021**, *11*, 1409. [CrossRef]
21. Cheng, R.; Zhang, J.; Zhang, H.; Qiu, Z.; Xie, Z. Ir-catalyzed enantioselective B–H alkenylation for asymmetric synthesis of chiral-at-cage o-carboranes. *Nat. Commun.* **2021**, *12*, 7146. [CrossRef]
22. Vaillant, J.F.; Guenther, K.J.; King, A.S.; Morel, P.; Schaffer, P.; Sogbein, O.O.; Stephenson, K. The medicinal chemistry of carboranes. *Coord. Chem. Rev.* **2002**, *232*, 173–230. [CrossRef]
23. Armstrong, A.F.; Vaillant, J.F. The bioinorganic and medicinal chemistry of carboranes: From new drug discovery to molecular imaging and therapy. *Dalton Trans.* **2007**, *38*, 4240–4251. [CrossRef]

24. Scholz, M.; Hey-Hawkins, E. Carbaboranes as Pharmacophores: Properties, Synthesis, and Application Strategies. *Chem. Rev.* **2011**, *111*, 7035–7062. [CrossRef]
25. Issa, F.; Kassiou, M.; Rendina, L.M. Boron in Drug Discovery: Carboranes as Unique Pharmacophores in Biologically Active Compounds. *Chem. Rev.* **2011**, *111*, 5701–5722. [CrossRef]
26. Stockmann, P.; Gozzi, M.; Kuhnert, R.; Sarosi, M.-B.; Hey-Hawkins, E. New keys for old locks: Carborane-containing drugs as platforms for mechanism-based therapies. *Chem. Soc. Rev.* **2019**, *48*, 3497–3512. [CrossRef]
27. Gozzi, M.; Schwarze, B.; Hey-Hawkins, E. Preparing (Metalla)carboranes for Nanomedicine. *ChemMedChem* **2021**, *16*, 1533–1565. [CrossRef]
28. Marfavi, A.; Kavianpour, P.; Rendina, L.M. Carboranes in drug discovery, chemical biology and molecular imaging. *Nat. Rev. Chem.* **2022**, *6*, 486–504. [CrossRef]
29. Chen, Y.; Du, F.; Tang, L.; Xu, J.; Zhao, Y.; Wu, X.; Li, M.; Shen, J.; Wen, Q.; Cho, C.H.; et al. Carboranes as unique pharmacophores in antitumor medicinal chemistry. *Mol. Ther. Oncolytics* **2022**, *24*, 400–416. [CrossRef]
30. Waddington, M.A.; Zheng, X.; Stauber, J.M.; Moully, E.H.; Montgomery, H.R.; Saleh, L.M.A.; Král, P.; Spokoyny, A.M. An Organometallic Strategy for Cysteine borylation. *J. Am. Chem. Soc.* **2021**, *143*, 8661–8668. [CrossRef]
31. Gazvoda, M.; Dhanjee, H.H.; Rodriguez, J.; Brown, J.S.; Farquhar, C.E.; Truex, N.L.; Loas, A.; Buchwald, S.L.; Pentelute, B.L. Palladium-Mediated Incorporation of Carboranes into Small Molecules, Peptides, and Proteins. *J. Am. Chem. Soc.* **2022**, *144*, 7852–7860. [CrossRef]
32. Ma, Y.-N.; Gao, Y.; Ma, Y.; Wang, Y.; Ren, H.; Chen, X. Palladium-Catalyzed Regioselective B(9)-Amination of *o*-Carboranes and *m*-Carboranes in HFIP with Broad Nitrogen sources. *J. Am. Chem. Soc.* **2022**, *144*, 8371–8378. [CrossRef]
33. Ma, Y.-N.; Ren, H.; Wu, Y.; Li, N.; Chen, F.; Chen, X. B(9)-OH-*o*-Carboranes: Synthesis, Mechanism, and Property exploration. *J. Am. Chem. Soc.* **2023**, *145*, 7331–7342. [CrossRef]
34. Ren, H.; Zhang, P.; Xu, J.; Ma, W.; Tu, D.; Lu, C.; Yan, H. Direct B–H Functionalization of icosahedral Carboranes via Hydrogen Atom Transfer. *J. Am. Chem. Soc.* **2023**, *145*, 7638–7647. [CrossRef]
35. Bregadze, V. Dicarba-*closo*-dodecaboranes $C_2B_{10}H_{12}$ and their derivatives. *Chem. Rev.* **1992**, *92*, 209–223. [CrossRef]
36. Junk, P.C.; Cole, M.L. Alkali-metal bis(aryl)formamidinates: A study of coordinative versatility. *Chem. Commun.* **2007**, *16*, 1579–1590. [CrossRef]
37. Edelmann, F.T. Chapter 3—Advances in the Coordination Chemistry of Amidinate and Guanidinate Ligands. *Adv. Organomet. Chem.* **2008**, *57*, 183–352.
38. Edelmann, F.T. Lanthanide amidinates and guanidinates in catalysis and materials science: A continuing success story. *Chem. Soc. Rev.* **2012**, *41*, 7657–7672. [CrossRef]
39. Deacon, G.B.; Hossain, M.E.; Junk, P.C.; Salehisaki, M. Rare-earth N,N′-diarylformamdinate complexes. *Coord. Chem. Rev.* **2017**, *340*, 247–265. [CrossRef]
40. Sengupta, D.; Gómez-Torres, A.; Fortier, S. Guanidinate, Amidinate, and Formamidinate Ligands. In *Comprehensive Coordination Chemistry*; University of Texas at El Paso: El Paso, TX, USA, 2021; Volume 3, pp. 366–405.
41. Edelmann, F.T. Carboranylamidinates. *Z. Anorg. Allg. Chem.* **2013**, *639*, 655–667. [CrossRef]
42. Yao, Z.-J.; Jin, G.-X. Transition metal complexes based on carboranyl ligands containing N, P, and S donors: Synthesis, reactivity and applications. *Coord. Chem. Rev.* **2013**, *257*, 2522–2535. [CrossRef]
43. Rädisch, T.; Harmgarth, N.; Liebing, P.; Beltrán-Leiva, M.J.; Páez-Hernández, D.; Arratia-Pérez, R.; Engelhardt, F.; Hilfert, L.; Oehler, F.; Busse, S.; et al. Three new types of transition metal carboranylamidinate complexes. *Dalton Trans.* **2018**, *47*, 6666–6671. [CrossRef]
44. Liebing, P.; Harmgarth, N.; Zörner, F.; Engelhardt, F.; Hilfert, L.; Busse, S.; Edelmann, F.T. Synthesis and Structural Characterization of Two New Main Group Element Carboranylamidinates. *Inorganics* **2019**, *7*, 41. [CrossRef]
45. Wang, H. Recent advances on carborane-based ligands in low-valent group 13 and group 14 elements chemistry. *Chin. Chem. Lett.* **2022**, *33*, 3672–3680. [CrossRef]
46. Harmgarth, N.; Hrib, C.G.; Lorenz, V.; Hilfert, L.; Edelmann, F.T. Unprecedented formation of polycyclic diazadiborepine derivatives through cage deboronation of *m*-carborane. *Chem. Commun.* **2014**, *50*, 13239–13242. [CrossRef]
47. Stalke, D.; Wedler, M.; Edelmann, F.T. Dimere Alkalimetallbenzamidinate: Einfluß des Metallions auf die Struktur. *J. Organomet. Chem.* **1992**, *431*, C1–C5. [CrossRef]
48. Schmidt, J.A.R.; Arnold, J. Synthesis and characterization of a series of sterically-hindered amidines and their lithium and magnesium complexes. *Dalton Trans.* **2002**, *14*, 2890–2899. [CrossRef]
49. Baker, R.J.; Jones, C. Synthesis and characterisation of sterically bulky lithium amidinate and bis-amidinate complexes. *J. Organomet. Chem.* **2006**, *691*, 65–71. [CrossRef]
50. Kahl, S.B.; Kasar, R.A. Simple, High-Yield Synthesis of Polyhedral Carborane Amino Acids. *J. Am. Chem. Soc.* **1996**, *118*, 1223–1224. [CrossRef]
51. Harmgarth, N.; Gräsing, D.; Dröse, P.; Hrib, C.G.; Jones, P.G.; Lorenz, V.; Hilfert, S.; Busse, S.; Edelmann, F.T. Novel inorganic heterocycles from dimetalated carboranylamidinates. *Dalton Trans.* **2014**, *43*, 5001–5013. [CrossRef]
52. Stoe & Cie. *X-Area and X-Red*; Stoe & Cie: Darmstadt, Germany, 2002.
53. Sheldrick, G.M. *SHELXT*—Integrated space-group and crystal-structure determination. *Acta Cryst.* **2015**, *A71*, 3–8. [CrossRef]
54. Sheldrick, G.M. Crystal structure refinement with SHELXL. *Acta Cryst.* **2015**, *71*, 3–8.

55. Farha, O.K.; Spokoyny, A.M.; Mulfort, K.L.; Hawthorne, M.F.; Mirkin, C.A.; Hupp, J.T. Synthesis and Hydrogen Sorption Properties of Carborane Based Metal-Organic Framework Materials. *J. Am. Chem. Soc.* **2007**, *129*, 12680–12681. [CrossRef]
56. Dash, B.P.; Satapathy, R.; Maguire, J.A.; Hosmane, N.S. Polyhedral boron clusters in materials science. *New J. Chem.* **2011**, *35*, 19551972. [CrossRef]

Article

DFT Surface Infers Ten-Vertex Cationic Carboranes from the Corresponding Neutral *closo* Ten-Vertex Family: The Computed Background Confirming Their Experimental Availability

Michael L. McKee [1], Jan Vrána [2], Josef Holub [3], Jindřich Fanfrlík [4] and Drahomír Hnyk [3,*]

[1] Department of Chemistry and Biochemistry, Auburn University, Auburn, AL 36849, USA; mckeeml@auburn.edu
[2] Faculty of Chemical Technology, University of Pardubice, CZ-532 10 Pardubice, Czech Republic; jan.vrana@upce.cz
[3] Institute of Inorganic Chemistry of the Czech Academy of Sciences, CZ-250 68 Husinec-Řež, Czech Republic; holub@iic.cas.cz
[4] Institute of Organic Chemistry and Biochemistry of the Czech Academy of Sciences, CZ-166 10 Praha 6, Czech Republic; jindrich.fanfrlik@uochb.cas.cz
* Correspondence: hnyk@iic.cas.cz

Abstract: Modern computational protocols based on the density functional theory (DFT) infer that polyhedral *closo* ten-vertex carboranes are key starting stationary states in obtaining ten-vertex cationic carboranes. The rearrangement of the bicapped square polyhedra into decaborane-like shapes with open hexagons in boat conformations is caused by attacks of N-heterocyclic carbenes (NHCs) on the *closo* motifs. Single-point computations on the stationary points found during computational examinations of the reaction pathways have clearly shown that taking the "experimental" NHCs into account requires the use of dispersion correction. Further examination has revealed that for the purposes of the description of reaction pathways in their entirety, i.e., together with all transition states and intermediates, a simplified model of NHCs is sufficient. Many of such transition states resemble in their shapes those that dictate Z-rearrangement among various isomers of *closo* ten-vertex carboranes. Computational results are in very good agreement with the experimental findings obtained earlier.

Keywords: carboranes; N-heterocyclic carbenes; cations; DFT; reaction pathways

Citation: McKee, M.L.; Vrána, J.; Holub, J.; Fanfrlík, J.; Hnyk, D. DFT Surface Infers Ten-Vertex Cationic Carboranes from the Corresponding Neutral *closo* Ten-Vertex Family: The Computed Background Confirming Their Experimental Availability. *Molecules* **2023**, *28*, 3645. https://doi.org/10.3390/molecules28083645

Academic Editors: Michael A. Beckett and Igor B. Sivaev

Received: 27 March 2023
Revised: 12 April 2023
Accepted: 15 April 2023
Published: 21 April 2023

1. Introduction

The exclusive chemistry of polyhedral carborane clusters is a result of the presence of delocalized electron-deficient bonding [1,2] based on the formation of multi-center–two-electron (mc-2e) bonds, which do not exist in classical organic chemistry, characterized by classical two-center–two-electron (2c-2e) bonds. When two carbons in a carborane cluster form the nearest-neighbor separation, such bonding is also of the 2c-2e nature [3]. The BH vertices in the carboranes are assembled in trigonal faces, which enables the creation of three-dimensional shapes, with octahedron [4], icosahedron, and bicapped square antiprism [2] being among the most prevalent, and are related to *closo* systems [1,2]. The preparation and subsequent reactivity of these clusters has been extensively studied by experiments [1,2].

In contrast to well-understood reaction mechanisms in organic chemistry, those in boron cluster chemistry can be very complex because there are very small energy differences between many possible intermediates and transition states. On that basis, the reaction of boron hydrides may involve many competing pathways [5]. Consequently, relatively little progress has been made so far in the understanding of the reaction mechanisms of boron hydrides and carboranes of various molecular shapes [6–8]. Apart from the reaction of the icosahedral carboranes [1,2], a few conversions of ten-vertex *closo* carboranes have already emerged.

The *closo*-$C_2B_8H_{10}$ molecular shape adopting a bicapped square antiprism (Scheme 1) exists in seven positional isomers, only three of which (1,2-, 1,6-, and 1,10-isomers) are known experimentally [2]. The most stable one is the 1,10-isomer and the least stable one is the 2,3-isomer [9]. The mutual isomerization of these carboranes has recently been interpreted as a result of Z-rearrangement, which is characterized by a transition state of the decaborane molecular shape between various pairs of these seven isomers. Therefore, upon reaction with hydroxides and amines 1,2-isomer is converted into open *arachno* systems: see [10,11], respectively. These reaction pathways have recently been searched computationally [12]; these attempts have shown that the products are negatively charged or neutral. The same applies to the reaction of this isomer with "wet" fluoride [13]. Such charge conditions are very common in boron cluster chemistry. However, when three isomers of *closo*-1,N-$C_2B_8H_{10}$ (N = 2,6,10) were allowed to react with N-heterocyclic carbenes (NHCs, Scheme 1), the cationic arrangements of decaborane shapes were obtained by HCl acidification of the neutral species of the decaborane shapes [14]. Note that such shapes were already spotted when these *closo* structural motifs originated by click chemistry [15], also including the aforementioned Z-rearrangement [16]. The existence of cationic ten-vertex carboranes, which retain the bicapped square antiprismatic arrangement only in the case of *closo*-1,10-$C_2B_8H_{10}$, prompted us to search the corresponding reaction pathways of their formation computationally by employing modern computational protocols.

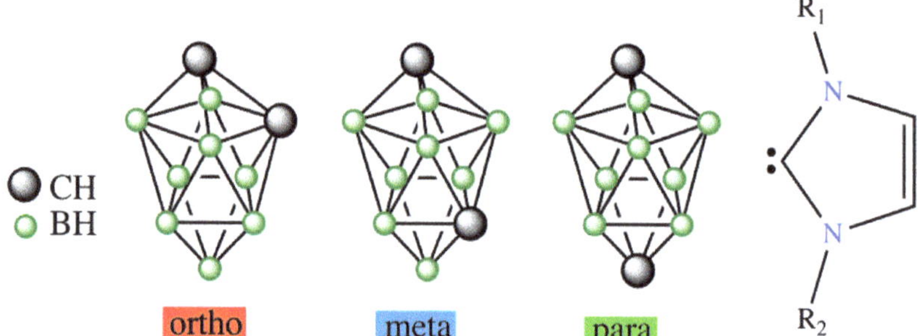

Scheme 1. Molecular diagrams of *closo*-1,N-$C_2B_8H_{10}$ (red: N = 2, blue: N = 6, green: N = 10) and N-heterocyclic carbenes.

2. Results and Discussion

2.1. The Reactions of closo-1,2-$C_2B_8H_{10}$ (o) with NHC

The reaction starts with the attack of a lone electron pair of a NHC on one of the most positive boron atom pairs within the cage, i.e., on B(3) [10–12], and following this proceeds very smoothly; **o-2**, conforming to the decaborane shape, is afforded (in line with Ref. [14]) via two transition states and one intermediate **1**, in which the opening of the square antiprismatic cage is marked: see Figure 1. The position of the C-C vector in **o-2** is consistent with that observed in the transition state of the isomerization of *closo*-1,2-$C_2B_8H_{10}$ to itself due to Z-isomerization, with the latter being experimentally observed in the R-substituted derivatives of *closo*-1,2-$C_2B_8H_{10}$ [16].

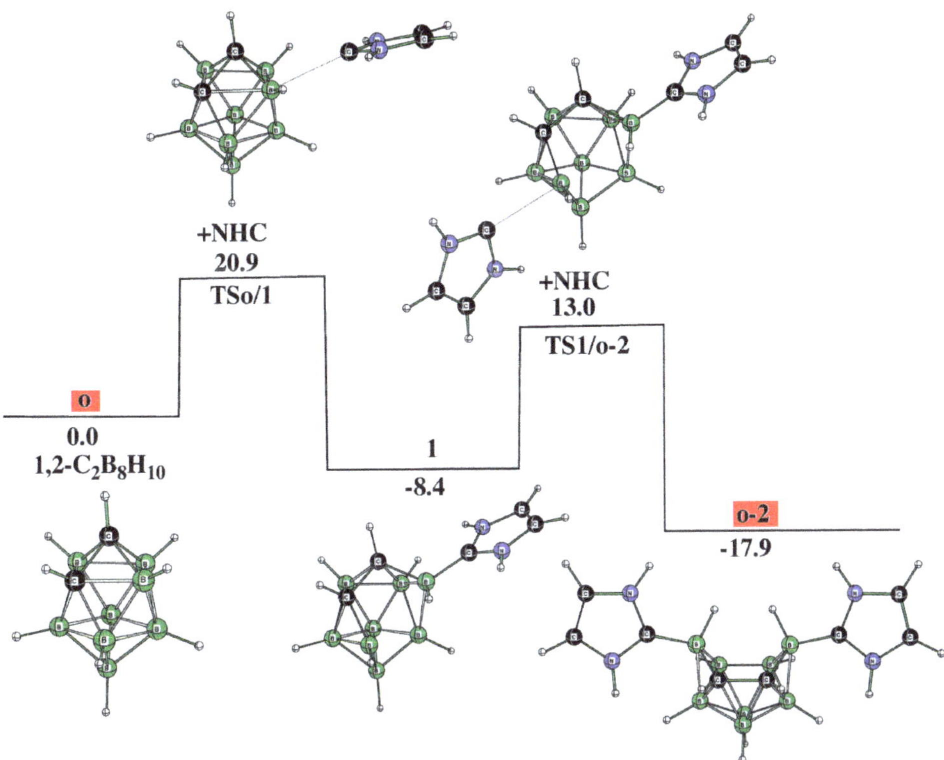

Figure 1. Relative free energies (kcal.mol^{-1}) of the individual stationary points on the potential energy surface (PES) of the formation of **o-2** (see Tables 1 and S1 and the Section 3 for details).

Table 1. Free energies (kcal.mol^{-1}) of the experimentally inferred products (characterized by ab initio/GIAO/NMR and X-ray diffraction structural tools [14]) relative to 1,2-C₂B₈H₁₀ (o) + one or two NHC(s).

Reactant to Product	B3LYP [a]	B3LYP [b]
o to o-1	−14.4	−21.8 (−2.7)
o to o-2	−17.9	−21.8 (24.6) [c]
m to m-2	−34.7	−47.4 (−4.0) [c]
p to p-2	−34.1	−49.0 (−6.1)
o-2 to o-2a	−33.2	−47.4 (0.2)
m-2 to m-2a	−42.3	−63.6 (−18.7)
p-2 to p-2a	−26.7	−50.1 (−5.8)
p-2a to 20a	−49.5	−61.2 (−41.6)

[a] SMD(DiethylEther)/B3LYP/6-311+G(2d,p)//B3LYP/6-31G(d), with NHC being employed for searching the reaction pathways as well as in single-point computations. [b] SMD(DiethylEther)/B3LYP/6-311+G(2d,p)+D3(BJ)//B3LYP/6-31G(d): single-point computations of the stationary points were carried out with Idip under the same reaction pathways as in (a). The values in parentheses result from not considering dispersion correction by Grimme [17] (D3(BJ)) in the single-point computations. [c] Of the two isomers differing in the mutual positions of the carbene moieties, only that with the lower energy is tabulated.

However, **1** appears to be an intermediate that brings about the formation of not only **o-2** but also other experimentally available decaborane-like shaped **o-1**, which can originate, as seen, from the kinetics depicted in Figure 2. Like **o-2**, **o-1** also occurs through two TSs and one intermediate, i.e., **2** in this case. This reaction path shows that the cage deformation, demonstrated by inspecting TS1/2, is achieved without using another NHC, with the latter being necessary for obtaining **o-2**. Both **o-1** and **o-2** have space for accommodating an extra hydrogen bridge to arrive at the cationic motifs **o-1a** and **o-2a**. The molecular shape of **o-1** is identical to the shape of the TS, through which the isomerization of 1,2- to 2,6-$C_2B_8H_{10}$ occurs due to Z-rearrangement [16].

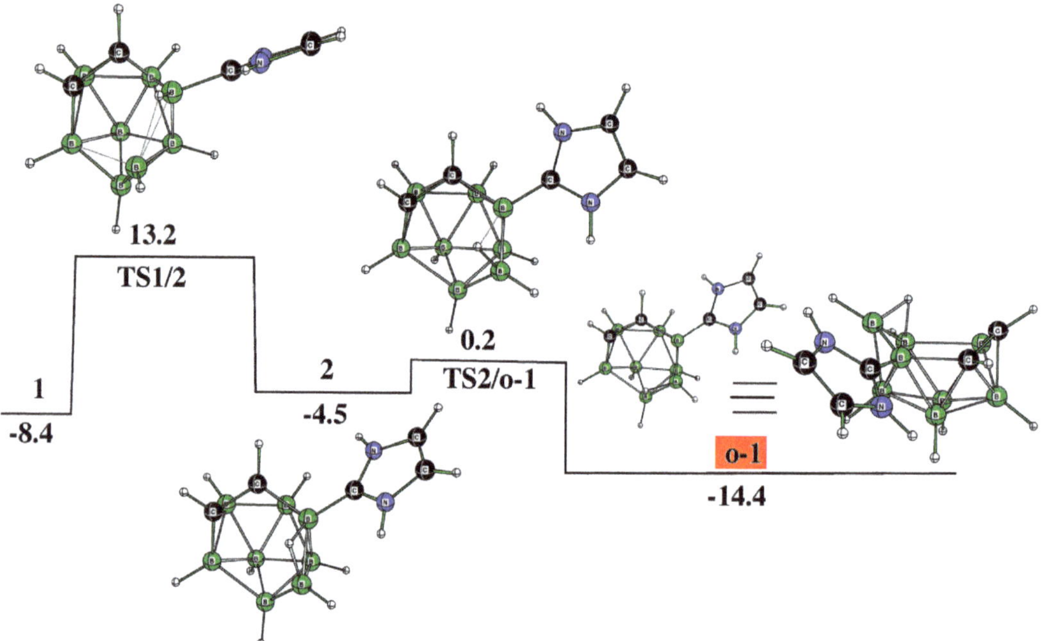

Figure 2. Relative free energies (kcal.mol^{-1}) of the individual stationary points on the potential energy surface (PES) of the formation of **o-1** (see Tables 1 and S1 and Section 3 for details).

It has already been mentioned that the individual species within the *closo*-$C_2B_8H_{10}$ family are prone to mutual isomerization. This process was also detected experimentally among the 1,2-(**o**), 1,6-(**m**), and 1,10-(**p**) isomers and was proposed to be explained in terms of Z-isomerization [16], although the previously suggested diamond-square-diamond (DSD) mechanism cannot be completely ruled out [18]. Guided by these results, we also examined whether such a linkage could exist among the corresponding $C_2B_8H_{10}$ architectures of the decaborane shape, i.e., among **o-2** (see Figure 1), **m-2** (see below), and **p-2** (see below). Indeed, such isomerizations have been computed to be feasible, as depicted in Figure 3. Seemingly, this overall process is intrinsically more demanding and occurs through a cascade of nine TSs and seven intermediates until the conversion from **o-2** to **p-2**, such as from **o** to **p**, has been accomplished: see below for the molecular diagrams of **m-2** and **p-2**.

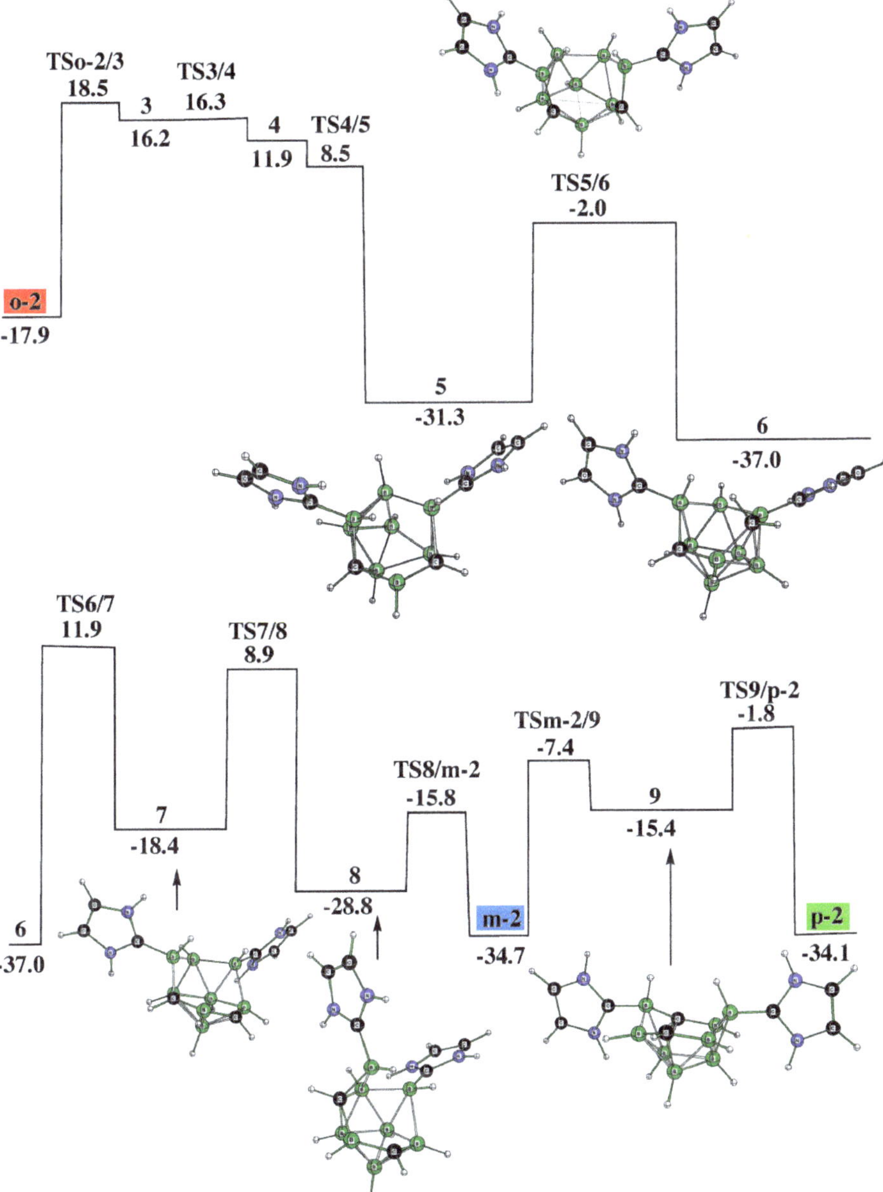

Figure 3. Relative free energies (kcal.mol^{-1}) of the individual stationary points on the potential energy surface (PES) of the formation of **m-2** and **p-2** from **o-2**. Only a few TSs and intermediates are shown for simplicity (see Tables 1 and S1 and Section 3 for details).

2.2. The Reactions of closo-1,6-C$_2$B$_8$H$_{10}$ (m) with NHC

Under favorable circumstances, **m-2** can directly originate from **o-2**; nevertheless, we tried to model the experimentally proved reaction [14] that transforms 1,6-C$_2$B$_8$H$_{10}$, **m**, to experimentally detected **m-2** [14]. **TSm/10** is a result of the NHC attack, see Figure 4, on the positive B(3) atom, which is in analogy with the formation of **o-2**. However, by continuing from this stationary point, the reaction pathway becomes quite complex as proven by locating many more TSs and intermediates on this surface (Figure 4).

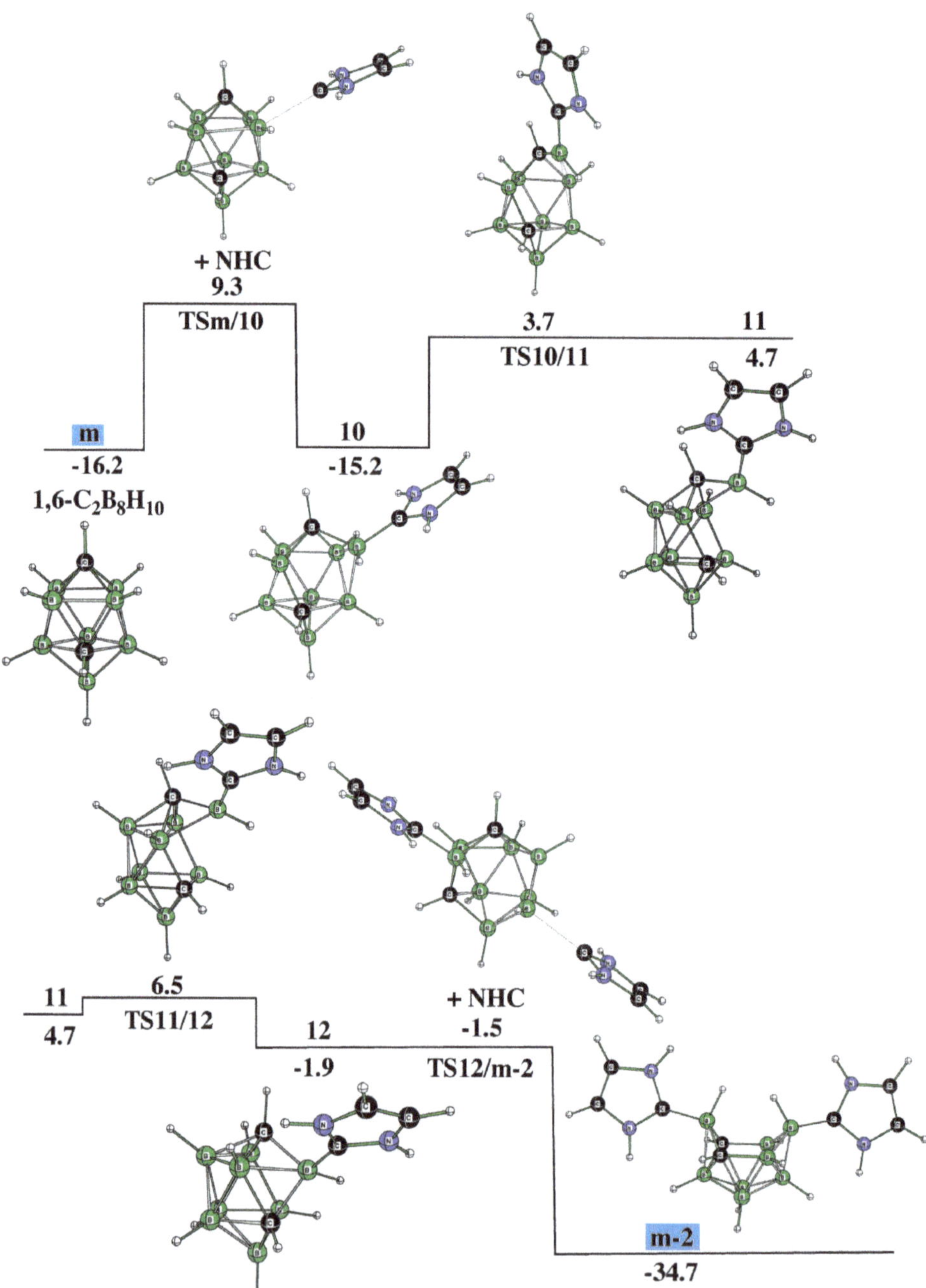

Figure 4. Relative free energies (kcal.mol^{-1}) of the individual stationary points on the potential energy surface (PES) of the formation of **m-2** from **m** (see Tables 1 and S1 and Section 3 for details).

2.3. The Reactions of closo-1,10-C₂B₈H₁₀ (p) with NHC

As in the **o**-to-**o-2** transformation, two TSs and one intermediate (i.e., **13**) are needed to obtain **p-2** from **p**. In other words, such a process is relatively easy to accomplish. As in the preceding two *closo* to electron-count-obeying *hypercloso*-like transformations (note that the NHC acts as an electron donor to the electron-deficient *closo* shape, with the latter appearing to be a slight electron acceptor experimentally [19]), it is the positively charged B(3)H vertex that accepts a lone pair of the C atoms within the carbene molecule. Intermediate **13** demonstrates the initiation of such a cage-opening process.

Interestingly, a feature of the experimentally known **p-2** bears a strong resemblance to the transition state of the 2,7- to 1,10-C₂B₈H₁₀ isomerization under the recently coined Z-rearrangement [16], see Figure 5.

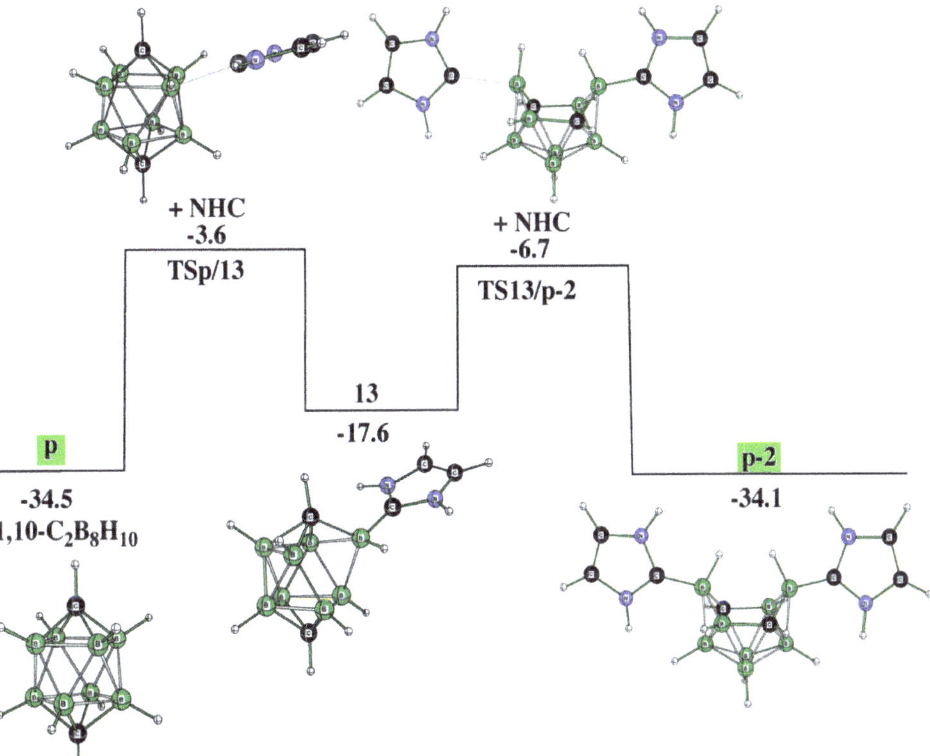

Figure 5. Relative free energies (kcal.mol⁻¹) of the individual stationary points on the potential energy surface (PES) of the formation of **p-2** from **p** (see Tables 1 and S1 and the Section 3 for details).

Synthetic routes [16] have also revealed the cage closure from both **o-2** and **m-2** to obtain **p**, with the latter being the starting stationary point to obtain **p-2** as described in [16] and computed in this work. In other words, the connection between **o-2**, **m-2**, and **p-2** examined experimentally has been confirmed computationally. Relatively stable intermediates **5** and **6**, located also in the pathway from **o-2** to **p-2** (see Figure 3), open the cage closure with the resulting **p** (Figure 6). This complex procedure is surmounted by six TSs, which proceed through four intermediates. The fifth intermediate, **13**, responsible for the elimination of the second NHC, links this pathway together with that describing the cage opening **p** to **p-2** (Figure 3). In essence, Figures 3, 5 and 6 illustrate the possible reversibility of the **p**-to-**p-2** and **p-2**-to-**p** processes.

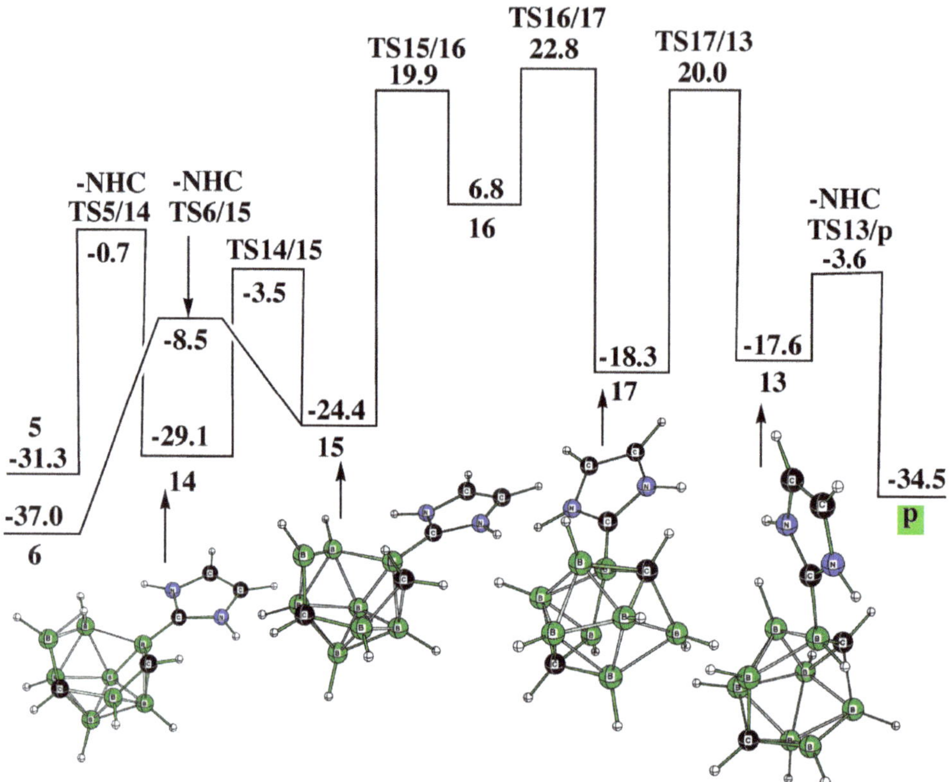

Figure 6. Relative free energies (kcal.mol^{-1}) of the individual stationary points on the potential energy surface (PES) of the reformation of **p** from both **o-2** and **m-2** (see Tables 1 and S1 and Section 3 for details).

In essence, both **o-2** and **m-2** are very easily protonated (with HCl) to provide the *cationic* structures **o-2a** and **m-2a**. These processes are exothermic, e.g., the energy gain of the first process is computed to be 15.3 kcal.mol^{-1}. In contrast, the protonation of **p-2** is mechanistically more complex: "**p-2a**" (not the final product of the protonization) with the same molecular shape as **o-2a** and **m-2a** is not the final product of this acidification. The presence of the extra hydrogen bridge in the "**p-2a**" arrangement results in the elimination of one NHC; through two cationic TSs and one intermediate, the cationic product **20a** is then formed as the final product, see Figure 7.

Table 1 shows that the energetic balance in both columns exhibits a very reasonable correlation (see also Figure 8). In other words, the necessity of including the dispersion correction when recomputing the energetic criteria with the carbene bearing 1,3-*i*Pr2-C$_6$H$_3$, i.e., with Idip (b) with respect to (a), indicates agreement with the fact that only the electron sextet on the carbene moiety justifies the modeling of the Idip with simpler NHCs in the search of the reaction pathways, to reduce the problem to more manageable dimensions. Note that for the cationic species, the necessity of considering dispersion energy is less crucial.

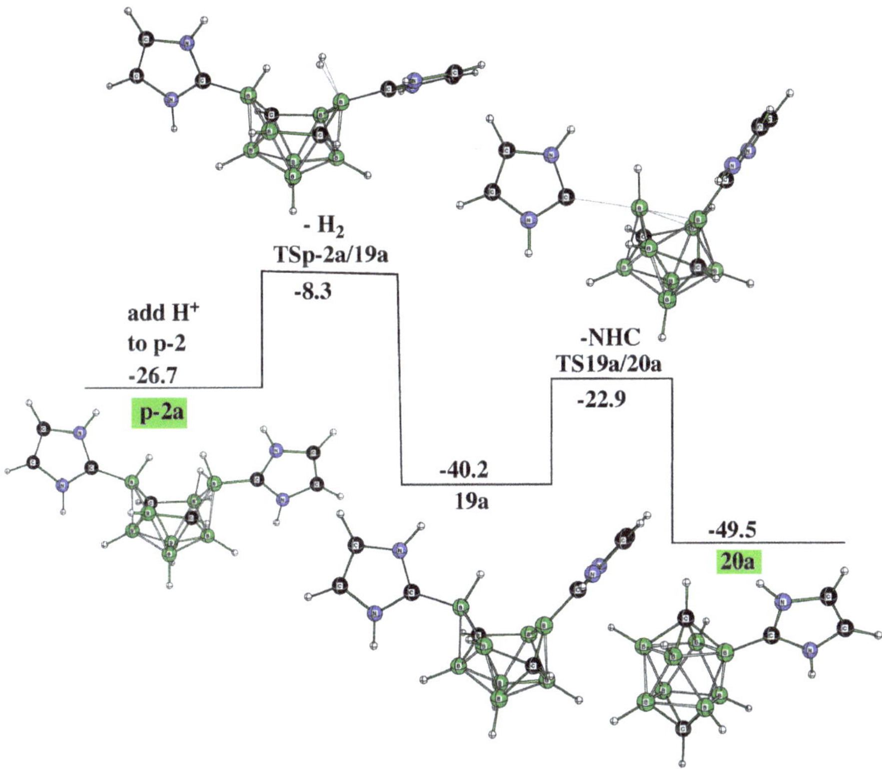

Figure 7. Relative free energies (kcal.mol^{-1}) of the individual stationary points on the potential energy surface (PES) of the entire protonation of **"p-2a"** (see Tables 1 and S1 and Section 3 for details).

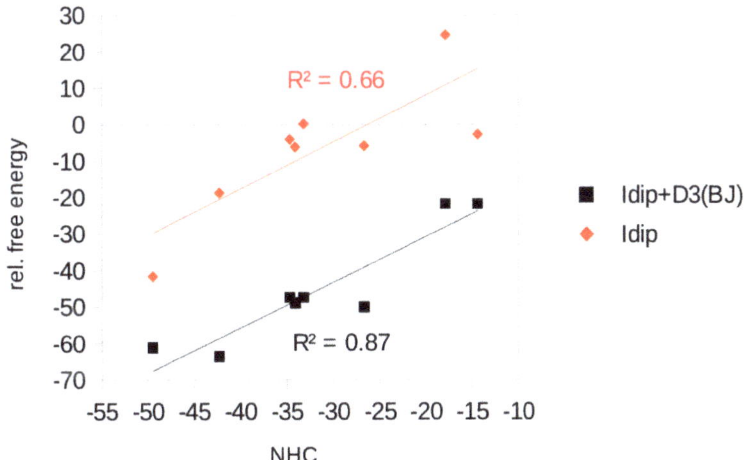

Figure 8. The correlation of the relative free energies of the existing species given in Table 1, optimized both with dispersion correction (black, Idip + D3(BJ)) and without it (red, Idip), with those related to stationary points in which Idip is simplified with NHC.

2.4. Further Possible Ten-Vertex Carbocations

o-2 is the most stable cation among those derived earlier. The very stable C-C bond in the carborane cage might be one of the driving forces to support this. The possible cleavage of this bond may generate another cation, **23a**, which resembles the TS responsible for the isomerization of the 2,3-$C_2B_8H_{10}$ to the 1,6-isomer (see Figure 9). The driving force of this pathway is the formation of CH_2, which is, like the C-C vector, a stable moiety. Moreover, such a pathway clearly goes through the stable but not isolable cation **22a**, resembling **o-2a** from which it differs in the position of the hydrogen bridge and the mutual position of the NHC-type pentagons.

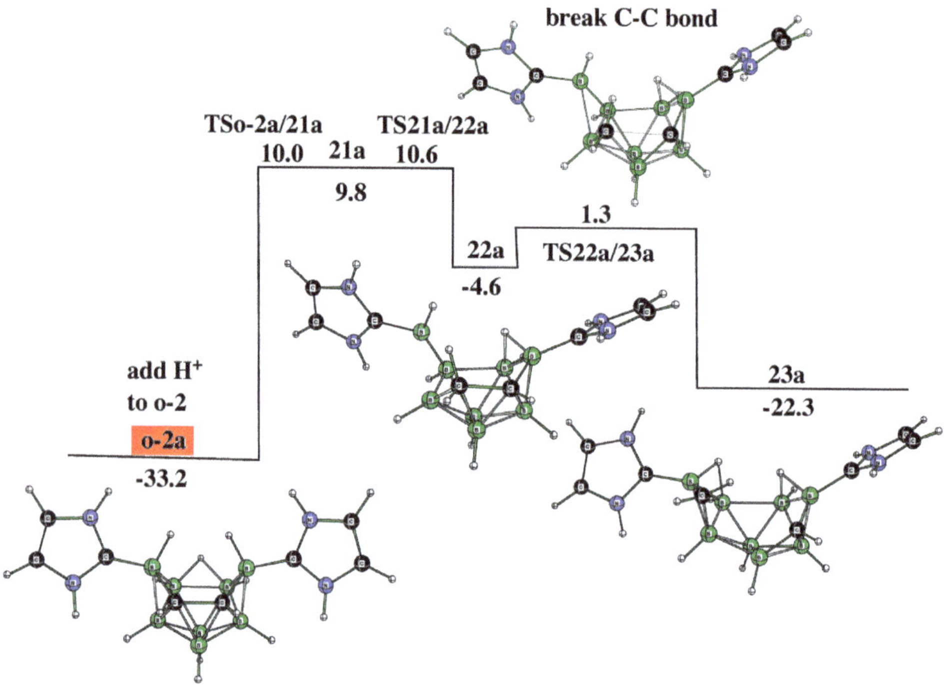

Figure 9. Relative free energies (kcal.mol^{-1}) of the individual stationary points on the potential energy surface (PES) of the formation of the cation, **23a**, through another cation, **22a** (see Tables 1 and S1 and Section 3 for details).

m-2a is more stable than "**p-2a**", on the basis of which it can serve as the more important intermediate in further pathway linkages. Figure 10 illustrates how this cation, by losing two hydrogens, can also be converted to **20a**, the latter being located also in the pathway depicted in Figure 7. Such a conversion occurs via the cation **25a**, which appears to be in analogy with **20a**. Interestingly, such a process is accomplished through two Z-rearrangements, as revealed by the last two TSs [16].

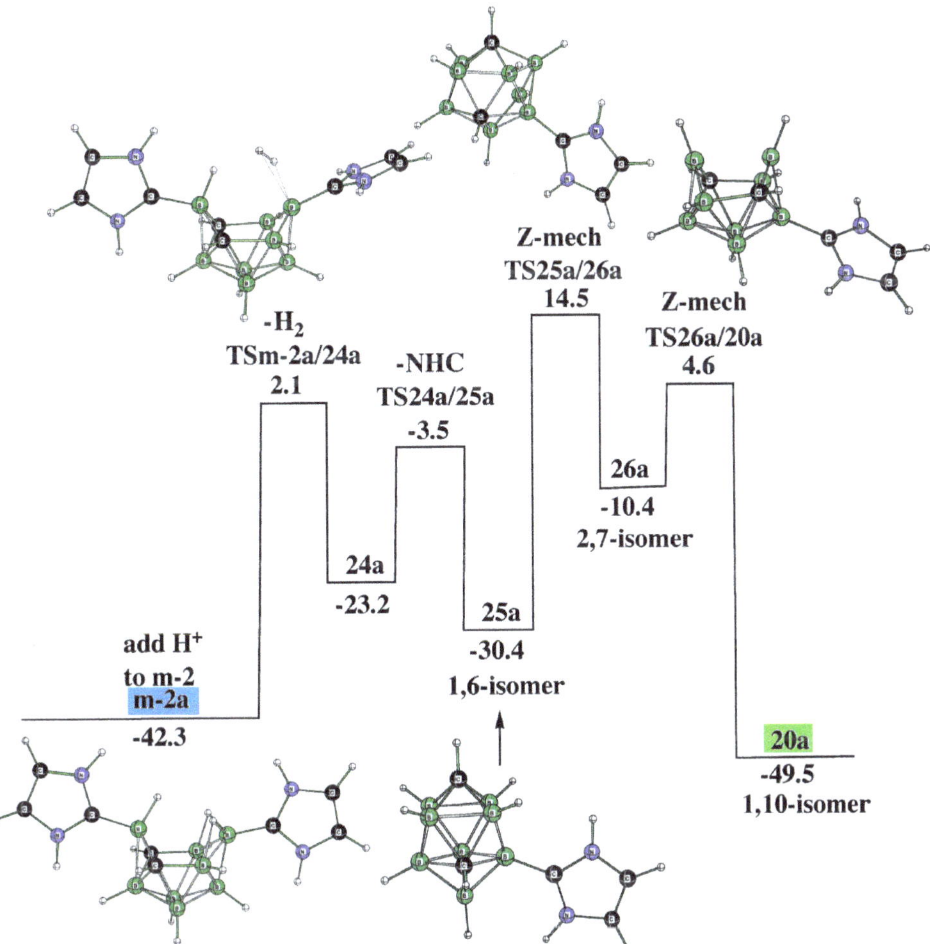

Figure 10. Relative free energies (kcal.mol^{-1}) of the individual stationary points on the potential energy surface (PES) of the transformation of **m-2a** through the cationic **25a** based on the 1,6-isomer and two Z-rearrangements to the final **20a** (see Tables 1 and S1 and Section 3 for details).

3. Methods

Free energies of all the stationary points appeared in the reactions of *closo*-1,2-, 1,6-, and 1,10-$C_2B_8H_{10}$ with a NHC in which both bulky Rs, represented in Ref. [14] with 1,3-iPr$_2$-C_6H_3 and abbreviated as Idip, were replaced with hydrogens during the pathway examinations (NHC in further notations below), and the SMD(DiethylEther)/B3LYP/6-311+G(2d,p)//B3LYP/6-31G(d) computational protocol was used. Seemingly, DiethylEther turned out to be a good approximation for CPME (not in the library of G16), in the latter protonizations that occurred [14]. The reactants and products were connected from the corresponding transition states in terms of computing the intrinsic reaction coordinates (IRC). However, such computations were not carried out in cases where the reactant and product were obvious from inspecting the corresponding transition vector, with the latter being revealed from the second derivative analysis. Such physically correct TSs are recognized as being responsible, e.g., for a simple addition of a NHC or a hydrogen movement. Due to the presence of benzene rings and iPr groups in Idip, it has many non-bonded distances that are below the sum of the van der Waals radii. Therefore, the importance of the inclusion of a dispersion correction was obvious. This was resolved by adding the D3(BJ) correction [17,20] to the B3LYP/6-311+G(2d,p) model chemistry in the single-point computations of all experimentally detectable species, as shown in Table 1 when Idip was used in the computations. The "a" indicates a cation, the integers are unobserved intermediates, and TSs ("reactant"/"products") refers to transition states; these notations are common for the reaction pathways below. All the computations were performed using Gaussian16, in which the above model chemistries and basis sets are incorporated [21]. *closo*-1,2-, 1,6-, and 1,10-$C_2B_8H_{10}$ and the corresponding stationary points are marked in red trace, blue trace, and green trace, respectively.

4. Conclusions

The computational protocol based on the density functional theory computational tool has revealed the formation of various cationic polyhedral structures of *nido* and *closo* molecular shapes, as a result of the reaction of ten-vertex *closo* dicarboranes with a N-heterocyclic carbene modeled with a simple $C_3N_2H_4$ pentagonal belt. Since previously obtained experimental [11]B chemical shifts of the cations were very successfully reproduced by high-level computations using this simple carbene, this approximation is justified. Based on that, all previously experimentally observed cations were computationally processed in the single-point computations in the same manner as the syntheses reported in Ref. [14], i.e., with a N-heterocyclic carbene in which nitrogen-bonded hydrogens were replaced with 1,3-*i*Pr-C_6H_3. Seemingly, the corresponding computed energetic balance was in line with that in which the simple carbene was considered only if the correction for dispersion energy was used in these more demanding computational efforts. Further computational experiments open the possibility of the existence of other cationic isomers of such nature, which offers experimenters further challenges.

Supplementary Materials: The following supporting information can be downloaded at: https://www.mdpi.com/article/10.3390/molecules28083645/s1, Table S1: Total Energies (hartrees) and Free Energies (kcal.mol^{-1}) Relative to 2NHC+1,2-$C_2B_8H_{10}$ or 2Idip+1,2-$C_2B_8H_{10}$, Table S2: Cartesian Coordinates of Stationary Points listed in Table S1 Optimized at the B3LYP/6-31G(d) Level.

Author Contributions: Computations, M.L.M., J.F. and D.H.; extension of the synthetic background, J.V. and J.H.; paper writing up, D.H. All authors have read and agreed to the published version of the manuscript.

Funding: We thank the Czech Science Foundation (project no. 22-03945S) for financial support. M.L.M. thanks Auburn University for access to the HOPPER computer.

Institutional Review Board Statement: Not applicable.

Informed Consent Statement: Not applicable.

Data Availability Statement: Data supporting reported results may be available on demand (to the corresponding author).

Conflicts of Interest: The authors declare no conflict of interest.

Sample Availability: Not applicable.

References

1. Hnyk, D.; Wann, D.A. Molecular Structures of Free Boron Clusters. In *Boron: The Fifth Element*; Challenges and Advances in Computational Chemistry and Physics; Hnyk, D., McKee, M., Eds.; Springer: Dordrecht, The Netherlands, 2016; Volume 20, pp. 17–48.
2. Grimes, R.N. *Carboranes*, 3rd ed.; Academic Press: Cambridge, MA, USA, 2016.
3. Melichar, P.; Hnyk, D.; Fanfrlík, J. A Systematic Examination of Classical and Multi-center Bonding in Heteroborane Clusters. *Phys. Chem. Chem. Phys.* **2018**, *20*, 4666–4675. [CrossRef] [PubMed]
4. Keller, W.; Hofmann, M.; Sárosi, M.B.; Fanfrlík, J.; Hnyk, D. Reactivity of Perhalogenated Octahedral Phospha- and Arsaboranes toward THF: A Joint Experimental/Computational Study. *Inorg. Chem.* **2022**, *61*, 16565–16572. [CrossRef] [PubMed]

5. McKee, M.L. Deconvoluting the Reaction Path from $B_{10}H_{14}$ Plus BH_4^- to $B_{12}H_{12}^{2-}$. Can Theory Make a Contribution? In *Boron: The Fifth Element*; Challenges and Advances in Computational Chemistry and Physics; Hnyk, D., McKee, M., Eds.; Springer: Dordrecht, The Netherlands, 2016; Volume 20, pp. 121–138.

6. McKay, D.; Macgregor, S.A.; Welch, A.J. Isomerisation of *nido*-$[C_2B_{10}H_{12}]^{2-}$ Dianions: Unprecedented Rearrangements and New Structural Motifs in Carborane Cluster Chemistry. *Chem. Sci.* **2015**, *6*, 3117–3128. [CrossRef] [PubMed]

7. Shameena, O.; Pathak, B.; Jemmis, E.D. Theoretical Study of the Reaction of $B_{20}H_{16}$ with McCN: *Closo/Closo* to *Closo/Nido* Conversion. *Inorg. Chem.* **2008**, *47*, 4375–4382. [CrossRef] [PubMed]

8. Štíbr, B.; Holub, J.; Bakardjiev, M.; Lane, P.D.; McKee, M.L.; Wann, D.A.; Hnyk, D. Unusual Cage Rearrangements in 10-Vertex *nido*-5,6-Dicarbaborane Derivatives: An Interplay between Theory and Experiment. *Inorg. Chem.* **2017**, *56*, 852–860. [CrossRef] [PubMed]

9. Schleyer, P.V.R.; Najafian, K. Stability and Three-Dimensional Aromaticity of closo-Monocarbaborane Anions, $CB_{n-1}H_{n-}$, and *closo*-Dicarboranes, $C_2B_{n-2}H_n$. *Inorg. Chem.* **1998**, *37*, 3454–3457. [CrossRef] [PubMed]

10. Hnyk, D.; Holub, J. Handles for the Dicarbadodecaborane Basket Based on $[arachno$-5,10-$C_2B_8H_{13}]^-$: Oxygen. *Dalton Trans.* **2006**, *22*, 2620–2622. [CrossRef] [PubMed]

11. Janoušek, Z.; Dostál, R.; Macháček, J.; Hnyk, D.; Štíbr, B. The First Member of the Eleven-Vertex Azadicarbaborane Series, 1,6,9-$NC_2B_8H_{13}$ and its N-alkyl Derivatives. *Dalton Trans.* **2006**, *39*, 4664–4671. [CrossRef] [PubMed]

12. Holub, J.; Fanfrlík, J.; McKee, M.L.; Hnyk, D. Reactions of Experimentally Known *Closo*-$C_2B_8H_{10}$ with Bases. A Computational Study. *Crystals* **2020**, *10*, 896. [CrossRef]

13. Holub, J.; Bakardjiev, M.; McKee, M.L. Ten-Vertex *Closo*-Carboranes React with "Wet" Fluoride: A Direct *Closo*-to-*Arachno* Transformation as a Result of a Hydride Transfer. *Inorg. Chim. Acta* **2022**, *538*, 12100. [CrossRef]

14. Vrána, J.; Holub, J.; Samsonov, M.A.; Růžičková, Z.; Cvačka, J.; McKee, M.L.; Fanfrlík, J.; Hnyk, D.; Růžička, A. Access to Cationic Polyhedral Carboranes via Dynamic Cage Surgery with *N*-heterocyclic Carbenes. *Nat. Commun.* **2021**, *12*, 4971. [CrossRef] [PubMed]

15. Tok, O.L.; Bakardjiev, M.; Štíbr, B.; Hnyk, D.; Holub, J.; Padělková, Z.; Růžička, A. Click Dehydrogenation of Carbon-Substituted *nido*-5,6-$C_2B_8H_{12}$ Carboranes: A General Route to *closo*-1,2-$C_2B_8H_{10}$ Derivatives. *Inorg. Chem.* **2016**, *55*, 8839–8843. [CrossRef] [PubMed]

16. Bakardjiev, M.; Holub, J.; Růžičková, Z.; Růžička, A.; Fanfrlík, J.; Štíbr, B.; McKee, M.L.; Hnyk, D. Transformation of Various Multicenter Bondings within Bicapped-Square Antiprismatic Motifs: Z-Rearrangement. *Dalton Trans.* **2021**, *50*, 12019–12334. [CrossRef] [PubMed]

17. Grimme, S.; Antony, J.; Ehrlich, S.; Krieg, H. A Consistent and Accurate Ab initio Parametrization of Density Functional Dispersion Correction (DFT-D) for the 94 Elements H–Pu. *J. Chem. Phys.* **2010**, *132*, 154104. [CrossRef] [PubMed]

18. Gimarc, B.M.; Ott, J.J. Isomerization of Carboranes $C_2B_6H_8$, $C_2B_8H_{10}$ and $C_2B_9H_{11}$ by the Diamond-Square-Diamond Rearrangement. *J. Am. Chem. Soc.* **1987**, *109*, 1388–1392. [CrossRef]

19. Hnyk, D.; Všetečka, V.; Drož, L.; Exner, O. Charge Distribution within 1,2-Dicarba-closo-dodecaborane: Dipole Moments of its Phenyl Derivatives. *Collect. Czech. Chem. Commun.* **2001**, *66*, 1375–1379. [CrossRef]

20. Maué, D.; Streber, P.H.; Bernhard, D.; Rösel, S.; Schreiner, P.R.; Gerhards, M. Dispersion-Bound Isolated Dimers in the Gas Phase: Observation of the Shortest Intermolecular C-H . . . H-C Distance via Stimulated Raman Spectroscopy. *Angew. Chem. Int. Ed.* **2021**, *60*, 11305–11309. [CrossRef] [PubMed]

21. Frisch, M.J.; Trucks, G.W.; Schlegel, H.B.; Scuseria, G.E.; Robb, M.A.; Cheeseman, J.R.; Scalmani, G.; Barone, V.; Petersson, G.A.; Nakatsuji, H.; et al. *Gaussian 16, Revision C.01*; Gaussian, Inc.: Wallingford, CT, USA, 2016.

Article

Synthesis of Novel Carborane-Containing Derivatives of RGD Peptide

Alexander V. Vakhrushev, Dmitry A. Gruzdev *, Alexander M. Demin, Galina L. Levit and Victor P. Krasnov *

Postovsky Institute of Organic Synthesis, Russian Academy of Sciences (Ural Branch),
620108 Ekaterinburg, Russia; avv@ios.uran.ru (A.V.V.); amd2002@mail.ru (A.M.D.); ca512@ios.uran.ru (G.L.L.)
* Correspondence: gruzdev-da@ios.uran.ru (D.A.G.); ca@ios.uran.ru (V.P.K.)

Abstract: Short peptides containing the Arg-Gly-Asp (RGD) fragment can selectively bind to integrins on the surface of tumor cells and are attractive transport molecules for the targeted delivery of therapeutic and diagnostic agents to tumors (for example, glioblastoma). We have demonstrated the possibility of obtaining the *N*- and *C*-protected RGD peptide containing 3-amino-*closo*-carborane and a glutaric acid residue as a linker fragment. The resulting carboranyl derivatives of the protected RGD peptide are of interest as starting compounds in the synthesis of unprotected or selectively protected peptides, as well as building blocks for preparation of boron-containing derivatives of the RGD peptide of a more complex structure.

Keywords: 3-amino-1,2-dicarba-*closo*-dodecaborane; RGD peptide; linker fragment; protecting groups

Citation: Vakhrushev, A.V.; Gruzdev, D.A.; Demin, A.M.; Levit, G.L.; Krasnov, V.P. Synthesis of Novel Carborane-Containing Derivatives of RGD Peptide. *Molecules* **2023**, *28*, 3467. https://doi.org/10.3390/molecules28083467

Academic Editors: Michael A. Beckett, Igor B. Sivaev and Vito Capriati

Received: 20 March 2023
Revised: 7 April 2023
Accepted: 12 April 2023
Published: 14 April 2023

1. Introduction

The search for efficient pharmaceuticals for the diagnostics and treatment of tumor diseases is one of the most urgent problems of medicinal chemistry. Currently, molecular vectors—namely, short peptides, antibodies, aptamers, and other compounds that provide targeted delivery of the functional part of the molecule—are widely used in the constructs of targeted therapy agents. The mechanism of their selective accumulation is based on the interaction of the vector with a target molecule, typically a receptor protein located on the surface of tumor cells.

Today, the RGD peptide (L-arginyl-glycyl-L-aspartic acid, Arg-Gly-Asp) and structurally similar peptides (Figure 1) are widely used as molecular vectors in the drug design of targeted agents for the diagnostics and therapy of tumor diseases [1–6]. The RGD amino acid sequence has a tropism for cell adhesion proteins, integrins, which are particularly overexpressed in tumor cells (namely, $\alpha v \beta_3$ and $\alpha v \beta_5$ integrins). Integrin inhibitors represent an important class of agents for the treatment of tumors, macular degeneration, acute coronary syndrome, and other diseases [7,8]. Among the derivatives and analogs of the RGD peptide, a number of integrin inhibitors have been found [9,10]. Cilengitide, a selective inhibitor of $\alpha v \beta_3$ and $\alpha v \beta_5$ integrins proposed for the treatment of recurrent glioblastoma [11,12], has not passed phase III clinical trials because of insufficient pharmacokinetic parameters [13]. At the same time, studies of a number of other integrin inhibitors related to the RGD peptide are currently ongoing [14–17].

Based on the RGD peptide, a wide range of conjugates containing isotopic [18–22], fluorescent [23–26], or magnetic contrast labels [27–29], residues of cytostatic molecules [30–34], as well as agents for photodynamic therapy [35–38] have been synthesized. For efficient binding of the RGD peptide-based compounds to integrins (for example, on the surface of tumor cells), it is preferrable that the guanidine fragment of arginine and the carboxyl group of aspartic acid remain unsubstituted [39].

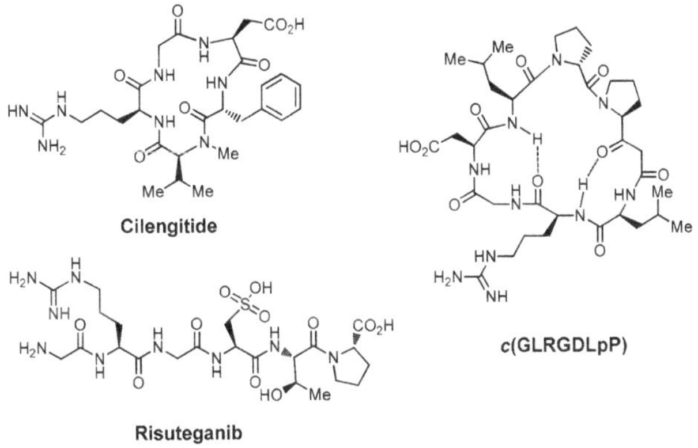

Cilengitide

c(GLRGDLpP)

Risuteganib

Figure 1. Biologically active compounds based on the RGD motif.

One of the emerging approaches to tumor treatment is boron neutron capture therapy (BNCT). This method is based on the ability of the ^{10}B isotope to interact with thermal neutrons with the emission of ^{4}He and ^{7}Li nuclei, which locally damage cells containing boron compounds [40–42]. A crucial condition for the application of BNCT is the selective accumulation of boron-containing molecules by tumor cells. The design of low-toxic boron-containing tumor-targeting compounds is an urgent task of modern medicinal chemistry [43–46]. An important group of potential boron delivery agents are derivatives of 1,2-dicarba-*closo*-dodecaborane (carborane), the molecule of which contains ten boron atoms and can be modified using various functional groups. Certain properties of carboranes such as stability under physiological conditions and low toxicity make them unique pharmacophores for the design of new biomimetics [47–49]. Carborane conjugates with natural amino acids and peptides are of particular interest from the point of view of drug design of BNCT agents, as well as theranostic agents [50]. In particular, carborane-containing derivatives of the c(RGDfK) peptide have been used for adhesion of cells expressing the $\alpha v \beta_3$ integrin receptors [51], as well as for boron delivery to tumor cells [52,53]. The boron-containing conjugate of the cyclic RGD peptide was able to selectively accumulate in murine SCCVII carcinoma cells but was highly toxic [53]. Boron-containing nanoparticles containing FITC-labeled RGD-K peptide residues [54] or internalizing RGD fragments [55,56] were selectively accumulated by ALTC1S1 glioma, GL261 glioma, and A549 adenocarcinoma cells. Modification of the sodium dodecaborate-loaded liposomes by c(RGDfK) [57,58] and c(RGDyC) [59] peptides made it possible to achieve their binding to human umbilical cord endothelial cells. The fact that RGD-functionalized *closo*-dodecaborate albumin conjugates are capable of accumulating in U87 MG xenografts has recently demonstrated the efficacy of BNCT in in vivo experiments [60]. The c(RGDfK) peptide-based theranostic agent containing both a dodecaborane residue and ^{67}Ga and ^{125}I isotope labels was highly stable and capable of accumulating in U87 MG glioblastoma cells [61].

Recently, we have demonstrated the possibility of obtaining carborane-containing derivatives and analogs of natural amino acids as a result of modifications of protected amino acids using classical methods of peptide chemistry (formation of an amide bond, selective introduction and removal of *N*- and *C*-protecting groups) [62–67].

The purpose of this work was to synthesize new *N*- and *C*-protected derivatives of the RGD peptide containing a *closo*-carborane residue linked to the arginine α-amino group via a short linker (compounds **1a–c**, Scheme 1). We used a glutaric acid residue as a linker, which makes it possible to obtain conjugates of the RGD peptide with readily available 3-amino-*ortho*-carborane with a high boron content. The choice of protecting groups was due to the possibility of either selective deblocking of the guanidino group in the arginine residue and carboxyl groups in the aspartate residue (compound **2a**), or removal of all protecting groups in one step (compounds **2b,c**).

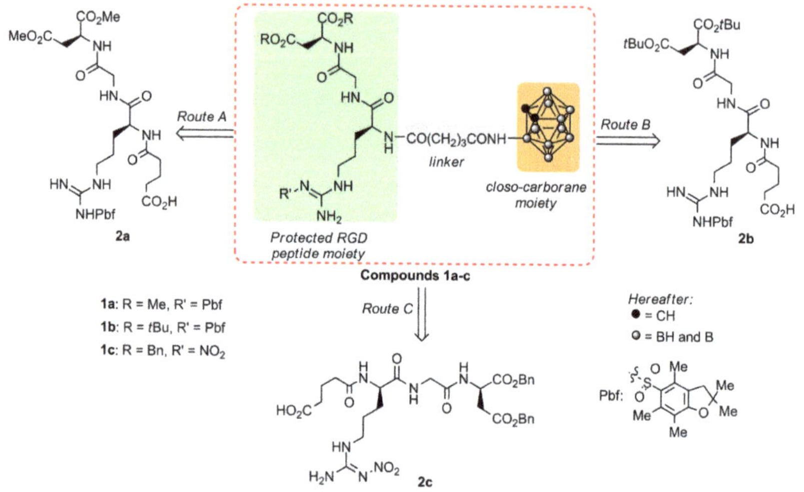

Scheme 1. Synthetic routes to protected *closo*-carboranyl RGD peptide derivatives **1a–c**.

2. Results and Discussion

We have carried out a comparative study of three synthetic routes for *closo*-carboranyl derivatives of the RGD peptide involving the use of different protecting groups.

The synthesis of peptides **1a,b** was carried out starting from dimethyl and di-*tert*-butyl esters **2a** and **2b**, which we had previously obtained, containing a 2,2,4,6,7-pentamethyldihydrobenzofuran-5-sulfonyl (Pbf) group in the arginine side chain and a glutaryl fragment at the arginine α-amino group [68–70]. The protecting groups of compounds **1a** and **2a** can be removed selectively: ester groups by alkaline hydrolysis; and the Pbf group by the action of an acid, for example, TFA. Removal of the three protecting groups in compounds **1b** and **2b** can be carried out in one step, by acid treatment.

To obtain conjugate **1c**, it was necessary to synthesize a glutaryl derivative **2c** of the protected RGD peptide containing a nitro group in the guanidine fragment and two benzyl ester groups, which can be simultaneously removed by hydrogenolysis. The synthesis of derivatives of the RGD peptide containing benzyl aspartate and a nitro group protecting the side chain of arginine has been described in the literature; however, information on the physicochemical characteristics of intermediate compounds is fragmentary [71–76].

We synthesized glutaryl derivative **2c** starting from dibenzyl (*S*)-aspartate (**3**) (Scheme 2). Coupling of amino ester **3** to *N*-Boc-glycine using *N,N'*-dicyclohexylcarbodiimide (DCC) as a coupling agent in the presence of *N*-hydroxysuccinimide (HOSu) and subsequent treatment of protected dipeptide **4** with hydrochloric acid in methanol led to amino ester **5** in moderate yield after chromatographic purification. Coupling of compound **5** to N^α-Boc-N^ω-nitro-(*S*)-arginine in the presence of TBTU gave the protected tripeptide **6**. Removal of the Boc group of compound **6** under acidic conditions and subsequent treatment of tripeptide **7** with glutaric anhydride gave compound **2c** containing a free carboxyl group.

At each stage of the synthesis of glutaryl tripeptide **2c**, the formation of side products was observed, so in order to obtain pure compounds **2c**, **4–7**, it was necessary to perform chromatographic purification. It is known that peptides containing an aspartic acid residue, including those in the RGD fragment, are prone to degradation, isomerization, and epimerization [77–81]. In our case, the total yield of compound **2c** (Scheme 2) was only 9.2% relative to the starting amino ester **3**. At the same time, the total yields of peptides **2a** and **2b** obtained from dimethyl and di-*tert*-butyl (*S*)-aspartates were about 20% [69].

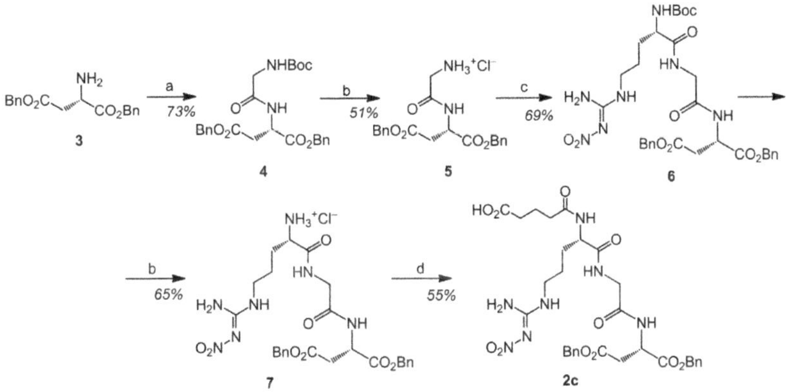

Scheme 2. Synthesis of compound **2c**. (a) *N*-Boc-Gly-OH, HOSu, DIPEA, DCC, CH$_2$Cl$_2$, rt, 24 h; (b) HCl conc., MeOH, rt, 15 min; (c) *N*-Boc-Arg(NO$_2$)-OH, DIPEA, TBTU, CH$_2$Cl$_2$, rt, 20 h; (d) glutaric anhydride, DIPEA, CH$_2$Cl$_2$, rt, 20 h.

Coupling of compounds **2a–c** to 3-amino-*ortho*-carborane (**8**) by the mixed anhydride method in the presence of ethyl chloroformate led to protected carboranyl peptides **1a–c** in moderate yields (Scheme 3). Attempts to implement an alternative approach consisting in the acylation of amine **8** with glutaric anhydride followed by coupling to peptide **7** failed because of the low nucleophilicity of 3-aminocarborane.

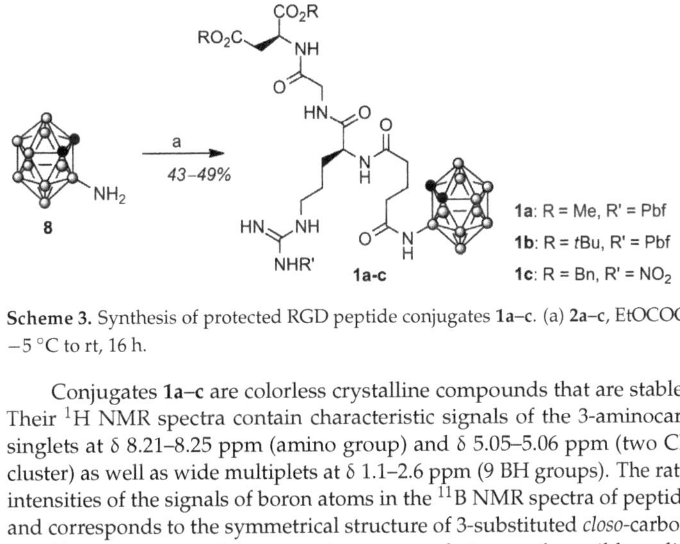

Scheme 3. Synthesis of protected RGD peptide conjugates **1a–c**. (a) **2a–c**, EtOCOCl, NMM, CH$_2$Cl$_2$, −5 °C to rt, 16 h.

Conjugates **1a–c** are colorless crystalline compounds that are stable during storage. Their ^{1}H NMR spectra contain characteristic signals of the 3-aminocarborane protons: singlets at δ 8.21–8.25 ppm (amino group) and δ 5.05–5.06 ppm (two CH groups in the cluster) as well as wide multiplets at δ 1.1–2.6 ppm (9 BH groups). The ratio of the integral intensities of the signals of boron atoms in the ^{11}B NMR spectra of peptides **1a–c** is 4:1:2:3 and corresponds to the symmetrical structure of 3-substituted *closo*-carborane.

To remove protecting groups in compounds **1a–c**, rather mild conditions are usually suitable, in which, as a rule, cleavage of peptide bonds or degradation of the *closo*-carborane residue do not occur. Thus, these derivatives can be considered as convenient starting compounds for further modifications.

3. Conclusions

Thus, we synthesized several protected derivatives of the RGD peptide containing 3-amino-*closo*-carborane and glutaryl residue as a linker. The structural motif of the RGD peptide can be considered as a basis for the synthesis of potential boron delivery agents for BNCT; at the same time, the preparation of compounds of this group requires careful selection of reaction conditions. The derivatives obtained by us differ in the structure of the protecting groups; their removal can be carried out both in one stage (by hydrogenolysis or acidic treatment) and separately. This opens up prospects for further modification of the peptide fragment and the synthesis of carborane-containing peptides of a more complex structure.

4. Materials and Methods

Dimethyl (S,S)-$(N^\alpha$-4-carboxybutanoyl-N^ω-Pbf-arginyl)-glycyl-aspartate (**2a**) [69], di-*tert*-butyl (S,S)-$(N^\alpha$-4-carboxybutanoyl-N^ω-Pbf-arginyl)-glycyl-aspartate (**2b**) [69], dibenzyl (S)-aspartate 4-toluenesulfonate (**3**) [82], and 3-amino-1,2-dicarba-*closo*-dodecaborane (**8**) [83] were obtained according to known procedures. Other reagents were commercially available and were purchased from Alfa Aesar (Heysham, UK). Solvents were purified according to traditional methods [84] and used freshly distilled. Melting points were obtained on a SMP3 apparatus (Barloworld Scientific, Staffordshire, UK). Optical rotations were measured on a Perkin Elmer M341 polarimeter (Perkin Elmer, Waltham, MA, USA). The ^{1}H, ^{11}B, and ^{13}C NMR spectra were recorded on a Bruker Avance 500 instrument (Bruker, Karlsruhe, Germany) with operating frequencies of 500, 160, and 126 MHz, respectively, at ambient temperature using TMS as an internal standard and $BF_3 \cdot Et_2O$ as an external standard. The NMR spectra of the compounds were obtained; see the Supplementary Materials, Figures S1–S19. CHN-Elemental analysis was performed using a Perkin Elmer 2400 II analyzer (Perkin Elmer, Waltham, MA, USA). Analytical TLC was performed using Sorbfil plates (Imid, Krasnodar, Russia). Flash column chromatography was performed using Silica gel 60 (230–400 mesh) (Alfa Aesar, Heysham, UK). The high-resolution mass spectra were obtained using a Bruker maXis Impact HD mass spectrometer (Bruker, Karlsruhe, Germany), with electrospray ionization at atmospheric pressure in positive or negative mode, with direct sample inlet (4 L/min flow rate). Analytical reversed-phase HPLC was carried out with an Agilent 1100 instrument (Agilent Technologies, Santa Clara, CA, USA) using a Kromasil 100-5-C18 column (Nouryon, Göteborg, Sweden) thermostated at 35 °C, with detection at 230 nm (compounds **1b**, **1c**, **2c**, **4**, **6**, and **7**) or 254 nm (compound **1a**), and a 0.8 mL/min flow rate; the mobile phases are indicated in each specific case. For the HPLC data for compounds **4**, **6**, **7**, **2c**, and **1a–c**, see the Supplementary Materials, Figures S20–S26.

Dibenzyl *N*-Boc-glycyl-(*S*)-aspartate (4). DCC (0.48 g, 2.33 mmol) and DIPEA (1.22 mL, 6.98 mmol) were added to a solution of *N*-Boc-glycine (0.41 g, 2.33 mmol), dibenzyl (S)-aspartate 4-toluenesulfonate (**3**) (1.13 g, 2.33 mmol) and *N*-hydroxysuccinimide (0.13 g, 1.16 mmol) in CH_2Cl_2 (10 mL). The reaction mixture was stirred at room temperature for 24 h, and then filtered. The filtrate was successively washed with 10% citric acid solution (2 × 8 mL), saturated aqueous NaCl solution (2 × 8 mL), 5% aqueous $NaHCO_3$ solution (2 × 8 mL), and saturated aqueous NaCl solution (8 mL). The organic layer was dried over Na_2SO_4 and evaporated to dryness under reduced pressure. The residue was purified by flash column chromatography (eluent benzene–EtOAc from 8:2 to 6:4). Yield, 0.79 g (73%). Colorless powder; m.p., 56 °C. $[\alpha]_D^{20}$ + 9.9 (*c* 1.0, $CHCl_3$). TLC (benzene–EtOAc 3:1): R_f, 0.44. RP-HPLC (MeCN–H_2O 1:1, 230 nm): τ, 4.7 min. ^{1}H NMR (DMSO-d_6) (*major conformer*) δ (ppm): 1.38 (s, 9H, *t*Bu), 2.80 (dd, *J* = 16.6, 6.8 Hz, H-3B Asp), 2.90 (dd, *J* = 16.6, 6.2 Hz, H-3A Asp), 3.55–3.57 (m, 2H, 2×H-2 Gly), 4.74–4.79 (m, H-2 Asp), 5.07 (s, 2H, Bn), 5.09 (s, 2H, Bn), 6.99 (t, *J* = 6.1 Hz, 1H, NH Gly), 7.31–7.37 (m, 10H, Ar), 8.35 (d, *J* = 8.0 Hz, 1H, NH Asp). ^{13}C NMR (DMSO-d_6) (*major conformer*) δ (ppm): 28.1 (3C), 35.8, 42.9, 48.5, 65.8, 66.2, 78.0, 127.6 (2C), 127.9 (2C), 128.0 (2C), 128.4 (2C), 128.4 (2C), 135.6, 135.7, 155.7, 169.4, 169.8, 170.3. Calcd (%) for $C_{25}H_{30}N_2O_7$: C, 63.82; H, 6.43; N, 5.95. Found (%): C, 63.89; H, 6.47; N, 5.99. HRMS (ESI) (*m/z*) [M+H]$^+$: calcd for $[C_{25}H_{31}N_2O_7]^+$: 471.2126; found: 471.2127.

Dibenzyl *N*-Glycyl-(*S*)-aspartate Hydrochloride (5). Concentrated HCl (2.0 mL, 24.0 mmol) was added to a solution of compound **4** (1.13 g, 2.4 mmol) in MeOH (10 mL). The reaction mixture was stirred at room temperature for 15 min, then evaporated to dryness under reduced pressure. The residue was purified by flash column chromatography on silica gel (eluent $CHCl_3$–EtOH from 100:0 to 1:1). Yield, 0.34 g (51%). Yellowish oil.

$[\alpha]_D^{20}$ +5.9 (*c* 1.0, CHCl$_3$). TLC (CHCl$_3$–EtOH 3:1): R_f, 0.69. ^{1}H NMR (DMSO-d_6) (*major conformer*) δ (ppm): 2.88 (dd, *J* = 16.8, 6.8 Hz, H-3B Asp), 2.93 (dd, *J* = 16.8, 5.7 Hz, H-3A Asp), 3.56–3.65 (m, 2H, H-2 Gly), 4.82–4.86 (m, H-2 Asp), 5.09 (s, 2H, Bn), 5.12 (s, 2H, Bn), 7.31–7.40 (m, 10H, Ar), 8.09 (s, 3H, NH$_3^+$), 9.00 (d, *J* = 7.9 Hz, 1H, NH Asp). ^{13}C NMR (DMSO-d_6) (*major conformer*) δ (ppm): 35.7, 43.7, 48.6, 66.0, 66.5, 127.8 (2C), 128.0 (2C), 128.1 (2C), 128.4 (4C), 135.6, 135.7, 166.4, 169.7, 170.0. HRMS (ESI) (*m/z*) [M+H]$^+$: calcd for [C$_{20}$H$_{23}$N$_2$O$_5$]$^+$: 371.1602; found: 371.1604.

Dibenzyl (*S,S*)-(*N*$^\alpha$-Boc-*N*$^\omega$-nitroarginyl)-glycyl-aspartate (6). TBTU (0.45 g, 1.41 mmol) and DIPEA (1.46 mL, 4.36 mmol) were added to a solution of amino ester hydrochloride **5** (0.57 g, 1.41 mmol) and *N*$^\alpha$-Boc-*N*$^\omega$-nitro-(*S*)-arginine (0.45 g, 1.41 mmol) in CH$_2$Cl$_2$ (20 mL). The reaction mixture was stirred at room temperature for 20 h then successively washed with 10% citric acid solution (2 × 15 mL), saturated aqueous NaCl solution (2 × 15 mL), 5% aqueous NaHCO$_3$ solution (2 × 15 mL) and saturated aqueous NaCl solution (10 mL). The organic layer was dried over Na$_2$SO$_4$ and evaporated to dryness under reduced pressure. The residue was purified by flash column chromatography on silica gel (eluent CHCl$_3$–EtOH from 10:0 to 8:2). Yield, 0.65 g (69%). Colorless powder; m.p., 112 °C (lit. m.p.: 98–99 °C [85], 99–102 °C [70]). $[\alpha]_D^{20}$ +4.0 (*c* 1.0, CHCl$_3$). TLC (CHCl$_3$–EtOH 3:1): R_f, 0.49. RP-HPLC (MeCN–H$_2$O–AcOH 80:20:0.0025, 230 nm): τ, 4.2 min. ^{1}H NMR (DMSO-d_6) (*major conformer*) δ (ppm): 1.37 (s, 9H, *t*Bu), 1.43–1.59 (m, 3H, H-3B and 2×H-4 Arg), 1.59–1.71 (m, 1H, H-3A Arg), 2.79 (dd, *J* = 16.6, 6.8 Hz, H-3B Asp), 2.90 (dd, *J* = 16.3, 6.4 Hz, H-3A Asp), 3.07–3.17 (m, 2H, 2×H-5 Arg), 3.71 (dd, *J* = 16.8, 5.6 Hz, 1H, H-2A Gly), 3.76 (dd, *J* = 16.8, 5.7 Hz, 1H, H-2A Gly), 3.91–3.95 (m, 1H, H-2 Arg), 4.74–4.78 (m, H-2 Asp), 5.07 (s, 2H, Bn), 5.09 (s, 2H, Bn), 6.96 (d, *J* = 7.8 Hz, 1H, N$^\alpha$H Arg), 7.31–7.38 (m, 10H, Ar), 7.55–8.25 (br. s, 2H, 2×N$^\omega$H Arg), 8.07 (dd, *J* = 5.7, 5.6 Hz, 1H, NH Gly), 8.42 (d, *J* = 7.9 Hz, 1H, NH Asp), 8.44–8.54 (br. s, 1H, N$^\omega$H Arg). ^{13}C NMR (DMSO-d_6) δ (ppm): 24.6, 28.2 (3C), 29.1, 35.8, 40.0, 41.6, 48.6, 53.9, 65.9, 66.3, 78.2, 127.7 (2C), 127.9 (2C), 128.0 (2C), 128.4 (4C), 135.7, 135.8, 155.4, 159.3, 168.8, 169.7, 170.3, 172.2. Calcd (%) for C$_{31}$H$_{41}$N$_7$O$_{10}$: C, 55.43; H, 6.15; N, 14.60. Found (%): C, 55.07; H, 6.26; N, 14.77. HRMS (ESI) (*m/z*) [M+H]$^+$: calcd for [C$_{31}$H$_{42}$N$_7$O$_{10}$]$^+$: 672.2988; found: 672.2983.

Dibenzyl (*S,S*)-(*N*$^\omega$-Nitroarginyl)-glycyl-aspartate Hydrochloride (7). Concentrated HCl (0.50 mL, 5.95 mmol) was added to a solution of compound **6** (0.20 g, 0.30 mmol) in MeOH (5 mL). The reaction mixture was stirred at room temperature for 15 min, then evaporated to dryness under reduced pressure. The residue was purified by flash column chromatography on silica gel (eluent CHCl$_3$–EtOH from 10:0 to 3:7). Yield, 0.12 g (65%). Yellowish powder; m.p., 61–64 °C. $[\alpha]_D^{20}$ +17.6 (*c* 1.0, CHCl$_3$). TLC (CHCl$_3$–EtOH 3:1): R_f, 0.28. RP-HPLC (MeCN–H$_2$O–CF$_3$CO$_2$H 70:30:0.01, 230 nm): τ, 5.1 min. ^{1}H NMR (DMSO-d_6) δ (ppm): 1.47–1.61 (m, 2H, 2×H-4 Arg), 1.67–1.77 (m, 2H, 2×H-3 Arg), 2.82 (dd, *J* = 16.6, 7.0 Hz, H-3B Asp), 2.91 (dd, *J* = 16.6, 6.1 Hz, H-3A Asp), 3.18 (br. s, 2H, 2×H-5 Arg), 3.82–3.88 (m, 2H, 2×H-2 Gly and H-2 Arg), 4.76–4.80 (m, H-2 Asp), 5.08 (s, 2H, Bn), 5.10 (s, 2H, Bn), 7.31–7.38 (m, 10H, Ar), 7.68–8.23 (br. s, 2H, NH$_2$Arg), 8.14 (s, 3H, NH$_3^+$), 8.47–8.63 (br. s, 1H, NH Arg), 8.65 (d, *J* = 7.9 Hz, 1H, NH Asp), 8.72 (t, *J* = 5.3 Hz, 1H, NH Gly). ^{13}C NMR (DMSO-d_6) δ (ppm): 24.4, 31.1, 35.8, 40.3, 41.5, 48.5, 53.6, 65.9, 66.3, 127.6 (2C), 127.9 (2C), 128.0 (2C), 128.4 (4C), 135.6, 135.7, 159.2, 168.8, 169.7, 170.3, 173.9. Calcd (%) for C$_{26}$H$_{33}$N$_7$O$_8$×1.5HCl: C, 49.86; H, 5.55; N, 15.66; Cl, 8.49. Found (%): C, 49.41; H, 5.54; N, 15.64; Cl, 8.24. HRMS (ESI) (*m/z*) [M+H]$^+$: calcd for [C$_{26}$H$_{34}$N$_7$O$_8$]$^+$: 572.2464; found: 572.2462.

Dibenzyl (*S,S*)-(*N*$^\alpha$-Glutaryl-*N*$^\omega$-nitroarginyl)-glycyl-aspartate (2c). A solution of compound **7** (0.30 g, 0.49 mmol), glutaric anhydride (0.056 g, 0.49 mmol) and DIPEA (0.13 mL, 0.74 mmol) in CH$_2$Cl$_2$ (5 mL) was stirred at room temperature for 20 h, then evaporated to dryness under reduced pressure. The residue was purified by flash column chromatography on silica gel (eluent CHCl$_3$–EtOH from 9:1 to 3:7). Yield, 0.185 g (55%). Colorless powder; m.p., 103–108 °C. $[\alpha]_D^{20}$ −1.7 (*c* 1.0, EtOH). TLC (CHCl$_3$–EtOH 1:1): R_f, 0.49. RP-HPLC (MeCN–H$_2$O–AcOH 60:40:0.005, 230 nm): τ, 6.6 min. ^{1}H NMR (DMSO-d_6) δ (ppm): 1.41–1.58 (m, 3H, H-3B and 2×H-4 Arg), 1.62–1.74 (m, 3H, CH$_2$ glutaryl and H-3A

Arg), 2.14–2.20 (m, 4H, 2×CH$_2$ glutaryl), 2.80 (dd, J = 16.6, 6.9 Hz, H-3B Asp), 2.90 (dd, J = 16.6, 6.3 Hz, H-3A Asp), 3.14 (br. s, 2H, 2×H-5 Arg), 3.71 (dd, J = 16.9, 5.7 Hz, 1H, H-2B Gly), 3.75 (dd, J = 16.9, 5.7 Hz, 1H, H-2A Gly), 4.19–4.23 (m, 1H, H-2 Arg), 4.74–4.78 (m, H-2 Asp), 5.07 (s, 2H, Bn), 5.09 (s, 2H, Bn), 7.31–7.38 (m, 10H, Ar), 8.16 (br. s, 2H, 2×N$^\omega$H Arg), 8.07 (d, J = 7.3 Hz, 1H, NH Asp), 8.24 (t, J = 5.7 Hz, 1H, NH Gly), 8.38 (d, J = 6.9 Hz, 1H, N$^\alpha$H Arg), 8.51 (br. s, 1H, N$^\omega$H Arg), 12.03 (s, 1H, CO$_2$H). ^{13}C NMR (DMSO-d_6) δ (ppm): 20.6, 24.7, 28.9, 33.0, 34.2, 35.8, 40.2, 41.6, 48.5, 52.4, 65.9, 66.3, 127.7 (2C), 127.9 (2C), 128.0 (2C), 128.4 (4C), 135.7, 135.8, 159.3, 168.8, 169.7, 170.3, 172.0, 172.1, 174.2. Calcd (%) for C$_{31}$H$_{39}$N$_7$O$_{11}$: C, 54.30; H, 5.73; N, 14.30. Found (%): C, 53.94; H, 5.65; N, 13.99. HRMS (ESI) (m/z) [M$-$H]$^-$: calcd for [C$_{31}$H$_{38}$N$_7$O$_{11}$]$^-$: 684.2684; found 684.2685.

General Procedure for the Synthesis of Carboranylaminoglutaryl Tripeptides 1a–c. Ethyl chloroformate (63 µL, 0.66 mmol) was added to a cold (-10 °C) solution of an appropriate compound **2a**, **2b** or **2c** (0.66 mmol) and *N*-methylmorpholine (145 µL, 1.32 mmol) in CH$_2$Cl$_2$ (10 mL). The mixture was stirred at -10 °C for 15 min; then, 3-aminocarborane (**8**) (0.11 g, 0.66 mmol) was added. The reaction mixture was stirred at room temperature for 16 h, then successively washed with 10% citric acid solution (2 × 8 mL), saturated aqueous NaCl solution (2 × 8 mL), 5% aqueous NaHCO$_3$ solution (2 × 8 mL) and saturated aqueous NaCl solution (8 mL). The organic layer was dried over Na$_2$SO$_4$ and evaporated to dryness under reduced pressure. The residue was purified by flash column chromatography (eluent CHCl$_3$–EtOH from 10:0 to 8:2).

Dimethyl (S,S)-{N$^\alpha$-[4-(1,2-Dicarba-*closo*-dodecaboran-3-yl)aminocarbonylbutanoyl]-N$^\omega$-Pbf-arginyl}-glycyl-aspartate (1a). Yield, 0.28 g (48%). Colorless powder; m.p., 120–122 °C. [α]$_D^{20}$ +7.0 (c 0.9, CHCl$_3$). TLC (CHCl$_3$–EtOH 7:1): R_f, 0.7. RP-HPLC (MeCN–H$_2$O 1:1, 254 nm): τ, 8.2 min. ^{1}H NMR (DMSO-d_6) (*major conformer*) δ (ppm): 1.1–2.6 (br. m, 9H, 9×BH), 1.36–1.50 (m, 2H, 2×H-4 Arg), 1.41 (s, 6H, Pbf), 1.59–1.72 (m, 2H, CH$_2$ glutaryl and 2×H-3 Arg), 2.01 (s, 3H, Pbf), 2.14 (t, J = 7.5 Hz, 2H, CH$_2$ glutaryl), 2.19 (t, J = 7.4 Hz, 2H, CH$_2$ glutaryl), 2.42 (s, 3H, Pbf), 2.48 (s, 3H, Pbf), 2.72 (dd, J = 16.6, 6.9 Hz, H-3B Asp), 2.80 (dd, J = 16.6, 6.2 Hz, H-3A Asp), 2.96 (s, 2H, Pbf), 3.00–3.04 (m, 2H, 2×H-5 Arg), 3.60 (s, 3H, CO$_2$Me), 3.62 (s, 3H, CO$_2$Me), 3.66–3.77 (m, 2H, 2×H-2 Gly), 4.17–4.21 (m, 1H, H-2 Arg), 4.65–4.69 (m, H-2 Asp), 5.06 (s, 2H, CH carborane), 6.37 (br. s, 1H, N$^\omega$H Arg), 6.56–7.12 (br. m, 2H, 2×N$^\omega$H Arg), 7.99 (d, J = 7.6 Hz, 1H, N$^\alpha$H Arg), 8.22 (t, J = 5.9 Hz, 1H, NH Gly), 8.23 (s, 1H, NH carborane), 8.28 (d, J = 7.8 Hz, 1H, NH Asp). ^{11}B NMR (DMSO-d_6) δ (ppm): -15.0 (br. s, 3B), -13.43 (2B), -10.69 (1B), -5.51 (4B). ^{13}C NMR (DMSO-d_6) (*major conformer*) δ (ppm): 12.2, 17.5, 18.8, 20.8, 25.4, 28.2 (2C), 29.0, 34.2, 35.6, 35.9, 41.5, 41.6, 42.4, 48.3, 51.6, 52.1, 57.1 (2C), 59.8, 86.2, 116.2, 124.2, 131.4, 134.1, 137.2, 156.0, 157.4, 168.6, 168.7, 170.2, 170.9, 172.0, 176.2. HRMS (ESI) (m/z) [M+H]$^+$: calcd for [C$_{34}$H$_{59}{}^{11}$B$_{10}$N$_7$O$_{11}$S]$^+$: 884.5044; found: 884.5045.

Di-*tert*-butyl (S,S)-{N$^\alpha$-[4-(1,2-Dicarba-*closo*-dodecaboran-3-yl)aminocarbonylbutanoyl]-N$^\omega$-Pbf-arginyl}-glycyl-aspartate (1b). Yield, 0.31 g (49%). Colorless powder; m.p., 126 °C. [α]$_D^{20}$ +5.5 (c 1.0, CHCl$_3$). TLC (CHCl$_3$–EtOH 7:1): R_f, 0.71. RP-HPLC (MeCN–H$_2$O–AcOH 40:60:0.0025, 230 nm): τ, 2.4 min. ^{1}H NMR (DMSO-d_6) (*major conformer*) δ (ppm): 1.1–2.6 (br. m, 9H, 9×BH), 1.32–1.51 (m, 2H, 2×H-4 Arg), 1.380 (s, 9H, *t*Bu), 1.384 (s, 9H, *t*Bu), 1.41 (s, 6H, Pbf), 1.58–1.66 (m, 1H, H-3B Arg), 1.66–1.76 (m, 3H, CH$_2$ glutaryl and H-3A Arg), 2.01 (s, 3H, Pbf), 2.12–2.16 (m, 2H, CH$_2$ glutaryl), 2.19 (t, J = 7.6 Hz, 2H, CH$_2$ glutaryl), 2.42 (s, 3H, Pbf), 2.47 (s, 3H, Pbf), 2.54 (dd, J = 16.3, 6.9 Hz, H-3B Asp), 2.64 (dd, J = 16.3, 6.1 Hz, H-3A Asp), 2.96 (s, 2H, Pbf), 3.00–3.04 (m, 2H, 2×H-5 Arg), 3.68–3.74 (m, 2H, 2×H-2 Gly), 4.16–4.24 (m, 1H, H-2 Arg), 4.47–4.52 (m, H-2 Asp), 5.06 (s, 2H, CH carborane), 6.18–7.28 (m, 3H, 3×N$^\omega$H Arg), 7.98 (d, J = 7.5 Hz, 1H, N$^\alpha$H Arg), 8.13 (d, J = 8.0 Hz, 1H, NH Asp), 8.20 (t, J = 6.2 Hz, 1H, NH Gly), 8.23 (s, 1H, NH carborane). ^{11}B NMR (DMSO-d_6) δ (ppm): -15.0 (br. s, 3B), -13.46 (2B), -10.69 (1B), -5.52 (4B). ^{13}C NMR (DMSO-d_6) (*major conformer*) δ (ppm): 12.2, 17.5, 18.8, 20.8, 25.4, 27.5 (3C), 27.6 (3C), 28.2 (2C), 29.1, 34.2, 35.9, 37.1, 40.0 (overlapped by DMSO-d_6 signal), 41.6, 42.4, 49.1, 52.3, 57.0 (2C), 80.4, 80.9, 86.2, 116.2, 124.2, 131.4, 134.1, 137.2, 156.0, 157.4, 168.5, 169.0, 169.5, 171.9, 172.0, 176.2. Calcd (%) for C$_{40}$H$_{71}$B$_{10}$N$_7$O$_{11}$S: C, 49.72; H, 7.41; N, 10.15. Found (%): C, 49.65; H, 7.30; N, 9.98. HRMS (ESI) (m/z) [M+H]$^+$: calcd for [C$_{40}$H$_{72}{}^{11}$B$_{10}$N$_7$O$_{11}$S]$^+$: 968.5988; found: 968.5972.

Dibenzyl (*S*,*S*)-{*N*$^\alpha$-[4-(1,2-Dicarba-*closo*-dodecaboran-3-yl)aminocarbonylbutanoyl]-*N*$^\omega$-nitroarginyl}-glycyl-aspartate (1c). Yield, 0.23 g (43%). Colorless powder; m.p., 94–98 °C. [α]$_D^{20}$ +2.0 (*c* 1.0, CHCl$_3$). TLC (CHCl$_3$–EtOH 7:1): R_f, 0.44. RP-HPLC (MeCN–0.06 M phosphate buffer solution (pH 7.0) 8:2, 230 nm): τ, 20.9 min. ^{1}H NMR (DMSO-d_6) (*major conformer*) δ (ppm): 1.2–2.6 (br. s, 9H, 9×BH), 1.42–1.57 (m, 3H, 2×H-4 and H-3B Arg), 1.62–1.75 (m, 3H, CH$_2$ glutaryl and H-3A Arg), 2.13–2.19 (m, 4H, 2×CH$_2$ glutaryl), 2.81 (dd, *J* = 16.6, 6.9 Hz, H-3B Asp), 2.90 (dd, *J* = 16.6, 6.3 Hz, H-3A Asp), 3.08–3.18 (m, 2H, 2×H-5 Arg), 3.68–3.78 (m, 2H, 2×H-2 Gly), 4.19–4.26 (m, 1H, H-2 Arg), 4.74–4.79 (m, H-2 Asp), 5.05 (s, 2H, 2×CH carborane), 5.07 (s, 2H, Bn), 5.09 (s, 2H, Bn), 7.31–7.38 (m, 10H, Ar), 7.52–8.30 (br. s, 2H, 2×N$^\omega$H Arg), 8.01 (d, *J* = 7.5 Hz, 1H, N$^\alpha$H Arg), 8.23 (m, 2H, NH carborane and NH Gly), 8.41 (d, *J* = 7.9 Hz, 1H, NH Asp), 8.51 (br. s, 1H, N$^\omega$H Arg). ^{11}B NMR (DMSO-d_6) δ (ppm): −15.0 (br. s, 3B), −13.42 (2B), −10.67 (1B), −5.51 (4B). ^{13}C NMR (DMSO-d_6) (*major conformer*) δ (ppm): 20.9, 24.7 (br. s), 29.0, 34.2, 35.7, 35.9, 40.2, 41.5, 48.5, 52.2, 57.1 (2C), 65.8, 66.3, 127.6 (2C), 127.8 (2C), 127.9, 128.0, 128.3 (4C), 135.6, 135.8, 159.2, 168.8, 169.6, 170.2, 171.9, 172.1, 176.2. HRMS (ESI) (*m/z*) [M+H]$^+$: calcd for [C$_{33}$H$_{51}$^{11}B$_{10}$N$_8$O$_{10}$]$^+$: 829.4699; found: 829.4694.

Supplementary Materials: The following supporting information can be downloaded at: https://www.mdpi.com/article/10.3390/molecules28083467/s1, Figures S1–S19: ^{1}H, ^{11}B, and ^{13}C NMR spectra of compounds **4–7**, **2c**, and **1a–c**; Figures S20–S26: HPLC data for compounds **4**, **6**, **7**, **2c**, and **1a–c**.

Author Contributions: Conceptualization and methodology, D.A.G. and V.P.K.; investigation, A.V.V. and A.M.D.; writing—original draft preparation, D.A.G.; writing—review and editing, G.L.L.; supervision, V.P.K.; project administration, D.A.G. All authors have read and agreed to the published version of the manuscript.

Funding: This research was funded by the Russian Science Foundation, Grant No. 21-73-10073.

Institutional Review Board Statement: Not applicable.

Informed Consent Statement: Not applicable.

Data Availability Statement: Not applicable.

Acknowledgments: Equipment from the Centre for Joint Use "Spectroscopy and Analysis of Organic Compounds" at the Postovsky Institute of Organic Synthesis was used.

Conflicts of Interest: The authors declare no conflict of interest.

Sample Availability: Not applicable.

References

1. Ruoslahti, E. RGD and other recognition sequences for integrins. *Annu. Rev. Cell Dev. Biol.* **1996**, *12*, 697–715. [CrossRef] [PubMed]
2. Park, J.; Singha, K.; Son, S.; Kim, J.; Namgung, R.; Yun, C.-O.; Kim, W.J. A review of RGD-functionalized nonviral gene delivery vectors for cancer therapy. *Cancer Gene Ther.* **2012**, *19*, 741–748. [CrossRef] [PubMed]
3. Danhier, F.; Le Breton, A.; Préat, V. RGD-Based Strategies to Target Alpha(v) Beta(3) Integrin in Cancer Therapy and Diagnosis. *Mol. Pharm.* **2012**, *9*, 2961–2973. [CrossRef] [PubMed]
4. Asati, S.; Pandey, V.; Soni, V. RGD Peptide as a Targeting Moiety for Theranostic Purpose: An Update Study. *Int. J. Pept. Res. Ther.* **2019**, *25*, 49–65. [CrossRef]
5. Ludwig, B.S.; Kessler, H.; Kossatz, S.; Reuning, U. RGD-Binding Integrins Revisited: How Recently Discovered Functions and Novel Synthetic Ligands (Re-)Shape an Ever-Evolving Field. *Cancers* **2021**, *13*, 1711. [CrossRef]
6. Battistini, L.; Bugatti, K.; Sartori, A.; Curti, C.; Zanardi, F. RGD Peptide-Drug Conjugates as Effective Dual Targeting Platforms: Recent Advances. *Eur. J. Org. Chem.* **2021**, *2021*, 2506–2528. [CrossRef]
7. Ley, K.; Rivera-Nieves, J.; Sandborn, W.J.; Shattil, S. Integrin-based therapeutics: Biological basis, clinical use and new drugs. *Nat. Rev. Drug Discov.* **2016**, *15*, 173–183. [CrossRef] [PubMed]
8. Slack, R.J.; Macdonald, S.J.F.; Roper, J.A.; Jenkins, R.G.; Hatley, R.J.D. Emerging therapeutic opportunities for integrin inhibitors. *Nat. Rev. Drug Discov.* **2022**, *21*, 60–78. [CrossRef]
9. Mas-Moruno, C.; Rechenmacher, F.; Kessler, H. Cilengitide: The First Anti-Angiogenic Small Molecule Drug Candidate. Design, Synthesis, and Clinical Evaluation. *Anti-Cancer Agents Med. Chem.* **2010**, *10*, 753–768. [CrossRef]
10. Kapp, T.G.; Rechenmacher, F.; Neubauer, S.; Maltsev, O.V.; Cavalcanti-Adam, E.A.; Zarka, R.; Reuning, U.; Notni, J.; Wester, H.-J.; Mas-Moruno, C.; et al. A Comprehensive Evaluation of the Activity and Selectivity Profile of Ligands for RGD-binding Integrins. *Sci. Rep.* **2017**, *7*, 39805. [CrossRef]

11. Stupp, R.; Hegi, M.E.; Gorlia, T.; Erridge, S.C.; Perry, J.; Hong, Y.-K.; Aldape, K.D.; Lhermitte, B.; Pietsch, T.; Grujicic, D.; et al. Cilengetide combined with standard treatment for patients with newly diagnosed glioblastoma with methylated *MGMT* promoter (CENTRIC EORTC 26071-22072 study): A multicentre, randomized, open-label, phase 3 trial. *Lancet Oncol.* **2014**, *15*, 1100–1108. [CrossRef] [PubMed]

12. Tucci, M.; Stucci, S.; Silvestris, F. Does cilengetide deserve another chance? *Lancet Oncol.* **2014**, *15*, e584–e585. [CrossRef] [PubMed]

13. Chinot, O.L. Cilengitide in glioblastoma: When did it fail? *Lancet Oncol.* **2014**, *15*, 1044–1045. [CrossRef]

14. Reichart, F.; Maltsev, O.V.; Kapp, T.G.; Räder, A.F.B.; Weinmüller, M.; Marelli, U.K.; Notni, J.; Wurzer, A.; Beck, R.; Wester, H.-J.; et al. Selective Targetting of Integrin $\alpha v \beta 8$ by a Highly Active Cyclic Peptide. *J. Med. Chem.* **2019**, *62*, 2024–2037. [CrossRef] [PubMed]

15. Shaw, L.T.; Mackin, A.; Shah, R.; Jain, S.; Jain, P.; Nayak, R.; Hariprasad, S.M. Risuteganib—A novel integrin inhibitor for the treatment of non-exudative (dry) age-related macular degeneration and diabetic macular edema. *Expert Opin. Investig. Drugs* **2020**, *29*, 547–554. [CrossRef] [PubMed]

16. Maturi, R.; Jaffe, G.J.; Ehlers, J.P.; Kaiser, P.K.; Boyer, D.S.; Heier, J.S.; Kornfield, J.A.; Kuppermann, B.D.; Quiroz-Mercado, H.; Aubel, J.; et al. Safety and Efficacy of Risuteganib in Intermediate Non-exudative Age-Related Macular Degeneration. *Investig. Ophthalmol. Vis. Sci.* **2020**, *61*, 1944.

17. Khanani, A.M.; Patel, S.S.; Gonzalez, V.H.; Moon, S.J.; Jaffe, G.J.; Wells, J.A.; Kozma, P.; Dugel, P.U.; Maturi, R.K. Phase 1 Study of THR-687, a Novel, Highly Potent Integrin Antagonist for the Treatment of Diabetic Macular Edema. *Ophthalmol. Sci.* **2021**, *1*, 100040. [CrossRef] [PubMed]

18. Liu, S. Radiolabeled Cyclic RGD Peptide Bioconjugates as Radiotracers Targeting Multiple Integrins. *Bioconjugate Chem.* **2015**, *26*, 1413–1438. [CrossRef]

19. Chakravarty, R.; Chakraborty, S.; Guleria, A.; Shukla, R.; Kumar, C.; Nair, K.V.V.; Sarma, H.D.; Tyagi, A.K.; Dash, A. Facile One-Pot Synthesis of Intrinsically Radiolabeled and Cyclic RGD Conjugated ^{199}Au Nanoparticles for Potential Use in Nanoscale Brachytherapy. *Ind. Eng. Chem. Res.* **2018**, *57*, 14337–14346. [CrossRef]

20. Shao, Y.; Liang, W.; Kang, F.; Yang, W.; Ma, X.; Li, G.; Zong, S.; Chen, K.; Wang, J. A direct comparison of tumor angiogenesis with ^{68}Ga-labeled NGR and RGD peptides in HT-1080 tumor xenografts using microPET imaging. *Amino Acids* **2014**, *46*, 2355–2364. [CrossRef]

21. Ramezanizadeh, M.; Masterifarahani, A.; Sadeghzadeh, N.; Abediankenari, S.; Mardanshahi, A.; Maleki, F. ^{99m}Tc-D(RGD): Molecular imaging probe for diagnosis of $\alpha v \beta 3$-positive tumors. *Nucl. Med. Commun.* **2020**, *41*, 104–109. [CrossRef] [PubMed]

22. Liolios, C.; Sachpekidis, C.; Kolocouris, A.; Dimitrakopoulou-Strauss, A.; Bouziotis, P. PET Diagnostic Molecules Utilizing Multimeric Cyclic RGD Peptide Analogs for Imaging Integrin $\alpha v \beta 3$ Receptors. *Molecules* **2021**, *26*, 1792. [CrossRef] [PubMed]

23. Choi, J.; Rustique, E.; Henry, M.; Guidetti, M.; Josserand, V.; Sancey, L.; Boutet, J.; Coll, J.-L. Targeting tumors with cyclic RGD-conjugated lipid nanoparticles loaded with an IR780 NIR dye: In vitro and in vivo evaluation. *Int. J. Pharm.* **2017**, *532*, 677–685. [CrossRef]

24. Wu, Y.; Wang, C.; Guo, J.; Carvalho, A.; Yao, Y.; Sun, P.; Fan, Q. An RGD modified water-soluble fluorophore probe for in vivo NIR-II imaging of thrombosis. *Biomater. Sci.* **2020**, *8*, 4438–4446. [CrossRef] [PubMed]

25. Li, M.; Liu, J.; Chen, X.; Dang, Y.; Shao, Y.; Xu, Z.; Zhang, W. An activatable and tumor-targeting NIR fluorescent probe for imaging of histone deacetylase 6 in cancer cells and in vivo. *Chem. Commun.* **2022**, *58*, 1938–1941. [CrossRef]

26. Yu, C.; Xiao, E.; Xu, P.; Lin, J.; Hu, L.; Zhang, J.; Dai, S.; Ding, Z.; Xiao, Y.; Chen, Z. Novel albumin-binding photothermal agent ICG-IBA-RGD for targeted fluorescent imaging and photothermal therapy of cancer. *RSC Adv.* **2021**, *11*, 7226–7230. [CrossRef]

27. Zheng, S.W.; Huang, M.; Hong, R.Y.; Deng, S.M.; Cheng, L.F.; Gao, B.; Badami, D. RGD-conjugated iron oxide magnetic nanoparticles for magnetic resonance imaging contrast enhancement and hyperthermia. *J. Biomater. Appl.* **2014**, *28*, 1051–1059. [CrossRef]

28. Melemenidis, S.; Jefferson, A.; Ruparelia, N.; Akhtar, A.M.; Xie, J.; Allen, D.; Hamilton, A.; Larkin, J.R.; Perez-Balderas, F.; Smart, S.C.; et al. Molecular Magnetic Resonance Imaging of Angiogenesis In Vivo using Polyvalent Cyclic RGD-Iron Oxide Microparticle Conjugates. *Theranostics* **2015**, *5*, 515–529. [CrossRef]

29. Arriortua, O.K.; Insausti, M.; Lezama, L.; Gil de Muro, I.; Garaio, E.; Martínez de la Fuente, J.; Fratila, R.M.; Morales, M.P.; Costa, R.; Eceiza, M.; et al. RGD-Functionalized Fe_3O_4 nanoparticles for magnetic hyperthermia. *Colloids Surf. B* **2018**, *165*, 315–324. [CrossRef]

30. Colombo, R.; Mingozzi, M.; Belvisi, L.; Arosio, D.; Piarulli, U.; Carenini, N.; Perego, P.; Zaffaroni, N.; De Cesare, M.; Castiglioni, V.; et al. Synthesis and Biological Evaluation (in Vitro and in Vivo) of Cyclic Arginine–Glycine–Aspartate (RGD) Peptidomimetic–Paclitaxel Conjugates Targeting Integrin $\alpha_v \beta_3$. *J. Med. Chem.* **2012**, *55*, 10460–10474. [CrossRef]

31. Hou, J.; Diao, Y.; Li, W.; Yang, Z.; Zhang, L.; Chen, Z.; Wu, Y. RGD peptide conjugation results in enhanced antitumor activity of PD0325901 against glioblastoma by both tumor-targeting delivery and combination therapy. *Int. J. Pharm.* **2016**, *505*, 329–340. [CrossRef] [PubMed]

32. Wang, X.; Qiao, X.; Shang, Y.; Zhang, S.; Li, Y.; He, H.; Chen, S. RGD and NGR modified TRAIL protein exhibited potent anti-metastasis effects on TRAIL-insensitive cancer cells in vitro and in vivo. *Amino Acids* **2017**, *49*, 931–941. [CrossRef] [PubMed]

33. Wang, G.; Wang, Z.; Li, C.; Duan, G.; Wang, K.; Li, Q.; Tao, T. RGD peptide-modified, paclitaxel prodrug-based, dual-drug loaded, and redox-sensitive lipid-polymer nanoparticles for the enhanced lung cancer therapy. *Biomed. Pharmacother.* **2018**, *106*, 275–284. [CrossRef] [PubMed]

34. Noh, G.J.; Oh, K.T.; Youn, Y.S.; Lee, E.S. Cyclic RGD-Conjugated Hyaluronate Dot Bearing Cleavable Doxorubicin for Multivalent Tumor Targeting. *Biomacromolecules* **2020**, *21*, 2525–2535. [CrossRef] [PubMed]

35. Li, M.-M.; Cao, J.; Yang, J.-C.; Shen, Y.-J.; Cai, X.-L.; Chen, Y.-W.; Qu, C.-Y.; Zhang, Y.; Shen, F.; Xu, L.-M. Effects of arginine–glycine–aspartic acid peptide-conjugated quantum dots-induced photodynamic therapy on pancreatic carcinoma in vivo. *Int. J. Nanomed.* **2017**, *12*, 2769–2779. [CrossRef]

36. Zhao, C.; Tong, Y.; Li, X.; Shao, L.; Chen, L.; Lu, J.; Deng, X.; Wang, X.; Wu, Y. Photosensitive Nanoparticles Combining Vascular-Independent Intratumor Distribution and On-Demand Oxygen-Depot Delivery for Enhanced Cancer Photodynamic Therapy. *Small* **2018**, *14*, 1703045. [CrossRef]

37. Wang, H.; Wang, Z.; Chen, W.; Wang, W.; Shi, W.; Chen, J.; Hang, Y.; Song, J.; Xiao, X.; Dai, Z. Self-assembly of photosensitive and radiotherapeutic peptide for combined photodynamic-radio cancer therapy with intracellular delivery of miRNA-139-5p. *Bioorg. Med. Chem.* **2021**, *44*, 116305. [CrossRef]

38. Li, R.; Zhou, Y.; Liu, Y.; Jiang, X.; Zeng, W.; Gong, Z.; Zheng, G.; Sun, D.; Dai, Z. Asymmetric, amphiphilic RGD conjugated phthalocyanine for targeted photodynamic therapy of triple negative breast cancer. *Signal Transduct. Target. Ther.* **2022**, *7*, 64. [CrossRef]

39. Dong, X.; Yu, Y.; Wang, Q.; Xi, Y.; Liu, Y. Interaction Mechanism and Clustering among RGD Peptides and Integrins. *Mol. Inform.* **2017**, *36*, 1600069. [CrossRef]

40. Dymova, M.A.; Taskaev, S.Y.; Richter, V.A.; Kuligina, E.V. Boron neutron capture therapy: Current status and future perspectives. *Cancer Commun.* **2020**, *40*, 406–421. [CrossRef]

41. Suzuki, M. Boron neutron capture therapy (BNCT): A unique role in radiotherapy with a view to entering the accelerator-based BNCT era. *Int. J. Clin. Oncol.* **2020**, *25*, 43–50. [CrossRef] [PubMed]

42. Malouff, T.D.; Senevirante, D.S.; Ebner, D.K.; Stross, W.C.; Waddle, M.R.; Trifiletti, D.M.; Krishnan, S. Boron Neutron Capture Therapy: A Review of Clinical Applications. *Front. Oncol.* **2021**, *11*, 601820. [CrossRef] [PubMed]

43. Xuan, S.; Vicente, M.D.G.H. *Boron-Based Compounds: Potential and Emerging Applications in Medicine*; Hey-Hawkins, E., Viñas Teixidor, C., Eds.; Wiley: Hoboken, NJ, USA, 2018; pp. 298–342.

44. Barth, R.F.; Mi, P.; Yang, W. Boron delivery agents for neutron capture therapy of cancer. *Cancer Commun.* **2018**, *38*, 35. [CrossRef]

45. Hu, K.; Yang, Z.; Zhang, L.; Xie, L.; Wang, L.; Xu, H.; Josephson, L.; Liang, S.H.; Zhang, M.-R. Boron agents for neutron capture therapy. *Coord. Chem. Rev.* **2020**, *405*, 213139. [CrossRef]

46. Sauerwein, W.A.G.; Sancey, L.; Hey-Hawkins, E.; Kellert, M.; Panza, L.; Imperio, D.; Balcerzyk, M.; Rizzo, G.; Scalco, E.; Herrmann, K.; et al. Theranostics in Boron Neutron capture Therapy. *Life* **2021**, *11*, 330. [CrossRef]

47. Lesnikowski, Z.J. Challenges and opportunities for the application of boron clusters in drug design. *J. Med. Chem.* **2016**, *59*, 7738–7758. [CrossRef]

48. Stockmann, P.; Gozzi, M.; Kuhnert, R.; Sárosi, M.B.; Hey-Hawkins, E. New keys for old locks: Carborane-containing drugs as platforms for mechanism-based therapies. *Chem. Soc. Rev.* **2019**, *48*, 3497–3512. [CrossRef]

49. Marfavi, A.; Kavianpour, P.; Rendina, L.M. Carboranes in drug discovery, chemical biology and molecular imaging. *Nat. Rev. Chem.* **2022**, *6*, 486–504. [CrossRef]

50. Gruzdev, D.A.; Levit, G.L.; Krasnov, V.P.; Charushin, V.N. Carborane-containing amino acids and peptides: Synthesis, properties, and applications. *Coord. Chem. Rev.* **2021**, *433*, 213753. [CrossRef]

51. Neirynck, P.; Schimer, J.; Jonkheijm, P.; Milroy, L.-G.; Cigler, P.; Brunsveld, L. Carborane–β-cyclodextrin complexes as a supramolecular connector for bioactive surfaces. *J. Mater. Chem. B* **2015**, *3*, 539–545. [CrossRef]

52. Kimura, S.; Masunaga, S.; Harada, T.; Kawamura, Y.; Ueda, S.; Okuda, K.; Nagasawa, H. Synthesis and evaluation of cyclic RGD-boron cluster conjugates to develop tumor-selective boron carriers for boron neutron capture therapy. *Bioorg. Med. Chem.* **2011**, *19*, 1721–1728. [CrossRef] [PubMed]

53. Masunaga, S.; Kimura, S.; Harada, T.; Okuda, K.; Sakurai, Y.; Tanaka, H.; Suzuki, M.; Kondo, N.; Maruhashi, A.; Nagasawa, H.; et al. Evaluating the Usefulness of a Novel ^{10}B-Carrier Conjugated with Cyclic RGD Peptide in Boron Neutron Capture Therapy. *World J. Oncol.* **2012**, *3*, 103–112. [CrossRef]

54. Kuthala, N.; Vankayala, R.; Li, Y.-N.; Chiang, C.-S.; Hwang, K.C. Engineering Novel Targeted Boron-10-Enriched Theranostic Nanomedicine to Combat against Murine Brain Tumors via MR Imaging-Guided Boron Neutron Capture Therapy. *Adv. Mater.* **2017**, *29*, 1700850. [CrossRef]

55. Chen, J.; Yang, Q.; Liu, M.; Lin, M.; Wang, T.; Zhang, Z.; Zhong, X.; Guo, N.; Lu, Y.; Xu, J.; et al. Remarkable Boron Delivery Of iRGD-Modified Polymeric Nanoparticles for Boron Neutron Capture Therapy. *Int. J. Nanomed.* **2019**, *14*, 8161–8177. [CrossRef] [PubMed]

56. Chen, J.; Dai, Q.; Yang, Q.Y.; Bao, X.; Zhou, Y.; Zhong, H.; Wu, L.; Wang, T.; Zhang, Z.; Lu, Y.; et al. Therapeutic nucleus-access BNCT drug combined CD47-targeting gene editing in gliblastoma. *J. Nanobiotechnol.* **2022**, *20*, 102. [CrossRef]

57. Koning, G.A.; Fretz, M.M.; Woroniecka, U.; Storm, G.; Krijger, G.C. Targeting liposomes to tumor endothelial cells for neutron capture therapy. *Appl. Radiat. Isot.* **2004**, *61*, 963–967. [CrossRef] [PubMed]

58. Krijger, G.C.; Fretz, M.M.; Woroniecka, U.D.; Steinebach, O.M.; Jiskoot, W.; Storm, G.; Koning, G.A. Tumor cell and tumor vasculature targeted liposomes for neutron capture therapy. *Radiochim. Acta* **2005**, *93*, 589–593. [CrossRef]

59. Kang, W.; Svirskis, D.; Sarojini, V.; McGregor, A.L.; Bevitt, J.; Wu, Z. Cyclic-RGDyC functionalized liposomes for dual-targeting of tumor vasculature and cancer cells in glioblastoma: An in vitro boron neutron capture therapy study. *Oncotarget* **2017**, *8*, 36614–36627. [CrossRef]

60. Kawai, K.; Nishimura, K.; Okada, S.; Sato, S.; Suzuki, M.; Takata, T.; Nakamura, H. Cyclic RGD-Functionalized *closo*-Dodecaborate Albumin Conjugates as Integrin Targeting Boron Carriers for Neutron Capture Therapy. *Mol. Pharm.* **2020**, *17*, 3740–3747. [CrossRef]

61. Mishiro, K.; Imai, S.; Ematsu, Y.; Hirose, K.; Fuchigami, T.; Munekane, M.; Kinuya, S.; Ogawa, K. RGD Peptide-Conjugated Dodecaborate with the Ga-DOTA Complex: A Preliminary Study for the Development of Theranostic Agents for Boron Neutron Capture Therapy and Its Companion Diagnostics. *J. Med. Chem.* **2022**, *65*, 16741–16753. [CrossRef]
62. Levit, G.L.; Krasnov, V.P.; Gruzdev, D.A.; Demin, A.M.; Bazhov, I.V.; Sadretdinova, L.S.h.; Olshevskaya, V.A.; Kalinin, V.N.; Cheong, C.S.; Chupakhin, O.N.; et al. Synthesis of *N*-[(3-amino-1,2-dicarba-*closo*-dodecaboran-1-yl)acetyl] derivatives of α-amino acids. *Collect. Czech. Chem. Commun.* **2007**, *72*, 1697–1706. [CrossRef]
63. Gruzdev, D.A.; Levit, G.L.; Bazhov, I.V.; Demin, A.M.; Sadretdinova, L.S.; Ol'shevskaya, V.A.; Kalinin, V.N.; Krasnov, V.P.; Chupakhin, O.N. Synthesis of novel carboranyl derivatives of α-amino acids. *Russ. Chem. Bull.* **2010**, *59*, 110–115. [CrossRef]
64. Gruzdev, D.A.; Levit, G.L.; Olshevskaya, V.A.; Krasnov, V.P. Synthesis of *ortho*-Carboranyl Derivatives of (*S*)-Asparagine and (*S*)-Glutamine. *Russ. J. Org. Chem.* **2017**, *53*, 769–776. [CrossRef]
65. Gruzdev, D.A.; Nuraeva, A.S.; Slepukhin, P.A.; Levit, G.L.; Zelenovskiy, P.S.; Shur, V.Y.; Krasnov, V.P. Piezoactive amino acid derivatives containing fragments of planar-chiral *ortho*-carboranes. *J. Mater. Chem. C* **2018**, *6*, 8638–8645. [CrossRef]
66. Gruzdev, D.A.; Telegina, A.A.; Levit, G.L.; Solovieva, O.I.; Gusel'nikova, T.Y.; Razumov, I.A.; Krasnov, V.P.; Charushin, V.N. Carborane-Containing Folic Acid *bis*-Amides: Synthesis and In Vitro Evaluation of Novel Promising Agents for Boron Delivery to Tumour Cells. *Int. J. Mol. Sci.* **2022**, *23*, 13726. [CrossRef]
67. Gruzdev, D.A.; Telegina, A.A.; Levit, G.L.; Krasnov, V.P. N-Aminoacyl-3-amino-*nido*-carboranes as a Group of Boron-Containing Derivatives of Natural Amino Acids. *J. Org. Chem.* **2022**, *87*, 5437–5441. [CrossRef]
68. Demin, A.M.; Vigorov, A.Y.; Nizova, I.A.; Uimin, M.A.; Shchegoleva, N.N.; Ermakov, A.E.; Krasnov, V.P.; Charushin, V.N. Functionalization of Fe_3O_4 magnetic nanoparticles with RGD peptide derivatives. *Mendeleev Commun.* **2014**, *24*, 20–22. [CrossRef]
69. Vigorov, A.Y.; Demin, A.M.; Nizova, I.A.; Krasnov, V.P. Synthesis of Derivatives of the RGD Peptide with the Residues of Glutaric and Adipic Acids. *Russ. J. Bioorg. Chem.* **2014**, *40*, 142–150. [CrossRef]
70. Demin, A.M.; Vakhrushev, A.V.; Tumashov, A.A.; Krasnov, V.P. Synthesis of glutaryl-containing derivatives of GRGD and KRGD peptides. *Russ. Chem. Bull.* **2019**, *68*, 2316–2324. [CrossRef]
71. Xin, M.; Xiang, H.; Si, W.; Zhao, W.; Xiao, H.; You, Q. Synthesis and antiangiogenic properties of 2-methoxestradiol-RGD peptide conjugates. *J. China Pharm. Univ.* **2011**, *42*, 198–205.
72. Jiang, B.; Cao, J.; Zhao, J.; He, D.; Pan, J.; Li, Y.; Guo, L. Dual-targeting delivery system for bone cancer: Synthesis and preliminary biological evaluation. *Drug Deliv.* **2012**, *19*, 317–326. [CrossRef]
73. Jiang, B.; Zhao, J.; Li, Y.; He, D.; Pan, J.; Cao, J.; Guo, L. Dual-targeting Janus Dendrimer Based Peptides for Bone Cancer: Synthesis and Preliminary Biological Evaluation. *Lett. Org. Chem.* **2013**, *10*, 594–601. [CrossRef]
74. Fu, Q.; Zhao, Y.; Yang, Z.; Yue, Q.; Xiao, W.; Chen, Y.; Yang, Y.; Guo, L.; Wu, Y. Liposomes actively recognizing the glucose transporter $GLUT_1$ and integrin $α_vβ_3$ for dual-targeting of glioma. *Arch. Pharm.* **2019**, *352*, e1800219. [CrossRef] [PubMed]
75. Zhao, Z.; Zhao, Y.; Xie, C.; Chen, C.; Lin, D.; Wang, S.; Lin, D.; Cui, X.; Guo, Z.; Zhou, J. Dual-active targeting liposomes drug delivery system for bone metastatic breast cancer: Synthesis and biological evaluation. *Chem. Phys. Lipids* **2019**, *223*, 104785. [CrossRef]
76. Pu, Y.; Zhang, H.; Peng, Y.; Fu, Q.; Yue, Q.; Zhao, Y.; Guo, L.; Wu, Y. Dual-targeting liposomes with active recognition of $GLUT_5$ and $α_vβ_3$ for triple-negative breast cancer. *Eur. J. Med. Chem.* **2019**, *183*, 111720. [CrossRef]
77. Inglis, A.S. Clevage at Aspartic Acid. *Meth. Enzymol.* **1983**, *91*, 324–332. [CrossRef]
78. Oliyai, C.; Borchardt, R.T. Chemical Pathways of Peptide Degradation. VI. Effect of the Primary Sequence on the Pathways of Degradation of Aspartyl Residues in Model Hexapeptides. *Pharm. Res.* **1994**, *11*, 751–758. [CrossRef]
79. Bogdanowich-Knipp, S.J.; Jois, D.S.S.; Siahaan, T.J. The effect of conformation on the solution stability of linear vs. cyclic RGD peptides. *J. Pept. Res.* **1999**, *53*, 523–529. [CrossRef]
80. Bogdanowich-Knipp, S.J.; Chakrabarti, S.; Williams, T.D.; Dillman, R.K.; Siahaan, T.J. Solution stability of linear vs. cyclic RGD peptides. *J. Pept. Res.* **1999**, *53*, 530–541. [CrossRef] [PubMed]
81. Hinterholzer, A.; Stanojlovic, V.; Regl, C.; Huber, C.G.; Cabrele, C.; Schubert, M. Detecting aspartate isomerization and backbone cleavage after aspartate in intact proteins by NMR spectroscopy. *J. Biomol. NMR* **2021**, *75*, 71–82. [CrossRef]
82. Bergmeier, S.C.; Cobás, A.A.; Rapoport, H. Chirospecific Synthesis of (1*S*,3*R*)-1-Amino-3-(hydroxymethyl)cyclopentane, Precursor for Carbocyclic Nucleoside Synthesis. Dieckmann Cyclization with an α-Amino Acid. *J. Org. Chem.* **1993**, *58*, 2369–2376. [CrossRef]
83. Zakharkin, L.I.; Kalinin, V.N.; Gedymin, V.V. Synthesis and some reactions of 3-amino-*o*-carboranes. *J. Organomet. Chem.* **1969**, *16*, 371–379. [CrossRef]
84. Armarego, W.L.F.; Chai, C.L.L. *Purification of Laboratory Chemicals*, 6th ed.; Butterworth Heinemann: Burlington, MA, USA, 2009; 743p.
85. Song, M.-M.; Ju, P.-P.; Cao, J.-J.; Shen, S.-B. Study on the synthesis of RGD tripeptide by chemical method. *Chin. J. Bioprocess Eng.* **2005**, *3*, 23–26.

 *molecules*

Article

Structural (XRD) Characterization and an Analysis of H-Bonding Motifs in Some Tetrahydroxidohexaoxidopentaborate(1-) Salts of *N*-Substituted Guanidinium Cations [†]

Michael A. Beckett [1,*], Simon J. Coles [2], Peter N. Horton [2] and Thomas A. Rixon [1]

[1] School of Natural Sciences, Bangor University, Bangor, Gwynedd LL57 2UW, UK
[2] Chemistry Department, University of Southampton, Southampton SO17 1BJ, UK
* Correspondence: m.a.beckett@bangor.ac.uk
[†] Dedicated to Professor John D. Kennedy on the occasion of his 80th birthday.

Abstract: The synthesis and characterization of six new substituted guanidium tetrahydroxido hexaoxidopentaborate(1-) salts are reported: $[C(NH_2)_2(NHMe)][B_5O_6(OH)_4]\cdot H_2O$ (**1**), $[C(NH_2)_2(NH\{NH_2\})][B_5O_6(OH)_4]$ (**2**), $[C(NH_2)_2(NMe_2)][B_5O_6(OH)_4]$ (**3**), $[C(NH_2)(NMe_2)_2][B_5O_6(OH)_4]$ (**4**), $[C(NHMe)(NMe_2)_2][B_5O_6(OH)_4]\cdot B(OH)_3$ (**5**), and $[TBDH][B_5O_6(OH)_4]$ (**6**) (TBD = 1,5,7-triazabicyclo [4.4.0]dec-5-ene). Compounds **1–6** were prepared as crystalline salts from basic aqueous solution via self-assembly processes from $B(OH)_3$ and the appropriate substituted cation. Compounds **1–6** were characterized by spectroscopic (NMR and IR) and by single-crystal XRD studies. A thermal (TGA) analysis on compounds **1–3** and **6** demonstrated that they thermally decomposed via a multistage process to B_2O_3 at >650 °C. The low temperature stage (<250 °C) was endothermic and corresponded to a loss of H_2O. Reactant stoichiometry, solid-state packing, and H-bonding interactions are all important in assembling these structures. An analysis of H-bonding motifs in known unsubstituted guanidinium salts $[C(NH_2)_3]_2[B_4O_5(OH)_4]\cdot 2H_2O$, $[C(NH_2)_3][B_5O_6(OH)_4]\cdot H_2O$, and $[C(NH_2)_3]_3[B_9O_{12}(OH)_6]$ and in compounds **1–6** revealed that two important H-bonding $R_2^2(8)$ motifs competed to stabilize the observed structures. The guanidinium cation formed charge-assisted pincer cation–anion H-bonded rings as a major motif in $[C(NH_2)_3]_2[B_4O_5(OH)_4]\cdot 2H_2O$ and $[C(NH_2)_3]_3[B_9O_{12}(OH)_6]$, whereas the anion–anion ring motif was dominant in $[C(NH_2)_3][B_5O_6(OH)_4]\cdot H_2O$ and in compounds **1–6**. This behaviour was consistent with the stoichiometry of the salt and packing effects also strongly influencing their solid-state structures.

Keywords: borate; guanidinium salts; H-bonding; oxidoborate; pentaborate(1-); $R_2^2(8)$ motifs; tetrahydroxidohexaoxidopentaborate(1-) salts; XRD

Citation: Beckett, M.A.; Coles, S.J.; Horton, P.N.; Rixon, T.A. Structural (XRD) Characterization and an Analysis of H-Bonding Motifs in Some Tetrahydroxidohexaoxidopen taborate(1-) Salts of *N*-Substituted Guanidinium Cations. *Molecules* **2023**, *28*, 3273. https://doi.org/10.3390/molecules28073273

Academic Editor: Vito Capriati

Received: 21 March 2023
Revised: 4 April 2023
Accepted: 5 April 2023
Published: 6 April 2023

1. Introduction

Poly(hydroxidooxidoborate) compounds are well-known and are represented by many naturally occurring minerals [1–6] and by synthetic analogues [7–11]. These compounds are structurally diverse and are comprised of hydroxidooxidoborate anions paired with various cationic units. Structural features include simple neutral transition-metal complexes [11], insular salts (main group cations, transition-metal complexes and organic cations) [8–11], and many more highly condensed species comprised of 2-D chains or 3-D networks [1–7]. Some of these compounds, e.g., $Na_2B_4O_5(OH)_4\cdot 8H_2O$ (borax) and the zinc borate $ZnB_3O_4(OH)_3$, are of great importance for many industrial applications [12–14]. Other compounds, e.g., β-BaB_2O_4 (BBO) have more specialized small-scale applications [4] and the physicochemical properties of others are currently being actively investigated for new potential applications [10,11].

The synthesis of new polyborate [15] salts is easily achieved through simple aqueous synthetic procedures or through solvothermal/hydrothermal methods [10,11]. The nitrogen-rich base guanidine, $(H_2N)_2C = NH$, is a strong base (pK_{aH} of its conjugate acid = ca. 13.6 [16]) and organic bases of this strength readily form, when protonated, polyborate salts. Usually, solid-state interactions in polyborate chemistry favour specific cation–anion combinations [10] and therefore it is surprising to find that *three* guanidinium polyborate salts of differing stoichiometries have been structurally characterized by single-crystal X-ray diffraction studies: $[C(NH_2)_3]_2[B_4O_5(OH)_4]\cdot 2H_2O$ [17], $[C(NH_2)_3][B_5O_6(OH)_4]\cdot H_2O$ [18], and $[C(NH_2)_3]_3[B_9O_{12}(OH)_6]$ [8,19]. The tetraborate(2-) salt, $[C(NH_2)_3]_2[B_4O_5(OH)_4]\cdot 2H_2O$, was first reported in 1921 [20] and crystallographically characterized in 1985 [17]. Likewise, pentaborate(1-) salts of various degrees of hydration, $[C(NH_2)_3][B_5O_6(OH)_4]\cdot xH_2O$ have been described in early literature [21] but $[C(NH_2)_3][B_5O_6(OH)_4]\cdot H_2O$ was crystallographically characterized in 2020 together with its spectral, optical, thermal, and third-order NLO properties [18]. The large-scale high temperature synthesis of a salt containing the structurally rare nonaborate(3-) anion, $[C(NH_2)_3]_3[B_9O_{12}(OH)_6]$, was reported in 2000 [19]. $[C(NH_2)_3]_3[B_9O_{12}(OH)_6]$ has been used as a precursor to BN powders [22].

The guanidinium cation is therefore particularly well-suited in stabilizing polyborate anions in the solid state, whereas other nonmetal cations are in general less adaptable. The small size, high symmetry (D_{3h}), and high H-bond donor capacity may be contributing factors for this [23,24]. In this manuscript, we investigated the synthesis of new polyborate salts by crystallization from aqueous solutions containing $B(OH)_3$ and the substituted guanidinium cations. Six new tetrahydroxidohexaoxidopentaborate(1-) salts were obtained (see Figure 1 for schematic structures) and herein we report their synthesis and their solid-state structures as determined by single-crystal XRD studies. Their solid-state H-bond interactions were analysed together with those found in the polyborate salts of the unsubstituted guanidinium cations as a potential means of accounting for this behaviour [25,26].

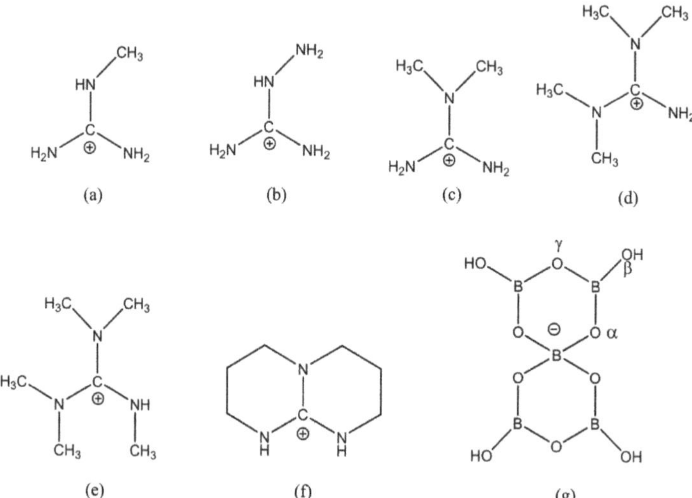

Figure 1. Schematic drawings of the cations and anions found in compounds **1–6**: (**a**) *N*-methyl guanidinium(+1) in **1**; (**b**) *N*-aminoguanidinium(+1) in **2**; (**c**) *N,N*-dimethylguanidinium(+1) in **3**; (**d**) *N,N,N′,N′*-tetramethylguanidinium(1+) in **4**; (**e**) *N,N,N′,N′,N″*-pentamethylguanidinium(1+) in **5**; (**f**) TBDH (TBD = 1,5,7-triazabicyclo[4.4.0]dec-5-ene) in **6**; (**g**) tetrahydroxidohexaoxidopentaborate(1-) found in compounds **1–6** with H-bond acceptor sites labelled as α, β, or γ.

2. Results and Discussion

2.1. Synthesis

The three structurally characterized guanidinium polyborate salts $[C(NH_2)_3]_2[B_4O_5(OH)_4]\cdot2H_2O$ [17], $[C(NH_2)_3][B_5O_6(OH)_4]\cdot H_2O$ [18], and $[C(NH_2)_3]_3[B_9O_{12}(OH)_6]$ [19] were all prepared by crystallization from aqueous solutions containing $[C(NH_2)_3]_2CO_3$ and $B(OH)_3$ at various temperatures and ratios [8]. Two alternative syntheses of $[C(NH_2)_3]_3$ $[B_9O_{12}(OH)_6]$ have been reported using either (i) $[C(NH_2)_3]Cl$, $Na_2B_4O_7\cdot5H_2O$ and $B(OH)_3$ or (ii) $[C(NH_2)_3]_2[B_4O_5(OH)_4]\cdot2H_2O$ and $B(OH)_3$ in appropriate ratios [19]. It is well-known that $B(OH)_3$, and other borate salts, exists in alkaline aqueous solution as equilibrium mixtures of numerous polyborate anions [27,28]. The guanidinium cation that is present templates the crystallization of specific products under the reaction conditions in what can be described as self-assembly processes [29,30].

The *N*-substituted guanidinium starting materials used in this study were all commercially available and were non-carbonate species. We have previously synthesized many nonmetal cation polyborate salts by a room temperature crystallization of aqueous solutions originally primed with an organic free base, or its protonated cation (prepared in situ as its $[OH]^-$ salt by a metathesis reaction) and $B(OH)_3$ [10,11]. Based on this strategy we prepared six new *N*-substituted guanidinium salts as shown in Scheme 1. Crude yields of these compounds ranged from 71% to near quantitative. The recrystallization of samples from H_2O gave crystals suitable for single-crystal X-ray diffraction studies.

$$[C(NH_2)_2(NHMe)]Cl \xrightarrow[\text{H}_2\text{O/RT}]{\substack{\text{(i) [OH]}^-\text{ ion exchange} \\ \text{(ii) 5 B(OH)}_3}} [C(NH_2)_2(NHMe)][B_5O_6(OH)_4] \quad (\mathbf{1}) \quad 71\%$$

$$[C(NH_2)_2(NH\{NH_2\})]_2SO_4 \xrightarrow[\text{H}_2\text{O/RT}]{\substack{\text{(i) Ba(OH)}_2\cdot8H_2O \\ \text{(ii) 10 B(OH)}_3}} 2\,[C(NH_2)_2(NH\{NH_2\})][B_5O_6(OH)_4] \quad (\mathbf{2}) \quad 95\%$$

$$[C(NH_2)_2(NMe_2)]_2SO_4 \xrightarrow[\text{H}_2\text{O/RT}]{\substack{\text{(i) Ba(OH)}_2\cdot8H2O \\ \text{(ii)10 B(OH)}_3}} 2\,[C(NH_2)_2(NMe_2)][B_5O_6(OH)_4] \quad (\mathbf{3}) \quad 87\%$$

$$C(NH)(NMe_2)_2 \xrightarrow[\text{H}_2\text{O/RT}]{5\,B(OH)_3} [C(NH_2)(NMe_2)_2][B_5O_6(OH)_4] \quad (\mathbf{4}) \quad 99\%$$

$$[C(NHMe)(NMe_2)_2]I \xrightarrow[\text{H}_2\text{O/RT}]{\substack{\text{(i) [OH]}^-\text{ ion exchange} \\ \text{(ii) 5 B(OH)}_3}} [C(NHMe)(NMe_2)_2][B_5O_6(OH)_4]\cdot B(OH)_3 \quad (\mathbf{5}) \quad 97\%$$

$$TBD \xrightarrow[\text{H}_2\text{O/RT}]{5\,B(OH)_3} [TBDH][B_5O_6(OH)_4] \quad (\mathbf{6}) \quad 93\,\%$$

Scheme 1. Synthetic procedure for the synthesis of compounds **1–6** (TBD = 1,5,7-triazabicyclo[4.4.0]dec-5-ene and the structure of TBDH is illustrated in Figure 1f).

Thermal and spectroscopic data (Section 2.2) and elemental analysis data (Sections 3.3–3.9) on the crude products **1–6** were consistent with formulating these materials as *N*-substituted guanidinium pentaborate salts and these formulations were confirmed by single-crystal XRD studies (Section 2.3). All compounds were colourless and stable in the solid state, insoluble in organic solvents but soluble in H_2O with decomposition.

2.2. Thermal and Spectroscopic Properties

Thermal gravimetric analysis (TGA) data (in air) for nonmetal cation polyborate salts are often reported with a thermal decomposition leading to B_2O_3 via multistage processes involving the loss of interstitial H_2O (100–200 °C), the condensation of hydroxy groups bound to boron with the loss of a further two H_2O molecules (250–400 °C), and then the oxidation of organics (400–700 °C) [10,31,32]. A TGA was undertaken on compounds **1–3** and **6** as representative examples of the new substituted guanidinium pentaborate(1-) salts. Compounds **2**, **3**, and **6** did not have interstitial H_2O and their TGA curves showed the expected loss of two H_2O molecules between 240 and 320 °C. Compound **1** followed this expected behaviour but there was no clear distinction between the two lower temperature processes that occurred (100–275 °C) with the loss of three H_2O molecules. The glassy solids that remained after heating to 700 °C for compounds **1–3** and **6** had residual masses consistent with 2.5 B_2O_3.

^1H and ^{13}C NMR were obtained for compounds **1** and **3–6** dissolved in D_2O and these all gave spectra consistent with the appropriate guanidinium cations being present, e.g., the ^1H spectrum of compound **4** gave a signal for the *N*-Me groups as a singlet at 2.85 ppm with exchangeable H atoms with HOD (4.7 ppm). The $^{13}C\{^1H\}$ spectrum of compound **4** gave two signals at 36.98 and 160.2 ppm for the CH_3 and CN_3 carbons, respectively, with the downfield signal being weak. The presence of boron in these compounds was confirmed by ^{11}B NMR (in D_2O) for all compounds. All ^{11}B spectra all showed three "signature" signals [31,33], arising through the decomposition of a pentaborate, with relative intensities appropriate for a sample that had attained polyborate/aqueous equilibrium [27]. Thus, compound **4** gave three signals at 1.0, 12.8, and 19.1 ppm with approximate relative intensities of 5%, 10%, and 85%, respectively. These signals have been previously assigned, moving downfield, to $[B_5O_6(OH)_4]^-$ (tetrahedral *B*), $[B_3O_4(OH)_4]^-$, and $B(OH)_3/[B(OH)_4]^-$ [27]. Three signals were observed as the sample was relatively concentrated rather than one signal at +16.1, which would have been expected at infinite dilution [33].

All samples were characterized by FTIR spectroscopy. Strong and potentially diagnostic absorptions were to be expected for the cation $\nu(NH)$ at ca. 3400 cm^{-1} and $\nu(CN)$ at ca. 1650 cm^{-1} [34] and the polyborate anion $\nu(O\text{-}H)$ at 3500 cm^{-1} and the $\nu(B\text{-}O)$ stretches grouped between 1450 and 740 cm^{-1} [35]. In particular, a strong adsorption at ca. 925 cm^{-1} in the $(B_{trig}\text{-}O)_{sym}$ stretching region has previously been described as diagnostic for the $[B_5O_6(OH)_4]^-$ anion [36]. This strong absorption, amongst other strong B-O stretches, together with the strong adsorptions associated with the cation (1671–1625 cm^{-1}), was present in all samples.

2.3. Single-Crystal XRD Studies

Compounds **1–6** were characterized by single-crystal XRD studies. All the compounds contained the expected insular *N*-substituted guanidinium(1+) cations and insular tetrahydroxidohexaoxidopentaborate(1-) anions within the asymmetric unit. Additionally, compound **1** had one H_2O of crystallization per cation/anion and compound **5** was cocrystallized with a molecule of $B(OH)_3$ per cation/anion. Crystallographically, compounds **1–5** were free of disorder, whilst compound **6** was modulated, but was only refined using the subcell with all atoms disordered over two positions of equal population. Brief crystallographic details for each compound are given in the experimental section, and atomic numbering schemes for compounds **1–6** are shown in Figure 2. The full crystallographic information is available as Supplementary Materials. Generally, the gross structures, bond angles, and bond distances within the guanidinium [17,37] and pentaborate [31–33,36] units were within the expected ranges for these ions and need no further comment. The *N*-substituted guanidinium cations had a variable number of potential H-bond donor sites (one to seven in this study) that were dependent on the extent of the substitution. Each pentaborate(1-) unit had four potential H-bonds donor sites and ten potential H-bond acceptor sites in locations described as α, β, or γ, as defined elsewhere [10,32], and they are illustrated in Figure 1. Since solid-state H-bonding is likely to be important in stabilizing these structures [23–26], the H-bond interactions in compounds **1–6** were analysed and are described in detail using an Etter graph set terminology [38].

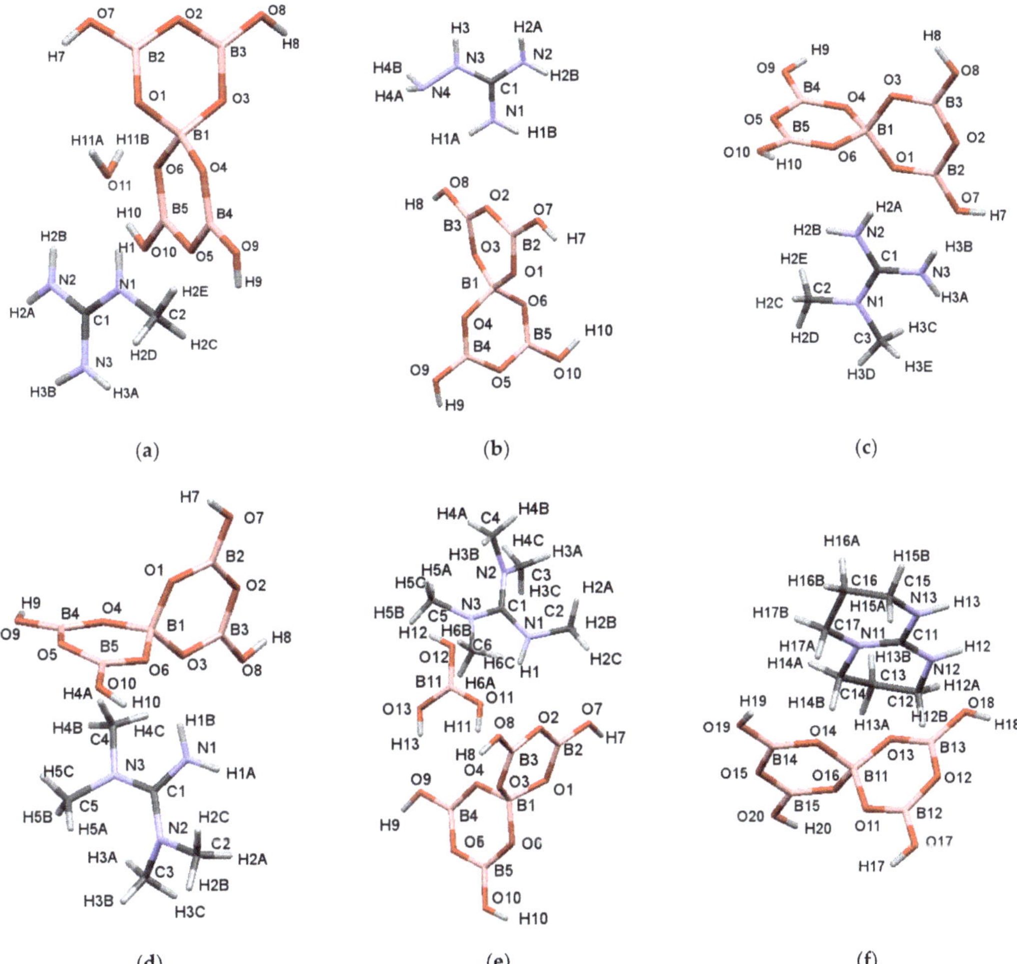

(a)

(b)

(c)

(d)

(e)

(f)

Figure 2. Drawings of the crystals structures of (**a**) [C(NH$_2$)$_2$(NHMe)][B$_5$O$_6$(OH)$_4$]·H$_2$O (**1**), (**b**) [C(NH$_2$)$_2$(NH{NH$_2$})][B$_5$O$_6$(OH)$_4$] (**2**), (**c**) [C(NH$_2$)$_2$(NMe$_2$)][B$_5$O$_6$(OH)$_4$] (**3**), (**d**) [C(NH$_2$) (NMe$_2$)$_2$][B$_5$O$_6$(OH)$_4$] (**4**), (**e**) [C(NHMe)(NMe$_2$)$_2$][B$_5$O$_6$(OH)$_4$]·B(OH)$_3$ (**5**), and (**f**) [TBDH][B$_5$O$_6$ (OH)$_4$] (**6**) (TBD = 1,5,7-triazabicyclo[4.4.0]dec-5-ene) showing atomic numbering. Compound **6** is modulated and just one component of the disordered anion and cation is shown in (**f**).

Compounds **2**, **3**, **4**, and **6** had the extended anion–anion H-bonded lattices that are found in many nonmetal cation pentaborate salts [10], with the cations filling the "voids" and "channels" and forming additional H-bonds to the anions. In these structures, each pentaborate(1-) anion donates four H-bond to four neighbouring pentaborate(1-) anions. The extended lattice structures are 3-D because the central B atom in each pentaborate(1-) is tetrahedral with its two associated boroxole rings perpendicular to each other (D_{2d} for heavy atoms).

Thus, in compound **4**, the H-bonds originating from each pentaborate were to three α and one β (or α,α,α,β) acceptor sites of its neighbours (see Figure 1 for acceptor site positions). This particular anion–anion giant structure is commonly found in many non-metal cation pentaborate salts [10] and has be described as a "brick wall" with three $R_2^2(8)$ [38] reciprocal-α interactions forming a "layer", resembling a brick wall, and the fourth interaction, at a β-site, linking these layers by infinite C(8) [38] chains. These $R_2^2(8)$ reciprocal interactions are known to be strong and are particularly common in polyborate chemistry [10,31–33,36,39]. $R_2^2(8)$ interactions are also prevalent elsewhere, e.g., carboxylic acid and boronic acid dimers [26]. In compound **4** these donors were O7-H7, O9-H9, and O10-H10 and the acceptors were O1, O4, and O6, respectively, on three different neighbouring pentaborate anions. The C(8) chain that linked layers arose through O8H8··· O10′. The $[C(NH_2)(NMe_2)_2]^+$ cations in compound **4** were in the channels of the brick-wall structure (when viewed along the b axis) and each formed two donor H-bonds from the amino group to the γ (N1H1A··· O2) and α (N1H1B··· O3) positions of two neighbouring pentaborate(1-) anions.

Compounds **2** and **3** had similar, and previously observed [10], anion–anion H-bond arrangements ($\alpha,\alpha,\alpha,\gamma$ acceptor sites) as three $R_2^2(8)$ reciprocal-α interactions and one $R_2^2(8)$ reciprocal-γ interaction. The reciprocal-α H-bond donor interactions originated from O7-H7, O8-H8, and O10-H10 in compound **2** and from O8-H8, O9-H9, and O10-H10 in compound **3**. The reciprocal-γ interactions for compound **2** were -O9-H9··· O5′-B4′-O9′-H9′··· O5-B4- and those for compound **3** were -O7-H7··· O2′-B2′-O7′-H7′··· O2-B2-. The $[C(NH_2)_2(NMe_2)]^+$ cation in compound **3** formed four H-bonds to three pentaborates of the anionic lattice. These interactions are shown in Figure 3a. The two interactions starting at N2-H2B and N3-H3A were both part of a dimeric centrosymmetric $R_4^4(20)$ ring comprised of two anions and two cations. The H-bonds originating from N2-H2A and N3-H3B were also of interest since these amino-hydrogens participated in a charge-assisted pincer $R_2^2(8)$ ring [23,40], as a double donor to an α-site (O1) and a β-site (O7) of a single pentaborate(1-) anion. Each $[C(NH_2)_2(NH\{NH_2\})]^+$ cation in compound **2** had seven potential H-bond donor sites and all were involved in H-bonds to five neighbouring pentaborate anions in a complex manner. However, the H-bond motifs described for compound **3** were also readily discerned in compound **2**. Thus, "simple" interactions involved N1-H1A and N4-H4A with O8 (β) and O4 (α), respectively. N4-H4B was involved in a bifurcated H-bond bridging two pentaborates at the O5 (γ) sites. The four remaining H-bond donors, N1-H1B, N2-H2B, N2-H2A, and N3-H3, participated in two pincer-type $R_2^2(8)$ rings to two neighbouring pentaborates at O9 (β)/O4 (α) and O2 (γ)/O7 (β), respectively.

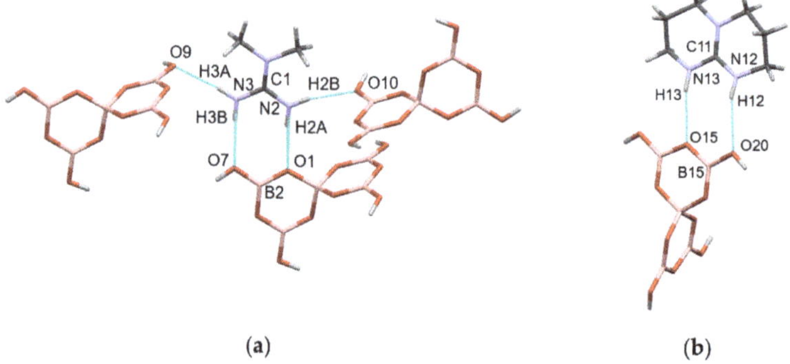

Figure 3. H-bonds originating from the *N*-substituted guanidinium centres present in compounds (a) **2**, and (b) **6**, illustrating the "pincer"-like $R_2^2(8)$ rings.

Compound **6** also had a previously observed anion–anion giant lattice with $\alpha,\alpha,\alpha,\beta$ acceptor sites (also observed in compound **4**) but now arranged as three reciprocal-α rings and one $R_2^2(12)$ reciprocal-β ring [10]. The reciprocal-α H-bonds donor originated from O17-H17, O19-H19, and O20-H20 and the $R_2^2(12)$ reciprocal-β ring involved O18-

H18$\cdots$O17$'$ and O18$'$-H18$'\cdots$O17. The bicyclic [TBDH]$^+$ cation in compound **6** had two H-bonds donor sites, N12-H12 and N13-H13. Both were involved in a pincer-type $R_2^2(8)$ ring at O20 (β) and O15 (γ) acceptor sites, respectively, of a single pentaborate anion, Figure 3b.

Compounds **1** and **5** had additional cocrystallized molecules with the cation and anion. Cocrystallized molecules in H-bonded anionic polyborate lattices usually function either by (i) helping to fill space within the areas where the cations sit or by (ii) expanding the anionic lattices to help accommodate larger cations [10,41].

Compound **1** had a pentaborate lattice structure that was very similar to compound **3** with $\alpha,\alpha,\alpha,\gamma$ acceptor sites involved in four $R_2^2(8)$ rings. In compound **1**, these donors were O7-H7, O8-H8, O9-H9, and O10-H10, and the acceptors were O3 (α), O1 (α), O5 (γ), and O6 (α) on four different neighbouring pentaborate anions. The structure of compound **1** differed from that of compound **3** in the detail of their H-bonded $R_2^2(8)$ rings with H-bonds originating from O10-H10 and O9-H9 reciprocal and those originating from O7-H7 and O8-H8 paired. The [C(NH$_2$)$_2$(NHMe)]$^+$ cation in compound **1** had five H-bond donor sites. N2-H2A and N3-H3B were both part of a pincer-like $R_2^2(8)$ ring to O4 (α) and O9 (β) acceptor sites, respectively. N3-H3A and N2-H2B both donated H-bonds to a β-sites (O7 and O8). The H-bond from N2-H2B had a N2-H2B-O8 angle of 141.1(13)$°$ and a H2B$\cdots$O8 distance of 2.322(15) Å. The H-bond arising from N2-H2B may be considered as bifurcated since O11 (H$_2$O) had a N2-H2B-O11 angle of 138.8(13)$°$ with a long, potentially H-bonding H2B$\cdots$O11 distance of 2.661(15) Å. This H$_2$O molecule was also involved with closer and stronger acceptor interaction from N1-H1 (H1$\cdots$O11, 1.943(15) Å and an angle of N1-H1$\cdots$O11 170.2(14)$°$. This interaction together with that from N2-H2B formed a pincer $R_2^1(6)$ ring. The H$_2$O molecule formed two donor bonds to two pentaborates at α,α (O1/O4, bifurcated) and β (O10) sites. The anion–anion lattice in compound **1** was largely unaffected by the presence of the cocrystallized H$_2$O molecule. Since the cation in compound **1** was relatively small, the H$_2$O positioned itself to fill space within the lattice reserved for the cations and further stabilized the structure by these H-bond interactions.

Compound **5** had B(OH)$_3$ cocrystallized with the pentaborate salt and consequentially had a unique anionic pentaborate/boric acid lattice. However, the $R_2^2(8)$ building motifs seen in most polyborate structures were also readily identified here. Thus, one "plane" was comprised of chains of pentaborates each linked via two reciprocal-α rings. These reciprocal-α interactions originated from O7-H7 and O8-H8 with O3 and O1 acceptors, respectively. The other two H-bond interactions originating from each pentaborate (O9-H9 and O10-H10) are shown in Figure 4. The interactions shown here were approximately perpendicular to those just described an originating from O7-H7 and O-H8. The "horizontal" chains in Figure 4 were comprised of alternating pentaborate(1-)/boric acid moieties with the H-atom positions of the B(OH)$_3$ acid arranged to maximise $R_2^2(8)$ interactions. Thus, O10-H10 on the pentaborate and O12-H12 on the B(OH)$_3$ were involved in a standard borate–borate interaction and O11-H11 and O13-H13 were involved in a "pincer" double H-bond to the pentaborate at O4 (α) and O9 (β) sites. Each B(OH)$_3$ formed three donor H-bonds to two α sites and one β site of neighbouring pentaborates. The fourth and final H-bond originating from each pentaborate is also shown in Figure 4. Here, O9-H9 formed an H-bond to a β-acceptor site (O8) of its neighbour, cross-linking the pentaborate(1-)/boric acid chains into the "plane" shown in Figure 4 by the formation of C(8) chains. The sterically demanding [C(NHMe)(NMe$_2$)$_2$]$^+$ cation in compound **5** had only one amino group (N1-H1) available for H-bonding and this was utilized in the H-bonding to O11 of the B(OH)$_3$ (Figure 4). The B(OH)$_3$ in this structure served to expand the anionic pentaborate lattice to enable the sterically demanding cation more space. This behaviour has been observed before in several polyborate [36,41–43] salts including [N(nPr)$_4$][B$_5$O$_6$(OH)$_4$][B(OH)$_3$]$_2$ and [HPS][B$_5$O$_6$(OH)$_4$]·B(OH)$_3$ (PS = proton sponge, 1,8-bis(dimethylamino)naphthalene).

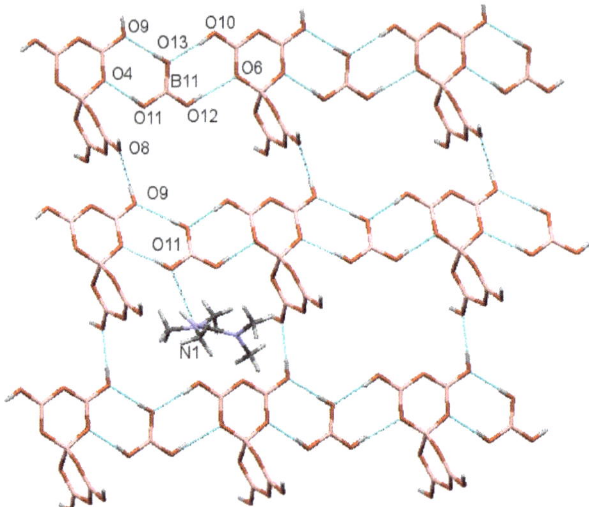

Figure 4. In [C(NHMe)(NMe$_2$)$_2$][B$_5$O$_6$(OH)$_4$]·B(OH)$_3$ (**5**) the "horizontal" chains (three partially shown) of alternating H-bonded R$_2^2$(8) pentaborate(1-)/boric acid units are cross-linked by H-bonded C(8) "vertical" chains (O9H9···O8′) into planes. For clarity, H atoms are not labelled but are numbered the same as the oxygen/nitrogen atom to which they are attached. The [C(NHMe)(NMe$_2$)$_2$]$^+$ cations (one shown) fill the "voids" and are H-bonded to the B(OH)$_3$ units (N1H1···O11).

2.4. Discussion on H-Bonding in Guanidinium Polyborates

The anions in polyborate salts, [B$_a$O$_b$(OH)$_c$]$^{n-}$, are invariably rich in H-bond donor sites (there are c donor BOH groups) and oxygen-atom H-bond acceptor sites with acceptor sites ($b + c$) outnumbering the donor sites and enabling the opportunity for H-bond donor cations to contribute to the H-bonding networks. R$_2^2$(8) are important stabilizing interactions [10,33] and the *maximum* number of borate–borate R$_2^2$(8) rings available to a polyborate anion is limited to the lower value of b or c, since bridging oxygen atoms are acceptors in these R$_2^2$(8) interactions. In actual structures, packing/steric effects may reduce further this maximum number.

The guanidinium (G) cation is well-known as a multi-H-bond donor in solid-state structures and several H-bond motifs are commonly observed [23,24,37,44–48]. These include a donor H-bond to an electronegative atom from a single N-H site (often this simple motif is incorporated into larger more complex rings and/or chains), and two pincer-type donor ring motifs to either a single acceptor site, R$_2^1$(6), or to two acceptor sites, R$_2^2$(8). The structure of GCl has Cl$^-$ ions bridging G cations as part of alternating anion–cation chains linked by H-bonded R$_2^1$(6) rings [44]. Oxoanions (e.g., borate esters [24], carboxylate [45,46], sulfonates [47,48], phosphonates [48]), are good R$_2^2$(8) pincer H-bond acceptors and these interactions are strong and have been described as "'charge-assisted" [45,49]. The structure of G[RSO$_3$] is self-assembled and templated into layers with each G cation forming three such pincer R$_2^2$(8) rings with three organosulfonate anions [47]. Cubic hydrogen-bonded borate networks containing four pincer R$_2^2$(8) rings per borate have been observed in G$_4$[B(OMe)$_4$]$_3$Cl·5H$_2$O [24].

As noted in the introduction, the G cation was found in three polyborate structures. These "oxoanions" are ideally set up to accept the pincer R$_2^2$(8) rings, and we examined, focussing on the polyborate anion, the extent of such interactions as they are in direct competition (see below) with borate–borate R$_2^2$(8) rings in stabilizing these structures. We also examined how *N*-substitution affected this R$_2^2$(8) balance and how this affected the observed structures.

When considering solid-state H-bonding interactions the generalised guanidinium (either N-substituted or unsubstituted) polyborate salt can be represented as $(xG)_n[B_aO_b(OH)_c]$ where x represents the *maximum* number of potential donor pincer $R_2^2(8)$ motifs available to the cation. Each unsubstituted G cation has the potential to form a maximum of three donor pincer $R_2^2(8)$ motifs (i.e., $x = 3$), whilst substituted G cations will have fewer opportunities ($x = 0$, 1, or 2) with the value of x dependent on the extent and position of the substitution. Within guanidinium polyborate salts, the maximum number of $R_2^2(8)$ acceptor interactions (both pincer and normal) is b, since bridging oxygen atoms generally can only be used once and are required for both types of these $R_2^2(8)$ interactions. Usually, borate–borate interactions are approximately coplanar with boroxole rings. Packing/steric effects may reduce the maximum number for both types of interactions.

This analysis focused on an analysis of two ratios. Firstly, a value of >1 for $(xGn + c)/b$ would indicate borate–borate and pincer $R_2^2(8)$ motifs were in direct competition for the available borate acceptor sites. Secondly, potential and actual (i.e., observed) $(xGn)/c$ ratios for each compound were also of interest since a difference in these values may indicate a structural preference for one or the other of these H-bonding motifs.

Using this approach, $G_2B_4O_5(OH)_4$, $GB_5O_6(OH)_4$, and $G_3B_9O_{12}(OH)_6$, had $(xGn + c)/b$ ratios of 2.0, 1.17, and 1.25, respectively, confirming that pincer/borate donor rings were in competition for the available acceptor sites. The potential donor ratios, $(xGn)/c$, for the three compounds were 1.5, 0.75, and 1.5.

The examination of the structure of $G_2[B_4O_5(OH)_4]\cdot2H_2O$ [17] revealed two independent G cations. Two molecules of H_2O within this formula unit also affected the structure but did not affect this analysis. All five bridging oxygen atoms of the tetraborate(2-) anion were involved in $R_2^2(8)$ rings with four cation–borate pincer interactions and only one borate–borate interaction. Thus, within this structure, 80% of these rings involved G cations and the observed $(xGn)/c$ ratio was 4.0. This value was much higher than would be expected (1.5) based on the number of potential donor sites. The higher-than-expected observed $(xGn)/c$ ratio would indicate that the G cations were either able to replace the strong borate–borate interactions with stronger pincer motifs or that additional factors were also important. Such factors may include the compound's stoichiometry and that the three-fold symmetry of G cations (D_{3d}) were a good packing match for the tetraborate(2-) anion (C_{2v}). H-bonding in tetraborate(2-) structures not containing the G cation have been recently reviewed [50] and their structural architectures are, as expected, dominated by borate–borate interactions with the majority of structures having at least two $R_2^2(8)$ rings.

Within the structure of $G_3[B_9O_{12}(OH)_6]$ [19], all twelve bridging oxygen acceptor sites were involved in eleven $R_2^2(8)$ H-bonded rings with seven pincer charge-assisted interactions involving the three G cations and four borate–borate interactions. Interestingly, the four bridging oxygen atoms joined to the central boron were involved, solely amongst themselves, in three pincer rings; the two oxygens that bridged two four-coordinate borons were both involved in two pincer rings. These rings were not coplanar with boroxole rings. In this salt, 64% of the $R_2^2(8)$ rings involved the G cations and 36% were borate–borate. This observed ratio of 1.78 was higher than that expected (1.5) based on total potential donor sites and would again indicate that strong cation–anion pincer interactions were able to replace the strong borate–borate interactions, but again, stoichiometry and packing considerations were also important.

The H-bonded structure of $G[B_5O_6(OH)_4]\cdot H_2O$ [18] utilizes only four of the six available acceptor bridging oxygen atoms in $R_2^2(8)$ interactions. Here, and in contrast to the tetraborate(2-) and nonaborate(3-) structures, 75% of the H-bond ring interactions were borate–borate $R_2^2(8)$ rings with the charge-assisted pincer rings only accounting for 25%. The observed $(xGn)/c$ ratio of 0.33 was lower than would be expected (0.75) from the number of potential donor sites and confirmed that the borate–borate interactions strongly influenced that structure and that the templating influence of G was minimal. These structures were strongly stabilized by these anion–anion interactions with any additional cation–anion interactions helping to further stabilize the anionic lattice [10]. Many non-metal cation pentaborate(1-) salts show three such borate–borate interactions [10]. This was also the likely situation in this salt since the three-fold symmetry of the cation did not pack well with the D_{2d} symmetry of the insular pentaborate(1-) anion and its perpendicularly H-bonded giant lattice.

Adding substituents to the G cations affects their structure directing potential in three ways: it reduces their ability to form H-bonds, increases their steric bulk, and disrupts their three-fold symmetry. The first factor also correspondingly lowers x, reduces the $(xGn)/c$ ratio, and hence encourages more borate–borate H-bonds. The structures of the *N*-substituted guanidinium cation pentaborate(1-) salts **1–6** were described in detail in Section 2.2, but further general comments based on the analysis above would now be appropriate. The numbers of potential pincer H-bond donor sites (x) available in compounds **1–6** were 2, 2, 1, 0, 0, and 1, and the $(xGn)/c$ ratios were lower than that for $G[B_5O_6(OH)_4]$ (0.75) at 0.5, 0.5, 0.25, 0, 0, and 0.25, respectively. This analysis would indicate borate–borate interactions should dominate the H-bond interactions in compounds **1–6**, with additional charge-assisted pincer H-bonds forming wherever possible. Unsurprisingly, three or four borate–borate $R_2^2(8)$ rings were observed in compounds **1–6** and these interactions were clearly important in stabilizing their solid-state structures. Pincer interactions were unavailable for compounds **4** and **5** and the amino-H atoms formed simple H-bonds to borate/boric acid acceptor sites. Compounds **3** and **6** both had the potential to form one pincer bond to stabilize the lattice and each had one such pincer bond. Compound **2** had the potential to form two pincer bonds, and both were formed. Compound **1** had potentially two pincer bonds to the borate but here, only one was observed. However, this was a direct consequence of the cocrystallized H_2O since the H_2O was part of an alternative $R_2^1(6)$ pincer ring motif. Compound **5** had a molecule of $B(OH)_3$ cocrystallized and no opportunity to form a pincer ring. As noted in Section 2.2, the $[C(NHMe)(NMe_2)_2]^+$ in compound **5** was relatively bulky and the $B(OH)_3$ served to extend the anionic lattice, by positioning itself between two pentaborate(1-) anions. It is interesting to note that the H-atom positions in $B(OH)_3$ were asymmetric (see Figure 4) and therefore allowed it to partake in an $R_2^2(8)$ donor pincer interaction to one pentaborate(1-) and a standard $R_2^2(8)$ borate interaction to the other.

3. Materials and Experimental Methods

3.1. General

Reagents were all obtained commercially. FTIR spectra were obtained as KBr pellets on a Perkin-Elmer 100FTIR spectrometer (Perkin-Elmer, Seer Green, UK). ^1H, ^{11}B, and $^{13}C\{^1H\}$ NMR spectra were obtained on a Bruker Ultrashield Plus 400 spectrometer (Bruker, Coventry, UK) on samples dissolved in D_2O at 400, 128, and 100 MHz, respectively. Chemical shifts are in ppm with positive values to the high frequency (downfield) of TMS (^1H, ^{13}C) or $BF_3 \cdot OEt_2$ (^{11}B). TGA and DSC were performed on an SDT Q600 V4 instrument (TA Instruments, New Castle, DE, USA) using Al_2O_3 crucibles with a temperature ramp-rate of 10 °C per minute (20–700 °C in air). X-ray crystallography was performed at the EP-SRC national crystallography service centre at Southampton University. Drawings in this manuscript have been generated using Mercury software (CCDC, Cambridge, UK). CHN analyses were obtained from OEA Laboratories (Callingham, Cornwall, UK).

3.2. X-ray Crystallography

Crystallographic data for compounds **1–3**, **5**, and **6** were obtained, with crystals kept at $T = 100(2)$ K during data collection, on either a Rigaku FRE+ diffractometer equipped with VHF Varimax confocal mirrors and an AFC12 goniometer and Hypix 6000 detector (**1, 3**) or a Rigaku 007HF equipped with HF Varimax confocal mirrors and an AFC11 goniometer and HyPix 6000 detector (**2, 5, 6**) using CrysAlisPro [51]. Structures (**1–3, 5, 6**) were solved with the ShelXT [52] structure solution programme using the dual-solution structure method (**1, 5**) or the intrinsic-phasing solution method (**2, 3, 6**).The crystallographic data collection for compound **4** was obtained at 120(2) K on a Bruker-Nonius FR591 diffractometer with a

Nonius KappaCCD detector using Collect [53] and Denzo [54] and *SHELXS97* [55] was used for the structure solution. By using Olex2 [56] as the graphical interface, all structure models were refined with ShelXL [57] using a least-squares minimisation. Structure refinement details and ORTEP diagrams with 50% probability displacement ellipsoids for compounds **1–6** are given in the Supplementary Materials and selected crystallographic data are given for each compound in Sections 3.3–3.9.

3.3. Preparation of $[C(NH_2)_2(NHMe)][B_5O_6(OH)_4]\cdot H_2O$ (**1**)

DOWEX 550A (OH)$^-$ ion exchange resin (18 g) was added to a solution of $[C(NH_2)_2(NHMe)]Cl$ (0.5 g, 4.6 mmol) in H_2O (25 mL) and the mixture was stirred for 18 h. The resin was removed by filtration and $B(OH)_3$ (1.4 g, 23 mmol) was added to the filtrate. The solution was stirred for a further 2 h, evaporated under reduced pressure to yield 1.1 g of a white powder as a crude product (71%). Recrystallization of 0.3 g of the product from warm H_2O afforded a few crystals suitable for XRD studies. TGA: 100–275 °C, condensation of pentaborate with loss of 3 H_2O 17.4% (17.4% calc.); 275–700 °C, oxidation of organic residue leaving residual 2.5 B_2O_3 55.9% (56.1% calc.). NMR/ppm: δ_H: 2.69 (s, CH_3) 4.70 (HOD); δ_B: 1.1 (3%), 13.2 (24%), 18.1 (73%); δ_C: 36.98 (CH_3). FTIR (KBr, cm^{-1}): 3435(s), 3400(s), 2958(m), 2924(m), 1664(s), 1439(m), 1347(m), 1103(m), 1025(m), 925(s), 782(m), 696(m). XRD crystallographic data: $C_2H_{14}B_5N_3O_{11}$, M_r = 310.21, monoclinic, $P2_1/c$ (no. 14), a = 9.9962(2) Å, b = 10.9047(2) Å, c = 11.7215(2) Å, β = 96.101(2), $\alpha = \gamma = 90°$, V = 1270.47(4) Å^3, T = 100(2) K, Z = 4, Z' = 1; 17,592 reflections were measured, 2893 unique (R_{int} = 0.0115), which were used in all calculations. The final wR_2 was 0.0743 (all data) and R_1 was 0.0267 (I > 2σ(I)).

3.4. Preparation of $[C(NH_2)_2(NH\{NH_2\})][B_5O_6(OH)_4]$ (**2**)

$[C(NH_2)_2(NH\{NH_2\})]_2SO_4$ (1 g, 4.0 mol) was dissolved in H_2O (25 mL) and to this was added $Ba(OH)_2\cdot8H_2O$ (1.3 g, 4.1 mmol) in H_2O (5 mL). The resulting cloudy solution was stirred for 1 h and the solid barium sulphate was filtered from the solution. $B(OH)_3$ (2.3 g, 37 mmol) was added to the clear solution, and stirred with gentle heating for 3 h. The solution was then evaporated under reduced pressure and dried in an oven to afford 2.0 g of a white powder as crude product (95%). A sample (0.3 g) of this solid was redissolved in 10 mL deionised water and left to crystallise over a few days to yield a few crystals suitable for XRD studies. $CH_{11}B_5N_4O_{10}$. Anal. Calc.: C = 4.1%, H = 3.8%, N = 19.1%. Found: C = 4.2%, H = 3.7%, N = 18.7%. TGA: 240–320 °C, condensation of pentaborate(1-) anions with loss of 2 H_2O 11.9% (12.3% calc.); 320–700 °C, oxidation of organic residue leaving residual 2.5 B_2O_3 59.3% (59.3% calc.). NMR/ppm: δ_B: 1.2 (10%), 13.3 (9%), 19.2 (81%). FTIR (KBr, cm^{-1}) 3451(s), 3372(s), 3295(s), 3235(m), 1672(s), 1657(s), 1437(s), 1361(s), 1252(m), 1196(m), 1103(s), 1026(s), 925(s), 782(s), 698(s). XRD crystallographic data: $CH_{11}B_5N_4O_{10}$, M_r = 293.19, triclinic, P-1 (no. 2), a = 7.4870(2) Å, b = 8.5076(2) Å, c = 9.6502(2) Å, α = 93.906(2)°, β = 98.470(2)°, γ = 96.457(2)°, V = 601.88(3) Å^3, T = 100(2) K, Z = 2, Z' = 1; 11,885 reflections were measured, 2142 unique (R_{int} = 0.0404), which were used in all calculations. The final wR_2 was 0.0827 (all data) and R_1 was 0.0294 (I > 2σ(I)).

3.5. Preparation of $[C(NH_2)_2(NMe_2)][B_5O_6(OH)_4]$ (**3**)

An aqueous solution (5 mL) containing $Ba(OH)_2\cdot8H_2O$ (1.2 g, 3.8 mmol) was added to an aqueous solution (25 mL) of $[C(NH_2)_2(NMe_2)]_2SO_4$ (1.0 g, 3.7 mmol). The mixture was stirred for 1 h and the $BaSO_4$ was removed from the solution by gravity filtration and washed with H_2O. $B(OH)_3$ (2.2 g, 36 mmol) was added to the filtrate and the solution was stirred under gentle warming for 1 h. The solvent was removed under reduced pressure to yield 2.0 g of a white powder as the crude product (87%). This solid was characterised by FT-IR and NMR (^1H, ^{11}B, ^{13}C) studies. A sample (0.3 g) of the crude product was redissolved in warm H_2O and left to recrystallize. A few white single crystals suitable for XRD studies were obtained. $C_3H_{14}B_5N_3O_{10}$. Anal. Calc.: C = 11.8%, H = 4.6%, N = 13.8%. Found: C = 11.9%, H = 4.6%, N = 13.5%. TGA: 240–275 °C, condensation of pentaborate(1-) anions with loss of 2 H_2O 12.3% (11.8% calc.); 275–700 °C, oxidation of organic residue leaving residual 2.5 B_2O_3 55.3 (56.8% calc.). NMR/ppm δ_H: 2.89 (s, CH_3N), 4.70 (HOD); δ_B: 1.0 (13%), 13.2 (11%), 18.3 (77%); δ_C: 37.34 (CH_3). FTIR (KBr, cm^{-1}): 3370(s), 3249(m), 3200(m), 1688(m), 1671(s), 1634(s), 1435(vs), 1393(s), 136 (s), 1313(s), 1114(s), 1033(s), 926(vs), 780(s), 699(m). XRD crystallographic data: $C_3H_{14}B_5N_3O_{10}$, Mr = 306.22, monoclinic, $P2_{1/c}$, a = 9.9747(2) Å, b = 11.2563(3) Å, c = 11.6174(3) Å, $\alpha = \gamma = 90°$, $\beta = 96.084(2)°$, V = 1297.03(5) Å^3, T = 100(2) K, Z = 4, Z' = 1; 18,168 reflections were measured, 2969 unique (R_{int} = 0.0191), which were used in calculations. The final wR_2 was 0.0811 (all data) and R_1 was 0.0280 (I > 2σ(I)).

3.6. XRD Data for [C(NH₂)(NMe₂)₂][B₅O₆(OH)₄] (4)

$B(OH)_3$ (6.2 g, 100 mmol) was dissolved in H_2O (100 mL) and [C(NH)(NMe₂)₂] (1.2 g, 20 mmol) was added with stirring. The solution was stirred at RT for 1 h and evaporated to dryness to yield 6.6 g (99%) of the crude product. NMR/ppm δ_H: 2.85 (s, CH_3N), 4.70 (HOD); δ_B: 1.0 (5%), 12.8 (10%), 19.1 (85%); δ_C: 38.77 (CH_3), 161.3 (CN_3). FTIR (KBr cm^{-1}): 3164(m), 1669(m), 1377(s), 1291(s), 1041(m), 917(s), 772(m), 725(m). A sample (0.3 g) of this product was recrystallized from H_2O to afford crystals suitable for sc-XRD studies. $C_5H_{18}B_5N_3O_{13}$, M_r = 334.27, triclinic, $P\text{-}1$ (no.2), a = 9.5035(3) Å, b = 9.5151(3)Å, c = 10.4386(3) Å, α = 65.3810(10)°, β = 69.049(2), γ = 88.603(2), V = 792.72(4) Å^3, T = 100(2) K, Z = 2; 16,517 reflections were measured, 3638 unique (R_{int} = 0.0406), which were used in all calculations. The final wR_2 was 0.0935 (all data) and R_1 was 0.0373 (I > 2σ(I)).

3.7. Preparation of [C(NHMe)(NMe₂)₂]I

C(NH)(NMe₂) (2.1 g, 2.3 mL, 18.2 mmol) was added to an Ar charged two-necked flask, and to this, CH_3CN (10 mL) was added. The sealed vessel was then cooled to 0 °C using an ice bath. MeI (2.9 g, 20.4 mmol) was then added to CH_3CN (8 mL) and this solution was added dropwise to the cold C(NH)(NMe₂) solution. The mixture was then stirred for 12 h and left to slowly equilibrate to room temperature. The solvents were then removed via rotary evaporation. The resulting oil was then washed with EtOAc (3 × 25 mL) and dried under vacuum at room temperature to yield a white crystalline product [C(NHMe)(NMe₂)₂]I (3.2 g, 63%). NMR/ppm: δ_H: 2.73 (3H, s, CH_3N), 2.80 (12H, s, CH_3N); δ_C: 38.91 (CH_3).

3.8. Preparation of [C(NHMe)(NMe₂)₂][B₅O₆(OH)₄]·B(OH)₃ (5)

[C(NHMe)(NMe₂)₂]I (1.0 g, 4.0 mmol) was dissolved in H_2O (25 mL) along with a DOWEX 550A (OH)$^-$ ion exchange resin (18 g) and stirred for 18 h. The resin was removed by filtration and the filtrate was added to $B(OH)_3$ (1.3 g, 21 mmol). The solution was stirred for 2 h and evaporated under reduced pressure to yield 1.4 g of a white powder as a crude product (97%). A 0.3 g sample of the product was redissolved in H_2O (15 mL) and white crystals suitable for X-ray diffraction studies were obtained after a few days. $C_6H_{23}B_6N_3O_{13}$. Anal. Calc.: C = 17.6%, H = 5.7%, N = 10.2%. Found: C = 17.7%, H = 5.5%, N = 10.1%. NMR/ppm: δ_H: 2.73 (3H, s, CH_3N), 2.80 (12H, s, CH_3N), 2.82 (1H, s), 4.70 (HOD); δ_C: 38.89 (CH_3). δ_B: 1.1 (9%), 13.2 (13%), 18.8 (78%); FTIR (KBr, cm^{-1}): 3370(s), 3236(s), 1625(s), 1592(m), 1426(s), 1398(m), 1367(s), 1324(m), 1156(m), 1022(m), 922(vs), 779(m), 709(m). XRD crystallographic data: $C_6H_{23}B_6N_3O_{13}$, M_r = 410.13, monoclinic, $P2_1/c$ (no. 14), a = 9.49570(10) Å, b = 11.44900(10) Å, c = 16.84590(10) Å, β = 98.0710(10), $\alpha = \gamma = 90$, V = 1813.28(3) Å^3, T = 100(2) K, Z = 4, Z' = 1; 20,493 reflections were measured, 3281 unique (R_{int} = 0.0196), which were used in all calculations. The final wR_2 was 0.0674 (all data) and R_1 was 0.0263 (I > 2σ(I)).

3.9. Preparation of [TBDH][B₅O₆(OH)₄] (6)

1,5,7-Triazabicyclo[4.4.0]dec-5-ene (TBD, 1.0 g, 7.2 mmol) was dissolved in H_2O (20 mL) and to this was added $B(OH)_3$ (2.2 g, 36 mmol). The resulting solution was stirred under gentle heating to fully dissolve the $B(OH)_3$ and left for 3 h. The solution was then evaporated under reduced pressure to yield a crude white product (2.4 g, 6.7 mmol, 93%).

A small sample of this solid (0.3 g) was redissolved in H_2O (20 mL) and left to recrystallize over 7 d to afford a small number of white crystals, suitable for X-ray diffraction studies. $C_7H_{18}B_5N_3O_{10}$. Anal. Calc.: C = 23.5%, H = 5.1%, N = 11.7%. Found: C = 23.7%, H = 5.0%, N = 11.5%. TGA: 245–295 °C, condensation of pentaborate with the loss of 2 H_2O, 11.0% (10.0% calc.); 295–700 °C, oxidation of organic residue leaving residual 2.5 B_2O_3 49.0% (48.5% calc.). NMR/ppm: δ_H: 1.84 (4H, quin, CH_2), 3.12 (4H, t, CH_2N), 3.20 (4H, t, CH_2N), 4.70 (HOD); δ_C: 20.12 (CH_2), 37.71 (CH_2), 46.38 (CH_2). δ_B: 0.8 (1%), 13.0 (29%), 17.7 (70%). FTIR (KBr/cm^{-1}): 3238(m), 1630(m),1408(m), 1294(s), 1144(m), 1086(m), 1015(m), 910(vs), 816(m), 772(s), 723(m), 706(s). XRD crystallographic data: $C_7H_{18}B_5N_3O_{10}$, M_r = 358.29, triclinic, P-1 (no. 2), a = 9.3096(6) Å, b = 9.3175(3) Å, c = 9.3733(6) Å, α = 76.598(5)°, β = 85.611(5)°, γ = 79.947(4)°, V = 778.22(8) Å^3, T = 100(2) K, Z = 2, Z' = 1; 13,382 reflections were measured, 2836 unique (R_{int} = 0.0443), which were used in all calculations. The final wR_2 was 0.1747 (all data) and R_1 was 0.0561 (I > 2σ(I)).

4. Conclusions

The synthesis and structural characterization of six *N*-substituted guanidinium pentaborate(1-) salts was accomplished. Reactant stoichiometry, solid-state packing, and H-bonding interactions are all important factors in the self-assembly of these structures. An analysis of the H-bonding motifs in these salts and in related unsubstituted guanidinium polyborates was undertaken to evaluate this contribution. This analysis indicated that strong charge-assisted pincer $R_2^2(8)$ motifs originating from the cation were in competition with the strong borate–borate $R_2^2(8)$ motifs for the available borate acceptor sites. The ratios of potential cationic pincer-type H-bond donors to borate H-bond donors were determined for all compounds and compared with the ratios observed in their solid-state structures. Differences in these values were indicative of a structural preference for one or the other of these H-bonding motifs. The salts $[C(NH_2)_3]_2[B_4O_5(OH)_4]\cdot 2H_2O$ and $[C(NH_2)_3]_3[B_9O_{12}(OH)_6]$ had potential ratios of >1 with the observed ratios higher than their potential ratios, indicating that the cation–anion pincer H-bond motifs were favoured. In contrast, $[C(NH_2)_3][B_5O_6(OH)_4]\cdot H_2O$ and the six new substituted guanidinium pentaborates described within this manuscript all had potential ratios of <1 and the observed ratios were generally lower than their potential ratios, indicating favourable borate–borate H-bond interactions. However, as both types of H-bond motifs were expected to be equally strong, other factors such as stoichiometry and crystal packing were strongly influential in determining the observed structures.

Supplementary Materials: The following supporting information can be downloaded at: https://www.mdpi.com/article/10.3390/molecules28073273/s1, Crystallographic data for compounds **1–6** are available as Supplementary Materials. CCDC 2249242-7 also contain the supplementary crystallographic data for this paper. These CCDC data can be obtained free of charge via http://www.ccdc.cam.ac.uk/conts/retrieving.html (or from CCDC, 12 Union Road, Cambridge, CB2 1EZ. Fax: +44 1223 336033; E-mail: deposit@ccdc.cam.ac.uk).

Author Contributions: M.A.B. conceived the research strategy and drafted this manuscript, supervision; T.A.R., experimental synthetic procedures and initial draft writing (Ph.D. thesis); P.N.H. and S.J.C., experimental crystallography and structure determinations; M.A.B., P.N.H. and T.A.R. manuscript review and editing; S.J.C., funding acquisition. All authors have read and agreed to the published version of the manuscript.

Funding: This research received no external funding.

Institutional Review Board Statement: Not applicable.

Informed Consent Statement: Not applicable.

Data Availability Statement: Crystallographic available from CCDC, UK, see Supplementary Materials for details. CIF files for published unsubstituted guanidinium polyborate salts were also obtained from CCDC: 146622 (WEHFAQ), 1132636 (CUGNAT), and 1824332 (QUFJAF).

Acknowledgments: Crystals of compound **4** were kindly provided by J. Morris. We thank the EPSRC for the NCS X-ray crystallographic service (Southampton).

Conflicts of Interest: The authors declare no conflict of interest.

Sample Availability: Samples of the compounds are not available from the authors.

References

1. Farmer, J.B. Metal borates. *Adv. Inorg. Chem Radiochem.* **1982**, *25*, 187–237.
2. Christ, C.L.; Clark, J.R. A crystal-chemical classification of borate structures with emphasis on hydrated borates. *Phys. Chem. Miner.* **1977**, *2*, 59–87. [CrossRef]
3. Heller, G. A survey of structural types of borates and polyborates. *Top. Curr. Chem.* **1986**, *131*, 39–98.
4. Becker, P. Borate materials in nonlinear optics. *Adv. Mater.* **1998**, *10*, 979–992. [CrossRef]
5. Burns, P.C.; Grice, J.D.; Hawthorne, F.C. Borate minerals I. Polyhedral clusters and fundamental building blocks. *Can. Mineral.* **1995**, *33*, 1131–1151.
6. Grice, J.D.; Burns, P.C.; Hawthorne, F.C. Borate minerals II. A hierarchy of structures based upon the borate fundamental building block. *Can. Mineral.* **1999**, *37*, 731–762.
7. Touboul, M.; Penin, N.; Nowogrocki, G. Borates: A survey of main trends concerning crystal chemistry, polymorphism and dehydration process of alkaline and pseudo-alkaline borates. *Solid State Sci.* **2003**, *5*, 1327–1342. [CrossRef]
8. Schubert, D.M.; Smith, R.A.; Visi, M.Z. Studies of crystalline non-metal borates. *Glass Technol.* **2003**, *44*, 63–70.
9. Schubert, D.M.; Knobler, C.B. Recent studies of polyborate anions. *Phys. Chem. Glasses Eur. J. Glass Sci. Technol. B* **2009**, *50*, 71–78.
10. Beckett, M.A. Recent Advances in crystalline hydrated borates with non-metal or transition-metal complex cations. *Coord. Chem. Rev.* **2016**, *323*, 2–14. [CrossRef]
11. Xin, S.-S.; Zhou, M.-H.; Beckett, M.A.; Pan, C.-Y. Recent advances in crystalline oxidopolyborate complexes of d-block or p-block metals: Structural aspects, synthesis, and physical properties. *Molecules* **2021**, *26*, 3815. [CrossRef]
12. Schubert, D.M. Borates in industrial use. *Struct. Bond.* **2003**, *105*, 1–40.
13. Schubert, D.M. Boron oxide, boric acid, and borates. In *Kirk-Othmer Encyclopedia of Chemical Technology*, 5th ed.; John and Wiley and Sons: Hoboken, NJ, USA, 2011; pp. 1–68.
14. Schubert, D.M. Hydrated zinc borates and their industrial use. *Molecules* **2019**, *24*, 2419. [CrossRef]
15. Beckett, M.A.; Brellocks, B.; Chizhevsky, I.T.; Damhus, T.; Hellwich, K.-H.; Kennedy, J.D.; Laitinen, R.; Powell, W.H.; Rabinovich, D.; Vinas, C.; et al. Nomenclature for boranes and related species (IUPAC Recommendations 2019). *Pure Appl. Chem.* **2020**, *92*, 355–381. [CrossRef]
16. Shaw, J.W.; Grayson, D.H.; Rozas, I. *Topics in Heterocyclic Chemistry*; Springer: Berlin/Heidelberg, Germany, 2015; pp. 1–51.
17. Weakley, T.J.R. Guanidinium tetraborate(2-) dihydrate, $(CH_6N_3)_2[B_4O_5(OH)_4]\cdot 2H_2O$. *Acta Cryst.* **1985**, *C41*, 377–399. [CrossRef]
18. Dhatchaiyini, M.K.; Rajasekar, G.; Mohideen, M.N.; Bhaskaran, A. Investigation on structural, spectral, optical, thermal studies and third order NLO properties of guanidinium pentaborate monohydrate single crystal. *J. Mol. Sci.* **2020**, *1210*, 128065. [CrossRef]
19. Schubert, D.M.; Visi, M.Z.; Knobler, C.B. Guanidinium and imidazolium borates containing the first examples of an isolated nonaborate oxoanion: $[B_9O_{12}(OH)_6]^{(3-)}$. *Inorg. Chem.* **2000**, *39*, 2250–2251. [CrossRef]
20. Rosenheim, A.; Lyser, F. Über Polyborate in wäßriger lösung. (Zur kenntnis der iso-und heteropolysäuren. XVII. Mitteilung). *Z. Anorg. Allg. Chem.* **1921**, *119*, 1–38. [CrossRef]
21. Bowden, G.H. *Supplement to Mellor's Comprehensive Treatise on Inorganic and Theoretical Chemistry*; Boron, Part A: Boron-Oxygen Compounds; Longman Group Ltd.: London, UK, 1980; Volume 5.
22. Wood, G.L.; Janik, J.F.; Visi, M.Z.; Schubert, D.M.; Paine, R.T. New borate precursors for boron nitride powder synthesis. *Chem Mater.* **2005**, *17*, 1855–1859. [CrossRef]
23. Blondeau, P.; Segura, M.; Perez-Fernandez, R.; de Mendoza, J. Molecular recognition of oxoanions based on guanidinium receptors. *Chem. Soc. Rev.* **2007**, *26*, 198–210. [CrossRef]
24. Abrahams, B.F.; Haywood, M.G.; Robson, R. Guanidinium ion as a symmetrical template in the formation of cubic hydrogen-bonded borate networks with the boracite topology. *J. Am. Chem. Soc.* **2005**, *127*, 816–817. [CrossRef] [PubMed]
25. Desiraju, G.R. Supramolecular synthons in crystal engineering—A new organic synthesis. *Angew. Chem. Int. Ed. Engl.* **1995**, *34*, 2311–2327. [CrossRef]
26. Dunitz, J.D.; Gavezzotti, A. Supramolecular synthons: Validation and ranking of intermolecular interaction energies. *Cryst. Growth Des.* **2012**, *12*, 5873–5877. [CrossRef]
27. Salentine, G. High-field ^{11}B NMR of alkali borate. Aqueous polyborate equilibria. *Inorg. Chem.* **1983**, *22*, 3920–3924. [CrossRef]
28. Anderson, J.L.; Eyring, E.M.; Whittaker, M.P. Temperature jump rate studies of polyborate formation in aqueous boric acid. *J. Phys. Chem.* **1964**, *68*, 1128–1132. [CrossRef]
29. Corbett, P.T.; Leclaire, J.; Vial, L.; West, K.R.; Wietor, J.-L.; Sanders, J.K.M.; Otto, S. Dynamic combinatorial chemistry. *Chem. Rev.* **2006**, *106*, 3652–3711. [CrossRef]
30. Sola, J.; Lafuente, M.; Atcher, J.; Alfonso, I. Constitutional self-selection from dynamic combinatorial libraries in aqueous solution through supramolecular interactions. *Chem. Commun.* **2014**, *50*, 4564–4566. [CrossRef]

31. Beckett, M.A.; Horton, P.N.; Hursthouse, M.B.; Knox, D.A.; Timmis, J.L. Structural (XRD) and thermal (DSC, TGA) and BET analysis of materials derived from non-metal cation pentaborate salts. *Dalton Trans.* **2010**, *39*, 3944–3951. [CrossRef] [PubMed]

32. Visi, M.Z.; Knobler, C.B.; Owen, J.J.; Khan, M.I.; Schubert, D.M. Structures of self-assembled nonmetal borates derived from α,ω-diaminoalkanes. *Cryst. Growth Des.* **2006**, *6*, 538–545. [CrossRef]

33. Beckett, M.A.; Coles, S.J.; Davies, R.A.; Horton, P.N.; Jones, C.L. Pentaborate(1−) salts templated by substituted pyrrolidinium cations: Synthesis, structural characterization, and modelling of solid-state H-bond interactions by DFT calculations. *Dalton Trans.* **2015**, *44*, 7032–7040. [CrossRef] [PubMed]

34. Bonner, O.D.; Jordan, C.F. The infrared and Raman spectra of guanidinium salts. *SpectroChim. Acta Part A Mol. Spectrosc.* **1976**, *32*, 1243–1246. [CrossRef]

35. Li, J.; Xia, S.; Gao, S. FT-IR and Raman spectroscopic study of hydrated borates. *Spectrochim. Acta* **1995**, *51*, 519–532.

36. Beckett, M.A.; Horton, P.N.; Coles, S.J.; Kose, D.A.; Kreuziger, A.-M. Structural and thermal studies of non-metal cation pentaborate salts with cations derived from 1,5-diazobicyclo[4.3.0]non-5-ene, 1,8-diazobicyclo[5.4.0]undec-7-ene and 1,8-bis(dimethylamino)naphthalene. *Polyhedron* **2012**, *38*, 157–161. [CrossRef]

37. Drozd, M. Molecular structure and infrared spectra of guanidinium cation a combined theoretical and spectroscopic study. *Mat. Sci. Eng. B* **2007**, *6*, 20–28. [CrossRef]

38. Etter, M.C. Encoding and decoding hydrogen-bond patterns of organic chemistry. *Acc. Chem. Res.* **1990**, *23*, 120–126. [CrossRef]

39. Durka, K.; Jarzembska, K.N.; Kaminski, R.; Lulinski, S.; Serwatowski, J.; Wozniak, K. Structure and energetic landscapes of fluorinated 1,4-phenylboronic acids. *Cryst. Growth Design* **2012**, *12*, 3720–3734. [CrossRef]

40. Heller, G. Die hydrolyse von borsäuretrimethylester in gegenwart organischer basen. *J. Inorg. Nucl. Chem.* **1968**, *30*, 2743–2754. [CrossRef]

41. Beckett, M.A.; Coles, S.J.; Horton, P.N.; Jones, C.L. Polyborate anions partnered with large non-metal cations: Triborate(1-), pentaborate(1-) and heptaborate(2-) salts. *Eur. J. Inorg. Chem.* **2017**, *2017*, 4510–4518. [CrossRef]

42. Freyhardt, C.C.; Wiebcke, M.; Felsche, J.; Englehardt, G. N(nPr$_4$)[B$_5$O$_6$(OH)$_4$][B(OH)$_3$]$_2$ and N(nBu$_4$)[B$_5$O$_6$(OH)$_4$][B(OH)$_3$]$_2$: Clathrates with a diamondoid arrangement of hydrogen bonded pentaborate anions. *J. Inclusion Phenom. Mol. Recogn. Chem.* **1994**, *18*, 161–175. [CrossRef]

43. Beckett, M.A.; Horton, P.N.; Hursthouse, M.B.; Timmis, J.L.; Varma, K.S. Templated heptaborate and pentaborate salts of cyclo-alkylammonium cations: Structural and thermal properties. *Dalton Trans.* **2012**, *41*, 4396–4403. [CrossRef]

44. Hass, D.J.; Harris, R.R.; Mills, H.H. The crystal structure of guanidinium chloride. *Acta Cryst.* **1965**, *19*, 676–679. [CrossRef] [PubMed]

45. Smith, G.; Wermuth, U.D. Three-dimensional hydrogen-bonded structures in guanidinium salts of the monocyclic dicarboxylic acids rac-trans-cyclohexane-1,2-dicarboxylic acid (2:1, anhydrous) isophthalic (1:1, monohydrate) and terephthalic acid (2:1, trihydrate). *Acta Cryst. Sect. C* **2010**, *C66*, o575–o580. [CrossRef]

46. Vadivel, S.; Sultan, B.A.; Samad, S.A.; Shunmuganarayanan, A.; Muthu, R. Synthesis, structural elucidation, thermal, mechanical, linear and nonlinear optical properties of hydrogen bonded organic single crystal guanidinium propionate for optoelectronic device application. *Chem Phys. Lett.* **2018**, *707*, 165–171. [CrossRef]

47. Russell, V.A.; Etter, M.C.; Ward, M.D. Layered materials by design: Structural enforcement by hydrogen bonding in guanidinium alkane- and arene-sulfonates. *J. Am. Chem Soc.* **1994**, *116*, 1941–1952. [CrossRef]

48. Schug, K.A.; Lindner, W. Noncovalent Binding between Guanidinium and Anionic Groups: Focus on Biological- and Synthetic-Based Arginine/Guanidinium Interactions with Phosph[on]ate and Sulf[on]ate Residues. *Chem. Rev.* **2005**, *105*, 67–113. [CrossRef]

49. Thomas, M.; Lagones, T.A.; Judd, M.; Morshedi, M.; White, N.G. Hydrogen bond-driven self-assembly between amidiniums and carboxylates: A combined molecular dynamics, NMR ppectroscopy, and single crystal X-ray diffraction study. *Chem. Asian J.* **2017**, *12*, 1587. [CrossRef] [PubMed]

50. Avdeeva, V.V.; Malinina, E.A.; Vologzhanina, A.V.; Sivaev, I.B.; Kuznetsov, N.T. Formation of oxidoborates in destruction of the [B$_{11}$H$_{14}$]$^-$ anion promoted by transition metals. *Inorg. Chim. Acta* **2020**, *509*, 119693. [CrossRef]

51. *CrysAlisPro Software System*; Rigaku Oxford Diffraction; Agilent Technologies UK Ltd: Yarnton, UK, 2019.

52. Sheldrick, G.M. *ShelXT*—intergrated space-group and crystal structure determination. *Acta Cryst.* **2015**, *A71*, 3–8. [CrossRef]

53. Hooft, R.; Nonius, B.V. *COLLECT, Data Collection Software*; Nonius BV: Delft, The Netherlands, 1998.

54. Otwinowski, Z.; Minor, W. Processing of X-ray diffraction data collected in oscillation mode. *Meth. Enzymol.* **1997**, *276*, 307–326.

55. Sheldrick, G.M. A short history of SHELX. *Acta Cryst. A* **2008**, *A64*, 112–122. [CrossRef]

56. Dolomanov, O.V.; Bourhis, L.J.; Gildea, R.J.; Howard, J.A.K.; Puschmann, H. *Olex2*: A complete structure solution, refinement and analysis program. *J. Appl. Cryst.* **2009**, *42*, 339–341. [CrossRef]

57. Sheldrick, G.M. Crystal structure refinement with ShelXL. *Acta Cryst.* **2015**, *C27*, 3–8.

 *molecules*

Article

Polyhedral Dicobaltadithiaboranes and Dicobaltdiselenaboranes as Examples of Bimetallic *Nido* Structures without Bridging Hydrogens

Amr A. A. Attia [1], Alexandru Lupan [1,*] and Robert Bruce King [2,*]

[1] Faculty of Chemistry and Chemical Engineering, Babeş-Bolyai University, 400347 Cluj-Napoca, Romania; amr.attia@ubbcluj.ro

[2] Department of Chemistry and Center for Computational Quantum Chemistry, University of Georgia, Athens, GA 30602, USA

* Correspondence: alexandru.lupan@ubbcluj.ro (A.L.); rbking@chem.uga.edu (R.B.K.)

Abstract: The geometries and energetics of the n-vertex polyhedral dicobaltadithiaboranes and dicobaltadiselenaboranes $Cp_2Co_2E_2B_{n-4}H_{n-4}$ (E = S, Se; n = 8 to 12) have been investigated via the density functional theory. Most of the lowest-energy structures in these systems are generated from the (n + 1)-vertex most spherical *closo* deltahedra by removal of a single vertex, leading to a tetragonal, pentagonal, or hexagonal face depending on the degree of the vertex removed. In all of these low-energy structures, the chalcogen atoms are located at the vertices of the non-triangular face. Alternatively, the central polyhedron in most of the 12-vertex structures can be derived from a $Co_2E_2B_8$ icosahedron with adjacent chalcogen (E) vertices by breaking the E–E edge and 1 or more E–B edges to create a hexagonal face. Examples of the *arachno* polyhedra with two tetragonal and/or pentagonal faces derived from the removal of two vertices from *isocloso* deltahedra were found among the set of lowest-energy $Cp_2Co_2E_2B_{n-4}H_{n-4}$ (E = S, Se; n = 8 and 12) structures.

Keywords: polyhedral boranes; sulfur; selenium; cobalt; density functional there

Citation: Attia, A.A.A.; Lupan, A.; King, R.B. Polyhedral Dicobaltadithiaboranes and Dicobaltadiselenaboranes as Examples of Bimetallic *Nido* Structures without Bridging Hydrogens. *Molecules* **2023**, *28*, 2988. https://doi.org/10.3390/molecules28072988

Academic Editors: Michael A. Beckett and Igor B. Sivaev

Received: 11 March 2023
Revised: 24 March 2023
Accepted: 26 March 2023
Published: 27 March 2023

1. Introduction

The structures of the polyhedral borane dianions $B_nH_n^{2-}$ as well as those of the isoelectronic carborane monoanions $CB_{n-1}H_n^-$ and neutral dicarbaboranes $C_2B_{n-2}H_n$ are based on the most spherical deltahedra, which are known as the *closo* deltahedra (Figure 1) [1,2]. Except for the 13-vertex systems, the most spherical deltahedra have exclusively triangular faces with vertex degrees as nearly similar as possible. The 11- and 13-vertex systems contain 1 and 2 degree-6 vertices, respectively, whereas the other *closo* polyhedra have exclusively degree-4 and -5 vertices, for which the degree of a vertex is defined as the number of edges meeting at the vertex in question. The chemical bonding in these polyhedral boranes and carboranes is based on $2n$ + 2 skeletal electrons for an n-vertex system, with each BH and CH vertex contributing 2 and 3 skeletal electrons, respectively, after providing an electron for the external B–H or C–H bond. This is a key aspect of the Wade–Mingos rules [3–5] relating polyhedral geometry to skeletal electron count. A reasonable chemical bonding model for these deltahedra consists of a resonance hybrid of canonical structures containing a single n-center bond composed of orbitals from each of the n-vertex atoms overlapping at the center of the polyhedron supplemented by n two-center, two-electron bonds on the surface of the polyhedron [6]. This structural model accounts for the $2n$ + 2 skeletal electrons in the stable *closo* deltahedral borane structures. The delocalization implicit in this bonding model, particularly the presence of the multicenter core bond, is consistent with the interpretation of these species as three-dimensional aromatic systems [7,8].

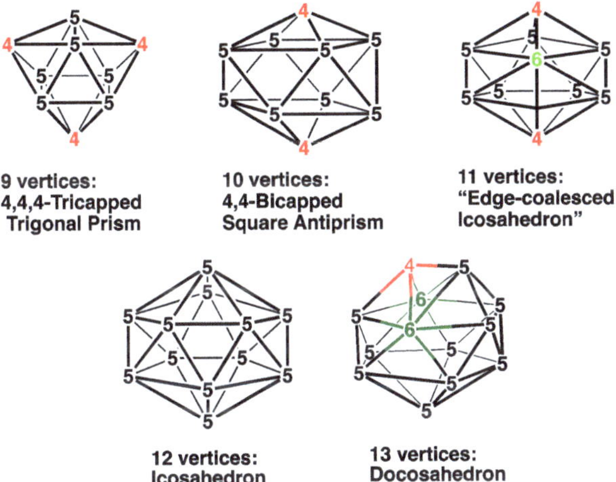

Figure 1. The most spherical *closo* deltahedra having 9 to 13 vertices from which the *nido* structures discussed in this paper are generated. Vertices of degrees 4, 5, and 6 are indicated in red, black, and green, respectively, in Figures 1, 2, and 8.

The research groups of Hawthorne [9] and Grimes [10] were the first to show that the boron vertices in these polyhedral boranes could be replaced by transition metal units. The cyclopentadienylcobalt unit (CpCo) was especially useful for this purpose due to the robustness of the linkage between the cyclopentadienyl ring and the cobalt atom. Furthermore, a CpCo vertex is a donor of two skeletal electrons like a BH vertex so that it can replace a BH unit in the polyhedral structures. The research of the Hawthorne group [9] focused on metallaboranes that can be obtained from decaborane ($B_{10}H_{14}$) as a precursor, whereas the research of the Grimes group [10] studied metallaboranes obtained from pentaborane (B_5H_9) as a precursor. The chemistry of polyhedral boranes and their transition metal derivatives remains of interest even after approximately half a century following the discovery of the original metallaboranes with the possibilities of applications in medicine [11,12] and catalysis [13].

Further development of metallaborane chemistry, especially by Kennedy and his co-workers, by using a variety of metal vertices, particularly those containing second- and third-row transition metals, resulted in the discovery of alternative so-called *isocloso* deltahedra in 9- and 10-vertex metallaborane structures with a degree-6 vertex for the metal atom (Figure 2) [14–17]. The 11-vertex *closo* deltahedron can also be an *isocloso* deltahedron since it necessarily has a degree 6 vertex. A reasonable chemical bonding model for an n-vertex *isocloso* metallaborane deltahedron consists exclusively of n 3-center, 2-electron (3c-2e) bonds in n of the $2n - 4$ faces of the deltahedron without the n-center core bond found in the *closo* chemical bonding [18]. Thus, the *isocloso* metallaboranes are $2n$ rather than $2n + 2$ skeletal electron systems.

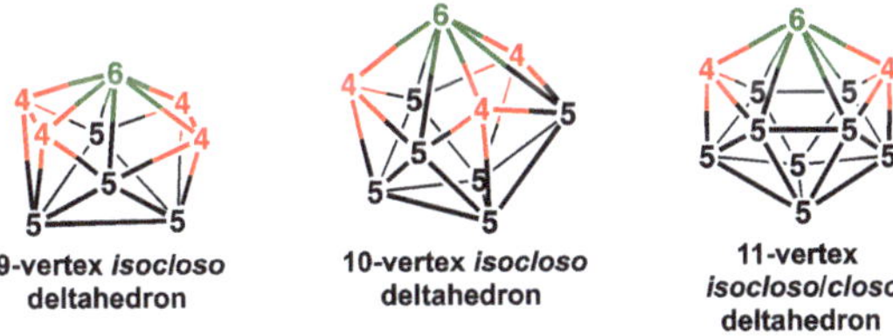

Figure 2. The *isocloso* deltahedra from which the *nido* polyhedra with hexagonal holes are generated.

Electron-richer polyhedral borane derivatives with *n*-vertices are obtained by removing a vertex from an (*n* + 1)-vertex *closo* or *isocloso* deltahedron to give a so-called *n*-vertex *nido* polyhedron. Such a vertex removal process creates a smaller polyhedron with a non-triangular face at the site of vertex removal. This non-triangular face can be regarded as a "hole" in an otherwise triangulated polyhedral surface. Williams [2] introduced a convenient notation for *nido* polyhedra as ni-*n*⟨*h*⟩, where *n* is the number of vertices, and *h* is the number of vertices in the non-triangular face as indicated by a Roman numeral. For example, by using this notation, the square pyramid can be designated as a ni-5⟨IV⟩ polyhedron. Since the original (*n* + 1)-vertex *closo* deltahedron has 2(*n* + 1) + 2 skeletal electrons, the *n*-vertex *nido* polyhedron derived from it has 2*n* + 4 skeletal electrons.

The discovery of neutral binary boranes B_nH_{n+4} (*n* = 2, 5, 6, 8, 9, 10) exhibiting *nido* structures, of which $B_{10}H_{14}$ is the most stable [19,20], predates that of the more stable *closo* boranes by going back to the original boron hydride syntheses of Stock. The structures of B_5H_9 and B_6H_{10} are derived by the removal of a degree-4 vertex from an octahedron and a degree-5 vertex from a pentagonal bipyramid, respectively, thereby generating a square and a pentagonal face, respectively (Figure 3). The structures of the larger B_nH_{n+4} (*n* = 8, 9, 10) boranes are derived by removal of the unique degree-6 vertex from the corresponding (*n* + 1) vertex *isocloso* deltahedron. The polyhedral frameworks of all of the B_nH_{n+4} boranes (*n* = 5, 6, 8, 9, 10) all have *n* BH vertices with the four "extra" hydrogen atoms bridging the B–B edges surrounding the non-triangular face. A relatively large non-triangular face such as a hexagonal hole versus a pentagonal or tetragonal face provides more space for the four hydrogen atoms bridging the hole B–B edges.

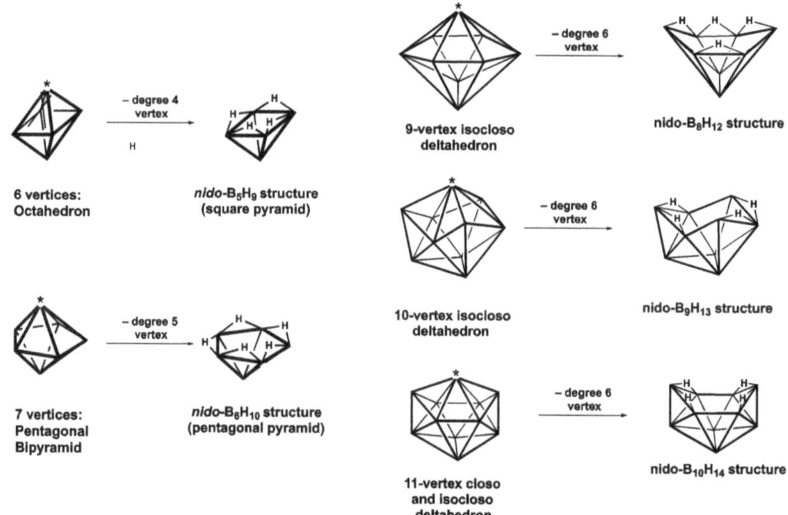

Figure 3. Generation of the binary B_nH_{n+4} borane structures from *closo* and *isocloso* deltahedra by the removal of the starred vertices.

A topic of interest is the design of viable *n*-vertex *nido* borane structures that have the necessary 2*n* + 4 skeletal electrons without the need for bridging hydrogen atoms. Tetracarbaborane structures of the type $C_4B_{n-4}H_n$ provide the required 2*n* + 4 skeletal electrons. In this connection, species of the type $C_4B_nR_n$ (R = H or alkyl) are known experimentally to have such structures with 6 [21,22], 8 [23,24], 10 [25–28], 11 [29], and 12 [30–36] vertices. Furthermore, the structures and energetics of the tetracarbaboranes have been studied by using modern density functional theory methods [37].

Generating a *nido* polyhedral borane with the requisite 2*n* + 4 skeletal electrons by using electron-richer carbon atoms in CH vertices to replace boron atoms in BH vertices requires four such carbon heteroatoms in the carborane polyhedron. Using bare sulfur or selenium vertex atoms as heteroatoms in polyhedral borane structures to provide the $2n + 4$ skeletal electrons for *nido* geometry requires only two heteroatoms in an $E_2B_{n-2}H_{n-2}$ (E = S, Se) structure since bare sulfur or selenium vertices are each donors of four skeletal electrons after diverting 2 of their 6 valence electrons to an external lone pair. In this connection, the 11-vertex *nido*-$B_9H_9E_2$ (E = S, Se) structures have been synthesized and shown via X-ray crystallography to have a central polyhedron obtained via the removal of one vertex from an icosahedron [38].

Dithiaboranes and diselenaboranes with one or two CpCo vertices have been synthesized and structurally characterized via X-ray crystallography (Figure 4). The 12-vertex cobaltadiselenaborane $CpCoSe_2B_9H_9$ has been synthesized and shown to have a ni-13$\langle VI \rangle$ structure derived via the removal of a degree-6 vertex from a docosahedron, which is the 13-vertex *closo* deltahedron [35]. This structure can also be derived from a central $CoSe_2B_9$ icosahedron with a Se–Se edge by breaking the Se–Se edge and an adjacent Se–B edge to generate an SeCoSeBB pentagonal face, leaving a degree-3 selenium vertex. This structure is shown by density functional theory to be 1 of the 3 lowest-energy structures lying within 1 kcal/mol of each other [39].

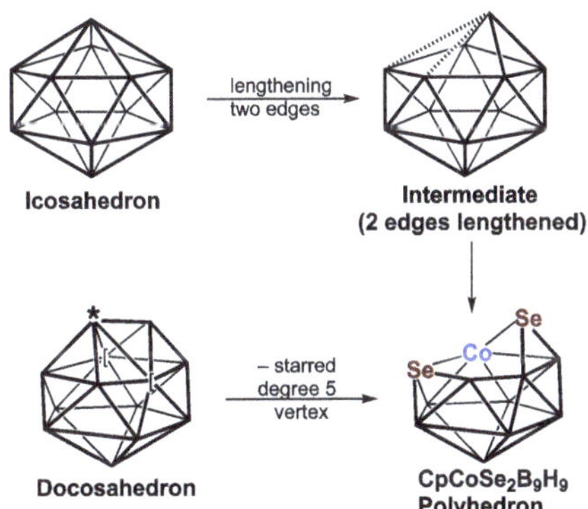

Figure 4. Alternative ways of generating the central 12-vertex polyhedron of the experimentally known $CpCoSe_2B_9H_9$ from the 12-vertex icosahedron by breaking edges and from the 13-vertex docosahedron by removing the starred degree-5 vertex.

In total, 3 isomeric structures of the 11-vertex dicobaltadithiaborane $Cp^*_2Co_2S_2B_7H_7$ with 2 pentamethylcyclopentadienylcobalt vertices (Cp^*Co) have been isolated by Kang and Sneddon [40] from the mixture that was obtained from the reaction between LiCp*, $NaS_2B_7H_8$, and $CoCl_2$ (Figure 5). Attempts to obtain definitive structural information on these species via X-ray crystallography were prevented by disorder problems. However, the poor X-ray data were sufficient to indicate the relative positions of the cobalt atom and 11-vertex $Co_2S_2B_7$ geometries obtained via removal from a vertex from a central icosahedron. Furthermore, 1 of the 3 structures appears to be an ultimate pyrolysis product at 300 °C. The structures were assigned on the basis of the ^{11}B NMR spectra. In addition, trace quantities of a 10-vertex $Cp^*_2Co_2S_2B_6H_6$ structure were obtained, which was suggested to have a decaborane-like structure on the basis of its ^{11}B NMR.

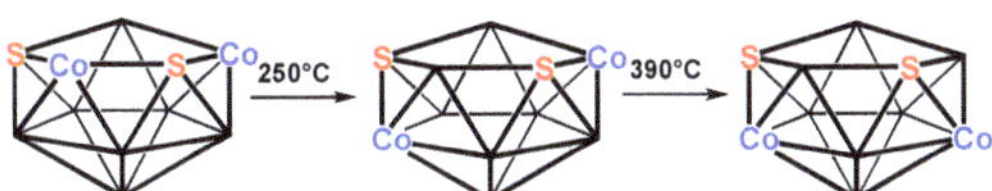

Figure 5. The 11-vertex frameworks suggested for the 3 $Cp^*_2Co_2S_2B_7H_7$ isomers isolated by Kang and Sneddon from the reaction between LiCp*, $NaS_2B_7H_9$, and $CoCl_2$.

In view of the difficulty in obtaining definitive structural information on the dicobaltadithiaboranes, we have undertaken density functional theory studies on the complete series of $Cp_2Co_2E_2B_{n-4}H_{n-4}$ (E = S, Se; n = 8, 9, 10, 11, 12) systems with unsubstituted cyclopentadienyl rings. In addition, we have included the $Cp^*_2Co_2S_2B_7H_7$ system with pentamethylcyclopentadienyl (Cp*) rings in this study for comparison with the reported experimental data. This theoretical study extends the previous study [36] of the dicobaltadiselenaboranes by introducing a second cobalt atom into the metallaborane polyhedron. This provides an opportunity to assess preferences for a pair of cobalt atoms in these structures to occupy adjacent bonding positions or to be as far removed from each other as possible or something in between. The dicobaltadithiaboranes and dicobaltadiselenaboranes are of interest in representing the first examples of *nido* structures without bridging hydrogen atoms containing two transition metal vertices in the underlying polyhedron.

2. Results and Discussion

2.1. The 11-Vertex Systems $Cp_2Co_2E_2B_7H_7$ (E = S, Se)

In total, 8 structures were found for each of the 11-vertex $Cp_2Co_2E_2B_7H_7$ (E = S, Se) systems within 15 kcal/mol of the lowest-energy structure (Figure 6 and Table 1). The central ni-11⟨V⟩ polyhedron in all of the eight structures can be derived from an icosahedron via the removal of one vertex, leaving a pentagonal face that is similar to the dicarbollide anion $C_2B_9H_{12}{}^-$ that was originally used by Hawthorne and co-workers [9] to synthesize a variety of transition metal complexes having a central MC_2B_9 icosahedron. For both $Cp_2Co_2E_2B_7H_7$ (E = S, Se) systems, the lowest-energy structures **B7Co2E2-1** (E = S, Se) have both cobalt atoms and both chalcogen atoms at the surface of the pentagonal face and, thus, at degree-4 vertices. This is the most prevalent example of generating an n-vertex *nido* polyhedral borane by the removal of a vertex from an (n + 1)-vertex *closo* deltahedral borane. Furthermore, in each of the 8 low-energy $Cp_2Co_2E_2B_7H_7$ structures, both chalcogen atoms are located at the degree-4 pentagonal face vertices rather than at interior degree-5 vertices. This is consistent with the preference of chalcogen atoms for lower degree vertices in polyhedral selena- and thiaboranes.

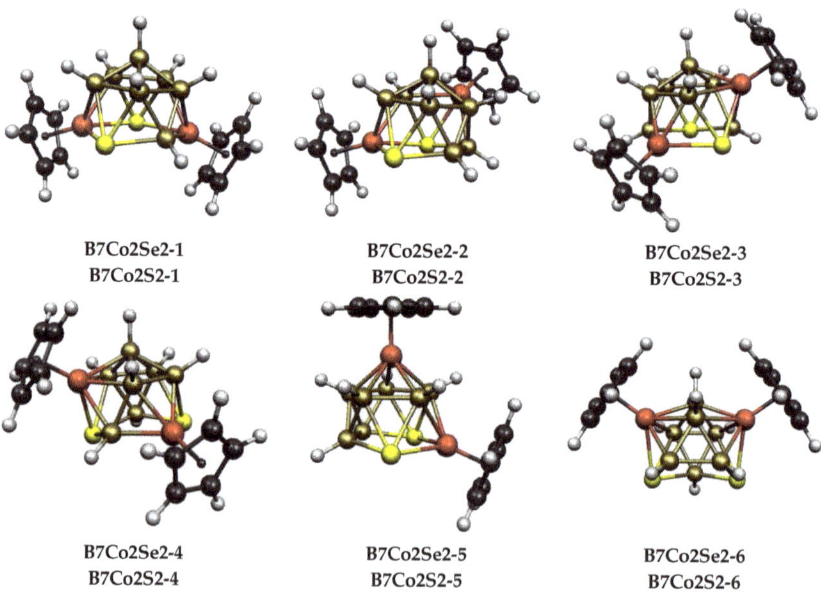

B7Co2Se2-1
B7Co2S2-1

B7Co2Se2-2
B7Co2S2-2

B7Co2Se2-3
B7Co2S2-3

B7Co2Se2-4
B7Co2S2-4

B7Co2Se2-5
B7Co2S2-5

B7Co2Se2-6
B7Co2S2-6

Figure 6. *Cont.*

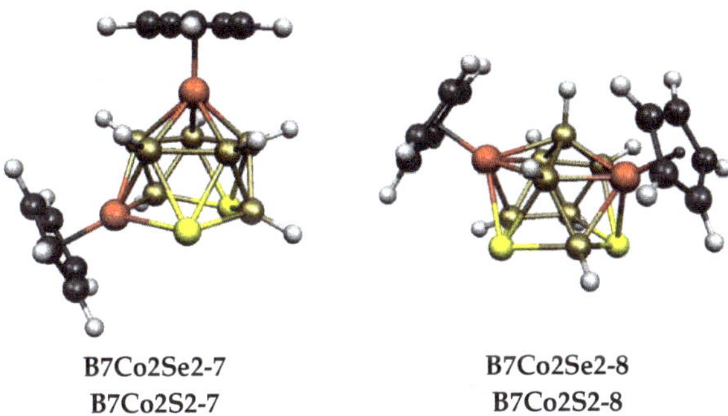

B7Co2Se2-7
B7Co2S2-7

B7Co2Se2-8
B7Co2S2-8

Figure 6. The eight lowest-energy $Cp_2Co_2E_2B_7H_7$ (E = S, Se) structures oriented to have the open pentagonal face at the bottom.

Table 1. Relative energies (kcal/mol) and geometries of the lowest-energy 11-vertex $Cp_2Co_2E_2B_7H_7$ (E = Se, S) structures. In all 8 structures, the central $Co_2E_2B_7$ polyhedron is derived from an 11-vertex polyhedron that is obtained via the removal of a vertex from the 12-vertex regular icosahedron, leaving a pentagonal face.

$Cp_2Co_2Se_2B_7H_7$ Structure (sym)	ΔE	$Cp_2Co_2S_2B_7H_7$ Structure	ΔE	Vertex Degrees S	Co	Co⋯S Edges	Co⋯Co (E = Se) Dist (Å)	WBI	Pentagonal Face Atoms
B7Co2Se2-1 (C_1)	0.0	B7Co2S2-1	0.0	4, 4	4, 4	3	3.91	0.10	SCoSBCo
B7Co2Se2-2 (C_1)	2.5	B7Co2S2-2	1.8	4, 4	4, 5	3	3.81	0.10	SCoSBB
B7Co2Se2-3 (C_1)	8.1	B7Co2S2-3	6.2	4, 4	4, 5	2	3.79	0.09	SCoBSB
B7Co2Se2-4 (C_1)	9.4	B7Co2S2-4	7.7	4, 4	4, 5	2	3.71	0.10	SCoBSB
B7Co2Se2-5 (C_s)	10.1	B7Co2S2-5	18.2	4, 4	4, 5	2	3.68	0.09	SCoSBB
B7Co2Se2-6 (C_s)	11.8	B7Co2S2-6	19.9	4, 4	5, 5	2	3.73	0.09	SBSBB
B7Co2Se2-7 (C_1)	13.5	B7Co2S2-7	10.7	4, 4	4, 5	1	3.67	0.08	SCoBSB
B4Co2Se2-8 (C_1)	14.7	B7Co2S2-8	12.6	4, 4	5, 5	2	3.73	0.08	SBSBB

Three isomeric pentamethylcyclopentadienyl $Cp^*_2Co_2S_2B_7H_7$ complexes have been isolated by Kang and Sneddon in small quantities from the reaction between LiCp*, $NaS_2B_7H_8$, and $CoCl_2$ which they designated by the Roman numerals III, IV, and V in their paper (Figure 5) [37]. Extensive disorder prevented complete X-ray structure determinations on these molecules beyond location of the cobalt atoms. On the basis of ^{11}B NMR and 2D ^{11}B-^{11}B COSY NMR, Kang and Sneddon assigned structures analogous to **B7Co2S-1**, **B7Co2S-2**, and **B7Co2S-3** to III, IV, and V, respectively. We found that complete substitution of hydrogen atoms with methyl groups did not affect the relative energy ordering of the 3 lowest-energy structures with the relative energies of the Cp* structures **B7Co2S-1***, **B7Co2S-2***, and **B7Co2S-3*** being 0.0, 0.6, and 6.7 kcal/mol, respectively. What is strange is the observation that the Kang/Sneddon isomer III, which has the assigned structure **B7Co2S2-1*** and has both cobalt atoms as well as both sulfur atoms located on pentagonal face vertices, is converted ultimately upon heating to 300 °C to the Kang/Sneddon isomer V (Figure 5), which was assigned the higher-energy structure **B7Co2S2-3***, as it has both cobalt atoms located at degree-5 interior vertices. This is contrary to expectation because pyrolysis, particularly to a temperature as high as 300 °C, would be expected to give a lower-energy isomer rather than a higher-energy isomer. Our theoretical studies cast some doubt about the structure assignments of III, IV, and V that were given by Kang and Sneddon in their 1988 study [37]. Note that the predicted Co⋯Co distances in **B7Co2S-1**, **B7Co2S-2**, and **B7Co2S-3** are all approximately 3.8 to 3.9 Å, so the determination of these distances by an otherwise incomplete X-ray crystallography study on an extensively disordered system would not be sufficient to distinguish between these three structures. The improvements in

X-ray crystallography methodology in the 35 years since the Kang/Sneddon report of the 3 $Cp^*_2Co_2S_2B_7H_7$ isomers might provide a resolution to this dilemma.

2.2. The 12-Vertex Systems $Cp_2Co_2E_2B_8H_8$ (E = S, Se)

In total, 9 structures were found for the 12-vertex $Cp_2Co_2E_2B_8H_8$ (E = S, Se) systems up to 9 kcal/mol (E = S) and 12 kcal/mol (E = Se) in energy (Figure 7 and Table 2). These structures are of three types. The polyhedra for the lowest-energy $Cp_2Co_2S_2B_8H_8$ structure **B8Co2S2-1**, as well as those for 4 of the 5 next higher-energy structures **B8Co2S2-2**, **B8Co2S2-3**, **B8Co2S2-5**, and **B8Co2S2-6**, lying 3.0, 3.1, 3.8, and 5.6 kcal/mol in energy above **B8Co2S2-1**, respectively, can be derived from a $Co_2E_2B_8$ icosahedron with an S–S edge by breaking the S–S edge and at least one S–B edge to create typically a gaping, bent hexagonal face. These five structures can be divided in two categories. In **B8Co2S2-1**, **B8Co2S2-5**, and **B8Co2S2-6**, the cobalt atoms are located in *meta* (non-adjacent, non-antipodal) positions of the original octahedron. However, in **B8Co2S2-2** and **B8Co2S2-4**, the cobalt atoms are located in *para* (antipodal) positions in the original icosahedron.

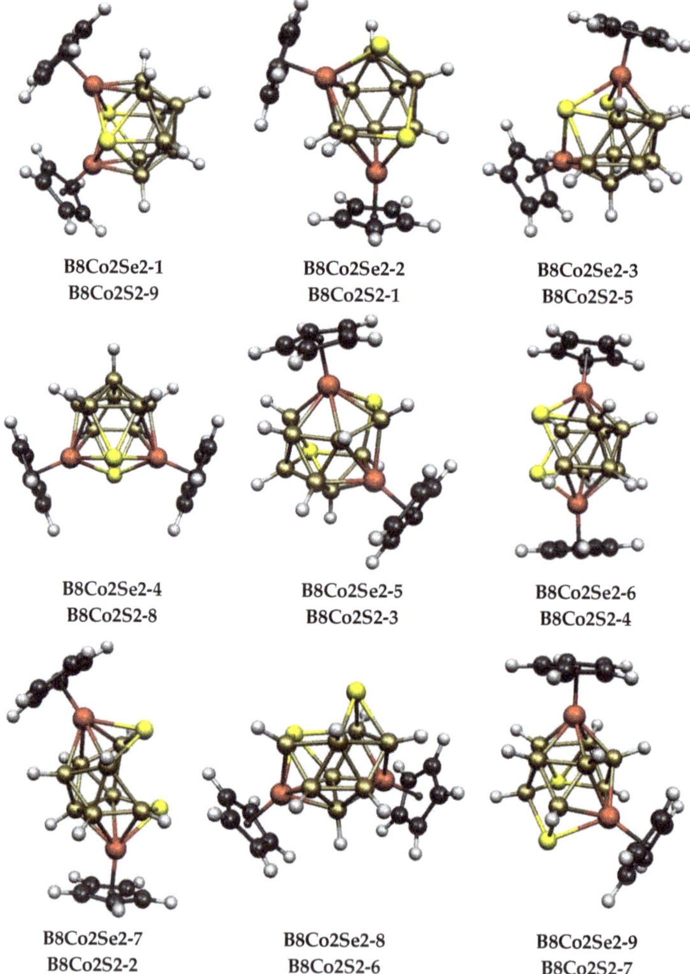

Figure 7. The nine lowest-energy $Cp_2Co_2E_2B_8H_8$ (E = S, Se) structures.

Table 2. Relative energies (kcal/mol) and geometries of the lowest-energy 12-vertex $Cp_2Co_2E_2B_8H_8$ (E = S, Se) structures.

$Cp_2Co_2Se_2B_8H_8$ Structure (sym)	ΔE	$Cp_2Co_2S_2B_8H_8$ Structure	ΔE	Vertex Degrees S	Co	Co $\cdots$ S Edges	Co $\cdots$ Co (E = Se) Dist (Å)	WBI	Polyhedron
B8Co2Se2-1 (C_{2v})	0.0	B8Co2S2-9	9.0	5, 5	4, 4	4	3.18	0.17	ni-12⟨IV⟩
B8Co2Se2-2 (C_1)	4.0	B8Co2S2-1	0.0	3, 3	5, 5	2	3.85	0.12	*meta* Co2 open icosahedron
B8Co2Se2-3 (C_1)	4.5	B8Co2S2-5	3.8	3, 4	4, 5	2	3.83	0.09	*meta* Co2 open icosahedron
B8Co2Se2-4 (C_{2v})	5.3	B8Co2S2-8	7.9	4, 4	5, 5	4	3.48	0.15	ni-12⟨IV⟩
B8Co2Se2-5 (C_1)	8.2	B8Co2S2-3	3.1	3, 4	5, 5	1	3.61	0.09	ar-12⟨IV,V⟩
B8Co2Se2-6 (C_2)	8.2	B8Co2S2-4	3.5	3, 3	5, 5	2	4.22	0.05	*para* Co2 open icosahedron
B8Co2Se2-7 (C_2)	9.3	B8Co2S2-2	3.0	3, 3	5, 5	2	4.82	0.08	*para* Co2 open icosahedron
B8Co2Se2-8 (C_1)	11.8	B8Co2S2-6	5.6	3, 4	5, 5	1	3.73	0.11	*meta* Co2 open icosahedron
B8Co2Se2-9 (C_1)	12.3	B8Co2S2-7	7.1	3, 4	5, 5	1	3.72	0.10	ar-12⟨IV,V⟩

Substituting selenium for sulfur in the 12-vertex system to give $Cp_2Co_2Se_2B_8H_8$ derivatives leads to a different energy ordering of the 9 lowest-energy structures (Table 2). The lowest-energy $Cp_2Co_2Se_2B_8H_8$ structure **B8Co2Se2-1** and the higher-energy structure **B8Co2Se2-4**, which lies 4.0 kcal/mol higher in energy, are exceptional among the complete series of $Cp_2Co_2E_2B_{n-4}H_{n-4}$ (n = 8 to 12) structures in exhibiting ideal C_{2v} symmetry, whereas all of the remaining structures of this type have the lower-symmetry point groups of C_1, C_2, or C_s. These 2 structures are ni-12⟨IV⟩ structures derived from the 13-vertex *closo* deltahedron, namely, the docosahedron (Figure 1), by the removal of the unique degree-4 vertex, thereby creating a tetragonal face with alternating degree-5 and degree-4 vertices. In both **B8Co2Se2-1** and **B8Co2Se-4**, the tetragonal face has alternating cobalt and sulfur vertices. The $Cp_2Co_2S_2B_8H_8$ structures corresponding to **B8Co2Se2-1** and **B8Co2Se2-4** are **B8Co2S-9** and **B8Co2S-8**, respectively, which lie 9.0 and 7.9 kcal/mol in energy above **B8Co2S-1**.

The remaining 2 of the 9 lowest-energy $Cp_2Co_2E_2B_8H_8$ (E = S, Se), structures, namely, **B8Co2S-3** and **B8Co2S-7** for E = S, which lie 3.1 and 7.1 kcal/mol, respectively, in energy above **B8Co2S-1**, and **B8Co2Se-5** and **B8Co2Se-9**, respectively, which lie 8.2 and 12.3 kcal/mol, respectively, in energy above **B8Co2Se2-1**, can be considered to be *arachno* ar-12⟨V,V⟩ structures that are obtained by removing 2 vertices from a 14-vertex deltahedron. However, the central 14-vertex deltahedron from which these structures are derived is not the 14-vertex *closo* deltahedron, namely, the D_{6d} bicapped hexagonal antiprism with 2 degree-6 vertices in antipodal positions, as well as 12 degree-5 vertices, but instead, it is a less symmetrical 14-vertex deltahedron with 3 degree-6 vertices, 10 degree-5 vertices, and 1 degree-4 vertex (Figure 8). This 14-vertex deltahedron is closely related to the 14-vertex polyhedron in experimentally known $Cp^*_2Ru_2C_2B_{10}H_{12}$ by lengthening an edge connecting a degree-6 vertex with a degree-5 vertex [41,42].

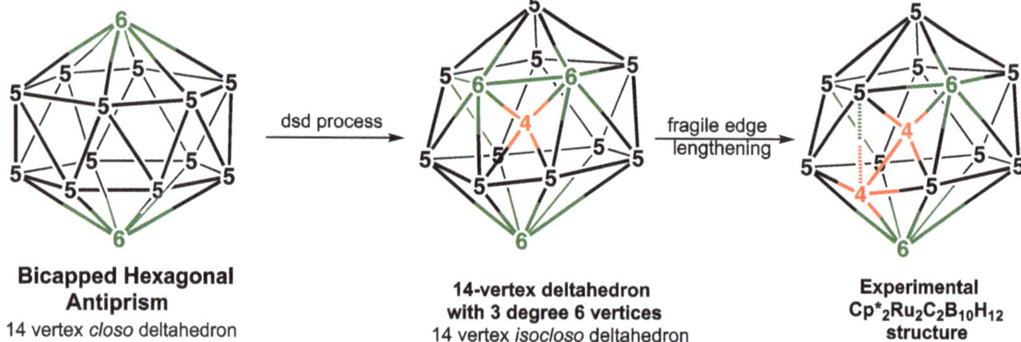

Figure 8. The D_{6d} bicapped hexagonal antiprism as the 14-vertex *closo* deltahedron and a 14-vertex *isocloso* deltahedron derived from it via a diamond-square-diamond process that is the polyhedron found in the experimentally known $Cp*_2Ru_2C_2B_{10}H_{12}$. The 12-vertex *arachno* $Cp_2Co_2E_2B_8H_8$ (E = S, Se) structures **B8Co2S2-3**, **B8Co2S2-7**, **B8Co2Se2-5**, and **B8Co2eS2-9** are obtained from this 14-vertex *isocloso* deltahedron by the removal of a degree-4 vertex and a degree-5 vertex, which are so situated that the resulting tetragonal and pentagonal faces share an edge.

2.3. The $Cp_2Co_2E_2B_{n-4}H_{n-4}$ (E = S, Se) Systems Having 8 to 10 Vertices

The central $Co_2E_2B_4$ polyhedra in 7 of the 8 lowest-energy structures of the 8-vertex $Cp_2Co_2E_2B_4H_4$ (E = S, Se) systems (Figure 9 and Table 3) are all derived by the removal of a vertex from the *closo* 9-vertex deltahedron, namely, the tricapped trigonal prism (Figure 1). Removal of a degree-4 vertex from the tricapped, trigonal prism gives the bicapped trigonal prism, which is found in the lowest-energy structures **B4Co2E2-1** (E = S, Se), as well as the higher-energy structures **B4C2S2-2** and **B4Co2Se2-5**, which lie ~6 kcal/mol in energy above the lowest-energy structures (Table 3). In the lowest-energy structure **B4Co2E2-1**, the atoms of the open tetragonal face are alternating cobalt and chalcogen atoms with all four boron atoms located at interior vertices. The other bicapped trigonal prismatic $Cp_2Co_2E_2B_4H_4$ (E = S, Se) structures have the two sulfur atoms as well as one of the cobalt atoms at the open tetragonal face, with the other cobalt atom at an interior vertex.

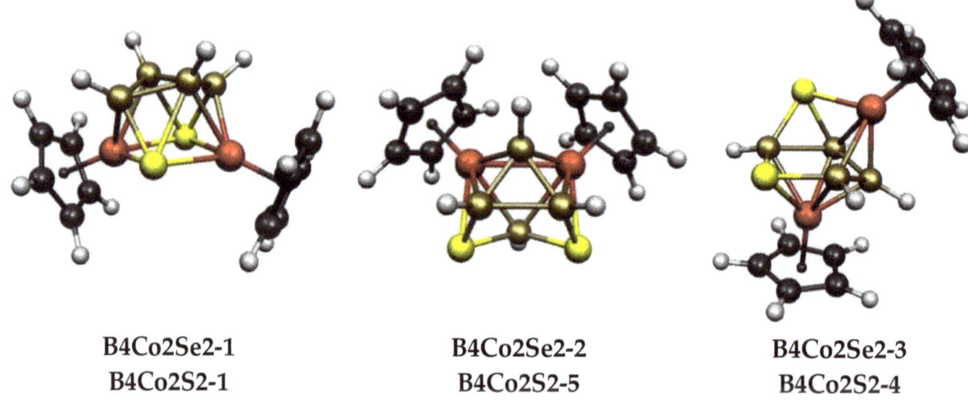

B4Co2Se2-1
B4Co2S2-1

B4Co2Se2-2
B4Co2S2-5

B4Co2Se2-3
B4Co2S2-4

Figure 9. *Cont.*

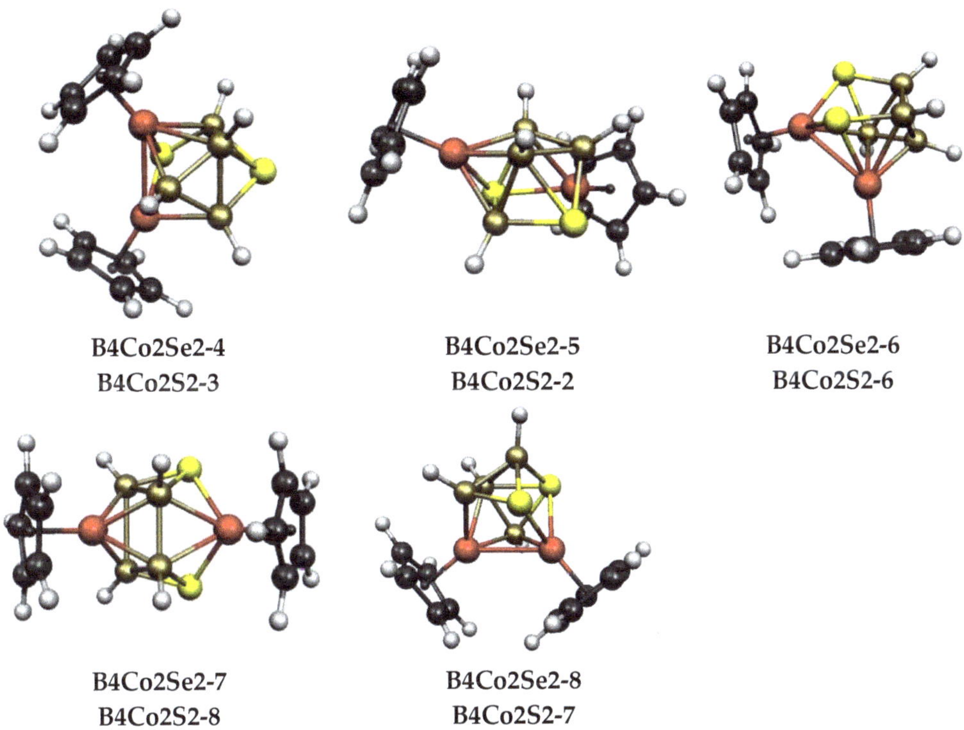

B4Co2Se2-4
B4Co2S2-3

B4Co2Se2-5
B4Co2S2-2

B4Co2Se2-6
B4Co2S2-6

B4Co2Se2-7
B4Co2S2-8

B4Co2Se2-8
B4Co2S2-7

Figure 9. The eight lowest-energy $Cp_2Co_2E_2B_4H_4$ (E = S, Se) structures.

Table 3. Relative energies (kcal/mol) and geometries of the lowest-energy $Cp_2Co_2E_2B_4H_4$ (E = S, Se) structures.

$Cp_2Co_2Se_2B_4H_4$ Structure (sym)	ΔE	$Cp_2Co_2S_2B_4H_4$ Structure	ΔE	Vertex Degrees S	Co	Co⋯S Edges	Co⋯Co (E = Se) Dist (Å)	WBI	Non-Triang Face Atoms	Polyhedron
B4Co2Se2-1 (C_2)	0.0	**B4Co2S2-1**	0.0	4, 4	4, 4	4	3.39	0.12	SCoSCo	bicap trig prism
B4Co2Se2-2 (C_s)	7.1	**B4Co2S2-5**	5.2	4, 4	4, 4	3	3.75	0.14	SCoSB	bicap trig prism
B4Co2Se2-3 (C_1)	11.1	**B4Co2S2-4**	1.9	3, 3	4, 5	2	2.55	0.38	SCoBBB	ni-8⟨V⟩
B4Co2Se2-4 (C_1)	11.4	**B4Co2S2-3**	0.7	3, 3	4, 5	2	3.44	0.10	SCoBSB	ni-8⟨V⟩
B4Co2Se2-5 (C_s)	12.5	**B4Co2S2-2**	0.1	3, 3	5, 5	2	2.64	0.24	SCoBSB	ni-8⟨V⟩
B4Co2Se2-6 (C_1)	13.5	**B4Co2S2-6**	6.5	3, 3	4, 5	3	2.59	0.40	SCoSBB	ni-8⟨V⟩
B4Co2Se2-7 (C_1)	14.9	**B4Co2S2-8**	10.3	3, 4	4, 4	2	2.65	0.44	2 × SCoSB	ar-8⟨IV,IV⟩
B4Co2Se2-8 (C_s)	15.3	**B4Co2S2-7**	8.3	3, 3	4, 4	2	2.62	0.09	SCoSBB	ni-8⟨V⟩

For the 8-vertex selenium derivative $Cp_2Co_2Se_2B_4H_4$, the 2 low-energy bicapped trigonal prismatic structures **B4Co2Se-1** and **B4Co2Se-2** are essentially isoenergetic, as they lie within ~1 kcal/mol of each other (Figure 9 and Table 3). The lowest-energy $Cp_2Co_2Se_2B_4H_4$ ni-8⟨V⟩ structure **B4Co2Se2-3**, which is derived from a tricapped trigonal prism by removing a degree-5 rather than a degree-4 vertex, lies 11.1 kcal/mol above **B4Co2Se-1**. In total, 4 more ni-8⟨V⟩ $Cp_2Co_2Se_2B_4H_4$ structures, namely, **B4Co2Se2-4**, **B4Co2Se2-5**, **B4Co2Se2-6**, and **B4Co2Se2-8,** lie in the energy range of 11 to 15 kcal/mol above **B4Co2Se2-1**. The potential energy surface of the corresponding 8-vertex sulfur system $Cp_2Co_2Se_2B_4H_4$ is significantly different since 3 of the 5 ni-8⟨V⟩ structures **B4Co2S2-2**, **B4Co2S2-3**, and **B4Co2S2-4** lie within ~2 kcal/mol of the lowest-energy structure **B4Co2S2-1** and below the higher-energy, bicapped trigonal prism isomer **B4Co2S2-5**. In all eight lowest-energy

$Cp_2Co_2E_2B_4H_4$ (E = S, Se) structures, both sulfur atoms lie on tetragonal or pentagonal face vertices, which is consistent with the preference of sulfur for lower degree vertices in borane polyhedra.

Of the 8 lowest-energy $Cp_2Co_2E_2B_4H_4$ (E = S, Se) structures, **B4Co2S2-8** and **B4Co2Se2-7**, which lie at 10.3 and 14.9 kcal/mol in energy, respectively, above the corresponding **B4Co2E2-1** structure, are not derived by removing a vertex from a tricapped trigonal prism (Figure 9 and Table 3). Instead, the central $Co_2S_2B_4$ polyhedron in these structures is generated via the removal of the 2 degree-4 vertices that are bridged by the unique degree-6 vertex from the 10-vertex *isocloso* deltahedron of ideal C_{3v} symmetry. This leads to an *arachno* 8-vertex ar-8⟨IV,IV⟩ structure with 2 tetragonal faces sharing a cobalt atom and a sulfur atom. The process of removing 2 degree-4 vertices from the 10-vertex *isocloso* deltahedron to give the 8-vertex $Cp_2Co_2E_2B_4H_4$ structures **B4Co2S2-8** and **B4Co2Se2-7** is analogous to the process of removing a degree-4 and a degree-5 vertex from a 14-vertex *isocloso* 14-vertex deltahedron (Figure 8) to give the 12-vertex ar-14⟨IV,V⟩ $Cp_2Co_2E_2B_8H_8$ structures **B8Co2S23** and **B8Co2S27** (Figure 7) that are discussed above.

The central polyhedra of the 6 lowest-energy 9-vertex $Cp_2Co_2E_2B_5H_5$ (E = S, Se) structures (Figure 10 and Table 4) are all derived from the 10-vertex *closo* deltahedron, namely, the bicapped square antiprism, by removing either a degree-4 vertex or a degree-5 vertex. Whether a degree-4 vertex is removed to give a capped square antiprism or a degree-5 vertex is removed to give a ni-9⟨V⟩ structure with a pentagonal face makes relatively little difference in energy since the 6 structures lying within 7 kcal/mol of the lowest-energy structures **B5Co2E2-1** (E = S, Se) include 2 representatives of the former type and 4 representatives of the latter type. Both chalcogen vertices are always located at a non-triangular face in all of the low-energy structures.

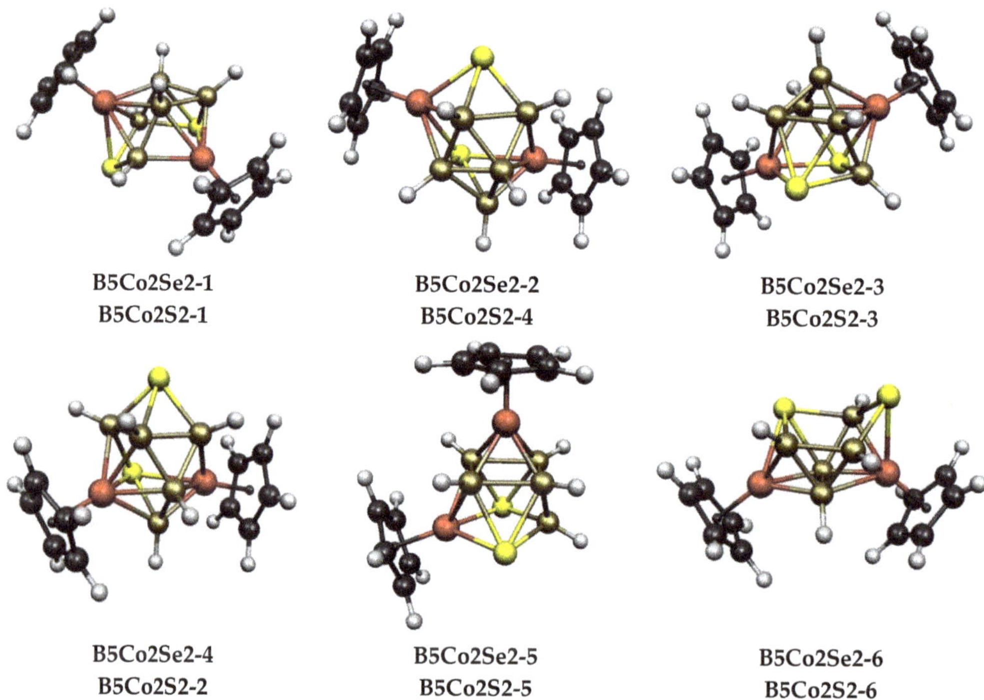

Figure 10. The six lowest-energy $Cp_2Co_2E_2B_5H_5$ (E = S, Se) structures.

Table 4. Relative energies (kcal/mol) and geometries of the lowest-energy 9-vertex $Cp_2Co_2E_2B_5H_5$ (E = S, Se) structures.

$Cp_2Co_2Se_2B_5H_5$ Structure (sym)	ΔE	$Cp_2Co_2S_2B_5H_5$ Structure	ΔE	Vertex Degrees S	Co	Co⋯S Edges	Co⋯Co (E = Se) Dist(Å)	WBI	Non-Triang Face Atoms	Polyhedron
B5Co2Se2-1 (C_1)	0.0	B5Co2S2-1	0.0	3, 4	4, 5	2	3.67	0.11	SCoBSB	ni-9⟨V⟩
B5Co2Se2-2 (C_1)	0.7	B5Co2S2-4	3.9	3, 4	4, 5	3	3.82	0.07	SCoBSCo	ni-9⟨V⟩
B5Co2Se2-3 (C_1)	0.9	B5Co2S2-3	3.6	4, 4	4, 5	3	3.77	0.12	SCoSB	capped square antiprism
B5Co2Se2-4 (C_1)	2.7	B5Co2S2-2	2.8	3, 4	4, 5	3	3.40	0.07	SCoBSB	ni-9⟨V⟩
B5Co2Se2-5 (C_s)	3.7	B5Co2S2-5	6.0	4, 4	4, 4	2	3.65	0.11	SCoSB	capped square antiprism
B5Co2Se2-6 (C_1)	6.9	B5Co2S2-6	6.2	3, 4	4, 5	2	3.72	0.12	SBBSB	ni-9⟨V⟩

The potential energy surfaces for the 10-vertex $Cp_2Co_2E_2B_6H_6$ (E = S, Se) systems are the simplest of all, with only 3 structures lying within 16 kcal/mol (E = S) or 21 kcal/mol (E = Se) of the lowest-energy structures **B6Co2E2-1** (Figure 11 and Table 5). The central $Co_2E_2B_6$ framework of all three structures is that of the very stable decaborane, $B_{10}H_{14}$, which is obtained via the removal of the unique degree-6 vertex of the *closo* 11-vertex deltahedron (sometimes called by the confusing name of "edge-coalesced icosahedron") to create a bent hexagonal face. The structures **B6Co2E2-1**, in which the hexagonal face has only boron and sulfur atoms with the sulfur atoms in opposite ("*para*") positions, is favored energetically over the next lowest-energy structures **B6Co2E2-1** by significant margins of ~11 kcal/mol (E = S) and ~12 kcal/mol (E = Se).

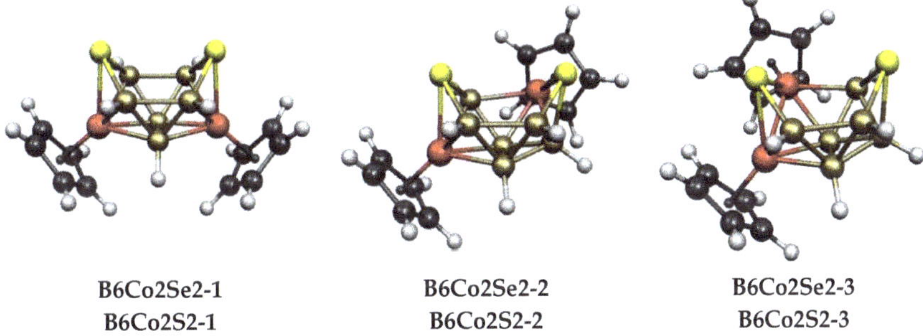

B6Co2Se2-1
B6Co2S2-1

B6Co2Se2-2
B6Co2S2-2

B6Co2Se2-3
B6Co2S2-3

Figure 11. The three lowest-energy $Cp_2Co_2E_2B_6H_6$ (E = S, Se) structures are oriented so that the bent hexagonal face is at the top.

Table 5. Relative energies (kcal/mol) and geometries of the lowest-energy 10-vertex $Cp_2Co_2E_2B_6H_6$ (E = S, Se) structures. In all 3 structures, the central $Co_2E_2B_6$ polyhedron has the same geometry as the B_{10} polyhedron in decaborane-14 with a hexagonal face.

$Cp_2Co_2Se_2B_6H_6$ Structure (sym)	ΔE	$Cp_2Co_2S_2B_6H_6$ Structure	ΔE	Vertex Degrees S	Co	Co⋯S Edges	Co⋯Co (E = Se) Dist (Å)	WBI	Hexagonal Face Atoms	Polyhedron
B6Co2Se2-1 (C_{2v})	0.0	B6Co2S2-1	0.0	3, 3	5, 5	2	3.77	0.10	SBBBBS	$B_{10}H_{14}$ framework
B6Co2Se2-2 (C_1)	10.6	B6Co2S2-2	12.4	3, 3	4, 5	2	3.78	0.09	SBBSCoBS	$B_{10}H_{14}$ framework
B6Co2Se2-3 (C_1)	15.7	B6Co2S2-3	18.1	3, 3	4, 5	2	2.49	0.41	SBBSCoBS	$B_{10}H_{14}$ framework

3. Theoretical Methods

The chemical models that were investigated in this study are based on various $B_nH_n^{2-}$ polyhedra, for which a systematic substitution of 2 BH vertices with 2 CpCo units, followed by the substitution of 2 BH vertices with 2 chalcogen atoms (sulfur or selenium) led to the generation of a total of 5389 different starting structures for each of the $Cp_2Co_2S_2B_{n-4}H_{n-4}$ and $Cp_2Co_2Se_2B_{n-4}H_{n-4}$ systems (n = 8 to 12) (see the Supporting Information).

Full geometry optimizations were carried out on all systems by using the PBE0 DFT functional [43], coupled with the def2-TZVP basis set [44], as implemented in the Gaussian 09 package [45]. The natures of the stationary points after optimization were checked via calculations of the harmonic vibrational frequencies to ensure genuine minima. Furthermore, single-point energy calculations were performed on the lowest-energy optimized structures by using the DLPNO-CCSD(T) method [46–59] coupled with the def2-QZVP basis set, as implemented in the ORCA 3.0.3 software package [60–68]. Zero-point corrections taken from the PBE0/def2-TZVP computations were then added to the final energies.

The polyhedral dicobaltadithiaborane and dicobaltadiselenaborane structures $Cp_2Co_2E_2B_{n4}H_{n-4}$ (E = S, Se; n = 8 to 12) are designated as **B(n−4)Co2E2-x** throughout the text, where n is the total number of polyhedral vertices, and **x** is the relative ordering of the structure on the energy scale. Only the lowest-energy and, thus, potentially chemically significant structures are considered in detail in this paper. More comprehensive structural information including higher-energy structures, connectivity information not readily seen in the figures, and orbital energies and HOMO-LUMO gaps are provided in the Supporting Information.

4. Summary

The central $Co_2E_2B_{n-4}$ polyhedra in the low-energy structures of the n-vertex polyhedral dicobaltadithiaboranes and dicobaltadiselenaboranes $Cp_2Co_2E_2B_{n-4}H_{n-4}$ (E = S, Se; n = 8 to 12) in general are generated from the (n + 1)-vertex most spherical *closo* deltahedra via the removal of a single vertex, leading to a tetragonal, pentagonal, or hexagonal face, depending on the degree of the vertex removed. In all of these low-energy structures, both chalcogen atoms are located on the non-triangular face vertices, reflecting the energetic preference of chalcogens for lower degree vertices. For the 8- and 9-vertex systems, the structures obtained via the removal of a degree-4 or degree-5 vertex from the corresponding (n + 1)-vertex *closo* deltahedra, namely, the tricapped trigonal prism and the bicapped square antiprism, have similar energies. The low-energy structures for the 10-vertex $Cp_2Co_2E_2B_6H_6$ (E = S, Se) systems all have the framework of the most stable B_nH_{n+4} borane, namely, $B_{10}H_{14}$ with a bent hexagonal face. The lowest-energy of these 10-vertex $Cp_2Co_2E_2B_6H_6$ structures by significant margins exceeding 10 kcal/mol has only boron and both sulfur atoms located at the 6 hexagonal face vertices. The central polyhedra of all of the 11-vertex $Cp_2Co_2E_2B_7H_7$ structures are similar to the polyhedron of the dicarbollide anion $C_2B_9H_{12}^-$ in being generated by loss of a vertex from a regular icosahedron to generate a pentagonal face.

In principle, the central polyhedra in most of the low-energy 12-vertex $Cp_2Co_2E_2B_8H_8$ (E = S, Se) structures can be derived via the removal of a vertex from the 13-vertex *closo* deltahedron, namely, the docosahedron. However, the central polyhedron in most of the 12-vertex structures can also be derived from a $Co_2E_2B_8$ icosahedron with adjacent chalcogen vertices by breaking the E–E edge and 1 or more E–B edges to create a hexagonal hole.

Two examples of *arachno* polyhedra were found among the set of lowest-energy $Cp_2Co_2E_2B_{n-4}H_{n-4}$ (E = S, Se; n = 8 to 12) structures. The central polyhedron of one structure within 15 kcal/mol of the lowest-energy structure in each of the 8-vertex $Cp_2Co_2E_2B_4H_4$ (E = S, Se) systems is an *arachno* polyhedron with 2 tetragonal faces sharing an edge that is derived from the 10-vertex *isocloso* deltahedron via the removal of the 2 degree-4 vertices bridged by the unique degree-6 vertex. In addition, 2 of the structures in each of the 12-vertex $Cp_2Co_2E_2B_8H_8$ (E = S, Se) systems are ar12⟨IV,V⟩ structures that are derived from the 14-vertex *isocloso* deltahedron that is found in the experimentally known $Cp^*_2Ru_2C_2B_{10}H_{12}$ by removing the unique degree-4 vertex as well as a degree-5 vertex.

Supplementary Materials: The following supporting information can be downloaded at: https://www.mdpi.com/article/10.3390/molecules28072988/s1, Initial structures, distance, and energy ranking tables, orbital energies and HOMO/LUMO gaps, complete Gaussian09 reference; concatenated .xyz file containing the Cartesian coordinates of the lowest-energy optimized structure. Table S1A: Initial 8-vertex starting structures. Table S1B: Distance matrices and energy rankings for the lowest energy $Cp_2Co_2S_2B_4H_4$ structures. Table S1C: Distance matrices and energy rankings for the lowest energy $Cp_2Co_2Se_2gB_4H_4$ structures. Table S2A: Initial 9-vertex starting structures. Table S2B: Distance matrices and energy rankings for the lowest energy $Cp_2Co_2S_2B_5H_5$ structures. Table S2C: Distance matrices and energy rankings for the lowest energy $Cp_2Co_2Se_5B_5H_5$ structures. Table S3A: Initial 10-vertex starting structures. Table S3B: Distance matrices and energy rankings for the lowest energy $Cp_2Co_2S_2B_6H_6$ structures. Table S3C: Distance matrices and energy rankings for the lowest energy $Cp_2Co_2Se_2B_6H_6$ structures. Table S4A: Initial 11-vertex starting structures. Table S4B: Distance matrices and energy rankings for the lowest energy $Cp_2Co_2S_2B_7H_7$ structures. Table S4C: Distance matrices and energy rankings for the lowest energy $Cp_2Co_2Se_2B_7H_7$ structures. Table S5A: Initial 12-vertex starting structures. Table S5B: Distance matrices and energy rankings for the lowest energy $Cp_2Co_2S_2B_8H_8$ structures. Table S5C: Distance matrices and energy rankings for the lowest energy $Cp_2Co_2Se_2B_8H_8$ structures. Table S6A: Distance matrices and energy rankings for the lowest energy permethylated $Cp^*_2Co_2S_2B_7H_7$ structures. Table S6B: Distance matrices and energy rankings for the lowest energy permethylated $Cp^*_2Co_2Se_2B_7H_7$ structures. Table S7: Orbital energies and HOMO-LUMO gaps for the lowest $Cp_2Co_2S_2B_{n-4}H_{n-4}$ (n = 8 to 12) structures. Table S8: Orbital energies and HOMO-LUMO gaps for the lowest $Cp_2Co_2Se_2B_{n-4}H_{n-4}$ (n = 8 to 12) structures. Table S9: Orbital energies and HOMO-LUMO gaps for the lowest permethylated $Cp^*_2Co_2E_2B_7H_7$ (E = S, Se) structures

Author Contributions: R.B.K. and A.L. conceived the project; A.A.A.A. performed the calculations and initially organized the results; A.L. supervised the computations and curated the data; R.B.K. generated the initial draft of the manuscript; R.B.K. and A.L. generated the final draft of the manuscript. All authors have read and agreed to the published version of the manuscript.

Funding: The computational facilities were provided by the Babeş-Bolyai University under project POC/398/1/1/124155, which is co-financed by the European Regional Development Fund (ERDF) through the Competitiveness Operational Program for Romania 2014–2020.

Data Availability Statement: The data presented in this study are available in supplementary material.

Conflicts of Interest: The authors declare no conflict of interest.

Sample Availability: Samples of the compounds are not available from the authors.

References

1. Muetterties, E.L. (Ed.) *Boron Hydride Chemistry*; Academic Press: New York, NY, USA, 1975.
2. Williams, R.E. The polyborane, carborane, carbocation continuum—Architectural patterns. *Chem. Rev.* **1992**, *92*, 177–207. [CrossRef]
3. Wade, K. The structural significance of the number of skeletal bonding electron-pairs in carboranes, the higher boranes and borane anions, and various transition-metal carbonyl cluster compounds. *J. Chem. Soc. D Chem. Commun.* **1971**, *15*, 792–793. [CrossRef]
4. Mingos, D.M.P. A General theory for cluster and ring compounds of the main group and transition elements. *Nat. Phys. Sci.* **1972**, *236*, 99–102. [CrossRef]
5. Mingos, D.M.P. Polyhedral skeletal electron pair approach. *Accts. Chem. Res.* **1984**, *17*, 311–319. [CrossRef]
6. King, R.B.; Rouvray, D.H. A graph-theoretical interpretation of the bonding topology in polyhedral boranes, carboranes, and metal clusters. *J. Am. Chem. Soc.* **1977**, *99*, 7834–7840. [CrossRef]
7. Aihara, J.-I. 3-Dimensional aromaticity of polyhedral boranes. *J. Am. Chem. Soc.* **1978**, *100*, 3339–3342.
8. King, R.B. Three-dimensional aromaticity in polyhedral boranes and related molecules. *Chem. Revs.* **2001**, *101*, 1119–1152. [CrossRef]
9. Callahan, K.P.; Hawthorne, M.F. Ten years of metallocarboranes. *Adv. Organometal. Chem.* **1976**, *14*, 145.
10. Grimes, R.N. The role of metals in borane clusters. *Accts. Chem. Res.* **1983**, *16*, 22–26. [CrossRef]

11. Kuan, H.; Yang, Z.; Zhang, L.; Xie, L.; Wang, L.; Xu, H.; Josephson, L.; Liang, S.H.; Zhang, M.-R. Boron agents for neutron capture therapy. *Coord. Chem. Revs.* **2020**, *405*, 213139.
12. Avdeeva, V.V.; Garaev, T.M.; Malinina, E.A.; Zhizhin, K.Y.; Kuznetsov, N.T. Physiologically active compounds based on membranotropic cage carriers—Derivatives of adamantane and polyhedral boron clusters. *Russ. J. Inorg. Chem.* **2022**, *67*, 33–53. [CrossRef]
13. Sivaev, I.B. Functional group directed B–H activation of polyhedral boron hydrides by transition metal complexes. *Russ. J. Inorg. Chem.* **2021**, *66*, 1192–1246. [CrossRef]
14. Bould, J.; Kennedy, J.D.; Thornton-Pett, M. Ten-vertex metallaborane chemistry. Aspects of the iridadecaborane *closo*→*isonido*→*isocloso* structural continuum. *J. Chem. Soc. Dalton* **1992**, *4*, 563–576. [CrossRef]
15. Kennedy, J.D.; Štibr, B. *Current Topics in the Chemistry of Boron*; Kabalka, G.W., Ed.; Royal Society of Chemistry: Cambridge, UK, 1994; pp. 285–292.
16. Kennedy, J.D. *The Borane-Carborane-Carbocation Continuum*; Casanova, J., Ed.; Wiley: New York, NY, USA, 1998; Chapter 3; pp. 85–116.
17. Štibr, B.; Kennedy, J.D.; Drdáková, E.; Thornton-Pett, M. Nine-vertex polyhedral iridamonocarbaborane chemistry. Products of thermolysis of [(CO)(PPh$_3$)$_2$IrCB$_7$H$_8$] and emerging alternative cluster-geometry patterns. *J. Chem. Soc. Dalton* **1994**, *2*, 229–236. [CrossRef]
18. King, R.B. The oblate deltahedra in dimetallaboranes: Geometry and chemical bonding. *Inorg. Chem.* **2006**, *45*, 8211–8216. [CrossRef]
19. Parry, R.W.; Walter, M.K.; Jolly, W.L. (Eds.) *Preparative Inorganic Reactions*; Interscience: New York, NY, USA, 1968; Volume 5, pp. 45–102.
20. Kasper, J.S.; Lucht, C.M.; Harker, D. The structure of the decaborane molecule. *J. Am. Chem. Soc.* **1948**, *70*, 881–882. [CrossRef]
21. Onak, T.P.; Wong, G.T.F. Preparation of the pentagonal pyramidal carborane, 2,3,4,5-tetracarba-nido-hexaborane(6). *J. Am. Chem. Soc.* **1970**, *92*, 5226. [CrossRef]
22. Pasinski, J.P.; Beaudet, R.A. Microwave spectrum, structure, and dipole moment of 2,3,4,5-tetracarbahexaborane(6). *J. Chem. Phys.* **1974**, *61*, 683–691. [CrossRef]
23. Wrackmeyer, B.; Schang, H.-J.; Hofmann, M.; Schleyer, P.v.R. A new carborane cage: Hexacarba-arachno-dodecaborane(12). *Angew. Chem. Int. Ed.* **1998**, *37*, 1245–1247. [CrossRef]
24. Wrackmeyer, B.; Schanz, H.-J.; Hofmann, M.; Schleyer, P.v.R.; Boese, R. The structures of a tetracarba-nido-octaborane(8) and a novel spiro derivative of a 2,3,5-tricarba-nido-hexaborane(7). *Eur. J. Inorg. Chem.* **1999**, *3*, 533–537. [CrossRef]
25. Köster, R.; Seidel, G.; Wrackmeyer, B.; Bläser, D.; Boese, R.; Bühl, M.; Schleyer, P.v.R. Decaethyl-2,6,8,10-tetracarba-nido-decaborane(10)—Preparation, structure in the solid state, and stability. *Chem. Ber.* **1991**, *124*, 2715–2724. [CrossRef]
26. Rayment, T.; Shearer, H.M.M. Crystal structure of 2,4,6,8,9,10-hexamethyl-2,4,6,8,9,10-hexabora-adamantane. *J. Chem. Soc. Dalton* **1977**, *2*, 136–138. [CrossRef]
27. Köster, R.; Seidel, G.; Wrackmeyer, B. Dimerization of a C$_2$B$_3$-*closo*-carbaborane(5) to the C$_4$B$_6$-adamantane and its 2Z/3Z valence isomerization to the C$_4$B$_6$-*nido*-carbaborane(10). *Angew. Chem. Int. Ed.* **1985**, *24*, 326–328. [CrossRef]
28. Wrackmeyer, B.; Schanz, H.-J. Tetracarba-*nido*-hexa-, octa-, and –decaborane derivatives. NMR study and DFT calculations. *J. Organomet. Chem.* **2015**, *798*, 268–273. [CrossRef]
29. Štíbr, B.; Jelínek, T.; Drdáková, E.; Heřmánek, S.; Plešek, J. A new family of stable parent *nido*-tetracarbaboranes 5,6,8,9-C$_4$B$_6$H$_{10}$, 2,7,8,11-C$_4$B$_7$H$_{11}$, and 7,8,9,10-C$_4$B$_7$H$_{11}$. *Polyhedron* **1988**, *7*, 669–670. [CrossRef]
30. Grimes, R.N. Carbon-rich carboranes and their metal derivatives. *Adv. Inorg. Chem. Radiochem.* **1983**, *26*, 55–117.
31. Maynard, R.B.; Grimes, R.N. Oxidative fusion of carborane ligands in iron and cobalt complexes—A systematic study. *J. Am. Chem. Soc.* **1982**, *104*, 5983–5986. [CrossRef]
32. Maxwell, W.M.; Miller, V.R.; Grimes, R.N. 4-Carbon carboranes—Synthesis of tetra-C-methyltetracarba-dodecaborane(12) and its metallocarborane derivatives. *J. Am. Chem. Soc.* **1974**, *99*, 7116–7117. [CrossRef]
33. Maxwell, W.M.; Miller, V.R.; Grimes, R.N. Iron-hydrogen and iron-cobalt metallocarboranes—Synthesis and chemistry of [(CH$_3$)$_2$C$_2$B$_4$H$_4$]$_2$Fe^{II}H$_2$ and a novel tetracarbon carborane system (CH$_3$)$_4$C$_4$B$_8$H$_8$. *Inorg. Chem.* **1976**, *15*, 1343–1348. [CrossRef]
34. Freyberg, D.P.; Weiss, R.; Sinn, E.; Grimes, R.N. Crystal and molecular structure of a tetracarbon carborane, (CH$_3$)$_4$C$_4$B$_8$H$_8$, a new type of *nido* cage system. *Inorg. Chem.* **1977**, *16*, 1847–1851. [CrossRef]
35. Venable, T.L.; Maynard, R.B.; Grimes, R.N. Crystal and molecular structure of a tetracarbon carborane, (CH$_3$)$_4$B$_8$H$_8$, a new type of *nido* cage system. *J. Am. Chem. Soc.* **1984**, *106*, 6187–6193. [CrossRef]
36. Spencer, J.T.; Pourlan, M.R.; Butcher, R.J.; Sinn, E.; Grimes, R.N. Pi-complexation of *nido*-(PhCH$_2$)$_2$C$_2$B$_4$H$_6$ at the C$_2$B$_3$ and C$_6$ rings—Synthesis and crystal structures of *nido*-2,3-(CO)$_3$Cr[(.eta.^{6}-C$_6$H$_5$)CH$_2$]$_2$-2,3,-C$_2$B$_4$H$_6$ and (PhCH$_2$)$_4$C$_4$B$_8$H$_8$, a nonfluxional C$_4$B$_8$ cluster. *Organometallics* **1987**, *6*, 335–343. [CrossRef]
37. Attia, A.A.A.; Lupan, A.; King, R.B. Tetracarbaboranes: *nido* structures without bridging hydrogens. *Dalton Trans.* **2016**, *45*, 18541–18551. [CrossRef] [PubMed]
38. Friesen, G.D.; Barriola, A.; Daluga, P.; Ragatz, P.; Huffman, J.; Todd, L. Chemistry of dithiaboranes, selenathiaboranes, and diselenaboranes. *Inorg. Chem.* **1980**, *19*, 458–462. [CrossRef]
39. Attia, A.A.A.; Lupan, A.; King, R.B. Polyhedral cobaltadiselenaboranes: Nido strucures without bridging hydrogens. *RSC Adv.* **2016**, *6*, 53635–53642. [CrossRef]

40. Kang, S.O.; Sneddon, L.G. Synthesis of new dithiacobaltaborane clusters derived from *arachno*-6,8-$S_2B_7H_9$. *Inorg. Chem.* **1988**, *27*, 3769–3772. [CrossRef]

41. Robertson, A.P.M.; Beattie, N.A.; Scott, C.; Man, W.Y.; Jones, J.J.; Macgregor, S.A.; Rosair, G.M.; Welch, A.J. 14-Vertex heteroboranes with 14 skeletal electron pairs: An experimental and computational study. *Angew. Chem. Int. Ed.* **2016**, *55*, 8706–8710. [CrossRef]

42. Szabolcs, J.; Lupan, A.; Kun, A.-Z.; King, R.B. *Isocloso* versus *closo* deltahedra in slightly hypoelectronic supraicosahedral 14-vertex dimetallaboranes with 28 skeletal electrons; Relationship to icosahedral dimetallaboranes. *New J. Chem.* **2020**, *44*, 16977–16984.

43. Adamo, C.; Barone, V. Toward reliable density functional methods without adjustable parameters: The PBE0 model. *J. Chem. Phys.* **1999**, *110*, 6158–6169. [CrossRef]

44. Weigend, F. Accurate Coulomb-fitting basis sets for H to Rn. *Phys. Chem. Chem. Phys.* **2006**, *8*, 1057–1065. [CrossRef]

45. Gaussian 09, Revision E.01; Gaussian, Inc.: Wallingford, CT, USA. 2016. Available online: https://gaussian.com/g09citation/ (accessed on 10 February 2023).

46. Schneider, W.B.; Bistoni, G.; Sparta, M.; Saitow, M.; Riplinger, C.; Auer, A.A.; Neese, F. Decomposition of intermolecular interaction energies within the local pair natural orbital coupled cluster framework. *J. Chem. Theory Comput.* **2016**, *12*, 4778–4792. [CrossRef] [PubMed]

47. Riplinger, C.; Pinski, P.; Becker, U.; Valeev, E.F.; Neese, F. Sparse maps—A systematic infrastructure for reduced-scaling electronic structure methods. II. Linear scaling domain based pair natural orbital coupled cluster theory. *J. Chem. Phys.* **2016**, *144*, 024109. [CrossRef]

48. Pavošević, F.; Pinski, P.; Riplinger, C.; Neese, F.; Valeev, E.F. SparseMaps—A systematic infrastructure for reduced-scaling electronic structure methods. IV. Linear-scaling second-order explicitly correlated energy with pair natural orbitals. *J. Chem. Phys.* **2016**, *144*, 144109. [CrossRef] [PubMed]

49. Kubas, A.; Berger, D.; Oberhofer, H.; Maganas, D.; Reuter, K.; Neese, F. Surface adsorption energetics studied with "gold standard" wave-function-based ab initio methods: Small-molecule binding to TiO_2 (110). *J. Phys. Chem. Lett.* **2016**, *7*, 4207–4212. [CrossRef] [PubMed]

50. Isegawa, M.; Neese, F.; Pantazis, D.A. Ionization energies and aqueous redox potentials of organic molecules: Comparison of DFT, correlated ab initio theory and pair natural orbital approaches. *J. Chem. Theory Comput.* **2016**, *12*, 2272–2284. [CrossRef]

51. Guo, Y.; Sivalingam, K.; Valeev, E.F.; Neese, F. SparseMaps—A systematic infrastructure for reduced-scaling electronic structure methods. III. Linear-scaling multireference domain-based pair natural orbital N-electron valence perturbation theory. *J. Chem. Phys.* **2016**, *144*, 094111. [CrossRef]

52. Dutta, A.K.; Neese, F.; Izsák, R. Towards a pair natural orbital coupled cluster method for excited states. *J. Chem. Phys.* **2016**, *145*, 034102. [CrossRef]

53. Datta, D.; Kossmann, S.; Neese, F. Analytic energy derivatives for the calculation of the first-order molecular properties using the domain-based local pair-natural orbital coupled-cluster theory. *J. Chem. Phys.* **2016**, *145*, 114101. [CrossRef]

54. Pinski, P.; Riplinger, C.; Valeev, E.F.; Neese, F. Sparse maps—A systematic infrastructure for reduced-scaling electronic structure methods. I. An efficient and simple linear scaling local MP2 method that uses an intermediate basis of pair natural orbitals. *J. Chem. Phys.* **2015**, *143*, 034108. [CrossRef]

55. Mondal, B.; Neese, F.; Ye, S. Control in the rate-determining step provides a promising strategy to develop new catalysts for CO_2 hydrogenation: A local pair natural orbital coupled cluster theory study. *Inorg. Chem.* **2015**, *54*, 7192–7198. [CrossRef]

56. Liakos, D.G.; Sparta, M.; Kesharwani, M.K.; Martin, J.M.L.; Neese, F. Exploring the accuracy limits of local pair natural orbital coupled-cluster theory. *J. Chem. Theory Comput.* **2015**, *11*, 1525–1539. [CrossRef] [PubMed]

57. Liakos, D.G.; Neese, F. Domain based pair natural orbital coupled cluster studies on linear and folded alkane chains. *J. Chem. Theory Comput.* **2015**, *11*, 2137–2143. [CrossRef] [PubMed]

58. Liakos, D.G.; Neese, F. Is it possible to obtain coupled cluster quality energies at near density functional theory cost? domain-based local pair natural orbital coupled cluster vs. modern density functional theory. *J. Chem. Theory Comput.* **2015**, *11*, 4054–4063. [CrossRef] [PubMed]

59. Demel, O.; Pittner, J.; Neese, F. A local pair natural orbital-based multireference Mukherjee's coupled cluster method. *J. Chem. Theory Comput.* **2015**, *11*, 3104–3114. [CrossRef]

60. Neese, F. The ORCA program system. *Wiley Interdiscip. Rev. Comput. Mol. Sci.* **2012**, *2*, 73–78. [CrossRef]

61. Izsák, R.; Neese, F. Speeding up spin-component-scaled third-order pertubation theory with the chain of spheres approximation: The COSX-SCS-MP3 method. *Mol. Phys.* **2013**, *111*, 1190–1195. [CrossRef]

62. Izsák, R.; Neese, F. An overlap fitted chain of spheres exchange method. *J. Chem. Phys.* **2011**, *135*, 144105. [CrossRef]

63. Kossmann, S.; Neese, F. Efficient structure optimization with second-order many-body perturbation theory: The RIJCOSX-MP2 Method. *J. Chem. Theory Comput.* **2010**, *6*, 2325–2338. [CrossRef]

64. Kossmann, S.; Neese, F. Comparison of two efficient approximate Hartee–Fock approaches. *Chem. Phys. Lett.* **2009**, *481*, 240–243. [CrossRef]

65. Neese, F.; Wennmohs, F.; Hansen, A.; Becker, U. Efficient, approximate and parallel Hartree–Fock and hybrid DFT calculations. A "chain-of-spheres" algorithm for the Hartree–Fock exchange. *Chem. Phys.* **2009**, *356*, 98–109. [CrossRef]

66. Neese, F. An improvement of the resolution of the identity approximation for the formation of the Coulomb matrix. *J. Comput. Chem.* **2003**, *24*, 1740–1747. [CrossRef] [PubMed]
67. Dutta, A.K.; Neese, F.; Izsák, R. Speeding up equation of motion coupled cluster theory with the chain of spheres approximation. *J. Chem. Phys.* **2016**, *144*, 034102. [CrossRef] [PubMed]
68. Christian, G.J.; Neese, F.; Ye, S. Unravelling the molecular origin of the regiospecificity in extradiol catechol dioxygenases. *Inorg. Chem.* **2016**, *55*, 3853–3864. [CrossRef] [PubMed]

Thermal Polymorphism in CsCB$_{11}$H$_{12}$

Radovan Černý [1,*], Matteo Brighi [1], Hui Wu [2], Wei Zhou [2], Mirjana Dimitrievska [2,3,4], Fabrizio Murgia [1], Valerio Gulino [5], Petra E. de Jongh [5], Benjamin A. Trump [2] and Terrence J. Udovic [2,6]

[1] Laboratory of Crystallography, Department of Quantum Matter Physics, University of Geneva, Quai Ernest-Ansermet 24, CH-1211 Geneva, Switzerland
[2] NIST Center for Neutron Research, National Institute of Standards and Technology, Gaithersburg, MD 20899-6102, USA
[3] National Renewable Energy Laboratory, Golden, CO 80401, USA
[4] Transport at Nanoscale Interfaces Laboratory, Swiss Federal Laboratories for Material Science and Technology (EMPA), Ueberlandstrasse 129, CH-8600 Duebendorf, Switzerland
[5] Materials Chemistry and Catalysis, Debye Institute for Nanomaterials Science, Utrecht University, 3584 CG Utrecht, The Netherlands
[6] Department of Materials Science and Engineering, University of Maryland, College Park, MD 20742, USA
* Correspondence: radovan.cerny@unige.ch

Abstract: Thermal polymorphism in the alkali-metal salts incorporating the icosohedral monocarba-hydridoborate anion, CB$_{11}$H$_{12}^{-}$, results in intriguing dynamical properties leading to superionic conductivity for the lightest alkali-metal analogues, LiCB$_{11}$H$_{12}$ and NaCB$_{11}$H$_{12}$. As such, these two have been the focus of most recent CB$_{11}$H$_{12}^{-}$ related studies, with less attention paid to the heavier alkali-metal salts, such as CsCB$_{11}$H$_{12}$. Nonetheless, it is of fundamental importance to compare the nature of the structural arrangements and interactions across the entire alkali-metal series. Thermal polymorphism in CsCB$_{11}$H$_{12}$ was investigated using a combination of techniques: X-ray powder diffraction; differential scanning calorimetry; Raman, infrared, and neutron spectroscopies; and ab initio calculations. The unexpected temperature-dependent structural behavior of anhydrous CsCB$_{11}$H$_{12}$ can be potentially justified assuming the existence of two polymorphs with similar free energies at room temperature: (*i*) a previously reported, ordered *R*3 polymorph stabilized upon drying and transforming first to *R*3*c* symmetry near 313 K and then to a similarly packed but disordered *I*$\bar{4}$3*d* polymorph near 353 K and (*ii*) a disordered *Fm*$\bar{3}$ polymorph that initially appears from the disordered *I*$\bar{4}$3*d* polymorph near 513 K along with another disordered high-temperature *P*6$_3$*mc* polymorph. Quasielastic neutron scattering results indicate that the CB$_{11}$H$_{12}^{-}$ anions in the disordered phase at 560 K are undergoing isotropic rotational diffusion, with a jump correlation frequency [1.19(9) $\times$ 10^{11} s^{-1}] in line with those for the lighter-metal analogues.

Keywords: monocarba-hydridoborate; polymorphism; crystal structure; anion dynamics

Citation: Černý, R.; Brighi, M.; Wu, H.; Zhou, W.; Dimitrievska, M.; Murgia, F.; Gulino, V.; de Jongh, P.E.; Trump, B.A.; Udovic, T.J. Thermal Polymorphism in CsCB$_{11}$H$_{12}$. *Molecules* **2023**, *28*, 2296. https://doi.org/10.3390/molecules28052296

Academic Editors: Michael A. Beckett and Igor B. Sivaev

Received: 20 January 2023
Revised: 22 February 2023
Accepted: 27 February 2023
Published: 1 March 2023

1. Introduction

Icosahedral hydridoborates M^{x+}(B$_{12}$H$_{12}$)$_{x/2}$ and their C-derivatives M^{x+}(CB$_{11}$H$_{12}$)$_x$ are extensively used in organic syntheses, medicine, nanoscale engineering, catalysis, metal recovery from radioactive waste, and recently as solid ionic conductors [1–3]. Their crystal structures are classified among so-called plastic (rotatory) crystals [4] as they show with temperature an order/disorder transition of dynamic nature into a state with orientationally disordered icosahedral anions B$_{12}$H$_{12}^{2-}$ or CB$_{11}$H$_{12}^{-}$. As shown by solid-state NMR, quasielastic neutron scattering (QENS) experiments and ab-initio calculations, the icosahedral anions undergo discrete symmetry-preserving reorientational jumps, even in their ordered state [5]. Upon transformation to the disordered state, the anion reorientational mobilities typically increase by several orders of magnitude, and the motions become more rotationally diffusive. While monocarba-hydridoborates of smaller alkali metals, such as Li and Na have been characterized in detail due to their potential importance as solid ionic conductors [2,6], the thermal polymorphism in $KCB_{11}H_{12}$ has been studied only recently [7]. No reports are available for $RbCB_{11}H_{12}$. Aside from the recent detailed room-temperature (*rt*) structural study of $CsCB_{11}H_{12}$ [8], only one study of thermal polymorphism in $CsCB_{11}H_{12}$ has been published, which was 18 years ago [9,10]. We have been motivated to understand more thoroughly the thermal polymorphism in $CsCB_{11}H_{12}$ in comparison with the monocarba-hydridoborate salts of the lighter alkali metals to get a broader insight into the nature of the anion–anion interaction and its effect on cation mobility. We will show that an understanding of the polymorphism in this compound is complicated by the presence of hydrated and metastable phases.

2. Results

2.1. Crystal Structures and Thermal Polymorphism

Structural characterizations were performed for the single-cation sample $CsCB_{11}H_{12}$, as well as for the Rb-doped $Cs_{0.93}Rb_{0.07}CB_{11}H_{12}$ salt.

2.1.1. $CsCB_{11}H_{12}$

All measured SR-XPD data (two typical temperature-dependent SR-XPD patterns are shown in Figures S1 and S2) show the presence of an anhydrous polymorph γ observed at *rt* and three other polymorphs (β, α, and α') observed on heating. The temperature region where the diffraction peaks of the polymorphs were observed depends on the heating/cooling rate. The list of the observed polymorphs is given in Table 1. The γ-polymorph has been characterized using single-crystal X-ray diffraction in refs. [9,10] as a hydrated phase with the composition $CsCB_{11}H_{12}\cdot1/3\ H_2O$. A recent study of anhydrous $CsCB_{11}H_{12}$ has also been performed using single-crystal X-ray diffraction [8]. The thermal stability of the various polymorphs is analyzed in the Discussion.

Table 1. Thermal polymorphs observed in anhydrous $CsCB_{11}H_{12}$ with their space group symmetry (*s.g.*), lattice parameters, cell volume, and the number of formula units (*f.u.*)/unit cell (Z). The naming of the polymorphs is according to ref. [9].

Polymorph	*s.g.*	*a* [Å]	*c* [Å]	*V* [Å³]	Z	T_{exp} [K]
γ	$R3$	20.9533 (1)	13.2400 (1)	5034.1 (1)	18	303
		20.7818 (2)	13.0935 (2)	4897.3 (1)		156
β	$I\bar{4}3d$	15.1493 (6)		3476.8 (4)	12	423
α	$Fm\bar{3}$	10.6491 (9)		1207.7 (3)	4	523
α'	$P6_3mc$	7.3987 (5)	12.5639 (3)	595.5 (2)	2	539

The γ-polymorph changes its symmetry reversibly from $R3$ to $R3c$ at 313 K. The structural change is very subtle and practically invisible in the DSC curves (Figure S3) but clearly detected in diffraction patterns (Figures S1, S2, and S4). It is understood as a degeneration of two independent $CB_{11}H_{12}{}^-$ anions in the asymmetric unit of *s.g.* $R3$ into one in *s.g.* $R3c$, as also discussed in ref. [8]. The γ-polymorph of $CsCB_{11}H_{12}$ is stable down to at least 100 K as observed using low-temperature (*lt*) SR-XPD. Its hydrated version, studied in [9], contains 1/3 of a water molecule/*f.u.* disordered in the channels running along the *c*-axis shown in Figure 1 in light blue. In the anhydrous sample, the channel is empty with the C-H bond pointing inside it [8].

At 353 K, the γ-polymorph transforms into the β-polymorph. It has the symmetry of the cubic space group $I\bar{4}3d$ and an unusual structural prototype of anti-Th_3P_4 [11] with Cs occupying 3/4 of the P positions, and $CB_{11}H_{12}$ localized on the Th positions. Its diffraction pattern corresponds to the phase called β in [9] where the water molecules (non-dried sample) are probably disordered on the remaining 1/4 of the P positions in the anti-Th_3P_4 prototype shown in light blue in Figure 1.

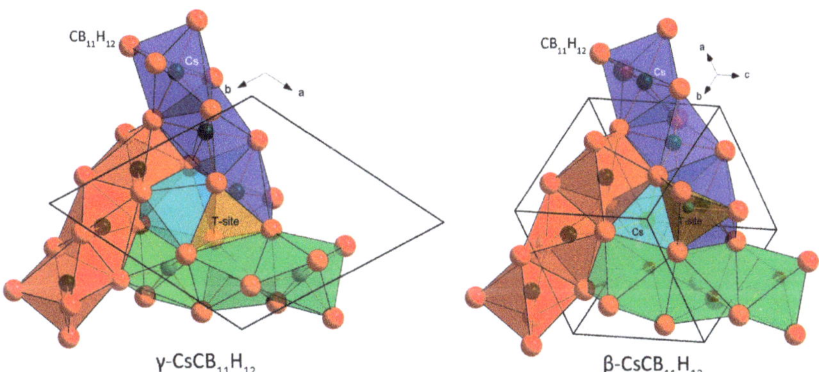

Figure 1. Comparison of the basic structural motif in the γ- ($R3c$) and β- ($I\bar{4}3d$) polymorphs of $CsCB_{11}H_{12}$. The crystal structures in both polymorphs can be constructed from three columns built from face-sharing $Cs(CB_{11}H_{12})_6$ octahedra ($CB_{11}H_{12}^-$ anions simplified as red spheres) and inter-connected by edges (red, blue, and green columns, respectively). The three columns then share triangular faces with the fourth column (light blue). In the γ-polymorph, this column is empty, but may be occupied by a water molecule in the Cs/H_2O ratio of 3/1. In the β-polymorph, Cs is randomly distributed inside all four columns with an occupancy of 0.75.

The anion packing in the γ- and β-polymorphs is very similar. In both structures, the anion–anion coordination number is CN = 12, but the packing is not a close packing (*ccp* or *hcp*). For *n* anions, the packing in both structures contains 4/3 *n* of octahedral O-sites and *n* tetrahedral T-sites. The distribution of O-sites is best described by three columns built from face-sharing $Cs(CB_{11}H_{12})_6$ octahedra and inter-connected by edges (red, blue, and green columns in Figure 1). The three columns then share octahedral faces with the fourth column (light blue). The size of shared triangular faces between the octahedra (O-O bottleneck) and between the octahedra and tetrahedra (O-T bottleneck) is similar and large enough (4.425 Å) in both structures to allow for cation mobility, which may be blocked by the presence of water molecules in hydrated samples. The Rietveld refinements suggest that the β-polymorph is a disordered version of γ with respect to both Cs distribution and anion rotational dynamics.

At 512–515 K, the diffraction peaks of β-polymorph disappear and peaks of α- and α′-polymorphs appear simultaneously, the latter disappearing at 533–598 K, depending on the heating rate. The α-polymorph is observed in diffraction data up to 623 K. Additional DSC-TGA curves measured up to 868 K (not shown here) indicate that the compound starts to lose mass (i.e., decomposes) noticeably by ~798 K.

The α-polymorph is built as a cubic close packing (*ccp*) of anions with NaCl structure type, and its crystal structure can be refined in $Fm\bar{3}m$ or $Fm\bar{3}$ symmetry without significant difference in the agreement factors (Figure 2). The powder pattern shown in Figure 4b of ref. [9] and attributed to a phase called α, corresponds in reality to a mixture of β- and α-polymorphs. The symmetry of the α′-polymorph is hexagonal ($P6_3mc$). It was not possible to refine its crystal structure, due to low data quality. The labeling as α- and α′-polymorphs is not unambiguous because they appear simultaneously at the same temperature, but the α′-polymorph exists only in a limited temperature range. They may correspond to two polymorphs with very close free energies, as has been recently observed in $CaB_{10}H_{10}$ [12].

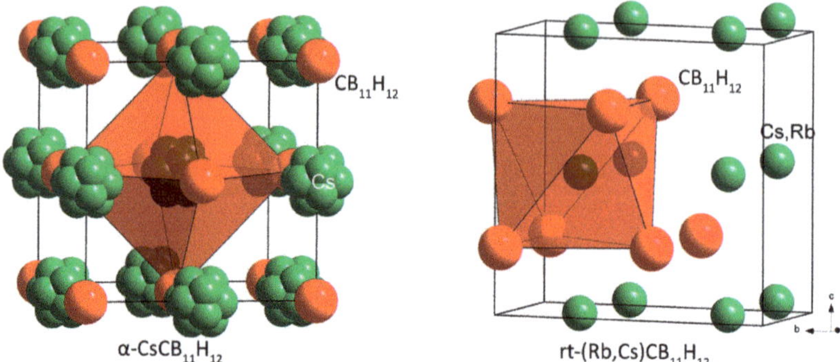

Figure 2. Structural representations of α-CsCB$_{11}$H$_{12}$ with anions *ccp*, and of mixed cation phase *rt*-(Rb,Cs)CB$_{11}$H$_{12}$ with anions hcp. The CB$_{11}$H$_{12}{}^-$ anions are simplified as red spheres.

2.1.2. Cs$_{0.93}$Rb$_{0.07}$CB$_{11}$H$_{12}$

In the *rt* diffraction data of a CsCB$_{11}$H$_{12}$ sample containing 7% Rb alkali-metal substitution, we have observed, besides the dominant peaks of γ-CsCB$_{11}$H$_{12}$, two other sets of diffraction peaks. The first one corresponds probably to the pure RbCB$_{11}$H$_{12}$ phase and disappears very rapidly. The second set has been successfully indexed on an orthorhombic cell and the structure solved in the space group *Pbcm* with the structural prototype of the high-pressure polymorph of NaOH, i.e., with anions *hcp* (Figure 2). As the fraction of this phase in the sample was very low (see Figure S9), the Rb/Cs ratio in this phase was not possible to refine.

The Rietveld plots of all refined structures are given in the Supplementary Materials (Figures S5–S9). The CIF files may be obtained from the Fachinformationszentrum Karlsruhe, 76344 Eggenstein-Leopoldshafen (Germany), quoting the depository numbers CSD-2169329-216934.

2.2. Anion Dynamics

In addition to the crystallographic results, the vibrational dynamics of both CsCB$_{11}$H$_{12}$ and Cs$_{0.93}$Rb$_{0.07}$CB$_{11}$H$_{12}$ samples at 4 K were characterized using NVS. The neutron vibrational spectrum for CsCB$_{11}$H$_{12}$ at 4 K is compared in Figure 3 with the simulated phonon density of states (PDOS) based on the DFT-optimized γ-CsCB$_{11}$H$_{12}$ *R*3-structure determined from the single crystal diffraction results. Due to the overwhelming large neutron scattering cross-section for H atoms with respect to other elements, the spectrum in this energy region is dominated by the normal modes involving the various possible CB$_{11}$H$_{12}{}^-$ anion deformations, all of which entail significant H-atom displacements. The measured spectral signature is typically sensitive to the particular crystal structure [13,14] and in this instance agrees well with the simulated PDOS for the trigonal arrangement of CB$_{11}$H$_{12}{}^-$ anions and Cs$^+$ cations, taking into account the additional minor contributions expected from two-phonon combination bands. The presence of CsCB$_{11}$H$_{12}$ lattice effects on the measured PDOS is made clearer by the discrepancies observed between the CsCB$_{11}$H$_{12}$ spectrum and that calculated for the isolated CB$_{11}$H$_{12}{}^-$ anion in Figure 3. We note that the spectrum for the CB$_{11}$H$_{12}{}^-$ anion deformations measured with the Cs$_{0.93}$Rb$_{0.07}$CB$_{11}$H$_{12}$ sample is nearly identical to that for CsCB$_{11}$H$_{12}$, as expected based on this mixed-cation salt's dominant γ-CsCB$_{11}$H$_{12}$-like structure and the lack of any significant vibrational perturbations from the relatively low concentration of Rb atoms. Further information about the characters and energies of the different CsCB$_{11}$H$_{12}$ phonon modes contributing to the simulated PDOS for the ordered *R*3 structure can be found in the animation file in the Supplementary Materials [15].

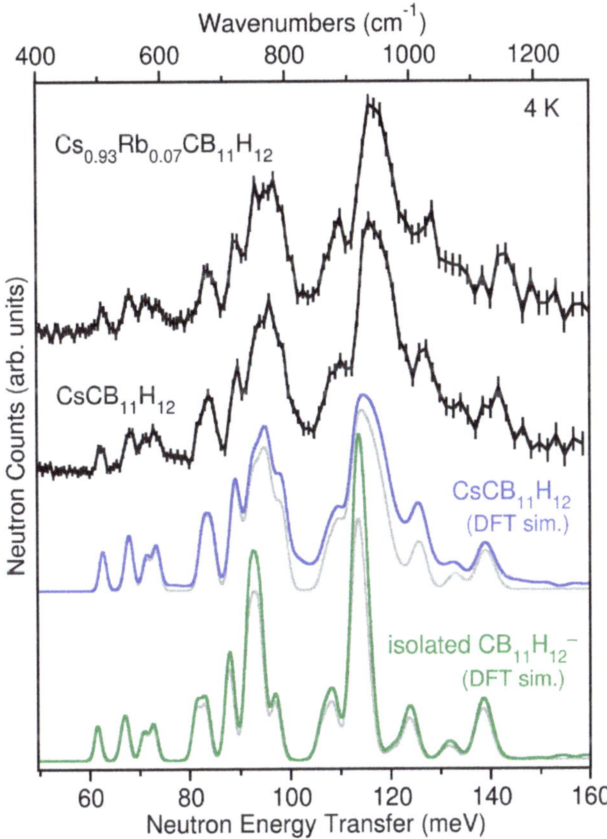

Figure 3. Neutron vibrational spectra (black) of CsCB$_{11}$H$_{12}$ and Cs$_{0.93}$Rb$_{0.07}$CB$_{11}$H$_{12}$ at 4 K compared to the simulated one + two-phonon densities of states from DFT phonon calculations of the optimized γ-CsCB$_{11}$H$_{12}$ R3-structure (blue) and the isolated CB$_{11}$H$_{12}{}^{-}$ anion (green, from ref. [2], using a $30 \times 30 \times 30$ supercell and full $C5_v$ molecular symmetry). The simulated one-phonon densities of states considering only the fundamental single-phonon modes are shown for comparison in grey. Vertical error bars represent ± 1 σ. (N.B., 1 meV $\approx$ 8.0655 cm^{-1}).

Quasielastic neutron scattering measurements were performed only on the anhydrous Cs$_{0.93}$Rb$_{0.07}$CB$_{11}$H$_{12}$ sample, although we expect the observed anion reorientational behavior to largely mimic that of pure CsCB$_{11}$H$_{12}$. At 560 K, the CB$_{11}$H$_{12}{}^{-}$ anions in the predominant disordered polymorph α displays a reorientational jump correlation frequency of $1.19(9) \times 10^{11}$ s^{-1}, a rather high reorientational mobility in general agreement with the values extrapolated from lower-temperature QENS data for the lighter-metal Li, Na, and K carba-hydridoborates [7,16] (see Figures S10 and S11). Moreover, this 560 K jump frequency is almost 20× higher than that of the B$_{12}$H$_{12}{}^{2-}$ anions in the symmetry-related Cs$_2$B$_{12}$H$_{12}$ [17,18], which is not unexpected since the B$_{12}$H$_{12}{}^{2-}$ anions in Cs$_2$B$_{12}$H$_{12}$ are relatively more "confined", i.e., they are surrounded by twice as many cations as the CB$_{11}$H$_{12}{}^{-}$ anions in the Rb-doped CsCB$_{11}$H$_{12}$ salt. The momentum transfer dependence of the elastic incoherent structure factor at 560 K (Figure S12) indicates that the CB$_{11}$H$_{12}{}^{-}$ anion reorientations at this high temperature are more akin to isotropic rotational diffusion, a behavior not yet fully reached by the B$_{12}$H$_{12}{}^{2-}$ anions in Cs$_2$B$_{12}$H$_{12}$ (by 530 K) [18] or by the CB$_{11}$H$_{12}{}^{-}$ anions in the Li, Na, and K carba-hydridoborates (by 473–480 K) [7,16].

We note that high $CB_{11}H_{12}{}^-$ anion rotational mobility can also significantly affect the translational mobility of the cations [16]. Although the Cs^+ cation diffusive mobility in disordered α-$CsCB_{11}H_{12}$ is certainly much lower than that for the lighter and smaller alkali-metal cations (Li^+, Na^+, and K^+) in their analogous *ht*-disordered $MCB_{11}H_{12}$ phases [2,7], the emergence of rotationally fluidic anions for this phase can still provide a more accommodating potential-energy landscape [16] for greatly enhancing the cation conductivity compared to that for the lower-temperature ordered structure. Indeed, increased anion rotational mobility may be at least partially responsible for the four-orders-of-magnitude increase in Cs^+ conductivity observed above ~483 K ($\sigma = 5.5 \times 10^{-5}$ S cm^{-1}) for the *ht*-disordered α-polymorph of the related Cs dodecahydro-7,8-dicarba-*nido*-undecaborate salt, Cs-7,8-$C_2B_9H_{12}$, as compared to its low- and medium-temperature ordered polymorphs [19], although the authors attributed this conductivity jump solely to favorable α-polymorph structural effects.

3. Discussion

3.1. Metastable and Stable Polymorphs

Understanding of $CsCB_{11}H_{12}$ thermal polymorphism is complicated by hydrated pristine samples. Contrary to hydrated $Cs_2B_{12}X_{12}$ (X = Cl, Br, and I) and alkali-metal salts $A_2B_{12}F_{12}$ where the water molecule coordinates via its lone pair with the cations [20,21], in hydrated $CsCB_{11}H_{12}$, the disordered water molecule coordinates with the $CB_{11}H_{12}{}^-$ anions. DFT structural optimization of $R3$ γ-$CsCB_{11}H_{12}\cdot1/3$ H_2O (with artificial H_2O site ordering) points to the formation of di-hydrogen bonds between the hydrogens of the water molecule and hydrogens from two different $CB_{11}H_{12}{}^-$ anions. As the water molecule is normally disordered in the channels, we were unable to get a more precise answer. The single crystal results from ref. [8] suggest the C-H bond to point inside the channels, i.e., towards the possible water molecules, and DFT calculations confirm that this particular anion orientation is energetically preferred.

The drying procedure of the as-purchased hydrated samples under dynamic vacuum at 593–693 K provides anhydrous powders verified using Raman and IR spectroscopies (Figures S13 and S14) where the characteristic broad O-H stretching and H-O-H bending bands are missing at 3200–3550 and 1600 cm^{-1}, respectively.

$CsCB_{11}H_{12}$ behaves unexpectedly with temperature and its polymorphism can be understood only by accepting similar *rt* free energies of the ordered γ- and disordered α-polymorphs. One possible scenario for $CsCB_{11}H_{12}$ is the metastability of ordered γ- and stability of disordered α-polymorphs at *rt*. In such a scenario, the as-purchased sample contains hydrated γ-phase. When dried, the channels occupied by water molecules (light blue in Figure 1) become empty, but the $CsCB_{11}H_{12}$ framework stays intact on fast cooling from the drying temperature to *rt*. The coordination of the $CB_{11}H_{12}{}^-$ polyanion in the γ-phase with $R3/R3c$ symmetry points to a metastable structure (Figure 4 left) called as unexpected in ref. [8]. Without the water molecules, which were located in the channels on the right side from the anion in Figure 4 left, the Coulombic forces between Cs^+ and $CB_{11}H_{12}{}^-$, even if balanced according to DFT calculations, can be easily pushed out of balance. On the other hand, the forces seem to be better balanced in the disordered β-polymorph (Figure 4 right). In the temperature-dependent SR-XPD data where the heating has been stopped before the appearance of α- and α'-phases, the β-polymorph does not transform back to the γ-polymorph on cooling and persists in the sample down to 100 K.

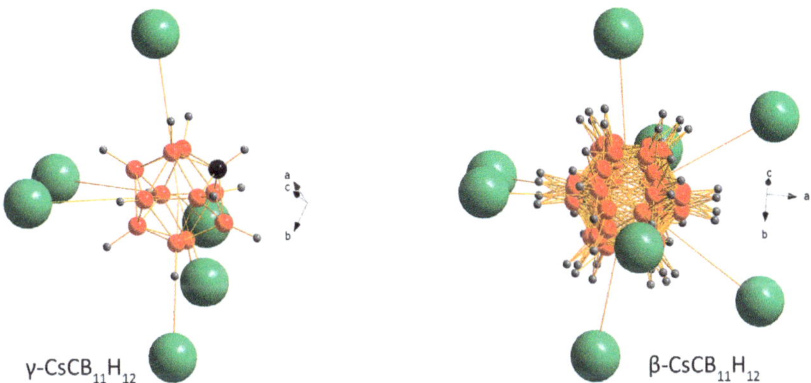

Figure 4. Coordination of the $CB_{11}H_{12}^-$ anion in the crystal structures of γ- and β-CsCB$_{11}$H$_{12}$. The γ-structure as obtained from single-crystal data [8] is optimized here with DFT also allowing carbon localization. While the anion is ordered in the former, it is rotationally disordered in the latter. The water molecules in hydrated phases occupy the channels on the right side from the anion in γ and share the positions with Cs atoms in the ratio 1:3 in β.

On heating, the γ-polymorph transforms to its disordered variant β, and at 512–515 K, the α- and α′-polymorphs crystallize from melted β. Please note the diffuse intensity signal starting from this temperature marking the presence of a melted sample fraction, which does not recrystallize into α- and α′-polymorphs due to the slow crystallization kinetics of the latter two polymorphs (Figures S1 and S2). The sample behavior on cooling depends on the cooling rate. During fast cooling (10 K/min) used in Figure S1 and during the fast DSC scans (Figure S3 left), the melted fraction of the sample does not crystallize as the stable α-polymorph, but rather as metastable β, which transforms upon further cooling into the γ-polymorph. The crystallization of β- and γ- instead of the α-polymorph on cooling can be explained by the Ostwald step rule [22] stating that the phase with the lower kinetic barrier forms first due to its faster crystallization kinetics. When slow cooling is used (2 K/min), Figure S2 does not show any recrystallization of β- and γ-polymorphs. The persisting diffuse scattering visible in both diffraction data sets, fast and slow heating/cooling rates, has its origin in the melted fraction of the sample, which does not transform into β- and γ-phases and stays glassy. In slow DSC scans (Figure S3 right), the β- and γ-recrystallization events are still visible on cooling, but with decreasing intensity compared to heating. The difference between the temperature-dependent SR-XPD data and the DSC curve may be understood as due to different crystallization kinetics between a closed system (capillary in the diffraction experiment) and an open system (sample holder in the DSC experiment). The two unexplained signals in the cooling part of DSC scans without an equivalent in the heating part are tentatively attributed to glassy transitions in the melted part of the sample.

We have suggested in the above discussion that the disordered α-polymorph may be the stable (i.e., thermodynamically preferred) structure at *rt*. Nonetheless, at first glance, it would seem that the γ-polymorph is the actual stable *rt*-form as it is supported by the crystallization of γ from the water solution in ref. [8]. Yet, we caution that this somewhat unexpected γ-crystallization may have been aided by the participation of water molecules in this case, a structural pathway which may not necessarily occur in an anhydrous crystallization environment.

3.2. Cation Coordination in Alkali-Metal Carba-Hydridoborates

We will now discuss the relation of Cs and other alkali-metal carba-hydridoborates. In all *rt*-structures, the alkali metal is located in O-sites. In the ordered phases of Li, Na, and K [2,7], the cation is displaced from the center of the O-site towards a triangular face of the octahedron $(CB_{11}H_{12})_6$ due to the size effect. The Cs is disordered around the octahedron center in the α-polymorph. The localization of the relatively smaller Li and Na in O-sites compared to hydridoborates $Li_2B_{12}H_{12}$ and $Na_2B_{12}H_{12}$ where it occupies T-sites may be explained by the repulsion between the cation and the least electronegative H bonded to the C atom pointing preferentially towards the T-site in carba-hydridoborate as recently shown by molecular dynamics simulations [16]. In disordered *ht*-phases of Li and Na carba-hydridoborates, the cation occupies both O- and T-sites [2], but in K [7] and Rb (structural results not shown here, manuscript under preparation), carba-hydridoborates the cation occupies only the O-site. The difference in the occupation of O- and T-sites by the cation between disordered carba-hydridoborates of Li and Na on one side and K, Rb, and Cs on the other side is visible as a difference in the volume/*f.u.* vs. cation size plot shown in Figure 5. The temperature of cation and anion disordering is also given in the figure showing its dependence on the cation coordination.

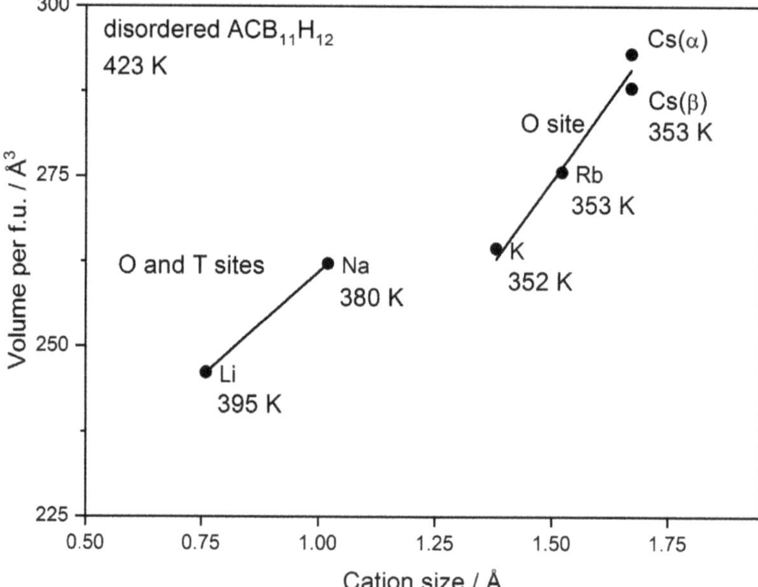

Figure 5. Dependence of volume per formula unit in $ACB_{11}H_{12}$ (A = Li, Na, K, Rb, and Cs) on alkali cation radius in octahedral coordination according to Shannon [23]. The temperature of cation- and anion-disordering is also given. The structure of all compounds is disordered with *ccp* or *hcp* anion packing with the exception of disordered β-$CsCB_{11}H_{12}$.

4. Materials and Methods

4.1. Samples Preparation

$CsCB_{11}H_{12}$ and $Cs_{0.93}Rb_{0.07}CB_{11}H_{12}$ were obtained from Katchem (The mention of all commercial suppliers in this paper is for clarity and does not imply the recommendation or endorsement of these suppliers by any involved institution). The as-purchased products were hydrated and were first superficially dried under a dynamic vacuum at 453 K for 12 h. Any remaining residual water was removed by evacuating at 603 K for 16 h in a quartz tube followed by the removal of the tube from the oven, resulting in a cooldown to *rt* over approximately 0.5 h.

4.2. Synchrotron Radiation X-ray Powder Diffraction (SR-XPD)

The data used for the crystal structure solution and refinement were collected at the Swiss Norwegian Beamlines of ESRF (European Synchrotron Radiation Facility, Grenoble, France) between 100 and 723 K. Temperature-dependent SR-XPD data were collected on the Dectris Pilatus M2 detector at wavelengths of 0.8212, 0.7399, and 0.64113 Å. The 2D images were integrated and treated with the locally written program Bubble. For all measurements, the samples were sealed into borosilicate capillaries of diameter 0.5 mm (under argon atmosphere), which were spun during data acquisition. The temperature above *rt* was controlled with a hot air blower calibrated with thermal dilatation of silver. The temperature below *rt* was controlled with a Cryostream 700 (Oxford Cryosystems, Long Hanborough, UK). The wavelength was calibrated with a NIST SRM640c Si standard. Additional SR-XPD patterns were collected for $CsCB_{11}H_{12}$ in a 1 mm diameter sealed quartz capillary (unspun) at the Advanced Photon Source at Argonne National Laboratory at a wavelength of 0.45246 Å using a two-dimensional amorphous Si-plate detector. The two-dimensional data were converted to one-dimensional data using GSAS-II [24]. Crystal structures were solved ab-initio using the software Fox [25] and refined with the Rietveld method using TOPAS [26] or GSAS [27]. The $CB_{11}H_{12}{}^-$ anion was modelled as a rigid body with an ideal icosahedral shape and with B-H and B-B distances of 1.81 and 1.16 Å, respectively. As carbon localization is largely hindered by the low X-ray scattering contrast between B and C atoms, the anion was modelled as a $B_{12}H_{12}$ cluster. The interatomic distances and coordination polyhedra were analyzed using the program DIAMOND [28].

4.3. Differential Scanning Calorimetry (DSC)

Differential scanning calorimetry (DSC) measurements for $CsCB_{11}H_{12}$ were performed using a Mettler Toledo HP DSC 1-STAR (Mettler Toledo, Greifensee, Switzerland). Approximately 5−10 mg of the sample were loaded in 40 mg aluminum crucibles, closed with a lid inside a glove box, and transferred to the DSC under an inert atmosphere. The measurements were performed under an Ar flow, using heating and cooling ramps of 2 or 10 K/min in the desired temperature range.

4.4. Raman Spectroscopy

The Raman spectra at *rt* were collected using the spectrometer Horiba (Kyoto, Japan), LabRAM HR Evolution using the same samples in the capillary that were used in the SR-XPD experiments.

4.5. Fourier Transformation Infrared Spectroscopy (FTIR)

Infrared spectra were collected in diffuse reflectance infrared Fourier transform spectroscopy (DRIFTS) mode, using a PerkinElmer 2000 spectrometer (PerkinElmer, Waltham, MA, USA). Thirty-two scans were accumulated in the $4000-900$ cm^{-1} range with a resolution of 4 cm^{-1}. KBr powder was used to collect the background. To prevent air contamination, the sample was placed in an airtight sample holder, with KBr windows, inside the glove box.

4.6. Neutron Scattering Measurements

Neutron scattering measurements for $CsCB_{11}H_{12}$ were performed at the NIST Center for Neutron Research (NCNR) using thin flat-plate-shaped samples in reflection to minimize neutron beam attenuation from the highly-neutron-absorbing ^{10}B present in natural boron. The H-weighted phonon density of states (PDOS) at 4 K was measured using neutron vibrational spectroscopy (NVS) on the Filter-Analyzer Neutron Spectrometer (FANS) [29] using the Cu(220) monochromator with pre- and post-collimations of 20' of arc, yielding a full-width-at-half-maximum (FWHM) energy resolution of about 3% of the neutron energy transfer. Quasielastic neutron scattering (QENS) spectra were measured at 100 K and 560 K on the Disc Chopper Spectrometer (DCS) [30], utilizing incident neutron wavelengths of 4.8 Å and 8 Å with respective resolutions of 56 μeV and 30 μeV FWHM. The instrumental resolution function was obtained from the purely elastic 100 K QENS spectrum. All neutron data analyses were done with the DAVE software package [31]. Quantitative elemental analysis of the $Cs_{1-x}Rb_xCB_{11}H_{12}$ mixed cation alloy was also performed at the NCNR using the NGD Cold-Neutron Prompt Gamma-Ray Activation Analysis Spectrometer [32]. The

Cs/Rb atomic ratio was determined after neutron irradiation of the sample from analysis of spectral peaks in the gamma-ray decay spectrum at the element-specific gamma-ray energies of 605 keV and 796 keV (for ^{134}Cs) and 1077 keV (for ^{86}Rb).

4.7. Density Functional Theory (DFT) Calculations

To complement the $CsCB_{11}H_{12}$ structural refinements and NVS measurement, first-principles calculations were performed within the plane-wave implementation of the generalized gradient approximation to Density Functional Theory (DFT) using a Vanderbilt-type ultrasoft potential with Perdew–Burke–Ernzerhof exchange-correlation and the program QUANTUM ESPRESSO [33]. A cutoff energy of 544 eV and a $2 \times 2 \times 2$ k-point mesh (generated using the Monkhorst-Pack scheme) were used and found to be enough for the total energy to converge within 0.01 meV/atom. For comparison with the NVS measurement, the PDOS was calculated for the 0 K DFT-optimized $CsCB_{11}H_{12}$ structure (as determined using SR-XPD) using the supercell method with finite displacements [34,35] and was appropriately weighted to take into account the H, Cs, B, and C total neutron scattering cross sections.

5. Conclusions

The various structural and dynamical results presented above for $CsCB_{11}H_{12}$ suggest a more complex thermal polymorphic behavior than expected from purely thermodynamic considerations, with observed structural transformations intimately dependent on the specific thermal history. In particular, a likely similarity in free energies between metastable and stable polymorphs combined with the sensitivity to the magnitude of the thermal ramping and cooling rates on the preferred order-disorder and recrystallization pathways may result in kinetically preferred rather than thermodynamically preferred structural configurations. The situation is made more complicated by the potential presence of H_2O molecules that must be removed from within the void spaces of the synthesized $CsCB_{11}H_{12}$ lattice, lest they adversely affect the preferred structural pathways of otherwise anhydrous $CsCB_{11}H_{12}$. Such trace H_2O is strongly bound and/or kinetically trapped and can only be removed using extensive thermal evacuation near 600 K.

The anion reorientational mobility of the $CB_{11}H_{12}{}^-$ anions in the disordered $Fm\bar{3}$-polymorph at 560 K ($>10^{11}$ jumps s^{-1}) is found to be in line with the temperature-dependent trend reported for the other disordered, lighter-alkali-metal analogues. At this high temperature, QENS results indicate that the anions undergo a reorientational mechanism akin to isotropic rotational diffusion. This rotational-liquid-like mobility likely has some synergistic influence on enhancing the translational diffusive mobility of the Cs^+ cations, such as typically observed for the cations in various other related disordered polyhedral hydridoborate and carba-hydridoborate salts.

It is still unclear which polymorph of $CsCB_{11}H_{12}$ is thermodynamically preferred at *rt*. The answer might be found by more extensive ab initio calculations comparing the relative stabilities of the *R3*- and $Fm\bar{3}$-polymorphs, although the latter polymorph is already disordered at *rt*, making a definitive theoretical assessment of its relative stability more problematic. Finding a minimal-energy ordered version of this *ccp* structure that might appear at a lower temperature will require a nontrivial computational search of *ccp*-like structures starting from lower crystal symmetries. Of course, experimentally observing such a low-temperature structure would be the ultimate goal. As it would be both beneficial to this study and of fundamental interest to understand the structural trends across the entire alkali-metal family of $MCB_{11}H_{12}$ salts, we are currently extending our investigations to characterize the thermal polymorphism in the still largely unexplored $RbCB_{11}H_{12}$ analogue.

Supplementary Materials: The following supporting information can be downloaded at: https://www.mdpi.com/article/10.3390/molecules28052296/s1: Figures S1 and S2: Temperature dependent powder patterns; Figure S3: DSC curves; Figure S4: The diffraction peak positions (blue lines) generated by R3 and R3c space groups compared to the respective 298 K and 353 K SR-XPD patterns measured for CsCB11H12; Figures S5–S9: Rietveld plots; Figure S10: QENS spectrum; Figure S11: Arrhenius plot of jump correlation frequency; Figure S12: Elastic incoherent structure factors as a function of Q, Figure S13: Raman spectrum of CsCB11H12 measured at room temperature; Figure S14: IR spectra of CsCB11H12 measured at room temperature, Phonon animation file.

Author Contributions: Conceptualization, R.Č. and T.J.U.; Formal Analysis, R.Č., T.J.U., M.B., H.W., W.Z., M.D., F.M., V.G. and B.A.T.; Investigation, M.B., H.W., W.Z., M.D., T.J.U., F.M., V.G. and B.A.T.; Resources, R.Č. and T.J.U.; Writing—Original Draft Preparation, R.Č. and T.J.U.; Writing—Review & Editing, R.Č., T.J.U., M.B., H.W., W.Z., M.D., F.M., V.G., B.A.T. and P.E.d.J.; Visualization, M.B., H.W., W.Z., M.D., F.M., V.G. and B.A.T.; Supervision, R.Č. and T.J.U.; Project Administration, R.Č. and T.J.U.; Funding Acquisition, R.Č., T.J.U. and P.E.d.J. All authors have read and agreed to the published version of the manuscript.

Funding: M. D. gratefully acknowledges support from the US DOE Office of Energy Efficiency and Renewable Energy, Fuel Cell Technologies Office, under Contract No. DE-AC36-08GO28308 and from H2020 through SMARTCELL Project (Proj. No. 101022257). This work utilized facilities that are supported by the National Science Foundation under Agreement No. DMR-1508249. Use of the Advanced Photon Source, an Office of Science User Facility operated for the US DOE Office of Science by Argonne National Laboratory, was supported by the US DOE under Contract No. DE-AC02-06CH11357.

Institutional Review Board Statement: Not applicable.

Informed Consent Statement: Not applicable.

Data Availability Statement: All data supporting reported results may be obtained upon the request from the authors.

Acknowledgments: Dmitry Chernyshov and Vadim Diadkin from the Swiss-Norwegian Beamlines, ESRF Grenoble, are acknowledged for collecting the synchrotron X-ray powder diffraction data. Jérémie Teyssier, DQMP, University of Geneva, is acknowledged for collecting the Raman spectra. Rick L. Paul, NIST, is acknowledged for collecting and analyzing the data from the NGD Cold-Neutron Prompt Gamma-Ray Activation Analysis Spectrometer.

Conflicts of Interest: The authors declare no conflict of interest.

Sample Availability: Samples of the compounds are not available from the authors.

References

1. Grimes, R.N. *Carboranes*; Elsevier: Burlington, VT, USA, 2011; pp. 1–1107.
2. Tang, W.S., Unemoto, A., Zhou, W., Stavila, V., Matsuo, M., Wu, H., Orimo, S.-I., Udovic, T.J. Unparalleled lithium and sodium superionic conduction in solid electrolytes with large monovalent cage-like anions. *Energy Environ. Sci.* **2015**, *8*, 3637–3645. [CrossRef]
3. Brighi, M.; Murgia, F.; Černý, R. *Closo*-hydroborate sodium salts: An emerging class of room-temperature solid electrolytes. *Cell Rep. Phys. Sci.* **2020**, *1*, 100217d. [CrossRef]
4. Timmermans, J. Plastic Crystals: A Historical Overview. *J. Phys. Chem. Solids* **1961**, *18*, 1–8. [CrossRef]
5. Skripov, A.V.; Soloninin, A.V.; Babanova, O.A.; Skoryunov, R.V. Anion and Cation Dynamics in Polyhydroborate Salts: NMR Studies. *Molecules* **2020**, *25*, 2940. [CrossRef] [PubMed]
6. Murgia, F.; Brighi, M.; Piveteau, L.; Avalos, C.E.; Gulino, V.; Nierstenhöfer, M.C.; Ngene, P.; de Jongh, P.; Černý, R. Enhanced Room-Temperature Ionic Conductivity of NaCB$_{11}$H$_{12}$ via High-Energy Mechanical Milling. *Appl. Mater. Interfaces* **2021**, *13*, 61346–61356. [CrossRef]
7. Dimitrievska, M.; Wu, H.; Stavila, V.; Babanova, O.A.; Skoryunov, R.V.; Soloninin, A.V.; Zhou, W.; Trump, B.A.; Andersson, M.S.; Skripov, A.V.; et al. Structural and Dynamical Properties of Potassium Dodecahydro-monocarba-closo-dodecaborate: KCB$_{11}$H$_{12}$. *J. Phys. Chem. C* **2020**, *124*, 17992–18002. [CrossRef]
8. Bareiss, K.U.; Friedly, A.; Schleid, T. The unexpected crystal structure of cesium dodecahydro-monocarba-*closo*-dodecaborate Cs[CB$_{11}$H$_{12}$]. *Z. Naturforsch.* **2020**, *75*, 1049–1059. [CrossRef]
9. Romerosa, A.M. Thermal, structural and possible ionic-conductor behaviour of CsB$_{10}$CH$_{13}$ and Cs B$_{11}$CH$_{12}$. *Thermochim. Acta* **1993**, *217*, 123–128. [CrossRef]
10. Romerosa, A.M. Thermal, Structural and Possible Ionic-Conductor Behaviour of CsB$_{10}$CH$_{13}$ and Cs B$_{11}$CH$_{12}$. Ph.D. Thesis, University of Barcelona, Barcelona, Spain, 1992.
11. Meisel, K. Kristallstrukturen von Thoriumphosphiden. *Z. Anorg. Allg. Chem.* **1939**, *240*, 300–312. [CrossRef]

12. Jorgensen, M.; Zhou, W.; Wu, H.; Udovic, T.J.; Paskevicius, M.; Černý, R.; Jensen, T.R.J. Polymorphism of Calcium Decahydrido-closo-decaborate and Characterisation of its Hydrates. *Inorg. Chem.* **2021**, *60*, 10943–10957. [CrossRef]

13. Her, J.H.; Wu, H.; Verdal, N.; Zhou, W.; Stavila, V.; Udovic, T.J. Structures of the Strontium and Barium Dodecahydro-closo-Dodecaborates. *J. Alloys Compd.* **2012**, *514*, 71–75. [CrossRef]

14. Verdal, N.; Zhou, W.; Stavila, V.; Her, J.-H.; Yousufuddin, M.; Yildirim, T.; Udovic, T.J. Alkali and Alkaline-Earth Metal Dodecahydro-*closo*-Dodecaborates: Probing Structural Variations via Neutron Vibrational Spectroscopy. *J. Alloys Compds.* **2011**, *509S*, S694–S697. [CrossRef]

15. V_Sim, Open-Source Software. Available online: https://gitlab.com/l_sim/v_sim-website/-/tree/master/download (accessed on 5 January 2023).

16. Dimitrievska, M.; Shea, P.; Kweon, K.E.; Bercx, M.; Varley, J.B.; Tang, W.S.; Skripov, A.V.; Stavila, V.; Udovic, T.J.; Wood, B.C. Carbon Incorporation and Anion Dynamics as Synergistic Drivers for Ultrafast Diffusion in Superionic LiCB$_{11}$H$_{12}$ and NaCB$_{11}$H$_{12}$. *Adv. Energy Mater.* **2018**, *8*, 1703422. [CrossRef]

17. Verdal, N.; Wu, H.; Udovic, T.J.; Stavila, V.; Zhou, W.; Rush, J.J. Evidence of a Transition to Reorientational Disorder in the Cubic Alkali-Metal Dodecahydro-closo-Dodecaborates. *J. Solid State Chem.* **2011**, *184*, 3110–3116. [CrossRef]

18. Verdal, N.; Udovic, T.J.; Rush, J.J.; Cappelletti, R.; Zhou, W. Reorientational Dynamics of the Dodecahydro-*Closo*-Dodecaborate Anion in Cs$_2$B$_{12}$H$_{12}$. *J. Phys. Chem. A* **2011**, *115*, 2933–2938. [CrossRef]

19. Rius, J.; Romerosa, A.; Teixidor, F.; Casabó, J.; Miravitllest, C. Phase Transitions in Cesium 7,8-Dicarbaundecaborate(12): A New One-Dimensional Cesium Solid Electrolyte at 210 °C. *Inorg. Chem.* **1991**, *30*, 1376–1379. [CrossRef]

20. Tiritiris, I.; Schleid, T. Die Kristallstrukturen der Dicaesium-Dodekahalogeno-*closo*-Dodekaborate Cs$_2$[B$_{12}$X$_{12}$] (X = Cl, Br, I) und ihrer Hydrate. *Z. Anorg. Allg. Chem.* **2004**, *630*, 1555–1563. [CrossRef]

21. Peryshkov, D.V.; Bukovsky, E.V.; Lacroix, M.R.; Wu, H.; Zhou, W.; Jones, W.M.; Lozinšek, M.; Folsom, T.C.; Heyliger, D.L.; Udovic, T.J.; et al. Latent Porosity in Alkali Metal M$_2$B$_{12}$F$_{12}$ Salts: Structures and Rapid Room-Temperature Hydration/Dehydration Cycles. *Inorg. Chem.* **2017**, *56*, 12023–12041. [CrossRef]

22. Ostwald, W. Studien über die Bildung und Umwandlung fester Körper. 1. Abhandlung: Übersättigung und Überkaltung. *Z. Phys. Chem.* **1879**, *22*, 289–330. [CrossRef]

23. Shannon, R.D. Revised effective ionic radii and systematic studies of interatomic distances in halides and chalcogenides. *Acta Crystallogr. A* **1976**, *32*, 751–767. [CrossRef]

24. Toby, B.H.; Von Dreele, R.B. GSAS-II: The Genesis of a Modern Open-Source All Purpose Crystallography Software Package. *J. Appl. Crystallogr.* **2013**, *46*, 544–549. [CrossRef]

25. Favre-Nicolin, V.; Černý, R. FOX, "Free Objects for Crystallography": A modular approach to ab initio structure determination from powder diffraction. *J Appl. Crystallogr.* **2002**, *35*, 734–743. [CrossRef]

26. Coelho, A.A. Whole-profile structure solution from powder diffraction data using simulated annealing. *J. Appl. Crystallogr.* **2000**, *33*, 899–908. [CrossRef]

27. Larson, A.C.; Von Dreele, R.B. *General Structure Analysis System*; Report LAUR 86–748; Los Alamos National Laboratory: Los Alamos, NM, USA, 1994.

28. Putz, H.; Brandenburg, K. (Eds.) *Diamond-Crystal and Molecular Structure Visualization*; Crystal Impact: Bonn, Germany.

29. Udovic, T.J.; Brown, C.M.; Leão, J.B.; Brand, P.C.; Jiggetts, R.D.; Zeitoun, R.; Pierce, T.A.; Peral, I.; Copley, J.R.D.; Huang, Q.; et al. The Design of a Bismuth-based Auxiliary Filter for the Removal of Spurious Background Scattering Associated with Filter-Analyzer Neutron Spectrometers. *Nucl. Instr. Meth. A* **2008**, *588*, 406–413. [CrossRef]

30. Copley, J.R.D.; Cook, J.C. The Disk Chopper Spectrometer at NIST: A New Instrument for Quasielastic Neutron Scattering Studies. *Chem. Phys.* **2003**, *292*, 477–485. [CrossRef]

31. Azuah, R.T.; Kneller, L.R.; Qiu, Y.; Tregenna-Piggott, P.L.W.; Brown, C.M.; Copley, J.R.D.; Dimeo, R.M. DAVE: A Comprehensive Software Suite for the Reduction, Visualization, and Analysis of Low Energy Neutron Spectroscopic Data. *J. Res. Natl. Inst. Stan.* **2009**, *114*, 341–358. [CrossRef]

32. Paul, R.L.; Sahin, D.; Cook, J.C.; Brocker, C.; Lindstrom, R.M.; O'Kelly, D.J. NGD Cold-Neutron Prompt Gamma-Ray Activation Analysis Spectrometer at NIST. *J. Radioanal. Nucl. Chem.* **2015**, *304*, 189–193. [CrossRef]

33. Giannozzi, P.; Baroni, S.; Bonini, N.; Calandra, M.; Car, R.; Cavazzoni, C.; Ceresoli, D.; Chiarotti, G.L.; Cococcioni, M.; Dabo, I.; et al. QUANTUM ESPRESSO: A Modular and Open-Source Software Project for Quantum Simulations of Materials. *J. Phys. Condens. Matter* **2009**, *21*, 395502. [CrossRef]

34. Kresse, G.; Furthmuller, J.; Hafner, J. Ab initio Force Constant Approach to Phonon Dispersion Relations of Diamond and Graphite. *Europhys. Lett.* **1995**, *32*, 729–734. [CrossRef]

35. Yildirim, T. Structure and Dynamics from Combined Neutron Scattering and First-Principles Studies. *Chem. Phys.* **2000**, *261*, 205–216. [CrossRef]

Article

Synthesis: Molecular Structure, Thermal-Calorimetric and Computational Analyses, of Three New Amine Borane Adducts

Kevin Turani-I-Belloto [1], Rodica Chiriac [2], François Toche [2], Eddy Petit [1], Pascal G. Yot [3], Johan G. Alauzun [3] and Umit B. Demirci [1,*]

1   Institut Europeen des Membranes, IEM–UMR 5635, ENSCM, CNRS, Universite de Montpellier, 34090 Montpellier, France
2   Laboratoire des Multimateriaux et Interfaces, UMR CNRS 5615, Université Claude Bernard Lyon 1, 69622 Villeurbanne, France
3   ICGM, Universite de Montpellier, CNRS, ENSCM, 34293 Montpellier, France
*   Correspondence: umit.demirci@umontpellier.fr

Abstract: Cyclopropylamine borane $C_3H_5NH_2BH_3$ (C3AB), 2-ethyl-1-hexylamine borane $CH_3(CH_2)_3$-$CH(C_2H_5)CH_2NH_2BH_3$ (C2C6AB) and didodecylamine borane $(C_{12}H_{25})_2NHBH_3$ ((C12)2AB) are three new amine borane adducts (ABAs). They are synthesized by reaction of the corresponding amines with a borane complex, the reaction being exothermic as shown by Calvet calorimetry. The successful synthesis of each has been demonstrated by FTIR, Raman and NMR. For instance, the ^{11}B NMR spectra show the presence of signals typical of the NBH_3 environment, thereby implying the formation of B–N bonds. The occurrence of dihydrogen bonds (DHBs) for each of the ABAs has been highlighted by DSC and FTIR, and supported by DFT calculations (via the Mulliken charges for example). When heated, the three ABAs behave differently: C3AB and C2C6AB decompose from 68 to 100 °C whereas (C12)2AB is relatively stable up to 173 °C. That means that these ABAs are not appropriate as hydrogen carriers, but the 'most' stable (C12)2AB could open perspectives for the synthesis of advanced materials.

Keywords: adduct; amine borane; boranes; boron chemistry; dihydrogen bonds

Citation: Turani-I-Belloto, K.; Chiriac, R.; Toche, F.; Petit, E.; Yot, P.G.; Alauzun, J.G.; Demirci, U.B. Synthesis: Molecular Structure, Thermal-Calorimetric and Computational Analyses, of Three New Amine Borane Adducts. *Molecules* **2023**, *28*, 1469. https://doi.org/10.3390/molecules28031469

Academic Editors: Michael A. Beckett and Igor B. Sivaev

Received: 3 January 2023
Revised: 30 January 2023
Accepted: 31 January 2023
Published: 3 February 2023

1. Introduction

Ammonia borane NH_3BH_3 and ethane C_2H_6 have one point in common—they are isoelectronic—otherwise they bear important differences, starting from their physical state. Ammonia borane is solid at ambient conditions whereas ethane is gaseous, because of dissimilarities in terms of electronegativity, partial charge of the hydrogen atoms, bond polarity and dipole moment [1]. The intermolecular interactions within the ammonia borane solid are driven by dihydrogen bonds (DHBs) that occur between the acidic hydrogen atoms $H^{\delta+}$ of the NH_3 group and the basic hydrogen atoms $H^{\delta-}$ of the BH_3 group [2]. Such DHBs are stronger than the van der Waals interactions that occur between the almost neutral hydrogen atoms H of CH_3 of the ethane molecules [3].

Ammonia borane is the 'lightest', the fully hydrogenated, representative of amine borane adducts (ABAs), and certainly the best known from the fact that it has been much investigated as a potential hydrogen carrier since the mid-2000s [4]. However, many other ABAs have been developed in parallel, not necessarily for being used as a hydrogen carrier, thus for various uses in chemistry. Examples are follows: ethylenediamine bisborane as hydrogen carrier [5]; diisopropylaminoborane for one-pot borylation [6]; morpholine borane as radical initiator [7]; pyridine borane as reducing agent [8]; ammonia borane as hydrogenation reagent [9]; 1,3,5-(*p*-aminophenyl)benzene-borane as monomer of borazine-linked polymer [10]; and, methylamine borane as precursor of boron-carbon-nitrogen layers [11]. Providing an exhaustive list of the ABAs reported so far and of their potential applications goes beyond the scope of the present article, and for further information, the reader is referred to appropriate review articles [12–20].

In the chemistry of the ABAs, DHBs arising between $H^{\delta+}$ of NH_x (x = 1, 2 or 3) of one molecule and $H^{\delta-}$ of BH_3 of another one play an important role [21]. The examples are as follows. With the aforementioned ammonia borane, the molecules have close intermolecular $N-H\cdots H-B$ contacts owing to DHBs, and resulting in this ABA being solid at ambient conditions [22,23] and that, under moderate heating (ca. 90–110 °C) conditions, $H^{\delta+}$ and $H^{\delta-}$ react and combine to release molecular hydrogen H_2 [24,25]. Diammoniate of diborane $[(NH_3)_2BH_2]^+[BH_4]^-$ is known as the reactive intermediate of ammonia borane [26], and on account of the existence of DHBs, it forms by transfer of $H^{\delta-}$ from BH_3 of one ammonia borane molecule to BH_3 of another one resulting in the formation of $[BH_4]^-$ and $[(NH_3)_2BH_2]^+$, and thus the formation of the ionic $[(NH_3)_2BH_2]^+[BH_4]^-$ [27]. α-Methylbenzylamine borane $C_6H_5CH(CH_3)NH_2BH_3$ is a chiral ABA that, in chloroform, exists as dimer owing to DHBs [28]. Dodecylamine borane $C_{12}H_{25}NH_2BH_3$, when solubilized in (anhydrous) tetrahydrofuran, forms stable core-shell aggregates consisting of five self-assembling, dihydrogen-bonded, stretched molecules [29]. Methylamine borane $CH_3NH_2BH_3$, in the presence of an Ir(III) pincer complex catalyst, dehydrocouples at 20 °C, the process mediated by DHBs and leading to the production of a polyaminoborane of high molecular weight ($M_w > 20,000$) [30]. It is thus clear that, with ABAs, DHBs are ubiquitous. However, Chen et al. [21] noticed that DHBs have not been used to their full potential yet, and they could be further developed in fields such as crystal engineering, self-assembly and synthesis of advanced materials.

Our current research belongs within the context described above. In a first stage, we have been synthesizing and acutely characterizing new ABAs. Recently, we released the results concerning eight alkylamine borane adducts $C_nH_{2n+1}NH_2BH_3$ (n = 4, 6, 8, 10, 12, 14, 16 or 18) [31,32]. All have shown higher melting points (by 20–40 °C) than the parent amine $C_nH_{2n+1}NH_2$, and such a gain of stability has been explained by the existence of DHBs. For the present study, we considered amines that are not based on monoalkylamines; consequently, we focused on three new ABAs such as cyclopropylamine borane $C_3H_5NH_2BH_3$ (C3AB) as a low molecular weight ABA and where the carbonaceous group is cyclic; 2-ethyl-1-hexylamine borane $CH_3(CH_2)_3CH(C_2H_5)CH_2NH_2BH_3$ (C2C6AB) where the carbonaceous chain is an isomer of the n-octyl chain of the $C_8H_{17}NH_2BH_3$ cited above, and didodecylamine borane $(C_{12}H_{25})_2NHBH_3$ ((C12)2AB) as the secondary amine analog of the aforementioned $C_{12}H_{25}NH_2BH_3$. These three new ABAs are synthesized by Lewis acid-base reaction using a commercial amine and a commercial borane complex (Figure 1). They were intensively analyzed by Calvet calorimetry, differential scanning calorimetry, thermogravimetric analysis, Fourier-transform infrared spectroscopy, Raman spectroscopy, ^1H and ^{11}B nuclear magnetic resonance spectroscopy and solid-state ^{11}B magic angle spinning magnetic resonance spectroscopy. Their molecular structures were also studied by density functional theory calculations. We had three objectives in mind: to show the occurrence of DHBs; to highlight the properties of ABAs; and to layout the beginnings for their prospect use.

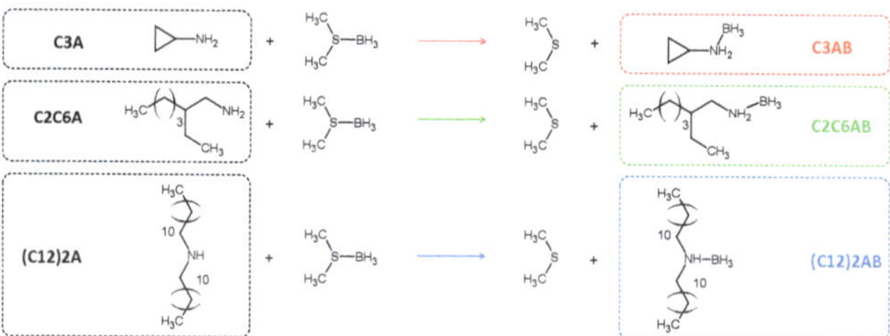

Figure 1. Reaction path for the synthesis of the ABAs in diethyl ether, under argon atmosphere and at ambient temperature: C3AB for cyclopropylamine borane C2C6AB for 2-ethyl-1-hexylamine borane, and (C12)2AB for didodecylamine borane.

2. Results and Discussion

2.1. Syntheses of the ABAs Driven by Calvet Calorimetry

The amines C3A, C2C6A or (C12)2A readily react with $(CH_3)_2S \cdot BH_3$. We determined the enthalpies of reaction by Calvet calorimetry (Figure 2). They are of -40.8, -39 and -56.3 kJ mol^{-1} for the syntheses of C3AB, C2C6AB and (C12)2AB, respectively. These enthalpies confirm the exothermic nature of the reaction, as observed during the preparation in the glove box.

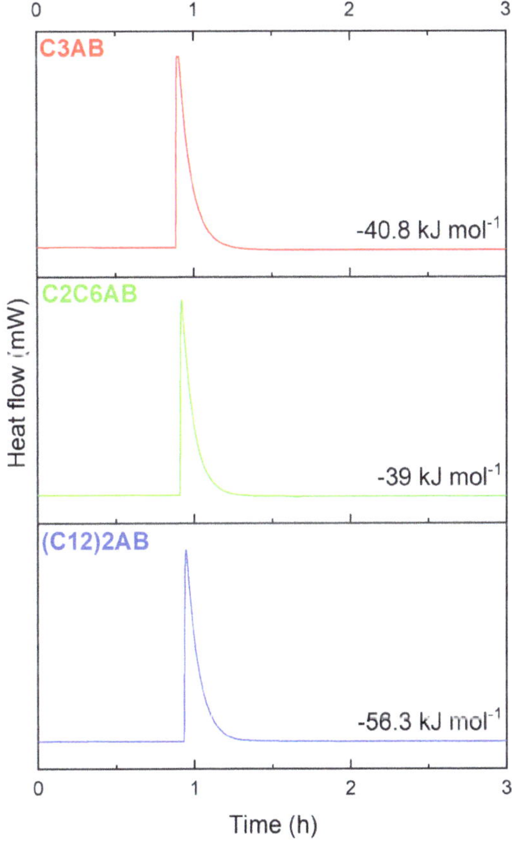

Figure 2. Determination of the enthalpy of reaction by Calvet calorimetry for C3AB, C2C6AB and (C12)2AB.

At room temperature and under argon atmosphere, C3AB is a pasty solid (X-ray diffraction pattern, not reported in this work, is characteristic of an amorphous solid). We determined the onset temperature of the melting event with the help of DSC (Figure 3). When heated from subzero temperatures up to 50–100 °C, an endothermic signal is observed, and it is counterbalanced by an exothermic signal on cooling. This evidences the melting and solidifying of C3AB. The onset temperature of the melting event is 39.9 °C. It is much higher (ΔT of ca. 85 °C) than the melting point of the amine reactant (-45 °C). This indicates additional intermolecular interactions in the ABA, that is, DHBs [21].

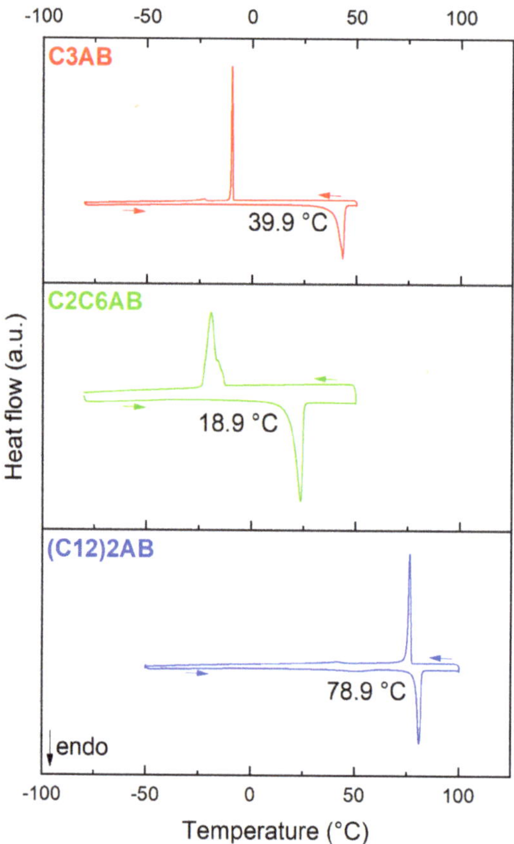

Figure 3. DSC curves of C3AB, C2C6AB and (C12)2AB with the onset temperature of the melting event mentioned.

C2C6AB is an oily liquid at room temperature. The onset temperature of its melting is 18.9 °C (Figure 3). It is much higher (ΔT of ca. 95 °C) than the melting point of the amine reactant (-76 °C), which is evidence of DHBs occurring between the C2C6AB molecules. It is worth mentioning that, at room temperature, $C_8H_{17}NH_2BH_3$, the n-alkyl analog of C2C6AB, is solid. The melting point of $C_8H_{17}NH_2BH_3$ is higher, with 34.7 °C [29], which can be explained by a lower symmetry (because of the ethyl side group) of the C2C6AB molecule, as is the case with the alkane analogs 2-ethyl-hexane (m.p. -118 °C) and n-octane (-95 °C) [33].

(C12)2AB is a powdery lowly crystalline solid (Figure S1 (Supplementary Materials)). At first glance and with the quality of the pattern, (C12)2AB was found to crystalize into the monoclinic system. The following lattice parameters were determined: $a = 5.260(1)$ Å, $b = 54.232(5)$ Å, $c = 4.621(5)$ Å and $\beta = 109.17(1)°$. The space group is possibly the $P2_1$ (No. 4) one. (C12)2A would then be isostructural with $C_8H_{17}NH_2BH_3$ [31]. The onset temperature of the melting event for (C12)2AB is 78.9 °C (Figure 3), and it is much higher (ΔT of ca. 51 °C) than that of the amine (C12)2A (28 °C). This is due to the occurrence of DHBs between the ABA molecules.

2.2. Molecular Characterization of C3AB

C3AB was analyzed by FTIR spectroscopy. The spectrum was compared to that of the amine reactant C3A (Figure 4). There are differences. The spectra were analyzed and the bands were assigned with the help of the references [34–37]. Bands due to B−H have appeared at 2600–2100 cm^{-1} (stretching) and 1300–1100 cm^{-1} (deformation). The N−H stretching bands (3350–3100 cm^{-1}) are more intense and sharper, and they are red-shifted, which is in good agreement with the formation of DHBs between the NH$_2$ and BH$_3$ groups. Similar changes are observed for the N−H deformation bands (around 1600 and 800 cm^{-1}). The bands within the ranges 1500–1300 (C−H deformation for example) and 1100–850 cm^{-1} (C−N and C−C stretching for example) are greater and sharper [37]. The B−N bond is featured by a stretching vibration at about 700 cm^{-1}. Otherwise, the spectrum of C3AB favorably compares with the one we predicted by DFT (Figure S2).

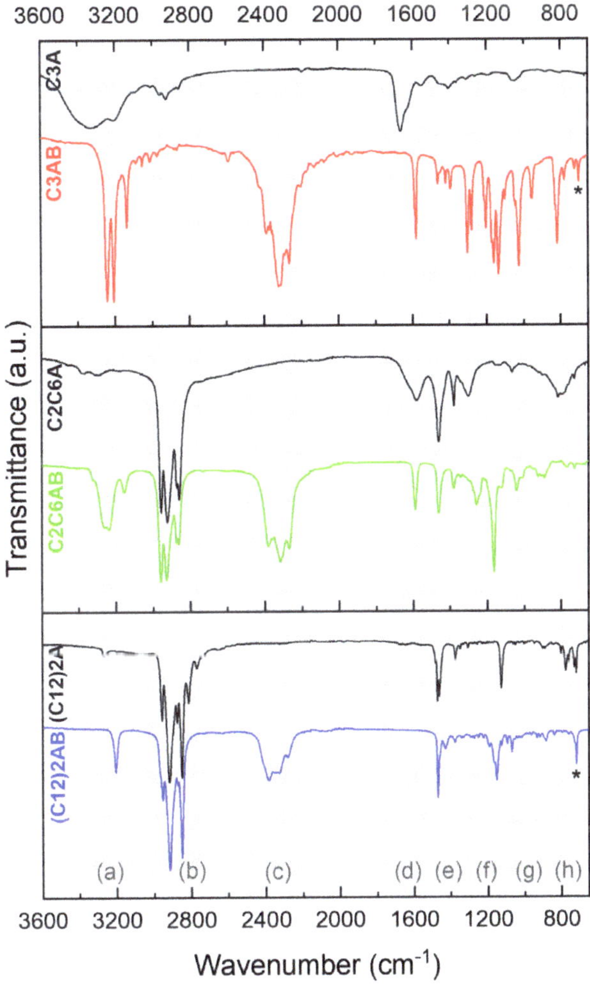

Figure 4. FTIR spectra of C3AB, C2C6AB and (C12)2AB. The spectra of the starting amines C3A, C2C6A and (C12)2A are given for comparison. The bands are attributed to the corresponding vibration modes as follows: (a) N−H stretching; (b) C−H stretching; (c) B−H stretching; (d) N−H deformation; (e) C−H and N−H deformation; (f) B−H deformation; (g) C−N and C−C stretching and C−H deformation; (h) N−H deformation; star for B−N.

C3AB was also analyzed by Raman spectroscopy (experimental in Figure 5 and predicted in Figure S3). The spectrum was exploited with the help of the reference [38]. The Raman results (experimental and predicted) are in line with the FTIR obtained, and both allow us to conclude the successful production of C3AB occurred.

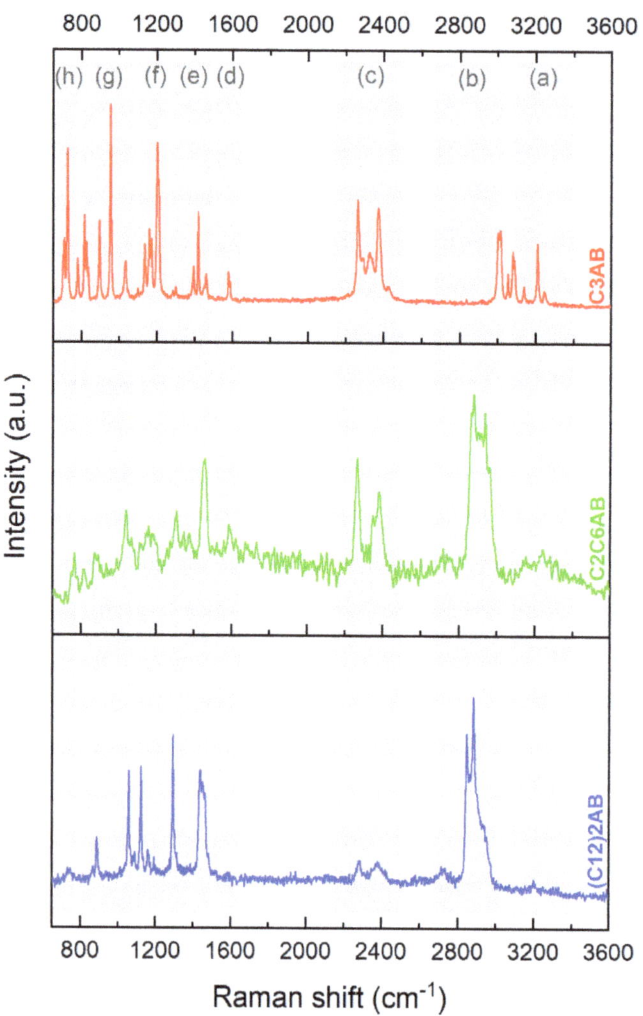

Figure 5. Raman spectra of C3AB, C2C6AB and (C12)2AB. The bands are assigned: (a) N−H stretching; (b) C−H stretching; (c) B−H stretching; (d) N−H deformation; (e) C−H deformation; (f) B−H deformation; (g) C−N and C−C stretching and C−H deformation; (h) B−N stretching.

C3AB was analyzed by NMR spectroscopy. The ^{1}H NMR spectrum (Figure S4) shows four remarkable signals, in good agreement with the molecular structure of C3AB and the chemical shifts predicted by DFT calculations (Figure S5): a quartet of normalized intensity 1:1.1:1.1:1 at 1–1.7 ppm ($^1J_{B-H}$ of 95.6 Hz) typical of BH_3; a signal at 4.05 ppm due to the 2 H of NH_2; a multiplet at 0.61 ppm due to the 4 H of the cyclopropyl's CH_2 groups; and, a septuplet at 2.25 ppm due to the H of $CHNH_2$. Based on these data, the purity of C3AB was found to be ≥98%.

The ^{11}B NMR spectrum (Figure 6) shows a quartet at −20.4 ppm, of normalized intensity 1:2.5:2.5:1, and with a coupling constant $^{1}J_{B-H}$ of 94.1 Hz [39,40]. It is favorably due to the NBH$_3$ environment of C3AB.

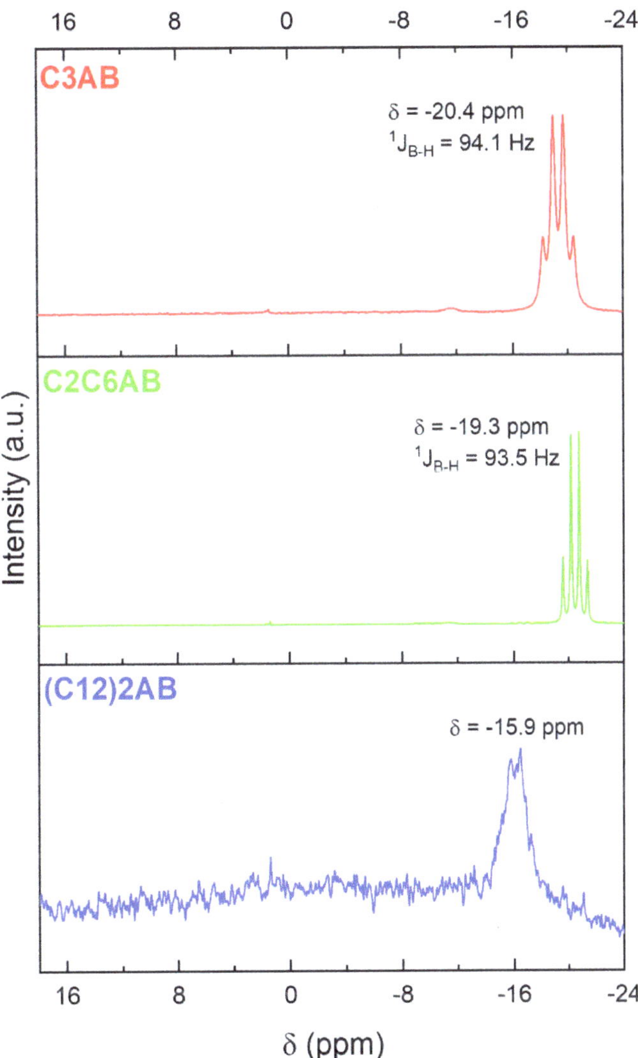

Figure 6. ^{11}B NMR spectra of C3AB, C2C6AB and (C12)2AB (dissolved in CD$_3$CN). The very small signal at 1.4 ppm is due to hydrolysis of the ABAs because of traces of water in CD$_3$CN. The very small signal at −11.5 ppm for C3AB is ascribed XBH$_3$ (with X = OH$^-$ from H$_2$O, CH$_3$CN from CD$_3$CN or (CH$_3$)$_2$S) (i.e., the borane used as reactant).

The ^{11}B MAS NMR spectrum of C3AB is typical of the spectrum of an ABA (Figure 7), showing a double-horned peak due to NBH$_3$ [41]. The signal is centered −23.5 ppm. There are other signals of much smaller intensity (at −16.9, 0.4 and 5–15 ppm). They are attributed to the N$_2$BH$_2$, BO$_x$, and B(III) environments [42,43]. They are likely to be due to dehydrocoupling and hydrolysis of C3AB, but to a very small extent. The former reaction may have taken place during analysis and rotor rotation, and the latter one because of a slight moisture contamination (when transferring the rotor from the lab to the NMR apparatus located in another building) given that ABAs are moisture sensitive [29].

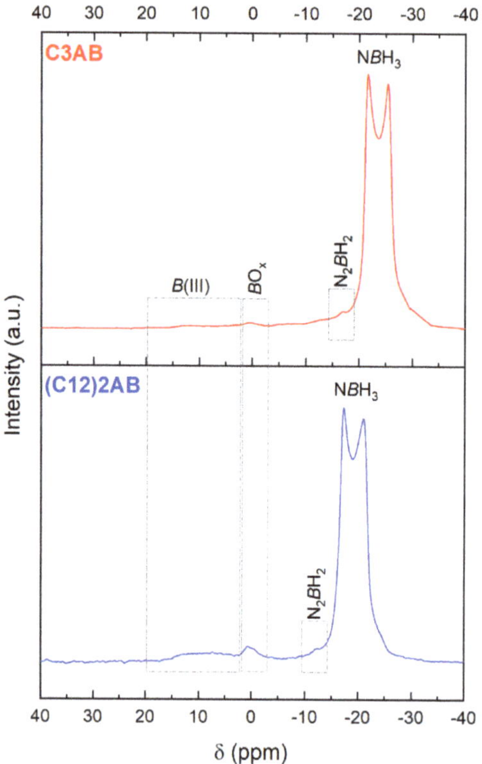

Figure 7. ^{11}B MAS NMR of the ABAs in solid state, i.e., of C3AB and (C12)2AB. The signals are assigned, and discussed in the main text.

2.3. Molecular Characterization of C2C6AB

The FTIR spectrum of C2C6AB (Figure 4) shows [34–37]: (i) the vibration modes of the B−H bonds (stretching between 2600 and 2100 cm^{-1} and deformation at 1200–1100 cm^{-1}); (ii) red shifted and more intense bands for the N−H stretching (3400–3100 cm^{-1}); (iii) less intense bands for the N−H deformation (around 800 cm^{-1}); and (iv) the vibration modes of the C−H, C−C and C−N bonds as for the amine C2C6A. The B−N stretching band cannot be properly assigned as overlapping the N−H deformation bands. The spectrum favorably compares to the predicted one (Figure S6).

C2C6AB was also analyzed by Raman spectroscopy (experimental in Figure 5 and predicted in Figure S7) [38]. The results are in line with the FTIR ones. These observations allow us to conclude the successful production of C2C6AB occurred.

C2C6AB was analyzed by NMR spectroscopy. The ^{1}H NMR spectrum (Figure S8) shows eight signals in good agreement with the molecular structure of the analyzed ABA and the chemical shifts predicted by DFT calculations (Figure S9): namely, a multiplet at 0.8–1.8 ppm due to the 3 H of BH₃; a triplet at 0.92 ppm due to the 3 H of CH₃ of the main alkyl chain that is partly overlapping a triplet at 0.88 ppm due to the 3 H of CH₃ of the ethyl chain; a singlet of high intensity at 1.28 ppm due to the CH₂ groups of the hexyl chain between CH₃ and the beta CH₂; a multiplet at 1.34 ppm due to the CH₂ of the ethyl chain; a multiplet at 1.56 ppm due to the H of CHCH₂N; a quintet at 2.57 ppm due to the 2 H of

CH_2N; a broad singlet of low intensity at 3.83 ppm due to the H of NH_2. Based on these data, the purity of C2C6AB was found to be $\geq$99%.

The ^{11}B NMR spectrum (Figure 6) shows the quartet due to NBH_3 of the C2C6AB [39,40]. It is located at -19.3 ppm, is of normalized intensity 1:2.6:2.6:1 and has a coupling constant $^{1}J_{B-H}$ of 93.5 Hz.

2.4. Molecular Characterization of (C12)2AB

(C12)2AB was analyzed by FTIR spectroscopy (Figure 4), and its spectrum favorably compares to the predicted spectrum (Figure S10). As for the aforementioned ABAs, there are differences between the spectrum of (C12)2AB and that of (C12)2A. The main features are as follows [34–37]. The N–H stretching band has red-shifted with the addition of the BH_3 group by 60 cm^{-1} (3207 cm^{-1} for (C12)2AB and 3267 cm^{-1} for the starting amine). Another remarkable change is for the N–H deformation mode between 800 and 700 cm^{-1}: the number of bands has decreased with the addition of BH_3. This can be explained by the presence of DHBs that impose constraints on the $N-H$ bond [35]. The band labelled by a star belongs to the $B-N$ bond.

Similar observations can be made from the Raman spectrum (experimental in Figure 5 and predicted in Figure S11). These results confirm the successful production of (C12)2AB.

(C12)2AB was analyzed by NMR spectroscopy. The ^{1}H NMR spectrum (Figure S12) shows six signals: a multiplet at 0.5–2 ppm due to the 3 H of BH_3; a triplet at 0.91 ppm due to the 6 H of the two CH_3; a singlet of high intensity at 1.31 ppm due to the eighteen CH_2 groups between CH_3 and the beta CH_2; a multiplet at 1.56 ppm due to the 4 H of $[CH_2CH_2]_2N$; a quintet at 2.61 ppm due to the 4 H of $[CH_2]_2N$; a broad singlet of low intensity at 3.89 ppm due to the H of NH. The shifts predicted by DFT calculations (Figure S13) are in line with the experimental data. Based on these data, the purity of (C12)2AB was found to be $\geq$98%.

The ^{11}B NMR spectrum (Figure 6) was systematically recorded with a low resolution because (C12)2AB is lowly soluble in CD_3CN. Nevertheless, the spectrum shows one main signal which shape suggests a quartet and that is centered at -15.9 ppm. It is due to NBH_3 of the ABA [44].

The ^{11}B MAS NMR spectrum (Figure 7) shows the two-horned peak at -19 ppm featuring NBH_3 [41]. Like for C3AB, the spectrum shows additional signals of small intensity indicating evolution of (C12)2AB to a small extent (possibly due to dehydrocoupling and hydrolysis because of contamination with moisture).

2.5. Further Molecular Analyses by DFT Calculations

The total energy of each of the ABA molecules was determined by DFT calculations. The C3AB molecule has a total energy of -199.97 Hartree. The total energy of the C2C6AB is -397.84 Hartree. It is, in absolute value, higher, which is due to the additional carbon and hydrogen atoms. With respect to the heavier (C12)2AB molecule, the total energy is -1027.04 Hartree.

The DFT calculations allowed extracting the Mulliken charges of the atoms of each of the ABA molecules (Figures S14–S16). The Mulliken charge of the elements B, N, H of BH_3, H of NH or NH_2, alpha C and beta C are listed in Table 1. The charges of the hydrogens of the BH_3 and NH_2/NH groups are, respectively, negative and positive, thereby confirming the presence of $H^{\delta+}$ and $H^{\delta-}$ and the occurrence of DHBs between the ABA molecules. The B element is almost neutral for the C3AB and C2C6AB molecules, whereas it has a positive charge in the (C12)2AB molecule. In contrast, the N element has a negative charge in each of the molecules. It is also interesting to mention that the charges of the alpha and beta C elements of the three ABAs are not comparable. For the C3AB molecule, the alpha C element is positively charged and the beta one negatively, but for the C2C6 molecule, it is the opposite and the charges (in absolute value) are bigger. Concerning the (C12)2AB molecule that has two C12 alkyl chains, the two alpha C elements are negatively charged while one of the beta C element has a positive charge and the other one is almost neutral.

These differences of charges are also illustrated by the mapped electrostatic potentials (Figures S17–S19).

Table 1. Mulliken charges of the elements B, N, H of BH_3, H of NH_2 or NH_2, alpha C and beta C, for the molecules C3AB, C2C6AB and (C12)2AB.

	C3AB	C2C6AB	(C12)2AB
B	−0.028	0.049	0.243
N	−0.449	−0.402	−0.278
H of BH_3	−0.092 to −0.086	−0.114 to −0.091	−0.131 to −0.109
H of NH_2 or NH	0.278 and 0.297	0.293 and 0.306	0.278
alpha C	0.156	−0.632	−0.496 and −0.743
beta C	−0.283	0.861	0.298 and −0.017

The distribution of the highest occupied molecular orbital (HOMO) and that of the lowest unoccupied molecular orbital (LUMO), over the three ABA molecules, were plotted (Figure 8). They are in line with the observations made above about the Mulliken charges. Indeed, the HOMO of the C3AB molecule is mainly localized on the N element as well as on the two beta C elements, and the LUMO is localized on the B element. With respect to the C2C6AB molecule, the HOMO is localized on the hydrogens of the BH_3 as well as, in a lesser extent, on the C−C−N bonds, and the LUMO is mainly localized on the B−N bond. Finally, the plots obtained for the (C12)2AB indicate a HOMO localized on the BH_3 group, with a smaller contribution on the N−C bonds, and a LUMO localized onto the C–C–C–C bonds of one alkyl chain bound to N. These observations confirm that the $NHBH_3$ and NH_2BH_3 groups should be reactive, which is typical of the ABAs [19,45,46], and suggest that the alpha and beta C elements, but not only these carbons in the case of (C12)2AB, are likely to be reactive.

2.6. Thermal Properties of the ABAs

Under heating, C3AB decomposes from 68 °C (Figures S20 and S21). Hydrogen is released in two steps over the temperature range 68–200 °C. At about 100 °C, C3AB shows some decomposition with the release and detection of C3A (Figure S22), indicating a breaking of the B–N bond. This could be related to the Mulliken charges of the BN and N elements as well as the localization of the HOMO mainly onto N. Above 150 °C, some NH_3 is released. The weight loss at 250 °C is of about 85 wt.%, which is higher than the hydrogen content of NH_2BH_3 (16.7 wt.%) of C3AB. In other words, pristine C3AB decomposes more than it dehydrogenates.

C2C6AB starts to dehydrogenate from 100.3 °C (Figures S23 and S24). The dehydrogenation is stepwise, with two events peaking at 128 and 179 °C. The weight loss at 200 °C is 8.2 wt.%, which is higher than the 3.5 wt.% of the H of the NH_2BH_3 group, thereby suggesting that C2C6AB decomposes more than it dehydrogenates. The main decomposition takes place between 230 and 500 °C. The total weight loss is 97.3 wt.% at 600 °C, because of the release of ammonia together with hydrocarbon fragments (methane and unsaturated C4–C7 species). This decomposition behavior is comparable to that of the previously reported $C_8H_{17}NH_2BH_3$ [31].

(C12)2AB starts to dehydrogenate from about 78 °C but the main decomposition shows an onset temperature of at 173.6 °C (Figures S25 and S26). Up to about 200 °C, the dehydrogenation is featured by two peaks at 154 and 193 °C, and the second dehydrogenation step is concomitant with the decomposition of the alkyl chain. At about 180 °C, (C12)2AB has lost 0.3 wt.%, which corresponds to the release of 0.5 equiv. H_2. Above 180 °C and up to about 550 °C, (C12)2AB predominantly decomposes though H_2 is released stepwise (with events peaking at 303, 359, 410 and 474 °C). A number of saturated and unsaturated C4 to C12 fragments (as well as methane), C13 and C14 hydrocarbons due to some fragments combinations and dodecylmethylamine were detected by microGC-MS and GC-MS (Table S1). At 600 °C, the weight loss is 93.3 wt.%, The remaining 6.7 wt.% well matches with the content of B and N (6.7 wt.%) in (C12)2AB, indicating the formation of a boron nitride-based solid.

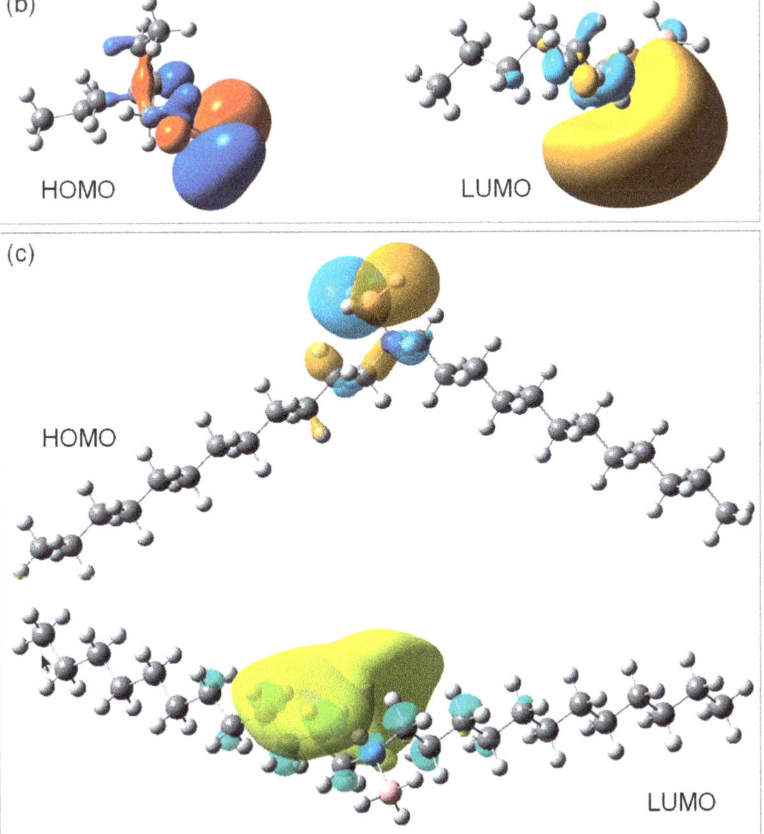

Figure 8. HOMO and LUMO of the (**a**) C3AB, (**b**) C2C6AB and (**c**) (C12)2AB molecules, as predicted by DFT calculations.

In the same way as pristine ammonia borane, C3AB and C2C6AB dehydrogenate first, but they then decompose over the temperature range 100–250 °C. Thus, both, in pristine state, cannot be considered as potential hydrogen carriers. They might however have a prospect in this field if combined to an amine, as performed elsewhere [5]. With respect to (C12)2AB, it is more stable. It dehydrogenates from about 100 °C, but the amount of H_2 released up to 180 °C remains low with 0.5 equiv. H_2 (over a maximum of 2.5 equiv. H_2 for the NH_2BH_3 group). Such properties are not compatible with those expected for chemical hydrogen storage.

The relative thermal stability of (C12)2AB below 200 °C is interesting when considering synthesis of advanced materials. For instance, polymers using (C12)2AB as monomer could be produced by dehydrocoupling, and such polymers could be used as precursors to produce BCN ceramics by pyrolysis. Another example would be to consider (C12)2AB as a possible soft template to produce nanostructured materials by self-assembly. These two possible applications might also be considered for C3AB and C2C6AB, but one has to keep in mind that they are likely to decompose from 50 to 60 °C.

3. Materials and Methods

The reactants, all from Merck, were cyclopropylamine $C_3H_5NH_2$ (98%; 57.09 g mol^{-1}; m.p. −45 °C; denoted C3A), 2-ethyl-1-hexylamine $CH_3(CH_2)_3CH(C_2H_5)CH_2NH_2$ (98%; 129.24 g mol^{-1}; m.p. −76 °C; denoted C2C6A), didodecylamine $(C_{12}H_{25})_2NH$ (≥97%; 353.67 g mol^{-1}; m.p. 28 °C; denoted (C12)2A), and borane dimethyl sulfide $(CH_3)_2S \cdot BH_3$ (5.0 M in diethyl ether; 75.97 g mol^{-1}). The solvent, from Merck also, was anhydrous diethyl ether (≥99.7%). They were all used as received.

The ABAs were synthesized by substitution of $(CH_3)_2S$ of the borane complex by one of the amines (Figure 1). Typically, 500 mg of an amine was dissolved in 8 mL of diethyl ether, under stirring (500 rpm) for 1 h, at ambient temperature and under argon atmosphere (glove box MBraun M200B, O_2 < 0.1 ppm, H_2O < 0.1 ppm). A slight excess of $(CH_3)_2S \cdot BH_3$ (1.1 mol versus 1 mol of amine) was added dropwise. Upon a 24-h stirring, the diethyl ether solvent (b.p. 34.6 °C) and the only reaction product $(CH_3)_2S$ (b.p. 37.3 °C) were extracted by vacuum distillation at 0 °C in 2 h. In doing so, cyclopropylamine borane $C_3H_5NH_2BH_3$ (70.92 g mol^{-1}; denoted C3AB), 2-ethyl-1-hexylamine borane $CH_3(CH_2)_3CH(C_2H_5)CH_2NH_2BH_3$ (143.07 g mol^{-1}; denoted C2C6AB) and didodecylamine borane $(C_{12}H_{25})_2NHBH_3$ (367.5 g mol^{-1}; denoted (C12)2AB) were produced.

Calorimetric and thermal analyses were performed using the following techniques. Calvet calorimetry (the C80 model from Setaram) was used to calculate the enthalpy of reaction for the three ABAs from the heat flow monitored against time. The calorimeter uses a reversal stainless-steel hermetic mixing cell with two separated chambers, used for pressures up to 5 bar. In the glove box, the cell was filled so that the 2.5-mL chamber contained the amine solution and the 2-mL chamber contained the borane complex. A reference cell that was kept empty was also used. Both cells were inserted in the calorimeter, and the reaction temperature was fixed at 28 °C. The reactants were mixed by turning the calorimeter, and thus the cells, upside down several times with a rotational motion. The motion was stopped when the reaction peak reached its maximum. The heat flow was monitored against time. It allows for the calculation of the enthalpy of reaction. Differential scanning calorimetry (DSC 1 from Mettler-Toledo) was used for determining the melting points and enthalpies of fusion of the ABA. Indium and mercury were used as standards to calibrate the temperature. Indium and zinc were used to calibrate the enthalpy. This resulted in errors of <1%. Aluminum crucibles (40 μL) were used. They are sealable, which allows preventing the ABAs from air contamination. Typically, the samples to analyze (4 to 7 mg) were prepared in the glove box, and the sealed crucibles were transferred into the DSC oven. The heating rate was 5 °C min^{-1}. The nitrogen flow rate was 30 mL min^{-1}. A full cycle consisting of a heating step and a cooling step was performed. Thermogravimetric (TG) analysis (TGA/DSC2 from Mettler Toledo) was used for studying the thermal stability and decomposition of the ABA. The analysis conditions were as follows: 15 to 20 mg of ABA; aluminum crucible (100 μL) having a pierced lid; heating of 5 °C min^{-1}; and nitrogen flow rate of 30 mL min^{-1}. The analyzer was coupled to gas chromatography (GC) and mass spectrometry (MS) detector (7890B GC/5977A MS from Agilent, Les Ulis, France). Two types of couplings can be used: TGA/microGC-MS (SRA Instruments, Marcy l'Etoile, France) to follow hydrogen and other small molecules, and TGA/Storage-Interface/GC-MS for heavier volatile products.

Molecular analyses were performed using the following techniques: Fourier-transform infrared spectroscopy (FTIR; IS50 Thermo Fisher Scientific; from 4000 to 650 cm^{-1}; 64 scans; resolution of 4 cm^{-1}); Raman spectroscopy (Horiba Jobin Yvon LabRAM 1B; laser Ar/Kr 100 mW 647.1 nm); ^{1}H nuclear magnetic resonance spectroscopy (^{1}H NMR; Bruker Avance-400 NMR; BBOF probe; CD3CN; 5-mm NMR tube); ^{11}B NMR spectroscopy (Bruker AVANCE-400; probe head BBFO; CD3CN; 5-mm tube; 128.378 MHz); and solid-state ^{11}B magic angle spinning (MAS) NMR spectroscopy (^{11}B MAS NMR; Varian VNMR4000; 128.378 MHz).

For the lowly crystalline (C12)2AB, the lattice parameters were refined by LeBail refinement from diffraction patterns collected at room temperature on a PANalytical X'PERT Pro multipurpose diffractometer (Cu-K$_{\alpha 1}$ radiation, λ = 1.54059 Å, 45 kV and 40 mA) equipped with an X'Celerator detector and using Scherrer geometry. The acquisition time was about 10 h. The corresponding powders were loaded into 0.5 mm borosilicate glass capillary tubes in an argon-filled glove box (Jacomex PBOX; O_2 < 1 ppm, and H_2O < 2 ppm), and sealed to prevent the samples from moist air contamination.

The molecular structures of C3AB, C2C6AB and (C12)2AB were also studied by density functional theory (DFT) calculations. A gas phase geometry optimization was performed using the DFT/B3LYP method with the 6–311++G (2d, p) basis set available in the Gaussian16 program, which is a good compromise between accuracy and cost. The optimized conformers were calculated at 298.15 K. The FTIR and Raman spectra, as well as the NMR shifts, were simulated and predicted. The Mulliken charges, the electrostatic potentials, the HOMO and the LUMO were calculated.

4. Conclusions

The reactions between the selected amines and the borane dimethyl sulfide complex are exothermic (thus spontaneous), as evidenced by Calvet calorimetry. This allowed us to produce three new ABAs, i.e., C3AB, C2C6AB and (C12)2AB, each of them being pure. The formation of a B–N between N of the amine and B of the borane has been confirmed by NMR, FTIR and Raman spectroscopy. For instance, the ^{11}B NMR spectra show signals belonging to the NBH$_3$ environment that is typical of ABAs.

ABAs are molecules of interest as they are able to interact with each other owing to DHBs. The onset melting temperature of each ABA has been determined by DSC. The values are much higher than the onset temperature of the amine reactants. This is typical of the occurrence of additional intermolecular interactions, that is, of DHBs. By FTIR spectroscopy, a red shift of the N–H stretching bands can be observed when the spectra of the amines and ABAs are compared. This is in good agreement with the existence of DHBs. DFT calculations have given, for each ABA molecule, the Mulliken charges, the electrostatic potentials, the HOMO and the LUMO, and all of these data suggest the occurrence of DHBs between the molecules.

Under heating, C3AB and C2C6AB start to dehydrogenate from 68 to 100 °C and decompose mainly above 100 °C. Unlike these two ABAs, (C12)2AB is more stable, decomposing from 173 °C. With such thermal behaviors, none of these ABAs appears to be appropriate as hydrogen carrier. However, the relative thermal stability of (C12)2AB below 200 °C may open up perspectives for the synthesis of advanced materials.

Supplementary Materials: The following supporting information can be downloaded at: https://www.mdpi.com/article/10.3390/molecules28031469/s1, Figure S1. Observed (black line) and calculated (red line) powder XRD profile for the LeBail refinement of (C12)2AB; Figure S2. FTIR spectra of C3AB: predicted versus experimental; Figure S3. Raman spectra of C3AB: predicted versus experimental; Figure S4. ^{1}H NMR spectrum of C3AB (in CD$_3$CN); Figure S5. Simulated ^{1}H NMR chemical shifts for C3AB; Figure S6. FTIR spectra of C2C6AB: predicted versus experimental; Figure S7. Raman spectra of C2C6AB: predicted versus experimental; Figure S8. ^{1}H NMR spectrum of C2C6AB (in CD$_3$CN); Figure S9. Simulated ^{1}H NMR chemical shifts for C2C6AB; Figure S10. FTIR spectra of (C12)2AB: predicted versus experimental; Figure S11. Raman spectra of (C12)2AB: predicted versus experimental; Figure S12. ^{1}H NMR spectrum of (C12)2AB (in CD$_3$CN); Figure

S13. Simulated ^{1}H NMR chemical shifts for (C12)2AB; Figure S14. C3AB: partial charges extracted from Gaussian calculations following the Mulliken scheme; Figure S15. C2C6AB: partial charges extracted from Gaussian calculations following the Mulliken scheme; Figure S16. (C12)2AB: partial charges extracted from Gaussian calculations following the Mulliken scheme; Figure S17. Electrostatic potentials of the C3AB molecule; Figure S18. Electrostatic potentials of the C2C6AB molecule; Figure S19. (Electrostatic potentials of the (C12)2AB molecule; Figure S20. TG curve of C3AB superimposed with the evolution of H_2.as obtained by TGA/microGC-MS; Figure S21. TG (left axis) and DTG (right axis) curves of C3AB; Figure S22. Thermal decomposition of C3AB: evolution of the peak area of the volatile products detected by TGA/microGC-MS coupling as a function of the sampling temperature; Figure S23. TG curve of C2C6AB, superimposed with the evolution of H2 as obtained by TGA/microGC-MS; Figure S24. TG (left axis) and DTG (right axis) curves of C2C6AB; Figure S25. TG curve of (C12)2AB, superimposed with the evolution of H2 as obtained by TGA/microGC-MS; Figure S26. TG (left axis) and DTG (right axis) curves of (C12)2AB; Tableau S1. Volatile organic compounds obtained by coupling TGA with micro GC-MS and GC-MS during the decomposition of (C12)2AB.

Author Contributions: Conceptualization, U.B.D. and J.G.A.; methodology, all of the authors; validation, all of the authors; formal analysis, K.T.-I.-B., R.C., F.T., E.P. and P.G.Y.; investigation, K.T.-I.-B.; writing—original draft preparation, U.B.D.; writing—review and editing, R.C., E.P., P.G.Y., J.G.A. and U.B.D.; supervision, J.G.A. and U.B.D.; project administration, U.B.D.; funding acquisition, U.B.D. All authors have read and agreed to the published version of the manuscript.

Funding: This research was funded by the AGENCE NATIONALE DE LA RECHERCHE, grant number ANR-18-CE05-0032.

Institutional Review Board Statement: Not applicable.

Informed Consent Statement: Not applicable.

Data Availability Statement: Data supporting reported results may be available on demand (to the corresponding author).

Conflicts of Interest: The authors declare no conflict of interest.

Sample Availability: Samples of the compounds are available from U.B.D., on request for sale on purchase order (after contracting with the CNRS Occitanie Est or the University of Montpellier).

References

1. Demirci, U.B. Ammonia borane, a material with exceptional properties for chemical hydrogen storage. *Int. J. Hydrogen Energy* **2017**, *42*, 9978–10013. [CrossRef]
2. Mitoraj, M.P. Bonding in ammonia borane: An analysis based on the natural orbitals for chemical valence and the extended transition state method (ETS-NOCV). *J. Phys. Chem. A* **2011**, *115*, 14708–17716. [CrossRef] [PubMed]
3. Bartell, L.S. On the effects of intramolecular van der Waals forces. *J. Chem. Phys.* **1960**, *32*, 827–831. [CrossRef]
4. Staubitz, A.; Robertson, A.P.M.; Sloan, M.E.; Manners, I. Amine- and phosphine-borane adducts: New interest in old molecules. *Chem. Rev.* **2010**, *110*, 4023–4078. [CrossRef] [PubMed]
5. Zhang, G.; Morrison, D.; Bao, G.; Yu, H.; Yoon, C.W.; Song, T.; Lee, J.; Ung, A.T.; Huang, Z. An amine-borane system featuring room-temperature dehydrogenation and regeneration. *Angew. Chem. Int. Ed.* **2021**, *60*, 11725–11729. [CrossRef]
6. Ramachandran, P.V.; Hamann, H.J.; Mishra, S. Aminoboranes via tandem iodination/dehydroiodination for one-pot borylation. *ACS Omega* **2022**, *7*, 14377–14389. [CrossRef]
7. Liautard, V.; Delgado, M.; Colin, B.; Chabaud, L.; Michaud, G.; Pucheault, M. In situ generation of radical initiators using amine-borane complexes for carbohalogenation of alkenes. *Chem. Commun.* **2022**, *58*, 2124–2127. [CrossRef]
8. Gurram, S.; Srivastava, G.; Badve, V.; Nandre, V.; Gundu, S.; Doshi, P. Pyridine borane as alternative reducing agent to sodium cyanoborohydride for PEGylation of L-asparaginase. *Appl. Biochem. Biotechnol.* **2022**, *194*, 827–847. [CrossRef]
9. Guo, X.; Unglaube, F.; Kragl, U.; Mejia, E. B(C_6F_5)$_3$-Catalyzed transfer hydrogenation of esters and organic carbonates towards alcohols with ammonia borane. *Chem. Commun.* **2022**, *58*, 6144–6147. [CrossRef]
10. Jackson, K.T.; Reich, T.E.; El-Kaderi, H.M. Targeted synthesis of a porous borazine-linked covalent organic framework. *Chem. Commun.* **2012**, *48*, 8823–8825. [CrossRef]
11. Leardini, F.; Flores, E.; Galvis, A.R.; Ferrer, I.J.; Ares, J.R.; Sánchez, C.; Molina, P.; van der Meulen, H.P.; Navarro, C.G.; López Polin, G.; et al. Chemical vapor deposition growth of boron–carbon–nitrogen layers from methylamine borane thermolysis products. *Nanotechnology* **2018**, *29*, 025603. [CrossRef]
12. Hutchins, R.O.; Learn, K.; Nazer, B.; Pytlewski, D.; Pelter, A. Amine boranes as selective reducing and hydroborating agents. A review. *Org. Prep. Proced. Int.* **1984**, *16*, 335–372. [CrossRef]

13. Burnham, B.S. Synthesis and pharmacological activities of amine-boranes. *Curr. Med. Chem.* **2005**, *12*, 1995–2010. [CrossRef]
14. Kalidindi, S.B.; Sanyal, U.; Jagirdar, B.R. Chemical synthesis of metal nanoparticles using amine-boranes. *Chem. Sus. Chem.* **2011**, *4*, 317–324. [CrossRef]
15. Rossin, A.; Peruzzini, M. Ammonia-borane and amine-borane dehydrogenation mediated by complex metal hydrides. *Chem. Rev.* **2016**, *116*, 8848–8872. [CrossRef]
16. Colebatch, A.L.; Weller, A.S. Amine-borane dehydropolymerization: Challenges and opportunities. *Chem. Eur. J.* **2019**, *25*, 1379–1390. [CrossRef]
17. Han, D.; Anke, F.; Trose, T.; Beweries, T. Recent advances in transition metal catalysed dehydropolymerisation of amine boranes and phosphine boranes. *Coord. Chem. Rev.* **2019**, *308*, 260–286. [CrossRef]
18. Faverio, C.; Boselli, M.F.; Medici, F.; Benaglia, M. Ammonia borane as a reducing agent in organic synthesis. *Org. Biomol. Chem.* **2020**, *18*, 7789–7813. [CrossRef]
19. Reddy, D.O. A short chronological review on the syntheses of amine-boranes. *Chem. Rev. Lett.* **2020**, *3*, 184–191.
20. Lau, S.; Gasperini, D.; Webster, R.L. Amine-boranes as transfer hydrogenation and hydrogenation reagents: A mechanistic perspective. *Angew. Chem. Int. Ed.* **2021**, *60*, 14272–14294. [CrossRef]
21. Chen, X.; Zhao, J.C.; Shore, S.G. The roles of dihydrogen bonds in amine borane chemistry. *Acc. Chem. Res.* **2013**, *46*, 2666–2675. [CrossRef] [PubMed]
22. Klooster, W.T.; Koetzle, T.F.; Siegbahn, P.E.M.; Richardson, T.B.; Crabtree, R.H. Study of the N-H$\cdots$H-B dihydrogen bond including the crystal structure of BH_3NH_3 by neutron diffraction. *J. Am. Chem. Soc.* **1999**, *121*, 6337–6343. [CrossRef]
23. Morrison, C.A.; Siddick, M.M. Dihydrogen bonds in solid BH_3NH_3. *Ang. Chem. Int. Ed.* **2004**, *43*, 4780–4782. [CrossRef]
24. Al-Kukhun, A.; Hwang, H.T.; Varma, A. Mechanistic studies of ammonia borane dehydrogenation. *Int. J. Hydrogen Energy* **2013**, *38*, 169–179. [CrossRef]
25. Tao, J.; Lv, N.; Wen, L.; Qi, Y.; Lv, X. Hydrogen-release mechanisms in $LiNH_2BH_3 \cdot NH_3BH_3$: A theoretical study. *J. Mol. Struct.* **2012**, *1081*, 437–442. [CrossRef]
26. Zhao, Q.; Li, J.; Hamilton, E.J.M.; Chen, X. The continuing story of the diammoniate of diborane. *J. Organomet. Chem.* **2015**, *798*, 24–29. [CrossRef]
27. Shore, S.G.; Parry, R.W. Chemical evidence for the structure of the "diammoniate of diborane". II. The preparation of ammonia-borane. *J. Am. Chem. Soc.* **1958**, *80*, 8–12. [CrossRef]
28. Merten, C.; Berger, C.J.; McDonald, R.; Xu, Y. Evidence of dihydrogen bonding of a chiral amine-borane complex in solution by VCD spectroscopy. *Angew. Chem. Int. Ed.* **2014**, *53*, 9940–9943. [CrossRef]
29. Theorodatou, A.; Turani-I-Belloto, K.; Petit, E.; Dourdain, S.; Alauzun, J.G.; Demirci, U.B. Synthesis of n-dodecylamine borane $C_{12}H_{25}NH_2BH_3$, its stability against hydrolysis, and its characterization in THF. *J. Mol. Struct.* **2022**, *1248*, 131484. [CrossRef]
30. Staubitz, A.; Sloan, M.E.; Robertson, A.P.M.; Friedrich, A.; Schneider, S.; Gates, P.J.; Schmedt auf der Grüne, J.; Manners, I. Catalytic dehydrocoupling/dehydrogenation of *N*-methylamine-borane and ammonia-borane: Synthesis and characterization of high molecular weight polyaminoboranes. *J. Am. Chem. Soc.* **2010**, *132*, 13332–13345. [CrossRef]
31. Turani-I-Belloto, K.; Valero-Pedraza, M.J.; Chiriac, R.; Toche, F.; Granier, D.; Cot, D.; Petit, E.; Yot, P.G.; Alauzun, J.G.; Demirci, U.B. A series of primary alkylamine borane adducts $C_xH_{2x+1}NH_2BH_3$: Synthesis and properties. *ChemistrySelect* **2021**, *6*, 9853–9860. [CrossRef]
32. Turani-I-Belloto, K.; Valero-Pedraza, M.J.; Petit, E.; Chiriac, R.; Toche, F.; Granier, D.; Yot, P.G.; Alauzun, J.G.; Demirci, U.B. Solid-state structures of primary long-chain alkylamine borane adducts—Synthesis, properties and computational analysis. *ChemistrySelect* **2022**, *7*, e202203533. [CrossRef]
33. Brown, R.J.C.; Brown, R.F.C. Melting point and molecular symmetry. *J. Chem. Educ.* **2000**, *77*, 724–731. [CrossRef]
34. Smith, J.; Seshadri, K.S.; White, D. Infrared spectra of matric isolated BH_3NH_3, BD_3ND_3, and BH_3ND_3. *J. Mol. Spectr.* **1973**, *45*, 327–337. [CrossRef]
35. Wolff, H.; Gamer, G. Hydrogen bonding and complex formation of dimethylamine. Infrared investigations on the NH stretching vibration bands. *J. Phys. Chem.* **1972**, *76*, 871–876. [CrossRef]
36. Vijay, A.; Sathyanarayana, D.N. Theoretical investigation of equilibrium structure, harmonic force field and vibrational spectra of borane diammine: Effects of basis set and electron correlation. *J. Mol. Struct.* **1996**, *375*, 127–141.
37. Krueger, P.J.; Jan, J. Infrared spectra and the molecular conformations of some aliphatic amines. *Can. J. Chem.* **1970**, *48*, 3229–3235. [CrossRef]
38. Durig, J.R.; Lindsay, N.E.; Hizer, T.J.; Odom, J.D. Infrared and raman spectra, conformational stability and normal coordinate analysis of ethyldimethylamine-borane. *J. Mol. Struct.* **1988**, *189*, 257–277. [CrossRef]
39. Eaton, G.R. NMR of boron compounds. *J. Chem Educ.* **1969**, *46*, 547–556. [CrossRef]
40. Flores-Parra, A.; Guadarrama-Pérez, C.; Galvez Ruiz, J.C.; Sanchez Ruiz, S.A.; Suarez-Moreno, G.V.; Contreras, R. Mono- and di-alkyl-[1,3,5]-dithiazinanes and their N–borane adducts revisited. Structural and theoretical study. *J. Mol. Struct.* **2013**, *1047*, 149–159. [CrossRef]
41. Kobayashi, T.; Gupta, S.; Caporini, M.A.; Pecharsky, V.K.; Pruski, M. Mechanism of solid-state thermolysis of ammonia borane: A ^{15}N NMR study using fast magic-angle spinning and dynamic nuclear polarization. *J. Phys. Chem. C* **2014**, *118*, 19548–19555. [CrossRef]

42. Sinton, S.W. Complexation chemistry of sodium borate with poly(vinyl alcohol) and small diols. A ^{11}B NMR study. *Macromolecules* **1987**, *20*, 2430–2441. [CrossRef]

43. Roy, B.; Pal, U.; Bishnoi, A.; O'Dell, L.A.; Sharma, P. Exploring the homopolar dehydrocoupling of ammonia borane by solid-state multinuclear NMR spectroscopy. *Chem. Commun.* **2021**, *57*, 1887–1890. [CrossRef] [PubMed]

44. Hermanek, S. Boron-11 NMR spectra of boranes, main-group heteroboranes, and substituted derivatives. Factors influencing chemical shifts of skeletal atoms. *Chem. Rev.* **1992**, *92*, 325–362. [CrossRef]

45. Carboni, B.; Monnier, L. Recent developments in the chemistry of amine- and phosphine-boranes. *Tetrahedron* **1999**, *55*, 1197–1248. [CrossRef]

46. Kumar, R.; Karkamkar, A.; Bowden, M.; Autrey, T. Solid-state hydrogen rich boron–nitrogen compounds for energy storage. *Chem. Soc. Rev.* **2019**, *48*, 5350–5380. [CrossRef]

 *molecules*

Article

Incorporation of a Boron–Nitrogen Covalent Bond Improves the Charge-Transport and Charge-Transfer Characteristics of Organoboron Small-Molecule Acceptors for Organic Solar Cells

Jie Yang [1], Wei-Lu Ding [2], Quan-Song Li [1,*] and Ze-Sheng Li [1,*]

[1] Key Laboratory of Cluster Science of Ministry of Education, Beijing Key Laboratory of Photoelectronic/Electrophotonic Conversion Materials, School of Chemistry and Chemical Engineering, Beijing Institute of Technology, Beijing 100081, China
[2] Beijing Key Laboratory of Ionic Liquids Clean Process, CAS Key Laboratory of Green Process and Engineering, State Key Laboratory of Multiphase Complex Systems, Institute of Process Engineering, Chinese Academy of Sciences, Beijing 100190, China
* Correspondence: liquansong@bit.edu.cn (Q.-S.L.); zeshengli@bit.edu.cn (Z.-S.L.)

Abstract: An organoboron small-molecular acceptor (OSMA) $M_{B\leftarrow N}$ containing a boron–nitrogen coordination bond (B←N) exhibits good light absorption in organic solar cells (OSCs). In this work, based on $M_{B\leftarrow N}$, OSMA $M_{B\text{-}N}$, with the incorporation of a boron–nitrogen covalent bond (B-N), was designed. We have systematically investigated the charge-transport properties and interfacial charge-transfer characteristics of $M_{B\text{-}N}$, along with $M_{B\leftarrow N}$, using the density functional theory (DFT) and the time-dependent density functional theory (TD-DFT). Theoretical calculations show that $M_{B\text{-}N}$ can simultaneously boost the open-circuit voltage (from 0.78 V to 0.85 V) and the short-circuit current due to its high-lying lowest unoccupied molecular orbital and the reduced energy gap. Moreover, its large dipole shortens stacking and greatly enhances electron mobility by up to 5.91×10^{-3} cm^2·V^{-1}·s^{-1}. Notably, the excellent interfacial properties of PTB7-Th/$M_{B\text{-}N}$, owing to more charge transfer states generated through the direct excitation process and the intermolecular electric field mechanism, are expected to improve OSCs performance. Together with the excellent properties of $M_{B\text{-}N}$, we demonstrate a new OSMA and develop a new organoboron building block with B-N units. The computations also shed light on the structure–property relationships and provide in-depth theoretical guidance for the application of organoboron photovoltaic materials.

Keywords: organoboron; non-fullerene acceptor-based organic solar cells; density functional theory; charge transport; charge transfer

Citation: Yang, J.; Ding, W.-L.; Li, Q.-S.; Li, Z.-S. Incorporation of a Boron–Nitrogen Covalent Bond Improves the Charge-Transport and Charge-Transfer Characteristics of Organoboron Small-Molecule Acceptors for Organic Solar Cells. *Molecules* **2023**, *28*, 811. https://doi.org/10.3390/molecules28020811

Academic Editors: Michael A. Beckett and Igor B. Sivaev

Received: 16 December 2022
Revised: 9 January 2023
Accepted: 11 January 2023
Published: 13 January 2023

1. Introduction

Organoboron plays a crucial role in the field of optoelectronic materials applications [1–3]. Over the past few years, the emerging strategy of the incorporation of boron-nitrogen (BN) units into organic structures has been widely applied in optoelectronic devices, attracting great attention due to their interesting electronic and optical properties [4,5]. In 2011, Nakamura et al. synthesized a series of BN-fused polycyclic aromatic compounds with high mobility, especially 4b-aza-12b-boradibenzo[g,p]chrysene, which predicted that BN-substituted aromatic hydrocarbons were potential candidates for organic electronic materials [6]. In 2013, Pei and Wang et al. reported two novel tetrathienonaphthalene derivatives incorporating a BN unit, namely BN-TTN-C3 and BN-TTN-C6, signifying that a boron nitride fused ring had been applied to organic electronic devices for the first time [7]. Later, they developed and studied many BN-embedded compounds with excellent properties which could be applied to organic semiconductor materials in the field of electronics [7–9]. In 2015, Liu et al. reported the first BN-based acceptor applied in the field of organic solar cells (OSCs) through replacement of a C-C bond by a boron–nitrogen coordination bond (B←N) [10], which sparked great research interest in the chemistry of organoboron acceptors. In 2022, Liu et al. reported an organoboron compound (SBN-1) based on N-B←N units with a balanced resonance hybrid of boron–nitrogen covalent bond (B-N) and B←N, which can be used as an effective building block to construct small band gap conjugated polymers for OSCs [11].

Organoboron provides a new idea to design optoelectronic materials [12–22]. However, the development of n-type organoboron small-molecular acceptors (OSMAs) with advantages of facile synthesis, synthetic versatility, and simplified purification lags far behind that of their polymer counterparts [23,24]. Therefore, it is necessary to further increase the research on OSMAs. The BN unit is one of effective building structures for the construction of OSMAs, which contain two kinds of chemical bonds between B and N, a boron–nitrogen coordination bond, B←N, and a boron–nitrogen covalent bond, B-N. The former has been widely applied to construct non-fullerene acceptors and has achieved great success. The typical representative form is M-BNBP4P-1, which exhibits superior sunlight harvesting capability because of its unique wide absorption spectrum, with two strong bands in the long-wavelength region (771 nm) and the short-wavelength region (502 nm) [25,26]. Besides, Piers et al. devoted their efforts to synthesizing a series of novel cores with BN units for use in optoelectronic devices [27,28]. Nevertheless, the latter is rarely used in OSCs. Recently, Duan et al. developed a novel OSMA called BNTT2F, which was the first reported B-N covalent bond-based electron acceptor for OSCs, achieving the highest power conversion efficiency (PCE) among OSMAs [29]. At this stage, there are not many computational studies regarding OSMAs at the atomistic level, especially concerning their homojunction interfacial charge-transport characteristics and heterojunction interfacial charge-transfer properties [25]. Inspired by the remarkable characteristics of M-BNBP4P-1 (called $M_{B←N}$ in this work) and the fact that the core of the acceptor could effectively regulate its performance, we introduced the B-N bonds into the core group to design a B-N-containing OSMA M_{B-N} and studied whether the derivative containing B-N bond would prove to be an excellent OSMA. The structures of B←N-based OSMA $M_{B←N}$, the B-N-containing compound (M_{B-N}), and their carbon-counterpart (M_{C-N}) are shown in Figure 1; density functional theory (DFT) and time-dependent density functional theory (TD-DFT) were used to systematically investigate their electronic, optical, and interfacial properties, particularly their charge-transport properties and charge-transfer characteristics. Our results show that the B-N-embedded compound M_{B-N} is regarded as an excellent OSMA, which is expected to improve open-circuit voltage (V_{OC}), short-circuit current (J_{SC}), and fill factor (FF) originated from the upshifted frontier molecular orbitals, as well as to exhibit superior charge-transport properties with enhanced electron mobility up to 5.91×10^{-3} cm^2·V^{-1}·s^{-1}, outstanding optical absorption, and sterling interfacial charge-transfer characteristics. This theoretical work will provide useful guidance for the application of OSMAs in OSCs.

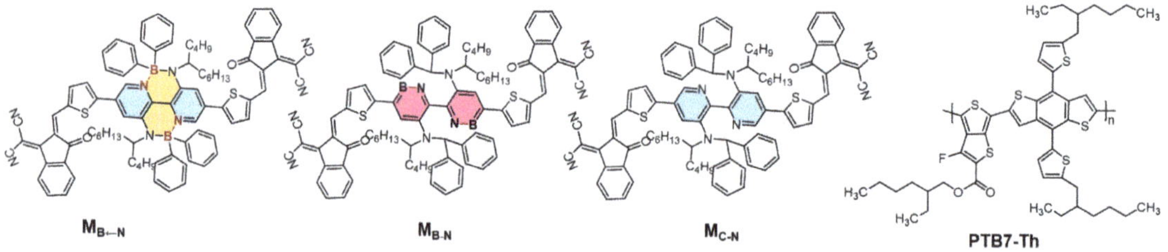

Figure 1. Molecular structures of the investigated acceptors and donor PTB7-Th.

2. Results and Discussion

2.1. Monomolecular Characteristics

In OSCs, determining the energy-level alignment of an acceptor is particularly important since the key photovoltaic parameter V_{OC} is dependent on the difference (ΔE_{DA})

between the LUMO energy level of the acceptor and the HOMO energy level of the donor ($\Delta E_{DA} = E_{LUMO}(A) - E_{HOMO}(D)$) [30,31], which is essential to drive devices [32]. As shown in Figure 2a, $M_{B\leftarrow N}$ and $M_{B\text{-}N}$ possess relatively higher-lying HOMO levels than that of $M_{C\text{-}N}$, indicating less energy loss (E_{loss}) in the boron-containing acceptors. Moreover, the empty 2p orbital of the boron atom adopts sp^2 hybridization in the tri-coordination B-N compounds and sp^3 hybridization in the tetra-coordination $B\leftarrow N$ molecules [15,33]. Therefore, the p-π^* conjugation with the π system of $M_{B\text{-}N}$ due to the empty 2p orbital of the boron atom upshifts the LUMO energy level (-3.85 eV) in comparison with $M_{B\leftarrow N}$ (-3.92 eV) and then increases ΔE_{DA} [33 35], which equates to a higher V_{OC} in OSCs devices. According to the formula evaluated, V_{OC} ($eV_{OC} = \Delta E_{DA} - E_{loss}$) and the experiment value of $M_{B\leftarrow N}$ (0.78 V), the E_{loss} is estimated as 0.52 eV. Therefore, $M_{B\text{-}N}$ yields a high V_{OC} of 0.85 V, which is enhanced by about 9% compared to the reference acceptor. The energy gap is related to the planarity of the geometry, and organoboron compounds shows a smaller dihedral angle off the plane. In comparison with $M_{C\text{-}N}$, the energy gaps of organoboron compounds greatly decrease because of the substantially upshifted HOMOs; thus, improved absorptions are expected [36]. In terms of $M_{B\text{-}N}$, on one hand, its relatively high-lying LUMO levels can result in high V_{OC} due to increased ΔE_{DA}; on the other hand, high J_{SC} may be realized because of the reduced energy gaps with upshifted HOMOs compared to those of $M_{C\text{-}N}$ [37]. The orbital delocalization index (*ODI*) can quantitatively investigate the degree of orbital delocalization. The smaller the value, the higher the degree of orbit delocalization. As displayed in Figure 2b, the *ODI* values of the LUMOs are smaller than those of the HOMOs, indicating that the studied molecules have good electron transport properties.

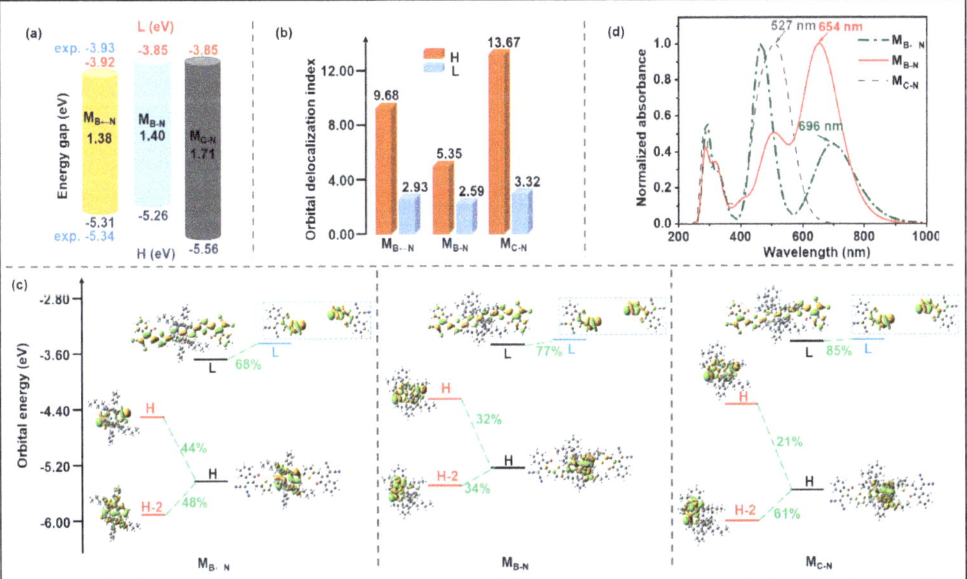

Figure 2. (a) Frontier molecular orbital energy levels, (b) orbital delocalization index, (c) orbital interaction diagram, and (d) absorption spectra of $M_{B\leftarrow N}$, $M_{B\text{-}N}$, and $M_{C\text{-}N}$ (H: the highest occupied molecular orbital; L: the lowest unoccupied molecular orbital; green texts indicate the major contribution of the molecular orbitals from the fragments to the orbitals of the complex).

Molecular orbital correlation (MOC) analysis has been proven to be an effective method for analyzing the contribution of each molecular orbital fragment to the entire molecular orbital and further exploring the intramolecular orbital interactions [38–40]. To study the influence of different forms of boron and nitrogen substitution on the distribution and energy of frontier molecular orbitals, we carried out MOC analysis on these molecules. The studied molecules can be divided into central cores and electron-withdrawing units on both sides. The results (Figure 2c) show that the contributions of the LUMO of the studied molecules are mainly from the electron-withdrawing groups, and the contributions of the HOMO are mainly from the central cores, which explains why the LUMO energy levels of these molecules show little change, but the HOMO exhibits a large shift. Compared with M_{C-N}, the introduction of B atoms into the system increases the contribution of the HOMO of the central cores to the whole molecular HOMO, and slightly reduces the contribution of the LUMO of the electron-withdrawing group to the overall LUMO. As an important photovoltaic parameter, the absorbance of materials in OSCs equals the power input, which significantly determines the performance of OSCs [41]. Thus the optical characteristics of molecules are widely impacted [42,43]. As depicted in Figure 2d, the maximum absorption wavelengths of $M_{B\leftarrow N}$, M_{B-N}, and M_{C-N} are calculated as 696 nm, 654 nm, and 527 nm, respectively. The absorption spectra of the organoboron acceptors $M_{B\leftarrow N}$ and M_{B-N} are red-shifted relative to that of their carbon-counterpart as a result of the reduced energy gaps mentioned previously. M_{C-N} exhibits the strongest absorption at 527 nm, which loses the energy of the long-wavelength region, thus hindering the increase in J_{SC}. $M_{B\leftarrow N}$ presents two peaks in the UV-Vis region, which is consistent with previously reported experimental results [25]. Similar to $M_{B\leftarrow N}$, there are two unique peaks from 400 nm to 800 nm in M_{B-N}. Different from $M_{B\leftarrow N}$, M_{B-N} exhibits stronger absorption at the long wavelength of about 650 nm, while the strongest peak of $M_{B\leftarrow N}$ is located at 450 nm. In terms of absorption spectra, $M_{B\leftarrow N}$ and M_{B-N} are expected to be excellent acceptors because they meet the absorption demands of long wavelengths [44,45]. Particularly, M_{B-N} not only exhibits a wide absorption range covering the entire visible region and extending to the near-infrared region with two unique absorption peaks in the visible region, but also presents relatively high absorption intensity, which is expected to improve J_{SC} [46].

2.2. Acceptor/Acceptor Charge Transport Properties

Electron mobility (μ_e) is the most important property for acceptors in OSCs, which reflects the electron transport behavior [47–49]. In this work, the semi-classical Marcus electron-transfer theory, combined with the Einstein relation (Table S1), was employed to assess μ_e based on the crystal structures predicted by the polymorph module in Materials Studio [50]. According to the Marcus theory, the one factor affecting the rate is reorganization energy (λ): a large λ value will lead to a decrease in mobility, subsequently, to poor electron transport [51,52]. As summarized in Figure 3a, introductions of the boron atoms increase λ values (0.16 eV for $M_{B\leftarrow N}$ and 0.26 eV for M_{B-N}). The larger λ values in organoboron compounds compared with those in M_{C-N} (0.13 eV) indicate larger energy barriers during the electron transfer process [53]. We have to admit that the introductions of boron atoms are disadvantageous in terms of λ. The other factor is the electron transfer integral v (see Table S1), which reflects the electronic coupling between two neighboring molecules and depends on the relative orientations of adjacent molecules. As shown in Figure 3d, the M_{C-N} molecular stacking includes face-to-face hopping pathways with long distances and edge-to-edge hopping pathways, resulting in relatively small v values. Meanwhile, all hopping pathways in $M_{B\leftarrow N}$ are face-to-edge stacking, but with a small dipole of 0.61 Debye, giving rise to moderate v values. The large dipole of the B-N unit is expected to shorten the π-π stacking and construct good molecular packing for the transport charge [54–56]. Owing to the large dipole of 3.51 Debye in M_{B-N} due to B-N substitution, the stacking exhibits a compact face-to-face π-stacking pattern, leading to large v values and thus, excellent charge transporting properties. As described in the Einstein formula, the maximum electron transfer integral (v_{max}) plays an important role in μ_e, which is attributed to its large weight in the investigated pathways. It can be seen from Figure 3b that v_{max} values of molecules $M_{B\leftarrow N}$ and M_{B-N} are one order of magnitude larger than those of M_{C-N}, with a v_{max} of 3.68×10^{-4} eV. M_{B-N} has the largest v_{max} of 5.36×10^{-3} eV, which is about four times that of $M_{B\leftarrow N}$ (1.23×10^{-3} eV, Figure 3a). Interestingly, despite a larger λ, $M_{B\text{-}N}$ possesses as high as μ_e of 5.91×10^{-3} cm$^2 \cdot$V$^{-1} \cdot$s^{-1} (Figure 3c), which is one order of magnitude larger than that of $M_{C\text{-}N}$ (1.03×10^{-4} cm$^2 \cdot$V$^{-1} \cdot$s^{-1}), as a result of its excellent v_{max}. In addition, $M_{B\leftarrow N}$ delivers a large μ_e of 2.43×10^{-3} cm$^2 \cdot$V$^{-1} \cdot$s^{-1}, also thanks to its large v_{max}. In conclusion, although the large λ values of $M_{B\leftarrow N}$ and $M_{B\text{-}N}$ are disadvantageous factors, the significant enhancements of v_{max} values help to achieve high μ_e values, especially for B-N-containing $M_{B\text{-}N}$ with the largest μ_e, which are conducive to charge transport and thus, are expected to improve the J_{SC} and FF parameters of the devices. The results show that μ_e of organoboron compounds are mainly determined by the v_{max} in this work. Although the active layer is amorphous, the small range of the ordered domains of the non-fullerene acceptor is the key to the charge transport characteristics [57]. Here, the information and calculation results are of significance and of reference value.

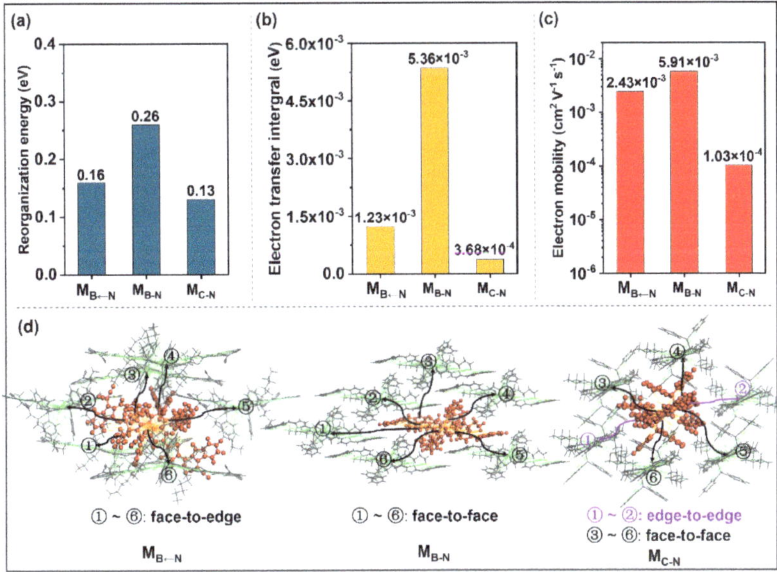

Figure 3. (a) Internal reorganization energy (λ), (b) maximum transfer integral (v_{max}), (c) electron mobility (μ_e), and (d) charge hopping pathways of the studied compounds.

2.3. Donor/Acceptor Interfacial Charge Transfer Performance

In OSCs, the excitons photogenerated within the donor or acceptor components dissociate at the donor/acceptor (D/A) interfaces, which significantly control the photocurrent and thus affect the PCE of the devices [58–61]. To gain a deeper insight into the interfacial optical properties of the organoboron acceptors, the interfaces between the donor PTB7-Th and the investigated organoboron molecules, namely PTB7-Th/$M_{B\leftarrow N}$ and PTB7-Th/$M_{B\text{-}N}$, were constructed, and the excited-state properties of the D/A interfaces were assessed. These excited states can be divided into three classes, namely, charge-transfer (CT) state, local-excitation (LE) state, and hybrid charge-transfer (HCT) state. Among these, the CT state between the donor and the acceptor at the D/A interface is a paramount intermediate state to realize charge separation [58,62,63], which is characterized by the fact that the two singly occupied molecular orbitals separately locate on the donor and the acceptor. There are generally three mechanisms agreed upon by scientists to form the CT state, namely, the direct excitation mechanism, the intermolecular electric field (IEF) mechanism, and the hot exciton (HE) mechanism [62,64,65]. The LE state refers to an electron excitation localized on the donor or the acceptor, with high transition probability, but which is difficult to separate. The HCT state is another important state which possesses a high exciton utilization, resulting from the CT state, and a large oscillator strength, originating from the LE component; it effectively becomes the CT state in the subsequent process, but only given the relevant mechanisms.

Given that the lower excited states are essential in photo-physical and photochemical processes [66], Figure 4 provides the crucial parameters for the three lowest excited states (S_1–S_3), including the transition density matrix (TDM), the charge difference density (CDD) map, the net transferred charges (Δq), D index, and the oscillator strength (f), which are widely used to evaluate the features of excited states. Among these, the TDM and CDD maps are usually used to visually study the spatial span and primary sites of electron transitions. A large TDM value in the off-diagonal term denotes that a strong electron-hole coherence presents between the donor and the acceptor, which corresponds to the CT state, while a large value in the diagonal region indicates a strong charge coherence within the donor or acceptor, which represents the LE component. Parameter Δq can quantitatively express electrons transferring from the donor to the acceptor [67]. The degree of charge separation can be represented by the D index, which is defined as the distance from the hole centroid to the electron centroid [68]. Parameter f indicates the transition probability; an excited state with a high f value means strong absorption. The direct excitation process is a paramount mechanism to generate a CT state, that is, charge carriers of D/A interface are directly excited into the CT state manifolds upon illumination [66]. According to Figure 4a, the TDM of the first excited state (S_1) shows that the photoexcitation of PTB7-Th/$M_{B\leftarrow N}$ is mainly distributed on the off-diagonal part. In addition, as shown in the CDD map (Figure 4b), the electron distributes on the $M_{B\leftarrow N}$, and the hole distributes on the PTB7-Th; therefore, the S_1 of PTB7-Th/$M_{B\leftarrow N}$ is the CT_1 state. However, the S_1 of PTB7-Th/M_{B-N} is regarded as the HCT_1 state, since the photoexcitation of interface PTB7-Th/M_{B-N} distributes not only in the off-diagonal region, but also in the diagonal part. It can be proven by the CDD map, in which the hole only locates in PTB7-Th, and the electron and hole locate in M_{B-N}. Note that, in the studied distributions of the second excited state (S_2), two interfaces are reserved to their S_1 states, according to the TDM and CDD maps; thus, the S_2 states are HCT_1 in the PTB7-Th/$M_{B\leftarrow N}$ interface and CT_1 in the PTB7-Th/M_{B-N} interface. PTB7-Th/$M_{B\leftarrow N}$ and PTB7-Th/M_{B-N} exhibit similar transition characteristics in the third excited states (S_3), and S_3 are CT_2 states in both of the two interfaces. For CT_1 and CT_2, the Δq of PTB7-Th/M_{B-N} are larger than those of PTB7-Th/$M_{B\leftarrow N}$. M_{B-N} transfers more charge to PTB7-Th compared to $M_{B\leftarrow N}$, reflecting stronger charge-separation ability. The calculated D indexes (Figure 4c) follow the order of PTB7-Th/$M_{B\leftarrow N}$ (2.37 Å) < PTB7-Th/M_{B-N} (3.14 Å) for the CT_1 states and PTB7-Th/$M_{B\leftarrow N}$ (2.58 Å) < PTB7-Th/M_{B-N} (2.81 Å) for the CT_2 states. The larger Δq and D index of PTB7-Th/M_{B-N} lead to stronger CT characteristics, which are favorable to the charge separation. In the case of HCT_1, PTB7-Th/M_{B-N} has the strongest light-harvesting ability, with an f of 1.17, which is about 1.5 times larger than that of PTB7-Th/$M_{B\leftarrow N}$. Our results show that the organoboron M_{B-N} has excellent charge-transfer characteristics of direct excitation into the CT state manifold.

In addition to the interfacial direct excitation mechanism discussed above, the intermolecular electric field (IEF) mechanism is one of the main ways, which separate exciton by producing more CT states [62,69,70]. Two key factors determine the realization of the IEF mechanism, namely energy difference and molecular electrostatic potential (ESP) [71]. The low energy state, with a small energy difference, can generate the CT state through the IEF mechanism driven by the difference in molecular ESP between the donor and the acceptor [72]. The energy difference and average ESP values of these molecules were calculated and summarized in Figure 5. The smaller energy difference of 0.05 eV between the HCT and CT state in PTB7-Th/M_{B-N} can be easier to overcome in comparison with PTB7-Th/$M_{B\leftarrow N}$, thus promoting interfacial exciton dissociation. In addition, compared to the value of 3.08 kcal/mol in $M_{B\leftarrow N}$, the average ESP values of M_{B-N} are enhanced to 3.28 kcal/mol. The greatly enhanced ESP of M_{B-N} is favorable for attracting negative electrons and improving charge separation, which is expected to improve J_{SC} and FF.

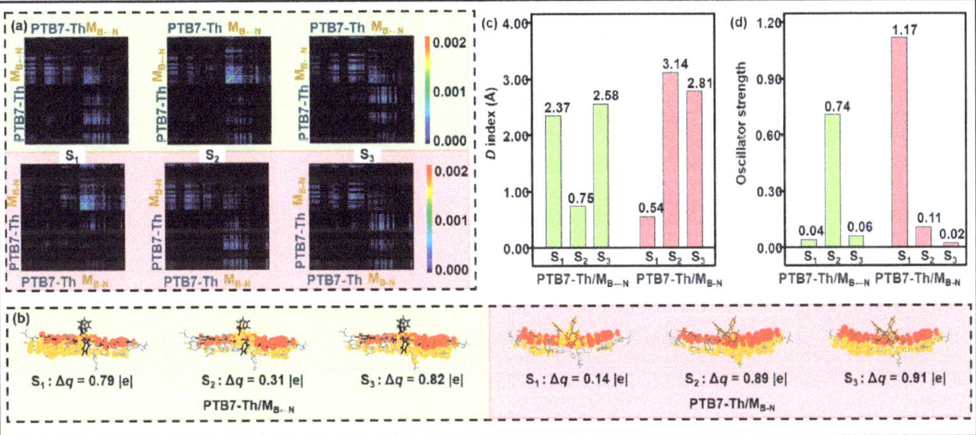

Figure 4. (**a**) 2D site representation of the transition density matrix (TDM), (**b**) charge difference density (CDD) maps with net transferred charge (Δq), (**c**) D index, and (**d**) oscillator strengths of the lowest three excited states (S_1–S_3) of the studied interfaces (CDD: red/yellow stands for accumulation/depletion of negative charges).

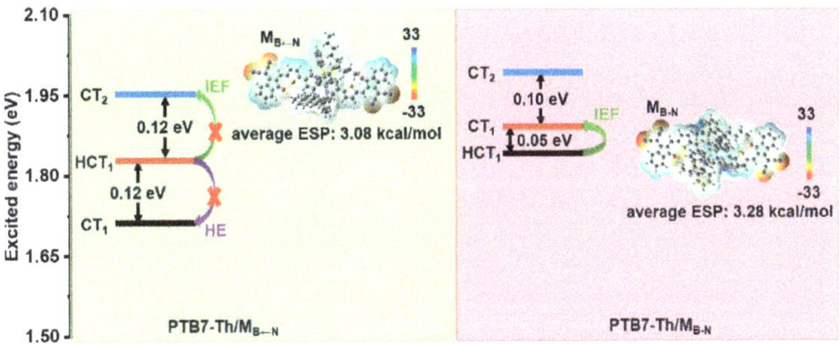

Figure 5. Charge-transfer mechanisms of the lowest three excited states (S_1–S_3) for the studied interfaces (IEF: intermolecular electric field mechanism; HE: hot exciton mechanism; ESP: molecular electrostatic potential).

Another frequently discussed mechanism producing the CT state—the hot exciton (HE) mechanism—can generate the CT state from higher-lying states with close energy resonance, which requires the energy difference between adjacent excited states to be as small as possible [43,73–75]. For PTB7-Th/$M_{B\leftarrow N}$ (Figure 5), the energy difference between the high-lying HCT_1 and the lower CT_1 state is 0.12 eV, which prevents the production of CT from HCT_1 through the HE mechanism due to the lack of close energy resonance.

3. Computational Methods

3.1. Computational Details

The energy minima geometries of the investigated molecules were performed at the B3LYP/6-31G (d, p) level [76], which was confirmed to be suitable for the geometrical parameters [77]. It is well known that the highest occupied molecular orbital (HOMO) eigenvalues are relatively sensitive to the fraction of the Hartree–Fock exchange in the exchange-correlation functional; seven functionals (MPWLYP1M, TPSSH, B3PW91, MPW1B95, PBE38, and M06-2X) with a broad range of Hartree–Fock exchange ratios (from 5% to 54%) were used to calculate the HOMO of $M_{B\leftarrow N}$ to obtain more accurate HOMO eigenvalues (Figure S1). Note that the computed HOMO energy level using the B3PW91 functional (−5.31 eV) was in very good agreement with the experimental value (−5.34 eV) [25]; thus, functional B3PW91 was chosen to calculate HOMOs in this work. Considering that the virtual orbitals are generally more difficult to describe theoretically than the occupied orbitals, the lowest unoccupied molecular orbital (LUMO) eigenvalues were obtained by adding the corrected HOMO energies to the TD-DFT HOMO-LUMO gap (E_1), namely $E_{LUMO} = E_{HOMO} + E_1$ [78]. The molecular packings of the acceptors were obtained from the crystal structure prediction at the DREIDING force field [79], which was considered to be a more appropriate force field for molecular crystal prediction [80]. Crystal structure predictions of the studied molecules were performed by using the polymorph predictor module in Materials Studio [50]. Electrostatic potential charges of all atoms were obtained by the DMol3 module, and the crystal structure prediction was then carried out by employing the Perdew–Burke–Ernzerhof (PBE) [81] exchange-correlation energy functional. Finally, we sorted the obtained crystal structures in terms of their total energies and selected crystal structures with the lowest energies for further DFT calculations regarding their electron mobilities. The M06-2X functional is a high nonlocality functional with a double amount of nonlocal exchange (2X), which provides a good description of the non-covalent interaction [82,83]. Thus, the M06-2X functional was employed to calculate the transfer integrals of all hopping pathways based on the direct coupling approach. To obtain a reliable method, the maximum absorption wavelength of $M_{B \leftarrow N}$ was calculated by functionals B3LYP, PBE33, PBE38, M06-2X, CAM-B3LYP, and wB97XD (Figure S2). The maximum absorption wavelength (696 nm) obtained by PBE38 is in good agreement with the experimental result of 698 nm [25]. Therefore, the excited-state properties of the investigated compounds were characterized at the TD-PBE38/6-31G (d, p) level. The empirical D3 dispersion corrections were included using the Becke–Johnson damping potential in DFT and TD-DFT calculations [84,85]. The polarizable continuum model was employed in the single molecules to consider the solvation effect in chlorobenzene [86]. The above calculations were performed with the Gaussian 16 code [87]. The large overlap between the donor and the acceptor enhances the interface interactions, thus reducing the overall energy [88]. Therefore, the donor PTB7-Th was stacked face-to-face, stacking $M_{B \leftarrow N}$ and M_{B-N}, respectively, at a distance of 3.5 Å to form the donor/acceptor (D/A) interfaces, which has been proven to obtain essential results [62,89,90]. Then, these D/A interfaces containing a donor/acceptor pair were simulated by molecular dynamics (Figure S3). The equilibrated simulation time was set to 10 ns, with an integration time of 1 fs, using a universal force field [91], which is suitable for organic molecules [92]. The NVE ensemble was performed at 298K using the Forcite module of the Materials Studio software [50]. After reaching equilibrium, the D/A interfacial configuration containing a donor/acceptor pair remained relatively stable, and the lowest energy configurations obtained by equilibrated dynamic simulations were fully optimized at the B3LYP-D3/6-31G (d, p) level with the Gaussian 16 code. Given the slight effect of side chains on the electronic properties, the long alky chains were substituted by the methyl groups in the subsequent computations in order to balance the time and accuracy. The analysis of optical properties was performed using the Multiwfn software [93].

3.2. Orbital Delocalization Index

The orbital delocalization index (*ODI*) can quantitatively investigate the degree of orbital delocalization, which is expressed as [93]:

$$ODI_i = 0.01 \times \sum_A (\Theta_{A,i})^2,$$

(1)

where $\Theta_{A,i}$ represents the composition of the A atom in the i orbital.

3.3. Electron Mobility

The Marcus theory, with the hopping model, was employed to describe the electron transport behavior [94,95]. The charge hopping rate (k) between two identical molecules is [96,97]:

$$k = \frac{2\pi}{h} v^2 \frac{1}{\sqrt{4\pi k_{\mathrm{B}} T}} \exp\left(\frac{-\lambda}{4 k_{\mathrm{B}} T}\right),\tag{2}$$

where k_{B}, T, and h are the Boltzmann constant, the temperature in Kelvin, and the Planck constant, respectively ($T = 300$ K in our work). λ denotes the reorganization energy, which is calculated using the adiabatic potential energy surface method. In this work, only internal reorganization energy, which mainly originates from the geometrical relaxation during the charge transfer process and reflects the barriers from one molecule to another, was considered. The reorganization energy can be expressed as follows [98]:

$$\lambda = (E_0^* - E_0) + (E_-^* - E_-),\tag{3}$$

where E_-^* and E_0 are the energies of neutral species in the anionic and neutral geometries, respectively. E_0^* and E_- represent the energies of the anionic species with the geometries of neutral and anionic molecules, respectively.

The transfer integral (v) is obtained by adopting a direct approach at the M06-2X/6-31G (d, p) level [30], which has been proven to be suitable in describing the non-covalent interaction. In our work, v can be calculated by [99]:

$$v = \left\langle \Psi_i^{\mathrm{LUMO}} \middle| SC\varepsilon C^{-1} \middle| \Psi_j^{\mathrm{LUMO}} \right\rangle,\tag{4}$$

where Ψ_i^{LUMO} and Ψ_j^{LUMO} represent the LUMOs of the isolated molecules i and j. The Kohn–Sham orbital C and eigenvalue ε are evaluated by diagonalizing the zeroth-order Fock matrix. S denotes the overlap matrix for the dimer. The electron mobility of the investigated molecules was calculated using the Einstein relation [99,100]:

$$\mu = \frac{1}{2d} \frac{e}{k_{\mathrm{B}} T} \sum r_i^2 k_i P_i,\tag{5}$$

where d represents the spatial dimensionality and is 3 in our work, i is a selected hopping pathway, and r_i and k_i are the charge hopping centroid-to-centroid distance and charge hopping rate, respectively. P_i is defined as the hopping probability, which can be obtained using:

$$P_i = \frac{k_i}{\sum k_i},\tag{6}$$

3.4. Net Transferred Charge

The net transferred electrons from donor (D) to acceptor (A) can be obtained by using the following formula [67,93]:

$$\Delta q = Q_{\mathrm{D,A}} - Q_{\mathrm{A,D}},\tag{7}$$

where $Q_{\mathrm{D,A}}$ ($Q_{\mathrm{A,D}}$) corresponds to the electron transfer from D (A) to A (D) during the excitation, which can be calculated from:

$$Q_{\mathrm{D,A}} = \sum_i^{\mathrm{OCC}} \sum_a^{\mathrm{vir}} \left[(w_i^a)^2 - (w_i'^a)^2 \right] \sum_{R \in D} \Theta_{R,i} \sum_{S \in A} \Theta_{S,a},\tag{8}$$

where w_i^a and $w_i'^a$ are the configuration coefficient of the excitation molecular orbital i to a and the de-excitation molecular orbital a to i, respectively; $\Theta_{R,i}$ ($\Theta_{S,a}$) is the contribution of atom R (S) to the molecular orbital i (a).

3.5. D Index

The distance from the hole centroid to the electron centroid can be expressed from the following equation [68]:

$$D = \sqrt{D_X^2 + D_Y^2 + D_Z^2}\,,\tag{9}$$

The charge transfer (CT) length in X/Y/Z can be measured by the centroid distances between the hole and the electron in corresponding directions:

$$D_{X/Y/Z} = |N_{ele} - N_{hole}|\,,\tag{10}$$

The electron centroid (N_{ele}) and hole centroid (N_{hole}) can be calculated from the following equation:

$$N_{ele/hole} = \int n\rho^{ele/hole}(\mathbf{r})d\mathbf{r}\,,\tag{11}$$

where n is the X (Y or Z) component of position vector $\mathbf{r}$. $\rho^{ele/hole}$ presents the spatial charge distribution.

4. Conclusions

Our studies on two organoboron small-molecular compounds and their carbon counterpart provide an in-depth understanding of the relationship between structures and their electronic, optical, and charge-transport characteristics, as well as their interfacial charge-transfer properties. In comparison with the carbon-counterpart M_{C-N}, the introduction of boron strongly lowers the optical gaps and concurrently, dramatically enhances electron mobility due to the unique characteristics originating from the presence of a vacant p orbital on the B atom. The results show that boron atoms are necessary, and that the M_{B-N} containing a boron–nitrogen covalent bond outperforms the $M_{B\leftarrow N}$ comprising the boron–nitrogen coordination bond due to increased V_{OC} (enhanced by about 9% compared to the $M_{B\leftarrow N}$) as the result of its high-lying LUMO, and its enhanced J_{SC} and FF because of excellent absorption and significantly increased electron mobility of up to 5.91×10^{-3} cm$^2 \cdot$V$^{-1} \cdot$s^{-1}. Further, more CT states originating from the direct excitation mechanism and the IEF process help to improve the interfacial charge-transfer properties of PTB7-Th/M_{B-N}; thus, M_{B-N}-based OSCs are expected to achieve a high V_{OC}, J_{SC}, and FF. Our results not only predict an excellent organoboron small-molecular acceptor M_{B-N} by explaining the internal mechanisms, but also provide a theoretical description of the structure–property relationships.

Supplementary Materials: The following supporting information can be downloaded at: https://www.mdpi.com/article/10.3390/molecules28020811/s1, Figure S1: The E_{HOMO} of $M_{B\leftarrow N}$ at the different functionals; Figure S2: Maximum absorption wavelength of $M_{B\leftarrow N}$ employed different functionals; Figure S3: The potential energy evolutions as a function of simulation time; Figure S4: The dihedral angles of studied molecules; Table S1: Electron transfer intervals of the studied molecules; Table S2: Net transferred charge and D index of the studied interfaces; Table S3: Charge difference density maps of the studied interfaces.

Author Contributions: Conceptualization, J.Y., Q.-S.L. and Z.-S.L.; methodology, J.Y., Q.-S.L. and Z.-S.L.; software, J.Y. and W.-L.D.; validation, J.Y., W.-L.D., Q.-S.L. and Z.-S.L.; formal analysis, J.Y. and W.-L.D.; investigation, J.Y. and Q.-S.L.; resources, Q.-S.L. and Z.-S.L.; data curation, J.Y., Q.-S.L. and Z.-S.L.; writing—original draft preparation, J.Y.; writing—review and editing, W.-L.D., Q.-S.L. and Z.-S.L.; visualization, J.Y. and Q.-S.L.; supervision, Q.-S.L. and Z.-S.L.; project administration, Q.-S.L. and Z.-S.L.; funding acquisition, Q.-S.L. and Z.-S.L. All authors have read and agreed to the published version of the manuscript.

Funding: This research was funded by National Natural Science Foundation of China (grant numbers 22173008) and the Beijing Key Laboratory for Chemical Power Source and Green Catalysis (grant number 2013CX02031).

Institutional Review Board Statement: Not applicable.

Informed Consent Statement: Not applicable.

Data Availability Statement: Data is contained within the article or Supplementary Material.

Acknowledgments: We thank Li-Jie Li for the code to predict crystal structures.

Conflicts of Interest: The authors declare no conflict of interest.

Sample Availability: Samples of the compounds are not available from the authors.

References

1. Kim, H.J.; Godumala, M.; Kim, S.K.; Yoon, J.; Kim, C.Y.; Park, H.; Kwon, J.H.; Cho, M.J.; Choi, D.H. Color-Tunable Boron-Based Emitters Exhibiting Aggregation-Induced Emission and Thermally Activated Delayed Fluorescence for Efficient Solution-Processable Nondoped Deep-Blue to Sky-Blue OLEDs. *Adv. Opt. Mater.* **2020**, *8*, 1902175. [CrossRef]
2. Lu, L.; He, J.; Wu, P.; Wu, Y.; Chao, Y.; Li, H.; Tao, D.; Fan, L.; Li, H.; Zhu, W. Taming electronic properties of boron nitride nanosheets as metal-free catalysts for aerobic oxidative desulfurization of fuels. *Green Chem.* **2018**, *20*, 4453–4460. [CrossRef]
3. Lu, H.; Chen, K.; Bobba, R.S.; Shi, J.; Li, M.; Wang, Y.; Xue, J.; Xue, P.; Zheng, X.; Thorn, K.E.; et al. Simultaneously Enhancing Exciton/Charge Transport in Organic Solar Cells by an Organoboron Additive. *Adv. Mater.* **2022**, *34*, 2205926. [CrossRef] [PubMed]
4. Kothavale, S.S.; Lee, J.Y. Three- and Four-Coordinate, Boron-Based, Thermally Activated Delayed Fluorescent Emitters. *Adv. Opt. Mater.* **2020**, *8*, 2000922. [CrossRef]
5. Ma, X.D.; Tian, Z.W.; Jia, R.; Bai, F.Q. B–N counterpart of biphenylene network: A theoretical investigation. *Appl. Surf. Sci.* **2022**, *598*, 153674. [CrossRef]
6. Hatakeyama, T.; Hashimoto, S.; Seki, S.; Nakamura, M. Synthesis of BN-fused polycyclic aromatics via tandem intramolecular electrophilic arene borylation. *J. Am. Chem. Soc.* **2011**, *133*, 18614–18617. [CrossRef] [PubMed]
7. Wang, X.Y.; Lin, H.R.; Lei, T.; Yang, D.C.; Zhuang, F.D.; Wang, J.Y.; Yuan, S.C.; Pei, J. Azaborine compounds for organic field-effect transistors: Efficient synthesis, remarkable stability, and BN dipole interactions. *Angew. Chem. Int. Ed.* **2013**, *52*, 3117–3120. [CrossRef]
8. Wang, J.Y.; Pei, J. BN-embedded aromatics for optoelectronic applications. *Chin. Chem. Lett.* **2016**, *27*, 1139–1146. [CrossRef]
9. Wang, X.Y.; Zhuang, F.D.; Zhou, X.; Yang, D.C.; Wang, J.Y.; Pei, J. Influence of alkyl chain length on the solid-state properties and transistor performance of BN-substituted tetrathienonaphthalenes. *J. Mater. Chem. C* **2014**, *2*, 8152–8161. [CrossRef]
10. Dou, C.; Ding, Z.; Zhang, Z.; Xie, Z.; Liu, J.; Wang, L. Developing conjugated polymers with high electron affinity by replacing a C-C unit with a B←N unit. *Angew. Chem. Int. Ed.* **2015**, *54*, 3648–3652. [CrossRef]
11. Shao, X.; Liu, M.; Liu, J.; Wang, L. A Resonating B, N Covalent Bond and Coordination Bond in Aromatic Compounds and Conjugated Polymers. *Angew. Chem. Int. Ed.* **2022**, *61*, e202205893. [CrossRef]
12. Grandl, M.; Schepper, J.; Maity, S.; Peukert, A.; von Hauff, E.; Pammer, F. N→B Ladder Polymers Prepared by Postfunctionalization: Tuning of Electron Affinity and Evaluation as Acceptors in All-Polymer Solar Cells. *Macromolecules* **2019**, *52*, 1013–1024. [CrossRef]
13. Zhao, R.; Liu, J.; Wang, L. Polymer Acceptors Containing B←N Units for Organic Photovoltaics. *Acc. Chem. Res.* **2020**, *53*, 1557–1567. [CrossRef]
14. de la Torre, G.; Bottari, G.; Torres, T. Phthalocyanines and Subphthalocyanines: Perfect Partners for Fullerenes and Carbon Nanotubes in Molecular Photovoltaics. *Adv. Energy Mater.* **2017**, *7*, 1601700. [CrossRef]
15. Miao, J.; Wang, Y.; Liu, J.; Wang, L. Organoboron molecules and polymers for organic solar cell applications. *Chem. Soc. Rev.* **2022**, *51*, 153–187. [CrossRef]
16. Ikeda, N.; Oda, S.; Matsumoto, R.; Yoshioka, M.; Fukushima, D.; Yoshiura, K.; Yasuda, N.; Hatakeyama, T. Solution-Processable Pure Green Thermally Activated Delayed Fluorescence Emitter Based on the Multiple Resonance Effect. *Adv. Mater.* **2020**, *32*, 2004072. [CrossRef] [PubMed]
17. Park, I.S.; Min, H.; Yasuda, T. Ultrafast Triplet-Singlet Exciton Interconversion in Narrowband Blue Organoboron Emitters Doped with Heavy Chalcogens. *Angew. Chem. Int. Ed.* **2022**, *61*, e202205684. [CrossRef] [PubMed]
18. Wang, X.; Zhang, Y.; Dai, H.; Li, G.; Liu, M.; Meng, G.; Zeng, X.; Huang, T.; Wang, L.; Peng, Q.; et al. Mesityl-Functionalized Multi-Resonance Organoboron Delayed Fluorescent Frameworks with Wide-Range Color Tunability for Narrowband OLEDs. *Angew. Chem. Int. Ed.* **2022**, *61*, e202206916.
19. Yan, Z.P.; Yuan, L.; Zhang, Y.; Mao, M.X.; Liao, X.J.; Ni, H.X.; Wang, Z.H.; An, Z.; Zheng, Y.X.; Zuo, J.L. A Chiral Dual-Core Organoboron Structure Realizes Dual-Channel Enhanced Ultrapure Blue Emission and Highly Efficient Circularly Polarized Electroluminescence. *Adv. Mater.* **2022**, *34*, 2204253. [CrossRef]
20. Cheon, H.J.; Woo, S.J.; Baek, S.H.; Lee, J.H.; Kim, Y.H. Dense Local Triplet States and Steric Shielding of Multi-Resonance TADF Emitter Enables High-Performance Deep Blue OLEDs. *Adv. Mater.* **2022**, *34*, 2207416. [CrossRef]
21. Gurubasavaraj, P.M.; Sajjan, V.P.; Munoz-Flores, B.M.; Jimenez Perez, V.M.; Hosmane, N.S. Recent Advances in BODIPY Compounds: Synthetic Methods, Optical and Nonlinear Optical Properties, and Their Medical Applications. *Molecules* **2022**, *27*, 1877. [CrossRef]

22. Nowicki, K.; Pacholak, P.; Lulinski, S. Heteroelement Analogues of Benzoxaborole and Related Ring Expanded Systems. *Molecules* **2021**, *26*, 5464. [CrossRef]

23. Dou, C.; Long, X.; Ding, Z.; Xie, Z.; Liu, J.; Wang, L. An Electron-Deficient Building Block Based on the B←N Unit: An Electron Acceptor for All-Polymer Solar Cells. *Angew. Chem. Int. Ed.* **2016**, *55*, 1436–1440. [CrossRef] [PubMed]

24. Zhao, R.; Wang, N.; Yu, Y.; Liu, J. Organoboron Polymer for 10% Efficiency All-Polymer Solar Cells. *Chem. Mater.* **2020**, *32*, 1308–1314. [CrossRef]

25. Liu, F.; Ding, Z.; Liu, J.; Wang, L. An organoboron compound with a wide absorption spectrum for solar cell applications. *Chem. Commun.* **2017**, *53*, 12213–12216. [CrossRef]

26. Liu, F.; Liu, J.; Wang, L. Effect of fluorine substitution in organoboron electron acceptors for photovoltaic application. *Org. Chem. Front.* **2019**, *6*, 1996–2003. [CrossRef]

27. Morgan, M.M.; Patrick, E.A.; Rautiainen, J.M.; Tuononen, H.M.; Piers, W.E.; Spasyuk, D.M. Zirconocene-Based Methods for the Preparation of BN-Indenes: Application to the Synthesis of 1,5-Dibora-4a,8a-diaza-1,2,3,5,6,7-hexaaryl-4,8-dimethyl-s-indacenes. *Organometallics* **2017**, *36*, 2541–2551. [CrossRef]

28. Morgan, M.M.; Nazari, M.; Pickl, T.; Rautiainen, J.M.; Tuononen, H.M.; Piers, W.E.; Welch, G.C.; Gelfand, B.S. Boron-nitrogen substituted dihydroindeno[1,2-b]fluorene derivatives as acceptors in organic solar cells. *Chem. Commun.* **2019**, *55*, 11095–11098. [CrossRef]

29. Liu, X.; Pang, S.; Zeng, L.; Deng, W.; Yang, M.; Yuan, X.; Li, J.; Duan, C.; Huang, F.; Cao, Y. An electron acceptor featuring a B-N covalent bond and small singlet-triplet gap for organic solar cells. *Chem. Commun.* **2022**, *58*, 8686–8689. [CrossRef]

30. Blom, P.W.M.; Mihailetchi, V.D.; Koster, L.J.A.; Markov, D.E. Device Physics of Polymer:Fullerene Bulk Heterojunction Solar Cells. *Adv. Mater.* **2007**, *19*, 1551–1566. [CrossRef]

31. Liu, Z.; Wang, N. Enhanced Performance and Stability of Ternary Organic Solar Cells Utilizing Two Similar Structure Blend Fullerene-Free Molecules as Electron Acceptor. *Adv. Opt. Mater.* **2019**, *7*, 1901241. [CrossRef]

32. Wang, Y.; Fan, Q.; Guo, X.; Li, W.; Guo, B.; Su, W.; Ou, X.; Zhang, M. High-performance nonfullerene polymer solar cells based on a fluorinated wide bandgap copolymer with a high open-circuit voltage of 1.04 V. *J. Mater. Chem. A* **2017**, *5*, 22180–22185. [CrossRef]

33. Yin, X.; Liu, J.; Jakle, F. Electron-Deficient Conjugated Materials via p-π* Conjugation with Boron: Extending Monomers to Oligomers, Macrocycles, and Polymers. *Chem. Eur. J.* **2021**, *27*, 2973–2986. [CrossRef]

34. Liu, Z.; Wang, N. Small Energy Loss and Broad Energy Levels Offsets Lead to Efficient Ternary Polymer Solar Cells from a Blend of Two Fullerene-Free Small Molecules as Electron Acceptors. *Adv. Opt. Mater.* **2019**, *7*, 1900913. [CrossRef]

35. Li, S.Y.; Sun, Z.B.; Zhao, C.H. Charge-Transfer Emitting Triarylborane pi-Electron Systems. *Inorg. Chem.* **2017**, *56*, 8705–8717. [CrossRef]

36. Li, C.; Song, J.; Cai, Y.; Han, G.; Zheng, W.; Yi, Y.; Ryu, H.S.; Woo, H.Y.; Sun, Y. Heteroatom substitution-induced asymmetric A–D–A type non-fullerene acceptor for efficient organic solar cells. *J. Energy Chem.* **2020**, *40*, 144–150. [CrossRef]

37. Jeon, S.J.; Kim, Y.H.; Kim, I.N.; Yang, N.G.; Yun, J.H.; Moon, D.K. Utilizing 3,4-ethylenedioxythiophene (EDOT)-bridged non-fullerene acceptors for efficient organic solar cells. *J. Energy Chem.* **2022**, *65*, 194–204. [CrossRef]

38. Wu, J.; Liao, Y.; Wu, S.X.; Li, H.B.; Su, Z.M. Phenylcarbazole and phosphine oxide/sulfide hybrids as host materials for blue phosphors: Effectively tuning the charge injection property without influencing the triplet energy. *Phys. Chem. Chem. Phys.* **2012**, *14*, 1685–1693. [CrossRef]

39. Wang, X.; Liu, Z.; Yan, X.; Lu, T.; Zheng, W.; Xiong, W. Bonding Character, Electron Delocalization, and Aromaticity of Cyclo[18]Carbon (C$_{18}$) Precursors, C$_{18}$-(CO)$_n$ (*n* = 6, 4, and 2): Focusing on the Effect of Carbonyl (-CO) Groups. *Chem. Eur. J.* **2022**, *28*, e202103815.

40. Wu, J.; Wu, S.; Geng, Y.; Yang, G.; Muhammad, S.; Jin, J.; Liao, Y.; Su, Z. Theoretical study on dithieno[3,2-b:2′,3′-d]phosphole derivatives: High-efficiency blue-emitting materials with ambipolar semiconductor behavior. *Theor. Chem. Acc.* **2010**, *127*, 419–427. [CrossRef]

41. Lin, H.; Wang, Q. Non-fullerene small molecule electron acceptors for high-performance organic solar cells. *J. Energy Chem.* **2018**, *27*, 990–1016. [CrossRef]

42. Aldrich, T.J.; Matta, M.; Zhu, W.; Swick, S.M.; Stern, C.L.; Schatz, G.C.; Facchetti, A.; Melkonyan, F.S.; Marks, T.J. Fluorination Effects on Indacenodithienothiophene Acceptor Packing and Electronic Structure, End-Group Redistribution, and Solar Cell Photovoltaic Response. *J. Am. Chem. Soc.* **2019**, *141*, 3274–3287. [CrossRef] [PubMed]

43. Liang, Y.J.; Zhao, Z.W.; Geng, Y.; Pan, Q.Q.; Gu, H.Y.; Zhao, L.; Zhang, M.; Wu, S.X.; Su, Z.M. Can we utilize the higher Frenkel exciton state in biazulene diimides-based non-fullerene acceptors to promote charge separation at the donor/acceptor interface? *New J. Chem.* **2020**, *44*, 9767–9774. [CrossRef]

44. Xia, K.; Li, Y.; Wang, Y.; Portilla, L.; Pecunia, V. Narrowband-Absorption-Type Organic Photodetectors for the Far-Red Range Based on Fullerene-Free Bulk Heterojunctions. *Adv. Opt. Mater.* **2020**, *8*, 1902056. [CrossRef]

45. Jia, Z.; Qin, S.; Meng, L.; Ma, Q.; Angunawela, I.; Zhang, J.; Li, X.; He, Y.; Lai, W.; Li, N.; et al. High performance tandem organic solar cells via a strongly infrared-absorbing narrow bandgap acceptor. *Nat. Commun.* **2021**, *12*, 178. [CrossRef]

46. Li, Z.; Zhu, C.; Yuan, J.; Zhou, L.; Liu, W.; Xia, X.; Hong, J.; Chen, H.; Wei, Q.; Lu, X.; et al. Optimizing side chains on different nitrogen aromatic rings achieving 17% efficiency for organic photovoltaics. *J. Energy Chem.* **2022**, *65*, 173–178. [CrossRef]

47. Guo, Y.; Han, G.; Duan, R.; Geng, H.; Yi, Y. Boosting the electron mobilities of dimeric perylenediimides by simultaneously enhancing intermolecular and intramolecular electronic interactions. *J. Mater. Chem. A* **2018**, *6*, 14224–14230. [CrossRef]

48. Koh, S.E.; Risko, C.; da Silva Filho, D.A.; Kwon, O.; Facchetti, A.; Brédas, J.L.; Marks, T.J.; Ratner, M.A. Modeling Electron and Hole Transport in Fluoroarene-Oligothiopene Semiconductors: Investigation of Geometric and Electronic Structure Properties. *Adv. Funct. Mater.* **2008**, *18*, 332–340. [CrossRef]

49. Kupgan, G.; Chen, X.K.; Brédas, J.L. Molecular packing of non-fullerene acceptors for organic solar cells: Distinctive local morphology in Y6 vs. ITIC derivatives. *Mater. Today Adv.* **2021**, *11*, 100154. [CrossRef]

50. Dassault Systemes. One Molecular Simulation Software. Available online: https://www.accelrys.com (accessed on 28 October 2017).

51. Shuai, Z.; Wang, L.; Li, Q. Evaluation of charge mobility in organic materials: From localized to delocalized descriptions at a first-principles level. *Adv. Mater.* **2011**, *23*, 1145–1153. [CrossRef]

52. Li, G.; Feng, L.W.; Mukherjee, S.; Jones, L.; Jacobberger, R.; Huang, W.; Young, R.M.; Pankow, R.M.; Zhu, W.; Lu, N.; et al. Non-Fullerene Acceptors with Direct and Indirect Hexa-fluorination Afford >17% Efficiency in Polymer Solar Cells. *Energy Environ. Sci.* **2022**, *15*, 645–659. [CrossRef]

53. Wang, L.; Nan, G.; Yang, X.; Peng, Q.; Li, Q.; Shuai, Z. Computational methods for design of organic materials with high charge mobility. *Chem. Soc. Rev.* **2010**, *39*, 423–434. [CrossRef] [PubMed]

54. Zhang, P.F.; Zeng, J.C.; Zhuang, F.D.; Zhao, K.X.; Sun, Z.H.; Yao, Z.F.; Lu, Y.; Wang, X.Y.; Wang, J.Y.; Pei, J. Parent B$_2$N$_2$-Perylenes with Different BN Orientations. *Angew. Chem. Int. Ed.* **2021**, *60*, 23313–23319. [CrossRef] [PubMed]

55. Zhao, K.; Yao, Z.F.; Wang, Z.Y.; Zeng, J.C.; Ding, L.; Xiong, M.; Wang, J.Y.; Pei, J. "Spine Surgery" of Perylene Diimides with Covalent B-N Bonds toward Electron-Deficient BN-Embedded Polycyclic Aromatic Hydrocarbons. *J. Am. Chem. Soc.* **2022**, *144*, 3091–3098. [CrossRef] [PubMed]

56. Li, W.; Du, C.Z.; Chen, X.Y.; Fu, L.; Gao, R.R.; Yao, Z.F.; Wang, J.Y.; Hu, W.; Pei, J.; Wang, X.Y. BN-Anthracene for High-Mobility Organic Optoelectronic Materials through Periphery Engineering. *Angew. Chem. Int. Ed.* **2022**, *61*, e202201464.

57. Oh, C.M.; Jang, S.; Lee, J.; Park, S.H.; Hwang, I.W. Versatile control of concentration gradients in non-fullerene acceptor-based bulk heterojunction films using solvent rinse treatments. *Green Energy Environ.* **2022**, *7*, 1102–1110. [CrossRef]

58. Cao, Z.; Yang, S.; Wang, B.; Shen, X.; Han, G.; Yi, Y. Multi-channel exciton dissociation in D18/Y6 complexes for high-efficiency organic photovoltaics. *J. Mater. Chem. A* **2020**, *8*, 20408–20413. [CrossRef]

59. Wang, J.; Jiang, X.; Wu, H.; Feng, G.; Wu, H.; Li, J.; Yi, Y.; Feng, X.; Ma, Z.; Li, W.; et al. Increasing donor-acceptor spacing for reduced voltage loss in organic solar cells. *Nat. Commun.* **2021**, *12*, 6679. [CrossRef]

60. Rahmanudin, A.; Yao, L.; Jeanbourquin, X.A.; Liu, Y.; Sekar, A.; Ripaud, E.; Sivula, K. Melt-processing of small molecule organic photovoltaics via bulk heterojunction compatibilization. *Green Chem.* **2018**, *20*, 2218–2224. [CrossRef]

61. Holmes, N.P.; Munday, H.; Barr, M.G.; Thomsen, L.; Marcus, M.A.; Kilcoyne, A.L.D.; Fahy, A.; van Stam, J.; Dastoor, P.C.; Moons, E. Unravelling donor–acceptor film morphology formation for environmentally-friendly OPV ink formulations. *Green Chem.* **2019**, *21*, 5090–5103. [CrossRef]

62. Zhao, Z.W.; Pan, Q.Q.; Geng, Y.; Wu, Y.; Zhao, L.; Zhang, M.; Su, Z.M. Theoretical Insight into Multiple Charge-Transfer Mechanisms at the P3HT/Nonfullerenes Interface in Organic Solar Cells. *ACS Sustain. Chem. Eng.* **2019**, *7*, 19699–19707. [CrossRef]

63. Zarrabi, N.; Sandberg, O.J.; Zeiske, S.; Li, W.; Riley, D.B.; Meredith, P.; Armin, A. Charge generating mid-gap trap states define the thermodynamic limit of organic photovoltaic devices. *Nat. Commun.* **2020**, *11*, 5567. [CrossRef]

64. Gao, F.; Inganas, O. Charge generation in polymer-fullerene bulk-heterojunction solar cells. *Phys. Chem. Chem. Phys.* **2014**, *16*, 20291–20304. [CrossRef] [PubMed]

65. Yang, J.; Li, Q.S.; Li, Z.S. End-capped group manipulation of indacenodithienothiophene-based non-fullerene small molecule acceptors for efficient organic solar cells. *Nanoscale* **2020**, *12*, 17795–17804. [CrossRef] [PubMed]

66. Vandewal, K.; Albrecht, S.; Hoke, E.T.; Graham, K.R.; Widmer, J.; Douglas, J.D.; Schubert, M.; Mateker, W.R.; Bloking, J.T.; Burkhard, G.F.; et al. Efficient charge generation by relaxed charge-transfer states at organic interfaces. *Nat. Mater.* **2013**, *13*, 63–68. [CrossRef]

67. Shi, X.; Yang, Y.; Wang, L.; Li, Y. Introducing Asymmetry Induced by Benzene Substitution in a Rigid Fused π Spacer of D−π−A-Type Solar Cells: A Computational Investigation. *J. Phys. Chem. C* **2019**, *123*, 4007–4021. [CrossRef]

68. Yao, C.; Yang, Y.; Li, L.; Bo, M.; Zhang, J.; Peng, C.; Huang, Z.; Wang, J. Elucidating the Key Role of the Cyano (−C≡N) Group to Construct Environmentally Friendly Fused-Ring Electron Acceptors. *J. Phys. Chem. C* **2020**, *124*, 23059–23068. [CrossRef]

69. Sobolewski, A.L. Organic photovoltaics without p-n junctions: A computational study of ferroelectric columnar molecular clusters. *Phys. Chem. Chem. Phys.* **2015**, *17*, 20580–20587. [CrossRef]

70. Schwarz, C.; Bassler, H.; Bauer, I.; Koenen, J.M.; Preis, E.; Scherf, U.; Kohler, A. Does conjugation help exciton dissociation? a study on poly(p-phenylene)s in planar heterojunctions with C$_{60}$ or TNF. *Adv. Mater.* **2012**, *24*, 922–925. [CrossRef]

71. Yao, H.; Cui, Y.; Qian, D.; Ponseca, C.S., Jr.; Honarfar, A.; Xu, Y.; Xin, J.; Chen, Z.; Hong, L.; Gao, B.; et al. 14.7% Efficiency Organic Photovoltaic Cells Enabled by Active Materials with a Large Electrostatic Potential Difference. *J. Am. Chem. Soc.* **2019**, *141*, 7743–7750. [CrossRef]

72. Yao, H.; Qian, D.; Zhang, H.; Qin, Y.; Xu, B.; Cui, Y.; Yu, R.; Gao, F.; Hou, J. Critical Role of Molecular Electrostatic Potential on Charge Generation in Organic Solar Cells. *Chin. J. Chem.* **2018**, *36*, 491–494. [CrossRef]

73. Zhu, X.Y.; Yang, Q.; Muntwiler, M. Charge-transfer excitons at organic semiconductor surfaces and interfaces. *Acc. Chem. Res.* **2009**, *42*, 1779–1787. [CrossRef]
74. Grancini, G.; Maiuri, M.; Fazzi, D.; Petrozza, A.; Egelhaaf, H.J.; Brida, D.; Cerullo, G.; Lanzani, G. Hot exciton dissociation in polymer solar cells. *Nat. Mater.* **2013**, *12*, 29–33. [CrossRef] [PubMed]
75. Iizuka, H.; Nakayama, T. Quantum process of exciton dissociation at organic semiconductor interfaces: Effects of interface roughness and hot exciton. *Jpn. J. Appl. Phys.* **2016**, *55*, 021601. [CrossRef]
76. Becke, A.D. Density-functional exchange-energy approximation with correct asymptotic behavior. *Phys. Rev. A* **1988**, *38*, 3098–3100. [CrossRef]
77. Grimme, S. Semiempirical GGA-type density functional constructed with a long-range dispersion correction. *J. Comput. Chem.* **2006**, *27*, 1787–1799. [CrossRef] [PubMed]
78. Zhang, G.; Musgrave, C.B. Comparison of DFT methods for molecular orbital eigenvalue calculations. *J. Phys. Chem. A* **2007**, *111*, 1554–1561. [CrossRef]
79. Mayo, S.L.; Olafson, B.D.; Goddard, W.A. DREIDING: A generic force field for molecular simulations. *J. Chem. Phys.* **1990**, *94*, 8897–8909. [CrossRef]
80. Sokolov, A.N.; Atahan-Evrenk, S.; Mondal, R.; Akkerman, H.B.; Sánchez-Carrera, R.S.; Granados-Focil, S.; Schrier, J.; Mannsfeld, S.C.B.; Zoombelt, A.P.; Bao, Z.; et al. From computational discovery to experimental characterization of a high hole mobility organic crystal. *Nat. Commun.* **2011**, *2*, 437. [CrossRef]
81. Ernzerhof, M.; Scuseria, G.E. Assessment of the Perdew–Burke–Ernzerhof exchange-correlation functional. *J. Chem. Phys.* **1999**, *110*, 5029–5036. [CrossRef]
82. Zhao, Y.; Truhlar, D.G. The M06 suite of density functionals for main group thermochemistry, thermochemical kinetics, noncovalent interactions, excited states, and transition elements: Two new functionals and systematic testing of four M06-class functionals and 12 other functionals. *Theor. Chem. Acc.* **2008**, *120*, 215–241.
83. Geng, Y.; Wu, S.X.; Li, H.B.; Tang, X.D.; Wu, Y.; Su, Z.M.; Liao, Y. A theoretical discussion on the relationships among molecular packings, intermolecular interactions, and electron transport properties for naphthalene tetracarboxylic diimide derivatives. *J. Mater. Chem.* **2011**, *21*, 15558–15566. [CrossRef]
84. Grimme, S.; Antony, J.; Ehrlich, S.; Krieg, H. A consistent and accurate ab initio parametrization of density functional dispersion correction (DFT-D) for the 94 elements H-Pu. *J. Chem. Phys.* **2010**, *132*, 154104. [CrossRef] [PubMed]
85. Becke, A.D.; Johnson, E.R. A density-functional model of the dispersion interaction. *J. Chem. Phys.* **2005**, *123*, 154101. [CrossRef]
86. Mennucci, B. Polarizable continuum model. *WIREs Comput. Mol. Sci.* **2012**, *2*, 386–404. [CrossRef]
87. Frisch, M.J.; Trucks, G.W.; Schlegel, H.B.; Scuseria, G.E.; Robb, M.A.; Cheeseman, J.R.; Scalmani, G.; Barone, V.; Petersson, G.A.; Nakatsuji, H.; et al. *Gaussian 16, Revision A.03*; Gaussian, Inc.: Wallingford, CT, USA, 2016; Volume 3.
88. Mahmood, A.; Irfan, A.; Wang, J.L. Molecular level understanding of the chalcogen atom effect on chalcogen-based polymers through electrostatic potential, non-covalent interactions, excited state behaviour, and radial distribution function. *Polym. Chem.* **2022**, *13*, 5993–6001. [CrossRef]
89. Pan, Q.Q.; Li, S.B.; Duan, Y.C.; Wu, Y.; Zhang, J.; Geng, Y.; Zhao, L.; Su, Z.M. Exploring what prompts ITIC to become a superior acceptor in organic solar cell by combining molecular dynamics simulation with quantum chemistry calculation. *Phys. Chem. Chem. Phys.* **2017**, *19*, 31227–31235. [CrossRef] [PubMed]
90. Cui, Y.; Li, P.; Song, C.; Zhang, H. Terminal Modulation of D−π–A Small Molecule for Organic Photovoltaic Materials: A Theoretical Molecular Design. *J. Phys. Chem. C* **2016**, *120*, 28939–28950. [CrossRef]
91. Rappe, A.K.; Casewit, C.J.; Colwell, K.S.; Goddard, W.A.; Skiff, W.M. UFF, a full periodic table force field for molecular mechanics and molecular dynamics simulations. *J. Am. Chem. Soc.* **2002**, *114*, 10024–10035. [CrossRef]
92. Casewit, C.J.; Colwell, K.S.; Rappe, A.K. Application of a universal force field to organic molecules. *J. Am. Chem. Soc.* **2002**, *114*, 10035–10046. [CrossRef]
93. Lu, T.; Chen, F.J. Multiwfn: A multifunctional wavefunction analyzer. *J. Comput. Chem.* **2012**, *33*, 580–592. [CrossRef] [PubMed]
94. Tender, L.; Carter, M.T.; Murray, R.W. Cyclic Voltammetric Analysis of Ferrocene Alkanethiol Monolayer Electrode Kinetics Based on Marcus Theory. *Anal. Chem.* **1994**, *66*, 3173–3181. [CrossRef]
95. Mayer, J.M. Understanding Hydrogen Atom Transfer: From Bond Strengths to Marcus Theory. *Acc. Chem. Res.* **2011**, *44*, 36–46. [CrossRef]
96. Cornil, J.; Brédas, J.L.; Zaumseil, J.; Sirringhaus, H. Ambipolar Transport in Organic Conjugated Materials. *Adv. Mater.* **2007**, *19*, 1791–1799. [CrossRef]
97. Marcus, R.A. Electron transfer reactions in chemistry. Theory and experiment. *Rev. Mod. Phys.* **1993**, *65*, 599–610. [CrossRef]
98. Chen, H.Y.; Chao, I. Effect of perfluorination on the charge-transport properties of organic semiconductors: Density functional theory study of perfluorinated pentacene and sexithiophene. *Chem. Phys. Lett.* **2005**, *401*, 539–545. [CrossRef]

99. Yang, X.; Li, Q.; Shuai, Z. Theoretical modelling of carrier transports in molecular semiconductors: Molecular design of triphenylamine dimer systems. *Nanotechnology* **2007**, *18*, 424029. [CrossRef]
100. Coropceanu, V.; Cornil, J.; da Silva Filho, D.A.; Olivier, Y.; Silbey, R.; Brédas, J.L. Charge Transport in Organic Semiconductors. *Chem. Rev.* **2007**, *107*, 926–952. [CrossRef]

MDPI

St. Alban-Anlage 66

4052 Basel

Switzerland

www.mdpi.com

Molecules Editorial Office

E-mail: molecules@mdpi.com

www.mdpi.com/journal/molecules